D0202643

WITHDRAWN

# Textbook of Work Physiology

## Physiological Bases of Exercise

Fourth Edition

Per-Olof Åstrand, MD
Professor Emeritus, The Swedish College of Physical Education, Stockholm, Sweden

Kaare Rodahl, MD
Professor Emeritus, The Norwegian University of Sport and Physical Education, Oslo, Norway
Former Director, The Norwegian Institute of Work Physiology

Hans A. Dahl, MD
Sigmund B. Strømme, PhD
Professors, The Norwegian University of Sport and Physical Education, Oslo, Norway

Human Kinetics

**Library of Congress Cataloging-in-Publication Data**

Textbook of work physiology / Per-Olof Åstrand ... [et al.]. -- 4th ed.
p. ; cm.
Rev. ed. of: Textbook of work physiology / Per-Olof Åstrand, Kaare Rodahl. c1986.
Includes bibliographical references and index.
ISBN 0-7360-0140-9 (hard cover)
1. Work--Physiological aspects. 2. Exercise--Physiological aspects.
[DNLM: 1. Exertion--physiology. 2. Exercise--physiology. 3. Physical Education and Training. WE 103 T355 2003] I. Åstrand, Per-Olof.
QP301 .A67 2003
612'.042--dc21

2002153149

ISBN: 0-7360-0140-9



This book is a revised edition of *Textbook of Work Physiology: Physiological Bases of Exercise, Third Edition,* published in 1986 by McGraw-Hill.

**Acquisitions Editor:** Michael S. Bahrke, PhD; **Developmental Editor:** Myles Schrag; **Assistant Editors:** Kathleen Bernard, Derek Campbell, Jennifer L. Davis; **Copyeditors:** Julie Anderson, Karen Bojda; **Proofreader:** Joanna Hatzopoulos Portman; **Indexer:** Craig Brown; **Permission Manager:** Dalene Reeder; **Graphic Designer:** Nancy Rasmus; **Graphic Artist:** Dawn Sills; **Photo Manager:** Leslie A. Woodrum; **Cover Designer:** Keith Blomberg; **Photographers (cover):** © Human Kinetics (left) and USDA Forest Service, Technology & Development Center, Missoula, MT (right); **Art Manager:** Kelly Hendren; **Illustrators:** Bjørn Norheim (medical art) and Craig Newsom; **Printer:** Transcontinental

Printed in Canada 10 9 8 7 6 5 4 3 2 1

**Human Kinetics**
Web site: www.HumanKinetics.com

*United States:* Human Kinetics
P.O. Box 5076
Champaign, IL 61825-5076
800-747-4457
e-mail: humank@hkusa.com

*Canada:* Human Kinetics
475 Devonshire Road Unit 100
Windsor, ON N8Y 2L5
800-465-7301 (in Canada only)
e-mail: orders@hkcanada.com

*Europe:* Human Kinetics
107 Bradford Road
Stanningley
Leeds LS28 6AT, United Kingdom
+44 (0) 113 255 5665
e-mail: hk@hkeurope.com

*Australia:* Human Kinetics
57A Price Avenue
Lower Mitcham, South Australia 5062
08 8277 1555
e-mail: liahka@senet.com.au

*New Zealand:* Human Kinetics
P.O. Box 105-231, Auckland Central
09-523-3462
e-mail: hkp@ihug.co.nz

# Contents

# PREFACE

The purpose of this revised and updated fourth edition of the *Textbook of Work Physiology* is the same as that of the original text: to bring together into one volume the various factors affecting human physical performance in a manner that is comprehensible to the physiologist, the physical educator, and the clinician. Contrary to most conventional physiology textbooks, which emphasize the regulation of the various functions of the body at rest, this book emphasizes the regulatory mechanisms that occur during physical activity. Some knowledge of elementary physics and chemistry as well as human anatomy and physiology is assumed of the reader. However, some basic physiology and biochemistry are included to facilitate the understanding of some of the physiological and biochemical events encountered during work stress and physical exercise.

In this text, an attempt has been made to meet the contemporary needs of the physical education student at both the graduate and the postgraduate levels. To make it easy for readers to get to the roots as well as to the cutting edge of our present knowledge, we have included more references than is customary in most textbooks. This is, however, a very dynamic field, and inevitably new developments have taken place since the submission of our manuscript. To acknowledge some of the people who have inspired us in our work, we have also pointed out examples of classical studies throughout the book.

We are aware that the curriculum in many physical education programs may not permit as comprehensive a study of physiology as this book entails. For this reason, each chapter has been written as a fairly complete entity, relatively independent from the rest of the book. With this arrangement, and with the extensive list of references, the book may also be useful to those students who wish to concentrate more deeply on a particular field or a limited area of study. Also toward this end, we have included "For Additional Reading" boxes throughout the book, which provide readers with references to pursue a particular subject beyond these pages. Our hope is that this text is useful not only in the teaching of physical education but also in the teaching of clinical and applied physiology. We also hope that it serves to stimulate the appreciation of the role of physical education for young and old, healthy and infirm.

We wish to let the reader know that Per-Olof Åstrand, who has been such a vital force in this book's previous three editions, for various reasons has been unable to participate actively in the revision of this fourth edition. His classic research and professionalism in the field of physiology are, however, still evident in these pages.

In gathering much of the unpublished material in this book, we have collaborated with our colleagues at the College of Physical Education in Stockholm and at the Institute of Occupational Health and the Norwegian University of Sport and Physical Education in Oslo. We gratefully acknowledge their kind cooperation. We have also benefited greatly from our personal associations and frequent discussions with our many colleagues in these and other institutions. We are especially indebted to Per Brodal, Arne T. Høstmark, Svein Linge, Terje Lømo, Eric Rinvik, Nina K. Vøllestad, and Ola Wærhaug for their valuable comments and expert suggestions during our work on the various chapters. We are also indebted to Hege Underthun, Grete Eggemoen, Anne Grethe Gabrielsen, Line Arneberg, Kirsti Lome, and Unni Lund for their help in the search for and retrieval of pertinent literature. We also gratefully acknowledge the technical assistance provided by Stein Hjeltnes, Gunnar F. Lothe, Carina Knudsen, and Tove Riise during the preparation of the manuscript. In particular, we are indebted to Joan Rodahl for her invaluable help with the references. Finally, we want to thank Michael S. Bahrke and Myles Schrag for their patient support and advice in the preparation of the book.

*Note:* Previous editions of this textbook have served a global audience; they have appeared in Italian, Japanese, Portuguese, Spanish, French, and Chinese.

# Chapter 1

# Our Biological Heritage

To remind ourselves that it has taken humanity a long time to develop into the beings we are today, let us start this book by presenting a brief sketch of our evolutionary history. Close to 100% of the biological existence of our species has been dominated by outdoor activity. Hunting and foraging for food and other necessities in the wild have been a condition of human life for millions of years. We are adapted to that style of life; this applies to our emotional and social lives and our intellectual skills. After a brief spell in agrarian culture, we have ended up in urbanized, highly technological societies. There is obviously no way to revert to our natural way of life, which, by the way, was not without problems. But with insight into our biological heritage, we may yet be able to modify our current lifestyle. Knowledge of the function of the body at rest as well as during exercise under various conditions is important as a basis to optimize our existence. Many of the biological processes, which are the basis of our present physical abilities dealt with in detail in this book, are in fact thousands of millions of years old. Thus, a comprehensive textbook of biochemistry written some 1,500 million years ago would no doubt still be up-to-date in its treatment of the cell.

### For Additional Reading

For an interesting review of the consequences of the conflict between the modern, sedentary lifestyle and our biological heritage, see Booth et al. (2000).

It is assumed that our solar system was created some 4,600 million years ago. At that time the atmosphere surrounding our planet did not contain oxygen. This was a prerequisite for the evolution of life from nonliving organic matter, for without atmospheric oxygen and thus without high-altitude ozone, the ultraviolet radiation from the sun could reach the surface of the earth. This radiation could then provide the energy for the photosynthesis of organic compounds from molecules such as water, carbon dioxide, and ammonia. The photosynthetic process that enabled living organisms to capture solar energy for the synthesis of organic molecules such as glucose can be clearly traced in fossils dated about 3,500 million years old. Anaerobic glycolysis is probably the oldest energy-extracting pathway found in life on earth.

Ancient organisms split water by photosynthesis and gradually released free oxygen into the atmosphere. It may have taken 2,000 million years to create an atmosphere in which one out of every five molecules was oxygen. Oxygen became toxic to many of the original oxygen producers, and new metabolic patterns developed, namely aerobic energy yield, which uses oxygen as a hydrogen acceptor. Another result of the production of oxygen was the formation of an ozone layer in the upper atmosphere. Nonbiological synthesis of organic matter ceased, because ultraviolet radiation now had to pass through the ozone layer in order to reach the surface of our planet, and it thus was markedly reduced. However, solar energy still manages to reach the earth in the visible wave lengths, and this energy can be used for biological photosynthesis.

## Development of Primitive Organisms

A new milestone in biological evolution occurred about 1,500 million years ago, when the unicellular organism with a nucleus (the eukaryote) developed (Vidal 1984). These primitive organisms, representing the simplest form of aerobic life, incorporated fundamental functions such as metabolism, excitability, locomotion, and reproduction, based on very elaborate and complicated biological phenomena. From this point of view, it is hardly fair to call them primitive. Thus, the energy-absorbing and energy-yielding processes typical of present cellular activity—such as the adenosine triphosphate–adenosine diphosphate system, the primary mode of transportation for chemical energy in every cellular reaction—are merely repetitions of events occurring thousands of millions of years ago (Schopf 1978).

Actually, **adenosine triphosphate (ATP)** is the principal medium for the storage and exchange of energy in almost all living organisms. A large amount of energy is released when ATP is hydrolyzed into **adenosine diphosphate (ADP)** and a phosphate ion. However, because it is a heavy fuel, the supply of ATP is very limited. Depending on how physically active an individual is, within 24 h she or he could expend an amount of energy equivalent to 50% to 100% more than her or his own body weight in ATP. Therefore, a very rapid resynthesis of ATP is essential; the *anaerobic processes,* several millions of years old, are supplemented by the *aerobic* energy yield taking place inside the mitochondria.

Calcium ions play a key role in the regulation of many processes in the body, including the activation of heart and skeletal muscles. A special protein (calmodulin) serves as an intracellular calcium receptor and mediates the calcium-regulatory functions. This protein is structurally conserved and functionally preserved throughout the plant and animal kingdoms. This is yet another example of how a mechanism that developed thousands of millions of years ago has proved to be very efficient and has survived the test of time.

A cell—microscopic in size, from a few micrometers up to a few millimeters—is able to transport nutrients, waste products, electrolytes, and dissolved gases both intra- and extracellularly using simple physical forces. Diffusion and osmosis (differences in cellular concentrations) are the main driving forces. In addition, energy-consuming biological processes assist in this exchange of matter and molecules.

Thus, it may be summarized that during thousands of millions of years of evolution, a unicellular living organism evolved. Through trial and error, the fundamental biological principles for maintaining life were developed; these processes are still in efficient operation.

## Appearance of Mammals

Evolution was then ready for the next major step, the development of larger animals, probably beginning some 700 million years ago (Valentine 1978). In the evolution of larger organisms it was impossible simply for the cell to increase in size, which might jeopardize its supply of oxygen and fuel. The living cell needs oxygen for its metabolism. As long as oxygen is available in the environment, it will diffuse toward the place of metabolism: the mitochondria. The distance that the oxygen molecules have to travel and the difference in oxygen tension between the solution extra- and intracellularly determine the rate of diffusion. It has been calculated that a hypothetical cell with a 10-mm radius and a reasonable metabolic rate would need an external oxygen pressure of 25 times the barometric pressure at sea level in order to secure an oxygen supply to the center of the cell by diffusion alone (Krogh 1941). This, of course, is out of the question. Furthermore, the available oxygen pressure in ambient air is only about one fifth of the barometric pressure.

Similarly, the transport of fuel into the cell by diffusion also limits the size of the individual cell. Suppose one million sugar molecules were placed on the bottom of a cylinder filled with water. After 1 h, half of them would have traveled 1 mm or more by diffusion but only 20 molecules would have covered a distance of 7 mm. It would take about 100 years to obtain an even concentration of the sugar molecules up to a level of 1 m. In other words, diffusion is an efficient transport mechanism over short distances but inefficient over long distances.

Consequently, in the evolution of larger animals, the individual cell retained its original size, that is, the same size as the unicellular organism living more than 1,000 million years ago. However, more of these single cells were piled together as a means of increasing the size of the organism. Special structures in the cell, the genes, were encoded with detailed instructions about the proliferation of the cell mass, guiding specialization in shape, structure, and function. Human beings have about 200 different types of cells. Certain cells aggregate to form tissues and organs with a relatively homogeneous composition. Some tissues became specialized for support (bone, cartilage, connective tissue), others developed the potential for motion (muscles), and still others specialized in dealing with excitability and conduction of information (sensory cells, nerve cells). All living cells have a certain metabolic capability, but certain groups of cells are capable of handling specific metabolic tasks, as in the case of liver cells and gastrointestinal tract cells. In order for the organism to survive as a whole, it is essential that its individual cells collaborate according to the principle of one for all and all for one.

As an inevitable consequence of piling thousands of millions of some 200 different types of cells together in one organism, the individual cell lost intimate contact with the external environment. Furthermore, throughout the course of evolution, for some organisms this environment changed from water to air. Transporting adequate supplies of building materials, fuel, and oxygen to each cell and removing waste products are two of the greatest challenges facing an organism as the number of cells increase. Both problems are solved by bathing each cell in water, that is, the interstitial fluid. Like the amoeba, each cell in our bodies (with some exceptions) is surrounded by fluid, basically similar in composition to that of the ancient oceans.

The organism brought the sea water with it, so to speak, in a bag made of skin. The distance between the cell interior and its external environment is so small that gases and substances are easily transferred. For optimal function, the cell needs an environment that is as stable as possible. Thus, the composition of the cell-bathing fluid must be kept fairly constant and prevented from large fluctuations. Its content of organic compounds such as fatty acids, glucose, hormones, and enzymes and of inorganic substances such as sodium, potassium, and calcium exerts a vital influence on the cell. The continuous supply of oxygen and removal of carbon dioxide are crucial. The tolerance for an increase in hydrogen ions, that is, a decrease in pH, during heavy exercise (pH less than 7 in the arterial blood) is remarkable, but there is a limit. The tolerance for changes in body temperature is also limited. All warm-blooded animals live their lives only a few degrees away from their thermal death points.

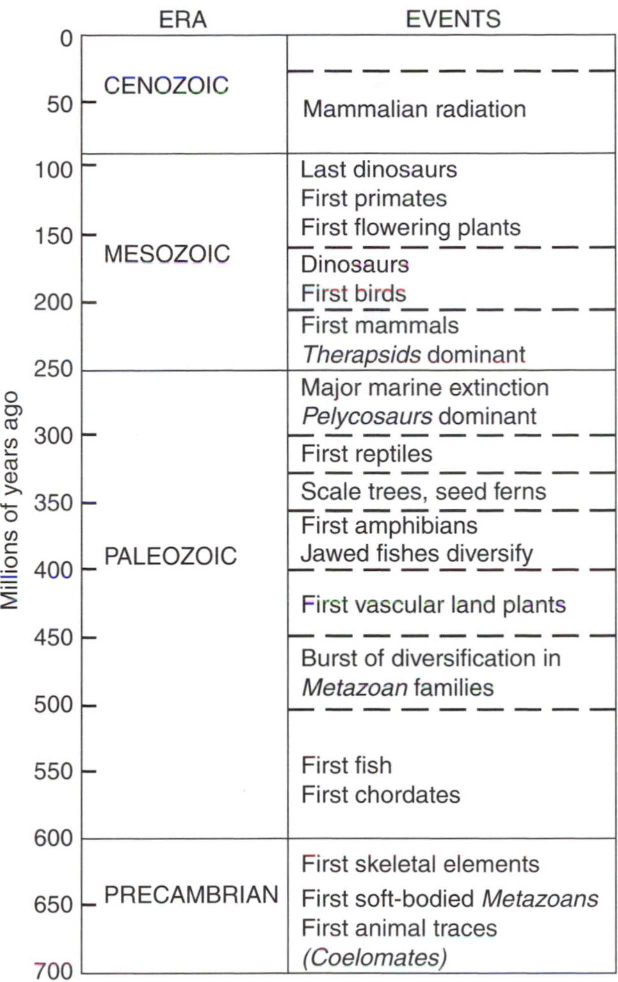


**Figure 1.1** Major events in the evolution of multicellular organisms.

Adapted from Valentine 1978.

## Differentiation of Life

In the course of diversification of multicellular organisms that took place over the last 700 million years, new types of organisms appeared and divergence occurred within already established groups. The earliest fossil traces of animal life are burrows that begin to appear in rocks younger than 700 million years (Valentine 1978). Figure 1.1 depicts the major events in the evolution of multicellular organisms. It should be noted that the history of the mammals covers the last 220 million years, if not more. The first primates (the order including human beings) can be traced to 60 to 70 million years ago, to a period when the dinosaurs still dominated the scene. With the extinction of the dinosaurs, there was a mammalian radiation into newly available niches. Sixty to 70 million years ago, an evolutionary explosion occurred with a radiation of flowering plants, birds, and mammals.

What then are the mechanisms that underlie the origin of species and the evolutionary relationships between them, that is, Darwinism? Lewin (1980) has summarized the views held by different researchers in this field. According to the modern synthesis, evolution is a consequence of the gradual accumulation of genetic differences due to point mutations and rearrangements in the chromosomes. The direction that an evolutionary change takes is then determined by natural selection, promoting those variants that are best adapted to their environment. However, the fact remains that, on the whole, the fossils do not document a smooth transition from old morphologies to new ones; this was also discussed by Darwin. For millions of years species remain unchanged in the fossil record, suddenly to

## Classic Study

Charles Darwin (1809–82) was the promulgator of *Darwinism,* the theory of the origin and perpetuation of new species of animals and plants, which holds that organisms tend to produce offspring varying slightly from their parents. This theory suggests that the process of natural selection tends to favor the survival of individuals whose peculiarities best adapt them to their environment and not only that new species have been and may still be produced chiefly by the continued operation of these factors, but that organisms of widely differing groups may have arisen from common ancestors.

be replaced by something that is substantially different but clearly related (Lewin 1980). It is indeed conceivable, however, that future fossil discoveries may fill many of the gaps and provide some of the missing links.

# Emergence of Primates

Maybe only 10 million or as much as 20 million years ago, the family tree of primates developed a branch called the hominids, which finally resulted in *Homo sapiens sapiens,* the only surviving hominid. Another branch led to the development of the anthropoid apes: the orangutan, gorilla, and chimpanzee. According to evolutionary theory, the variants that survive are those best adapted to their particular environment. These individuals reached adulthood and produced viable offspring. The weaker variants failed to survive. In the past million years, there were periods of tropical climate in large areas, as well as four glacial periods.

Somewhere along the line a prototypic anthropoid ape abandoned life in the trees and started to forage and hunt on the ground. A species related to these creatures may have been *Ramapithecus.* According to some experts this transition occurred some 14 million years ago; others believe it happened several million years later. Humankind thereby initiated a bipedal adaptation to terrestrial life, first at the margins of forests and then gradually out on the savanna, forming bands of hunters and gatherers. Valentine (1978) suggested that the final rise of the human species was associated with a further shift toward big game hunting, increasing the value of cunning, intelligence, and cooperation.

There is a puzzling lack of fossil hominids for a long period of time after *Ramapithecus.* Not until about four million years ago do the African fossils reveal the presence of the hominid genus *Australopithecus.* The pelvis permitted upright posture with bipedal gait and free arms. Its brain size was 450 to 550 $cm^3$, which is the same size as a gorilla brain. The body height was 110 to 120 cm. Archaeological records of tools—pebble "choppers" and small stones—are probably more than three million years old (Lewin 1981). Thus, toolmaking was established before there was a marked brain expansion in the hominid stock. Although a few species of *Australopithecus* have been identified, it was a relatively homogeneous genus that survived for more than two million years. The last representatives of the Australopithecines became extinct less than two million years ago. A few years ago a 3.6-million-year-old complete skeleton of a 1.2-m-tall hominid, *Australopithecus africanus,* was found at the outskirts of Johannesburg, South Africa.

Another member of the hominid family tree has been named *Homo habilis* by some authorities and existed from 2.3 to 1.5 million years ago, with a brain volume ranging from 600 to 800 $cm^3$. The first well-identified true human in the fossil record is *Homo erectus.* By this time the brain had nearly doubled in size, to an average of 1,050 $cm^3$. "In the struggle for survival through technology, selection for bigger and more efficient brains would seem to have been at work. For tools to be greatly improved and diversified, the appropriate brain capacity had to evolve" (Weiner 1971). *Homo erectus* had a truly modern pelvis and moved with a striding gait. These hominids lived as hunters and gatherers with a wide geographic range. Their body height was probably 150 to 160 cm. They made use of fire, as evidenced by a hominid occupation site 1.4 million years old (Gowlett et al. 1981).

The general public is probably most familiar with the Neanderthal *(Homo sapiens neanderthalensis),* who, from archaeological findings, appears to have been well established some 200,000 years ago (Stringer 1990). Neanderthals were skilled hunters of both large and small game, forming bands similar to those of more recent hunting people, and were probably linked into tribal groupings, or at least into groups with a common language. They formed a human population complex extending from Gibraltar across Europe into East Asia. The Neanderthal population was as homogeneous as the human population is today. The brain encased in the Neanderthal skull, however, was on the average slightly larger than the brain of modern humans. According to Trinkhaus and Howells (1979), this anatomical feature is undoubtedly related to the fact that the musculature of the Neanderthals was more substantial than that of modern humans. Neanderthals had apparently the same postural abilities, manual dexterity, and range and character of movement that are typical for modern humans. But they had more massive limb bones and more muscular mass and power. They had broad shoulders; a flat, protruding forehead; and a broad nose. A number of caves that had been occupied by the Neanderthals have been excavated in Gibraltar. So far some 30 skeletons of Neanderthals have been found. The departure of the Neanderthals occurred some 35,000 years ago; when they disappeared from the scene, anatomically modern humans, *Homo sapiens sapiens,* were already in existence. No one knows why humans of the modern type came to the fore and the Neanderthals disappeared (Stringer 1990). One hypothesis is that modern humans evolved in Africa and then spread throughout the world, developing racial features in the process. Modern humans and Neanderthals could be distinct lines that diverged from a common ancestor more than 200,000 years ago in Africa and Europe, respectively. At a later stage they spread, and in some parts of the world they shared the environment.

An alternative hypothesis is a "gene-flow" model: The genetic contribution varied from region to region, and the rate of intermixture gradually increased as modern humans evolved. Stringer (1990) pointed out that in the gene-flow model racial features preceded the appearance of modern humans, whereas the African model reverses the order. He supports the African model with dispersal of early modern humans from Africa within the past 100,000 years. However, the dating of our origin as modern humans is controversial.

## Modern Human Evolution

Most likely, a human being living 50,000 years ago had the same potential for physical and intellectual performance, such as playing a piano or constructing a computer, as anyone living today. As mentioned, the Neanderthals had larger brains and greater muscle mass than modern humans, but these findings do not suggest any difference in intellectual or behavioral capacities (Trinkhaus and Howells 1979). From all indications, *Homo sapiens sapiens* has remained biologically unchanged during at least the last 50,000 years. By 30,000 years ago, modern humans had spread to nearly all parts of the world. It was not until some 10,000 years ago that the transition from a roaming hunter and gatherer to a stationary farmer began.

To illustrate the evolutionary time scale, let us compare the 4,600 million years our planet has existed with a 460-km journey (figure 1.2). Life began after the first 100 km of the trip had been covered. It took another 200 km before the unicellular organism with a nucleus was born. Multicellular animals were living by the 400-km mark. An evolutionary radiation of the mammalian stock started somewhere around the 453-km milepost. The first hominid probably appeared approximately 6 km further on. *Australopithecus* joined the journey about 400 to 200 m from the end, and the Neanderthals disappeared just about 3.5 m from the finishing line,

OUR BIOLOGICAL EVOLUTION

| Start | | | | |
|---|---|---|---|---|
| 0 | km | The earth is created | 4,600,000,000 | years ago |
| 110 | km | Biochemical processes are developed that can trap solar energy | 3,500,000,000 | years ago |
| 310 | km | Single-celled organisms with a nucleus appear | 1,500,000,000 | years ago |
| 390 | km | Multicellular organisms appear | 700,000,000 | years ago |
| 453 | km | Modern mammals appear | 70,000,000 | years ago |
| 459 | km | Hominid branch is being developed | 10,000,000 | years ago |
| 459.6 | km | *Australopithecus* appears | 4,000,000 | years ago |
| 459.99 | km | Neanderthal man appears | 100,000 | years ago |
| 459.999 | km | Agriculture is introduced | 10,000 | years ago |
| 459.99999 | km | Present 100-year-old man is born | 100 | years ago |

**Figure 1.2** We can compare the 4,600 million years our planet has existed with a 460-km journey. Life began after the first 100 km of the trip had been covered. A person 100 years old today has covered a distance of merely 10 mm.

where they were replaced by modern humans. The cultivation of land and keeping of livestock occurred 1 m from our present position. A person 100 years old today has covered a distance of merely 10 mm of the entire 460-km journey.

The purpose of this very brief summary of an ongoing evolution is to provide an outline of our genetic background. Many structures and functions are common to different species in the animal kingdom. For instance, there appears to be no fundamental difference in structure, chemistry, or function between the neurons and synapses of a human and those of a squid, a snail, or a leech (Kandel 1979). We can therefore learn a great deal from studying different species. It is remarkable that all living organisms have a genetic code based on the same principle. Data indicate, for instance, that humans and chimpanzees share more than 98% of their genetic material (Washburn 1978). However, minimal genetic changes can effect major morphological modification. Consequently, one should be careful when extrapolating findings from one species to another, including humans. Over millions of years many species underwent minor or major modifications in physical and other characteristics. In general, however, evolution is a very conservative process. All vertebrates, including the hominids, have backbones. Backbones are complicated in design, yet they are quite similar in all animals that have them. This finding supports the hypothesis that backbones have evolved only once. In other words, it appears that all vertebrates share a common ancestor with a backbone. At an early stage the human embryo starts to develop gills, even though it is going to breathe with lungs.

Other examples of structures and functions shared by many species can be taken from the mammals: All mammals have three separate bones in the middle ear; the females have milk-producing glands, although the composition of the milk varies markedly among different species.

Among vertebrates, locomotion is generally genetically programmed. Fish are able to swim as soon as they are born; birds may walk as soon as they are hatched. Many species of mammals are well developed at birth. Some are even able to run as soon as they are born, and some of them may attain a speed of up to 35 km $\cdot$ $h^{-1}$ when they are only a few days old. Ultimately their survival may depend on their ability to get away. In the case of human beings, who are utterly helpless at birth and entirely dependent on parental care, it may be to their advantage not to be able to remove themselves very far from their parents until they are mature enough to stand on their own two feet.

The evolutionary process continues, and more recent mammalian history has seen a wave of extinction, particularly severe for large mammals, including the hominids. Extinctions are a measure of the success of evolution in adapting organisms for particular environmental conditions. New forms have their chance when their adaptations provide entry into a relatively empty niche. In the balance between existence and extinction, the odds are not too favorable: It has been estimated that 2,000 million species have appeared on earth during the last 700 million years, but the number of multicellular species now living is in the order of two million; that is, only 0.1% have survived.

The cortex of the human brain mirrors humans' evolutionary success. Just as the proportions of the human hand—with its large, opposable, and muscular thumb—reflect a successful arboreal and later tool-using adaptation, so does the anatomy of the human brain reflect a successful adaptation for manual and intellectual skills. Adults of most vertebrate species allocate 2% to 8% of their basal metabolism for maintenance of the central nervous system (CNS; Mink, Blumenschine, and Adams 1981). These authors hypothesize that "an optimal functional relationship between the energy requirements of an animal's executor system (muscle metabolism) and its control system (CNS metabolism) was established early in vertebrate evolution." One important exception is *Homo sapiens*, whose CNS consumes 20% of its basal metabolism.

Just as upright walking and toolmaking were the unique adaptations of early phases of human evolution, the physiological capacity for speech was the biological basis for the later stages. Indeed, it is through language that human social systems are mediated. Speech is the form of behavior that differentiates humans from other animals more than any other behavior. It is the passing down of knowledge and experience by language from one generation to the next that has enabled humans, biologically unchanged for tens of thousands of years, to accelerate progress so dramatically and to apply their endowed intellectual resource in a technical revolution leading to entirely new and complex tools, weapons, shelters, boats, wheeled locomotion, exploratory voyages, and the attainment of the seemingly impossible: landing on the moon. And yet, in the midst of these splendid achievements, there are those who now wonder whether the evolution of the human brain has gone too far. While its ability to conceive, invent, create, and construct is astonishing, it remains to be seen whether the human brain has retained or developed equally well its capability for ethical conduct or the responsible application of its endowed potential. At the time when our ancestors roamed around in small bands, any destructive consequence of their activity was quite limited. But today, because of social developments and technical innovations, basically the same brain is capable of turning the human being into a self-destructive monster.

## The Mobile Human

Humans, like all higher animals, are basically designed for mobility. Consequently, our locomotive apparatus and service organs constitute the majority of our total body mass. The shape and dimensions of the human skeleton and musculature are such that the human body cannot compete with a gazelle in speed or with an elephant in sturdiness, but in diversity human beings are indeed outstanding.

The basic instrument of mobility is the muscle. It is a very old tissue. As already mentioned, the earliest animal fossils were the burrowers living some 700 million years ago. Evidently, using muscle force, these animals could dig in the sea bed. Muscles have retained the metabolic pathways developed when the air had no oxygen, that is, the anaerobic energy yield. The pyruvic acid formed in our muscles under anaerobic conditions is removed by the formation of **lactic acid.** One old-fashioned alternative could have been the transformation of the pyruvate into ethyl alcohol. There may be those among us who regret that skeletal muscles did not select this alternative route; if they had, producing pyruvate by exercising to exhaustion or running uphill might have been a very popular endeavor!

Skeletal muscle is unique in that it can vary its metabolic rate to a greater degree than any other tissue. In fact, active skeletal muscles can increase their oxidative processes to more than 50 times the resting level. Such an enormous variation in metabolic rate must necessarily create serious problems for the muscle cell because when the consumption of fuel and oxygen increases 50-fold, the rate of removal of heat, carbon dioxide, water, and waste products must be similarly increased. To maintain the chemical and physical equilibrium of the cell, there must also be a tremendous increase in the exchange of molecules between intra- and extracellular fluid; fresh fluid must continuously flush the exercising cell. When muscles are thrown into vigorous activity, the ability to maintain the internal equilibria necessary to continue the exercise entirely depends on those organs that service the muscles. This dependence is especially true in the case of respiration and circulation, but certainly food intake, digestion and handling of substrates, kidney function, and water balance are also affected by variations in the metabolic rate.

## Summary

The purpose of this brief review of our biological heritage is to provide some basic facts that may be useful in understanding the complicated interplay between all the biological processes that form the basis for our existence and performance.

Humans are made to be physically active. Almost all of the biological existence of our species has been based on outdoor activity, but we have ended up in an urbanized, highly technological society. With insight into our biological heritage, we may yet be able to modify our current lifestyle. Knowledge of the function of our bodies is important for optimizing our existence.

### For Additional Reading

For a review of human evolution, the British Museum publication *Man's Place in Evolution* (1980) is recommended.

### For Additional Viewing

An excellent review of the evolution of primates is presented in the video *Evolution: Primates* (1997). A fascinating picture of the genetic role in organic development is presented in the video *The Hopeful Monsters* (1997).

# Chapter 2

# The Cell and Its Regulatory Mechanisms

The last decades of the 20th century brought us a great deal of new information about the cell and its regulatory mechanisms. Many of the "black boxes" have been opened and their contents revealed, some of them in great detail. The different classes of receptors for intercellular signaling have become part of most biological curricula, and new knowledge about cells' **downstream** signaling pathways helps us understand how cells react in health and disease.

### For Additional Reading

For a personal account of the discovery of the structure of DNA, see Watson (1968).

## The Cell

Among the major advances of the 20th century were the discoveries of the structure of DNA and the genetic code. At the turn of the century, the entire human genome has been revealed, and we are about to enter what has been named the "era of proteomics," that is, the era when our scientific focus will be on how the about 30,000 different genes in our genome affect our lives through the proteins they encode. It seems safe to foresee that our acquisition of knowledge in the area of proteomics will be much faster than the almost 50 years that passed between the initial descriptions of the double helix and the sequencing of the human genome, thus improving our understanding of how our bodies adapt to different challenges.

The aim of this chapter is to give a short overview of the basic structure of a mammalian cell and its possible strategies for adapting to new challenges.

### For Additional Reading

A more detailed account of cell biology is found in Alberts et al. (1998).

Let us begin by summarizing the general design of a mammalian cell (figure 2.1, a and b). Every cell in the human body is completely surrounded by a cell membrane, irrespective of the size or shape of the cell. Sometimes the cell membrane is called a plasma membrane, plasmalemma, or, in the case of a muscle cell, a sarcolemma (see chapter 3). The cell interior consists of organelles, small bodies of different sizes and shapes, suspended in a viscous fluid called **cytosol.** The different structures in the cell serve different functions, which we outline in the following sections, starting with the cell membrane.

### Classic Studies

By unveiling the structure of DNA, two classic papers by Watson and Crick (1953a, 1953b) prepared the way for the development of modern cell biology. In addition to showing the three-dimensional structure of the double DNA helix, they also introduced the law of base-pairing and suggested how DNA could replicate itself faithfully. In 1962, Watson and Crick were awarded a Nobel Prize for this work.

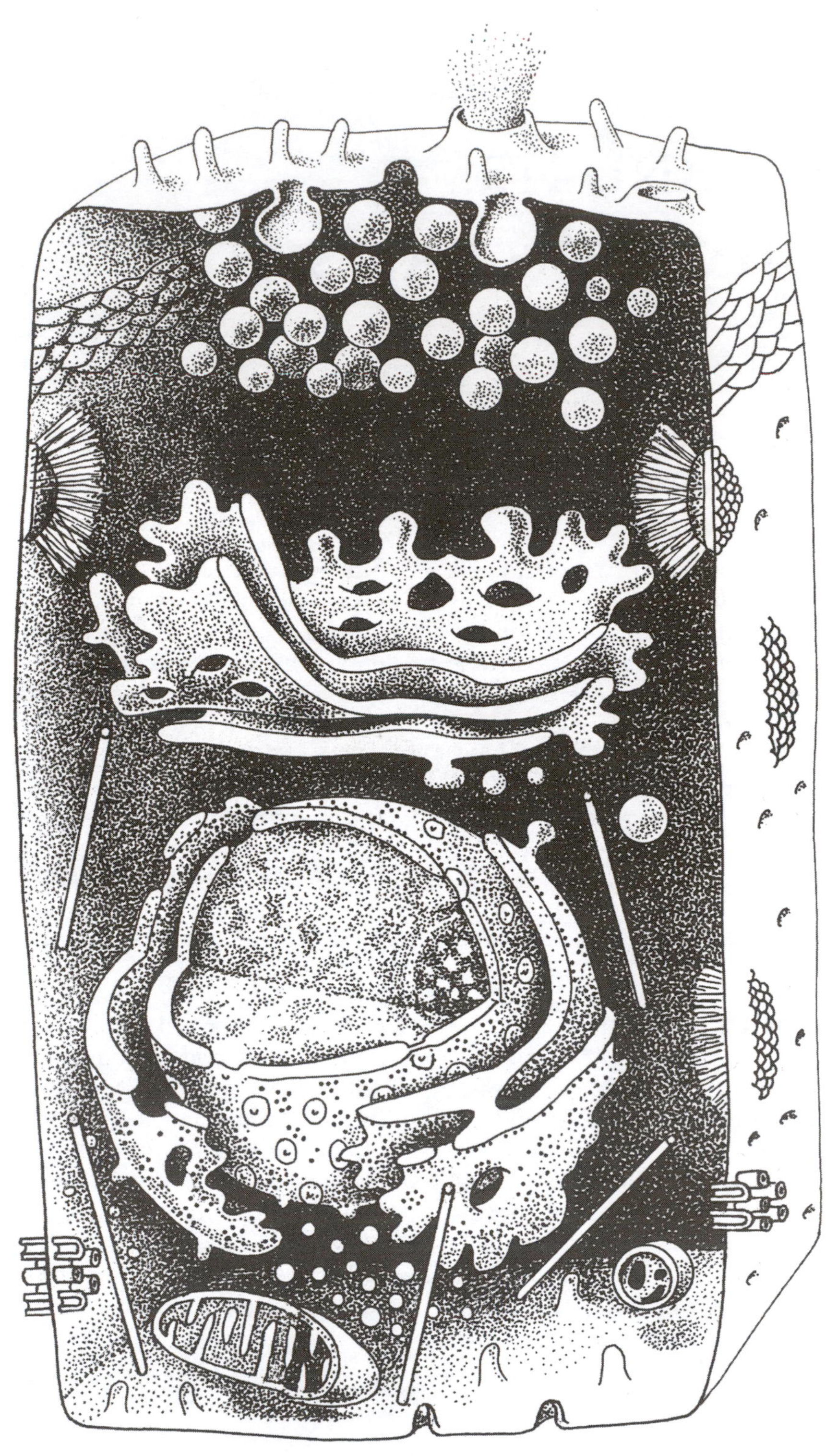

**Figure 2.1** *(a)* Simplified three-dimensional picture of a columnar epithelial cell. Although the shape of a cell and its content of organelles may vary within wide limits, this specialized cell can serve as an example of the basic design of a mammalian cell.

Reprinted, by permission, from P. Brodal, H.A. Dahl, and S. Fossum, 1990, *Menneskets anatomi og fysiologi* (Oslo, Norway: J.W. Cappelens Forlag).

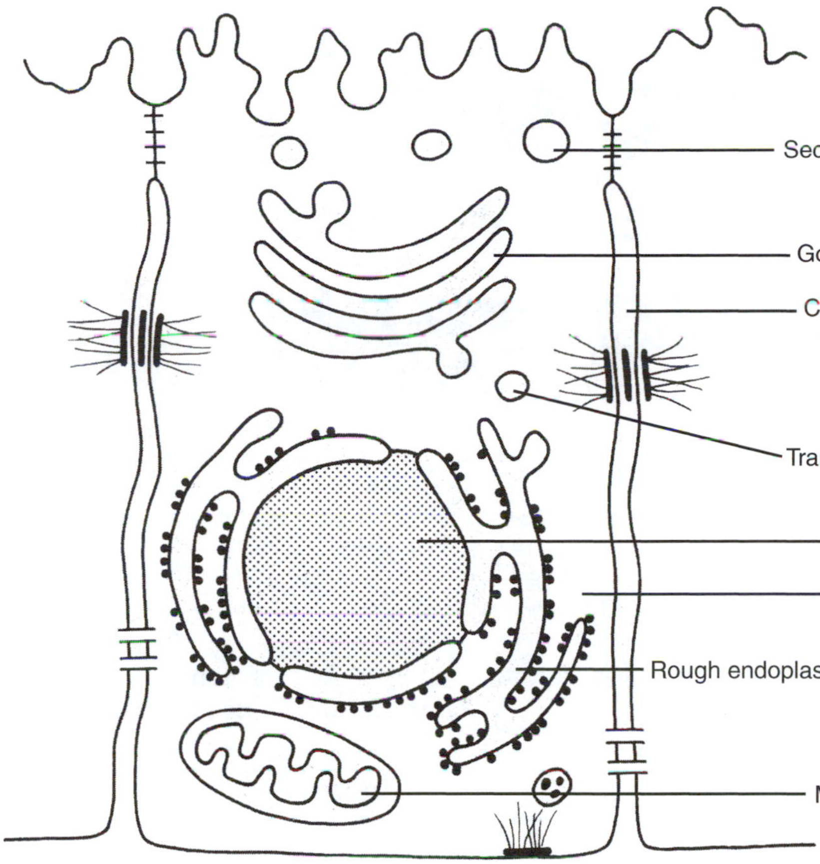

**Figure 2.1** *(b)* The same cell as it would appear in a microscopic section.

Reprinted, by permission, from P. Brodal, H.A. Dahl, and S. Fossum, 1990, *Menneskets anatomi og fysiologi* (Oslo, Norway: J.W. Cappelens Forlag).

## The Cell Membrane

The structural framework of the cell membrane, approximately 5 nm thick, is a double layer of lipid molecules. The individual lipid molecule has a head and two tails. **Polar** (i.e., **hydrophilic,** or soluble in water) heads form the outer and inner membrane surfaces, while **apolar** (i.e., **hydrophobic,** or insoluble in water) tails meet in the membrane interior. This structure serves as an anchor for other components of the membrane, such as proteins and glycoproteins (figure 2.2). Actually, what make one cell membrane different from another are the specific proteins attached to or built into the membrane in one way or another. The membrane proteins serve different purposes: as pumps, channels, receptors, enzymes, or structural proteins. Individual membrane proteins may serve more than one function; for example, a protein can be at the same time a receptor and a channel or an enzyme and a pump. As receptors, membrane proteins can transfer information and instructions from the environment into the cell, using hormones or other signal molecules as triggers. Some molecules, such as oxygen, carbon dioxide ($CO_2$), nitric oxide (NO), and molecules of a lipid nature, can diffuse through the membrane, but for water-soluble molecules, the lipid core of the membrane forms a barrier to molecular **diffusion.** Accordingly, most biologically important molecules and ions have specific transport mechanisms for their flux between the environment and the cell interior (figure 2.3).

Many of the lipids in mammalian cell membranes are unsaturated and liquid at body temperature. Therefore, the membrane has the consistency of light oil and forms a sheetlike structure, much like the wall of a soap bubble. To a certain extent, this makes the different lipid and protein molecules free to move about within the membrane but still maintains the functional diversity of different membrane domains (see chapter 3).

Protein molecules may lie close to either membrane surface, or they may penetrate the membrane completely. In either case, part of the protein molecule may stick out from the inner or outer surface of the membrane. In the case of receptor proteins,

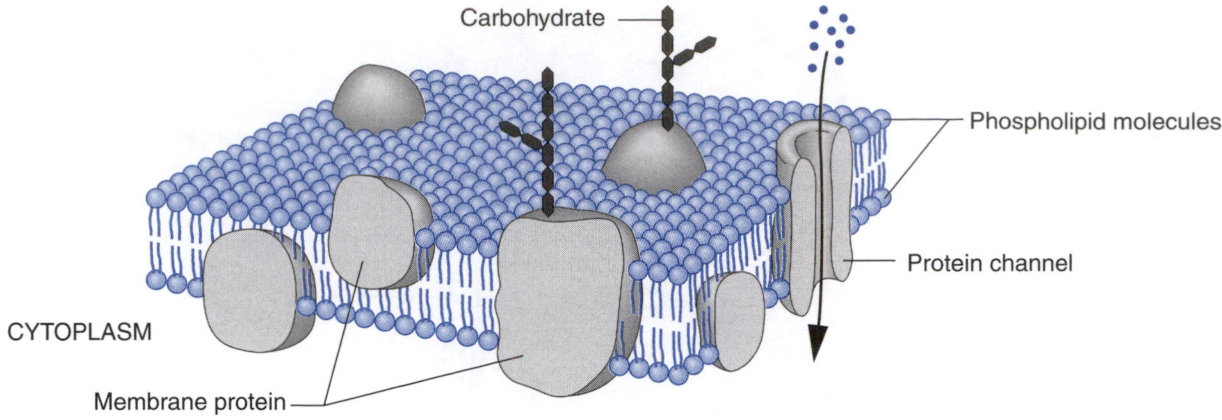


**Figure 2.2** Schematic drawing of a cell membrane. Its framework consists of a double layer of phospholipid molecules, the hydrophilic (polar) heads of which face the outer and inner surfaces of the membrane, while their hydrophobic (apolar) tails are hidden in the middle of the cell membrane. Embedded in the lipid bilayer are proteins and glycoproteins (proteins with carbohydrates attached) serving different functions.

Adapted, by permission, from H.A. Dahl and E. Rinvik, 1999, *Menneskets funksjonelle anatomi* (Oslo, Norway: J.W. Cappelens Forlag).

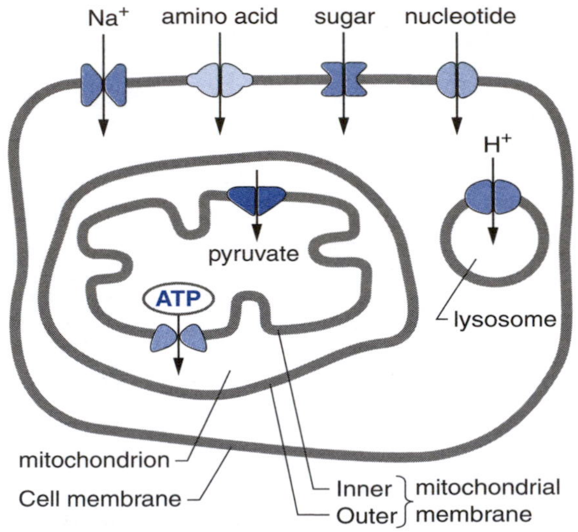


**Figure 2.3** Some examples of substances transported across various types of cell membranes—surface or intracellular membranes—by carrier proteins. Each type of cell membrane has its own characteristic set of carrier proteins, and each carrier transports one particular kind of molecule or ion.



the part of the molecule protruding from the outside contains a docking site for its specific signal molecule, while the part protruding inside may serve as a transducer that generates a new signal to the interior of the cell.

Because the cell membrane is able to control almost all transport between a cell and its surroundings, including the transfer of intercellular signals, its integrity is obviously essential both for a cell's survival and for its proper function.

## Transport Mechanisms Through the Cell Membrane

Transport through cell membranes can be active or passive, depending on whether it consumes energy or not (figure 2.4). The energy necessary to drive **active transport** comes from the hydrolysis of adenosine triphosphate (ATP) to adenosine diphosphate (ADP) and inorganic phosphate. **Passive transport** is driven by differences in concentration or electric charge, together named the **electrochemical gradient,** and can be simple diffusion or **facilitated diffusion.**

Simple diffusion means that the molecules in question diffuse through the membrane like ghosts through the wall. This is possible only for simple molecules such as oxygen, $CO_2$, and NO and for molecules that are lipid in nature, notably the steroid hormones. In facilitated diffusion the molecules have to pass through channels formed by protein molecules (channel-mediated passive transport) or by means of carrier-mediated passive transport. Such protein channels and carriers are highly specific and admit only one type of molecule or one that is closely related to it. Examples are ion channels; glucose transporters (**GLUTs;** Bell et al. 1990); **monocarboxylate transporters (MCTs),** which transport lactate and so are also known as lactate transporters (C.K. Garcia et al. 1994); and water channels, also known as **aquaporins** (Beitz 1999). Each of

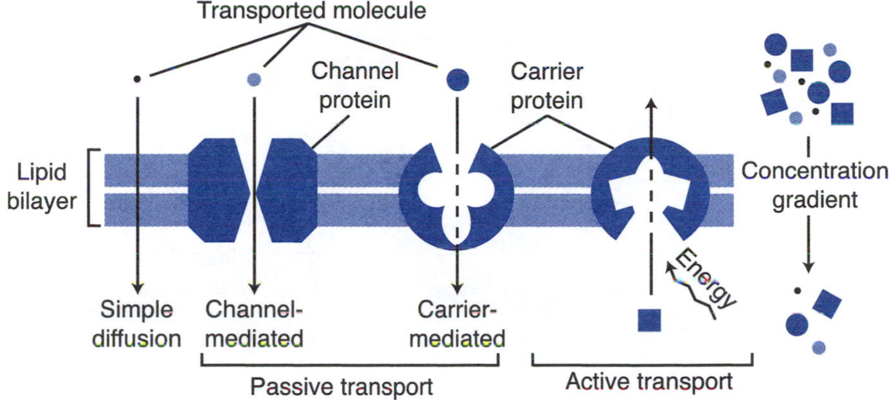


**Figure 2.4** Different modes of transport through a cell membrane. If uncharged molecules are small enough, they can move down the concentration gradient directly across the lipid bilayer by simple diffusion. Most solutes, however, can cross the membrane only if there is a carrier protein or a channel protein for that particular solute. Passive transport of a molecule in the same direction as its concentration gradient occurs spontaneously. Transport against a concentration gradient (active transport), on the other hand, requires an input of energy. Only carrier proteins can carry out active transport, but both carrier proteins and channel proteins can carry out passive transport.



these can exist in different isoforms, depending on tissue and cell type.

## For Additional Reading

For an in-depth review of aquaporin water channels, see Agre et al. (2002).

Active transport also takes place through protein channels, but in addition to permitting passage, these channel proteins have the ability to hydrolyze ATP to provide the energy necessary to transport the molecules against the electrochemical gradient for that particular type of molecule. In other words, the protein serves as an enzyme, more specifically, as an ATP-splitting enzyme, or **ATPase.** In line with this nomenclature, proteins that transport sodium ions ($Na^+$) and potassium ions ($K^+$) are called **$Na^+$–$K^+$ ATPases** (sodium–potassium ATPases), and those that transport calcium ions ($Ca^{2+}$) are called **$Ca^{2+}$ ATPases** (calcium ATPases). In many cases the same transport molecule transports two different ions or molecules at the same time, either in the same direction, in which case the transporter is called a **symport,** or in the opposite direction, an **antiport.** This can be seen as an energy-saving mechanism, much like a truck carrying two types of cargo at the same time or two people passing through the same revolving door in opposite directions. In the case of the $Na^+$–$K^+$ ATPase, the two ions are transported in opposite directions, potassium into the cell and sodium out of it, thus creating a concentration gradient for these two ions between the cytosol and the surrounding tissue fluid. This concentration gradient is the very reason for the existence of a membrane potential, so vital for the electrical activities of nerve and muscle cells. We shall return to the membrane potential and its importance in excitable cells in chapter 3.

Even larger particles or molecules that do not seem to have specific transport mechanisms of their own may pass into or out of a cell. Such material is transported concealed in vesicles, much like in a Trojan horse. This transport mechanism is called **endocytosis** when the direction is into the cell and **exocytosis** when the direction is out of the cell (figure 2.5*a*). In cells that specialize in endocytosis, such as white blood cells, the process is also known as **phagocytosis** or cellular eating (literally "cell swallowing"). When fluid or larger molecules are endocytosed, the process is called **pinocytosis.** Such vesicle transport is, however, not only a transport into or out of the cell. It is part of an extensive transport system both between the different compartments inside the cell (endoplasmic reticulum, **Golgi apparatus**) and between the cell interior and the surrounding tissue fluid. A survey of intracellular trafficking can be found in Alberts et al. (1994, 1998).

At first sight, endocytic processes may seem unspecific, like a swimmer swallowing a mouthful of water irrespective of what it may contain, but that is not always the case. In a process called **receptor-mediated endocytosis,** specific macromolecules in the tissue fluid are captured by receptors on the cell surface (figure 2.5*b*). When loaded with their cargo,

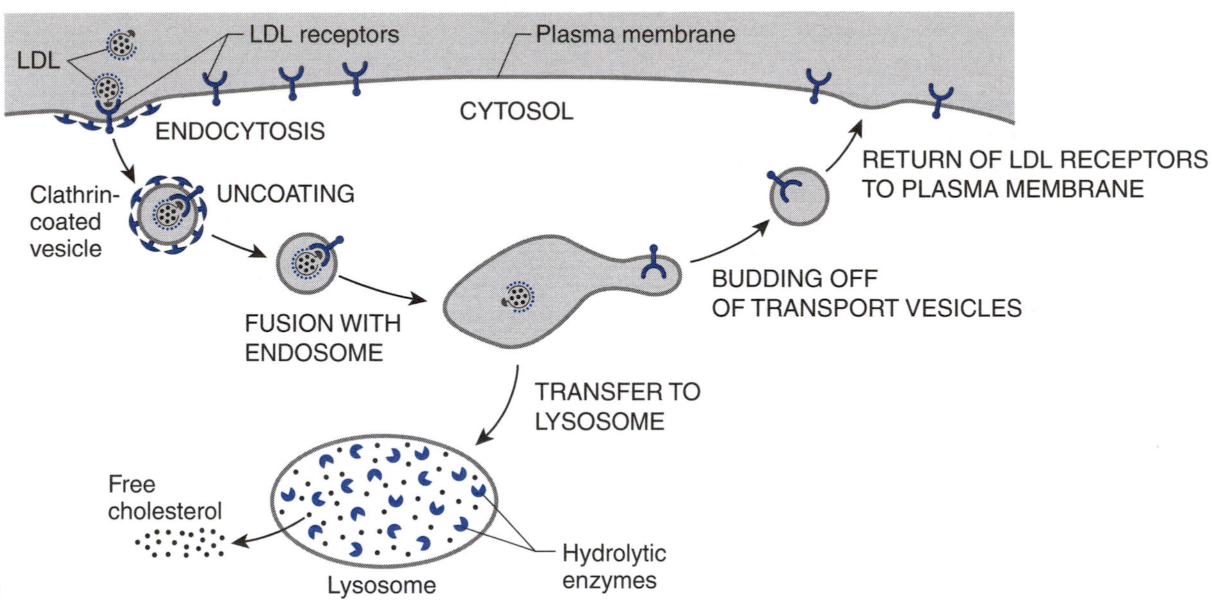


**Figure 2.5** *(a)* The processes by which cells import and export molecules that are unable to pass through the plasma membrane by means of a carrier protein or a channel protein are called endocytosis and exocytosis, respectively. In endocytosis, vesicles containing extracellular material are pinched off from the plasma membrane. Inside the cell these vesicles fuse with other intracellular vesicles, where their content is degraded or processed. In exocytosis, the material to be exported is segregated in vesicles, which ultimately fuse with the cell membrane and release their content to the exterior. *(b)* In receptor-mediated endocytosis, the material to be imported (in this case, low-density lipoprotein [LDL] particles) is captured by specific receptors exposed on the outer surface of the cell membrane. The receptor with its cargo is internalized in clathrin-coated vesicles. For details about clathrin-coated vesicles, see Alberts et al. (2002). Inside the cell the vesicle loses its coating and fuses with an **endosome,** inside which the LDL is released from its receptor. The receptor is returned to the cell surface by an exocytotic vesicle, which fuses with the cell membrane. The LDL particle ends up in a lysosome, where it is degraded by hydrolytic enzymes to release free cholesterol. For simplicity, only one LDL receptor is shown entering and leaving the cell interior. In fact, each vesicle contains a large number of specific receptors, which constantly cycle between the cell surface and the cell interior. Typically, one round trip from the cell surface into the cell and back again takes 10 min.

the receptors and the part of the cell membrane in which they are embedded are pinched off and taken into the cytosol as a so-called **coated vesicle.** Inside the cell the endocytic vesicle fuses with another type of cytoplasmic vesicle called an endosome, where the vesicle content is processed further. This mechanism enables the cell to take up specific molecules from the surrounding fluid even though their concentration is very low.

In some cells, especially in specialized phagocytes, the endocytic activity may be very intense, amounting to engulfing vesicles with a combined surface area equal to its own surface area in less than one hour. Since we know that most cells have a fairly constant volume and surface area, this means that a corresponding volume of vesicles must be exocytosed during the same time. For cells or parts of cells that primarily engage in exocytosis (secretory cells and synaptic terminals), the same amount of surface membrane has to be endocytosed again to keep the size constant.

In addition to providing a means of transport into and out of the cell, exocytosis and endocytosis also enable the cell to vary the type and number of specific receptors and transporters displayed on the cell surface. In the kidney, vasopressin increases the number of aquaporin water channels in the apical cell membrane of cells in the collecting ducts, thereby increasing the uptake of water from the urine (Frokiaer et al. 1998). In a muscle cell, both contractile activity and insulin increase the number of **GLUT4** glucose transporters in the cell membrane (Hirshman et al. 1988; Klip et al. 1987). The increased number of transporters enables the cell to take up glucose more efficiently, thereby providing fresh energy substrate for the cell and lowering the blood glucose level. Obviously, both the cycling of aquaporin and the cycling of glucose transporters between the cell surface and an intracellular pool are important homeostatic mechanisms during work and physical activity.

### *Summary*

The cell membrane is an extremely thin but tenaciously stable film of lipid and protein molecules. By virtue of the barrier properties of its lipids and the selective transport mechanisms provided by its protein molecules, it controls the trafficking of ingredients between the cell interior and its surroundings. By means of endocytosis and exocytosis, cells are able to import and export material that lacks specific channel or transport proteins. In addition, exocytosis and endocytosis enable the cell to increase or decrease, respectively, the number of specific receptors or transporters exposed on the cell surface.

## The Cell Interior

Encapsulated within the boundary of the cell membrane is the cytoplasm. It contains a number of formed elements called organelles (see figure 2.1), suspended in a "soup" mainly consisting of proteins, nutrients, and ions dissolved in water. This soup is called cytosol and contains, among other things, enzymes that support the metabolic processes for the anaerobic regeneration of ATP in the cell. Anaerobic means literally "without air," which means that these processes can take place in the absence of oxygen. As we shall see later, however, the initial steps of the breakdown of glucose always take place anaerobically, even in the presence of oxygen.

Membranes very similar in structure to the membrane covering the cell surround most organelles. This similarity is not surprising if we remember the lively shuttling of vesicles between the cell membrane and the cell interior described earlier. Actually, cytoplasmic vesicles form an intricate and well-regulated trafficking system between different organelles. Segregating different enzymes and functions in vesicles is an important way to enable the cell to perform many different chemical tasks at the same time without interference between the tasks.

The **endoplasmic reticulum (ER)** is an elaborate system of flattened vesicles that may or may not be covered on the outside by **ribosomes.** ER with ribosomes attached is called **rough endoplasmic reticulum.** Without ribosomes it is called smooth endoplasmic reticulum. Ribosomes are small ribonucleoprotein particles that are able to join amino acids in a linear fashion to peptides and proteins of different sizes in the process called translation. Future intracellular proteins are synthesized by polyribosomes, ribosomes attached one after the other like beads on a string to a molecule of **messenger ribonucleic acid (mRNA)** that codes for that particular protein. Proteins destined for export from the cell, on the other hand, are synthesized by membrane-bound ribosomes and at the same time funneled into the lumen (i.e., the hollow inside) of the ER. While inside the ER, the protein may be modified in different ways. Such modifications are collectively named **posttranslational modifications** and are extremely important for the function of the protein. The **collagen** molecule can serve as an example here. Inside the ER the early collagen molecule, called procollagen, is hydroxylated and glycosylated, making the resulting mature collagen fibers stronger. These modification steps require the presence of vitamin C, and one of the characteristics of scurvy, the result of a vitamin C deprivation, is collagen fibers of inferior quality.

From the ER the molecules to be exported are transferred inside vesicles to the Golgi apparatus for final preparation and eventually to the extracellular space by exocytosis. Even though protein molecules are far too large to pass through the cell membrane by any known mechanism, they can be exported by means of exocytosis, provided that the protein molecules have been transferred to the ER lumen as described previously. A leakage to the tissue fluid and eventually to the blood of intracellular proteins that are not intended for export is always taken as a sign of cell injury.

Intracellular vesicle trafficking is a highly regulated process. A small part of each protein molecule serves as an address tag, identifying the destination of that particular molecule, much like a baggage tag in an airport. And the analogy goes further. Like the baggage system in air travel, this system sometimes fails, but unlike the baggage, which may turn up some time later, a failure in intracellular trafficking may cause permanent disease.

The mitochondria are rod-shaped bodies variable in size from 0.5 μm to 12 μm in length, separated from the cytosol by a double membrane. The inner membrane forms more or less regular ridges, the mitochondrial cristae, which protrude into the interior of the mitochondrion (figure 2.6). Many of the enzyme proteins of the mitochondrion combine to form supramolecular aggregates that are organized in an orderly array along the inner membrane.

The mitochondria are the "powerhouses" of the cell, responsible for the aerobic part of its energy metabolism. A skeletal muscle fiber can contain several thousand mitochondria, but as we shall see in later chapters, the number may vary substantially, depending on the muscle fiber type and training status. Since aerobic resynthesis of ATP from ADP and inorganic phosphate is the only form of ATP resynthesis that can go on for a long time without causing fatigue, it is easy to understand that the number of mitochondria in a muscle cell is a good indicator of its endurance.

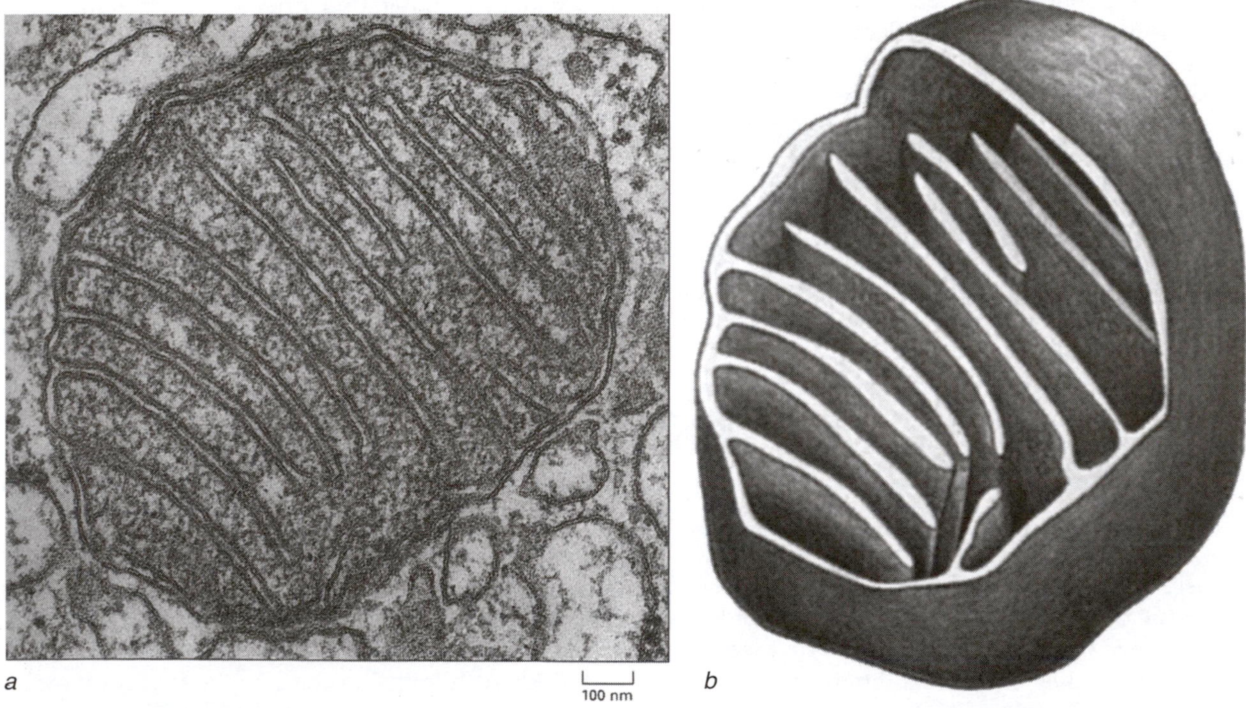


**Figure 2.6** *(a)* A mitochondrion seen in cross section with an electron microscope. *(b)* A three-dimensional schematic drawing of a mitochondrion. Note the ridges—the mitochondrial cristae—which are formed by the inner membrane of the mitochondrion and protrude into its interior, vastly increasing the total surface of the inner membrane. The inner membrane contains most of the proteins responsible for the oxidative activity of the mitochondrion.

The nucleus is the largest organelle in the cell and contains its genetic material, the chromosomes, consisting of deoxyribonucleic acid **(DNA)** and proteins. The genetic information is stored as a linear sequence of purine and pyrimidine bases along the sugar–phosphate backbone of the DNA molecule. During cell division (mitosis) the individual chromosomes become clearly visible in the microscope as small rods of different sizes, 46 in number. At the time they become visible (in the metaphase of the mitosis), the replication of DNA has already taken place, and each of the 46 chromosomes has its own conjoined, identical twin. We can call it a double chromosome or a two-chromatid chromosome (figure 2.7a). Later in the mitosis, in the anaphase, the two twins separate and are segregated, one to each of the two daughter cells (figure 2.7b).

With very few exceptions all cells in the human body carry the same genetic information. This is a logical consequence of the fact that they all derive from the same parent cell, the fertilized ovum, and that the type of cell division that makes a human being out of one cell, mitosis, is a conservative type of cell division. This ensures that the genetic material is faithfully copied and transferred to all its descendants. An important exception to this rule is the mature sex cell, the gamete, which has only half the number of chromosomes.

Storing genetic information in the form of DNA is extremely efficient. The efficiency has been illustrated in a variety of ways: It would take 3,500 volumes, each 500 pages long, to spell out all the genetic material in the human genome in clear text, while in the cell it is contained in 46 chromosomes, which theoretically could fit into a cube less than 2 μm on each side. Equally astonishing is the fact that most of the genetic material is very old, some of it millions of years, and most likely very little has been added to the human genome during the past 10,000 years. It has been calculated that the genomes of humans and chimpanzees are 98% identical, and some of the genes found in humans can be found in all types of cells, both eukaryotic and prokaryotic.

For a long time the valid definition of a gene was a stretch of DNA that encodes one polypeptide chain. It has turned out, however, that the coding sequence of most genes is discontinuous. Interspersed between short coding stretches, called **exons,** are long noncoding sequences called **introns** (see figure 2.14 later in this chapter). During transcription, the process whereby the gene is transcribed into a molecule of mRNA, both exons and introns are copied, but the introns have to be removed before the mRNA molecule is allowed to leave the nucleus. This process is called **splicing,** but the exons are not always joined together in the same sequence. If the sequence is changed or if one exon is omitted in what is now commonly called **alternative splicing,** the result is a protein with slightly different properties. As a result of this, the definition of a gene has been changed to "any DNA sequence that is transcribed as a single unit and encodes one set of closely related polypeptide chains (protein isoforms)" (Alberts et al. 2002, p. 437).

Cell activity includes several hundreds of simultaneous chemical reactions, each catalyzed by highly specific enzymes. Specific means that each type of enzyme is able to catalyze only one particular step in a sequence of chemical transformations. Consequently, such a sequence needs to be catalyzed by a corresponding selection of different enzymes, without which that particular sequence of chemical transformations cannot take place. It is the selection of enzymes in a cell and their instantaneous activity that determine the functional properties of a cell at any given moment.

The chemical reactions of a cell serve different purposes. Some are involved in the making of new building blocks that are needed to replace old ones in the continuous rebuilding that makes the cell able to adapt to changing demands. Another important group of enzymes is involved in the regeneration of ATP to provide energy for the cell's different tasks. Other enzymes are involved in signal transduction inside the cell following the interaction of an extracellular signal molecule with a receptor on the cell surface. In the following sections we take a closer look at the latter two types of reactions, starting with energy metabolism.

## *Summary*

The cell interior consists of organelles, small structures of different shapes and functions, suspended in the cytosol. The cytosol is a viscous solution of proteins, carbohydrates, ions, and nutrients suspended in water. The various organelles serve different functions in the cell, much like the different organs in our body. Consequently, the selection and number of organelles in a cell reveal the functional properties of that particular cell. The nucleus, the largest of the organelles, contains DNA (deoxyribonucleic acid), which is the carrier of genetic information. A gene is a stretch of DNA that codes for one particular protein or some closely related variants (isoforms) of the same protein.

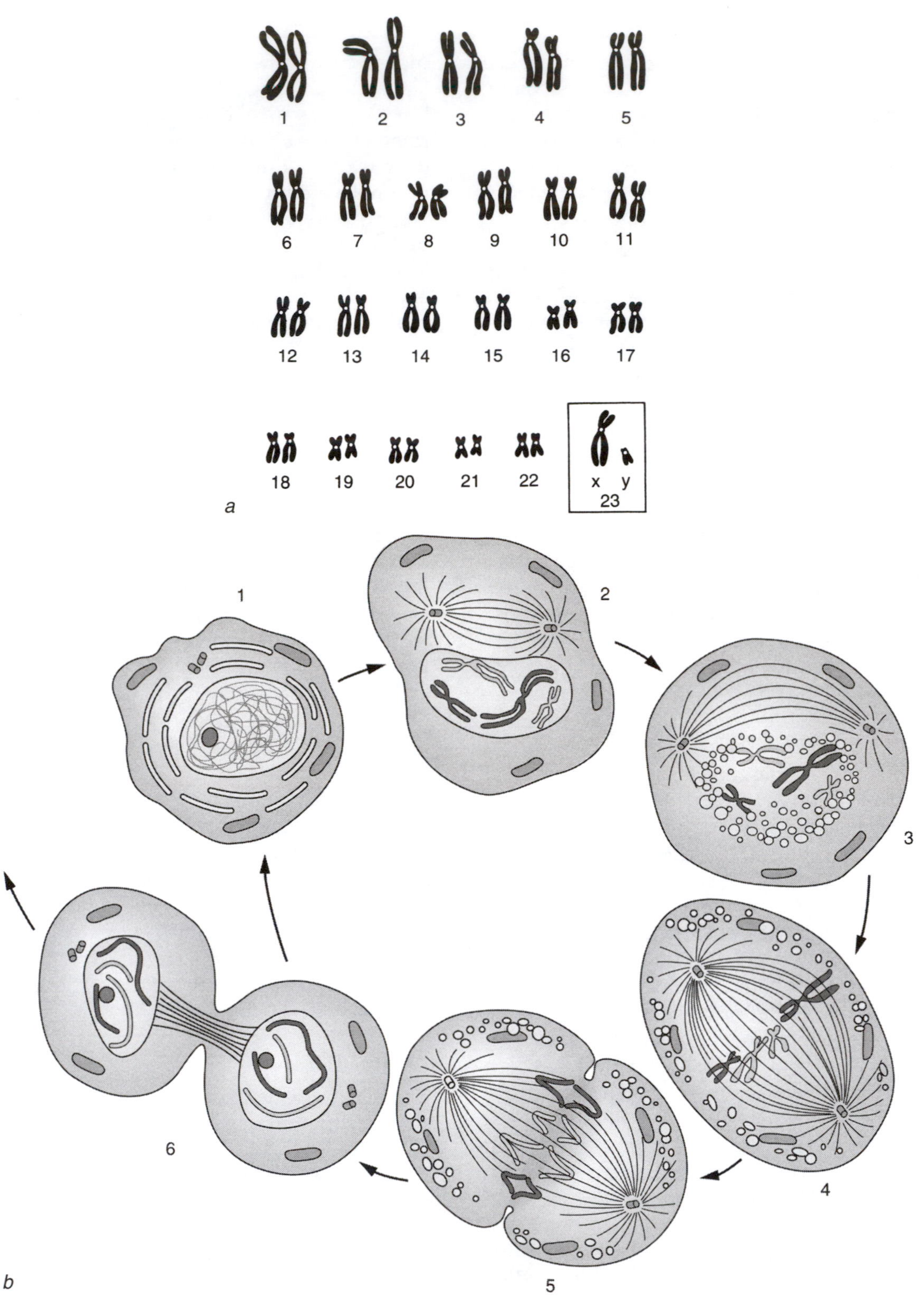


**Figure 2.7** *(a)* The 23 pairs of chromosomes from a human male, as they look during metaphase. Note the gradual reduction in size from number 1 to number 22, and note that the Y chromosome is much smaller than the X chromosome. *(b)* Mitosis: 1. interphase, 2. prophase, 3. prometaphase, 4. metaphase, 5. anaphase, 6. telophase (cytokinesis). Only 2 of the 23 pairs of chromosomes are shown. In each pair one chromosome (dark gray) comes from the individual's father and the other (light gray) from her or his mother. Somewhere in between stage 6 and stage 2 (in the S phase of the cell cycle), each chromosome has been replicated, making the Siamese-twin-like double chromosomes shown in part *a*.

Adapted, by permission, from H.A. Dahl and E. Rinvik, 1999, *Menneskets funksjonelle anatomi* (Oslo, Norway: Cappelen Akademisk Forlag A/S).

## Energy Metabolism

Our basic source of energy is the food that we eat, but the energy in our food is too slowly accessible to be used directly for cellular activity. Therefore, the energy present in the chemical bonds of our nutrients has to be transferred to a common energy "currency," ATP, in a process we name energy metabolism (figure 2.8).

Many different processes in the cell need a supply of ATP to take place, and one of the most obvious of these processes is muscle contraction. Muscle contractile work is basically a question of transforming chemical energy into kinetic (mechanical) energy, much in the same way as the combustion engine in a car. In the combustion engine, gasoline and air are introduced into the cylinder. The spark from the spark plug initiates the explosive combustion of the gas mixture. The expansion of the gas forces the piston to move; the chemical energy in the gasoline has been converted into kinetic energy and heat. The engine has to be cooled by fluid or air to prevent overheating, and the waste products are expelled with the exhaust.

Some of the energy used for muscle contraction is also converted to heat, which is taken up by the blood and carried to the skin, where it is dissipated to the environment. The $CO_2$ resulting from muscle metabolism is transported to the lungs and expired, while the extra water may be removed by sweating or lost in the urine, depending primarily on the relative intensity of the muscular work and the ambient temperature and humidity. Last but not least, both the combustion engine and the muscle need a supply of fuel to work properly, but there is one important difference. While the combustion engine is able to empty its stores of fuel completely, that is never the case with a muscle. Even in a state of complete exhaustion, there is a tiny amount of ATP left in a muscle cell, sufficient to keep it alive.

### Glucose Is the Major Energy Source for Most Cells

Glucose, the major energy source for most cells, is the only one that can be broken down anaerobically. Lipids and proteins, to the extent that the latter is used for energy purposes, can be broken down only aerobically (see figure 2.8). The first steps in glucose **catabolism** (breakdown) from glucose to pyruvic acid—collectively called **glycolysis**—are always anaerobic; that is, they take place in the same manner in the presence of oxygen as in its absence. The only difference is the way the cell deals with the molecules of the reduced coenzyme nicotinamide adenine dinucleotide **(NADH)** resulting from some of the steps in the glycolytic pathway. In the presence of oxygen, NADH is reoxidized to **NAD** by the mitochondria. In its absence, it is reoxidized by the enzyme lactic acid dehydrogenase **(LDH)** in a chemical reaction that turns pyruvic acid into lactic acid while restoring the level of NAD necessary to keep the glycolysis going (figure 2.9). Although the glycolytic pathway enables the cell to function under anaerobic conditions, it remains an inefficient solution energetically. Only two molecules of ATP are the yield from the breakdown of each molecule of glucose, and two molecules of lactic acid accompany it. In comparison, the complete breakdown of one molecule of glucose yields 36 molecules of ATP in addition to water ($H_2O$) and $CO_2$ but, most important, no lactic acid.

Lactic acid has, however, gained an undeservedly bad reputation, especially among laypeople. This is understandable in view of its putative role in muscle fatigue. It seems appropriate at this point, however, to modify this all-too-common view of lactic acid. First of all, it should be called a metabolic intermediate rather than a waste product, simply because most of the energy originating from the glucose molecule is still intact. Secondly, since intracellular glucose and all the metabolic intermediates of the glycolytic pathway are unable to traverse the muscle cell membrane because they are phosphorylated, lactic acid becomes important as transferable energy. It can be released by muscle fibers working anaerobically and later taken up and metabolized by other fibers working aerobically (Chatham 2002). This is the only way by which energy can be transferred from one muscle cell to another, or from any cell to any other cell, for that matter (figure 2.10). Thus, it has even been hypothesized that neurons metabolize glia-derived lactate rather than glucose during activity (for discussion, see Chih, Lipton, and Roberts 2001). There has also been postulated an intracellular lactate shuttle, but its existence has been a matter of debate (Brooks 2000 and 2002; Sahlin et al. 2002).

### Lipids and Proteins Can Also Be Used As Sources of Energy

Although glucose is the most readily available source of energy for muscle fibers, both lipids and, to a lesser degree, proteins can be used as energy sources for cellular activity.

Lipids emerge as an important source of energy both for muscle fibers and other cells, not least

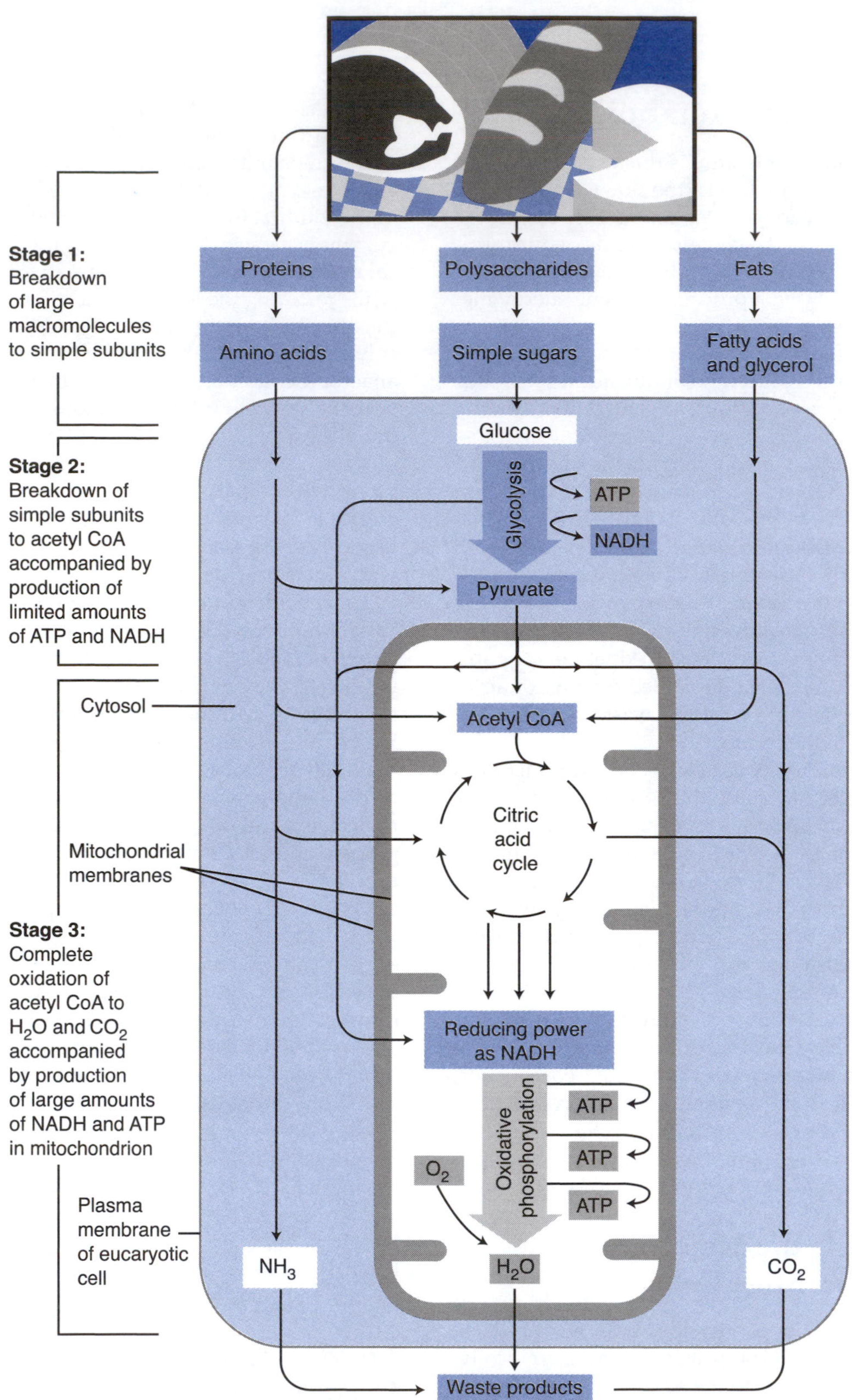


**Figure 2.8** Simplified diagram of the three stages of breakdown that lead from food to waste products and provide energy in the form of adenosine triphosphate (ATP) for animal cells. Stage 1 takes place in your kitchen while you cook the meal and in your gastrointestinal tract once you have eaten it. No energy is obtained from this stage. Stage 2 occurs mainly in the cytosol, providing limited amounts of ATP. No oxygen is consumed during this stage. Stage 2 may therefore take place even in the absence of oxygen, that is, under anaerobic conditions. Stage 3 takes place in the mitochondria, provided that oxygen is available. CoA = coenzyme A; NADH = nicotinamide adenine dinucleotide, reduced form.

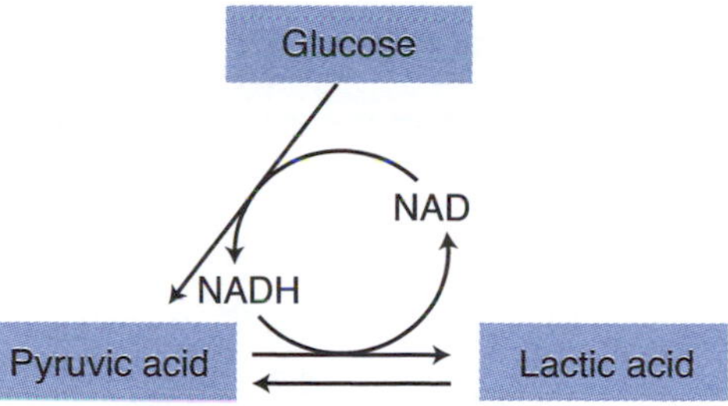


**Figure 2.9** Glycolysis. Under aerobic conditions, the reduced nicotinamide adenine dinucleotide (NADH) generated during glycolysis is reoxidized to NAD by the mitochondria. In the absence of oxygen, the mitochondria are unable to do that. Since the cell needs NAD to keep the glycolytic pathway going and NAD is in short supply in the cell, NADH has to be reoxidized to NAD in a reaction that converts pyruvic acid to lactic acid. The conversion of pyruvate to lactate is, however, a blind alley, and to avoid undue accumulation of lactate, the cell has to export it or convert it back to pyruvic acid and funnel it into the **citric acid cycle** when oxygen is again available in the cell.

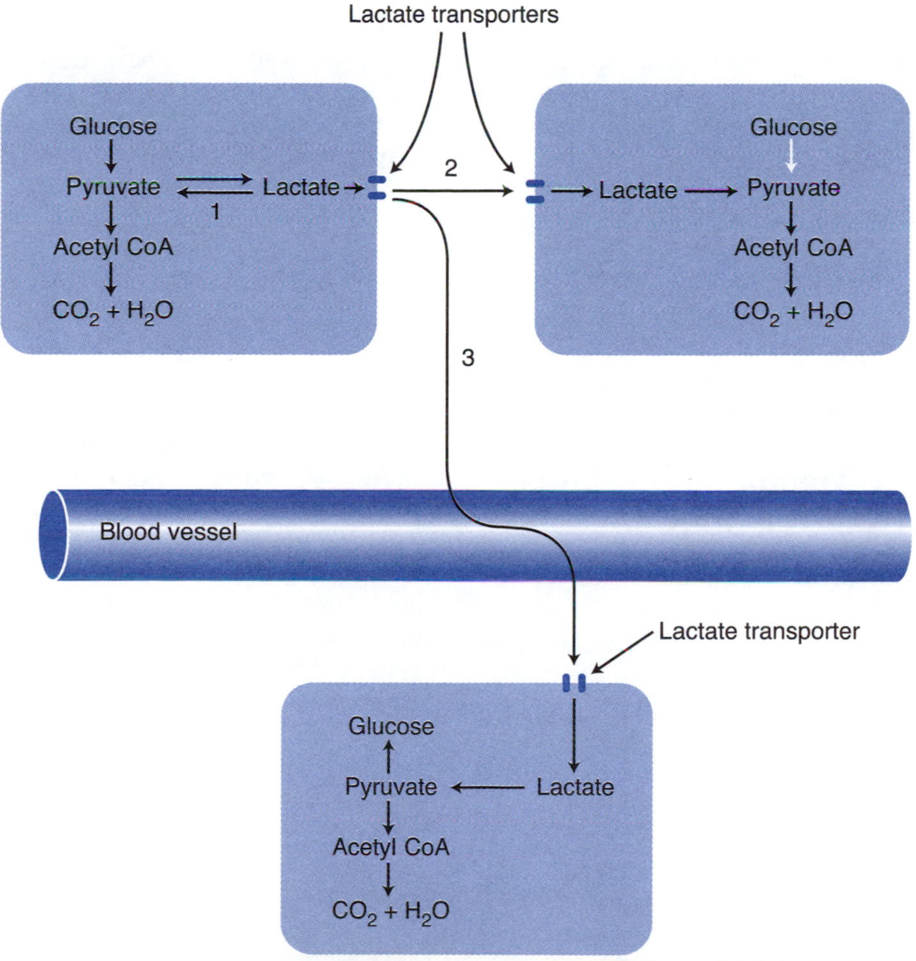


**Figure 2.10** Three possible fates of lactate. 1. Lactate may be reoxidized to pyruvate and funneled into the citric acid cycle in the cell where it was produced. 2. Lactate may be exported from the cell that produced it and imported and metabolized by a neighboring cell. 3. Lactate may be exported and transported by the bloodstream to some distant cell, which—depending on need and metabolic capacity—may use it as a source of energy or turn it into glucose by the process called gluconeogenesis.

because, unlike carbohydrates, lipids can be stored in almost unlimited amounts in the body. In addition, lipids are a very efficient way of storing energy. Lipids storing the same amount of energy as carbohydrates would increase the weight of the energy stores by a factor of five. It must be considered a disadvantage, however, that the energy stored as lipid is not as readily accessible as energy stored as carbohydrates. The main lipid stores are in subcutaneous fat tissue as **triacylglycerol,** which must be broken down to glycerol and free fatty acids **(FFA)** by the enzyme triacylglycerol lipase. In the blood, the FFAs are transported bound to **albumin,** and even though the enzyme is located in an entirely different tissue, the triacylglycerol lipase reaction is the **flux-generating reaction** for fatty-acid oxidation in muscle. In spite of lipids' abundance, they are unable to cover the energy demand during high-intensity aerobic activity. Therefore, when the carbohydrate stores are empty, lipid oxidation alone can provide only the energy necessary for activity at about 50% of $\dot{V}O_2$max.

Proteins, on the other hand, are not a preferred fuel for muscle cells, and when amino acids are used for energy purposes, the amino group must first be removed. This occurs in the liver. The resulting deaminated carbon skeleton can be converted to acetyl coenzyme A **(acetyl CoA)** and can enter the citric acid cycle or be converted to glucose in the process called **gluconeogenesis.**

## How Energy Metabolism Adapts to the Needs of the Body

The energy requirement of the human body varies within wide limits, both among individuals and from one situation to another in the same individual. This is primarily due to variations in physical activity. Consequently, it is of utmost importance that the energy metabolism be adjusted to the actual demand. This is achieved by means of a handful of basic biochemical mechanisms.

Energy metabolism consists of a sequence of chemical transformations, coupled in series, that stepwise degrade a glucose molecule to $H_2O$ and $CO_2$ (see figure 2.8). Some of the chemical steps are what are commonly called equilibrium reactions, that is, reactions that can go in both directions, depending on the relative concentrations of substrate and reaction product. Since coupling in series means that the product of one chemical reaction is the substrate for the next, the flux through a series of equilibrium reactions is determined by the flux through the first reaction in the series. Other steps are nonequilibrium reactions, where the flux in one direction far exceeds the flux in the opposite direction. The nonequilibrium reactions make sure that the reactions proceed in the direction necessary to turn the glucose into $H_2O$ and $CO_2$, at the same time covering the energy requirements of the cell.

The long sequence of chemical transformations starting with glucose and ending with $H_2O$ and $CO_2$ can be divided not only into an anaerobic and an aerobic part but also into shorter sequences of reactions, each starting with a nonequilibrium reaction. Such sequences can be called **metabolic pathways** (Newsholme and Leech 1984). The enzymes of the nonequilibrium reactions have the additional property that molecules binding to a regulator site on the enzyme molecule can modulate their activity. Since this regulator site is different from the **active site** where the substrate binds, the regulator site is also called an allosteric site (**allos** = other). Such a regulatory molecule may be a downstream reaction product acting back on the first reaction in that particular sequence in a feedback mechanism, or it may be some other relevant molecule.

The flux through the equilibrium reactions adjusts to the amount of reaction product coming out of the preceding nonequilibrium reaction. Consequently, the flux through the initial nonequilibrium reaction determines the flux through that particular part of the metabolic pathway. In this way **allosteric regulation** of nonequilibrium reactions makes it possible to tune the metabolic activity of the cell to its actual energy requirements.

At rest the energy requirements of the muscle cell are modest, but when the cell faces an immediate danger, the energy demand rises sharply. Such acute changes are due to the action of **adrenaline,** released from the adrenal medulla as a result of sudden increased activity in the sympathetic nervous system. Through the action of **cyclic adenosine monophosphate (cAMP)** and a **kinase** cascade, adrenaline increases the release of glucose from intracellular stores of glycogen, thus providing an increased supply of substrate for the glycolytic pathway.

# Intercellular Communication

Like multicellular organisms, an individual cell needs to communicate with its environment. As part of a multicellular organism, it has to coordinate its behavior in a number of different ways, partly to play its role in the orchestrated activities of the cells of the body, partly to adjust to the changing needs of the body as a whole.

Communication between cells takes place through chemical signals, signal molecules, which are produced by a signaling cell and released into the tissue fluid. Such signal molecules may act locally on neighboring cells or even on the signaling cell itself, or they may be transported to more distant parts of the body by the blood (figure 2.11). The latter case is called **endocrine signaling,** which

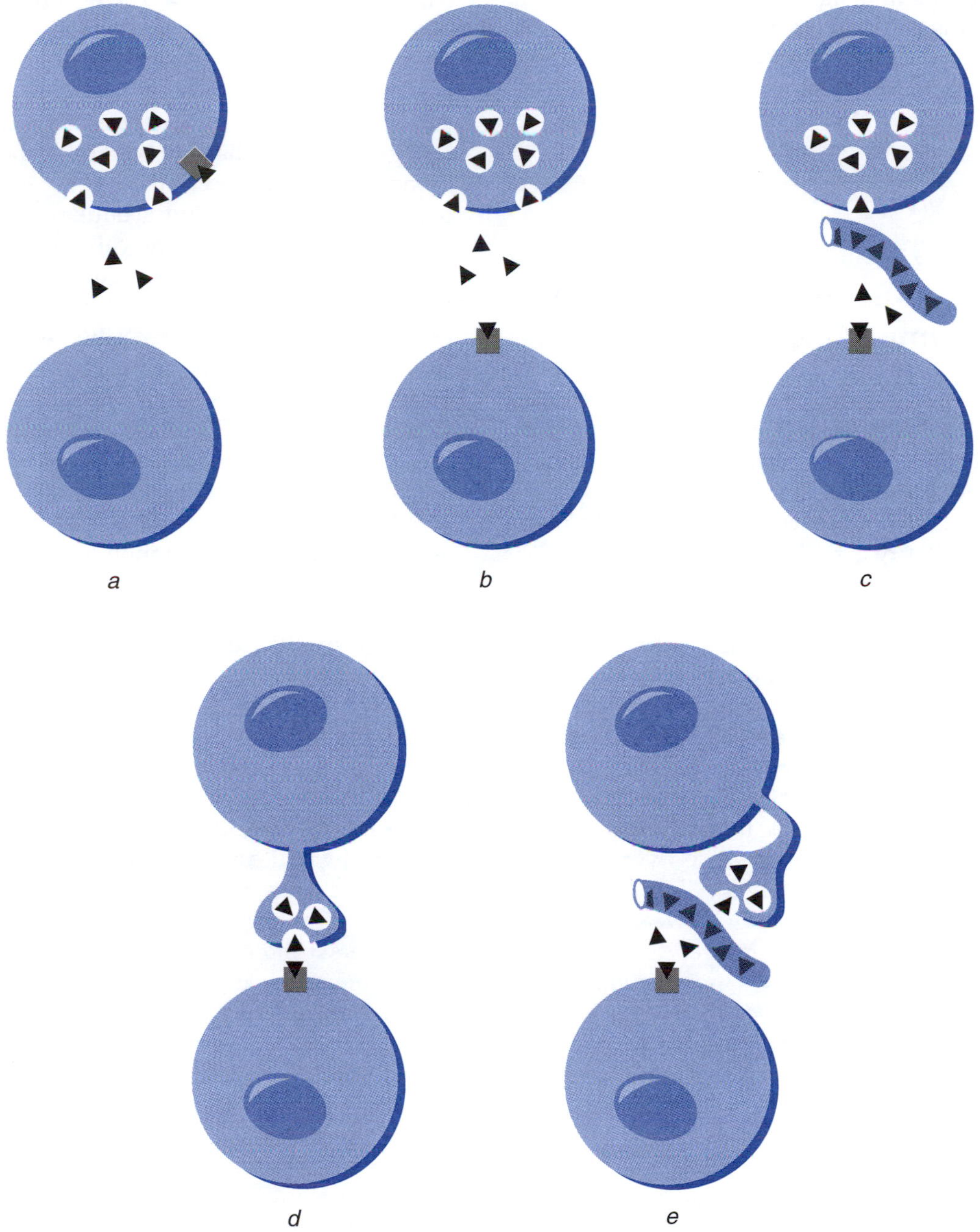


**Figure 2.11** Different forms of cell signaling. *(a)* In autocrine signaling, the signaling molecules act on the cell itself. *(b)* In paracrine signaling, the signaling molecules act on neighboring cells. *(c)* In classic endocrine signaling, the signaling molecules—the hormones—are taken up by the blood and distributed to cells all over the body. Only cells with the appropriate receptors are able to respond to the signal. In this figure only a cell surface receptor is shown. In the case of steroid hormones and thyroxin, which are able to traverse the cell membrane, the appropriate receptors are found inside the cell. *(d)* In neural signaling, the signaling molecules—the transmitter substance—are liberated from nerve terminals lying closely apposed to the cell membrane of the target cell. *(e)* Neuroendocrine signaling is a combination of endocrine and neural signaling. Instead of being delivered directly to the target cell, the signaling molecules are released to the bloodstream and distributed accordingly.

Reprinted, by permission, from P. Brodal, H.A. Dahl, and S. Fossum, 1990, *Menneskets anatomi og fysiologi* (Oslo, Norway: J.W. Cappelens Forlag).

has been known for a long time in the hormone system. If the target cell is a neighbor of the signaling cell, it is called **paracrine signaling.** If the signal affects the signaling cell itself, it is called **autocrine signaling.** The first signal molecules to be identified were the classical hormones, described in the first decades of the 20th century. In more recent years, the number of known signal molecules has exploded, and more important, their modes of action have been revealed in increasing detail (Alberts et al. 1994).

## Synaptic Transmission and Hormone Signals Are Based on the Same Mechanism

The increased knowledge about signal molecules has revealed that neuronal signaling, also known as synaptic transmission, is basically the same mechanism as endocrine signaling, although these signals govern different kinds of cellular activity. The main point of similarity is that in both cases signal molecules are released from one cell and detected by another by means of specific receptor molecules. The main difference is the number of cells that can be affected by the signal. Paracrine or endocrine signaling is like shouting a message to every cell in the neighborhood or in the whole body. Neuronal signaling, on the other hand, is more like whispering in the ear of the recipient because the signal molecules, long known as neurotransmitters or transmitter substances, are released very close to the cell membrane of the target cell. In both cases, the specificity of the interaction between the signal molecule and its receptor ensures that only the correct cells react to the signal. Understandably, this is of fundamental importance to the precision of endocrine signaling. When the production of female sex hormones increases during puberty, for example, only the cells that have the correct receptors respond by developing into breasts. In the absence of such specificity, breasts would have developed all over the body's surface.

A major advantage of neuronal signaling is its speed of action. When activated by signals from other nerve cells or from the environment, a neuron sends electrical impulses down its axon at a speed of more than 100 m per second, releasing transmitter molecules from its axon terminals. This makes the nervous system able, for example, to withdraw a hand from a painful stimulus in a fraction of a second.

## Signal Molecules Differ Chemically

The signal molecules belong to several different chemical families. For practical purposes, however, they can be divided into two categories: those that are lipid soluble and those that are not. The latter type is unable to penetrate the cell membrane and consequently has to be captured by receptor molecules exposed on the outer surface of the cell membrane. The signal–receptor complex then initiates the generation of an intracellular signal, which changes the behavior of the cell in a specific way. The lipid-soluble signal molecules, which include the steroid hormones and thyroxin, are able to pass through the lipid barrier of the cell membrane. Inside the cell they are captured by one of a family of nuclear receptors specific for each of the lipid-soluble hormones. The nuclear receptors serve as transcription factors and are activated by the binding of their specific signal molecule. When activated, they are able to bind to DNA in the nucleus and influence the transcription of certain genes.

The water-soluble signal molecules, which, by the way, are the more numerous, have to be captured by specific receptors exposed on the outer surface of the cell membrane. These receptors belong to one of three different families of receptors: **ion channel–linked receptors, G-protein-linked receptors,** and **enzyme-linked receptors** (figure 2.12).

The ion channel–linked receptors are also known as **ligand-gated,** or transmitter-gated, **ion channels,** since they are the ones that are found in neuronal signaling. We shall return to ion channels in chapters 3 and 4.

G-protein-linked receptors are the largest family of cell surface receptors, comprising hundreds of different types and responding in a specific way to a vast number of different signal molecules. The receptor itself is a large transmembrane protein, which upon contact with the appropriate signal molecule interacts with the GTP-binding protein **(G-protein)** attached to the inner surface of the cell membrane. In spite of differences in function, all G-proteins have a remarkably similar basic structure, being composed of three subunits called $\alpha$, $\beta$, and $\gamma$. The interaction between the transmembrane receptor and the G-protein leads to a dissociation of the G-protein subunits, giving rise to an activated $\alpha$-subunit and an activated $\beta\gamma$-complex. Each of them can initiate further intracellular signals by interacting with enzymes attached to the cytoplas-

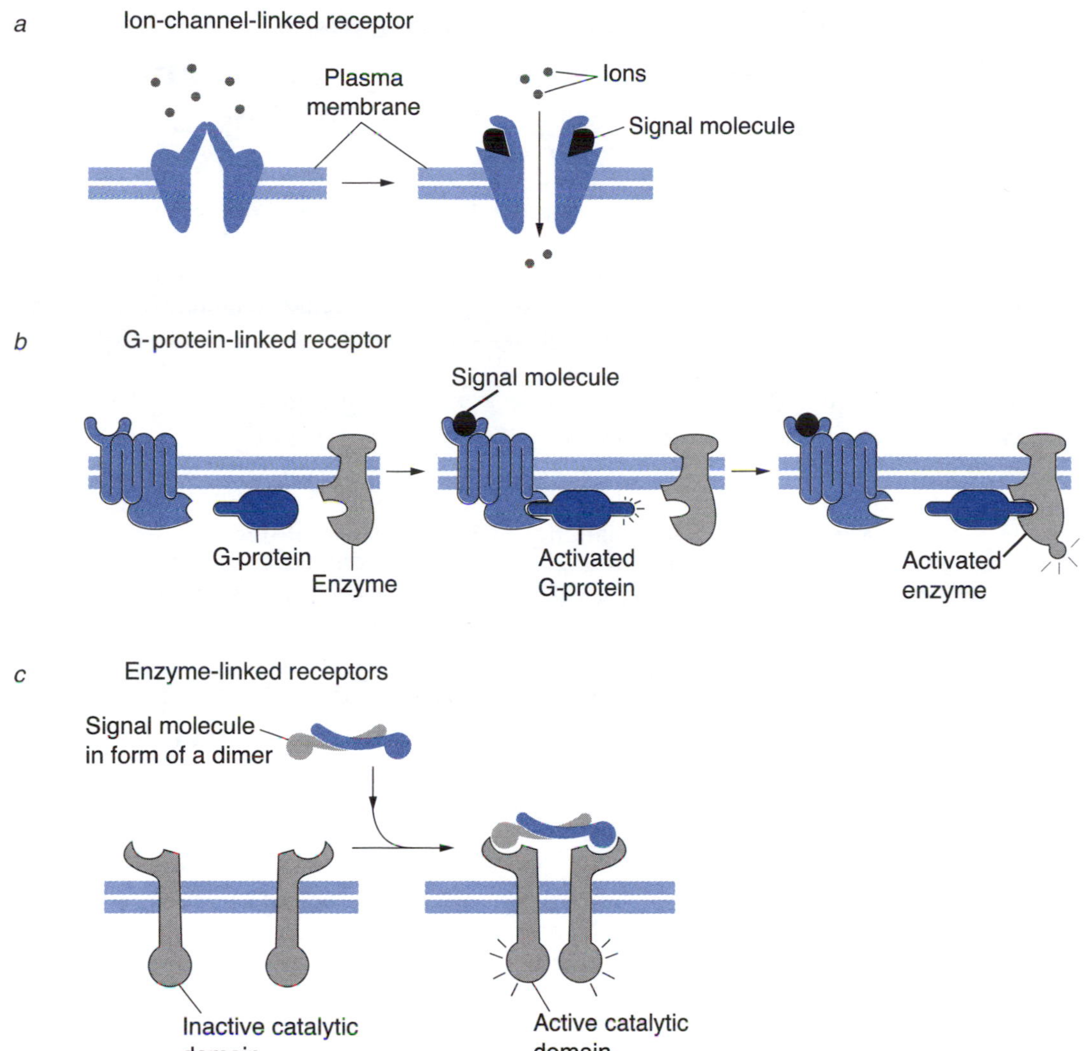


**Figure 2.12** Three classes of cell surface receptors. *(a)* Ion channel–linked receptors open in response to binding of their ligand, the signaling molecule. In other instances it may have the opposite effect (not shown). *(b)* G-protein-linked receptors activate a G-protein, which can activate an enzyme, which in turn gives rise to an intracellular signal. *(c)* Enzyme-linked receptors are in fact enzymes themselves. The receptor site is exposed on the cell surface, while the cytoplasmic part of the molecule is an inactive enzyme. Binding of the signal molecule in the form of a dimer brings two receptors together, making them able to activate each other and relay the signal into the cell interior.



mic surface of the cell membrane or with ion channels.

When adrenaline is bound to its G-protein-linked receptor in the cell membrane of a muscle cell, the α-subunit activates the enzyme adenylate cyclase. In turn, the cyclase activates a sequence of kinases (a kinase cascade), resulting in an increased breakdown of intracellular glycogen to glucose. This provides the necessary energy for a muscular reaction to the situation, be it fight or flight.

The enzyme-linked receptors serve many different signal molecules, including both growth factors and classical hormones, the best known of which is insulin. When stimulated by the appropriate signal molecule attached to the extracellular part of the receptor, enzyme activity is switched on at the cytoplasmic end of the receptor. This generates a series of further signals, depending on the nature of the signal and its receptor. (For a review of intercellular signaling, see Alberts et al. 1994.)

## Sensitivity of the Cell Adapts to the Level of the Stimulus

It is obvious from the preceding sections that cellular activity can vary enormously from one situation to another as a response to changing levels of intercellular signal molecules. The level of a particular kind of signal molecule may also vary within wide limits. It is important, therefore, that the sensitivity of the signal transduction system of the target cell is constantly adjusted in a way that "makes the cell responsive both to messages that are whispered and to those that are shouted" (Alberts et al. 1998, p. 503). Such adjustments can occur in different ways.

One important way is up- and down-regulation of the number of relevant receptors on the cell surface. As a response to sustained stimulation, receptors may be internalized and transiently sequestered inside the cell by endocytosis. The sensitivity of receptors can also be modulated by changes in their level of phosphorylation, or the sensitivity of the downstream signal transduction system inside the cell can be temporarily increased or decreased. (For a review of target cell adaptation, see Alberts et al. 1994.)

## Summary

To ensure proper cooperation between the cells of the body in different situations, a host of different signal molecules serve to adjust the activity of individual cells to the need of the body as a whole. These chemical messengers are released from one type of cell and influence the activity of others. Sometimes they influence cells in the close vicinity, other times cells in distant parts of the body. The signal molecules are transported from the cells of origin to their target cells in the tissue fluid or the blood, or they are deposited close to the membrane of the target cell in the case of neuronal signaling.

Chemically, the signal molecules are very different, but for the sake of simplicity they can be divided in two groups: those that are lipid soluble and those that are not. The lipid-soluble signal molecules can diffuse through the cell membrane and attach to their specific receptor inside the cell. The water-soluble signal molecules are unable to traverse the cell membrane and must be captured by specific receptors exposed on the outer surface of the cell membrane to exert their effect. These surface receptors belong to three main classes, ion channel–linked receptors, G-protein-linked receptors, and enzyme-linked receptors, each of which in its own way is able to initiate intracellular signals that change the behavior of the target cell.

# Common Cellular Strategies for Adaptation to New Demands

We shall return to the adaptations that occur in the muscle cells as a result of changes in the level of physical activity in later chapters. It is already necessary, however, at this stage to review briefly the general strategies available to the cell to attain these adaptations.

We have already stated that it is the selection of enzymes in a given cell and their instantaneous activity that determine the functional properties of that cell at any given moment. As a logical consequence of this, changes in functional properties must be a result of qualitative or quantitative changes in the selection of enzymes present and in the activity of these enzymes. Such changes take place on two different timescales.

## Short-Term and Long-Term Adaptations Serve Different Purposes

First of all, in an emergency situation, muscle cells are able to increase their activity from a state of total rest to maximal activity in a matter of seconds. This is too short a time to allow any new synthesis of enzyme proteins. The synthesis of a new protein molecule of average size (400 amino acids) takes about 20 s (Alberts et al. 1994). Therefore, more abrupt changes must be due to changes in the activity of enzyme molecules already present in the cell. Such switching on and off of enzyme activity is caused by covalent modifications, most often phosphorylation and dephosphorylation of the enzyme molecule in question (figure 2.13), and is the result of the action of a hormone or some other intercellular signal molecule.

On a longer timescale, adaptations take place through processes involving both degradation and new synthesis of protein molecules. Such turnover goes on all the time, even when no adaptation is necessary. In fact, it can be considered the very basis for long-term cellular adaptability. In theory, the amount of a given protein in a cell can be monitored by changes in its synthesis, by changes in its degradation, or by a combination of the two processes. The most obvious result of this adaptation is a change in the amount of the protein in question, but in addition qualitative changes can take place. Instead of replacing old protein molecules with identical ones, the cell may switch to another isoform more able to meet the challenges of the new situation. As we shall see in chapter 3, this is exactly what happens when a muscle fiber changes from one fiber type to another.

By and large, there are more possible sites for control of protein synthesis than for control of protein degradation. To be able to comment on the different possibilities for control, however, we need to present a brief overview of the main steps in protein synthesis (figure 2.14). During transcription the relevant piece of DNA, the gene, is copied into a molecule of mRNA. The genetic information in DNA is not continuous, however; rather, bits and pieces of information, the exons, are separated by long, noncoding sequences called introns. These introns must be removed and the exons joined together in the process called splicing before the mRNA molecule is allowed to leave the nucleus. This splicing is only one of the changes that have to take place. Other

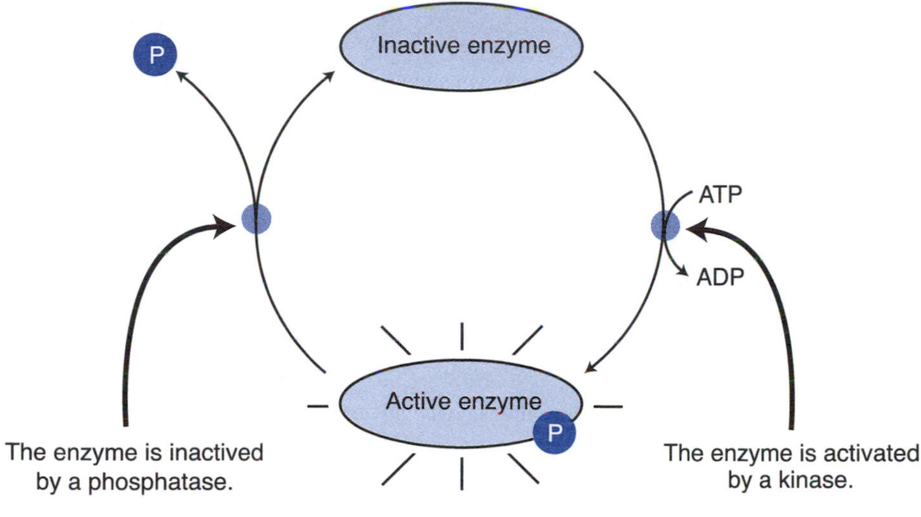


**Figure 2.13** Enzymes can alternate between an active and an inactive state depending on the presence or absence of a phosphate group. In most instances, as in this figure, the enzyme activity is turned on when the enzyme molecule is phosphorylated by a kinase and inactivated when the phosphate group is removed by a phosphatase, but for some enzymes it is the opposite.

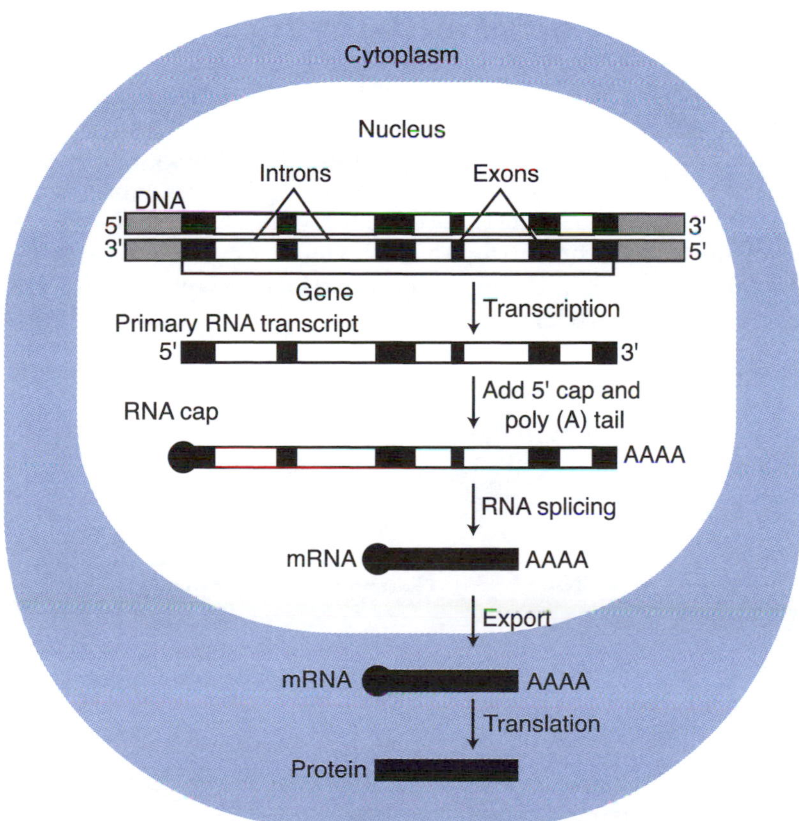


**Figure 2.14** An overview of the steps leading from gene to protein. The nucleotide sequence of one deoxyribonucleic (DNA) strand serves as a template for a messenger ribonucleic acid (mRNA) strand in the process called transcription. This gives rise to a primary RNA transcript containing both exons and introns. During several steps, collectively known as RNA processing, the introns are removed and a 5' cap (discussed later in this chapter) and a poly(A) tail are added before the mRNA molecule is allowed to exit from the nucleus. In the cytosol, the mRNA molecule serves as the recipe for protein synthesis. Note that the two complementary DNA strands are antiparallel. The 5' end of one strand faces the 3' end of the other.

modifications determine the lifetime of the actual mRNA molecule or function as an address tag telling its final destination in the cell. Together these different modifications are called RNA processing.

In the cytoplasm the mRNA molecules can be used as a template for protein synthesis. The ribosome subunits assemble on one end of the mRNA molecule and translate the genetic code into the appropriate sequence of amino acids. Both the number of mRNA molecules for one specific protein and the rate at which they are translated are subject to regulation. The number of each particular type of mRNA molecule depends on its rate of production **(transcriptional control)** and its rate of degradation. The rate at which the different mRNAs are used in actual protein synthesis is known as **translational control.** Evidently, transcriptional control and alternative splicing are the only possible means of changing the type of protein produced by the cell. All other types of control can influence only the amount of types of protein already present in the cell (figure 2.15).

## Control of Gene Expression

We shall return to what is actually known about the adaptive changes taking place in muscle cells in later chapters. Before we leave the topic, however, we shall take a closer look at the control of gene expression, that is, the type of control that determines which genes the cell is going to use as a basis for its protein synthesis. Like many other scientific disciplines, the field of molecular genetics has developed its own nomenclature, some of which need to be mentioned before we continue.

The two complementary strands of polynucleotides in DNA are antiparallel. One end is called 5', the other 3' (see Alberts et al. 2002 for further information). Consequently, *antiparallel* means that the 5' end of one DNA strand faces the 3' end of the other (see figure 2.14). The important thing for our discussion is that synthesis of a new polynucleotide, be it DNA or RNA, can take place in only one direction: from the 5' end of the new polynucleotide toward the 3' end. This means that the synthesis starts in the 3' end of the part of the original template DNA strand to be copied and progresses toward its 5' end. The template DNA strand thus has been likened to a river running in one direction; **upstream** means in the direction of the 3' end of a gene, and **downstream** means in the direction of its 5' end.

A few more concepts are necessary to understand the story. Just upstream of the start of the gene lies an important regulatory portion of DNA, called the **promoter** of the gene (figure 2.16). For each particular gene to be transcribed, a complex of general transcription factors must bind to its promoter and activate the enzyme **RNA polymerase,** which synthesizes the **primary RNA transcript** (compare figure 2.14). In addition, several other portions of the DNA strand, both upstream and downstream of the gene itself, are important as **gene regulatory sequences.** Some of these sequences bind factors called **enhancers** that facilitate or enhance transcription of that particular gene. Other regulatory sequences bind transcription factors that repress the transcription of the gene. Details of the mechanism controlling the expression of different genes are just beginning to be unveiled, but one thing seems reasonably clear even at this early stage. Most if not all genes in a complex organism like a human being are controlled by a combination of gene regulatory factors in what is now commonly known as **combinatorial control** (figure 2.16). A more comprehensive review of this topic can be found in Alberts et al. (1994, 1998, 2002).

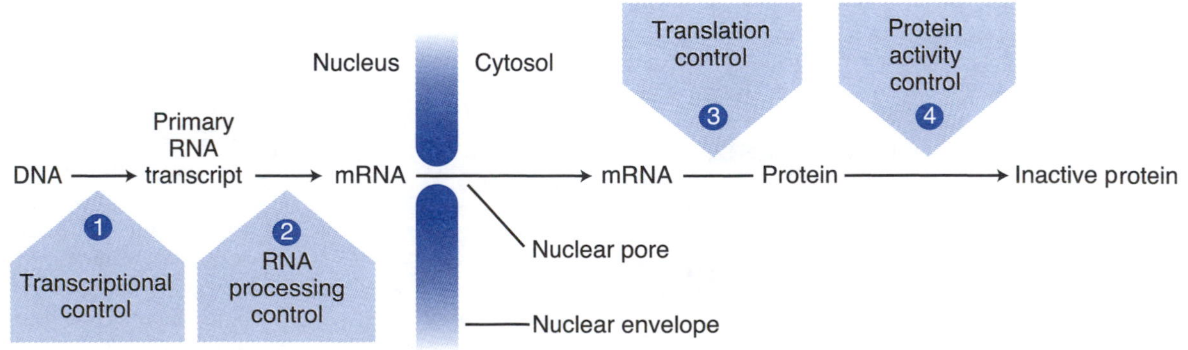


**Figure 2.15** Schematic overview of possible control points during protein synthesis.

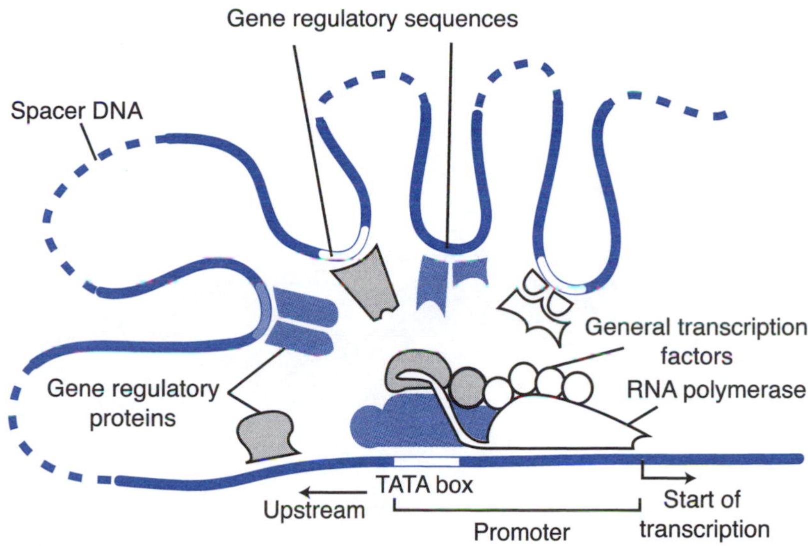


**Figure 2.16** For transcription of a gene to start, a complex consisting of general transcription factors and ribonucleic acid (RNA) polymerase must attach to a binding site called the promoter of the gene, situated just upstream of transcription start point. The binding of the complex is modulated by a group of **gene regulatory proteins (transcription factors)** attaching to specific gene regulatory sequences located both upstream and downstream of the gene. While the general transcription factors are necessary for any transcription to take place, it is the gene regulatory proteins that govern which proteins the cell is going to make. The term *combinatorial control* implies that it is the combination of gene regulatory proteins rather than individual proteins that is important. DNA = deoxyribonucleic acid; TATA = a sequence of DNA consisting primarily of T and A nucleotides.



## Summary

This chapter provides an overview of cell biology as a basis for understanding how cells and organs of the body react to different challenges in connection with work and physical activity. A major objective of this chapter has been to help the reader understand that work and exercise physiology are not separate kinds of physiology, different from human physiology in general, but that they are based on fundamental properties of all cells, properties that we need to know to fully understand how our bodies react, and sometimes do not react as expected, to different kinds of physical activity.

CHAPTER 3

# THE MUSCLE AND ITS CONTRACTION

Our knowledge of muscle biology increased significantly over the last two decades of the 20th century. A substantial part of this is due to a better understanding of muscle fiber type and its molecular basis. In parallel with this increased knowledge, however, is an increasing awareness of the limitations inherent in the concept of muscle fiber type. This awareness, together with the emerging knowledge of the factors that govern the development of the different muscle fiber types and of how the muscle fibers adapt to new challenges, will undoubtedly make the concept of muscle fiber type a more adequate tool for all who deal with muscle functional properties. This chapter attempts to recapitulate the main points in our understanding of muscle and its constituent muscle fibers shortly after the turn of the century.

### FOR ADDITIONAL READING

For a more comprehensive account of the scientific basis of myology, see Engel and Franzini-Armstrong (1994).

## Skeletal Muscle Architecture

The muscles, known by their Latin names such as *brachialis* or *biceps*, are bundles of contractile muscle fibers that join into a tendon at each end. Each bundle or fascicle consists of thousands of muscle fibers, individually wrapped in a thin layer of connective tissue, the **endomysium** (figure 3.1). A layer of connective tissue called the **perimysium** surrounds each fascicle. These fascicles, usually about the thickness of a match or a toothpick, are easy to distinguish in fresh meat. The muscle itself is wrapped in a sheath of connective tissue called the **epimysium.** In most muscles the epimysium forms a smooth surface, allowing the muscle to move freely in relation to neighboring muscles or other structures during contraction. In such cases the epimysium is called a **fascia.**

### FOR ADDITIONAL READING

For a survey of the different collagen isoforms found in the intramuscular extracellular matrix, see Kovanen (2002).

What is commonly called a muscle fiber is in fact a multinucleated muscle cell, which shows all the basic characteristics of other cells, apart from the fact that it is no longer able to divide mitotically. It is a **postmitotic cell.** The reason that it is often called a fiber is its impressive length in relation to its diameter. A muscle fiber 15 cm long and 50 μm in diameter has a length/diameter ratio of 3,000! This ratio is often called the **aspect ratio** of the muscle fiber (Trotter, Richmond, and Purslow 1995).

These highly elongated cells arise in fetal life by fusion of mononucleated **myoblasts.** Each of the resulting multinucleated muscle cells or muscle fibers is surrounded by its own basal lamina, bordering directly on the endomysial connective tissue mentioned earlier. Remaining myoblasts that do not fuse with the muscle fiber are found in adult life as **satellite cells** between the muscle fiber membrane—the **sarcolemma**—and the basal lamina (figure 3.2). The satellite cells represent a myoblast reserve (Bischoff 1994). They are the only cells in the **myogenic** cell line that retain their ability to reenter the cell cycle and divide (Moss and Leblond 1971), thereby giving rise to myoblasts that are able to fuse with the muscle fiber during **hypertrophy** (Roy, Monke et al. 1999) or repair.

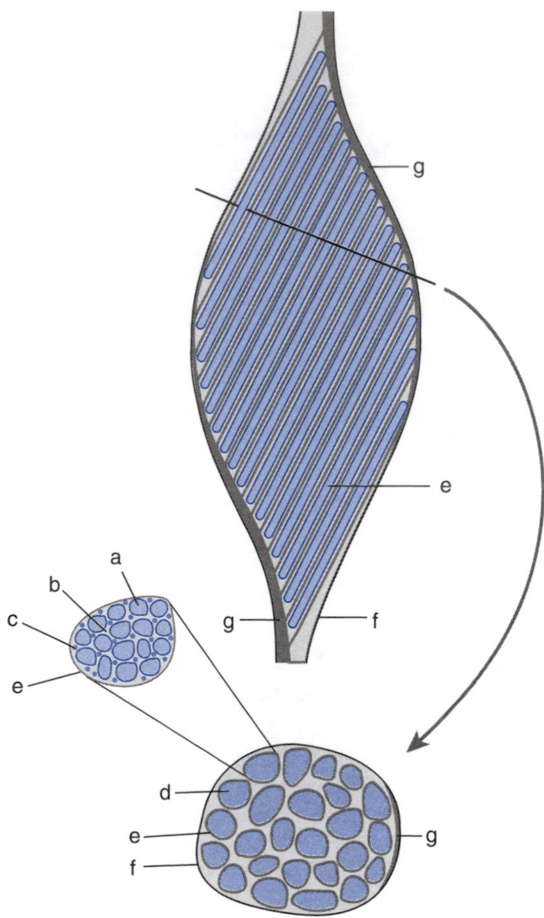


**Figure 3.1** A schematic picture of a unipennate muscle as it is usually seen in textbooks. Individual muscle fibers *(a)* are surrounded by a thin layer of connective tissue called the endomysium *(b)* containing capillaries *(c)*. Bundles of muscle fibers *(d)* are wrapped in a layer of connective tissue called perimysium *(e)*, and the whole muscle is covered by a layer of connective tissue called the epimysium *(f)*. In muscles that usually move freely in relation to their surroundings (e.g., the biceps brachii muscle), the epimysium appears as a smooth sliding surface and is usually called a fascia. Note the thickening *(g)* of the fascia where the connective tissue of the perimysium blends with the surface connective tissue, i.e., the epimysium *(f)*, appearing as a continuation of the tendon alongside the muscle belly *(g)*. Note that individual muscle fibers *(a)* are visible only in the more detailed cross section in the lower part of the figure. The structures *(d)* running from one side of the muscle belly to the other are bundles of muscle fibers. Some of the fibers within a bundle may be shorter than the bundle in which they lie and may be arranged in series with other fibers.

Adapted, by permission, from H.A. Dahl and E. Rinvik, 1999, *Menneskets funksjonelle anatomi* (Oslo, Norway: J.W. Cappelens Forlag).

The textbook picture of a skeletal muscle fiber is that of an almost cylindrical fiber, with nearly constant diameter from one end to the other, extending all the way from the tendon of origin to the tendon of insertion (see figure 3.1). That is not always true. Reports of muscle fibers that are considerably shorter than the fascicle in which they lie have appeared sporadically since the 19th century but have not had any impact on our everyday picture of a muscle fiber until quite recently.

## For Additional Reading

For review of the functional morphology of series-fibered muscles, see Trotter (1993) and Trotter, Richmond, and Purslow (1995).

When the muscle fibers are shorter than the fascicle in which they lie, the distance between the tendon of origin and the tendon of insertion has to be spanned by several muscle fibers in series. Consequently, such muscles are called **series-fibered muscles** (Trotter, Richmond, and Purslow 1995). In addition to being shorter than the bundle of which they are a part, they also deviate significantly from the cylindrical form by having tapered ends. We shall return to series-fibered muscles when dealing with force transmission later in this chapter. The amount of connective tissue (endomysium, perimysium, and epimysium) varies between different muscles and between different animal species. In fusiform muscles part of the epimysium may be visibly thickened and reinforced, appearing as a continuation of the tendon along one side of the muscle belly.

Whether the number of muscle fibers in human muscle is finally established shortly after birth (Gollnick et al. 1981; MacCallum 1898) or increases later in life as a result of hyperplasia (Antonio and Gonyea 1993) is still a matter of debate. We shall return to this discussion in chapter 11. The thickness of a muscle fiber is, however, undoubtedly subject to variation during an individual's lifetime. In adult life this is mainly due to hypertrophy or atrophy, depending first and foremost on changes in the level of physical activity (see chapter 11). Even within a single muscle the diameter may vary; type II muscle fibers usually have a larger diameter than type I fibers. The normal range of fiber diameter is often said to be 30 to 100 μm (Dubowitz and Brooke

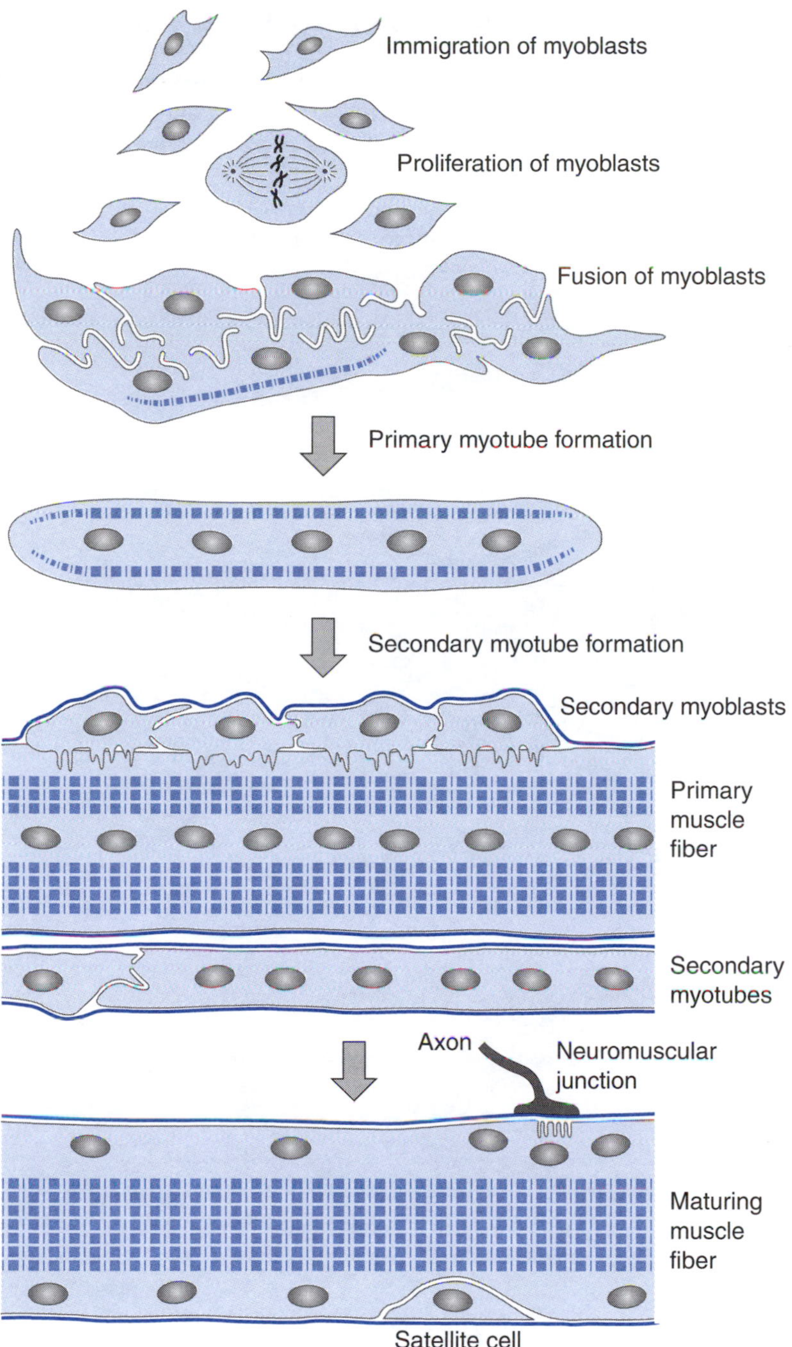


**Figure 3.2** Stages in the formation of a skeletal muscle fiber. Mononucleated myoblasts fuse to form multinucleate primary myotubes, characterized initially by central nuclei. A second clone of myoblasts fuses alongside the primary myotubes, forming secondary myotubes. As the contractile apparatus is assembled, the proliferating myofibrils occupy the center of the fiber, and the nuclei move to the periphery. The cross striations become visible, and the neuromuscular junction starts to develop. Later, small adult-type myoblasts—satellite cells—can be seen lying between the basement membrane and the cell membrane of the muscle fiber.

Reprinted from *Gray's Anatomy*, S. Salmons, Muscle, 737-900, 1995, by permission of the publisher Churchill Livingstone.

1973). Since the diameter is measured in cross sections of the muscle, however, the value of such a range is seriously confounded by the presence of serially arranged fibers with tapered ends (Heron and Richmond 1993).

It should be emphasized already at this early stage that although a muscle is an anatomical entity, it is not always a single unit functionally. Especially in large muscles such as the deltoid and the adductor magnus, different parts of the muscle may have quite different functions. This is possible because a muscle consists of individually working groups of muscle fibers called **motor units.** By definition a motor unit is one motoneuron in the spinal cord, its axon and axonal branches, and all the muscle fibers that are innervated by these branches. The number of muscle fibers in one particular motor unit is always constant under normal circumstances, but the number in different units can vary from around ten to maybe a couple of thousands. The name *motor* *unit* implies that the muscle fibers work as a team, always contracting at the same time and in the same manner, more or less independent of other motor units in the same muscle. When different parts of one muscle serve different functions, it simply means that the motor units in one part are able to work independently of motor units in the other part. This is part of a concept called **muscle regionalization** (Kernell 1998).

## The Muscle Cell Membrane

The cell membrane of the muscle fiber possesses all the characteristics of cell membranes in general, including the ability to create and maintain a membrane potential so vital to its contractile function. Structurally, it is closely apposed to the basement membrane surrounding each muscle fiber. Together, the two are called the sarcolemma, although some-

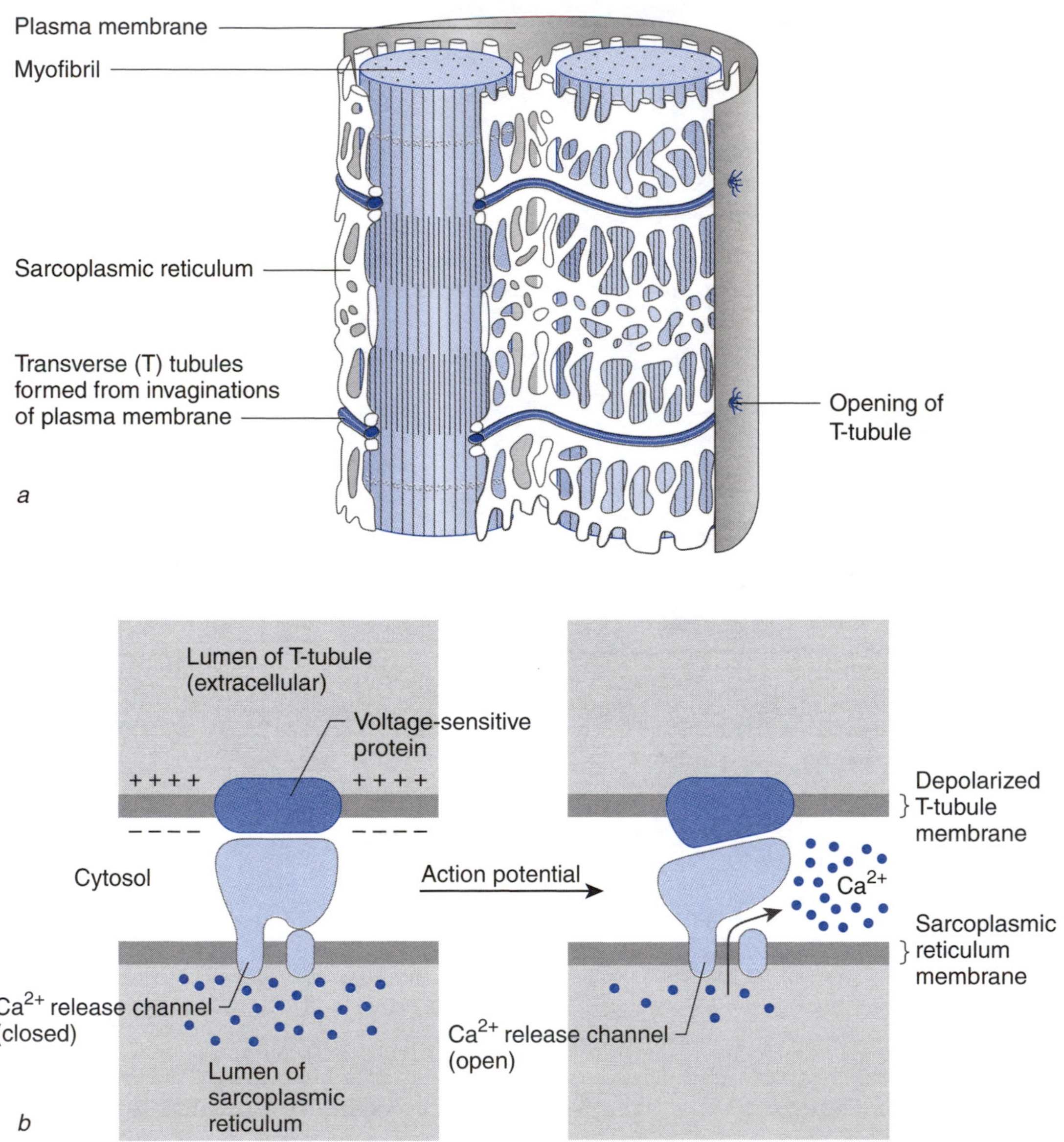


**Figure 3.3** *(a)* T-tubules and sarcoplasmic reticulum form the morphological basis for the coupling between the muscle fiber action potential and the contractile activity of the **myofibrils.** In human and other mammalian skeletal muscle fibers, there is one T-tubule at each A-I junction. *(b)* Schematic diagram showing how a $Ca^{2+}$ release channel (ryanodine receptor) in the sarcoplasmic reticulum membrane is thought to be opened by a voltage-sensitive dihydropyridine receptor in the adjacent T-tubule membrane.

times this designation is used for the cell membrane alone. Functionally, the muscle cell membrane is a mosaic of functional domains (Horwitz, Schotland, and Franzini-Armstrong 1994). Examples are the neuromuscular and myotendinous junctions and the transverse tubules. We shall return to the myotendinous junctions later in this chapter and to the neuromuscular junction in chapter 4, but we need to take a closer look at the transverse tubules at this stage.

The transverse tubules, or **T-tubules,** are tubular invaginations of the muscle cell membrane, running more or less perpendicularly to the long axis of the muscle fiber, hence the name (figure 3.3). T-tubules are found at regular intervals along the length of the muscle fiber. In mammalian skeletal muscle they are found at each A-I junction (figure 3.3). Since a T-tubule is a true invagination of the surface membrane, it follows that its lumen is a continuation of the extracellular space. This has turned out to be an important point when discussing muscle fatigue. We shall return to this question in chapter 15.

## The Interior of the Muscle Cell

Now it is time to turn to the interior of the muscle cell. We have already mentioned that one muscle cell may contain many nuclei, sometimes as many as several thousands (Roy, Monke et al. 1999). In relation to training adaptations (see chapter 11), it is important to realize that each nucleus seems to govern a part of the muscle cell cytoplasm surrounding it, called a **nuclear domain** (Pavlath et al. 1989). Among other things, this implies that the number of muscle cell nuclei has to increase if the volume of the muscle cell increases by hypertrophy to keep the nucleus/cytoplasm ratio constant. This happens by recruitment and fusion of satellite cells with the hypertrophying muscle fiber.

The interior of a skeletal muscle cell is dominated by the contractile elements, the myofibrils, running parallel to the long axis of the muscle cell. Each myofibril consists of regularly repeating units, called **sarcomeres,** arranged in series (figure 3.4). The most striking feature, however, is the regularity of the arrangement of sarcomeres in neighboring myofibrils, which gives rise to the transverse striations so typical of skeletal muscle cells. The traditional view is that each sarcomere is composed of two types of protein filaments, **actin** filaments and **myosin** filaments, often collectively named **myofilaments.** In addition to actin molecules, the actin filaments contain two types of regulatory proteins, troponin and tropomyosin. As we shall see, this is an oversimplification, but let us start by taking a look at the traditional picture. Despite the fact that neither actin nor myosin is able to shorten, both are commonly referred to as contractile proteins. The contractility of the myofibril is now known to be a function of an interaction between its two types of myofilaments rather than a property of each individual myofilament (R. Craig 1994).

The myofibrils, each 1 to 3 µm thick, are characterized by the alternating dark and light bands constituting the sarcomeres. The dark bands are called **A-bands** and the light ones **I-bands** (see figure 3.4). The letters *A* and *I* stand for anisotropic and isotropic, designations originally given to the bands because of their optical properties. Today this explanation has mainly historical interest. In the electron microscope a thin line can be seen in the middle of each I-band. This line is often called the Z-line, but it should rather be called the **Z-disk** because it crosses the whole more or less circular cross section of the myofibril. The letter *Z* comes from German *zwischen,* which means "between," because the Z-disks divide the myofibril in the sections called sarcomeres. One sarcomere can therefore be said to consist of one A-band and two halves of I-bands, one on each side of the A-band (see figure 3.4). The myosin filaments, each 1.6 µm long, extend from one end of the A-band to the other. The actin filaments, on the other hand, extend from the Z-disk through the half I-band and partly into the A-band, leaving a zone in the middle of the A-band devoid of actin filaments. This zone appears lighter than the rest of the A-band and is called the **H-band** after Hensen, who first described it in the light microscope in 1868 (see Häggquist 1931). In the center of the H-band, some thin lines cross the myofibril. They are called **M-lines** (from German *Mittelinie,* midline).

Two other proteins have been shown to play significant roles in sarcomere assembly and function (figure 3.5). **Nebulin,** so called because its function for a long time seemed nebulous, is now regarded primarily as a ruler, lying alongside the actin filament, assisting its assembly from **G-actin** monomers, and securing its correct length (Krüger, Wright, and Wang 1991). **Titin,** also named **connectin,** is the largest protein known, with a molecular mass of 3,000 kDa. It spans the entire half sarcomere, from the Z-disk to the M-lines in the middle of the A-band. Titin is an elastic protein, especially the part located in the I-band (R. Craig 1994; Patel and Lieber 1997). It is therefore believed that titin confers a substantial fraction of the passive

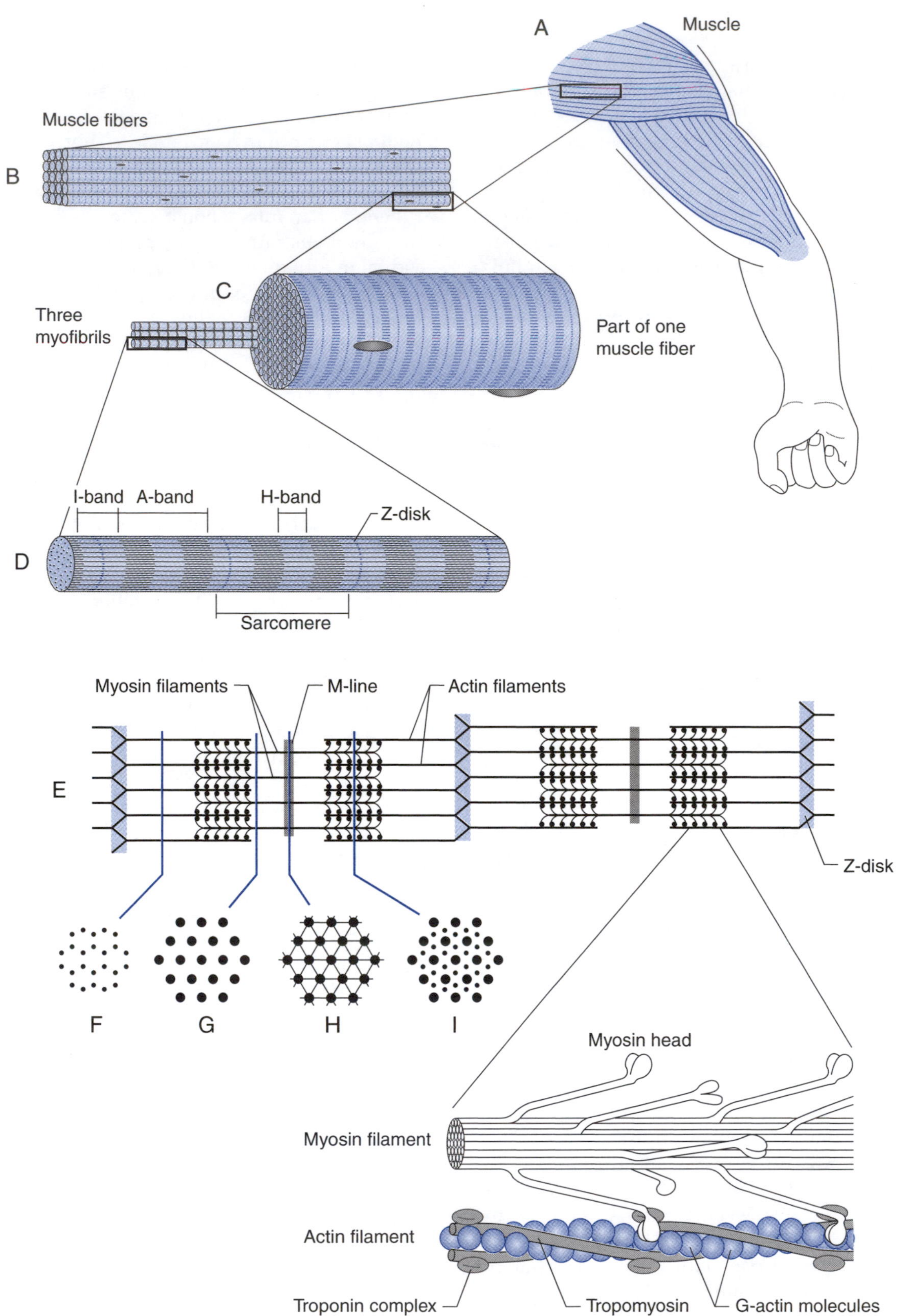


**Figure 3.4** The macroscopic, microscopic, and molecular organization of a skeletal muscle. Of the sarcomeric proteins, only actin, myosin, tropomyosin, and troponin are shown. For the location of titin and nebulin, see figure 3.5. *F, G, H,* and *I* show cross sections through different parts of the sarcomere as indicated. Note that the actin filaments are arranged in a hexagonal lattice *(F, I).*

Adapted, from W. Bloom and D.W. Fawcett, 1994, *A Textbook of Histology* (London: Chapman and Hall), 279. By permission of Arnold Publishing.

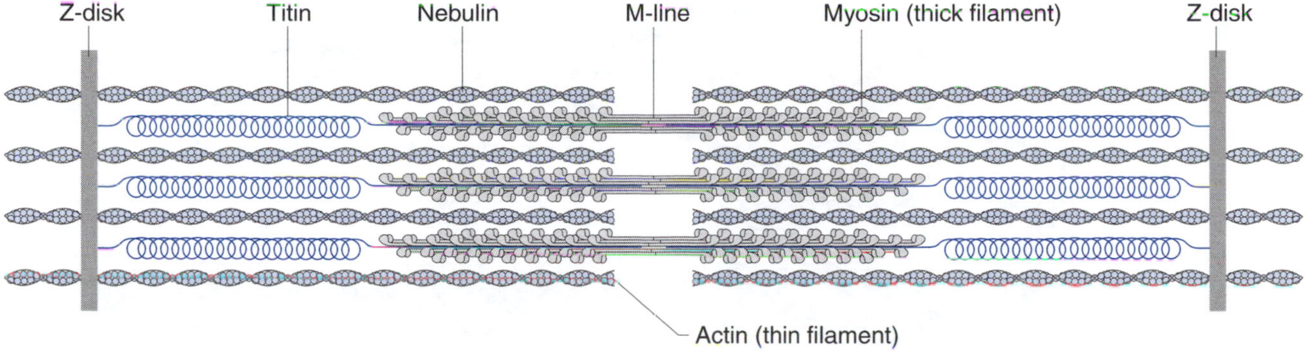


**Figure 3.5** Location of titin and nebulin in a skeletal muscle sarcomere. Each giant titin molecule extends from the Z-disk to the M-line. Within the A-band the titin molecule is tightly associated with the myosin filament. The rest of the molecule, lying in the I-band, is highly elastic and changes length as the myofibril contracts and relaxes. Each nebulin filament extends from the Z-disk to the tip of the actin filament and is believed to act as a molecular ruler that determines the length of the actin filament.



elasticity of the muscle fiber, thereby contributing to what is commonly known as its resting tension (Patel and Lieber 1997). One of its main functions may be to keep the myosin filaments in the center of the sarcomere (R. Craig 1994), thus ensuring equal force production in each half sarcomere (compare with the force–length relationship, discussed later in this chapter). In addition, experiments indicate that it also has a function in serial force transmission from the myosin filament to the Z-disk (Patel and Lieber 1997). We shall return to the question of force transmission in a later section of this chapter. Force transmission is another area that has turned out to be not quite as simple as once thought.

## For Additional Reading

For discussion of the possible role of titin for the shortening velocity, see Minajeva et al. (2002).

As mentioned earlier, one of the most striking features of a striated muscle cell is the regularity with which its myofibrils are aligned with regard to A- and I-bands (see figure 3.4), giving rise to the well-known cross striations. It has turned out that the myofibrils are not independent of each other but on the contrary are closely interconnected by so-called **intermediary filaments,** in particular **desmin.** These filaments extend from Z-disk to Z-disk and from the Z-disks of the most peripheral myofibrils to the sarcolemma (figure 3.6). In addition to ensuring correct alignment of neighboring myofibrils, these intermediary filaments probably assist in lateral force transmission, and it has even been claimed that they take part in the control of gene expression (Traub and Shoeman 1994).

In accordance with the frequent use of the prefix *sarco-* in connection with structures in the muscle cell, the term **sarcoplasm** is often used for its cytoplasm. As in other cell types, it consists of several types of organelles suspended in a cytosol, a viscous solution of ions, proteins, carbohydrates, and other molecules in water. Apart from the myofibrils, which by far dominate the muscle cell interior, the most important organelles for the specific functions of the muscle cell are the mitochondria and the **sarcoplasmic reticulum (SR).**

The mitochondria are often called the powerhouses of the cell, alluding to the fact that the major (i.e., aerobic) part of the energy metabolism takes place in the mitochondria (see chapter 2). In muscle fibers most mitochondria are found in clusters beneath the cell membrane (figure 3.7), more and larger clusters in some muscle fibers than in others, depending both on muscle fiber type and training status.

The SR is the endoplasmic reticulum of the muscle fiber and functions as an intracellular calcium store. As we shall see later, calcium ions ($Ca^{2+}$) are the ultimate intracellular signal in the sequence of events connecting the nerve impulse to muscle contraction. The SR consists of flattened vesicles wrapped around the myofibrils like lace cuffs or sleeves (see figure 3.3). Before we continue with the description of the SR, we need to return to the T-tubules for a moment.

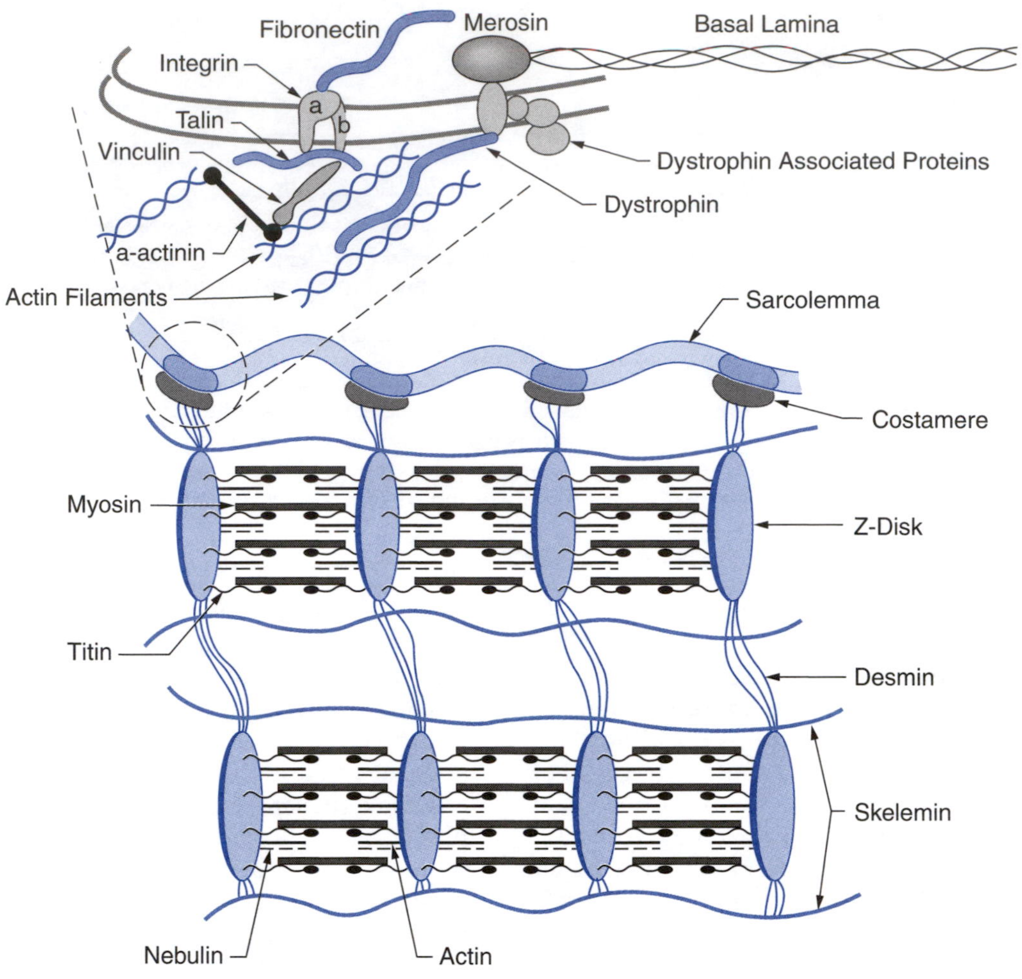


**Figure 3.6** Diagram of the muscle fiber cytoskeletal lattice. Some of the details about the relationship between the different proteins involved are hypothesized, based on indirect evidence. The variety of connections between structures provides a myriad of possible routes of force transmission from the sarcomere to the extracellular matrix.

Reprinted, by permission, from T.J. Patel and R.L. Lieber, 1997, "Force transmission in skeletal muscle: from actomyosin to external tendons," *Exercise and Sport Sciences Reviews* 25:321-363.

In mammalian skeletal muscle there is one T-tubule at each A-I junction (Franzini-Armstrong 1994). On each side of a T-tubule is a slightly dilated part of the SR called the **terminal cisterna.** The membrane of the terminal cisterna and the T-tubule membrane are very close together, and with high resolution electron microscopy it is possible to see small, electron-dense spots representing material bridging the gap between the two membranes. Together the T-tubule and the two abutting terminal cisternae are called a **triad** (Franzini-Armstrong 1994). The close apposition of the T-tubule, which represents an extension of the sarcolemma into the interior of the muscle fiber, and the SR is the morphological basis for the coupling of a muscle fiber **action potential** to the release of calcium ions in the process called **excitation–contraction coupling,** or **E–C coupling.**

## Myofibrillar Fine Structure

In spite of the discovery of new components of the myofibril, the myosin and actin filaments still play the leading parts in muscle contraction. The myosin filament, 1.6 μm long, is a multimolecular aggregate of myosin molecules, each 150 nm long (figure 3.8*a*). The form of the myosin molecule has been compared to that of a golf club, with a shaft forming part of the stem of the myosin filament and a head sticking out toward a neighboring actin filament. The myosin head contains the site for attachment to the actin filament and for adenosine triphosphatase (ATPase) activity, splitting adenosine triphosphate (ATP) to provide energy for contractile work. Several hundred myosin molecules are packed in a sheaf, with their heads pointed in one direction along half the filament and in the opposite direction

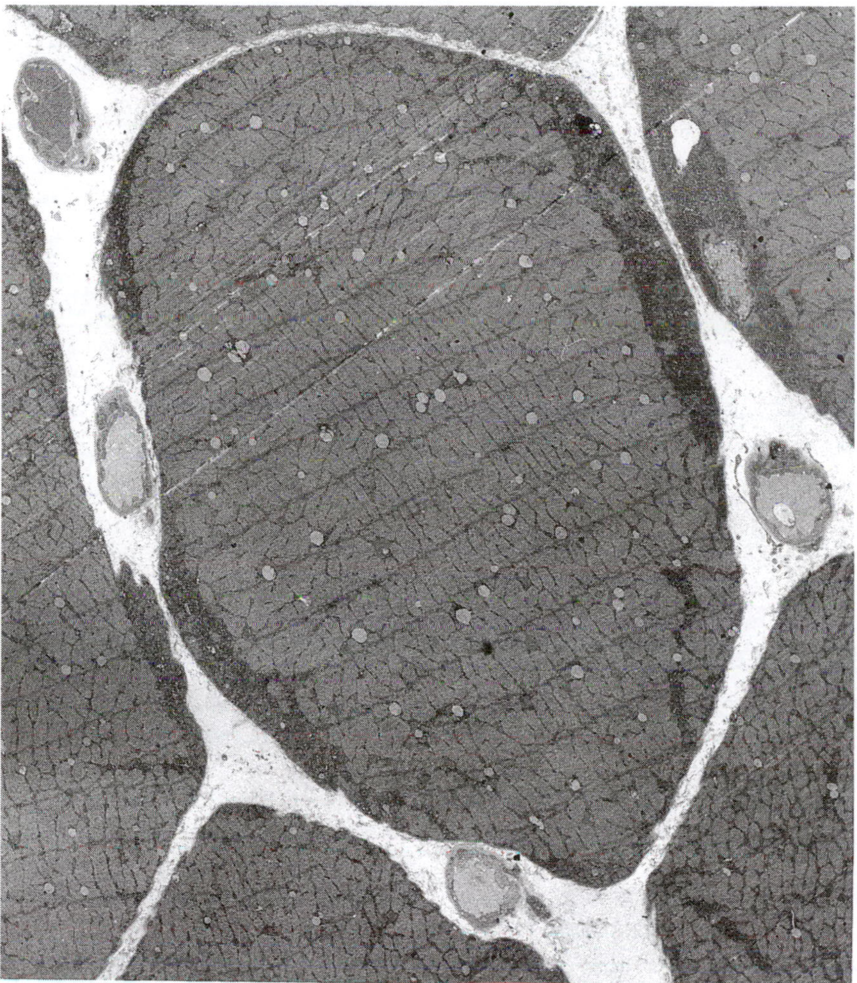

**Figure 3.7** Electron micrograph of a cross section of a type I muscle fiber from the vastus lateralis muscle of a trained individual. Clusters of mitochondria are seen beneath the cell membrane.

Reprinted, by permission, from H.A. Dahl and E. Rinvik, 1999, *Menneskets funksjonelle anatomi* (Oslo, Norway: J.W. Cappelens Forlag).

along the other half, leaving a zone devoid of protruding heads midway along the filament's length (figure 3.8*b*).

The myosin heads are arranged in pairs, which means that for each head sticking out from the myosin filament, another one is sticking out in the same plane on the opposite side of the filament. As we move along the filament, each pair is rotated 60° in relation to the preceding pair, meaning that every third pair is in the same plane. The distance between two neighboring heads in the same plane is 43 nm.

A cross section through the part of the A-band region that contains actin filaments reveals a very regular and characteristic pattern. Each myosin filament is surrounded by six actin filaments in a hexagonal lattice, corresponding exactly to the myosin heads protruding from the myosin filament (see figure 3.4). This places the myosin heads in an ideal position to interact with a neighboring actin filament during contraction.

Like actin filaments in general, the actin myofilament is composed of globular actin monomers, called G-actin, which attach to each other in a linear fashion like beads on a string. Such filamentous actin is called **F-actin.** In spite of their globular appearance, the actin monomers have a certain polarity, which is transferred to the filament as well. Two such strings are twisted around each other to make the stem of the thin filament. The polarity of the actin filaments in one half sarcomere is opposite to the polarity of the actin filaments in the other half sarcomere.

The two other components of the thin filament, troponin and tropomyosin, are located in the grooves

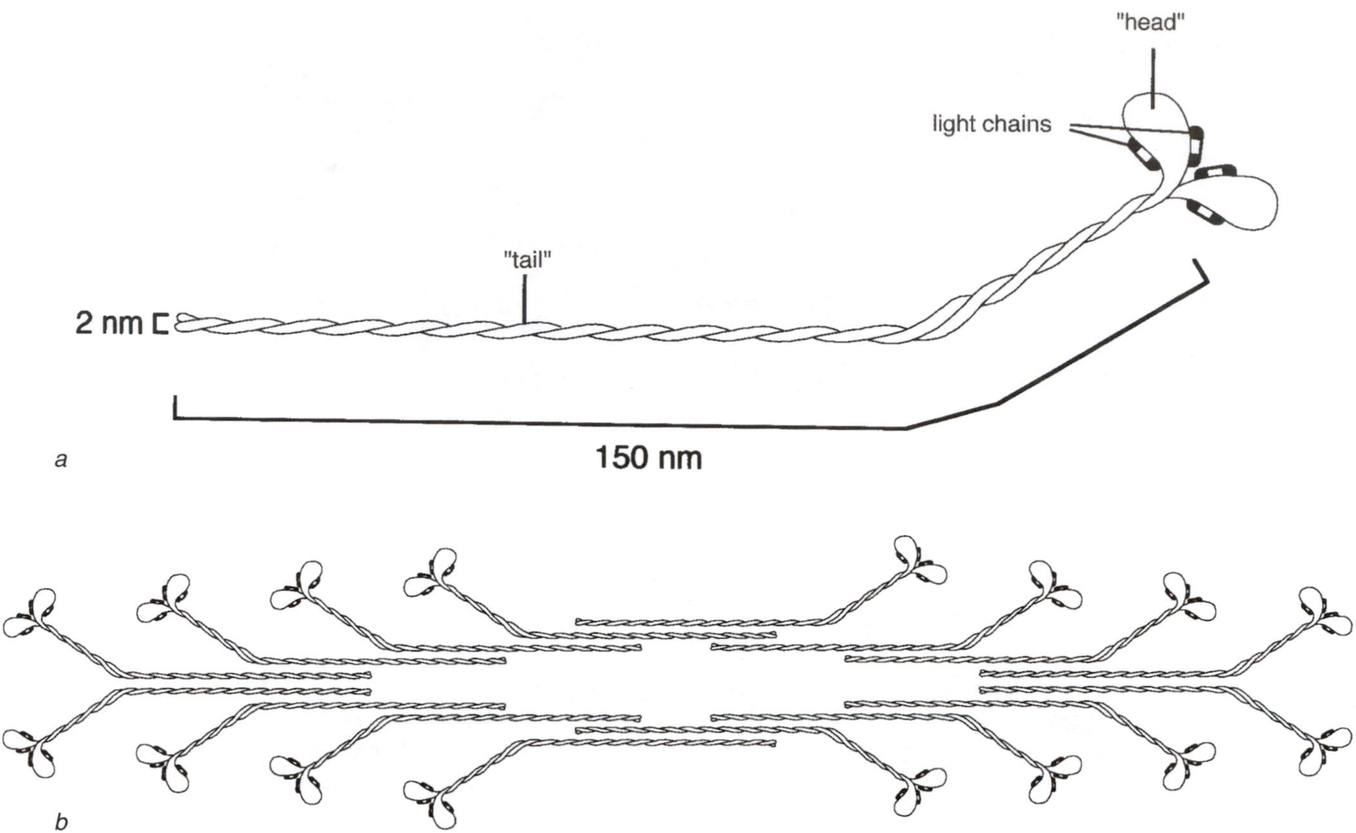


**Figure 3.8** *(a)* A myosin molecule consists of six subunits, two intertwined **heavy chains** and four **light chains.** The adenosine triphosphatase (ATPase) activity is located in the heads. *(b)* In a myosin filament (here shown exploded schematically), the tails of individual myosin molecules form the shaft with the heads protruding like barbs. Note that the myosin molecules in the two halves of the filament have opposite polarity, leaving a bare zone in the middle devoid of heads. Only myosin heads protruding in one plane are shown. Compare figure 3.4.

formed by the twisted strings. Tropomyosins are 40-nm-long rod-shaped molecules located in each of the two grooves formed by the intertwined F-actin strings. Each tropomyosin molecule is a dimer, either an αα or a ββ homodimer or an αβ heterodimer. The tropomyosin molecules are arranged in a head-to-tail fashion with a slight overlap. Each tropomyosin dimer spans seven actin monomers. Close to the end of each tropomyosin molecule sits one troponin molecule, which is composed of three subunits, troponin C (TnC), troponin I (TnI), and troponin T (TnT). TnC is a calcium-binding protein, TnI is the inhibitory subunit, able to inhibit the interaction between actin and myosin, and TnT is the subunit that confers binding to tropomyosin. TnC belongs to a **superfamily** of calcium-binding proteins that also includes calmodulin, myosin light chains, and parvalbumin. All subunits of troponin and tropomyosin exist in several isoforms (Schiaffino and Reggiani 1996).

For a long time it was puzzling that sarcomeric actin filaments are of very constant length, while the lengths of actin filaments in other cells are highly variable. The solution to the problem was nebulin. Nebulin is a filamentous protein that lies alongside the actin filament, acting as a molecular ruler. Evidence for this comes from various studies summarized by Patel and Lieber (1997). It has, for example, been shown that nebulin isoform size correlates with actin filament length (Krüger, Wright, and Wang 1991).

The muscle myosin molecule belongs to the myosin II subfamily of myosins (Schiaffino and Reggiani 1996). It is a **hexamer,** that is, it is composed of six subunits, two of which are **myosin heavy chains (MyHC).** The other four are **myosin light chains (MyLC).** As mentioned earlier, each heavy chain resembles a golf club with a shaft and a globular end. In the native myosin molecule, the tails of the two heavy chains are twisted around each other, while the two globular ends together make the myosin head. It is the twisted shafts of the heavy chains that assemble as the stem of the myosin filament, while the heads protrude from the stem like barbs.

The MyLCs are located close to the myosin head. Two light chains are associated with each of the heavy chains (see figure 3.8*a*). One of these is always LC2, the other either LC1 or LC3. LC2 can be phosphorylated and is called a regulatory or phosphorylatable light chain, while LC1 and LC3 are called essential, or alkali, light chains. We shall return to the importance of the light chains when we deal with muscle fiber types later. Figures 3.5 and 3.6 summarize schematically the structure of myofibrils and the location of their constituent proteins.

## Summary

A muscle cell is a long, multinucleated cell resulting from the fusion of mononucleated myoblasts in fetal life. Due to its considerable length in relation to its diameter, a muscle cell is commonly known as a muscle fiber. Each muscle fiber is surrounded by a basement membrane, which in turn borders directly on the connective tissue of the endomysium. The muscle fiber itself is a postmitotic cell, but some myoblasts persist as satellite cells in adult life. The satellite cells are found between the muscle fiber membrane and the basement membrane. They are able to divide and fuse with the muscle fiber during hypertrophy or repair.

A certain number of muscle fibers are contacted by branches from the same motoneuron and contract simultaneously and in the same manner when the motoneuron fires. Consequently, a motoneuron, its axon branches, and all the muscle fibers contacted by these branches are collectively called a motor unit. A motor unit may be likened to a work crew, all members of which always work at the same time and with equal intensity.

The interior of the muscle fiber is dominated by the myofibrils, the main components of which are actin and myosin myofilaments. In addition to actin molecules, the thin actin filaments contain two types of regulatory proteins, troponin and tropomyosin. The length of the actin filaments is fairly constant in skeletal muscle due to a molecule named nebulin, which acts as a ruler. The myosin filaments are thicker and consist of myosin molecules. Individual myosin molecules are hexamers, consisting of two MyHCs and four MyLCs. Most myofibrillar proteins exist in several isoforms.

# Interaction of Actin and Myosin

There can be no doubt any longer that contraction of a skeletal muscle fiber takes place by a sliding movement of its actin and myosin filaments relative to each other (R. Craig 1994; Huxley 1974). During this sliding movement the actin filaments in each half sarcomere are pulled toward the center of the sarcomere by the myosin heads, resulting in a progressive narrowing of the I-bands but leaving the A-bands unchanged. Because the actin filaments are attached to the Z-disks, they are drawn together too, and the whole sarcomere shortens.

### FOR ADDITIONAL READING

For a tribute to Hugh E. Huxley and his sliding filament model, see Weber and Franzini-Armstrong (2002).

This, of course, refers to a concentric contraction, that is, a contraction that leads to shortening of the muscle fiber. During an isometric contraction, which causes no change in muscle length, the length of the I-band remains constant (however, see comments on isometric contractions of whole muscle in the section "Force Transmission From Muscle Fiber to Tendon" on page 45). During an eccentric contraction the actin filaments are pulled out of the A-band by an external force acting on the muscle, despite attempts by the myosin heads to pull them in the opposite direction, and the I-band broadens. Irrespective of the mode of contraction, however, neither of the myofilaments changes its length.

## From Nerve Signal to Muscle Contraction: Excitation–Contraction Coupling

During rest there is no electrical activity in the muscle (see chapter 4), and there is no contractile activity. A muscle's resting tension is entirely due to passive elements in the muscle, especially the titin filaments (Horowits 1992). In this relaxed state the myosin heads are located close to the actin filaments but are unable to attach because of the blocking action of the regulatory proteins.

Under normal circumstances only a nerve impulse can initiate a contraction in a skeletal muscle. The contact between the motor nerve fiber and the muscle fiber is called a motor endplate or a **neuromuscular synapse.** A fundamental characteristic of the motor endplate is that each endplate contains a cluster of synaptic endings (figure 3.9*b*), all of which are activated simultaneously by an action potential in the parent nerve fiber. By means of spatial summation of their individual postsynaptic potentials, this simultaneous activation ensures that every action potential traveling down the motor axon results in the generation of an action potential in the muscle fiber. (For further details, see chapter 4.) On this basis, we now consider the effect of one single action potential arriving at the motor endplate.

When the action potential reaches one of the synaptic endings, synaptic vesicles within the terminal release their content of the neuromuscular transmitter substance **acetylcholine (ACh).**

## FOR ADDITIONAL READING

For details regarding the steps leading to the release of neurotransmitters, see Schiavo and Stenbeck (1998).

The ACh molecules are captured by specific ion channel receptors (see chapters 2 and 4), whereupon the channels open, allowing sodium ions to diffuse down their electrochemical gradient from the extracellular fluid into the muscle cell. Just as in other excitatory synapses, this leads to a depolarization of the postsynaptic cell. Spatial summation of the individual depolarizations results in an endplate potential that exceeds the firing threshold, giving rise to a new action potential in the muscle fiber.

The muscle action potential spreads from the motor endplate, usually located somewhere in the middle third of the muscle fiber, toward both ends, at the same time penetrating, by means of the T-tubules, into the interior of the muscle fiber. Embedded in the T-tubule membrane, opposite the termi-

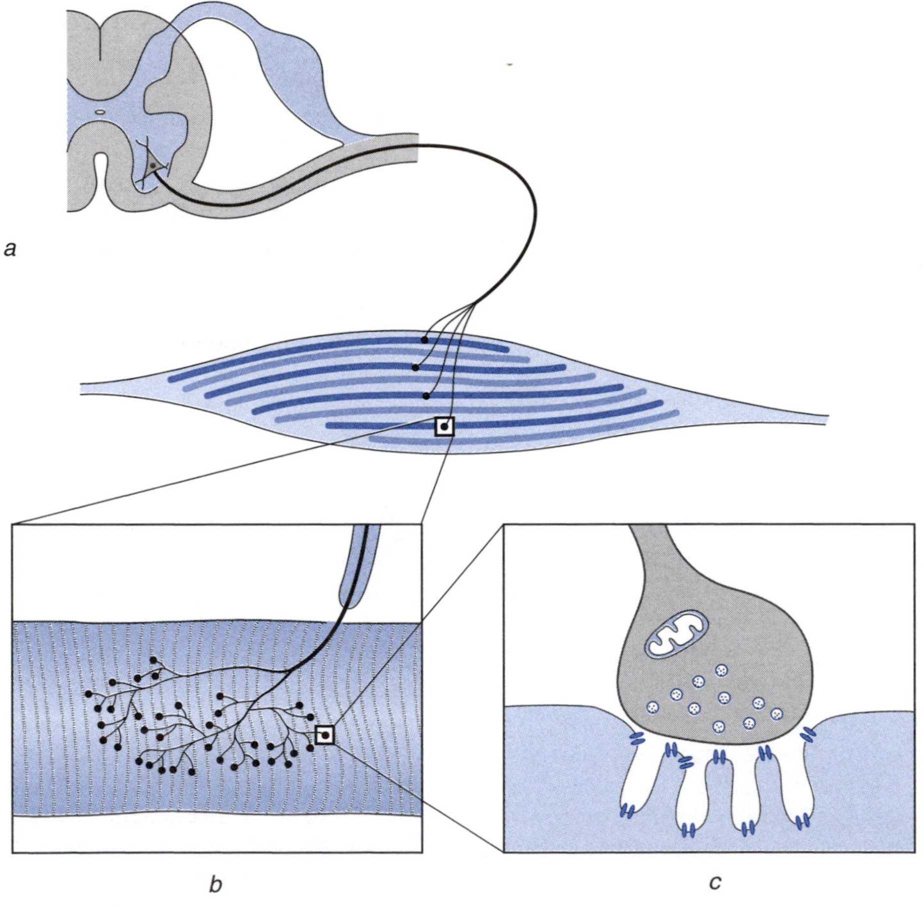


**Figure 3.9** *(a)* A motor unit consists of a motoneuron in the ventral horn of the spinal cord, its axon and axon branches, and all the muscle fibers contacted by these branches. *(b)* As each axon branch approaches the surface of the muscle fiber, it divides into smaller branches, each ending in a synaptic terminal. *(c)* Each synaptic terminal is in close contact with the muscle cell membrane, which is highly folded under the terminal. The terminal contains synaptic vesicles with acetylcholine (ACh), and the muscle cell membrane is equipped with many ion channels: ACh-gated ion channels at the top of the junctional folds, close to the sites where ACh is released, and voltage-gated ion channels at the bottom.

Adapted, by permission, from H.A. Dahl and E. Rinvik, 1999, *Menneskets funksjonelle anatomi* (Oslo, Norway: J.W. Cappelens Forlag).

nal cisternae of the SR, are voltage-sensitive protein molecules called **dihydropyridine receptors,** which in turn are in close contact with proteins named **ryanodine receptors** in the SR membrane. Both the voltage-sensitive dihydropyridine receptor and the ryanodine receptor are in fact calcium channels, and when an action potential travels past the voltage-sensitive T-tubule protein, the latter interacts in some way with the SR calcium channel, resulting in a release of calcium ions from the SR lumen into the cytosol (see figure 3.3). The calcium ions diffuse into the myofibrillar lattice, where they are captured by troponin C. This initiates a conformational change in the troponin–tropomyosin complex, which exposes actin's myosin-binding sites, allowing actin and myosin to interact. In a way, the troponin–tropomyosin complex can be regarded as a chaperone that prevents untimely interaction between actin and myosin. When the complex is "distracted" by calcium ions, however, its chaperone function is temporarily discontinued and remains so until the level of free calcium ions in the cytosol is back to resting level.

If the rise in calcium concentration is the result of one single action potential, the calcium level returns to resting level very quickly, due to active calcium reuptake mechanisms in the SR membrane. The energy necessary to transport calcium ions against their electrochemical gradient comes from ATP. Consequently, these pumps are known as $Ca^{2+}$ ATPases. The brief rise in calcium concentration, often called the **calcium transient,** allows actin and myosin to interact for an equally short time, resulting in a very brief contraction, a twitch. The duration of a single twitch is, however, too short to result in a visible movement of the muscle. A surprisingly high percentage of single action potentials has, however, been observed during normal ambulatory activity in rat muscle (Hennig and Lømo 1985), possibly serving as corrective impulses during the course of a movement. In general, however, the signal adequate for muscle contraction is not a single action potential but a series of action potentials known as an impulse train. It is not the individual action potential that is the main signal, neither in muscle nor in the central nervous system. It is the impulse frequency in a train of impulses. In other words, the signal is frequency modulated, like FM (i.e., frequency-modulated) radio transmission signals.

The instantaneous concentration of calcium ions in the cytosol depends both on their rate of release from the SR and on their reuptake into the SR by means of $Ca^{2+}$ ATPases. In the present context, the rate of uptake into the SR can be regarded as constant in a given fiber. This leaves the frequency of the impulse train more or less alone in charge of the instantaneous cytosolic calcium concentration. A low frequency gives a low calcium concentration and a low contraction force. Increasing the impulse frequency progressively increases calcium concentration and force, until a maximal force is obtained (figure 3.10), possibly because all TnC calcium-binding sites are occupied.

Depending on the isoform, TnC has three or four calcium-binding sites: two high-affinity sites and one or two low-affinity sites. The high-affinity sites are saturated even at resting calcium levels. It is the low-affinity sites that bind calcium ions at high levels and release them again when the level declines (Schiaffino and Reggiani 1996); they thus

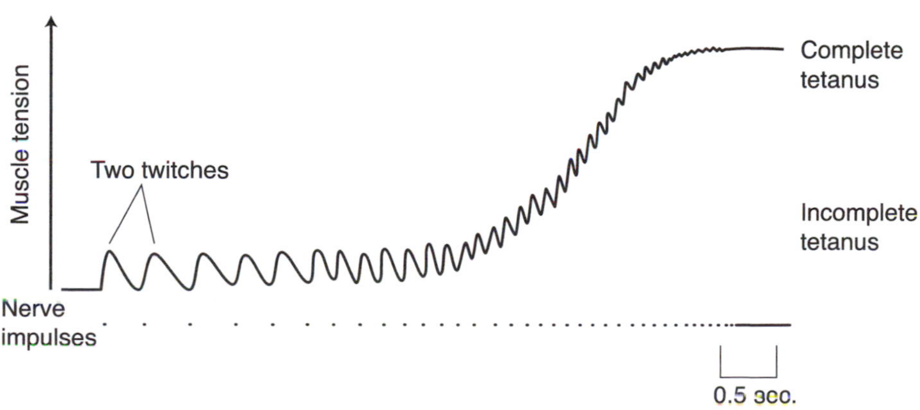


**Figure 3.10** Rate coding of muscle force. The curve shows isometric force development of a muscle fiber during electrical stimulation at a progressively increasing rate. At low rates, single twitches are evoked. Depending on the type of motor unit, individual twitches start to fuse at higher rates, at first to incomplete tetanus, then at even higher rates to complete tetanus.

represent the switch that turns the contractile process on and off.

The rate of binding of calcium to the low-affinity sites is, however, low in relation to the duration of the twitch calcium transient. Therefore, the transient is too brief to allow equilibration between cytosolic and TnC-bound calcium, and the resulting twitch force is lower than expected from the cytosolic calcium concentration. In other words, it is the amount of calcium ions bound to TnC, rather than the cytosolic concentration per se, that determines the contractile force.

It has been debated whether the calcium concentration obtained in the cytosol during a twitch is of the same magnitude as the concentration during a tetanic concentration (see for example J.C. Rüegg 1992). In an **in vitro** experiment with skinned fibers, it can be shown beyond doubt that the contractile force is positively correlated to the calcium concentration in the bathing solution. Today, there is general agreement that the cytosolic calcium concentration is higher during a high-frequency impulse train than following a single stimulus (Sculptoreanu, Scheuer, and Catterall 1993). This may be due to the following mechanisms. There is evidence that it may be due to potentiation of the activity of the dihydropyridine receptors (DHPR) by voltage-dependent phosphorylation, leading to increased $Ca^{2+}$ current through the DHPR, which is a calcium channel in its own right (Sculptoreanu, Scheuer, and Catterall 1993; Catterall 1997). In turn, this increases the calcium release from the sarcoplasmic reticulum, possibly by calcium-induced calcium release (figure 3.11), which to a large extent may occur through ryanodine receptors not directly coupled to DHPRs (Shoshan-Barmatz and Ashley 1998; Sorrentino and Reggiani 1999).

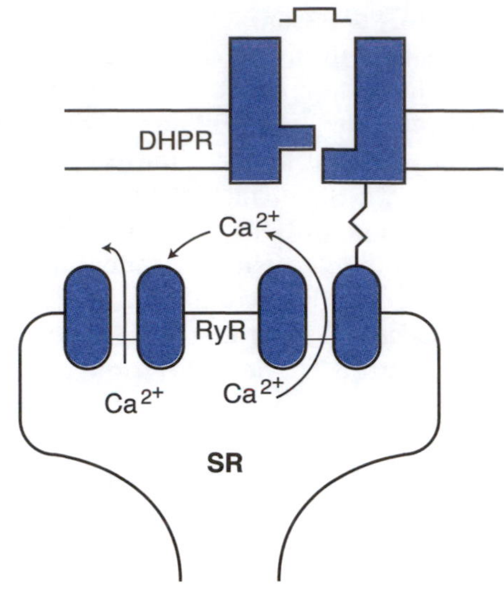


**Figure 3.11** Model of the coupling between the voltage-sensitive dihydropyridine receptor (DHPR) in the T-tubule membrane and the ryanodine receptor (RyR) in the sarcoplasmic reticulum (SR). The action potential activates the DHPR calcium channel, which in turn activates the RyR, allowing calcium ions ($Ca^{2+}$) to escape from the SR. In turn, $Ca^{2+}$ activates neighboring RyRs, thereby increasing the cytosolic concentration of $Ca^{2+}$.

Reprinted, by permission, from J.A. Wasserstrom, 1998, "New evidence for similarities in excitation-contraction coupling in skeletal and cardiac muscle," *Acta Physiologica Scandinavica* 162:247-252.

### For Additional Reading

For a discussion of the possible effect of nitric oxide on calcium release and cross-bridge kinetics, see Heunks, Cody, Geiger et al. (2001) and Heunks, Machiels, Dekhuijzen et al. (2001).

## Relaxation Is As Important As Contraction in Normal Movements

Switching between an active and an inactive state is a common feature to most biological mechanisms, and in every case, switching off is just as important as switching on. In this regard, muscle contraction is no exception. Imagine a man running as fast as he can. Not only must he be able to move his legs with sufficient speed, he must also be able to switch his hip flexors and hip extensors on and off in a reciprocal manner, the flexors relaxing when the extensors contract, and vice versa. Even though running can be a most demanding activity with regard to the speed and correct timing of the relaxation, a concerted collaboration between agonists and antagonists is a fundamental requirement in most movements.

Because muscle contraction is caused by an increase in the cytosolic calcium concentration, the logical key to relaxation is a lowering of the calcium concentration. For this to happen, the train of impulses from the central nervous system that cause and maintain the outpouring of calcium ions from the SR must stop, and the calcium ions already in the cytosol must be removed. The former is, of course, the task of the nervous system. The latter is taken care of by the $Ca^{2+}$ ATPases in the SR membrane.

Inside the SR, especially in the lateral sacs, free calcium ions are in equilibrium, with calcium ions bound to a protein named **calsequestrin** (Franzini-Armstrong and Jorgensen 1994). In fast muscle fibers of small animals such as the rat, **parvalbumin,** a calcium-binding protein in the cytosol, has been claimed to help in the lowering of free $Ca^{2+}$, thereby removing calcium ions from TnC and increasing the speed of relaxation. The prolonged relaxation time found in parvalbumin knockout mice supports this assumption (Schwaller et al. 1999). In larger organisms such as humans, where rates of relaxation are lower, only negligible amounts of parvalbumin are found, even in muscles with a high percentage of fast muscle fibers (Perry 1994).

## Summary

Contraction of a skeletal muscle fiber takes place by movement of the actin and myosin filaments in relation to each other. Neither of the filaments changes length during contraction. Under normal circumstances, contractile activity of a muscle fiber is initiated by a sequence of nerve impulses—an impulse train—in its motor nerve fiber. Each nerve impulse initiates the generation of an action potential in the muscle cell membrane. These action potentials spread out from the motor endplate toward both ends of the muscle fiber, at the same time penetrating into the interior of the muscle cell by means of the T-tubules. The T-tubules are in close contact with the lateral sacs of the SR, forming the structural link between the action potential and the resulting release of calcium ions from the SR. The calcium ions bind to troponin C, thereby removing the inhibition imposed by the regulatory proteins on the contractile process. Relaxation of the muscle fiber is the result of a lowering of the calcium concentration in the cytosol due to active relocation of the calcium to the SR by calcium ATPases in the SR membrane.

# Force Transmission From Muscle Fiber to Tendon

We now take a closer look at the transmission of force from the muscle fiber to the tendons of attachment. In fact, we need to start by looking at the transmission from the myofibrils to the muscle fiber membrane, realizing that unless the myofibrils are firmly anchored in the sarcolemma, no force transmission can take place.

The traditional view holds that force is transmitted serially from the end of individual myofibrils to their attachment to the sarcolemma in the end of the muscle fiber, adjacent to the **myotendinous junction (MTJ).** In electron microscope pictures of a myotendinous junction, fingerlike extensions of the muscle fiber are seen to interdigitate with similar extensions of the connective tissue of the tendon (figure 3.12). With the recognition of series-fibered muscles, the traditional view cannot any longer be regarded as the only mode of force transmission. There are two major reasons for this.

First of all, serially arranged muscle fibers lack MTJs, at least in the end terminating intrafascicularly (Trotter, Richmond, and Purslow 1995). Second, serially arranged muscle fibers have tapered ends (figure 3.13), leaving room for fewer myofibrils in the ends than in the middle part of the fiber. Consequently, for the shorter myofibrils to contribute to the overall force of the muscle fiber, their force has to be transmitted laterally to neighboring myofibrils and eventually to the sarcolemma. A number of experiments have shown that such lateral transmission of force really occurs.

### For Additional Reading

For discussion of lateral force transmission in muscle fibers, see Trotter, Richmond, and Purslow (1995), Patel and Lieber (1997), and Monti et al. (1999).

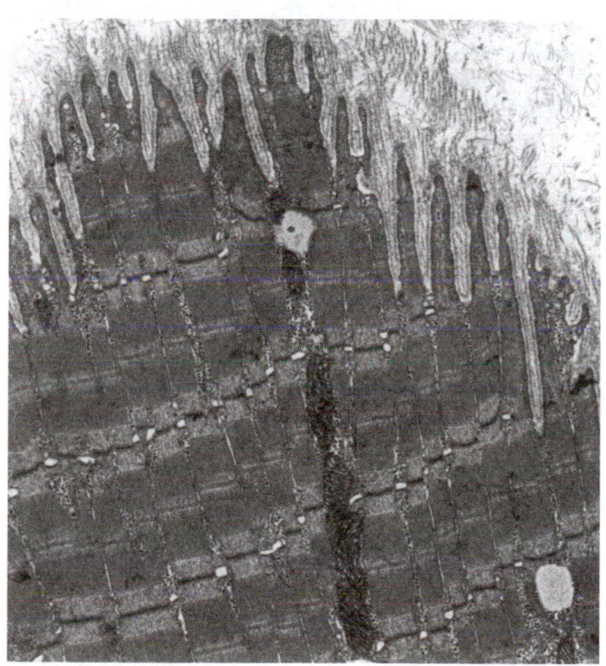

**Figure 3.12** Electron micrograph showing a longitudinal section through the myotendinous junction of a frog muscle fiber. The muscle cell membrane is deeply invaginated, forming what appear as fingerlike extensions interdigitating with the connective tissue of the tendon. Note the near-perfect alignment of the sarcomeres in neighboring myofibrils.

From Horwitz, et al., 1994. In *Myology*, edited by A.G. Engel and C. Franzini-Armstrong (New York: McGraw-Hill), 213. By permission of McGraw-Hill.

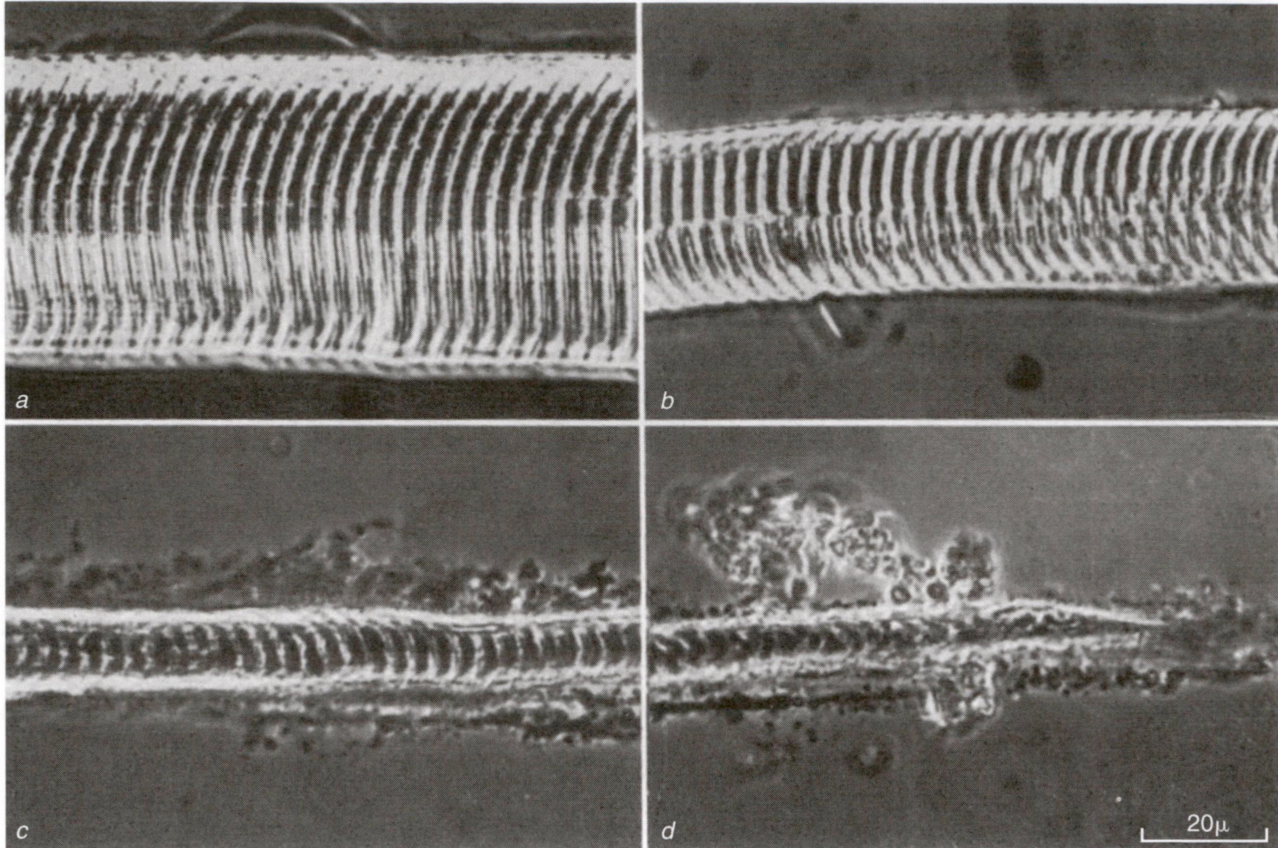


**Figure 3.13** Phase contrast micrographs of a tapering muscle fiber from the human sartorius. Pictures are taken at .5-cm intervals.

B. Barrett, 1962, "The length and mode of termination of individual muscle fibers in the human sartorius and posterior femoral muscles," *Acta Anatomica* 48:251. Reprinted, by permission, from Karger, Basel.

The lattice of intermediary filaments may account for such lateral force transmission. In addition to ensuring correct alignment of A- and I-bands of neighboring myofibrils, intermediary filaments are now believed to transmit contractile forces not only between myofibrils, but also between the most peripheral myofibrils and the sarcolemma (see figure 3.6). The intermediary filaments are attached to the sarcolemma through a complex of proteins assembled in structures called **costameres.** This name derives from their riblike appearance after they are labeled with antibodies against vinculin, one of their constituent proteins (Pardo, Siliciano, and Craig 1983).

How then is the force transmitted from the sarcolemma to the endomysium and eventually to the tendon? Most probably this occurs through shearing forces from the sarcolemma to the endomysium and from there through other muscle fibers and the perimysium to the tendon. For details, see Trotter, Richmond, and Purslow (1995) and Patel and Lieber (1997). Shearing forces are the same kind of forces that allow you to ski uphill classic style or keep your car on the right track when making a turn.

This mechanism of lateral force transmission through shearing forces in no way invalidates the traditional view of force transmission through MTJs. The two mechanisms should be regarded as complementary rather than mutually exclusive. The ultrastructural changes in MTJs after hind-limb unweighting and the accompanying increased tendency to rupture (Tidball 1991b) support this assumption. Such complementary routes of force transmission may ensure some continued function when a muscle fiber is ruptured (Tidball 1991a), an ability that may have conferred increased chances of survival to our ancient predecessors.

## Force Transmission Through Passive Tissue Structures

One more question needs to be addressed in this section, and that is the efficiency of force transmission through passive tissue structures. This question is of paramount importance in relation to the function of series-fibered muscles. Admittedly, we do not know for sure how individual fibers in a series-fibered motor unit are placed in the muscle, but chances are that they do not span the entire distance between muscle origin and attachment (Trotter, Richmond, and Purslow 1995). If that is the case, forces may have to be transmitted through passive structures such as unrecruited muscle fibers and connective tissue to the tendon of attachment. Intuitively, this seems questionable, but calculations have shown that it is entirely feasible (Trotter, Richmond, and Purslow 1995). In animal muscles, it has been shown that type I muscle fibers more often span the entire distance from tendon to tendon than type II fibers (Ounjian et al. 1991). If that is the case in human muscle as well, the problem of force transmission through passive structures at low force levels may be rather theoretical. At high force levels the percentage of recruited motor units may be assumed to be sufficiently high to secure effective force transmission to the tendon.

As one can easily observe during contraction of the quadriceps muscle with the knee extended, such "isometric" contractions give rise to considerable movement of the patella. This observation goes back to A.V. Hill, but recently it has been visualized by means of ultrasound at the muscle fascicle level. Fukunaga et al. (1997) have shown that the compliance of passive structures during isometric contractions of the human quadriceps muscle may lead to fascicle shortening amounting to more than 20% when the knee is extended. Even in the flexed position, when the tendinous structures are at their maximal physiological length, the elongation was 5% (compare to figure 3.27). The most obvious consequence of this is that some sarcomere shortening probably takes place initially even in isometric contractions. Less obvious are the consequences for muscular function at the whole-muscle level.

## Summary

Force transmission from the muscle fibers to the tendon takes place both through specialized myotendinous junctions in the ends of the muscle fibers and through the endomysial connective tissue surrounding each fiber. In the latter case, force transmission takes place through shearing forces. This type of force transmission may be the only one in muscle fibers that are shorter than the fascicles in which they lie. The connective tissue of a muscle is, however, rather compliant, making contractions that appear to be isometric on the whole-muscle level at least partly concentric on the muscle fiber level.

# Muscle Fiber Types and Their Molecular Basis

From a functional point of view, muscle is not a homogeneous tissue. Most muscles contain muscle fibers with different contractile and metabolic properties. In some species such differences also manifest themselves as visible color differences. This is the case in the rabbit, where Ranvier (1873) was able to show that red muscle had a slower contraction than white muscle. Although the color differences are less conspicuous in many species, including humans, the names *red muscle* and *white muscle* were widely used until the middle of the 20th century. With the advent of muscle histochemistry, the ability to distinguish muscle fiber types on the basis of their content of various enzymes increased significantly. Not until it was shown that the ATPase activity of a muscle fiber correlates with its speed of shortening (Bárány 1967), however, was a more meaningful basis for classification established.

## Strategies for Fiber Typing

In animal experiments it is possible to measure the contractile properties of single motor units and afterward perform experiments to reveal their histochemical profile. Such experiments rely heavily on a safe method to identify the motor unit that has been characterized physiologically. The glycogen depletion method, introduced by Edström and Kugelberg (1968), fulfills this need. By extending the findings of Edström and Kugelberg (1968), Burke et al. (1973) established a basis for classification of muscle fiber types (figure 3.14). As we shall see in a later section of this chapter, the methods of classification have been substantially improved since then, but the main conclusions in that paper are still valid.

Before discussing muscle fiber types in greater detail, we need to define what we mean by this concept. A useful definition could be "a group of muscle fibers with similar functional properties," but as we shall see, this definition has serious limitations. The most important limitation is the rather large variation in functional properties that can be found within groups of muscle fibers with the same fiber-type designation. The obvious explanation for this is the plasticity of muscle fibers, always adapting to changes in the level of activity, be it training or inactivity. Under physiological

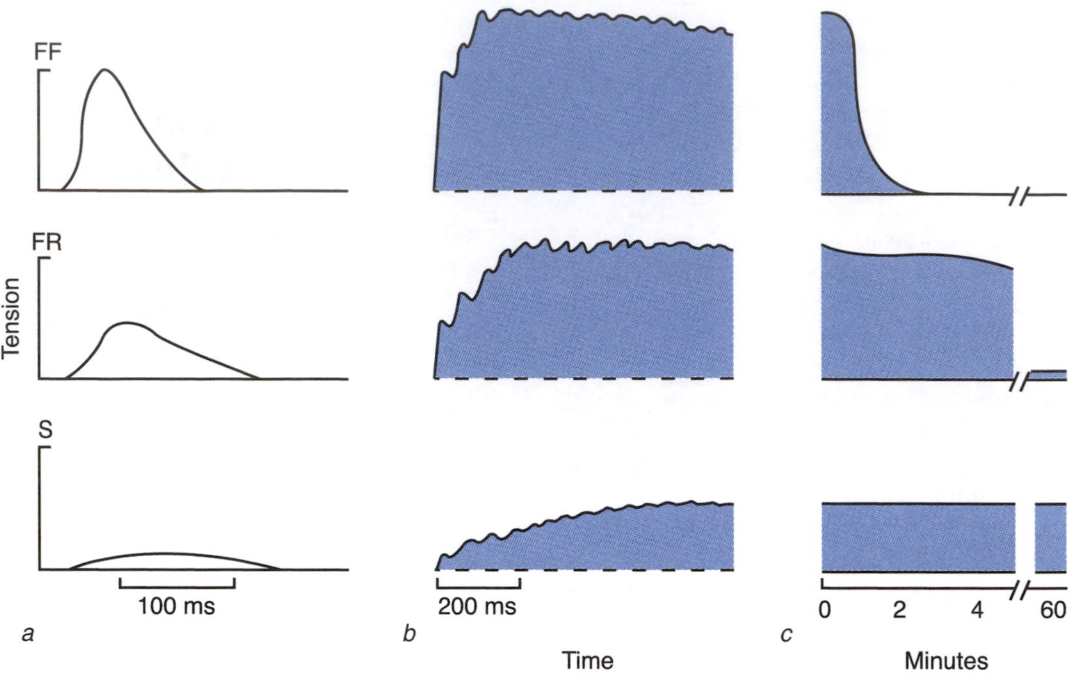


**Figure 3.14** Graphs of typical isometric tension development of the three types of motor units found in the cat gastrocnemius muscle. *(a)* Twitch tension. *(b)* Unfused tetanus evoked by stimulus trains of 20 Hz. *(c)* Curve of isometric maxima obtained by trains of stimulation at 40 Hz lasting 330 ms, repeated every 1 s. Note the large twitch force in the FF unit in *a*, the slow buildup of tetanic force in the S unit in *b*. The rapid decline of tension in the FF unit gives rise to its designation FF, for fast fatigue. The S (slow) unit is clearly the most fatigue resistant of the three, showing no force decay after 60 min, while the FR (fatigue resistant) unit behaves in an intermediate fashion.

Reprinted from SCIENCE 174:709-712, R.E. Burke, et al., 1971, "Mammalian motor units: Physiological-histochemical correlation in three types in cat gastrocnemius."

## CLASSIC STUDIES

Two papers, Edström and Kugelberg (1968) and Burke et al. (1971), were the first to show the correlation between physiological parameters and histochemistry in single motor units. By introducing the glycogen depletion technique to identify muscle fibers belonging to one motor unit, Edström and Kugelberg (1968) paved the way for the more detailed physiological characterization by Burke et al. (1971), who introduced the terms S (slow), FR (fatigue resistant), and FF (fast contraction, fast fatigue) for the three types of motor units in cat gastrocnemius. See figure 3.14.

circumstances, such adaptations regularly affect the metabolic properties of a muscle fiber long before any fiber-type transformation takes place, although under experimental conditions in animals, changes in MyHC expression can occur quite rapidly. This relative stability of the MyHC expression pattern means that the same "muscle fiber types" can be found in world-class athletes and in old people in nursing homes, but obviously their functional properties are vastly different. This does not mean, however, that the concept is useless. It only means that the fiber-type concept should be used with caution, with its strengths and weaknesses always in mind.

The heart of the problem regarding the concept of muscle fiber type lies in the basis of classification used. In animal experiments, such as the ones performed by Burke et al. (1971, 1973), it is easy to distinguish between slow and fast muscle fiber types on the basis of their contractile characteristics (see figure 3.14). Although it is possible to do the same thing using single fibers isolated from human biopsies (Larsson and Moss 1993), it is not useful for routine fiber-type classification.

Over the years, different enzyme histochemical techniques have been used for classification. One of the earliest methods was based on oxidative capacity by incubating frozen cross sections of muscle tissue for succinic dehydrogenase (SDH) activity (Dubowitz and Pearse 1960), but following the report of the proportional relationship between myosin ATPase activity and speed of shortening (Bárány 1967), myosin-based classification methods quickly appeared.

The oldest myosin-based classification system was based on the ATPase activity of the myosin molecule (mATPase) and separated muscle fibers with low ATPase activity, called type I muscle fibers, from muscle fibers with high ATPase activity, called type II muscle fibers. Guth and Samaha (1970) were the first to separate the two by exploiting the differential susceptibility of slow and fast myosin to alkaline and acid buffers in slide histochemistry. Their procedure was based on the pioneering method of Padykula and Herman (1955a, 1955b). During the following years several modifications were launched, the most widely used of which turned out to be the method presented by Brooke and Kaiser (1970b). In addition to separating type I muscle fibers from type II muscle fibers, this method exploits the different pH sensitivities of type II myosin isoforms to separate subgroups of type II muscle fibers, now known as type IIA and type IIX (figure 3.15).

The usefulness of myosin-based classification systems is due to the relative stability of the myosin isoform content in a given muscle fiber. Most parameters that can be used for classification are affected by changes in the level of activity, but some are affected more easily and sooner than others. The intracellular level of energy stores (glycogen and lipid) is most easily affected. Even one bout of exercise can make changes that invalidate their use for classification purposes, and consequently such methods have not been used since the early days of muscle fiber typing. PAS staining for glycogen may be very useful, however, to identify muscle fibers that have been active during studies of contractile properties (Edström and Kugelberg 1968) like the one shown in figure 3.14. Training or inactivity more readily affects the level of metabolic enzymes than the myosin isoform (see chapter 11), leaving myosin isoform as the most suitable basis for fiber-type classification.

The most recent classification systems are based on immunohistochemistry, treating cryostat sections of muscle tissue with **monoclonal antibodies (Mabs)** against the various myosin isoforms (figure 3.16). This work was pioneered by Schiaffino et al. (1986) in Italy and subsequently refined both by Schiaffino's group and others. A major advantage of muscle histochemistry is that it is possible to subject consecutive, serial cross sections of muscle fibers to different histochemical treatments, creating what has become known as the histochemical profile of the different muscle fibers. Creating such a profile for muscle fibers from a frail old individual and a world-class athlete would show very clearly that muscle fibers containing the same myosin isoform may differ widely with regard to other parameters such as oxidative capacity. Such differences are easy to comprehend, but they are not the only ones. Even within one individual, corresponding differences may be found, albeit usually not to the same degree.

| Muscle fiber type | 1 | 2A | 2B | 2C |
|---|---|---|---|---|
| Routine ATPase | 1+ | 3+ | 3+ | 3+ |
| ATPase preincubated pH 4.6 | 3+ | 0 | 3+ | 3+ |
| ATPase preincubated pH 4.3 | 3+ | 0 | 0 | 2+ |
| NADH-TR | 3+ | 2+ | 1+ | 2+ |
| SDH | 3+ | 2+ | 1+ | 2+ |
| α-glycerophosphate-menadione linked | 0 | 2+ | 2+ | 1+ |
| PAS | 1+ + 2+ | 3+ | 2+ | 2+ |
| Phosphorylase | 1+ + 0 | 3+ | 3+ | 3+ |

**Figure 3.15** Relative intensity of key histochemical reactions of major human muscle fiber types. During preincubation, sections of muscle tissue are subjected to a buffer adjusted to a certain pH value. The NADH-TR and SDH reactions show the oxidative capacity of the different fiber types, while the αglycerophosphate reaction is an indication of their glycolytic capacity. PAS staining is in fact a staining for glycogen, the level of which reflects the recent contractile activity of the muscle fiber in question. Glycogen phosphorylase is the key enzyme in the breakdown of glycogen to glucose-1-phosphate. Type 2B corresponds to what is now called Type IIX, and type 2C represents hybrid fiber types.

Reprinted, by permission, from V. Dubowitz and M.H. Brooke, 1973, *Muscle biopsy: A modern approach* (Philadelphia, PA: W.B. Saunders), 51.

Upper body muscles may differ significantly from lower body muscles in the same person simply because of differences in the training status between different parts of the body (Larsson and Ansved 1985). It seems reasonable to say, therefore, that the value of the concept of muscle fiber type is first and foremost the information it gives about the adaptive possibilities of a particular muscle or individual. It can also tell a lot about the functional characteristics of one particular muscle, but less about other muscles in the same individual.

## For Additional Reading

For a basic update on antibodies and immunological techniques, see Alberts et al. (1998).

So far, we have been dealing with fiber-typing techniques based on serial cross sections of muscle tissue. Another powerful technique, primarily used for molecular dissection of different muscle fiber types, is **polyacrylamide gel electrophoresis (PAGE).** Although it can be used to characterize individual muscle fibers both in humans (J.L. Andersen, Terzis, and Kryger 1999; Larsson and Moss 1993; Sant'ana Pereira et al. 1995) and experimental animals (Schiaffino and Reggiani 1996), this approach is too laborious for routine purposes. It is, however, very useful in what we could call molecular dissection of individual fibers, especially in combination with other biochemical methods (Pette, Peuker, and Staron 1999). An overview

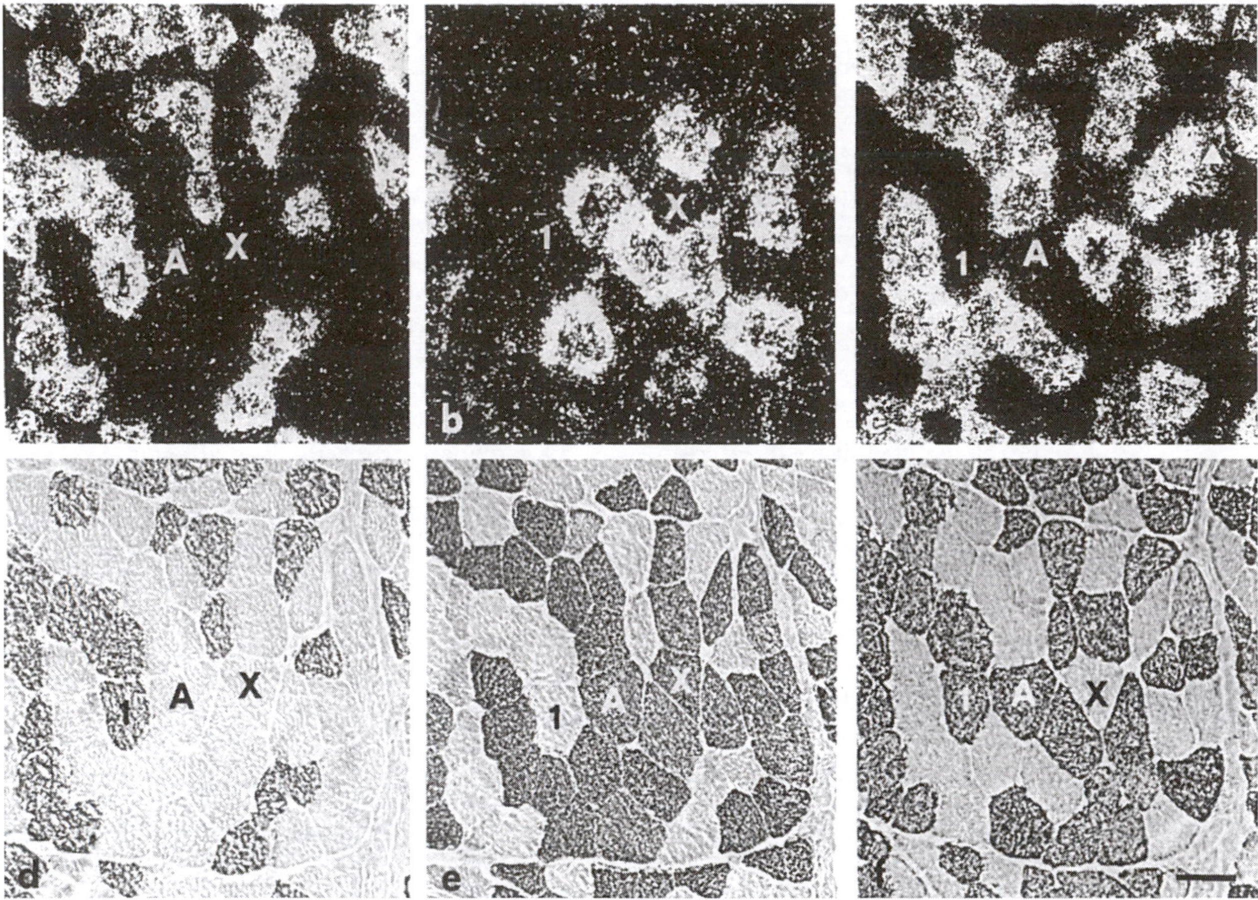


**Figure 3.16** *(a–c)* In situ hybridization correlated with *(d–f)* immunohistochemical staining of fiber types from human adult trapezius muscle. Serial cryosections were hybridized with $^{35}$S-labeled cRNA probes complementary to 3'-UTR of *(a)* β/slow (type I), *(b)* IIA, and *(c)* IIX MyHC transcripts (mRNA molecules), followed by autoradiography and dark-field microscopy. Most fibers contain exclusively either *(1)* β/slow (type I), *(A)* IIA, or *(X)* IIX MyHC mRNA. Some fibers coexpress IIA and IIX mRNA. Serial sections were incubated with anti-MyHC **antibody** reactive with *(d)* β/slow (type I) MyHC, *(e)* all type II MyHC, or *(f)* an antibody that in the rat reacts with all MyHCs except IIX. Bound antibody was revealed by immunoperoxidase staining. Note that fibers reactive for IIX MyHC mRNA correspond to fibers unlabeled in *(c)*. Bar = 30 μm. For further details, see original publication.

Adapted, by permission, from V. Smerdu et al., 1994, "Type IIx myosin heavy chain transcripts are expressed in type IIb fibers of human skeletal muscle," *American Journal of Physiology: Cell Physiology* 267:C1723-C1728.

of potential combinations of methods is shown in figure 3.17.

## FOR ADDITIONAL READING

For the basic principles of gel electrophoresis, see Alberts et al. (1998).

It may also be appropriate to mention that a rough estimate of type II muscle fibers or MyHC type II content can be obtained indirectly by means of **isokinetic equipment** (Aagaard and Andersen 1998; Thorstensson, Grimby, and Karlsson 1976). In addition, Wretling, Gerdle, and Henriksson-Larsen (1987) demonstrated a significant correlation between the percentage of type I muscle fibers in the vastus lateralis muscle and the mean power frequency of electromyograms during maximal isokinetic knee extensions.

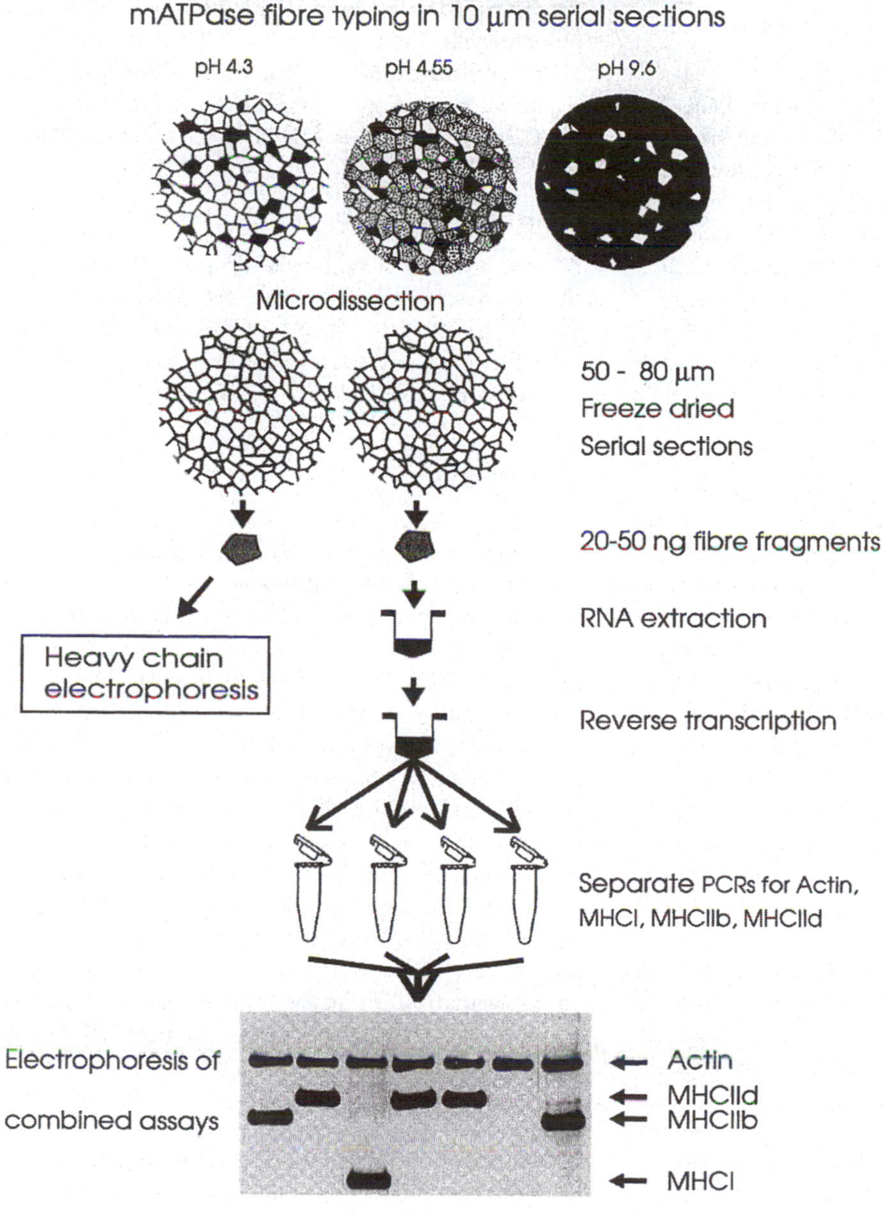


**Figure 3.17** Overview of possible combination of methods for molecular dissection of muscle fiber types. See source for details.

Reprinted, by permission, from D. Pette, H. Peuker, and R.S. Staron, 1999, "The impact of biochemical methods for single muscle fibre analysis," *Acta Physiologica Scandinavica* 166:270.

## Muscle Fiber Types: How Many and What Kind?

"Muscle Fiber Types: How Many and What Kind?" was the title of an article by Brooke and Kaiser (1970a), illustrating that this question has been a matter of debate since the early days of muscle histochemistry. Romanul (1964) was able to discriminate between eight different "muscle fiber types" in rat muscle, and as we shall see in a later section of this chapter, today even more muscle fiber types can be discerned with modern techniques. The question today should therefore not be how many different muscle fiber types we are able to identify, but rather how many are useful in our discussions about an individual's performance capacity and adaptability to new demands.

This textbook is primarily about humans, but since many of the milestones in muscle biology have been reached in animal experiments, it is necessary to include some of those, partly to illustrate basic phenomena, and partly to make clear the main differences between human muscle and muscles from experimental animals.

In the rat, one of the most common experimental animals in muscle biology, it is now customary to separate four different muscle fiber types: type I, type IIA, type IIB, and type IIX. The latter type was described independently by two groups: one in Germany (Bär and Pette 1988) and one in Italy (Schiaffino et al. 1989). Pette's group called it type IID since they found it in the diaphragm, while Schiaffino's group called it type IIX. Both names are used, often in combination (IID/X), but IIX alone is encountered more often than IID alone. Figure 3.18 illustrates rabbit muscle fiber types.

In humans, three major muscle fiber types are distinguished. Originally, they were called type I, type IIA, and type IIB, in accordance with the nomenclature used for experimental animals prior to the discovery of type IIX. More refined methods have revealed that transcripts of the IIX gene are expressed in what was previously called type IIB fibers (figure 3.19), and that the human type "IIB" gene is very homologous to the rat type IIX gene (Smerdu et al. 1994). Therefore, and since there seems to be no isoform homologous to the rat IIB in humans, more and more authors now recommend the use of the designation IIX instead of IIB in human muscle (Ennion et al. 1995). Unfortunately, this change of nomenclature may lead to some confusion since some authors continue to use the old name. We shall return to this question next.

Based on a combination of myosin-based and other techniques, type I, type IIA, and type IIB in both rats and humans have been called slow oxidative (SO), fast oxidative glycolytic (FOG), and fast glycolytic (FG), or slow (S), fast fatigue resistant (FR), and fast fatigable (FF), respectively. Other designations such as fast twitch a (FTa) and fast twitch b (FTb) have also been used for type IIA and type IIB, respectively. It seems, however, to be a good idea to use fiber type designations that show directly the technique used for muscle fiber typing (Dubowitz and Brooke 1973). For example, the names *slow* and *fast* should be preferred when twitch characteristics have actually been measured, and type I and type II with subgroups when myosin-based methods have been used. In humans, type I fibers are the more oxidative. Type IIX fibers (formerly named IIB) are the least oxidative, relying to a large extent on glycolysis, while type IIA are intermediary in their metabolic characteristics. This can be seen as a logical consequence of what is supposed to be their relative position in the recruitment hierarchy (see chapter 4). Equally logical is the differential distribution of lactate transporter isoforms in the various muscle fiber types. Contrary to earlier belief, lactate needs a specific transporter molecule, a monocarboxylate transporter (MCT) to traverse the cell membrane.

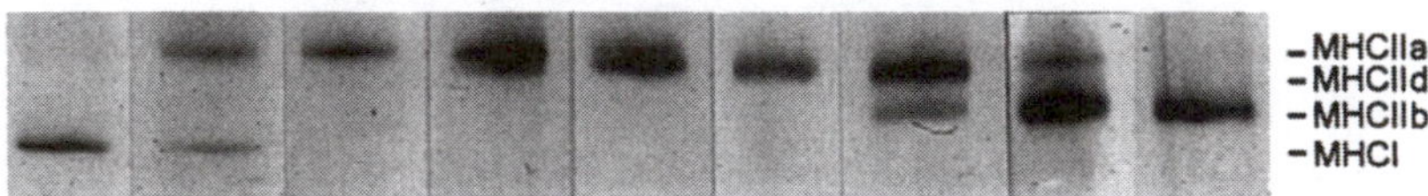


**Figure 3.18** Electrophoretic separation of myosin heavy chain (MyHC) isoforms in pure and hybrid single fibers from rabbit muscle. The fiber type continuum is indicated by the ordered sequence. ↔ indicates the possible directions of altered MyHC isoform expression. Apart from IIC, which here has been used for a hybrid fiber containing type I and type IIA MyHC, a single letter designates a pure fiber type, while double letters (AD, DA, DB, and BD) designate hybrid fibers, with the first letter indicating the major isoform present. Note that fiber type IID is the same as IIX.

Reprinted, by permission, from D. Pette, H. Peuker, and R.S. Staron, 1999, "The impact of biochemical methods for single muscle fibre analysis," *Acta Physiologica Scandinavica* 166:263.

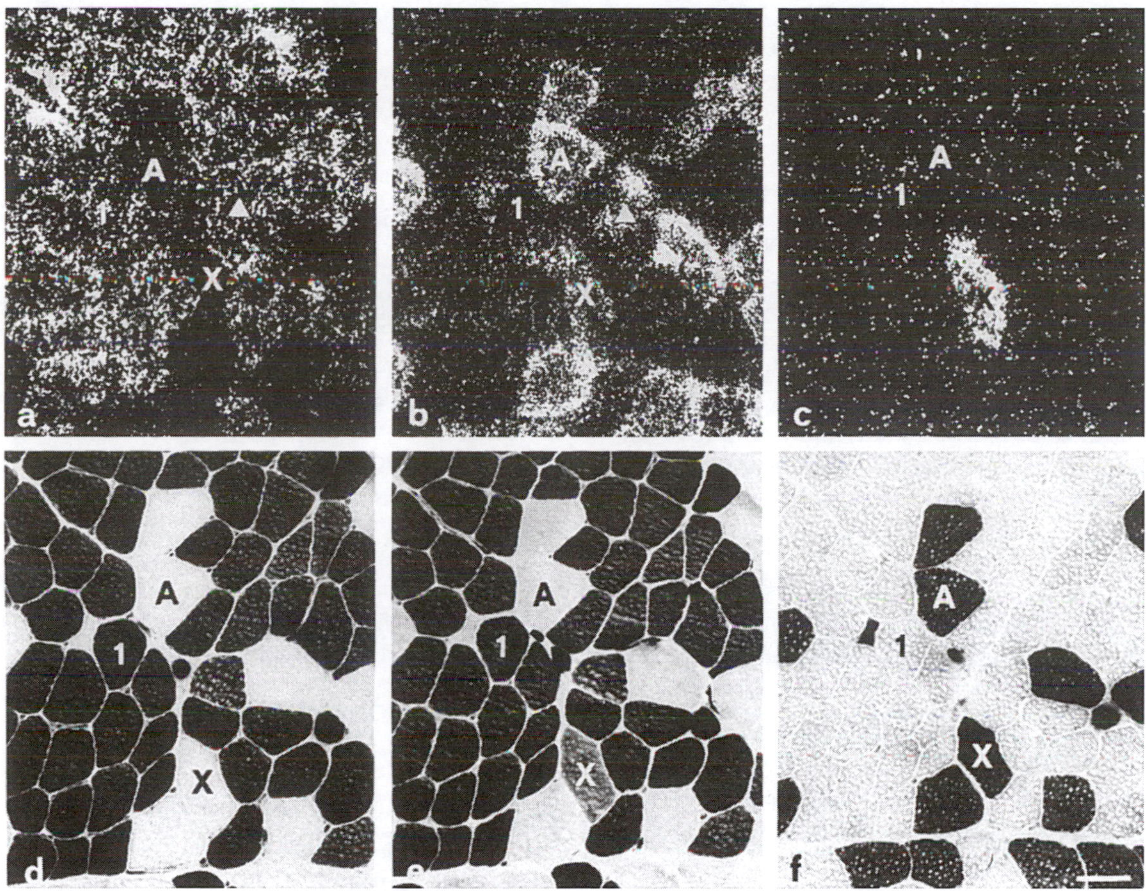


**Figure 3.19** *(a–c)* In situ hybridization correlated with *(d–f)* histochemical adenosine triphosphatase (ATPase) staining of fiber types from human adult adductor hallucis muscle. Serial cryosections were hybridized with probes specific for *(a)* I, *(b)* IIA, and *(c)* IIX myosin heavy chain (MyHC) transcripts or were processed for myosin ATPase histochemistry after preincubation at pH *(d)* 4.3, *(e)* 4.6, and *(f)* 10.4. Note that fiber *X*, strongly reactive for type IIX MyHC transcripts and weakly reactive for type IIA transcripts, shows a strong reaction after preincubation at pH 4.6. Based on ATPase histochemistry alone, this fiber would have been classified as type IIB. Bar = 30 μm. For details, see source.

Adapted, by permission, from V. Smerdu et al., 1994, "Type IIx myosin heavy chain transcripts are expressed in type IIb fibers of human skeletal muscle," *American Journal of Physiology: Cell Physiology* 267:C1723-C1728.

Type II muscle fibers, the more glycolytic muscle fiber type, have been shown to express the monocarboxylate transporter isoform MCT4, which is known to export lactic acid in skeletal muscles (M.C. Wilson et al. 1998). In the article by M.C. Wilson et al. (1998), this isoform was originally called MCT3 but later was renamed MCT4. Type I muscle fibers, on the other hand, express MCT1, which in this type of cell has been shown to import lactic acid (S.K. Baker, McCullagh, and Bonen 1998).

One final word of warning seems to be appropriate regarding fiber-type designations. Unfortunately, some of the names in common use today were earlier assigned to other properties of muscle fibers. Type I and type II (or 1 and 2), for example, have been used for muscle fibers of different length (Barrett 1962), as well as for fibers with different SDH activity (Dubowitz and Pearse 1960). In addition, the fiber type designation IIA has been used in combination with numbers (IIA1, IIA2, IIA3) to indicate different susceptibility to acid buffers (Gollnick, Parsons, and Oakley 1983). Comments on intermediary and hybrid fiber types are found in the next section about the molecular basis of muscle fiber types.

## Muscle Fiber Types at the Molecular Level

Underlying the rather restricted number of major muscle fiber types is a high degree of molecular variability, due to the existence of multiple isoforms of each myofibrillar component (Schiaffino and Reggiani 1996). As in every other aspect of muscle biology, however, more details have been revealed in experimental animals, especially rodents, than in humans. We shall therefore start with a description of rat muscle fiber types at the molecular level before we return to human muscle.

## *Rat Muscle*

Adult rat limb muscles contain four major fiber types, one slow (type I) and three fast types (types IIA, IIX, and IIB). Each of these muscle fiber types corresponds to one type of MyHC, encoded by a separate gene (Vikstrom et al. 1997). In addition, there are two developmental isoforms (MyHC embryonic and MyHC perinatal) and a couple of special isoforms expressed in some head and neck muscles (MyHC extraocular and MyHC mandibular). The latter isoforms fall outside the scope of this text.

### FOR ADDITIONAL READING

More details about myosin isoforms can be found in an excellent review by Schiaffino and Reggiani (1996).

Since myosin-based techniques are the basis of state-of-the-art muscle fiber-type identification, it may seem unnecessary to state that the MyHC is the sole determinant of muscle fiber type. It is also the major determinant of its ATPase activity (Schiaffino and Reggiani 1996), making the ATPase reaction a useful method to separate muscle fiber types in cryostat sections. There is, however, increasing evidence that mature mammalian muscle fibers very often coexpress variable amounts of different MyHC isoforms (Staron and Pette 1993). This may be a problem in section histochemistry since the mATPase staining profile of such hybrid fibers has been claimed to assume the characteristics of the dominant isoform (Klitgaard et al. 1990). Using computerized image analysis, however, Sant'ana Pereira et al. (1995) were able to show a highly linear relationship between percentage content of MyHC IIB (which now should be called IIX) and mATPase staining intensity in single human skeletal muscle fibers.

The role of the MyLCs has been more enigmatic, but studies of single muscle fibers have improved the picture. As mentioned earlier, each hexameric myosin molecule contains four MyLCs (see figure 3.8). Two of these are always MyLC2, one associated with each MyHC. The other two may be either MyLC1 or MyLC3. One of the difficulties of speaking about MyLCs is the different names given to them. MyLC1 and MyLC3 are also called **essential** or **alkali light chains,** and MyLC2 is called the **regulatory, phosphorylatable,** or **DTNB light chain.** In addition, all of these exist in more than one isoform. Apart from MyLC1f and MyLC3f, which originate from a single gene by alternative splicing, the MyLCs are encoded by separate genes, which are distributed throughout the genome (Barton and Buckingham 1985).

Apparently, both MyHC and MyLC isoforms influence the maximal velocity of shortening in the rat (Bottinelli et al. 1994), while MyHCs alone seem to determine the ATP consumption and isometric tension cost (Schiaffino and Reggiani 1994; figure 3.20). With a given fast-type MyHC content, however, the relative proportion of MyLC3f and MyLC1f affects the unloaded shortening velocity in a significant way, especially in fibers containing type IIB MyHC (Bottinelli et al. 1994). In human muscle fibers a corresponding relationship has been difficult to find, possibly because fibers with the same MyHC type but various MyLC contents have been hard to find (Larsson and Moss 1993).

### FOR ADDITIONAL READING

For a discussion of the importance of postactivation potentiation, thought to be due to phosphorylation of MyLC, in human performance, see Sale (2002).

## *Human Muscle*

Only three different isoforms of MyHC are found in adult human locomotor muscles, one slow and two fast. The slow isoform is called MyHC I, as in rat muscle, but as mentioned earlier, the nomenclature for fast fiber types has been changed. What used to be called type IIB has recently been renamed IIX based on sequence homology with the corresponding rat gene (Smerdu et al. 1994). In accordance with this new nomenclature, we have chosen to use the name IIX throughout this book for the human muscle fiber type formerly called IIB, even when the authors of the original publications used IIB. The main reason for doing this is the existence of an apparently "mute" human MyHC IIB gene, located with the other MyHC genes on chromosome number 17 (Vikstrom et al. 1997). So far, no expression of this gene has been found in human locomotor muscles, but real MyHC IIB transcripts may turn out to be present in more specialized human muscles.

Muscle fiber types previously designated IC or IIC have now been shown to represent hybrid fiber types, expressing more than one, usually two, MyHC isoforms (Sant'ana Pereira et al. 1995). Especially in light of the high number of possible combinations of MyHCs and MyLCs in human muscle (Wada et al.

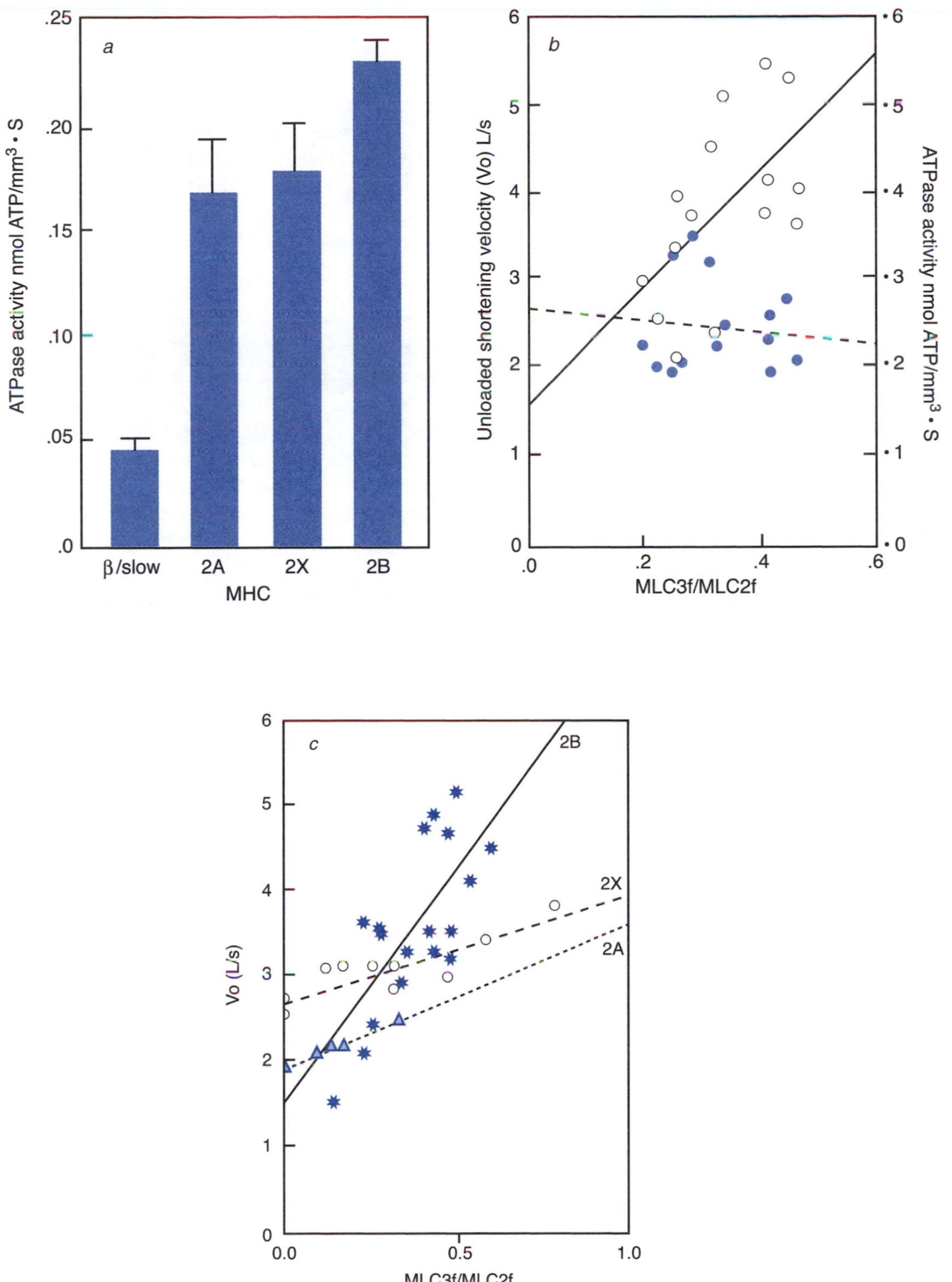


**Figure 3.20** Influence of myosin heavy chain (MyHC) and myosin light chain (MyLC) isoform composition on adenosine triphosphatase (ATPase) activity and unloaded shortening velocity (V0) in rat skeletal muscle fibers. *(a)* Differences in adenosine triphosphate (ATP) hydrolysis during isometric contraction between four sets of fibers, each containing one single MyHC isoform. *(b)* $V_0$ and ATPase activity in rat type IIB fibers with different MyLC isoform contents. Since the hexameric myosin molecule always contains two MyLC2 molecules, the fraction MyLC3/MyLC2 can be used as an indicator of the presence of MyLC1, MyLC3, or both. When the fraction has the value of 0, the myosin molecule contains two MyLC1, no MyLC3, and two MyLC2. When the value is 1, the MyLC isoform content is no MyLC1, two MyLC3, and two MyLC2. The regression lines show that $V_0$ is significantly related to MyLC3 content (solid line), whereas ATPase activity (dashed line) is not. *(c)* Relationship between $V_0$ and relative content of MyLC1 and MyLC3 in subtypes of fast fibers. Regression lines show a significant impact of MyLCs on $V_0$ for all three fiber types.

Adapted, by permission, from S. Schiaffino and C. Reggiani, 1994, "Myosin isoforms in mammalian skeletal muscle," *Journal of Applied Physiology* 77:493-501.

1996), it may be a good idea to try to refrain from designations other than those telling directly which MyHC isoforms the fiber in question contains. This may serve to keep the muscle fiber-type designations as descriptive as possible and their number within reasonable limits.

Hybrid muscle fiber types seem to be more frequent in sedentary and endurance-trained individuals than in those moderately active (Klitgaard et al. 1990). Fibers coexpressing MyHC IIA and MyHC IIX are more common than fibers coexpressing MyHC I and MyHC IIA, but the prevalence of the latter increases significantly in old age. This makes it difficult or even impossible to distinguish between type I and type II muscle fibers in old age (J.L. Andersen, Terzis, and Kryger 1999). Hybrid muscle fibers have often been regarded as fibers in transition (Kugelberg 1973) or as primitive fibers capable of differentiating into one of the major fiber types (Dubowitz and Brooke 1973). More recently, however, the question has been raised whether some human muscle fibers may be in a stable hybrid state, thereby securing a wide range of maximal shortening velocities (Kelly and Rubinstein 1994; Sant'ana Pereira et al. 1995).

The specific tension (i.e., the tension per unit cross-sectional area) of the different muscle fiber types has been a matter of debate for a long time and still is. This may at least in part be due to differences in the experimental approach. Thus, Larsson and Moss (1993) found no difference in specific tension in chemically skinned human muscle fibers, while in freeze-dried muscle fibers the specific tension varied according to MyHC type. In whole muscle, the question of specific tension is further complicated by the presence of other tissue components in addition to the muscle fibers. For further discussion of specific tension, see Larsson and Moss (1993) and J.A. Taylor and Kandarian (1994).

## Species Differences Versus Differences Between Muscles

A major question in muscle biology is to what extent the results from animal experiments are valid for human muscle as well. Apparently, some results are, while others are more uncertain. Molecular biology has shown that many of the proteins involved in the contractile process are **highly conserved**, which means that the proteins in question have changed very little during the course of evolution and only those most acquainted with the field are able to tell an electron micrograph of rat muscle from one of human muscle. When it comes to other levels of organization and functional characteristics, the situation may be different. In spite of this, results from animal experiments like those reported by Burke et al. (1973) are more often than not used without reservation when describing human muscle properties. For discussion, see Bigland-Ritchie, Fuglevand, and Thomas (1998).

The application of results in animals to humans may be excusable, however, in light of the lack of directly comparable results from human muscle. With the advent of intraneural motor axon stimulation (R.N. Lemon, Johansson, and Westling 1995; Westling et al. 1990), this situation has changed (Bigland-Ritchie, Fuglevand, and Thomas 1998). So far, however, mostly intrinsic hand muscles have been investigated in humans. The results show that, as in animal muscle, human motor units are recruited in an orderly sequence from weak to strong and from the more to the less fatigue resistant. They fail, however, to show the correlation between unit contractile speed and force found in studies of animal muscle (Bigland-Ritchie, Fuglevand, and Thomas 1998). Whether this is due to differences between hand muscles and other muscles or to differences between humans and experimental animals is so far unknown.

## Muscle Biopsies

Much of our knowledge concerning the fiber types and their characteristics in human muscle is based on the analysis of tissue samples obtained by muscle **biopsy.** Muscle biopsies can be obtained by an open biopsy method, where a piece of muscle is cut out under visual guidance, but more often a biopsy needle (J. Bergström 1975) or a conchotome is used (Dietrichson et al. 1987). The question is, however, how representative the biopsy sample is. From animal muscle it is known that the deeper part of a muscle often contains a higher percentage of type I muscle fibers than more superficial parts. In the chicken, where color differences between fiber types are prominent, this is easy to recognize with the naked eye. Comparable differences in fiber-type distribution have been demonstrated in the human quadriceps muscle (Lexell, Henriksson-Larsén, and Sjöström 1983), but they are not macroscopically visible.

Blomstrand and Ekblom (1982) have reported that the percentage of type I muscle fibers varied 6% between duplicate biopsies taken from one leg. When the biopsies were taken from both legs, the variation was 12%. The variation in the percentage of type IIA fibers was 4.4% and 7.3%, respectively. Evidently, data obtained from the analysis of single biopsies should be evaluated critically. Ideally, multiple muscle biopsies should be performed, and the depth from which the biopsies are taken should be defined (Lexell, Henriksson-Larsén, and Sjöström 1983).

## Large Individual Variations in Fiber Types

The human vastus lateralis muscle is the one most often used for biopsy studies, at least outside a clinical setting. In a biopsy from this muscle, the percentage of type I muscle fibers can be anywhere from 10% to 95% (figure 3.21). Within one individual there is also an almost systematic variation, with a tendency for more type I muscle fibers in **antigravity muscles,** the soleus muscle being a typical example (figure 3.22).

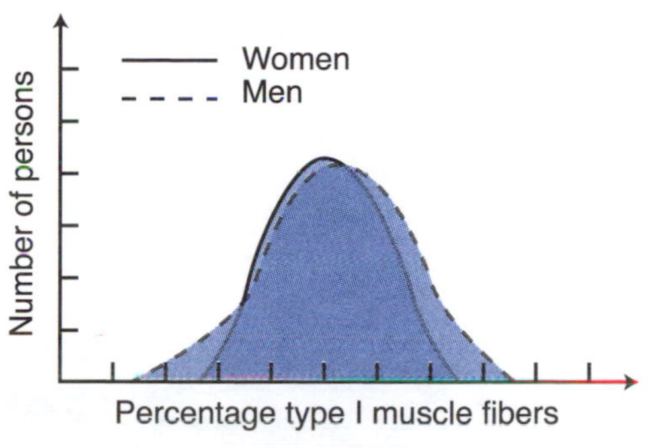


**Figure 3.21** Percentage of type I muscle fibers in male and female vastus lateralis muscle.

Adapted, by permission, from H.A. Dahl and E. Rinvik, 1999, *Menneskets funksjonelle anatomi* (Oslo, Norway: J.W. Cappelens Forlag).

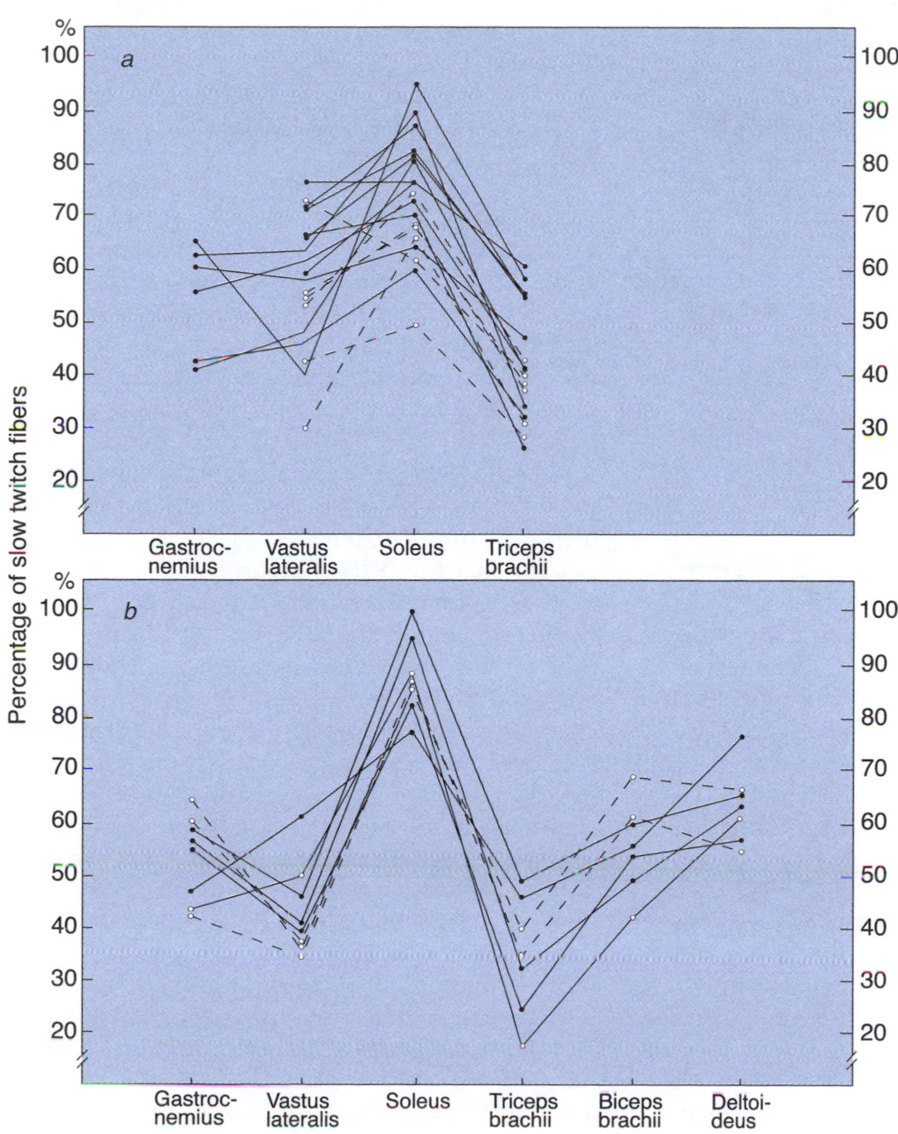


**Figure 3.22** Relative occurrence of type I fibers in some muscles of the human body. *(a)* Muscle samples obtained by needle biopsy. *(b)* Muscle samples obtained postmortem within 24 h of death.

From HANDBOOK OF PHYSIOLOGY, SECTION 10, edited by Peachey, copyright 1988 by American Physiological Society. Used by permission of Oxford University Press, Inc.

When comparing the muscle fiber-type distribution in athletes engaged in different kinds of sports, a quite typical picture emerges (figure 3.23). Athletes engaged in endurance events typically have a higher percentage of type I muscle fibers in their thigh muscles than those engaged in more explosive-type events such as sprinting or weightlifting. The question then is whether the fiber-type distribution is a consequence of their type of training or whether their choice of sport is a consequence of their inherited muscle fiber-type distribution. The answer is that the influence most probably goes both ways: The inherited muscle fiber-type pattern possibly initially determines success or failure in the type of sport chosen, and the training regime followed enhances the differences in the long run.

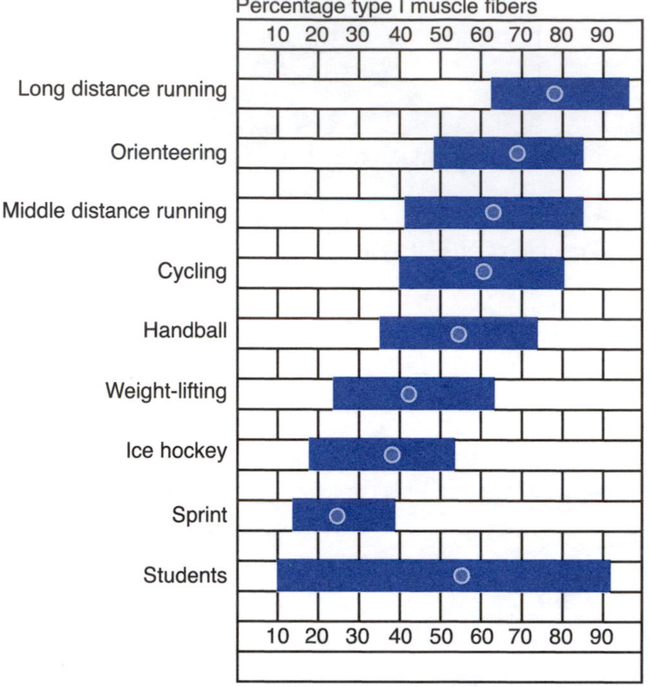


**Figure 3.23** Percentage of type I fibers in the vastus lateralis muscle in different kinds of athletes and in randomly selected young students. There is a clear tendency toward a higher percentage of type I fibers in athletes engaged in endurance activities than in those engaged in more explosive types of activities.

Adapted, by permission, from H.A. Dahl and E. Rinvik, 1999, *Menneskets funksjonelle anatomi* (Oslo, Norway: J.W. Cappelens Forlag).

A few studies support the notion that muscle fiber-type distribution influences the type and level of physical activity chosen by an individual. In a study originally designed to reveal whether the number of capillaries in the muscles increases during endurance training (Ingjer and Brodal 1978), half of the experimental subjects dropped out after 7 weeks of endurance training. Based on biopsy material obtained from the vastus lateralis before the training started, Ingjer and Dahl (1979) were able to show a significantly lower percentage of type I muscle fibers in the dropout group than in the group that completed the training period. A similar correlation between the percentage of type I fibers and leisure physical activity was demonstrated by Glenmark (1994). It has been shown that running and race-walking performance is significantly related to the economy of motion (P.A. Farrell et al. 1979; J.M. Hagberg and Coyle 1983). Since it is quite possible that muscle fiber type influences contractile economy (Horowitz, Sidossis, and Coyle 1994), muscle fiber type's having an effect on leisure physical activity seems not at all farfetched.

## For Additional Reading

For more information on genetics and physical performance traits, see Rankinen, Pérusse, Rauramaa et al. (2001).

# Factors That Determine Muscle Fiber Type

What factors determine whether a muscle fiber becomes type I or type II? This question can be addressed in several ways. It can be addressed in terms of genetics versus environmental factors or in terms of local factors versus influence from the nervous system. It can also be addressed in terms of molecular genetics; that is, which factors activate the expression of the various MyHC genes?

The question of the relative importance of genetic factors versus training and other environmental factors has been the topic of many scientific papers since the 1970s. An early study by Komi et al. (1977), comparing identical and nonidentical twins, concluded that the proportion of type I muscle fibers was almost entirely genetically determined. Since then, several papers have reported changes in the proportion of type I fibers after changes in the level of physical activity (see chapter 11), and more recent heritability estimates are significantly lower. Even though the relative importance of genetic and environmental factors remains to be settled definitely, the most recent evidence indicates that about 45% of the variance in the proportion of type I fibers is due to genetic factors (figure 3.24). Sampling and technical factors are claimed to account for 15%, while the remaining 40% have to be accounted for by training and other environmental factors (Simoneau and Bouchard 1995). It seems reasonable to assume that the importance of environmental factors, train-

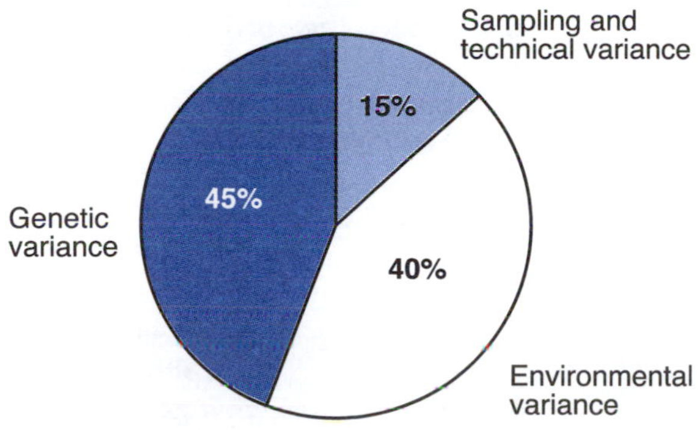


**Figure 3.24** Estimates of genetic and other determinants of the proportion of type I fibers in human skeletal muscle.

Reprinted, by permission, from J.A. Simoneau and C. Bouchard, 1995, "Genetic determinism of fiber type proportion in human skeletal muscle," *FASEB* 9:1094.

ing particularly, increases with the amount and type of training performed.

How then is this genetic influence exerted? It could be a local genetic effect in the individual muscle fiber or its progenitor cells, or it could be more indirect through an influence from the motoneuron. Let us start by taking a look at the importance of the latter.

## *The Motoneuron and Muscle Fiber-Type Diversity*

There are many indications that the matching of the motoneuron and its **muscle unit** is not incidental. There is, for example, a direct correlation between the size of the motor unit as measured by its twitch force and the size of the motoneuron as measured by the conduction velocity in its axon or by its threshold force (Dengler, Stein, and Thomas 1988; Milner-Brown, Stein, and Yemm 1973). A priori, this ensures that the smaller motoneurons, which are easier to recruit (Henneman, Somjen, and Carpenter 1965) and which innervate small muscle units with low force output, are the ones recruited first in most movements.

Most of the evidence concerning the importance of neural factors comes from animal studies, starting with the classic cross-reinnervation experiment by Buller, Eccles, and Eccles (1960). In that paper they showed that a slow muscle becomes faster when reinnervated by a nerve that originally innervated a fast muscle, and a fast muscle becomes slower when reinnervated by a "slow nerve." They were unable to tell, however, whether this effect is due to the impulse pattern imposed on the muscle fiber or to some unknown "trophic factor." This question has been addressed by stimulating denervated muscles through implanted platinum electrodes, thereby avoiding any contact between the muscle fibers and nervous tissue that could release trophic factors. Such studies show that it is possible to modify the contractile and histochemical properties of muscle fibers by electrical stimulation with a new impulse pattern (Lømo, Westgaard, and Dahl 1974), although they do not entirely exclude the possible influence of some unknown factor.

The adaptive possibilities are not unlimited, however. Animal experiments have shown that while transformation between some rat muscle fiber types can be rather easily induced, transformation between other types is far more difficult, maybe impossible, to obtain. This has led to the introduction of the concept of adaptive range, which indicates the adaptive possibilities at hand for each muscle fiber type (Westgaard and Lømo 1988). There are, however, indications that the adaptive range may be wider than the early results suggested (Gundersen 1998; Windisch et al. 1998). A corresponding range of possibilities in human muscle has not been established yet, but undoubtedly some transformations are easier to obtain than others. We shall return to this question when dealing with changes due to training in chapter 11.

### FOR ADDITIONAL READING

For a discussion of the role of myoblast diversity for the adaptive capacity of muscle, see Parry (2001).

## *What Local Genetic Factors Influence Muscle Fiber Type?*

The most obvious candidates for a genetic influence at the local level are the myogenic cells. There is evidence that some kind of specialization occurs as early as in the myoblast, giving rise to two generations of myotubes (Kelly and Rubinstein 1994), at least in small experimental animals with short gestational times. The primary generation is believed to give rise to mostly slow muscle fibers, while the secondary generation gives rise primarily to fast muscle fibers. Experimental cross transplantation of satellite cells between cat temporalis and limb muscles indicates that satellite cells are type specific. The unique MyHC isoform present in cat temporalis and masseter muscles was expressed even when satellite cells were transferred from temporalis to limb muscles, but not when limb-muscle satellite cells were transferred to the temporalis (Hoh, Hughes, and Hoy 1988). If this also holds for satellite cells in general, it implies that satellite cells are predetermined to end up as a more or less specific muscle fiber type with a certain adaptive range. Apart from this we have no reason so far to believe that local genetic factors are important.

## Transcriptional Control of Muscle Fiber Type

Although we are able to describe muscle fiber types in molecular terms, we do not yet have a clear picture of how they are regulated. Piece by piece, however, new information is added. It is well established that the MyHC genes belong to what is now known to be a multigene family. In humans six different MyHC genes are located within a 500-kb interval on the short arm of chromosome 17. In addition, the slow/type I/β cardiac MyHC gene is located on chromosome 14. The MyLC genes, on the other hand, are distributed on several chromosomes, at least in the mouse (Barton and Buckingham 1985). Although the clustering of the MyHC genes on chromosome 17 has been seen as suggestive of some regulatory significance (Yoon et al. 1992), their sequence in the chromosome (figure 3.25) does not support this assumption (Vikstrom et al. 1997).

It has been known for some time that a group of transcription factors known as **myogenic determination factors (MDFs),** myogenic regulatory factors (MRFs), or simply myogenic factors are necessary to commit naive cells in the somite to a myogenic fate. Among the members of the MDF family—MyoD, myf5, mrf4, and **myogenin**—two have been shown to accumulate in muscles of different fiber type: MyoD in fast-twitch muscle and myogenin in slow-twitch muscle. The significance of this selectivity has only recently started to be unveiled. Experimental mutations in the promoter of the MyHC IIB gene prevent binding of MyoD and myogenin to this element, and a time-course analysis of rat soleus muscles subjected to hind-limb suspension shows that an increase in MyoD messenger ribonucleic acid (mRNA) precedes the increase in de novo expression of MyHC IIB mRNA. Taken together, these findings suggest a possible causative role for MyoD in the up-regulation of MyHC IIB taking place in unweighted soleus muscle (Wheeler et al. 1999). Myogenin, on the other hand, may have relevance for the metabolic qualities of the muscle fiber types. When myogenin is overexpressed in transgenic animals, it causes the muscle fibers in question to become more oxidative and less glycolytic without any detectable change in myosin isoform (S.M. Hughes et al. 1999).

### For Additional Reading

A survey of the cellular and molecular embryology of vertebrate skeletal muscle can be found in review articles by Hauschka (1994) and Brand-Saberi and Christ (1999).

Hormones are obvious candidates whenever we are looking for reasons for changes in cell function. Apart from thyroxin and possibly anabolic steroids (Czesla et al. 1997; Noirez and Ferry 2000), however, no hormone has been shown to affect locomotor skeletal muscle fiber **phenotype.** The effect of sex hormones on sexually dimorphic muscles, on the other hand, is another story. However, although thyroxin may induce profound changes in fiber type when its level is manipulated experimentally (Izumo, Nadal-Ginard, and Mahdavi 1986), it is unable to account for muscle fiber diversity in the euthyroid state (Chin et al. 1998).

Since it is known that changes in the neural stimulation pattern can induce a shift in muscle fiber type, we have to look for possible determining factors among the changes in the muscle fiber invoked by the impulse pattern. Of course, this means

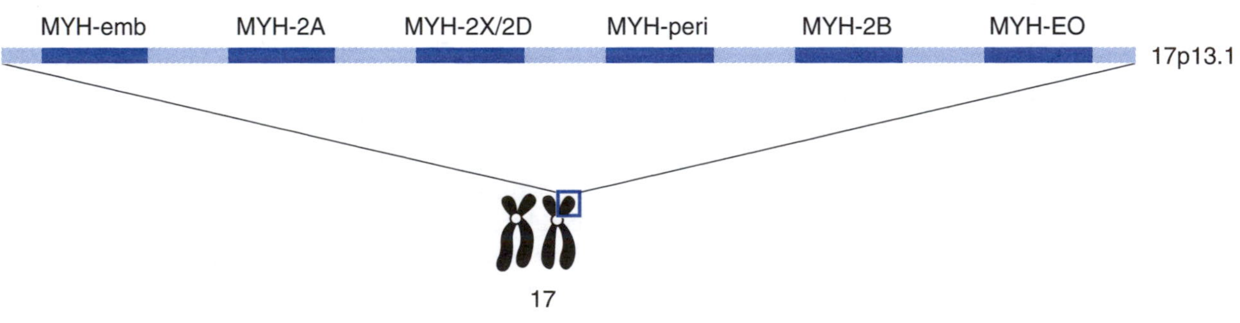


**Figure 3.25** The organization of six sarcomeric myosin heavy chain genes on human chromosome 17. The distances between the various genes as well as their transcriptional orientation with respect to each other (head/tail, head/head, or tail/tail) have yet to be determined.

Adapted from Vikstrom et al. 1997.

that there are many candidates, but one of the most obvious ones is the intracellular calcium ion concentration, $[Ca^{2+}]_i$. This was the basis for an experiment by Chin et al. (1998), which we outline later.

**Calcineurin (CaN)** is a serine–threonine phosphatase that is abundant in skeletal muscle (Hoey et al. 1995). It has been shown to be involved in the activation of **cytokine** gene expression in lymphocytes (Rao, Luo, and Hogan 1997). It is thought to act in the following way: Binding of calcium to a calmodulin–CaN complex activates a serine–threonine phosphatase, which dephosphorylates a transcription factor called NFAT (nuclear factor of activated T-cells), allowing it to translocate to the nucleus and stimulate target genes. It has been shown that the calmodulin–CaN complex is sensitive to sustained increases in $[Ca^{2+}]_i$, but insensitive to brief calcium transients, irrespective of their amplitude (Dolmetsch et al. 1997). Such differences in calcium sensitivity make this or similar complexes promising candidates as mechanisms that differentiate between the slow- and fast-activity patterns that characterize the two major fiber types (Hennig and Lømo 1985) and their calcium transients (Westerblad and Allen 1991).

So far this is little more than an attractive theory. It turns out, however, that the immunosuppressant cyclosporin A is a specific inhibitor of CaN (J. Liu et al. 1991), making it possible to block this signaling pathway under experimental conditions that otherwise would result in slow fibers. This is exactly what happens, making the CaN-dependent transcriptional pathway a possible link between neural activity and muscle fiber type (Chin et al. 1998). Seemingly, this CaN-dependent pathway not only switches on the expression of the slow-fiber program. It also switches off the fast program (figure 3.26), but it is not able to account for how the subgroups of muscle fiber types come about. There are indications that CaN may also be involved in other types of muscle adaptations, notably hypertrophy (S.E. Dunn, Burns, and Michel 1999; Semsarian, Wu, et al. 1999).

### For Additional Reading

For a discussion of possible strategies aimed at identifying the signal transduction pathways mediating the effects of nerve activity on muscle phenotype, see Schiaffino et al. (1999).

## Age Effects on the Expression of Myosin Heavy Chains

Age affects the biological systems of the body in different ways, and it should be no surprise that it also affects the expression of MyHCs. At the level of the whole organism, advanced age results in a generalized slowing of movements (Kirkendall and Garrett 1998). At the muscle fiber level, the prevalence of type II muscle fiber atrophy increases with

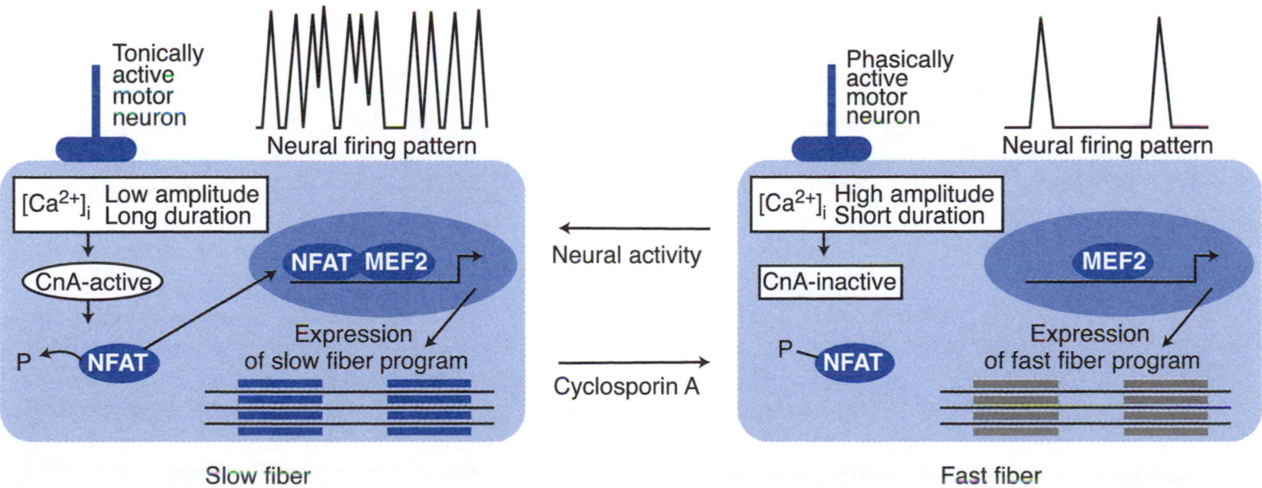


**Figure 3.26** Model of a possible calcineurin (CaN)-dependent pathway linking specific patterns of motoneuron activity to distinct programs of gene expression that establish phenotypic differences between slow and fast myofibers. **MEF2 (myocyte enhancer–binding factor 2,** a transcription factor of the **MADS** family) is shown to represent the requirement for collaboration between activated NFAT proteins and muscle-restricted transcription factors in slow-fiber-specific gene transcription, but other proteins (not shown) also are likely to participate.

Reprinted, by permission, from Chin, et al., 1995, "A calcineurin-dependent transcriptional pathway controls skeletal muscle fiber type," *Genes and Development* 12:2505.

advancing age (Larsson, Sjödin, and Karlsson 1978; Tomonaga 1977). The percentage of type I muscle fibers, on the other hand, has been reported to increase as well as decrease with age (Glenmark, Hedberg, and Jansson 1992). A possible explanation for this discrepancy can be found at the molecular level, since it has been shown that an increasing number of muscle fibers from old people express more than one MyHC isoform (J.L. Andersen, Terzis, and Kryger 1999). Most often fibers containing MyHC I and IIA or MyHC IIA and IIX are found, but even fibers coexpressing MyHC I and IIX or all three major MyHC isoforms have been reported (J.L. Andersen, Terzis, and Kryger 1999). Understandably, the existence of more than one MyHC isoform in a single fiber makes the myofibrillar ATPase method less reliable for fiber-typing purposes (J.L. Andersen, Terzis, and Kryger 1999; Klitgaard et al. 1990). Moreover, as stated by J.L. Andersen, Terzis, and Kryger (1999), what we see "is not so much a change in the ratio between type 1 and type 2 fibers in the classical sense but more an obfuscation of the border between type 1 and type 2 fibers" (p. 453).

Two important questions arise from this observation. First, what is the mechanism behind this increased prevalence of hybrid fibers? And second, what are the functional consequences? One of the explanations that have been forwarded originates from the assumption that there may be an increased frequency of **denervation–reinnervation** processes in aging muscles because of neuropathic changes that are known to occur (J.L. Andersen, Terzis, and Kryger 1999). Hybrid fibers are, however, also found in young people, both after endurance training (Klitgaard et al. 1990) and after 5 weeks of experimental bed rest (J.L. Andersen, Gruschy-Knudsen, et al. 1999). A possible unifying explanation could therefore be that changes in the workload sufficient to induce fiber-type changes tend to lead to the coexistence of two different MyHC isoforms in single fibers. This is based on the fact that the turnover of a MyHC isoform at the protein level is slower than the turnover at the mRNA level (J.L. Andersen and Schiaffino 1997). An increased number of fibers with a mismatch between MyHC isoform and MyHC mRNA after bed rest in young people (J.L. Andersen, Gruschy-Knudsen, et al. 1999) can be taken as support for this. As to the functional consequences of hybrid fibers, nothing is known so far (J.L. Andersen, Terzis, and Kryger 1999).

## Summary

A skeletal muscle contains different kinds of muscle fibers. Historically, these muscle fiber types have been given many different names, depending on the methods available to separate them. Today, the state-of-the-art methods are based on the different isoforms of MyHCs. In humans, three different types of skeletal muscle fiber types are found in locomotor muscles: one slow-twitch, called type I, and two types of fast-twitch fibers, called type IIA and IIX. Type IIX was formerly called IIB. Type I muscle fibers are the more economical and oxidative muscle fiber type, making them highly suitable for sustained work. Among the fast fibers, type IIA is the more oxidative. After periods of changed levels of physical activity and in old age, the prevalence of hybrid muscle fibers containing more than one isoform of MyHC increases. This tends to make the borders between the different muscle fiber types less distinct.

The fiber-type composition of an individual's muscles has significant influence on the muscles' functional characteristics. A high percentage of type II fibers predisposes for strength and explosive action, while a high percentage of type I fibers predisposes for endurance activities. There are also indications that muscle fiber type has an impact on leisure-time physical activity.

Both genetic and environmental factors determine the fiber-type composition of our muscles, the genetic influence being the single most important factor. Among the environmental factors, training seems to be the most important. Much of the genetic influence seems to be mediated by the impulse pattern in the nerve; long trains of low-frequency impulses result in type I fiber characteristics, while short bursts of high-frequency impulses give the muscle fiber type II characteristics. The link between impulse pattern and fiber type seems to be the resulting intracellular concentration of calcium ions.

# Types of Muscle Activation

Depending on the magnitude of the contractile force in relation to the imposed workload, the result of a muscle contraction varies. If the contractile force equals the force resisting the movement, no movement takes place, and the contraction is called **isometric** (i.e., with the same length), or **static.** If the contractile force exceeds the resistance at hand, the muscle shortens, creating a contraction in the true meaning of the word, commonly known as a **concentric contraction.** If the contractile force is less than the opposing force, a lengthening contraction occurs, and the muscle acts like a brake, reducing the effect of the external force. This is known as an **eccentric contraction.** Together, concentric and eccentric contractions are called **dynamic contractions.** Both types occur regularly during normal movements in daily life. When you lift a cup to have a sip of coffee, your biceps works concentrically. When you put the cup back on the saucer, it works eccentrically to brake the movement and prevent breaking the cup.

Theoretically, a contraction can take place totally unresisted, in which case a shortening but no rise in tension take place. This has been called an **isotonic contraction.** It should be emphasized, however, that this situation occurs very rarely, if ever, during normal movements. It is hard to imagine a contraction that is totally unopposed. In any case, the muscle has to overcome the inertia of its own mass during shortening. Even if the term *isotonic* is restricted to contractions with constant force development during the movement, it is doubtful whether it really occurs outside of the laboratory. Even if the external load is kept constant, the force developed by the muscle varies as the lever arms become shorter or longer.

According to physical laws, work is the product of force (expressed in newtons) and the distance (in meters) through which it is expressed. Consequently, the unit for work is newton-meter, N · m. Contrary to all experience this means that static work is no work at all, since the distance is zero. It cannot be disputed, however, that isometric activity demands energy and can be very fatiguing. From a physiological point of view, the "work" in this case is related to the force times contraction time rather than to force times displacement. One of the reasons that isometric or static contractions are very fatiguing is that they tend to compromise the circulation in the muscle. The shortening and resulting thickening of individual muscle fibers compress and temporarily stop circulation in the endomysial capillaries surrounding the fibers. Understandably, this becomes a crucial point in ergonomics. As mentioned earlier in this chapter (in the section "Force Transmission From Muscle Fiber to Tendon" on page 45), a contraction that is isometric at the whole-muscle level is not necessarily so at the muscle fiber level (Ito et al. 1998), at least not in the initial phases of the contraction (figure 3.27).

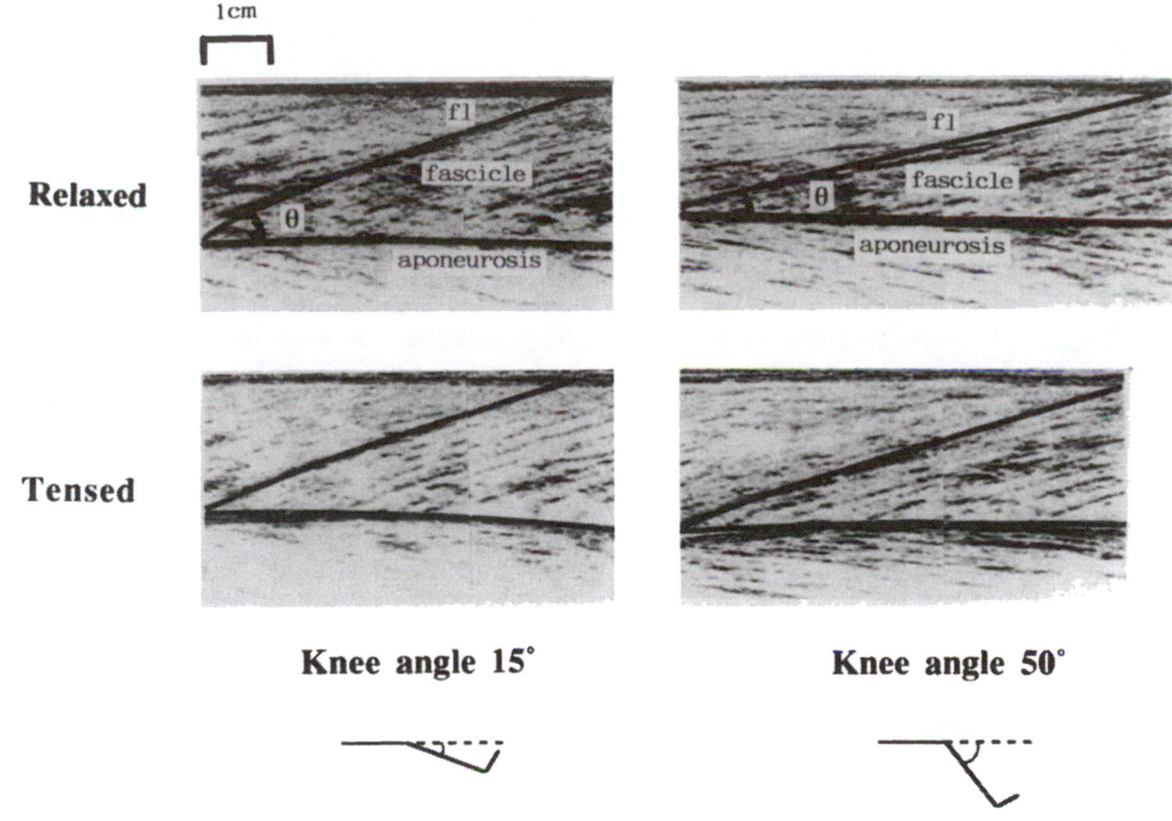


**Figure 3.27** Ultrasonic image obtained from human vastus lateralis muscle at two different joint angles during relaxed and tensed conditions at 10% of maximal static strength. The fascicle length (fl) was determined as the length of a line drawn along the ultrasonic echo parallel to fascicles. The fascicle angle (θ) was determined as the angle between the echo obtained from the fascicles and the deep aponeurosis in the ultrasonic image.

## Force in Relation to Sarcomere Length: Force–Length Relationship

The isometric force exerted by a muscle depends heavily on the actual length of the muscle or, more correctly, on the actual length of its muscle fibers and their constituent sarcomeres. The obvious explanation for this is that the force developed is related to the degree of overlap between the actin and myosin filaments (figure 3.28).

When an unstimulated muscle is stretched, its passive elastic components create a force that increases exponentially with increasing length (see figure 3.28, curve A). If, in addition, the muscle is stimulated maximally at a certain constant length, the resulting force measured is the sum of the passive tension at that particular length and the contractile force developed (see figure 3.28, curve B). Consequently, the active isometric force at each different constant length can be found by subtracting the passive force from the total force measured (see figure 3.28, curve C).

When the sarcomeres are stretched until there is no longer any overlap between actin and myosin filaments, no active force can be developed. Each human skeletal muscle actin filament is approximately 1.27 μm long (Rassier, MacIntosh, and Herzog 1999), while the myosin filament is 1.6 μm long. Consequently, this happens at sarcomere lengths larger than 4 μm. At shorter sarcomere lengths the active isometric force is proportional to the overlap between the interacting myofilaments. A maximum is reached when all cross-bridges in each half sarcomere are in a position to interact with the appropriate actin filaments. At shorter sarcomere lengths, the isometric force remains at the same level in the beginning but soon starts to decline as the sarcomere becomes successively shorter. This is more difficult to explain than the force deficit at longer sarcomere lengths, but several possible explanations are available. First of all, when an actin filament enters the cross-bridge region of the opposite half sarcomere, it may at least interfere with proper cross-bridge cycling in that region. It is also possible that the increased distance between the actin and myosin filaments as the muscle shortens and becomes thicker interferes with force development (Rassier, MacIntosh, and Herzog 1999). Furthermore, both the actin filaments and myosin filaments in each half sarcomere have opposite polarity, theoretically creating a possibility for more active counteracting force production at short sarcomere lengths.

The practical consequences of this force–length relationship have been a matter of discussion. One obvious confounding issue is the pennate organization of many muscles. Both theoretical considerations and practical experiments indicate that the **pennation angle** influences the width of the length–tension curve (Kaufman, An, and Chao 1989). Increasing the angle of pennation makes the curve narrower, indicating that a pennate muscle is able to develop maximal force over a smaller range of lengths than a muscle whose fibers are parallel to its direction of pull. As compensation, its maximal force is higher because of a larger physiological cross-sectional area. Increasing the angle of pennation tends to decrease the length of individual muscle fibers in relation to total muscle length. This relationship between muscle fiber length and muscle length has been named index of architecture, $i_a$ (figure 3.29; Kaufman, An, and Chao 1989). To a certain extent this concept fits neatly into our perception of how certain muscles are used in real

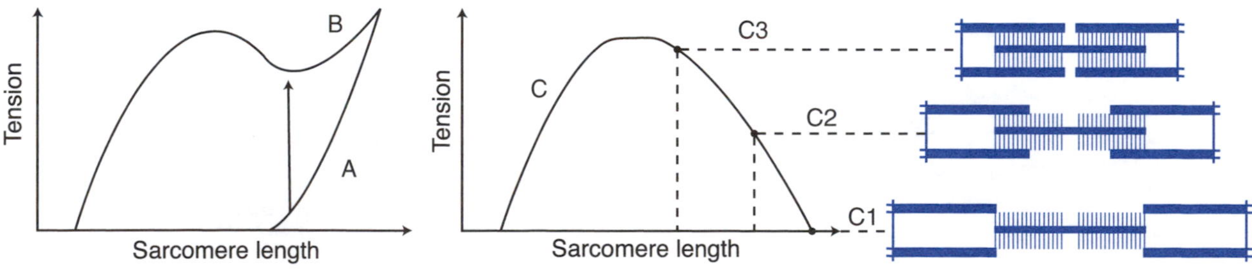


**Figure 3.28** The force–length relationship. Curve A shows an increase in passive tension with increasing muscle length. Curve B shows the sum of passive tension (curve A) and contractile force during isometric contractions at different sarcomere lengths. Curve C, the difference between curves B and A, shows the isometric force at different sarcomere lengths. Point C1 corresponds to a situation with no overlap between actin and myosin filaments. Point C3 corresponds to a maximal overlap between actin filaments and the myosin heads.

Adapted, by permission, from H.A. Dahl and E. Rinvik, 1999, *Menneskets funksjonelle anatomi* (Oslo, Norway: J.W. Cappelens Forlag).

movements. The two heads of the biceps femoris muscle can serve as an example here (Roy and Edgerton 1992). The long head is a biarticular muscle, has an index of architecture of .25, and undergoes only minor changes in length during normal ambulatory activity because of the linkage between hip and knee movements. The short head, on the other hand, is a monoarticular muscle, has an index of architecture of .52, and is subject to the full range of motion taking place in the knee.

Unfortunately, there is one very common misconception regarding the force–length relationship. All too often it is taken as a description of force development during a dynamic contraction. That is not the case. The force–length relationship is a static property of skeletal muscle and does not predict the consequences of dynamic contractions (Rassier, MacIntosh, and Herzog 1999). This misconception may, however, be the result of the way we draw the curve and the names we use to describe it. Rather than being drawn as a continuous line, it ought to be shown as separate points, each representing the active force at one particular sarcomere length. Furthermore, we routinely use the terms *ascending limb* and *descending limb* for the left and right part of the curve, respectively, more or less indicating by our nomenclature the possibility of some sort of movement along the curve. Of course, this does not mean that the muscle is unable to change its length. It only means that the curve we draw does not represent such a movement.

Considering the force–length relationship as a continuous, dynamic property has led to the notion of instability in sarcomere length on the descending limb of the force–length curve (Rassier, MacIntosh, and Herzog 1999). Although the question of sarcomere instability is still a matter of debate (e.g., see Allinger, Epstein, and Herzog 1996 and Zahalak 1997), it is beyond the scope of this chapter. More relevant is the question of the actual range of muscle lengths used by a muscle in its everyday movements. Intuitively, one might assume that a muscle preferably operates essentially at or around the plateau where it is strongest. However, this assumption is not correct (Rassier, MacIntosh, and Herzog 1999). Most muscles whose in situ force–length properties have been determined—albeit mostly animal muscles—have been found to operate primarily on the ascending or on the descending limb of the force–length relationship, reaching the plateau toward the end of the range of joint motion (Rassier, MacIntosh, and Herzog 1999). It must be realized that muscles very often operate in a way that combines more than one of the basic modes of contraction found in textbooks. Muscles operating in stretch–shortening cycles can serve as an example of this. We return to this later.

## Importance of the Lever Arm

The ability of a muscle to create a movement is dependent not only on the initial length of the engaged muscles. It is the resulting **torque** that matters. As the product of the contractile force and the lever arm, the torque depends heavily on the latter. This is clearly illustrated in the performance of a chin-up. Intuitively, everybody prefers to do this with supinated rather than pronated forearm, simply because it feels easier. The reason for this is found in the way the biceps tendon attaches to the tuberosity of the radius.

When the elbow is at 90° and the forearm is supinated (i.e., ulna and radius are parallel and the palm of the hand faces upward), it is easy to feel the

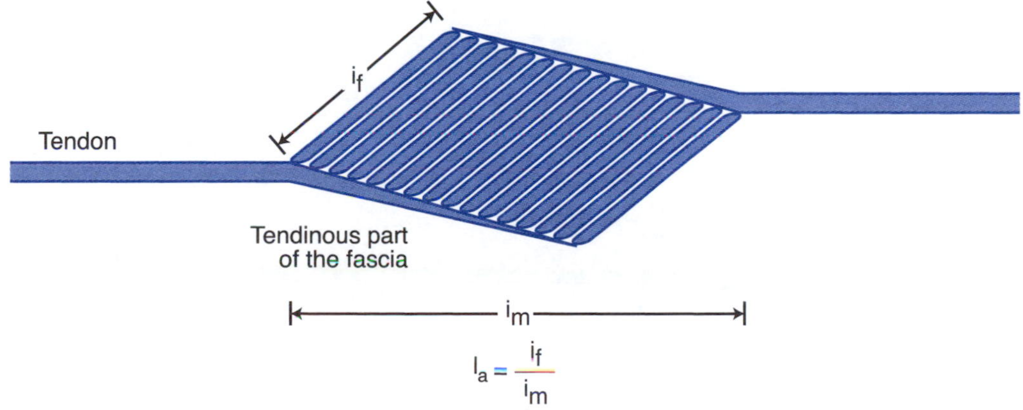


**Figure 3.29** The index of architecture ($i_a$) of a muscle is the relationship between the length of muscle fascicles and that of the muscle itself.

Adapted, by permission, from *Pflügers Archive: European Journal of Physiology,* Influence of muscle architecture on the length-force diagram. A model and its verification, R.D. Woittiez, P.A. Huijing, and R.H. Rozendal, 397, 73-74 1983, © Springer-Verlag.

biceps tendon as a string on the ventral side of the joint. When the forearm is pronated from this position, the tendon retracts and assumes a position closer to the joint, reducing the lever arm and the resulting torque considerably.

Even when the forearm is supinated, the elbow flexors are at some sort of disadvantage regarding the torque they are supposed to counteract. Holding a 10-kg (22-lb) weight in the palm of the hand does not usually represent a problem for a healthy person. The lever arm of the weight, however, is approximately 10 times larger than the lever arm of the muscles keeping the elbow joint at 90°. Consequently, the combined contractile forces of the elbow flexors must be in the range of 100 kg (220 lb) to resist the weight (figure 3.30).

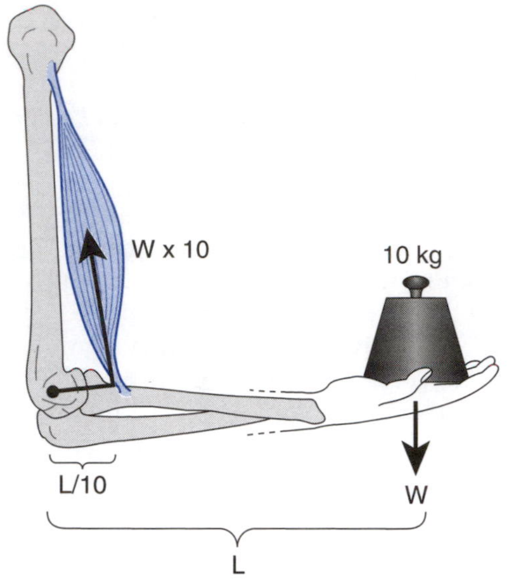


**Figure 3.30** To hold a weight of 10 kg (22 lb) in the hand in the position shown in the figure, the elbow flexors must exert a force 10 times larger, since the lever arm of the weight is 10 times longer than the lever arm of the elbow flexors. This example does not take the weight of the forearm itself into account.

## The Importance of the Pennation Angle

In pennate muscles the line of pull of each individual muscle fiber is at an angle to that of the muscle as a whole. According to simple vector rules, the maximal force of the muscle is less than the combined maximal forces of its constituent muscle fibers. In fact, the force is reduced to a percentage equal to the cosine of the angle of pennation. If the angle of pennation is 25°, the force is reduced by 10% since the cosine of 25° is approximately 0.90. For larger angles of pennation, the reduction is larger. It has been shown that the angle of pennation tends to increase with increasing muscle thickness (figure 3.31), implying that muscle hypertrophy may lead to increased pennation (Fukunaga et al. 1997). The somewhat paradoxical consequence of this is that when strength training results in hypertrophy of a pennate muscle, the increased angle of pennation tends to reduce the effective increase in muscle force (Fukunaga et al. 1997).

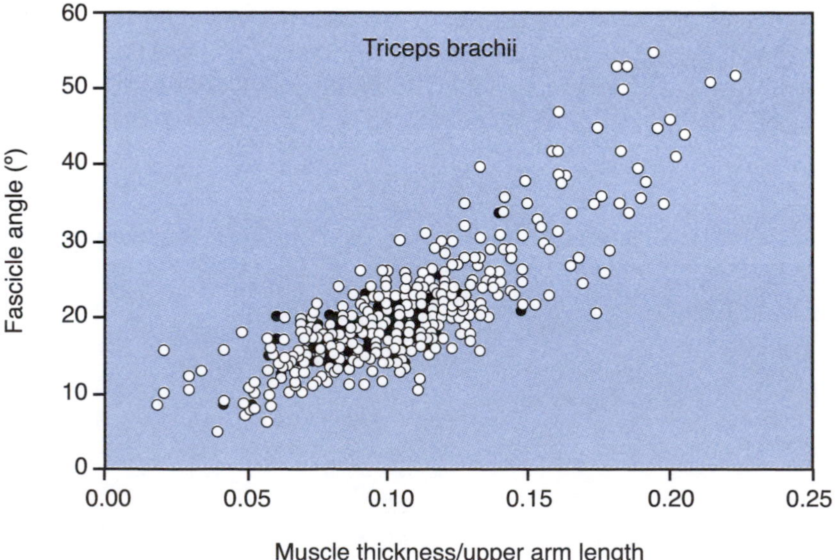


**Figure 3.31** Relationship between fascicle angles of the triceps brachii (long head) and the ratio of muscle thickness to upper-arm length for 537 male and female subjects ages 10 to 60 years. $r = .81$; $p < .001$.

Reprinted from Journal of Biomechanics, 30, T. Fukunaga, et al., Muscle architecture and function in humans, 462, Copyright 1997, with permission from Elsevier Science.

## Force in Relation to Speed of Shortening: Force–Velocity Relationship

The force of a muscular contraction depends heavily on the speed of shortening. This was shown experimentally by Fenn and Marsh (1935) and expressed in a mathematical formula by A.V. Hill (1938). The curve showing the relationship between force and speed of shortening has since become known as the Hill curve (figure 3.32). It is very often not acknowledged, however, that the curve does not describe the force development during an accelerated movement. In fact, it should not be presented as a continuous curve at all, being as it is composed of a series of discrete points, each representing the force obtained during a different constant shortening velocity.

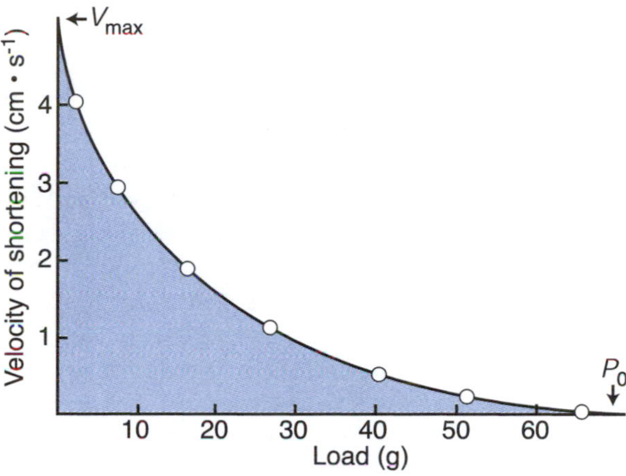


**Figure 3.32** Relationship between force and velocity of shortening measured in a whole sartorius muscle of the frog.

Reprinted, by permission, from A.V. Hill, 1938, "The heat of shortening and the dynamic constants of muscle," *Proceedings of the Royal Society of London* B126:136-195.

The curve shows very clearly that the force during an isometric contraction, that is, at zero shortening velocity, is higher than at any positive shortening velocity. On the other hand, the force is even higher during an eccentric contraction, that is, with negative shortening velocity. The reason for this is elastic elements in series with the contractile machinery, which are stretched during the contraction, adding passive tension to the active contractile force. Experiments with frog muscle show that the curve is very steep around zero shortening velocity (Edman 1992). If this holds true for human muscle as well, it may mean that during the initial phase of an eccentric movement, the force increases considerably without an accompanying increase in the speed of lengthening of the muscle. This may prevent the muscle from yielding untimely to the external force acting on it (e.g., when landing after a jump or when walking downhill carrying a heavy weight).

The ability of a muscle fiber to create force at high shortening velocities is closely related to its speed of cross-bridge cycling. Consequently, the slope of the curve is different for slow and fast muscle fibers, slow fibers losing force more rapidly with increased speed of shortening (figure 3.33). Intuitively, this

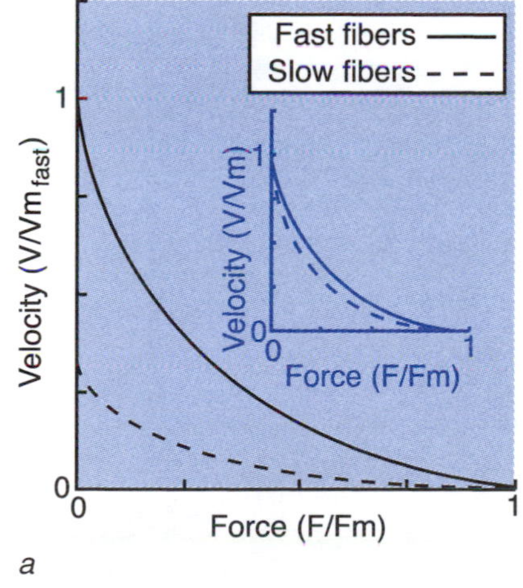


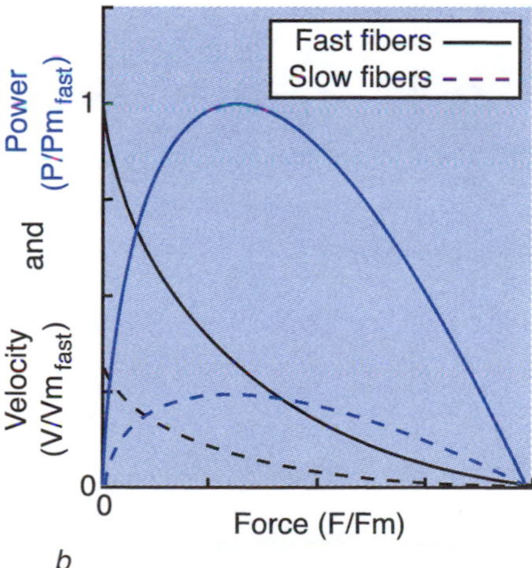


**Figure 3.33** Velocity of shortening and power output as a function of force for fast and slow human muscle fibers. Force is normalized by maximal isometric tetanic force. *(a)* Velocity of shortening as a function of force for fast and slow fibers. Velocities are normalized by the maximal shortening velocity of the fast fibers. Insert illustrates the difference in curvature of the force–velocity curves of the fast and slow fibers. Velocities are normalized by their respective maximal shortening velocity. *(b)* Power output as a function of force added to the curve in *a*. Power is normalized by the maximal power output of the fast fibers.

Reprinted, by permission, from J.A. Faulkner, D.R. Claflin, and K.K. McCully, 1986, Power output of fast and slow fibers from human skeletal muscles. In *Human muscle power*, edited by N.L. Jones, N. McCartney, and A.J. McComas (Champaign, IL: Human Kinetics), 84.

makes slow muscle fibers unfit for contractions at high shortening velocities, and as we shall see later, the consequences for the resulting power are severe.

In all kinds of motors, what matters is the power developed rather than the force. In the case of a muscle, the power is the product of the speed of shortening and the force developed, or in other words, it is equal to the area of a rectangle with sides corresponding to the coordinates of the corresponding force–velocity point (figure 3.34). Understandably, the power is zero when either the velocity or the force is zero, and it turns out to be maximal at a shortening velocity that is approximately one third of maximal shortening velocity, $V_{max}$ (figure 3.35; Edgerton et al. 1986). Since $V_{max}$ is considerably lower for slow muscle fibers than for fast fibers, the speed of shortening at which maximal power occurs in slow fibers is correspondingly low. It has been calculated that a muscle with 50% slow and 50% fast fibers has a peak power output that is only 55% of the peak power of a muscle composed exclusively of fast fibers (Faulkner, Claflin, and McCully 1986). Furthermore, the contribution from slow fibers is restricted to low shortening velocities (see figure 3.35). The consequences of this in everyday movements are intriguing, considering the role of slow fibers in sustained physical activity. Maybe we tend to overestimate the speeds of shortening used in daily life and fail to realize that much of human movement is performed at shortening speeds within the range of slow fibers (see discussion following the presentation by Faulkner, Claflin, and McCully 1986). This contention is supported by the fact that the optimal speed of shortening of type IIB fibers is lower than the $V_{max}$ of slow fibers, ensuring that slow fibers do not go slack but contribute to power generation (Bottinelli et al. 1999).

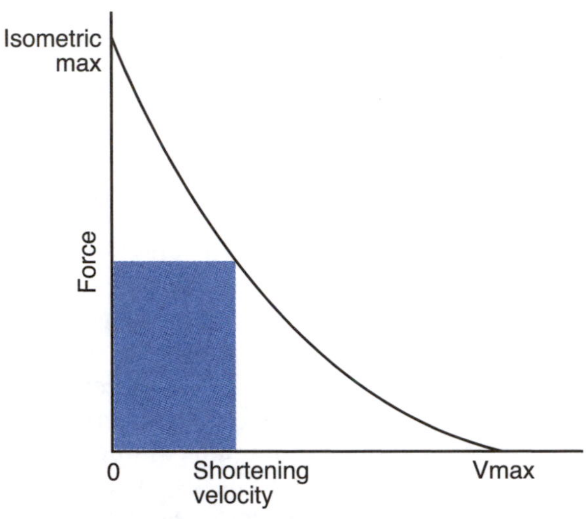


**Figure 3.34** The power output at a certain velocity of shortening equals the area of the rectangle under the force–velocity curve for that particular velocity.

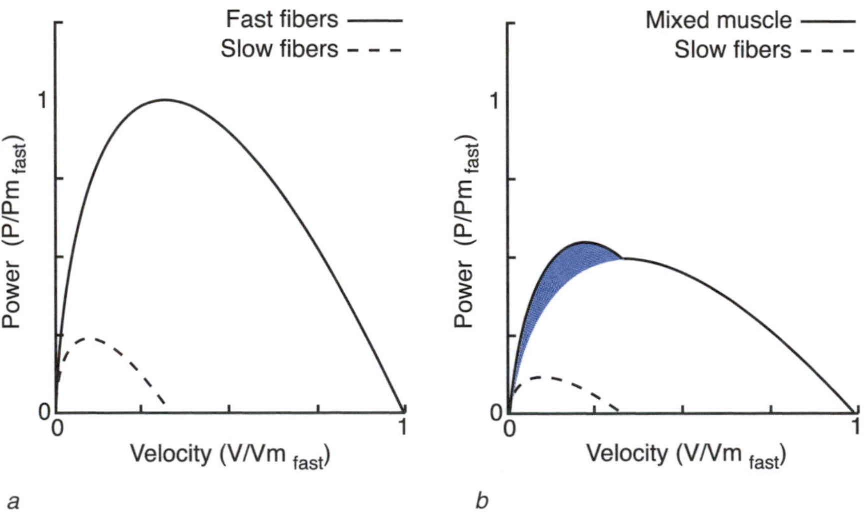


**Figure 3.35** Power output as a function of shortening velocity *(a)* for pure fast and pure slow muscle and *(b)* for mixed muscle with 50% fast and 50% slow muscle fibers. The shaded portion of the curve in *b* represents the contribution of the slow fibers to the power output of the mixed muscle. Values for power and velocity are normalized to those of pure fast fibers.

Reprinted, by permission, from J.A. Faulkner, D.R. Claflin, and K.K. McCully, 1986, Power output of fast and slow fibers from human skeletal muscles. In *Human muscle power*, edited by N.L. Jones, N. McCartney, and A.J. McComas (Champaign, IL: Human Kinetics), 85.

## Mechanical Efficiency of Muscle Contraction

The **mechanical efficiency** of muscular work is the ratio of external work performed, expressed in energy units, to the extra energy consumption incurred by the work. The extra energy consumed is calculated as total energy consumed *(E)* minus the resting metabolic energy consumption *(e)*. To express the mechanical efficiency (ME) as a percentage, this ratio is multiplied by 100%:

$$ME = 100\% \times W(E\text{-}e)^{-1} \quad (1)$$

It has been claimed that the human body may perform work with a mechanical efficiency of 20% to 30% under favorable conditions (Robinson 1980). This makes the human body an efficient machine, far better than a steam engine (10%) and comparable to a gasoline engine. The efficiency of work may, however, vary considerably with speed, load, fatigue, training, and fuel used (Robinson 1980). It is also influenced in a significant way by the type of contraction involved, resulting in values that have been claimed to reach 60% in eccentric contractions. In a stretch–shortening cycle (discussed later) the ME of the concentric phase is influenced by the level of prestretch and may approach 50% for high prestretch levels (Komi 1992).

During work performed in the upright position, a basic energy cost is incurred at all speeds by muscles supporting the body weight. At very low speeds, therefore, the efficiency is low because this basic energy cost is relatively high compared with the energy cost of the work itself. With increasing speed, the efficiency increases, until a maximum is attained at intermediate speeds. Observations of whole muscle as well as of single fibers indicate that optimal efficiency is reached at a speed of shortening close to $V_{opt}$. $V_{opt}$ is defined as the shortening velocity at which maximum power is obtained. Therefore, shortening at optimal velocity would optimize not only power but also efficiency (Bottinelli et al. 1999). Increasing the speed beyond this optimum leads to a progressive loss of efficiency (figure 3.36; Robinson 1980).

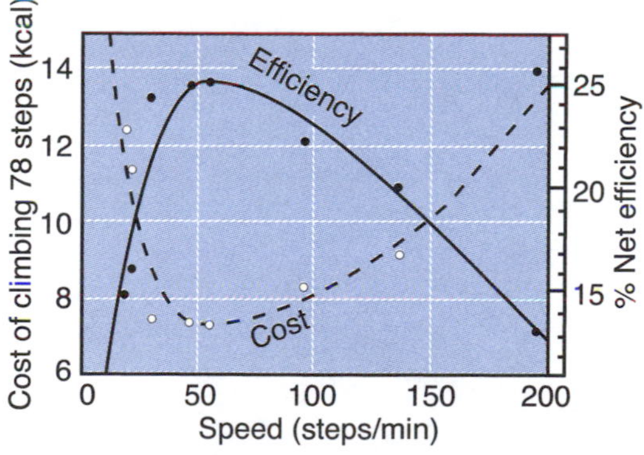


**Figure 3.36** Influence of speed of climbing stairs on energy cost (dashed curve) and efficiency (solid curve). Cost was lowest and efficiency highest at intermediate speed.

Reprinted, by permission, from S. Robinson, 1980, Physiology of muscular exercise. In *Medical physiology*, edited by V.B. Mountcastle (Philadelphia, PA: Mosby), 1399.

## Storage and Utilization of Elastic Energy

In tutorial texts it is usual to describe concentric, isometric, and eccentric muscle activity as separate events. In most daily activities, however, they are not. It has even been shown that what is perceived as an isometric contraction on the whole-muscle level is far from isometric on the muscle fiber level due to compliance of tendinous structures (Ito et al. 1998). In most daily activities, alternating sequences of eccentric and concentric work occur. During walking and running the quadriceps muscle starts out with an eccentric phase on ground contact, followed by a transitory isometric phase and a concentric phase ending at toe-off. Such frequently occurring sequences of muscular activity are known as stretch–shortening cycles (Komi 1992). During the eccentric phase the muscle acts like a brake, and part of the kinetic energy absorbed is converted to heat (thermal energy). Some of the kinetic energy is, however, absorbed by elastic elements in series with the contractile apparatus, called series elastic elements or **series elastic components (SECs).** This elastic energy is released again during the concentric phase, making it more powerful.

The major part of the SEC is believed to be located in the tendinous tissues of the muscle, while some may reside in the cross-bridges themselves (for discussion, see Huijing 1992).

To regard the increased performance during the eccentric phase as purely due to elastic components may, however, be an oversimplification. Several lines of evidence indicate that changes in the reflex activity may play a role as well. The importance of modulations of stretch and cutaneous reflexes for the maintenance of stability in ambulatory movements is well documented (Kearney, Lortie, and Stein 1999; Zehr and Stein 1999). Furthermore, it has been shown that runners adjust leg stiffness for their first step on a surface with a different compliance (Ferris, Liang, and Farley 1999). In spite of a 25-fold change in running surface compliance, their center of mass followed nearly the same path before and after the change.

# Chapter 4

# Motor Function

The purpose of this chapter is to give an overview of the basic structure and mechanisms of the nervous system and to provide an idea of how sensory information and volition are integrated to create adequate movements.

The central nervous system (CNS) receives information concerning the outside world via exteroceptors as they react to light, sound, touch, temperature, or chemical agents and via interoceptors that are stimulated by changes within the body (figure 4.1). The latter include **proprioceptors** (such as muscle spindles, Golgi tendon organs, joint receptors, and vestibular receptors), **chemoreceptors,** and visceroceptors. The CNS is equipped to receive, interpret, and handle information and then eventually to transform the result into appropriate reactions. Even the process of feeding information to the CNS may involve some muscular activity. For example, the eye muscles are almost continuously active, especially while the individual is awake. Reactions to various stimuli include gestures, speech, writing, and sometimes very vigorous muscular contractions, as in the event of an approaching danger. There is scarcely any stimulus that does not, through reflexes or voluntary action, affect smooth muscle, heart muscle, or skeletal muscle.

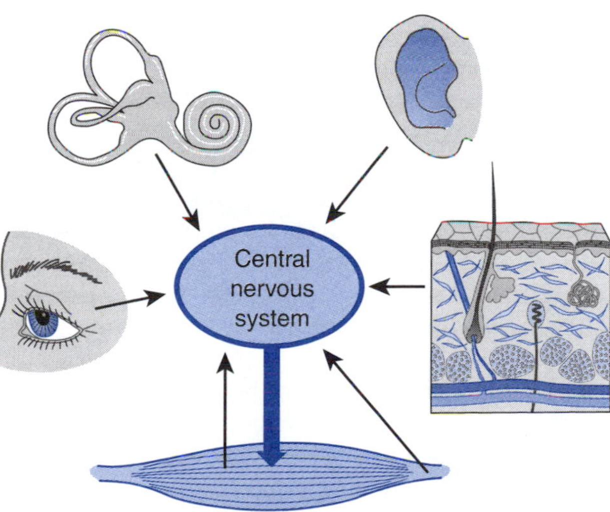


**Figure 4.1** Different kinds of receptors provide the central nervous system (CNS) with information about the surrounding world as well as about the body itself. Different kinds of stimuli are transduced by specific receptors into nerve impulses in afferent nerves. Appropriate motor commands are created by the CNS and transmitted to the muscles through peripheral motor axons, which are the only available route to the skeletal muscles.

A presentation of the physiology of work and exercise would justify a very detailed discussion of the function of the CNS. In fact, it is difficult to decide which aspects are more and which are less essential for the understanding of muscular movements in work and other types of physical activity. We shall, however, limit this discussion mainly to the parts of the nervous system more directly involved in the activation of skeletal muscles, including a brief survey of the **autonomic nervous system.** For a more comprehensive description of the structure and function of the nervous system, the reader should consult basic neurobiological textbooks and reviews (e.g., Brodal 1998; Kandel, Schwartz, and Jessell 2000). In the following discussion, it is assumed that the reader is familiar with the basic structure of the nervous system.

Correct timing and smooth and targeted execution of a movement are the result of a close cooperation between sensory and motor systems. Of course, the final executors of the movement are the muscles,

## CLASSIC STUDY

In his seminal work, Sherrington (1906) introduced a series of new concepts, which not only have survived the passage of time but still rank among the fundamental principles of neurobiology. He had already coined the term *synapse* in 1897, and now he introduced the well-known phrase "final common path" for the motoneuron and its axon to the muscle. He also saw the motoneuron as an integrator of nervous information, a very anticipatory view considering that neither he nor any of his contemporaries had any idea of chemical transmission yet.

but normal, rested muscles obey their commanding motoneurons completely. Every time a group of motoneurons launches their impulse trains, the corresponding muscle units react in a predictable way. A muscle unit is defined as the muscle fibers belonging to one motor unit.

The motoneuron is the site of final integration of all the different neural influences affecting the movement, for example, through descending pathways from the brain and from sensory sources relevant for the movement in question. It is this role of the motoneuron and its motor axon as the sole communicator from the nervous system to the muscle that Sherrington (1906) epitomized as "the final common path." Accordingly, the motoneurons and their obedient muscle fibers, collectively known as motor units, take center stage in this chapter. We start by reviewing the basic structure and function of neurons in general before we take a closer look at the role of sensory systems in motor function.

### FOR ADDITIONAL READING

For a fuller account of Sherrington's contribution to neurobiology, see G.M. Shepherd and Erulkar (1997).

## Basic Structure and Function of Neurons

The brain is made up of nerve cells, which are the communicating elements, and the assisting and nourishing glial cells. The number of nerve cells in our body is somewhere around a hundred billion ($10^{11}$), give or take a factor of 10; it is about the same as the number of stars in our galaxy! There are approximately 1,000 different types of nerve cells, or neurons. Each has a diameter from 5 to 100 μm and consists of four morphological regions: (1) the cell body (soma or **perikaryon**), which is the "heart" of the nerve cell, (2) the dendrites, a set of short, fine, arborizing processes radiating from the cell body, (3) the axon, from less than 1 to 20 μm in diameter and from 1 mm to 1 m long, and (4) the **axon terminals** (figure 4.2). Each region has distinctive functions, which are described in more detail later. A membrane, 5 nm thick and basically of the same type as cell membranes in general, covers the soma and the processes. The abundance of mitochondria in the cell indicates a high level of metabolic activity. In fact, the oxygen uptake of nerve cells per unit of weight is probably of the same order as that of skeletal muscles during maximal contraction. If we take a power of 80 W as being a reasonable figure for a resting adult person, 15 to 20 W is the portion of this power used by the nervous system. This level is, however, relatively constant whether the person is sleeping or concentrating very hard on an intricate intellectual problem. This contrasts with the metabolic rate of the skeletal muscles, which in the case of top athletes can increase from 20 W at rest to more than 2,000 W during maximal exercise.

Generally, the dendrites and the cell body receive signals, while the axon hillock—the initial segment of the axon—combines and integrates them and "decides" whether or not a signal will be sent through the axon to the terminals. The axon may branch off near its beginning, but more often the branching takes place closer to its end. In most cases, communication with other cells is based on chemical signals released from the nerve terminal that act on the target cell in a synapse. In the context of this book, all synaptic transmission can be considered chemical.

A typical nerve cell can have anywhere from 1,000 to 10,000 synapses on its surface, which relay information from something like 1,000 other neurons. The task of the single nerve cell is to react to the signals it receives and, if appropriate, to generate a new nerve impulse, which in turn influences its own target cells. Such target cells can be other nerve cells, muscle cells, or gland cells. With regard to the human nervous system, it is important to realize that there is no such thing as "impulse transmission." Each nerve impulse is restricted to one nerve cell and ends up in the release of signal molecules,

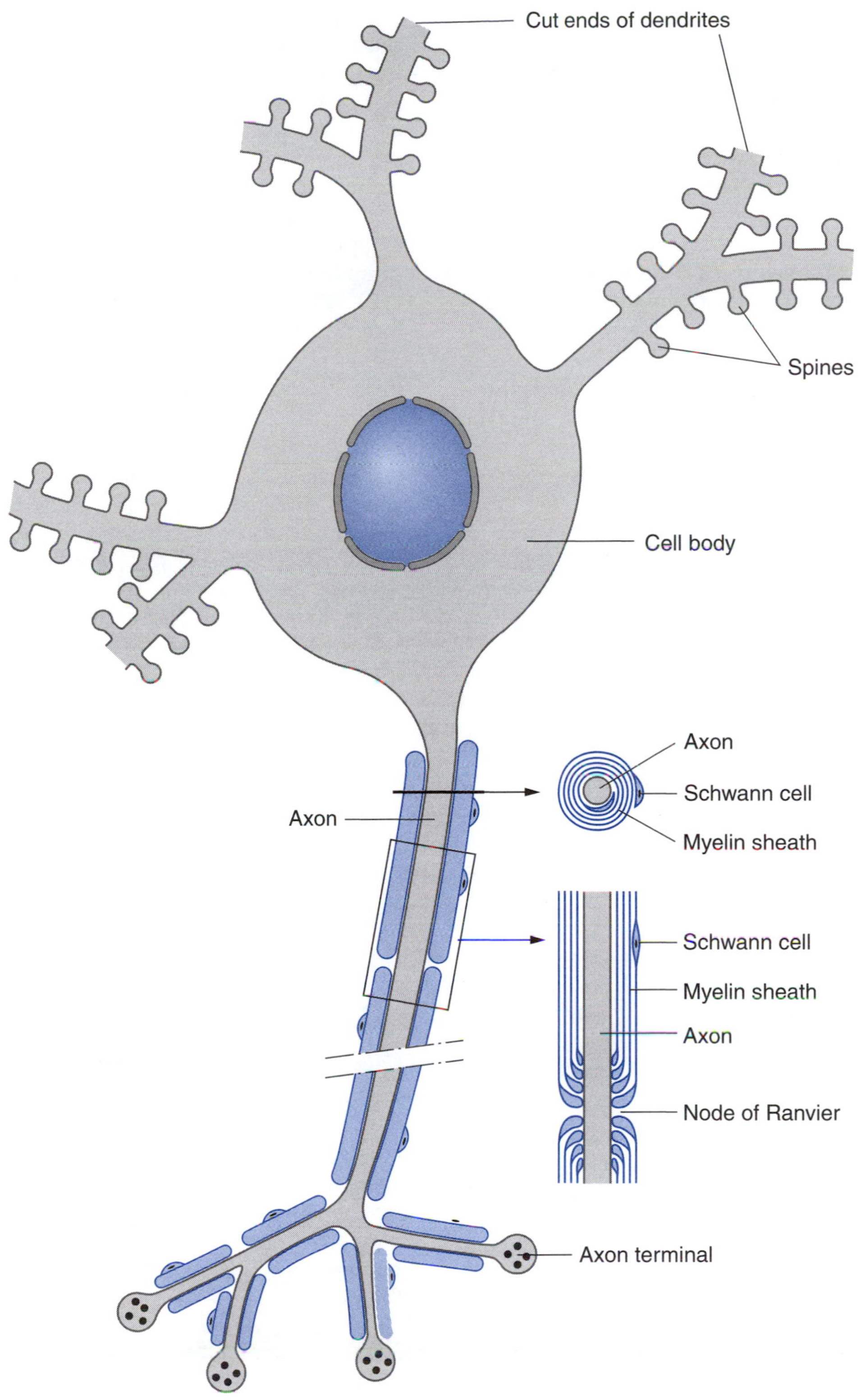


**Figure 4.2** This neuron shows clearly the four different morphological regions of a typical nerve cell: the cell body (also known as the soma or perikaryon), the dendrites with dendritic spines (the small, lollipop-like protrusions), the axon, and the axon terminals.

traditionally called transmitter substance or neurotransmitter, from the nerve terminal. In a way, a nerve impulse can be likened to the first marathon runner, who, according to history, ran from Marathon to Athens in 490 B.C. to tell the Athenians that they had conquered the Persians. After having delivered his message, he fell to the ground dead.

In addition, the concept of "transmission" of impulses from one cell to another becomes untenable, or at least too easily misunderstood, when we realize that some synapses are inhibitory. Instead of promoting the creation of a new impulse, such synapses serve to prevent this from happening in the **postsynaptic cell** (figure 4.3). Among the thousands of synapses acting on a nerve cell, some are excitatory while others are inhibitory. For all practical purposes it can be stated that individual synapses never change from one kind to the other. What matters at any given moment is the relative activity in the **excitatory** and **inhibitory synapses** acting on a particular nerve cell. If excitatory activity prevails, the chances are that the postsynaptic cell fires. If not, the cell remains silent. We return to this in greater detail later in this chapter.

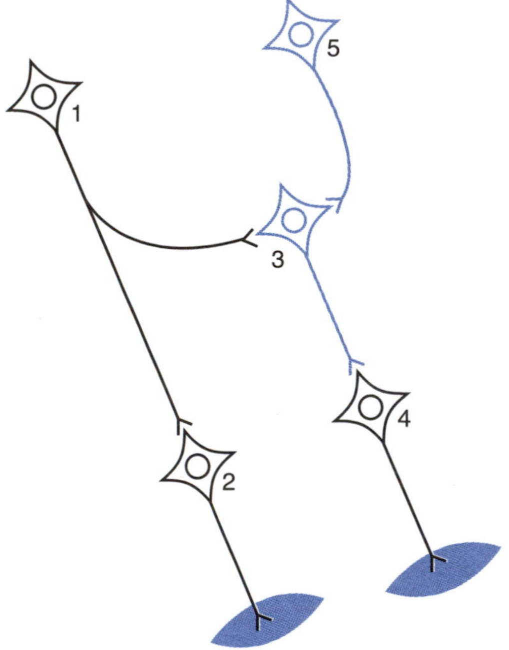


**Figure 4.3** Inhibitory neurons have important roles in the central nervous system. In this example, cells 2 and 4 can be thought to be motoneurons innervating antagonistic muscles. For cell 1 to be able to excite cell 2 and at the same time prevent excitation of cell 4 (which may be counterproductive), cell 1 has to excite an inhibitory interneuron, cell 3 (blue), which in turn inhibits cell 4. In other instances, cell 3 may be inhibited (rather than excited) by another inhibitory neuron (cell 5, blue), thus relieving the inhibition of cell 4. This is called disinhibition. See also figure 4.24.

Eccles (1982) points out that chemical transmission was an essential evolutionary advance. It made possible the design of brains with all of the infinite complexity required for higher nervous functions. The chemical transmission of information provided an opportunity for the evolution of the mechanism of inhibition, a process essential to the function of the CNS, and one that would not be possible with electrical transmission alone.

As stated earlier, the entire nerve cell with all its processes is covered by a cell membrane. The basic composition of this membrane is that of cell membranes in general, as outlined in chapter 2. Of special interest for the function of nerve cells are the protein molecules that are embedded in the cell membrane, forming, among other things, various kinds of ion channels and ion transporters (see the next section). In addition to mitochondria, the cytoplasm of the nerve cell body is characterized by large amounts of granular endoplasmic reticulum and free ribosomes, serving the requirements of the entire cell, including its processes, for protein synthesis. In the nerve branches different kinds of filaments and tubules run parallel to the long axis of the process. They serve partly as a cytoskeleton, important for creating and maintaining the shape of the cell and its processes, and partly as transport routes between the cell body and the distant parts of the processes.

## For Additional Reading

For details about intracellular transport, see Alberts et al. (1994).

In fact, the processes are arteries for busy molecular traffic moving up and down between the cell body and its most distant ramifications. This traffic is essential for the survival of the dendrites and the axon. In the axon, there is a slow outgoing flow at a rate of 0.5 to 3 mm per day but also a fast transport, mainly of organelles, running at approximately 400 mm per day. Such axoplasmic flow is essential for the function of the nerve cell. As we shall see, the nerve cells can be regarded as secretory cells, and fast transport of the secretory products from the cell body to the tip of its processes is important. Just think of the phrenic nerve of a giraffe! Its motoneurons are located in the upper cervical segments of the spinal cord, and their axons extend all the way down to the diaphragm. In the axons, there is also a retrograde (backward) flow of organelles and molecules. Not least, the axon and the parent neuron are critically influenced by factors released by the target cells. Some of these neurotrophic factors are essential to the survival of the neuron. They act through various types of receptors in the axon terminal membrane, many of which are enzyme-linked receptors (see chapter 2) of the tyrosine kinase type.

### For Additional Reading

For a general discussion of cell surface receptors, see Alberts et al. (1998). The role of neurotrophic factors in the survival of nerve cells is discussed in Kandel, Schwartz, and Jessell (2000, chapter 53).

Even though this axoplasmic transport is important, it is the creation and propagation of nerve impulses that take center stage when it comes to nerve cell function. The impulses travel down the axon at speeds up to more than 100 m · $s^{-1}$ (328 ft · $s^{-1}$) or 360 km · $h^{-1}$ (224 mph)! The speed of conduction depends on the diameter of the axon (the thicker the axon, the faster the speed) and on whether it is myelinated or not. Consequently, thick, myelinated axons like the ones belonging to the motoneurons in the spinal cord are the fastest. Thin, unmyelinated sensory axons can have conduction velocities as low as 1 m · $s^{-1}$ or even less, while thick, myelinated sensory axons can be just as fast as the fastest motor axons.

The myelin sheath is in fact layers of cell membrane wrapped around the axon. In the central nervous system the cell membrane is part of oligodendroglial cells, while in the peripheral nervous system it is part of Schwann cells (see figure 4.2). The myelin sheath serves as insulation, preventing the flux of ions between the axoplasm and the surrounding tissue fluid. At certain intervals, however, the myelin sheath is interrupted by so-called **nodes of Ranvier,** which the advent of the electron microscope in the middle of the 20th century revealed to be intervals between two neighboring myelin segments, leaving a short piece of the axolemma naked. The impulse jumps from node to node in what is now commonly known as **saltatory conduction,** thus significantly enhancing the velocity of the impulse. Since the long internodal sections remain passive, the ionic fluxes are restricted to less than 1% of the surface area of the nerve fiber, relieving the internodal sections of the metabolic burden of running the sodium–potassium pump.

Axon terminals (also called boutons or synaptic endings) of other nerves end at the surface of the soma and the dendrites (see figure 4.2). Actually, a large percentage of the surface of the soma, the dendrites, and even the axon hillock can be covered by synaptic endings from thousands of other nerve cells. The space between the membranes of the synaptic ending and the contacted nerve cell (known as the presynaptic and postsynaptic membranes, respectively) is called the **synaptic cleft.** It is about 20 to 30 nm wide, and there is no cytoplasmic continuity between the two cells.

## Resting Membrane Potential

In nerve cells, as in other cells, the chemical composition of the fluid on either side of the selectively permeable cell membrane is different. In relation to the membrane potential, the major difference is in the ionic composition, the external fluid in essence having a higher concentration of sodium and chloride and a lower concentration of potassium than the internal cytosol. These differences in the concentration of ions would cause diffusion of the ions through the membrane if permeability and electrical gradients could permit this, but the membrane does not allow a free passage of ions. On the contrary, it acts as a physical barrier to their diffusion.

As mentioned in chapter 2, the cell membrane contains proteins with different functions. Some of the proteins form ionic channels, most of which are gated, that is, able to switch between an open and a closed position. Apparently, slight changes of shape in critically placed parts of the channel protein molecule play a decisive role in the transition between the open and closed situation and thus in the membrane conductance for specific ions. In one type of ion channel, a change in the potential across the membrane can open or close the gate. We call these **voltage-gated ion channels.** Another type of channel is chemically gated, which means that the channel opens when a particular molecule, a neurotransmitter, binds to its receptor region. These are called ligand-gated ion channels or chemically gated ion channels. A third type of channel, found in sensory cells, is opened by specific physical stimuli, such as light (eye), vibrations (ear), pressure, temperature (skin), or stretch. There is also specificity between different ions—potassium ion ($K^+$), sodium ion ($Na^+$), chloride ion ($Cl^-$), and calcium ion ($Ca^{2+}$)—and their respective channels. This specificity rests on an interaction between the ion in question and parts of the channel protein. The density of any particular type of channel on a cell ranges from zero up to some 10,000 per square micrometer, and there are significant differences in the types of ion channels found in the various parts of the nerve cell membrane. The axon hillock, for example, has the highest density of voltage-gated $Na^+$ channels, making it the part of the cell with the lowest threshold for generating action potentials.

Important characteristics of the membrane and its proteins account for the nerve impulse and other complex features of neuron function. We now discuss some of the electric and chemical events that take place on both sides of the nerve cell membrane, as well as inside the membrane at rest and when active. Before considering the events underlying the nerve impulse, we have to take a closer look at the nerve cell at rest.

Because of their respective electrochemical gradients, $K^+$, $Na^+$, and $Cl^-$ all are able to get through the membrane by means of their respective ion channels, provided the channels are open. The main intracellular anions, on the other hand, are proteins that are far too large to escape through the membrane into the extracellular fluid, even though the extracellular concentration of the protein is lower.

At rest, the gated ion channels are closed, but nongated ion channels allow potassium ions to leak out into the extracellular fluid because the concentration of potassium is higher inside. This creates a relative deficit in positive charges inside the cell, making the inside of the cell membrane electrically negative in relation to the outside. Electric charges of opposite signs attract each other, and the increasing negative charge inside the cell therefore attracts $K^+$ ions, pulling them back into the cell, so to speak. At a certain point, called the **equilibrium potential** for potassium, this internal negative charge is able to balance the outward leak of potassium driven by the difference in concentration.

Even though the membrane is much less permeable to $Na^+$ than to $K^+$, an inward leak of sodium driven by the electrochemical gradient for this ion would eventually eliminate the differences across the membrane. At the resting membrane potential, both the concentration gradient and the charge difference tend to drive $Na^+$ into the cell. Consequently, to maintain the difference in concentration of ions across the resting cell membrane, there must be active transport of $Na^+$ outward and $K^+$ inward virtually equaling the diffusion in the opposite directions (figure 4.4). This is taken care of by the **sodium–potassium pump,** more often simply called the **sodium pump.** The energy required to transport these ions against their electrochemical gradient comes from adenosine triphosphate (ATP). In accordance with this, the sodium–potassium pump is also known as a sodium–potassium ATPase. To convey a rough idea of the capacity of the sodium pump, when operating at its maximal rate, each pump can transport some 200 $Na^+$ and 130 $K^+ \cdot s^{-1}$ across the membrane. The reason for the difference between the two ions is that the pump expels three sodium ions for every two potassium ions taken in. One neuron may have as many as one million pumps, with a potential to move approximately 200 million $Na^+ \cdot s^{-1}$ (C.F. Stevens 1979).

The net effect of these membrane characteristics is that the inside of the cell membrane is electrically negative compared with the outside, the difference being in the range of 60 to 70 mV. This difference is,

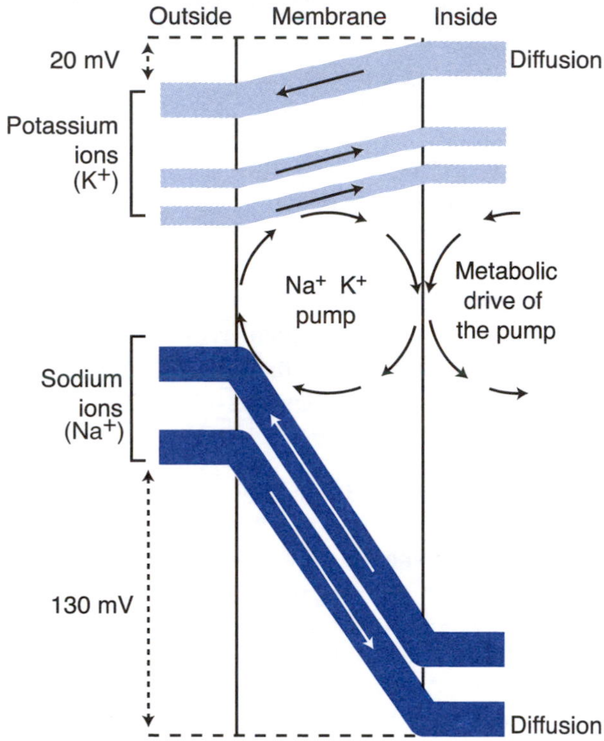


**Figure 4.4** Schematic diagram showing the potassium ion ($K^+$) and sodium ion ($Na^+$) fluxes through the surface membrane of a nerve cell in the resting state. The slopes of the flux channels across the membrane represent the respective electrochemical gradients of the ions. The voltage difference across the membrane is in the range of 45-75 mV, usually 70 mV, the inside being negative in relation to the outside. Under resting conditions, chloride ions ($Cl^-$) diffuse inward and outward at equal rates. The concentration of sodium and potassium must, however, be maintained by some sort of active transport, called the "$Na^+$ $K^+$ pump," driven by ATP. Because of a slow inward leak of $Na^+$ ions, the potential inside the nerve cell is less negative than the equilibrium potential for potassium ions, resulting in an outward leak of $K^+$ ions. Therefore, the pump has to counteract this leak by transporting $K^+$ ions back into the cell. For sodium ions, both the concentration gradient and the electrical gradient tend to drive sodium ions into the cell. To keep the sodium out, therefore, a very active pump is needed.

Modified from Eccles 1965.

however, restricted to the part of the cytosol and the external tissue fluid immediately bordering on the cell membrane (figure 4.5). By convention, the outside is said to be zero, giving the inside a charge of –60 to –70 mV, called the resting membrane potential. In some cells the resting membrane potential can be as high as –75 mV; in others it can be as low as –45 mV.

## *Summary*

Nongated ion channels allow $K^+$ to leak out from the cell, giving the cell interior a charge of –60 to –70 mV in relation to the outside. At this resting membrane potential, the outward leak of $K^+$ is balanced by the internal negative charge pulling $K^+$ back into the cell. For $Na^+$ the situation is different. Both the concentration gradient and the charge difference tend to drive $Na^+$ into the cell, eventually reducing the differences. As long as the gated $Na^+$ channels are closed, however, only a minor inward leak of $Na^+$ occurs, and this is electrically balanced by a compensatory outward leak of $K^+$. The resulting ionic perturbation is taken care of by the sodium–potassium pump.

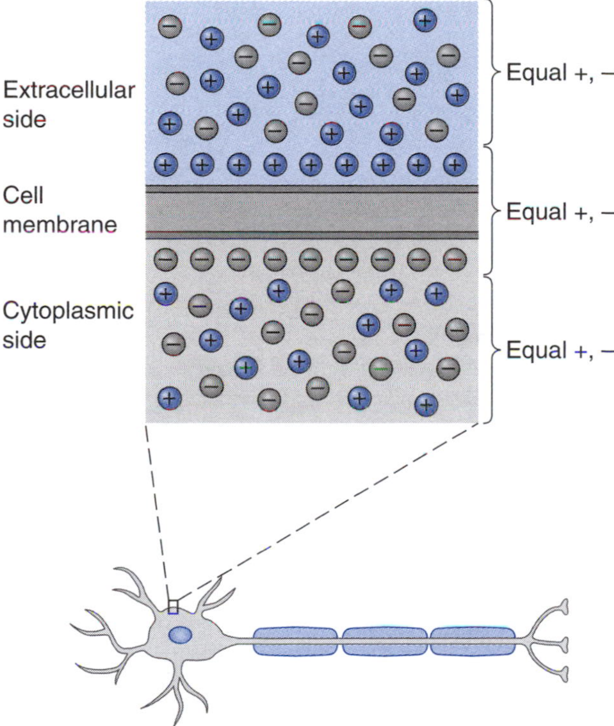


**Figure 4.5** The membrane potential of a cell is the result of an inequality of electric charges on the cytoplasmic and extracellular sides of the cell membrane. Note that the inequality is restricted to a narrow zone bordering on either side of the membrane. In the bulk of the cytosol and extracellular fluid, no such differences exist.

Adapted, by permission, from E.R. Kandel, J.H. Schwartz, and T.M. Jessell, 2000, *Principles of Neural Science*, 4th ed. (New York: McGraw-Hill), 29.

## Excitation and Inhibition

When a nerve impulse arrives at a synaptic terminal, voltage-gated $Ca^{2+}$ channels in the terminal open, allowing $Ca^{2+}$ from the outside into the terminal. This initiates a series of events (summarized in Schiavo and Stenbeck 1998) culminating with a number of synaptic vesicles emptying their contents into the synaptic cleft. The result of this depends on the nature of the transmitter or signal molecules contained in the synaptic vesicles and the type of ion channel on which they act. Both transmitters and ion channels exist in many varieties, but nevertheless, as a rule they can be dichotomously divided into excitatory and inhibitory ones. In both cases the effect of the transmitter molecules is very short-lived because the transmitter molecules are almost instantly removed or inactivated in some way or another. (See, however, comments on neuromodulators later in this chapter.) Even though we routinely speak of single impulses when explaining the synaptic events, bear in mind that impulses normally occur in series, known as impulse trains.

In excitatory synapses, the transmitter molecules bind to specific binding sites on ligand-gated $Na^+$ channels and open the channels, allowing $Na^+$ into the cell. This makes the local area under the postsynaptic membrane less negative. Such local changes in the membrane potential are called **synaptic potentials** or **postsynaptic potentials;** in the case of excitatory synapses, they are more specifically called **excitatory postsynaptic potentials (EPSPs).** Since the difference between the inside and the outside charges (the polarization) in this case has been reduced, this change can also be called a depolarization (figure 4.6). It is, however, important to note that a synaptic potential is restricted to the area of the postsynaptic cell just underlying the synapse. It is a local phenomenon and does not alone affect the rest of the cell to a significant degree. In other words, one single impulse is not sufficient to change the behavior of the postsynaptic cell, but if the same or a neighboring excitatory synapse fires again before the synaptic potential has faded away, the combined depolarizations are added. If the sum of such depolarizations reaches a certain firing threshold (around –55 mV), voltage-gated $Na^+$ channels open, allowing a sufficient number of $Na^+$ into the cell to reverse the membrane potential.

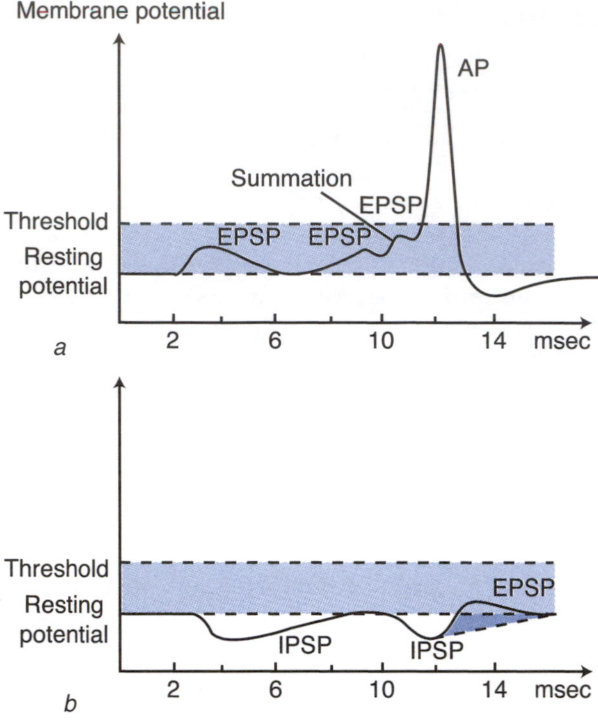


**Figure 4.6** Synaptic potentials are local and transient effects. *(a)* In general, a single excitatory postsynaptic potential (EPSP), shown in the left part of the figure, is not sufficient to depolarize the local area of the membrane to the firing threshold, and the membrane potential returns to its resting level. In the right part of the figure, a second EPSP occurs before a previous one has disappeared. Together they reach the firing threshold, and an action potential (AP) ensues. *(b)* In contrast to an EPSP, an inhibitory postsynaptic potential (IPSP) removes the membrane potential from the firing threshold. If an EPSP occurs during an IPSP (right part of the figure) they are summed up; from a hyperpolarized state, more EPSPs are needed to reach the firing threshold. Since the EPSP and the IPSP are caused by different synapses, this is an example of spatial summation. Since the EPSP starts a little later than the IPSP, it is an example of temporal summation as well.

From Brodal 1998.

Soon, however, the conductance for $Na^+$ returns to its resting value because the voltage-gated $Na^+$ channels are inactivated by the persisting depolarization. Furthermore, because the inside now has become positive in relation to the outside, both the concentration gradient and the charge drive $K^+$ out of the cell through voltage-gated $K^+$ channels, which after a short delay also open because of the depolarization (figure 4.7). The outward movement of $K^+$ brings the membrane potential back toward its resting level. It does not stop at the resting level, however, since it takes a few milliseconds for all the voltage-gated $K^+$ channels to close, allowing more $K^+$ to escape from the cell. This makes the cell temporarily hyperpolarized. Together, the sequential opening and closing of voltage-gated $Na^+$ and $K^+$ channels and the ensuing fluxes of ions result in a rapid and characteristic change in the membrane potential, called an action potential or, in everyday language, a nerve impulse (see figure 4.7). An action potential is an all-or-none phenomenon. Increasing the depolarization beyond the firing threshold does not increase the amplitude of the action potential. As mentioned earlier, action potentials are able to travel along the axon at very high speed with no decay in energy. In relation to the high speed of conduction in the axon, however, the signal is delayed about .5 ms in the synapse. "It used to be thought that one of the distinguishing properties of synapses was fatigue, but this is belied by the performance of synaptic

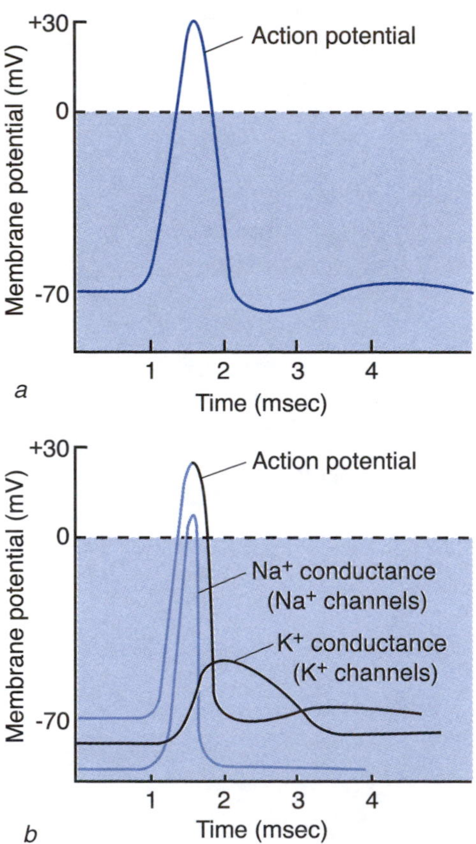


**Figure 4.7** The action potential (AP) is due to the sequential opening of voltage-gated $Na^+$ and $K^+$ channels. The first part of the AP (the depolarization, shown in blue) is due to opening of $Na^+$ channels (the permeability for $Na^+$ is also shown in blue), while the second part (the repolarization phase) is due to opening of $K^+$ channels.

Adapted, by permission, from P. Brodal, H.A. Dahl, and S. Fossum, 1990, *Menneskets anatomi og fysiologi* (Oslo, Norway: J.W. Cappelens Forlag).

mechanisms in the normal functioning of the brain" (Eccles 1973b). Actually, synapses and neurons of the CNS are well adapted metabolically to respond continuously at frequencies as high as 100 impulses per second with peaks even higher. We shall return to the propagation of action potentials later. Suffice it to say here that excitatory neurons are the true information carriers in the nervous system.

In inhibitory synapses, the first steps in the sequence of events are just the same as in excitatory ones. The arrival of the impulse at the synaptic terminal, the inflow of $Ca^{2+}$ into the terminal through voltage-gated ion channels, and the resulting emptying of the contents in the synaptic vesicles into the synaptic cleft are all the same. The difference lies in the nature of the transmitter molecules that are liberated and the type of ion channel that is affected. Inhibitory transmitter molecules bind to ligand-gated $K^+$ or $Cl^-$ channels, allowing $K^+$ to move out of the cell or $Cl^-$ into the cell, provided that the membrane potential is less negative than the equilibrium potentials for the two ions. Naturally, this makes the interior of the cell even more negative. It is a hyperpolarization, which is called an **inhibitory postsynaptic potential (IPSP).** Like EPSPs, IPSPs are local and transient events, but unlike the latter, they do not bring the membrane potential closer to the firing threshold. On the contrary, they move the membrane potential away from the firing threshold, making it less likely that the cell will fire an action potential.

The effect of inhibitory synapses is to reduce the depolarizing effect of excitatory synapses. In other words, EPSPs and IPSPs occurring at the same time can be summed up: If the combined potentials result in a depolarization still sufficient to reach the firing threshold, an action potential will ensue. If not, nothing happens. The postsynaptic cell is inhibited. An inhibitory effect may be seen even if the electrochemical gradient at the outset is such that no ion flux takes place when the inhibitory ion channels open. If excitatory synapses become active while the chloride channels are open, however, $Cl^-$ will start to flow into the cell as soon as the membrane is depolarized, in a way short-circuiting the depolarization. This makes it much less likely that the depolarization will reach the firing threshold. Many inhibitory neurons are so-called interneurons, that is, neurons with rather short axons that do not extend outside the area of the CNS where the cell body is located. Their main function is to modulate the effect of excitatory neurons. According to Eccles (1973b) inhibition can be likened to a sculpting process in which the inhibition "chisels away at the diffuse and rather amorphous mass of excitatory action and gives a more specific form to the neuronal performance at every stage of synaptic relay." Inhibitory neurons may, however, also modulate the effect of other inhibitory neurons. When an inhibitory neuron is inhibited, its inhibition of its target cell is reduced, and we call the process **disinhibition.**

The excitatory synapses tend to be peripherally located on the dendrites, while the inhibitory synapses are likely to be concentrated on the soma or the axon hillock. As mentioned earlier, the initial segment of the axon usually has the lowest threshold for the initiation of an action potential; it is a trigger zone for action. This is an example of good strategic design, because the inhibitory synapses located on the soma have final control over whether impulses will arise and be allowed to propagate down the axon (Eccles 1973b). As a result of the longer latency for inhibitory impulses than for excitatory impulses (about 1 ms), the former are most effective when they precede the excitatory volley. The majority of excitatory synapses end on dendritic spines, small lollipop-like protrusions on the dendrites (figure 4.8; see also figure 4.2). There is strong evidence that such spines are involved in learning and memory. We shall return to that topic both later in this chapter and when we deal with the effect of training in chapter 11.

During the course of the action potential itself, the cell is unable to respond to any stimulus. This is called an absolute refractory period. A hyperpolarization, either because of an IPSP or following an action potential (the after-hyperpolarization phase), makes it less likely that the cell will fire a new action potential. Therefore, the hyperpolarization phase is called a relative refractory period. Different nerve cells have after-hyperpolarization phases of different durations. Small motoneurons, for example, have longer after-hyperpolarization phases than larger motoneurons. This is a major reason that the maximal firing frequency of large motoneurons is higher than that of smaller motoneurons.

So far we have been dealing with axosomatic and axodendritic synapses. Synapses are, however, also found on axons, notably on their terminals. They are called axoaxonic synapses and serve to modulate the effect of their target terminals. Most often they act to reduce or inhibit the effect of axon potentials arriving at the terminal, so-called **presynaptic inhibition.** Presynaptic inhibition is a prominent feature in sensory pathways. In other instances axoaxonic synapses enhance the effect of the target terminal.

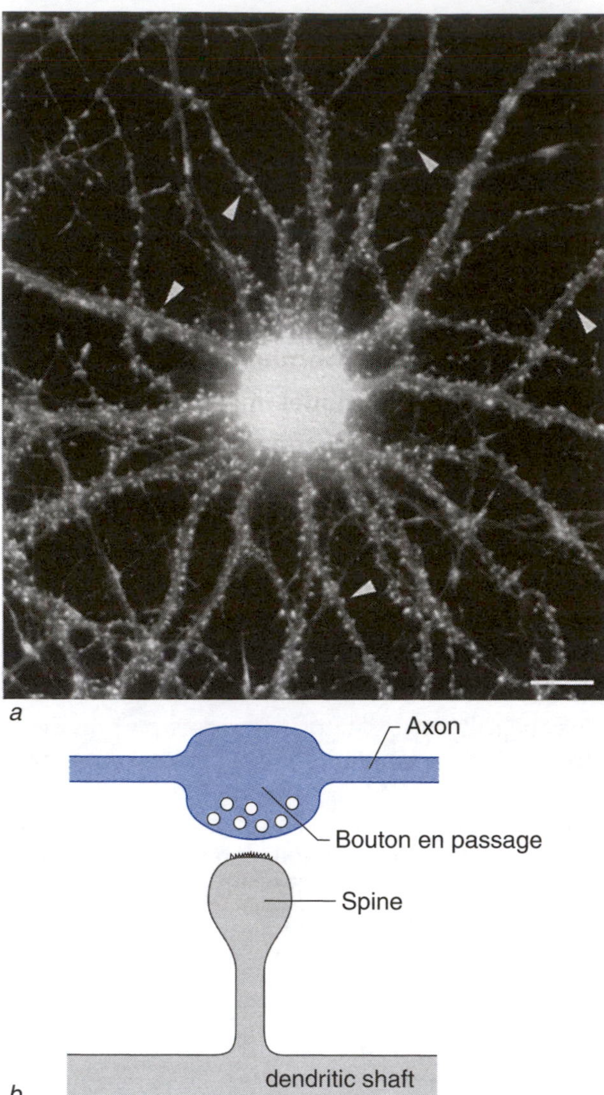


**Figure 4.8** *(a)* Hippocampal neuron grown in culture and stained for F-actin. Dendritic spines are seen in this digital image as intensely labeled dots along the surface of the dendrites (arrowheads). Scale bar, 10 μm. *(b)* Schematic drawing of a synapse between a *bouton en passage* and a dendritic spine.

*(a)* Reprinted from Trends in Neuroscience, 23, S. Halpain, Actin and the agile spine: how and why do dendritic spines dance?, 142, Copyright 2000, with permission from Elsevier Science.

## For Additional Reading

For further details concerning structure and function of synapses, see Brodal (1998) or Kandel, Schwartz, and Jessell (2000, chapters 10–15).

# Temporal and Spatial Summation

As stated earlier, summation of synaptic potentials is necessary to evoke an action potential. This is one of the reasons why single impulses are usually not regarded as proper signals in the nervous system. To result in an additive effect, synapse potentials must occur sufficiently close in time for the effect of the ionic currents to be combined. If the second synapse potential occurs after the transient ionic perturbation underlying the first synapse potential has faded away, no summation is possible. If the time lag between two or more excitatory potentials is gradually reduced, the combined synaptic potential is increased correspondingly, eventually reaching the firing threshold and resulting in an action potential.

This may be achieved in two different ways, either separately or in combination. Since the duration of a synaptic potential is longer than the duration of the action potential causing it, sequential action potentials arriving at the same terminal can be added. This is called **temporal summation.** But even synaptic potentials caused by different synaptic terminals can be added together, provided that they are located sufficiently close. This is called **spatial summation** and is most clearly seen in neuromuscular synapses, which are also known as motor endplates. With a few exceptions, which are outside of the scope of this book, all terminals in a motor endplate belong to the same parent axon. Consequently, they are all active at the same time, and their combined endplate potential normally exceeds the firing threshold by good margin, resulting in an action potential and a twitch contraction in the muscle cell.

When the postsynaptic cell is another nerve cell, it is almost impossible to decide whether two or more presynaptic terminals belong to the same parent axon or not. On the other hand, most postsynaptic nerve cells are contacted by a large number of synaptic terminals from different sources, the synaptic potentials of which can be added together. Such summation is always spatial but very often also temporal, since the probability of simultaneous firing by two or more different synaptic terminals is low unless the terminals belong to the same parent axon.

It has been emphasized that the action potential is an all-or-none response to effective stimuli, which means that the amplitude of this potential has a standard size, being stereotypic and similar in all axons (110 mV and typically with a duration of 1 ms). Whether or not a stimulus is effective is evidently determined by the algebraic summation of the impact of the EPSPs and IPSPs at the integrative, decision-making point, the axon hillock, which serves as a trigger zone. Since action potentials are "all-or-nothing" phenomena, how then is the intensity and nature of the stimulus reported and interpreted? In the individual neuron, the intensity of the stimulus is conveyed by changes in the number of spikes in a "train" or volley and by the interval between the spikes. It is a frequency code. But a population code, based on varying the number of cells participating in roughly the same type of operation, is also of major informational importance. In the case of sensory signals, the nature of the stimulus and its behavioral end results are determined by what part of the brain and which sets of neurons receive the signals. Later in the text we shall see how various sensory surfaces of our body are represented in a topographical manner in the brain and how the same holds true for muscles and movements. These two areas of the cerebral cortex, one for sensory perception and the other for motor commands, are extensively interconnected.

## Neurotransmitters and Neuromodulators

Thousands of millions of neurons communicate with each other and with other cells by means of their transmitter substances, the neurotransmitters. When the transmitter diffuses across the synaptic cleft, its effect depends on the ability of the receptor on the postsynaptic membrane to recognize and quickly combine with the transmitter. Like a key the neurotransmitter must fit into a lock and open specific ligand-gated ion channels. Biochemically, transmitters are quite heterogeneous, but it may be very difficult to prove whether a substance found in the CNS really is a true transmitter. In the range of 10 different transmitter substances have been identified as true transmitters, but there are some 30 more substances that are possible candidates. Whether a molecule is called a neurotransmitter is at least partly a question of definition, as discussed, for example, by Brodal (1998) and Kandel, Schwartz, and Jessell (2000). To properly deserve the designation *neurotransmitter,* the molecule should meet certain criteria. If not, it should rather be named transmitter candidate or **putative transmitter** until it is shown to meet the criteria.

It used to be thought that the mature neuron is specialized and capable of producing only one kind of transmitter, which is released from its synaptic terminals. That view cannot be upheld any longer. We now know for sure that more than one neurotransmitter, often one classical transmitter and one or several so-called neuropeptides, can be colocalized in the same synaptic terminal, even in the same synaptic vesicle. In such cases it is difficult to tell which of them is responsible for the observed effect.

It is the type of receptor and the ionic channels that it influences that determine the synaptic effect of the transmitter, not only whether the synaptic response is going to be an EPSP or IPSP, but also whether it will be a fast or slow synaptic effect. EPSPs and IPSPs are classical fast synaptic responses, resulting from neurotransmitter binding to receptor sites on the ion channel protein itself. In addition to such **ionotropic events,** a much slower type of effect has been shown to occur. In such cases, the receptor may be somewhere else on the surface of the postsynaptic membrane, away from the ion channel, implying that its effect on ion channels is indirect, through some kind of intracellular messenger pathway. This kind of synaptic influence is not only slower than a classical postsynaptic potential, but its effect lasts considerably longer as well, in a way modulating the effect of other synaptic influences on the ion channels in question or affecting the metabolism of the postsynaptic cell. Thus, while fast synaptic responses are very precise both in time and space, the slow responses are not. Such slow synaptic events are often called **metabotropic,** and the molecules that cause them are called **neuromodulators.** So far, we can safely say that less is known about the effects of neuromodulators than about the effects of classical neurotransmitters. One complicating fact is that the same signal molecule may act as a classical neurotransmitter in one place and as a neuromodulator in another.

It was emphasized earlier that the effect of a given neurotransmitter depends on the type of receptor and ion channel it influences. This is most clearly seen in **adrenergic** synapses. Different classes of adrenergic receptors, called $\alpha$ and $\beta$ with subgroups, account for the large variations in responses due to adrenaline (epinephrine) and **noradrenaline** (norepinephrine). We shall return to this in the section about the autonomic nervous system later in this chapter.

Substances with opiate-like properties, collectively called **opioids,** are widely distributed in the brain and may be involved in the perception of pain, emotion, and pleasure (Hökfelt et al. 1980). Like morphine, they may act as painkillers. In fact, the reason that morphine is such an effective painkiller is that it binds to and exerts its effect through opioid receptors. It has been reported that physical activity can enhance the release of opioids (M. Allen 1983; Harber and Sutton 1984). This may help to explain why minor cuts and bruises inflicted during physical activity may remain undiscovered until after the activity. Like morphine, opioids are believed to cause euphoria, that is, an undue elevation of mood, and the question has been raised whether some people are addicted to running in the same way as one can be addicted to morphine. Along the same lines of thought, the pleasure experienced when the demand for activity has been satisfied has been called "runner's high."

When a classical transmitter has performed its job on the postsynaptic membrane, its action must be rapidly terminated; otherwise, its effect would last too long, and the precision control of synaptic traffic would be lost (figure 4.9). The removal of the transmitter can be achieved by diffusion, but obviously widespread diffusion would jeopardize the precision of neural signaling.

## For Additional Reading

For a discussion of the effect of transmitter diffusion on synaptic cross talk, see Barbour and Häusser (1997) and Rusakov, Kullmann, and Stewart (1999). A more focused survey of transporters for glutamate, the major excitatory transmitter in the CNS, including discussion of the role of glial cells, is presented by Danbolt (2001).

It is assumed that glial cells, surrounding the nerve cell, function as chemical insulators by means of specific uptake mechanisms, thus preventing diffusion of transmitters away from their sites of release. The glial cells may metabolize the neurotransmitters, thereby keeping the extracellular spaces of the CNS free from disturbance by free-floating transmitters. Inactivation of the transmitter molecules may also take place by enzymatic degradation, as in the motor endplate (discussed later), or they may be recaptured back into the presynaptic nerve terminal by a very fast and efficient transport process. For a general review of neurotransmitter transporters, see Reith (1997).

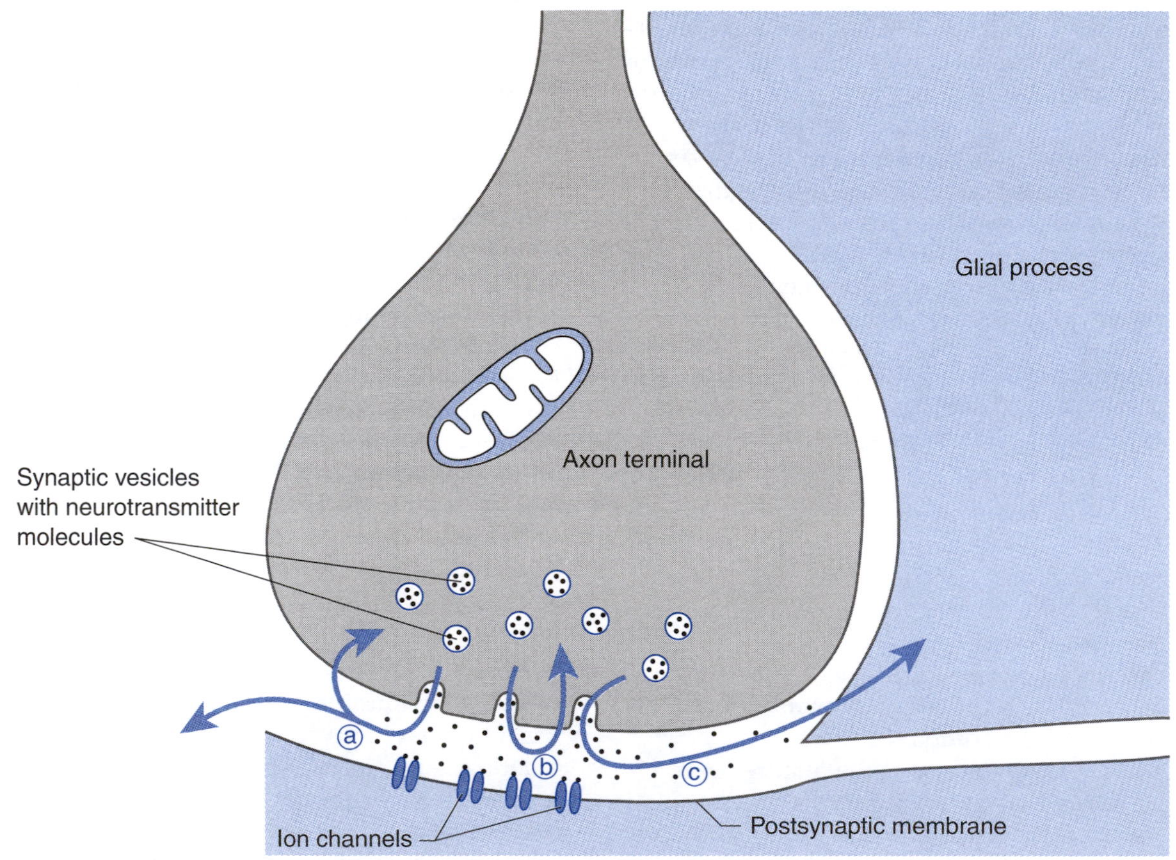


**Figure 4.9** Strategies used by nervous tissue to prevent diffusion of neurotransmitter molecules. *(a)* The neurotransmitter may be inactivated by enzymatic cleavage. In the case of acetylcholine, the acetate moiety is taken up by the synaptic terminal and used for resynthesis. *(b)* Serotonin (5-hydroxytryptamine) is an example of a transmitter inactivated by reuptake into the synaptic terminal. *(c)* Glutamate, the major excitatory transmitter in the central nervous system, is removed from the synaptic cleft by uptake both into neighboring glial cells and back into the synaptic terminal. Glial processes may surround the synapse and effectively insulate the synaptic cleft from the rest of the nervous tissue.

## *Summary*

The synapse is a unique structure capable of transferring information from one nerve cell to another or to other cell types. When an action potential reaches the synaptic terminal, signal molecules, traditionally known as neurotransmitters or transmitter substance, are released from the terminal. The release is due to an influx of $Ca^{2+}$ into the terminal through voltage-gated ion channels. The classical transmitter molecules bind to specific receptor sites on ligand-gated ion channels, permitting a flux of ions through the channel and creating a local postsynaptic potential. The exact type of ion and the nature of the synaptic potential depend on the type of transmitter used by the actual synapse and the nature of the ion channel affected by it.

In excitatory synapses, the transmitter molecules bind to $Na^+$ channels, allowing a flux of $Na^+$ down its steep electrochemical gradient into the postsynaptic cell and giving rise to an EPSP. In general, the depolarization of one EPSP is insufficient to reach the firing threshold, but several EPSPs occurring more or less at the same time or place may sum up and bring the cell to its firing threshold. In inhibitory synapses, the inhibitory transmitter molecules bind to ligand-gated $Cl^-$ or $K^+$ channels, resulting in a hyperpolarization of the postsynaptic membrane, an IPSP.

Repeated or concurrent impulses, both excitatory and inhibitory, can be summed temporally or spatially, the result depending on whether or not the postsynaptic cell is depolarized to its firing threshold. If the threshold is reached, voltage-gated ion channels open, and an action potential ensues. After fulfilling their task, the transmitter molecules are inactivated. The exact way this happens varies from one type of transmitter to another.

In addition to this classic fast (ionotropic) synaptic response, a slower, metabotropic response can occur. The metabotropic signal molecules affect the ion channels indirectly, through intracellular pathways, and modulate the effect of the classical neurotransmitters.

# Motor Units, Effectors of the Motor System

As defined in chapter 3, a motor unit consists of a motoneuron, all its axonal branches, and all muscle fibers contacted by those branches. An action potential arriving at a motor endplate always results in a synaptic potential, or more correctly an endplate potential, well above the firing threshold. As we shall see later, this is due to the anatomy of the motor endplate. Since action potentials in a rested, normal motoneuron always propagate down every branch of its axon (however, see the discussion of possible branch failure during fatigue in chapter 15), all muscle fibers of one particular motor unit must be active at the same time and to the same degree.

As the axon of the motoneuron approaches the muscle fiber, it loses its myelin sheath and divides into a large number of terminal branches, each of which ends in a synaptic terminal in close contact with the sarcolemma of the muscle fiber (see figure 3.9). Each motoneuron supplies from fewer than ten (in the eye muscles) up to possibly several thousand muscle fibers (in the proximal limb muscles), referred to as small and large motor units, respectively. Irrespective of the size of the motor unit, its constituent muscle fibers all work at the same time and with the same intensity since an impulse in the axon activates all the fibers of the unit simultaneously. The fibers in a motor unit are scattered and intermingled with fibers of other units, and they can be spread over an approximately circular region with an average diameter of 5 mm (Buchthal and Schmalbruch 1980). In any case, muscle fibers belonging to one particular motor unit are located within defined areas (compartments) of the muscle (English, Wolf, and Segal 1993), compatible with the functional compartmentalization of muscle described in chapter 3. All muscle fibers of adult human locomotor muscles have only one endplate, which is supplied by a single nerve fiber. The terminal branches of the axon end near the middle of the muscle fiber.

## The Motor Endplate

The motor endplate, the synapse between the motoneuron and its "slave," the skeletal muscle fiber, is the most intensively studied type of synaptic connection, and acetylcholine (ACh) has long since been identified as its transmitter substance.

The arrival of an action potential at the terminal initiates an inflow of $Ca^{2+}$, causing an almost synchronous ejection of 100 to 200 quanta of ACh. Within microseconds, transmitter molecules bind to ACh receptors (AChR) in the junctional folds of the endplate (figure 4.10; see also figure 3.9), leading to the entry of $Na^+$ into the muscle fiber. The resulting change in membrane potential for each of the synaptic terminals of the endplate is subliminal, but since all terminals of the endplate discharge simultaneously, allowing spatial summation to take place, the combined endplate potential is well above the

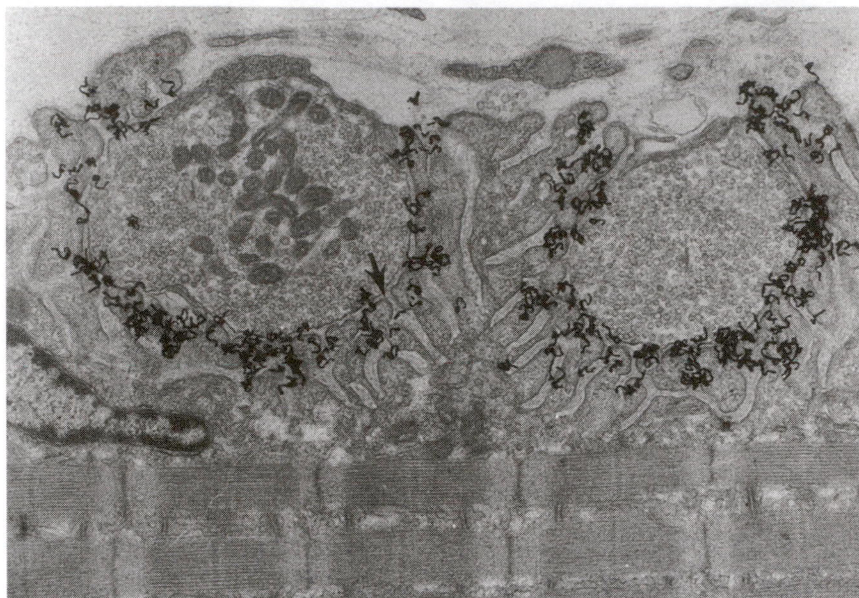

**Figure 4.10** Electron microscope autoradiograph of a vertebrate neuromuscular junction, showing the location of acetylcholine (ACh) receptors (black silver grains). Compare to figure 3.9.

Reprinted, by permission, from E.R. Kandel, J.H. Schwartz, and T.M. Jessell, 2000, *Principles of Neural Science*, 4th ed. (New York: McGraw-Hill), 189.

firing threshold. The degree of depolarization by which the threshold is exceeded is referred to as the **safety factor** of the endplate, and it means that an action potential in the nerve always leads to an action potential in the muscle fiber under normal, nonfatigued circumstances. In spite of this, a single muscle fiber action potential is of little use since it elicits only a twitch contraction in the muscle fiber, too brief to overcome the inertia of the actual body part.

At the firing threshold, voltage-gated $Na^+$ channels in the muscle cell membrane open, and after a short delay, so do the $K^+$ channels as well, allowing $K^+$ to escape out of the muscle cell in the same way as during action potentials in nerve cells. A single channel is open for about 1 ms on the average. During this time, some 10,000 $Na^+$ and $K^+$ can pass through the membrane. As many as 250,000 ion channels may open as the result of one single nerve impulse (Peper, Bradley, and Dreyer 1982). The resulting action potential spreads along the muscle fiber membrane, at the same time penetrating into the interior of the muscle fiber by means of the T-tubules and triggering a contraction as outlined in chapter 3. The enzyme acetylcholinesterase, present in the postsynaptic membrane, inactivates ACh almost instantly by enzymatic breakdown into choline and acetic acid. Much of the choline thus formed is recaptured from the synaptic cleft and brought back into the nerve terminal. In this way, it becomes available for the resynthesis of ACh, which will be packed into vesicles, ready for reuse. Actually, a release of ACh accelerates its synthesis, so that under optimal conditions, the level of ACh in the endplate remains relatively constant (Hebb 1972). Even the vesicle membranes are retrieved and brought back to the terminal for reuse. In fact, this is the same process as exo- and endocytosis, described in chapter 2.

ACh is also a transmitter substance in the CNS, but its receptors and mechanism of action are different from those present in the motor endplate. The motor endplate receptor is of the **nicotinic** type, while most cholinergic receptors in the CNS are **muscarinic.** The nicotinic receptor at the neuromuscular junction is of the direct, excitatory type, while the muscarinic synapses of the CNS work indirectly through G-proteins; that is, ACh works as a neuromodulator in the CNS. In the autonomic nervous system, ACh is the transmitter in both parasympathetic and sympathetic preganglionic neurons and in parasympathetic postganglionic neurons. We shall return to this in the section about the autonomic nervous system later in this chapter.

Recall that within the CNS, the algebraic sum of the electrochemical effects of impulses arriving at excitatory and inhibitory synaptic terminals in contact with a particular neuron determines whether or not it will fire an action potential. This is different at the neuromuscular junction. All synaptic terminals in a mammalian motor endplate are of the same kind, and they are all excitatory. Consequently, the only physiological way to promote relaxation of mammalian muscle fibers is to decrease or stop the discharge of their motoneurons.

## Regulation of Contractile Force

A long time ago it was believed that the motor unit also obeys the all-or-none law, meaning that it either contracts maximally or not at all. That is not the case. Admittedly, the creation of an action potential in the muscle fiber in response to a nerve impulse arriving at its motor endplate is an all-or-none phenomenon, but the resulting contraction is not. The contractile force relies heavily on the frequency of the impulse train (see figure 3.10). In addition, the force depends on a number of other factors, such as the muscle fiber's length, speed of shortening, temperature, and oxygen supply. Some of these factors were discussed in chapter 3; others will be discussed in later chapters. The action potentials of the muscle fibers can be recorded with needle electrodes inserted into the muscle or with surface electrodes attached to the skin over the active muscle. Depending on the technique, the obtained electromyogram reveals either action potentials in single fibers or the sum of motor unit activity.

### For Additional Reading

For further details about electromyography and electrodiagnosis, see Kandel, Schwartz, and Jessell (2000, chapter 34).

We can proceed to discuss the control of contractile force. In a completely relaxed muscle, no motor units are active, and the muscle is electrically silent, although so-called miniature endplate potentials, probably caused by the spontaneous ejection of a few quanta of ACh from the presynaptic terminal, can be recorded (Kandel, Schwartz, and Jessell 2000). The resulting postsynaptic depolarization is subliminal, however, and no contraction ensues. Nevertheless, there is a certain tension (tonus) even in the relaxed muscle due to the elasticity of the myofibrils and fibrous tissues. As discussed in chapter 3, this resting tonus is now believed to be due mainly to the elastic titin filaments, and differences in resting tonus among muscles are believed to be due to different isoforms of titin.

The functional division of a muscle into separately working muscle units (the muscle fibers of a motor unit), which can be recruited sequentially (figure 4.11), and the **rate coding** of force in each of them represent two more or less parallel mechanisms of force regulation at the whole-muscle level. Conceptually, we can regard the sequential **recruitment** of increasingly large motor units as the basic mechanism of force regulation. The gain in muscular tension with the activation of one additional motor unit depends on the number of muscle fibers in the unit. The fewer there are, the finer the grading of the contractile force can be. Still, if this were the only mechanism, it would lead to a stepwise increase in force incompatible with finer motor control. Combined with the frequency-coded regulation of force in individual motor units, however, a smooth increase or decrease in force is ensured, as demanded by the motor task at hand.

Figures 4.11 and 4.12 illustrate an important principle in the normal orderly recruitment of motor units (Freund 1983; Henneman, Somjen, and Carpenter 1965). The low-threshold motoneurons, recruited when the demand for muscular force is low, usually activate motor units with relatively few muscle fibers per unit. The selective mobilization of varying numbers of small motor units at low force levels provides a means for achieving precisely controlled and finely graded movements. When additional force is demanded, motor units with increasingly higher thresholds are gradually recruited. These motor units contain a larger number of muscle fibers. The largest motor units in the human calf muscle can develop 200 times more tension than the smallest ones. Therefore, when such units are made active, progressively greater increments of force are added. As illustrated in figure 4.11 (and figure 3.10), increasing the firing rate of the motoneurons also produces a greater muscular force. The sequential recruitment of motor units combined with rate coding of contractile force in each unit was nicely illustrated by De Luca and Erim (1994), shown on page 86 as figure 4.12.

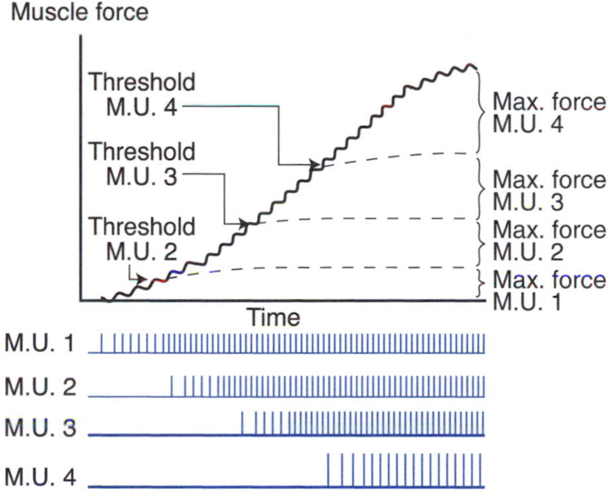


**Figure 4.11** The size principle of recruitment of motor units (M.U.). Smaller motor units (M.U. 1) are recruited first; successively larger units (M.U. 2–4) begin firing at increasing force levels. Newly recruited units start at a base frequency and then increase to a maximum.

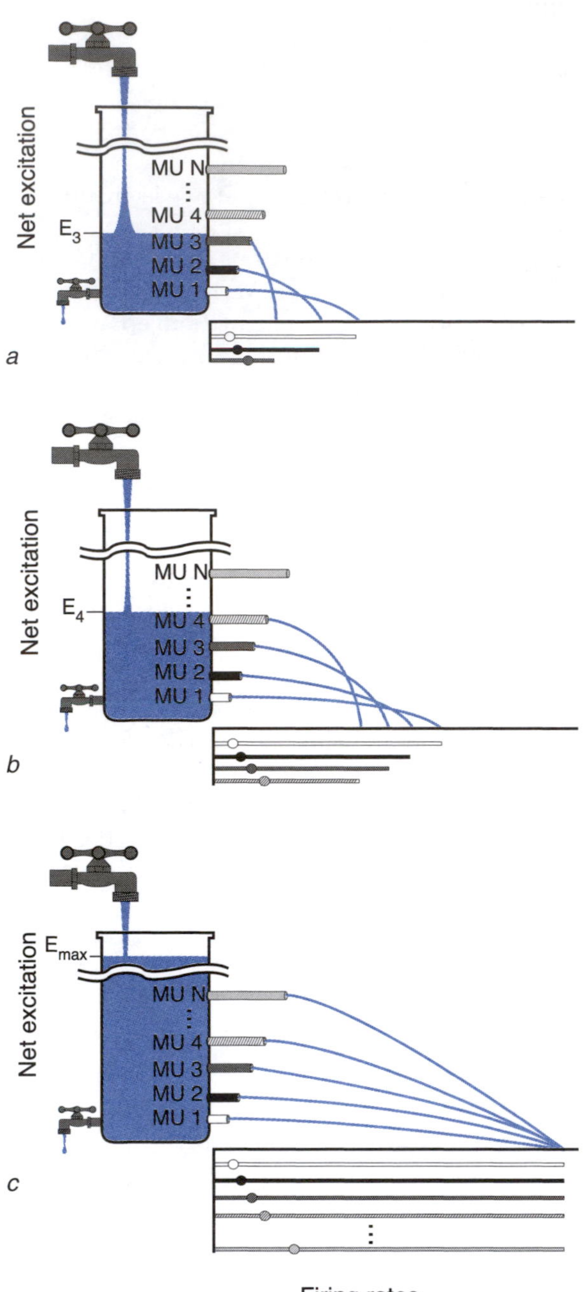


**Figure 4.12** A simple hydraulic model to summarize the rules governing the regulation of motor units in muscle-force production. The water flow into the tank corresponds to the excitatory drive directed at the motoneuron pool, while the outflow from the individual spouts and the distance each outflow travels (indicated by horizontal lines) correspond to the recruitment of a given motor unit and its firing rate. The length of each spout represents the initial firing rate and corresponds to the circular dot on the line representing the firing rate, below which the motor unit cannot fire. The outlet valve on the bottom left represents the inhibition to the pool. The net accumulation of water in the vat corresponds to the resulting net drive (excitation minus inhibition). Broken lines are used to show that vat height is greater than shown in the figure. *(a)* The behavior of firing rates when the drive is only enough to recruit three motor units. *(b)* The recruitment of a new motor unit and the increase in the firing rates of already active motor units as the drive to the pool increases further. *(c)* The convergence of the firing rates to the same maximal value in the case of an extreme drive (water height) at which the differences between the individual spout heights become negligible compared with the water level.



It turns out that the frequency of the first impulses in an impulse train is particularly important, since an initial high-frequency doublet or triplet results in a contractile force that is higher than would be predicted from the frequency of the rest of the train (figure 4.13). This is known as the **catch property** of muscle (Lee, Becker, and Binder-Macleod 1999). Hennig and Lømo (1985) found initial doublets to be a regular event during continuous registration of the electrical activity in muscles of free-moving rats.

As in most other areas of biology, basic knowledge about motor recruitment derives from animal experiments. Questions have been raised, however, regarding the validity of the recruitment pattern found in animal experiments for human muscles (e.g., see Bigland-Ritchie, Fuglevand, and Thomas 1998). We know that the number of major fiber types is different between small laboratory animals and humans (see chapter 3) and that obvious differences exist in the types of motor tasks undertaken by humans and animals. However, differences also exist among various human muscles with regard to motor tasks and degree of motor control. This may make the recruitment pattern less stereotyped in humans than in experimental animals.

Data accumulated by means of electromyography and electroneurogram illustrate the recruitment pattern of different motor units in human muscles activated in voluntary contraction. Hannerz (1974) and L. Grimby and Hannerz (1976) mapped out the discharge pattern and **recruitment order** of single motor units in voluntary contractions of human anterior tibial muscle. Low-threshold units were recruited in mild, sustained contractions starting at a frequency of 5 to 10 Hz. During such activity, when the exerted tension reached a certain value, a particular motor unit started to contract and continued its activity until the tension again dropped

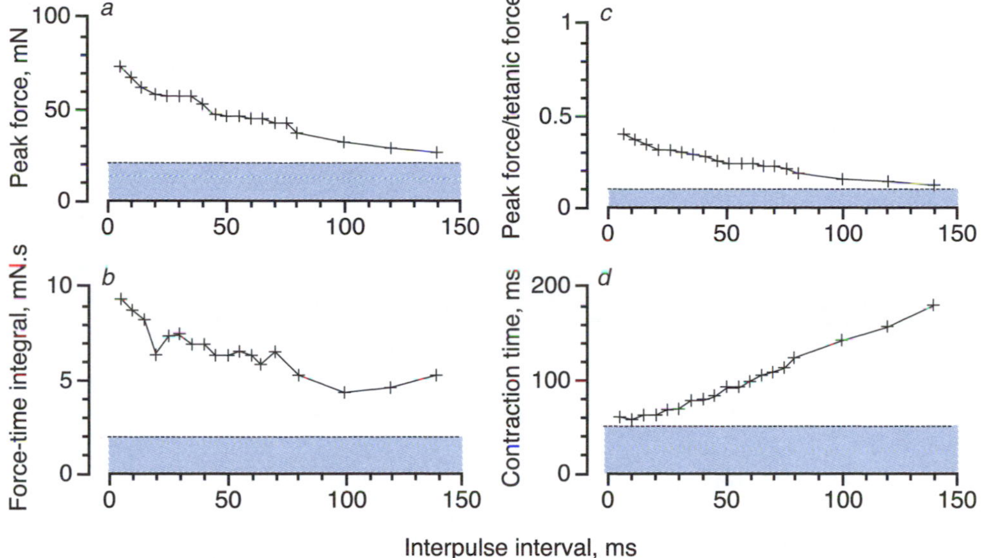


**Figure 4.13** The effect of various interpulse intervals on contractile force: *(a)* peak force, *(b)* force–time integral, *(c)* peak force normalized to tetanic force, and *(d)* time to peak force. Dotted lines indicate initial twitch values.

Reprinted, by permission, from C.K. Thomas, R.S. Johansson, and B. Bigland-Ritchie, 1999, "Pattern of pulses that maximize force output from single human thenar motor units," *Journal of Neurophysiology* 82:3190.

below the threshold level. The same unit usually started its activity at about the same force level and at a regular rate. With increasing force, the units increased their discharge rate up to a relatively low maximum (around 25 Hz). New motor units were recruited, but they started at a higher frequency (in some cases exceeding 30 Hz), and they attained a higher maximum (up to 65 Hz). It was noted that new motor units were recruited at all levels of tension up to maximal effort. High-threshold units, recruited at higher force levels, tended to exhibit a discontinuous discharge, particularly when the force exceeded 80% of the maximum. As an example, a motor unit recruited at a sustained contraction with 15% of maximal isometric strength had a minimal frequency of about 10 Hz and reached 30 Hz at maximal voluntary tension. A unit starting at 60% load at about 20 Hz increased its rate up to 45 Hz, and a 90% unit began at 35 Hz and increased to about 60 Hz.

Slow-twitch fibers are activated by motoneuron activity at a frequency range from about 5 Hz up to 20 or 30 Hz. The fast-twitch fibers are driven by a frequency range from approximately 30 Hz up to 60 or 65 Hz.

During steady contractions, the firing range extends from approximately 8 to 30 Hz. It is possible to produce high-frequency discharges (80 to 120 Hz) only for very short periods of time (about 100 ms) and then mainly for the generation of ballistic contractions (defined later in this chapter). Firing rates exceeding 30 Hz are used exclusively for variations in the speed of the muscle contraction. Due to the inherent time course of the muscle twitch contraction, the fastest possible alternating human finger or hand movements cannot exceed rates of about 7 to 8 Hz (Freund 1983).

We can assume that the low-threshold units are slow-twitch fibers innervated by small motoneurons and that the high-threshold motor units are composed of fast-twitch fibers with axons from large α motoneurons. From experiments in which they manipulated the proprioceptive afferent input, L. Grimby and Hannerz (1976) concluded that the low-frequency motor units have more support by afferent feedback than the high-threshold units. The recruitment order can also be modified or reversed by various procedures that change the afferent feedback, as shown by Garnett and Stephens (1981), who reported a reversal of recruitment order in the first dorsal interosseus muscle during stimulation of cutaneous afferents.

Different muscles might exhibit different orders of motoneuron recruitment, but the recruitment pattern according to the size principle seems to be the basic one, at least for concentric and isometric contractions at increasing force levels in monofunctional muscles. Muscles serving joints with more than one degree of freedom, on the other hand, may have different orders of recruitment for different movements. Thus, the order of recruitment in the first dorsal interosseus muscle, which abducts the index finger or contributes to its flexion at the metacarpophalangeal joint, has been reported to be different for the two directions of movement (Desmedt and Godaux 1981). Such rank deordering may be explainable on the basis of the position of the muscle units relative to the respective axis of movement.

The same kind of reasoning applies to the task-specific recruitment of motor units in the long head of the biceps brachii muscle. Motor units located medially were preferentially recruited during isometric **supination** (figure 4.14) and may have a slightly more favorable lever arm during supination than the more lateral motor units recruited first during flexion (Haar Romeny, van der Gon, and Gielen 1984).

By recording the discharge properties of single motor units in the short extensors of the toes, L. Grimby (1984) observed that the single motor units were recruited in the same order during slowly increasing speed of locomotion as they were during slowly increasing voluntary tension. Low-threshold units fired 5 to 10 times at 80- to 40-ms intervals in each step cycle during walking at normal speed. The number of units recruited, as well as their firing rates, increased with the speed of locomotion. At a given speed the variations in the pattern from one step cycle to the next were small. In rapid walking or running, high-threshold motor units were also recruited, and they fired only in short, high-frequency bursts. Some of them fired only during rapid acceleration, rapid changes of direction, or other rapid corrective movements.

Increasing the shortening velocity of concentric contractions lowers the force threshold level for individual motor units. This means that a particular motor unit is recruited at lower force levels in fast ballistic contractions than in isometric contractions or ramp movements (Desmedt and Godaux 1978). If we suppose that the recruitment order is maintained, this can be explained by the force–velocity relationship alone, since each motor unit contributes less force at high shortening velocities. Whether high-threshold units really can be recruited before low-threshold units at higher velocities, as discussed by Burke (1981), remains an unsettled question, although some reports explicitly deny such rank deordering (Desmedt and Godaux 1981). Theoretically, changes in recruitment order can be brought about by means of **recurrent inhibition** (discussed later), driven by high-threshold motor units (Hultborn, Katz, and Mackel 1988). Still, there may be differences between muscles in this respect. The question can be regarded as rather theoretical, however, since slow, low-threshold units probably contribute only marginally to power in ballistic movements (Faulkner, Claflin, and McCully 1986).

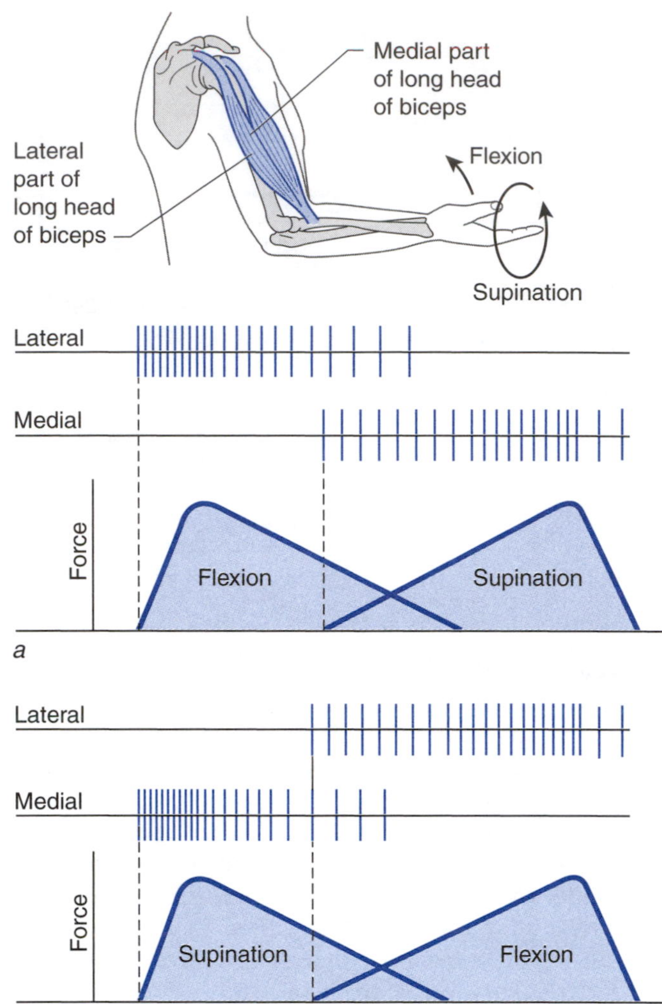


**Figure 4.14** The biceps brachii is a multifunctional muscle acting on several joints. In the elbow it acts on the humeroulnar joint (elbow flexion) and the radioulnar joint (supination). The sequence in which its motor units are recruited has turned out to be different in these two movements. If a subject starts to exert an isometric flexion force from the position shown, units in the lateral part of the long head are preferentially recruited. *(a)* If the flexion force is discontinued and the subject starts to exert an isometric supination force, units in the medial part of the long head are recruited instead. *(b)* The same pattern is seen if the subject starts by exerting a supination force and later changes to a flexion force.

Adapted, by permission, from D.G. Sale, 1992, Neural adaptation to strength training. In *Strength and power in sport*, edited by P.V. Komi (Oxford: Blackwell Science Ltd.), 260.

The reversal of the recruitment order in eccentric movements as reported by Nardone, Romano, and Schieppati (1989; figure 4.15) is less easily explained. Teleologically, we can speculate that the faster cross-bridge cycling of fast muscle fibers makes them less vulnerable to tear during eccentric contractions.

Apparently, muscles also differ somewhat with regard to the relative importance of recruitment and rate coding at low and high force levels, although

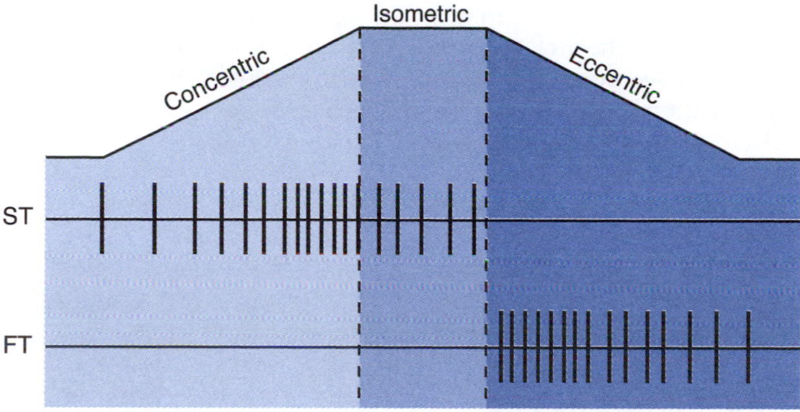


**Figure 4.15** Task-specific recruitment of motor units. During plantar flexion in the ankle until standing on the toes and while maintaining that position by means of isometric contraction of the gastrocnemius muscle and the other plantar flexors, slow-twitch (ST) motor units are preferentially recruited. On the other hand, during the eccentric phase, when returning to normal standing, fast-twitch (FT) units are the more active.

Adapted, by permission, from D.G. Sale, 1992, Neural adaptation to strength training. In *Strength and power in sport*, edited by P.V. Komi (Oxford: Blackwell Science Ltd.), 249-265.

Milner-Brown, Stein, and Yemm (1973) state that recruitment is generally the more important. Stein (1974) points out that rate coding would be less useful than recruitment for maintaining a steady force because there is only a limited range of rates at which motor units can produce a reasonable steady force without fatigue, that is, at low force levels. Rate coding becomes more prominent when higher tension is demanded, however.

It has been known for a long time that invertebrate motoneurons show so-called bistable behavior, which means that the membrane potential shifts between two quasi-stable levels, one around resting level and another at a more depolarized level, called a plateau potential. It has now been shown that rodent motoneurons (Eken 1999; Eken and Kiehn 1989) and even humans show plateau potentials (D.F. Collins, Burke, and Gandevia 2002; Hounsgaard 2002). Such bistable behavior may serve as a basis for movement pattern generators, and it may also explain the rotation of activity between motor units reported during sustained, low-force contractions (Fallentin, Jørgensen, and Simonsen 1993) as well as the unexpected strength sometimes experienced by individuals in critical situations (D.F. Collins, Burke, and Gandevia 2002; Hounsgaard 2002).

## *Summary*

An action potential elicited in a motoneuron propagates along its axon and ends up by releasing ACh from the axon terminals of the motor endplate. The ACh molecules attach to and open ligand-gated sodium channels in the sarcolemma, resulting in a depolarization that by good margin exceeds the firing threshold and gives rise to an action potential in the muscle fiber in the same way as in synapses between nerve cells. The muscle action potential propagates along the membrane, at the same time penetrating into the interior of the muscle fiber by means of the T-tubules, and initiates a series of events culminating in the interaction between myosin and actin described in chapter 3. The resulting muscle tension depends on the number of motor units activated and on the frequency with which each of them is stimulated.

Both in reflex and voluntary contractions, small, slow-twitch motor units are recruited first in activities with low demands on force, initially with a relatively low frequency. With increasing demand for force, these low-threshold units increase their discharge rate, and in addition, new motor units are recruited. The fast-contracting motor units gradually start their activity at a relatively high frequency. Varying the number of active motor units (recruitment) and their frequency of excitation (rate coding, or frequency modulation) brings about the gradation of force of a muscle contraction. The recruitment of slow-twitch fibers combined with a rotation of activity between motor units in sustained contractions of low tension is a wise arrangement, since these fibers are aerobic and fatigue resistant. Recruiting new motor units and activating many muscle fibers when more tension is demanded keeps the metabolic load on the individual muscle fiber low.

The relatively stereotyped recruitment order of motor units applies primarily to monofunctional muscles. In muscles serving more than one joint or a joint with more than one degree of freedom, the recruitment order may depend on the direction and type of movement. Such task dependency must be a consequence of central pattern generators at the level of the spinal cord and certainly at higher levels of the CNS (innate or learned patterns). These centers seem capable of selecting in advance the appropriate recruitment order for the task intended. The continuous interaction of central commands with sensory feedback can modify the number of motor units recruited and their frequency of contraction.

# The Role of Sensory Systems and Reflexes in Motor Function

Some human behavior is innate and follows a stereotypic pattern, basically identical in all individuals. Examples of such behavior patterns are swallowing, coughing, breathing, coordination of eye movements, blinking, vomiting, orgasm, and startling when taken by surprise. Central programs in the nervous system can coordinate the motoneurons involved in such activities, and once initiated, the motoneurons do not require additional incoming (afferent) sensory feedback for the continuation of their essential pattern, even if a dozen or more muscle groups are involved. Even more complex movement combinations such as walking and running are basically genetically programmed. However, sensory inputs are essential for the adaptation and modification of these programs to changing environmental conditions, including the avoidance of obstacles. Many universal emotional expressions, such as anger, fear, sorrow, disgust, or joy, have rather fixed and rigid components, but they can be modified and controlled by will.

During childhood, we learn new movements, and an ongoing process seeks to modify behavior as a result of experience. For an understanding of some basic principles underlying the function of the CNS, it is useful to examine how relatively simple reflexes are brought about and how they are modified, for the reflex is an elementary model of behavior. Before we do so, however, we take a brief look at the basic properties of sensory receptors.

## General Properties of Sensory Receptors

All sensory receptors in the body, regardless of type, share one basic property: They translate the physical or chemical signals acting on them into nerve signals. Each type of sensory receptor reacts most easily to one particular type of stimulus (e.g., light in the case of the eye or sound in the case of the ear), but if they are strong enough, even inappropriate stimuli can elicit nerve signals. A blow to the eye, for example, can make you see flashes of light. In much the same way, mechanical stimulation of a nerve can give rise to such false signals. A blow on the ulnar nerve in the elbow region is a common and well-known cause of tingling pain radiating down the forearm to the ulnar side of the hand. In these examples it is easy to see the relationship between cause and effect, but even more subtle and covert influences, more difficult to identify, can give rise to symptoms such as tinnitus or vertigo, sometimes even with incapacitating results.

## Receptor Potentials and Adaptation

A **receptor potential** in a sensory cell is the equivalent of a synaptic potential in a postsynaptic cell. It is not due to ligand-gated ion channels, but to ion channels sensitive to specific physical stimuli. Most sensory systems relevant for motor control react to mechanical stimuli such as stretch or vibration. If the receptor potential, or **generator potential** as it is also called, reaches the firing threshold, an action potential is generated in the same way as for excitatory synapses.

An important property of all sensory receptors is that they adapt to constant stimulation, meaning that the receptor potential gradually decreases in amplitude during a constant, sustained stimulus. Of course, this makes it less likely that the stimulus will give rise to action potentials. At first, the frequency of action potentials decreases. Later, they may cease completely. Such adaptation can be rapid or slow, separating receptors into **slowly adapting receptors** and **rapidly adapting receptors.** Cutaneous receptors responding to touch are often used as examples of rapidly adapting receptors. It is a common experience that the touch of a piece of clothing or jewelry is unnoticeable after some time unless it is in constant motion. Pain receptors, on the other hand, adapt rather slowly. Teleologically, this seems to be advantageous. Pain is a sign of danger and ought to be persistent until, if at all possible, we take steps to avoid the situation causing it.

## The CNS Can Enhance or Suppress Sensory Information

If the skin is touched by an object, touch and eventually pain receptors are stimulated. The most strongly stimulated receptors may, through collaterals from the afferent fibers, stimulate inhibitory interneurons. In turn, these inhibitory interneurons may abolish the weaker afferent nerve impulses by making the nerve cells that receive signals from adjacent receptors sufficiently hyperpolarized, a strategy that can functionally isolate cells that are anatomically adjacent. The most activated neurons will be surrounded by an inhibitory zone of less active or silent neurons. This general strategy is called lateral inhibition. Functionally, this strategy is advantageous when there are several synaptic relays for afferent impulses from the various receptors. At every synaptic transmission, there is a possibility of an inhibitory action that can sharpen the neuronal signals by eliminating the weaker excitatory actions, which are elicited, for example, when an object touches the skin. In this way, the touch stimulus can be more precisely located and evaluated by the cerebral cortex (figure 4.16).

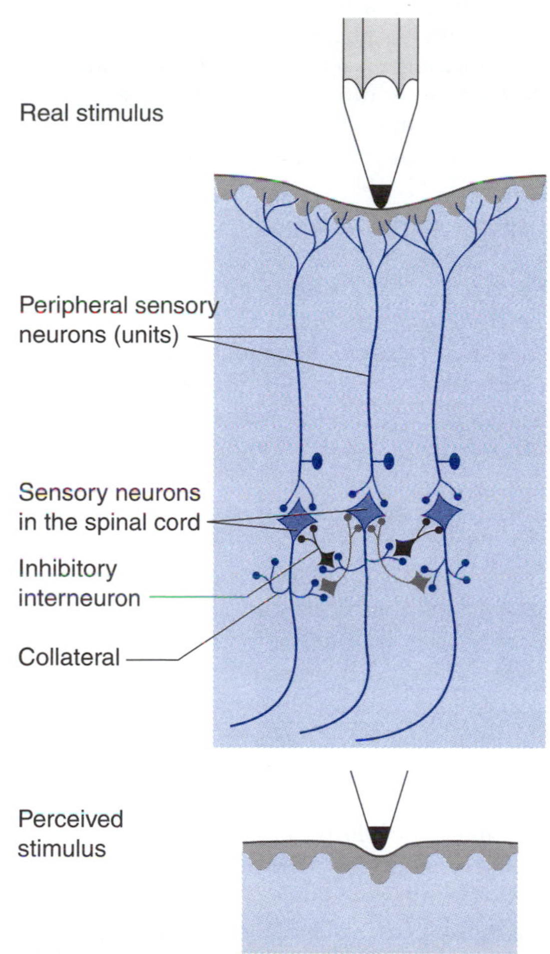


**Figure 4.16** Simplified presentation of how inhibitory interneurons in the CNS can improve the precision of the sensory information reaching consciousness by means of lateral inhibition. Cells shown in black and gray are inhibitory.

Reprinted, by permission, from P. Brodal, 1998, *The central nervous system: Structure and function*, 2nd ed. (New York: Oxford University Press), 193.

If all sensory receptors that are active at the same time demanded full conscious attention, the result would be confusing. Fortunately, several strategies exist that limit the amount of information reaching the CNS, in particular the cerebral cortex and hence consciousness, at any given moment. First of all, as discussed earlier in this chapter, sensory receptors adapt to continuous stimulation by reducing their response, that is, by lowering their impulse frequency. Obviously, this serves to give new sensory information priority over old sensations, but inhibiting the relay of information within the CNS can to a certain extent eliminate even new information. According to Eccles (1973b) this enables the cortex to "protect itself from being bothered by stimuli that can be neglected." Most notably, this can happen to impulses from pain receptors, which are slow adapters and thus scarcely affected by the duration of the painful stimulus. Such inhibition can take place at the synapses between the primary afferent neuron and the next link in the pathway or at more centrally located synapses. Part of such inhibition derives from descending pathways from the cerebral cortex that inhibit information at the level of dorsal horn cells through relay nuclei in the brain stem. Experiments, in both humans and animals, have shown that this analgesic effect is mediated by both opioid and non-opioid mechanisms. It can be elicited by various kinds of stress, be it due to real danger, shortage of time to catch a plane, or a mere competitive situation in the sport field.

## For Additional Reading

For further details and discussion of pain perception, see Brodal (1998) or Kandel, Schwartz, and Jessell (2000, chapter 24).

Even though the pathways conveying cutaneous impulses from the periphery to the cerebral cortex usually and correctly are treated as separate tracts, this does not mean that they are totally independent. Several lines of evidence indicate that the sensation of pain may be modulated by activity in other cutaneous afferents, exemplified by the well-known soothing effect of blowing at, shaking, or rubbing your finger after hitting it with a hammer. In the 1960s Melzack and Wall (1965) put forward their gate control theory to explain this effect (figure 4.17).

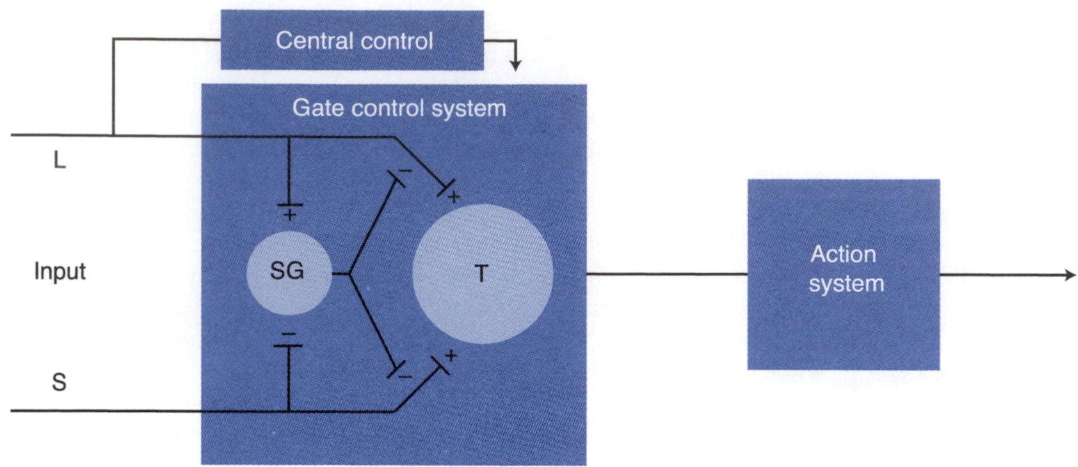


**Figure 4.17** Schematic diagram of the gate control theory, as presented by Melzack and Wall (1965). Large-diameter (L) and small-diameter (S) fibers project to the substantia gelatinosa (SG, the dorsalmost part of the dorsal horn) and the first central transmission cells (T). The inhibitory effect exerted by the SG on the afferent fiber terminals is increased by activity in the L fibers and decreased by activity in the S fibers. The central control trigger is represented by a line running from the large-fiber system to the central control mechanisms; these mechanisms, in turn, project back to the gate control system. The T cells project to the entry cells of the action system. + = excitation; – = inhibition.



## *Summary*

Motor function depends heavily on sensory information. Part of this information reaches consciousness and may serve as a basis for voluntary movements, but most of it takes part in reflexes. Irrespective of the type of sensory information, all sensory receptors share one basic property: They translate specific physical or chemical stimuli into nerve signals.

All sensory receptors adapt to constant stimulation by gradually decreasing their response. Such adaptation is either rapid or slow, depending on the type of receptor involved, and can be regarded as a strategy to reduce the flow of information to the CNS. The same adaptive result can be obtained by means of inhibitory interneurons acting at synapses within the CNS. Inhibitory interneurons can also serve to improve the quality of sensory information. By means of so-called lateral inhibition, sensory information can be more precisely localized.

# General Properties of Reflexes

In reflexes relevant for motor function, there are at least two neurons in the reflex chain: the afferent (receptor) neuron and the efferent (effector) neuron. The cell body (perikaryon) of the afferent neuron is in a dorsal root **ganglion** or an equivalent ganglion of a cranial nerve, and it conveys cutaneous, muscular, or special sense information. The cell body of the efferent motoneuron is located in the ventral horn of the spinal cord (or a motor nucleus of a cranial nerve). The afferent fibers entering the dorsal root of the spinal cord may thus end monosynaptically around motoneurons, but in general several connecting interneurons intervene between the afferent and efferent neurons (figure 4.18).

To a large extent, neural control of skeletal muscles is reflexive in nature. The membrane potential of the motoneurons is increased or decreased, depending on the sum of the excitatory and inhibitory activity in the synaptic terminals acting on the motoneuron. If the muscles are the slaves of the motoneurons, the motoneurons themselves are, if not slaves, nevertheless heavily influenced by spinal and supraspinal mechanisms. The reason that motoneurons can be considered less slavelike than the muscles is that their reaction is more integrative. A slave has only one master, which it obeys more or less always in the same way. The motoneuron is influenced from several sources, the relative importance of which varies with the type of movement in question.

Even in humans, most of the descending tracts from **supraspinal levels,** including the **pyramidal tract,** terminate on interneurons. According to Eccles (1957) "it can be stated that by establishing synaptic connections with interneurons rather than with motoneurons, the pyramidal tract and the other descending tracts are able to operate through the coordinative mechanisms at the segmental levels of the spinal cord." Some of these interneurons are inhibitory, which enables the pyramidal tract to inhibit motoneurons that innervate muscles that are

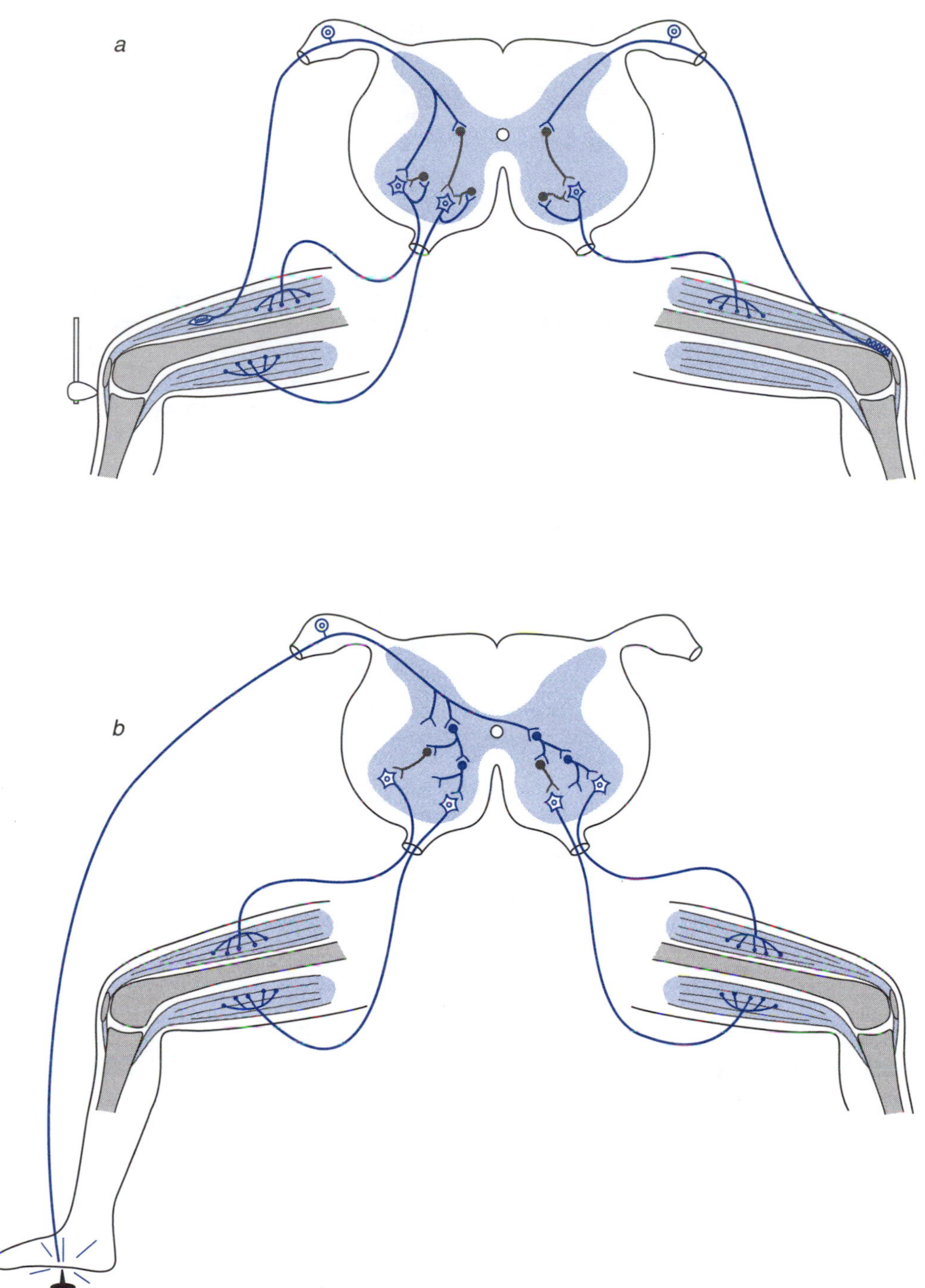


**Figure 4.18** Examples of spinal reflexes. Excitatory neurons are shown in blue, inhibitory neurons in black. *(a)* The left part of the figure shows a monosynaptic stretch reflex (the knee jerk) with concomitant inhibition (via an inhibitory interneuron) of motoneurons to antagonistic muscles. The right part of the figure shows the inhibitory effect of tendon organ afferents (via inhibitory interneurons) on motoneurons belonging to the same muscle. Both sides also show recurrent collaterals to inhibitory Renshaw interneurons (discussed later in this chapter). *(b)* The polysynaptic flexor reflex is nociceptive (reacting to painful and potentially harmful stimuli). The stronger the stimulus, the brisker the reflex answer and the shorter the latency. This is a typical feature of polysynaptic reflexes. If sufficiently strong, the stimulus also evokes a contralateral extensor reflex to provide the necessary support for the body when the stimulated leg is withdrawn.

Reprinted, by permission, from H.A. Dahl and E. Rinvik, 1999, *Menneskets funksjonelle anatomi* (Oslo, Norway: J.W. Cappelens Forlag).

antagonistic to the intended movement. In addition, the interneurons allow the pyramidal tract to have a modulating effect on reflex activity at the **spinal level.** The descending impulses from the higher levels of the brain, including those voluntarily evoked, will necessarily be modified on the basis of coincident information from all kinds of receptors. It also seems important that information based on past experiences permits modification of responses via interneuronal circuits. Thus, the impulse traffic in the final common path, the axons of the motoneurons, reflect an integration of synaptic activity based on both past and present experience.

Either the synaptic endings of the afferent fiber are located in the same segment of the spinal cord that the fiber enters, or collaterals pass up or down the spinal cord, eventually reaching the brain. Theoretically, there are innumerable pathways that an impulse can travel via nerve fibers, but some tracts are preferred, partially on an anatomical basis. One axon may branch and terminate with several synaptic endings on the same nerve cell and its dendrites.

Unfortunately, reflex nomenclature is rather unsystematic and potentially confusing. A reflex may be named after the location of the reflex center, such as spinal reflex or bulbar reflex (bulbus is an old name for the brain stem), or it may be named after the nature of the stimulus (e.g., stretch reflex, nociceptive reflex) or after the effector of the reflex, that is, the muscles that perform the reflex movement (e.g., flexor reflex, extensor reflex). In still other cases it may be named after the part of the body where the stimulus is applied (corneal reflex, tendon reflex).

## The Muscle Spindle and the Gamma Motor System

From the skeletal muscles, afferent nerves report to the CNS about the muscles' tension, length, and position and about changes in these parameters. These nerves are activated by special receptors, one of which is the **muscle spindle.** The axons of the motoneurons so far discussed are of the so-called **alpha (α)** type (12 to 20 μm in diameter), and they supply the skeletal muscle fibers that make up the bulk of skeletal muscles. In the ventral horn of the spinal **gray matter,** there are also other types of motoneurons giving rise to axons that are thinner and supplying the special kind of muscle fibers situated inside the muscle spindle. These motoneurons are of the **gamma (γ)** type. To distinguish these special muscle fibers within the muscle spindle from the ordinary ones outside, they are called **intrafusal muscle fibers** (*intra* = inside; *fusus* = spindle). Consequently, the ordinary muscle fibers outside the muscle spindle are known as **extrafusal muscle fibers.**

The muscle spindle consists of a small number of intrafusal muscle fibers (usually 5–10) enclosed in an elongated, fusiform connective tissue capsule, hence the name. The long axis of the spindle, only a few millimeters long, is parallel to the extrafusal muscle fibers, but more important, the spindle is coupled in parallel to the extrafusal fibers as well (figure 4.19), meaning that the muscle spindle is stretched when the extrafusal fibers are and is relaxed when the extrafusal muscle fibers shorten. In fact, the muscle spindle is a stretch receptor, providing the CNS with proprioceptive information about the length of the muscle, changes in the length of the muscle, and the speed of these changes. The ends of the muscle spindle are attached to the connective tissue within the muscle and thus indirectly to its tendons, enabling it to monitor the actual length of the muscle.

The intrafusal muscle fibers are of two major types but share one important feature. Except in the middle part, both types show cross striations because of their content of contractile, myofibrillar material. It is these striated parts that are innervated by the γ motoneurons. The central, unstriated part, on the other hand, is the main sensory region, innervated by thick, myelinated fibers belonging to group I and II afferent fibers.

### For Additional Reading

For more information about the nomenclature of peripheral nerve fibers, see Brodal (1998) or Kandel, Schwartz, and Jessell (2000, chapters 22 and 36).

The importance of the efferent γ **innervation** of the intrafusal muscle fibers can be summarized as follows: It adjusts the length of the intrafusal muscle fibers to the actual length of the extrafusal ones and can change the sensitivity of the muscle spindle to external stretch. We shall return to this later in this chapter.

The two types of intrafusal muscle fibers are different in many respects, but the morphological differences are most clearly visible in their central parts. In addition to being thicker all along its length, one type has a dilated middle part containing a cluster of nuclei (see figure 4.19). This has given rise to the name **nuclear bag fiber** for this type. The more slender type lacks this dilatation, and its nuclei are arranged in a row all along its length. Consequently, fibers of this type are called **nuclear chain fibers.**

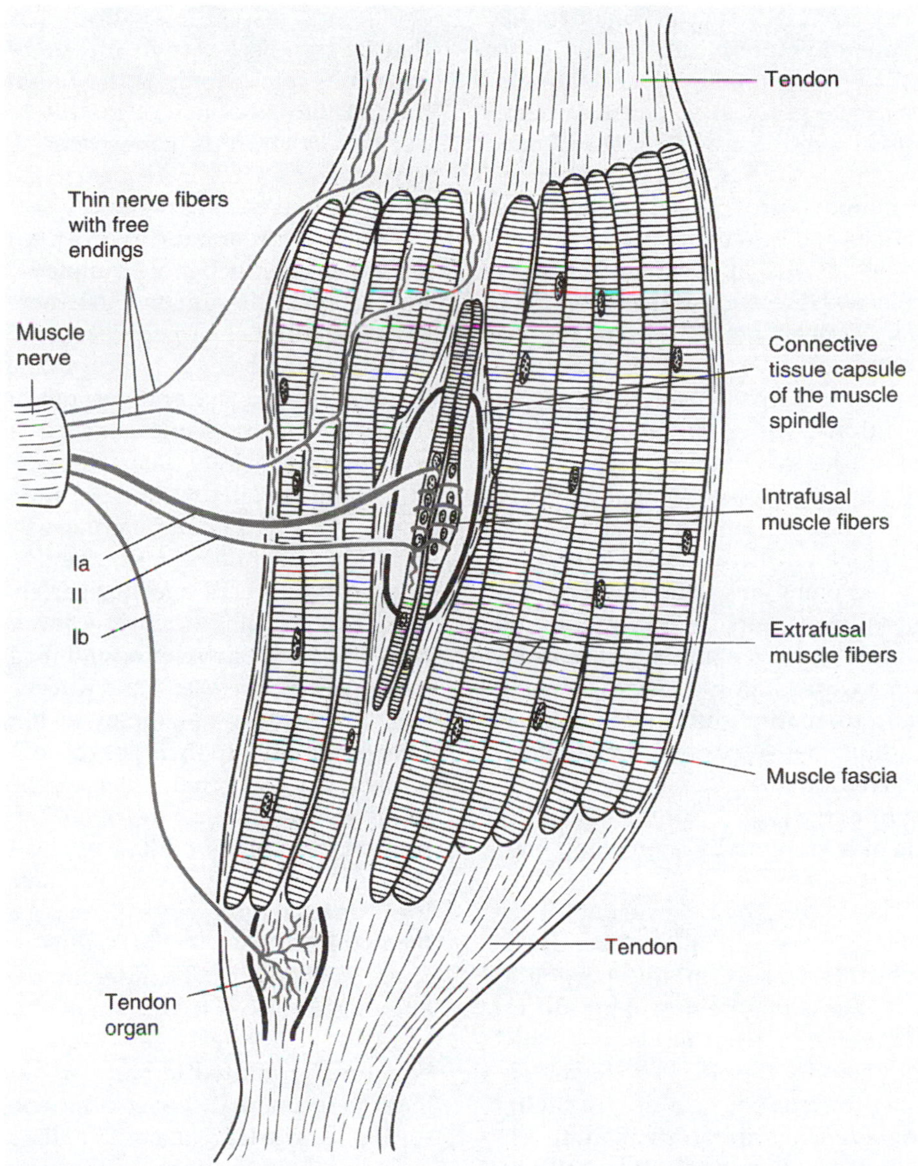


**Figure 4.19** Highly schematic drawing of a muscle and its sensory innervation. The size of both the receptors and the individual extrafusal muscle fibers is exaggerated in relation to the muscle as a whole. Only two intrafusal fibers are shown, a nuclear chain fiber to the left and a nuclear bag fiber to the right (these types are discussed later in the text). The muscle spindle is attached to the intramuscular connective tissue and thus indirectly to the tendons of the muscle. It is stretched whenever the muscle as a whole is stretched but not when the muscle contracts and becomes shorter. The tendon organ (discussed later in the text), on the other hand, is activated by both passive stretching and active contraction of the muscle but is more sensitive to the latter. Many of the free nerve endings are nociceptors.

Reprinted, by permission, from P. Brodal, 1998, *The central nervous system: Structure and function*, 2nd ed. (New York: Oxford University Press), 196.

The muscle spindle is the sense organ involved in stretch reflexes, the most widely known example of which is the knee jerk, or patellar reflex. In this reflex a tap on the quadriceps tendon (patellar ligament) results in a contraction of the quadriceps muscle and a sudden and brief extension of the knee. It is obvious to anyone who has experienced it, however, that such a blow to the quadriceps tendon is a very unphysiological stimulus and that nothing like it is ever felt during normal movements. Although useful in routine medical examinations, this phasic reflex response has little relevance for normal motor control. It demonstrates, however, that one of the consequences of a train of impulses in the spindle afferents is monosynaptic excitation of homonymous α motoneurons (**homonymous** means literally "with the same name," i.e., α motoneurons belonging to the same muscle as the spindles). As stated earlier, a major role of muscle spindles is a more "silent" one, namely, to provide the CNS with proprioceptive information for motor control. This means that the spindle afferents make other synaptic connections in addition to the ones involved in the knee jerk.

Lying in parallel with the extrafusal muscle fibers, the intrafusal muscle fibers are destined to become slack when the extrafusal fibers shorten, unless they too shorten to the same degree. This is taken care of by the γ motoneurons. The γ motoneurons for a particular muscle are located within the same motoneuron pool as the α motoneurons. Actually, most of the fiber tracts descending from the higher levels of the CNS that impinge on the α motoneurons also stimulate the γ motoneurons. This is called **α-γ coactivation.** The γ motoneurons can adjust the length of the muscle spindle so that it maintains its sensitivity to stretch over a wide range of changes in length of the extrafusal musculature during voluntary as well as reflex contractions. A passive stretch of the muscle, irrespective of its length, can therefore increase the firing from the muscle spindle. By reflex, the same muscle may respond by contracting, thus counteracting the stretching.

At constant muscle length, a shortening of the contractile ends of an intrafusal muscle fiber due to γ activity stretches its sensory middle portion in much the same way as a lengthening of the muscle and the spindle as a whole. If the muscle is stretched during a train of γ activity, the effects of the stretching and γ activity are added, resulting in higher frequency in the afferent nerve than a similar stretching without γ activity. In other words, the sensitivity of the spindle to external stretch is increased.

The muscle spindle is a slowly adapting receptor; that is, the discharge in the afferent nerve fiber continues for as long as the muscle is stretched, though the frequency of the discharge gradually declines. A stimulus that increases slowly in intensity produces a lower frequency of impulses than a stimulus that rises very rapidly to the same level.

The structure and function of the muscle spindles are, however, much more complex than revealed in the preceding discussion. First, there are two different kinds of nuclear bag fibers: the dynamic $bag_1$ and the static $bag_2$. A typical mammalian muscle spindle contains two or three nuclear bag fibers and usually about five nuclear chain fibers (Kandel, Schwartz, and Jessell 2000). There are also two types of γ efferents, called γ-static ($\gamma_s$) and γ-dynamic ($\gamma_d$), respectively. The latter innervates dynamic nuclear $bag_1$ fibers, while γ-static fibers innervate static nuclear $bag_2$ fibers and nuclear chain fibers. The $bag_1$ fibers contain mainly slow developmental (slow tonic) myosin heavy chain, while $bag_2$ fibers express α cardiac myosin (Walro and Kucera 1999). Nuclear $bag_1$ fibers are very similar to tonic (nontwitch) fibers and build up their state of contraction slowly, while the nuclear chain fibers are fast-twitch fibers like extrafusal fibers. $Bag_2$ fibers have intermediate properties (A. Taylor, Ellaway, and Durbaba 1999). It is only the γ-static fibers that adapt the length of the intrafusal muscle fibers during a contraction of the extrafusal muscle fibers, thus preventing them from slackening. The afferent nerves from the spindles are also of two different kinds: the primary, fast-conducting nerve fibers (group Ia), which are equatorially located in both bag and chain fibers, and the secondary, slower-conducting nerve fibers (group II). The large majority of the secondaries are found on the chain fibers, with a minority on the bag fibers (figure 4.20).

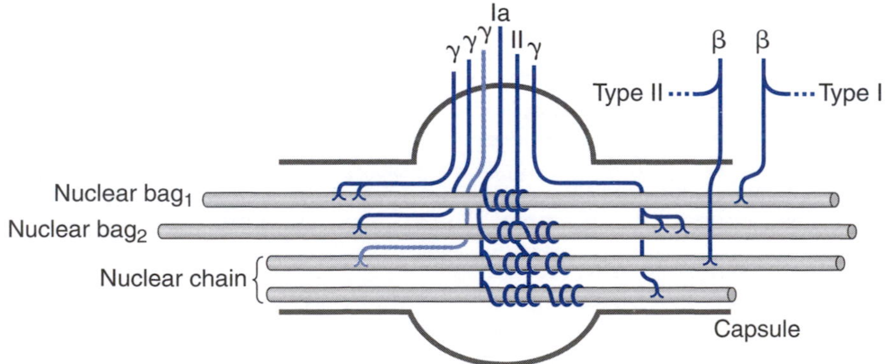


**Figure 4.20** The innervation of a mammalian muscle spindle. The β fibers shown to the right are collaterals of α motoneurons that innervate intrafusal muscle fibers.

Figure 4.21 illustrates the different properties of the responses of primary (Ia) and secondary (II) nerve endings to various stimuli. The primaries are particularly sensitive to stretch. They cease to fire when the stretch is released. The secondaries, on the other hand, are better adapted for signaling the actual length of the muscle, and they do not show a period of silence after release of pull. The primaries respond strongly to a tap on the tendon of the muscle, in contrast to the secondary nerve endings. If the muscle or its tendon is subjected to sinusoidal stretches or high-frequency vibrations, it is only the primaries that vary their action-potential frequency to a marked degree. When recording from single afferent nerve fibers, this characteristic of the primaries can be used for identification. It should be pointed out that γ activity during the release of pull can adapt the length of the muscle spindles, and the pause noticed in firing from the primary nerve endings in figure 4.21 may be filled out. Both static and dynamic γ fibers can be activated or inhibited in voluntary muscle activities.

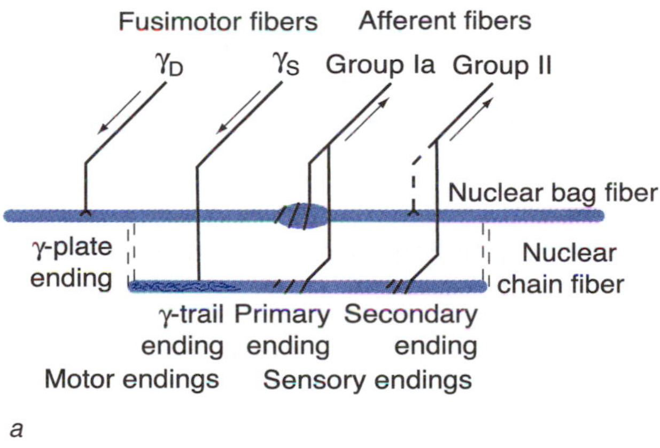


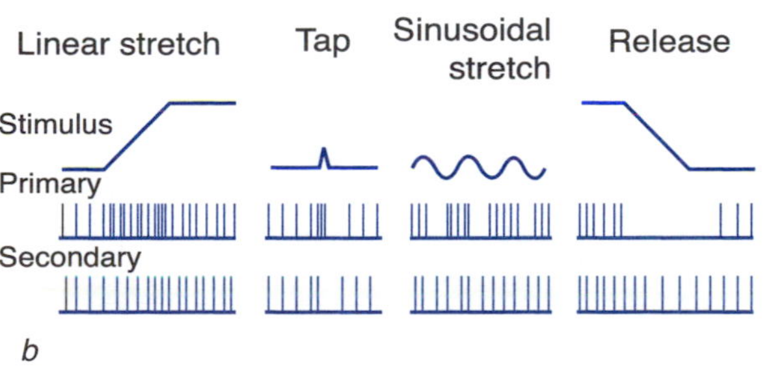


**Figure 4.21** A simplified version of *(a)* muscle spindle innervation and *(b)* the pattern of afferent impulses induced by typical stimuli. The responses of primaries and secondaries are drawn as if the muscle were under moderate initial stretch and as if there were no fusimotor activity.

Part *(a)* from Rudjord 1972, part *(b)* from Matthews 1964.

The primary nerve fibers can monosynaptically activate the motoneurons supplying their own and synergistic muscles. In fact, a primary afferent fiber makes contact with virtually all homonymous motoneurons, and via interneurons it inhibits those of the antagonistic muscles. This is the function of the muscle spindles, as seen in a classic stretch reflex. We can conclude by stating that, in general, muscle spindle primaries and secondaries have been found to behave similarly, but the primary nerve endings are more sensitive to the velocity at which a muscle is being stretched (dynamic conditions), while the secondary nerve endings more precisely report the muscle length (static conditions). At a subconscious level, they provide the CNS with essential information about the state of the muscle.

## Golgi Tendon Organs

The **Golgi tendon organs,** a few millimeters in size, are connected in series with extrafusal muscle fibers and inserted between the muscles and their tendons (see figure 4.19). Each Golgi tendon organ is responsive to contraction of about 10 to 20 single muscle fibers, each belonging to a separate motor unit (Desmedt 1981; Jami 1992). Its afferent nerve fiber has many branches twisted within the braids of collagen fiber bundles. A stretch of the Golgi tendon organ gives rise to action potentials that are conducted centrally in relatively thick group Ib nerve fibers. For mechanical reasons, the receptor is particularly sensitive to active tension caused by a contraction of the extrafusal motor units that are attached in series to the portion of the tendon in which the receptor is located. Passive tension created by pulling the muscle is a much less effective stimulus on the tendon organ because the force is exerted on the whole cross section of the tendon.

When stimulated, the afferent nerve fibers from Golgi tendon organs have been found to cause an inhibition of their corresponding muscle, elicited via interneurons. This led to the belief that the function of the tendon organ was to prevent the development of dangerously high tension in the muscle, a belief that is now largely abandoned. First, the effect of afferent impulses from the tendon organ is not always inhibition but may be excitation of homonymous α motoneurons, depending on the type and phase of the actual movement. During walking, the effect is inhibition of extensor motoneurons to the lower extremity during the swing phase, but during the stance phase, it changes to excitation, thus providing a positive feedback and improving the support of body weight.

Moreover, an at least equally important task of the tendon organ is to provide the CNS with continuous information regarding the active tension in the muscle as part of the proprioceptive information needed for motor control.

## Joint Receptors

The ligaments and the capsules of the joints contain different kinds of receptors, resembling Ruffini and Pacinian receptors of the skin and Golgi tendon organs. Some of these proprioceptors are specialized to respond to movement of the joint; others show an impulse discharge that varies with the exact position of the joint but are less sensitive to movement (see chapter 7). The information from joint receptors reaches neurons at all levels of the CNS, including the cerebellum and the somatosensory cortex, as do the impulses originating in other peripheral receptors. These signals can modulate the activity of neurons in the motor cortex, cerebellum, and other areas of the CNS. There are indications that joint afferents do not play a crucial role in the sense of the static limb position, but they can signal joint movement and position and thus duplicate the kinesthetic input from muscle spindle endings (Ferrell, Gandevia, and McCloskey 1987).

As we consider sensory information as a basis for motor control, it may be appropriate to draw attention to the apparent redundancy of information available to the CNS. Joint position and movements can be monitored both by means of joint receptors and by the muscle spindles, since muscle length changes more or less in parallel with movement of the actual joint. There is some uncertainty, however, about whether the CNS is able to substitute one kind of information for the other. In acute experiments with anesthesia of finger nerves that carry afferent impulses from finger joints, the position sense is reduced, and the researchers concluded that full proprioceptive acuity depends on the combined contribution of receptors in muscles, skin, and joints (Gandevia et al. 1983). On the other hand, patients with hip joint replacement, who have lost at least the major part of their hip joint receptors, seem to have an adequate sense of hip joint position. Of course, there is a difference between acute experiments and the chronic situation after an operation, and in addition there may be large differences in the demand for position sense in the fingers and the hip. Thus, the proprioceptive sense of finger joints has been reported to be superior to that of more proximal joints (Hall and McCloskey 1983).

### *Summary*

Most neural control of movements is reflexive in nature. Even descending impulses from supraspinal levels, including the cerebral cortex, to a large extent act through their influence on spinal reflex circuits. Unfortunately, reflex nomenclature is unsystematic and rather confusing to those not yet aware of this fact.

The muscle spindle is the sensory organ for the most familiar of all motor reflexes, the knee jerk. Actually, the knee jerk is just one example of a family of reflexes known as stretch reflexes because their adequate stimulus is stretching of muscle spindles.

A typical muscle spindle consists of 5 to 10 tiny intrafusal fibers enclosed in a fusiform connective tissue capsule. There are three types of intrafusal muscle fibers: nuclear chain fibers and two different nuclear bag fibers. The nuclear chain fibers and one of the nuclear bag types show a static response with little adaptation and are able to monitor the length of the muscle at all times. The other bag type has a dynamic response, reacting to a sudden stretch with a burst of activity as in the knee jerk, and it rapidly adapts to the new length.

Golgi tendon organs are inserted between a muscle and its tendon. Their afferent nerves were once thought to act like a fuse, preventing dangerously high tension in the muscle by inhibiting its motoneurons. Their effect is not always inhibitory; it can be inhibitory or excitatory, depending on the phase of the movement in question. In addition, and probably more important, they provide the CNS with information about the current tension of the muscle at all times.

Different kinds of receptors in the capsule and ligaments around a joint monitor the position of the joint, changes in this position, and the speed of these changes. To a large extent, this is the same information as provided by the muscle spindles, since there is a causal relationship between muscle length and joint position. The relative importance of the two sensory systems that signal joint position may vary from one joint to another, possibly in relation to the demand for precise position sense of the joint in question.

# Motor Control

Motor control as a research area has caught the interest of people from various professions, and their ways of approaching it vary accordingly. This is not the time or place to elaborate on different theories of motor control; more information can be found in more specialized books.

### For Additional Reading

More information about theories of motor control can be found in Schmidt and Lee (1999) or Shumway-Cook and Woollacott (1995).

The muscles, the true effectors of the neuromuscular system, are composed of muscle units. As defined earlier, *muscle unit* is the name assigned to the collective muscle fibers belonging to one motor unit. The muscle units are more or less independently controlled by the CNS. Therefore, the degree of motor control is determined by the recruitment of the appropriate muscle units, as well as by their level of recruitment (rate coding), and the sequence in which they are brought into and removed from action. Since an action potential from the motoneuron under normal circumstances always leads to an appropriate reaction in the muscle fibers innervated by it, our attention can focus on the motoneuron and its level of excitation. In other words, the degree of motor control can be said to be determined by the level of control of motoneuron excitation. In turn, this level of excitation is determined by the combined action of afferents from supraspinal and peripheral sources, as well as from local spinal circuitry.

## Functional Organization of the Spinal Cord

Before we go into any more detail concerning motor control, it may be wise to take a closer look at the structure and functional organization of the spinal cord. As in many other instances in biology, structure is a reflection of function.

The gross anatomy of the spinal cord is of minor interest in this respect, apart from the fact that the segments supplying the upper and lower extremities contain more nerve cells (gray matter) than the segments supplying the rest of the body. This makes the spinal cord thicker in the cervical and lumbar regions. Of course, a major reason for the higher number of nerve cells in these regions is that the muscles of the extremities, the more distal ones in particular, have smaller motor units than most other muscles. Consequently, the number of motor units is rather high, and a correspondingly high number of motoneurons are necessary to control them.

In a cross section of the spinal cord, the gray matter occupies a butterfly-like zone in the center, surrounded by **white matter** (figure 4.22). The gray matter consists mainly of nerve cell bodies and their dendrites, including the nerve cells of the local spinal circuitry, the axons of which are more or less confined to the gray matter as well. The white

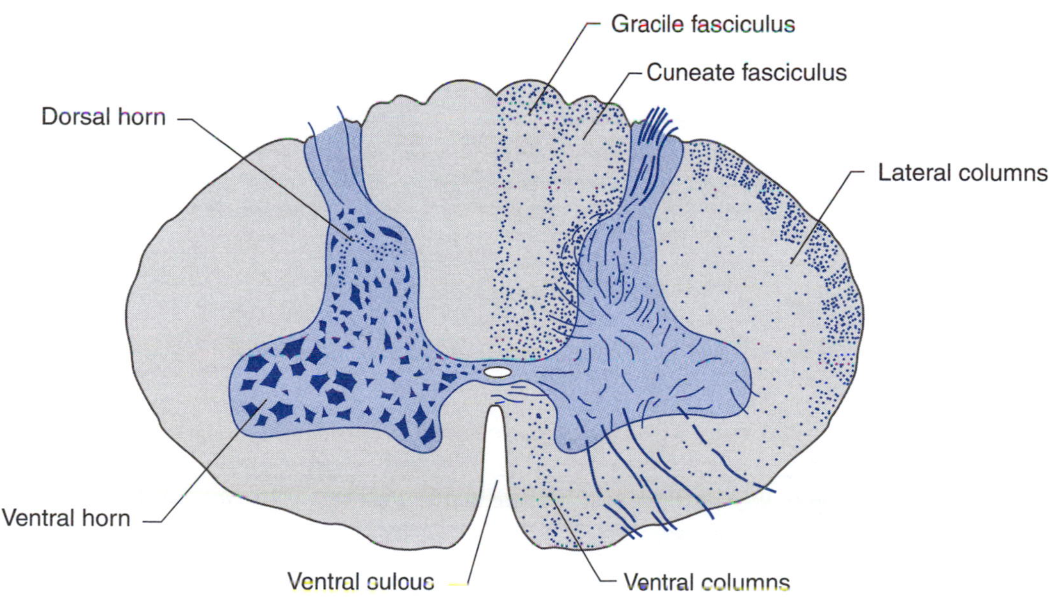


**Figure 4.22** Schematic drawing of a cross section of the spinal cord. The left part is shown as stained for neuronal cell bodies; the right part, for myelinated fibers. The gracile and cuneate fasciculi are collectively called the dorsal column.

Reprinted, by permission, from E.R. Kandel, J.H. Schwartz, and T.M. Jessell, 2000, *Principles of Neural Science*, 4th ed. (New York: McGraw-Hill), 338.

matter, on the other hand, consists of longer axons running up or down the spinal cord, connecting the nerve cells of the various segments with each other as well as with supraspinal parts of the CNS.

Within the gray matter, the motoneurons are found in the ventral horns (see figure 4.22). The α and γ motoneurons belonging to the same muscle are located together, forming clusters of nerve cell bodies called **motoneuron pools.** Motoneuron pools belonging to distal extremity muscles are typically found in the lateral part of the ventral horn, while motoneuron pools belonging to proximal and axial muscles are found more medially (figure 4.23*a*).

In the white matter, axons connecting different segments of the spinal cord, the **propriospinal fibers,** are found in a zone closely surrounding the gray matter (figure 4.23*b*). This is as may be expected from the general and orderly arrangement of axons in the CNS. "Short-distance" propriospinal fibers, connecting neighboring segments of the spinal cord, are found in the lateral fascicle of the white matter, ideally placed to connect, for example, antagonistic muscles in the extremities. "Long-distance" propriospinal fibers, on the other hand, connecting motoneuron pools belonging to more proximal muscles and trunk muscles in different parts of the body are found medial to the ventral horns.

## Renshaw Cells and Recurrent Inhibition

Motoneurons give off collateral branches on their way to a ventral root. They form excitatory synaptic contacts with interneurons located in the ventromedial region of the ventral horn. The axons of these **Renshaw cells** establish inhibitory synaptic contacts with the same and other motoneurons and interneurons (Ia inhibitory interneurons, among others) in an overlapping and diffuse fashion (figure 4.24). Since the Renshaw cells project back to the same motoneurons, which excite them, this is called **recurrent inhibition.** Renshaw cells may also be excited by descending fibers from supraspinal levels (figure 4.25). The Renshaw cells provide a feedback, and a single volley in the axon of the motoneuron can evoke a repetitive discharge of the Renshaw cell with the consequent tendency to dampen and stabilize the motoneural activity. The influence from supraspinal levels adapts the spinal circuitry to the motor task at hand. Through the Renshaw cells' inhibition of agonistic motoneurons and simulta-

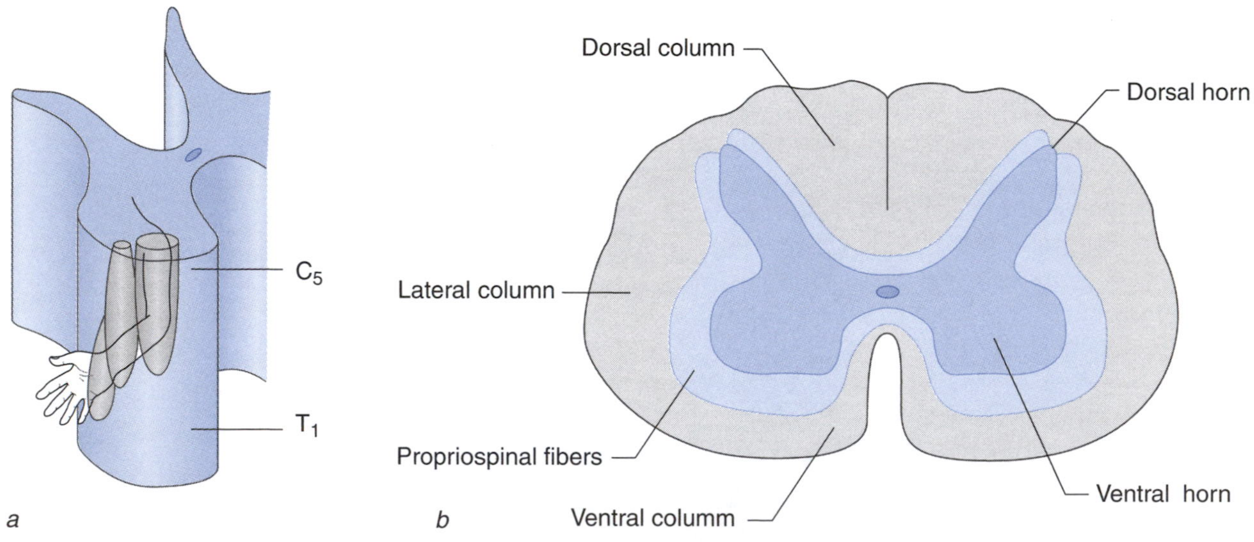


**Figure 4.23** *(a)* A thick section of the spinal cord with white matter removed. C5 = fifth cervical segment; T1 = first thoracic segment. The motoneurons that supply muscles in the distal part of the extremities are located more laterally and caudally in the ventral horn than motoneurons supplying more proximal and axial muscles. Part of the right arm is inserted into the figure to show the location of the motoneurons that innervate the muscles in the different parts of the arm. *(b)* The propriospinal fibers, which connect motoneurons in different spinal segments, occupy a zone surrounding the gray matter of the spinal cord. Propriospinal fibers connecting motoneurons belonging to distal muscles are found lateral to the ventral horns, while the longer propriospinal fibers connecting more proximal and axial muscles are found medial to the ventral horns. Compare the location of motoneurons to distal and proximal muscles in part *a*.

Reprinted, by permission, from P. Brodal, 1998, *The central nervous system: Structure and function*, 2nd ed. (New York: Oxford University Press), 313.

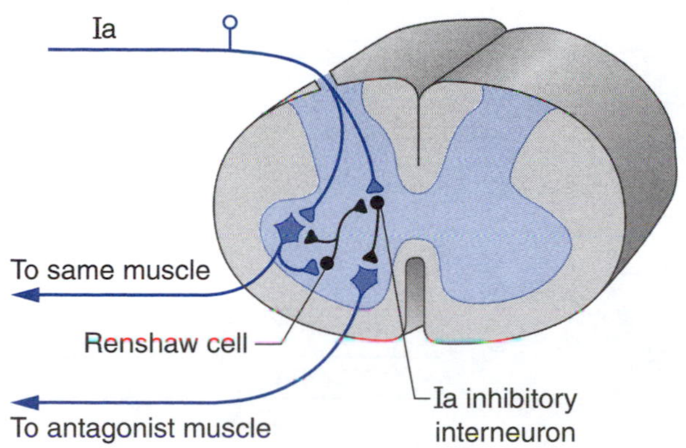


**Figure 4.24** Motoneurons give off branches, called recurrent collaterals, before their axons leave the gray matter of the spinal cord. These collaterals make excitatory synapses on inhibitory interneurons, called Renshaw cells. In turn, Renshaw cells synapse on motoneurons, including the one that gave rise to its input. In addition, they make synaptic contacts with other inhibitory interneurons, resulting in inhibition of inhibition, that is, disinhibition. Inhibitory neurons are shown in black. Compare to figure 4.3.

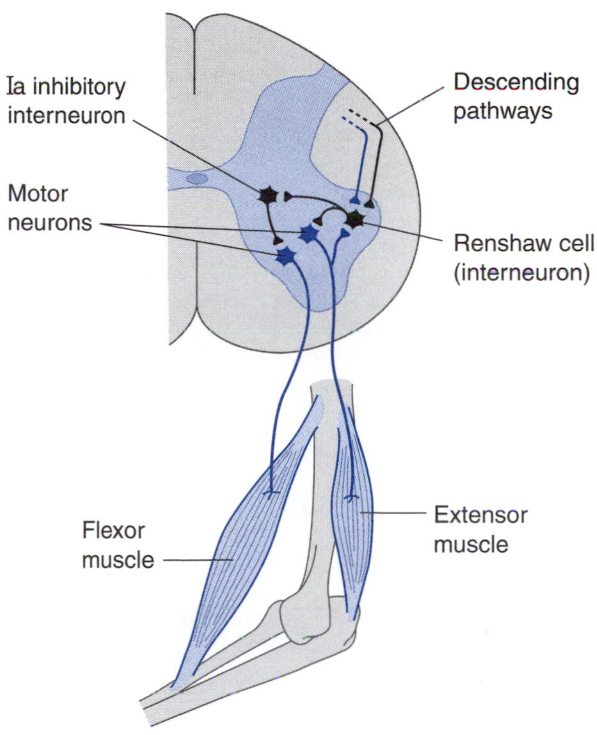


**Figure 4.25** In addition to being excited by recurrent motoneuron collaterals, Renshaw cells are influenced by both excitatory and inhibitory (blue) descending fibers from supraspinal levels. Thus, descending inputs that modulate the excitability of Renshaw cells are able to adjust the excitability of all the motoneurons supplying the muscles around a joint. Inhibitory neurons and their axons are shown in black.

Adapted, by permission, from E.R. Kandel, J.H. Schwartz, and T.M. Jessell, 2000, *Principles of Neural Science*, 4th ed. (New York: McGraw-Hill), 722.

neous disinhibition of antagonistic motoneurons (through inhibition of Ia inhibitory interneurons, figure 4.25), Renshaw cells may contribute to the generation of rhythmic movements (e.g., during walking).

## Classification of Movements

Movements can be classified in many different ways. One obvious way is a division into voluntary movements and automatic movements. It has turned out, however, that there seems to be a gradual transition from voluntary to nonvoluntary movements, rather than a clear dichotomous division between the two. Another useful division is **ballistic movements** and **ramp movements.** In short, ballistic movements are high-speed movements of short duration, too short for any correction to take place during the course of the movement, while ramp movements are slower, allowing correction of possible perturbations. The term *ballistic* derives from military language, where *ballistics* means the science of the motion of projectiles such as bullets or shells. Typically, all kinetic energy is supplied early in the course of a ballistic movement, as when you throw a ball or hit a nail with a hammer. On the contrary, kinetic energy is supplied more or less throughout the course of a ramp movement, as when you follow a moving target with your finger.

Operationally, the possibility for any correction to take place during a movement is the basis of distinction between the two types, but it is far from easy to tell where the border is between ballistic and ramp movements. In general, however, movements lasting less than 0.2 s are difficult, maybe impossible, to correct. They are said to take place in an **open-loop mode,** while ramp movements take place in a **closed-loop mode.**

## Motor Areas in the Cerebral Cortex

The part of the cerebral cortex called neocortex consists of six layers of nerve cells. Some of the layers are primarily concerned with afferent impulses from other areas of the CNS. Others are mainly efferent. It is not surprising, therefore, that differences in the various layers exist among cortical areas, reflecting the different functional roles of the cortical areas in question. These differences form the basis for the so-called cytoarchitectonic maps of the cerebral cortex (figure 4.26*a*).

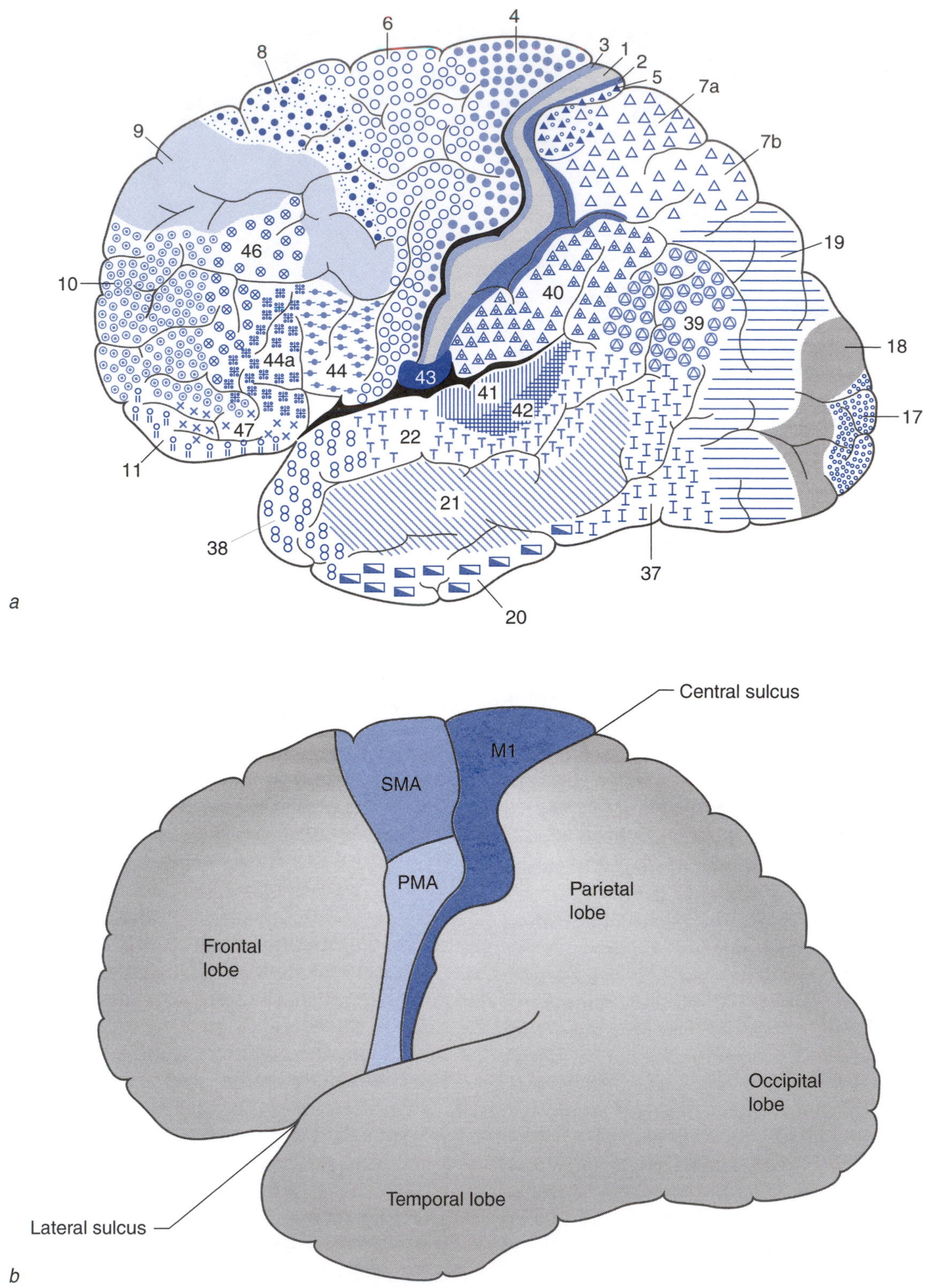


**Figure 4.26** *(a)* Brodmann's cytoarchitectonic map of the human cerebral cortex as viewed from the lateral side of the hemisphere. This map—from the early part of the 20th century—is based on variations in the arrangement of the different layers of the cerebral cortex. The various areas are labeled with different symbols and numbers. *(b)* Some of the major functional divisions of the cerebral cortex relevant for motor control. The primary motor cortex (M1) corresponds to Brodmann's area 4 and borders directly on the primary sensory cortex along the central sulcus.; PMA = premotor area; SMA = supplementary motor area.

The region of the cerebral cortex just anterior to the **central sulcus** has a key role in the voluntary control of muscle activity. It is not the initiator of movement, but it is the final relay station, receiving instructions from widely dispersed areas in the brain and putting them into effect via the motoneurons. The area with the most direct and powerful influence on motoneuronal excitability is the **precentral gyrus,** immediately bordering on the central sulcus. Therefore, this area has been named the **primary motor cortex,** or **M1** (see figure 4.26b). It corresponds to area 4 in the cytoarchitectonic map in figure 4.26a. All descending axons that make monosynaptic contact with motoneurons come from M1.

The primary motor cortex is organized in a somatotopic manner; that is, each region of the body is represented by a specific part of the M1. The part of the cortex governing the muscles of the head is found in the most lateral (lower) part of the precentral gyrus, close to the **lateral cerebral fissure of Sylvius.** The part governing lower extremity muscles is found in the most medial part, toward and partly facing the interhemispheric fissure. The rest of the body is represented in the intervening part of the precentral gyrus, in an anatomically correct sequence (figure 4.27).

One additional feature deserves mentioning, and that is the size of the cortical areas representing the different parts of the body, which reflects the level of motor control, not the actual size of the body part in question. Specifically, this means that the parts of the primary motor cortex governing, for example, muscles of the hand, lips, and tongue are far larger than predicted by the size of these body parts. In a way, the size of the cortical representation of a body part is inversely related to the size of the motor units of its muscles.

The area of the motor cortex just anterior to M1 is usually subdivided into the **premotor area (PMA)** and **supplementary motor area** (**SMA;** see figure 4.26b). Both of these also give rise to descending axons to motor areas in the brain stem and spinal cord, but their main influence on motor control is due to their connections to M1. This is why they are often placed above M1 in a hierarchical organization of motor control. The SMA is situated most medially, close to the interhemispheric fissure, and seems to be important for the planning of complex movements, while the execution of the movement is taken care of by M1. This is nicely illustrated by the fact that merely imagining a complex movement gives rise to activity in the SMA, while there is little or no activity in M1. The PMA occupies the major part of area 6, just anterior to M1. It seems to be particularly important for movements under visual guidance. It is necessary to emphasize that this represents a rough and highly schematic view only. In fact, these areas also serve other functions, and other areas not mentioned also participate in these tasks. The anterior

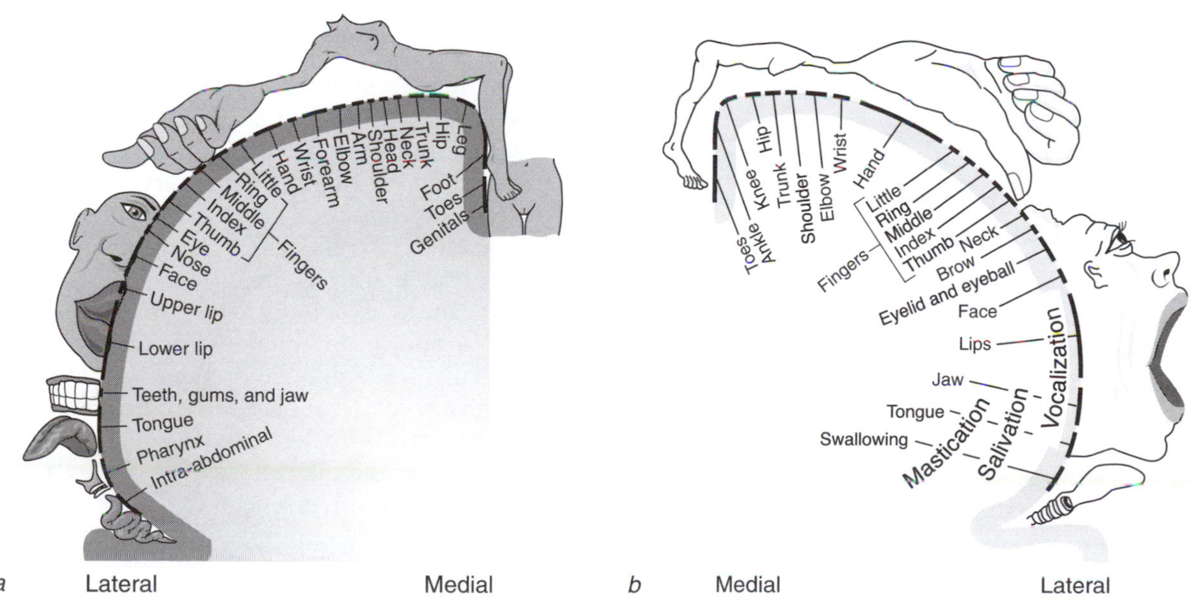


**Figure 4.27** Grotesque figures like these are used to illustrate the relative size of the cortical areas representing different parts of the body in *(a)* the primary sensory cortex and *(b)* the primary motor cortex. Homunculus means "little man."

Reprinted, by permission, from E.R. Kandel, J.H. Schwartz, and T.M. Jessell, 2000, *Principles of Neural Science*, 4th ed. (New York: McGraw-Hill), 344.

speech area (of Broca), for example, is found in the lower part of the PMA, just anterior to the mouth area of M1. In addition, the SMA has recently been subdivided into a pre-SMA and the SMA proper (for discussion of functional significance, see Yazawa et al. 2000). The visual control of hand action requires a 3-D perception of the shape and position of the object to be handled in relation to the body's own "egocentric position." The parietal association cortex has been claimed to play an essential role in this important function (Sakata et al. 1997).

These cortical areas are not the only parts of the cerebral cortex of importance for motor control. First of all, it is important to realize that motor control without sensory information is significantly impaired. Imagine what happens if in the middle of an activity you suddenly feel an urge to scratch your nose and decide to yield to that urge. You just do it, without any need to consider how you have to move your hand to do it. That information is already present in the CNS as a result of afferent impulses from proprioceptors in your arm and shoulder. This is one of the reasons why the sharp distinction long upheld between motor and sensory areas of the brain has now been abandoned.

In addition, especially in relation to voluntary movements, widespread activity occurs over large parts of the cerebral cortex long before any activity can be found in the motor cortex in the narrow sense. This so-called **readiness potential** starts as early as 800 ms before the activity in M1 and about 850 ms before the movement itself. In the intervening time a large number of other brain areas show increased activity, demonstrating that many parts of the brain cooperate in the preparation of a movement.

## Supraspinal Control of Motoneurons

When dealing with motor control, it is customary to speak about spinal and supraspinal levels of the CNS. The spinal level comprises the areas of the CNS where we find motoneurons that make direct contact with the muscle fibers by means of their axons. In the broadest sense, this makes the motoneurons of the brain stem part of the spinal level, which may be confusing, but functionally they are on the same level as the spinal motoneurons since they send their axons to muscle fibers. Supraspinal levels are all other areas of the CNS that are able to participate in motor control but have to do it through their influence on the motoneurons. As mentioned earlier, this is what makes the motoneurons deserve the designation "final common path."

The supraspinal control of the motoneurons takes place through descending pathways from the cerebral cortex and the brain stem. In addition to direct connections to the spinal cord, the cerebral cortex also has connections to the nuclei in the brain stem, which in turn give rise to descending axons to the spinal cord. Traditionally, descending axons from supraspinal levels to the spinal cord have been said to belong to the pyramidal tract or to extrapyramidal tracts. The pyramidal tract is a well-defined entity, while the term *extrapyramidal tract* has now been abandoned because of lack of a precise meaning. Of course, a large number of descending tracts other than the pyramidal tract exist, but they should rather be treated as separate entities, as we shall do later. Suffice it to say at this stage that the pyramidal tract is the only direct connection between the cerebral cortex and the spinal cord. The other descending tracts start from motor nuclei in the brain stem, which in turn receive connections both from motor areas in the cortex and other parts of the CNS. Thus, we can say that they are indirect connections between the cortex and the spinal cord.

The descending motor tracts most often mentioned in addition to the pyramidal tract are the rubrospinal tract, the reticulospinal tract, the tectospinal tract, and the vestibulospinal tract. We shall return to these tracts in the section "Various Nuclei Involved in Movement" on page 109 in this chapter.

It is estimated that the pyramidal tract in humans contains about 1.2 million fibers. This is far more than in experimental animals. The cat has fewer than 200,000 pyramidal fibers. In the higher primates it is estimated that some 10% of the pyramidal endings in the spinal cord form monosynaptic contacts with α and γ motoneurons. These contacts are excitatory, but pyramidal fibers may also inhibit motoneurons by exciting inhibitory interneurons. Not least important is the effect of pyramidal fibers on local reflex circuits in the spinal cord. This gives the pyramidal tract control over spinal reflexes, adapting the reflex activity to the overall motor task.

Both α and γ motoneurons are affected by the pyramidal tract, as previously mentioned in connection with α-γ coactivation. Such coactivation is seen in all movements, except maybe in very fast ones, which at least initially are carried out by pure α activity. In most movements the sequence is as follows: (1) The skeletal muscle is activated by the α motoneuron discharge; (2) the thinner γ fibers conduct impulses more slowly than the α fibers, and therefore the muscle spindle contraction begins somewhat later than the extrafusal contraction; and (3) the afferent discharge from the muscle spindles follows, with the triple role of directly supporting the α activity to the same muscle, inhibiting the antagonists (via inhibitory interneurons), and reporting to the brain about the progress of the initiated movement.

Our present knowledge of how supraspinal levels influence spinal motoneurons in general is rather scarce. Evidence is accumulating, however, that synchronization of neuronal activity is a general mechanism linking different cortical areas during voluntary movement (J.F. Marsden et al. 2000; McCormick 1995). Such synchronization is also evident in the rhythmic and pulsatile fashion in which supraspinal levels activate the appropriate motoneurons during slow finger and wrist movements (Kakuda, Nagaoka, and Wessberg 1999).

## For Additional Reading

For a discussion of the role of motor unit synchronization for neuromuscular performance, see Semmler (2001).

One important feature of the pathways from the cerebral motor cortex is their extensive collateral branching to brain stem nuclei (figure 4.28). Some of these nuclei project to the cerebellum and provide it with "blueprints" of the motor commands to the spinal cord.

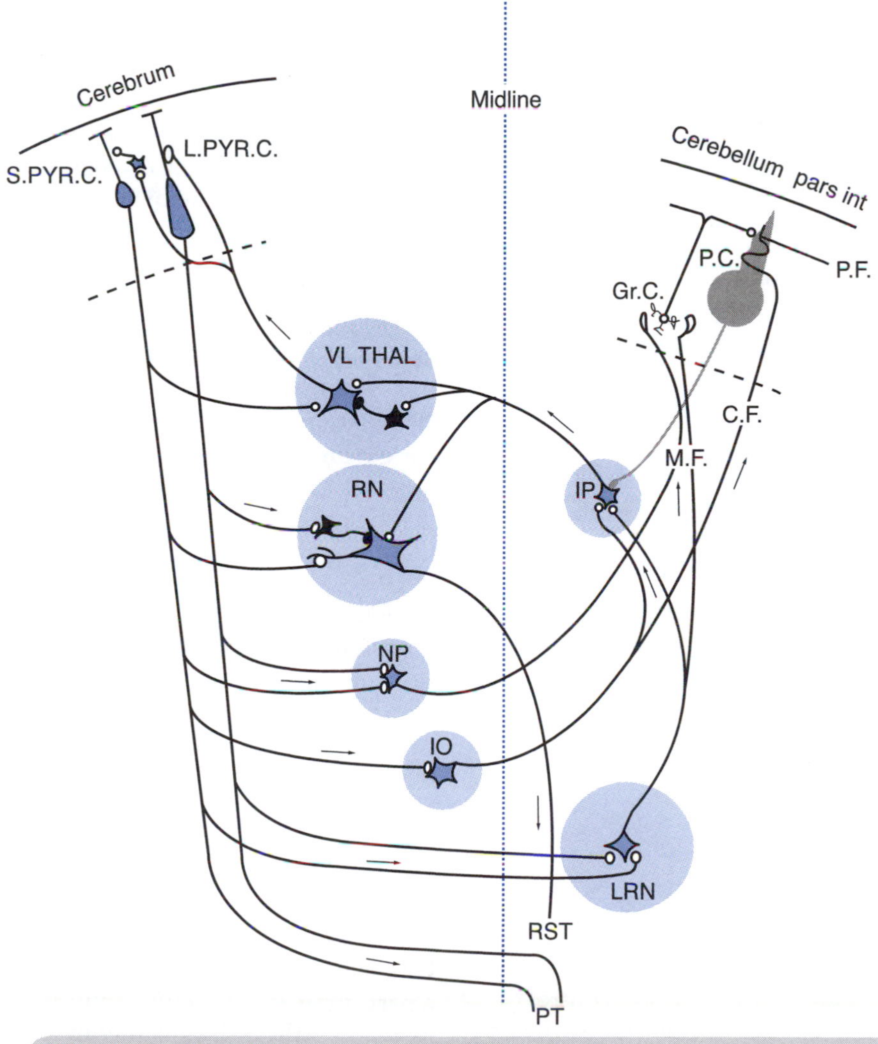


**Figure 4.28** Pathways linking the sensorimotor areas of the cerebral cortex with the cerebellum, motor nuclei in the brain stem, and the spinal cord. Inhibitory neurons are shown in black.

Adapted, by permission, from J.C. Eccles, M. Ito, and J. Szentágothai, 1967, *Cerebellum as a neuronal machine*. (New York: Springer-Verlag).

## *Summary*

The motoneurons are in charge of the ultimate command of muscle contraction. Irrespective of whether a reflex or a voluntary contraction is called for, the motor command to the muscles has to be conveyed by motoneurons, earning for the latter the name "final common path." In turn, the motoneurons are influenced from several sources, including afferents from the periphery and descending tracts from supraspinal levels as well as local spinal circuitry.

The term *supraspinal* alludes to a hierarchical organization of motor control and comprises all areas of the CNS that contribute to motor control but have to do it through their influence on motoneurons only. Supraspinal motor areas of the CNS include motor areas of the cerebral cortex, the cerebellum, and various nuclei in the brain and brain stem. Some of these give rise to descending tracts that affect the motoneurons directly or—more often—indirectly through interneurons.

The area of the cerebral cortex immediately anterior to the central sulcus has the most powerful influence on the motoneurons, but it is not the place where voluntary movements are initiated. As early as 800 ms before any activity can be found in the primary motor cortex, there is widespread activity in the cerebral cortex, called the readiness potential.

# Cerebellum

In spite of rather detailed knowledge of the neuronal circuitry of the cerebellum, including its afferent and efferent connections, surprisingly little is known about the way in which it functions. This presentation is therefore rather general.

### FOR ADDITIONAL READING

For a more comprehensive discussion of the cerebellum and its function, see, for example, Brodal (1998), Kandel, Schwartz, and Jessell (2000), or the special September 1998 issues about the cerebellum in *Trends in Neurosciences* and *Trends in Cognitive Sciences*, where the possible role of the cerebellum in cognitive processes also is discussed.

Of course, it is known that the cerebellum has a key function in the smooth and efficient control of movement. It integrates and organizes information arriving from peripheral proprioceptive receptors and other somatosensory receptors as well as from other parts of the CNS. The proprioceptive information from the peripheral receptors reaches the cerebellum through spinocerebellar tracts. The main afferents from other parts of the CNS come from the **vestibular nuclei,** relaying information about posture and equilibrium, and from the cerebral cortex via pontine nuclei, informing the cerebellum about motor tasks at hand. The cerebellum takes part in several impulse circuits, integrating sensory information and motor commands to provide purposeful movements. The afferents to the cerebellum from the spinal cord, the cerebral cortex, and the vestibular nuclei end in different parts of the cerebellar cortex, often called spinocerebellum, cerebrocerebellum, and vestibulocerebellum, respectively. In turn, these areas largely send their efferent signals, directly or indirectly, back to the same parts of the CNS from which they received afferents. In spite of the apparent differences in the kinds of information handled by the different parts of the cerebellar cortex, its structure is remarkably similar all over. Another striking feature is the modest number of fibers leaving the cerebellar cortex in relation to the number of afferent fibers. In humans, the afferent fibers to the cerebellum outnumber the efferent ones by a factor of 40, illustrating a high degree of convergence of information. One can safely say that the cerebellum listens far more than it speaks.

The afferents to the cerebellum are of two kinds, **mossy fibers** and **climbing fibers.** Both are excitatory, albeit very different in their behavior, as we shall see later. The efferent fibers from the cerebellar cortex, on the other hand, are all axons of Purkinje cells. It came as a major surprise, therefore, when it was discovered that the Purkinje cells are inhibitory (Fonnum, Storm-Mathisen, and Walberg 1970; Ito et al. 1970). Most Purkinje cells project to the cerebellar nuclei, which in turn give rise to the major part of the efferent fibers from the cerebellum. Collaterals from mossy fibers and climbing fibers provide excitatory drive to the cerebellar nuclei (figure 4.29).

The Purkinje cells are among the most conspicuous neurons in the CNS. They have extensively branched dendritic trees, but the branches are confined to one plane only, making the cells look like espalier trees clinched against a wall. All Purkinje cells are oriented in the same direction, with the plane of their dendritic trees perpendicular to the long axis of the cerebellar folium of which they are a part (figure 4.30). *Folium* (plural *folia*) is the name assigned to the rather narrow, transversely oriented cerebellar "gyri." The climbing fibers, originating in the inferior olive in the ventral part of the **medulla oblongata,** make multiple synapses with the Purkinje cells while climbing through the dendritic tree like ivy or Virginia creeper. The inferior olive receives fibers from the spinal cord, from other

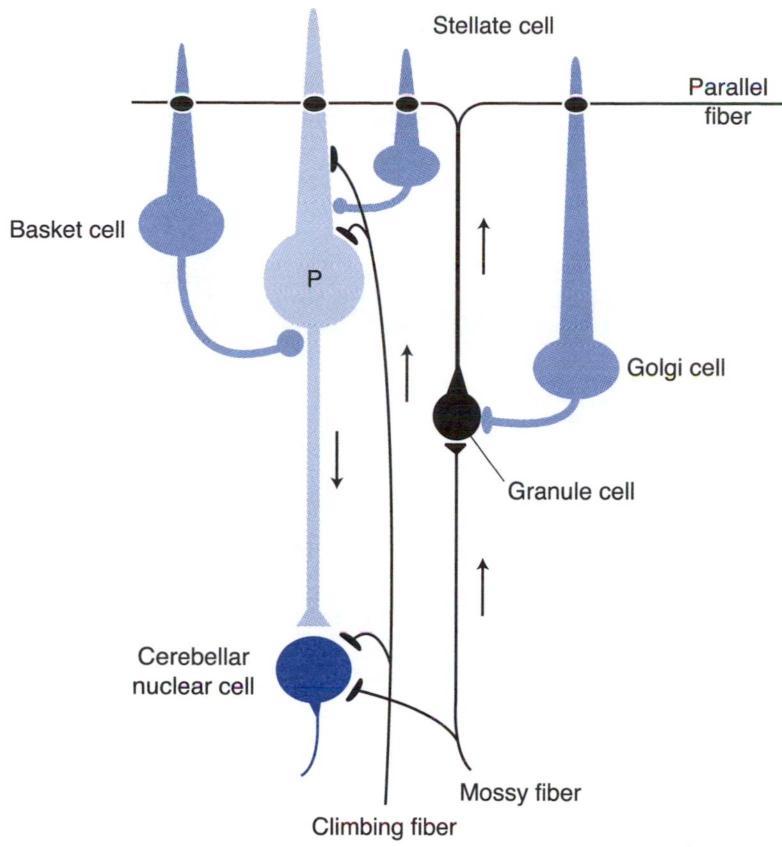


**Figure 4.29** Basic features of the cerebellar circuitry. Cells shown in different shades of blue are inhibitory. Cells and terminals drawn in black are excitatory. Note that collaterals of climbing fibers and mossy fibers provide excitatory drive to the deep cerebellar nuclei.

Reprinted, by permission, from P. Brodal, 1998, *The central nervous system: Structure and function*, 2nd ed. (New York: Oxford University Press), 404.

nuclei in the brain stem, and to a small extent, also from the cerebral cortex. The climbing fibers exert a very powerful excitation on the Purkinje cells, but they do so at a very low frequency, often less than one impulse per second. On the other hand, one impulse from the climbing fiber gives rise to a large excitatory potential in the Purkinje cell, resulting in a series of action potentials associated with a large influx of $Ca^{2+}$ into the Purkinje cell dendrite. It is believed that the climbing fibers convey error signals and that they have an important role in motor learning. There are also indications that the inferior olive may contribute through climbing fibers to the synchronization of neuronal activity mentioned earlier (McCormick 1995; J.P. Welsh et al. 1995).

The mossy fibers are different and convey impulses to the cerebellum from all sources other than the inferior olive. Each mossy fiber branches extensively and makes synaptic contacts with a large number of granule cells, the most numerous type of neuron in the cerebellar cortex. It has been estimated that there are about 1,000 million granule cells for each Purkinje cell in the human cerebellum (Eccles 1973a). Each granule cell sends its axon out into the outermost layer of the cerebellar cortex, the molecular layer, where it bifurcates and gives rise to the so-called parallel fibers, which run parallel to the long axis of the cerebellar folium. In fact, the major part of the molecular layer consists of bundles of parallel fibers running perpendicularly to the plane of the Purkinje cell dendrites, making excitatory contacts with one Purkinje cell after another. Thus, each granule cell makes contact with a large number of Purkinje cells, and each Purkinje cell is contacted by a large number of different granule cell axons (see figure 4.30). Judged from the resulting effect on the Purkinje cells, however, the excitatory action of the mossy fibers is rather weak. Many mossy fibers must be active at the same time to fire a Purkinje cell, possibly because the parallel fibers excite different types of inhibitory interneurons in addition to the Purkinje cells—interneurons that inhibit both granule cells and Purkinje cells. The impulse frequency in a mossy fiber is typically

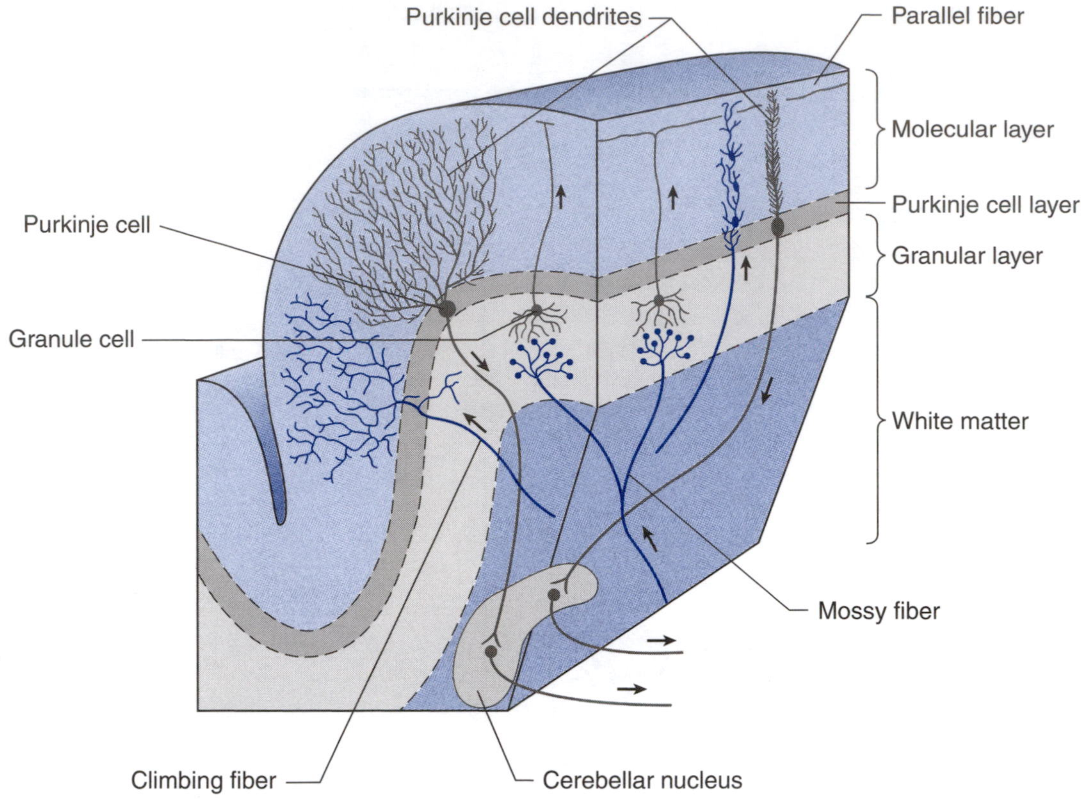


**Figure 4.30** The structure of the cerebellar cortex. The three layers and their main cell types are shown in a piece of one cerebellar folium. The cerebellar folia are oriented transversely. Consequently, the left part of the figure shows a sagittal section, while the right part shows a transversal section, parallel to the long axis of the folium. The Purkinje cell dendrites are oriented in a plane perpendicular to the long axis of the folium and to the parallel fibers. Therefore, the Purkinje cell dendrites are seen head on in the sagittal plane. For clarity, the terminal arborizations of the climbing fibers are shown alongside the Purkinje cell dendrites instead of superimposed on them.

Adapted, by permission, from P. Brodal, 1998, *The central nervous system: Structure and function*, 2nd ed. (New York: Oxford University Press), 404.

rather high. As stated earlier, most Purkinje cells project to the cerebellar nuclei, located in the deep white matter of the cerebellum. In turn, these nuclei project to other motor nuclei, the **red nucleus (nucleus ruber)** and **thalamus** in particular.

Because all the neurons of the cerebellar cortex except the granule cells are inhibitory, the inhibition has dominance. Eccles (1973b) pointed out that "there can be no prolonged chattering in chains of excitatory neurons. . . . Within 0.1 s after some computation, that area of the cerebellar cortex is 'clean,' ready for the next computation. This automatic cleansing is very important in giving reliable performance during quick movements" (p. 125).

To get a rough idea of one of the main tasks of the cerebellum, just try to imagine what kinds of motor commands are needed to make your hand draw a figure eight in the air. The direction of movement changes all the time, and so does the selection of motor units active at different phases of the movement. In fact, it can be said that the figure-eight movement is a sequence of different movements rather than one single movement. The individual pieces of movement in the sequence have to be joined together smoothly, so to speak, to conceal its individual parts. This is done by the cerebellum, and patients with cerebellar lesions are unable to do this in a smooth manner. Instead, it becomes all too apparent that the figure eight is a sequence of movements. In these patients the cerebellum may also be unable to tailor the individual pieces of the sequence to measure, another of its important tasks, making the whole figure out of shape as well as discontinuous and jerky. Such loss of function shows clearly why the cerebellum has been named the coordinator of movements.

The cerebellum also plays an important part in the maintenance of body equilibrium and posture. As stated earlier, it both receives input from and projects back to the vestibular nuclei in the brain stem, which in turn give rise to the descending vestibulospinal tract, among other things. In this way, the cerebellum becomes a major link between the equilibrium sense organs in the inner ear and the locomotor apparatus in charge of the maintenance of equilibrium.

There are apparently "memory stores" in the cerebellar nuclei related to specific innate as well as learned movements. The cerebellum continuously receives messages from peripheral receptors about joint positions, muscle length and tension, movements, environment, and so on. When a call for movement is reported to the cerebellum, it has all the necessary requirements for execution and control of the movement. At the time the motor command descends to the motoneurons, the cerebellum updates the intended movement on the basis of the somatosensory description of body position and velocity on which the movement is to be superimposed. Within a fraction of a second, the return circuit from the cerebellum to the cerebral cortex can modify the discharges down the descending tracts to the spinal cord.

The muscle spindles and other proprioceptors are part of a mechanism for checking the execution of movement in relation to a command. The cerebellum has the potential to compute and integrate the total sum of information about a movement and to execute a follow-up correction. In taking such compensatory measures, the cerebellum has a key function. If the muscle spindle's own contraction is set for a shortening that is suddenly prevented by an unexpected increase in the load, increased afferent impulse traffic from the spindles driving the motoneuron pool automatically compensates for the extra load. This drive stops when the intrafusal and extrafusal lengths are functionally equal. In this way, the muscle spindles can serve as quick error detectors. We can see how a movement planned within the cortex in close cooperation with the cerebellum and **basal ganglia** is finally executed by the "upper motoneurons" in the cerebral motor cortex via the "lower motoneurons" located in the spinal cord. There is a plan-ahead activity based on previous experience (memories) or memory fragments of fixed movement patterns, modified by a continuous updating of the motor command at its beginning and throughout its duration. Some kind of preprogram is always available. This is particularly important in very rapid (ballistic) movements, for they cannot be adequately updated once they begin. Therefore, in learning an exercise, one must first execute it slowly; the cerebral cortex continually intervenes, but the cerebellum participates and is capable of learning. With practice, the movement becomes preprogrammed and can eventually be executed more and more rapidly. This means that the feedback from the periphery becomes less essential for a successful execution. The cerebellum's way of functioning has been compared with the servomechanisms commonly used in modern technology (automatic pilot, industrial control systems, antirobot weapons, etc.).

## Various Nuclei Involved in Movement

Several nuclei in the brain and the brain stem have important roles in motor control. Some of them give rise to descending axons to the spinal cord. Others do not. Two large nuclear complexes in the brain, the thalamus and the basal ganglia, serve central roles in motor control without sending any fibers to the spinal cord. The thalamus is a large, ovoid mass of gray matter, constituting the major part of the diencephalon. It consists of several distinct nuclei and is an important integrating relay station, handling both sensory information from the spinal cord and information related to motor control from the cerebellum and the basal ganglia. It reports to both motor and sensory areas of the cerebral cortex. It also receives numerous axons from the cerebral cortex. The basal ganglia are a group of nuclei in the depth of each cerebral hemisphere. The larger of them, the lentiform nucleus, is separated from the lateral part of the thalamus by the internal capsule (figure 4.31).

Both the thalamus and the basal ganglia are involved in supraspinal motor circuits (figure 4.32). One such circuit involves the cerebral cortex, the pontine nuclei, the cerebellum, and the thalamus, which in turn projects back to the cortex (figure 4.32*a*). Another circuit goes from the cerebral cortex via the basal ganglia and thalamus back to the cerebral cortex (figure 4.32*b*). The latter circuit also receives important input from the substantia nigra in the brain stem, the loss of which results in Parkinson's disease. It is believed that these supraspinal motor circuits are involved, among other things, in the translation of a movement idea into a patterned sequence of motor unit recruitment, that is, in the neural activity taking place between the readiness potential and the activity in the primary motor area.

The nuclei, which project to the spinal cord, take part in automatic movements rather than in conscious, voluntary movements. Among the more conspicuous of these nuclei is the red nucleus (nucleus ruber), which is situated in the part of the brain stem called the **mesencephalon** (figure 4.33). In animals, the rubrospinal tract appears to supplement the pyramidal tract in the control of movements, but its func-

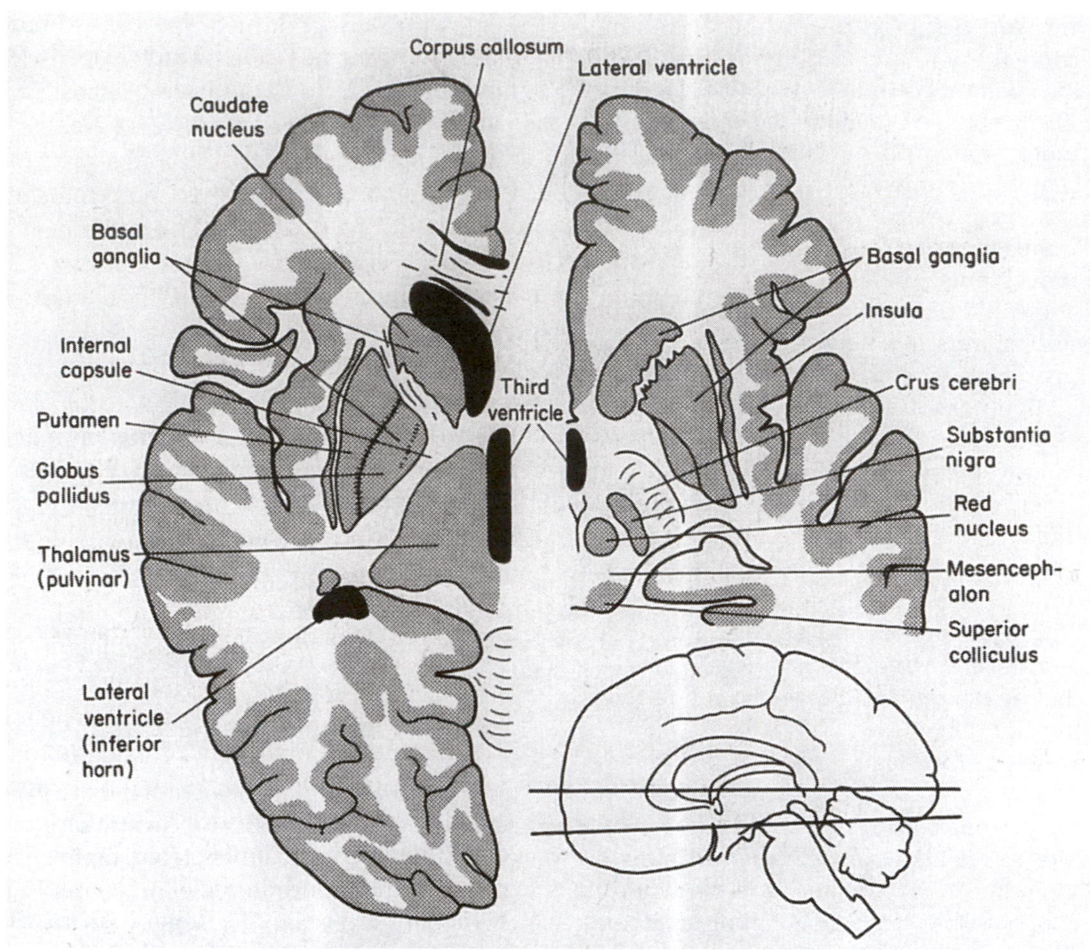


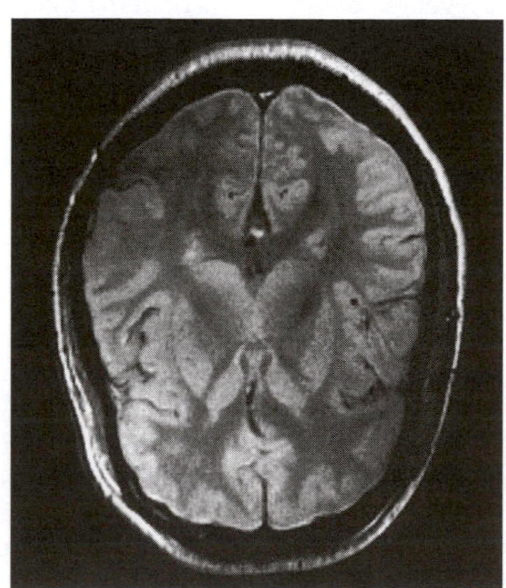


**Figure 4.31** The internal structure of the cerebral hemispheres. *(a)* Drawings of two horizontal sections at different levels (indicated in the inset). The section to the right is the lower one. *(b)* Magnetic resonance image (MRI) in the same direction and level of sectioning as the left part of *a*.

Reprinted, by permission, from P. Brodal, 1998, *The central nervous system: Structure and function*, 2nd ed. (New York: Oxford University Press), 99.

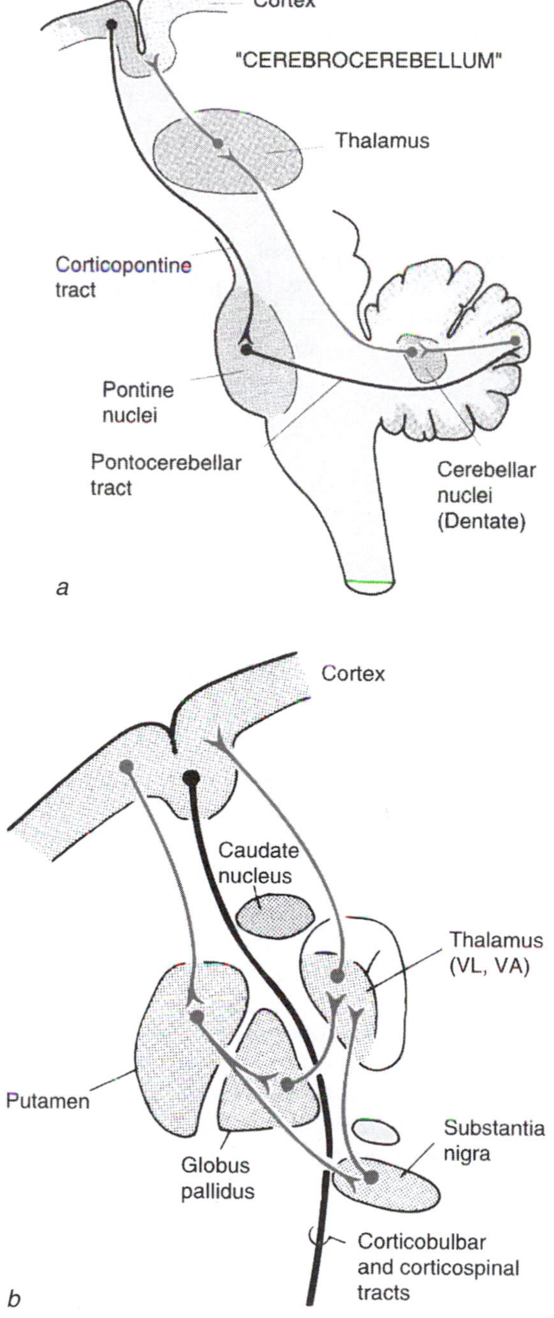


**Figure 4.32** Supraspinal motor circuits. *(a)* The motor circuit from the cerebral cortex via pontine nuclei to the cerebellum and back to the cerebral cortex via thalamus. *(b)* The main connections of the basal ganglia. For further details, see Brodal (1998).

Reprinted, by permission, from P. Brodal, 1998, *The central nervous system: Structure and function*, 2nd ed. (New York: Oxford University Press), 373, 400.

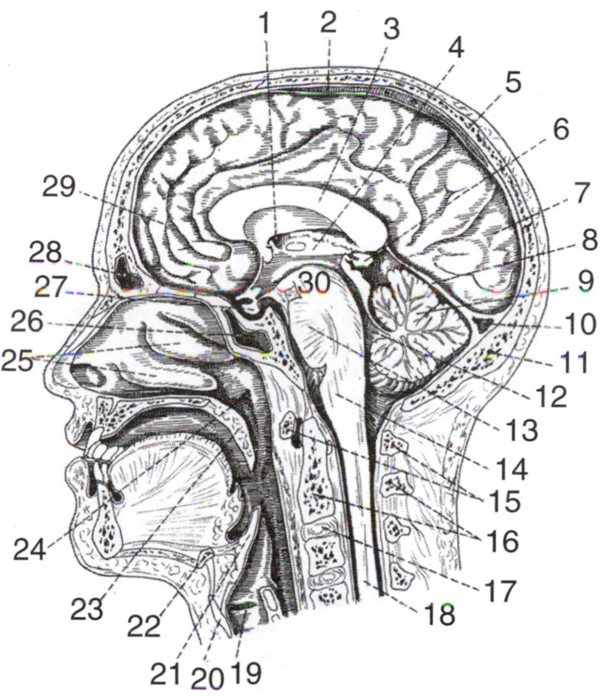


**Figure 4.33** A midline section through the head showing the brain, brain stem, and the upper part of the spinal cord and their respective bony cages. (1) The interventricular foramen (the opening between the lateral ventricle and the third ventricle); (2) superior sagittal sinus (a venous sinus); (3) corpus callosum, containing the major part of the commissural fibers, which connect the two hemispheres; (4) third ventricle; (5) parietal bone; (6) pineal gland; (7) occipital lobe of the brain; (8) tentorium cerebelli, an almost horizontal fold of dura mater between the occipital lobe and the cerebellum; (9) cerebellum; (10) transversal sinus (another venous sinus); (11) occipital bone; (12) fourth ventricle; (13) pons; (14) medulla oblongata; (15) atlas (first cervical vertebra); (16) axis (second cervical vertebra); (17) intervertebral disk; (18) spinal cord; (19) the vocal cord; (20) thyroid cartilage; (21) epiglottis; (22) hyoid bone; (23) the soft palate; (24) opening of the auditory tube; (25) nasal conchae; (26) sphenoid sinus; (27) hypophysis; (28) frontal sinus; (29) frontal lobe of the brain.

Adapted, by permission, from H.A. Dahl and E. Rinvik, 1999, *Menneskets funksjonelle anatomi* (Oslo, Norway: J.W. Cappelens Forlag).

tion in humans is now considered more dubious, partly since the human pyramidal tract is much more developed than in laboratory animals, partly because the part of the red nucleus, which gives rise to the descending fibers, is poorly developed in man. This does not mean that the red nucleus is considered to be without importance for human motor function, but its effect is rather via the cerebellum and motor cortex.

The **reticular formation** is a rather heterogeneous mass of gray matter extending through the central parts of the brain stem from the medulla oblongata to the mesencephalon. It appears as a meshwork of different nuclei and fibers, hence the name. Most neurons that give rise to reticulospinal fibers are located in the reticular formation of the pons and medulla oblongata. Primarily, they seem to be important for the automatic maintenance of the upright position (postural reflexes) and for rather crude movements involving proximal muscles. Such movements are under cortical control and are mediated by a corticoreticulospinal pathway. Monoaminergic fibers from the brain stem to the spinal cord are often included among the reticulospinal fibers. They are thought to mediate a more general facilitation to the neurons of the spinal cord and may be responsible for the increased excitability of the spinal motoneurons reported under conditions of stress.

The vestibulospinal tract originates from the vestibular nuclei, a group of nuclei in the lateral part of the brain stem at the level of the pons and medulla oblongata. The vestibular nuclei receive primary afferent fibers through the vestibulocochlear nerve (the eighth cranial nerve), which conveys information about the position of the head and changes in rotational movements of the head. Note that the information concerns the head only. In addition, information about the position of the head in relation to the body comes from muscles and joints in the neck. The majority of vestibulospinal fibers affect motoneurons to antigravity muscles, that is, muscles that help us maintain an upright position.

The **superior colliculus,** situated on the dorsal side of the mesencephalon (the part called the tectum), is an important center for visual reflexes. It also gives rise to descending fibers to the spinal cord, named tectospinal fibers. Most of them end at cervical levels in accordance with their role in coordinating movements of the eyes and the head. The superior colliculus also receives auditory input, establishing a reflex pathway responsible for turning the head toward unexpected sounds.

### *Summary*

The cerebellum receives nerve impulses from peripheral receptors and other parts of the brain via mossy fibers and climbing fibers. The climbing fibers convey impulses from the inferior olivary nucleus in the medulla oblongata, while the mossy fibers convey impulses from all other sources that send axons to the cerebellum. The climbing fibers make direct synaptic contacts with the Purkinje cells, while the mossy fibers do so via the granule cells. The granule cells are the only excitatory ones in the cerebellar cortex. On their way to the cerebellar cortex, both climbing and mossy fibers send off excitatory collaterals to the cerebellar nuclei. The axons of the inhibitory Purkinje cells are the only output channel from the cerebellar cortex. Thus, the effect of the Purkinje cells is to modulate the activity of the cerebellar nuclei, which give rise to most of the axons leaving the cerebellum.

In spite of our rather extensive knowledge of cerebellar circuitry, our understanding of how the cerebellum performs its functions is surprisingly deficient. Its main tasks are, however, well known. It has a pivotal role in the coordination of movements, not least in the process of making one smooth complex movement out of a sequence of smaller movements, and in tailoring the movements to measure. It is also important for the maintenance of equilibrium and posture. Neither the cerebellum, the thalamus, nor the basal ganglia sends axons to the spinal cord, but rather they exert their influence on motor control through their connections to other supraspinal areas that do.

## Integration of Neuronal Activity in Movement

From this simplified presentation of motor control, one might get the impression that only the cerebral cortex, cerebellum, and other major relay stations are able to integrate the various kinds of information and commands relevant for motor control. That is not so, however. Most, if not all, synaptic events in the CNS are part of an integrative process in which excitatory and inhibitory influences are combined. As mentioned earlier the ultimate nervous integration in motor control takes place at the level of the motoneuron.

The principle of reciprocal inhibition is one example of such integration. This important design in locomotion can be seen clearly in a simple stretch reflex, whereby impulses are evoked in the muscle spindle afferents, as mentioned earlier. These fibers excite the motoneurons of both the same and synergistic muscles monosynaptically. Their collaterals, releasing the same transmitter, at the same time excite inhibitory interneurons, which in turn inhibit the motoneurons to antagonistic muscles. It is certainly functionally economic that a synergistic team of muscles, when activated, is not faced with the resistance of their antagonists. Not least, this is important in fast movements, such as the reflex mentioned earlier, and in fast, alternating movements, such as running. The reciprocal inhibition is not obligatory, however. In many situations synergists and antagonists are active simultaneously to provide stability to a joint. This is a good illustration of the fact that spinal reflex activity, rather than being stereotyped, is modulated to the need of the motor task at hand.

In a knee jerk, the stretch reflex may seem rather stereotypic, but in a more physiological situation that is not so. One reason for this is that another stretch reflex with longer latency seems to be the more physiological one. There has been a long dispute whether this long-latency reflex is transcortical or not. The weight of the evidence now indicates that it is (Matthews 1991). Consequently, the long-latency reflex is also called the **functional stretch reflex** or **long-loop stretch reflex.** This reflex is polysynaptic and highly adaptable to the actual situation. If a relaxed muscle is stretched, no reflex is elicited at all. If a person is asked to resist a sudden stretch, the reflex is stronger than if the instruction was to relax and let go. It is also modifiable during learning of a new motor task.

The main reason that the monosynaptic stretch reflex appears stereotypic is its constant latency. On the other hand, the reflex commonly shows substantial individual variation, indicating a possible influence from other parts of the CNS. In support of this influence, recent experiments in animals have shown that even the monosynaptic stretch reflex and its electrically evoked counterpart, the H-reflex, can be modified by training (Wolpaw 1997).

Theoretically, such training modifications are due to changes in the interneuronal circuitry at the spinal level, notably a decrease in the segmental inhibition of the reflex. The Jendrassik maneuver is widely used to increase the response of the knee jerk in clinical settings, but the mechanism behind its effect has been elusive. Zehr and Stein (1999), however, offered experimental evidence that it may act through a reduced segmental presynaptic inhibition of the reflex.

In a now classic experiment, Nashner (1976) showed that if a person is standing on a movable platform and the platform is unexpectedly moved backward, the plantar flexors on the dorsal side of the leg are stretched, eliciting a stretch reflex. This leads to a contraction of the plantar flexors, primarily the gastrocnemius and soleus muscles, which counteracts the stretch and prevents a forward fall (figure 4.34). If the platform is tilted backward, on

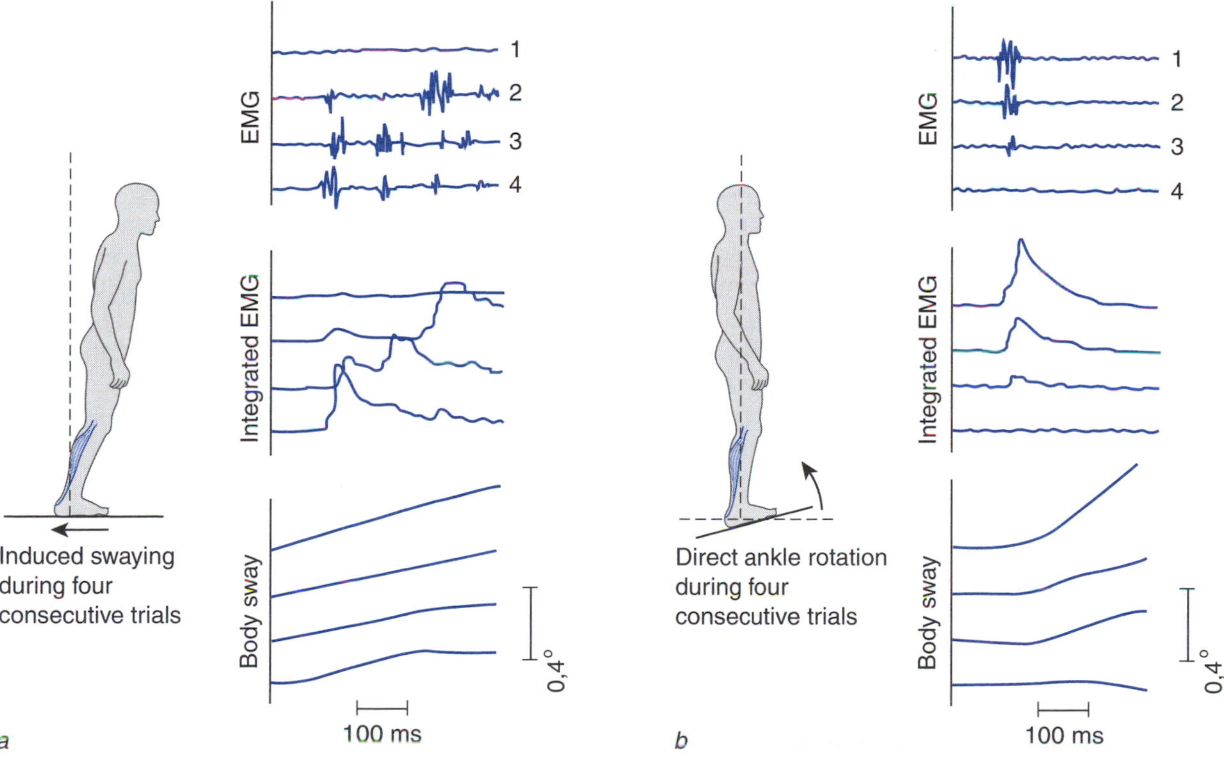


**Figure 4.34** Appropriate anticipatory responses to postural disturbances can be learned. *(a)* Backward movement of a sliding platform tilts the body forward, calling for countervailing action in the stretched gastrocnemius to maintain balance. In successive trials the muscular response is enhanced and its latency reduced. *(b)* When the platform is tilted backward, action by the gastrocnemius would worsen the backward body tilt. Accordingly, in successive trials the muscle's response is decreased, with a corresponding decrease of backward sway.

Reprinted, by permission, from E.R. Kandel, J.H. Schwartz, and T.M. Jessell, 2000, *Principles of Neural Science*, 4th ed. (New York: McGraw-Hill), 820, and L.M. Nashner, 1976, "Adapting reflexes controlling the human posture," *Experimental Brain Research* 26:59-72.

the other hand, the plantar flexors are stretched in the same way, but now a reflex contraction would add to the instability instead of counteracting it. In the first few trials, such a destabilizing reflex contraction is exactly what happens, but after a while the CNS is able to change the reflex response. The reflex contraction of the plantar flexors disappears. Instead, there is a reflex contraction of the ankle dorsiflexors, which is more appropriate in relation to the overall goal of maintaining stability. A similar type of reflex adaptation is seen almost momentarily if the person voluntarily tilts the platform on which he or she is standing.

Imagine also the kind of integration and coordination of motor commands that it takes to extend your arm to grab an object. The overall action consists of simultaneous movements at several joints (at least the shoulder, elbow, and wrist joints). The hand follows a direct trajectory toward the object in question from wherever its starting position might be, and the relevant joints contribute to the movement more or less as they would if they were to comply to a passive movement of the hand along the same trajectory. The main problems to be solved by the CNS in such tasks are to localize the target, to identify the present state and position of the limb in question, and to determine a suitable movement trajectory (Desmurget et al. 1998). Although both the premotor area of the cerebral cortex and the cerebellum most certainly are important in this kind of movement, the exact circuitry involved is so far unknown.

The traditional point of view has been that the cerebral cortex reigns at the highest level in the brain's hierarchical organization for the motor function. However, Evarts (1973) argued that the cerebral motor cortex is at a rather low level of the motor control system, close to the muscular apparatus itself. The cerebellum and basal ganglia are at a higher functional level in the neural chain of command that initiates and controls movement, but there is no evidence that they initiate the command. The primary function of the cerebral motor cortex may not be volition, but rather the refined control of motor activity. It has been claimed that the cerebellum is particularly involved in rapid, ballistic movements, whereas the basal ganglia are preferentially active in slow movements, but this may be a too simplistic view. More recent data seem to indicate that the basal ganglia are important in motor learning, especially in learning by repetition, and that they have a major role in situations where it is important to change motor behavior. A malfunction of the basal ganglia causes postural disturbances, muscular tremor at rest, increased tonus, muscular rigidity, and difficulty in the initiation of movement. Such symptoms are typical of Parkinson's disease. Muscular tremor is also seen after damage to the cerebellum, but this tremor is more severe during voluntary movement. The person's movements are clumsy and slow, carried out in a more or less disorderly fashion. The person also suffers from poor equilibrium. Higher primates with the cerebral motor cortex destroyed can still perform many activities, but there is a loss of fine motor control, particularly of the fingers, and decreased voluntary movement activity. Although the left cerebral hemisphere is dominant with regard to speech in most individuals, the idea of a dichotomous division of function between the two hemispheres, which was so common in the 1970s and 1980s, has now largely been abandoned. Instead, it is now generally accepted that both hemispheres take part in most functions.

## The Role of Reflex Activity in Motor Control

Depending on the degree of voluntary control, the execution of movement is governed by the spinal cord, the brain stem, or the cerebral cortex. The more automatic movements are pure reflex movements involving local circuits in the spinal cord only, while the least automatic voluntary movements are initiated and controlled by the cerebral cortex. Even basically automatic sequences of movements such as walking are initiated and terminated from the cerebral cortex, but the movements are based on reflex activity during their execution, with voluntary influences restricted to necessary corrections only. On the other hand, various types of reflexes are active during voluntary movements to adapt the movement to the situation at hand and to make necessary adjustments in postural muscles to maintain equilibrium.

### For Additional Reading

For further discussion of the role of reflexes in motor control, see Kandel, Schwartz, and Jessell (2000, chapter 36).

The basic neural machinery essential for the generation of reflex action is located in the spinal cord. These local spinal circuits, however, can be opened and closed by descending axons from higher brain areas as well as by other regions within the spinal cord. In addition, descending axons can influence the motoneurons both directly and indirectly.

At various levels of the nervous system, there are "generators" that can initiate and coordinate rhythmic movements. Reciprocal innervation is a basic feature of such movements, and as we have seen, it is established all the way down to the spinal level. If a person steps on a sharp stone while barefoot, pain receptors are stimulated, and the limb is withdrawn from the source of the pain even before any sensation of pain has been experienced. The response is executed by means of excitatory connections in the spinal cord between the afferent fibers whose nerve endings were stimulated and the motoneurons of the flexor muscles of the extremity. It is a polysynaptic reflex with several interneurons included in the reflex arc. Some of the intercalated interneurons have an inhibitory influence on the motoneurons of the extensors of the same extremity. This is an example of an ipsilateral flexor reflex with a simultaneous inhibition of ipsilateral extensor motoneurons. If the **noxious** stimulation is strong enough, the extensor motoneurons to the opposite (contralateral) limb are activated almost simultaneously with inhibition of the flexor motoneurons, resulting in an extension of the contralateral limb—a contralateral extensor reflex—to provide continued support for the body (see figure 4.18).

Examples of muscular activities that are normally executed and controlled subconsciously are breathing, swallowing, blinking, and coughing. At any moment one can voluntarily interfere with most such normally subconscious muscle actions. Two parallel pathways can activate motoneurons that innervate the respiratory muscles, an automatic one serving the metabolic function of breathing and a voluntary behavioral one originating in the cerebral cortex. The latter is used during speech or singing, for example.

Grillner (1975) reported that cats with chronic spinal transections were able to generate the essential and basic features of the step cycle as well as coordinate the limbs and shift from one type of gait to another. This also holds true for deafferented animals (whose dorsal roots have been cut), meaning that locomotor patterns can be generated in the spinal cord by central pattern generators without any sensory input. There are several structures in the brain stem from which locomotion can be initiated, and these patterns are activated via pathways descending from the brain stem. Continuous electrical stimulation of an area called the mesencephalic locomotor region can elicit active walking movements—that is, rhythmic activity—in decerebrated cats. When the strength of the stimulation was increased, cats changed stride frequency and pattern to a trot and eventually to a gallop (Grillner 1981). The afferent input is important when, for one reason or another, the locomotor movements are disturbed. Grillner and his group (1981) noticed that a light touch on the dorsum of a spinal kitten's paw during the flexion phase in walking gave an additional flexion. However, the same stimulus during the extension phase produced no response in the flexor muscles but enhanced the extensor activity. In other words, there was a reversal of the reflex. In one position, the afferent pathway to the flexor motoneurons was wide open; in a different position, it was shut off. Receptors influenced by the hip joint position can apparently modify the response to a given stimulus. Anesthesia of the skin abolished these reflexes.

Deafferented higher primates, such as monkeys, can develop motor skill; they can walk and run, and the timing of their muscles can be normal, showing that afferent signals from a limb are not indispensable for movement. The eyes can take over part of the information input to the CNS. Patients with nonfunctioning dorsal roots, however, show a retardation and prolongation of the execution of voluntary movements, particularly the slower movements. Faster movements are the least influenced. This can be explained by the basic importance of central programming, especially of the fast movements. It should be emphasized that the level of muscle recruitment in a given movement depends on whether gravity works for or against the movement, in addition to the weight of the external load. Numerous sense organs of different kinds provide feedback from the periphery, and their proper function is therefore essential for very efficient and skilled movements.

Forssberg and Svartengren (1983) studied the locomotor output of the gastrocnemius muscle in cats after a transposition that gave the muscle a dorsiflexing action around the ankle. The old extensor pattern was retained in all six cats throughout the study period, more than one year after surgery. The researchers concluded that neither the spinal locomotor network controlling the gastrocnemius muscle nor the supraspinal circuits influencing the network exhibit a high degree of plasticity in response to locating the muscle in a position antagonistic to its original position. They point out that in humans most transposed muscles can be voluntarily controlled in their new positions, while only some muscles seem to convert their activity pattern during locomotion.

Stretch reflexes have been regarded as important for motor control for a long time, but, partly due to the long unresolved question of spinal versus long-loop stretch reflexes, their exact role has been a matter of discussion. The present view can be summarized as follows: Both spinal and long-loop stretch reflexes are active during normal movements; the relative contribution of each depends on the task and the muscles involved (Kandel, Schwartz, and Jessell 2000). Long-loop reflexes seem to be particularly important for movements involving distal muscles, which are under rather direct cortical control (J.F. Marsden et al. 1999).

By nature, a stretch reflex is a kind of servomechanism counteracting the unexpected stretch of a muscle. In this way, it can also act to compensate for at least a minor force deficit during an attempt to lift an object slightly heavier than expected. Everyday experience shows, however, that it is unable to compensate for more significant weight differences.

Stretch reflexes may also serve to adjust the stiffness of muscles during eccentric work (Houk 1979). Anyone who has driven a delivery van has experienced the vastly different compliance of its springs when the van is empty as compared to when it is loaded with cargo. The compliance of the extensor muscles of the lower limb during human walking is, on the other hand, surprisingly little affected by the carrying of even a substantial additional weight, due to the reflexively adjusted stiffness of the extensor muscles.

### *Summary*

Central pattern generators control stereotypic locomotor movements such as walking and running. The neural networks are to a significant degree located within the spinal cord. Supraspinal regions can activate the relevant spinal programs as well as control and modify these programs. Similarly, powerful signals from peripheral receptors can control the central pattern generators on the spinal level or less directly via loops passing through higher levels of the CNS. The receptors reporting the hip position, as well as the load on a limb, can control and modify a step cycle. The central program does not require afferent feedback for its essential pattern or maintenance but becomes functionally more efficient and adaptable to unexpected events when fed afferent signals. Faster movements are less dependent on external cues than slower ones. Both spinal and long-loop (transcortical) stretch reflexes can be active during normal movements, but their level of activity is adjusted to suit the actual purpose of the movement.

## Motor Learning

Unfamiliar movements are performed clumsily and with difficulty, but we know that with proper practice they become smooth and easy. We are, however, far from a complete understanding of the reasons for this in physiological terms.

Like other tissues, the CNS is characterized by plasticity in its structural, biochemical, and functional properties. The gross neuronal connections are established before birth, and the mechanisms governing these processes are beginning to be revealed. In postnatal life, activity has turned out to be necessary for neurons and their synapses to develop and function optimally. In kittens blinded shortly after birth, the visual cortex fails to develop properly. On the other hand, the size of an area of the motor cortex engaged in a particular activity can increase if that particular activity is repeated extensively. Learning can also make certain combinations of neural connections more efficient, so that movements can be quickly repeated with remarkable precision and even be retrieved in spite of lack of practice for years. Today, scientists distinguish between different forms of learning and memory. A major distinction is between explicit or declarative memory and implicit or nondeclarative memory. Motor learning belongs to the latter category (figure 4.35).

#### FOR ADDITIONAL READING

For a recent update on CNS development and plasticity, see Kandel, Schwartz, and Jessell (2000, chapters 38, 52, 53, and 56) and Jurata, Thomas, and Pfaff (2000).

There is now strong evidence that learning involves changes in synaptic efficacy called **long-term potentiation (LTP)** and **long-term depression (LTD).** Bliss and Lømo (1973) first demonstrated LTP in the hippocampus, a region of the brain long known to be involved in memory. At that time, nothing was known about the underlying mechanism, but over the years many details have been revealed.

#### FOR ADDITIONAL READING

More information about LTP and LTD can be found in Daniel, Levenes, and Crepel (1998) and Kandel, Schwartz, and Jessell (2000, chapter 63).

Suffice it to say here that an elevated calcium concentration in the postsynaptic cell is a key element in both instances, emphasizing the importance of calcium signals in general. It should be mentioned, however, that the involvement of cer-

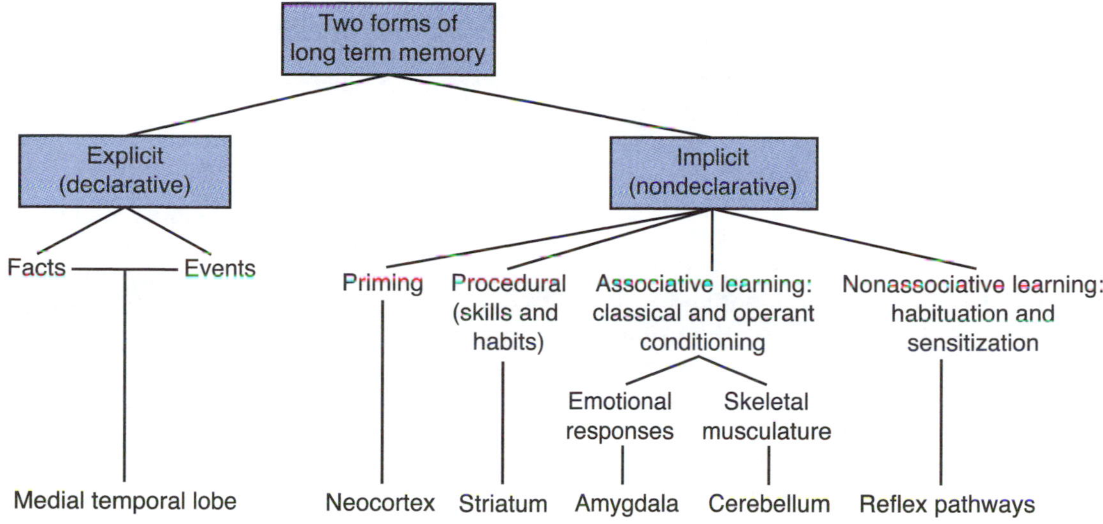


**Figure 4.35** Various forms of memory. Motor memory is implicit or nondeclarative.

Reprinted, by permission, from E.R. Kandel, J.H. Schwartz, and T.M. Jessell, 2000, *Principles of Neural Science*, 4th ed. (New York: McGraw-Hill), 1231.

ebellar LTD in actual motor learning has not yet been proven directly (Brodal 1998), although it is well supported by experiments on transgenic mice (Daniel, Levenes, and Crepel 1998).

In adults as well as children, a structural plasticity is evident at the microlevel; that is, some synapses may develop, while others regress. High-frequency synaptic activity has been reported to change the number and shape of dendritic spines (Halpain 2000). Such changes in spine morphology can take place in a matter of seconds or minutes, and they may persist for hours to days. Indeed, the rapid changes in spine morphology have led to the notion of "spine dancing." Admittedly, the functional significance of spine motility is so far only a subject of speculation, but it has been discussed in relation to retrograde signaling and rapid changes in synaptic efficacy.

## For Additional Reading

For further details about dendritic spines and list of Web sites displaying dynamic images of spines, see Halpain (2000).

In one way or another, repetition of a movement can be coded and memory engrams stored and played back with great accuracy and precision. Although many questions remain unanswered, available results suggest that the cerebellum is directly involved in the learning of motor actions and that it does so by means of the modifying effect (LTD) of the climbing fibers on the efficiency of the parallel fiber–Purkinje cell synapses. In this mechanism it is believed that the climbing fibers detect differences between expected and actual afferent input rather than simply monitoring afferent information (Kandel, Schwartz, and Jessell 2000, chapter 42).

Even simple, monosynaptic reflexes can be changed during a learning process, as evidenced by animal experiments involving operant conditioning of the H-reflex, the electrically evoked counterpart of the spinal stretch reflex. Changes were found in both the firing threshold and conduction velocity of the motoneuron and in its afferent connections, and there is evidence that the pyramidal tract has an essential role in producing this plasticity. It is believed that this plasticity contributes to motor development in childhood as well as to the learning of motor skills later in life (Wolpaw 1997). When we discuss the role of stretch-reflex circuits in motor behavior, we should remember that the spindle afferents, like other proprioceptive afferents, are key components in motor control, even in situations where no overt reflex activity takes place.

In view of the sharply defined projection of impulses within and between areas of the brain, it is difficult to understand how a general movement pattern may function. The organization does not appear to favor a transfer effect; that is, learning by practicing a certain movement pattern does not in itself enhance the performance of another movement pattern, not even one that is relatively similar. However, the technique of learning new tasks can be improved. One can learn and memorize specific activities that can then be woven together in different combinations. The pianist, having practiced many hours at the keyboard, has the potential to learn new pieces of music quickly. Whether it should be called a transfer effect is a matter of definition; however, the pianist can learn to play a piece of music slowly and softly, but once he or she has learned it, the pianist can play it fast and loud just as well. If someone is asked to write his or her name in letters a meter high, his or her normal signature can easily be identified in the greatly exaggerated letters. Apparently, a central program for the shape and pattern of movements is developed, but there is no rigid and formal program. A certain selection of motor units is activated at each stage of the movement, but obviously a different selection of motor units must be activated for the normal signature and the exaggerated one. Central programs must conduct many of the movements learned because they can be executed without peripheral feedback, whether visual, auditory, or factual. It is possible, for instance, to write one's signature with closed eyes.

Learned movements can be easily performed subconsciously, but the motor cortex is still engaged in the control of the movement, regardless of whether the movement is innate or learned. The same nerve cells seem to control the pattern of muscle contraction regardless of the circumstances or context of the movement (Evarts 1973).

Little by little, a "sense" is developed, based on both endowed and acquired capabilities. Granit (1972) reported old experiments showing that a person, when lifting two equally heavy spheres, one 4 cm (1.6 in.) and the other 10 cm (3.9 in.) in diameter, feels the larger sphere to be lighter. Subjects who lifted equally heavy objects (112 g, or 4 oz) with volumes ranging from 10 ml (.3 fl oz) to 2,000 ml (67.6 fl oz) felt that the largest ones were also lighter; they were then asked to add weights to the apparently lighter (2,000-ml) object until it felt as heavy as the 10-ml one. It turned out that the amount required averaged about 112 g (4 oz)!

Granit and Burke (1973) quoted Hagbarth's observation: "If, in a completely darkened room, a normal subject swings his arm in an arc around the elbow joint and the arm is stroboscopically illuminated only when it is at 90°, the subject has a strong sensation that he is not really moving his arm at all. On closing his eyes, this illusion immediately disappears." No doubt visual input has a relatively dominant influence for the determination of position. It can also, as previously mentioned, partially compensate for defects in other afferent input.

#### For Additional Reading

For a recent review of central mechanisms of motor skill learning, see Hikosaka et al. (2002).

### *Summary*

Today, motor learning is categorized as one kind of implicit or nondeclarative memory. Along with other types of learning, motor learning is believed to involve changes in synaptic efficacy, LTP and LTD. Available results indicate that the cerebellum is directly involved in motor learning. Although it is well supported by experiments on transgenic mice, however, the role of cerebellar LTD in motor learning has not yet been directly proven. Rapid changes in dendritic spine morphology have also been observed in connection with learning, but their functional significance is so far only a subject of speculation.

## Coordination

In fast, ballistic movements, an initial spurt of activity in the agonist produces momentum and kinetic energy in the segment, and then the muscle relaxes as the limb proceeds by its own momentum. By reciprocal inhibition the antagonist relaxes completely, except perhaps at the end of a movement or when the movement is stopped by the limits of the joint or an external force. In slow movements in some activities, discrete bursts of neural activity are observed in agonists and antagonists to accelerate and decelerate the segment (Hubbard 1960). An integration of central programs and the feedback loops from the proprioceptors are probably responsible for the periodicity and modulation of the motoneural activity. Proprioceptor and visual stimuli are probably relayed too late to correct a misdirected fast movement but in time to make adjustments in succeeding movements. In slow movements, continuous close control is possible.

Movements of the hand toward an object that last less than 250 ms are programmed in advance. With movements of longer duration, the first phase is programmed, but then there is also visual control as well as guidance from peripheral proprioceptors. Such feedback is essential for accuracy and the learning of new exercises.

Hubbard (1960) also points out that skilled pitching, shot putting, and discus throwing are excellent examples of developing moments serially to stretch agonists successively. In terms of efficient production, the important factor is to develop tension under conditions as like isometric (or even eccentric) as possible and to maintain this condition as long as possible. This can be accomplished by moving the proximal segment ahead of the distal segment so that the agonist develops tension while lengthening or remaining at the same length as long as possible. The difference between a good discus thrower and a poor one is that the poor one uncoils while spinning across the circle, but the good one stays coiled until ready to throw.

## Posture

Life began in water. In that environment, it is much less of a problem to maintain a given posture than it is for an organism living on land. The density of animal tissues is close to that of water but quite different from that of air. In air, appropriate muscles have to be activated at sufficient force, and the body's center of gravity has to be kept within the area of support. In a two-legged posture, most of the corrective movements take place in the ankle joints, to a lesser degree in the knees and the hips. For control of posture (and in walking and many other movements), three types of information are essential: information from (1) the proprioceptive system (receptors in joints, muscles, tendons, skin), (2) the visual system, and (3) the vestibular system. The proprioceptive afferent inflow from the ankle region and afferent impulses from the sole of the foot are of special importance.

The upright position is maintained by muscles acting against the force of gravity. In theory, the myotatic stretch reflex may be an important factor in maintaining an adequate posture. The muscles acting against the pull of gravity are stretched; therefore, the muscle spindles in those muscles are also stretched. Afferent impulses are evoked, and the muscles contract so that the pull of gravity is counterbalanced. Since the intrafusal muscle fibers of the muscle spindle can be activated from higher centers via the $\gamma$ fibers, its receptor may be more or less prone to respond to a stretch. A feeling of happiness, alertness, or attention can increase the $\gamma$ activity, whereas unhappiness, drowsiness, or lack of attention can reduce the activity. In this way, part of the noticeable relationship between an individual's mood and posture may be explained. Descending control of spinal mechanisms is, however, important in the maintenance of posture, just as it is in the control of locomotion.

In many joints it is the ligaments, not the muscles, that normally maintain the integrity of the joint. For example, muscles that are able to support the arches in the foot are generally inactive when standing at rest. When a person sits upright with the arms hanging in the relaxed, neutral position, the muscles that cross the shoulder joint and the elbow joint are not active in preventing dislocation of these joints by a heavy downward pull applied to the arm (Basmajian and De Luca 1985). It is an interesting observation that during activity with forward flexion of the spinal column, there is a marked muscular activity until flexion is extreme. At that point, ligament structures assume the forces, and the electromyographic-recorded discharge from the trunk muscles ceases (Basmajian and De Luca 1985; Floyd and Silver 1955).

With the center of gravity of the head and trunk very close to the supporting spinal column, the human being has the most economical antigravity mechanism among the mammals once the upright posture is attained. Oxygen uptake is only slightly higher in the erect position (0.30–0.35 $L \cdot min^{-1}$) than in the supine position (0.25 $L \cdot min^{-1}$). Many antigravity muscles have a high percentage of slow-twitch fibers; they are more affected by the $\gamma$ loop than the fast-twitch fibers are. Stronger afferent impulse traffic recruits more motoneurons rather than increasing the rate of discharge in those units already active. Electromyographic studies have revealed that the antigravity muscles are activated at a frequency of 5 to 20 Hz unless they are under special stress.

Three types of responses to stabilize posture and to prevent falling are observed: an immediate muscle stiffness; activation of a long-latency stretch reflex, with an approximately 120-ms latency; and vestibular activation, with a latency longer than 180 ms. The myotatic stretch reflex, acting after a 45- to 50-ms delay, seems to be less significant.

The long-latency reflex (latency about 120 ms, also called the postural synergy) involves many muscle groups, starting in the peripheral muscles. It seems to involve supraspinal, probably transcortical, and spinal circuits (Brodal 1998; Matthews 1991). In some experimental subjects this reflex dominated. As mentioned earlier, rotational stimulation of the ankle (platform tilting backward) initially led to an inappropriate response from the gastrocnemius muscle, but the responses evoked by succeeding stimuli became progressively weaker, and the sway therefore became less pronounced.

In other experimental subjects, there was a delay in the compensatory responses, and apparently visual or vestibular inputs, or both, were involved (with 180- to 200-ms latency). If subjects in similar experiments used the arms for additional support, the reflex response was switched off. These examples illustrate how higher levels of the CNS evaluate whether a reflex response is appropriate or not and whether it can be extensively modified to optimize the response. Descending influences can gate basic reflex loops by opening them (facilitation) or closing them (inhibition).

The so-called antigravity muscles have the important functions of producing the powerful movements necessary for changing from a supine to a sitting or standing position and of providing a firm but flexible foundation for the variety of muscular activity of everyday life. Actually, posture is the basis of movement. All movements start from and end in a posture. Accordingly, not only do peripheral factors automatically evoke postural responses, but voluntary movements (e.g., of an arm) have been shown to initiate activation of the postural muscles. The sequence of events in such activation is similar to that typical of an automatic postural reaction. Certainly, any movement disturbs the position of the body to some degree. It has been observed that changes in the activity of postural muscles may precede the actual start of voluntary movement, particularly if that movement moves the body's center of gravity. It appears that principal postural components are included in the composition of the central program of an impending movement, but the exact pattern is determined by the state of the locomotor organs at that particular time. (For further discussion, see Kandel, Schwartz, and Jessell 2000, chapter 41.)

The free normal posture is characterized by a postural sway, so that the center of gravity varies with respect to its projection on the ground with a frequency of 5 to 6 cycles per minute. With the eyes closed (or in the dark), the swaying is about 50% more pronounced, but the projection of the center of gravity is still within the area of support. During walking the projection of the center of gravity alternately moves out of and back into the area of support. It is inside, giving the body a relatively stable equilibrium, only during the double stance phase. When it is outside during the swing phase, the body is in a state of labile equilibrium, which can be maintained for only a very limited time. If at the end of the swing phase the leg is unable to find new support, a fall is almost inevitable. As soon as one begins to fall, reflex compensatory muscular programs that restore the state of equilibrium are activated. Any of the muscles of the trunk and legs act as antigravity muscles.

### *Summary*

For the maintenance of a balanced body position in standing, in locomotion, and in any kind of exercise, an integrated coordination of the proprioceptive, visual, and vestibular systems is essential. The proprioceptive and visual systems can jointly manage most of the neuromuscular interactions necessary to secure an optimal situation. The vestibular system seems to function more as a reference system, controlling which adaptive modifications should be performed in the corrections elicited by the proprioceptive and visual systems. Stretch reflexes are important, but there also seem to be central pattern generators for posture.

## Motor Function in a Lifelong Perspective

Uncoordinated movements of many parts of the body are characteristic of the newborn, but gradually, more coordinated reflexes develop (postural reflexes, tonic neck reflexes, walking, and so on). The CNS gradually matures, allowing the movements to become more complex. Nevertheless, when beginning school, the child usually fails in the performance of more complicated, "artificial" movements. It is not until puberty that the child becomes capable of the fine coordinated movements necessary for the most precise actions, which are based on the integration of nervous activity from various levels of the CNS and the impact from all peripheral receptors. In particular, changing body dimensions during adolescence necessitate continuous modification of the interpretation of impulses exchanged between muscles and CNS to ensure correct motor unit activation patterns for given tasks. Apparently, there are definite anatomic and physiological limitations to the complex movements that can be performed during early adolescence.

A child only a few weeks old already possesses visual depth perception. Over time this perception is further developed, which is essential for optimal postural control. Children who have just learned to stand have acquired supporting reflexes, and their proprioceptive function is well developed, albeit far from perfect. In experiments using platforms that can be rotated or moved backward and forward (as discussed earlier in this chapter), the response in children under the age of 7 or 8 is much more variable than in older individuals. When facing conflicting information from the visual and proprioceptive systems, small children usually fail to stand. Evidently, the three systems of key importance for postural balance are not yet integrated. Over the course of evolution, the locomotor activity generated in the CNS has, in important aspects, been modified to make bipedal walking possible. The stepping movements performed by a newborn child, if supported and moved across the floor, is not the typical pattern of *Homo sapiens,* but is more related to quadruped locomotion. The generators of these primitive locomotive movements are probably located in the spinal cord. Gradually, a modification in the impulse traffic to the muscles engaged in walking takes place from the higher brain areas. Most likely other generators within the CNS take over the coordination of muscles. Three requirements for independent walking are sequentially met in a child. A stepping pattern is present already at birth or shortly after, while the control of balance—the major limiting factor—appears much later, when the child is about 1 year old. The third requirement, ability to adapt the walking movements to the actual environment, is the last to appear (Shumway-Cook and Woollacott 1995).

To a large extent, therefore, walking is a question of balance control and of motivation (courage) when the child starts to walk without support. A child who cannot walk without support may stand and walk if offered a stick to grab. If the child notices that no one holds the other end of the stick, the child immediately sits down. After the child has learned to walk, there is a further optimizing of the sequence of muscle activation, and as previously pointed out, at the age of approximately 8 years, the proprioceptive, vestibular, and visual systems are fully integrated. This enhances the child's choice of exercises and improves the flexibility and modifiability of the locomotor system so that new demands, whether expected or unexpected, can be handled adequately.

As soon as a stable gait pattern is established during the first decade of life, it usually stays that way for many decades, often with remarkable similarities between parent and offspring, unless perturbed by accident or disease. In old age, however, the gait pattern may gradually change, for a host of different reasons. Typically, the speed of walking is reduced, the stride length is reduced, but the stride width is increased. The latter may serve to increase the base of support and compensate for decreased balance capabilities. Many elderly people complain of dizziness, often resulting in a fear of falling, which clearly influences the gait pattern. It has been discussed whether the gait pattern in old age is a reappearance of an infantile pattern. It is rather a question of lack of balance in both cases, albeit for different reasons in the young and the old, but they use similar strategies to handle it (Shumway-Cook and Woollacott 1995).

In addition to decreased balance capabilities, decreased muscular function may be an additional limiting factor to mobility in the elderly. In general, muscle strength is a limiting factor for independent living. Admittedly, there seems to be an almost inevitable decrease in muscle function with age, but it can be counteracted significantly by appropriate physical activity and training (G. Grimby et al. 1992; Porter, Vandervoort, and Lexell 1995). The loss of strength is primarily due to a loss of muscle tissue—muscle atrophy or **sarcopenia**—which has been calculated to amount to 1% or more per year after the age of 50 (Frontera et al. 2000). The annual decline in strength has been reported to be in the range of 1.4% to 5.4% per year and tends to be higher in longitudinal than in cross-sectional studies, indicating a possible underestimation of the changes in the latter studies (Frontera et al. 2000). In addition, changes in explosive strength, important for the speed of walking and climbing of stairs, tend to be more severe than changes in static strength. When differences in muscle strength are taken into account, force fluctuations during a contraction are more pronounced in old individuals than in younger ones, especially at low force levels (Laidlaw, Bilodeau, and Enoka 2000). Such fluctuations may seriously reduce manual dexterity, but strength training may improve steadiness, possibly by improving CNS control over motor unit discharge rates.

## For Additional Reading

The role of insulin-like growth factor-1 in the prevention of muscle atrophy in old age is discussed by Hameed, Harridge and Goldspink (2002).

Reports about changes in the percentage of different muscle fiber types with age are ambiguous (Frontera et al. 2000). One obvious reason for this ambiguity is the increased prevalence of hybrid muscle fibers in old age and the inability of the myofibrillar ATPase reaction to reveal it (see chapter 3). There are several possible reasons for the appearance of hybrid fibers. One is the denervation–reinnervation processes known to occur in old age, since a different impulse pattern in the reinnervating motor axon is bound to change the myosin phenotype of the muscle fiber in question (Lømo, Westgaard, and Dahl 1974; see also chapter 3). Another consequence of these denervation–reinnervation processes is that the motor units of a muscle become fewer but larger (Brown, Strong, and Snow 1988). Whether manual dexterity and precision movements are affected by the increased size of the motor units involved is not known yet, but it is tempting to assume that they are affected in a negative way.

The loss of α motoneurons after the age of 60 appears to be one of the reasons for declining motor function in old age, and several authors have speculated about possible reasons for this neuronal death (surveys are given by F.W. Booth, Weeden, and Tseng 1994 and by Faulkner and Brooks 1995). **Ciliary neurotrophic factor (CNTF)** is a protein synthesized by Schwann cells that promotes the differentiation and survival of many cell types in the nervous system. In animal experiments, the production of CNTF has been shown to decline in parallel with the decline in muscular strength, and an exogenous supply of CNTF was able to improve strength in aging animals (Guillet et al. 1999).

## Degeneration and Regeneration of Nerves

The classic view holds that, in adults, neurons in the CNS are no longer capable of cell division; they are postmitotic cells. More recent findings in the human hippocampus seem to indicate that this is not entirely true, however, but the practical implications of this are so far unknown. Regarding the question of regeneration in general, intriguing theoretical possibilities emerge from the increasing evidence that stem cells, both in the CNS and elsewhere, retain a broader developmental potential than previously thought (Weissman 2000; Kondo and Raff 2000).

After damage to an axon, the loss of the normal metabolic connection with the cell body causes a degeneration of the axon and myelin sheath distal to the site of injury. But there are also retrograde changes, which are often severe and may include presynaptic nerves. Glial cells and Schwann cells play an important role in healing. A proliferation of glial cells can give rise to scars around the zone of the trauma, which can effectively block restoration of synaptic connections within the brain and spinal cord. The conditions for regeneration of peripheral nerves are, on the other hand, well developed. Regeneration is facilitated by the presence of uninterrupted endoneurial tubes. The regenerating axons grow into the peripheral endoneurial tubes at a rate of several millimeters per day. How the regenerating nerve finds its way is now beginning to be revealed. It is believed that regenerating axons are guided by the same mechanisms that guide embryonic axons (Kandel, Schwartz, and Jessell 2000).

A few days after a partial denervation of a muscle, the remaining axons begin sprouting at the terminals and the last node of Ranvier (collateral sprouting), but the capacity of this process decreases in old age (Aoyagi and Shephard 1992). Denervation leads to extrajunctional expression of AChR molecules and the muscle-specific **receptor tyrosine kinase** MuSK, paving the way for synapse formation outside the original endplate area, but the original synaptic site retains certain qualities that make it a preferred site of reinnervation (Bennett 1999). If extrajunctional reinnervation has taken place, it will be eliminated once a junction is reestablished in the old site.

### For Additional Reading

For details regarding synapse formation and regeneration, see M.A. Rüegg and Bixby (1998) or Kandel, Schwartz, and Jessell (2000, chapter 55).

The synaptic sites on a muscle fiber are unable to distinguish between its original axon terminals and sprouts that belong to other motoneurons. It is possible that such indiscriminate reinnervation may perturb the original directional preference of the reinnervated muscle fibers (Herrmann and Flanders 1998), thus contributing to the deterioration of finer motor control, especially in the small hand muscles.

Motoneurons that fail to form appropriate synapses ultimately die (Kandel, Schwartz, and Jessell 2000). Apparently, a factor within the muscle is necessary for the survival of the motoneuron. With age, there is a loss in muscle mass (sarcopenia), mainly due to a loss of muscle fibers. Although available evidence indicates that the reason for this is a primary death of motoneurons, a primary degeneration of muscle fibers cannot be ruled out. It should be noted that there are normally no signs of degenerated fibers in aging muscle. On the other hand, the muscle fibers per motor unit have been reported to increase in number, which may be interpreted as a reduction in motoneurons, partly compensated for by peripheral sprouting and reinnervation of the muscle fibers that lost their original motoneurons. To what extent changes in habitual physical activity can prevent this development is largely unknown (see also chapter 11).

### *Summary*

Functional regeneration of damaged nerves within the CNS is very limited, at least in adults. Plasticity within the CNS can often restore normal function, but the potential is limited. There are, however, indications that some neurons may retain their ability to reenter the mitotic cycle and that stem cells in the CNS may have a broader developmental potential than previously thought. So far, however, it is unknown whether and how these experimental results translate into restoration of function after injury.

In contrast, the prognosis for a regeneration of peripheral nerves is good in the presence of uninterrupted endoneurial tubes. In addition, denervated skeletal muscle fibers can be reinnervated by collateral sprouting from neighboring axons. Only those motoneurons that make successful contacts with their target muscle fibers survive.

## Autonomic Nervous System

The autonomic nervous system is the part of the nervous system that serves internal organs—that is, smooth muscle, heart muscle, and glands—a fact that has earned it the name **visceral nervous system** as well. The name *autonomic* derives from the fact that it is not controlled by conscious will. To a large extent it is a reflex system. In spite of that, the focus in many tutorial texts on the autonomic nervous system is on the efferent side of its peripheral part, largely ignoring the afferent link. The obvious reason for this is the anatomical design of the efferent link, with the typical differences between the **sympathetic** and **parasympathetic** parts. There has also been a tendency to regard the autonomic and somatic nervous system as totally separate entities, but although the peripheral parts of the two are easy to distinguish, their central parts are highly integrated. We start by looking at the peripheral part of the autonomic nervous system.

In spite of typical differences between the efferent parts of the sympathetic and the parasympathetic parts of the autonomic nervous system (the visceral efferent fibers), they share one feature that distinguishes them from the somatic efferent fibers. In the somatic system the spinal motoneurons send their axons all the way to their final targets, the muscle fibers. In the autonomic system, in contrast, the distance from the CNS to the target cells is spanned by a relay of two neurons with an intervening synapse. Groups of such synapses are found as more or less well-defined ganglia in different places around the body.

The differences between the sympathetic and the parasympathetic nervous system are most evident in their peripheral connections, particularly on the efferent side. In the sympathetic nervous system the ganglia are typically located rather close to the spinal cord, while in the parasympathetic system they are found close to the target (figure 4.36). Thus, the **preganglionic axon,** spanning the distance from the CNS to the ganglion, is rather short in the sympathetic system but long in the parasympathetic system. Consequently, the sympathetic **postganglionic axon** is longer than the parasympathetic one. There is also a considerable divergence, i.e., a spreading of impulses, in the autonomic ganglia, notably in the sympathetic ganglia, where the ratio of preganglionic to postganglionic neurons is about 1:20. Together with the length of the postganglionic axons, this accounts for the widespread effects so typical of the sympathetic nervous system.

The parasympathetic system accounts for the daily routines of the internal organs, while the sympathetic system is in charge of more urgent effects, typically flight-or-fight reactions. These systems' different effects on the target organs are due to the different signal molecules (transmitter substances) they use. The parasympathetic postganglionic neurons are **cholinergic;** that is, they use ACh as transmitter. In contrast, the sympathetic postganglionic neurons use noradrenaline. They are **noradrenergic.** The preganglionic neurons in both systems, on the other hand, use the same transmitter, ACh. This does not invalidate the differential effects of the two systems, however, since ACh in the autonomic ganglia serves as an excitatory transmitter between pre- and postganglionic neurons only and is unable to affect the final target organs.

So far, the story is a very simplistic one, but a few more features need to be added. First of all, we have omitted the role of the sympathetic system in the daily activities of the body. Second, the effect of sympathetic activity is fine-tuned by the differential distribution of the various types of adrenergic receptors, notably α and β with subtypes, in the internal organs. In addition, various types of neuropeptides colocalize with noradrenaline and acetylcholine in autonomic ganglia.

One of the major functions of the sympathetic system is to ensure sufficient blood pressure to provide high-priority organs such as the brain and the heart with adequate blood flow. This is done by a direct influence on the smooth muscle cells in the wall of arterioles. Increased impulse activity increases the vascular tone, resulting in a **vasoconstriction** and increased blood pressure. In addition, a sympathetically caused tachycardia increases cardiac output, also contributing to increased blood pressure. This is exactly what happens in a stressful situation, but in addition the vascular tone needs to be adjusted every time there is a change of body position, for example, when you rise from a supine to a standing position. Unless the vascular tone in the lower body is increased, a fall in blood pressure

The sympathetic nervous system

Ganglion

The central nervous system

Effector

Short preganglionic fiber with many collaterals

Long postganglionic fibers

The parasympathetic nervous system

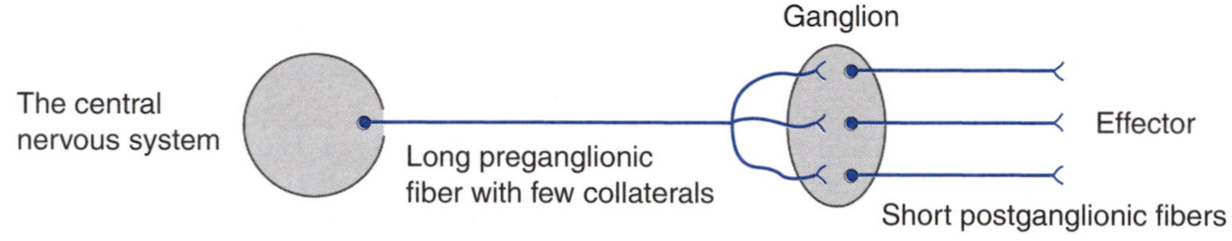


**Figure 4.36** Some major differences between the sympathetic and parasympathetic parts of the autonomic nervous system. In contrast to the somatic nervous system, the distance between the CNS and the effector is spanned by two neurons. The preganglionic neurons make synaptic contacts with the postganglionic neurons in autonomic ganglia. The preganglionic sympathetic neurons are rather short with many collaterals. In general, the sympathetic ganglia are located close to the CNS. The preganglionic parasympathetic neurons, on the other hand, are longer and have fewer collaterals, and the ganglia are located closer to the effector. Due to the high number of collaterals of the preganglionic fibers and the great length of the postganglionic ones, the effect of sympathetic nervous activity is widespread.

From Brodal, Dahl, and Fossum 1990.

(orthostatic hypotension) ensues, with dizziness and possibly fainting as a result.

In a fight-or-flight reaction, increased sympathetic activity causes widespread vasoconstriction and increased blood pressure as just outlined. Vasoconstriction in the muscles, however, is inappropriate since the increased muscular activity involved necessitates increased, not decreased, blood flow. The cells of adrenal medulla, which in fact are analogous to postganglionic sympathetic neurons, secrete adrenaline, which in turn can promote **vasodilation** in skeletal muscles. More important, however, local factors related to the metabolic activity in the working muscles contribute so powerfully to the vasodilation that they totally override any neural control of **vasomotor** tone.

The cell bodies of the preganglionic sympathetic neurons are found in the lateral horn of the spinal cord in the segments T1–L2, while those of the preganglionic parasympathetic neurons are located in cranial nerve nuclei in the brain stem as well as in sacral segments of the spinal cord. Because of the long preganglionic axons, the parasympathetic ganglia are typically found close to the target organs. In contrast, most sympathetic ganglia are found close to the spinal column.

The preganglionic sympathetic axons (the visceral efferent axons) leave the spinal cord through the ventral roots but part with the somatic efferent axons (the $\alpha$ and $\gamma$ fibers to the skeletal muscle fibers) soon after passing through the intervertebral foramen to enter a ganglion in the **sympathetic trunk.**

What is commonly called the sympathetic trunk is in fact two chains of ganglia, one on each side of the spinal column, extending almost all the way from the base of the skull to the pelvis (figure 4.37). The string connecting the ganglia consists of preganglionic axons running up or down the trunk before

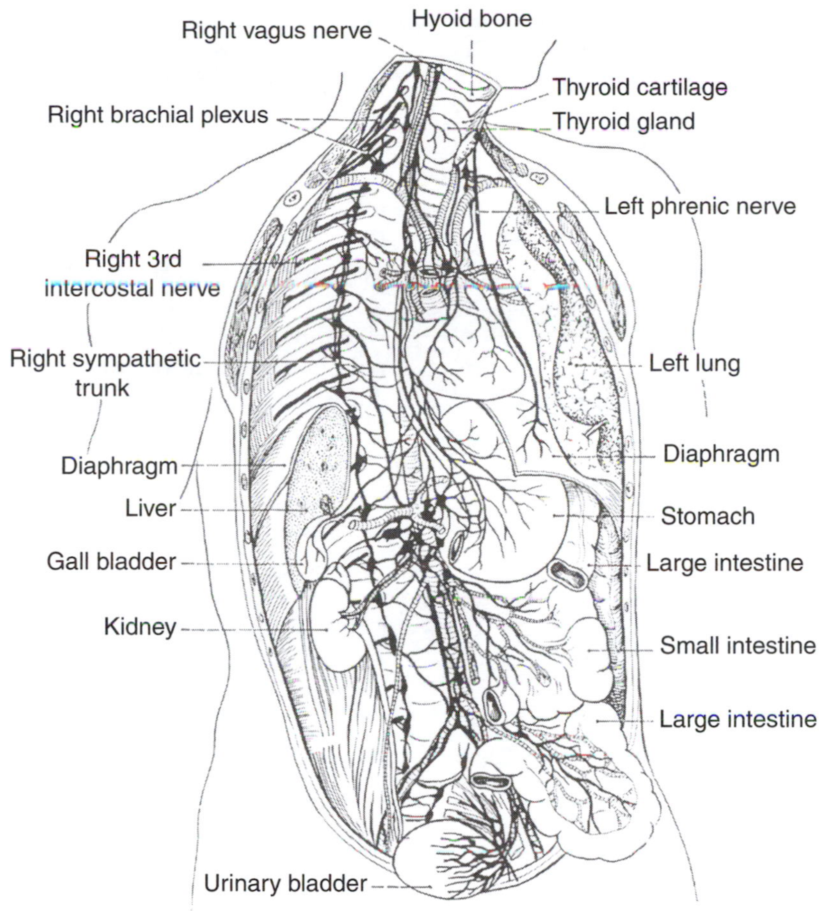


**Figure 4.37** The sympathetic trunk on the right side of body. The thorax and abdominal cavity are opened, the right lung has been removed, and the left pulled aside. Part of the diaphragm and the liver as well as most of the intestines are removed.

Reprinted, by permission, from A. Schreiner, 1966, *Menneskets anatomi og fysiologi* (Oslo, Norway: J.W. Cappelens Forlag).

establishing synaptic contact with postganglionic neurons. Such ascending and descending connections are necessary to ensure sympathetic innervation to the parts of the body innervated by spinal segments above T1 and below L2.

From the sympathetic ganglia the postganglionic fibers may take several routes to their target organs. Some, mainly fibers supplying the skin, rejoin the ventral root and follow branches of the spinal nerve to their final destinations. Others, mainly those in charge of vascular tone, form a plexus around arterial branches on their way to the resistance vessels (see chapter 5). Not all the preganglionic sympathetic fibers make synapses in the sympathetic trunk, however. Some just pass through to make synapses in so-called **prevertebral ganglia,** mainly in the thorax and the abdominal cavity.

In addition to the sympathetic and parasympathetic parts of the autonomic nervous system, there is a third part, called the **enteric system.** As indicated by its name, this part is found in the intestines. Groups of ganglion cells are found between the different layers of the intestinal wall. These ganglion cells were once regarded as postganglionic parasympathetic neurons, but now they are viewed as an integral part of the enteric system, which has emerged as a subdivision of the autonomic nervous system in its own right, largely governed by local reflexes. Admittedly, the classical parts of the autonomic nervous system still have some influence on the enteric system, as evidenced by changes in bowel function during periods of stress.

Another fact that challenges the traditional dichotomous division of the autonomic system into adrenergic sympathetic and cholinergic parasympathetic fibers is the finding of autonomic fibers that use neither adrenaline nor acetylcholine (Ach) transmitter. These fibers, mainly found in the enteric system, are called **nonadrenergic-noncholinergic (NANC)** and use other signal molecules, such as ATP and nitric oxide (NO).

The afferent autonomic fibers (visceral afferent fibers) enter the CNS through the dorsal roots together with the somatic afferent fibers from skin and the musculoskeletal system. Their cell bodies are found in the spinal ganglia or in corresponding ganglia in the cranial nerves. The distinction between sympathetic and parasympathetic afferent fibers is not nearly as apparent as on the efferent side, and there is good reason to regard visceral afferent fibers as serving both parts of the autonomic nervous system alike.

As mentioned earlier, the central part of the autonomic nervous system is difficult to separate from the somatic parts. Basically, the autonomic nervous system is based on reflexes, but the reflex responses are modulated by other parts of the CNS, including the cerebral cortex, the hypothalamus, and the periaqueductal gray. The hypothalamus serves a special role by establishing a link between the two signaling systems that regulate the activity of internal organs: the autonomic nervous system and the endocrine system.

CHAPTER 5

# BODY FLUIDS, BLOOD, AND CIRCULATION

The function of individual cells within the body depends on the constancy of their internal and surrounding environment. Claude Bernard (1813–78) recognized that an evolution of higher forms of organisms could not have taken place without the establishment of a stable **milieu interne,** its composition being guarded by regulatory mechanisms. His concept was later elaborated by Walter B. Cannon (1871–1945) using the term **homeostasis.**

The muscle cell is unique in regard to its ability to increase its metabolic rate. The maintenance of a constant *milieu interne* in the cell during the transition from rest to vigorous exercise necessarily represents, at times, a tremendous challenge to the circulation. The result can be that the muscle cell must cease its contractions or slow down its rate of activity because the changed composition and property of the fluid within or surrounding the cell interfere with the various processes that are necessary for the cell to perform under optimal conditions. In this chapter we present a general framework for understanding basic circulatory physiology and how the cardiovascular system contributes to the sustenance of optimal homeostasis during rest and exercise.

### FOR ADDITIONAL READING

For a more comprehensive and detailed treatment of cardiovascular control in health and disease, the reader is referred to Rowell (1993) and Saltin, Boushel et al. (2000).

## Body Fluids

In normal young adult men and women, the average **total body water (TBW)** accounts for 60% and 50% of the body mass, respectively. Because adipose tissue contains very little water, the TBW percentage varies inversely with the body's fat content. The relationship between TBW and fat-free body weight (lean body mass) is fairly constant; in an adult, TBW is about 72% of the lean body mass.

The TBW is distributed between two major compartments: the **intracellular fluid (ICF)** and the **extracellular fluid (ECF),** which are separated by the cell membrane. The extracellular fluid is subdivided into **interstitial fluid** and the liquid portion of the blood, the **plasma,** separated by the capillary endothelium (figure 5.1).

ICF accounts for about 40% of the body mass, and ECF about 20%. The interstitial fluid, the fluid surrounding the cells in the various tissues, makes up about three fourths of the ECF. Thus, the plasma amounts to about one fourth of the ECF, i.e., 5% of body mass.

ECF has an ionic composition claimed to be similar to that of ancient sea water. The ions in ECF and modern sea water are found in approximately the same relative proportions, although the total ionic concentration of sea water today is several times that of ECF. It has been postulated that this similarity suggests that ECF was originally derived from the ancient oceans, which were more dilute than those of today. The composition and the volume of water in the ICF and ECF compartments are under precise physiological control because they are of decisive importance for the optimal functioning of the cells.

## Blood

The blood and the lymph transport material between different cells or tissues. The blood

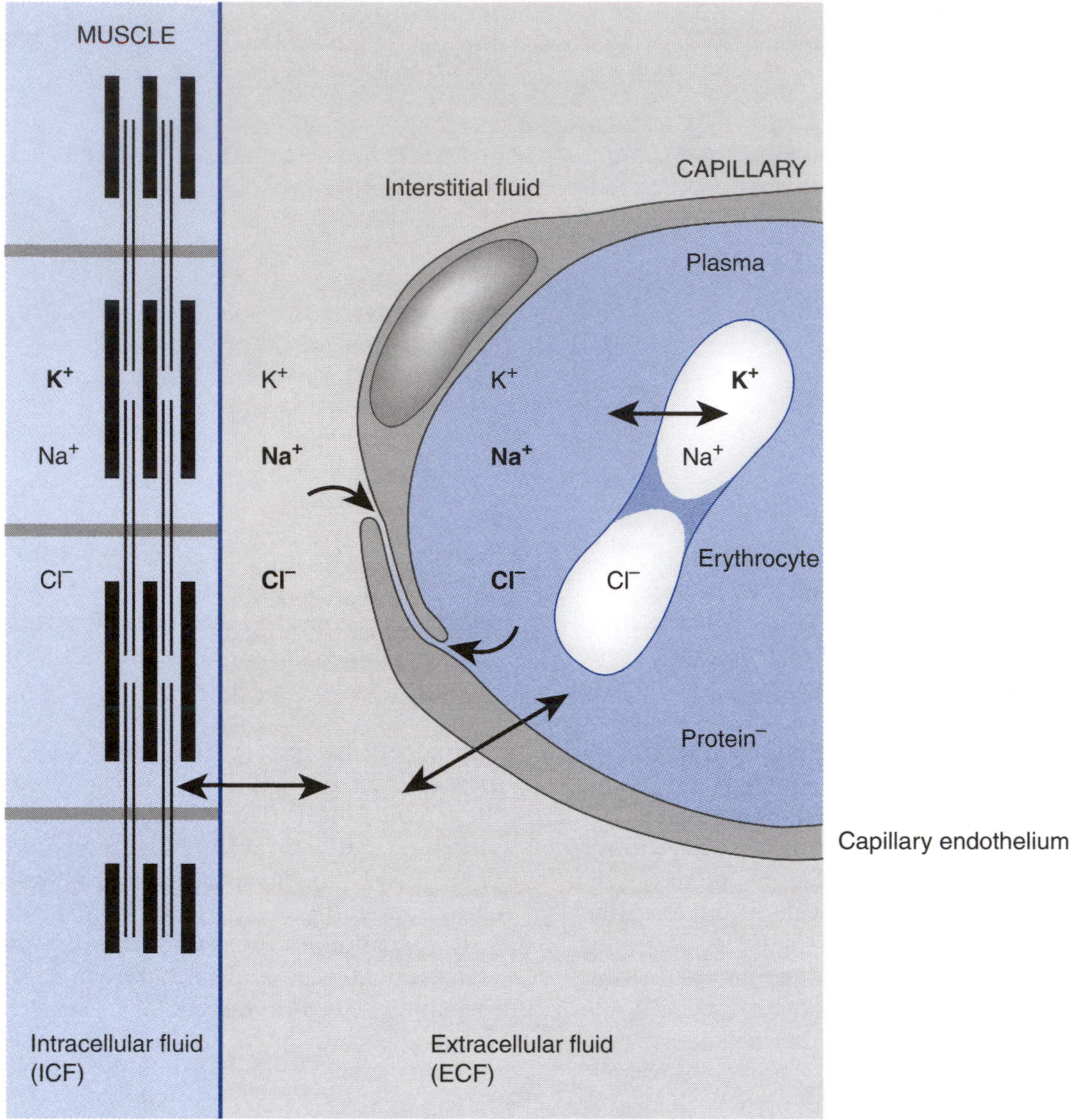


**Figure 5.1** Schematic drawing of capillary and muscle cell walls. There is a high concentration of sodium ion ($Na^+$) and chloride ion ($Cl^-$) outside the cells (boldface letters) and a high concentration of potassium ion ($K^+$) inside the cells (including the red cell).

brings nutrients from the digestive tract to the cells for catabolism, for synthesis of molecules in tissue structures, or for storage in depots that are later mobilized and redistributed. Heat and the chemical products of catabolism are removed, carbon dioxide is expelled through the lungs, heat is dissipated through the skin, and metabolites are transported to the liver and kidneys for further processing. Blood circulation has a key position in maintaining a proper water balance and fluid distribution. As a carrier of hormones and other active chemical agents, the blood can modify the function of cells and tissues in various ways.

## Volume

The most common procedure used to determine the blood volume in the human body is to inject into a vein a measured amount of a tracer substance that does not easily escape from the blood vessels. After some time, when complete mixing has taken place, a blood sample is secured to determine the concentration of the tracer substance, and the subject's volume of blood thus can be calculated. The total red cell volume can be estimated by reinfusion of a given volume of the subject's red cells labeled with radioactive $^{51}Cr$ or of plasma proteins labeled with $^{131}I$.

The individual variations in blood volume are large. A volume of 5 to 6 L for men and 4 to 4.5 L for women can be considered normal (about 75 and 65 ml · $kg^{-1}$ of body weight for men and women, respectively, and 60 ml · $kg^{-1}$ of body weight for children). The blood volume varies with the level of physical fitness, being larger in well-trained endurance athletes.

During heavy exercise there is a slight reduction in blood volume (discussed later in this chapter). The blood volume also can be reduced when the individual is dehydrated (e.g., because of diarrhea or profuse sweating and limited fluid intake).

## Cells

The solid elements of the blood that are visible under the ordinary microscope are red cells, or **erythrocytes;** white cells, or **leukocytes;** and platelets, or **thrombocytes.** When a blood sample in a tube containing an anticoagulant is centrifuged, these elements are separated from the plasma. The percentage of corpuscles in blood is known as the **hematocrit.** For men the hematocrit averages 47% (42%–54%), and for women and children it averages 42% (38%–46%).

The erythrocyte is a biconcave circular disk without a nucleus. It has an average diameter of 7.3 μm and a thickness of 1 μm in the center and 2.4 μm near the edge (see figure 5.1). Their number in an adult man averages 5.4 (4.6–6.2) million per 1 μl of blood; in women and children it averages 4.8 (4.2–5.4) million per 1 μl of blood. The life span of a red cell is about 4 months. Red cells are formed in the bone marrow at a speed that normally matches their rate of destruction. It can be estimated that the formation of erythrocytes proceeds at a rate of 2 to 3 million per second. The production of the red cells (**erythropoiesis**) is regulated by the hormone **erythropoietin (EPO),** which is secreted into the blood mainly by a special group of connective-tissue cells in the kidneys. EPO stimulates the production of erythrocytes in the bone marrow. The usual stimulus for EPO secretion is **hypoxia,** but the secretion of the hormone also can be stimulated by androgens and possibly other hormones. Besides the nutrients needed to synthesize any cell, the production of normal erythrocyte numbers requires the presence of iron and certain growth factors, including folic acid and vitamin $B_{12}$. The increase in hemoglobin concentration in the blood of people acclimatized to high altitude is a consequence of increased production of EPO.

The color of the red cell (and hence of the blood) is attributable to its content of **hemoglobin (Hb).** Hb is a globular molecule made up of four subunits. Each subunit consists of a polypeptide with a molecular group called **heme** attached to it. The polypeptides are referred to collectively as the **globin** portion of the Hb. The heme groups contain one atom of iron each, to which one molecule of oxygen binds. Thus, a single Hb molecule can combine with four oxygen molecules. This combination is not a chemical oxidation but is rather loose and reversible. In an oxygen-free medium, the Hb is reduced, and with oxygen available, it forms **oxyhemoglobin ($HbO_2$).** Another property of Hb is its affinity for carbon monoxide (CO), which is about 250 times greater than its affinity for oxygen. Exposure to CO results in a proportionally decreased capacity to take up oxygen. This phenomenon explains the high toxicity of CO.

In adult men, the average Hb content is 15.8 g per 100 ml of blood; in women it is 13.9 g per 100 ml. From a statistical point of view, a normal range (within which 95% of all people fall) is from 14.0 to 18.0 g for men and 11.5 to 16.0 g for women per 100 ml of blood. Each gram of Hb can maximally combine with 1.34 ml of oxygen. With 15.0 g of Hb per 100 ml of blood, 1 L of fully saturated blood can carry 0.201 L of oxygen plus some 0.003 L of oxygen dissolved in the plasma (an oxygen pressure of 100 mmHg, or 13.3 kPa).

The concentration of Hb in blood can be determined by two methods. The method used for routine measurements is based on the typical and specific light-absorption spectra of Hb and its derivatives. In the second method, the blood sample is saturated with oxygen, and the content of oxygen in a measured volume of blood is determined. The volume of dissolved oxygen can be calculated, and because the maximal oxygen-combining power of 1 g of Hb is fixed at 1.34 ml, the concentration of Hb in the sample easily can be calculated. In addition, a rough estimate of the Hb content of blood can be obtained from the hematocrit value, or from blood counts, with the presumption that each red cell has a normal Hb content. (For further discussion, see Shaskey and Green 2000.)

## Plasma

The plasma occupies about 55% of the total blood volume. It contains about 7% solutes and 93% water. The concentration of protein in the plasma is 60 to 80 g · $L^{-1}$. Three broad groups of proteins are present: albumins and **globulins,** which have many overlapping functions, and **fibrinogen.** The albumins are the most abundant of the three plasma protein groups. If blood is permitted to clot, the fibrinogen, in the presence of **thrombin,** forms **fibrin.** The fluid remaining when fibrinogen and other proteins have been removed as a result of blood clotting is called **serum.**

The plasma proteins have several important functions: Fibrinogen is necessary for blood coagulation, and among the globulins the gamma globulins serve a vital function as antibodies. All fractions, but especially the albumins, have the important transport function of carrying molecules to sites of need or elimination. The plasma proteins are active in buffer action and, in a way, constitute a mobile reserve store of amino acids. Finally, they play an important role in plasma volume and tissue-fluid balance.

All plasma proteins can pass through the capillary endothelium in small amounts, the albumins most readily. However, because most of the protein molecules are kept within the capillaries, they exert an osmotic pressure of about 25 mmHg (3.3 kPa). This so-called **colloid-osmotic pressure** is an important factor in the exchange of fluid between the intravascular and interstitial spaces.

The total amount of electrolytes in the plasma is 9 g · $L^{-1}$. The main ions are sodium ion ($Na^+$) and chloride ion ($Cl^-$). The plasma normally contains about 5 mmol · $L^{-1}$ (100 mg per 100 ml) of glucose. It also contains free fatty acids, amino acids, hormones, various enzymes, and about 25 different electrolytes in varying amounts.

## Viscosity

**Viscosity** can be defined as the resistance that a liquid exhibits to the flow of one layer over another. In tubes such as blood vessels, laminar layers of the fluid move at different speeds, causing a velocity gradient in a direction perpendicular to the wall of the tube. This velocity gradient is called the rate of shear. The faster layer tends to drag along and to be held back by the slower layer. The viscosity of blood is not a consequence of friction between the blood and the vessel wall but is attributable to the friction between adjacent laminae in the fluid. In a homogeneous fluid, the viscosity is constant at different rates of laminar flow. However, blood is definitely of a heterogeneous composition, and its viscosity varies depending on flow rates. Its viscosity depends mainly on the plasma proteins and particularly on the corpuscular components. At moderate or high shear rates, blood with a normal hematocrit of 45%, measured in vitro, is four to five times as viscous as water. An increase in hematocrit increases the blood viscosity. This is important in relation to blood doping. The use of the hormone EPO to improve **maximal aerobic power** can bring the viscosity to dangerous levels, increasing the risk of **thrombosis.**

The plasma viscosity in healthy subjects is about 2.2, relative to the viscosity of water, which is 1. However, blood has an anomalous viscosity. The red cells tend to accumulate along the axis of the blood vessels, which leaves a zone near the wall relatively free of cells. This axial accumulation of the erythrocytes can result in small side branches of a blood vessel containing a volume of red cells that is considerably less than that for mixed blood. This phenomenon is called **plasma skimming.** Another effect is a lower blood viscosity than expected. The axial accumulation of cells is more pronounced with an increase in velocity of the blood. However, in the physiological range of flow, the axial accumulation is complete, and therefore the effective viscosity is relatively constant. In capillaries (diameter about 6 µm), the highly deformable red cells (diameter about 7 µm) can be squeezed through single file, with a bulletlike configuration. Therefore, varying quantities of red cells can pass through the capillary per unit time with little effect on blood viscosity.

A third factor influences the effective viscosity of blood. In the very narrow vessels (arterioles and capillaries), the blood behaves as if the viscosity were reduced (called the Fåhraeus-Lindquist effect), which also contributes to reducing the demand on the heart to produce a driving force. This effect is of great importance during heavy muscular activity.

When measured in vivo, the effective viscosity of the blood passing through capillaries is, for the reasons just mentioned, approximately 50% of that in large vessels and is reduced to about the same as that of plasma, that is, 2.2 relative units.

The viscosity of the blood varies with its temperature. At 0 °C (32 °F), the viscosity is 2.5 to 3 times higher than at 37 °C (98.6 °F). This contributes to reduced circulation in tissues exposed to cold, as in the case of frostbite.

## Buffer Action, Blood pH, and Carbon Dioxide Transport

Apart from its many other functions, the blood also acts as a **buffer.** A buffer solution contains an acid or base that is only slightly ionized (weak) and a highly ionized salt of the same acid or base. A weak acid (HA) gives an equilibrium:

$$HA \leftrightarrow H^+ + A^- \quad (1)$$

If a strong acid is added to the solution, the hydrogen concentration increases, pushing the reaction to the left. As long as the buffer salt can provide $A^-$ ions, thereby forming undissociated HA, a change in the pH of the solution is prevented, or buffered.

At rest, the pH of arterial blood is 7.40 and of mixed venous blood approximately 7.37. (At a pH of 7.40, the hydrogen concentration is 0.00004 mM.) In the catabolism of the cells, carbon dioxide ($CO_2$) is formed, and in the case of anaerobic metabolism, there is a net production of hydrogen ion ($H^+$), with the hydrolysis of adenosine triphosphate (ATP) as the dominant source (see Busa and Nuccitelli 1984). The oxidation of phosphorus and sulfur in protein leads to the formation of phosphoric and sulfuric acids. The predominantly acid nature of the metabolites could easily explain the shift in pH as blood passes the tissue's capillary beds.

The **isoelectric points** of the proteins in the blood are on the acid side of the blood pH. Thus, the plasma proteins ionize as acids and form negatively charged anions ($NH_2$-R-COOH $\leftrightarrow$ $NH_2$-R-$COO^-$ + $H^+$, where R symbolizes the protein radical; for simplicity we can write protein$^-$ for the anion). The proteins therefore act as hydrogen acceptors in the blood. Proteins ionized as weak acids offer a buffer system: H-protein $\leftrightarrow$ $H^+$ + protein$^-$. The number of ionizable groups in the protein is large, and because whole blood contains about 190 g of protein per liter (with about 70 g in the plasma), its capacity to accept hydrogen without pH changes is considerable. In this respect, the potency of the plasma proteins is only one sixth that of Hb.

There is another reaction of considerable physiological importance. H-$HbO_2$ is a stronger acid than reduced H-Hb; that is, it dissociates more completely than does H-Hb. Hence, when blood is giving off oxygen, the $H^+$ concentration of the blood falls and pH rises.

Before we discuss the hydrogen exchange in the capillaries, we must draw attention to the second important buffer system of the blood. The $CO_2$ formed in the cells is dissolved and diffuses freely into the erythrocytes. Catalyzed by **carbonic anhydrase** present in the red cells, $CO_2$ with water forms $H_2CO_3$. This weak acid is dissociated as follows:

$$CO_2 + H_2O \leftrightarrow H_2CO_3 \leftrightarrow H^+ + HCO_3^- \quad (2)$$

The equilibrium of the reactions in this equation is determined by the concentration of the various molecules and ions. If free $H^+$ can be removed from the system, the reaction is pushed to the right. Potentially, this reaction gives place for more $CO_2$ in the solution without a change in the pH. Although more $H_2CO_3$ will be formed, this is only an intermediate step in the formation of $H^+$ and $HCO_3^-$. In this sense, the supply of a hydrogen acceptor actually determines the final equilibrium. Protein anions can serve as such a hydrogen acceptor. The $CO_2$ is produced in the tissue and diffuses into the red cell. At the same time, oxygen diffuses out to the tissue, where the oxygen concentration is lower than in the capillaries ($HbO_2 \rightarrow Hb + O_2$). Reduced Hb is a weaker acid than $HbO_2$, and when reduced, it binds some of the dissociated $H^+$ ions ($CO_2 + H_2O + Hb^- \rightarrow HCO_3^- +$ H-Hb).

We can now summarize the main functions involved in $CO_2$ transport from the muscle tissue to the lungs (figure 5.2):

1. The most important reactions in the oxygen and $CO_2$ exchange between blood and tissue are (a) the formation of a weak acid, reduced Hb, from the stronger one, $HbO_2$, and (b) the formation of $H^+ + HCO_3^-$ from $CO_2 + H_2O$, with carbonic anhydrase serving as an enzyme in the intermediate formation of $H_2CO_3$. The interplay between reactions (a) and (b) can be illustrated quantitatively by the following example: For each millimole of $HbO_2$ reduced, about 0.7 mmol of $H^+$ can be taken up, and consequently 0.7 mmol of $CO_2$ can enter the blood without changing pH.

   If the $CO_2$ were derived only from fat combustion, 0.7 mol of $CO_2$ would be produced per mole of oxygen used [**respiratory quotient (RQ)** = 0.7]. This means that the formation of reduced Hb from the oxygenated $HbO_2$ would completely buffer the $CO_2$ uptake. However, at rest, 0.82 to 0.85 mol of $CO_2$ is formed per mole of oxygen used. During heavy exercise, this figure is close to 1.00.
2. Some of the remaining $CO_2$ can combine directly with Hb, forming carbaminohemoglobin ($CO_2 + HbNH_2 \leftrightarrow HbNHCOOH$), and the simultaneous reduction of $HbO_2$ greatly favors this formation.
3. There is an increase in the volume of dissolved $CO_2$, so that the venous blood at rest contains about 0.003 L of $CO_2$ per liter more than the arterial blood.

Together, these three factors explain how the $CO_2$ produced at rest can be transported with the very small change in pH of 0.03 to the acid side.

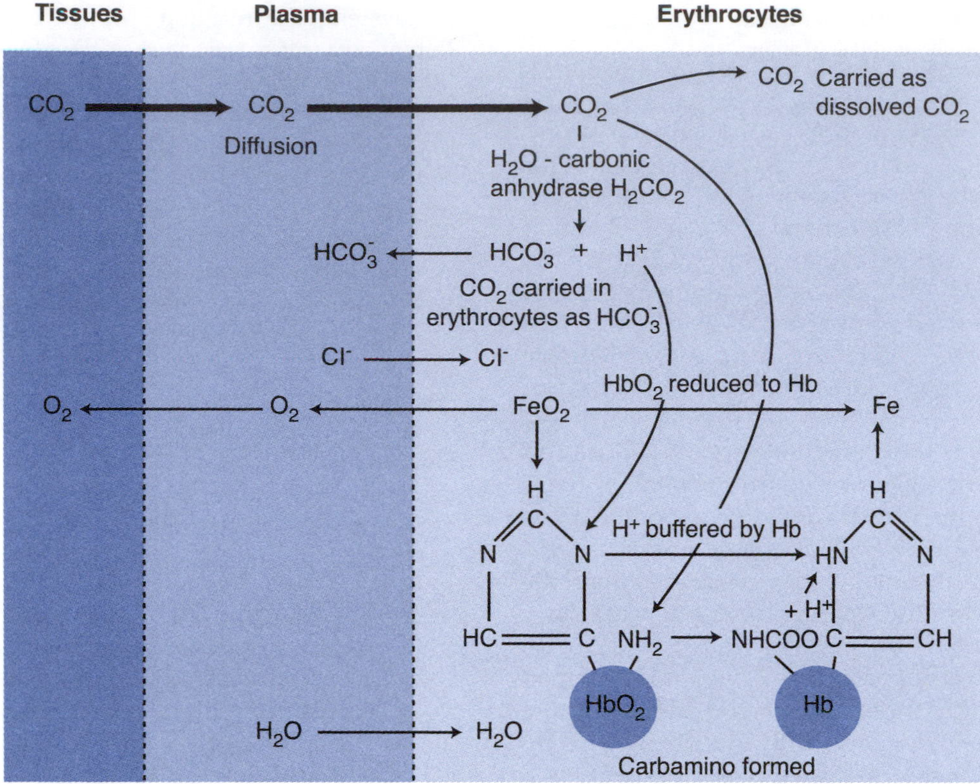


**Figure 5.2** Schematic presentation of the processes occurring when $CO_2$ passes from the tissues into the erythrocytes. At the bottom, the effect of oxygenation and reduction on the buffering action of hemoglobin (Hb) is illustrated. An increase in the acidity of the blood drives the reaction to the right, and oxygen is given off. (A decrease in the acidity of the solution would drive the reaction to the left, and oxygen would be taken up by reduced Hb.) In other words, the reduction of oxyhemoglobin ($HbO_2$) causes the Hb to become a weaker acid and to take up hydrogen ions from the solution.

From Davenport 1969.

The net result of the reaction under item 1 is an increase in hydrogen carbonate, or bicarbonate ions ($HCO_3^-$) in the red cells. $HCO_3^-$ can freely diffuse across the cell membrane and enter the plasma. Actually, there is a simultaneous decrease in the concentration of the anions as undissociated H-Hb is formed. The consequent excess of cations within the cell and lack of positive ions in the plasma cannot be compensated by a simple diffusion of a cation (in this case $K^+$) out to the plasma. The cellular activity prohibits an exchange of metallic cations. To restore the electrochemical equilibrium (Donnan equilibrium), $Cl^-$ ions diffuse into the cell, thereby replacing its loss of anions. This **chloride shift** allows about 70% of the formed $HCO_3^-$ to be transported in the plasma.

If these reactions are understood, the events that take place in the lung capillaries should be easily comprehensible. The reactions run "backward," and in an ingenious way the oxygen uptake and formation of the relatively strong acid $HbO_2$ facilitate the exclusion of $CO_2$. In brief, the increase in free $H^+$ with the dissociation of $HbO_2$ forces the reaction in equation 1 to the left. As the concentration of bicarbonate within the cell decreases, the plasma is ready to send in more $HCO_3^-$ ions. In exchange, the plasma gets back the $Cl^-$. At the same time, carbaminohemoglobin gives up some of its $CO_2$. The degree of these reactions depends on the oxygen and $CO_2$ tension of the alveolar air.

At rest, about 0.2 L of $CO_2$ is produced per minute in the tissues, and the cardiac output to transport this volume is about 5 L · min$^{-1}$. Thus 1 L of blood transports 0.040 L of this $CO_2$. The $CO_2$ content of 1 L of arterial blood is still about 0.5 L after delivery of $CO_2$ to the lungs. About 0.025 L is dissolved in the plasma, which is in equilibrium with a $CO_2$ tension of 40 mmHg (5.3 kPa). Most of the remaining $CO_2$ is in the plasma in the form of bicarbonate.

The buffer system composed of $CO_2$ and $HCO_3^-$ is, in itself, of relatively low capacity. However, because $CO_2$ can be expired and actually stimulates the respiration, the physiological importance is significant. For simplicity, the buffer action can be illustrated with the following example. In heavy muscular exercise, lactic acid (HLa) is formed. When HLa diffuses into the blood, the reaction

$$Na^+ + HCO_3^- + H^+ + La^- \leftrightarrow Na^+ + La^- + H_2CO_3 \quad (\leftrightarrow H_2O + CO_2) \qquad (3)$$

goes to the right because $H_2CO_3$ is a weaker acid (a stronger acceptor of $H^+$) than lactic acid. The increase in free $CO_2$ and the decrease in the pH of the blood stimulate increased ventilation, thereby decreasing the $CO_2$ content of the blood and body. This causes the buffer base to decrease because of the release of "volatile" carbonic acid anhydride. Because the lactic acid is later oxidized or transformed into glycogen, the blood becomes more alkaline.

It was mentioned that the pH of the arterial blood is close to 7.4 at rest. In the muscle cell, however, it is much lower, approximately 7.00. After exhaustive exercise, the arterial blood pH can fall below 7.00, and inside the exercising muscle it can decrease to 6.4 (Sahlin et al. 1978; Sahlin and Henriksson 1984). It is mainly the anaerobic breakdown of glycogen and glucose that causes this pH change. The intracellular bicarbonate concentration has been reported to decline from 10.2 $mmol \cdot L^{-1}$ to about 3 $mmol \cdot L^{-1}$. The intracellular pH was back to normal after 20 min of recovery, whereas the bicarbonate concentration was still well below normal (Sahlin et al. 1978). Exactly what factors cause exhaustion in exercise of only a few minutes duration are not fully known. Possible explanations are discussed in chapter 15.

Various proteins play an important role in the powerful buffer capacity of the blood. Because the protein content of many tissues is high, we should expect that such tissues have a good potential for contributing to the acid-base equilibrium. In fact, it appears that about five times as much acid is neutralized by other tissues than by the blood. Actually, the intracellular proteins can take a 50% share in buffering acids. The body of an average man might neutralize one equivalent (in other words, 1 L of a 1.0 M solution) of a strong acid, such as HCl, before the blood pH would fall below 7.0, close to the lowest pH value compatible with life. Without buffers, the pH would be about 1.6!

The buffer capacity of blood (and tissues) must be considered as a "first aid" service not capable of maintaining a constant acid-base equilibrium for long periods. It is no problem to eliminate the potential acid produced in the cell respiration; in one day, 12 mol of $CO_2$ is produced, with the potential to yield 12 mol of **protons,** but the $CO_2$ easily can be removed by expiration.

There is, however, the metabolic production of "nonvolatile" acids and the formation of phosphoric and sulfuric acids in the degradation of proteins. Depending on the diet, the production of protons can be 40 to 80 mmol daily. With these protons remaining in the body, the blood pH would fall to about 2.0 in one day. (A very protein-rich diet can contribute to the formation of 150 mmol per day.) Here the kidney function stands as a final guard over the body pH. This is accomplished by several mechanisms:

1. In the tubuli and the collecting ducts, a certain amount of $H^+$ is secreted into the lumen, increasing the acidity of the urine. This increase may cause the pH of the urine to drop to as low as 4.5. If the $CO_2$ content of the body fluid increases, more $H_2CO_3$ is formed in the kidneys and in turn is dissociated into $H^+ + HCO_3^-$. The hydrogen ions diffuse through the tubuli wall into the tubular fluid; $HCO_3^-$ diffuses more or less back into the blood.
2. In the tubuli there is also a secretion of ammonia, $NH_3$, which together with $H^+$ may form $NH_4^+$, thereby trapping hydrogen ions. (Ammonia actually is generated from the amide glutamine, produced in the muscle tissue.) If this proton were to be eliminated as, for instance, $H^+ + Cl^-$, the acidity would be intolerable. But as $NH_3 + H^+ + Cl^-$, the elimination takes place in the form of a neutral salt $NH_4Cl$.
3. In the blood there are phosphates, like $Na^+ + Na^+ + HPO_4^{2-}$, which, during the passage through the tubuli, pick up hydrogen ions, forming $Na^+ + H_2PO_4^-$. $Na_2HPO_4$ is more alkaline than $NaH_2PO_4$, and in the tubuli, the ratio $Na_2HPO_4/NaH_2PO_4$ may be varied according to the need for the elimination of $H^+$ (figure 5.3).

The formation and excretion of acidic phosphate, as well as the other two processes, help the body conserve sodium. On the other hand, excess alkali can be eliminated with the urine as bicarbonate or basic phosphate. If the amount of nonvolatile acids should increase, because of diet or metabolic disturbances, normal kidneys have the capacity to increase their rate of $H^+$ elimination. A reduction in arterial blood pH is called **acidosis;** an abnormally high arterial blood pH is called **alkalosis** (below or above 7.35 and 7.45, respectively).

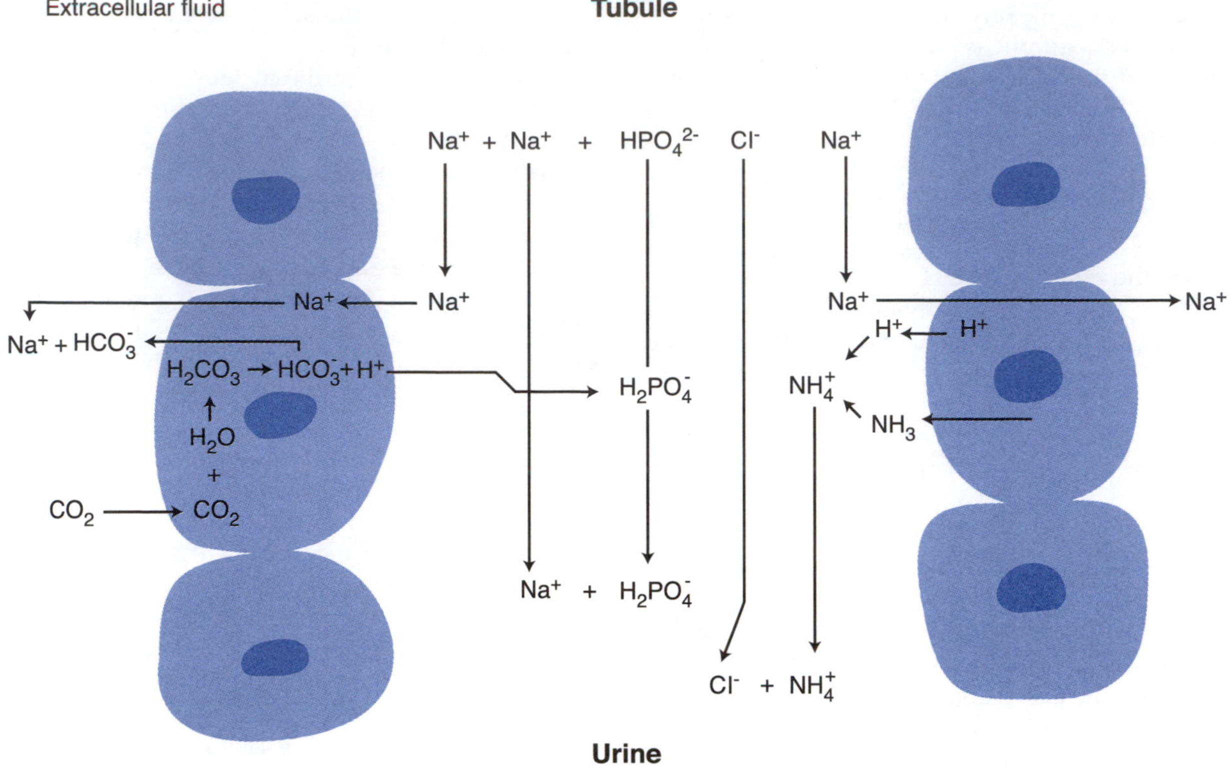


**Figure 5.3** To the left is illustrated how one $H^+$ is "trapped" from $H_2CO_3$ by the formation of $H_2PO_4^-$ from $HPO_4^{2-}$. In exchange another cation, $Na^+$, is leaving the $Na_2HPO_4$ complex, crossing the tubular cells into the extracellular fluid. The net effect is an ongoing passage of $Na^+ + H_2PO_4^-$ in the tubule. To the right is a simplified illustration of how $NH_3$ is secreted by the tubular epithelial cells with the formation of $NH_4^+$ by the inclusion of $H^+$. In this case, $Na^+$ also can be transferred to the extracellular fluid and eventually back to the blood.

Ingestion of sodium bicarbonate ($NaHCO_3$) may produce a significant **ergogenic** effect under specific exercise conditions, in particular during maximal workouts lasting from 1 to 7 min. This effect could be attributable to increased efflux of lactate and $H^+$ from the intracellular compartment of the working muscle to the extracellular fluid. However, gastrointestinal discomfort caused by $NaHCO_3$ ingestion can reduce performance ability (for further discussion, see Linderman and Gosselink 1994).

## *Summary*

The blood concentration of $H^+$ is low, with a pH around 7.40 at rest. The metabolic rate of $H^+$ production is high, but buffer systems can keep pH fluctuations at a very moderate level. In the healthy individual, vigorous muscular activity can markedly disturb the "normal" pH, and the intracellular pH can decrease to 6.4 and the arterial blood pH to levels below 7.0. An important aspect of the buffer systems is the dissociation of the weak carbonic acid $H_2CO_3$ into $CO_2$ and $H_2O$. An increase in arterial $CO_2$ and $H^+$ concentrations stimulates an increased pulmonary ventilation. Certainly, one cannot expire hydrogen ions, but the expiration of $CO_2$ has a similar effect, that is, it keeps the blood $H^+$ concentration low.

The proteins in the blood and tissues form powerful buffer systems. The difference in acidity between Hb and $HbO_2$ minimizes the variation in the pH of the blood as it passes the tissue, takes up $CO_2$, and subsequently releases it through the lungs. In the kidneys, excess $H^+$ produced in the metabolism of "nonvolatile" acids can be excreted as free ions or ions bound to $NH_4^+$ and as $NaH_2PO_4$. In other words, there are three strategies or lines of defense against increasing acidity: buffers, respiration, and renal excretion of protons.

# Heart

The heart is a four-chambered pump composed of a special type of muscle called myocardium and enclosed in a fibrous sac, the pericardium. To recall the basic anatomy of the heart, see figure 5.4a.

## Cardiac Muscle

Under a microscope, the myocardium shows transverse striations essentially similar to those of the skeletal muscle, but the nuclei are placed centrally within the cell; the fibers give off branches that anastomose with adjacent fibers, giving an impression of a syncytial continuity of the muscle fibers. However, electron microscope studies give a clearer picture of the structure. Apparently, a heart muscle fiber is composed of individual mononucleated muscle cells that are joined together in a linear array. The muscle cells are attached to each other by the so-called **intercalated disks,** which traverse the muscle fiber with around 10-μm intervals. Thus, the heart muscle does not represent a true syncytium, albeit a functional one, because gap junctions are important components of the intercalated disks, providing ionic continuity between adjacent muscle cells.

The cell mass of the left ventricle is approximately three times that of the right ventricle. Basically, the mechanism behind the contraction of the heart muscle fiber is similar to the events described for the skeletal muscle (chapter 3). There are, however, important differences in the way contractions are initiated. Whereas a skeletal muscle fiber needs impulses from the nervous system, a cardiac muscle cell is spontaneously active. The cells in the sinus node have a higher frequency than cells in other parts of the heart, allowing them to take command and act as the pacemaker of the heart. The conductive system of the heart and the effect of the refractory period after an excitation give the heart its rhythmic activity, unceasing from early embryonic life until death.

### For Additional Reading

For details about similarities and differences in excitation–contraction coupling in skeletal and cardiac muscle, see Wasserstrom (1998).

## Blood Flow

Normally the blood vessels in the heart do not form anastomoses if the diameter is larger than 40 μm (the size of small arterioles). Therefore, in experiments on animals, a sudden occlusion of an artery will be followed by an infarction of the tissue supplied by that artery. On the other hand, a gradual narrowing and final occlusion of a coronary artery will promote the development of a very rich network of anastomotic vessels. Studies indicate that effective collateral vessels develop more profusely if the animals are exercised, whereas training without concomitant coronary

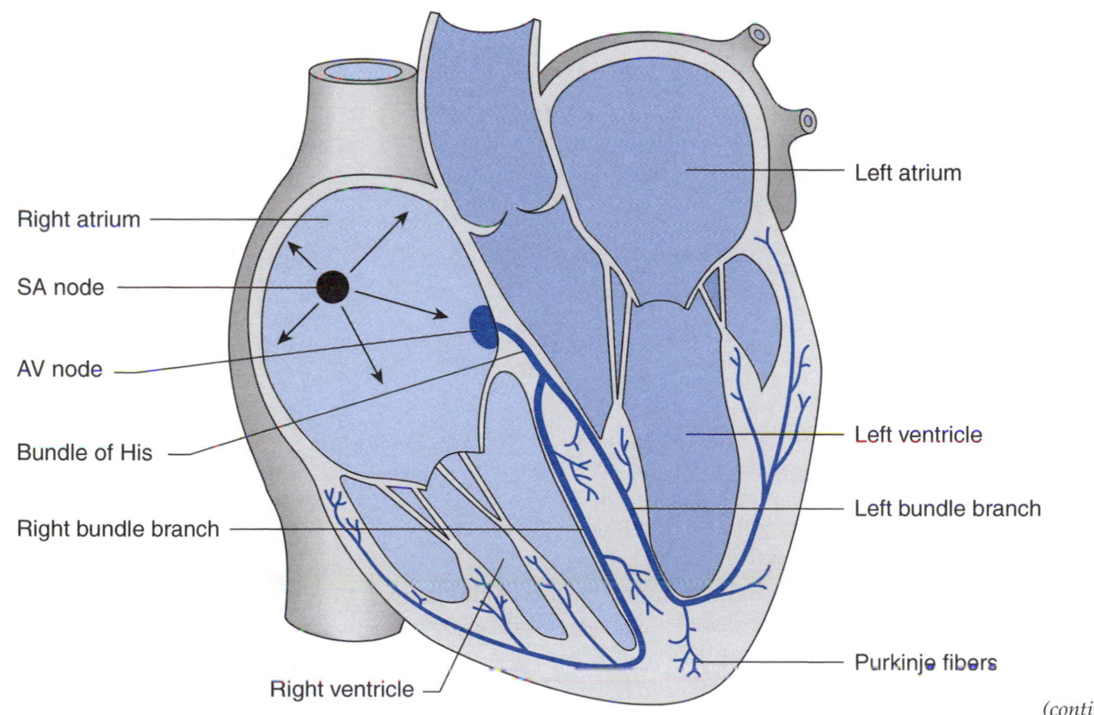


(continued)

**Figure 5.4** *(a)* Electrical impulses that cause regular, rhythmic contractions of the heart muscle originate in the sinoatrial (SA) node. The impulses spread along the atria and are conducted to the ventricles by way of the atrioventricular (AV) node, the bundle of His, the left and right bundle branches, and the Purkinje fibers.

artery constriction apparently does not improve collateral blood flow (Feigl 1983). Clinical experience shows that a collateral circulation can develop in cardiac patients (Gregg 1974). The capillary network in the heart is extraordinarily rich, with 2,500 to 3,000 capillaries per square millimeter. The capillary density increases by training (see chapter 11).

The contraction of the heart muscle interferes mechanically with the coronary blood flow. In a dog at rest, the left ventricle completely stops during the isometric contraction and a backflow actually is established, whereas a peak flow is reached when the systolic pressure is at its maximum, during the early ejection phase. A second peak flow is reached during early diastole, after which it decreases gradually to about 70% of maximum just at the end of diastole (see figure 5.4b). In the right ventricle, where pressure is lower during systole, the fluctuations are not so pronounced as in the left ventricle.

At rest, the coronary artery flow during systole is less than 20% of the total blood flow per cardiac cycle. During strenuous exercise, however, up to 40% of the total coronary blood flow takes place during systole, because systoles account for a higher fraction of the cardiac cycle.

The coronary blood flow increases a few heartbeats after the onset of exercise. Oxygen deficiency has a strong dilating effect on the arterioles of the heart. One causative factor for the vasodilation may be the nucleoside adenosine (Knabb et al. 1983). During the contraction of the heart muscle, ATP breaks down to adenosine diphosphate (ADP) as well as adenosine monophosphate (AMP). Some AMP is enzymatically catalyzed to adenosine, which can leave the cardiac cell, diffuse through the interstitial space, and reach the arterioles to cause vasodilation. Other factors may contribute, (e.g., $CO_2$ produced in the aerobic metabolism). $K^+$ diffusing out of the depolarized muscle cell can have a transient vasodilating effect at the onset of a **tachycardia** (increase in heart rate). Furthermore, there is increasing evidence that endothelium-derived **nitric oxide (NO)** influences vascular tone in the periods between exercise bouts. In animal studies, reactivity to stimuli that mediate their effects via NO is increased by training in both skeletal muscle and the coronary circulation. Human data confirm that rest-

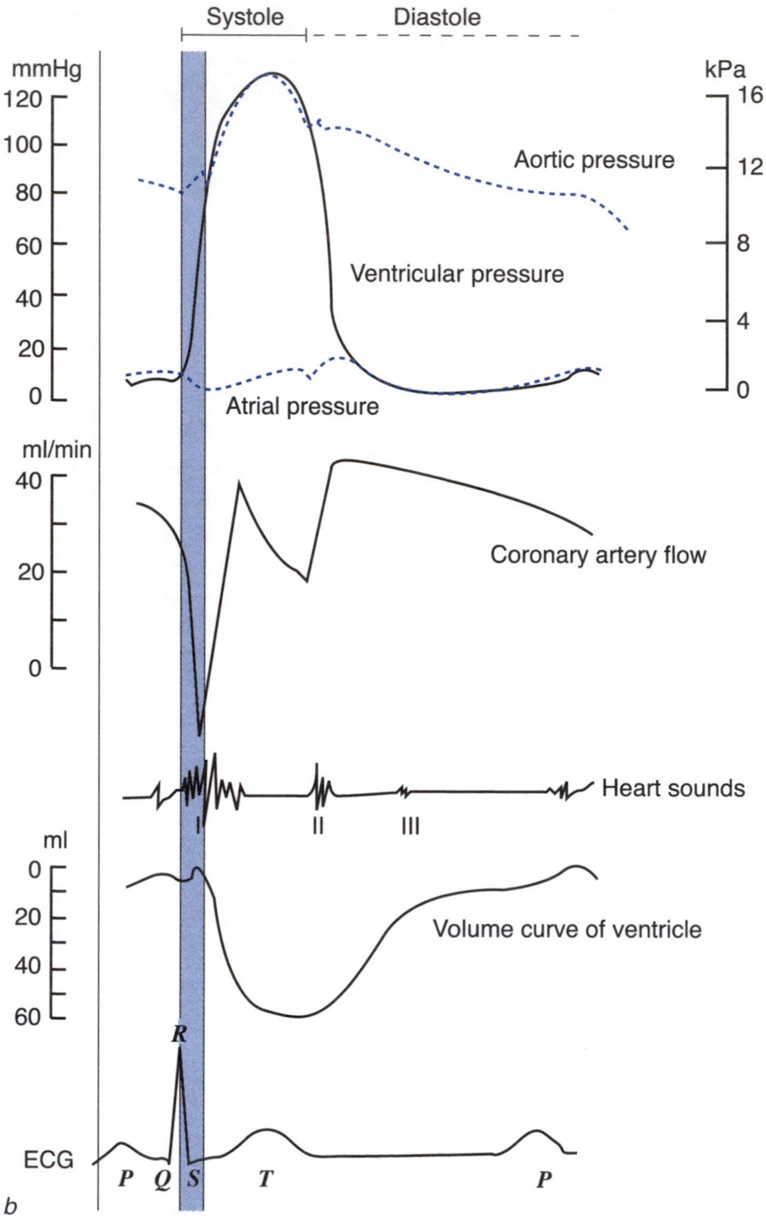

**Figure 5.4** *(b)* Pressure variations in the left heart chambers and aorta during the cardiac cycle. The heart sounds can be objectively analyzed under various pathological conditions by phonocardiography. Closure of the atrioventricular valves contributes largely to the occurrence of the first sound (I), but so do vibrations in the tissues and turbulence of the blood flow; closure of the aortic valves is of prime importance for the occurrence of the second heart sound (II); vibrations of the chamber walls caused by movement of blood into the relaxed ventricle produce the third sound (III).

ing NO production is increased during the early phases of training adaptation and that endothelium dependent dilator reserve is elevated in the peripheral circulation of highly trained athletes (Kingwell and Jennings 1998). Normally, there is an excellent match between myocardial metabolism and the coronary perfusion. Even at rest, a relatively large part of the oxygen offered to the tissue, or 60% to 70% of the arterial content, is used in the mitochondria; during maximal exercise, the arteriovenous oxygen difference is further increased by 10% to 20%. There is no evidence for an exhaustion of the coronary vasodilation reserve in healthy people even when exercising at maximal rate.

Studies of trained male subjects have shown that the myocardial oxygen demand and coronary blood flow are relatively low at rest as well as during a given work rate. A reduced heart rate could be o ne explanation of these findings (Jensen-Urstad et al. 1997).

### *Summary*

In healthy individuals, coronary blood flow and cardiac metabolic demands are well matched. The heart rate is well correlated with coronary blood flow. Adenosine and other factors like $CO_2$, $K^+$, and endothelium-derived NO act as local metabolic vasodilating transmitters between the cardiac cells and the coronary vascular smooth muscle cells. There is evidence that resting NO release is increased by physical training.

## Pressure Variations During a Cardiac Cycle

Figure 5.4b illustrates the pressure variations in the left ventricle as well as in the aorta. The systole starts with an isometric contraction, because the mitral valves close rapidly as the pressure in the ventricle exceeds that of the atrium (phase within the vertical lines). Within about .05 s, the ventricular pressure is increased above the level in the aorta, and the aortic valves open. The isotonic contraction increases the pressure further. The peripheral resistance does not permit the same volume of blood to escape from the aorta as is ejected into it. Part of this volume is "stored" in the distended aorta and its large branches. Then, as the pressure decreases in the ventricle during the relaxation of the muscle, the aortic valves close, and the elastic property of the aortic wall can propel the stored blood out into the arterial tree. The intermittent energy outbursts of the heart would give an intermittent flow if the vessels were rigid tubes. However, part of the potential energy is taken up by the elastic arterial wall and then released during diastole of the heart, keeping the hydraulic energy level close to the heart continuously high.

During the **systole,** blood returns to the large veins close to the heart and the atrium. It is possible that the atrium passively lengthens as the ventricle contracts, which may facilitate the filling of the atrium. During **diastole** there is a period of rapid filling of the ventricle after the opening of the mitral valves. The period of diastasis follows, during which the filling is much less rapid. The next cycle begins with the atrial contraction, which more or less empties the atrium. At rest, the atrial contraction is responsible for approximately 20% of the total ventricular filling.

The physical events in the right heart are essentially similar to those just described, but the ventricular and pulmonary artery pressures during systole are about one fifth of those in the left ventricle and aorta, respectively. In the aorta of normotensive persons, the pressure at rest varies between 120 mmHg (16 kPa) during systole and 80 mmHg (10.5 kPa) at the end of diastole. In the pulmonary artery, the values are 25 and 7 mmHg (3.3 and 0.9 kPa), respective1y. The sphygmomanometer cuff technique does not always give the same value as measurement via a catheter in the artery. This holds true particularly during exercise. Excess subcutaneous tissue also can make the sphygmomanometer technique unreliable.

## Innervation of the Heart

The heart receives a rich supply of sympathetic and parasympathetic nerve fibers. Parasympathetic fibers, terminating in the region of the pacemaker (sinoatrial node), release primarily acetylcholine (Ach), which slows the heart rate. Sympathetic postganglionic fibers releasing primarily noradrenaline are distributed not only to the conducting system but to the entire myocardium. A sympathetic activation of the adrenal medulla will release the chemically related adrenaline and also some noradrenaline. As mentioned in previous chapters, the effects of transmitter substances are determined by the membrane receptors of the target cells.

On the basis of purely pharmacological criteria, we can distinguish two main groups of receptors, α- and β-adrenergic receptors (with subgroups $\alpha_1$ and $\alpha_2$, $\beta_1$ and $\beta_2$). The α receptors are abundant in the cell membrane of all vascular smooth muscle cells, and throughout the body α-adrenergic activity causes vasoconstriction. In the heart, however, the α receptors are relatively sparse, making noradrenaline and adrenaline to activate mainly β receptors. This increases the heart rate by increasing the firing rate of the sinoatrial node (sinus node) and increases the conduction velocity in the atria, the atrioventricular node, and the Purkinje system. The effect is therefore the opposite of stimulation of the parasympathetic nerve and is called **chronotropic action.** In addition, the $\beta_1$-adrenergic activity increases the myocardial contractility, that is, the strength of contraction at any given **end-diastolic volume.** This increases **stroke volume** at the expense of a reduced end-systolic volume. This effect, exerted by the catecholamines, is known as a positive **inotropic action.** Certainly, the venous filling of the heart determines the size of the consecutive stroke volumes, as discussed later.

An increased sympathetic drive will, as just mentioned, elevate the heart rate, and the heartbeat becomes more forceful. This will increase the myocardial oxygen uptake and coronary blood flow. Therefore, the net effect will be a dilation of the coronary vessels. The inherent rate of beating can be highly modified; the healthy heart in a young, fully grown individual can cover a range from about 40 beats · $min^{-1}$ at rest to 200 beats · $min^{-1}$ during heavy exercise.

$\beta_2$-adrenergic receptors are present in vascular smooth muscles, most abundantly in resistance vessels in skeletal muscles, and also in the myocardium. An activation will dilate the vessels. The $\beta_2$ receptors are sensitive to adrenaline but insensitive to noradrenaline.

# Hemodynamics

Chemical energy is transformed into mechanical energy (external work) plus heat by the contraction of the heart muscle. The mechanical efficiency of the heart, that is, the external work divided by the total energy exchange, is rather low, or only some 10% at rest.

The external work is calculated from data on force times distance moved or, for a fluid, pressure times volume moved. The pressure can create kinetic energy and a flow of the fluid. The higher the velocity of the fluid, the higher the kinetic energy (related to velocity squared). On the other hand, the fluid pressure against the wall is smallest in the part of a vessel where the velocity is highest, because total hydraulic energy = pressure energy + kinetic energy (Bernoulli's principle). If the pressure is measured in a vessel with a catheter or tube connected to a manometer with the opening against the flow, the kinetic energy of the flow is reconverted to pressure (end pressure) and the total energy is measured. With only a side hole in the catheter, the side or lateral pressure is measured, and it will be less than the end pressure by an amount equivalent to the kinetic energy of the flow, according to the formula given previously. At rest, the kinetic factor in the aorta flow is only about 3% of the total work of the heart, but during exercise with a cardiac output five times the resting level, the aorta flow may be an important part of the total work of the heart, possibly as much as 30%. This means that a simultaneous measurement of end pressure and side pressure in the aorta may show a difference of some 75 mmHg (10 kPa), the end pressure being highest (see subsequent discussion). In the other arteries and in the smaller vessels, the kinetic energy factor is normally negligible, at least at rest. It should be pointed out that the sphygmomanometer cuff method for the measurement of blood pressure only measures the side pressure.

## Hydrostatic Pressure

The pressure in a vessel is created by continuous bombardment by the molecules of the fluid against the inner surface of the vessel. The pressure is equal at all points lying in the same horizontal plane in a static liquid. The hydrostatic pressure in a fluid at rest, under the influence of gravity, increases uniformly with depth under the free surface (Pascal's law). It is evident that in the supine position, the hydrostatic pressure is of similar magnitude in all parts of the body, but in the standing person it is higher (perhaps as high as 100 mmHg or 13.3 kPa) in the arteries of the feet than in the head. In the veins of the lower extremities, the hydrostatic factor in a standing individual is modified by the action of the valves. Arterial pressure should be measured at the horizontal level of the heart, or the value should be corrected to correspond to heart level. For practical purposes, measuring blood pressure with a sphygmomanometer cuff around the arm of a seated subject gives satisfactory values, provided there is not too much subcutaneous fat.

When a person is tilted from a supine position to an erect position with the feet down, the veins, and to some extent the other vessels below heart level, will dilate passively as a result of the hydrostatic force. Temporarily the blood is pooled, but when the vessels are filled, the flow continues unhindered.

## Tension

The pressure within the heart and vessels will more or less distend the walls (figure 5.5). The resistance to this force, which produces tension, depends on the thickness of the wall and its content of elastic and collagen fibers and active smooth muscle fibers. The elastic tissue can balance the pressure without any energy output, but the contraction of the muscles requires a continuous expenditure of energy. The tension $T$ is roughly proportional to the pressure difference $P$ between the inside and outside of the wall (**transmural pressure**) and the radius $r$ of the vessel, or $T = aPr$, where $a$ is a constant (Laplace's law). Therefore, a given pressure applied in a small vessel does not produce the same tension as in a vessel with a larger radius (see figure 5.5). The very thin membrane of the capillary, essential for the exchange of materials, can withstand the capillary blood pressure because its radius is so small.

The same law can be applied, with some reservations, to the heart. With a given tension developed by the heart muscle, a lower pressure is produced if the heart is dilated (i.e., radii of curvature are increased). If the diameter of the heart is doubled, the tension per unit length of ventricular wall must be about twice as great to produce the same pressure. The energy cost, that is, the oxygen uptake, of the heart is related to the tension that must be developed and the time this tension is maintained. Therefore, an increase in the size of the heart increases the load on the heart.

Heart muscle follows the same law as the skeletal muscle: Its ability to produce tension increases with length. By applying Laplace's law, one finds that the heart has to pay more to maintain a given pressure if a reduction of muscular strength is compensated for by a dilation (increase in length of the fibers).

The maximal active force developed by the cardiac muscle is attained at a sarcomere length of about 2.2 μm, but the maximal active force decreases rapidly at sarcomere lengths both below and beyond this optimum length. Skeletal muscle fibers have more of a force plateau, with changes in sarcomere length of approximately 2.0 to 2.2 μm. The cardiac muscle is also relatively stiff and tension increases extremely rapidly as the muscle is stretched over a sarcomere length range for which resting tension is minimal in skeletal muscle (Sonnenblick and Skelton 1974).

## Flow and Resistance

Because the difference in the hydraulic energy of a fluid in two parts of a system cannot be resisted by the fluid, it flows with a rate proportional to the energy difference or pressure load. The resistance to flow of blood results from the inner friction or viscosity of the blood. A cohesive force between the blood and the wall of the vessel "retards" the flow of the layers close to the wall. The nearer the center of the vessel, the higher the speed of each lamina of fluid, and this "friction" phenomenon results in a maximal speed in the very center of the vessel. The hydraulic energy provided to the blood by the contracting heart muscle is gradually spent and transformed to heat. Fluid flow can be **laminar** (streamlined); that is, each lamina of

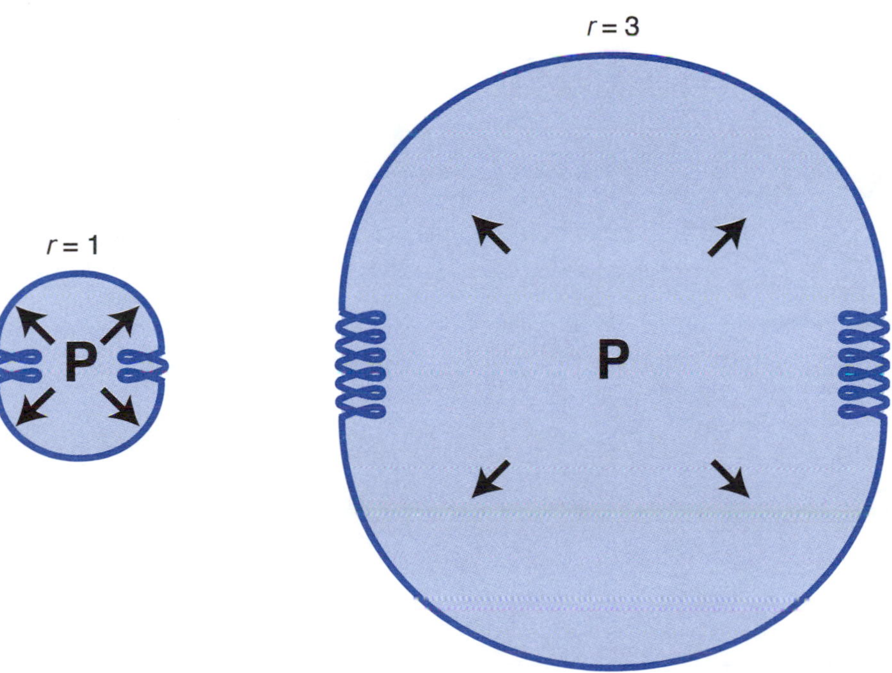


**Figure 5.5** The vascular tension at a given transmural pressure $P$ is roughly proportional to the radius of the vessel, $r$. (This is a simplification neglecting the thickness and properties of the wall.) Thus, if the radius is tripled, three times more tension will be required to maintain the same pressure.

liquid slips over adjacent laminae without mixture or interchange of fluid. At a critical velocity, the laminae of fluid move irregularly and start to mix with one another; the flow becomes **turbulent.** During turbulence the energy loss is larger, and for a given pressure gradient the flow is lower (figure 5.6). The velocity of flow in the ventricles of the heart and aorta is normally turbulent during the early phase of contraction. This turbulence produces vibrations, causing high-pitched sounds that can be heard easily by listening over the heart (see figure 5.4b). If the velocity is abnormal, as in mitral stenosis, the heart sounds are abnormal. Similarly, a shunt between the aorta and the pulmonary artery (open **Ductus arteriosus**) can give rise to a turbulent flow in that shunt vessel. When blood pressure is measured by applying a measured pressure over the brachial artery (with the sphygmomanometer cuff), the peak blood pressure suddenly overcomes the resistance of the gradually reduced compression. The blood flows with a high velocity through the narrowed artery, and a turbulence gives rise to sounds detectable by the stethoscope.

The radius of a tube is a deciding factor in the flow through it; the resistance to flow is a reverse function of its radius to the fourth power. A decrease to half the radius, other things being equal, will actually decrease the flow to one sixteenth of the original value. A dilation of 10% increases the blood flow about 50% compared with the flow through the same vessel constricted. Consequently, the variation in activity of the smooth muscles of the vessels is a very sensitive instrument for the control of blood flow.

The factors of importance for flow can be described in more detail by presenting the Poiseuille-Hagen formula, although the equation is only applicable for nonpulsatile flow of homogeneous fluids. Actually, flow in the blood vessels is normally laminar except in and close to the heart. The flow ($F$) through a tube is actually proportional to the driving force ($\Delta P$), inverse to the viscosity ($\pi$), proportional to the radius of the vessel raised to the fourth power ($r^4$), and inverse to the length of the vessel ($l$):

$$F = \Delta P \times \pi/8 \times 1/\eta \times r^4/l \quad (4)$$

(Poiseuille-Hagen formula)

The resistance to laminar flow ($R$) is proportional to pressure gradient per rate of flow.

The complex composition of blood and the distensible blood vessels complicates the situation in some ways. If a given pressure gradient ("driving pressure") is maintained between artery and vein, the flow is not, as expected, constant at varying pressure levels. At low pressures the flow becomes zero, even though there is still a considerable driving pressure. The pressure at which the vessels close completely is called the critical closing pressure. The decisive factor here is the transmural pressure. When the pressure outside the inner layer of the wall (tissue pressure and active muscular contraction in the vessel wall) exceeds the intravascular pressure, a collapsible vessel collapses (this happens "artificially" in the routine blood pressure measurement commented on earlier). At pressures well above the critical closing pressure, the flow is essentially proportional to the driving pressure.

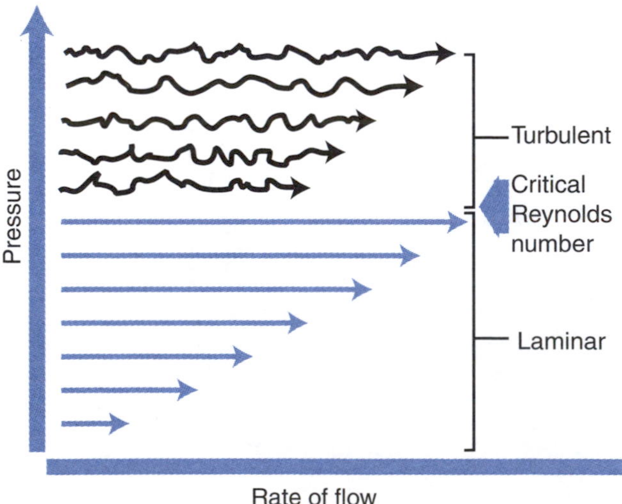


**Figure 5.6** In turbulent flow, the energy loss is greater and the rate of flow slower than in laminar flow at a given pressure gradient. Length of arrows indicates rate of flow. From Haynes and Rodbard 1962.

### *Summary*

The pressure–flow curve obtained in studies of the hemodynamics of circulation is not linear. The effect of driving pressure can be modified by the transmural pressure, especially in the arterioles and capillaries, and at low intravascular pressure. The resistance to flow can be profoundly altered by active and passive variations of the radius of the blood vessels, the flow being directly proportional to the fourth power of the radius ($r^4$).

## Blood Vessels

The blood vessels are classified into arteries, arterioles, capillaries, and veins. The exchange of water, electrolytes, and gases can occur only across the thin wall of the capillaries and, to some degree, in the postcapillary venules; the other vessels are only transport channels. The amount of elastic fibers, collagen fibers, and smooth muscles varies in the different vessels. In the big arteries, the walls are thick and elastic fibers dominate in the **tunica media.** In veins of the same size, the walls are thin, with few elastic and muscle fibers. A single layer of flattened endothelial cells covers the inner surface of the vessels. The vascular system adapts to the forces acting on the walls, modifying their mechanical property.

It has been emphasized that the local blood flow is determined mainly by the pressure load and the diameter of the actual vessels. The capacity of the different vessels to influence vascular resistance, blood flow, and blood distribution is, in a functional way, described by the following subdivision: **Windkessel vessels** (main arteries), resistance vessels (arterioles are the most important, but capillary and postcapillary sections are also of significance), **precapillary sphincters** (determining the functioning capillary surface area), shunt vessels (e.g., arteriovenous anastomoses), capacitance vessels (veins), and exchange vessels (capillaries). Most of this terminology is illustrated in figure 5.7.

## Arteries

The arteries are elastic, and forward flow is continuous because of the recoil of the vessel walls that have been stretched during the systole. For this reason the arteries are sometimes referred to as Windkessel vessels (from the German word for "elastic reservoir").

The expansion of the arterial wall during the ejection of blood causes a pressure wave that travels along the peripheral blood vessels at a speed of 5 to 9 $m \cdot s^{-1}$. In comparison, the velocity of the blood in the aorta at rest is .5 $m \cdot s^{-1}$. The more elastic the arterial wall, the lower the speed of the pulse wave. In practical application, the frequency of this wave is counted as the pulse rate, because it easily can be felt over the radial or carotid arteries.

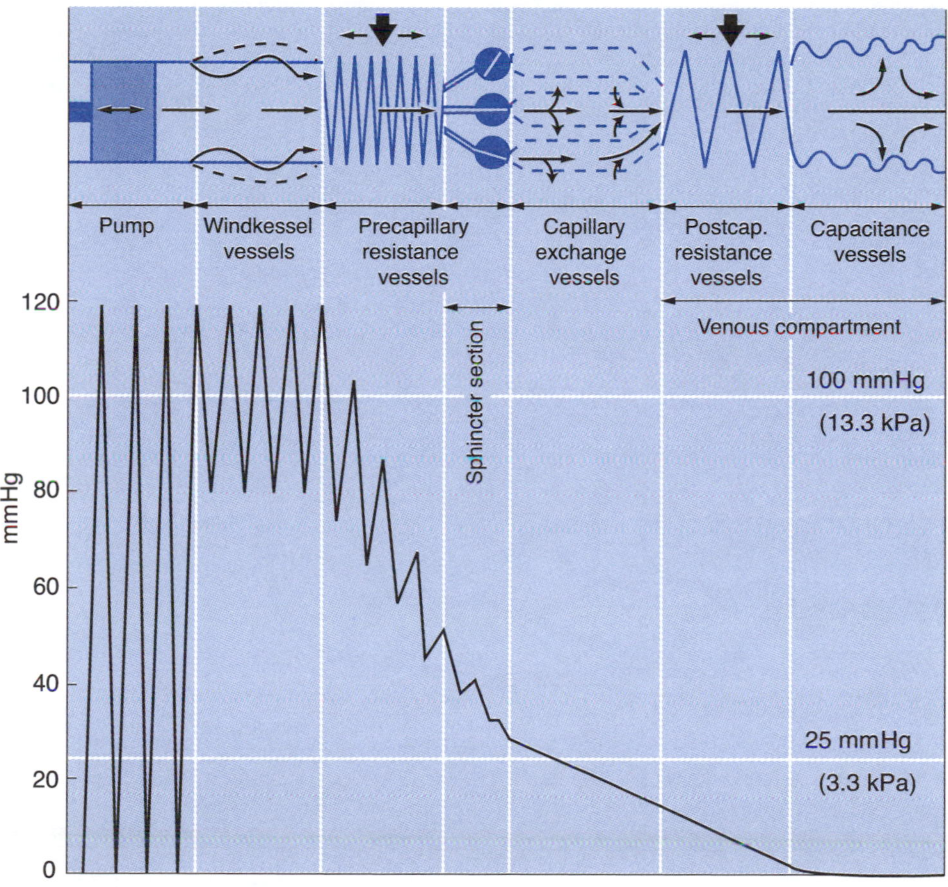


**Figure 5.7** A schematic illustration of the functionally different, consecutive sections of the vascular bed related to the blood pressure decrease along the circuit (ordinate blood pressure in mmHg). Note the marked pressure decrease and absorption of the pulse amplitudes in the precapillary resistance vessels. The line at 25 mmHg (3.3 kPa) represents the plasma protein osmotic pressure (colloid-osmotic pressure). A similar illustration can be made of the sections of the vascular bed in the "low pressure system" (pulmonary circulation), but the pressure in the right pump at rest alternates between about 25 and 0 mmHg (3.3 and 0 kPa) and in the pulmonary artery (Windkessel vessels) between 25 and 10 mmHg (3.3 and 1.3 kPa).

Adapted from Folkow and Neil 1971.

## Arterioles

In arterioles the peripheral resistance is high, markedly decreasing pressure. The velocity is still high because the total cross section of the arterioles is not large. The individual vessels are narrow, from .1 mm to 60 to 100 μm. In the wall are transversely oriented smooth muscle fibers that can be activated by their nerves or by local chemical factors. Figure 5.8(a-c) shows schematically various degrees of con-

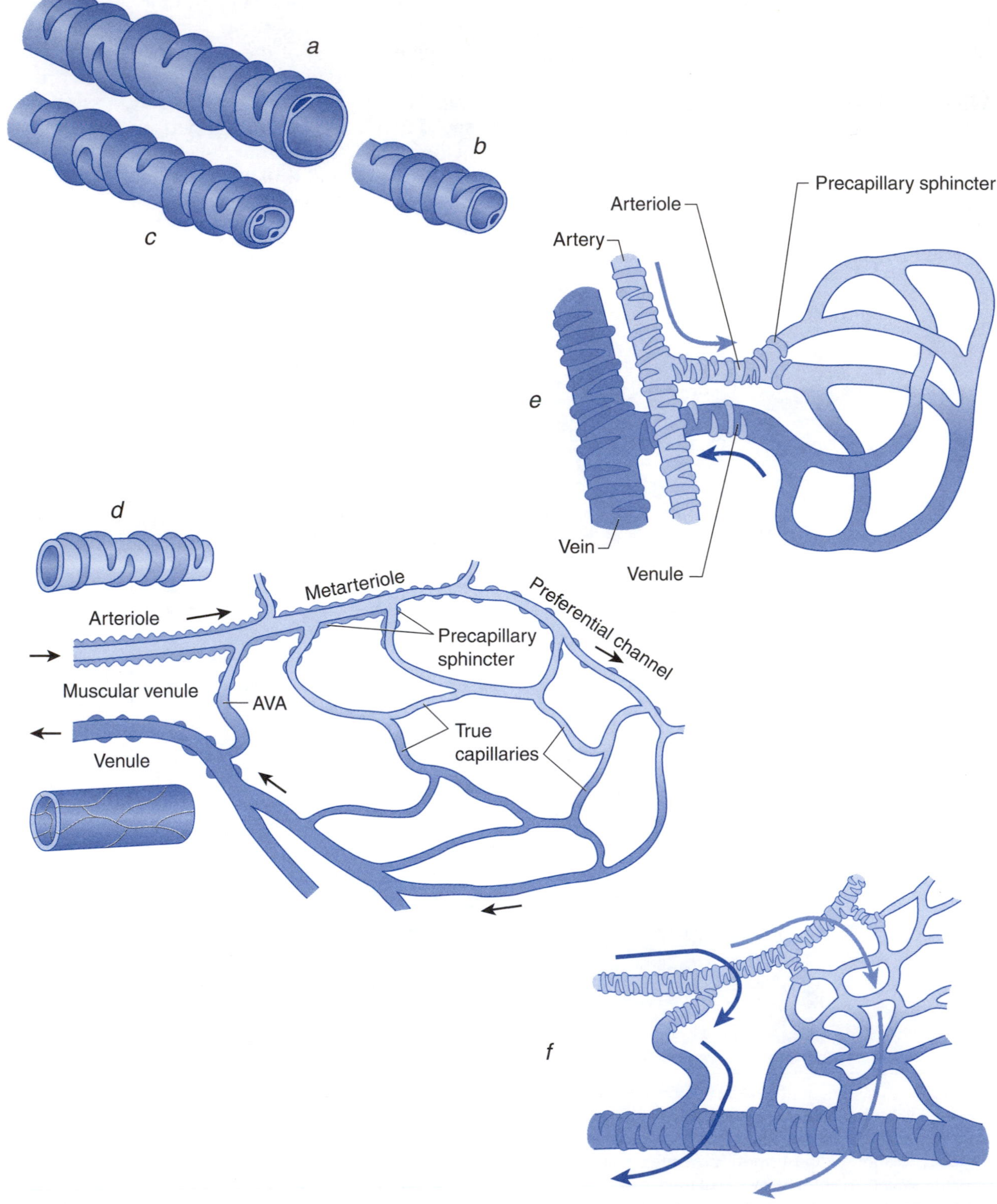


**Figure 5.8** A schematic representation of the structural pattern of the capillary bed. The distribution of smooth muscle is indicated in the vessel wall; see text for discussion. AVA = arteriovenous anastomose.

striction of an arteriole caused by contraction of the smooth muscle fiber. It should be stressed that the vessels do not have muscles that can actively dilate them.

Figure 5.9 illustrates how the arterioles (and capillaries) are arranged in parallel-coupled circuits between the arteries and the veins. They can effectively alter the total resistance against the outflow of blood from the arteries and thereby the arterial blood pressure and work of the heart; they markedly influence the distribution of blood flow to the various organs. An increase in caliber of the arterioles in a muscle will decrease the resistance in that area and hence increase the flow. If this local vasodilation is not compensated for by a vasoconstriction in another area or an increase in cardiac output, the arterial blood pressure inevitably will decrease.

The arterioles are arranged in a series of parallel channels joining the capillary bed (and veins) with the arterial side, effectively regulating the blood flow through the organs and tissues. The pressure is still high when the blood enters the arterioles but then drops to 30 to 40 mmHg (4–5.3 kPa). The systolic and diastolic pressure variations usually disappear before the blood reaches the capillaries (see figure 5.7).

## Capillaries

The microarchitecture of the capillary network is schematically illustrated in figure 5.8. In general, the capillary arrangements are as follows: The arterioles branch off into terminal or metarterioles. There the smooth muscle fibers are arranged in circular or

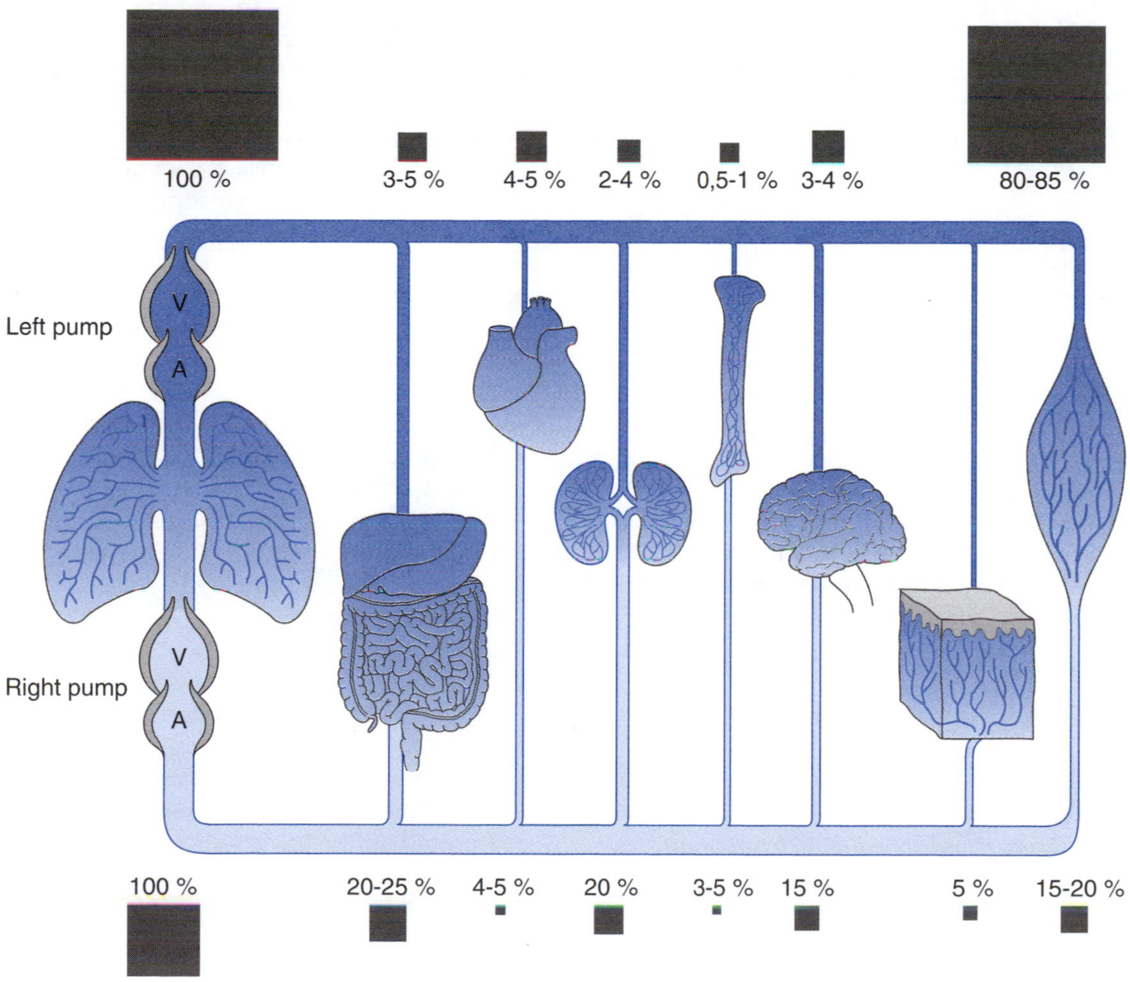


**Figure 5.9** Schematic drawing showing how the arterioles and capillaries are arranged in parallel-coupled circuits between the arteries (top) and the veins. The cardiac output may be increased fivefold when changing from rest to strenuous exercise. The figure indicates the relative distribution of the blood to the various organs at rest (lower scale) and during exercise (upper scale). During exercise, the circulating blood is diverted primarily to the muscles. The area of the black squares is roughly proportional to the minute volume of blood flow. Not included is an estimated blood flow of 5% to 10% to fatty tissues at rest, about 1% during heavy work.

spiral alignment and constitute a single layer. The metarteriole ends in a capillary-like channel or **thoroughfare channel** (preferential channel), from which the capillaries branch (see figure 5.8d). In the proximal wall portions of the thoroughfare channel is a discontinuous coat of thin muscle cells. Frequently a group of muscle fibers form precapillary sphincters, especially at the origin of the capillary. The preferential channel finally loses its muscle coat completely but can be differentiated from adjoining capillaries by the presence of a somewhat thicker supporting connective tissue coat. The true capillaries are thin-walled endothelial tubes with no muscle or connective tissue. In skeletal muscles, the true capillaries are 8 to 10 times as numerous as the preferential channels. In other tissues, there are relatively fewer capillaries. The capillaries and the preferential channels drain into venules, which in turn drain into the larger veins.

Another pattern of the microcirculation may be a net of arterioles that divide into anastomosing true capillaries (see figure 5.8e). Precapillary sphincters can be seen. The capillaries converge into venules, and no vessels resembling a thoroughfare channel are found.

Arteriovenous anastomoses or shunts are typical for the vascular bed in the skin but are also present in other tissues. Anatomically, this may be a vessel with a relatively thick muscle coat running from an artery or arteriole to a vein or venule (see figure 5.8f). The diameter can, in the latter case, be as small as 15 to 20 μm. Similar shunts are located proximally to the terminal arterioles or form a direct continuation from one of the branches of an arteriole to the venous circulation. In both cases, this shunt provides a good means of maintaining blood flow and low resistance, but the capillary vascular bed is bypassed (dark arrows in figure 5.8f).

### *Summary*

The circulation of blood from arteries to veins can take various routes via thoroughfare channels, true capillaries, and arteriovenous or arteriolovenous shunts. The pressure gradient and the activity of the smooth muscle cells in the walls or the precapillary sphincters decide which route the blood flow will take.

## Capillary Structure and Transport Mechanisms

Let us now take a closer look at the architecture of the capillary. Its length is less than 1 mm (as low as 0.4 mm) and its inner diameter ranges from 3 to 20 μm; therefore, there is sometimes hardly room for an erythrocyte (diameter 7.2 μm), which can stop the flow temporarily and be squeezed out of shape as it is forced through. The capillary wall is made of very thin, simple **squamous epithelial cells,** the endothelium. Numerous pinocytotic vesicles are seen in the endothelial cell cytoplasm. The cells are linked to one another by tight junctions, sealing the space between adjacent cells. On the outer side of the endothelium is a basement membrane or basal lamina. In most tissues, including muscle, the capillaries are of the so-called continuous type, but in endocrine glands and sites specialized for metabolite and fluid absorption, like the intestine, the capillaries are fenestrated. The **fenestrations** (*fenestra* = "window") are multiple, circular near-perforations, 80 to 100 nm in diameter, of the endothelial cell. The fenestrations are "closed" by a very thin diaphragm of material only, possibly representing remnants of the **glycocalyx** covering the surface of the endothelial cell.

The glycocalyx is a layer appearing in electron micrographs as a "fuzzy coat" on the extracellular face of cell membranes. It consists mainly of the carbohydrate moiety of glycoproteins and glycolipids in the cell membrane. It is believed to be important in cell recognition and to protect the cell surface. Carbohydrates are very diverse molecules, and contrary to some belief, they are very well suited as specific recognition molecules (Alberts et al. 1994).

The capillary wall has no smooth muscle cells or other contractile elements but has elastic properties. The thin-walled venules following the capillaries have walls, which in most places are similar to those of the capillaries, except that pericytes are more common. The postcapillary venules are important sites of action of vasoactive agents like histamine and serotonin, resulting in extravasation of fluid during allergy or inflammation.

The most important processes for the exchange of substances across the capillary membrane are filtration and diffusion. Isotope and other tracer studies have shown that gases, water, molecules in water, and lipids can rapidly diffuse back and forth in the direction of the concentration gradients through the capillary wall. Part of this transport has been shown to take place in pinocytotic vesicles.

Lipid-soluble materials can pass through the cell membrane by diffusion. Therefore, oxygen and $CO_2$ can use the entire surface area of the capillary wall and easily use the transendothelial route. Lipid-insoluble materials must use different mechanisms for passage. Theoretically, the interendothelial junction is one possibility for the passage of solutes through systems of "pores" or "slits." There is evidence of small pores with a diameter less than 10 nm, which water and small- to moderate-size molecules can use as pathways (chapter 3). For transport of larger molecules, like albumin and other proteins, large pores or "leak" pathways, with a diameter size of about 60 nm and more, are possible routes.

The large pores seem to correspond to the pinocytotic vesicle route mentioned previously, whereas the intercellular junction is the equivalent of the small pores. According to some estimates, only .02% of the capillary surface area is devoted to pores, and the ratio between small and large pores is approximately 100:1.

### *Summary*

The very thin walls of the capillaries (and venules) have a simple endothelium cell layer resting on a basal lamina. Lipid-soluble substances, which include gases, can diffuse freely through the endothelial cells. For the filtration and diffusion of water-soluble particles and larger molecules, the pathways are thought to be pores or other types of interruptions of the continuous lining of endothelial cells. In morphological studies, pinocytotic vesicles have been shown to constitute a discontinuous route of passage through endothelial cells for large molecules.

### For Additional Reading

For details concerning morphological and morphometrical analyses of muscle biopsy samples, see Lexell (1997).

## Filtration and Osmosis

Figure 5.10 shows the driving forces in a process outlined by Starling (1896). When blood flows continuously through a capillary, the hydrostatic pressure is normally well above the pressure in the interstitial fluid (ISF). Because the plasma proteins cannot pass through the capillary wall, they will exert a colloid-osmotic pressure much higher than that of the ISF. This statement, however, is more didactic than it is related to the mechanism of exchange of plasma and ISF. For simplicity, the pressures outside the capillary are disregarded in figure 5.10. The gradual change (less than 2%) in colloid-osmotic pressure as water filtrates to ISF or returns to the vessel also is ignored.

When blood enters the capillary, the hydrostatic pressure may be 40 mmHg (5.3 kPa), and this pressure decreases gradually to 10 mmHg (1.3 kPa) at the venous end of the capillary, in humans at rest.

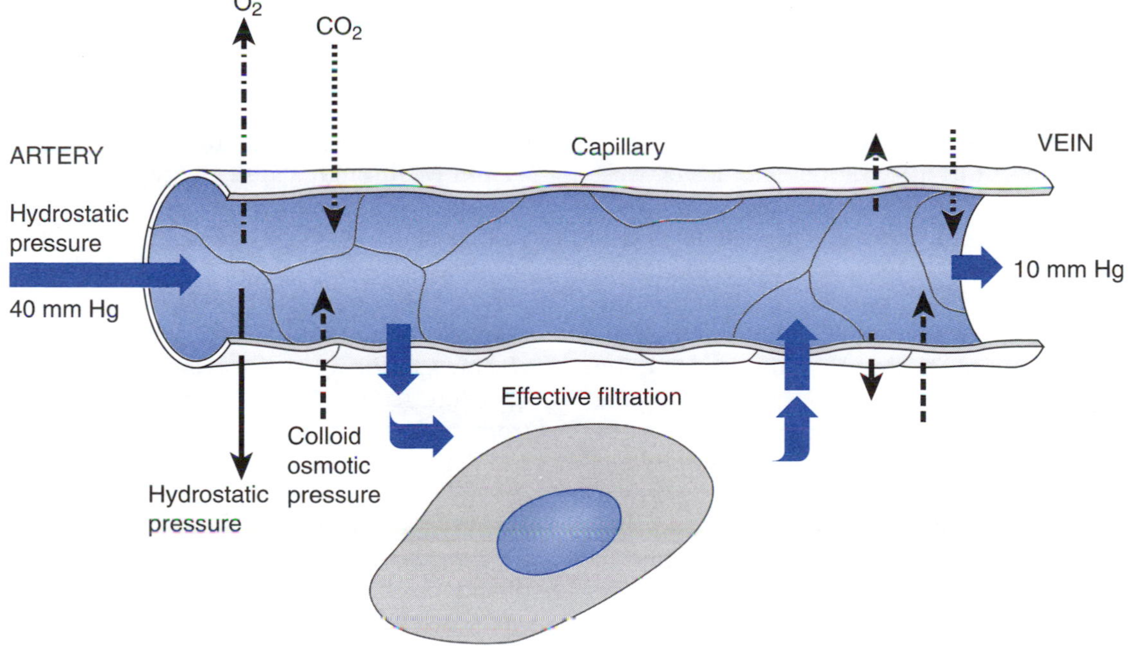


**Figure 5.10** Driving force in capillaries. The hydrostatic pressure (h.p.) depends on the degree of vasoconstriction in precapillary vessels. In a resting skeletal muscle, hydrostatic pressure may be only 15 mmHg (2.0 kPa) at the beginning of the capillary. For details, see text.

The colloid-osmotic pressure is about 25 mmHg (3.3 kPa, see figure 5.7). According to the driving forces in Starling's hypothesis, the filtration pressure is then 40 minus 25 mmHg (5.3–3.3 kPa) "proximally," giving an outflow of fluid. "Distally" the filtration pressure is 10 minus 25 mmHg (1.3–3.3 kPa), giving a negative pressure difference of 15 mmHg (2.0 kPa). This means that fluid is returned with a force of 15 mmHg (2.0 kPa). It is easy to see how small variations in pressures can profoundly alter the exchanges of fluid. Three examples are given: First, if the arterial pressure falls, for instance during blood loss, capillary pressure also is reduced. Thereby the normal balance between the hydrostatic pressure and colloid-osmotic pressure becomes disturbed, causing a greater retention of fluid in the capillary. Second, a decrease in the concentration of plasma protein (prolonged starvation or loss attributable to kidney disease causing albuminuria) lowers the colloid-osmotic pressure and the return of fluid from the tissue. An increase in ISF follows (edema). Third, an increase of the venous pressure increases the mean capillary pressure and hence the filtration of fluid into the tissue. At the same time, the return of fluid at the venous side of the capillary is reduced, contributing to the edema.

For the osmotic force in Starling's hypothesis to return ISF to the venous end of a capillary, the colloid-osmotic pressure of plasma must equal the hydrostatic pressure midway along the capillary (figure 5.10). In strenuous exercise, the hydrostatic pressure increases at the arterial end whereas the colloid-osmotic pressure of plasma remains constant. According to the Starling hypothesis, filtration of plasma fluid into the ISF should increase and the return of ISF should decrease, necessarily resulting in edema. But there is no edema associated with exercise. Recently, another explanation of the exchange of fluid between plasma and ISF has been proposed (Hammel, 1995, 1999, 2001). The problem in the Starling hypothesis is that the nearly constant colloid-osmotic pressure of the plasma does not have an effect on the exchange of plasma and ISF when blood flow is continuous and constant. Physiologists claim that proteins lower the water concentration in the plasma and that water diffuses from a higher water concentration in the ISF to the lower water concentration in the plasma. Contrary to this theory of osmosis long held by physiologists, Hammel (1999) has shown that water concentration is unrelated to the osmotic pressure in a solution and does not cause osmosis. Therefore, when the rate of flow of plasma through the capillary and its colloid-osmotic pressure are both constant, the plasma proteins have no osmotic effect. Hammel (1994) acclaims Hulett's theory of osmosis as the only valid explanation of osmosis. He extends this theory to account for the osmotic effect that does determine the exchange of plasma and ISF and that also depends on the metabolic rate of the tissue, as in strenuous exercise.

Hulett (1903) showed that the solute molecules contained within an aqueous solution exert a solute pressure against the boundaries of the solution. When no counter-pressure is applied to the solution, the pressure exerted by the solute molecules distends the boundaries of the solution and thereby alters the internal tension in the forces that bond the water molecules in its liquid form. The altered internal tension of the water by the solute accounts for all of the altered properties of the water in the solution, including its osmotic pressure and vapor pressure. Hammel (1995, 1999, 2001) extends Hulett's theory to apply to plasma flowing through a capillary. When solute molecules in a solution diffuse from a higher to a lower solute concentration within the solution, they have an osmotic effect on the water in the solution equal to the difference in the osmotic pressures of the higher and lower concentrations.

In humans at rest, the osmolality of the plasma is necessarily higher in venous blood than in arterial blood that flows through systemic capillaries. At rest, the difference is about 1.7 milliosmol (McKenna et al. 1997); the increase is due primarily to the increase in carbon dioxide and bicarbonate ion concentrations in the plasma. This amounts to an increase of 32 mmHg in osmotic pressure of the plasma as it flows through a systemic capillary. This is the osmotic force that accounts for the exchange of plasma and ISF at rest. In strenuous exercise, the increase in plasma osmolality is as much as 500 mmHg (McKenna et al. 1997). Thus, although the arterial pressure is substantially increased in skeletal muscle during exercise, the increased diffusion of solute (carbon dioxide and bicarbonate ions) from end to end in the plasma in the systemic capillaries prevents edema in the metabolizing tissue. Likewise, in lung capillaries, the osmotic pressure of the plasma at the arterial end becomes proportionally less at the venous end as carbon dioxide and bicarbonate ion concentrations decrease at the venous end.

## For Additional Reading

For further insight into the osmotic effect of solute diffusing through a solution, see Hammel (2001).

## Veins

The collecting venules have a supporting coat of connective tissue and irregularly spaced smooth muscle cells (see figure 5.8, e and f). In the distal part of the venules, a well-defined muscle layer, innervated by sympathetic neurons, appears. Also, the larger veins have muscles in their walls that can constrict the lumen of the vessel. Valves are frequent in the veins of the body extremities.

The veins within a skeletal muscle will be compressed mechanically during the muscular contraction. Blood is squeezed from the veins, and because of the higher resistance on the capillary side and the design of the valves, the blood return to the heart is facilitated by this "skeletal muscle pump." During muscular relaxation, the venous pressure decreases, and therefore blood can flow from the superficial veins, anastomosing the deep veins. The blood flow to an active muscle is higher during the period of relaxation than during contraction. The venous pressure has its minimum at the time of relaxation, which increases the arteriovenous pressure difference and thus the perfusion.

Normally, the venous systems contain some 60% to 70% of the total blood volume. The veins therefore are often referred to as capacitance vessels (see figure 5.7). The pulmonary veins account for about 10% to 15% of the total blood volume. At rest, the blood in the capillaries is estimated to be only 5% of the total (the spleen plays only a minor role in the human as a blood or red cell depot).

The resistance to flow in veins larger than 0.5 mm is very modest compared with the resistance to blood flow in the preceding vessels, perhaps only 10% of the total under normal conditions.

In an inactive individual, approximately 8,000 L of blood flows through the capillaries per day. Only about 24 L of this fluid is filtered from the capillaries (except for about 170 L filtered across the glomeruli of the kidneys). At the distal end of the capillaries, some 20 to 22 L of fluid is reabsorbed. The remaining 2 to 4 L returns to the blood via the lymphatic system.

## Vascularization of Skeletal Muscles

Nerves and vessels enter the muscle at a neuromuscular hilus often located at half-length of the muscle. The artery enters the muscle belly and branches freely in its course along the perimysium (figure 5.11). By anastomoses, a primary arterial network is established. Finer arteries arise and create a secondary network infiltrating the muscle tissue. The smallest arteries and terminal arterioles branch off, usually transversely to the long axis of the muscle fibers and at fairly regular intervals of 1 mm. The arterioles then supply the capillary network oriented parallel to the individual muscle fibers but also form frequent transverse links over or under the intervening fibers, thereby forming a delicate oblong mesh in the endomysium. The capillary network is especially well developed around the motor

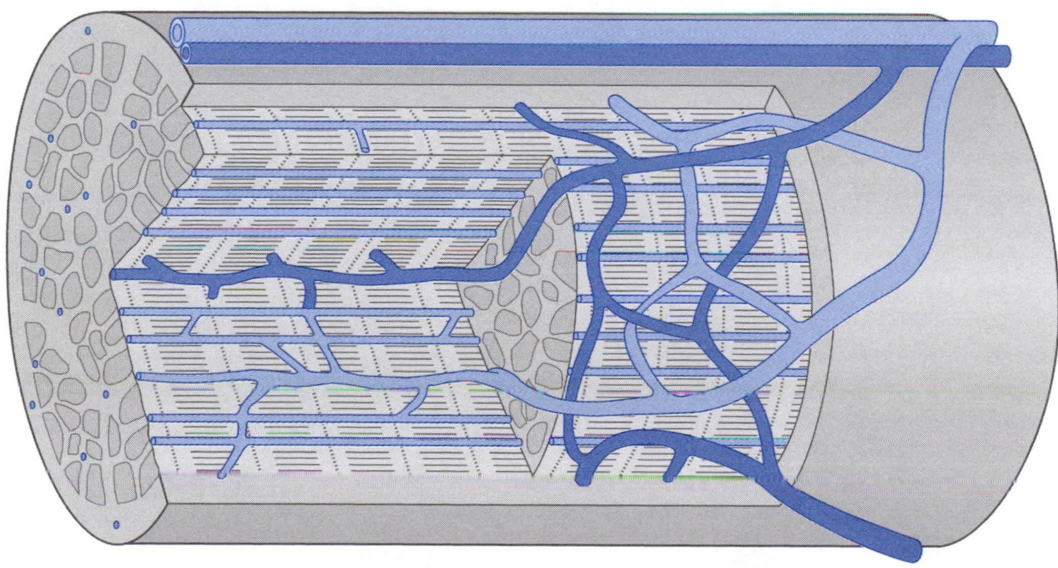

**Figure 5.11** Vascularization of skeletal muscle. Capillaries run parallel to the fibrils that make up the muscle fibers.

endplate. The veins have valves (from a caliber of 40 μm or larger) directing the blood flow toward the heart, and they follow the course of the arterioles and arteries.

The exact capillary density in skeletal muscles in the human body is not known. With the aid of the electron microscope, Brodal, Ingjer, and Hermansen (1977) counted the number of capillaries per square millimeter in cross sections of needle biopsies taken from the vastus lateralis muscle from 12 untrained and 11 endurance-trained individuals. The mean number of capillaries per square millimeter was 585 ± 40 in the untrained and 821 ± 28 in the well-trained individuals. Correcting for shrinkage and artificial spaces between the fibers, the investigators arrived at a mean number of capillaries per square millimeter of 305 for the untrained and 425 for the trained group, respectively. Thus, up to 500 capillaries/$mm^2$ may be open during heavy muscular activity. The total surface area of the capillary bed would then be about 200 $mm^2$ in an individual with 30 kg of muscles. With an average capillary diameter of 6 μm, the area of the capillaries would be close to 2% of the total tissue cross-sectional area; the distance from a capillary to the most remote mitochondria would be about 13 μm if the capillaries are evenly distributed. The length of the capillaries varies markedly; as a mean length, 1 mm is a reasonable figure. The arterioles can close the inlet to the capillary, and at rest the number of open capillaries is reduced markedly. The opening or closing is supposed to be operated by local chemical factors of a hypoxic or metabolic nature and only to a smaller degree by nerve activity (see subsequent discussion). At a certain level of metabolic activity in the tissue, the number of working capillaries is fairly constant, but the individual capillary can be intermittently open or closed (see also P. Andersen 1975; Ingjer 1979a. The human skeletal muscle is composed of both slow-twitch fibers (type I) and fast-twitch fibers (type II), and fibers of the different motor units are intermingled, forming a checkerboard pattern (see chapter 3). Therefore, two different fiber types often share common capillaries. On the average, the capillary density around the slow-twitch fibers is higher (three to four capillaries per fiber) than around the fast-twitch fiber (two to three capillaries per fiber). In the untrained muscle, the average tissue area that a capillary supplies is 20% to 30% larger for type IIX and 10% to 20% more for type IIA fibers compared with the type I fibers. As mentioned, training increases the capillary density; disuse of a muscle will reduce the density (Klausen, Andersen, and Pelle 1981).

### FOR ADDITIONAL READING

For further discussion concerning various issues of physical activity and the vascularization of skeletal muscle, see Coggan et al. (1992), Hepple et al. (1997), Chilibeck et al. (1997), and H. Green et al. (1999).

The abundance of capillaries in the muscles provides good facilities for supplying oxygen and nutritive materials to the cells and for removing products from metabolic activity. Krogh (1929) calculated that with 100 open capillaries per square millimeter, a gradient in oxygen tension of 12 mmHg (1.6 kPa) from a capillary to the most distant point would be sufficient during resting conditions to provide those distant parts with oxygen by diffusion. With all capillaries open, a lower oxygen gradient would be enough to drive a diffusion; the mitochondria function with an oxygen pressure of 1 mmHg (0.13 kPa) or even less.

# Regulation of Circulation at Rest

Any change in cellular activity should be met by a corresponding variation in local blood flow through the capillary bed. If an individual cell, in one way or another, could control its environment by varying the blood supply in balance with the actual nutritional demand, that cell would benefit. But other cells or tissues might suffer if some cells take more than their share. Hence, coordinative mechanisms are essential if the distribution of blood is to be balanced properly. An active regulatory mechanism ensures that more active and less active cells, as well as more susceptible and less susceptible organs, are supplied according to their need and to the capacity of the whole circulation.

## Arterial Blood Pressure and Vasomotor Tone

The blood pressure in the aorta is maintained by an integration of the following factors: (1) cardiac output, (2) peripheral resistance, (3) elasticity of the main arteries, (4) viscosity of the blood, and (5) blood volume. Evidently, a regulation of the arterial blood pressure can use factors 1 and 2, because factors 3, 4, and 5 are normally not at its disposal for rapid modifications.

The local blood flow is determined mainly by the pressure load and the diameter of the actual vessels. The smooth muscles of the arterioles and veins in many regions continuously receive nerve impulses that keep the lumen of the vessels more or less constricted. This vasomotor tone is provided by the sympathetic vasoconstrictor fibers driven from the vasomotor area in the medulla oblongata. The transmitter substance is noradrenaline. As mentioned, the membrane receptors of α-adrenergic type have a fairly general distribution in the vascular tree, and the effect on the smooth muscles by noradrenaline is a constriction of the vessels. In most tissues, the existence of β-adrenergic receptors on vascular smooth muscle is of minor importance. However, an important exception is the precapillary resistance sections in skeletal muscle where circulating adrenaline can relax the smooth muscles. The heart and brain receive few vasomotor fibers; the supply to the abdominal organs (by **splanchnic** nerves) and skin is very rich. The muscles have an intermediate portion. The vasomotor tone at rest can be demonstrated by sectioning or blocking the sympathetic nerve fibers in an animal. The arterioles dilate, and the arterial blood pressure decreases. The effect of such an inhibition of vasomotor tone is very marked in the skin but less pronounced in skeletal muscles. The basal blood flow of 3 to 5 ml · min$^{-1}$ per 100 ml of muscle tissue can double . During exercise, however, the blood flow can amount to 100 ml · min$^{-1}$ or more. The smooth muscles in precapillary vessels can exhibit spontaneous rhythmic contractions creating a basal vascular tone. The intravascular pressure may be a stimulating factor. Some of the smooth muscle fibers, predominantly those localized in the most narrow vessels, actually can serve as stretch receptors and pacemakers and thereby as triggers for the neighboring cells.

The vasomotor tone is important in keeping arterial blood pressure and cardiac output at an economical level. The splanchnic area could contain the whole blood volume after maximal dilation of the vessels. Also, the vascular bed of skin and muscles has a similarly large capacity. Fainting (vasovagal syncope) may be the result of a central inhibition of the vasomotor efferent impulses. The cardiac output certainly cannot exceed the venous return. A constriction of the postcapillary vessels (the capacitance vessels) with their large content of blood will increase the blood flow toward the heart, enabling an increase in cardiac output.

A decrease (inhibition) in the vasomotor tone, which actually relaxes the smooth muscles in the vessel wall and therefore causes vasodilation, can be obtained principally in two ways. In the first way, the smooth muscles can be affected by chemical substances liberated locally from neighboring cells or delivered from the blood. Known dilating agents are hypoxia, hyperosmolarity, lowered pH, an excess of $CO_2$ and lactic acid, adenosine compounds, an increase in extracellular potassium, inorganic phosphate, endothelium-derived NO, prostacyclin, and endothelin-1. A second way that vasodilation can be caused is by a decreased discharge in the sympathetic vasomotor nerves.

Regular physical activity results in peripheral vascular adaptations that enhance perfusion and flow capacity. Such adaptations may arise from structural modifications of the vasculature and alterations in the control of vascular tone (Delp 1995; Kingwell and Jennings 1998).

### *Summary*

The peripheral resistance to blood flow is determined by the vasomotor tone. The degree of contraction of the smooth muscles in the arterioles is of special importance for the local blood flow as well as the total resistance. In some tissues (such as the muscles), this tone is probably partly spontaneous, the smooth muscle contractions (myogenic activity) being triggered by mechanical stretch induced by the intravascular pressure, but a sympathetic vasomotor tone is superimposed. In other regions (such as the skin), this sympathetic vasomotor tone is dominating. A vasodilation can occur when the smooth muscles are affected by various dilating agents liberated locally or delivered from the blood. Vasodilation also can be caused by a decreased discharge in the sympathetic vasomotor nerves. Regular physical activity results in vascular adaptations that enhance perfusion and flow capacity.

### For Additional Reading

For a molecular approach on NO release during cyclic exercise, see Stefano et al. (2001).

## The Heart and the Effect of Nerve Impulses

The heart has its own pacemaker, the sinoatrial node, initiating about 110 impulses · min$^{-1}$ if left alone. Normally, however, it is under the influence of nervous activity, and at rest inhibitory impulses will dominate. Both sympathetic and parasympathetic nerve impulses can modify the heart rate. The parasympathetic activity from a cardioinhibitory center via the vagus nerve (and acetylcholine) slows the heart rate (**bradycardia**), and the sympathetic cardiac nerves (and noradrenaline) as well as blood-borne adrenaline can increase heart rate (tachycardia). In a resting subject, a blocking of the vagus nerve will cause the heart to beat faster, because this removes a predominating parasympathetic tone.

There are indications that the slowing of the heart rate during a standard rate of exercise after a period of physical training is induced by an increase in vagus tone and a reduction of the sympathetic drive (Ekblom, Kilbom, and Soltysiak 1973; Jensen-Urstad et al. 1997). The sympathetic nerves can increase the contractile force of the heart muscle fibers, but sympathetic nervous control of the vasomotor tone in the heart vessels is probably insignificant. As mentioned, the blood vessels of the heart dilate willingly when affected by, for example, metabolites, endothelium-derived NO, and hypoxia.

## Control and Effects Exerted by the Central Nervous System

Now that we have analyzed the tools available for a variation and redistribution of blood flow within the body, the actual control and regulating mechanisms can be discussed.

Important and essential nuclei are located in the brain stem, particularly in the medulla oblongata. Traditionally, textbooks have discussed a medullary vasomotor area, divided into a vasoconstrictor center and a vasodepressor area, the latter operating through inhibition of the sympathetic vasoconstrictor outflow. Keele, Neil, and Joels (1982) suggested the term *medullary cardiovascular centers*. Anatomically, they are found in the reticular formation. Such terms may be convenient but they give a false impression of well-defined collections of neurons with precise effects. This is not so. There is no clear anatomical separation between the pressor and depressor areas in the brain stem but rather an overlap in the distribution of nuclei that can excite sympathetic nerve fibers to the heart and blood vessels and neurons that can inhibit these effects.

The activators presumably release noradrenaline as a transmitter substance in their communication with preganglionic sympathetic neurons in the spinal cord (adrenergic neurons). Another class of neurons in the brain stem belonging to the sympathetic system also send their axons down toward preganglionic neurons, but presumably they release **serotonin (5-hydroxytryptamine).** According to one theory, these serotonergic neurons can presynaptically inhibit the adrenergic neurons. The balance in activity between these two systems provides the major control of vasomotor tone by increasing or decreasing the impulse traffic in the sympathetic nerves leaving the spinal cord (J.T. Shepherd and Vanhoutte 1979). As mentioned, the adrenergic nerve activity constricts most all vascular beds via α receptors (if not modified by other factors). In addition, it increases the heart rate and contractility (β receptors), whereas the transmitter substance, acetylcholine, reduces the heart rate by inhibiting the sinus node and the atrioventricular node (by hyperpolarization). There is an interesting interaction between the sympathetic and parasympathetic systems in organs that they both innervate. On the presynaptic endings of sympathetic fibers there may be specific receptors. When exposed to Ach, the release of noradrenaline is reduced. Conversely, catecholamines can reduce the release of acetylcholine. This effect is mediated by α-adrenergic receptors in the presynaptic region of the parasympathetic nerve fibers. This is an example of a reciprocal innervation at the presynaptic level. A high concentration of noradrenaline released into the synaptic cleft by an adrenergic fiber may limit the subsequent release mediated by α receptors on the presynaptic nerve terminal (a negative feedback mechanism).

Understanding the interaction between sympathetic and parasympathetic nerves is complicated by their use of signal molecules that are neither adrenaline nor acetylcholine, so-called nonadrenergic, noncholinergic (NANC) transmission. Such NANC signal molecules may be ATP or NO, or peptides like vasoactive intestinal peptide, somatostatin, cholecystokinin, or others.

The neurogenic vasomotor tone of the blood vessels originates essentially in the brain stem, and a continuous, somewhat rhythmic discharge synchronous with the pulse can be recorded in some areas. This discharge is probably mediated via **baroreceptors.** The nuclei eliciting vasodilation are essentially relay stations without spontaneous activity, but they can be activated by incoming impulses from peripheral receptors and other areas in the central nervous system (CNS).

In higher levels of the CNS are areas, especially in the cerebral cortex and diencephalon, from which cardiovascular reactions can be elicited. Although these higher centers do not contribute to the continuous vasomotor tone, many adjustments are initiated primarily from the brain above the level of the medullary centers. When these centers are stimulated, the combined effect is a vasodilation of precapillary resistance vessels in the skeletal muscles and vasoconstriction of the vessels of the abdominal organs and skin. With very little shift in arterial pressure, this automatic activation pattern leads to a remarkable and instantaneous redistribution of cardiac output to favor skeletal muscles. Simultaneously, accelerator fibers to the heart also are stimulated, and the medulla of the suprarenals liberate adrenaline. This hormone dilates the resistance vessels of the skeletal muscles and excites the smooth muscles of the capacitance vessels; noradrenaline strongly contracts both resistance and capacitance vessels. These various responses are consequences of the different receptors, $\alpha$ and $\beta$.

## Baroreceptors in Systemic Arteries

The afferent input is far from being completely mapped out. Important afferent fibers come from baroreceptors (also called mechanoreceptors) in blood vessels and in the heart. The systemic arterial receptors are located in the tissue of the carotid sinus, aortic arch, right subclavian artery, and common carotid artery (Heymans and Neil 1958; Kirchheim 1976). Mechanical deformation, or stretch, of the walls of the vessels is the normal stimulus of the receptors. They respond to the rate of blood pressure increase as well as to the amplitude of pulse pressure with a discharge transmitted to the CNS. If the wall of the vessel where the stretch receptors are located becomes less distensible, caused by increased activity of smooth muscles in the wall or a progressive structural change (in a hypertensive patient or aging person), a given pressure would induce less deformation of the receptors and a reduced impulse output. In fact, pathways from suprabulbar areas are probably involved in a central resetting of the baroreceptor reflex during increased activity of other receptor groups (e.g., during physical activity). This can augment or depress the reflex responses mediated through baroreceptors. Thus, during exercise the arterial baroreceptors are reset upward as exercise begins, causing them to respond as though arterial pressure had decreased. There is also evidence that the baroreflex sensitivity is augmented in highly fit subjects (Barney et al. 1988). Recent investigations have demonstrated that at the onset of low- to maximal-intensity dynamic exercise, the carotid baroreflex is reset in direct relationship to the intensity of exercise (see K.H. Norton, Boushel, et al. 1999). Furthermore, it has been shown that the progressive resetting of the carotid baroreflex and the shift of the reflex operating point render the carotid–cardiac reflex ineffectual in counteracting the continued decrement in mean arterial pressure that occurs during prolonged exercise (K.H. Norton, Gallagher, et al. 1999).

Pulsatile pressure about a given mean blood pressure is more effective than a steady mean pressure to set up an impulse traffic in the afferent nerves (sinus branch of the glossopharyngeal nerve and afferent fibers of the vagus). The threshold at which a stimulus is effective varies for the different receptors. A recording of activity in the sinus nerve normally reveals a continuous discharge and a variation in the impulse traffic with each pulse beat.

The baroreceptors can report a decrease as well as an increase in blood pressure to the cardiovascular centers, primarily the medullary vasomotor areas. At rest, the baroreceptors exert a restraining influence on the cardiovascular system, causing a reflex bradycardia and reflex inhibition of the medullary vasomotor center.

## Posture

The physiological interplay of the various factors involved in maintaining adequate arterial blood pressure can be elucidated by the following experiment: On a tilting table, a subject is tilted from supine to a head-up position (about a 60° angle to the horizontal). Because of the force of gravity, blood is pooled in the parts of the body below the heart level. Thus, the venous return to the heart is temporarily reduced. Consequently, the cardiac output decreases and so does the arterial blood pressure. The strain exerted on the baroreceptors is reduced, and fewer impulses are transmitted from them to the CNS. The impulse output from the parasympathetic neurons is diminished (which increases heart rate), and from the adrenergic sympathetic neurons in the brain stem there is an increased impulse traffic (the effect is a vasoconstriction in resistance vessels and capacitance vessels, especially in the splanchnic area, and an increase in heart rate). The precapillary vessels in skeletal muscles are also important targets for this baroreceptor reflex (see Rowell 1993). Thus, the peripheral resistance becomes higher, the cardiac output can be restored to an adequate level, and the arterial blood pressure can increase. The variation in heart rate in a subject passively tilted to different body positions is illustrated in figure 5.12. If blood pressure cuffs are placed around the upper parts of the thighs and a pressure of about 200 mmHg (26.6 kPa) is applied when the subject is in a horizontal or head-down position, the heart rate response to the head-up position is less pronounced. The hydrostatic forces are acting on a shorter "column," because the blood is prevented from circulating to the legs. If the pressure within the cuffs is suddenly released, the decrease in arterial blood pressure can be very pronounced, and eventually the subject faints as the blood, and thereby the oxygen supply to the brain, become inadequate. The capacity of the legs to retain blood has increased, because its arterioles

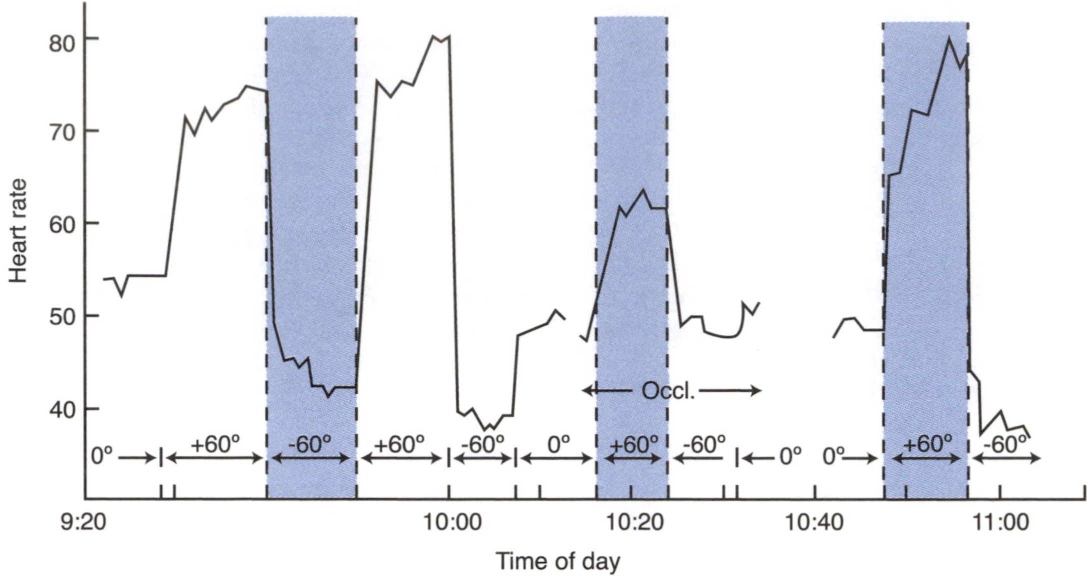


**Figure 5.12** Variation in heart rate (ordinate) in a subject passively tilted to different body positions. +60° = tilting to a head-up position; –60° = head-down position; Occl. = experiments in which inflated blood pressure cuffs were placed around the upper parts of the thighs. For further details, see text.

From Asmussen et al. 1939.

dilate as anaerobic metabolites accumulate during the period of circulatory occlusion. Tilting the subject to a head-down position will quickly restore the circulation and consciousness as the legs are drained. Figure 5.13 shows how the heart rate can be lowered some 10 beats · $min^{-1}$ if a subject in a head-up position (on a tilting table) contracts the leg muscles. The massaging effect of the repeatedly contracting muscles on the capillaries and veins enhances the venous return to the heart, and the heart rate is lowered. Most likely, the nerves from the cardiovascular baroreceptors form an important link in the reflex chain (see also Goldman et al. 1985).

The beneficial effect of the skeletal muscle pump on the venous return certainly should be stressed for people who work in a fixed sitting or standing position. Bandaged legs can partially reduce the hydrostatic shift of fluid to the legs in the upright position, and thereby the circulation is facilitated. A sudden standstill after prolonged exercise, particularly in a hot environment, can cause fainting, as the blood pools in the dilated vessels in exercised legs and in the skin.

Two important factors counteract edema formation in the legs in the erect position: The first is a pressure-induced facilitation of the rate of myogenic contractions of the smooth muscles of the arterioles and precapillary sphincters, decreasing the surface area of the capillary bed available for filtration. The second factor is an effective reabsorp-

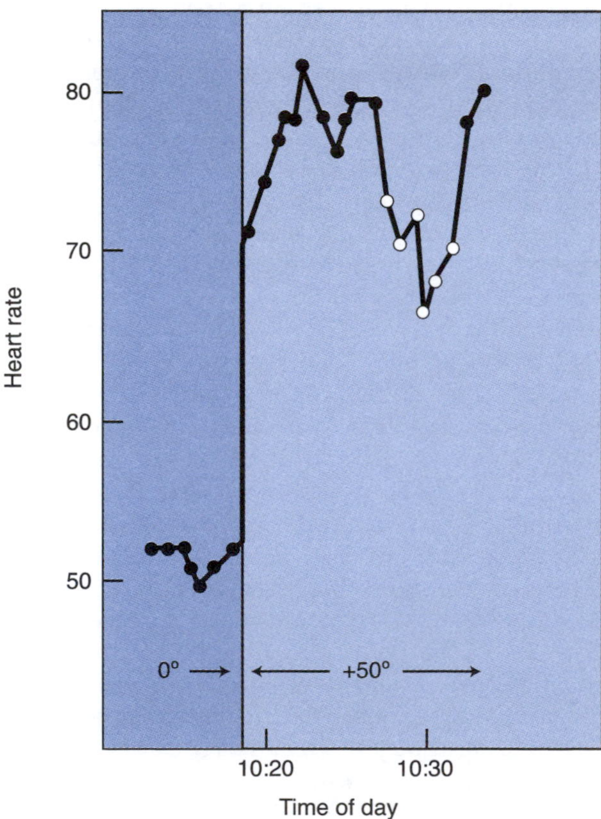


**Figure 5.13** Effect on heart rate when a subject in a head-up position (on the tilting table) voluntarily contracts the leg muscles. ● = passive standing; ○ = voluntary contraction of leg muscles.

From Asmussen et al. 1939.

tion of fluid via the red cells squeezed in the narrow capillaries. The close contact between red cells and tissue fluid will promote an uptake of water in the capillaries as well as the exchange of gases, as discussed earlier in this chapter.

Bjurstedt et al. (1983) reported a marked impairment of the orthostatic tolerance during the first 30 min of recovery after exhaustive exercise. A reduced plasma volume may be one of several factors behind a change in the reaction to a 70° head-up tilt test. It can take 50 to 60 min before the plasma volume is restored after such exercise (Harrison, Edwards, and Leitch 1975).

### For Additional Reading

For details concerning orthostatic intolerance, see Rowell (1993, pp. 118–61).

### *Summary*

A change in body position inevitably will affect the circulation as long as the individual stays under the influence of gravity. A head-up position primarily will increase the blood volume in the legs and decrease the central blood volume and cardiac output. Secondary variations in arterial blood pressure are reported from the baroreceptors in some arteries to the cardiovascular centers in the brain. The activity in sympathetic and parasympathetic nerves varies by reciprocal innervation in such a way that the arterial blood pressure and cardiac output return to a level fairly close to the one typical for the individual in a supine position.

## Other Receptors

In the walls of the pulmonary artery there are baroreceptors with reflex effects on the systemic circulation and heart similar to those caused by the systemic arterial baroreceptors. A third group of baroreceptors is located in the walls of the atria and ventricles of the heart. When stimulated, they cause by reflex a vasodilation, bradycardia, and systemic hypotension, so that the load on the heart diminishes. The other cardiovascular reflexes are of a buffer, depressor, or negative feedback nature. Stretch receptors also serve as one sensory mechanism in a reflex regulation of blood volume by control of urine output. The filling volume of the cardiac atria, related to the circulating or thoracic blood volume, seems to be the appropriate stimulus; a variation in the production of **antidiuretic hormone (ADH)** from the hypophysis is the tool.

Increased atrial distention also leads to the secretion of a peptide hormone known as **atrial natriuretic peptide (ANP),** also called atrial natriuretic factor, which is synthesized and secreted by cells in the cardiac atria. ANP inhibits sodium reabsorption and acts on the renal blood vessels to increase glomerular filtration rate. ANP concentration is increased concomitantly with increased cardiac filling (increased preload), but exercise-induced increases in heart rate and sympathetic nervous activity seem to be an even stronger stimulus for ANP release (Kanstrup, Marving, and Høilund-Carlsen 1992).

In this context, it is of interest that one of the primary adjustments that take place during a dive is circulatory in nature (Scholander et al. 1962). Most animals, including humans, display a diving bradycardia, which usually develops gradually. Thus, during the dive, the frequency may drop to one half of normal in some species and to one tenth in others. As a rule, the bradycardia stays with the animal whether it exercises or not during the dive. Blood pressure is maintained at a normal, or even an elevated, level as a result of peripheral vasoconstriction. In the diving animal, the reduction and selective redistribution of the circulation can save oxygen for vital organs such as the brain and heart. In the human, the heart rate increases immediately at the start of a dive and slows down very markedly as the dive progresses, even if the diver exercises vigorously (Strømme, Kerem, and Elsner 1970). Prompt return to normal sinus rhythm occurs with the first breath. This bradycardia during diving is not a result of apnea only, because it is reinforced by water submersion. Skin receptors, wetting of the nose, and asphyxial reflexes may contribute to the development of the bradycardia. The decline in heart rate is very rapid. However, it may be that the chemoreceptors gradually contribute to the bradycardia by a progressive hypoxic drive, because the decrease in alveolar oxygen tension is very marked, at least during exercise combined with breath holding.

### For Additional Reading

For further discussion about apneic heart rate responses in humans, see Manley (1990).

Afferent impulses to the cardiovascular centers from receptors in the skin and joints have been suggested but have not as yet been shown to exist. Receptors in the skeletal muscles can be stimulated by stretch, and they respond to muscular contraction. Group III and some group IV afferent fibers transmit the impulse traffic. Free nerve endings in the muscles serve as receptors that respond to pain (nociceptors). There is evidence that they can be activated by chemical factors such as serotonin,

**bradykinin,** potassium, hyperosmotic lactate, and phosphate. As is discussed subsequently, they are stimulated by metabolites and thus could be of some importance in regulating cardiovascular function during exercise.

## Regulation of Circulation During Exercise

The conclusions of the preceding discussions are as follows: The blood flow in the precapillary vessels of the muscles is regulated locally, for the most part by the level of metabolism in the muscle cells. The actual nutritional demand is the determining factor and is superimposed on any nervous influence. The blood flow in the splanchnic area, on the other hand, is under control of the CNS. The flow is varied so that the systemic arterial blood pressure is maintained at an adequate level to supply brain, heart, and other vital organs with blood. In this case, neural vasoconstrictor activity can be superimposed on the local dilator control. The vessels of the skin also subserve centrally controlled mechanisms. Impulses from the medullary cardiovascular area are important, but the final control probably is exerted by the temperature-regulating centers, at least while the subject is at rest or during submaximal exercise.

At rest, the kidneys receive about 25% of the cardiac output. There are no tonic impulses from the CNS to renal blood vessels, but electrical stimulation of the renal nerves causes intense renal vessel constriction with associated changes in blood flow (down to 250 ml · min$^{-1}$) and in excretion of water and electrolytes (Pappenheimer 1960). Exercise, postural changes, and circulatory stress in general can profoundly alter renal function, mediated through the hemodynamic effects of renal nerves.

This summarizes the main tools available for regulation of the circulatory system during exercise. The approximate blood distribution to the various organs is illustrated in figure 5.9.

The changes in heart function and circulation, from the moment muscular exercise begins (or even before), are initiated from the brain levels above the medullary area (probably the cerebral cortex and diencephalon). This is a central command, and it is speculated that the central activity that recruits the motor units also stimulates the medullary and spinal neuronal circuits that elicit the cardiovascular responses to exercise. In addition, the mechanical and chemical events in the exercising muscle are thought to contribute by reflex mechanisms to the cardiovascular adjustment by some sort of peripheral command (Mitchell, Kaufman, and Iwamoto 1983). By a reciprocal innervation, there is simultaneous increase in the sympathetic activity and decrease in the parasympathetic impulse traffic.

The skeletal muscle cells receive an increased share of the cardiac output because of vasodilation caused by local metabolic activity as well as circulating adrenaline acting on the large number of β-adrenergic receptors found in the arterioles of skeletal muscle. On the other hand, adrenaline, like noradrenaline released from sympathetic nerves, binds to α-adrenergic receptors, causing vasoconstriction and reduced blood flow to skin, kidneys, and splanchnic area (figure 5.14). Rowell (1993) calculated that at maximal vasoconstriction of the splanchnic and renal blood vessels, about 2.2 L · min$^{-1}$ can be redistributed to the activated muscles. This could increase their oxygen uptake by about 0.5 L · min$^{-1}$ without any additional increase in the cardiac output. It should be emphasized that a vasoconstriction of precapillary vessels will secondarily reduce the transmural (distending) pressure in the postcapillary vessels. The previously more or less distended veins collapse easily because of the passive recoil of the venous wall (Rothe 1983). The reduction in blood flow to the splanchnic tissues and kidneys (as well as the increase in heart rate) during exercise is more related to the severity of the exercise in relation to the individual's maximal oxygen uptake than to the absolute rate of oxygen uptake (Rowell 1993). Rowell noted a linear decline in the splanchnic blood flow down to about 30% of the flow at rest, during exercise or heat stress. Actually, the splanchnic blood flow in humans is inversely related to the heart rate under a wide variety of stresses (about 0.6% reduced flow per heartbeat increase). One may speculate that the sympathetic branches that control the heart rate and splanchnic vascular resistance are parallel. The splanchnic bed and the kidneys are particularly suited for short-term adjustments of the systemic vascular resistance because their blood supply greatly exceeds the metabolic requirements of their tissues. Mainly passive constriction of the veins (capacitance vessels), pumping action of the exercising muscles, and forced respiratory movements assist the venous return to the heart, causing what can be termed as an increased **preload.** Preload refers to the extent of filling of the ventricles at the end of diastole. Increasing preload leads to an increased stroke volume (the Starling effect). Because the heart escapes from the vagal inhibition, and sympathetic impulses can increase the force and frequency of the cardiac contractions, the heart gains capacity to take care of the increased inflow of blood and, if necessary, pumps it out against an elevated resistance, the **afterload.** The volume of blood in the

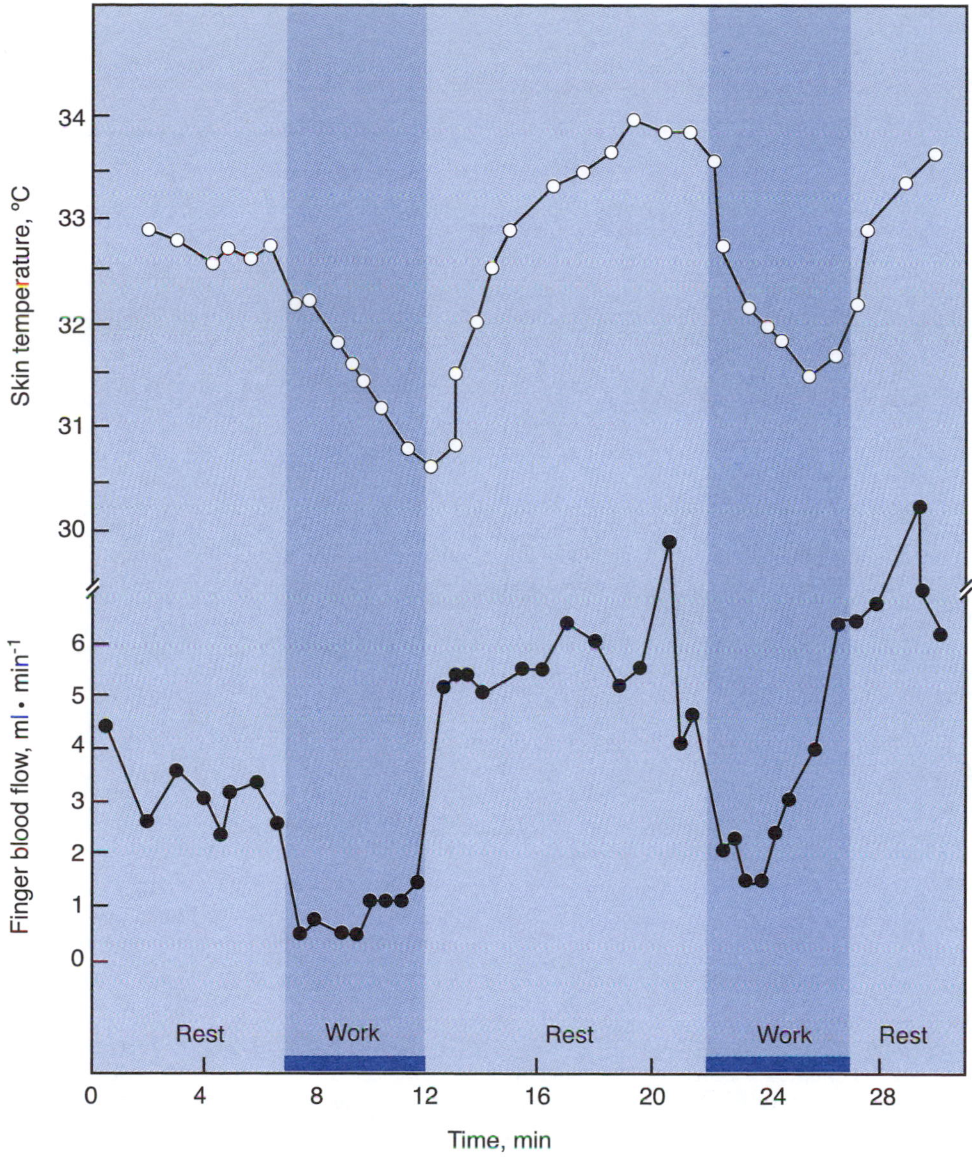


**Figure 5.14** Blood flow in the index finger (measured by means of a plethysmographic method) and skin temperature on the same finger at rest and during two periods of work on a cycle ergometer (1,080 kpm · $min^{-1}$, or 180 W).

From Christensen and Nielsen 1942.

heart, the end-diastolic volume, increases more during exercise in the supine position than when the individual is upright (Poliner et al. 1980).

Figure 5.15 illustrates the changes in the composition of the interstitial fluid in the skeletal muscle from a state of rest to that of exercise. Substances released by the activated skeletal muscle cells interact with specific vascular receptors, causing the smooth muscle cells of the resistance vessels to be hyperpolarized, which means relaxed. This local control of blood flow is a very important factor in securing an efficient blood supply to the exercising muscles. To the changes in the interstitial fluid that are caused by exercise, as illustrated in figure 5.15, can be added release of adenosine and some of the **prostaglandins** with a potential to produce vasodilation (Faber, Harris, and Miller 1982). Thus, Marshall (1995) suggested that the partial pressure of oxygen ($PO_2$) in muscle tissue affects the production of adenosine. This, in turn, is supposed to open an ATP-sensitive $K^+$ channel with interstitial $K^+$ concentration modulating the degree of vasodilation. However, earlier evidence shows that an increase in potassium concentration and hyperosmolarity only contribute to the early phase of vasodilation (Mohrman 1982; Olsson 1981). More

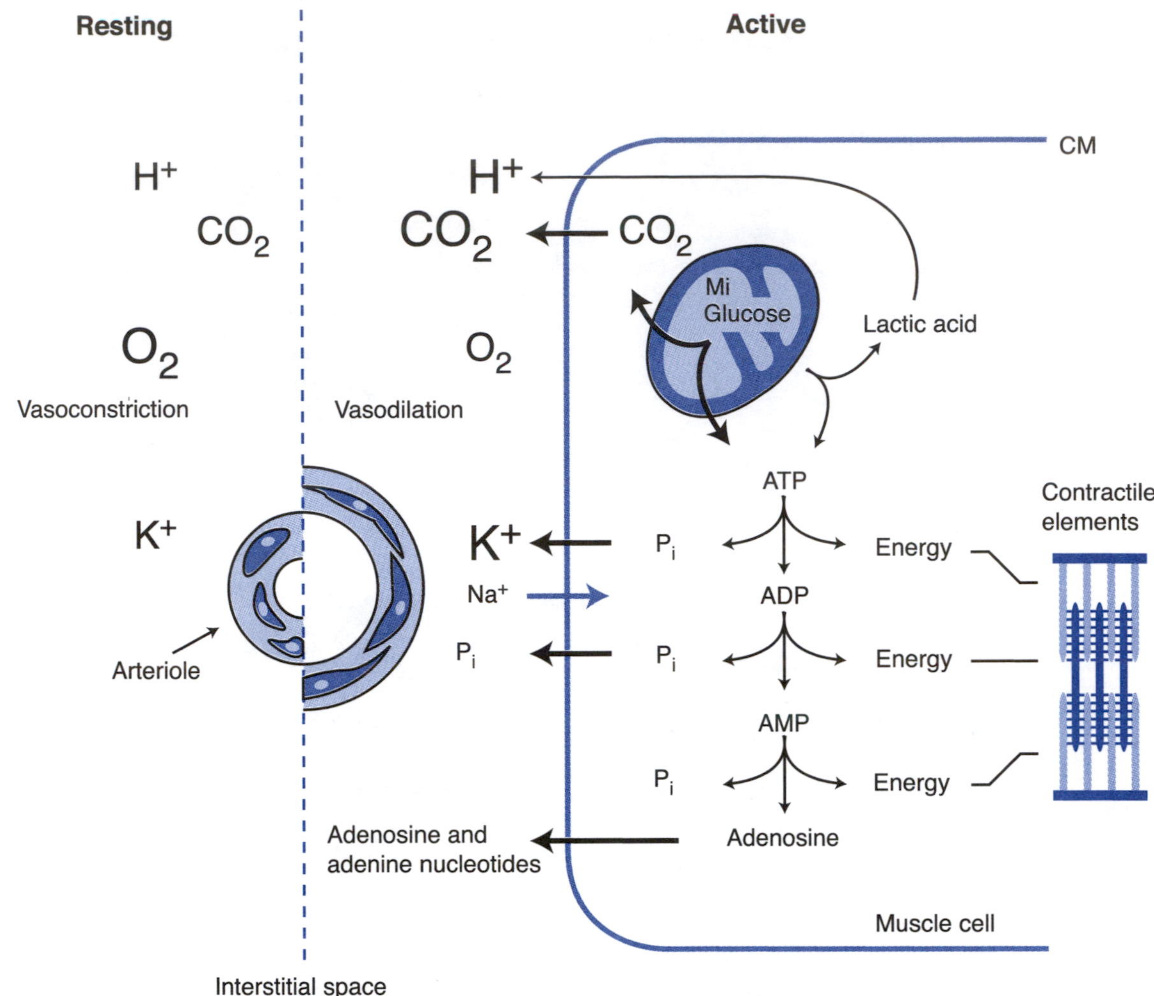


**Figure 5.15** Major changes in the composition of the interstitial fluid during contraction of muscle cells. When the muscle is inactive (left), the arterioles are constricted, the concentration of metabolites and carbon dioxide in the interstitial fluid is low, and little oxygen is used. When the muscles become active (right), (1) the depolarization of the cell membrane (CM) increases the potassium ion ($K^+$) concentration in the extracellular space; (2) the regeneration of adenosine triphosphate (ATP) by the mitochondria (Mi) augments the production of carbon dioxide, which diffuses to the extracellular space; (3) the anaerobic production of ATP in the cytoplasm results in the formation of lactic acid, which slowly diffuses out of the cell; (4) the increased amounts of lactic acid and carbon dioxide increase hydrogen ion ($H^+$) concentration of the extracellular fluid and thus decrease pH; (5) the breakdown of ATP to adenosine diphosphate (ADP) and adenosine monophosphate (AMP) and to adenosine, with liberation of inorganic phosphate (Pi), augments the concentration of adenosine and adenine nucleotides in the extracellular space; and (6) the osmolarity of the extracellular fluid increases. Each of these changes can relax contracted smooth muscle cells, and it is likely that their combination is responsible for the adjustment of the blood flow to the metabolic needs of the tissues. (Increased concentrations or osmolarity are symbolized by larger letters.) $Na^+$ = sodium ion; $CO_2$ = carbon dioxide. For further details, see text.

From J.T. Shepherd and Vanhoutte 1979.

recently, it has been suggested that a major part of the adenosine-mediated component of hypoxia-induced vasodilation is caused by adenosine acting on the endothelium and releasing NO. This is in line with the evidence that hypoxia also causes the release of ATP from the epithelium, which can be degraded to AMP and then broken down to adenosine. Last, it may be mentioned that the $PO_2$-sensitive enzyme $P_{450}$ (known from the kidney) also appears to be operative in skeletal muscle. This enzyme catalyzes the formation of vasoactive metabolites from arachnoid acid. The lower the $PO_2$, the higher the enzyme activity (Saltin, Boushel et al. 2000).

As the exercise proceeds, the blood vessels of the skin, especially the arteriovenous anastomoses, dilate, so that the produced heat can be transported to the surface of the body. The heavier the exercise and the higher the environmental temperature, the more pronounced is this secondary vasodilation in the skin. Indirectly, impulses in sympathetic nerve fibers are partly behind this dilation, and the temperature-regulating center in the hypothalamus guides the impulse traffic. These nerve fibers stimulate sweat production by means of acetylcholine. The local skin temperature also affects the lumen of the vessels. (Circulatory aspects of heat stress are discussed in chapter 13; see also Rowell 1993.)

During exercise, the integrated effect of neural and chemical factors (including hormones) gives a cardiac output that can be markedly higher and with quite a different distribution than when the subject is at rest (see figure 5.9). The reflex control of the resistance and capacitance vessels acts within seconds to move blood from the splanchnic region to the thorax. The blood flow through the exercising muscles increases by a capillary recruitment and an increase in flow velocity (Honig and Odoroff 1981).

The increased capillary pressure leads to a net outward filtration of fluid, the flow of which is facilitated by a simultaneous increase of the capillary area. The mean capillary pressure can increase from 15–20 to 25–35 mmHg (2.0–2.7 to 3.3–4.7 kPa) in activated skeletal muscles. In addition, the muscular contractions will result in an accumulation of metabolic end products inside the cell, causing an osmotic gradient that leads to a net uptake of water into the cell. At the same time, metabolites and potassium are transferred from the cell to the interstitium. Consequently, the interstitial water becomes hypertonic compared with blood, which shifts water from the blood to the interstitium (Coyle and Hamilton 1990). The plasma volume can decrease immediately by some 10% after the start of exercise, but this decrease will slowly return to 3% to 5% unless dehydration takes place (Sawka 1988). The muscle volume increases during exercise are most pronounced during anaerobic exercise, which causes large intracellular lactic acid accumulation.

Conversely, in tissues where the precapillary vessels are constricted, the mean capillary pressure decreases. This factor, plus an increased arterial osmolarity, favors a mobilization of extravascular fluid, so that the plasma volume can be maintained relatively well. There is a linear relationship between the bulk of plasma volume lost from the vascular system and the exercise intensity. At a given oxygen uptake, a relatively larger percentage decrease in plasma volume is noted for arm compared with leg exercise (D.S. Miles et al. 1983). During arm exercise, the mean arterial blood pressure is higher than in leg exercise and, as a consequence, the net outward filtration in the dilated capillaries is enhanced. On the other hand, the total capillary surface area is smaller in the exercising arm muscles than in leg muscles.

## Summary

We can schematically describe the circulatory response to exercise by considering four stages. First, at rest the skeletal muscles receive only some 15% of the minute blood flow, and their arterioles are constricted by a continuous vasoconstrictor activity and spontaneous vascular tone. Few capillaries are open, but the individual capillaries open and close alternatively. The heart rate is kept down by a parasympathetic outflow via the vagal nerve. In the second state, when (or even before) exercise begins, parasympathetic activity is inhibited and sympathetic impulse traffic is increased. According to one hypothesis, signals from higher levels of the CNS formulate a central command that initiates these autonomic nervous responses. Receptors in the activated skeletal muscles send afferent impulses to the medullary cardiovascular area, acting in concert with the central command. The heart is freed from its inhibition and, supported by the adrenergic activity, it beats faster and with increased cardiac contractility. Local metabolic activity and adrenaline acting on β-adrenergic receptors on vascular smooth muscle in the precapillary resistance section dilate arterioles in the muscles, thereby increasing their blood flow. On the other hand, sympathetic adrenergic vasoconstrictor nerves act on the vessels of abdominal organs and skin so that a decreasing share of the cardiac output flows through those tissues. The veins become constricted, both passively and by activity in the constrictor fibers. This constriction of veins, together with the pumping action of the working muscles and the forced respiratory movements, facilitates the blood return to the heart, enabling an increased cardiac output. In the third stage, the appropriate adjustment of the circulation occurs. In the activated muscles, the increased metabolism causes changes in the environment that locally dilate arterioles and open capillaries. Hormones, including circulating catecholamines, contribute to the constriction of vessels in nonactive areas. During the fourth stage, for temperature balance within the body, the produced heat is transported to the skin, because skin vessels become dilated.

# Cardiac Output and the Transportation of Oxygen

The cardiac output at rest is approximately 4.5 to 6 L · $min^{-1}$ depending on body size. With increasing oxygen uptake, there is an approximately rectilinear increase in cardiac output. Thus, in individuals with maximal oxygen uptake of 3.0 to 3.5 L · $min^{-1}$, cardiac output increases to 18 to 23 L · $min^{-1}$. In elite endurance athletes, values as high as 42 L · $min^{-1}$ at a maximal oxygen uptake of 6.2 L · $min^{-1}$ have been reported (Kanstrup and Ekblom 1978).

## Efficiency of the Heart

From studies of cardiac output and the energy output and work efficiency of the heart the following factors seem of special interest:

> Factor 1: A given stroke volume can be ejected with a minimum of myocardial shortening if the contraction starts at a larger volume.
>
> Factor 2: Energy losses in the form of friction and tension developed within the heart wall are also at a minimum in a dilated heart.
>
> Factor 3: The stretched muscle fiber can, within limits, provide a higher tension than the unstretched one.
>
> Factor 4: Loss of energy is larger when the contraction occurs rapidly, that is, with a high heart rate as compared with a slower contraction rate.
>
> Factor 5: On the other hand, the greater the volume of the heart, the higher the tension of the myocardial fibers necessary to sustain a particular intraventricular pressure (as a consequence of Laplace's law, as discussed earlier in this chapter). The energy need for a contraction is closely related to the tension that has to be developed.

The individual with a high capacity for oxygen transport because of natural endowment or training has a large stroke volume and a slow heart rate. Cardiac output and systolic/diastolic blood pressures at given rates of exercise or at rest are not noticeably different from normal. Of the factors listed previously, the first four tend to act in the individual's favor as far as each single heartbeat is concerned, whereas factor 5 acts against the person. However, because relatively few contractions are performed per minute, the total energy cost to maintain a given work level (flow times pressure) may be relatively low and the efficiency high. At maximal work rates, the person with the large diastolic heart volume and stroke volume may have as high a heart rate as the individual with a small diastolic filling. In that situation, factors 1 through 4 still favor the fit person, but factor 5 tends to decrease the efficiency when performance per unit time is considered.

A rich capillary network in the myocardium would provide the capacity to meet the demand of an increased metabolism. For the exercising skeletal muscles, however, it is essential that the increased demand for blood flow through the myocardium does not consume too many of the extra liters of blood that the heart eventually can manage to pump out. Apparently the energy requirement of the heart muscle is normally highly correlated with the cardiac output. The coronary blood flow is 4% to 5% of the cardiac output at rest as well as during vigorous exercise (see figure 5.9). There is only a slight increase in the arteriovenous oxygen difference ($(a\text{-}\bar{v})O_2$ difference), in the heart with increasing cardiac output. As mentioned, this $(a\text{-}\bar{v})O_2$ difference is as high as 16% to 17% by volume at rest. During exercise, it can increase to 18% to 19% by volume (oxygen content of the arterial blood = 20% by volume or higher).

From a general viewpoint, it is considered an advantage if a given level of cardiac output can be maintained with a low heart rate, that is, a large stroke volume. The reason why a training with "overload" apparently is necessary to improve the efficiency of the heart is presently unknown.

For the cardiac patient, the situation is different if a large diastolic filling is combined with a small stroke volume and a high heart rate. The dilation of the heart muscle improves the ability of the muscle fibers to produce tension as long as they are stretched, as in factor 3. Factors 1 and 2 help keep the efficiency high. Factors 4 and especially 5 tend to reduce efficiency. Undoubtedly, the heart has to pay for compensating the basically reduced myocardial strength. Eventually a critical equation for energy requirement and energy supply of the heart will come into play, and the vascular bed in the myocardium is the key in that formula. At rest, the capillary blood flow may cover the need, but during even mild exercise, a discrepancy arises, and the patient is forced to stop because of symptoms such as angina pectoris, which is a pain probably caused by hypoxia in the myocardium. With a high heart rate, the tension time, that is, the time during which the heart muscle is contracted per minute, is prolonged, and the blood flow through the heart muscle therefore is reduced (see figure 5.4b).

## For Additional Reading

For further discussion concerning functional and structural characteristics of the "athletic heart," see K.P. George, Wolfe, and Burggraf (1991).

# Venous Blood Return

Certainly, the cardiac output cannot exceed the flow of blood returning to the heart. When a person is in the supine position, the hydrostatic pressure on the venous side of most of the capillaries is about 10 mmHg (1.3 kPa), but the pressure gradient to the right atrium is increased by the negative intrathoracic pressure. At rest, this pressure is about 5 mmHg (0.7 kPa) less than the ambient barometric pressure because the elastic tissue of the lungs is expanded to the size of the thoracic cavity and the recoil of the tissue exerts a tension on the thin-walled vessels within the thorax. The inspiration increases this pulling force, and blood is sucked into the thorax. At the same time, the abdominal veins are compressed as the diaphragm contracts. The variations in intrathoracic and intra-abdominal pressures with breathing will significantly enhance the venous return, because a backflow in the veins is hindered by the capillary resistance and venous valves. During an expiratory effort with the **glottis** closed (**Valsalva maneuver**), the intrathoracic pressure increases, impairing both the venous return and the cardiac output; this is usually the case in weightlifting. **Hyperventilation** followed by the Valsalva maneuver can cause fainting (for an explanation, see chapter 6). There is evidence that the ventricular action contributes to the subsequent filling of the ventricle by exerting a suction force on the atrium (Folkow and Neil 1971).

A third important factor improving the venous return is dynamic activity of the skeletal muscles (figure 5.16; see also figure 5.13). As pointed out earlier in this chapter, the skeletal muscle pump is very effective in propelling blood toward the heart. When the leg muscles contract rhythmically, the blood volume of the legs decreases, indicating the emptying of blood. The action of the muscle pump

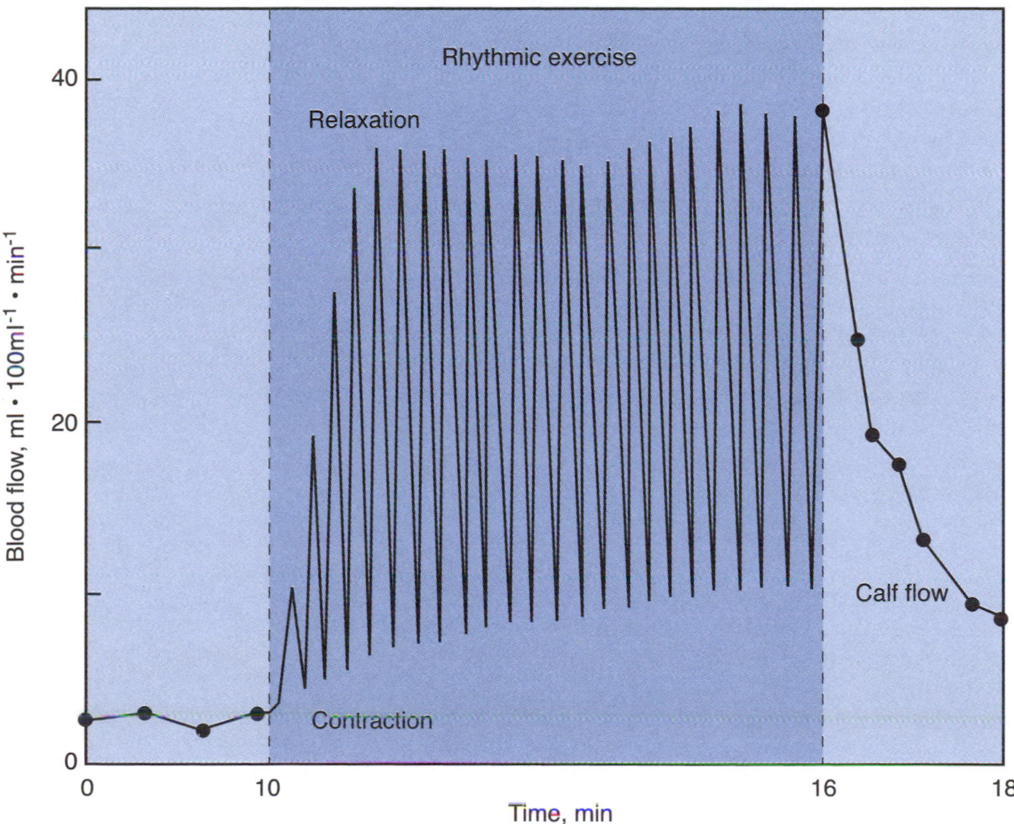


**Figure 5.16** Schematic representation of changes in blood flow through the calf muscle during strong rhythmic contractions. During contraction, the blood within the muscles is emptied toward the heart, but at the same time the inflow is greatly reduced.

From Barcroft and Swan 1953.

is especially important when a person is standing erect. If exercise is started (walking), the pressure in the veins of a foot can decrease from 100 mmHg (13.3 kPa) in passive standing to about 20 mmHg (2.7 kPa).

The increase in venous return to the heart as exercise starts will enable an immediate increase in cardiac output. An increased filling pressure in the heart (increased preload) may stimulate the stretch receptors in the myocardium and in a reflex manner accelerate the heart and improve its stroke. The larger the muscle mass involved, the more pronounced is this effect. As pointed out previously, when the exercise stops, the blood stays temporarily in the dilated vascular bed, and the decrease in venous return to the heart can cause such a decrease in cardiac output and arterial blood pressure that the person faints. This transient pressure decrease is more likely to occur if the skin vessels are dilated to secure heat elimination and the person remains motionless in a standing position.

At rest, the venous system (capacitance vessels) contains about 65% to 70% of the total blood volume. By constriction of the venules and veins, close to half of that blood volume can be mobilized and emptied toward the heart. Vasomotor activity in the skeletal muscles, the largest interstitial fluid depot in the body, is especially effective as a tool for a reflex control of the blood volume. Variations in blood volume by blood loss, dehydration, or prolonged inactivity can influence the filling of the heart.

To complete the picture, it should be emphasized that the normal heart has the capacity to pump all the blood that is returned to the heart into the arteries. Increases in heart rate and force of contraction thus can keep the atrial pressure low even during heavy exercise. Failure of the left or right ventricle would cause an increase in the central venous pressure and severe disturbances in the capillary fluid exchange (congestive heart failure).

## *Summary*

The venous return to the heart is determined by the balance between the filling pressure and the distensibility of the heart, that is, the intraventricular pressure minus the intrathoracic pressure. The filling is enhanced by (1) the variation in intrathoracic and intra-abdominal pressures during the respiratory cycle, (2) the effect of the muscle pump during muscular movements, and (3) a vasoconstriction in the postcapillary vessels. Changes in body position will, at least temporarily, affect the volume of blood in central veins (for further discussion, see Rowell 1993).

## Cardiac Output and Oxygen Uptake

At rest in the supine position, the cardiac output is 4 to 6 L · min$^{-1}$ with an extraction of 40 to 50 ml of oxygen per liter of blood and a total oxygen uptake of 0.2 to 0.3 L · min$^{-1}$. When a subject strapped to a tilting table is tilted from the horizontal to the feet-down position, the cardiac output can decrease from 5 to 4 L · min$^{-1}$. This decrease is attributable to the previously discussed venous pooling. The stroke volume is reduced and the heart rate is usually increased. Activation of the skeletal muscle pump propels the blood toward the heart, and the heart rate may even decrease as the stroke volume increases (see figure 5.13). In the passive feet-down position, the oxygen uptake is unchanged, and hence the $(a-\bar{v})O_2$ difference is increased.

During exercise, the cardiac output increases with the increase in oxygen uptake, but not quite linearly, if a range from rest value up to maximal is considered (figure 5.17). A very physically fit man can increase his oxygen uptake from 0.25 to 5.00 L · min$^{-1}$ or more when exercising on a cycle ergometer or treadmill. This increase is, let us assume, met by an

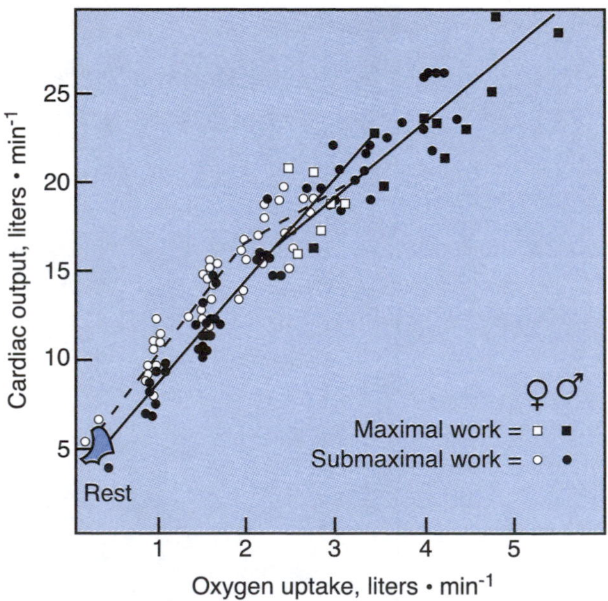


**Figure 5.17** Individual values on cardiac output in relation to oxygen at rest, during submaximal exercise, and during maximal exercise on 23 subjects using a cycle ergometer. Regression lines (broken lines for women) were calculated for experiments where the oxygen uptake was below 70% and above 70% of the individual's maximum.

From P.-O. Åstrand et al. 1964.

increase in heart rate from 50 to 200 beats · min$^{-1}$ and in stroke volume from 100 to 150 ml, which means that during maximal exercise the cardiac output has increased from 5.0 to 30.0 L · min$^{-1}$, or six times the resting level. Because the oxygen uptake increased 20 times, the (a-$\bar{v}$)$O_2$ difference must have changed from 50 to 165 ml per liter of blood, or 3.3 times the resting level (and 3.3 × 6 is close to 20). This better use of the oxygen transported by the blood is reached principally in two ways: First, the blood flow is redistributed during exercise so that skeletal muscles, with their pronounced ability to extract oxygen, receive 80% to 85% of the cardiac output, compared with some 15% at rest. Second, the oxygen dissociation curve is shifted so that more $HbO_2$ is reduced than normal at a given pressure for oxygen, that is, the percentage of saturation is less. This so-called Bohr effect is easier to understand if one studies figure 5.18. In the active muscles, the temperature can exceed 40 °C and the pH can be lower than 7.0, so there is really not a fixed relation between oxygen tension and oxygen saturation of the Hb.

At normal pH (7.40), $CO_2$ tension (40 mmHg), and blood temperature (37 °C), the blood keeps about 33% $HbO_2$ at an oxygen tension of 20 mmHg (5.3 kPa). At a pH of 7.2 and temperature of 39 °C, the percentage of $HbO_2$ at the same oxygen tension is reduced to about 17, which means that 1 L of blood with an oxygen capacity of 200 ml can deliver 26 ml more oxygen without changes in the pressure gradient for oxygen between capillary and muscle cell. With 24 L of blood circulating the working muscles per minute, this extra oxygen delivery amounts to about .6 L · min$^{-1}$ or 12% of the total uptake, in this example. However, with pH 7.0 and muscle temperature 40 °C, the extra oxygen uptake would not be increased further because the oxygen content of the arterial blood would be affected negatively by the low pH and high temperature. (The effect of diphosphoglycerate on the oxygen dissociation curve is discussed in chapter 6.)

With CO present, carboxyhemoglobin (HbCO) is formed. Such a conversion affects the $HbO_2$ dissociation curve with a shift to the left. The effect of

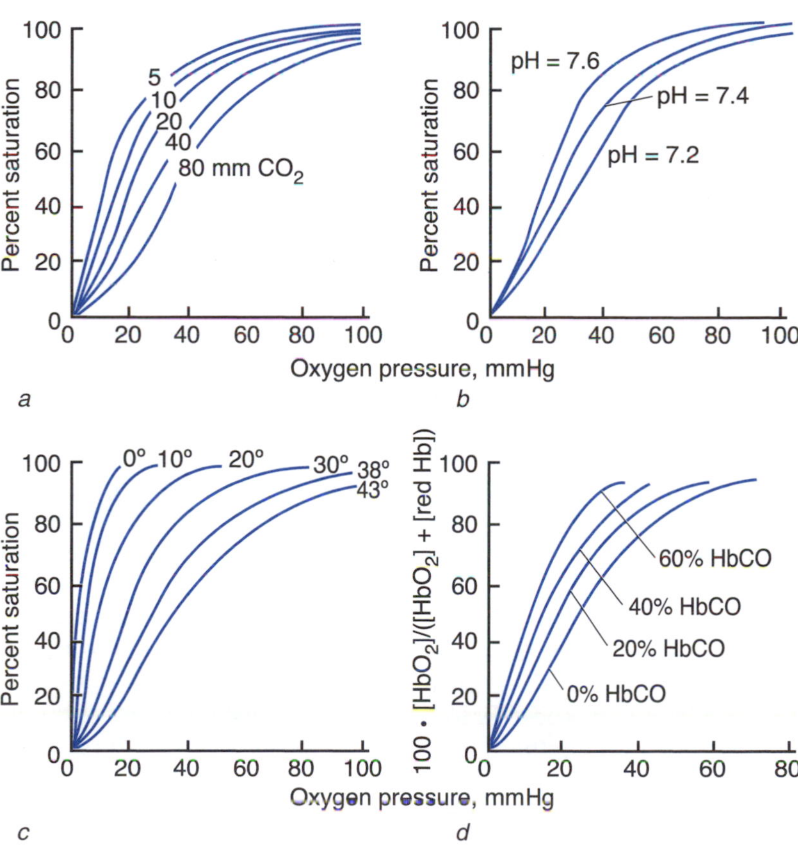


**Figure 5.18** Effects of *(a)* carbon dioxide ($CO_2$), *(b)* pH, *(c)* temperature (°C), and *(d)* carbon monoxide (CO) on the oxygen dissociation curve of the blood. Hb = hemoglobin; HbCO = carboxyhemoglobin; percent saturation = percent of oxyhemoglobin.

CO on oxygen transport is therefore twofold: It reduces the amount of Hb available for oxygen transport, and it interferes with the unloading of oxygen in the tissues. For smokers, the effect of the resulting CO on the Hb becomes a real handicap during exercise.

The shift in the dissociation curve is a result of the heat production by the exercising muscle cells and the formation of $CO_2$ and free protons during the heavy exercise. The effect of $CO_2$ in releasing oxygen from the blood is actually twofold: $CO_2$ lowers the pH of the blood and, by combining with Hb, reduces its affinity for oxygen.

By two mechanisms based on a regulation of the circulation and an inherent characteristic of Hb, the oxygen uptake can be elevated 20 times, but the cardiac output has to increase to only 30 $L \cdot min^{-1}$, not to $20 \times 5 = 100$ $L \cdot min^{-1}$.

## Oxygen Content of Arterial and Mixed Venous Blood

During exercise, there is a **hemoconcentration** of the blood, which is explained partly by the mentioned withdrawal of fluid to the active muscle cells and by the interstitial fluid (receiving the metabolites produced in the cells). Thus, the osmotic pressure is highest within and close to the metabolically active cells. The increased capillary pressure and surface area also increase outward filtration. This hemoconcentration makes the blood more viscous, but it also increases the transport capacity per liter of blood for both oxygen and carbon dioxide.

There is evidently a discrepancy between the increase in oxygen-binding capacity of the blood and the extra oxygen actually taken up during strenuous exercise. In other words, saturation of the arterial blood is slightly reduced during maximal exercise, despite a normal or even elevated oxygen tension in the lung alveoli. The arterial pH, however, may be below 7.2 and the blood temperature may be markedly elevated, and therefore the shift in the oxygen dissociation curve to the right gives a noticeable effect on the oxygen saturation even at a high oxygen tension in the blood. There is always an admixture of venous blood from the heart muscle and lung tissue to the arterial blood. These factors taken together can reduce the oxygen saturation of arterial blood from, say, 97% at rest to about 90% during maximal exercise. This shift is a disadvantage in the lungs, but the overall effect is, as discussed previously, an improved oxygen delivery attributable to the advantage at the tissue level, both in active and nonactive areas.

The increased extraction of oxygen from the arterial blood as exercise becomes heavier is illustrated in figure 5.19. During maximal exercise, the venous blood leaving the muscles has a very low oxygen content. In this study, the calculated oxygen content in mixed venous blood averaged about 20 $ml \cdot L^{-1}$ of blood for both women and men. At rest, the blood flow through most tissues is luxurious as far as the oxygen need is concerned, because other functions determine the flow distribution (e.g., through the vascular bed in kidneys, intestines, and skin; for the oxygen supply to the kidneys, about 50 ml of blood per minute would be enough, but at rest the actual flow exceeds 1 $L \cdot min^{-1}$). As emphasized in previous discussions, the blood flow during exercise is redistributed with the primary object of supplying metabolically active tissues with oxygen and removing the produced carbon dioxide. This blood flow has the potential to handle the transport of substrates, hormones, waste products, and the heat produced by metabolism.

It is reasonable to assume that some correlation should exist between the oxygen content of the arterial blood and the cardiac output at a given oxygen uptake. Women have about 10% lower concentration of Hb in the blood than men. P.-O. Åstrand et al. (1964) observed that the cardiac output required to transport 1.0 L of oxygen was 9.0 L for women (oxygen content in arterial blood: 167 $ml \cdot L^{-1}$) and 8.0 L for men (oxygen content: 192 $ml \cdot L^{-1}$) during submaximal exercise with an oxygen uptake of 1.5 $L \cdot min^{-1}$. The cardiac output in males is more effective in its oxygen-transporting function than in women, and this difference can be explained by the Hb content of the blood.

The relationship between oxygen uptake during maximal exercise and the oxygen content of arterial blood has been further analyzed. Subjects were exposed to acute hypoxia by reducing the ambient pressure of the inspired air to simulate an altitude of 4,000 m (barometric pressure = 460 mmHg; 61.2 kPa). The cardiac output during submaximal exercise was higher at high altitude than at sea level, but during maximal effort, no difference in cardiac output was observed. The oxygen uptake during maximal exercise, however, was reduced in proportion to the decrease in oxygen content of the arterial blood, or to about 70% of what it was at sea level (Hartley, Vogel, and Landowne 1973; Stenberg, Ekblom, and Messin 1966).

With part of the Hb blocked by CO (up to 20%), the oxygen transport at a given submaximal rate of work can be maintained. The heart rate increases

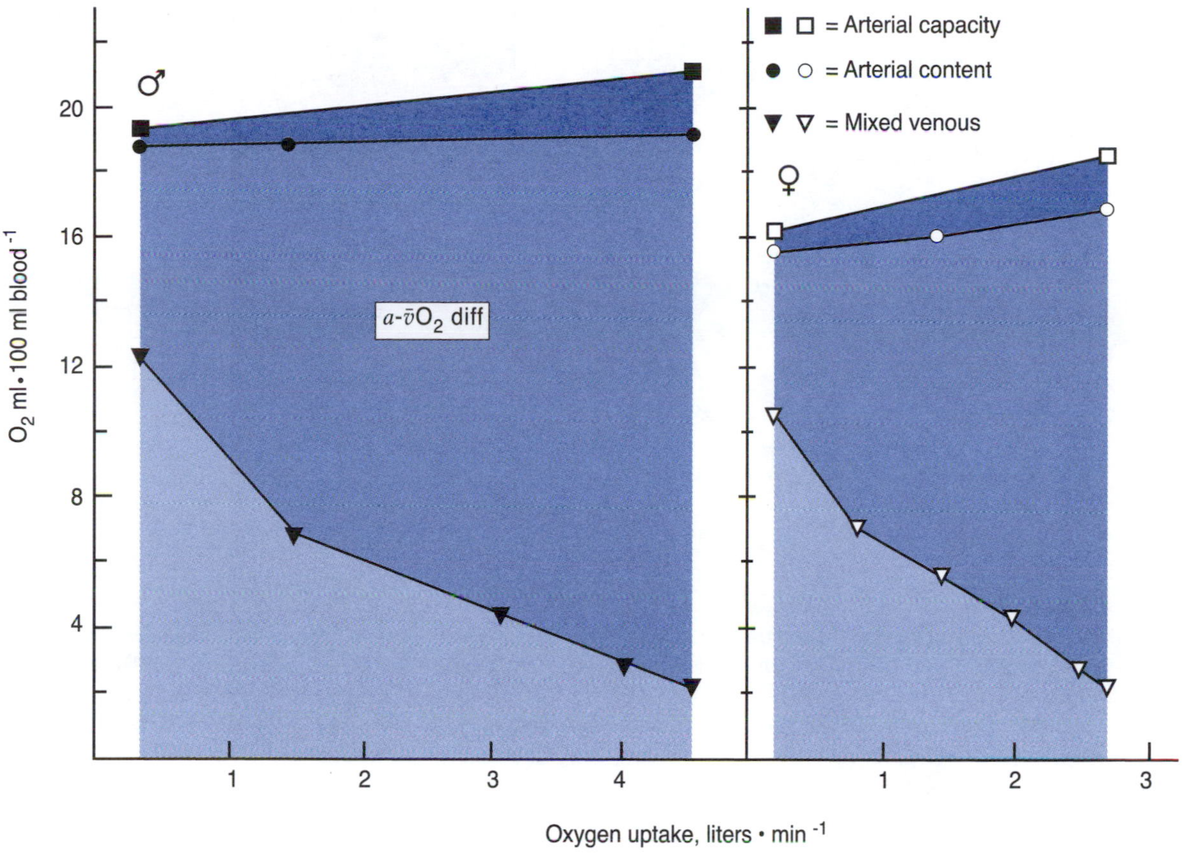


**Figure 5.19** Oxygen-binding capacity and measured oxygen content of arterial blood; calculated oxygen content of mixed venous blood at rest and during exercise up to maximum on cycle ergometer. Mean values for five female (right) and five male subjects (left), 20 to 30 years of age and with high maximal aerobic power (P.-O. Åstrand et al. 1964). During maximal exercise, the arterial saturation is about 92% compared with 97% to 98% at rest, and the venous oxygen content is very low and similar for women and men. $a\text{-}\bar{v}O_2$ diff = arteriovenous oxygen difference.

and the cardiac output is at control level or somewhat higher. During maximal exercise, the oxygen uptake is reduced more or less in proportion to the varied oxygen content of the arterial blood. However, Ekblom et al. (1975) reported that with 15% HbCO, the cardiac output averaged 5% lower than in a control experiment.

An increased oxygen tension in the inspired air will increase maximal oxygen uptake and improve performance. In the studies by Ekblom et al. (1975), eight subjects breathing 50% oxygen in nitrogen at sea level showed an average 12% increase in maximal aerobic power (in uphill running; figure 5.20a). During maximal running, the cardiac output was similar in both hyperoxia and in control groups.

With controlled blood loss and reinfusion of red cells, the effect of acute variations in hematocrit can be studied. The effect of blood loss is a deterioration of physical performance, which is related to the reduced maximal oxygen uptake. A reinfusion of red cells (equivalent to 800 ml of blood) in subjects who had recovered after blood loss could dramatically (overnight) improve these subjects' maximal oxygen uptake and performance to supernormal values (on average, an increase in maximal oxygen uptake of 9%; see figure 5.20b). In five subjects running at maximal speed, which could be maintained for about 5 min, the oxygen content of the arterial blood was on average 16% higher after reinfusion of red cells compared with the situation after blood loss. The difference in maximal oxygen uptake was actually about 14%, but the individual variations were large. The maximal heart rate and stroke volume were more or less identical in the different experiments (Ekblom, Wilson, and Åstrand 1976).

The positive effect on maximal oxygen uptake and performance of such an illegal "blood doping" has been confirmed repeatedly if the reinfused blood volume is 800 ml or higher (or the equivalent in

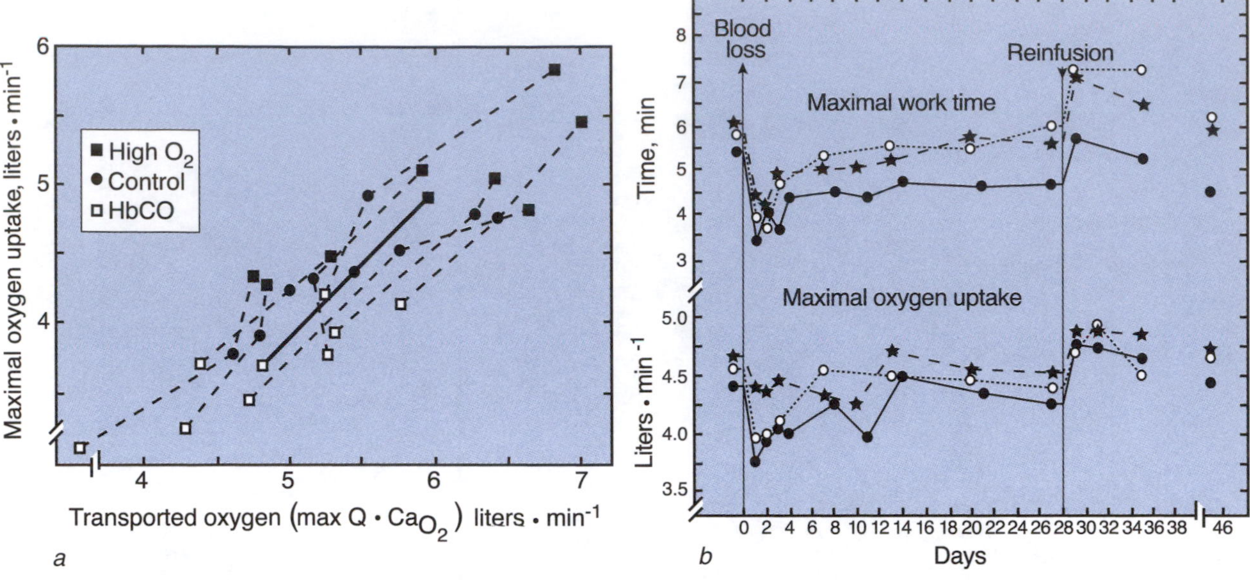


**Figure 5.20** *(a)* Relation between oxygen uptake and transported oxygen: Cardiac output (Q) · oxygen content in arterial blood ($Ca_{O_2}$) during maximal running. Individual values on eight subjects (broken lines) and means (solid lines) at hypoxia induced by about 15% carboxyhemoglobin (HbCO), control experiments, and hyperoxia with the subjects inhaling 50% oxygen in nitrogen, respectively. *(b)* Exercise time at a standard maximal run (maximal work time) and maximal oxygen uptake during control (day 0), after 800 ml of blood loss, and after reinfusion of the packed red cells (day 28) in three subjects.

*(a)* From Ekblom et al. 1975. *(b)* From Ekblom, Goldbarg, and Gullbring 1972.

packed red blood cells). Thomson et al. (1982) reported an increase in maximal oxygen uptake from 4.0 to 4.5 L · $min^{-1}$ after an **autologous blood reinfusion** that elevated the hematocrit from 42.4% to 46.2% (average of four subjects). During submaximal exercises the cardiac output was no different from the control level, but during maximal exercise the researchers found a small increase in cardiac output which, together with the increased content of oxygen in the arterial blood (about 20 ml · $L^{-1}$), explained the improved maximal aerobic power.

An induced **erythrocythemia** also can increase the hypoxia tolerance during physical exercise. Thus, an increase in the concentration of EPO, which is one of the hormones that stimulate red blood cell production, has been documented in athletes living at simulated altitude of between 2,500 and 3,000 m and training near sea level (Ashenden, Gore, Dobson, and Hahn 1999).

A plasma expansion can increase the stroke volume and cardiac output (Coyle and Hamilton 1990; Fortney et al. 1981a; Kanstrup, Marving, and Høilund-Carlsen 1992; Spriet et al. 1980). Kanstrup and Ekblom (1982) found that plasma expansion by infusion of dextran (a 700-ml increase in plasma volume, on average) increased the maximal stroke volume and cardiac output, just compensating for the concomitantly reduced Hb concentration so that the normal maximal aerobic power could be attained. An increase in blood volume is the only factor that so far has been shown to make a "supranormal" cardiac output possible, probably because of an enhanced filling of the heart. The total amount of Hb seems to be decisive for the maximal oxygen uptake (Kanstrup and Ekblom 1984).

The purpose of this brief summary is to illustrate that the maximal oxygen uptake (maximal aerobic power) in exercise engaging large muscle groups is apparently not limited by the capacity of the muscle mitochondria to consume oxygen. Slight variations in the volume of oxygen offered to the tissue (cardiac output (Q) · oxygen content in arterial blood ($Ca_{O_2}$)) will produce almost proportional changes in the volume of oxygen consumed. Exercise with the arms (in swimming) as well as with one leg (bicycling) includes muscle groups that are also engaged in normal swimming and two-leg work, respectively. It is remarkable, however, that the combined exercise does not dramatically increase the maximal oxygen uptake (Blomqvist and Saltin 1983).

A period of physical conditioning will increase the volume of mitochondria in trained muscles, increasing their aerobic energy potential. The crucial question is whether there are enzymes in the skeletal muscles that serve as a bottleneck for maximal aerobic energy yield. So far, the central circulation and the capillary bed available for perfusion have been considered to be the limitations for an individual's maximal oxygen uptake. Saltin and Gollnick (1983) estimated that the potential of enzyme systems of the skeletal muscles to consume oxygen by far exceeds the maximum actually attained.

## Stroke Volume

Two factors affecting the stroke volume are the venous return to the heart and the distensibility of the ventricles. The degree of diastolic filling has an anatomical limitation, but a range of various factors, some of which were discussed previously, affect the stretching of the muscle fibers. The final factor determining the stroke volume is the force of contraction in relation to the pressure in the artery (aorta or pulmonary artery).

The heart adjusts itself to changing conditions by an inherent self-regulatory mechanism. Starling (1896), using his famous lung–heart preparations, found that the normal heart tended to empty itself almost completely. It was distended to a greater diastolic volume in response to either a greater venous return or an increase of arterial pressure. In the latter case, stroke volume decreased transiently, but as the force of contraction increased with a greater initial length of the muscle fibers, the stroke volume and cardiac output became normal. By stimulation of the sympathetic cardiac nerves, the contraction force increased from the same initial length, and the arterial resistance could be overcome despite a greater extent of myocardial shortening.

It has been emphasized repeatedly that the increase in length of a muscle fiber (within limits) will improve its force-generating potential. In addition, catecholamines will elicit a similar effect plus an increase in the heart rate. What, then, is the mechanism behind these effects? In the resting muscle, troponin exerts an inhibitory effect on actin via tropomyosin. Thus, the actin and myosin filaments cannot interact and no cross-bridges are formed. A depolarization of the sarcolemma will cause a release of calcium from the sarcoplasmic reticulum, and as a result the inhibitory effect on actin of the troponin–tropomysin complex will be removed and the cross-bridges are formed. $Ca^{2+}$ also will activate adenosine triphosphatase (ATPase), which will break down ATP, thereby providing the energy yield for the muscle contraction. Then $Ca^{2+}$ is taken up again by the sarcoplasmic reticulum, the muscle relaxes, and the troponin–tropomyosin complex once more hinders contraction. The same events that take place when the skeletal muscle contracts and relaxes occur in the heart muscle. Findings suggest that catecholamines can increase the rate of calcium release and also the rate at which calcium is removed from troponin. In addition, the amount of calcium stored in the sarcoplasmic reticulum might increase. If so, more calcium is available for delivery to the contractile proteins in subsequent contractions. Cyclic AMP is involved as a mediator. Such mechanisms can explain the effects known to be produced by catecholamines, namely enhanced rate of tension increase, augmentation of contractility, and elevated heart rate.

The results from Starling's studies of the isolated heart also were considered applicable in the intact animal. Most earlier textbooks concluded that the diastolic volume of the heart was smaller at rest but increased during exercise, when the venous return increased and the arterial pressure was elevated. The same end-systolic volume could be maintained or was dependent on the strength of the heart muscle. The well-trained person was characterized by a small residual volume of blood in the heart after systole at rest as well as during exercise. The net effect was a substantial increase in stroke volume during exercise, according to these standard texts.

The present concept of stroke volume in the human during exercise can be summarized as follows: When a person's position changes from supine to standing or sitting, the end-diastolic size of the heart decreases, as does stroke volume. If muscular exercise is then performed, the stroke volume increases to approximately the same size or to a higher level as obtained in the recumbent position. At moderate exercise levels, the increase in stroke volume is mainly attributable to an enlarged left ventricular end-diastolic volume as a consequence of augmented venous return. Higher exercise levels result in a progressive decrease in left ventricular end-systolic volume because of increased cardiac contractility: in other words, a more complete emptying of the heart during systole.

### FOR ADDITIONAL READING

For further details about left ventricular hemodynamics during exercise, see Poliner et al. (1980), K.F. Adams et al. (1992), or Kanstrup et al. (1995).

## CLASSIC STUDY

The importance of the central blood volume for the stroke volume was demonstrated in 1939 by Asmussen and Christensen. Subjects in the sitting position exercised with their arms. In some experiments, the subjects lay down with their legs elevated for about 10 min before the exercise started. The circulation to the legs then was arrested by pressure cuffs around the thighs. When the subjects assumed a sitting position following this procedure, there was approximately 600 ml of blood less in the legs compared with when subjects sat without occlusion of the blood flow to the legs. The authors noted that the cardiac output was about 30% higher when the legs were "blood free," that is, when the central blood volume was high, compared with experiments with blood pooling in the legs. The high cardiac output was attributable to a high stroke volume, because the heart rate was actually lower than in exercise with reduced central blood volume (and low cardiac output) (Asmussen and Christenson 1939).

It has long been believed that the stroke volume reaches a plateau when oxygen uptake exceeds 40% to 50% of maximal aerobic power. Thus, in an early study, P.-O. Åstrand et al. (1964) concluded that during exercise in the sitting position, the stroke volume reached an "optimum" when oxygen uptake exceeded 40% of maximal aerobic power (at a heart rate between 110 and 120). No tendency toward a decrease in stroke volume at the peak load was observed. However, later studies showed that the training state of the subject is important for the maintenance of a maximal stroke volume during exercise. Both Higginbotham et al. (1986) and Christie et al. (1987) demonstrated that the stroke volume declines as exercise intensity is increased from submaximal to maximal exercise in young sedentary subjects. Spina et al. (1992) did the same, but they also showed that exercise training could prevent such a decline. Near half of the training-induced improvement in stroke volume at maximal exercise in their experiments was attributable to prevention of the decline in stroke volume that occurs in the untrained state (figure 5.21). Furthermore, B. Zhou et al. (2001) found that the stroke volume of elite male distance runners, in contrast to untrained university students, did not plateau but increased continuously with increasing intensity over the full range of an incremental exercise test. Augmented left ventricular fill-

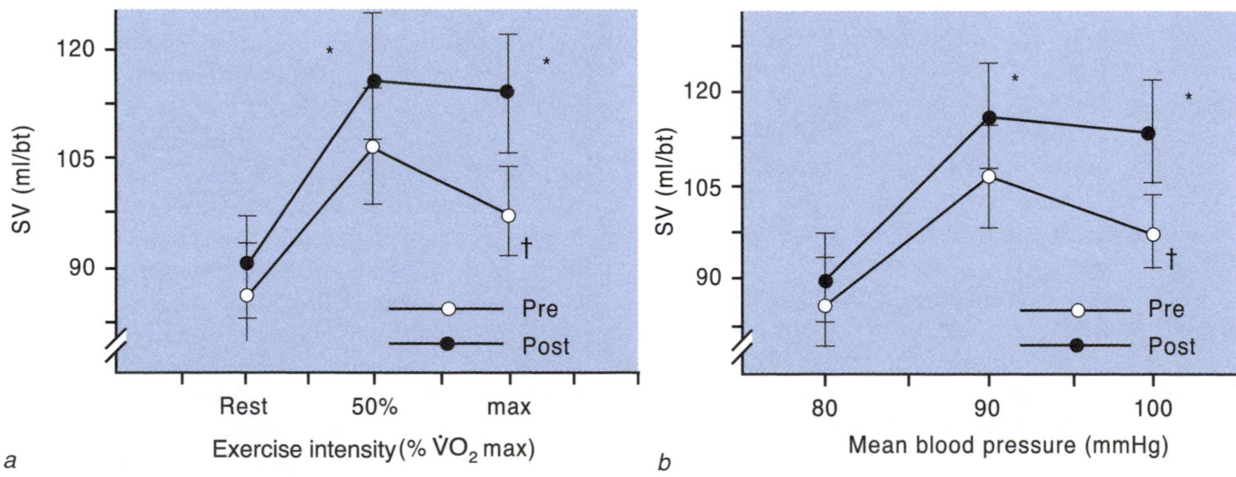


**Figure 5.21** *(a)* Stroke volume (SV) plotted as a function of exercise intensity before and after training. Values are mean ± standard error for 12 subjects. *Significantly higher after than before training ($p < .05$); †Significant decline from 50% to 100% maximal oxygen uptake ($p < .05$). *(b)* SV plotted as a function of mean blood pressure (MBP) before and after training. Values are mean ± standard error for 10 subjects. *At comparable levels of MBP, SV was significantly higher after than before training ($p < .01$). †Significant decline from 50% to 100% maximal oxygen uptake ($p < .05$).

Reprinted, by permission, from J.R. Spina et al., 1992, "Exercise training prevents decline in stroke volume during exercise in young healthy subjects," *Journal of Applied Physiology* 72 (6):2458-2462.

ing mediated by increased blood volume, enhanced venomotor tone, altered diastolic properties of the myocardium, or a combination of these factors could explain the lack of decline in exercise stroke volume after training.

Normally active male and female subjects with no history of training occasionally demonstrate high maximal aerobic power. Martino et al. (2002) have suggested that the primary explanation for the high $\dot{V}O_2max$ is a naturally occurring high blood volume that brings about a high maximal stroke volume and maximal cardiac output. Thus, in a comparison of two groups, each group consisting of six young age-matched and weight-matched men with no history of training, all subjects having $\dot{V}O_2max$ below 49.0 or above 62.5 mL · kg$^{-1}$ · min$^{-1}$, respectively, Martino and coworkers observed that the stroke volume did not plateau, but continued to increase throughout incremental work rates to maximum in both groups. Possibly, part of the explanation could be that a greater portion of the subjects' blood volume is hemodynamically active.

## Heart Rate

In many types of exercise, the increase in heart rate is linear with the increase in rate of exercise. There are exceptions, and those exceptions are perhaps more frequent among untrained subjects. When the subject performs very heavy exercise, the $(a-\bar{v})O_2$ difference may increase so that the oxygen uptake increases relatively more than the cardiac output. In most test procedures, the evaluation, from submaximal rates of exercise, of an individual's maximal oxygen uptake or capacity to perform work involves measuring the heart rate during steady state and then extrapolating to a fixed heart rate or to an assumed maximal heart rate. There are many pitfalls in this method. Here, the following should be noted: First, the standard deviation for maximal heart rate during exercise is ±10 beats · min$^{-1}$. Hence, for 25-year-old individuals, women or men, the maximal heart rate for 5 of 95 subjects may be below 175 or above 215, because the maximum is about 195 on an average. Second, there is a gradual decline in maximal heart rate with age, so that the 10-year-old attains 210 whereas the 65-year-old attains only about 165 beats · min$^{-1}$. Furthermore, longitudinal studies have shown a wide individual scatter in the decline in maximal heart rate with age (I. Åstrand et al. 1973).

When 50% of the maximal aerobic power is used, the heart rate in the 25-year-old man is about 130, but the same relative work rate and feeling of strain are experienced at a heart rate of 110 for the 65-year-old man (figure 5.22). For women, the 50% oxygen uptake is attained at a heart rate of about 140 beats · min$^{-1}$ at the age of 25.

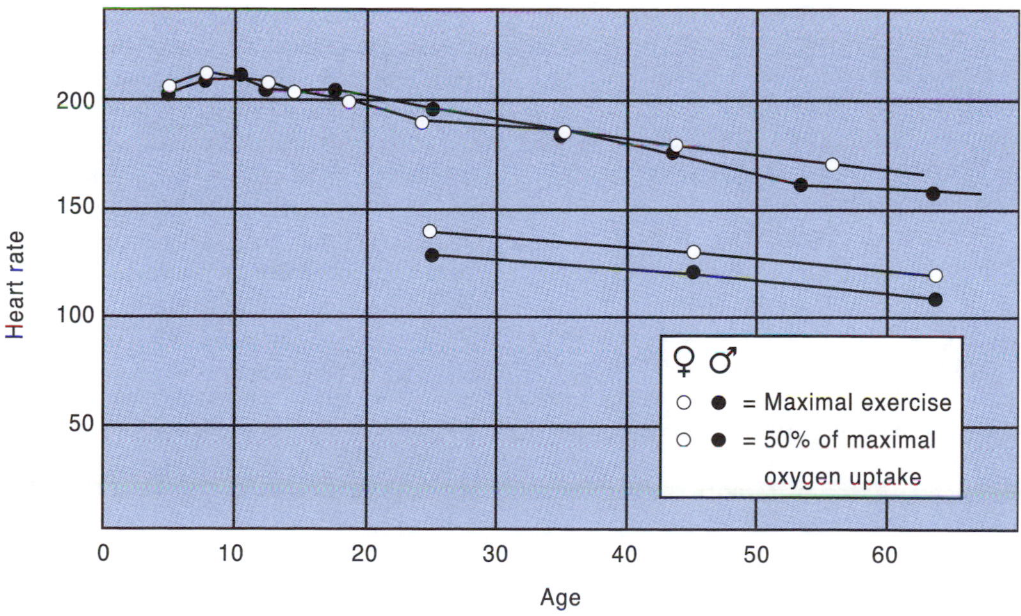


**Figure 5.22** The decline in maximal heart rate with age, and heart rate during a submaximal work rate. Mean values from studies on 350 subjects. The standard deviation in maximal heart rate is about ±10 beats · min$^{-1}$ in all age groups.

From P.-O. Åstrand and Christensen 1964.

Prolonged exercise in a hot environment causes a higher heart rate than exercise at a low ambient temperature. Emotional factors, nervousness, and apprehension also can affect the heart rate at rest and during exercise of light and moderate intensity. However, during repeated maximal exercise, the heart rate is remarkably similar under various conditions, with a standard deviation of ±3 beats · min$^{-1}$.

The heart rate at a given oxygen uptake is higher when the exercise is performed with the arms than with the legs (Christensen 1931a; Stenberg et al. 1967; Vokac et al. 1975). Static (isometric) exercise also increases the heart rate above the value expected from the metabolic rate (see subsequent discussion under "Blood Pressure"). The mechanism for these differences in heart rate response to exercise is not understood. However, the elevated heart rate usually is accompanied by a decreased stroke volume. It is known that a variation in heart rate at a given oxygen uptake at rest and during submaximal exercise often alters stroke volume, so that the cardiac output is maintained at an appropriate level. This information is based on studies in patients with artificial pacemakers or irregular heart rate and in subjects taking various drugs that influence the heart rate (Bevegård and Shepherd 1967; Braunwald et al. 1967).

Figure 5.23 presents data on subjects submitted to submaximal (cycle ergometer) and maximal (treadmill) exercise four times: (1) control; (2) after infusion of 10 mg of propranolol blocking the β-adrenergic receptors in the heart; (3) after infusion of 2 mg of atropine blocking the parasympathetic impulse traffic; and (4) after double blockade (propranolol and atropine). At a given oxygen uptake the heart rate varied about 40 beats · min$^{-1}$, taking the extremes, but the cardiac output was almost similar in the four situations, because the stroke volume compensated for the changes in heart rate. In the propranolol experiments, there was on average a 1.5 to 2 L · min$^{-1}$ reduction in cardiac output. The subjects reached their normal maximal oxygen uptake despite a reduction in maximal heart rate from 195 to about 160 beats · min$^{-1}$. The performance time was significantly shorter after β-blockade and after the intra-arterial blood pressure was reduced (Ekblom, Goldbarg, Kilbom, and Åstrand 1972). Sable et al. (1982) also reported a "normal" maximal aerobic power after β-blockade. In other studies, however, a reduction in maximal oxygen uptake as a consequence of β-blockade has been reported (see Hughson, Russel, and Marshall 1984).

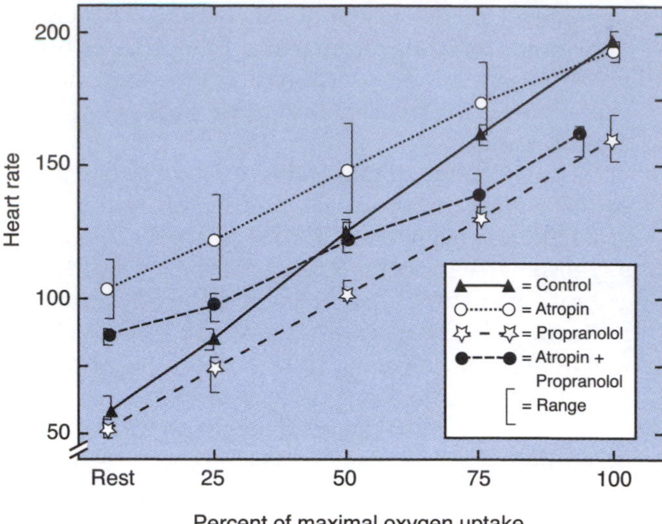


**Figure 5.23** Relationship between heart rate (means and ranges) and relative oxygen uptake in five sets of experiments (four subjects) during normal conditions (control) and after blockade of receptors. Resting heart rate was recorded in the sitting position.

From Ekblom, Goldbarg, Kilbom, and Åstrand 1972.

The regulation of the circulation in exercise is probably guided primarily by factors sensitive to an adequate cardiac output to secure the oxygen supply to the exercising skeletal muscles. There is a remarkable constancy in the relationship between oxygen uptake and cardiac output. If, for some reason, the stroke volume of the heart is reduced, heart rate increases in compensation. One exception is hypoxic conditions when the cardiac output at a given oxygen uptake is elevated. Heart rate and stroke volume are the variables, and the stroke volume is more likely to be directly influenced by such factors as venous return or peripheral vascular resistance to secure an adequate oxygen supply to the exercising skeletal muscles.

## Blood Pressure

There are some methodological aspects to be considered when defining blood pressure, especially arterial blood pressure. The total energy of the blood is the sum of the kinetic energy and the pressure energy. The side-hole catheter measures only the pressure energy (side pressure), but the catheter with the opening directed upstream also includes the kinetic energy. Because the kinetic energy factor of the blood in the aorta is high during exercise with a pronounced increase in cardiac output, the two catheters should give quite different pressure readings. The baroreceptors in the walls of arteries cannot sense the kinetic energy factor of the passing blood. The lateral distending pressure gives some information about the stretch to which these receptors are exposed, but a sympathetic vasoconstrictor activity involving the vessel wall will make the wall stiffer and less deformed by a given intravascular pulsatile pressure.

Simultaneous measurement of intra-arterial blood pressure in a peripheral artery and in the aorta during exercise gives a significantly higher systolic end pressure in the peripheral artery, but the mean and diastolic pressures are about the same as in the aorta. The systolic pressure in a peripheral artery is higher in a resting than in an exercising limb (P.-O. Åstrand et al. 1965). The progressive increase in systolic pressure (and pulse pressure) along an artery is attributable, at least in part, to a distortion in the transmission because of summation of the centrifugal wave and the reflected waves from the periphery. The importance of the wave reflection increases when the peripheral resistance is high, as is the case in a resting limb.

As a result of the vasodilation in the vascular bed in the active muscles, the peripheral resistance to blood flow is reduced during exercise, but the elevation in cardiac output causes the blood pressure to rise. If the cardiac output (Q) during the exercise is four times the resting value, giving a 25% increase in arterial mean pressure ($P_{mean}$), it follows that the resistance to flow (R) is reduced to more than one third of what it was at rest, because $Q \cdot R = P_{mean}$. The blood pressure obtained in a peripheral artery at rest, 120 mmHg (16 kPa) during systole and 80 mmHg (10.6 kPa) during diastole, can exceed 175 and 100 mmHg (23.3 and 13.3 kPa), respectively, during heavy exercise. The diastolic blood pressure, when measured with a blood pressure cuff, is constant or decreases slightly with an increasing rate of exercise (Hollmann and Hettinger 1980). In this case, the cuff method apparently gives a slightly different picture than the intravascular recording of the diastolic pressure.

Arterial blood pressure is significantly higher in arm exercise than in leg exercise (figure 5.24). The high blood pressure at a given cardiac output, when the work is performed by the arms, increases the stroke work of the heart. Therefore, for untrained individuals or for cardiac patients, it may be hazardous to exercise hard with the arms (e.g., to shovel snow, dig in the garden, or carry heavy bags). The relatively high blood pressure in exercise with small muscle groups is probably attributable to a vasoconstriction in the inactive muscles. The larger the activated muscle groups, the more pronounced is the dilation of the resistance vessels. The lower peripheral resistance is reflected in a lower blood pressure (because $Q \cdot R = P_{mean}$).

In isometric exercise, we have a similar situation with a considerable ventricular afterload. The arterial blood pressure, both systolic and diastolic, and heart rate increase abruptly with sustained isometric effort at 15% of the maximal voluntary contraction or higher. Usually there is no steady state in these functions but rather a gradual increase until the end of contraction. The blood pressures increase more or less linearly with the developed force in a given muscle group. The larger the muscle mass involved, the more pronounced the pressure and heart rate response (Kilbom and Persson 1981; Mitchell et al. 1981; D.R. Seals et al. 1983). With large muscle groups contracting at great force, the systolic pressure may well exceed 300 mmHg (40 kPa), and the diastolic pressure can rise beyond 150 mmHg (26.6 kPa). The stroke volume appears to remain constant, despite the increase in afterload. The increase in heart rate elevates cardiac output (Kilbom and Persson 1981). The reason why the arterial blood pressure at a given cardiac output is higher during isometric contraction than during dynamic exercise is not known. From a teleological viewpoint, one might say that a higher pressure facilitates the blood flow through a muscle with a high intramuscular pressure. The greater mental effort involved in maintaining the isometric force, and the accumulation of trapped metabolites in the muscle, can contribute to an exaggerated sympathetic drive.

If during recovery from isometric contractions, the blood flow through the engaged muscles is stopped by an inflated cuff, the heart rate returns to control level. Blood pressure also decreases but remains above the precontraction level until the cuff is released (Mitchell et al. 1981). Bonde-Petersen and Suzuki (1982) reported a similar pattern during recovery after dynamic exercise; that is, the mean arterial pressure remains elevated during occluded recovery, whereas the heart rate tends to recover at the same time as in the control without occlusion.

These findings support the previously discussed hypothesis that during exercise, the reflex control of the cardiovascular function, including the arterial blood pressure, is triggered by at least two more or less independent mechanisms. It has been

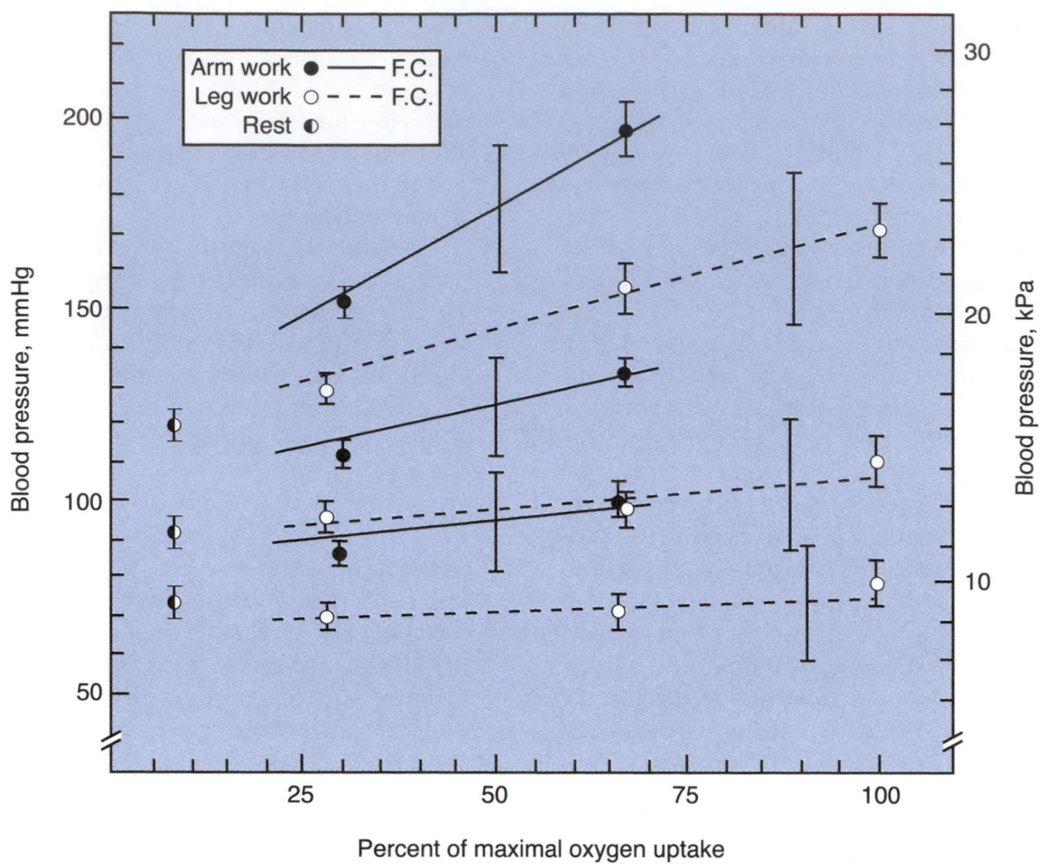


**Figure 5.24** Effect of exercise on blood pressure (end pressure). Regression lines of arterial systolic, mean, and diastolic blood pressures, respectively, in relation to oxygen uptake (in percentage of the maximum) during arm and leg exercise in the sitting position for 13 subjects. F.C. = femoral artery catheter. The figure summarizes data from 23 submaximal and 13 maximal work rates with arm work (cranking) and 44 and 13 experiments, respectively, with leg work. The vertical heavy lines represent ±1 standard deviation (SD) around the regression line. The dots and thin lines represent the mean ±1 standard error of the mean (SEM) for three groups of values at different levels of oxygen consumption. (The systolic pressure measured in a peripheral artery is higher than in the aorta, but the mean and diastolic pressures are similar in the two vessels.)

From P.-O. Åstrand et al. 1965.

hypothesized that a "central command" is related to the activation of motor units and that a peripheral control mechanism mediated by muscle afferents reports the metabolic changes in the contracting muscles.

In the control of arterial blood pressure, the baroreceptors in some artery walls play, as mentioned, a key role in that they perform some sort of a "buffer" function. If, at rest, there is a pressure decrease, the reduced activity (e.g., of the carotid sinus receptors) will activate the sympathetic system with reciprocal inhibition of the parasympathetic antagonist. The reaction to an increase in arterial blood pressure is the reverse. How do these receptors react to the "hypertension" induced by exercise? The answer is that they do respond and have a reflex buffer function even during exercise that will prevent marked deviations in arterial pressure from normal values. The maximal firing rate is reached around 180 mmHg (24 kPa), and with a further increase in blood pressure, the activity of the receptors in the carotid sinus does not change (J.T. Shepherd and Vanhoutte 1979). There is strong evidence suggesting that the control of the blood pressure by reflexes elicited from the arterial baroreceptors is not modified by either isometric or dynamic exercise. In other words, the carotid baroreflex finely balances the opposing effects of sympathetic vasoconstriction and metabolic vasodilation (Ludbrook 1983; Walgenbach and Donald 1983).

I. Åstrand, Guharay, and Wahren (1968) studied carpenters using hammers to nail at different heights. When they were hammering into a ceiling, their heart rate and intra-arterially measured blood pressures were significantly higher than when the subjects hammered at bench level. These pressures were also higher than during leg exercise on a cycle ergometer with a similar level of oxygen uptake.

In 1960, Reindell et al. reported that when they compared the arterial blood pressure response to exercise in subjects of different ages, older men had consistently higher systolic and diastolic pressures than younger ones. At rest, the 25-year-olds averaged 125/75 mmHg (16.0/10.0 kPa), and during exercise at a rate of 100 W (oxygen uptake about 1.5 L · min$^{-1}$) the pressures were 160 and 80 mmHg (21.3 and 10.6 kPa) in systole and diastole, respectively. For the 55-year-old group, the increase was from 140/85 mmHg (18.6/11.3 kPa) at rest up to 180/90 mmHg (23.9/12.0 kPa), the work rate being the same. Similar results were reported by Gerstenblith, Lakatta, and Weisfeldt (1976) in their review.

## Type of Exercise

The cardiac output at a given oxygen uptake is similar in many types of exercise, (e.g., in exercise with arms, cycling with one or two legs, combined arm and leg exercise, walking, running, and swimming) (Clausen 1976; C.T.M. Davies and Sargent 1974; Hermansen, Ekblom, and Saltin 1970; Holmér 1974a; Stenberg et al. 1967). The cardiac output at a given oxygen uptake is consistently 1 to 2 L less per minute in the erect position than when the subject is recumbent; the heart rate is about the same (Poliner et al. 1980). The compensation for the lower cardiac output must, by definition, be an increased (a-$\bar{v}$)$O_2$ difference when the person is erect. Peak oxygen uptake during cycling in the supine position is lower than during exercise in the sitting position on a cycle ergometer (figure 5.25). Similarly, cardiac output is somewhat lower during maximal exercise with the legs in the supine position. This difference in response may be explained in the following way: Rhythmic muscular contractions squeeze out blood from the veins, lowering the average venous pressure considerably and hence raising the effective perfusion pressure of flow. This is evident in exercise in the upright position, as the arterial pressure at the calf level is raised by some 70 to 80 mmHg (9.3–10.6 kPa) compared with the supine position, that is, in proportion to the distance from the heart because of the increase in hydrostatic pressure. However,

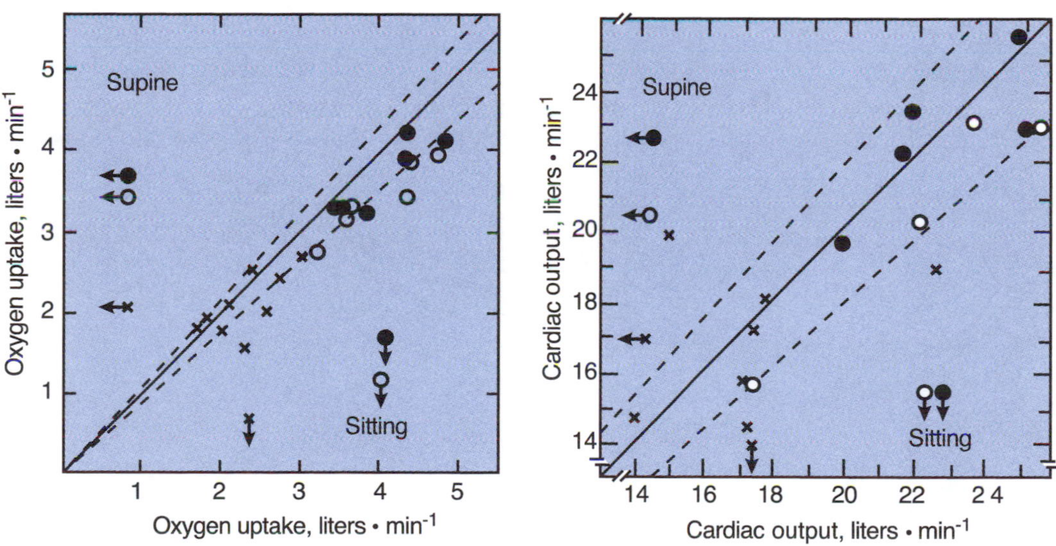


**Figure 5.25** Oxygen uptake (left) and cardiac output (right); comparison between the highest individual values attained in the sitting (abscissa) and supine (ordinate) position for arm exercise (**x**), leg exercise ( ○), and combined arm and leg exercise ( ●). Line of identity and lines corresponding to 10% deviation are drawn. Symbols with arrows give the mean of the different groups. Note that leg exercise in the supine position did not bring oxygen uptake and cardiac output to a maximum, but when the arm muscles also were exercised, the oxygen uptake and cardiac output increased to the same level as in leg exercise or in combined arm and leg exercise in the sitting position.

From Stenberg et al. 1967.

the pressure in the calf veins is maintained at a low level by the "milking" action of the muscle pump. Folkow et al. (1971) noted that the calf blood flow in humans could be 50% to 60% larger when a standard heavy rhythmic exercise was performed in the upright position compared with the reclining position. Combined arm and leg exercise in the sitting or supine position reveals almost the same values for maximal oxygen uptake, heart rate, and cardiac output as exercise in the sitting position with only the leg muscles (see figure 5.25). This emphasizes that the central circulation seems to limit the maximal cardiac output and thereby also the maximal oxygen uptake.

In prolonged exercise in a neutral thermal environment, the cardiac output is normally well maintained, but there is a progressive increase in heart rate and decrease in stroke volume. Furthermore, there is a gradual reduction in blood pressures (systemic arteries, pulmonary artery, right ventricular end-diastolic pressures; Rowell 1983).

## Heart Volume

Dimensional analyses of the heart can be carried out by echocardiography, Doppler technique, radionuclide ventriculography, and other methods. A high correlation has been established between heart volume and various parameters, such as blood volume, total amount of Hb, and stroke volume in healthy younger individuals (Rowell 1993). The difference between the athlete's heart and the dilated heart of the cardiac patient is not only the configuration of the heart but also the disproportion between heart size and maximal aerobic power and the total Hb content. Figure 5.26a shows data for a group of 30 young girls who had a calculated heart volume that in many cases was much larger than expected for their body size. The girls were some of the best Swedish swimmers and were not very likely to suffer from any heart disease. When the heart volume is related to the girls' maximal aerobic power (see figure 5.26b), the findings make func-

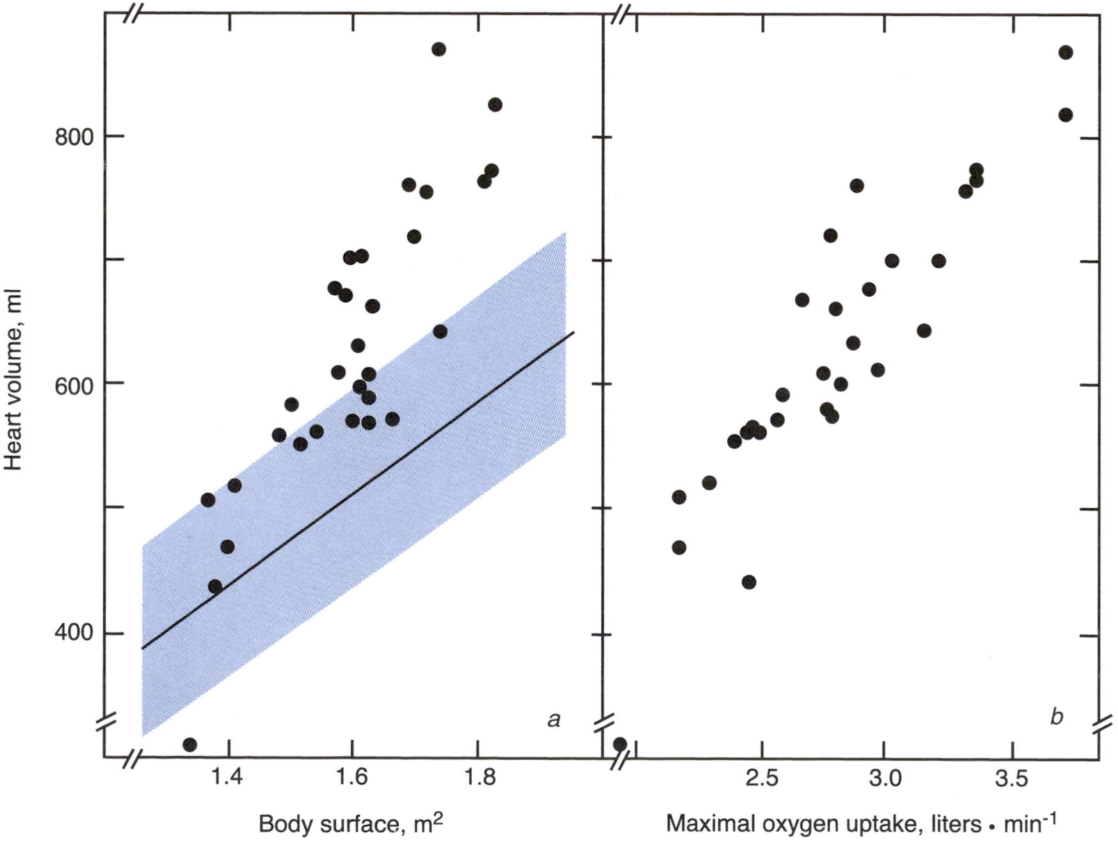


**Figure 5.26** Relationship between heart volume and *(a)* calculated body surface area and *(b)* maximal oxygen uptake in 30 young well-trained girl swimmers. Shadowed area gives the 95% range for "normal" girls.

Adapted from P.-O. Åstrand, Engström et al. 1963.

tional sense. The girl with the highest oxygen uptake and the largest heart was actually also the best swimmer; she was second in the 400-m freestyle in the 1960 Olympic Games (P.-O. Åstrand, Engström et al. 1963). These girls were reexamined in the 1970s, and most of them were then habitually quite inactive. They were healthy, but their maximal oxygen uptake was down to levels typical for "normal" women of the same age. However, they still had a large heart size (B.O. Eriksson, Lundin, and Saltin 1975).

In active athletes, there is a correlation between the demand placed on the oxygen transport system by the event in question and the estimated heart volume. On average, the heart volume has been found to be largest (above 900 ml) for well-trained athletes engaged in events calling for endurance (bicyclists, canoeists, cross-country skiers, long-distance runners). The average for middle-distance runners, swimmers, soccer players, and tennis players was between 800 and 900 ml and for boxers, fencers, gymnasts, jumpers, sprinters, throwers, and untrained controls below 800 ml. The individual variations are large, however (see Mitchell 1992; Rost and Hollmann 1983; Spirito et al. 1994).

Data from longitudinal studies of former endurance athletes who later become relatively inactive confirm the results from the study of the girl swimmers. Many of these previous athletes still had a large heart (Blomqvist and Saltin 1983). There are exceptions, however. Rost and Hollmann (1983) reported a heart volume of 1,700 ml in a professional bicyclist; his maximal aerobic power was above 6 L · $min^{-1}$. Four years after cessation of training, his heart volume was reduced to 980 ml. In former endurance-trained athletes, the correlation between maximal oxygen uptake and heart volume is low. This may be explained, at least in part, by their different levels of habitual physical activity.

Endurance training that demands a high oxygen uptake (and therefore inevitably a large stroke volume) generally is accompanied by an increase in left ventricular end-diastolic volume. The ventricular wall thickness may not change or it may increase slightly. Apparently, the weight of the heart increases. In athletes engaged in strength training that involves isometric effort, an increase in wall thickness is noted without any change in ventricular volume. As discussed previously under "Blood Pressure," isometric exercise engaging large muscle groups exposes the heart to high afterloads without change in stroke volume. Thus, there is definitely a difference in the demands on the heart when comparing strength training with endurance training. The weight of the athlete's heart does not generally exceed 500 g (18 oz), which is regarded as the limit of physiological hypertrophy. In patients with myocardial disease, the heart weight can exceed 1,000 g (35 oz).

### For Additional Reading

For further details concerning sport-specific adaptations and differentiation of the athlete's heart, see Perrault and Turcotte (1994), Gustafsson et al. (1996), K.P. George et al. (1999), and Urhausen and Kindermann (1999).

## Age

The heart rate reached during maximal exercise decreases with age. The value typical for a 10-year-old girl or boy is 210, for a 25-year-old 195, and for a 50-year-old 175 beats · $min^{-1}$ (see figure 5.22). The decrease in circulatory capacity in the old individual is more marked than predicted from heart rate, stroke volume, and cardiac output observed during submaximal exercise if the norms are the same as when evaluating young individuals. The old person has a larger heart volume, calculated from roentgenograms taken in supine position, than does the young person; blood volume may be slightly less, whereas the total amount of Hb is not different in young and old subjects (Timiras and Brownstein 1987). These findings should be related to the decrease in maximal stroke volume, cardiac output, and maximal aerobic power in older subjects.

Whether the decrease in maximal heart rate with age is a consequence of arteriosclerosis in the vessels of the heart is not known. The oxygen cost of a cardiac performance involving a high heart rate is great, and therefore the load on the heart is reduced by a lowered ceiling for heart rate. The lower heart rate during maximal exercise in the old individual is probably not a direct response to hypoxia, because breathing pure oxygen instead of room air during the exercise does not further elevate the heart rate (I. Åstrand, Åstrand, and Rodahl 1959).

Age changes in the sinus node and conductive system in the heart are possible explanations for the gradual decline in heart rate with age. The individual variation in maximal heart rate is of the same order of magnitude in old as in young people, that is, standard deviation of ±10 beats · $min^{-1}$.

Unfortunately, our knowledge of the cardiovascular response to exercise in the aged is rather scanty. One of the reasons for this is the problem of selecting a random sample. It is relatively simple to define a "normal healthy subject" when dealing with a young population. But, who can be called a "normal healthy subject" when approaching the age of 70? Such a group of older individuals is already a very selected group of subjects compared with younger counterparts.

### FOR ADDITIONAL READING

For further reading about age-related changes in cardiovascular function and aerobic fitness, see J.S. Green and Crouse (1993), Spirduso (1995), Bemben (1998), Paterson et al. (1999), and Cobb et al. (2000).

## Training and Cardiac Output

It was observed long ago that the heart rate at rest and during standard exercise, as well as during recovery, is lowered with training of the oxygen-transporting system. Published reports on the cardiac output during standard exercise repeated during a course of training indicate that cardiac output is maintained at the same level or is slightly reduced, that is, the $(a-\bar{v})O_2$ difference is increased (Rowell 1993). (For further discussion of the training effects, see chapter 11.)

Top athletes in endurance events are characterized by a very high maximal oxygen uptake. Their maximal values for circulatory parameters therefore must be high compared with those of less athletic individuals. Ekblom and Hermansen (1968) collected data obtained during maximal exercise on the treadmill in athletes, by using the dye dilution technique. Some data are presented in table 5.1. Both an intensive training and superb natural endowments contribute to the remarkable circulatory capacity for oxygen transportation in these subjects. The highest reported figure on maximal oxygen uptake, as far as we know, is 7.4 $L \cdot min^{-1}$. This subject's cardiac output was not measured, but if his $(a-\bar{v})O_2$ difference was 153 ml per liter of blood (table 5.1), the cardiac output would be 48.4 $L \cdot min^{-1}$.

Figure 5.27 summarizes data on 32 subjects with maximal aerobic power ranging from 2.8 to 6.2 $L \cdot min^{-1}$ of oxygen uptake. This figure shows a clear relationship between maximal cardiac output and oxygen uptake. It also shows that the stroke volume to a large extent determines maximal cardiac output. The most pronounced difference between the sexes is the smaller stroke volume and higher heart rate during exercise of a given severity for women compared with men. Actually, a similar difference in stroke volume and heart rate usually is observed when comparing individuals of the same age but with a low and high performance capacity (figure 5.28).

Of primary importance for the regulatory mechanisms is the relation between volume of oxygen supplied to the metabolically active tissue (cardiac output times the oxygen content of arterial blood) and the oxygen demand of the tissue. Within limits, other demands can be met by compensatory mechanisms; for example, if the stroke volume is reduced, the heart rate can increase so that an adequate cardiac output is still maintained.

**Table 5.1** Data Obtained During Maximal Exercise on the Treadmill Using the Dye Dilution Technique

| | Maximal values | | | | |
|---|---|---|---|---|---|
| **Subject** | $\dot{V}O_2$, $L \cdot min^{-1}$ | **Cardiac output,** $L \cdot min^{-1}$ | **Heart rate** | **Stroke volume, ml** | $a-\bar{v}O_2$ **difference,** $ml \cdot l^{-1}$ |
| G.P. | 6.00 | 39.8 | 188 | 212 | 151 |
| C.R. | 5.77 | 37.8 | 188 | 201 | 153 |
| A.H. | 5.60 | 34.4 | 189 | 182 | 163 |
| C.S. | 5.50 | 36.2 | 198 | 183 | 152 |
| B.T. | 5.64 | 38.0 | 193 | 197 | 148 |
| L.R. | 6.24 | 42.3 | 206 | 205 | 148 |
| Mean | 5.79 | 38.1 | 194 | 197 | 153 |

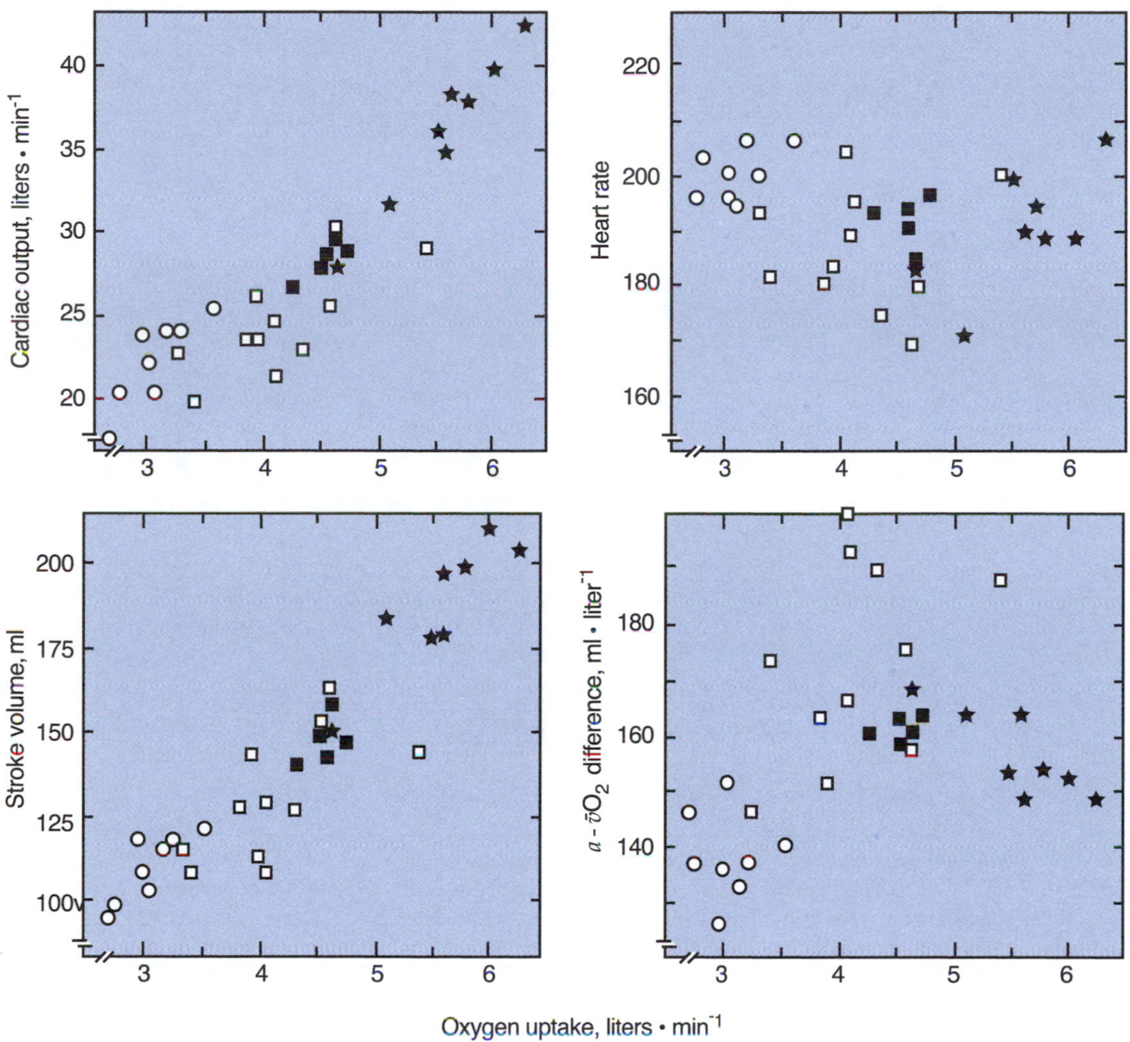


**Figure 5.27** Cardiac output, heart rate, stroke volume, and arteriovenous oxygen difference (a-$\bar{v}O_2$ difference) during maximal exercise in relation to maximal oxygen uptake in top athletes who were very successful in endurance events (★), well-trained but less successful athletes (■), and 25-year-old habitually sedentary subjects (O). Also included are maximal values on the male subjects presented in figure 5.17 (□).

From Ekblom 1969.

During maximal exercise, however, the cardiac output, oxygen uptake, and heart rate are remarkably fixed to values typical for the individual even if the performance is made under adverse conditions. In this situation, apparently all circulatory functions of decisive importance for a maximal oxygen supply to the active muscles are actually devoted to this task. Irrespective of environment, external and internal maximal vasoconstriction occurs in the blood vessels of the viscera and skin, so that practically the entire cardiac output is diverted to the vigorously contracting muscles. Maximal exercise involving large muscle groups creates an emergency reaction in the circulatory adjustment that favors the exercising muscles, including the heart, at the expense of all other tissues with exception of the central nervous system.

## For Additional Reading

For details concerning cardiovascular adaptations to physical training, see Wagner (1991), J.R. Sutton (1992), and Spina (1999).

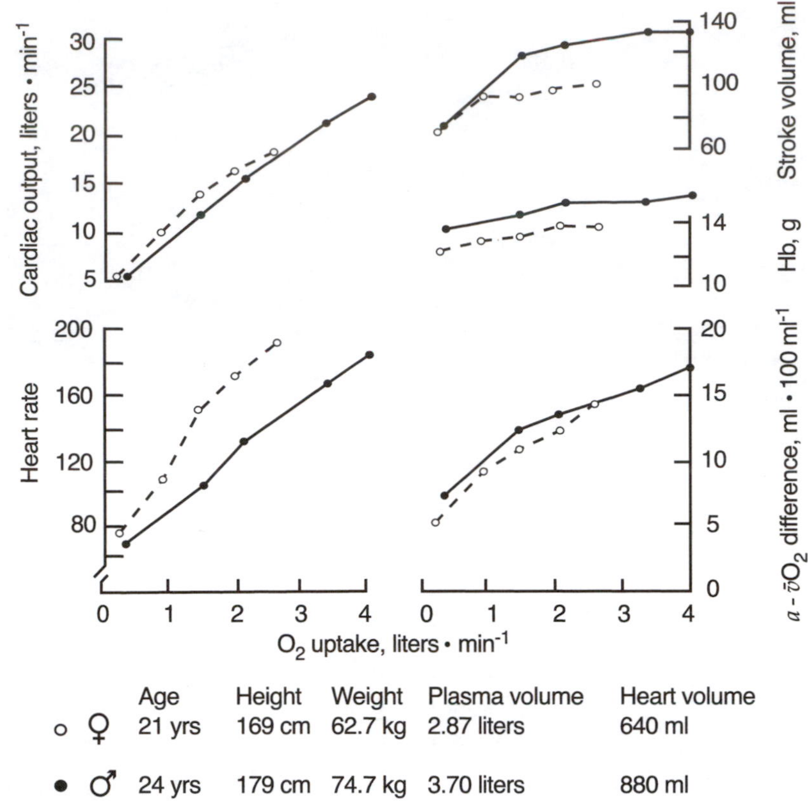


| | | Age | Height | Weight | Plasma volume | Heart volume |
|---|---|---|---|---|---|---|
| o | ♀ | 21 yrs | 169 cm | 62.7 kg | 2.87 liters | 640 ml |
| • | ♂ | 24 yrs | 179 cm | 74.7 kg | 3.70 liters | 880 ml |

**Figure 5.28** The figure is based on average values from measurements on 11 women and 12 men, all of them relatively well-trained and working on a cycle ergometer in the sitting position (P.-O. Åstrand et al. 1964.) Because the abscissa gives the oxygen uptake in absolute values, the calculated mean curves can be misleading. The less fit subjects have both a low maximal oxygen uptake and low stroke volume. Those with a high capacity for oxygen uptake also have a larger stroke volume. A man with a maximal aerobic power of 5 L · min$^{-1}$ eventually attains maximal stroke volume first at a work rate giving an oxygen uptake of 2 L · min$^{-1}$. The one with a maximal oxygen uptake of 3.5 L · min$^{-1}$ reaches his plateau for stroke volume when the oxygen uptake exceeds 1.3 L · min$^{-1}$. a-$\bar{v}O_2$ difference = arteriovenous oxygen difference.

# Chapter 6

# Respiration

The cells use oxygen for their metabolism, and in the process, carbon dioxide ($CO_2$) is produced. This lowers the concentration of oxygen within the cells, and oxygen will tend to diffuse toward the place of combustion. Similarly, $CO_2$, the major end product of oxidative metabolism, will tend to diffuse away. The exchange of oxygen and $CO_2$ depends on the distance the molecules have to travel and the pressure gradient. In the single-cell organism, the surface can be used effectively for the respiratory exchange. By diffusion, oxygen can easily reach every point within the cell, and $CO_2$ can be eliminated. The classic calculations of Krogh (1941) indicated that when metabolism is fairly high, diffusion can provide sufficient oxygen only to organisms with a diameter of 1 mm or less.

In multicell animals, the problem of gas transport is solved in various ways; for example, numerous airways, like tracheae in insects, or specialized organs, like lungs or gills, are developed, exposing an enlarged respiratory surface to the external medium to effect exchange of oxygen and $CO_2$. The spongelike structure of the lungs provides an enormous contact surface between air and blood. This air–blood interface of the average adult human lung is estimated to be some 70 to 90 $m^2$ (i.e., roughly the size of a tennis court), with a thickness of the tissue separating air and blood varying from 0.2 μm to several micrometers (average, some 0.7 μm). Fresh air and "new" blood must be supplied continuously to the many million gas-exchange units, because the store of oxygen in the human body is very limited and the central nervous system (CNS) and heart muscle do not tolerate any lag in the supply of oxygen.

In this chapter we first look at the basic pulmonary structure and function and then discuss respiration during physical activity and how breathing is regulated.

### For Additional Reading

For a comprehensive discussion of the many aspects of respiratory physiology, the reader is referred to Crystal and West (1997), Marcial and Slutsky (1997), and Altose and Kawakami (1999).

## Anatomy and Histology

The ideal gas exchanger involves four factors.

1. It should provide a large contact area between the air and blood with a very thin membrane separating the two media, because diffusion is directly proportional to the area but inversely related to the thickness of the membrane. The membrane should cause a minimal resistance to gas flow.
2. The inspired air must become saturated with water vapor and heated to tissue temperature to protect the delicate membranes from injury; any particles and agents in the air that are harmful should be removed during the passage through the airways and, if introduced, should be expelled.
3. Oxygen and $CO_2$ concentrations in the blood leaving the lungs should vary only within small limits, and therefore the distribution of gas and blood in the many exchange units should be closely matched.

4. The gas exchange, and therefore the perfusion of the units, must be proportional to the uptake of oxygen and production of $CO_2$ by the cells. This demands some sort of regulative mechanism linking the needs of the distant cells with the external respiration.

Point 4 of these factors is of special interest in this connection, because muscular exercise can increase the oxygen uptake of the body more than 20 times the resting level, with a similar increase in $CO_2$ production. The respiration also plays an important role in maintaining blood pH at normal levels. Hydrogen ions ($H^+$) cannot be exchanged between air and blood, but the acid-base equilibrium of the blood (and tissue) is closely associated with the $CO_2$ content and pressure in the blood (see chapter 5). During heavy exercise, there is, in the anaerobic metabolism, a high rate of proton ($H^+$) production and more $CO_2$ is driven out from the hydrogencarbonate ions ($HCO_3^-$), which together with the excess proton concentration in the blood stimulates respiration. Elimination of $CO_2$ by the increased ventilation limits the decline in the blood pH.

## Airways

Figures 6.1 and 6.2 illustrate the general architecture of the airways. Via nose or mouth, the inspired gases pass through the **pharynx** (throat), **larynx,** and **trachea** into the bronchial tree. Between the trachea and the alveolar sacs, the airways divide 23 times, each branching resulting in shorter, narrower, and more numerous tubes. The first 16 generations of branchings form the conductive zone of the airways. They include the bronchi, bronchioles, and terminal bronchioles. The remaining generations form the respiratory zone where gases are exchanged with the blood; these generations include the respiratory bronchioles, alveolar ducts and sacs, and alveoli.

The respiratory bronchioles have a diameter of about 1 mm and subdivide to produce about 1 million alveolar ducts. Whereas the cylindrical surface of the respiratory bronchioles bears a smaller number of variously spaced alveoli, the alveolar ducts and sacs are fully alveolated. Therefore, they lack proper walls but open out on all sides into alveoli, some 300 million altogether in the adult. Obviously, there are considerable normal variations in that number—from 200 million up to 600 million—related to the individual's body height, but the alveoli have similar dimensions in large and small adult human lungs. The transitional and respiratory zone, including the alveoli, amount to about 90% of the lung volume; about 65% of the air in the lungs is in the alveoli at three fourths the maximal inflation of the lungs (Weibel 1984).

The surface of the alveoli is not smooth but is corrugated by the capillaries and by various subcellular structures, like nuclei, bulging into the alveoli (figure 6.3). The alveoli share their walls with their neighbors and form the spongelike texture of the lungs.

The bronchi and bronchioles that are more than about 1 mm wide have a discontinuous cartilaginous support in the wall. Muscle fibers in circular or crisscrossing bundles are incorporated into a complex connective tissue framework of collagenous reticular and elastic fibers. The inner surface is covered with a ciliated epithelium (figure 6.4). **Goblet cells** occur singly or in groups between the epithelial cells and produce a mucus secretion. In the finest bronchioles, the mucus-secreting elements become sparse and finally absent. They lack cartilage, and the ciliated cells also disappear gradually. Muscle fibers as well as elastic, collagenous, and reticular fibers provide the supporting latticework of the interalveolar septa.

In the fetus, the alveoli contain no air, but the first influx of air in the newborn child provides a force that stretches the original cuboidal epithelium lining of the alveoli into an extremely thin layer of squamous cells. The lung of the newborn is not fully developed. The airways have subdivided into only some 17 generations of branchings, and the number of alveoli is less than one tenth of that found in the adult. As time goes on, additional branches are added and many more alveoli are formed as new ramifications grow out, so that before approximately 10 years of age, the adult number is reached.

The fluid lining the alveoli contains a lipid surface-tension-lowering agent called **surfactant**. It is produced in the alveoli and keeps the alveoli open and free from transudate from the blood by lowering the surface tension. This function is especially important as the volume decreases, such as during forced expiration, which would otherwise empty the small alveoli (see Marin, 1994; Rooney, Young, and Medelsohn, 1994).

## Blood Vessels and Nerves

The pulmonary arteries enter the lungs with the stem bronchi and provide arterial partners to the airways as they subdivide toward the respiratory zones of the lungs. The arterioles follow the bronchioles, alveolar ducts, and sacs and provide short twigs to capillary networks enveloping the alveoli that surround the particular airway terminal and to any other alveoli in the immediate vicinity. Each alveolus may be covered by a capillary network consisting of almost 2,000 segments; the capillary networks in the lungs are the richest in the body and are more or less continuous throughout large parts of the lungs (see figure 6.1b). The air–blood "barrier" is formed by the continuous alveolar epithelial and capillary endothelial cells with a tenuous interstitium with fibrous elements in between the two cell layers. The thickness of the barrier can vary from about 0.2 μm to several micrometers, and the variations are caused by various structures scattered throughout the continuous cell layers (see figure 6.3). The capillary network should be regarded as a sheet of blood floating along the alveolar surface, the sheet merely connecting the two walls. The capillary surface area is, in fact, about the size as the alveolar surface area. Thus, most of the space in the interalveolar septa is occupied by capillaries, and their blood flow is intercalated between two air-filled alveoli. The blood is separated from the air on either side only by a very thin tissue barrier.

Vagal efferent fibers go to the smooth muscles of the bronchial tree as far as the termination of the alveolar ducts and sacs and to the bronchial mucous glands. Nerve impulses stimulate the

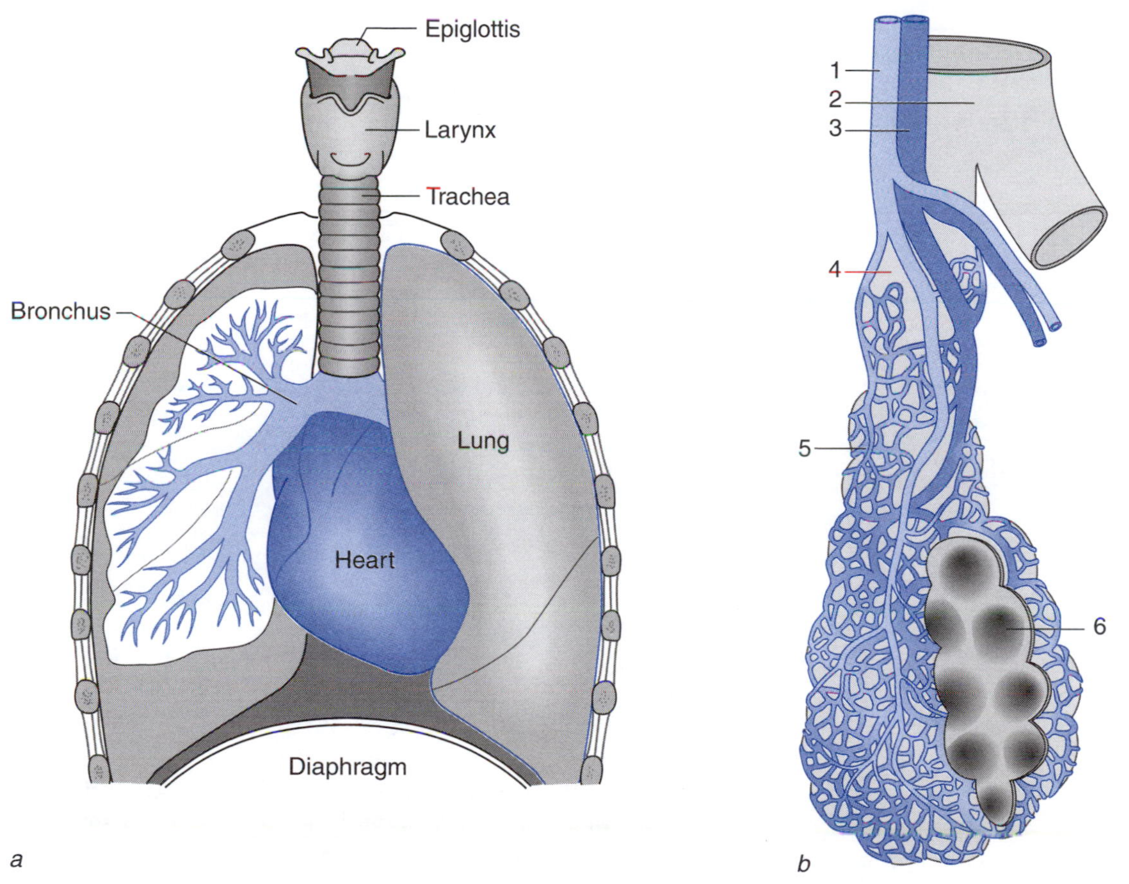


**Figure 6.1** *(a)* Principal organs of breathing. In this drawing, the ribs, the large arteries from the heart, and part of one lung have been cut away. *(b)* Schematic drawing of a bronchiole and its alveolar sac and alveoli. 1 = pulmonary artery, 2 = bronchiole, 3 = pulmonary vein, 4 = respiratory bronchiole, 5 = capillary network, 6 = alveoli.

Adapted from Dahl and Rinvik 1999.

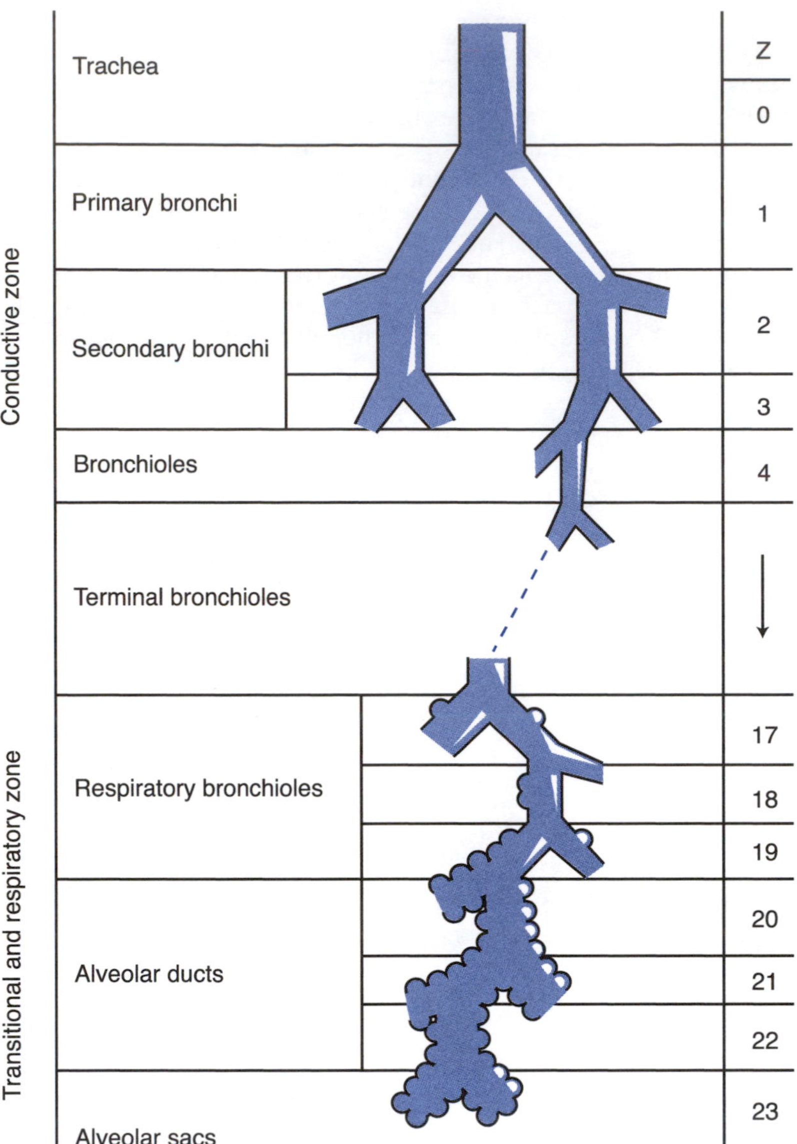


**Figure 6.2** General architecture of conductive/transitional and respiratory airways. The *z* column designates the order of generations of branching.

smooth muscles to contract and activate the glands. Vagal afferent fibers carry impulses from special stretch receptors scattered in lungs and **pleura.** Sympathetic fibers act as bronchodilators.

## Condition of Air

The inspired air may be cold or hot, dry or moist, but because of the rich blood supply of mucous membranes in the nose, the mouth, and the pharynx, the air becomes adjusted to body temperature and moistened before reaching the alveoli. When a person is inhaling large volumes of cold air, the point at which the air reaches body temperature moves progressively deeper into the lungs. Under extreme conditions (i.e., when a person is inspiring air with a temperature of –40° C), thermal transfer can be measured in airway branches less than 2 mm in diameter (McFadden 1983). Air saturated with water vapor at 37° C has a partial pressure of water vapor, $P_{H_2O}$, which equals 47 mmHg (6.3 kPa); the content of water is then 44 g/m³. At low temperatures, the water content in the air is low. Even if saturated, the air at 0° C contains only 5 g of $H_2O/m^3$. In a normal climate, about 10% of the total heat loss of the body at rest or during exercise takes place through the respiratory tracts by the "air conditioning" of the inspired air. At –15 to –20° C, the percentage is about

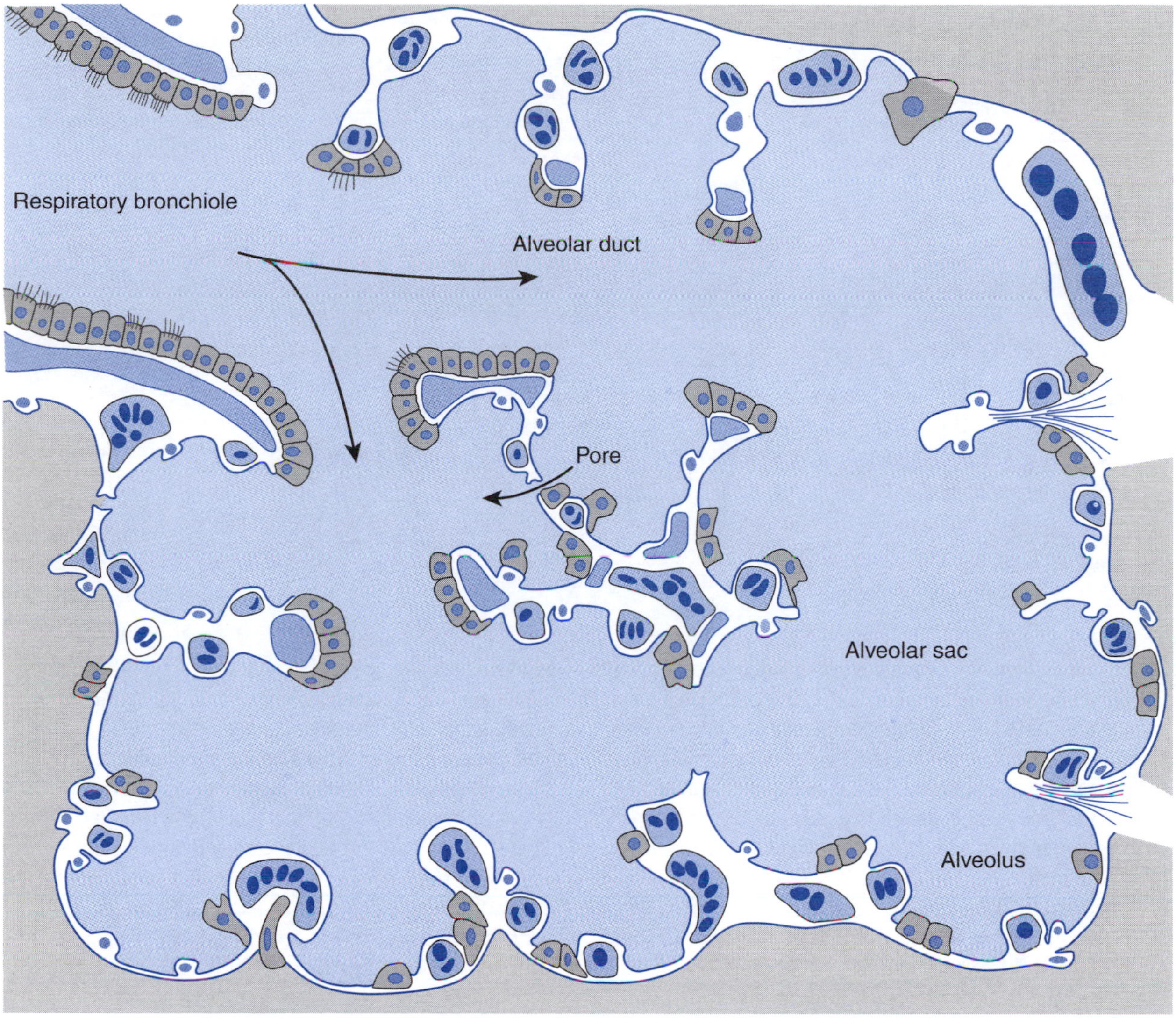

Respiratory bronchiole
Alveolar duct
Pore
Alveolar sac
Alveolus

*a*

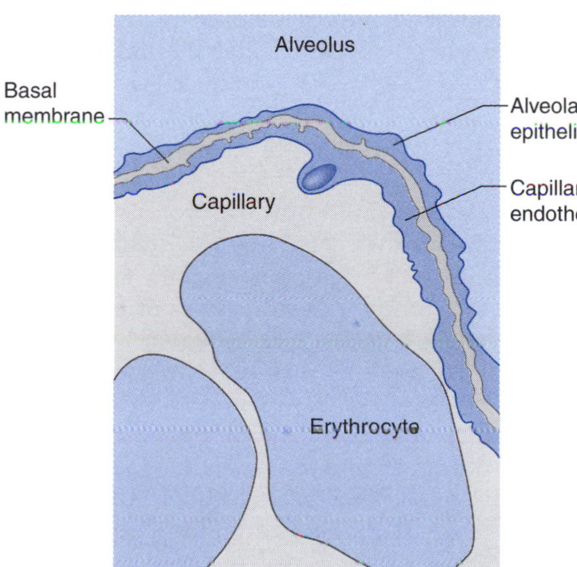


*b*

**Figure 6.3** *(a)* An idealized schematic representation of a respiratory unit of the lung, showing the epithelium and smooth muscle of the wall of the alveolar ducts. *(b)* Drawing of an electron micrograph (magnification × 30,000) showing a capillary with two erythrocytes and the capillary–alveolar septum, revealing the air–blood barrier composed of alveolar epithelium, basal membrane, and capillary endothelium.

*(a)* Adapted, by permission, from W. Bloom and D.W. Fawcett, 1994, *A Textbook of Histology* (London: Chapman and Hall), 715. *(b)* Adapted from Dahl and Rinvik 1999.

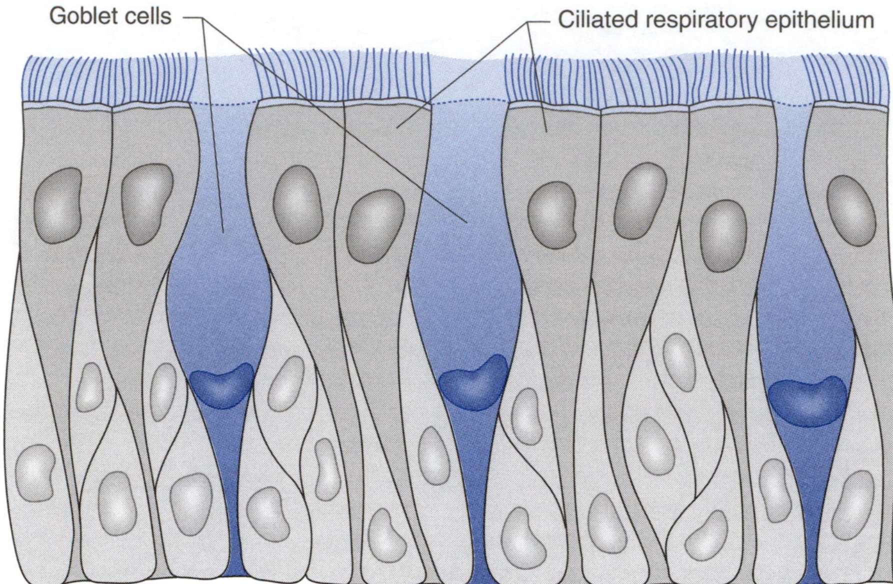


**Figure 6.4** A schematic representation of the ciliated epithelium of the airways. Between the epithelial cells are shown goblet cells, which may occur singly or in groups and produce a mucus secretion.

From Dahl and Rinvik 1999.

25. The respiratory tract serves as a regenerative system: The heating and humidifying of inspired air cool the mucosa. But during expiration, some of the heat and water are recovered by the mucosa from the passing alveolar air. Body heat and water are conserved. On a very cold day, this condensation of water vapor may result in excessive accumulation of water in the nostrils, leading to a runny nose! A cross-country skier breathing 100 L · min$^{-1}$ of air at –20° C must, in 1 hr, add about 250 ml of water to this air. Not all of this water volume is expired, however, thanks to the regenerative system (see Ferrus et al., 1984).

## Filtration and Cleansing Mechanisms

If living in a city, we may inhale thousands of millions of particles of foreign matter every day. Particles larger than about 10 μm are effectively removed from the inspired air in the nose, where they are trapped by the hair or the moist mucous membranes. The particles that escape these obstacles usually settle on the walls of the trachea, the bronchi, and the bronchioles. Therefore, only a few very small particles are likely to reach the alveoli, and this part of the lung is practically sterile. Alveolar macrophages perform a vital function of maintaining the alveoli clean and sterile. They are of hematogenous origin, but thanks to their extraordinary **amoeboid mobility,** they can pass into the alveoli and move freely over the airspace surfaces. They are able to phagocytose large quantities of foreign material, such as dust particles and bacteria. The macrophages and the phagocytosed remains can be cleared into the digestive tract via the airways or removed via the blood flow or lymphatics. The phagocytic capacity as well as the cleansing ability can be diminished by certain influences such as smoking (Weibel 1984). Actually, the lungs have a high metabolic activity. They can remove substances like serotonin, bradykinin, and certain prostaglandins, and they can synthesize and release a variety of substances, such as prostaglandins, peptides, and enzymes (Said 1982).

As mentioned, the epithelium of the airways within the lungs consists of ciliated cells (see figure 6.4). In the conductive zone, each cell carries up to 300 cilia about 6 to 7 μm in length at the free cell surface. The cilia of many thousands of cells beat in an organized, whiplike fashion in strokes, like oars of a boat, with a rapid upward propulsive stroke followed by a slower recovery downward stroke. Unlike the oars of a boat, however, the cilia do not move in synchrony. Instead, waves of movement sweep over the ciliated surface, regularly and continuously, day and night. The cilia are covered by a continuous surface of watery mucus. By the ciliary activity, this fluid carpet with all the entrapped particles moves toward the larynx at a speed of well over 1 cm · min$^{-1}$. This mucus is either expectorated or swallowed. The ciliary escalator is remarkably resistant to noxious influences. However, cigarette smoke has a deleterious effect on the ciliary function. The cilia slow down or stop beating when exposed to smoke. In long-time smokers, they even disappear.

From time to time we may sneeze or cough, and with our explosive blast during which the air moves with a speed approaching the speed of sound, foreign particles can be expelled.

# Mechanics of Breathing

The inner wall of the thoracic cavity is covered by the very thin *parietal pleura* and the lungs by the **visceral pleura.** These very thin membranes of single layers of flat epithelial cells on fibrous connective sheets continue uninterrupted from one pleural surface to the other across the pulmonary hilus. The two pleural surfaces are held close together with a thin fluid film in between (by adhesive force), providing smooth lubricated surfaces. If the thorax is opened so that the atmospheric pressure prevails in the intrapleural space, the elastic recoil of the lungs causes them to collapse (pneumothorax), and the chest expands a little, because a retractive force of the lungs normally is counterbalanced by an outward spring of the chest cage. Such an injury, of course, makes the lung involved incapable of any respiratory function. Normally, however, the pleurae are in close, but friction-free, contact with each other. Any volume changes in the thorax are completely transmitted to the lungs. The two pleural surfaces can be compared with two flat sheets of glass placed face to face with a thin layer of water between the two opposing surfaces. Although the two sheets of glass easily can be slid back and forth, a great force is required to move the two sheets away from one another by forces acting perpendicularly to the glass surfaces.

## Respiratory Muscles

Figure 6.5 shows the contours of the thorax and lungs at the end of an expiration and an inspiration, respectively. During quiet breathing at rest, the diaphragm is the principal muscle driving the inspiratory pump: The abdominal muscles relax, the abdomen protrudes, the thoracic volume increases, and the lungs expand. The contraction of the diaphragm causes its dome to descend some 1.5 cm, and the intra-abdominal pressure increases. During deep breathing, the vertical movements of the diaphragm can exceed 10 cm.

The role of the intercostal muscles has been a matter of discussion over the years (see Basmajian and De Luca 1985). Judged from their anatomy, the external intercostals have been assigned a role in inspiration, especially during exercise, whereas the internals may assist in expiration. Both of them may confer sufficient stiffness to the soft tissue of the intercostal spaces to prevent it from being sucked in during inspiration. Mm. scaleni, earlier believed to be active during forced inspiration, and mm. levatores costarum have turned out to be more important during quiet breathing than previously acknowledged.

When the inspiratory muscles relax during quiet breathing, the elastic recoil forces in the lung tissue, thoracic wall, and abdomen restore the chest to the resting position without any help from the expiratory muscles. During exercise or forced breathing at rest, with a ventilation exceeding two or three times the resting value, these recoil pressures are supplemented by activity of the expiratory muscles. The muscles of the abdominal wall are essentially expiratory muscles, as is the latissimus dorsi, but they

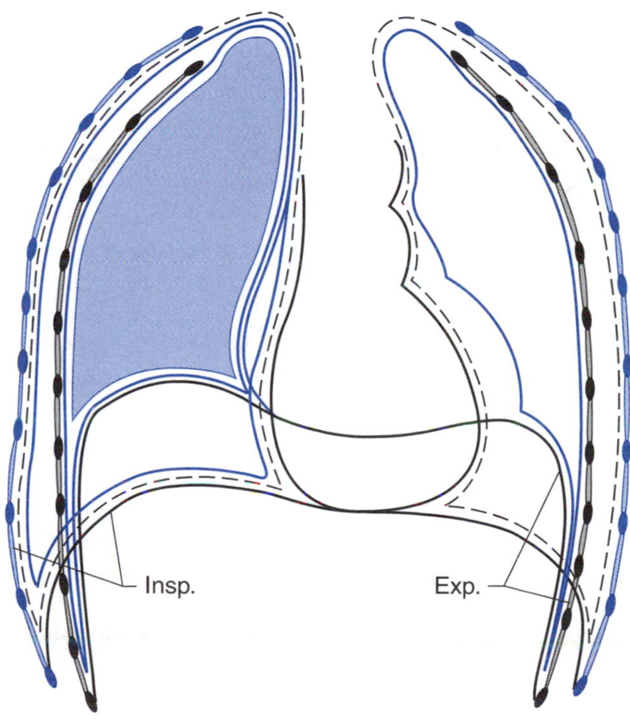


**Figure 6.5** Diagram of frontal section of thorax, based on roentgenograms, showing changes in the size of thorax and position of heart and diaphragm with respiration.

do not become engaged forcefully until the pulmonary ventilation reaches high levels, or during coughing, sneezing, or intense laughter.

At a ventilation exceeding 50 L · min$^{-1}$ and especially at very high ventilation, **accessory** inspiratory muscles may assist. The sternocleidomastoids and the pectoral muscles are the most important ones during forced inspiration. When the athlete grasps for support after an exhausting spurt, this posture may facilitate the action of the respiratory muscles.

The activity of the inspiratory muscles increases progressively throughout inspiration, and they actually continue to contract while being stretched during the early part of expiration. Part of the work done during inspiration is "stored" in the elastic structures of the system and is then available to supply part of the power for the expiration. If an expiration decreases the lung volume below the resting level of the system, the chest wall recoils outwardly, causing a passive inspiration back to the resting volume. During exercise, the inspiratory and expiratory muscles are activated reciprocally, especially the expiratory muscles in the last part of expiration.

## Resistance to Breathing

The respiratory muscles work mainly against an airway resistance and a pulmonary tissue and chest wall resistance. Most of the tissue resistance is offered by elastic forces. But the collagen fibers, which provide the supporting framework for the delicate structures of the lungs, also contribute to the resistance when a volume change occurs. Thanks to the soft, yielding tissues of the lungs, the resistance is low. Of the total pulmonary resistance, only about 20% is a tissue resistance and 80% is airway resistance. The resistance of the upper airways is about half that of the entire respiratory system (Bartlett 1979).

At high flow velocities, as during heavy exercise, the air flow is turbulent in the trachea and the main bronchi, giving a high flow resistance. Because of the large total cross-sectional area of the finest air tubes, the air flow in this region is low and therefore laminar. In fact, the greatest part of resistance to airflow within the lungs lies in airways greater than 2 mm internal diameter and particularly in the medium-sized bronchi (the 4$^{th}$–10th airway generations, see figure 6.2). An airflow of 1 L · s$^{-1}$ requires a pressure decrease along the airways of less than 2 cm $H_2O$.

At rest, the oxygen cost of breathing is only a small fraction of the total resting energy turnover: It has been estimated to be about .5 to 1.0 ml · L$^{-1}$ of moved air. With pulmonary ventilation of 6 L · min$^{-1}$, the oxygen uptake of the respiratory muscles would be up to 6 ml, compared with a total resting oxygen uptake of the body as a whole of about 250 to 300 ml. With the high pulmonary ventilation during heavy exercise, the energy cost per liter of ventilation becomes progressively greater, and the oxygen cost of breathing may be up to 8 ml · L$^{-1}$ when the pulmonary ventilation exceeds 100 L · min$^{-1}$. The air resistance when breathing through the nose is two to three times greater than that obtained by breathing through the mouth. It is therefore natural for almost everyone to breathe through the mouth (or oronasally) when performing heavy exercise. At a high pulmonary ventilation, approaching 100 L · min$^{-1}$, the oral minute volume accounts for more than 50% of the total ventilation (Niinimaa 1983).

In addition to overcoming the elastic resistance of the respiratory system, part of the energy of the respiratory muscles has to be applied to overcome two types of nonelastic resistance: a tissue viscous resistance attributable to friction, and a resistance to the movement of air in the air passages. This airway resistance can be doubled by bronchial smooth muscle contraction or reduced to half the normal resistance by bronchodilation. The airway resistance also can be increased by mucous edema or by intraluminal secretion. The factors causing this bronchoconstriction may be local or they may be a reflex response to inhaled fine, inert particles, smoke, dust, or noxious gases, or to the action of the parasympathetic system. The effect of the sympathetic system and adrenaline on bronchial tone is to dilate the airways. The increased sympathetic activity during muscular effort thus tends to lower the airway resistance. Individuals with reactive airways may develop an "exercise-induced airway constriction." Drying and cooling the inspired air will increase the response, whereas humidification and warming of the air can prevent the constriction. Exercise per se is not essential for the development of obstruction, and a heat loss and concomitant airway cooling relate directly and probably causally to both the occurrence and severity of the airway constriction (Maggi 1995; McFadden and Ingram 1983). Inhalation of cigarette smoke within seconds causes a two- to threefold increase in airway resistance that can last 10 to 30 min. (For further details concerning the effect of smoking on airway resistance and working capacity, see chapter 14.)

### FOR ADDITIONAL READING

For a thorough discussion of respiratory muscle activity and ventilatory support, see Basmajian and De Luca (1985) and Marcial and Slutsky (1997).

## *Summary*

When the respiratory muscles are relaxed (resting volume), the chest wall is retracted by the elastic recoil of the lungs. The beginning of an inspiration is assisted by the recoil of the chest wall, but the lung tissue is further stretched. During deeper inspiration, there is a retractive force from both the chest wall and the lungs. Part of the energy provided by the inspiratory muscles, mainly the diaphragm and the rib lifting muscles, is "stored" in the elastic structures and is used during the expiration. An expiration below the resting volume increases the outward recoil of the chest wall. In any volume position, the lungs and the chest cage behave like opposing springs, and at the resting volume, the forces exerted exactly counterbalance each other. The respiratory muscles mainly are devoted to doing elastic work and to overcoming the airway resistance.

# Volume Changes

Before we discuss pulmonary ventilation at rest and during exercise, a brief outline of the terminology and methods used for determining "static" and "dynamic" lung volumes, and the effect of age and sex on respiratory parameters, is presented.

## Terminology and Methods for the Determining Static Volumes

Figure 6.6 should be consulted for the terminology. When the respiratory muscles are relaxed, there is air still left in the lungs. This air volume is the functional residual capacity (FRC). A forced maximal expiration brings the volume down to the residual volume (RV) by expiration of the expiratory reserve volume. The limit of a maximal expiration is not only the capacity of the expiratory muscles to compress the thoracic cage. Many small airways become occluded during the forced expiration, and the lungs, with trapped gas, also are compressed. A maximal inspiration from the FRC adds the inspiratory capacity, and the gas volume contained in the lungs is then the total lung capacity (TLC). The maximal volume of gas that can be expelled from the lungs following a maximal inspiration is called the vital capacity (VC). It follows that VC plus the RV constitute the TLC. The volume of gas moved during each respiratory cycle is the tidal volume ($V_T$).

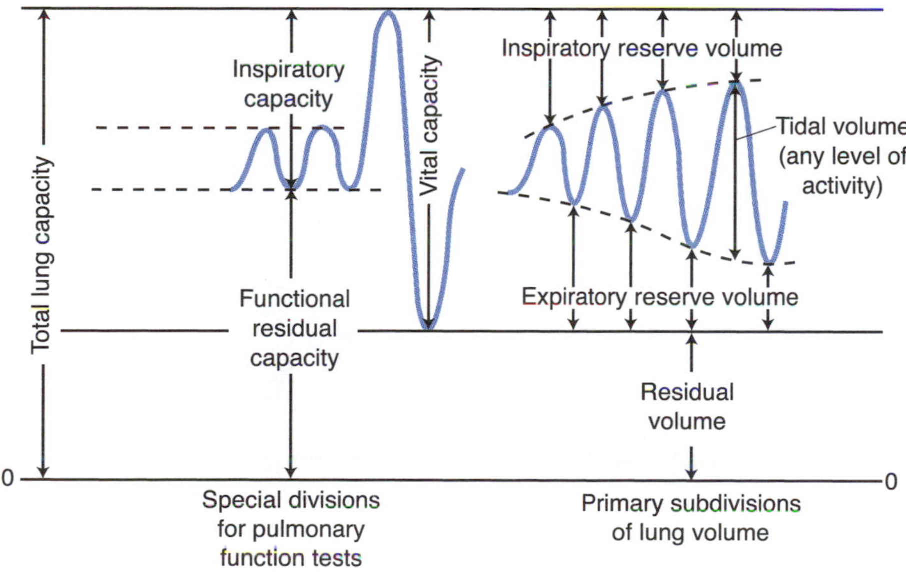


**Figure 6.6** Diagram of lung volumes and capacities.

The VC and its subdivisions commonly are measured with a **spirometer.** With the subject connected to the spirometer, any change in lung volume is reflected in a volume displacement in the spirometer. A calibration factor translates this displacement into liters.

The FRC can be measured with the closed-circuit methods (gas-dilution method). A closed spirometer contains a small, known amount of helium (or hydrogen). After a normal expiration, the subject is connected to the spirometer and rebreathes from the system. The expired $CO_2$ is absorbed by soda lime. Oxygen is added to the circuit at a rate to keep the volume at the end of expiration at a constant level. This refilling can be adjusted automatically. The concentration of the indicator gas falls in the spirometer and rises in the lungs. The final concentration is a simple function of the added gas volume, that is, the FRC. The principle for this method is clarified by figure 6.7. The concentration of the indicator is analyzed continuously, and a constant reading for about 2 min indicates a complete mixing. If the subject then performs a maximal expiration, followed by a maximal inspiration to TLC, the recordings permit calculation of RV and the subdivisions discussed previously (see figure 6.6). Normally, about 5 min of rebreathing is enough for complete mixing of the indicator gas within spirometer lungs, but in persons with an impaired lung function, up to 20 min of rebreathing may be necessary. The reason for using helium or hydrogen as the indicator gas is that these gases are absorbed by the lung tissues and blood to only a negligible degree.

If $He_1$ and $He_2$ are the initial and final concentrations, respectively, of helium, and $V_s$ is the volume of gas in the spirometer to the point of the subject's mouth, the functional residual capacity, $V_{FRC}$, can be calculated from the following formula:

$$V_s \cdot He_1 = (V_{FRC} + V_s) \cdot He_2 \text{ or } V_{FRC} = V_s \cdot (He_1 - He_2)/He_2 \quad (1)$$

The lung volumes are expressed at BTPS, that is, gas volume at body temperature and ambient pressure ($P_B$), saturated with water vapor ($P_{H_2O}$ = 47 mmHg, or 6.3 kPa), and therefore the gas volumes recorded must be recalculated. For a spirometer temperature of $t$ °C with a water pressure of $P_{H_2O}$, we have

$$FRC = V_{FRC} \cdot (310/273 + t) \cdot (P_B - P_{H_2O}/P_B - 47) \text{ L (BTPS)}. \quad (2)$$

If the subject is connected with the spirometer after a maximal expiration and then rebreathes deeply three times before being disconnected, the RV after a maximal expiration can be determined directly with the same formula.

Besides the gas dilution method, there are other methods of obtaining fairly accurate measurements of the absolute gas volumes and airspaces in the airways: the gas washout and the body **plethysmography** methods. The results obtained by these methods are closely comparable and reproducible

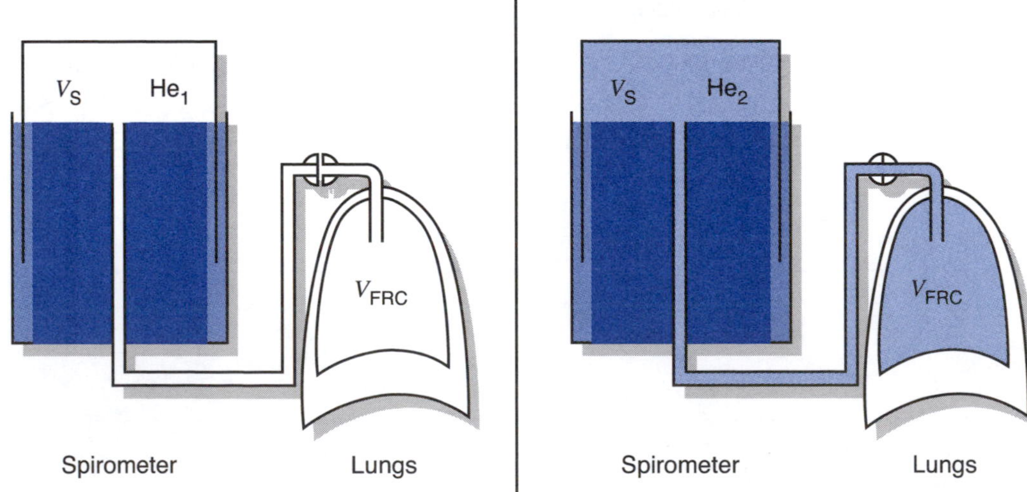


**Figure 6.7** A spirometer with a measured volume of gas ($V_s$) contains helium in a small, analyzed concentration ($He_1$). After a normal expiration (lung volume = functional residual capacity = $V_{FRC}$), the subject rebreathes from the spirometer until a homogeneous gas mixture is attained with a new and lower helium concentration ($He_2$) because of its dilution with the air in the lungs. $V_{FRC}$ can now be calculated (see text).

with a coefficient of variation of roughly ±5%. (For methods, see J.B. West 1994.)

## Effect of Age and Sex on Respiratory Parameters

Table 6.1 presents data on some of the lung volumes in liters obtained from fairly well-trained physical education students, about 24 years old (P.-O. Åstrand 1952), who were reinvestigated 21 and 32 years later (I. Åstrand, P.-O. Åstrand et al. 1973; Viljanen 1982).

Data on lung volume (liters) in 27 female and 26 male former physical education students (in 1949) were reinvestigated in 1970 and 1982. In 1949, the average age was 22 and 26 years, respectively. Maximal tidal volume and pulmonary ventilation were measured during exercise.

The data were obtained while subjects were in the standing position. During tilting from standing to supine position, the TLC and VC are reduced 5% to 10% because of a shift of blood to the thoracic cavity from the lower part of the body. This illustrates the effect of gravity on the blood distribution within the body (see chapter 5).

VC, RV, and TLC are related to body size and vary approximately as the cube of a linear dimension, such as body height, up to the age of about 25.

The individual dimensions are, however, not exclusively decisive for the size of the lung volumes. The lung volumes are about 10% smaller in women than in men of the same age and size. For the average person, the VC is up to 20% smaller than the values listed in table 6.1. Training during adolescence eventually will increase the VC and TLC. After the age of about 30, the RV and FRC increase and the VC usually decreases. However, there are observations that well-trained individuals attained the same VC at the age of 40 to 45 as 21 years earlier (I. Åstrand, P.-O. Åstrand et al. 1973). Twelve years later, a slight reduction in VC was observed (table 6.1). The ratio of RV/TLC · 100 in the young individual is about 20%, but for the 50- to 60-year-old individual, this ratio increases to about 40%, an increase that can be accounted for almost entirely by changes in lung elasticity with age (Viljanen 1982).

Athletes have similar or slightly higher values for VC and TLC compared with the data in table 6.1. The highest recorded value for VC is 9.0 L for a Danish rower, recorded by Secher and Jackson (unpublished data).

In a group of about 190 individuals 7 to 30 years of age, a significant correlation was found between VC and maximal oxygen uptake (figure 6.8). A closer examination of the individual figures reveals, however, that individuals with a VC of approximately 4 L may have a maximal oxygen uptake from about 2.0 to 3.5 L · $min^{-1}$. From this and similar studies, it is evident that vital capacities of 6.0 L are associated with oxygen uptake capacities varying from about 3.5 to 5.5 L · $min^{-1}$. This example shows that one function can appear closely related to another if the data are derived from persons of greatly different size.

However, the scattering of the data may still be considerable and sufficiently large to make any prediction of an individual's maximal oxygen uptake from such parameters as VC rather unreliable. The conclusion can be drawn, however, that an oxygen uptake of 4.0 L · $min^{-1}$ or more requires a VC of at least 4.5 L.

Measurement of the VC as part of a larger test battery can yield valuable information, especially concerning the distensibility of the respiratory system. Certain pathological conditions are associated with a reduced VC.

## "Dynamic" Volumes

Dynamic spirometry, that is, the determination of ventilatory capacity per unit time, also is used to assess an individual's respiratory function. The subject breathes into a low-resistance spirometer, and its displacements are recorded.

To determine forced expiratory volume (FEV), the subject first takes a deep breath and inspires

**Table 6.1** Lung Volumes From Well-Trained Physical Education Students

| | Females | | | Males | | |
|---|---|---|---|---|---|---|
| **Function** | **1949** | **1970** | **1982** | **1949** | **1970** | **1982** |
| Vital capacity | 4.26 | 4.25 | 4.05 | 5.55 | 5.39 | 4.92 |
| Residual volume | 1.10 | 1.71 | 1.64 | 1.45 | 2.04 | 2.12 |
| Total lung capacity | 5.36 | 5.96 | 5.69 | 7.00 | 7.43 | 7.04 |
| Maximal tidal volume | 2.24 | 2.26 | 2.23 | 3.38 | 3.31 | 3.23 |
| Maximal $\dot{V}_E$ L · $min^{-1}$ | 92.5 | 90 | 88.5 | 121 | 121 | 114 |

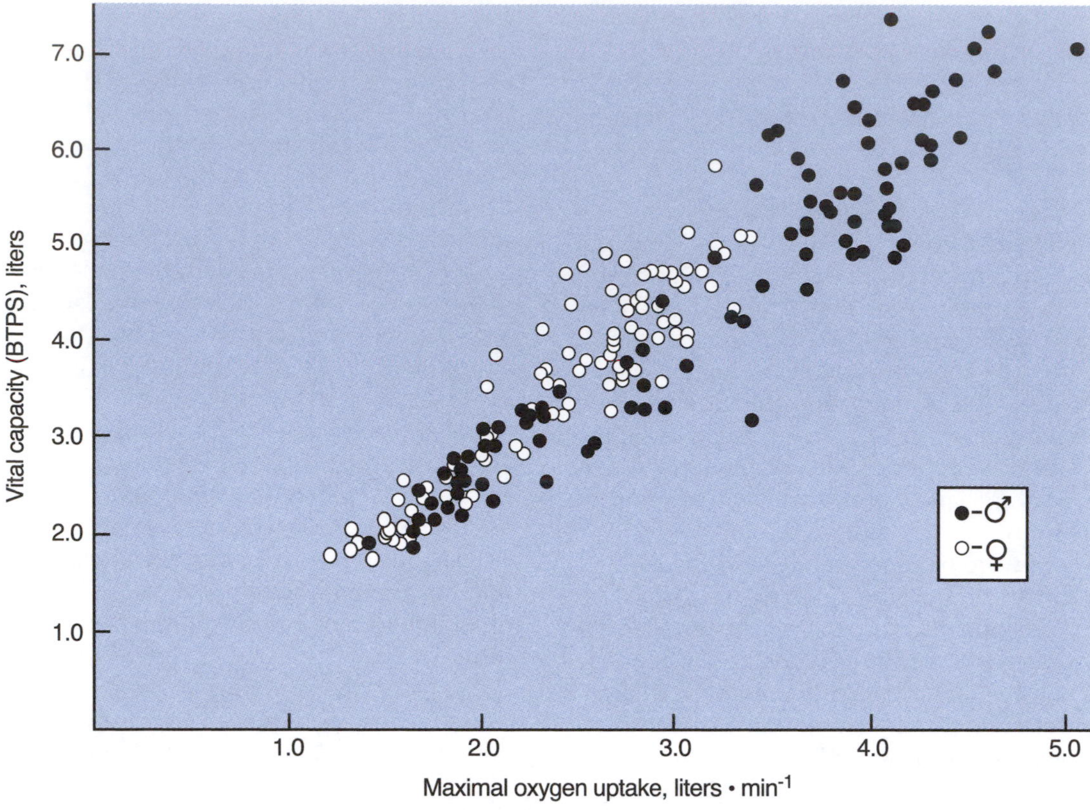


**Figure 6.8** Individual data on vital capacity measured in standing position in relation to maximal oxygen uptake during running or cycling in 190 subjects from 7 to 30 years of age.

From P.-O. Åstrand 1952.

maximally. The subject then exhales as forcefully and completely as possible. In this way the tester determines how much of the person's VC can be exhaled in the course of 1 s ($FEV_{1.0}$), and this volume is expressed as a percentage of the individual's entire VC. A normal figure for a 25-year-old individual is about 80%. The maximal flow is limited by the rate by which the muscles are able to transform chemical energy into mechanical energy and also by a rising flow resistance. Thus, $FEV_{1.0}$ is reduced in persons who have any airway obstructions.

The mechanical properties of the lungs and the chest wall can be evaluated by determining the maximal voluntary ventilation (MVV; also referred to as maximal breathing capacity). The subject is asked to breathe as rapidly and as deeply as possible during a given time interval, usually 15 s. The individual differences in MVV are large. For healthy 25-year-old men, the mean value is about 140 L · min$^{-1}$, with a range from 100 to 180 L · min$^{-1}$. For women, the normal values range from about 70 to 120 L · min$^{-1}$. The pulmonary ventilation during maximal work is somewhat lower than that obtained during the determination of MVV.

## Compliance

The lungs and the thorax are made partly of elastic tissue. During inspiration, these tissues are stretched. Because of their elastic nature, they return to their resting position as soon as the inspiratory muscles are relaxed. The more rigid these tissues are, the greater muscular force must be applied to achieve a given change in volume. The relation between force and stretch or between pressure and volume can be measured, which gives a measure of the tissue's elastic resistance to distension. With the aid of a balloon placed in the intrathoracic esophagus, the pressure is measured at the end of a normal expiration and again after the subject has inhaled a known volume of gas. These measurements can be repeated at different volume changes. The volume changes in liters produced by a unit of pressure change in centimeters of $H_2O$ give the lung compliance. If a pressure change of 5 cm $H_2O$ produces a change in lung volume of 1 L, the lung compliance is 1.0 L per 5 cm of $H_2O$, or .2 L per cm of $H_2O$, which is the normal value at quiet breathing. With a respiratory depth of about .5 L, the pressure variations in overcoming the resistance are a few centimeters of water. At lung volumes closer to maximal inspiration, or maximal expiration, a greater pressure is required for a given volume change; that is, the compliance is reduced. If the lungs are more rigid and less distensible (because of pathological changes, such as interstitial or pleural fibrosis), the compliance also is reduced and the respiratory work is increased.

# Pulmonary Ventilation at Rest and During Exercise

The pulmonary ventilation is the mass movement of gas in and out of the lungs. The pulmonary ventilation is regulated mainly to provide the gaseous exchange required for aerobic energy metabolism. Some gaseous exchange takes place through the skin, and some gas is lost in the urine and other secretions, but the volume of gas thus exchanged is negligible.

## Methods

Gas volumes can be measured very accurately by the classic way of using a water-filled spirometer. Air is inhaled through a respiratory valve, and the expired air is collected in the spirometer. The volume also can be measured by other means, such as with a gas meter or flow meter. The amounts of inhaled and exhaled air are usually not exactly equal, because the volume of inspired oxygen in most situations is larger than the volume of $CO_2$ expired. Pulmonary ventilation usually refers to the volume of air that is exhaled per minute. It is exceedingly important that the mouthpiece, respiratory valve, tubes, and stopcocks are so constructed that they cause a minimum of increased airway resistance during heavy physical exertion. The diameter of the tubes and all openings should be about 30 mm or wider.

## Pulmonary Ventilation during Exercise

Figure 6.9 shows how ventilation ($\dot{V}_E$) increases during increasing rates of exercise up to the maximal level. From a resting value of about 6.0 L · min$^{-1}$, the ventilation increases to 100, 150, and, in extreme cases, 200 L · min$^{-1}$ (Saltin and Åstrand 1967). The increase is semilinear, with a relatively larger increase at the heavier exercise intensities. (The explanation for this is discussed later in this chapter in the section "Regulation of Breathing" on page 202).

Figure 6.10 presents data on maximal pulmonary ventilation for about 225 subjects, from 4 to about 30 years of age, collected during maximal running for about 5 min. A positive correlation exists between maximal $V_E$ and $\dot{V}O_2$, but it is evident that maximal ventilation cannot be used to predict maximal oxygen uptake. Maximal pulmonary ventilation is actually not a well-defined parameter. Figure 6.9 illustrates a marked increase in ventilation during heavy exercise without any further increase in oxygen uptake. A well-motivated subject can continue to exercise at very high rates of exercise despite strain (as judged from blood lactic acid concentration), and the person attains a high ventilation; the less motivated one just quits exercise when still working at a submaximal level.

If pulmonary ventilation is expressed in relation to the magnitude of oxygen uptake, it is 20 to 25 L per liter of oxygen at rest and during moderately heavy exercise, but it increases to 30 to 40 L per liter of oxygen during maximal exercise. In children under 10 years of age, the values are about 30 L during light exercise and up to 40 L per liter of oxygen uptake during maximal exercise (see figure 6.10).

Figure 6.11 presents mean values for maximal pulmonary ventilation during exercise (running or cycling) in different age groups. The lower ventilation in the older individuals is associated with a reduced maximal oxygen uptake. In the subjects followed for a 33-year period, the maximal pulmonary ventilation declined only about 5%, on average (table 6.1).

## Dead Space

Only a part of the inhaled volume of air reaches the alveoli where the gaseous exchange can take place. This part is known as the effective tidal volume, or volume of alveolar gas ($V_A$). Part of the inspired tidal volume ($V_T$) occupies the conducting airways. This part is called **dead space** volume ($V_D$) because it does not take part in the gaseous exchange between alveolar air and blood. During expiration, this dead space component is exhaled first. It has a composition similar to moist inspired air. Then comes the alveolar component, which has a relatively high concentration of $CO_2$ and a low oxygen concentration. The total expired gas is therefore a mixture of dead space and alveolar gas, or $V_T = V_A + V_D$.

From a functional standpoint, this dead space is not merely the result of the anatomic features of the respiratory tract. In addition to the air

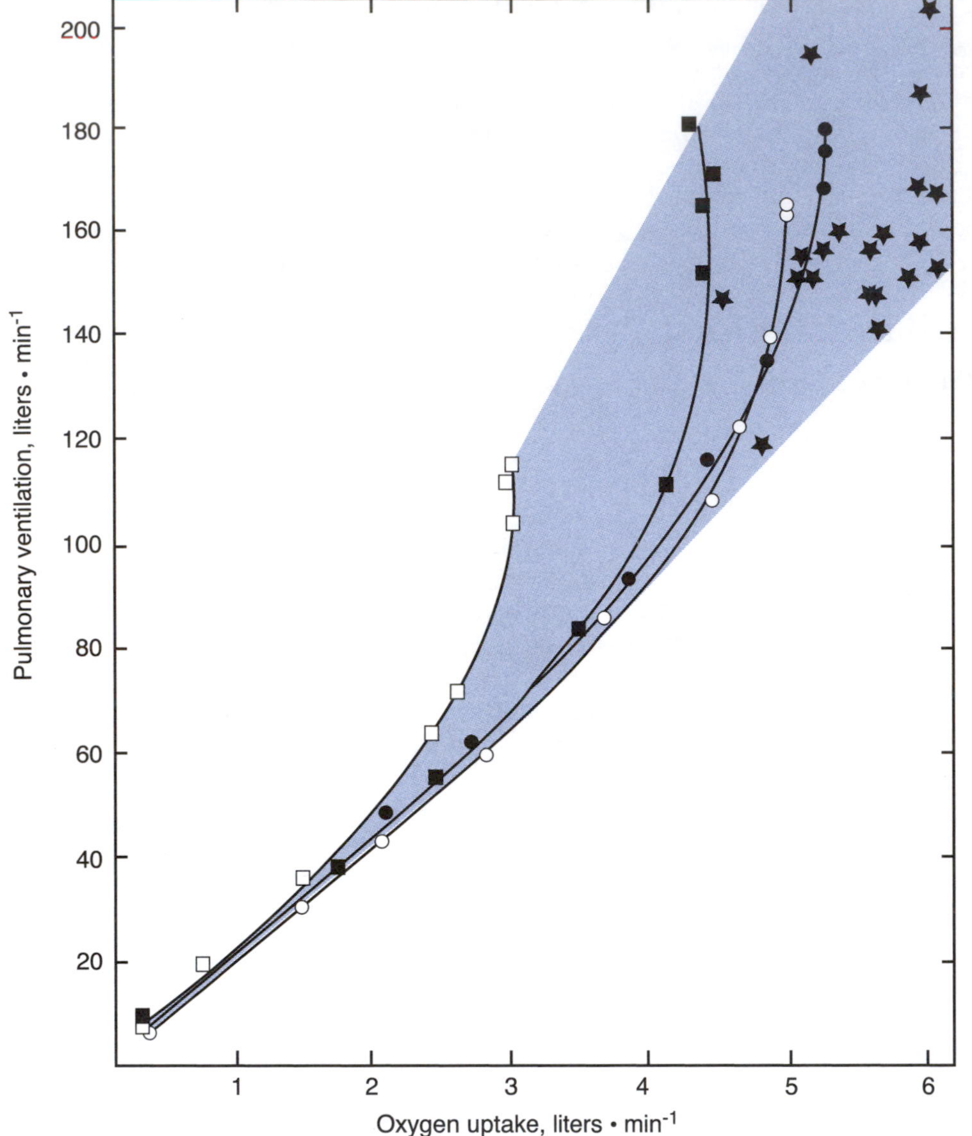


**Figure 6.9** Pulmonary ventilation at rest and during 2 to 6 min of exercise (running or cycling). Four individual curves are presented. Several rates of exercise gave the same maximal oxygen uptake. Individuals with maximal oxygen uptake of 3 L · $min^{-1}$ or higher usually fall within the shadowed area. Note the wide scattering at high oxygen uptakes. ★ = individual values for top athletes measured when maximal oxygen uptake was attained.

Data from Saltin and P.-O. Åstrand 1967.

volume that remains stagnant in the conductive airways, some air eventually reaches alveoli that are poorly perfused by capillary blood or not perfused at all. This reduces the gaseous exchange. In patients suffering from pulmonary disease, an unfavorable relationship between ventilation and perfusion may increase the physiological dead space.

The volume of the dead space can be estimated with the aid of Bohr's formula, which is based on the fact that the expired volume of oxygen at each respiration ($V_T \cdot F_EO_2$) is equal to the sum of the volume of oxygen contained in the dead space compartment ($V_D \cdot F_IO_2$) and the volume of oxygen coming from the alveolar air ($V_A \cdot F_AO_2$), where F = fraction of oxygen in the expired, inspired, and alveolar air, respectively. We therefore arrive at the following formula:

$$V_T \cdot F_EO_2 = V_D \cdot F_IO_2 + V_A \cdot F_AO_2 \quad (3)$$

Because $V_A = V_T - V_D$, the formula may be simplified as follows:

$$V_D = V_T \cdot F_EO_2 - F_AO_2 / F_IO_2 - F_AO_2 \quad (4)$$

If the oxygen content of the inspired air is 21%, the oxygen content of the expired air is 16%, the oxygen content of the alveolar air is 14%, and the depth of respiration ($V_T$) is 500 ml:

$$V_D = 500 \cdot 16 - 14/21 - 14 = = 143 \text{ ml.} \quad (5)$$

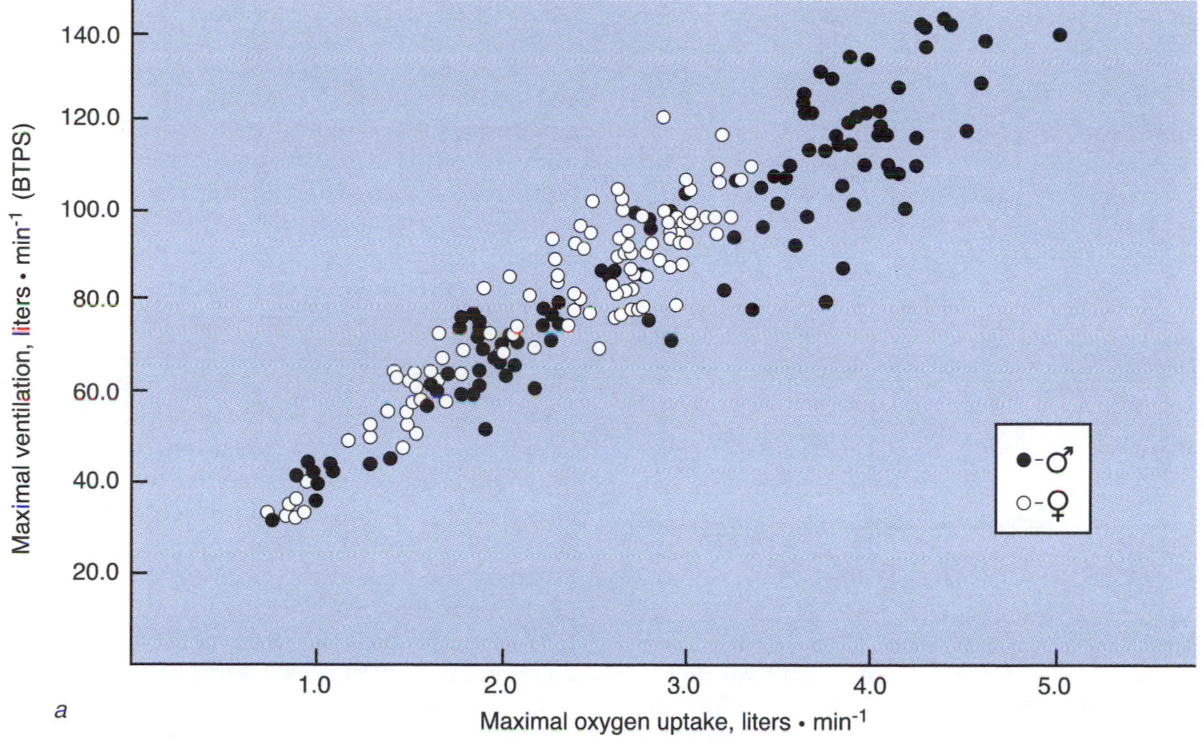


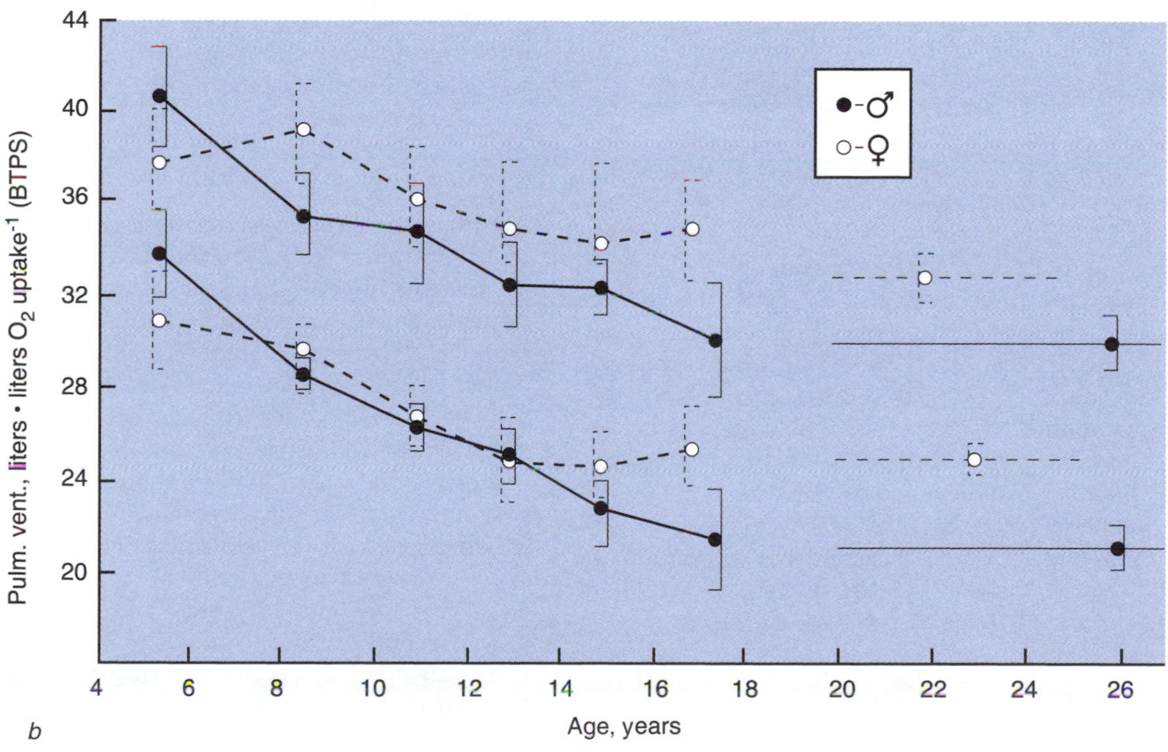


**Figure 6.10** Data on 225 subjects from 4 to about 30 years of age. *(a)* Maximal pulmonary ventilation in relation to maximal oxygen uptake was measured during running on a motor-driven treadmill for about 5 min. *(b)* Average values of ventilation per liter of oxygen uptake in relation to age. The upper curves show maximal values (attained during running), and the lower ones show submaximal values during running or cycling, with an oxygen uptake that was 60% to 70% of the subjects' maximal aerobic power [same subjects as in *(a)*]. Vertical lines denote ± 2 SE (standard error of the mean). BTPS = gas volume at normal body temperature and ambient barometric pressure, saturated with water vapor.

From P.-O. Åstrand 1952.

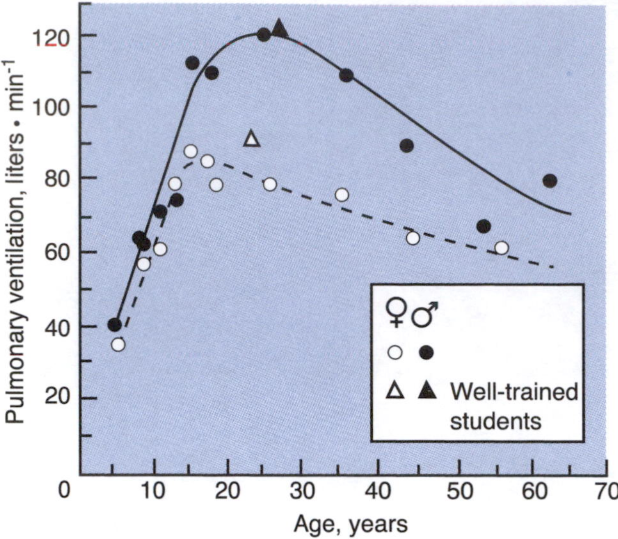


**Figure 6.11** Pulmonary ventilation measured after about 5 min exercise with a rate of exercise that brought the oxygen uptake to the individual's maximum. Mean values on 350 women and men and about 80 well-trained subjects; exercise on motor-driven treadmill or cycle ergometer.

Based on data from P.-O. Åstrand 1952 and I. Åstrand 1960.

The same calculation can be made on the basis of $CO_2$.

With a depth of respiration of 500 ml at rest, the dead space constitutes approximately 150 ml. The rest of the tidal volume reaches the alveoli. However, the first portion of the air reaching the alveoli during inspiration is, in reality, the air that remained in the dead space compartment from the previous respiratory cycle. The "fresh" air that is pulled down into the alveoli is diluted into a relatively large volume, that is, the FRC. The variations in the gas concentration are therefore relatively small in the alveoli during rest and normal breathing.

Relatively speaking, the dead space is reduced with increasing tidal volume. If, at a ventilation of 6.0 L · min$^{-1}$, the respiratory frequency is 10 and the dead space .15 L, the alveolar ventilation is

$$6.0 - .15 \cdot 10 = 4.5 \text{ L} \cdot \text{min}^{-1}. \quad (6)$$

If the respiratory rate, on the other hand, is 20, and the gross ventilation and dead space are assumed to be unchanged, the alveolar ventilation is only

$$6.0 - .15 \cdot 20 = 3.0 \text{ L} \cdot \text{min}^{-1}. \quad (7)$$

From the pulmonary ventilation data that are given in figure 6.9, part of the volume does not participate in the gas exchange. Hiding submerged in water and breathing through a snorkel, often described in adventure stories, represents a considerable complication of the gas exchange. The tube (snorkel) represents an extension of the respiratory dead space, and the tidal volume has to be increased by an amount equal to the volume of the tube if the alveolar ventilation is to be maintained unchanged. The breathing therefore can become very laborious.

A second complication of this scenario is the increased load on the inspiratory muscles. The lungs experience the same pressure as at the water surface, that is, atmospheric pressure. The outside of the thorax, however, is subjected to atmospheric pressure plus the pressure of the column of water above the diver. At a depth of 1.0 m, this extra pressure will be .1 atm, or about 76 mmHg (10.1 kPa). The highest pressure the inspiratory muscles can overcome is just above 70 mmHg (9.3 kPa), and, therefore, a depth of 1.0 m is the maximum that can be tolerated even if the problem of extra dead space could be solved by a system of valves.

## Tidal Volume–Respiratory Frequency

By definition, the pulmonary ventilation equals the frequency of breathing multiplied by the mean expired tidal volume, or

$$V_E = f \cdot V_T \quad (8)$$

At rest, the respiratory frequency is between 10 and 20. Inspiration occupies less than half the total cycle, the increase in flow being more abrupt than the decrease. During exercise at low intensity, primarily the tidal volume is increased. Naturally, the VC limits the tidal volume, but the latter rarely amounts to more than 60% of the VC (figure 6.12). The respiratory frequency also is increased, especially in the case of heavy exercise. Children about 5 years of age may have a respiratory frequency of about 70 at maximal exercise, 12-year-old children about 55, and 25-year-old individuals 40 to 45 (see figure 6.12). In well-trained athletes with high aerobic power, respiratory frequencies up to about 60 breaths · min$^{-1}$ are not unusual (J.M. Clark, Hagerman, and Gelfand 1983).

The increase in tidal volume is brought about through the use of both the inspiratory reserve volume and the expiratory reserve volume (see figure 6.6). The depth of respiration and respiratory frequency apparently is balanced in such a way that a certain ventilation takes place at optimal efficiency, that is, with the use of a minimum of energy

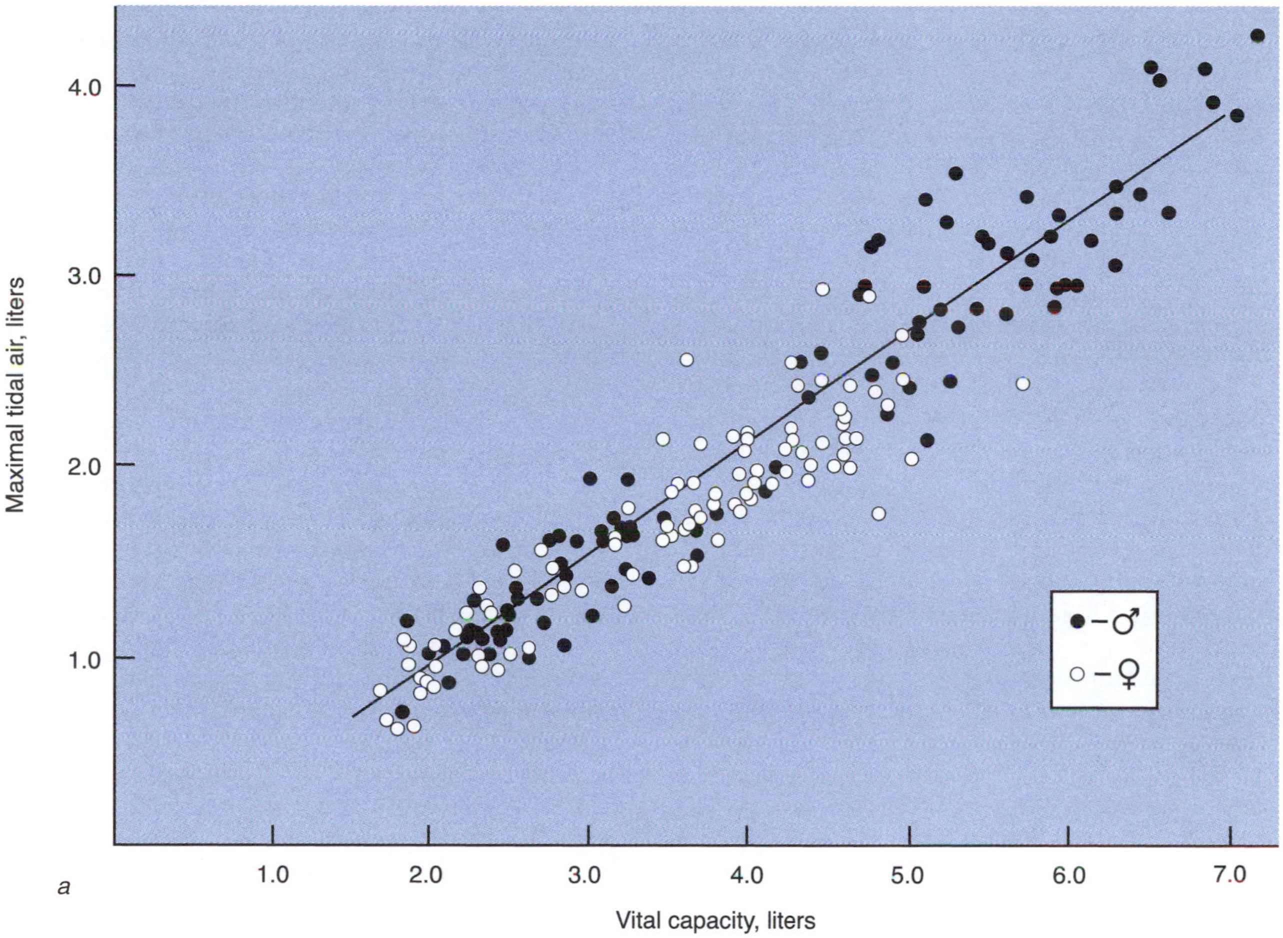

Maximal tidal air, liters

4.0

3.0

2.0

1.0

●–♂

○–♀

a

1.0 2.0 3.0 4.0 5.0 6.0 7.0

Vital capacity, liters

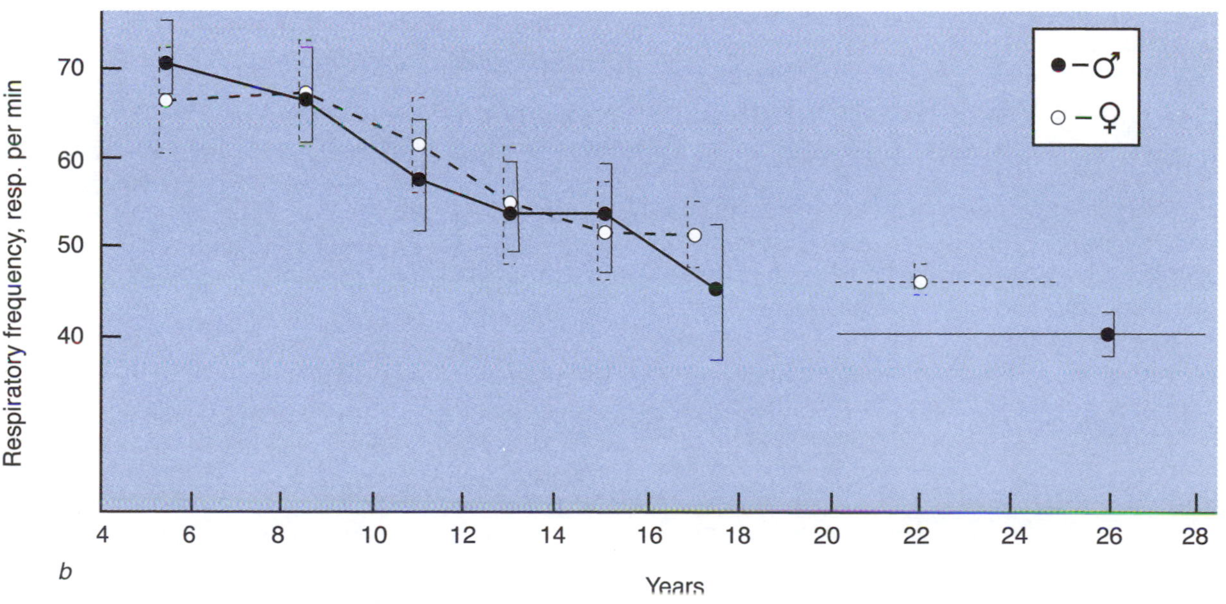


**Figure 6.12** *(a)* Highest tidal volume measured during running at submaximal and maximal speed (exercise time about 5 min) related to the individual's vital capacity measured in standing position. The data include 190 subjects from 7 to 30 years of age. *(b)* Respiratory frequency during running at a speed that brings the oxygen uptake up to maximum; average values (±2 standard errors) for 225 subjects from 4 to 10 years of age.

From P.-O. Åstrand 1952.

by the respiratory muscles. In athletic performance, it is therefore advisable to allow the athlete to assume the respiratory pattern that seems natural for him or her. In many types of physical exercise, the respiratory frequency tends to become fixed to the exercise rhythm (Jasinskas, Wilson, and Hoare 1980). There seems to be a strict locomotor–respiratory coupling, especially in exercise where the stress of locomotion tends to deform the thoracic complex (Bramble and Carrier 1983). Needless to say, this certainly holds for crawl swimming, but it also holds for such activities as bicycle riding, sculling, and running. Therefore, the ventilatory pattern is not guided exclusively by demand for minimal energy expenditure of the respiratory muscles. In swimming, instruction in the breathing technique may be necessary, but in other cases, the respiratory pattern should be allowed to follow its natural pattern. The phase-locked patterns can vary greatly in one type of exercise. Work on the cycle ergometer with a pedaling frequency of 50 usually gives a respiratory frequency related to this pedaling frequency. It often increases stepwise from 12.5 to 16.6, to 25.0, to 33.0, up to 50 at maximal work. It is important to keep this in mind if one is studying the mechanics and regulation of breathing during exercise, using only the cycle ergometer as a means of providing the workload.

### For Additional Reading

For further discussion about coordination-related changes in the rhythms of breathing, see Rassler and Kohl (2000).

## Respiratory Work (Respiration as a Limiting Factor in Exercise)

The work of the respiratory muscles consists primarily of stretching the elastic tissues of the chest wall and lungs and overcoming the flow-resistive forces. At rest, the respiratory muscles require from .5 to 1.0 ml of oxygen per liter of ventilation. During exercise, the oxygen cost per unit ventilation increases markedly. Once it was thought that the oxygen consumption of respiratory muscles increased disproportionately at high levels of ventilation, but later estimates show that the energy cost of breathing in normal individuals rarely represents more than 3% of the total energy expenditure (Casan et al. 1997).

A question of considerable importance is whether hyperventilation limits the body's total oxygen uptake. The answer is probably negative for the following four reasons. First, after the maximal oxygen uptake is reached, it is still possible for the subject to continue to exercise at higher rates because of the anaerobic processes. At the same time, the pulmonary ventilation is increased markedly, without any distinct ceiling being reached except in some extreme cases. The "normal" pattern is illustrated in figure 6.9. Second, during very heavy exercise that can be tolerated for only a few minutes at the most, the pulmonary ventilation is greater than at a somewhat lower but still maximal load that can be tolerated for about 6 min. The oxygen uptake is nevertheless the same in both cases (P.-O. Åstrand and Saltin 1961a). Third, at maximal exercise, it is possible to voluntarily increase the ventilation further, showing that the ability of the respiratory muscles to ventilate the lungs evidently is not exhausted during spontaneous respiration. One example: Five well-trained young subjects exercised on a cycle ergometer at, or very close to, maximal oxygen uptake for 10 min ($\dot{V}O_2$max 4.46 L · min$^{-1}$, 63 ml · kg$^{-1}$ · min$^{-1}$. Their MVV (30 s) was 211 L · min$^{-1}$ and their pulmonary ventilation during maximal cycling was, when they were breathing spontaneously, on the average 159 L · min$^{-1}$ or 76% of the MVV. Following another protocol, they hyperventilated voluntarily during the 5th min of the 10-min standard exercise. This increased the pulmonary ventilation to a mean value of 179 L · min$^{-1}$, or 85% of MVV. In other words, a reserve capacity was available. It should be noted that the hyperventilation increased the oxygen uptake (+.10 L · min$^{-1}$), and the blood lactate concentration increased from 11.1 to 12.7 mmol · L$^{-1}$, both changes being significant ($p < .01$;. Hartley, Lacour, and P.-O. Åstrand, unpublished data). Well-trained and very fit athletes can apparently use some 95% of this MVV during exercise, whereas less fit subjects normally only attain 60% to 70% of their MVV (Folinsbee et al. 1982). Fourth, at heavy rates of exercise, the pulmonary ventilation depends on the volume of $CO_2$ produced. The alveolar oxygen tension increases and the $CO_2$ tension decreases, a fact that indicates an effective gas exchange in the lungs (figure 6.13). The oxygen tension of the arterial blood is maintained or only slightly reduced except in highly trained individuals with high maximal aerobic power (Dempsey, Hanson, and Henderson 1984). These authors actually suggested that the failure to increase pulmonary ventilation adequately, and the very rapid transit time of red cells through the pulmonary capillary bed, could result in inefficient diffusion of oxygen from air to blood. The size of the contact area also may be critical when large volumes of oxygen are needed (see also Powers 1988; Weibel and Taylor 1981; J.H. Williams, Powers, and Stuart 1985).

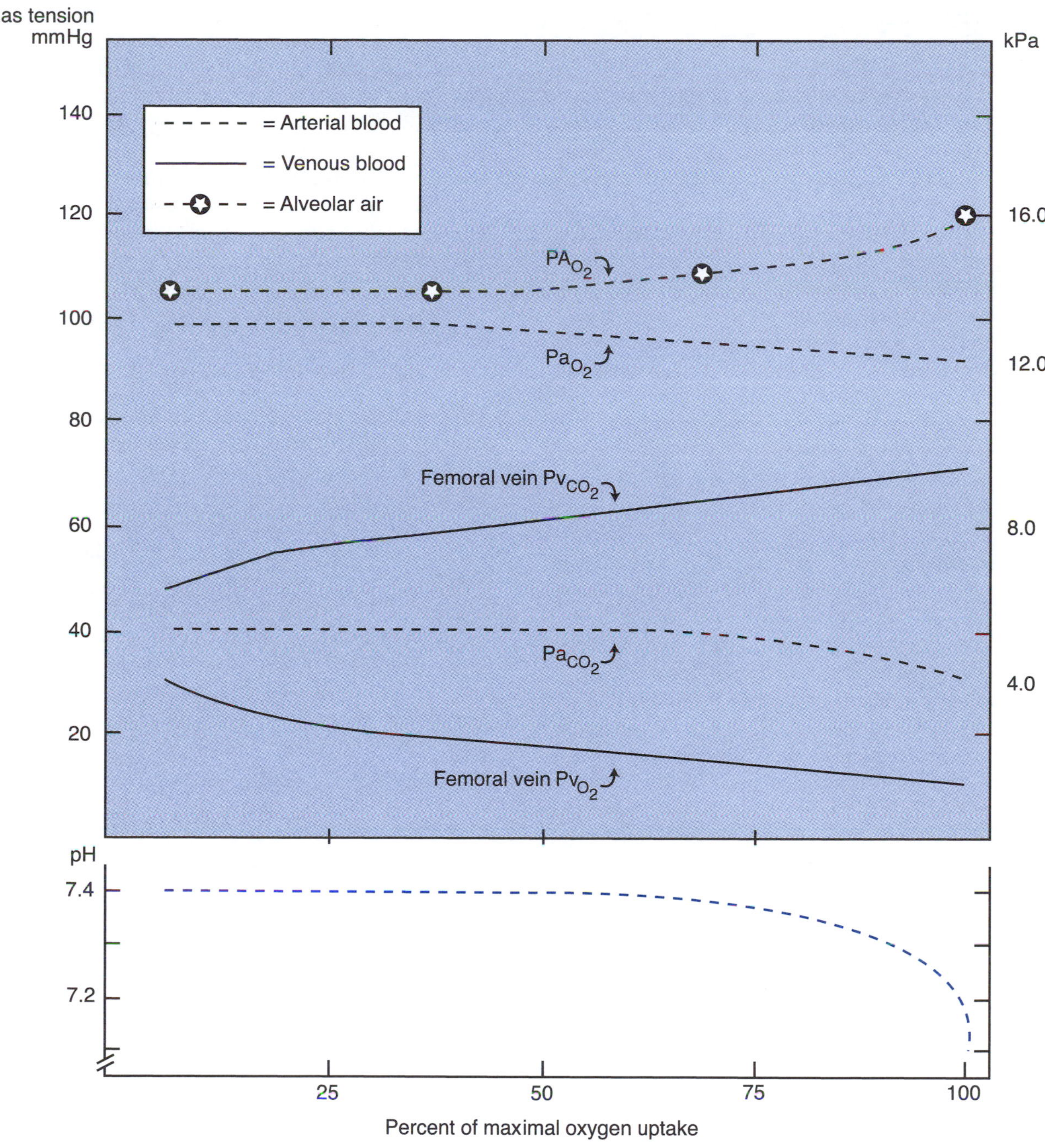


**Figure 6.13** Oxygen and carbon dioxide tensions in blood and alveolar air at rest and during various levels of exercise up to and exceeding the load necessary to reach the individual's maximal oxygen uptake (= 100%). At bottom, arterial pH. Curves are based on data from different authors and unpublished studies. $PA_{O_2}$ = alveolar partial pressure of oxygen; $Pa_{O_2}$ = arterial partial pressure of oxygen; $Pv_{CO_2}$ = venous partial pressure of carbon dioxide; $Pa_{CO_2}$ = arterial partial pressure of catbon dioxide; $Pv_{O_2}$ = venous partial pressure of oxygen.

To some extent, the pulmonary ventilation may be a limiting factor even though the maximal capacity of the respiratory muscles is not fully taxed. Thus, if the energy demand of the respiratory muscles required to increase the pulmonary ventilation necessitates such a marked increase in oxygen consumption that all the achieved increase in oxygen content of the alveolar air (and increase in oxygen content of the arterial blood) is entirely used by the respiratory muscles themselves, then none of this extra oxygen will benefit the rest of the active muscles of the body. In other words, an increase in pulmonary ventilation beyond a certain point is not physiologically useful, because all the additional oxygen thus gained would be required for the work of breathing (Bye, Farkas, and Roussos 1983). It is even conceivable that the oxygen use by the respiratory muscles may become so great that the oxygen supply to other tissues is reduced. However, normal individuals probably do not reach such a critical limit. It is likely that the blood flow in the vessels of the respiratory muscles is maximal even at a ventilation below the maximal ceiling and that the oxygen content of the blood is almost completely extracted. A further increase in ventilation beyond this point is probably met by anaerobic processes. A point in favor of this view is the fact that the oxygen uptake reaches a distinct plateau during extremely heavy exercise, even if the rate of exercise and the pulmonary ventilation are further increased.

Respiratory muscle fatigue invariably occurs with extreme ventilatory demands (e.g., during resistance breathing) (Bye, Farkas, and Roussos 1983). The question is whether this occurs during exercise. Keeping the energy cost of the respiratory muscles at a level acceptable to the muscles directly involved in the external task to be performed has to be balanced against the inevitable consequence of reducing oxygen tension in the arterial blood.

A substitution of air with a 79% helium/21% oxygen mixture (with reduced density) increases the maximal pulmonary ventilation and oxygen uptake significantly (Brice and Welch 1983). At present it is difficult to explain such findings. Loke, Mahler, and Virgulto (1982) reported that respiratory muscle fatigue is evidently a consequence of marathon running (both the strength of the respiratory muscles and the MVV declined following the race). B. Martin, Heintzelman, and Hsiun-ing Chen (1982) noticed that short-term maximal running performance decreased after 150 min of isocapnic hyperventilation at two thirds of the subjects' MVV. They concluded that a reduced ventilatory muscle performance explained their finding. Bye et al. (1984) also presented data indicating that diaphragmatic fatigue (evaluated from recorded electromyogram) was established in such subjects performing exhausting exercise at 80% of their maximal power output on a cycle ergometer. However, later studies that used phrenic nerve stimulation to test the force produced by the diaphragm showed that $CO_2$ retention (**hypoventilation**) and voluntary cessation of loading occur before the muscles become overtly fatigued (Gandevia, Allen et al. 1998).

### *Summary*

Under normal conditions, with the possible exception of individuals with very high aerobic power and therefore exceptional demands on the respiratory function, the potential of the respiratory muscles to provide sufficient ventilation does not seem to limit the maximal oxygen uptake. One complicating factor is that the improved oxygenation of the blood passing the lungs by an exaggerated pulmonary ventilation also leads to a proportionally increased oxygen demand by the respiratory muscles. Therefore, the net effect of an extra respiratory effort can be questioned.

## Diffusion in Lung Tissues, Gas Pressures

The pulmonary gas exchange between the capillary blood and the alveolar air is accomplished by diffusion. (For details, see R.E. Forster and Crandell 1976; J.B. West 1994.)

The diffusion takes place as a movement of gas molecules from a region of higher **partial pressure** of the gas to one of lower. The normal pressure of oxygen, $CO_2$, and nitrogen in atmospheric air ($P_B$ = 760 mmHg, or 101 kPa), in alveolar air, and in mixed venous blood and arterial blood at rest is given in figure 6.14. If, for example, the oxygen concentration in the alveolar air is 15% of the dry gas, the partial pressure of oxygen ($PO_2$) is

$$PO_2 = 15/100\ (760 - 47) = 107 \text{ mmHg or}$$
$$PO_2 = 15/100\ (101.1 - 6.3) = 14.2 \text{ kPa} \quad (9)$$

because the partial pressure of water vapor is 47 mmHg (6.3 kPa).

Blood flow through a tissue is not always determined by the metabolic activity in the tissue in question. Thus, the oxygen uptake of tissues such as the kidneys and skin is small compared with the magnitude of the blood flow through these tissues. For this reason, the $PO_2$ in the venous blood remains high and the $CO_2$ pressure relatively low. Comparatively more oxygen is used in the muscles, and here a greater amount of $CO_2$ is produced, so that the partial pressures of these gases in the venous blood from working muscles are different from those of the previously mentioned organs.

The total gas pressure in venous blood is considerably lower than in arterial blood (706 mmHg vs. 760 mmHg, or 94 vs. 101 kPa, respectively; see figure 6.14). In this way, accumulation of gas in the intrapleural space in the thorax is avoided, despite the opposed recoil of the lungs and the chest wall. If gas is trapped behind an occlusion of an airway, it becomes absorbed into the pulmonary circulation because of this subatmospheric gas pressure of

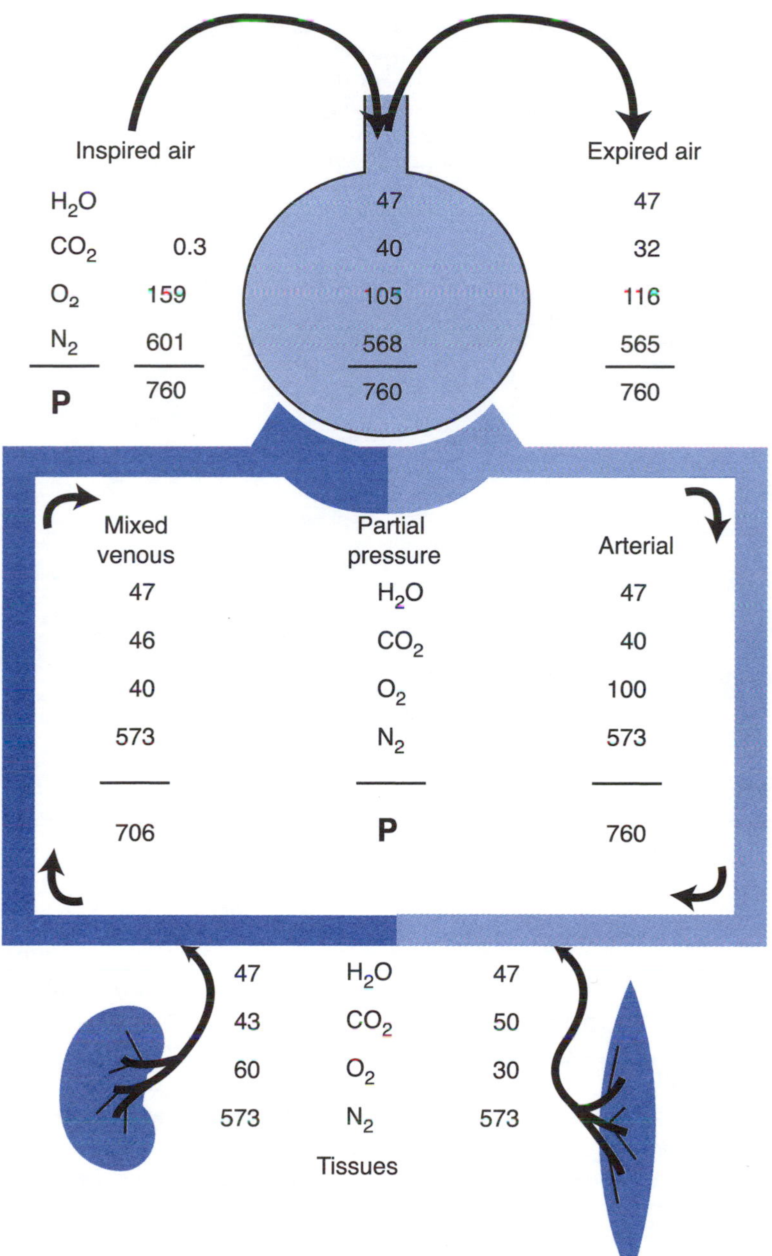


**Figure 6.14** Typical values of gas tensions in inspired air, alveolar air (encircled), expired air, and blood, at rest. Barometric pressure, 760 mmHg (100 mmHg = 13.3 kPa); for simplicity, the inspired air is considered free from water (dry). Tension of oxygen and carbon dioxide varies markedly in venous blood from different organs. In this figure, gas tensions in venous blood from the kidney and muscle are presented.

venous blood. Under normal conditions, the diffusion is so rapid that the gases in the blood leaving the pulmonary capillaries are approximately in equilibrium with the gases in the alveoli.

An analysis of the gas concentrations and pressures in the expired air at the end of the expiration gives an approximate idea of the gas pressure in the arterial blood. One can simply collect the last portion of the expiratory air volume, either at the end of a single forced expiration (Haldane-Priestley method) or from several repeated respiratory cycles (end-tidal sampling technique), to analyze the gaseous composition of expired air. With the aid of modern analytic and registration techniques, it is also possible to follow variations in gas concentrations and pressures continuously during one or several successive respirations (e.g., with the aid of a mass spectrometer). Figure 6.15 gives examples of the variations in $CO_2$ and oxygen pressures in the air in the trachea and in the alveoli during a single respiration. At rest, the variations in the composition of the gases in the alveoli are small because the inhaled air volume is diluted into a relatively large gas volume, the functional residual volume. During exercise, when the depth of respiration is increased, the variations become considerably larger.

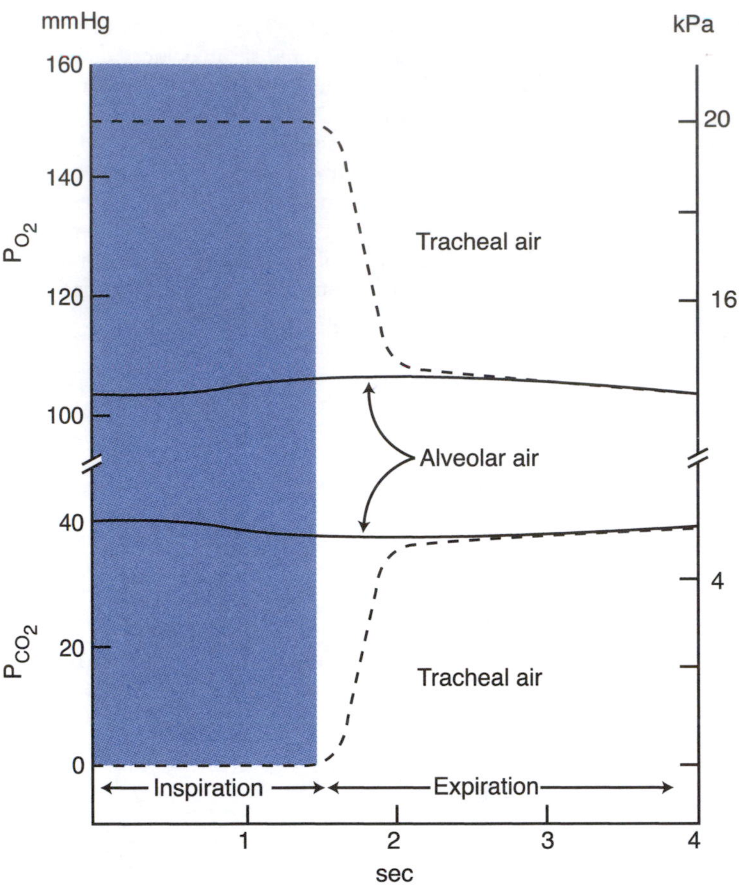


**Figure 6.15** Variations in oxygen and carbon dioxide tensions in tracheal air and alveolar air during one single breath at rest. Note the very small fluctuations in gas tensions of the alveolar air. $P_{O_2}$ = partial pressure of oxygen; $P_{CO_2}$ = partial pressure of carbon dioxide.

Because of the length of the respiratory tract, the gas movement during respiration can be considered as a mass movement of gas flow. For distribution within the small lung units, a molecular diffusion is the main determinant. Because of the small dimensions of the alveoli, a complete mixing within the alveolus probably occurs in less than .01 s. The rapid equalization of the gas pressures during the passage of the blood around the alveoli is evident from figure 6.16. At rest, the time it takes for the blood to pass the capillary is somewhat less than 1 s, but after .1 s the diffusion of the $CO_2$ already has reached an equilibrium. After a further few tenths of a second, the oxygen also has reached an equilibrium. Thus, during normal resting conditions, the blood in the pulmonary capillaries is almost completely equilibrated with the alveolar oxygen and $CO_2$ pressures. The size of the $CO_2$ molecule is larger than that of the oxygen molecule, which actually slows its rate of diffusion. On the other hand, the $CO_2$ is about 25 times more soluble in liquids than the oxygen, so that the net effect is that the $CO_2$ diffuses about 20 times more rapidly in aqueous liquids than does oxygen. Both $CO_2$ and oxygen are carried by the blood mainly in reversible chemical combinations. Hemoglobin (Hb) plays an overwhelming role in this transportation. In the exchange of respiratory gases with the blood in the lungs, the primary chemical reactions of these gases occur within the red cell. The barriers that have to be passed are the alveolar epithelium and basement membrane, the interstitial tissue, the capillary basement membrane and epithelium, the plasma, and the red cell membrane (see figure 6.3). However, the process is very rapid, as shown in figure 6.16. Even during vigorous exercise, when the transit time in the capillaries may be only .5 s or even less, it may be assumed that a gaseous equilibrium has been reached. In the narrow capillaries, there is hardly any plasma between the red cell and the endothelium, a situation that facilitates diffusion.

Even though the diffusion rate is sufficiently high, chemical processes are still essential if an adequate volume of gas is to pass the pulmonary membrane. Carbonic anhydrase plays an important role in the exchange of $CO_2$. If this were not present, the blood would have to remain in the capillaries for almost 4 min for all the $CO_2$ to be given off. Actually, this enzyme is present both in the pulmonary capillary endothelium and in the capillary endothelium

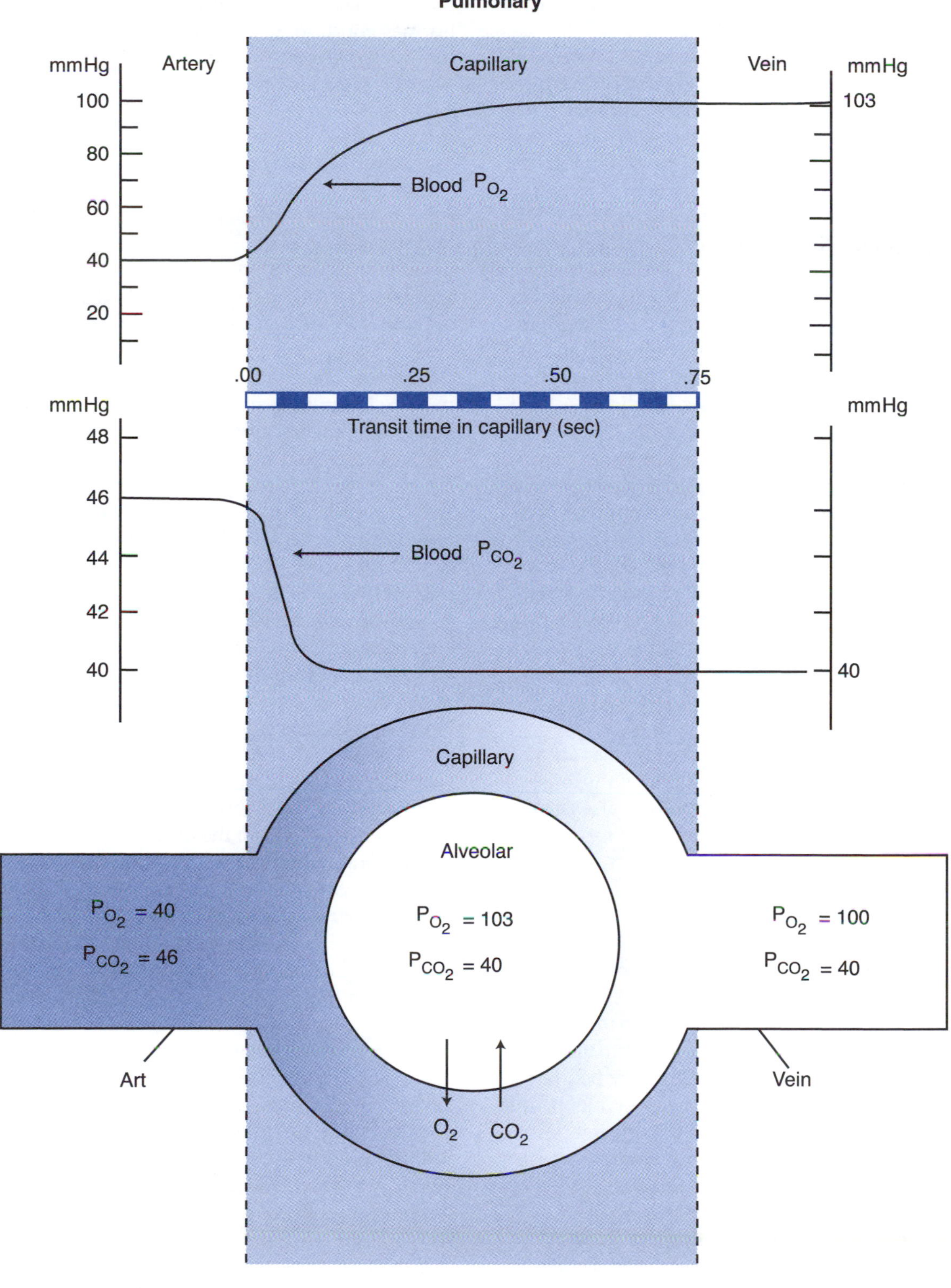


**Figure 6.16** Change in the partial pressures of oxygen and carbon dioxide as blood passes along the pulmonary capillary. Note that in the first part of the capillary, the blood is already equilibrated with the alveolar gas. $P_{O_2}$ = partial pressure of oxygen; $P_{CO_2}$ = partial pressure of carbon dioxide.

of skeletal muscles (Effros, Mason, and Silverman 1981; Lönnerholm 1982).

The diffusing capacity of the lung ($D_L$) is defined as the number of milliliters of a gas at STPD (which stands for standard temperature 0° C, pressure 760 mmHg = 101 kPA, dry) diffusing across the pulmonary membrane per minute and per millimeter of mercury of partial pressure difference between the alveolar air and the pulmonary capillary blood. The diffusion path includes the blood as stated previously. The diffusing capacity of the lung for the respiratory gases provides an index of the dimensions of the pulmonary capillary bed (the capillary wall, the interstitium, and the alveolar wall) and the overall efficiency of the system in the exchange of respiratory gases. Certain technical difficulties limit the possibility for an exact estimation. Usually the $D_LO_2$ is calculated from studies using carbon monoxide. From a resting value of 20 to 30 ml of oxygen per minute and a millimeter of mercury mean pressure difference between the alveolar oxygen pressure and the pulmonary capillary oxygen pressure, the $D_LO_2$ increases toward 75 ml in individuals with maximal oxygen uptake of about 5 L · min$^{-1}$ (J.B. West 1994).

## Ventilation and Perfusion

The difference in $PO_2$ between the alveolar air and the blood entering the left atrium depends on several factors: The membrane component plays a role, a certain amount of admixture of blood from bronchial veins to the pulmonary veins occurs, and there is the effect of the passage of some blood through poorly ventilated alveoli.

Because $CO_2$ diffuses about 20 times more rapidly than does oxygen, one cannot speak of any diffusion obstacle for $CO_2$. The inhaled air is not distributed equally to all the alveoli, and the composition of the gases is therefore not uniform throughout the lungs. The pulmonary capillary bed has a common blood supply, the mixed venous blood, but different areas of the lungs have an uneven perfusion. The composition of the gas in various parts of the alveolar space depends on the ventilation as well as on the blood flow, or the ratio $V_A/Q$.

Under extreme conditions, the $V_A/Q$ ratio may vary from zero (when there is perfusion but no ventilation) to infinity (when there is ventilation but no perfusion). When the ratio is zero, the tensions of oxygen and $CO_2$ of the arterial blood are equal to those in mixed venous blood, because there is no net gas exchange in the capillaries. In the latter case, no modification of the inspired air takes place. Although these extreme situations rarely occur under normal conditions, the various parts of the lung have a wide range of ventilation–perfusion ratio. The "alveolar air" actually represents various contributions from several hundred million alveoli, each possibly having slightly different exchange ratios and gas composition.

In other words, the alveolar gas tensions vary from moment to moment and from place to place within the lungs because of regional inhomogeneity of ventilation and blood perfusion; the supply of air and blood is not perfectly matched in the lungs, even in a healthy individual in any posture (J.B. West 1994).

There are mechanisms that to some extent compensate for an uneven ventilation in relation to the blood flow in the capillary bed of the alveoli, First, inadequately ventilated alveoli have a low $PO_2$, which in turn causes an alveolar vasoconstriction and reduced blood flow. Second, reduced blood flow reduces the alveolar partial pressure of $CO_2$ ($P_ACO_2$), which constricts the bronchioles and, therefore, reduces gas flow.

In the lungs, the blood flow is greatly affected by body position. Thus, the VC increases in the standing position, compared with the lying position, because of the reduction in blood volume in the thorax in the upright position. The effect of gravity on the distribution of the blood, as well as on perfusion, is such that in the upright position, the perfusion per unit lung volume is about five times greater at the base than at the apex of the lung. The ventilation per unit of lung volume changes in the same direction, but only slightly. As a result, the $V_A/Q$ ratio becomes much higher in the upper lobes than in the lower lobes in the erect posture, or above 3 at the top and below 1 at the bottom of the lungs (J.B. West 1994). It is apparent that the lung cannot be considered as a homogeneous unit. Overventilated alveoli cannot completely compensate for a desaturation of arterial blood from underventilated alveoli. The shape of the oxygen dissociation curve is such that an increase in arterial $PO_2$ ($P_aO_2$) above the "normal" 100 mmHg (13.3 kPa) adds little to the oxygen content of the blood; however, a reduction in $PO_2$ will affect the oxygen content considerably more.

When the position of the body is changed from upright to supine, the perfusion of the upper lung zone increases markedly at the expense of the lower zone. In the recumbent position, the calculated $V_A/Q$ therefore becomes quite uniform in the different pulmonary lobes (Amis, Jones, and Hughes 1984).

Even during light exercise in the sitting or standing position, the $V_A/Q$ ratio also becomes more uniform throughout the lungs. Certainly both upper and lower zone blood flows increase, but the former increases relatively more. The slight increase in pulmonary artery pressure that accompanies exercise is one factor that changes the balance between arterial, capillary, and venous pressures on the one side and the pressure outside the vessels on the other, favoring the perfusion of the upper zones of the lungs. The pressure within the pulmonary artery varies between 7 and 20 mmHg (.9 and 2.7 kPa) during a cardiac cycle at rest but between 15 and 35 mmHg (2.0 and 4.7 kPa) during exercise with an oxygen uptake of 2.0 L · min$^{-1}$. The systolic pulmonary pressure can exceed 50 mmHg (6.6 kPa) during maximal exercise. These pressure variations actually mean that the distribution of the blood flow to different parts of the lungs will vary not only with body position but also with the cardiac cycle. Even during heavy exercise, some parts of the lungs may be underperfused during part of the diastole.

### Summary

Considerable regional inequality in ventilation of the alveoli and in blood perfusion of the capillaries exists in the lungs, causing differences in the gas exchange in different parts of the lungs. When a person is at rest in the supine or prone position, however, the ventilation distribution is rather well adjusted to follow this perfusion. In the erect posture, hydrostatic forces cause a progressive decrease in perfusion from the bottom to the top of the lung without corresponding variations in the ventilation. Therefore, the upper lobes are relatively underperfused. During exercise, the $V_A/Q$ ratio becomes more uniform, and the increase in pulmonary arterial pressure is at least one important contributor to this change.

## Oxygen Pressure and Oxygen-Binding Capacity of the Blood

Figure 5.18 illustrates how the oxygen saturation of Hb is affected by the $PO_2$ in the blood. At $PO_2$ = 100 mmHg (13.3 kPa), 98% of the Hb is normally saturated with oxygen. The oxygen saturation curve is such that about half the Hb is in the form of oxyhemoglobin ($HbO_2$) and half is in the form of reduced Hb at $PO_2$ in the order of 26 mmHg (3.5 kPa; 37.0 °C, pH 7.40). The $HbO_2$ dissociation curve of the blood is affected by the $CO_2$ pressure, the pH, and blood temperature (see figure 5.18). $PCO_2$ affects pH and thereby the oxygen saturation curve, but it also has a specific effect in that $CO_2$ combines with Hb to form **carbamino compounds** ($HbNH_2 + CO_2 \leftrightarrow$ HbNHCOOH), thereby reducing the capacity of Hb to bind oxygen (see figure 5.2). During muscular exercise the $CO_2$ increases, as does the temperature locally in the muscle, while at the same time the pH is lowered. This causes the liberation of oxygen to increase at a given oxygen tension. Because of this feature of Hb, an effective oxygen diffusion gradient can be maintained between the capillaries and the oxygen-using cell. With regard to the effect of an increase in temperature, the diffusion increases about 2%/°C. These factors aid in the unloading of oxygen to the active muscles, but are, however, of less quantitative importance than the opening of additional capillaries in the muscles. In this manner, the distance between the capillary and the muscle cell is reduced, which shortens the diffusion distance. Because of the shape of the oxygen saturation curve, a change in pH, $PCO_2$, and blood temperature plays a relatively small role at a $PO_2$ around 100 mmHg (13.3 kPa). However, during very strenuous exercise, the oxygen saturation of arterial blood can be reduced below 95% without a corresponding decrease in $PO_2$ (see figure 5.19). At high altitude, where the alveolar $PO_2$ ($P_AO_2$) is lower, the arterial saturation is still more affected in a negative direction by a lowered pH and increased $PCO_2$ and temperature.

An organic phosphate compound, **diphosphoglycerate,** or **2,3-DPG,** is normally present in the red cells. It shifts the oxygen dissociation curve to the right, thus assisting the unloading of oxygen to peripheral tissues. Prolonged exercise training causes a slightly decreased Hb affinity for oxygen, which may be attributable to an increase in the erythrocyte 2,3-DPG concentration. Mairbäurl et al. (1983) noticed an increase in the oxygen tension with 1.3 mmHg (0.17 kPa) at 50% oxygen saturation of Hb in their fit subjects. However, they explained the reduced $HbO_2$ affinity after training as a shift in the average age of erythrocytes toward a younger range.

Factors affecting the concentration of 2,3-DPG in the red cell include pH. Because acidosis inhibits red cell glycolysis, the 2,3-DPG concentration decreases when pH is low. Ascent to high altitude increases 2,3-DPG concentration, with a consequent increase in the availability of oxygen to tissues. The increase in 2,3-DPG is secondary to the decrease in blood pH. Upon return to sea level, the 2,3-DPG concentration returns to normal level (Ganong 1999).

From an evolutionary standpoint, the shift of the oxygen dissociation curve to the right was beneficial, no doubt, under conditions prevailing at or near sea level. Unfortunately, at high altitudes, this shift may create disadvantages, in that a diffusion limitation is a determining factor in exercise tolerance (Bencowitz, Wagner, and West 1982). In this situation, a rightward shift of the oxygen dissociation curve inevitably becomes a handicap when the venous blood passes the alveoli, exposed to a relatively low oxygen tension.

In patients with severe iron-deficient anemia, oxygen delivery is assisted by a shift of the oxygen dissociation curve to the right, attributable in part to 2,3-DPG (Ohira et al. 1983). In other words, there are many stress situations that, by various mechanisms, affect the oxygen dissociation curve and thus enhance the oxygen yield at a given oxygen pressure.

Each Hb molecule has four iron atoms. In reality, the ratio between Hb and $HbO_2$ may vary as follows: $Hb_4$, $Hb_4O_2$, $Hb_4O_4$, $Hb_4O_6$, $Hb_4O_8$, in which $Hb_4$ is the completely reduced Hb. Only in the case of $Hb_4O_8$ is the molecule 100% saturated. Under normal conditions, these combinations are mixed in proportions that are determined by such factors as $PO_2$. Concerning the volume of $CO_2$ bound to Hb in the carbaminohemoglobin form, it varies with the degree of oxygen saturation and is greater at low oxygen saturation and less at high oxygen saturation.

The transport of $CO_2$ and oxygen was also discussed in chapter 5. It should be recalled that reduced Hb is a weaker acid than $HbO_2$, and, when reduced, it mops up $H^+$ more readily and thereby helps to prevent too large a reduction in pH through the formation of acid metabolites. Myoglobin facilitates the oxygen diffusion within both heart and skeletal muscle to approximately twice the level observed when this protein is inactivated. Favorable for this diffusion is a low intracellular oxygen tension, certainly present during vigorous exercise, as well as a gradient of the myoglobin–oxygen complex to provide a driving force for a facilitated diffusion, and a mobility within the cell of the myoglobin–oxygen to permit diffusion of the oxygen carrier (Cole et al. 1982; Livingston, LaMar, and Brown 1983; Wittenberg and Wittenberg 1981). The diffusion of oxygen is accelerated by an elevation of the tissue temperature. This temperature effect is greatly enhanced in the presence of myoglobin (Stevens and Carey 1981).

From an oxygen content of 20% of volume in arterial blood, it may eventually drop during heavy exercise to below 1% of volume because of the abundance of capillaries in muscle tissue with short diffusion distances, a pH around 7.0, and a temperature exceeding 40 °C. The mitochondria apparently function efficiently even if the intracellular $PO_2$ is less than 1 mmHg (0.13 kPa); at the end of a capillary in a vigorously contracting muscle, the oxygen tension probably is less than 10 mmHg (1.3 kPa; Chance, Schoener, and Schindler 1964).

# Regulation of Breathing

Because the object of breathing is to ensure the exchange of oxygen and $CO_2$ between the blood and atmospheric air, it would appear logical if these gases helped regulate breathing. Such is actually the case. A change in the $PO_2$, $PCO_2$, and $H^+$ concentration in the arterial blood alters ventilation in such a manner as to moderate the primary change (negative feedback). Stretch receptors in the lungs and respiratory muscles, as well as a number of other factors, affect respiration. As in the case of changes in the circulation under different conditions, the question arises: What is the regulating effect, and what is the disturbing effect? In respect to $PO_2$, $PCO_2$, and $H^+$ concentration in the blood, a change in pulmonary ventilation may influence these factors. The temperature of the blood affects respiration, but a change in pulmonary ventilation does not substantially change the body temperature of the human. In some animals, such as the dog, respiration also plays a part in temperature regulation. For dogs, the effect of an increased blood temperature has a much more pronounced effect on respiration than in the case of human beings. Catecholamines affect respiration, but a change in ventilation certainly does not affect the content of these substances in the blood. Also, emotion affects respiration as evidenced by the rapid breathing of an excited person. A number of different theories have been advanced concerning the regulation of breathing, but none of them has fully explained how the respiratory volume is adjusted to meet the demand at rest and during physical activity.

### For Additional Reading

For more extensive discussions concerning the control of breathing during rest and exercise, the reader is referred to Whipp (1983), Dempsey, Vidruk, and Mitchell (1985), Wasserman (1994), Altose and Kawakami (1999), and Caruana-Montaldo, Gleeson, and Zwillich (2000).

## Rest

In the CNS, a large number of physical, chemical, and nervous variables are integrated. $PO_2$, $PCO_2$, and $H^+$ concentration, or chemical changes related to them, appear to be the most prominently controlled chemical variables. A lowering of the $P_aO_2$ may stimulate the breathing via peripheral chemoreceptors in the carotid and aortic bodies. An increase in $PCO_2$ and $H^+$ concentration also increases ventilation, but this effect is primarily elicited from medullary chemosensitive receptors (usually called central chemoreceptors), located on the ventral surface of the medulla oblongata, separated from the blood by the blood–brain barrier.

The medulla contains widely dispersed clusters of neurons. The existence of anatomically well-localized concentrations of expiratory and inspiratory neurons has not been demonstrated, however. A respiratory center as such cannot be defined; it is rather a question of a central pattern generator guiding the respiration (von Euler 1983). Some neurons have an inherent rhythmicity, but so far it is unknown where this rhythmicity originates. However, there are powerful inputs from nuclear areas in rostral pons. The inspiratory activity is characterized by a sudden onset, followed by a slowly augmenting increase until it is "switched off." During the early phase of the expiration, there is again activity in the inspiratory motor neurons, apparently controlling the events of the early steps of the expiratory phase. Only then is the expiratory motor activity recruited, and only if the pulmonary function is under some sort of stress. During rest, the expiration is passive. The stronger the demand on an active expiration, as during exercise, the earlier the expiratory motor activity is brought into play. Apparently, the activity in neurons that causes an inspiration also activates some kind of an off-switch mechanism. When the inspiration has lasted for an adequate period of time to produce a sufficient tidal volume, a threshold is reached and the inspiration is terminated. Various inputs can adjust the off-switch threshold up or down, and thus the depth and rate of breathing can be regulated. As we will see, an increase in the $CO_2$ and $H^+$ concentrations in the blood and a reduction of its oxygen tension will raise the threshold that will delay the off-switch mechanism, thereby increasing tidal volume. There are powerful inputs from nuclear areas in the rostral pons. Furthermore, when the lung tissues are stretched in the course of inspiration, stretch receptors in the bronchioles are stimulated and, via afferent vagus fibers, inhibit the inspiratory activity, and the inspiration is stopped (the **Hering-Breuer reflex**). If the vagus is cut in an animal, the respiratory frequency slows and the respiratory depth is increased. These events are schematically illustrated in figure 6.17.

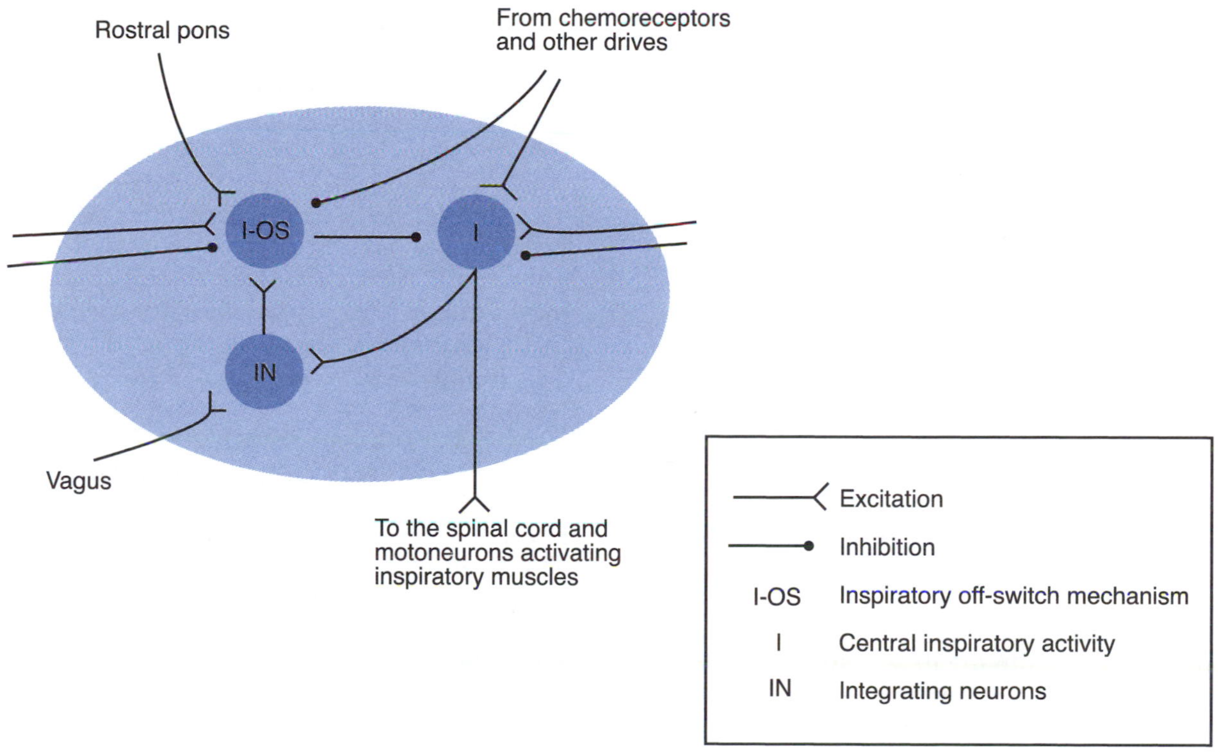


**Figure 6.17** Diagram illustrating one hypothesis for an inspiratory off-switch mechanism of the central pattern generator in the medulla. There is an input to the off-switch neurons from pulmonary stretch receptors (via vagus), from the rostral pons, and from a recurrent feedback from central inspiratory activators relayed over integrator neurons. The inspiratory off-switch mechanism and central inspiratory activators also receive other impulses, both excitatory and inhibitory.

Modified from von Euler 1983.

This report on the volume situation in the lungs to the respiratory central pattern generator in the brainstem is apparently not needed in humans until the tidal volume is 1 to 1.5 L or higher. During the expiratory phase, an inhibitory activity suppresses the inspiration. The duration and strength of this inhibition control the expiratory length.

Numerous neurons discharging with respiratory modulations can be found throughout the brainstem from caudal medulla to rostral pons.

The respiratory muscles, especially the intercostal muscles, are amply supplied with muscle spindles. Central respiratory drive descends by a common path to the spinal level, being then distributed to both α and γ motoneurons. Thereby we have the tool for coactivation of the α–γ system. The respiratory muscles can be volitionally controlled, and they are, in almost all of our daily activities, engaged in many "nonrespiratory" activities. We are all aware of how speech-controlling mechanisms take over the control of duration and rate of expiration to fit the requirements for phrasing, loudness, and articulation (see Doust and Patrick 1981). On the spinal level, different descending and afferent inputs interact and the interneuronal networks can exert reciprocal effects on the antagonistic muscles. This is the place for final integration; many situations require an instantaneous adjustment to trunk movements and postural changes to enable an immediate compensation for any changes in muscle length, direction of muscle forces, or other factors of chest wall mechanics.

## Central and Peripheral Chemoreceptors

The pulmonary ventilation at rest is regulated chiefly by the chemical state of the blood, particularly its $CO_2$ tension, which is highly related to the $H^+$ concentration. An increase in the $H^+$ and $CO_2$ concentration (**hypercapnia**) in the arterial blood increases ventilation. This response is correlated closely with changes in the local extracellular fluid $H^+$ concentration, as measured on the ventrolateral surface of the fourth ventricle and on, or slightly below, the medullary surface (Dempsey and Forster 1982). The exchange of $CO_2$ and $HCO_3^-$ between the blood and this tissue is very rapid, whereas the exchange is slow into the cerebrospinal fluid. The nature of the receptors and their contacts is poorly understood. It is assumed that $CO_2$ does not have an independent effect per se but acts via $H^+$.

The peripheral chemoreceptors are located at the bifurcation of the common carotid arteries and on the arch of the aorta. For the stimulation of respiration by hypoxia, a stimulation of the peripheral chemoreceptors is required.

The chemoreceptors in the carotid and aortic bodies consist of epithelioid cells with a rich innervation and blood supply. It has been shown conclusively that the decisive factor in the hypoxic stimulation of the chemoreceptors is a lowering of $PO_2$, not a reduced oxygen content as such. Thus, in laboratory animals, up to 80% of the Hb may be bound to carbon monoxide without the respiration being affected, providing the $PO_2$ is kept at a normal level (Lahiri, Mulligan, et al. 1981).

The discharge of the chemoreceptors increases if the blood flow through them is reduced. Stimulation of sympathetic fibers to the carotid bodies as well as blood-borne adrenaline and noradrenaline reduces the blood flow to the chemoreceptor tissue. Therefore, chemoreceptors send an increasing number of nerve impulses centrally, despite an unaltered oxygen tension of the arterial blood, if there is an increased sympathetic activity, as after hemorrhage or during exercise.

There are considerable individual variations in the reaction to hypoxia. Some persons become unconscious without pulmonary ventilation being significantly affected. In others, the pulmonary ventilation is increased to double or more during hypoxia.

What actually stimulates these peripheral chemoreceptors is not clear. If the oxygen tension in the arterial blood is high, as is the case when a person is breathing 95% $O_2$, the arterial $PCO_2$ ($P_aCO_2$; caused by breathing $CO_2$) may increase and its pH may decrease markedly without the chemoreceptors being stimulated. So, if a hypoxic drive does not exist but the oxygen saturation is adequate, these chemoreceptors seem rather insensitive to changes in $PCO_2$ and $H^+$ concentration. At low oxygen tension, however, the discharge from the chemoreceptors is increased further if $PCO_2$ is high and the pH is low at the same time. Thus, hypercapnia potentiates the effect of hypoxia, possibly by changing the intracellular $H^+$ concentration or reducing the blood flow to the area where the chemoreceptors are located (Eyzaguirre and Fidone 1980). The aortic chemoreceptors are much less responsive to $CO_2$ stimulus than are the carotid chemoreceptors (Lahiri, Mokashi, et al. 1981).

Normally, the $PO_2$ in persons breathing ordinary air at sea level is low enough to contribute a small but significant tonic ventilatory stimulus from the peripheral chemoreceptors (Dejours 1981). Most subjects exhibit a sustained increase in pulmonary ventilation only when breathing 16% oxygen and a marked increase in ventilation only when the inspired oxygen has dropped to about 10% or less. Thus, the chemoreceptors have different thresholds, some of which are not stimulated markedly until the $PO_2$ has decreased to 40 or 50 mmHg (5.3 or 6.6 kPa; 8–10% of oxygen inhaled). The chemoreceptors are very resistant to anoxia and therefore can maintain breathing reflexively for long periods (Dempsey and Forster 1982).

Because a hypoxic drive via the peripheral chemoreceptors increases ventilation, more $CO_2$ will be expired than is produced. The result of this hyperventilation is that the $P_aCO_2$ decreases (**hypocapnia**) and the pH increases. This means a reduced respiratory drive via $P_aCO_2$ and $H^+$ concentration. If this hypoxic stimulus is eliminated by the breathing of oxygen-rich air, the ventilation is immediately reduced. The $CO_2$ thereby is accumulated, which tends to offset the effect of interrupting the $PO_2$-dependent stimulation.

In many normal and pathological situations, a change in arterial $H^+$ concentration is attributable to causes other than a primary change in $PCO_2$. For example, during strenuous exercise there is an increase in arterial $H^+$ concentration caused by the production and release of lactic acid into the blood. The stimulation of the peripheral chemoreceptors by this change in $H^+$ concentration is responsible, in part, for the hyperventilation of severe exercise (see also Akiyama and Kawakami 1999).

### *Summary*

The central and peripheral inputs are integrated in the respiratory central pattern generators so that $PO_2$, $PCO_2$, the acid-base characteristics of the blood, and the internal environment of the body are maintained at constant levels. The inspiratory and expiratory generators in the medulla oblongata are inherently rhythmic and, in part, reciprocally coordinate the inspiratory and expiratory muscles that vary the rate and depth of breathing. Normal respiration is maintained by an interaction between this generator and neurons in the pons. Inspiration is interrupted by a dampening or cessation of the stimulating effect from chemical or other driving forces on the inspiratory generator and by an off-switch mechanism in the medulla. Stretch receptors in the lungs and muscle spindles in the respiratory muscles may participate in this switch from inspiration to expiration.

The $H^+$ concentration and, especially, $PCO_2$ in the blood continuously affect respiratory activity by sending drive inputs to cells in the ventrolateral parts of the medulla. In other words, changes in carbonic acid concentration in blood and other body fluids ameliorate the acid-base imbalance caused by the excess or deficit of nonvolatile acids. Deviation from a normal $P_aCO_2$ of 40 mmHg (5.3 kPa) results in a ventilatory response that minimizes the deviation. During hypoxia, ventilation increases because of discharge from peripheral chemoreceptors in the carotid and, to a smaller extent, aortic bodies. A number of structures in the medulla oblongata influence respiration. Activity of the reticular formation also increases ventilation. This effect forms a link between the proprioceptors in the muscles and joints, the cortex of the brain, and different centers of importance for the motor function, as well as the respiratory neurons. Emotion, voluntary actions, and reflexes such as swallowing can easily alter the pattern of respiration.

## Exercise

The mechanisms responsible for the control of breathing during exercise have attracted respiratory physiologists for a century but remain a topic of research. Scientists have been searching for a "work factor" that could explain exercise **hyperpnea,** which is an increase in depth and rate of respiration. Efforts have been made to quantitate various chemical and nervous factors with regard to their participation in the regulation of breathing during exercise. So far, these efforts have not been promising. Here we discuss the idea that a neurogenic factor is the primary activator of the respiratory muscles during exercise, with a secondary feedback mechanism of chemical nature that regulates and adjusts the respiratory volume mainly according to the composition of the arterial blood.

The idea is based on the fact that the pulmonary ventilation increases during muscular exercise almost rectilinearly with the increase in oxygen uptake up to a certain level, after which the increase in ventilation becomes steeper. Over the whole range, pulmonary ventilation is actually more related to the $CO_2$ volume exhaled than to the oxygen uptake (see figures 6.9 and 6.13). During submaximal physical activity, the $P_aCO_2$, $PO_2$, and $H^+$ concentrations are roughly at the same level as at rest (see figure 6.13). During very heavy exercise, the anaerobic contribution to the energy yield inevitably is coupled with the production of $H^+$. Thus, pH decreases and may be as low as 7.0 in the arterial blood. The relative hyperventilation that follows elevates the $P_AO_2$, but the $P_aO_2$ actually drops toward 90 to 85 mmHg (12.0–

## CLASSIC STUDY

In 1913, the Danish physiologists August Krogh (1874–1949) and Johannes Lindhard (1870–1947) published an article in *The Journal of Physiology (London)* describing the changes that take place in ventilation, respiratory exchange rate, alveolar $CO_2$ tension, blood flow, and pulse rate during the first few minutes of light or heavy muscular work on a bicycle ergometer. They also attempted to explain the mechanism behind these changes. The authors concluded as follows: "Evidence is brought forward to show that the rise in ventilation like the increase in heart rate is not produced reflexly *[sic]* but by irradiation of impulses from the motor cortex. In the case of ventilation these impulses are found to act indirectly through the respiratory centre by suddenly increasing its excitability towards hydrogen ions. The mechanisms disclosed provide a very rapid though not instantaneous adaptation of the respiratory and circulatory systems to sudden muscular exertions. Without it such exertions could not be kept up by the organism for more than a fraction of a minute."

Krogh and Lindhard 1913.

11.3 kPa) compared with the normal value of 95 to 100 mmHg (12.6–13.3 kPa). $P_aCO_2$ drops toward 35 mmHg (4.6 kPa) or even lower values (see figure 6.13). A similar decrease in $P_aO_2$ and pH at rest causes a rather moderate increase in ventilation, and a lowering of $P_aCO_2$ in itself will reduce respiratory activity. Thus, the chemical changes in the composition of the arterial blood cannot explain per se the ventilation of 100 to 250 L · min$^{-1}$ observed during heavy muscular exercise.

The threshold at which the ventilation increases proportionally more than the oxygen uptake actually varies from person to person. The individual's potential to supply the exercising muscles with oxygen is of decisive importance. A person who has a low maximal aerobic power reaches this threshold value at a lower oxygen uptake than does a person who has a high maximal $\dot{V}O_2$ (see figure 6.9). During exercise with small muscle groups (e.g., the arms), the ventilation at a given oxygen uptake is greater than it is when larger muscle groups are engaged (e.g., with exercise involving the leg muscles). For example, during cycle ergometer exercise with the arms, the ventilation was 89 L · min$^{-1}$ at an oxygen uptake of 2.8 L · min$^{-1}$ but was only 67 L · min$^{-1}$ when the same work was performed with the legs. The lactate concentration of the blood was 11 and 4 mmol · L$^{-1}$, respectively. The heart rate was 176 and 154, respectively.

## A Cerebral Drive?

For a long time researchers have discussed the possibility that impulses from motor centers in the cerebral cortex or from the active muscles and joints involved in exercise reach the respiratory generators (Asmussen 1967; Crystal and West 1997; Krogh and Lindhard 1913; Whipp 1983). Such impulses conceivably could alter the threshold or set point for the sensitivity of the respiratory neuron pools for $CO_2$, $H^+$, and oxygen mediated by peripheral chemoreceptors.

The cerebral hemispheres normally contribute an important component to the volume of breathing during wakefulness, and this cerebral drive maintains the rhythm of respiration even when metabolic stimuli are temporarily removed (e.g., after a period of hyperventilation). There might be a voluntary and behavioral system connected with the act of breathing located in the somatomotor and limbic forebrain structures, which adapts humans to vocalization and conditions us for the expected metabolic demands of exercise (Plum 1970). As mentioned, the efferent impulses can descend directly to the motoneurons of the respiratory muscles, but they also can make connections with the reticula formation. At this lower level of the brain, neural systems subserve the metabolic needs of the body. As a third system, we could consider the effects on the respiratory muscles by the postural demands governed by the cerebellum.

## The Role of Muscle Spindles

Campbell (1964), von Euler (1974), and other investigators have emphasized the importance of the muscle spindles for the activity of the respiratory muscles. Campbell pointed out that the respiratory muscles are voluntary muscles subject to all the spinal and supraspinal mechanisms that affect tone, posture, and movement. In the anterior horn cells, respiratory and nonrespiratory drives are integrated. Muscle spindles are numerous in the intercostal muscles but much less numerous and less evenly distributed in the diaphragm. An increased activity in the fusimotor γ fibers produces a contraction of the muscle spindle's intrafusal muscle fibers, and if the extrafusal muscle fibers of the parent muscle are not activated at the same time via the muscle's α fibers, the muscle spindle afferents send impulses via the dorsal roots to the spinal cord. These afferent impulses stimulate the α motoneuron, the entire muscle is activated, and the difference in length between the intrafusal and extrafusal muscle fibers is reduced sufficiently so that the afferent signals cease. This γ motor spindle system represents a "follow-up length servo" system. The importance of this mechanism for ventilation is shown by the fact that adding a resistance on the inspiratory side increases the force of contraction of the inspiratory muscles, thanks to an augmentation of their motoneuron discharge by the stretch reflex. In other words, the afferent nerves convey feedback signals from the muscle spindles, providing information about how successfully the respiratory muscles have accomplished their task.

## Combined Effects of Neurogenic and Chemical Stimuli

The system of γ and α motoneurons, linked together via the afferent nerves of the muscle spindles and affected by supraspinal mechanisms, especially the reticular formation, forms the basis for the ventilatory regulation. In connection with exercise, the activity in the reticular formation is increased, partially through impulses from the cortex cerebra and other higher centers, and partially through afferent impulses from the muscle spindles in the activated muscles. It would be natural if the respiratory movements and respiratory frequency were adjusted according to the exercise rhythm, which usually is the case: As mentioned, the frequency of respiration is adjusted in accordance with stride frequency or pedal frequency in running or cycling, for example. A change in posture and rhythm can change the pattern of breathing with reference to respiratory frequency and depth. The momentarily increased ventilation noticed in connection with the onset of exercise then must be adjusted according to the requirement for the elimination of $CO_2$ and uptake of oxygen.

Here, chemical regulation enters into the picture. The follow-up length servo system possibly has a tendency to produce hyperventilation, especially in the beginning of the exercise until the produced $CO_2$ has reached the lungs. Through a negative feedback elicited from the respiratory centers, if $P_aCO_2$ tends to decrease, the γ and α activity driving the respiratory muscles can be inhibited by the respiratory generators. When the $CO_2$ reaches the lungs without being eliminated in sufficient quantity, the $PCO_2$ of the arterial blood will increase and the inhibition is diminished.

To some extent this is a hypothetical discussion, which postulates that in connection with muscular exercise, α–γ activity to the respiratory muscles increases. However, the system has every possibility to produce a rhythmic, coordinated switching between inspiration and expiration, partly determined by the exercise rhythm. Recall that the respiratory muscles are involved in most activities not only to provide adequate pulmonary ventilation but also to stabilize the trunk in posture or movements. A synchronization between the respiratory movements and exercise rhythm is therefore important. The α–γ system provides the possibility for such a coordination.

Figure 6.18 is a simplified diagram presenting components involved in exercise hyperpnea. The neurogenic factors cannot be considered as a regulating stimulus but act as an activator. The chemical stimuli are of decisive importance for the finer adjustment of the ventilation. If one exercises very hard for 5 s, rests for 5 s, and so forth, the ventilation eventually may increase to 100 $L \cdot min^{-1}$, but no difference in ventilation will be detected between the periods of rest and exercise. A variation in afferent impulses from the exercising muscles is, in this case, of less importance than the input of $CO_2$ to the lungs, which is continuously high. The greater the $CO_2$ production, the greater the independence of the respiratory muscles.

There are many intriguing experimental findings to consider when analyzing the neurogenic respiratory drive during exercise. In eccentric exercise, the force developed by the contracting muscles can be five to seven times greater than in concentric exercise, but the pulmonary ventilation per liter oxygen uptake is similar in the two types of activities (Asmussen 1967). Prolonged static effort with maintained high muscular force does not represent an especially strong respiratory drive. The very moderate ventilatory response to the first seconds of maximal effort (e.g., a sprint, or running up a flight of stairs at maximal speed) also should be mentioned.

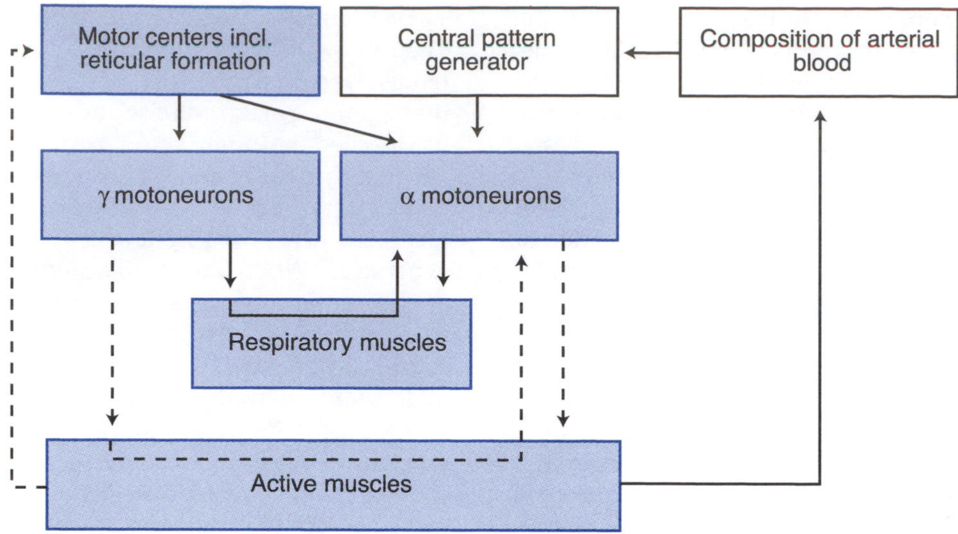


**Figure 6.18** Schematic summary of the regulation of breathing during exercise. The respiratory muscles are activated via their γ and α motoneurons (filled-line arrows). In a similar way, the other exercising muscles are activated (dotted-line arrows). The central pattern generator is influenced directly or indirectly by the chemical composition of the arterial blood, mainly its partial pressure of carbon dioxide, partial pressure of oxygen, and pH. These centers then can facilitate or inhibit the motoneurons of the respiratory muscles, depending on the effectiveness of the gas exchange in the lungs. Particularly critical is the carbon dioxide refill from the muscles and carbon dioxide output from the lungs. (The afferent nerve impulses to motor centers include impulses from receptors located in tendons, muscle spindles, and joints.)

However, the afferent impulses from exercising limbs and the stimulation from various parts of the brain attributable to the increased motor activity must be considered as coordinated activators of the respiratory muscles, but their motoneurons are subjected to various degrees of inhibition from the central pattern generator, depending on the chemical composition of the blood and the phase of the respiratory cycle.

Any change in respiratory frequency that adapts to the rhythm of movements is, within limits, automatically followed by a change in tidal air to provide an adequate alveolar ventilation. The regulation of cardiac output is similar: At rest and during moderate exercise, a change in heart rate is matched by a variation in stroke volume so that the cardiac output is maintained constant.

## Hypoxic Drive During Exercise

It has already been mentioned that the chemoreceptors have a relatively large blood flow and high oxygen uptake. They are innervated by sympathetic nerve fibers, which supply the arterioles of the carotid and aortic bodies with vasoconstrictor fibers. When these fibers are activated, blood flow through the chemoreceptor areas is reduced, possibly by the blood being diverted through adjacent arteriovenous anastomoses. It is possible that a change in the $PCO_2$ and $H^+$ concentration and hypoxia of the arterial blood contributes to modify the blood flow to the epithelioid cells of the chemosensitive areas. Actually, the oxygen tension inside specific tissue zones in the carotid body is much lower than that predicted from the $(a\text{-}\bar{v})O_2$ difference in the carotid body (Acker and Lübbers 1980).

The heavier the exercise in relation to the individual's performance capacity, the greater the sympathetic activity. Thereby, the chemoreceptor drive also can increase, despite the elevated perfusion pressure. This mechanism can contribute to the increase in ventilation during pronounced emotion as well as during exercise involving small muscle groups, when the ventilation is relatively high at a given oxygen uptake. Recall that during hypoxic conditions, the effect of $CO_2$ and $H^+$ concentration is a marked increase of the discharge from the peripheral chemoreceptors.

The following is a short summary of the time course of the increase in pulmonary ventilation from the start of exercise, with some comments on possible mechanisms in the regulation of ventilation. As mentioned, an abrupt increase in ventilation takes place at the first respiratory cycle following the transition from rest to exercise, and the increase is relatively independent of the rate of exercise. It lasts for some 15 to 20 s. This is phase 1, and everything speaks in favor of a neural influence from higher levels of the brain (Whipp 1983). Eldridge, Millhorn, and Waldrop (1981), who conducted experiments on walking, unanesthetized, decorticated cats, suggested that there are **hypothalamic** command signals primarily responsible for driving both locomotion and respiration during exercise. DiMarco et al. (1983), using high-level decerebrated cats, and Favier et al. (1983), using dogs, also supported the existence of a neurally mediated factor during the early stages of exercise hyperpnea.

The subsequent exponential increase in pulmonary ventilation is related to the severity of the exercise. This is phase 2 of the increase in ventilation. Its delay in onset can be explained if peripheral chemoreceptors act as important stimulators. The vascular transit time for metabolites from the exercising muscles to these receptors supports this theory. As mentioned, a local hypoxia can develop in critical areas of the chemoreceptors. During heavy exercise, the anaerobic metabolism produces protons that, via the blood, reach the carotid body and strengthen the hypoxic stimulus. Support for the importance of the carotid body during phase 2 is provided by the finding that patients with resected carotid bodies had substantially slower ventilatory kinetics than controls (Honda et al. 1979; Whipp 1983). Second, induced hypoxic subjects have a faster response in phase 2, whereas subjects breathing 100% oxygen, which functionally inactivates the carotid bodies, have a slower response (Whipp 1983).

Phase 3 is the steady-state condition. During exercise at an intensity that does not continuously lower the blood pH (i.e., below the anaerobic threshold), the $CO_2$ tension in the arterial blood seems to be more closely regulated than the oxygen tension. But, even in this steady-state condition, pulmonary ventilation often gradually increases during prolonged exercise despite no change in arterial pH, $Pco_2$, lactate, or $\dot{V}_{CO_2}$ (B. Martin et al. 1981).

Recall that stretch receptors in the lungs are supposed to activate the off-switch mechanism with a potential to shorten the inspiratory duration and therefore to increase the respiratory frequency. With an increase in tidal volume, this mechanism may be more active.

During heavy exercise, there is a steeper increase in pulmonary ventilation, as illustrated in figure 6.9. The ventilatory compensation for the metabolic acidosis becomes progressively inadequate. Both central and peripheral chemoreceptors probably drive the central pattern generator. There may also be additional effects of circulating catecholamines, blood temperature, and blood osmotic pressure. Simon et al. (1983) reported a study that suggested that exercise hyperventilation is not necessarily proportional to the increase in plasma lactate. There are also interesting reports that patients who cannot produce lactic acid (McArdle's disease) exhibit a normal hyperventilatory response during a progressive exercise test (J.M. Hagberg et al. 1982).

# Breathlessness (Dyspnea)

**Dyspnea** is difficult, labored, uncomfortable breathing (Comroe 1966). From experience we learn what to expect or experience in a certain situation, and the respiration itself, which normally proceeds unconsciously, can give rise to distress when it requires a conscious modification. Light activity at high altitude requires a ventilation that at sea level is associated with heavy exercise, and this difference can be experienced as a distress. A person unaccustomed to exercise may experience a ventilation of 75 L · min$^{-1}$ as unpleasant, but after suitable training this may even become a pleasant sensation (see Casaburi 1995).

The reason why respiration becomes consciously troublesome is not clear. The magnitude of the ventilation is not the determining factor: A patient can experience a pulmonary ventilation of 10 L · min$^{-1}$ as extremely disturbing, whereas the athlete is not consciously troubled by a ventilation as high as 200 L · min$^{-1}$. Impulses from muscle spindles, tendon organs, and thoracic joint receptors appear to give rise to conscious awareness and distress. It may be a question of length/tension, that is, tension appropriateness. Altered afferent signals from the muscle spindles and receptors in the chest wall reaching subcortical and cortical levels may cause unusual sensations. Gandevia (1998) suggested that the sense of dyspnea represents the conscious appreciation of the central outgoing command to the respiratory muscles.

### FOR ADDITIONAL READING

For further details concerning dyspnea and exercise, see Killian and Campbell (1983) and Killian, Jones, and Campbell (1999).

# Second Wind

During the first minutes of exercise, the load may appear very strenuous. One may experience dyspnea, but this distress eventually subsides, and one experiences a **second wind.** The factors eliciting the distress may be an accumulation of metabolites in the activated muscles and in the blood because the oxygen transport is inadequate to satisfy the requirement.

By what mechanism this changed environment is brought to consciousness is not known. During heavy exercise, there is actually a hypoventilation at the commencement of the activity caused by a time lag in the chemical regulation of the respiration. The issue then becomes a matter of a length/tension inappropriateness in the intercostal muscles. When the second wind occurs, the respiration is increased and adjusted according to requirement.

It appears that the respiratory muscles are forced to work anaerobically during the initial phases of the exercise if there is a time lag in the redistribution of blood. A **stitch** in the side can then develop. This is an exercise-related transient abdominal pain, which probably is caused by diaphragmatic **ischemia** and stress on the visceral ligaments. It is most common in untrained persons and is particularly apt to occur if heavy exercise is performed shortly after a large meal, when the circulatory adjustment at the commencement of exercise is slower (Morton and Callister 2000). As the blood supply to the respiratory muscles is improved, the pain disappears. This theory is not entirely satisfactory. The stitch is more common when running than in cycling and swimming. An alternative trigger of this stitch could be a mechanical stimulation of pain receptors in the abdominal region. A bouncing effect on the abdominal organs is certainly evident in jogging and running. However, it is puzzling that this type of problem does not follow a strict and reproducible course. It was previously believed that the pain was caused by an emptying of the blood depots in the spleen and the contractions taking place in the spleen. In humans, the spleen serves no such depot function, however. Furthermore, persons who have had their spleen removed can still experience such pain. Well-trained athletes who have warmed up adequately before a muscular effort seldom experience such pain.

# High Air Pressures and Breath Holding—Diving

Although human beings can become acclimatized to low air pressures, there is no way to become acclimatized to high air pressures such as are encountered in deep sea diving and during escape from a submarine when the survivor attempts to get from the inside of the craft where the pressure is normal to the surface through the sea where the air pressure is higher. For every 10 m (33 ft) of seawater the diver descends, an additional pressure of 1 atm acts on the body. As the pressure increases, more gases can be taken up by the body and dissolved in the various tissues. At a depth of about 10 m, twice as much gas will be dissolved in the diver's blood and tissue fluid as at sea surface. This is apt to give the diver trouble, mainly because of nitrogen. Nitrogen diffuses into the various tissues of the body very slowly and, once dissolved, it also leaves the body very slowly when the pressure is reduced to the normal atmospheric pressure. This is especially bad when the pressure is suddenly reduced from several atmospheres, as may be the case during submarine escape or deep sea diving. Then the nitrogen in the blood, and the nitrogen being released from the tissues into the blood, form insoluble gas bubbles. These bubbles congregate in the small blood vessels, where they obstruct the flow of blood. This gives rise to symptoms such as pains in the muscles and joints, and even paralysis can develop if the bubbles become trapped in the CNS. These symptoms are known as the **bends**. Obviously, the severity of the symptoms depends on the magnitude of the pressure, which means the depth to which the person has descended, the length of time spent at that depth, and the speed of ascent to the surface.

The bends can be avoided to a large extent by a slow return to normal pressure to allow time for the bodily fluids to get rid of their excess nitrogen without the formation of bubbles. Another way to avoid the bends is to prevent the formation of these bubbles by replacing atmospheric nitrogen with helium, which is less easily dissolved in the body. This is done by having the diver breathe a helium–oxygen gas mixture. Another advantage is that this practice is more apt to prevent the so-called nitrogen narcosis, which occurs when air is breathed at 3 atm or more and which results in an onset of euphoria and impaired mental activity with lack of ability to concentrate. With increasing pressures, the individual is progressively handicapped and may be rendered helpless at 10 atm.

### For Additional Reading

For further reading related to diving medicine, see Bove and Davis (1990).

## Breath Holding—Diving

The body's ability to store oxygen is extremely limited. The blood contains up to about 1.0 L of oxygen. In the lungs after a normal inspiration, there may be about .5 L of oxygen; after a maximal inspiration, it may be about 1.0 L. (A total capacity of 7.0 L and a concentration of oxygen of 15% = $15/100 \cdot 7.0 = 1.05$ L.)

The oxygen bound to myoglobin can amount to as much as .5 L, but it can be considered only as a local store and cannot be used by other tissues. The hyperbolic slope of the oxygen–myoglobin dissociation curve binds the oxygen effectively until its partial pressure drops to extremely low values.

When one holds one's breath at rest, a total of about 600 ml of oxygen is available and can be used. This is enough to last about 2 min. "Normal" maximal breath-holding time is 30 to 60 s. The $P_aO_2$ then decreases to about 75 to 50 mmHg (10–6.6 kPa) and the $PCO_2$ increases to 45 to 50 mmHg (6.0–6.6 kPa). This elevated $CO_2$ pressure plays a greater role in forcing the individual to discontinue the breath holding than does the reduced oxygen pressure. The hypoxic drive is also an important factor, as demonstrated by the fact that breath holding may be extended if it is preceded by oxygen breathing. Under these conditions, a further increase of $P_aCO_2$ of 5 to 10 mmHg (.65–1.30 kPa) can be tolerated.

Other factors also affect the capacity for breath holding. First, if one holds one's breath with a lung volume near the TLC (maximally inflated), the respiratory standstill may be extended until the $PCO_2$ has reached a value about 10 mmHg (1.3 kPa) higher than with a lung volume near the residual level. This is even the case with oxygen breathing; for this reason, a different hypoxic drive can be excluded. The reason is probably that afferent impulses from the receptors in the lungs and thoracic cage produce a greater stimulus to inspiration in the expiratory position. The second factor is that breath holding can be prolonged by swallowing; swallowing constitutes a momentary inhibition in inspiration. Third, single or repeated breaths after the breaking point, without change in alveolar air composition, make a new breath holding possible, and higher $P_aCO_2$ and lower $P_aO_2$ are obtained compared with the first trial. The fourth factor is that total paralysis of the muscles by **curare** prolonged the breath-holding time, and it totally abolished any sensation of breathlessness in the subjects despite a highly elevated $P_aCO_2$ (Godfrey and Campbell 1970). Apparently, a disproportion between the force developed in the respiratory muscles and a motor effect is in some way transmitted to sensation by afferent impulses from the muscles and the chest wall (Whitelaw et al. 1981).

## For Additional Reading

For further details regarding the trainability of underwater breath-holding time, see Hentsch and Ulmer (1984).

If a subject voluntarily hyperventilates forcefully for about a minute, the pulmonary air is exchanged more often and the composition of its gases more closely approaches that of the inspired air. In this manner, the $P_AO_2$ can increase to about 135 mmHg (18 kPa), the $PCO_2$ can drop below 20 mmHg (2.7 kPa), and the arterial blood will assume similar gas pressure. Because of the shape of the oxygen saturation curve (see figure 5.20), however, the oxygen content of the arterial blood is hardly affected.

A more important effect of hyperventilation in this connection is the reduction in the $CO_2$ content of the body. Thus, breath holding can be extended to several minutes by providing more room for $CO_2$. If a maximal breath holding is performed during muscular exercise, the breath-holding time is shorter than at rest, but because oxygen uptake and $CO_2$ production occur at a higher rate, the composition of the arterial blood and the pulmonary air will change more rapidly. At the breaking point, the $P_ACO_2$ may be as high as 75 mmHg (10 kPa) and the $PO_2$ may drop toward 40 mmHg (5.3 kPa). The arterial gas pressures are probably at the same level. This is still within the safe range of critical oxygen pressure for the function of the nervous system, which in the case of the oxygen pressure of the arterial blood is about 25 to 30 mmHg (3.3–4 kPa).

However, if a forceful hyperventilation precedes breath holding during heavy exercise, breath holding can be prolonged until convulsions or even fainting occurs (P.-O. Åstrand 1960). The $P_AO_2$ then drops toward 20 mmHg (2.7 kPa), which explains the symptoms.

If a dive is performed following a marked hyperventilation, one may apparently hold the breath until unconsciousness occurs. The total pressure of the alveolar air at sea level is about 760 mmHg (101 kPa), but at a depth of 10 m it is doubled (to twice the atmospheric pressure) by the pressure of the water on the thorax. An oxygen percentage which at sea level corresponds to an alveolar oxygen tension of about 25 mmHg or 3.3 kPa (3.5% by volume) produces at a depth of 2 m (total pressure = 912 mmHg, or 121 kPa) a pressure of 30 mmHg or 4 kPa [$3 \cdot 5/100 \cdot (912 - 47) = 30$]. As long as the diver remains at a depth of 2 m, the oxygen tension is adequate to meet the requirement of the nervous system. But when the diver approaches the surface, the oxygen tension can drop below the critical level. The danger is greater if the ascent is slow, in that the oxygen utilization then causes the oxygen pressure to decrease further. Several cases have been described of divers who were attempting to beat a record, or who for various other reasons took their time during prolonged diving, having lost their lives or having been rescued in the nick of time (A.B. Craig 1976). Often, hyperventilation had proceeded the diving: The diver then exhibited a strange behavior or simply ceased to swim and sank. Cases have been described when the swimmer had had no major difficulty holding his or her breath but then suddenly lost consciousness.

It is thus justifiable to warn against prolonged underwater swimming and diving without effective supervision. The diver should avoid hyperventilation before the dive, taking, at the most, five deep respirations. In this connection, another effect of hyperventilation should be stressed. The washing out of $CO_2$ in connection with the hyperventilation increases the pH. The effect of this alkalosis and reduced $P_aCO_2$ is a vasoconstriction, including the vessels in the brain. Dizziness and cramps may be the result. If one holds one's breath after a hyperventilation against a closed glottis and at the same time contracts the abdominal muscles (Valsalva maneuver), the cardiac output is reduced. This in combination with the vasoconstriction in the cerebral blood vessels can produce an oxygen deprivation in the CNS sufficient to cause a transitory loss of consciousness.

# Chapter 7

# Skeletal System

The purpose of this chapter is to summarize the structure and functions of bone as they pertain to physical performance, work stress, and exercise and to outline the mechanisms that enable bone tissue to adapt to changing demands.

### For Additional Reading

For a more extensive survey of bone physiology and the influence of physical activity on bone health, see Frost (2001) and Khan et al. (2001).

The human skeletal system provides the mechanical levers that enable the muscles to move the body. The skeleton is the supporting framework that prevents the entire body from collapsing into a heap of soft tissue, and it is the protecting shell or casing for such vital, viable organs as the brain, lungs, and heart. In addition, bone tissue is the major calcium and phosphorus reserve of the body, constantly being drawn on or added to, and the cavities found inside bones are filled with bone marrow, responsible for the lifelong production of blood cells.

## The Tissues of the Human Skeletal System

The human skeleton consists primarily of three types of connective tissue: bone, cartilage, and connective tissue proper. In the adult, bone is by far the more abundant, but in the fetus most "bones" are formed initially as cartilaginous primordia. Gradually, this cartilage is replaced by bone during fetal and early postnatal life. The connective tissue proper of the skeleton connects individual bones, partly in **synovial joints** that allow movements between the bones and partly in more rigid connections.

All three types of connective tissue consist of cells embedded in an extracellular matrix that is produced by the cells themselves. Despite a very close developmental relationship between the three main cell types—the **fibroblast** of connective tissue proper, the **chondroblast** of cartilage, and the **osteoblast** of bone —the three types of tissue differ considerably in their nature and appearance. All three types of extracellular matrix consist of a ground substance reinforced by collagen fibers, but there are major differences in both the ground substance and the type of collagen fibers.

### For Additional Reading

For a survey of the different isoforms of collagen, see Józsa and Kannus (1997).

The matrix of connective tissue proper is like a soft gel, and the wide variation in the mechanical properties of its different subtypes depends primarily on the amount of collagen fibers in relation to ground substance and cells. In ligaments and tendons—the major connective tissue components of the musculoskeletal system—the amount of collagen fibers by far outweighs the amount of cells and ground substance. Moreover, the major portion of the collagen fibers are oriented parallel to the major direction of tension, giving the structures in question high tensile strength.

The extracellular matrix of cartilage is different. The ground substance consists of large proteoglycan complexes forming a meshwork throughout the matrix, which, together with the collagen fibers reinforcing it, results in the firm, elastic consistency of cartilage.

The extracellular matrix of bone, as it is produced by the bone cells, has a consistency very much like that of cartilage, but there are major differences between the two, the more important of which is the ability of bone extracellular matrix to be mineralized by the deposition of calcium phosphate.

## The Structure and Function of Connective Tissue Proper

In accordance with the general scheme of connective tissue components, the extracellular matrix of connective tissue proper is a ground substance containing varying amounts of collagen fibers. The isoform of collagen found in connective tissue proper is collagen type I. The relative amounts of cells, fibers, and ground substance, and the resulting functional properties of the tissue, however, may vary. At one extreme, there is the loose connective tissue of the intestinal mucosa, at the other, the dense connective tissue found in ligaments and tendons.

The dense connective tissue of ligaments and tendons is by far dominated by collagen fibers. The ground substance is barely recognizable, and the cells are squeezed between bundles of collagen fibers. The major cell type in dense connective tissue is the fibroblast, the producer of the extracellular matrix—both the ground substance and the collagen fibers. This is in contrast to loose connective tissue, where other cell types like blood cells outnumber the fibroblasts.

Collagen fibers are very strong. Their tensile strength has been compared with steel, but the strength of a bundle of collagen fibers is in the longitudinal direction of the fibers only. Consequently, a tendon with almost parallel collagen fibers is very strong along its ordinary line of pull but yields far more easily to attempts to pull the fibers apart.

## The Structure and Function of Cartilage

Cartilage is a barely translucent tissue consisting of cartilage cells (chondrocytes) embedded in an extracellular matrix, largely produced by the cartilage cells themselves. The proteoglycan meshwork of the matrix is reinforced by collagen fibers of the type II isoform. Cartilage is a firm yet elastic tissue, flexible and capable of rather surprising modes of growth. Such growth may take place not only at the surface of the cartilage (appositional growth) but also in the middle of the tissue (interstitial growth). The latter is extremely important for the lengthening of the long bones of the skeleton and thus for growth during childhood and adolescence.

The lack of blood vessels is a characteristic and important feature of cartilage. In particular, it makes cartilage an ideal tissue to cover articular surfaces, because there are no blood vessels that can be pinched off during loading of the joint. Instead, part of the interstitial fluid of the articular cartilage is squeezed out of and sucked back into the tissue during loading and unloading of the joint, thus carrying nutrients from the synovial fluid to the cartilage cells, in addition to providing poroelastic properties (Fischer 2000).

Although cartilage is ideal on the articular surfaces, it is unable to fulfill the mechanical requirements of the adult human skeleton. In certain aquatic vertebrates, however, it is sufficient, because the stress is moderate. In the shark, for example, cartilage is the only skeletal material. On the other hand, in land vertebrates, such as the adult human being, cartilage plays only a limited part in the skeletal makeup.

The type of cartilage found in fetal bones, and consequently also in the **epiphyseal disks** and on most articular surfaces, is the more common one, called hyaline cartilage, because of its almost translucent appearance. The cartilage of articular disks, articular lips, and menisci is of another type called **fibrocartilage.** Actually, fibrocartilage can be regarded as a hybrid between connective tissue proper and hyaline cartilage. For further details about different forms of cartilage and their properties, see standard histology textbooks.

### *Summary*

All types of connective tissue consist of cells embedded in an extracellular matrix produced by the cells themselves. The three main cell types—the fibroblast of connective tissue proper, the chondroblast of cartilage, and the osteoblast of bone—are closely related, but the types of extracellular matrix they produce are different. The type of connective tissue proper found in the skeletal system consists of densely packed collagen fibers (type I) with few cells and very scarce ground substance. The major type of cartilage is the bluish-white hyaline cartilage found in fetal bones as well as on most articular surfaces.

## Bone Is a Calcified Tissue

Like cartilage and connective tissue proper, bone consists of cells embedded in an extracellular matrix produced by the cells themselves. The cells of bone are called **osteocytes.** The matrix is reinforced by type I collagen fibers in a systematic way, much like armoring iron in a concrete wall. Shortly after the matrix has been produced by the osteocytes (which actually are called osteoblasts during this early, matrix-producing phase of their life), the matrix is mineralized by the deposition of calcium phosphate in the form of hydroxyapatite. If this mineralization is suboptimal and delayed, the amount of unmineralized bone substance becomes too large to withstand the pull of the muscles and the weight of the body, resulting in typical skeletal abnormalities.

In mature bone, the osteocytes and the mineralized extracellular matrix are found as alternate layers where rows of osteocytes are interspersed between lamellae of mineralized bone matrix. In compact bone, as found in the shafts of the long bones, these lamellae are tightly apposed, hence the name (figures 7.1 and 7.2). In **trabecular bone,** the structure is more spongelike with irregular cavities containing bone marrow. Obviously, this means that the combined surface of all bone lamellae facing a marrow cavity is far larger in trabecular bone than in compact bone. This is an important fact, which we shall return to when discussing bone

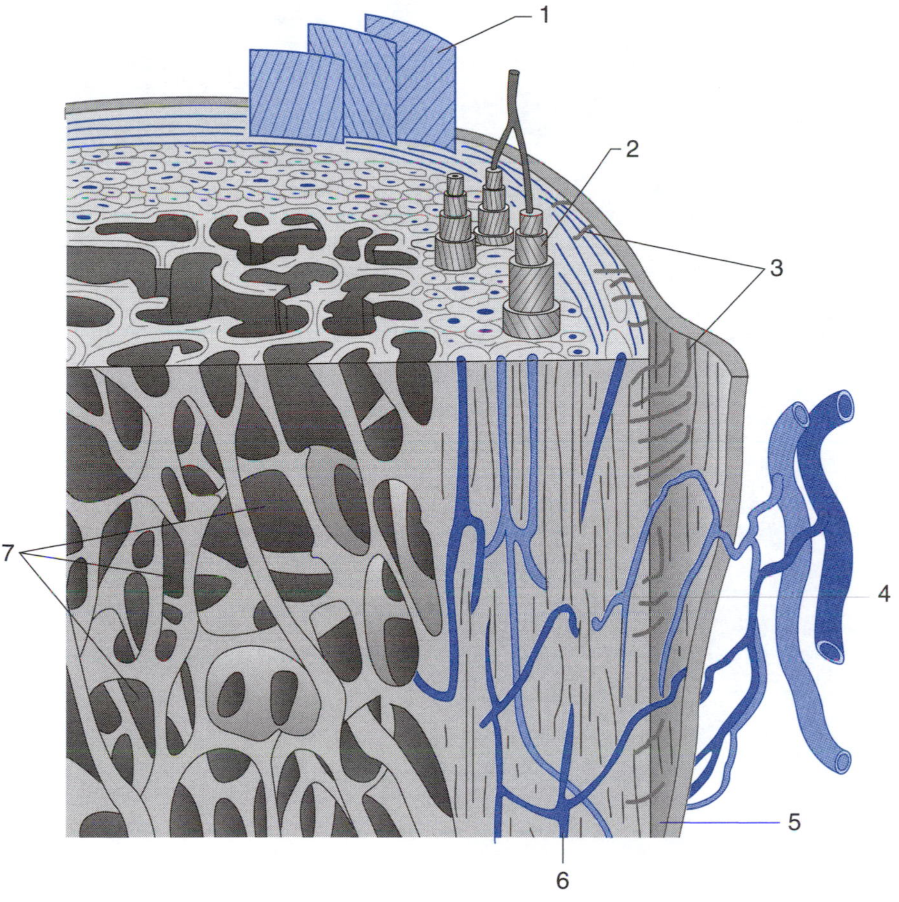


**Figure 7.1** Part of the shaft of a long bone. The solid outer part is made of compact bone, whereas the more spongelike trabecular bone is found toward the interior. The basic structure of compact bone is made of concentric bone lamellae (1), interspersed with haversian systems (2). Note the alternate directions of collagen fibers in neighboring lamellae. The outer surface of the bone is covered by a layer of connective tissue called periost (5). Blood vessels from the periost penetrate the layers of compact bone (4) and excavate canals, which subsequently become lined with concentric, haversian lamellae (6 and 2). At **osteotendinous junctions,** collagen fibers from the tendon merge with the periost and are inserted deeply into the bone matrix of the compact bone (3). The spaces (7) between the beams of trabecular bone are filled with bone marrow.

Adapted, by permission, from A. Benninghoff and K. Goertler, 1964, *Lehrbuch der Anatomie des Menschen* (Berlin, Germany: Urban und Schwarzenberg-München), 65.

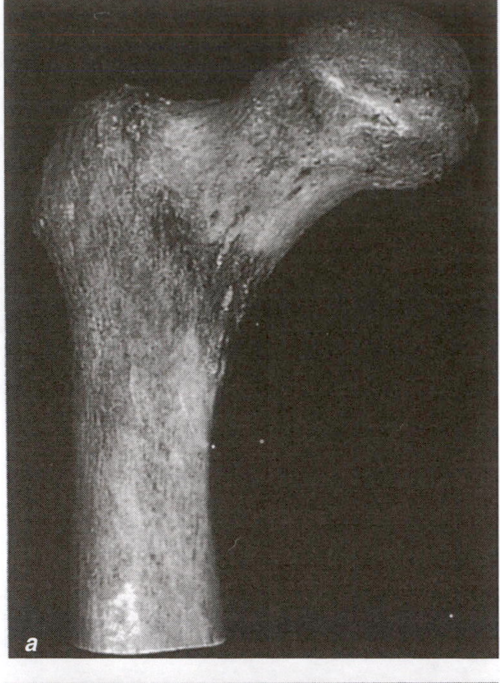


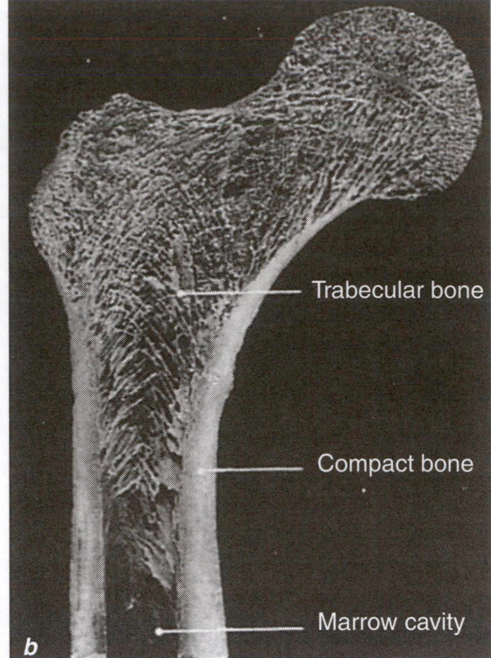


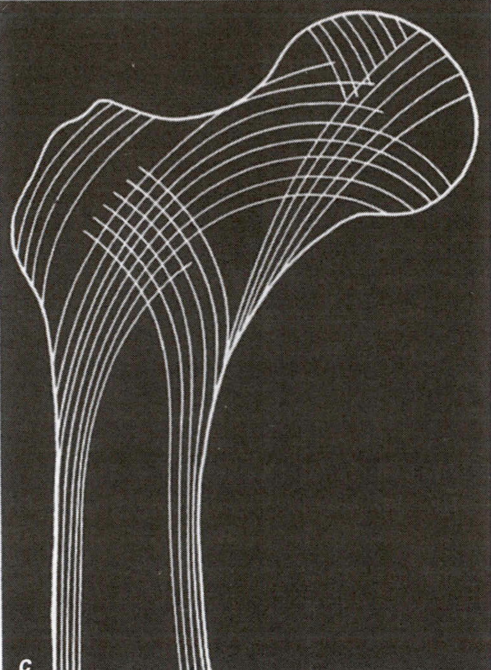


**Figure 7.2** The upper end of the right femur seen from the ventral side *(a)* and sectioned longitudinally *(b)*. In the diaphysis (lower part of the picture), a thick layer of compact bone surrounds the marrow cavity. In the metaphysis and epiphysis (upper part of the picture), the layer of compact bone becomes progressively thinner, and the interior is filled by trabecular bone. *(c)* The main directions of the bone beams seen in *(b)*.

Reprinted, by permission, from H.A. Dahl and E. Rinvik, 1999, *Menneskets funksjonelle anatomi* (Oslo, Norway: J.W. Cappelens Forlag).

remodeling. The trabeculae of trabecular bone are arranged in a way that is reminiscent of steel beams in a bridge (figure 7.2). Apparently, they are oriented to provide maximal strength with a minimum of material.

There are four different cell types in bone: **osteogenic cells** (also called **osteoprogenitor cells**—the stem cells of bone), osteoblasts, osteocytes, and **osteoclasts.** The first three of these are in fact developmental stages of the same cell type, whereas the osteoclasts are specialized bone macrophages, formed by fusion of mononucleated progenitors of the monocyte/macrophage family (Teitelbaum 2000). Together, these cell types build and rebuild bone tissue throughout life. As stated previously, the mature osteocytes are found trapped between layers of mineralized bone extracellular matrix. The other cell types are found at or near bone surfaces facing marrow cavities, in the marrow itself, or in the deep layers of the periost, the layer of connective tissue on the outer surface of bones. The osteoblasts (the suffix *blast* indicates a young, productive cell) are found as an almost epithelioid layer on the surface of bone lamellae. The osteoclasts are multinucleated cells interspersed between the osteoblasts, often situated in small excavations of the bone surface, created by their own "bone-eating" activity. Neither osteoblasts nor osteoclasts are long-lived cells. The lifespan of an osteoclast is about 2 weeks, whereas that of an osteoblast is about 3 months. Both cell types die by apoptosis (programmed cell death). Some osteoblasts are turned into osteocytes, which live longer than their parental cell type, but even they are not immortal (Manolagas 2000). Others end up as so-called lining cells, a kind of inactive cells lining resting bone surfaces.

**Osteoclast progenitor cells** express at least two membrane receptors important for their differentiation into mature, functioning osteoclasts (figure 7.3). One is a receptor for **macrophage colony-stimulating factor (M-CSF)**. The other is called **RANK** (which stands for receptor for activation of nuclear factor κ-B), a cytokine receptor of the tumor necrosis factor (TNF) receptor superfamily. Osteoblasts and marrow stromal cells are important for osteoclastogenesis because they express the appropriate ligands for both these receptors, called M-CSF and **RANK ligand (RANK-L),** respectively. Somewhat astonishingly, however, they also express and secrete a soluble cytokine receptor called **osteoprotegerin** (**OPG;** i.e., "the one that protects bone"), which is able to bind RANK-L and thus inhibit its interaction with the RANK receptor on

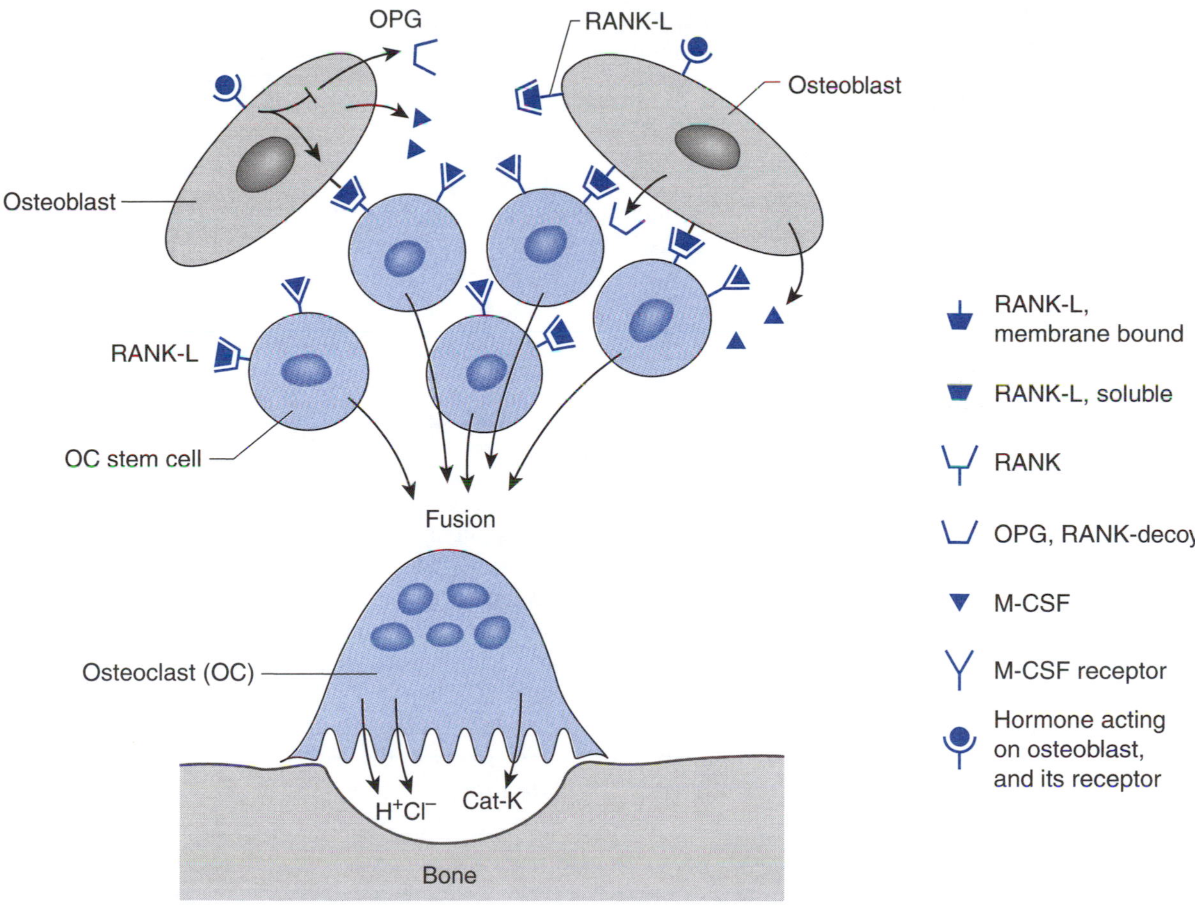


**Figure 7.3** Osteoclasts (OC) develop from progenitor cells of the monocyte/macrophage family. The progenitor cells have at least two different membrane receptors, called RANK (receptor for activation of nuclear factor κ-B) and macrophage colony-stimulating factor (M-CSF) receptor. The natural ligands for these receptors, RANK-ligand (RANK-L) and M-CSF, are expressed by osteoblasts and stromal cells, but they also express and secrete a soluble decoy receptor for RANK-L, called osteoprotegerin (OPG). When present, OPG binds to RANK-L and inhibits the interaction of the latter with its receptor on osteoclast progenitor cells. When stimulated by the appropriate ligands, osteoclast progenitor cells fuse to form osteoclasts, attach to the surface of bone, and secrete HCl and cathepsin K to digest bone matrix.

the surface of the osteoclast. In short, the balance between RANK-L and its soluble decoy receptor OPG determines the number of osteoclasts and thus the quantity of bone resorbed at any point (Teitelbaum 2000). Several of these molecules are known under different names. RANK-L, for example, has also been called **OPG-ligand (OPG-L;** because it binds OPG) and **osteoclast differentiation factor (ODF).**

## FOR ADDITIONAL READING

For further details about these cytokines and their receptors, see Hofbauer et al. (1999), Kong et al. (1999), and Manolagas (2000).

It has been known for a long time that parathyroid hormone (PTH), 1,25-dihydroxy-vitamin $D_3$ (the active vitamin $D_3$ hormone), calcitonin, and estrogen influence the rate of bone resorption, the former two increasing bone resorption and the latter two reducing it. PTH is now known to exert its osteoclast-stimulating effect by stimulating the expression of RANK-L by osteoblasts and stromal cells, as outlined previously. The same possibly holds true also for the vitamin $D_3$ hormone (Teitelbaum 2000). Estrogen (as well as selective estrogen receptor modulators, or SERM) inhibits bone resorption by blocking the stimulatory effect of interleukin (IL)-1 on osteoclast differentiation, whereas calcitonin exerts a direct inhibitory effect on osteoclast activity (Rodan and Martin 2000). The pancreatic polypeptide amylin and β-CGRP also inhibit osteoclast function, but in a way different from that of calcitonin (Alam et al. 1993). Other cytokines, in particular IL-6 and TNF-α, also are involved in the regulation of bone metabolism (Salamone et al. 1998), and polymorphism of the IL-6 gene has been shown to be an independent predictor of bone mineral density (Lorentzon et al. 2000).

Physical activity has been shown to induce an inflammatory cytokine response, especially if the activity leads to muscle soreness (Nieman et al. 2001), but how this affects bone metabolism is not easy to see. The increase in IL-1 receptor antagonist presumably shifts the balance toward less bone resorption, whereas an increase in IL-6 should work the other way. Because the pattern of cytokine secretion apparently differs between types and levels of physical activity, their combined effect on bone metabolism may differ too.

How, then, does the osteoclast digest bone? In much the same way as low pH in the mouth is harmful to the teeth, any acid is able to dissolve and remove the inorganic calcium salts in bone. The osteoclast attaches to the bone surface and creates a more or less sealed microenvironment between itself and the bone surface, into which it secretes hydrochloric acid, thereby dissolving the inorganic components of the underlying bone (see figure 7.3). The organic matrix is digested next by means of a proteolytic enzyme, cathepsin K, also secreted by the osteoclast (Teitelbaum 2000).

A typical long bone, such as the femur, consists of a shaft (the **diaphysis**) and two enlarged ends (the **epiphyses**). The part of the shaft that is next to the epiphysis is called the metaphysis. The surface of the bone is covered by a layer of connective tissue—the periost—the deeper layers of which contain osteogenic cells. The shaft of a long bone is like a hollow tube. The wall of the tube is a thick layer of compact bone surrounding a central marrow cavity. The metaphysis and epiphysis, on the other hand, contain trabecular bone covered by a thinner layer of compact bone (see figure 7.2). The joint surface of the epiphysis is covered with smooth, glistening, white hyaline cartilage. The articular cartilage has to absorb the shocks transmitted to the joint during motion and reflects the wear and tear of the joint.

The compact bone found in the shaft of the long bones consists basically of closely apposed lamellae of mineralized bone extracellular matrix, arranged concentrically around the marrow cavity, and with layers of osteocytes in between (see figure 7.1). The direction of the collagen fibers reinforcing the matrix alternate in neighboring lamellae, giving the whole structure maximal strength. The osteocytes have thin, branching processes penetrating into tiny canaliculi in the bony matrix of the neighboring lamellae. These canaliculi serve as routes for supply of nutrients to the osteocytes and enable the osteocyte processes to establish cell-to-cell contacts via

## CLASSIC STUDY

In a series of bed rest studies at the Lankenau Hospital in Philadelphia 40 years ago, K. Rodahl et al. found that the long bones of the body have to be loaded by the weight of the body in a standing position for at least 3 h · $day^{-1}$ to avoid loss of calcium from the bones.

K. Rodahl et al. 1966.

gap junctions with osteocytes in neighboring lamellae. This is part of an extensive cellular network comprising osteoblasts, bone marrow cells, and endothelial cells in addition to osteocytes, forming a functional syncytium (Burger and Klein-Nulend 1999; Manolagas 2000).

In the large bones of the human skeleton, however, an additional pattern of lamellar organization is seen. Small blood vessels accompanied by osteoclasts penetrate the bone matrix of compact bone and excavate rather roomy canals parallel to the long axis of the bone. As they are formed, these canals are filled with osteogenic cells, which start producing new bone lamellae lining the walls of the canal, in the end leaving only room for the tiny blood vessel in the middle. The blood vessel with its surrounding bone lamellae is called a **haversian system** (figure 7.4; see also figure 7.1). Such haversian systems are not permanent structures, because blood vessels excavate new canals all the time, destroying old haversian systems and making new ones (figure 7.5). This is part of the constant remodeling of bone.

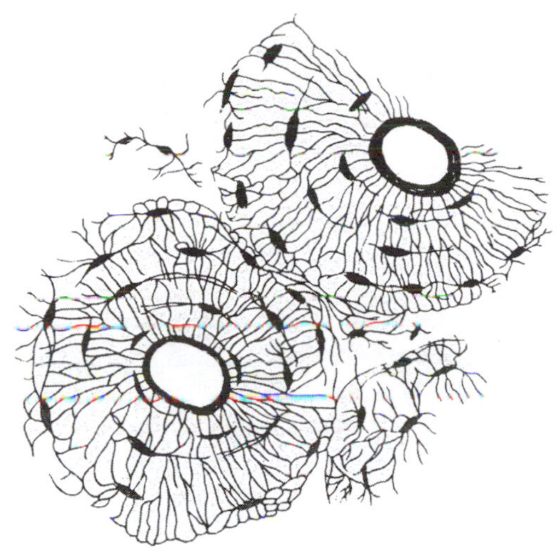

**Figure 7.4** Two haversian systems in cross section. Osteocytes (black, spiderlike profiles) with long, branching processes are interspersed between layers of mineralized bone matrix. The bone matrix is situated between the concentric layers of osteocytes and is not visible in this figure. The osteocyte processes are thought to establish cell-to-cell contacts with gap junctions between neighboring cells.

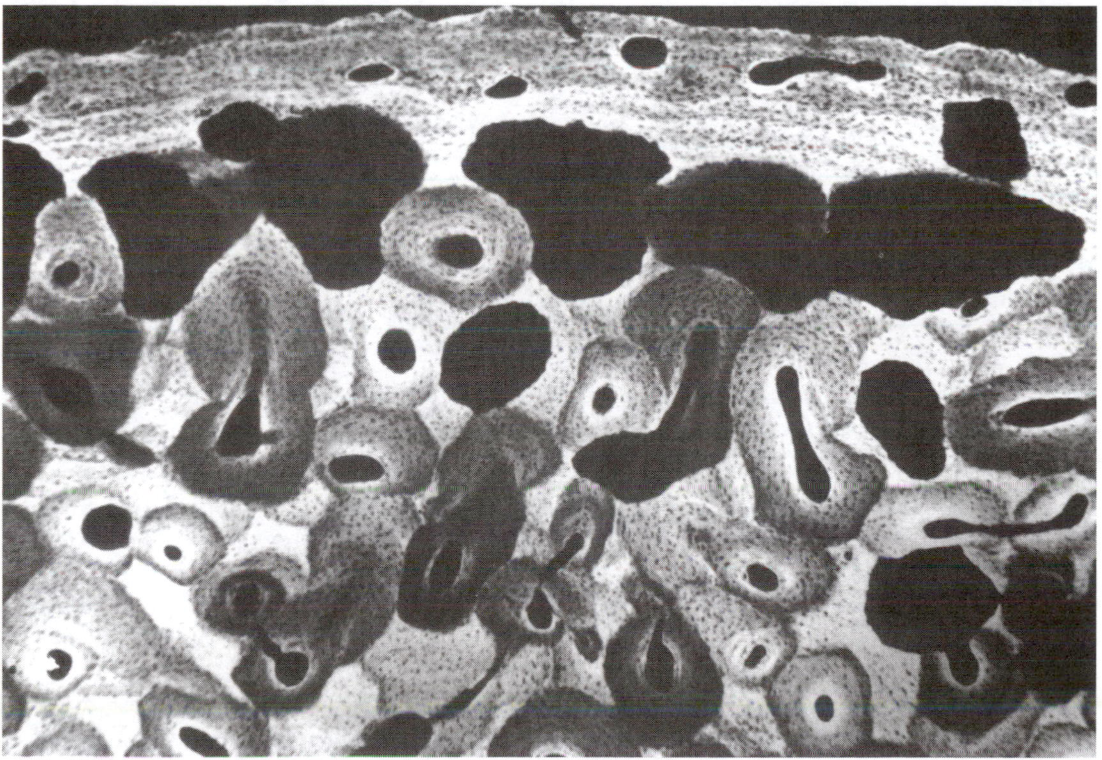

**Figure 7.5** Microradiograph of a 200-μm-thick cross section of bone from a normal 19-year-old male showing several generations of haversian systems. The older ones are very light because they are highly mineralized, whereas the newer ones are darker because of lower mineral content. Black areas represent soft tissue—newly formed excavations, not yet lined by bone lamellae, and the central canal of fully formed haversian systems. Note the large excavations in the upper part of the picture, close to the periosteal surface (top), representing the initial stage of the formation of new haversian systems.

Reprinted, by permission, from M.H. Ross, L.J. Romrell, and G.I. Kaye, 1995, *Histology, A Text and Atlas*, 3rd ed. (Philadelphia, PA: Lippincott, Williams, and Wilkins), 168.

## *Summary*

In contrast to the other two types of connective tissue, bone is a calcified tissue. Shortly after its production by the osteoblasts, the bone extracellular matrix is mineralized by the deposition of calcium phosphate. The minerals of bone serve two different purposes: They make the bones mechanically stronger, and they serve as a mineral bank, pivotal in the calcium and phosphate homeostasis of the rest of the body.

The removal of existing bone is taken care of by osteoclasts, multinuclear bone-eating cells of monocyte/macrophage descent. The differentiation of osteoclasts from the progenitor cells is governed by molecular signals from osteoblasts/stromal cells. Three kinds of signals seem to be particularly important, two of which—M-CSF and RANK-L—stimulate osteoclastogenesis, whereas the third—OPG—inhibits osteoclastogenesis by acting as a decoy receptor for RANK-L. Most hormones acting on bone do so via the osteoblasts by influencing the balance between RANK-L and OPG, but calcitonin directly inhibits osteoclast activity.

# Bone Formation

In the human fetus, most of the bones of the skeleton are cartilaginous, but the cartilage gradually is removed and replaced by bone, sparing the cartilage only in the epiphyseal disks, on the articular surfaces, and a few other places where elasticity is more important than stiffness, such as the sternal ends of the ribs, which are subjected to dynamic stress during breathing. The sequence in which the cartilage is replaced by newly formed bone is typical both in individual bones (figure 7.6) and between bones. The latter sequence can be used to determine the so-called skeletal age of a child. For further details about the stages in bone formation, see standard anatomy textbooks.

A series of in vitro studies have long since shown that the development of bone and cartilage is affected by environmental factors. C.A. Basset (1962) showed that the formation of normal bone from a tissue culture of primitive fibroblasts depended on adequate oxygenation. When high oxygen pressure was introduced in addition to compacting of the culture, bone formation resulted. With low oxygenation, cartilage was formed from the primitive fibroblasts.The molecular mechanism behind this change of behavior is unknown, but genetic-based studies have provided insight into how osteoblast differentiation is controlled through growth and transcription factors. Basically, the osteoblast can be regarded as a sophisticated fibroblast (Ducy, Schinke, and Karsenty 2000). In fact, all the genes expressed in fibroblasts also are expressed in osteoblasts, and only two osteoblast-specific transcripts have been identified, one of which is the transcription factor Cbfa1. The other is **osteocalcin,** a secreted molecule that inhibits osteoclast function (Ducy et al. 1996). Cbfa1 and its gene *Cbfa1* have emerged as important factors in osteoblast differentiation. In the embryo, Cbfa1 is expressed in mesenchymal cells destined to become chondroblasts or osteoblasts. Later, the expression becomes limited to osteoblasts, its transcription apparently being stimulated by **bone morphogenic proteins** (**BMPs;** Manolagas 2000). It seems, however, that members of all the major families of growth factors are involved in osteoblast differentiation. In turn, Cbfa1 activates osteoblast-specific genes. In accordance with this, Cbfa1-binding sites are found in the regulatory sequences of most genes required for the production of bone extracellular matrix (Manolagas 2000).

Bone resorption has been known for a long time to be under endocrine control, but evidence for hormonal control of bone formation is more recent. Interestingly, this seems to be yet another function—albeit indirect—of the multifunctional hormone **leptin** (figure 7.7). Both animals and humans deficient in leptin signaling are massively obese, but they also have a higher bone mass than normal individuals. Osteoblasts lack leptin receptors, and consequently, leptin has no effect on osteoblasts in culture. On the other hand, intraventricular injection of leptin in leptin-deficient animals normalizes their bone phenotype. In **wild type animals,** intraventricular administration of leptin causes bone loss. Together, this indicates that leptin acts through receptors located in the brain, most likely in the hypothalamus, and that it is a physiologically relevant pathway for inhibitory control of bone formation. It seems unlikely, however, that leptin is the only hormone that regulates bone formation. We have to wait for others to be revealed.

### For Additional Reading

For further details about leptin and bone, see Ducy, Schinke, and Karsenty (2000).

# The Bone Remodeling Cycle

Although from a macroscopic point of view the bones do not seem to change after reaching adult size, remodeling is continuous throughout life. It is easy to underestimate the extent of this remodeling process, however, because macroscopically, bones do not seem to change after reaching adulthood. But

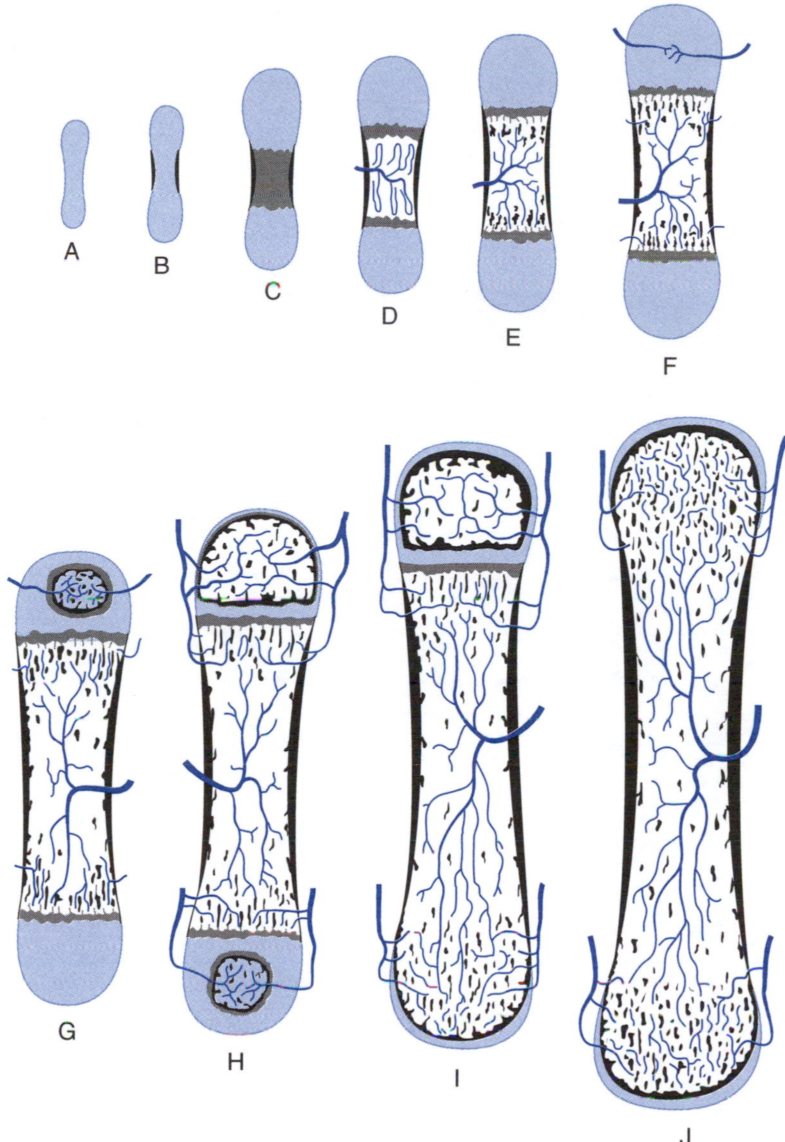


**Figure 7.6** Typical stages in the development from a cartilaginous "bone" to an adult, long bone. Blue color denotes cartilage, medium dark gray calcified (degenerate) cartilage, and dark gray (black) bone. *(a)* At first, the "bone" consists of pure cartilage. *(b)* Formation of a bone sleeve around the shaft of the bone. *(c)* Degenerative changes in the cartilage—deposition of calcium salts. *(d)* Blood vessels from the periost penetrate into and excavate the calcified cartilage. *(e)* Osteoblasts make new bone on the surface of remnants of calcified cartilage. *(f)* and *(g)* Blood vessels grow into the cartilage of the epiphysis and replace it with bone *(h)*, leaving the cartilage only in the epiphyseal disks *(h)* and *(i)* and on the articular surfaces *(h–j)*. In the end, bone formation overtakes the growth of the epiphyseal discs. With the disappearance of the epiphyseal discs *(j)*, any further longitudinal growth becomes impossible.

Adapted, from W. Bloom and D.W. Fawcett, 1994, *A Textbook of Histology* (London: Chapman and Hall), 216. By permission of Arnold Publishing.

they do change. By some estimates, the remodeling process is so extensive that an amount of bone corresponding to the entire adult skeleton is replaced by new bone every 10 years (Manolagas 2000).

Bone remodeling is performed by individual, independent bone remodeling units consisting of bone-resorbing osteoclasts and bone-forming osteoblasts (Snow-Harter and Marcus 1991), also known as **basic multicellular units (BMUs).** In healthy human adults, 3 to 4 million BMUs are initiated per year, and about 1 million of them are active at any given moment (Manolagas 2000). Each remodeling cycle starts by a group of osteoclasts digging a cavity in the bone surface (figure 7.8b). Next, osteogenic cells proliferate and line the cavity

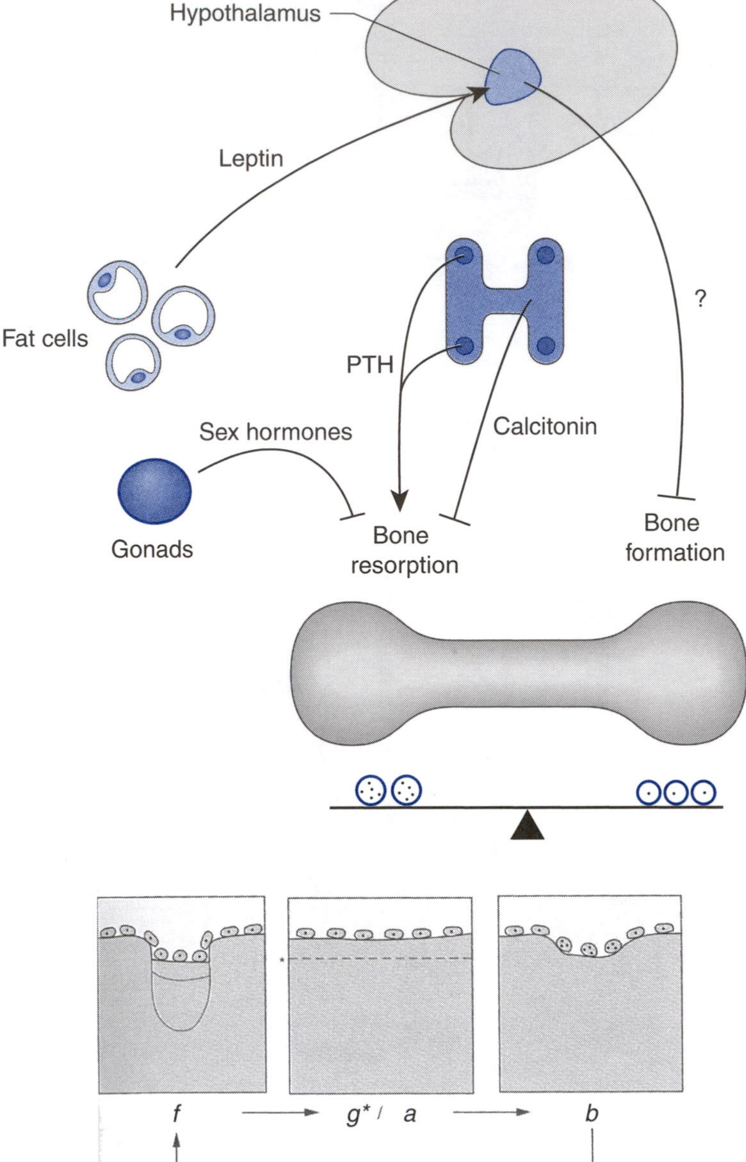


**Figure 7.7** Bone mass is the result of the balance between bone resorption and bone formation. Bone resorption has been known for a long time to be under endocrine control, with parathyroid hormone and vitamin $D_3$ stimulating bone resorption, and calcitonin and sex hormones decreasing it. Now leptin has been shown to have a negative effect on bone formation, but this effect is exerted via the hypothalamus, and the molecular signal from the hypothalamus to bone is unknown.

**Figure 7.8** The bone remodeling cycle: *(a)* resting trabecular surface; *(b)* multinucleated osteoclasts dig a cavity in the bone surface; *(c)* bone removal is completed by mononuclear phagocytes, and osteoblasts invade the resorption cavity and start secretion of new bone matrix *(d)*; *(e)* secretion of new bone matrix continues, with mineralization lagging a little behind; *(f)* mineralization is complete, and the bone returns to a quiescent state with a small deficit (*) in bone mass (*g*).



(figure 7.8, c and d), starting to produce new bone matrix (figure 7.8e), which after a small delay starts to become mineralized (figure 7.8f). At the end of the remodeling cycle, when the matrix is fully mineralized (figure 7.8g), it is noticeable that the amount of newly formed bone is somewhat less than that which was removed. The result is a minute loss of bone tissue. Over the years, these accumulated deficits result in the well-known age-associated bone loss.

The fact that bone remodeling cycles take place on bone surfaces makes it obvious that more remodeling cycles are found in trabecular bone than in compact bone, simply because there is more surface in the former than in the latter. Because each remodeling cycle ends with a little less bone than before, this means that the age-related bone loss is more pronounced in trabecular bone, like in the femoral neck (see figure 7.2) or in the vertebral bodies (figure 7.9), places well-known to be sites of osteoporotic fractures.

The strength of bone depends not only on the amount of bone tissue but also on how it is organized (figure 7.10). Most importantly, the reduced number of cross-linking bone trabeculae in osteoporotic bone (figure 7.11) weakens the structure more than would be expected from the percentage of bone loss.

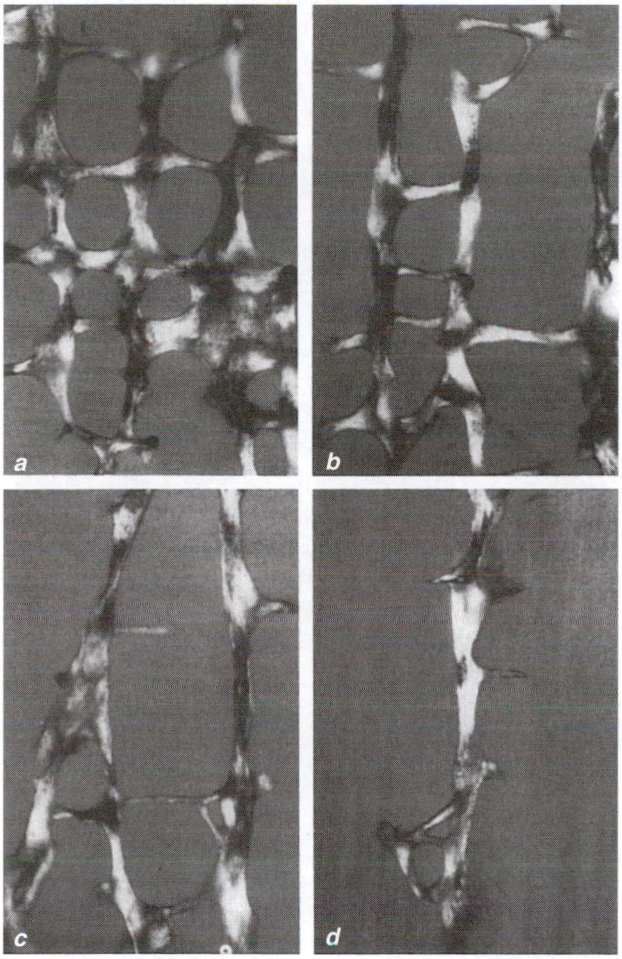

**Figure 7.9** Loss of trabecular bone with age. *(a)* A 50-year-old man with an almost perfect, continuous trabecular network. *(b)* A 58-year-old man with discernible thinning of horizontal trabeculae and some loss of continuity. *(c)* A 76-year-old man with continued thinning of horizontal trabeculae and wider separation of the vertical structures. *(d)* An 87-year-old woman who had advanced breakdown of the whole network with unsupported vertical trabeculae.

Reprinted from Bone, 9, L. Mosekilde, Age-related changes in vertebral trabecular bone architecture-assessed by a new method, 247-250, Copyright 1988, with permission from Elsevier Science.

## FOR ADDITIONAL READING

For further details about the importance of cross-linking between bone lamellae, see Snow-Harter and Marcus (1991).

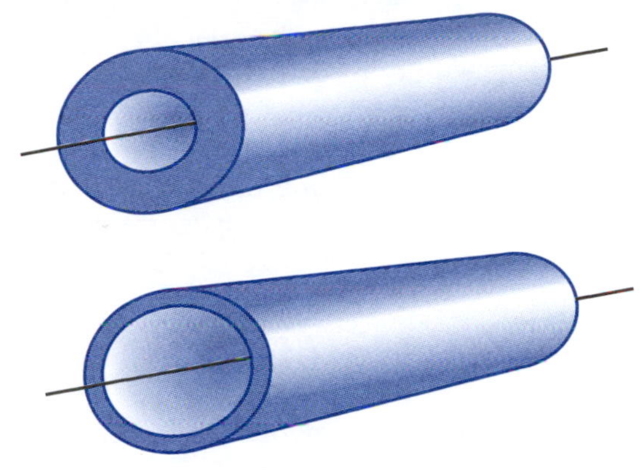

**Figure 7.10** The effect of mass distribution on strength. The two cylinders have equal mass, but in the lower one the mass is further away from the bending axis, resulting in substantially increased resistance to bending.

Reprinted, by permission, from C. Snow-Harter and R. Marcus, 2000, "Exercise, bone mineral density, and osteoporosis," *Exercise and Sport Sciences Reviews* 19:351-388.

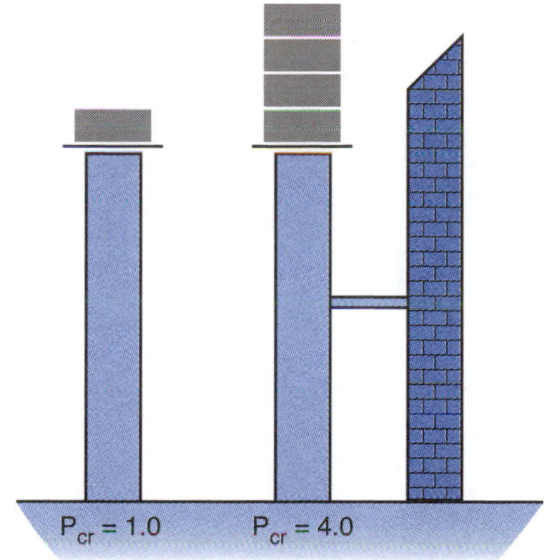


**Figure 7.11** The importance of horizontal trabeculae. A horizontal connection to a neighboring column increases the resistance to buckling by a factor of four (cf. figure 7.9). $P_{cr}$ = critical buckling load.

Adapted, by permission, from C. Snow-Harter and R. Marcus, 2000, "Exercise, bone mineral density, and osteoporosis," *Exercise and Sport Sciences Reviews* 19:351-388.

## *Summary*

In fetal life, bone formation may start in a sheet of connective tissue proper, but in most instances it takes place in cartilaginous primordia, where cartilage gradually is removed and replaced by bone tissue. After its initial formation, bone is constantly remodeled by independent bone remodeling units, called basic multicellular units. Each of the approximately 1 million of such units active at any given point consists of osteoclasts, which remove bone tissue, and osteoblasts, which replace lost bone with new bone. The amount of new bone is, however, slightly less than that which was removed. The result is a minute loss of bone for each remodeling cycle, over the years adding up to what may be a significant weakening of bone structure, especially of trabecular bone.

# Joints

Joints are formed where two or more bones of the skeleton meet one another (figure 7.12). In areas such as the skull, it is important that no movement occur between contiguous bones. In the vertebral column, on the other hand, some mobility is desirable, provided that it can be obtained without loss of strength or sturdiness and without compromising the integrity of the spinal cord. In other places, such as in the limbs, a wide range of movement is essential. In the latter case, synovial joints serve the purpose.

Synovial joints are what people usually mean by a joint. In a synovial joint, the ends of the articulating bones are covered by cartilage and only are separated by a potential cleft, often misleadingly called the joint cavity. In most joints, the articular cartilage is hyaline. A cuff of dense connective tissue forms a fibrous capsule around the joint and often is reinforced by ligaments. The synovial membrane is a layer of delicate, highly vascularized connective tissue lining the whole of the interior of the joint, with the exception of the cartilage-covered ends of the articulating bones. This membrane is a typical and most important feature of a synovial joint, hence the name of the joint. The synovial membrane produces synovial fluid, which lubricates and nourishes the articular cartilage. Not all synovial joints offer the same freedom of movement. The ball-and-socket joints at the shoulder and hip have the larger ranges of movement, whereas the iliosacral joints (between the hip bone and the sacrum) allow very little movement. In some synovial joints, a cartilaginous articular disk is inserted between the two bones. Such articular disks, made of fibrocartilage, act as shock-reducing structures and ensure perfect contact between the moving surfaces in any position of the joint. In some joints, such as the knee joint, fat pads covered by synovial membrane fill out any free space within the joint, constantly adjusting to the space available during movements of the joint.

In joints other than synovial joints, the bones are joined together by connective tissue or cartilage.

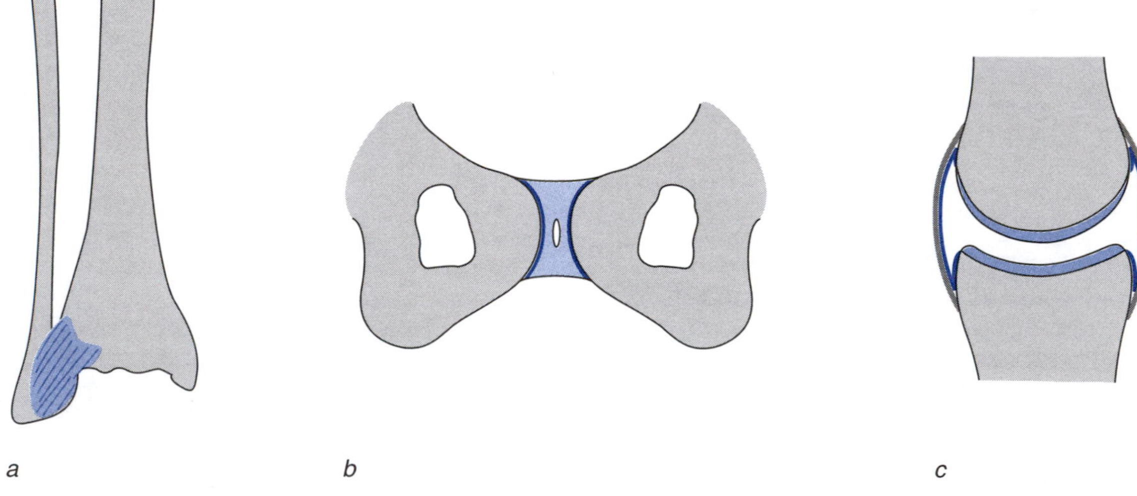


**Figure 7.12** Different kinds of joints. *(a)* Fibrous joints in the lower part of the leg. The lower ends of tibia and fibula are joined by bundles of dense connective tissue in a syndesmosis. In addition, the connective tissue membrane joining the two bones along their length—the interosseous membrane—is also a syndesmosis. *(b)* Cartilaginous joint—a symphysis—between the two pubic bones. Each bone is covered by a layer of hyaline cartilage, with a disk of fibrocartilage filling out the intervening space. *(c)* Synovial joint, the characteristic feature of which is the synovial membrane lining the inner surface of the fibrous capsule.

Such joints are capable of a very limited range of movement. In fibrous joints, the bones are fastened together by intervening fibrous connective tissue. Fibrous joints can be very tight, allowing no appreciable motion. Tight fibrous joints are found, for example, between the bones of the skull (sutures) and between the lower ends of the bones of the leg (syndesmosis).

In cartilaginous joints, the bones are joined by means of cartilage, either as a synchondrosis, where the bones are joined by solid hyaline cartilage, or as a symphysis, where the major component is a disk of fibrocartilage. The most familiar type of symphysis is the pubic symphysis, but the intervertebral disks also are regarded as symphyses.

### For Additional Reading

For further details about joints, see a standard textbook of anatomy, for example, *Gray's Anatomy* (P.L. Williams 1995).

The individual bones of the spinal column—the vertebrae—are joined together by three different kinds of connections: the intervertebral disks, small synovial joints, and a system of long and short ligaments. The intervertebral disks are located between the bodies of neighboring vertebrae, whereas the synovial joints, which are of the plane variety, are found between paired, small articular processes that protrude from the bony arc surrounding the vertebral canal, which contains the spinal cord. Long ligaments running the length of the vertebral column are found on the ventral and dorsal side of the vertebral bodies, respectively, as well as along the tips of the dorsal processes, the spines. The range of movement between neighboring vertebrae is normally rather small but varies between the different regions of the vertebral column, mainly because of differences in the orientation of the synovial joints, the fibrous capsules of which are rather lax, allowing small gliding movements.

The upper and lower aspect of adjacent vertebral bodies that face the intervening intervertebral disk are covered by hyaline cartilage, whereas the disk itself consists of fibrocartilage. In a way, the organization of the intervertebral disk resembles that of a haversian system (see figure 7.1) in having a lamellar structure. The lamellae are oriented parallel to the outer circumference of the disk. Together they form a ring, the **anulus fibrosus,** surrounding a central, soft part known as the **nucleus pulposus.** This soft nucleus can escape through a disrupted anulus fibrosus and compromise a spinal nerve root as it passes through the intervertebral foramen. This is what is commonly known as a disk hernia. The resemblance with a haversian system lies also in the alternate orientation of the collagen fibers in neighboring lamellae, giving the whole structure maximal strength. During lifting and in other situations imposing mechanical stress on the back, torsion of the vertebral column is apt to stress the outer lamellae of the anulus fibrosus more than the more central ones (Farfan 1973). This easily can lead to a progressive weakening and disruption of the integrity of the disk.

## Joint Movements

Movements in a synovial joint can be gliding, rotation, or a combination of the two, and the rotational movement can take place around one or several axes. Pure gliding movements are found in plane joints, whereas rotational movements are found in joints where one of the articulating surfaces is more or less convex and fits into a corresponding concave surface on the other bone. It is usual to speak about joints with one, two, or multiple axes. This is a simplified but nevertheless useful classification. The modified hinge joint at the elbow is an example of a one-axis joint, whereas the ball-and-socket joints at the shoulder and the hip are as examples of multiaxis joints.

When speaking about the mobility of a joint, one has to distinguish between its number of axes (degrees of freedom) and the range of movement in each possible direction. The number of possible movement axes in a joint is determined by the shape of the articulating surfaces and by the ligaments reinforcing the joint capsule. The range of movement possible in each plane depends on several factors such as tightness of the joint capsule or length of the muscles crossing the joint. Especially muscles crossing more than one joint can rather severely restrict movement in a joint, depending on the position of the other joint. The best known example of this is an inability to touch the floor with the fingers while standing with straight knees, because the muscles on the dorsal side of the thigh (the "hamstrings") are too short and restrict the flexion of the hip joint. In such cases, stretching the muscles in question can improve joint mobility. Of course, excess soft tissue, be it muscle or fat, also will restrict movements.

## Innervation of Bones and Joints

Periost, the connective tissue dressing on the outer surface of bones, is richly endowed with sensory and autonomic nerve fibers, and the bone cells have receptors for several neuropeptides present in the nerves (Bjurholm et al. 1992). The pain fibers of the periost are the immediate source of pain accompanying fracture or other kinds of trauma to bones.

Joint innervation is an integrated part of the proprioceptive sense, providing information about joint position and movement. In addition, the pattern of innervation suggests other possible roles. Synovial joints are innervated by the nerves that supply the muscles that act on them. It is reasonable to assume that this arrangement establishes local reflex arcs that ensure stability. The part of the joint capsule that is rendered taut by the contraction of a given muscle or group of muscles is innervated by the nerve or nerves supplying their antagonists. This condition prevents overstretching or tearing of the ligaments. Thus, Skoglund (1956) showed that afferent impulses from the joints affect the activity of the α motoneurons in a reciprocal manner; that is, during passive extension, the extensors are inhibited whereas the flexors are facilitated.

Studies of different joints in humans and animals have shown that nerve fibers to a joint arrive via special branches from motor nerves, periosteal nerves, and often cutaneous nerves (Brodal 1972). The contribution of these various fibers to the overall nerve supply of a joint varies widely from one joint to another. Thus, the anterior aspect of the knee joints is mainly supplied by nerve fibers coming from nerve branches supplying the adjacent muscles, whereas the ankle joint receives relatively few such fibers. Similarly, the total number of nerve fibers supplying a joint can vary within wide limits. Thus, the nerve supply to the knee joint is much greater than to the shoulder joint. The reason for this may be that the muscle mass surrounding the shoulder joint plays a role in the perception of the joint position, thus augmenting or partially replacing the function of the afferent nerve fibers of the joint (Brodal 1972).

In general, there are four types of nerve endings in the joints (figure 7.13). Three of them are nerves ending in specialized end organs, as in skin receptors, often surrounded by a capsule of connective tissue, whereas one of them ends as free-branching nerve endings. Of the former, one type is primarily specialized to give information about changes in joint position (type 1) and a second type to give information about the speed of movement in the joint (type 2). Both types are located in the parts of the joint capsule where bending and stretching take place. The type 3 receptor appears to be particularly adjusted to register the actual position of the joint, and it is located in the ligaments of the joints. The type 4 receptor consists of free-branching nerve endings, which are pain-sensitive fibers similar to those found elsewhere in the body. The synovial membrane has no nerve fibers of its own, as is also true of the joint cartilage.

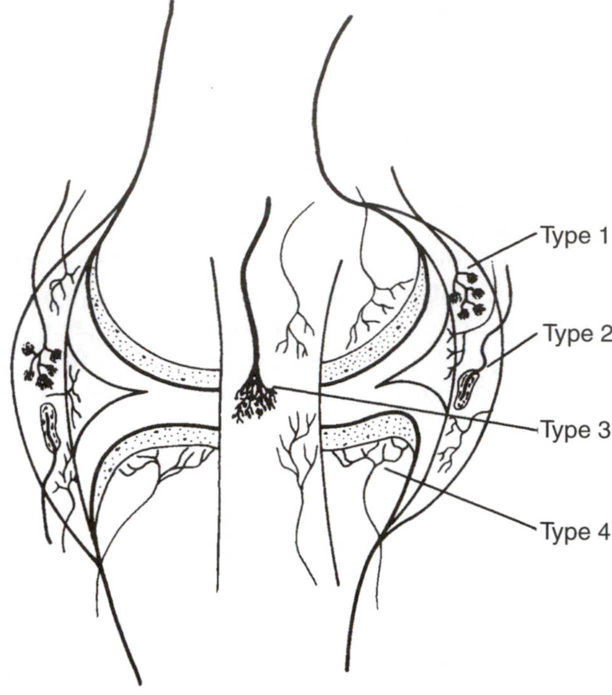


**Figure 7.13** The four different types of nerve endings in a joint.

Courtesy of P. Brodal.

Afferent impulses from the joints reach the somatosensory region of the cortex. For this reason, it has been postulated that joint innervation is of major importance for the conscious perception of joint position and motions, whereas afferent impulses from the muscles, which are not consciously perceived, are significant for the subconscious, reflex control of motor activity (Brodal 1972). Skoglund (1956) provided evidence that the joint receptors are able to provide information pertaining to the resistance against a movement.

## *Summary*

Joint is the general designation of a connection between two bones. Three main types of such connections exist: fibrous joints, cartilaginous joints, and synovial joints. Only the latter corresponds to what is generally perceived as a joint by laypeople. The kinds of movements that are possible in any particular synovial joint depend first and foremost on the geometry of the articular surfaces. The range of movement in each possible direction of movement depends on different factors, such as tightness of the joint capsule or of the muscles crossing the joint.

Joints are richly endowed with sensory nerve fibers. Different kinds of nerve endings are located in the joint capsule and in adjacent ligaments, and they provide information about joint position, movements in the joint, and the speed of such movements. Pain-sensitive free nerve endings are also present.

# Adaptive Capacity of the Skeletal System

The performance of the musculoskeletal system as a whole depends on the combined properties and behavior of its component tissues. Although the mechanical properties of muscle, bone, cartilage, and tendons are different, they share one important property: They adapt to long-term changes in the loads they are subjected to (Fischer 2000; Maffulli and King, 1992). In the following we outline such adaptive changes in bone, cartilage, ligaments, and tendons. The adaptive changes in muscles are covered in chapter 11.

## Bone Adaptation, Growth, and Repair

Bone adaptation, growth, and repair are inseparable from bone remodeling because the rebuilding phase of the remodeling cycles is influenced by the actual mechanical stress imposed on the bone. Remodeling is simply the basis of skeletal adaptation. Scientists have only recently started to unveil the mechanisms by which bone responds to mechanical loading by forming or removing tissue, and these mechanisms seem to be based on the osteocytes as the **mechanosensory cells** of bone (Burger and Klein-Nulend 1999; Manolagas 2000). Irrespective of the signaling pathways leading to bone adaptation, however, the end result depends on the balance between osteoclast and osteoblast activity during the remodeling cycles.

Bone growth is a unique physiological process, consisting of a constant construction and reconstruction that not only continue beyond the point of actual growth and development but also are far more than simply growth in length and width. Imagine, for example, the femur, which not only has to double its length between the age of 2 and adulthood but also maintains its outer shape. Figure 7.14 shows that growth of a long bone is far more than growth in length and breadth.

Following a bone fracture, the repair process starts with the formation of a cufflike mass of bone material around the fractured site, known as **callus**. Initially, the callus consists of unmineralized bone tissue, but subsequently it is mineralized just like during bone formation in general. The bridging of a fracture by new bone occurs in a fashion resembling a fixed-arch bridge (figure 7.15). When mineralized,

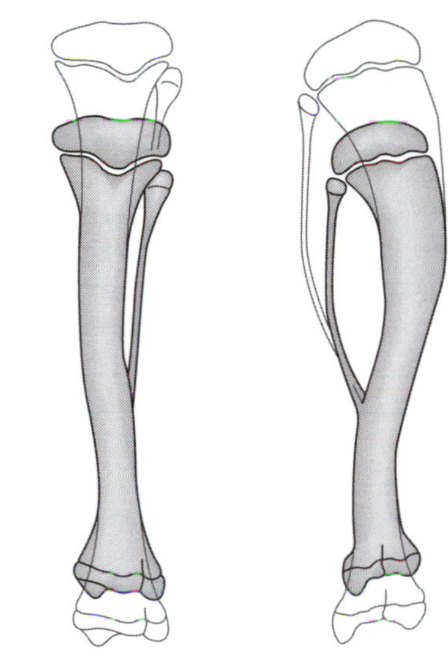

**Figure 7.14** Remodeling of rat tibia and fibula during growth, as seen from the anterior aspect (left) and in profile (right).

Adapted, from W. Bloom and D.W. Fawcett, 1994, *A Textbook of Histology* (London: Chapman and Hall), 223. By permission of Arnold Publishing.

**Figure 7.15** The similarity between the process of fracture healing and the construction of a fixed-arch bridge.

Adapted from McLean and Urist 1955.

the callus is easily seen on x-ray. It also is possible to feel the callus through the skin, or even see it, as in the case of a fractured clavicle. This callus is just a temporary fixation of the two bone fragments. When after some time the interior structure of the bone is restored, the callus gradually disappears. In the case of a fracture of a long bone in a young individual, this process is usually completed within about 3 to 6 months.

## The Role of Mechanical Factors

The fact that prolonged bedrest or immobility causes rapid and marked reduction in bone mineral density has been known since the 17th century (Judex et al. 1997) and was confirmed more recently (Chesnut, 1993). K. Rodahl et al. (1966) showed that the resulting bone loss and accompanying increase in urinary calcium excretion are not attributable to inactivity per se but to the absence of longitudinal pressure on the long bones. In their study, the increased urinary calcium excretion was unaffected by heavy cycle ergometer exercise performed in the supine position for 1 to 4 h daily and by 8 h of inactive sitting in a wheelchair. But 3 h of standing per day in addition to recumbent bedrest caused the urinary calcium excretion to return toward normal values (figure 7.16). When a pressure equivalent to the body weight of the individual confined to horizontal bedrest was exerted along the longitudinal axis of his body by heavy springs fixed to a shoulder harness and the foot of the bed for 3 h · day$^{-1}$, the urinary calcium elimination also returned toward normal values in one of two subjects.

If urinary calcium elimination indicates bone mineralization, it appears that gravitational stress on the long bones is essential for normal bone growth. This is supported by studies of calcium metabolism in astronauts during prolonged space flights. Thus, Whedon et al. (1975), studying mineral metabolism during the United States Skylab flights lasting from 28 to 84 days, observed that urinary calcium excretion increased as in bed rest studies by 80% to 100%. The urinary calcium excre-

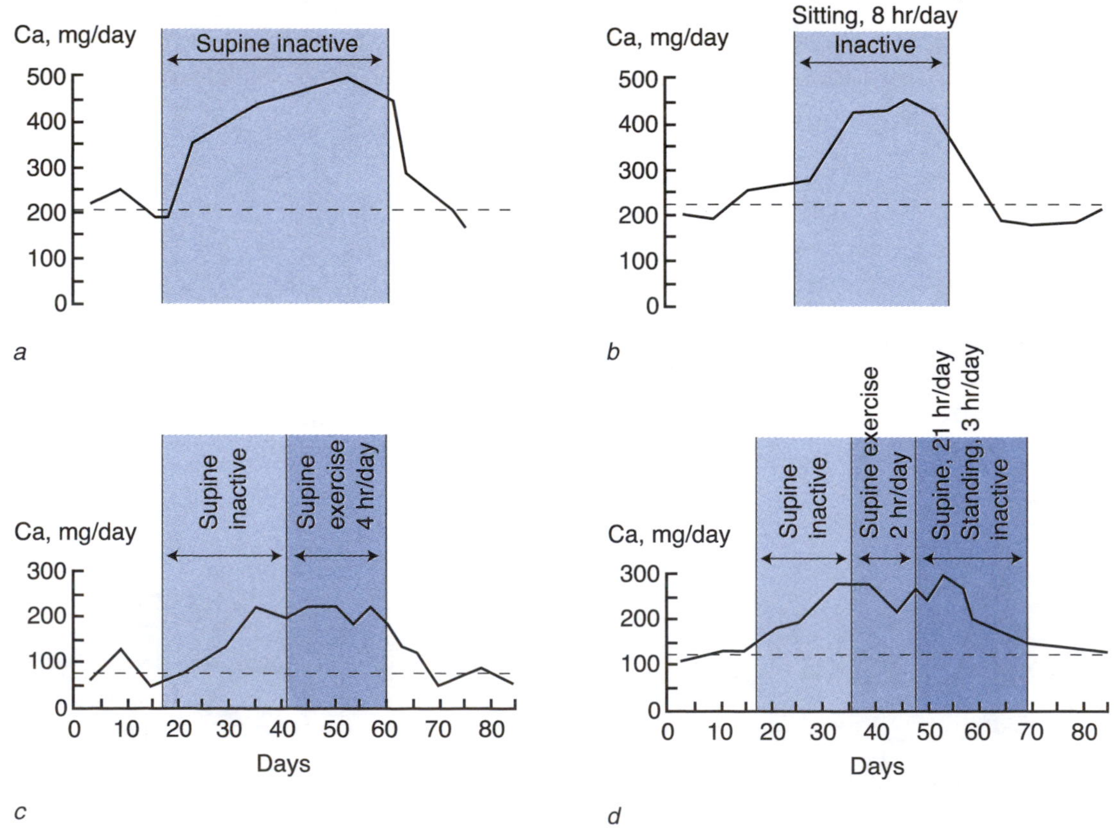


**Figure 7.16** Urinary output of calcium during *(a)* prolonged inactive bedrest; *(c)* bedrest combined with 4 h · day$^{-1}$ supine exercise; *(b)* bedrest combined with 8 h · day$^{-1}$ quiet sitting; and *(d)* bedrest combined with 3 h · day$^{-1}$ standing.
From K. Rodahl et al. 1966.

tion gradually increased during the first 3 weeks of the flight, after which it remained constant at the elevated level for the duration of the space flight, whether it lasted 28 or 84 days. Specific knowledge about the effect of gravity-free environment on the human skeleton is, however, scarce. For discussion, see Turner (2000). On the other hand, many studies have shown that resistance training and weight-bearing exercise positively affect bone mineral density (Bailey, Faulkner, and McKay 1996; Davee, Rosen, and Adler 1990; E.A. Krall and Dawson-Hughes 1994; Layne and Nelson 1999; Marcus et al. 1992; Marken Lichtenbelt et al. 1995; K. Rodahl et al. 1966; Snow-Harter and Marcus 1991).

Cross-sectional studies have shown that athletes, especially those who are strength trained, have greater bone mineral densities than nonathletes (Chilibeck, Sale, and Webber, 1995), whereas aerobic power per se is not consistently associated with bone density (Block et al. 1989). Stewart and Hannan (2000) even found that competitive cyclists can have a mild decrease in bone density in the spine, and they warned against cycling as the only type of physical activity, particularly in young, competitive athletes during the years of bone acquisition. This, together with the lower bone mineral density values found in swimmers, is seen as an additional indication that low bone density is attributable to the lack of weight bearing. In a review, Bailey, Faulkner, and McKay (1996) cited several studies providing evidence for a positive effect of physical activity on bone mineral accrual. They even cited studies showing that high-impact loading offsets, at least to some degree, the deleterious effects of menstrual disturbances in young, female athletes, whereas activities involving lower impact loads, such as running, do not. Evidently, it is through its load-bearing effect on the skeleton that physical activity is the most important influence on bone density and bone architecture (Lanyon 1996), and if physical activity is undertaken during the growing years, it has an impact beyond genetic factors. However, the control of bone mass is localized, and mechanical strain and its effect on bone can be very different from one part of the skeleton to another. It is even possible that net bone loss and net bone gain occur at the same time in two different places in the same bone (Bailey, Faulkner, and McKay 1996).

In a study of men and women between 50 and 73 years of age involved in 1-year fitness classes incorporating high-impact exercise, Welsh and Rutherford (1996) found that such exercise could increase hip bone mineral density in this age group. The fitness classes took place two to three times a week and included step and jumping exercises specifically to load the proximal femur and spine. The increased hip bone mineral density can be mediated by longitudinal weight bearing on the skeleton (Issekutz et al. 1966), which could be important for reducing hip fracture risk.

Montoye et al. (1980) examined 61 male veteran tennis players age 55 and over (mean age, 64), who had played tennis regularly for an average of 40 years. The volume and circumference of the hand and forearm were significantly greater in the dominant playing limb. Bone width and mineral content of the ulna, radius, and humerus were greater in the dominant side. The length, total area, and cortical cross-sectional area of the metacarpal bones were also significantly larger on the dominant side.

The beneficial effect of physical exercise as a prophylactic measure against vertebral bone loss in middle-aged women was emphasized by Krølner et al. (1982). The results of a study by M.K. White and associates (1984) support the hypothesis that mechanical loading caused by exercise might prevent postmenopausal **osteoporosis.** They used photon absorptiometry of the distal radius to compare the effect of walking and aerobic dancing on the bones in 73 recently postmenopausal women with a control group who did not exercise. The period of observation was 6 months. The control group and the walking group lost statistically significant amounts of bone mineral, but the dancing group did not. The dancing and walking groups both showed a significant increase in bone width, whereas the control group did not. Plasma estrogen levels were not influenced by the exercise.

Over the years, models have been developed to explain the relationship between mechanical loading and changes in bone mass and the way mechanical loading is translated into bone cell behavior. According to Frost's mechanostat theory (Frost 1987), mechanical usage windows can be defined as shown in figure 7.17. According to this model, there is a physiological loading zone where the load is sufficient to prevent bone loss but not to increase bone mass. To do that, the load must increase to values in the overload zone, preferably rather by high-impact loading of short duration than by more moderate loading of longer duration (Bailey, Faulkner, and McKay, 1996). This type of model leaves us, however, with at least two questions: What—in molecular terms—is this bone acquisition signal, and how is the signal translated into the correct arrangement of bone trabeculae? Huiskes et al. (2000) focused on the role of osteocytes, postulating that they react to loading in their local environments by producing a biochemical messenger in proportion to the strain imposed on them.

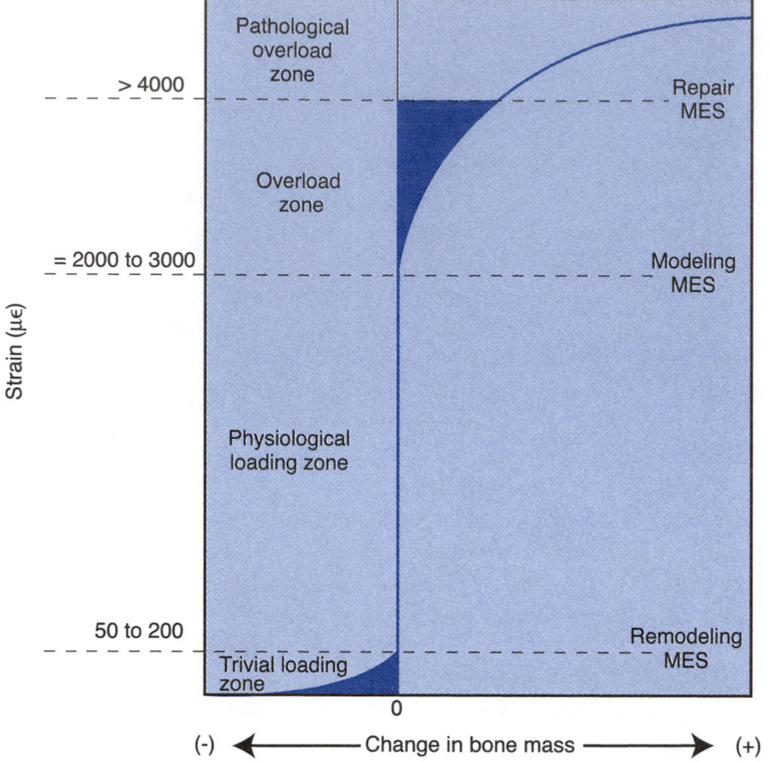


**Figure 7.17** Mechanical usage windows, as defined by Frost's mechanostat theory (Frost 1987). In the trivial loading zone, strains are low, resulting in increased remodeling and subsequent loss of bone, as in immobilization or disuse. In the physiological loading zone, strains are sufficient to maintain bone, remodeling is in a steady state, and bone is neither gained nor lost. In the overload zone, modeling is stimulated to add more bone and to organize it to respond to high strains. At very high strains, bone enters a repair mode, in which unorganized bone is added to meet a severe and immediate need. με = microstrain; MES = minimum effective strain.

Reprinted, by permission, from D.A. Bailey, R.A. Faulkner, H.A. McKay, 1996, "Growth, physical activity and bone mineral acquisition," *Exercise and Sport Sciences Reviews* 24:233-266.

### For Additional Reading

For a review of the life and death of bone cells, see Manolagas (2000).

The nature of this signal is unknown, but the vasoconstrictor peptide endothelin (Alam et al. 1992), prostaglandins (M. Li, Jee, et al. 1995), and nitric oxide (Zaman et al. 1999) all have been implicated in mediating the mechanical loading effect of bone, possibly released by endothelial cells because of changes in the fluid flow within bone during loading (Judex et al. 1997). For further discussion, see Burger and Klein-Nulend (1999).

It is thus clear that bone is strengthened by subjecting it to increased pressure. This process may be slower than the increase in muscle strength resulting from a program of weightlifting. In such training, the training intensity or the increase in training load should be sufficiently gradual to allow development of skeletal strength to keep pace with the increase in muscle strength. It is also noteworthy that bone mineral accrual lags behind longitudinal skeletal growth, resulting in a transient period of relative bone weakness and increased fracture risk during adolescence (Bailey, Faulkner, and McKay 1996; Blimkie et al. 1993).

## The Role of Hormones and the Importance of the Calcium Balance

Classical hormones are important signal molecules in the regulation of bone mineral density, but the relative importance of the hormones in question varies from one stage of life to another (Bailey, Faulkner, and McKay 1996). The prepubertal increase in bone mineral content seems to be largely attributable to the growth hormone, whereas sex hormones (especially estrogens) enter the picture during puberty and remain important throughout life, either through their inhibitory effect on osteoclast activity or (as in postmenopausal women) because of their absence. The effect of sex hormones on bone metabolism may be mediated, at least in part, by cytokines (Lorentzon, Lorentzon, and Nordstrom 2000). Other signal molecules involved in the control of bone mineral content include thyroid hormones and **insulin-like growth factor (IGF)**-1.

Calcium is a truly ubiquitous ion, serving purposes as diverse as mineralization of bone tissue and intracellular signaling, the most widely known example of which is the initiation of muscular contraction. In addition to providing the internal framework for the human body, therefore, bone tissue also serves as a readily mobilized source of calcium ions. Isotope studies using $^{32}P$ or $^{45}Ca$ indicate that more than 20% of the calcium and phosphate ions of bone are involved in a fairly rapid exchange in adults (LeBlond and Greulich 1956).

Calcium balance is negative during space flight, and increased bone resorption may occur (Zernicke et al. 1990). All changes in function of a bone are attended by alterations in its internal structure (Wolff's law). Because pressure will stimulate the appositional bone growth, increased weight bearing will increase the thickness of the bone and density of the shaft (Lanyon and O'Connor 1980). Atkinson, Weatherell, and Weidmann (1962) examined the density of the femur in physically active and sedentary subjects after 50 years of age. The bone density was slightly higher in the active group. Eisenberg and Gordan (1961) found, by measuring the dynamics of growth in the human with nonradioactive strontium, that muscular exercise accelerated the rate of bone deposition. Conversely, elimination of the "normal" effects of stress and strain led to a loss of bone tissue.

On the other side of the coin is the importance of maintaining calcium homeostasis in the blood and other extracellular fluids. In a context of whole-body biology, this part of calcium homeostasis has priority over bone mineralization and is taken care of mainly by PTH and the vitamin $D_3$ hormone.

In a review of the literature on osteoporosis, Marcus et al. (1992) discussed the possible effects of sex, age, diet, physical activity, and differences in hormonal function. They pointed out that young amenorrheic athletes with estradiol levels similar to those of postmenopausal women have significantly lower vertebral bone density than comparable athletes with normal menstrual cycles, and they questioned the effectiveness of exercise preventing osteoporosis in the absence of adequate estrogen stimulation. As pointed out earlier, however, this is a matter of discussion and may depend on the type of physical activity chosen (see Bailey, Faulkner, and McKay 1996).

There is reason to emphasize that the need for calcium may exceed even an optimal intake during the adolescent period of intense bone acquisition (Bailey, Faulkner, and McKay 1996). In this context it should be a matter of concern that teenagers often prefer soft drinks and junk food instead of dairy products, which are high in dietary calcium. Individuals working with young, especially female, athletes should make every attempt to ensure that their calcium intake is sufficient (P.W.R. Lemon 2000).

## Adaptation in Cartilage

Cartilage is not among the tissues most widely known for their ability to adapt to mechanical stimuli, but it too has the ability both to adapt to increased loading and, to a limited extent, to repair itself after injury (Fischer 2000).

### For Additional Reading

For an update on tissue culture and gene transfer techniques in cartilage repair at the turn of the century, see Yoo et al. (2000), Wroble (2000), and Mason et al. (2000).

It is common knowledge that normal cartilage structure and properties are optimized for load-bearing function, and it has become increasingly evident that mechanical forces have great influence on the synthesis and rate of turnover of articular cartilage molecules such as proteoglycans (Arokoski et al. 2000).

In fact, mechanical forces seem to be more important than motion, because movements in the absence of loading result in atrophic changes in the cartilage (Palmoski and Brandt 1981), but important differences exist. Regular cyclic loading of a joint increases proteoglycan synthesis and makes the cartilage stiffer, whereas continuous loading diminishes proteoglycan synthesis and damages the tissue through necrosis (Arokoski et al. 2000).

The finding of a beneficial effect of cyclic activity is in accordance with earlier reports. Holmdahl and Ingelmark (1948) showed that in trained animals, the articular cartilage was thicker than in untrained ones. The active animals had an increase in both the cellular and extracellular components of the cartilage. Animals living in small cages that restricted their opportunity to move around had, in general, thinner cartilages of the knee joints than animals provided with more spacious cages.

Articular cartilage also can respond to loading in a more acute way. Figure 7.18 illustrates how the articular cartilage in the knee joint can vary in thickness in a matter of minutes, depending on whether the animal is active or inactive. After 10 min of running, the increase in thickness was 12% to 13% compared with the thickness after 60 min of immobilization (Ingelmark and Ekholm 1948). Similar results have been obtained in humans. The explanation given for the rapid increase in thickness of the cartilage of an activated joint is that fluid seeps into the cartilage from the surroundings, when the cartilage is alternately compressed and decompressed.

One consequence of the increased fluid content of the cartilage is a change in its compressibility, diminishing the incongruence between the articulating surfaces. This change will increase the area of the contact surface of the joint in question and reduce the pressure per unit of area of the articular surface during compression with a given force (Ingelmark and Ekholm 1948). Warm-up activities before vigorous exercise therefore should include joints that may be stressed during the activity, to reduce their susceptibility to trauma.

Another advantage of regular dynamic motion of the joints is that the associated increase in fluid supply to the articular cartilage also will provide nutrients to the cartilage. It is a long-established clinical observation that a joint suffers from nutritional disturbances when kept inactive. It is not

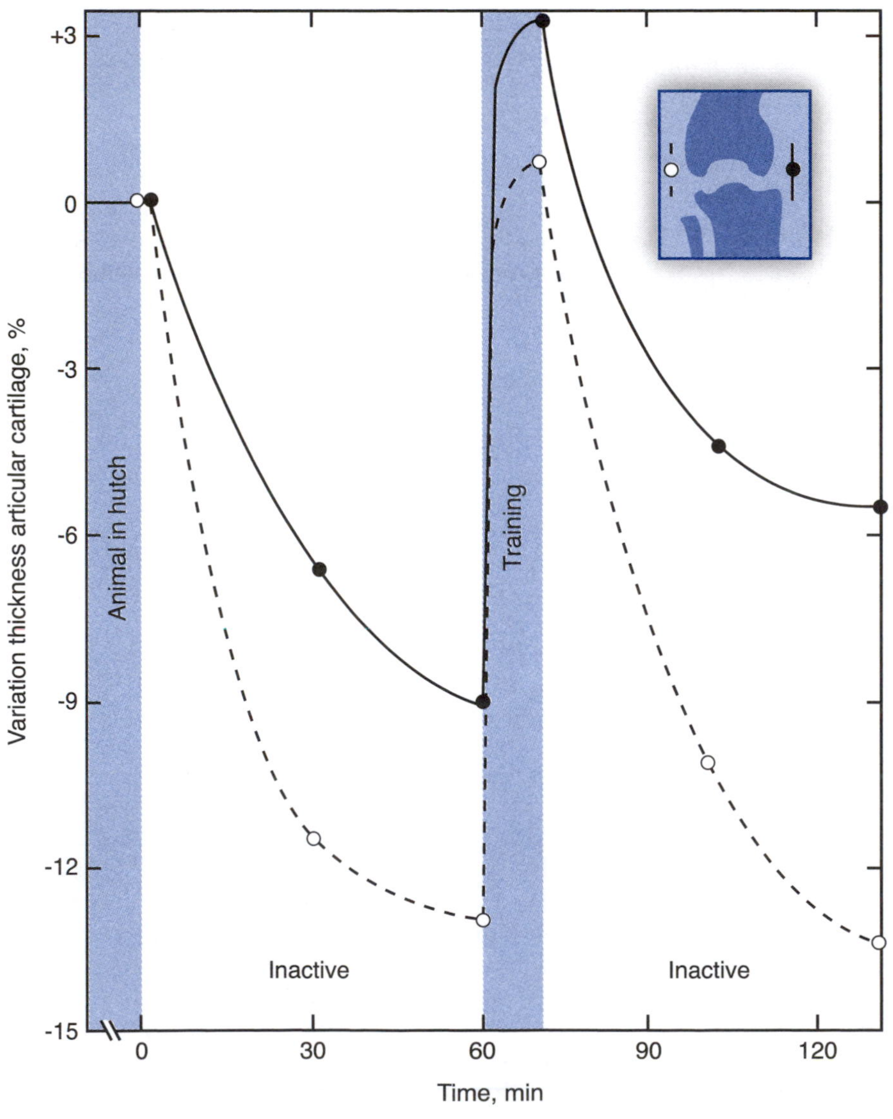


**Figure 7.18** Rabbits were taken from their spacious hutches where they could run freely. At 0 min, x-rays were taken to measure the thickness of the articular cartilage of lateral (○) and medial (●) parts of the knee joint. Measurements were repeated after 30 and 60 min of rest, during which the articular cartilage was unloaded. Then the animals were exercised on a motor-driven treadmill for 10 min at a speed of 40 m · min$^{-1}$. Measurements were repeated after exercise and a further period of rest. The figures represent the mean of about 50 animals.

Adapted from Ingelmark and Ekholm 1948.

known, however, how frequently such activities should be performed to provide optimal nutrition for the articular cartilage.

It is an obvious disadvantage that most studies of the adaptive capacity of cartilage are based on animals, especially young animals. Apart from the difficulty inherent in extrapolating across species, one cannot exclude a beneficial effect of growth factors still high in young animals but less so in humans beyond their 20s. Cadaver studies show, however, a correlation between adult cartilage thickness and presumed mechanical stress, but this might reflect a developmental trait (Fischer 2000).

A high prevalence of degenerative joint disease—**osteoarthritis**—in certain families may indicate hereditary differences in cartilage adaptability and resistance to wear and tear, but other hereditary factors, like joint malalignment or other variations in joint anatomy, also can contribute to such development.

## Adaptation in Ligaments and Tendons

Basically, ligaments and tendons are built in the same way, with bundles of more or less parallel collagen fibers in a sparse proteoglycan and glycosaminoglycan matrix. Morphological and biochemical evidence indicates, however, that ligaments are metabolically more active than tendons. This may mean that ligaments have a greater potential for adaptation than tendons (Fischer 2000).

### For Additional Reading

For a more detailed account of tendon anatomy and physiology, see Józsa and Kannus (1997). An overview of muscle and tendon interaction during human movements is offered by Fukunaga et al. (2002).

As in most other areas of biology, the major part of our knowledge about adaptation in connective tissues derives from animal experiments, and as usual, a certain degree of caution is warranted when extrapolating to humans. There are, however, results from human experiments, especially from Fukunaga's group, using real-time ultrasonography. Thus, Kubo, Kanehisa, Kawakami, and Fukunaga (2001) showed that repetitive muscle contractions caused the tendon structures to be more compliant, and that the changes in elasticity depend more on the duration of the action than on action mode or force level. On the other hand, Kubo, Kanehisa, Ito, and Fukunaga (2001) showed that 12 weeks of isometric training increased tendon stiffness, thus increasing the rate of torque development and shortening the electromechanical delay. The mechanisms leading to these changes are so far unknown, but Langberg et al. (2002) have shown that substantial amounts of the pleiotropic cytokine interleukin (IL)-6 are released from peritendinous tissue during prolonged exercise in humans, but indications of its possible effects on tendon tissue are so far lacking.

When speaking about adaptation in tendons, it is appropriate to take a systemic approach, that is, to consider the muscle–tendon unit as a whole. Knowing that no chain is stronger than its weakest link, this means that more or less concerted adaptations must take place in all components of the muscle–tendon unit. However, if the metabolic activity of its tissues indicate their adaptive capacity, then the passive components of the unit (the tendon and the myotendinous and osteotendinous junctions) are less able to adapt quickly to new demands than the muscle. For this reason, any increase in the load imposed on a muscle–tendon unit should be gradual to avoid imbalance in the tensile strength of its components with overuse injuries as a result. However, the adaptation in the passive components can be considerable if extended over sufficiently long time (Józsa and Kannus 1997). The metabolic activity in tendon tissue and the turnover of collagen have been shown to increase after exercise. Tendon diameter can increase, and it may decrease with aging and inactivity. It is a matter of concern, however, that an apparent weakening of tendinous structures has been reported in the initial phases of increased tensile loads (Archambault, Wiley, and Bray 1995; Fischer 2000), more so if the training is too vigorous than if it is more gradual (Józsa and Kannus 1997).

Very little is known about the effects of remobilization on musculoskeletal structures after injury, but as shown in figure 7.19, remobilization and rehabilitation of these structures require far more time than the time needed to cause the disuse atrophy (Józsa and Kannus 1997). Again, most experimental evidence is from animal studies. In wild primates, Noyes et al. (1974) showed that 8 weeks of immobilization significantly impaired the functional capacity of ligaments. After 20 weeks of resumed activity following immobilization, ligament strength recovered only partially.

Quite dramatic changes have been described in myotendinous junctions (MTJs) after immobilization. As little as 11 days of immobilization in a rat model decreased the breaking strength of MTJs by almost 50% (Tidball, 1991b). Changes in MTJs after immobilization of human muscle include reduced

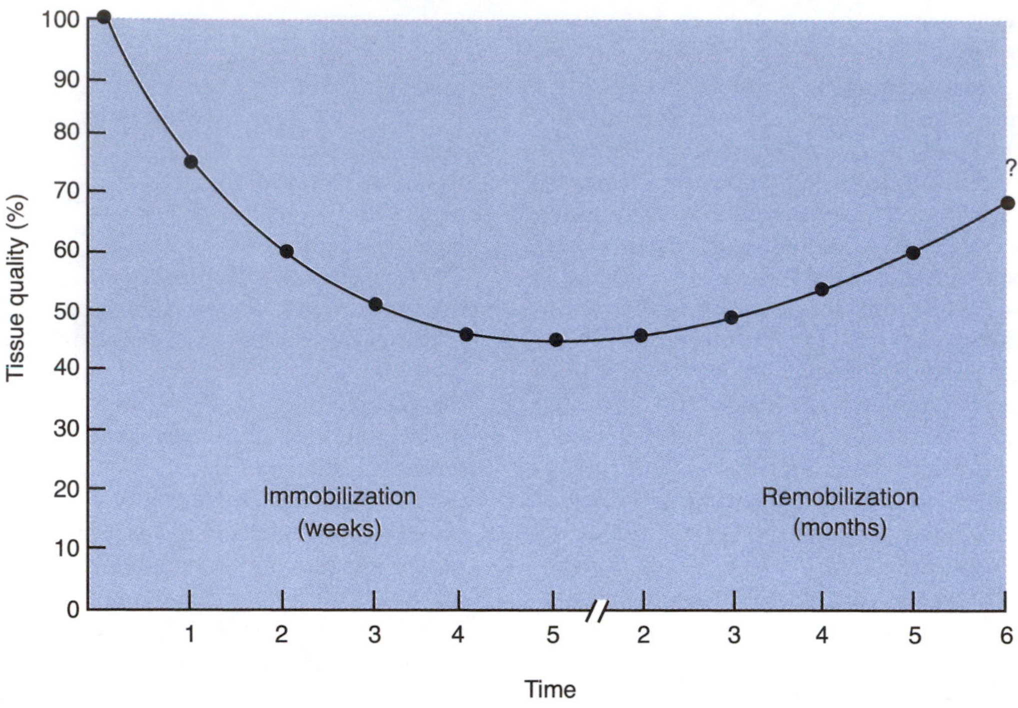


**Figure 7.19** The effects of immobilization and remobilization on the quality of musculoskeletal tissues. The theoretical and practical limits of remobilization are not known.

Reprinted, by permission, from L. Józsa and P. Kannus, 1997, *Human tendons: Anatomy, physiology and pathology* (Champaign, IL: Human Kinetics), 140.

membrane folding and increased angle of loading; an increased amount of type III collagen, which is weaker than type II collagen usually found at these sites; and a decreased amount of sulfated glycosaminoglycans, which serve as a glue (Józsa and Kannus 1997). This seems to be in accordance with the frequent finding of MTJs as a primary site of muscle tears (Tidball 1991b).

Tipton et al. (1975) showed that the mechanical stress produced by regular, prolonged exercise or training increases the strength of the junctions between ligaments or tendons and bones as well as the strength of ligaments that have been repaired. These researchers also found that the strength of the junction is closely related to both the duration of training and the type of exercise. Interestingly, the microscopic morphology of ligaments and tendons changes as it approaches the osseous insertion. Its cells become more rounded, and the tissue changes from dense connective tissue to fibrocartilage. The area with the roundest cells corresponds to the area with the highest compressive stress (Fischer 2000), a type of local environment known to promote development of fibrocartilage in tendons (Giori, Beaupre, and Carter 1993).

Contemporary exercise biology always looks for molecular signals that translate mechanical or other factors into changes in cell phenotype. As to the adaptive processes in ligaments and tendons, very little is known. It is known, however, that exercise changes the levels of IGF-1 in tendon fibroblasts (H.A. Hansson et al. 1988) and that IGF-1 stimulates collagen synthesis and cell replication (Fischer 2000). Thus, IGF-1 may emerge as a common extracellular signal in the muscle–tendon unit, acting in an autocrine or paracrine way. In tendons or regions of tendons subjected to increased compressive loading, an increased synthesis of **transforming growth factor (TGF)-β** has been reported to be a likely cause of the formation of fibrocartilage (Kjær et al. 2000). Because TGF-β possibly mediates chronic inflammation, this may provide clues about some of the overuse afflictions of the Achilles tendon, a tendon often subjected to compressive forces by footwear (Kjær et al. 2000).

## Summary

Bone remodeling is the basis for skeletal adaptation, influenced by the mechanical stress imposed on the bone. The osteocytes have emerged as the putative mechanosensory cells of bone, but the molecular signals connecting this sensory information to the activity of bone remodeling units are only starting to be revealed. It has been proven beyond doubt, however, that resistance training and weight-bearing exercise positively affect bone mineral density.

Even cartilage and connective tissue proper are, to a certain extent, able to adapt to new mechanical demands, but regarding adaptation, it is advisable to regard the muscle–tendon unit as a whole. It is noteworthy that the adaptive capacity of the tendon, including its myotendinous and osteotendinous junctions, is lower and slower than that of the muscle itself, raising the possibility of overuse injuries.

## Strategies to Reduce the Prevalence of Osteoporosis

Because the degree of osteoporosis in an aged individual depends on the amount and organization of his or her remaining bone tissue, the logical strategy to reduce the prevalence of osteoporosis is to increase the level of bone formation, especially in early years, and to reduce the bone loss, especially in later years (figure 7.20). As always, hereditary factors are important, but they are less easily modifiable than lifestyle factors.

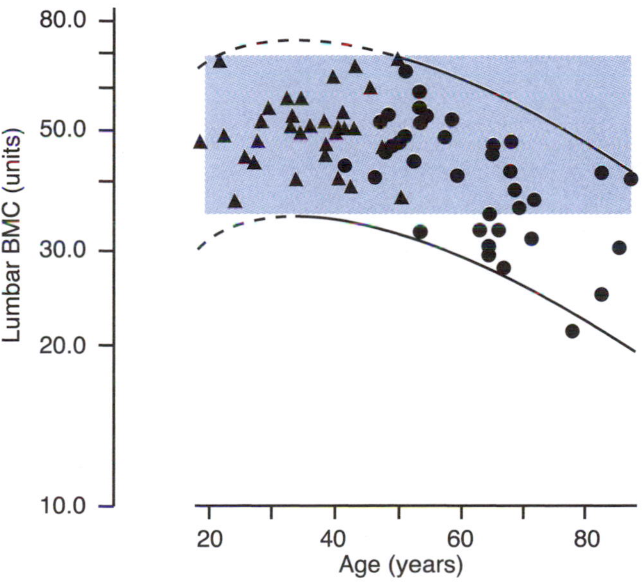


**Figure 7.20** Lumbar spine bone mineral content (BMC) in 70 normal women of different ages. ▲ = premenopausal women; ● = postmenopausal women. Significant bone loss does not occur until after the age of 50.

Adapted, by permission, from C. Snow-Harter and R. Marcus, 2000, "Exercise, bone mineral density, and osteoporosis," *Exercise and Sport Sciences Reviews* 19:351-388.

Although many lifestyle factors are known to influence bone mass throughout adult life, the single most important modifiable factor that determines bone mass in senior years seems to be bone mass acquisition during adolescence. The good thing about this is that a healthy lifestyle can promote bone formation during this period of life: a high, but not too high, level of physical activity, preferably of the weight-bearing type (Lehtonen-Veromaa et al. 2000), adequate nutrition, and no tobacco use (Bailey, Faulkner, and McKay 1996; Krall and Dawson-Hughes 1999). The bad thing about this advice is that the bone-related result of the adolescent lifestyle lies too many decades ahead to promote a change in a rather hedonistic behavior. Nevertheless, a substantial amount of scientific evidence supports the advice given here, and furthermore, these recommendations are, of course, valid for people of all ages.

Last, hormone replacement therapy and other types of pharmacological interventions are effective strategies to reduce bone loss in postmenopausal women. As more specific estrogen agonists, without or with risk of unwanted side effects, become available, this may become the pharmacological treatment of choice.

### *Summary*

The degree of osteoporosis in an aged individual depends on the amount and organization of the remaining bone tissue. The logical strategy to reduce the prevalence of osteoporosis is to increase bone acquisition in early years and reduce bone loss later in life. Fortunately, one can obtain this by having a healthy lifestyle with sufficient physical activity and no tobacco use. Unfortunately, however, there is a significant time lag between cause and effect, making a young person less likely to change lifestyle to prevent osteoporosis many decades ahead. Both hormone replacement therapy and other types of pharmacological intervention are available strategies to reduce bone loss in later years.

# Chapter 8

# Physical Performance

Athletic competition represents the classic test of physical fitness or performance capacity. Competitive performance can be measured objectively in centimeters or seconds, or it can be judged subjectively, as in gymnastics, figure skating, or diving. The individual's performance is the combined result of the coordinated exertion and integration of a variety of functions. The purpose of this chapter is to discuss the importance of some of these functions.

## Demand Versus Capability

The demands of the event must be perfectly matched by the individual's capabilities to achieve top performance and championship. It is impossible to present one formula that takes into account all aspects of a person's maximal performance, because the demands set by different types of activities vary greatly. However, the following factors can serve as a frame of reference for our discussion.

*Natural endowment* (genetic factors) undoubtedly plays a major role in a person's performance capacity, at least for those aspiring to the levels required for the attainment of Olympic medals. The individual's response to training also is associated with an endowed **genotype.** Thus, it appears that up to 70% of an individual's maximal force, power, or capacity is a matter of genetics (Bouchard and Malina 1983). Because the possible genetic combinations are astronomical in number, it is an interesting question whether a country must have a population of 100,000, 1 million, 10 million, or more, to "breed" an individual with proper endowment for top results. The more popular an event, the greater is the chance that an individual with the suitable constitution will participate and thus discover his or her ability. Obviously, the environment and geographic location are also important. If an individual with the perfect endowment for skiing grows up in a place where skiing is impossible, this endowment is wasted from an athletic standpoint. The fact that an increasingly large number of naturally endowed persons enter the ranks of competitive athletes has probably contributed to the gradual improvement of athletic records.

Granted the endowment, however, definite improvement in performance is achieved by training, and all factors listed in figure 8.1 as contributing to physical performance capacity can be modified. The very intense training programs used in many fields of athletic performance contribute greatly to the improved results. Another factor explaining the gradual improvement in athletic achievement over the years is the better techniques applied and the superior equipment that is becoming available through technical progress.

Athletes are mainly concerned with improving their ability to cut off seconds or add centimeters to their records. The scientist is interested in analyzing why the results improve or vary from time to time. Therefore, the scientific objective is (1) to evaluate quantitatively the influence of the various factors on the performance capacity in different tasks (performance requirements); (2) to examine how these factors vary with sex, age, and body size (capacity profile); and (3) to study the effect of such factors as training and environment. It is realistic to conclude that scientists have merely begun a systematic research on the performance capacity and the many factors involved. The most advanced information concerns the energy output by aerobic

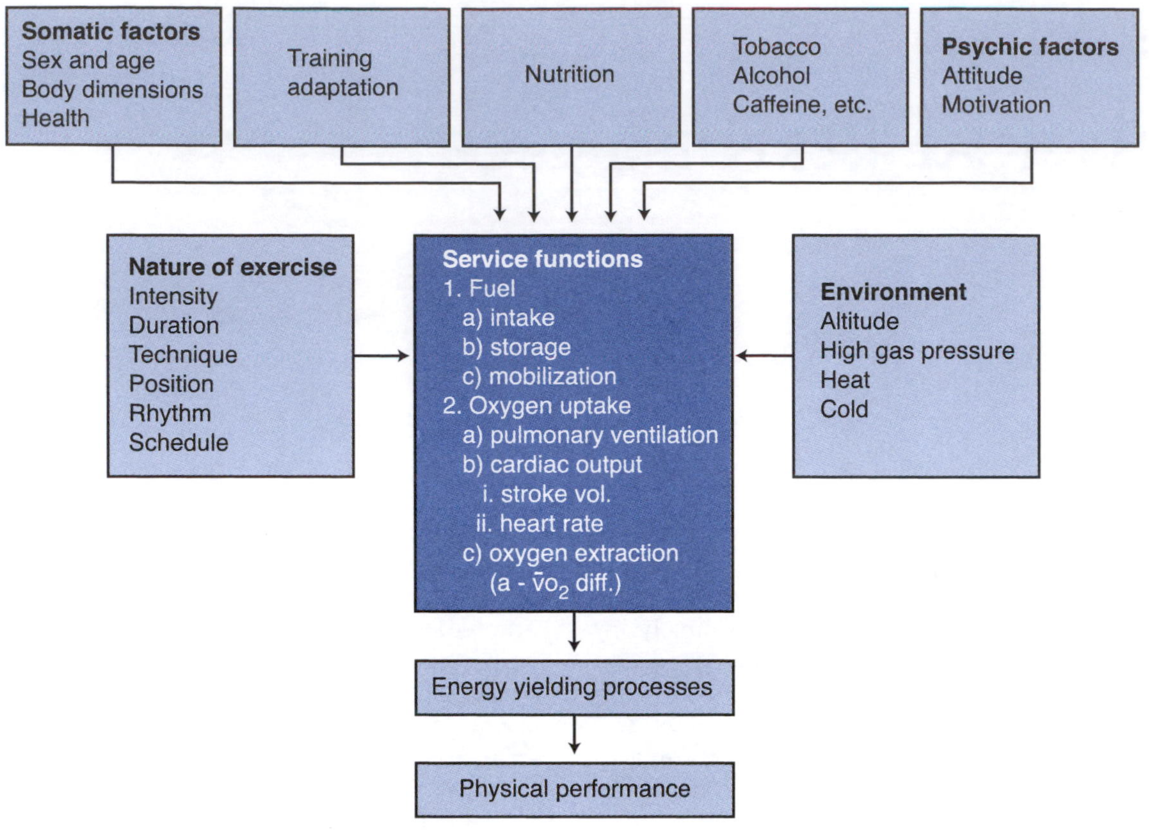


**Figure 8.1** Factors influencing physical performance ability.

processes. This can be explained by the fact that methods for quantitative measurements of energy output by the human combustion engine have long been available, ever since Lavoisier demonstrated in 1775 that animal life is a process of oxidation.

We therefore begin the more detailed analysis of physical performance with a discussion of the oxygen uptake during submaximal and maximal exercise and the **maximal aerobic power** (the individual's maximal oxygen uptake).

**Capacity denotes total energy available, and power means energy per unit of time.**

## Aerobic Processes

For each liter of oxygen consumed, about 20 kJ (range 19.7–21.2 kJ; or 5 kcal, range 4.7–5.05 kcal) will be delivered; hence, the higher the oxygen uptake, the higher the aerobic energy output. The oxygen uptake during exercise can be measured with an accuracy of ± 0.04 $L \cdot min^{-1}$ ($\dot{V}O_2$ >1 $L \cdot min^{-1}$). Figure

### Classic Study

Antoine L. Lavoisier (1743–94) demonstrated in 1775 that animal life is a process of oxidation. In the course of extensive studies that overthrew the phlogiston theory, Lavoisier showed that respiration involves only a portion of the air, the oxygen, and that the remainder, the nitrogen, is unaltered. In 1780, Lavoisier and Laplace established the quantitative relationship between respiration and animal heat production, and in 1789, Lavoisier and Seguin correlated physical activity and respiratory activity as estimated by oxygen uptake and carbonic acid output. This was the beginning of quantitative measurements of energy output by the human combustion engine.

Lavoisier 1775.

8.2 gives examples of how the classic, but today less used, **Douglas bag** method can be applied when studying the aerobic energy output during exercise. Today's equipment allows for breath-to-breath analysis from the first second.

## Intensity and Duration of Exercise

Figure 8.3 shows how oxygen uptake increases during the first minutes of exercise to a steady state, where the oxygen uptake corresponds to the demands of the tissues. When the exercise stops, the oxygen uptake gradually decreases to the resting level, and the so-called oxygen debt is paid off.

The slow increase in oxygen uptake at the beginning of exercise is explained by the sluggish adjustment of respiration and circulation, that is, the sluggish adjustment of the oxygen-transporting systems to exercise. Consequently, during the first 2 to 3 min there is an oxygen deficit. The attainment of the steady state coincides roughly with the adaptation of cardiac output, heart rate, and pulmonary

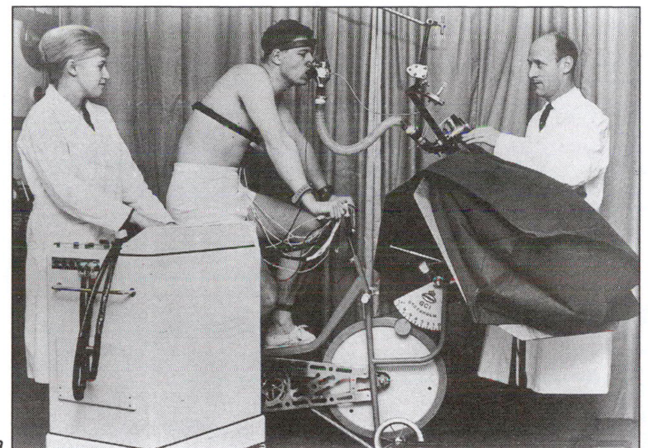
a

b

c

**Figure 8.2** Application of the classic Douglas bag method for measuring aerobic energy output during different types of exercise. The skier shown in *(c)* carries a three-way stopcock and a stopwatch on his chest to record the time during which the expired air is collected in the Douglas bag. The stopwatch automatically starts and stops when the stopcock is turned.

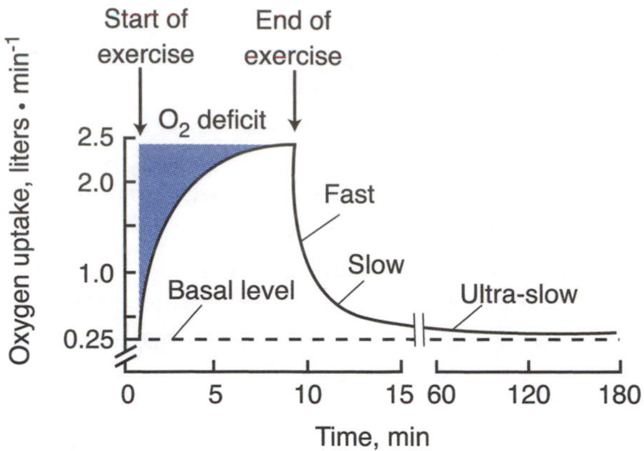


**Figure 8.3** During the first minutes of exercise there is an oxygen deficiency while the oxygen uptake increases to a level adequate to meet the oxygen demand of the tissue. At the cessation of exercise, oxygen uptake gradually decreases and several components can be identified. Note the change in time scale.

Modified from Newsholme and Leech 1984.

ventilation. A steady-state condition denotes a work situation where oxygen uptake equals the oxygen requirement of the tissues. Consequently, lactic acid does not accumulate in the body. At steady state, heart rate, cardiac output, and pulmonary ventilation have attained fairly constant levels.

Even in the steady-state situation, however, heart rate, cardiac output, and pulmonary ventilation can increase gradually, which can be a consequence of an increasing body temperature and an increase in the use of free fatty acids as a substrate. During prolonged exercise, the body's stores of energy and water decrease if not replaced. However, everything considered, the steady-state concept as defined here is very useful when discussing the energy-yielding processes during exercise.

In light exercise, the energy output during the first minutes of exercise can be delivered aerobically, because oxygen is stored in the muscles bound to myoglobin and in the blood perfusing the muscles. During more severe exercise, anaerobic processes must supply part of the energy during the early phase of exercise.

The breakdown of energy-rich phosphate compounds, adenosine triphosphate (ATP) and **phosphocreatine (PC)**, is certainly anaerobic and essential, but their quantitative role is limited. Therefore, the anaerobic breakdown of glycogen (**glycogenolysis**) and glucose (**glycolysis**) to lactic acid has an important potential to support the aerobic processes when they cannot provide enough energy for ATP production. In exercise that engages large muscle groups (e.g., running, which demands an oxygen uptake higher than 50% of the individual's maximum) and that is performed for some minutes, lactate produced in the activated muscles escapes and appears in the blood. The well-trained person can exercise at a higher intensity without an increase in blood lactate concentration than an untrained individual. The heavier the exercise, the more important the anaerobic energy yield and the higher the muscle and blood lactate concentrations. As the exercise becomes more strenuous, a decrease in the body's pH affects muscular tissues, respiration, and other functions.

From a methodological viewpoint, maximal oxygen uptake is attained at rates of exercise that are not necessarily maximal. Thus, an all-out test is not necessary to assess an individual's maximal aerobic power. On the condition that large muscle groups are involved in the exercise and that the exercise time is 4 to 5 minutes, three main criteria are used to show that the subject's maximal aerobic power is reached. First, there is no further increase in oxygen uptake despite further increase in the rate of exercise, and second, postexercise blood lactate concentration exceeds 8 to 9 mM. (In children and old subjects, it can be difficult to attain such high values, however.) The third criterion is that the respiratory exchange ratio, or respiratory quotient (R) should be above 1.15. (The R value has been shown to vary with the subjects' training status and age.) Thus, subjects should not be expected to meet all the criteria on any single test (Howley, Bassett, and Welch 1995).

Most individuals attain a slightly higher oxygen uptake (on average 6%) in a maximal running test up a grade compared with cycling, indicating that the type of exercise influences the values obtained. Therefore, one must be careful when evaluating "maximal oxygen uptake" data.

The heavier the rate of exercise, the steeper is the increase in oxygen uptake (and heart rate). This is illustrated by figure 8.4. After a 10-min period of exercise at 50% of maximal oxygen uptake, intensities of 300 to 450 W were applied until exhaustion. The tolerated exercise time varied from 6 min (300 W) to less than 2 min (heaviest load). The oxygen uptake at the end was the same in all experiments, about 4.1 L · $min^{-1}$. However, after 1 min of extremely heavy exercise, the oxygen uptake was 4.0 L · $min^{-1}$ at the "supramaximal" intensity but only 3.0 L · $min^{-1}$ during the less extreme but still heavy

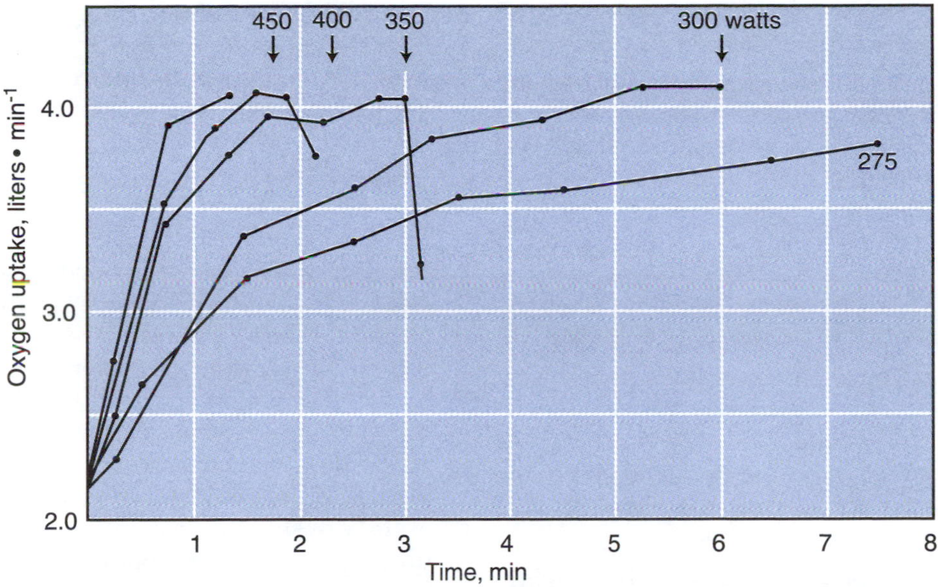


**Figure 8.4** Curves showing increase in oxygen uptake during heavy exercise following a 10 min warm-up period. Arrows indicate time when the subject had to stop because of exhaustion. Figures indicate workload on the cycle ergometer. The subject could continue the power of 275 W (1,650 kpm · $min^{-1}$) for more than 8 min.

From P.-O. Åstrand and Saltin 1961b.

work rate of 300 W, which could not be tolerated for more than 6 min (Åstrand and Saltin 1961b).

The kinetics of the increase in oxygen uptake during the first minutes of an exercise that lead to a steady-state situation have a time constant of about 30 s. If a given moderate exercise is preceded by a warm-up period, this value is reduced to about 20 s. The kinetics of the oxygen uptake at the beginning of exercise and during recovery were discussed by di Prampero (1981) and di Prampero et al. (1983).

## Recovery

It may take 60 min or more before the oxygen uptake and rate of aerobic metabolism return to the preexercise level. After relatively heavy exercise, one may identify three phases:

1. First is a fast exponential component in the decline in oxygen uptake with a half-time of about 30 s. Most likely this rapid phase is associated with aerobic replenishment of the ATP and PC stores and a refilling of the oxygen stores (myoglobin and hemoglobin). During the first minutes of recovery, there is a rapid resynthesis of the energy-rich phosphates. There is a close relationship between the reduction in ATP and PC on the one hand and both oxygen deficit and the fast component of the "oxygen debt" on the other (Knuttgen and Saltin 1973). A realistic figure for refilling depleted oxygen stores is about .5 L, and the oxygen cost for the ATP and PC production is up to 1.5 L (body weight about 70 kg, nonobese person).
2. Then comes a more complex slow component, which, after supramaximal exercise, has a half-time of about 15 min. In the classic literature, this component has been attributed to the energy cost of the elimination of the lactate produced, that is, a payment of a lactic oxygen debt. This concept has been criticized by Brooks and his group, who suggested that the term should be replaced by **excess postexercise oxygen consumption (EPOC),** a term that is commonly used today (Bahr and Sejersted 1991; Brooks and Fahey 1984; Trost, Wilcox, and Gillis 1997).

Inevitably, most of the energy yielded in the metabolism is converted to heat and the tissue temperature increases, which in turn elevates the metabolic rate by about 13% per degree centigrade. The increase in sympathetic nerve activity will also stimulate the metabolism, via adrenaline and noradrenaline. An elevated oxygen demand of the activated respiratory muscles and heart also contributes to the increased metabolism.

As mentioned, when the oxygen demand during exercise exceeds the oxygen supply, the breakdown of glycogen to lactate will support the metabolism. In very heavy exercise lasting for minutes or more, substantial amounts of glycogen are used. During recovery, some of that lactate is reconverted to glycogen, a process that demands energy.

Altogether, the oxygen uptake during recovery from maximal exercise of some 5 min duration can, in extreme cases, amount to almost 40 L during the following 60 min. At rest, during the same period of time, the individual would consume about 18 L of oxygen.

3. After sustained exercise there is a slightly elevated metabolism lasting for several hours, eventually for at least 24 h. It has been suggested that the oxygen consumption during this ultra-slow phase is attributable to a stimulation of substrate cycles (Newsholme and Leech 1984; Trost et al. 1997).

## Summary

In many types of exercise, the oxygen uptake increases roughly linearly with an increase in the rate of exercise. The maximal oxygen uptake or maximal aerobic power is defined as the highest oxygen uptake the individual can attain during exercise engaging large muscle groups while breathing air at sea level, exercise time 2 to 6 min, depending on the type of exercise.

During heavy exercise, anaerobic processes contribute to the energy yield not only at the beginning of exercise but continuously throughout the exercise period. An accumulation of metabolites eventually necessitates the termination of the exercise.

In very heavy exercise, the maximal oxygen uptake and maximal heart rate can be attained within 1 min, providing that a sufficient warm-up period precedes the maximal effort.

During recovery, there is a gradual decline in the oxygen uptake with a fast component that evidently is associated with the energy cost to replenish the ATP and PC pool and to refill the body's oxygen stores. Then follows a slow component and eventually an ultra-slow phase, the duration of which depends on the severity and duration of the exercise. The slow component is caused, at least partly, by an elevation in tissue temperature and circulating catecholamines, and if the exercise has been highly anaerobic, aerobic processes are involved in a resynthesis of glycogen from lactate.

## Intermittent Exercise

Muscular work in industrial or recreational activities is very seldom maintained for very long at a steady rate. For this reason, a steady state, as discussed earlier, is rarely attained. The classic laboratory studies, with subjects exercising continuously for 5 min or longer on the treadmill or cycle ergometer, in many ways represent artificial situations. Nevertheless, such procedures have distinct advantages when one is studying the physiology of exercise or studying patients, for these procedures provide standardized conditions and permit comparisons to be made on repeated occasions. They also can simulate the demands placed on the body in many sporting events. However, from both a practical and a theoretical point of view, it is equally important to study the effect of intermittent exercise, which better mirrors the type of muscular activities encountered in industry or at home and in most types of ordinary exercise or recreational activity. Furthermore, intermittent exercise at a high intensity level is an activity pattern of many types of sports, such as football, soccer, hockey, and tennis, where periods of intense exercise are interspersed with periods of active or passive recovery. Balsom (1995) attempted to evaluate how the different energy systems are used in this type of intermittent exercise with short periods (up to 10 s) of very high intensity (>200% $\dot{V}O_2$max) and which metabolic factors limit performance in a group of highly motivated, physically active males. The ability to maintain a high target power output during consecutive work periods was found to be greatly influenced by small changes in the exercise duration and intensity and in the duration of the intervening recovery periods. Low preexercise muscle glycogen concentration impaired performance, as was the case when oxygen availability to the working muscles was reduced. Following a regimen of creatine supplementation, which was shown to increase the total creatine concentration in m. vastus lateralis at rest, performance was enhanced.

Next some of the most important principles are discussed on the basis of a few classic experiments concerning intermittent exercise (I. Åstrand et al. 1960a, 1960b; Christensen et al. 1960; Essén 1978):

1. A subject whose maximal oxygen uptake was 4.6 L · min$^{-1}$ could exercise at 350 W for about 8 min. Because the oxygen need was approximately 5.2 L · min$^{-1}$, the anaerobic processes had to provide part of the energy. When the rate of exercise was reduced to 175 W, the exercise easily could be prolonged to 60 min, the final heart rate was 135, oxygen uptake was 2.45 L · min$^{-1}$, and the blood lactate concentration did not increase above resting level. The total oxygen uptake during the hour was 145 L.
2. In another experiment with the same subject, the rate of exercise was again 350 W, but now exercise periods of 3 min were alternated with 3-min rest periods. The subject could proceed with great difficulty for 1 hr, and the same total amount of work was performed as in experiment 1. The oxygen uptake and heart rate were now maximal, as was the peak blood lactate concentration (13.2 mM). The total energy output during the second experiment was about 10% higher than in the first one.
3. When the heavy exercise periods were shortened by introducing more frequent rest periods, the total oxygen uptake over the hour was not markedly reduced. The subjective feeling of strain was less severe, however, and peak oxygen uptake, heart rate, and blood lactate concentration were lower. Hence, with intermittent exercise and rest for 30 s, the heart rate did not exceed 150, the blood lactate was only 2.2 mM, and the total oxygen uptake was 154 L during the hour. The subject's maximal heart rate was 190.

Figure 8.5a illustrates another set of experiments with the same subject. He exercised on the cycle ergometer with an extremely heavy intensity of 412 W. When exercising continuously at this rate, he became exhausted within about 3 min (not shown). When exercising intermittently for 1 min and resting for 2 min, he could continue for 24 min before being totally exhausted, and the blood lactate concentration rose to 15.7 mM. In another experiment, the periods of exercise were reduced to 10 s and the rest periods to 20 s. Now the subjects could complete the intended production of 247 kJ (59 kcal) within 30 min with no severe feeling of strain, and his blood lactate concentration did not exceed 2 mM, indicating an almost balanced oxygen supply to his heavily stressed muscles. With periods of exercise and rest of 30 and 60 s, respectively, intermediate results were obtained.

Prolonging the rest periods, with the ratio between exercise and rest changed to 1:4, decreased total work output, of course, but had scarcely any beneficial effect on the subject's fatigue. The critical factor was the length of the exercise periods, and the duration of the rest pauses and the total time spent resting during the 30-min period were only of secondary importance.

Figure 8.5b attempts to explain these findings. When a person exercises intermittently for short periods (in this case 10 s) at an extremely high energy output, the aerobic metabolism is apparently adequate despite an insufficient transport of oxygen during the burst of activity. At least, blood lactate concentration does not increase continuously. Gradually there is a vasodilation in the active muscles, which will improve blood supply. In addition, an oxygen store in the myoglobin can be consumed during the bout of exercise. During the following period of rest, this depot is refilled with oxygen. Saltin, Essen, and Pederson (1976) and Essen (1978) repeated this type of intermittent exercise protocol, supplemented with studies of the metabolic events in the activated muscles. (Samples were secured with needle biopsy technique.) During the exercise (5–20 s), there was a reduction in the ATP and PC concentration, which, however, was restored during the period of rest, evidently by aerobic processes. In intermittent exercise at the same work rate as in continuous exercise, less glycogen is used and the lactate concentration in the muscle is much lower. Approximately 13 times more ATP can be replenished when glycogen is metabolized aerobically compared with the efficiency in anaerobic breakdown of glycogen to lactate. In intermittent exercise, lipids also contribute more to the energy yield than in continuous exercise at the same intensity, which also spares glycogen.

Apparently, during intermittent exercise with short exercise periods, one can endure very high rates of exercise aerobically and therefore experience little lactate production. However, it is essential that the exercise periods are kept sufficiently brief (around 15 s) to prevent the oxygen supply from being exhausted and the anaerobic lactate production from being too great. By spacing the exercise so that running periods lasted for 10 s and resting for 5 s, a subject could prolong the total exercise plus rest period to 30 min without undue fatigue at a speed that normally exhausted him after about 4 min of continuous running (Christensen,

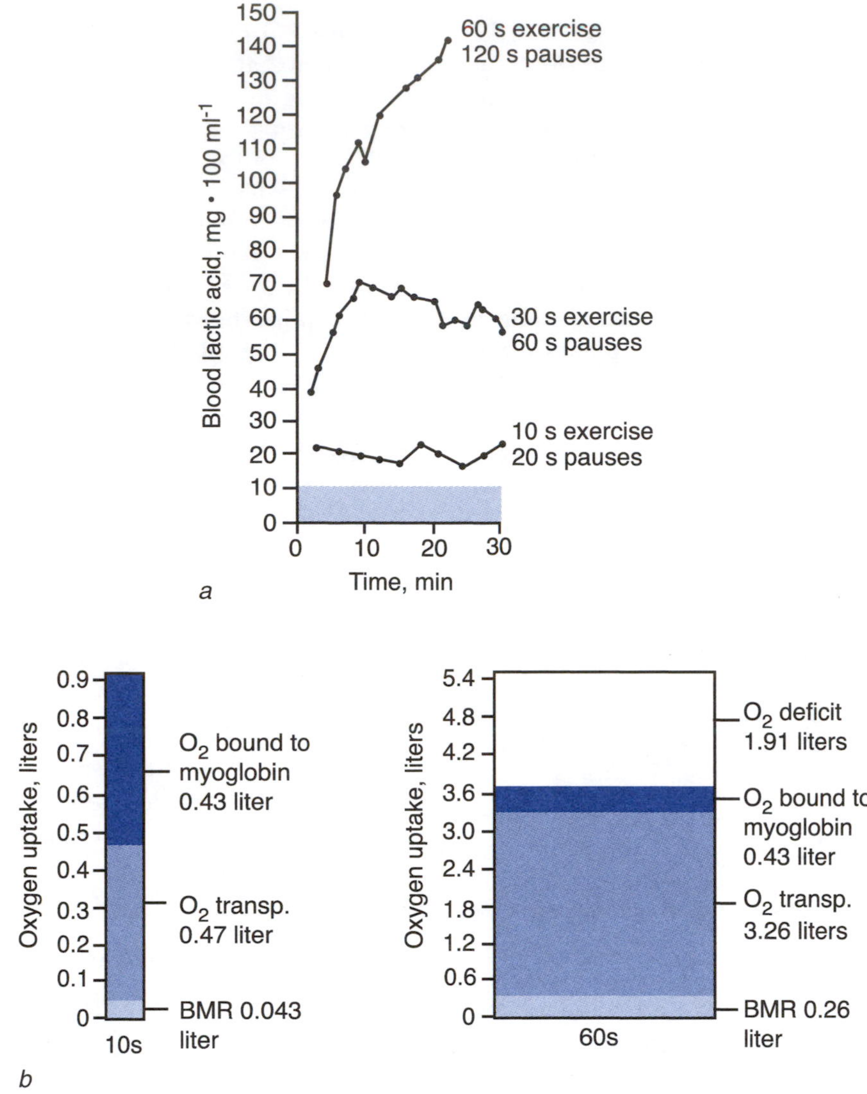


**Figure 8.5** *(a)* The blood lactate concentration in a total work production of 247 kJ (25,200 kpm) in 30 min. The exercise was accomplished with a work rate of 412 W (2520 kpm · $min^{-1}$), the exercise periods being 10, 30, and 60 s and the corresponding rest periods 20, 60, and 120 s, respectively. *(b)* The oxygen requirement for 10 and 60 s power of 412 W. The schematic drawing indicates the **basal metabolic rate (BMR),** the calculated fraction of oxygen bound to myoglobin, transported by the blood, and the oxygen deficit.

From I. Åstrand et al. 1960b.

Hedman, and Saltin 1960). This type of periodic activity is important in athletic training but also has applications in manual labor and many leisure activities.

Dynamic exercise is certainly an intermittent type of activity, and its superiority over static exercise as an endurance exercise can be explained partly on the basis of the muscle pump and the alternating emptying and filling of the oxygen stores during alternating muscle contraction and relaxation.

## Summary

The buffering effect of an oxygen store means that a great amount of exercise can be performed at an extremely heavy rate, with a relatively low peak demand on the circulation and respiration, by the introduction of properly spaced, short exercise and rest periods ("micropauses"). The heavier the work rate, the shorter should be the exercise periods. This physiological concept has at least two important applications:

1. It may explain why older or physically disabled individuals, despite a reduced maximal aerobic power, can remain in jobs involving heavy manual labor such as forestry, farming, and construction, or can enjoy physically demanding hobbies. As long as they are free to choose the optimal length of the exercise and rest periods, the acute loads on the respiration and circulation do not exceed the limits of their reduced capacity. However, if the pace is determined by a machine, even a less heavy peak load, but with relatively long activity periods, can overtax the capacity of the worker whose physical performance capacity is limited.
2. If the aim of a training program is to increase muscle strength, the highest load on the muscle fibers will be obtained within a given period of time if periods of rest are frequently interspersed between activity periods of 5 to 10 s. On the other hand, training the oxygen-transporting system will be easier if the exercise periods are prolonged to at least 2 to 3 min. This type of exercise also adapts the tissues to high lactate concentrations, provided the exercise is severe.

## Prolonged Exercise

Moderately well-trained individuals may walk or run for about 1 h with an oxygen uptake around 50% of the $\dot{V}O_2max$, maintaining the oxygen uptake, heart rate, and cardiac output at approximately the same level as attained after about 5 min of exercise. The lactate concentration in active muscles and in the arterial blood is not elevated, indicating a steady state (figure 8.6a). When the exercise time is further prolonged, oxygen uptake and heart rate progressively increase, and the subject becomes fatigued. Figure 8.6b illustrates an experiment in which exercise was performed continuously for seven 50-min periods at an oxygen uptake of 50% of the subject's maximal oxygen uptake. The subjects rested for 10 min in between, and after 4 h of exercise (cycling, alternating with running) they had a 1-h break for lunch. The most fit subject, with a maximal oxygen uptake of 5.60 $L \cdot min^{-1}$, exercised with an average oxygen uptake of 2.75 $L \cdot min^{-1}$; one subject with a maximal aerobic power of 2.25 $L \cdot min^{-1}$ exercised with a work rate requiring an oxygen uptake of 1.15 $L \cdot min^{-1}$. The four subjects participating in the experiments could fulfill the task, but they were fatigued. It appears that a 50% demand on the aerobic power is too high for a steady state if the physical activity is continuous for an entire working day. In manual labor, a 40% ceiling is advisable.

The well-trained individual can maintain steady state (as defined previously) at a higher relative work rate than 50%, indicating more efficient oxygen transport and oxygen and substrate utilization in the active muscles. Elite cross-country skiers can exercise at 85% of their maximal aerobic power at least for 1 h (P.-O. Åstrand, Hallbäck et al. 1963; Ingjer 1992).

Well-trained athletes, including marathon runners, can exercise for hours with an oxygen uptake around 75% to 85% of their maximum with little or no increase in blood lactate concentration (Costill 1970; Maron et al. 1976). C.T.M. Davies and Thompson (1979) studied "ultradistance" runners who were running for 24 h. The researchers noticed a gradual decline in the engagement of the aerobic potential. It dropped from about 90% at the beginning to slightly below 50% of the maximal aerobic power at the end. The energy output was estimated to be close to 78 MJ (18,600 kcal).

Chapter 15 discusses muscular fatigue, particularly in activities demanding maximal or close to maximal forces. In prolonged activities, a relatively small fraction of maximal strength is usually applied. Still, a gradually declining power output, despite efforts to continue at a high intensity, can be caused by disturbances in neuromuscular function, but other factors also are involved. During prolonged heavy exercise, the water balance can be disturbed and the stores of available energy, particularly glycogen, can be critically low. Therefore, the individual's ability to transport oxygen from the air to the active muscles may not always be the limiting factor. It has been observed that the subjective feeling of fatigue during heavy, prolonged exercise coincides with a decrease in blood glucose concentration, a depletion of the glycogen depots in the exercising muscles, or both. For a high rate of exercise, carbohydrate is an essential substrate. One explanation for the well-trained individual's ability to exercise for a long period of time at a high aerobic power in relation to $\dot{V}O_2max$ is that with endurance training, the ability to use fatty acids as a substrate is enhanced and this spares glycogen. In other words, it takes longer to deplete the glycogen stores. There is a price to pay for a higher energy yield from lipids with a reduced demand on the carbohydrate contribution. The oxygen required for a given energy yield is up to 7% higher with

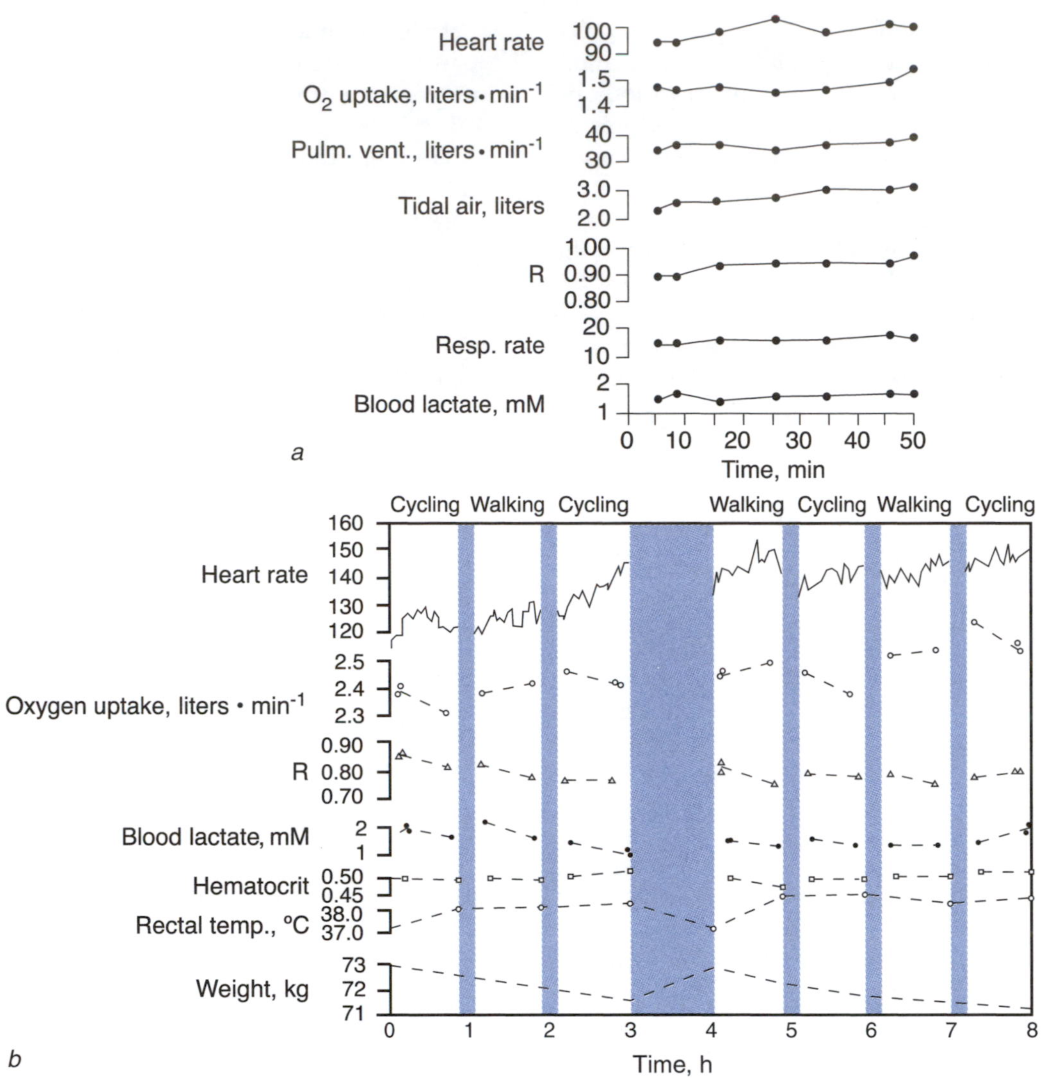


**Figure 8.6** *(a)* Metabolic parameters during 1 h of exercise in a subject exercising at an intensity demanding close to 50% of his maximal aerobic power (1.5 vs. 2.94 L of oxygen per minute). *(b)* Metabolic parameters in one subject during an experiment consisting of seven exercise periods of 50 min each. The shaded columns represent rest periods. The subject's maximal aerobic power was 4.6 L of oxygen per minute.

*(a)* Data from I. Åstrand, P.-O. Åstrand, and Rodahl 1959. *(b)* from I. Åstrand 1960.

lipids as a fuel compared with carbohydrates as a substrate. During prolonged exercise there is a gradual increase in the use of lipids in the muscle metabolism when a person is exercising at an oxygen uptake below about 70% of his or her maximum. Inevitably, this event will demand that more oxygen be delivered per unit of time. This may explain, at least in part, the slight increase in the oxygen uptake during exercise at a given rate and, as a consequence, an increase in the heart rate. In addition, as exercise proceeds, stroke volume often is reduced and heart rate increases in compensation to maintain an adequate cardiac output. The increase in heart rate and reduced stroke volume at a given oxygen uptake are particularly evident in exercise in a hot environment. If dehydration, a decrease in the blood sugar concentration, and depletion of the body's glycogen stores are prevented by proper supply of fluid and sugar, a high level of physical performance is better maintained during prolonged exercise.

The limiting factors in prolonged exercise may vary from individual to individual. It is conceivable that the electrolyte balance, the ratios of potassium and sodium ions, for example, across the muscular cell membrane are disturbed during prolonged exercise, and that the activity of key enzyme systems is hampered by the decreased pH and/or accumulation of metabolites. In fact, in some experiments involving prolonged severe exercise, none of the physiological parameters studied (e.g., blood sugar concentration, maximal oxygen uptake and cardiac output, and blood lactic acid level) correlated well with the subject's feeling of fatigue or reduction in performance capacity (Rowell 1983; Saltin 1964).

Motivation is undoubtedly an important factor determining endurance during heavy exercise. Well-trained, highly motivated subjects can maintain their oxygen uptake at a maximal level for at least 15 min, although most individuals feel an urge to stop after 4 to 5 min at a work rate that taxes the oxygen-transporting systems to a maximum. Furthermore, physical performance may be subject to diurnal (Bernard et al. 1998; D.W. Hill et al. 1992; A. Rodahl, O'Brien, and Firth 1976) as well as seasonal variations (J. Erikssen and Rodahl 1979; Shephard 1984a). Performance even has been reported to be influenced by the social class of the subject (Krombholz 1997). On the other hand, the contributions of the aerobic and anaerobic energy systems to high-intensity exercise performance show no changes following the loss of one night's sleep (D.W. Hill et al. 1994).

### *Summary*

It is obvious that an individual's maximal aerobic power plays a decisive role in his or her physical performance. If a given task demands an oxygen uptake of 2.0 $L \cdot min^{-1}$, the person with a maximal oxygen uptake of 4.0 $L \cdot min^{-1}$ has a satisfactory safety margin, but the 2.5 $L \cdot min^{-1}$ individual must exercise close to his or her maximum, and consequently the internal equilibrium becomes much more disturbed. In prolonged exercise, motivation, state of training, water balance, and depots of available energy are important for performance capacity. A technique and efficiency factor is of decisive importance for the energy cost of a given task, more so in activities that demand skill, such as swimming and cross-country skiing.

## Muscular Mass Involved in Exercise

The demand on the oxygen-transporting functions varies with the size of the active muscles. Since isometric contractions hinder the local blood flow and dynamic exercise facilitates the circulation, it follows that a greater oxygen uptake can be obtained during dynamic exercise. Usually, exercise involves both static and dynamic muscle contractions. Static exercise produces a relatively high heart rate and arterial blood pressure. This may complicate a task evaluation based on the measurements of heart rate and blood pressure (chapter 5).

In maximal work on a cycle ergometer in the supine position, the oxygen uptake is only about 85% of the value obtained in the sitting position. But, if the subject exercises with both legs and arms simultaneously in the supine position, the oxygen uptake, cardiac output, and heart rate increase to the values typical for maximal exercise in the upright position (Stenberg et al. 1967). One plausible explanation for the lower aerobic power for maximal cycling in the supine position, despite an optimal venous return to the heart, is the less favorable position, because the body weight cannot be used during the critical stages of pedaling. Second, the blood perfusion of the activated leg muscles is enhanced when a person exercises in the upright position.

In arm exercise, the maximal oxygen uptake is about 70% of what is attained in leg exercise. The intra-arterial blood pressure during arm exercise is higher than in leg exercise at a given oxygen uptake or cardiac output (see figure 5.24, page 170), and the heart rate is also higher. The consequence is a heavier load on the heart. For patients with heart disease or for completely untrained older individuals, heavy exercise with the arms (such as digging, shoveling snow) can be hazardous. This may be attributable, in part, to the Valsalva effect during such maneuvers. The subject is apt to hold his or her breath while lifting the load, increasing the intrathoracic pressure, which in turn hinders the normal venous return to the heart.

When arm and leg exercise (cranking and cycling) are combined, the highest oxygen uptake that can be attained depends on the relative load on the arms. Bergh, Kanstrup, and Ekblom (1976) noticed that the oxygen uptake was the same in maximal running as in arm plus leg exercise when the arm work rate was 20% to 30% of the total rate of work, and the total oxygen requirement exceeded the subject's maximal oxygen uptake. Subjects with strong arm and shoulder muscles could be submitted to relatively heavier arm exercise and still reach the maximum attained during uphill running. Otherwise, a typical finding was that the maximal oxygen uptake was reduced to 90% of the maximum during running, when the arm work rate was 40% of the total rate of work. However, the difference in oxygen uptake in leg exercise and arm plus leg exercise is much smaller than expected from the difference in mass of active muscles in the two procedures (see discussion in chapter 5). The central circulation may in some way limit aerobic power. There are advantages when a large muscle mass is activated. Let us look at figure 8.7. A work rate of 350 W could be tolerated for about 3 min if only the leg muscles were involved. However, with 100 W for the arms and 250 W for the legs, the exercise time could be prolonged to 6 min, even if the oxygen uptake (and cardiac output) did not increase further (P.-O. Åstrand and Saltin 1961a; Stenberg et al. 1967). Evidently the organism could tolerate a prolongation of the exercise period when a larger mass of skeletal muscles was activated. The subjective feeling of strain is related more to the metabolic rate per square area of muscle than to the total metabolism. Therefore, training the oxygen-transporting system is more efficient and is psychologically less strenuous, the larger the muscular mass involved in dynamic activities.

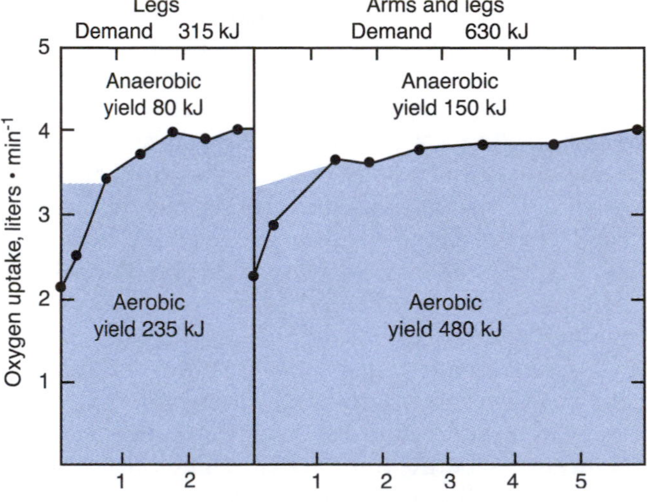


**Figure 8.7** Increase in oxygen uptake at start of exhausting exercise following a 10-min warm-up period: leg exercise only (left) and exercise with the same external power but with both arms and legs involved (right). This given exercise could be tolerated twice as long with arms and legs activated. Calculations of energy demands and yield are explained in text.

Data from P.-O. Åstrand and Saltin 1961a.

# Anaerobic Processes

During light exercise, the required energy is provided almost exclusively by aerobic processes, but during more severe exercise, anaerobic processes are brought into play as well. Anaerobic, energy-yielding metabolic processes play an increasingly greater role as the severity of the exercise increases. As discussed previously, the aerobic power during exercise can be followed quite accurately by measuring the oxygen uptake. Because we lack a similar tool for directly measuring the anaerobic power, indirect methods have to be applied when studying the kinetics, power, and capacity of these processes.

The energy yield from the breakdown of ATP and PC is indispensable, but, quantitatively, the available stores of these high-energy phosphates alone can only cover the energy requirement for less than 10 s during maximal effort. These processes of breakdown and resynthesis occur very quickly. Actually, processes of key importance in the anaerobic energy yield occur within fractions of a second, but with the available methods in human experiments (using muscle biopsies), seconds may elapse between sampling and stopping the biochemical events in the sample. The rate of turnover of ATP in a sprinting human is approximately 2.7 mmol $\cdot$ $s^{-1}$ per kilogram of muscle; in a high jump it may be as high as 7 mmol $\cdot$ $s^{-1} \cdot kg^{-1}$. With only 5 mmol $\cdot$ $kg^{-1}$ of ATP available, it obviously has to be quickly resynthesized, as in the case of a 100-m race (Newsholme and Leech 1984). PC is the initial source of energy in this process. For an analysis of anaerobic biochemistry, it may be convenient to discuss the breakdown of high-energy phosphate (~P) separately from the anaerobic glycogen breakdown to lactate (often named alactate or **alactic** period, and lactate or lactic period, respectively). There is an overlap in these processes as well as in the kinetics of the aerobic energy yield, but with an exercise period of about 5 s the **alactic power** will dominate. Reports indicate that there is a delay of up to 6 s before glycogenolysis sets in (see di Prampero, 1981). However, Hultman and Sjöholm (1983) claimed that during a near-maximal muscle contraction lasting 1.26 s, 20% of the energy yield comes from a degradation of glycogen to lactate. In the following, the calculations of the alactic power are based on a nonlactate contribution in maximal exercise lasting some 6 s, not preceded by warm-up.

## Power and Capacity for High-Energy Phosphate Breakdown

In some activities, the developed (external) power can be measured. C.T.M. Davies and Rennie (1968) had their subjects perform a high jump from a force platform. From the time course of the vertical velocity of the center of gravity of the body, the researchers could calculate the average peak power output to be 3,900 W for their male and 2,350 W for their female subjects. This peak power, developed during .2 s in the jump, is actually 15 times larger than the subject's maximal aerobic power, developed for 5 to 8 min during cycling. However, the mechanical efficiency in a jump is not known, and therefore its metabolic cost cannot be estimated.

A method by which one can estimate the **maximal anaerobic power** output has been developed (see di Prampero 1981). The subjects climb a normal flight of stairs at maximal speed taking two to three steps at a time. The peak speed is attained within 2 to 3 s and can be maintained up to the 6th second; from then on it declines. From speed, vertical distance climbed, and body weight, the power output can be calculated. By this method, values that are about 25% of the peak "external" power reported by C.T.M. Davies and Rennie (1968) have been obtained. The variation can be explained by the difference in the two types of exercise: In the high jump, it is a matter of one muscular contraction of the two legs simultaneously; in stair climbing, the values obtained represent an average involving a series of contractions using one leg at a time.

Assuming that the mechanical efficiency is 25% when a person is climbing stairs, an "external" power of 1,000 W would require a power of 4,000 W from the energy-yielding processes. With an exercise period lasting only a few seconds, the anaerobic breakdown of the high-energy phosphate compounds will dominate, and the stair-climbing test measures mainly the alactic anaerobic power. One crucial problem is that it is impossible to measure the mechanical efficiency accurately in maximal exercise. Therefore, calculating the power of the metabolic processes for a given external power output is rather hazardous. Using the elastic properties of a muscle can greatly improve the efficiency (e.g., when running at high speed), and therefore an extrapolation of energy demands at maximal exercise from data on submaximal steady-state conditions has its limitations. It has been estimated, however, that the energy cost of a 100-m sprint is about 33 kJ (8 kcal), and if the time is 10 s, the power is 3,300 W. It is estimated that approximately 85% of the energy comes from anaerobic processes and the remaining 15% from aerobic sources.

Another approach to calculating alactic power is to calculate the total energy bound in the ATP and PC compounds. If the concentrations are known, one can arrive at the maximal anaerobic alactic energy yield (capacity) per kilogram muscle if these phosphate compounds were completely exhausted. With an ATP concentration of 5 mmol $\cdot$ kg$^{-1}$ in fresh muscle and a PC concentration of 17 mmol $\cdot$ kg$^{-1}$, the potential total energy yield is approximately .9 kJ (.2 kcal). In 20 kg of muscle, the total energy content is 18 kJ (4.3 kcal). The equivalent oxygen requirement for an aerobic energy yield of that order is just .8 L. In other words, only a small oxygen deficit can be covered by the breakdown of ATP and PC. Actually, it is less than the theoretical value. During maximal exercise, PC exponentially declines in proportion to the severity of the exercise, down to values as low as about 2 mmol $\cdot$ kg$^{-1}$. The change in ATP concentration is, as mentioned, difficult to study in vivo. It may be reduced to 40% of the level in a resting muscle.

## Power and Capacity of Anaerobic Glycogen Breakdown

Figure 8.7 summarizes experiments performed on a cycle ergometer and illustrates one model used to estimate the size of the anaerobic energy yield during "supramaximal" exercise. With a correction for the alactic contribution, we can calculate the contribution from the anaerobic breakdown of carbohydrates.

By extrapolating from the steady-state oxygen uptake during different rates of submaximal exercise, we can estimate the oxygen and thereby the energy demand of the supramaximal rate of exercise. Consider the same subject presented in figure 8.4. The work rate of 350 W required 5.0 L $\cdot$ min$^{-1}$ and was maintained for 3 min. Therefore, the total energy demand during this period was $5 \cdot 21 \cdot 3 = 315$ kJ (75 kcal), the energy yield per liter of oxygen taken up in the mitochondria being about 21 kJ (5 kcal).

Let us now try to analyze how the energy demand of 315 kJ (75 kcal) was covered. The oxygen uptake was measured continuously during the 3 min and was found to be 10.7 L. An additional .5 L was used from stores, bound to myoglobin and hemoglobin, and refilled after the exercise. Thus, the aerobic energy yield can be estimated to be 235 kJ ($11.2 \times 21$). The deficit was therefore $315 - 235 = 80$ kJ (19 kcal), and this energy must have been derived anaerobically. A breakdown of ATP and PC yields 20 kJ (4.7 kcal) at the most (i.e., it may substitute for about 1 L of oxygen). The remaining deficit of about 60 kJ (14 kcal) must have been provided by glycogenolysis and glycolysis with a formation of lactic acid.

Because about 220 kJ (52 kcal) is released for each six-carbon unit of glycogen converted into lactate, a production of 2 mol or 180 g of lactate should yield 220 kJ. For a release of 60 kJ, the lactate production must then be about .55 mol (50 g).

The subject also performed the same work rate with both arms and legs. Under these conditions, the exercise could be prolonged to 6 min before exhaustion (see figure 8.7). At submaximal exercise, the mechanical efficiency is not significantly different from ordinary cycling. If anything, the oxygen uptake tends to be higher in exercise performed with the arms and legs than in leg exercise alone; it is also conceivable that the mechanical efficiency decreases during very heavy exercise because muscles that are at a mechanical disadvantage have to contribute. Therefore, the calculated energy demand is probably a minimal figure. The energy requirement is therefore assumed to be 105 kJ (25 kcal) per minute, or 630 kJ (150 kcal) altogether, during the 6 min. The measured oxygen uptake of 22.3 L, supplemented by .5 L from oxygen stores within the body, covers 480 kJ (114 kcal), leaving 150 kJ (36 kcal) for the anaerobic processes. Subtracting 30 kJ (7 kcal) as a contribution from high-energy phosphate compounds leaves 120 kJ (28.5 kcal) from glycogenolysis, that is, a formation of 1.1 mol or 100 g of lactic acid.

The glycogen concentration in the human muscle in individuals on a mixed diet is 80 mmol of glucosyl per kilogram of wet weight. Assuming that 25 kg of muscles are involved in arm and leg exercise, the available glycogen content in those muscles will be 2 mol (360 g). If so, approximately 25% of the available glycogen store was used anaerobically in the experiment depicted in figure 8.7, because a breakdown of .55 mol of glycogen ends up forming 1.1 mol of lactate.

Calculating the average breakdown of glycogen gives about 90 mmol · min$^{-1}$ or 3.6 mmol per kilogram of fresh muscle per minute. A degradation rate of approximately 2.5 mmol glucosyl units · kg$^{-1}$ in exercise at 90% to 100% of the fit individual's maximal aerobic power has been noted (Hultman and Sjöholm 1983). The total metabolic power was 1,750 W with a share of about 330 W to glycogenolysis. In figure 8.7, the anaerobic energy yield during the 5 to 6 min interval is about 20% of the total.

Figure 8.8 presents data from the study of a subject who exercised for 2.63 min with an exter-

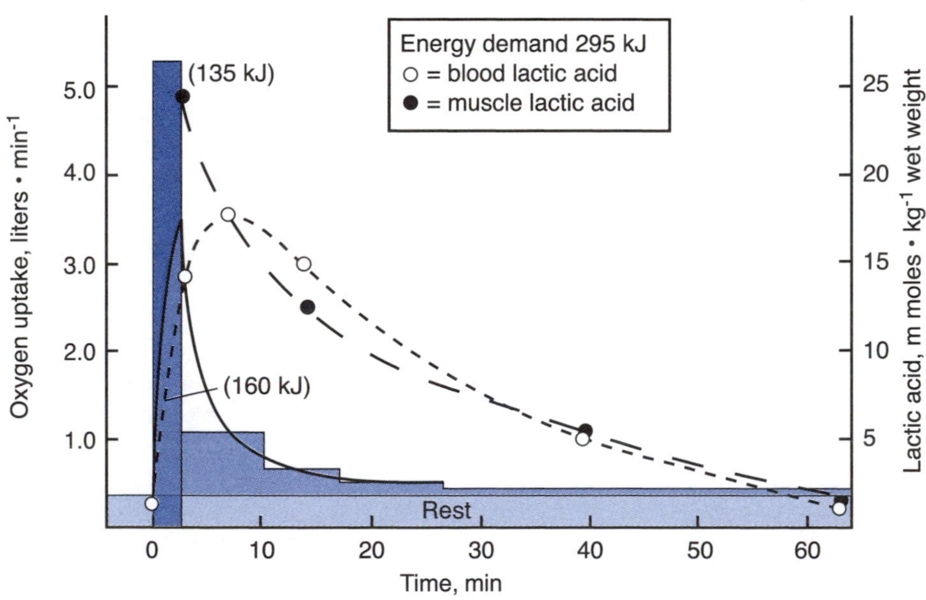


**Figure 8.8** Calculated energy requirement for 2.63 min of exercise on a cycle ergometer (column represents 295 kJ, 70 kcal) and measured oxygen uptake during exercise and during 60 min of recovery. Horizontal lines denote the level of oxygen uptake measured at rest before exercise. Calculated aerobic energy yield during exercise: 160 kJ (38 kcal); anaerobic energy yield: 135 kJ (32 kcal) (dotted area). Lactate concentration was analyzed in blood samples and pieces of skeletal muscle obtained by needle biopsy.

From B. Diamant, K. Karlsson, and B. Saltin (unpublished data).

nal power close to 400 W with a calculated energy demand of 295 kJ (70 kcal). The total oxygen uptake during cycling was 7 L. If .5 L is added from oxygen stores, the aerobic energy yield will be 160 kJ (38 kcal). Therefore, the anaerobic contribution was 135 kJ (32 kcal). Assuming that a degradation of glycogen yielded 105 kJ (25 kcal), the production of lactate would have been about .92 mol (83 g). In this case, the average anaerobic power was about 650 W. The maximal degradation rate of glycogen can be calculated to be 175 mmol · $min^{-1}$.

In maximal exercise of short duration, the rate of glycogen breakdown is higher. In his review, di Prampero (1981) presented the maximal power for the "lactic mechanism" for an average individual. The oxygen equivalent is 75 ml · $kg^{-1}$ ·$min^{-1}$. With a body weight of 75 kg, the total will be 5.6 L · $min^{-1}$ and the power as high as 1,950 W.

A degradation of one glucosyl unit to lactate yields energy that can resynthesize 3 ATP. The equivalent amount of glucose only covers the formation of 2 ATP. A second advantage of glycogen is that it is stored in the muscle fibers. During short-term exercise, there is not time to transport substrates from the liver and the fatty tissues; the muscles must be able to function on their own resources.

The blood lactate concentration has been used to evaluate anaerobic power and capacity. An increase in concentration means that the uptake of lactate by the blood exceeds the lactate removal. The problem is that we lack information about the total water pool in the body available for lactate uptake. There is also an uneven distribution of lactate between the extra- and intracellular water. Therefore, the weighted mean concentration at equilibrium of this water-soluble molecule is not the same throughout the different compartments. Di Prampero (1981) proposed a formula for calculating lactate production from the peak lactate concentration and the subject's body weight, although realizing the limitations of such a formula. Another missing piece of information is the rate by which lactate is removed during exercise and recovery. In the experiments presented in figure 8.7, the peak blood lactate concentration after 3 min of maximal exercise was almost identical with the peak value attained during recovery after the 6-min ride (17.1 and 17.4 mM, respectively). The estimated lactate production was, however, very different (.55 and 1.1 mol, respectively). This illustrates the poor correlation between peak blood lactate concentration and quantity of lactate produced.

With data on muscle lactate kinetics, one comes closer to the actual metabolic events. However, we cannot estimate how large a muscle mass is exercising and, second, we don't know whether the muscle sample is representative of all of the muscles that are activated. At high rates of exercise, the type II fibers (fast-twitch fibers) are recruited and their metabolic profile is different from that of type I fibers (slow-twitch fibers).

As pointed out by Halestrap and Price (1999), monocarboxylates such as lactate and pyruvate play a central role in cellular metabolism and metabolic communication between tissues. This is based on their rapid and specific transport across the plasma membrane, which is catalyzed by a family of proton-linked monocarboxylate transporters (MCTs). The isoform MCT1 is especially prominent in heart and red muscle, the latter containing a high percentage of type I fibers, and the level of MCT1 is upregulated in response to increased work. MCT4 is most evident in white muscle and other cells that have a high glycolytic rate. MCT2 has a 10-fold higher affinity for substrates than MCT1 and MCT4 and is found in cells where rapid uptake at low substrate concentrations is required, including the proximal kidney tubules, neurons, and sperm tails. MCT3 is uniquely expressed in the retinal pigment epithelium (Bonen 2000).

## FOR ADDITIONAL READING

For an extensive review of the structure, function, and regulation of the proton-linked MCT family, see Halestrap and Price (1999).

## *Summary*

Under highly standardized conditions, as in the experiments just discussed, one can estimate the rate of energy output. With the aerobic energy yield calculated from the oxygen uptake, the anaerobic contribution can be computed. It can be divided into an alactic part (energy from the breakdown of ATP and PC) and a lactic part (energy from the breakdown of glycogen). An increase in the lactate concentration in an active muscle shows that the lactate formation rate exceeds the lactate removal rate. However, the change in concentration may differ within the muscle itself. Calculating the lactate production is also impossible if one does not know the muscle mass involved and the amount of lactate removed to the blood or by chemical processes within the muscle itself before collection of the muscle tissue sample. Similarly, determinations of blood lactate concentration may give some information about whether the glycolysis contributes to the ATP resynthesis. However, an accurate estimation of the rate of lactate production and its magnitude is not possible.

In maximal exercises of a few seconds duration, the developed metabolic power can exceed 4,000 W, with the energy-rich phosphate compounds being the source of the energy. This is the alactic phase. With some seconds delay, the breakdown of carbohydrates to lactate, with glycogen as the dominating substrate, can produce a power that may come close to 2,000 W. As the exercise time is prolonged, the power of this anaerobic process gradually declines. The capacity of the energy-rich phosphates is very limited, about 0.9 kJ (0.2 kcal) · $kg^{-1}$ muscle. With 25 kg of muscles in maximal exercise, the potential is then 22 kJ (5.2 kcal), equivalent to 1 L of oxygen, providing all ATP and PC are used, which is not realistic. The capacity of the anaerobic carbohydrate metabolism is difficult to estimate. In athletes specialized in events demanding maximal effort of a few minutes duration, it may be up to 200 kJ (45 kcal).

The weak point in the calculations is that we do not have exact figures for mechanical efficiency during maximal cycling.

# Lactate Production, Distribution, and Disappearance

It has been known for a long time that skeletal muscle is a major producer of lactic acid in the body, but the way we regard lactate changed dramatically during the last quarter of the 20th century. We used to see it as a dead-end waste product during anaerobic conditions, responsible for such unwanted effects as muscle soreness and fatigue. Today, lactate has emerged as a normal metabolic intermediate, even under aerobic conditions. In fact, as much as 50% of the glucose metabolized under fully oxygenated conditions is converted to lactate (Connett, Gayeski, and Honig 1984; Gladden 2000).

Even our view of how lactate is transported across cell membranes has changed. We used to believe that it was by simple diffusion, but it has turned out to be facilitated diffusion by means of specific transporter molecules in the cell membrane (see chapter 2). As previously mentioned, these lactate transporter molecules belong to a family of proton-linked monocarboxylate transporters (MCTs) that plays an important role in the pH regulation of skeletal muscle. Skeletal muscle contains both the MCT1 and MCT4 isoforms. The amount of MCT1 is correlated with the aerobic capacity of the muscle fiber, possibly because MCT1 has been found in mitochondrial membranes (Brooks 2000). Consequently, there is more MCT1 in type I than in type II muscle fibers. MCT4, on the other hand, is found in all fiber types (Juel and Halestrap 1999). These monocarboxylate transporters are bidirectional, meaning that they are able to transport lactate out of or into muscle cells depending on the hydrogen ion ($H^+$) and lactate gradient (Donovan and Pagliassotti 2000). This makes them ideal for energy transfer between cells as described by the "lactate shuttle" hypothesis (Brooks 2000).

Our understanding of the role of lactate in energy metabolism has changed so much in recent years that it is appropriate to speak about a shift of paradigm. Furthermore, new evidence continues to accumulate at a rapid pace, partly supporting and partly challenging existing theories. We hope this means that our understanding of this important area of metabolism will continue to improve. At the moment, however, there is still some controversy regarding these phenomena, although at least some of it may be of more semantic nature. Alternative terms have been introduced to try to resolve the controversy. In addition to anaerobic or aerobic threshold, the terms **onset of blood lactate accumulation (OBLA)** and **lactate threshold** have been introduced. The following is an attempt to outline the situation in the new millennium.

## Time Course of the Blood Lactate Concentration

Blood lactate concentration is relatively simple to determine. Understandably, this has made it a frequently measured parameter. Even the salivary lactate concentration serves as a relevant indicator of blood lactate concentration (Ohkuwa et al. 1995). A semantic piece of warning is appropriate at this point. The terms *lactate concentration* and *lactate production* are often used synonymously. As will be evident in the following, they are not the same. The lactate concentration, wherever it is measured, reflects the difference between lactate's rate of appearance and its rate of removal (Brooks 1985).

The blood lactate concentration is an index of anaerobic metabolism but it does not inform us about the anaerobic power. Nor is it an adequate measure of the anaerobic energy release during exercise (Medbø, Mohn, and Tabata 1998). As mentioned previously, an increased lactate level in muscle and blood traditionally has been taken as an indication of anaerobic supplement to the aerobic production of ATP. Now it is seen rather as the result of a shift in the balance between pyruvate production by glycolysis and pyruvate consumption by the mitochondria (Gladden 2000). Because of the near-equilibrium nature of the lactate dehydrogenase reaction, increased pyruvate levels mean increased lactate levels. Not surprisingly, under hypoxic conditions, such as at high altitude, the oxidative disposal of pyruvate is at a disadvantage, and the blood lactate concentration is higher at a given work rate compared with normoxic conditions. When a person is exercising under hyperoxic conditions, the picture is reversed (P.-O. Åstrand 1954; di Prampero 1981; Knuttgen and Saltin 1973; P.K. Pedersen, 1983). Figure 8.3 illustrates how the peak blood lactate concentration starts to increase when the exercise rate increases. The events can be summarized as follows:

1. During light exercise, the demand for ATP is small and the oxidative removal of pyruvate keeps pace with its production. Most ordinary daily activities belong to this category.
2. During exercise of moderate intensity, the demand for ATP initially surpasses the supply by oxidative metabolism, and the glycolytic rate is increased temporarily to cover the demand until the aerobic oxidation can take over and completely cover the energy demand. Produced lactate diffuses out of the muscle fibers by means of lactate transporters (MCTs) in the cell membrane (facilitated diffusion) and can be traced in the venous blood draining the muscle. As the exercise proceeds, the blood lactate concentration decreases again, and the exercise can be continued for hours.
3. During heavier exercise, the imbalance between pyruvate production by glycolysis and its removal by the mitochondria is more prolonged but eventually can reach a steady state, depending on the intensity of the exercise. In such cases, the blood lactate concentration can remain constant and high throughout the exercise period. The length of time that the work rate can be endured will depend, to some extent, on the subject's motivation.
4. During very high-intensity exercise, there is a continuously increasing imbalance and a corresponding increase in the lactate content of the blood. As a rule, the exercise cannot be continued for more than a few minutes, because the subject's muscles can no longer function. The work rate has exceeded some sort of a threshold, and the condition is no longer one of a steady state.

Figure 8.9 illustrates how the arterial lactic acid concentration increases during and after severe exercise, followed by a slow decline back to the resting level. The lactate is produced in the muscles during the actual exercise, but there is a time lag for the diffusion from the active muscles and redistribution within the body. To determine peak lactate concentration in the blood, samples must be taken at intervals during the first 5 to 10 min of the recovery period. It takes up to 60 min or even longer before the resting level is again reached, so that if the effect of a stepwise increasing work rate is studied, the samples secured at the end of the last exercise period not only reflect the anaerobic component of the last exercise intensity but also are affected by the preceding work rates.

Figure 8.8 illustrates how the lactate concentration at the end of exercise is much higher in the active muscles than in the blood, but after some 5 min of recovery the concentrations in the two compartments run parallel. After exhausting exercise, the peak lactate concentration is attained after 5 to 8 min. The lactate is assumed to have reached equilibrium in 85% of the total body water. The elimination of lactate from the blood then has a half-time of 15 min if the person is resting during recovery, independent of its peak concentration, at least within the range 4 to 16 mM (di Prampero 1981).

With lactate concentrations as high as 30 mM in the exercising muscle, the peak blood concentration was around 20 mM. The fast-twitch fibers, particularly the type IIX fibers, have the highest potential for an anaerobic breakdown of carbohydrates. Elite cross-country skiers are characterized by a high maximal oxygen uptake (80–95 ml · kg$^{-1}$ · min$^{-1}$) and a high percentage of slow-twitch fibers (around 75%). Typical for endurance athletes, they can attain very high blood lactate concentrations in a 5- to 10-min maximal exercise on the treadmill, or on the average 15 mM. Apparently, the slow-twitch fibers also can exercise anaerobically quite effectively. Endurance athletes have a relatively low activity potential for the enzyme lactate dehydrogenase in their fibers, but this enzyme is not rate limiting. The highest values on blood lactate reported so far are in samples drawn from well-trained athletes during recovery from competitive events of 1 to 2 min duration.

Reports show that lactate actually is released from resting muscles (arms) during heavy, prolonged exercise with the legs. We have no reason to believe that these muscles are anaerobic, and the mechanism behind this release is probably a glycogenolysis induced by the increased sympathetic

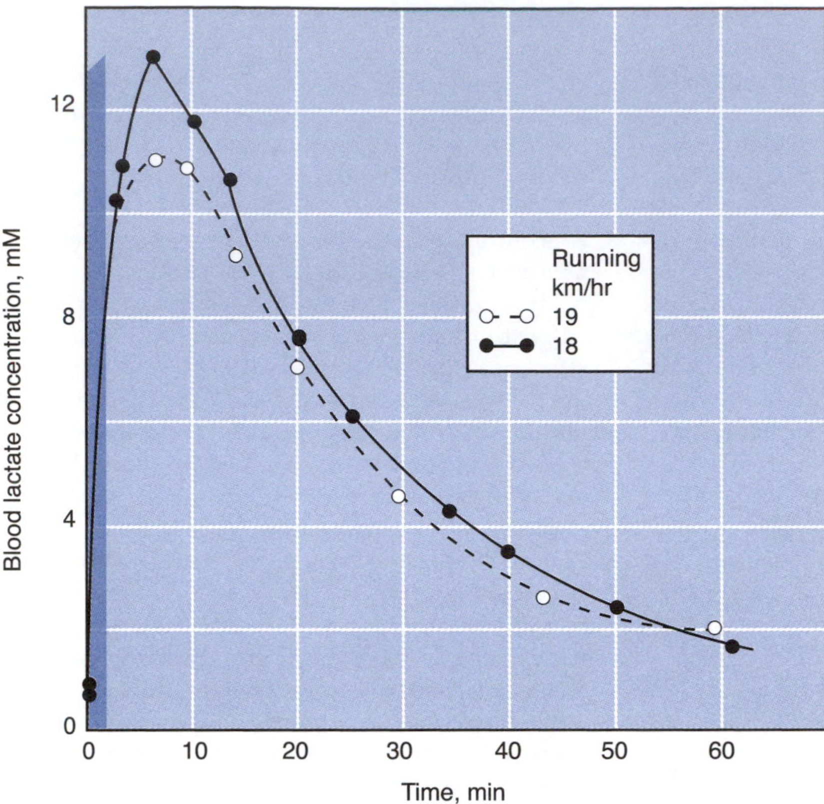


**Figure 8.9** Blood lactate concentration after severe exercise of 2 min duration (shaded column) in two subjects. Peak values occur several minutes following the cessation of exercise.

From I. Åstrand 1960.

activity. This lactate can be used as a substrate in the exercising muscles or as a precursor for a glucose formation in the liver (Ahlborg, Hendler, and Felig 1981).

## What Is the Fate of the Lactate Produced?

In an aerobic metabolism of glycogen down to carbon dioxide ($CO_2$) and water ($H_2O$), the energy yield for ATP production is 2,813 kJ (672 kcal) per six-carbon unit. Of this potential, only about 8% is available in the anaerobic breakdown to lactate. However, the produced lactate is not wasted. In addition to being an important metabolic intermediate for the muscle cell itself, it is a very important molecule for the transfer of energy substrates from one muscle cell to another. Once phosphorylated, a glucose molecule is trapped inside the muscle cell, and it is not liberated from the cell until it is converted to pyruvate or lactate. Without any loss of energy, the process of pyruvate transformation to lactate can be operated in reverse. From that point, there are two alternative routes: (1) The pyruvate can be oxidized, or (2) it can be a substrate for a synthesis of glucose and glycogen. When oxidized, it yields the remaining 92% as energy, and the heart muscle, kidney cortex (Keul, Keppler, and Doll 1967; E.V. Newman et al. 1937; Newsholme and Leech 1984), and skeletal muscles (both resting and exercising) can use lactate as a substrate (Ahlborg, Hangenfeldt, and Wahren 1976; Brooks and Fahey 1984; Gladden 2000).

It has been well established that lactate produced during exercise can be used for resynthesis of glycogen in the liver. It has been a matter of debate, however, to what extent such a synthesis (glyconeogenesis) can take place directly in mammalian muscles (Krebs 1964). Previously, it was believed that several of the essential enzymes for these pathways did not exist in the muscle. However, it has been shown that a set of key enzymes for one of several pathways does exist in the required amounts in the muscle.

Hermansen and Vaage (1977) reported from studies on humans that most of the lactate produced during repeated 1 min maximal running periods interspersed with 4 min rest periods was resynthesized to glycogen during recovery at rest, evidently directly in the muscles. The uptake of glucose in the muscle studied could not explain the increase in its glycogen content. According to the classic concept, about 20% of the lactate produced during exercise is reoxidized to pyruvate and then disseminated to

$CO_2$ and $H_2O$, and the remaining lactate is taken up by the liver and turned into glucose, which can be reconverted into glycogen or delivered to the blood. The muscles can then use this glucose in its glycogenesis to restore their glycogen depots (Krebs 1964). From the calculations by Hermansen and Vaage, approximately 75% of the lactate was reconverted into glycogen, but the pathway via the liver (the so-called **Cori cycle**) was not used.

With the oxygen uptake measured, not only during the period of exercise, as illustrated in figure 8.7, but also during 60 min of recovery, one can calculate the total energy yield. With lactate as the only substrate for all tissues in their aerobic metabolism during recovery, at the most about 50% of the lactate is removed by the **Krebs cycle** and respiratory chain. During recovery, the muscle's glycogen content increases significantly and this process is energy demanding (i.e., in a way, there is a payment of a debt).

From what we know about differences between muscle fiber types, the fate of lactate produced during a bout of exercise depends on several factors, the more important of which appear to be the muscle fiber type pattern of the muscles involved and their recruitment pattern, and the training status of the individual in relation to the intensity of the exercise. According to Donovan and Pagliassotti (2000), all fiber types in their rabbit muscle preparations released lactate when perfused with 1 mM lactate, which is the resting lactate concentration. As the lactate concentration in their perfusate was increased, all muscle fiber types showed a transition from net lactate release to net lactate uptake, but this transition took place at lower lactate concentrations in type I muscle fibers than in type II. In type I muscle fibers, lactate is disposed of primarily by oxidative degradation, whereas in type II fibers glyconeogenesis seems to be more important. The exact pathway for this glyconeogenesis has not yet been settled, but there are indications that it is via an extramitochondrial path (Donovan and Pagliassotti 2000).

## How Does Continued Light Exercise Influence the Lactate Removal?

E.V. Newman et al. (1937) noticed that the removal of lactate after exhausting exercise was enhanced if the subject continued to exercise, but at a lower intensity that normally did not produce any lactate. This observation was confirmed by several investigators (Belcastro and Bonen 1975; Gisolfi, Robinson, and Turrell 1966; Hermansen and Stensvold 1972; McMaster, Lawler, and Lee 1989). It is logical that this is attributable to lactate being used as a substrate for oxidative metabolism in the active muscles, partly replacing glycogen, glucose, and free fatty acids (Gladden 1991). Hermansen and Stensvold (1972) reported that the optimal removal rate of lactate occurred when the intensity of the active recovery exercise was around 60% of maximal oxygen uptake. Other investigators have recommended exercise intensities that vary between 30% and 50% of $\dot{V}O_2max$ (Belcastro and Bonen 1975; Dodd et al. 1984).

Because of the increased oxidative capacity of skeletal muscle observed with endurance training, it has been thought that trained subjects exhibit a faster decline in blood lactic acid after a maximal exercise bout. However, both B.W. Evans and Cureton (1983) and D.R. Bassett et al. (1991) reported no differences in blood lactic acid disappearance between trained and untrained individuals during resting recovery from intense exercise.

Observations show that if a standardized exhaustive exercise is performed with blood and muscle lactate concentrations elevated in advance (because other muscle groups previously have been exercised to exhaustion), then the performance time is reduced (Karlsson et al. 1975). An athlete should keep this in mind when warming up before an athletic event. At least from a theoretical point of view, it appears that if interval training is conducted with the aim of elevating the lactate concentration to very high values, one should rest between bouts of exercise. On the other hand, if the main purpose of the training is to attain a high oxygen uptake and cardiac output, one can tolerate more repetition if the lactate level is relatively low. Evidently, light activity during the "resting" period provides the optimal effect. The logical recovery for an ice hockey player when off the ice during the game, for example, might be to exercise on a cycle ergometer at low intensity.

## Effects of Metabolism on Tissue and Blood pH

The hydrolysis of one ATP produces about one proton at the muscle cell pH. However, during the oxidative phosphorylation, all the products of ATP hydrolysis—adenosine diphosphate (ADP), inorganic phosphate (Pi) and $H^+$—are reutilized. This means that there is no net accumulation of $H^+$ in the aerobic metabolism.

From their review, Busa and Nuccitelli (1984) concluded that lactic acid accumulation is not the source of the intracellular pH decrease during anaerobic glycolysis. This is interesting and important from a theoretical point of view. However, the result of the sequence ATP hydrolysis plus rephosphorylation by energy from the breakdown of glycogen is a constant production of 2 mol $H^+$ for each mole of glucosyl (because of the opposite pH dependencies of $H^+$ production by glycolysis and by ATP hydrolysis). In a way, the net effect is what the traditional formula reveals:

$$\text{glycogen or glucose} = 2 \text{ lactate} + 2 \text{ H}^+ \quad (1)$$

It is not surprising that the muscle pH is reduced during anaerobic exercise, from about 7.0 to 6.5 or even lower. Secondarily, the arterial blood pH can decrease from 7.4 to below 7.0. As discussed previously, the rate of the glycolysis is determined by the need to resynthesize ATP. Actually, the breakdown products of ATP—ADP, adenosine monophosphate, and Pi—are potent stimulators of key glycolytic enzymes, whereas ATP itself is inhibitory. In this way, the rates of ATP hydrolysis and lactate formation are coupled. It is not surprising that there is a high correlation between lactate concentration and pH values in blood samples taken at rest as well as during and after exercise (steady state as well as maximal; Keul, Keppler, and Doll 1967). These authors pointed out that because of the buffer systems of the blood, a 10-fold increase in lactate concentration causes only a 1.42-fold increase in $H^+$ concentration.

As discussed in chapter 6, one effect of the change in blood pH is a hyperventilation, that is, when the pulmonary ventilation per liter of oxygen uptake increases. Beyond a certain point, pulmonary ventilation does not increase linearly with the oxygen uptake but rather increases exponentially. More $CO_2$ is exhaled than produced in the aerobic metabolism. During recovery, the situation is the reverse.

It is suggested that an accumulation of protons is the factor that causes fatigue and failure to continue exercises involving the anaerobic metabolism. It is tempting to investigate whether manipulating the tissue and blood pH before an exhausting exercise of one or a few minutes duration would modify the exercise at the given work rate. A metabolic alkalosis can be induced by ingestion of sodium bicarbonate and an acidosis by ammonium chloride. In the 1930s, positive effects of an intake of sodium bicarbonate were reported, but later results were equivocal (Gerrard and Hollings 1992; Gledhill 1984). Some failed to find any performance improvements after induction of an alkalosis before standard exercise (McCartney, Heigenhauser, and Jones 1983, exercise time 30 s; A. Katz et al., 1984, exercise time 100 s). On the other hand, Gledhill (1984) reported a significantly faster 800-m race after "soda loading" (control, 2:09.9; sodium bicarbonate, 2:05.8). In his review, Gledhill (1984) quoted studies indicating that bicarbonate ingestion can improve performance in anaerobic events such as 400- to 1,500-m races. There is an ethical problem, however: Do sodium bicarbonate or chemicals with a similar effect on the body's pH in various compartments fall under the general definition of doping agents?

# Interaction Between Aerobic and Anaerobic Energy Yield

This is an attempt to summarize parts of the preceding discussion. Table 8.1 presents the contribution to energy output from aerobic and anaerobic processes in maximal efforts in exercise involving large muscle groups. The individual's maximal aerobic power is set to 5 L · $min^{-1}$, equivalent to about 100 kJ (24 kcal) · $min^{-1}$ and maximal anaerobic capacity to 200 kJ (48 kcal) · $min^{-1}$, equivalent to 9 L of oxygen uptake in aerobic exercise. It is assumed that 100% of the maximal oxygen uptake can be maintained for 10 min, 95% for 30 min, 85% for 60, and 80% for 120 min. For nonathletes, the figures are roughly 50% of the values listed.

For exercise periods up to 2 min, the anaerobic power is more important than the aerobic contribution; at about 2 min there is a 50:50 ratio, and with prolonged exercise the aerobic power becomes gradually more dominating. This is illustrated in figure 8.10.

It is very rare that an individual possesses top power for both aerobic and anaerobic processes. Therefore, the analysis in table 8.1 should not be interpreted and applied too literally. A maximal aerobic power of 100 kJ (24 kcal) · $min^{-1}$ can be coupled with a maximal anaerobic yield of 100 kJ. For this individual, the proportional participation of anaerobic and aerobic processes will be different compared with the tabulated data.

An analysis of the energetic demands of different sporting events and the athlete's capabilities to fulfill these requirements may help him or her both in training and in selecting suitable events. One factor to consider is that endurance athletes have skeletal muscles with a high proportion of slow-twitch fibers, and sprinters are characterized by a dominance of fast-twitch fibers. However, in many

**Table 8.1** The Contribution to Energy Output From Aerobic and Anaerobic Processes in Exercise Involving Large Muscle Groups

| | Exercise time, maximal effort | | | | | | | |
|---|---|---|---|---|---|---|---|---|
| **Process** | **10 sec** | **1 min** | **2 min** | **4 min** | **10 min** | **30 min** | **60 min** | **120 min** |
| *Anaerobic* | | | | | | | | |
| kJ | 100 | 170 | 200 | 200 | 150 | 125 | 80 | 65 |
| kcal | 25 | 40 | 45 | 45 | 35 | 30 | 20 | 15 |
| % | 85 | 65–70 | 50 | 30 | 10–15 | 5 | 2 | 1 |
| *Aerobic* | | | | | | | | |
| kJ | 20 | 80 | 200 | 420 | 1,000 | 3,000 | 5,500 | 10,000 |
| kcal | 5 | 20 | 45 | 100 | 250 | 700 | 1,300 | 2,400 |
| % | 15 | 30–35 | 50 | 70 | 85–90 | 95 | 98 | 99 |
| Total | | | | | | | | |
| kJ | 120 | 250 | 400 | 620 | 1,150 | 3,125 | 5,580 | 10,065 |
| kcal | 30 | 60 | 90 | 145 | 285 | 730 | 1,320 | 2,415 |

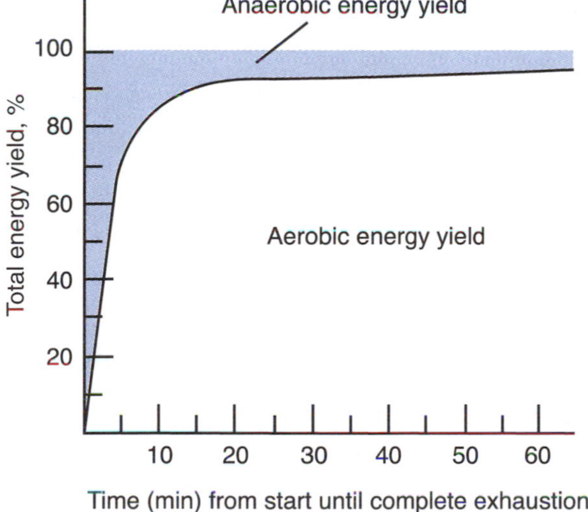


**Figure 8.10** Relative contribution in percentage of total energy yield from aerobic and anaerobic processes during maximal efforts of up to 60 min duration for an individual with high maximal power for both processes. Note that at a 2-min time period maximal effort hits the 50% mark, meaning that both processes are equally important for success.

events there is not such a strict pattern in fiber composition (Costill et al. 1976; Saltin and Gollnick 1983). So far, muscle fiber typing is not a good instrument for selecting potential athletes.

A trained individual can exercise at a relatively high oxygen uptake in relation to the $\dot{V}O_2$max without any continuous elevation in blood lactate concentration. For an untrained person, the critical point may be at 50% of the $\dot{V}O_2$max. For endurance-trained, top athletes, this relative work rate may be as high as 85% of the maximal oxygen uptake (see following section on "The Anaerobic Threshold Concept").

From a teleological point of view, it is very efficient that low-threshold, slow-twitch fibers are recruited during exercise of moderate intensity. They are enzymatically equipped for aerobic metabolism, and the myoglobin enhances the oxygen diffusion within the cells. The fast-twitch fibers, on the other hand, are well equipped for anaerobic intensive exercise. From a mechanical point of view, they are specialized for very intensive exercise. When recruitment of these fast-twitch fibers is required, as in the case of very vigorous exercise, their oxidative capacity is limited. With training, however, they can increased their number of mitochondria and improve their aerobic potential.

The conversion of the reduced form of nicotinamide dinucleotide **(NADH)** to the nonreduced form of nicotinamide dinucleotide **(NAD)** by the conversion of pyruvate to lactate is indeed a practical way of keeping the glycolysis going. Lactate can diffuse out of the lactate-producing muscle cells, and it is an excellent substrate in the aerobic energy yield of neighboring fibers as well as for the glucose and glycogen resynthesis. With the anaerobic processes running at a high rate, an inevitable consequence is the accumulation of protons that can hinder physical performance.

## The Anaerobic Threshold Concept

An anaerobic threshold concept was introduced to define the point when metabolic acidosis and associated changes in gas exchange in the lungs occur during graded exercise (Wasserman et al. 1973). Already in 1936, Bang had pointed out that there was a metabolic phase in which the exercising person would encounter an accumulation of lactate in the blood as the exercise proceeded. Since then, considerable efforts have been made to establish the oxygen uptake in relation to the person's maximal aerobic power when the blood lactate concentration gradually starts to increase during continuous exercise. The work rate at this "breaking point" has been named **anaerobic threshold**, lactate threshold, or OBLA (N. Jones and Ehrsam 1982; Karlsson and Jacobs 1982; Weltman et al. 1990).

This concept poses several problems, not least because lactic acid is a normal metabolic intermediate of aerobic metabolism. In the first place, there is a problem of terminology: Are we talking about a threshold for a single muscle fiber, a whole muscle group, or a regulatory system involving chemoreceptors; or are we talking about the onset of blood lactate accumulation, or some other lactate effects, or a pH-dependent function?

Definitely, some muscle fibers use an anaerobic metabolism for the ATP resynthesis even before the muscle, as a whole, has reached its maximal oxygen uptake. In fact, lactate concentration may be higher in the white part of a muscle than in the red part, with the arterial concentration somewhere in between (Baldwin, Campbell, and Cooke 1977). The capillary density varies within the muscle and so do the myoglobin concentration and the amount of mitochondria. In particular, the type IIX fibers are at a disadvantage. An increased sympathetic activity accompanying intense exercise can induce a lactate formation even in resting muscles. There is a work rate at which the lactate uptake in the blood exceeds the removal so that the lactate concentration increases gradually. From this point, only a relatively small increase in the intensity of exercise can reduce the time to exhaustion from, say, 1 h to 15 min. A threshold value often used is an arterial lactate concentration of 4 mM. However, there are marked individual variations in this threshold value. Stegmann and Kindermann (1982) reported one example of 19 rowing athletes, 15 of whom failed to endure the work rate on the cycle ergometer that in the first test trial gave a lactate concentration of 4 mM. They had to stop after 14.4 min, work rate 242 W, lactate 9.6 mM. The individual anaerobic thresholds also were established and the mean value was 193 W. The blood lactate value in this "threshold test" was 2.3 mM. The work rate 190 W could now be continued for the intended time, 50 min, with a lactate concentration of 3.75 mM at the end. The perceived exertion was rated by the Borg scale (Borg 1982). The values were 18 (close to *very, very heavy*) at the end of the 4.0 mM experiment and 14.0 (between *somewhat heavy* and *heavy*) after the 2.3 mM ride. The established individual threshold was 4.0 mM lactate for three subjects and 6.1 mM for the remaining subject.

Roberg et al. (1990) investigated the differences between lactate thresholds determined from venous and arterial blood and found that arterial blood lactate concentrations were significantly higher than venous blood at 350 W workload, at maximal exercise, and throughout recovery. Arterial lactate concentration was significantly higher than venous blood lactate concentration at the OBLA. Drawing on these findings, they suggested that differences between venous and arterial blood lactate need to be considered when we compare different anaerobic threshold determinations.

## Problems in Determining Anaerobic Threshold

There are different patterns in the kinetics of lactate production and disappearance, between individuals, between muscles within an individual, and even between different muscle fiber types within a muscle. It is therefore not surprising that the increase in blood lactate concentration with a stepwise or ramp-wise increasing work rate is rather smooth, making it difficult to define a threshold, and that the variability of the threshold value for a given subject is large (average range 16% in the study reported by Yeh et al. 1983). The threshold value found also is modified by such factors as the athlete's nutritional state and speed of movement (E.F. Hughes, Turner, and Brooks 1982).

In some laboratories, 2.5 mM is used as a threshold value, whereas others use something in between this value and 4 mM. Different protocols are used to determine the threshold. Some use a stepwise increase in work rate, whereas others use a continuous increase and with different rates of increase in metabolic demands. Such variations make it difficult to interpret and compare data from different laboratories.

As already mentioned, an alternative noninvasive method to establish the anaerobic or lactate threshold has been used, that is, by determining at what rate of exercise/oxygen uptake a nonlinear increase in pulmonary ventilation occurs. The theoretical basis for this is that the lactate accumulation in the blood should reduce its pH and thereby increases the chemoreceptor drive to the central respiratory generators. There are, however, similar methodological problems in defining a reproducible respiratory threshold, as with the lactate threshold. Second, other factors than pH can contribute to the nonlinear increase in pulmonary ventilation with increasing oxygen uptake. Local hypoxia in the peripheral chemoreceptor area may be more pronounced in severe exercise. This will stimulate the peripheral chemoreceptors and increase pulmonary ventilation, yet it has nothing to do with the blood lactate level. A change in the impulse traffic from higher brain centers also can modify the ventilation, more so if the exercise requires mental effort. It is interesting that patients with McArdle's syndrome, who, because of a lack of phosphorylase activity, cannot use glycogen and form lactate, nonetheless react with an "abrupt" increase in pulmonary ventilation, similar to an anaerobic threshold (Hagberg et al. 1982).

In the study by Stegmann and Kindermann (1982), the perceived exertion was, as mentioned, around 14 at the anaerobic threshold, determined individually for each person. Purvis and Cureton (1981) found similar mean values, of 13 to 14 *(somewhat hard)* for their female and male subjects at anaerobic threshold, ranging from 11 to 16. For both groups, the percentage of the maximal oxygen uptake was around 60 ± ~8. This relatively wide scatter is not surprising. The endurance-trained person can exercise closer to her or his maximal oxygen uptake than someone who is untrained. For example, Hurley et al. (1984) reported a 26% increase in maximal aerobic power in their eight subjects after a 12-week training program. A blood lactate of 2.5 mM was attained at 68 ± 4% of the maximal oxygen uptake before and at 75 ± 3% after training. In eight competitive runners, the relative oxygen uptake was even higher, or 83 ± 2% at the blood lactate concentration 2.5 mM. (See also Denis et al. 1982.) Several factors may have contributed to this improvement. An increased oxidative capacity of the muscle fibers in question may have shifted the pyruvate balance toward removal, thereby lowering the level of both pyruvate and lactate. A change in the capillary density and enzyme pattern in the muscles may have contributed, and it is also possible that an increased lactate transporter capacity in oxidative fibers, attributable to training, lowered the extracellular lactate concentration (S.K. Baker, McCullagh, and Bonen 1998).

### FOR ADDITIONAL READING

For a brief review about the concept of intracellular pH threshold in relation to an anaerobic threshold, see Iwanaga et al. (1996).

### *Summary*

The concept of anaerobic threshold (lactate threshold) or OBLA is based on an exponential increase in blood lactate concentration when a person exceeds a certain rate of exercise/oxygen uptake. Anaerobic threshold usually is determined during incremental exercise to establish the point mentioned. Indirectly, a similar breaking point of pulmonary ventilation versus oxygen uptake has been applied, under the assumption that these two points are highly correlated. In many laboratories, testing has been standardized with the goal of finding the rate of exercise or oxygen uptake at which the blood lactate concentration reaches a value between 2.5 and 4 mM.

It is evident that the theoretical considerations behind this concept need further clarification, to say the least. It is difficult to establish a well-defined "point," even for one individual, mainly because the balance between the production and removal of pyruvate differs both between muscle and between muscle fiber types in the same muscle. The work rate at which a nonlinear increase in ventilation occurs need not be the same exercise rate at which the lactate concentration increases. There are individual variations in the highest lactate level that can be tolerated during prolonged exercise.

These critical notes do not mean that it is of no interest to study, in normal subjects as well as in patients, how intensively they can exercise without an accumulation of lactate, or to follow the lactate response in subjects submitted to a given rate of exercise or oxygen uptake. On the contrary, this measurement provides important information about the individual's aerobic potential and about the effect of training. The threshold concept, as such, however, rests on an unstable scientific foundation. At any rate, experienced endurance athletes know quite well themselves what rate of speed they can tolerate without fatigue caused by lactate accumulations. Thus, there is hardly any need for the coach to tell them on the basis of blood lactate analyses.

## Maximal Aerobic Power—Age and Sex

Chronological age is not a very good reference point when analyzing biological data, particularly for children and teenagers. It is an inevitable evolutionary consequence that individuals within a species are different in many ways. In this aspect, people are not born equal. Tanner, one of the pioneers in this field, established the general framework of the biological age (Tanner 1989). By regularly measuring physical characteristics such as height and weight at least twice a year and observing the development of secondary sex characteristics (pubic hair and development of external genitalia in boys; breasts and time of first menstruation in girls), scientists can follow the maturity of the young individual (figure 8.11). In addition, x-ray analysis of the carpal bones provides information about the so-called skeletal age (Tanner et al. 1983). It is evident that a longitudinal recording of the child's height, which is certainly easy to do, gives a good picture of the onset of puberty. During the first years in life the child grows rapidly, followed by a somewhat slower rate of growth for about 10 years. Then comes a second spurt, when height can increase, on average, about 7 cm in 1 year for girls and 10 cm for boys (based on data from Northern European and North American subjects; figure 8.12). The girls' first menstruation normally occurs closely after this accelerated growth period, with a time lag of about 1 year. As mentioned, the development of sex characteristics in both girls and boys is also related to this adolescent growth spurt, which often is referred to as the peak height velocity. A peak weight velocity is usually observed some 6 months later.

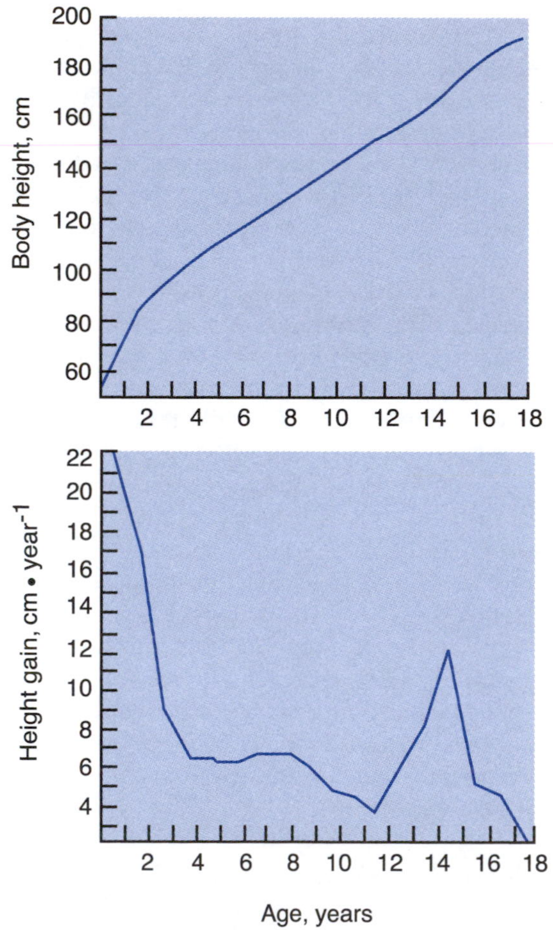


**Figure 8.11** The growth in height of a French boy over an 18-year period. The upper curve shows the gradual gain, the lower curve the height gain each year. Note the accelerated growth around the age of 14 to 15 years (peak height velocity).

From Tanner 1962.

To illustrate the fact that neither men nor women are born equal, figure 8.12 shows how the peak height velocity can occur in girls as early as the age of 9-1/2 years for one girl but not until the age of 15 years for another. In boys, one had his adolescent growth spurt at the age of 11 and another at age 17. Fifty % of the girls experienced their menarche between the ages of 12-1/2 and 13-3/4 years. With regard to sexual maturity, the girls are approximately 2 years ahead of the boys. At the age of about 13 years, the majority of the girls are already young ladies, while the boys are still kids. The pattern of the hormonal activity is the trigger mechanism for this development.

Although it is quite practical to characterize an individual on the basis of an age scale, it is biologically unsound. Yet it is not easy to find an alternative for such a classification. At any rate it is important to be aware of this problem. Because the adolescent growth spurt has a profound effect on physical performance, it is not only unfair but also possibly harmful to group children athletically in classes according to age. One early developed individual may be at the top of his or her class athletically for a period of time, causing the parents and coach to overestimate his or her athletic talent. In time, his or her classmates may catch up only to show that the early success was not attributable to talent but simply a matter of an early maturation.

In the aging individual, the genetic code may have more of an impact on the function of systems of key importance for physical performance than do environment and lifestyle. However, a change in lifestyle definitely can modify the "biological age," both upward and downward, at almost any chronological age.

## For Additional Reading

For further discussion about developmental aspects of physical performance parameters, see Krahenbuhl, Skinner, and Kohrt (1985), Ingjer (1992), Rowland (1996), and Armstrong and Welsman (1997).

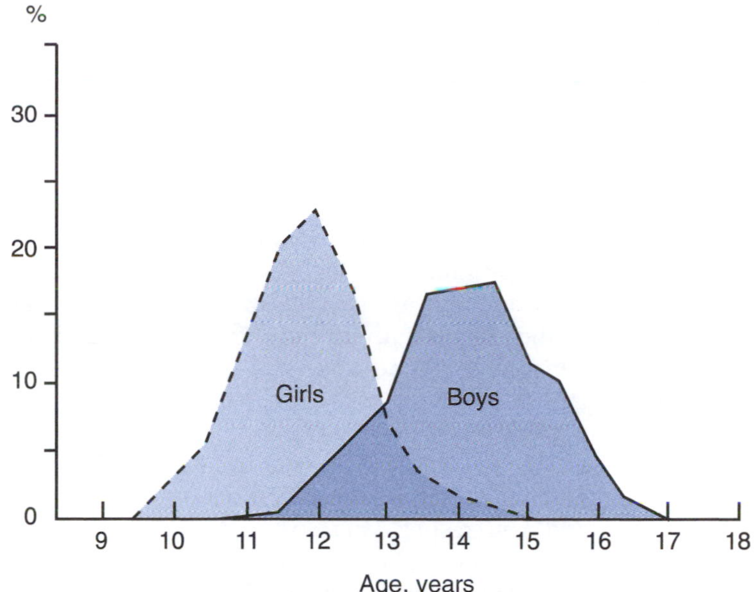


**Figure 8.12** Distribution of age at peak height velocity age for girls ($n$ = 358) and boys ($n$ = 373).

Adapted from Lindgren 1978.

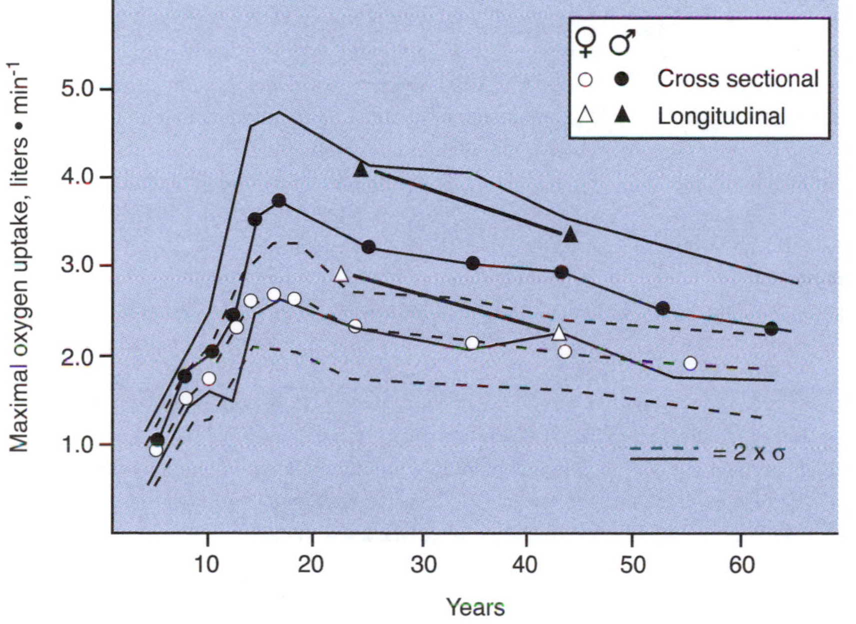


**Figure 8.13** Mean values for maximal oxygen uptake (maximal aerobic power) measured during exercise on treadmill or cycle ergometer in 350 female and male subjects 4 to 65 years of age. Included are values from a group of 86 students trained in physical education (from P.-O. Åstrand and Christensen 1964) and data from a follow-up study of 35 female and 31 male students belonging to the same group.

From I. Åstrand, P.-O. Åstrand et al. 1973.

## Maximal Aerobic Power—Absolute Values

The information provided by the assessment of maximal oxygen uptake is a measure of (1) the maximal energy output by aerobic processes and (2) the functional capacity of the circulation, because there is a high correlation between maximal cardiac output (and stroke volume) and maximal oxygen uptake. Note that the high correlation between these parameters is only valid when they are measured in liters per minute. The maximal oxygen uptake in milliliters per kilogram of body weight per minute is quite useless in this evaluation.

Direct measurements of the maximal oxygen uptake in 350 individuals ranging in age from 4 to 65 years are presented in figure 8.13. All subjects were healthy and moderately well trained; none was an athlete. It is almost impossible to present "normal" material, because it is very difficult to define what is normal. This material has been selected, but the age and sex factors that modify maximal aerobic power should be fairly evident in this homogeneous group of subjects.

Before puberty, girls and boys show no significant difference in maximal aerobic power. Thereafter, the woman's power is, on average, 65% to 75% of the man's. In a 3-year study of 34 girls, starting at ages 11 to 12 years, Laaneots, Karelson, and Viru (1996) observed age-related increases in maximal oxygen uptake, independent of the sexual maturation stage. Maximal aerobic power remained constant during the final stages of sexual maturation.

In both sexes, there is a peak in the maximal oxygen uptake at 18 to 20 years of age, followed by a gradual decline. At the age of 65, the mean value is about 70% of what it is for a 25-year-old individual. The maximal oxygen uptake for the 65-year old man (average) is the same as that typical for a 25-year old woman.

The individual variations should be noticed. Many older subjects have a maximal power that is higher than that found in many much younger individuals. In figure 8.13, the "±2 standard deviation line" for male subjects coincides closely with the average values for women, and the 95% range is actually ±20% to 30% of the mean value at a given age.

Similar data have been reported from other countries. Sex differences and changes in maximal oxygen uptake with age are very similar to those just described, but the absolute values vary. This can be explained by the different methods used to select subjects and variations in the average body size (e.g., in Sweden compared with Japan). A random sample is difficult to study successfully, particularly in older age groups. For subjects with different occupations, the mean values tend to vary with the nature of the occupation. It is not surprising that forestry workers have higher maximal oxygen uptakes than white-collar employees, for instance (I. Åstrand 1967b). Similarly, kibbutz dwellers, mostly engaged in agricultural activities, have a higher aerobic power than city dwellers (Epstein et al. 1981).

These differences in maximal aerobic power, as well as in many other parameters, are perhaps partly attributable to selection, because those who are endowed with a strong constitution are probably overrepresented in occupations that require physically demanding tasks. Furthermore, such jobs may in themselves train the oxygen transport system. The more mechanized the society, the less pronounced such differences may be between individuals from different occupations (I. Åstrand 1967a; Drinkwater, Horvath, and Welles 1975; Hermansen 1973; Hollmann and Hettinger 1980; Kobayashi 1982; R.R. Pate and Kriska 1984; Robinson 1938).

Figure 8.13 includes data originally obtained from a group of 86 physical education students. Their mean values are definitely higher than those for the average physically active woman or man, but the difference between the female and male students in maximal power is of the same magnitude as in other groups (females, 70% of the male maximum). Twenty-one years later, 66 of these students were restudied and without exception the maximal oxygen uptake was now reduced, on average, by about 20%. Thirteen years later they were again tested at maximal exercise, running and cycling. Of the entire group, 53 were able to participate in this third test. Their mean age was now 55 and 59 years for women and men, respectively. Surprisingly, over the intervening 13 years, the women had maintained their maximal aerobic power (mean value), but for the men there was an average decline of 4%. There were marked individual variations, however, in both groups. For some of the subjects who were still very active physically, the same maximum was now recorded as had been recorded 33 years before when they were physical education students (P.-O. Åstrand, Bergh, and Kilbom 1997).

It is not surprising that the maximal oxygen uptake increases during childhood and adolescence. During this period there is growth in all tissues of importance for strength and power. The sex differences are discussed later. But what about the gradual decline in maximal aerobic power beyond the age of 21?

There are many age-induced changes in tissue and organ functions that may explain this decline. A decline in maximal heart rate is quite evident (see figure 5.22). The lower maximal heart rate with higher age certainly must reduce the maximal cardiac output and hence the oxygen-transporting potential. In one of the few longitudinal studies just mentioned that included subjects with a relatively high level of habitual physical activity, the maximal heart rate declined from an average of 195 when subjects were in their mid-20s, to 176, 33 years later, a trend that fits figure 5.22. There is another consequence of a lower maximal heart rate. Studies on 33 building workers (bricklayers, carpenters, laborers) from 30 to 70 years of age showed that the mean heart rate during occupational activity was correlated with the individual's maximal heart rate (figure 8.14). For subjects with a maximal heart rate of 185 beats $\cdot$ min$^{-1}$, mostly the younger workers, the mean heart rate during occupational activity was

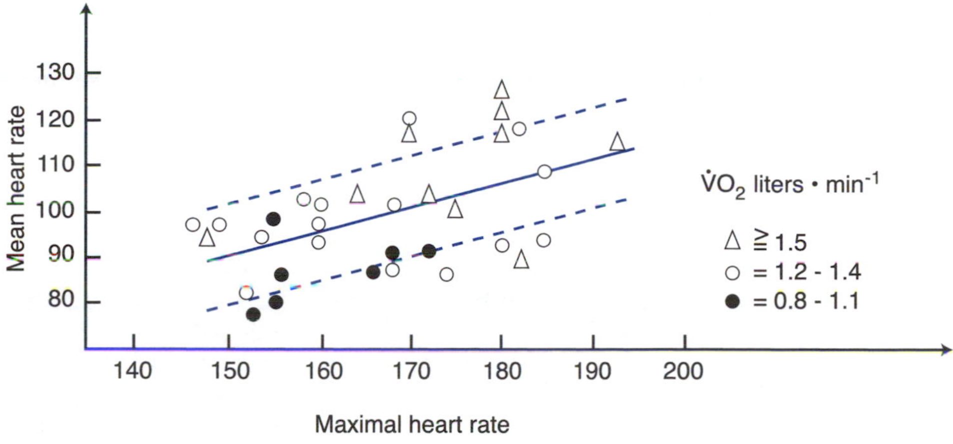


**Figure 8.14** Individual values for the relationship between mean heart rate during occupational activity (building) and maximal heart rate attained during exercise on a cycle ergometer. The heart rate was recorded by **telemetry.** The estimated oxygen uptakes during occupational work are presented on the right, together with their symbols.

From I. Åstrand 1967a.

110, and those with a maximum of 150 had a mean of 90 beats · min$^{-1}$. The maximal oxygen uptake ranged from 2.2 to 3.6 L · min$^{-1}$. In general, the worker used the same percentage of his maximum in the work operations irrespective of maximal oxygen uptake. In other words, the older worker with a lower maximal aerobic power keeps a slower tempo than the younger one, but the relative load is the same for the two workers, about 40%. The person with a high maximal heart rate can do a day's work at a higher mean heart rate than a person with a low maximum, but the relative strain may be the same on the two persons (I. Åstrand 1967b).

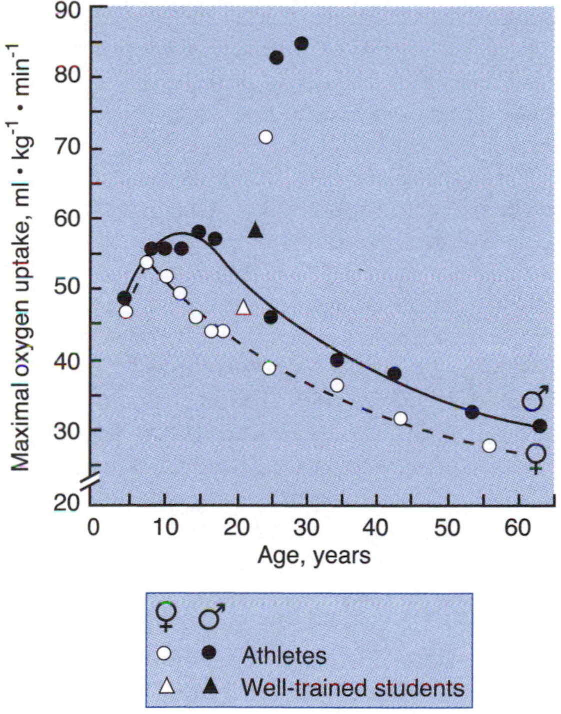


**Figure 8.15** Mean values for maximal oxygen uptake expressed in ml · kg$^{-1}$ · min$^{-1}$. Same subjects as presented in figure 7.13. The standard deviation is between 2.5 and 5 ml · kg$^{-1}$ · min$^{-1}$.

## Maximal Aerobic Power—Related to Body Size

According to Wilmore (1979), the average North American female at full maturity is approximately 13 cm shorter than the average male, 15 to 18 kg lighter in total weight, 18 to 23 lighter in lean body weight, and considerably fatter, i.e., 25% vs. 15% relative body fat.

Obviously, differences in body size are important, as well as how much of the sex differences in maximal aerobic power can be explained by differences in body size. Furthermore, in work and exercise where the body is lifted (as in walking, running, and climbing), the oxygen uptake potential related to body weight is certainly more relevant than the liter per minute figure. Figure 8.15 presents the data from the same study as in figure 8.13, but the maximal aerobic power is expressed in milliliters of oxygen per kilogram of gross body weight [ml · kg$^{-1}$ · min$^{-1}$, which also can be written as ml/(kg · min)]. (Too frequently one finds in the literature "ml/kg/min" or "ml/kg · min", both expressions being mathematically incorrect.) Also there is a general agreement in data reported from different countries regarding the slopes and relative positions of the curves, but the levels may differ. For young girls and boys there is no significant difference in the maximal aerobic power until the age of approximately 10 years, but from then on the females reach, on average, 75% to 85% of the males' maximum. Again, the variation in this sex difference can be explained by the sampling procedure.

The highest maximal aerobic powers reported so far are 94 ml · $kg^{-1}$ · $min^{-1}$ for a male and 77 ml · $kg^{-1}$ · $min^{-1}$ for a female cross-country skier, a sex difference in the same range as previously described.

Another way to express the individual's maximal oxygen uptake is to relate it to the dimensions of the body or to various organs. It may be of both theoretical and clinical value to examine whether the maximal oxygen uptake is proportional to heart size, muscular mass, or lung volume. Because fatty tissue is metabolically fairly inert, at least in this context, but can constitute a large proportion of the body weight, it may be important to exclude fatty tissue when evaluating the oxygen-transporting capacity.

When the weight of adipose tissue, estimated on the basis of **hydrostatic weighing,** is subtracted from the gross body weight of the well-trained students in figure 8.13 (12 kg or 20.3% of the body weight for the women and 8 kg or 10.6% for the men), the maximal oxygen uptake per kilogram of fat-free body weight (lean body mass, LBM) can be calculated. The average figure is the same for both groups: 71 ml · kg of $LBM^{-1}$. Other studies have reached similar results (Sparling and Cureton 1983; Wilmore 1979). However, because the aerobic metabolic rate at rest or during maximal exercise should vary with the body weight raised to the 2/3 power, it becomes evident that women should have a higher aerobic power per kilogram of lean body mass than men. The lower maximal oxygen uptake than expected may be attributable to the lower hemoglobin concentration found in women. Therefore, the lower maximal aerobic power of women might be natural, because their dimensions are different from those of men and the oxygen-binding capacity of the blood is lower. The relative increase in body fat content in women starts at puberty. In reality, a woman has to carry not only her LBM but also her fatty tissue; therefore, the maximal oxygen uptake per kilogram of gross body weight is a better expression of her potential to move her body than is the maximum related to free fat body weight.

Welsman et al. (1997) examined the relationship between thigh muscle volume and aerobic and anaerobic performance in 10-year-old children. They used a treadmill running test to exhaustion to determine peak oxygen uptake and a Wingate anaerobic test to determine peak power and mean power. The volume of the right thigh muscle was determined by **magnetic resonance imaging (MRI).** They found that when body size is appropriately accounted for by using **allometric scaling,** the thigh muscle volume is unrelated to aerobic and anaerobic power in 10-year-old children (see also Janz et al. 1998).

## A High Maximal Oxygen Uptake Does Not Guarantee Top Performance

An interesting question is why the best performance, in events demanding a high aerobic power, is usually obtained by athletes 20 to 30 years of age, when the highest maximal oxygen uptake is usually reached before the age of 20 (see figures 8.13 and 8.15). There are, however, other factors to be considered. In general, physical activity is more regular and vigorous for those below than for those above 20 years of age, at least as long as physical education is compulsory in schools. This may explain the results presented in the figures mentioned. On the other hand, if training is continued, the maximal aerobic power certainly can be maintained or even further increased for another 10-year period. Finally, performance also depends on technique, tactics, motivation, and other factors, and intensive training and experience over the years lead to gradual improvement.

Physical performance is related to the maximal oxygen uptake in exercises with large muscle groups vigorously involved for some minutes or longer. No one can attain top results in such exercises without a high aerobic power. On the other hand, a high power does not guarantee a good performance, because technique, state of training, and psychological factors have a positive or negative influence. It is tempting to look at how accurately one may predict a person's performance in an all-out run that taxes the oxygen transport system maximally, on the basis of the measured maximal oxygen uptake (ml · $kg^{-1}$ · $min^{-1}$). Many such studies have been made. Shephard (1984b) reviewed 37 such reports and noted coefficients of correlation from .04 to .90.

By choosing subjects with a wide range in $\dot{V}O_2max$, it is easier to obtain a high correlation.

A reliable prediction in the individual case, however, demands a correlation well above .90. A "significant" correlation is in itself useless in this context. An important conclusion is that data on maximal oxygen uptake do not reveal the person's potential to perform well in events that demand aerobic power.

For females, the peak maximal oxygen uptake in milliliters per kilogram per minute is attained at a very young age. They are not very good endurance runners, however. They have, like boys, a relatively poor running efficiency, which may explain the fact that their running performance is unrelated to their relative maximal aerobic power.

At the other end of the age scale, the gradual decline in maximal aerobic power has its consequences. To walk at the modest speed of 4 km · $h^{-1}$ on a level surface costs approximately 20 kJ (5 kcal) for a person weighing 70 kg. To cover that energy cost aerobically demands an oxygen uptake of 1 L · $min^{-1}$. There are many individuals 60 and over, particularly women, with maximal oxygen uptake well below 2 L · $min^{-1}$. (Note that the subjects represented in figure 8.13 were selected and relatively well trained.) These individuals would experience such walking as being quite strenuous. Any overweight will increase the energy cost in proportion to the additional number of kilograms. The oxygen requirement for a 100-kg person walking is about 1.5 L · $min^{-1}$; in this case, the walk may end up being an all-out effort. It is easy to understand that aging obese persons will gradually reduce their habitual physical activity with a resulting decrease in fitness, in a typical **circulus vitiosus** pattern. There are two solutions to this problem: the obese person (1) can devote more effort to losing weight while still middle-aged and (2) can increase the maximal oxygen uptake by training, thus expanding the safety margin between the oxygen requirement of daily routine activities and the maximal capacity.

## *Summary*

Particularly when discussing physical performance in relation to age, it is essential to realize that chronological age is a poor reference scale. For instance, in a class of 12- to 13-year-old children, a girl may be biologically 16 years old and a boy just 9. The adolescent growth spurt may start as early as at the age of 9 or as late as at the age of 17, with the average girl maturing about 2 years earlier than the average boy.

The maximal oxygen uptake expressed in liters per minute informs us about a person's potential for activity requiring a high aerobic power. It is highly correlated with the individual's maximal stroke volume and cardiac output (which limits the maximal aerobic power). The maximal oxygen uptake in milliliters per kilogram of body weight per minute indicates a person's potential to move the body during such activities as running or climbing stairs for several minutes or longer. The actual oxygen uptake that can be maintained is at a certain percentage of the $\dot{V}O_2max$, this percentage decreasing the longer the exercise is carried on. In many events, aerobic performance can be predicted by a certain minimal requirement, but other factors such as technique, motivation, and, in prolonged activity, muscle enzyme functions, muscle capillary density, and the supply of substrates play important roles in determining performance.

The maximal oxygen uptake (maximal aerobic power) increases with age up to 16 to 17 years in females and 18 to 20 years in males. After these ages there is a gradual decline, so that the 60-year-old individual attains about 70% of the $\dot{V}O_2max$ that he or she had at age 25. Before the age of 10, there is no significant difference between girls and boys; thereafter, the average difference in maximal oxygen uptake between women and men amounts to 25% to 35%. Related to body weight, the sex difference in maximal aerobic power after puberty is 15% to 25%. Calculating the maximal oxygen uptake per kilogram of fat-free body weight, or related to muscle mass, blood volume, or other such parameters, makes it possible to analyze dimensions versus function. When expressed per kilogram of fat-free body weight, the maximal aerobic power for women and men is very similar, particularly when they are well trained.

Training and lack of regular physical activity can significantly modify the maximal oxygen uptake. Top athletes in endurance events have a maximal aerobic power that is about twice as high as that of an average person. The basis for such high levels is a combination of constitution and effective training. The gradual decline in maximal oxygen uptake with age beyond 20 is attributable, at least in part, to a reduction in maximal heart rate. Physical inactivity, however, is the most important factor for the decrease in functional range of the oxygen transporting system. Inactivity reduces the stroke volume and perhaps the efficiency of the regulation of the circulation during exercise. The margin between the aerobic demands of various daily activities and the maximal aerobic power becomes narrower, and eventually this will create problems. Particularly vulnerable is the obese person, because any excess weight increases the energy cost of moving the body.

## Anaerobic Power—Age and Sex

Earlier in this chapter it was emphasized that we lack reliable methods to measure anaerobic power. According to di Prampero (1981), the average man is 15% to 30% superior to the average woman in maximal alactic anaerobic power (calculated per kilogram of body weight). The 60-year-old person has an alactic anaerobic power that is 60% of the value for one who is 20 years old. Applying the Margaria staircase sprint, di Prampero and Cerritelli (1969) calculated the anaerobic energy output in African natives and reported it as equivalent to oxygen delivery per kilogram of body weight. Their figure was 88 ml · $kg^{-1}$ · $min^{-1}$ for 8-year-old girls and boys. It increased to 143 and 165 ml · $kg^{-1}$ · $min^{-1}$ for boys and girls, respectively, at the age of 20. This power reflects mainly the potential of the high-energy phosphate compounds. Because children have a similar concentration of ATP and PC in their muscles as do adults, this improvement in a sprint lasting some 6 s is surprising. It may perhaps essentially reflect an improved technique with age, including a greater ability to use the elastic properties of the muscles. In a 30-s all-out test, teenage girls produce 20% to 25% less work than do boys. Part of this difference is attributable to the small muscle mass of the girls.

Prasad (1996) examined the relationship between aerobic and anaerobic exercise capacities in prepubertal children. He observed that the children who do best aerobically also do best anaerobically in the prepubertal age.

Chamari et al. (1995) showed that in endurance-trained athletes, the age-related difference in anaerobic peak power output was significantly greater than that of peak aerobic power. Pankey, Bacharach, and Gaugler (1996), on the other hand, found in two groups of physically active women, 18 to 29 and 30 to 42 years of age, that average and peak anaerobic power neither declined nor increased with increasing age in fit women who participated regularly in aerobic activities.

As mentioned, the peak blood lactate concentrations in the muscles and the blood do not reflect anaerobic power or capacity. However, they provide information about the extracellular environment after exhausting exercise. In most studies, children and older individuals do not attain as high blood lactate values as young adults. For 68 boys and girls, the peak recovery lactate value after about 5 min maximal treadmill exercise was, on average, 9 mM, which is lower than a "normal" maximum (P.-O. Åstrand 1952). Robinson (1938), in his classic study of male subjects, made a similar observation. B.O. Eriksson (1972) reported a lower muscle lactate concentration in 11-year-old boys (about 10 mM · $kg^{-1}$ of wet muscle) than usually noted in older subjects after an all-out effort (20 mM or higher). He observed that the concentration of a rate-limiting enzyme in the glycolysis, phosphofructokinase (PFK), was lower in children than in adults. This may be one factor explaining the children's lower peak lactate concentration. Actually, after a period of training, the PFK concentration increased and so did the lactate concentration after maximal exercise. In all such tests, motivation always plays a decisive role, and children in general might be more reluctant to push themselves to their "anaerobic limit." However, high lactate values in children have been observed (Cumming et al. 1980; Shephard 1982). There is no significant sex difference in peak lactate concentrations.

### For Additional Reading

For further discussions about definitions, limitations, and unsolved problems concerning measurement of anaerobic capacities, see S. Green and Dawson (1993).

## Muscle Strength—Age and Sex

In any test of muscular strength, the day-to-day variation is usually on the order of ±10%. The correlation between the strength of different muscle groups in the same individual is low, moderate, or fairly high, depending on which muscle groups are compared.

### Age

Figure 8.16 presents curves for maximal muscle strength in relation to age for males and females. A peak is usually reached at the age of 20 for men and a few years earlier for women. The strength of the 65-year-old person is, on average, 75% to 80% of that attained between the ages of 20 to 30, with a further decline to about 60% in leg and back muscles and to

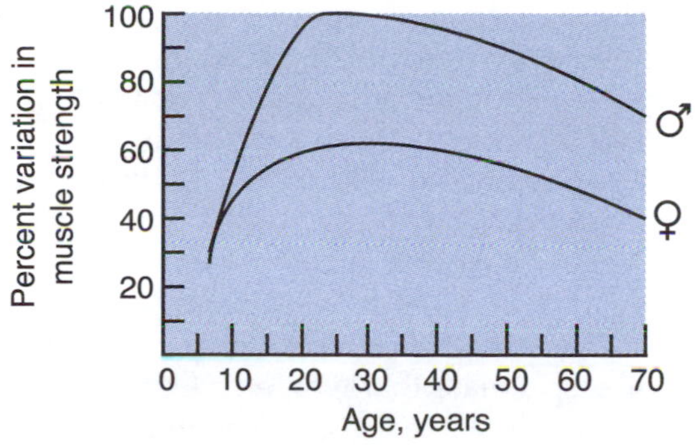


**Figure 8.16** Changes in maximal isometric strength with age in women and men.

Courtesy of E. Asmussen.

70% in arm muscles from 30 to 80 years of age. In other words, the rate of the decline with age in the strength of the leg and trunk muscles is in both sexes greater than the decline in the strength of arm muscles (Grimby and Saltin 1983; Hollmann and Hettinger 1980). Some studies have provided strength curves that are flatter than the ones presented in figure 8.16 up to the age of 50 (Larsson, Grimby, and Karlsson 1979). As in the case of maximal aerobic power, training can improve muscular performance by increasing muscle strength and by increasing the engagement of muscle synergists in the activities of everyday life. This greatly influences the test results.

The decline in maximal muscle strength with age seems to parallel the reduction in muscle mass. The individual muscle fibers have a relatively "normal" cross-sectional area during adult life, up to the age of 80 years, with some diminution in the area of the fast-twitch fibers (type II; Grimby and Saltin 1983). The reason for the reduced muscle mass, and hence the reduced strength in older persons, is a loss of muscle fibers, perhaps down to some 60% of the initial number. A loss of motoneurons is most likely the cause of this type of degeneration. It is not surprising that older individuals have a relatively poor power potential for performing a vertical jump (C.T.M. Davies, White, and Young 1983).

In the case of children, age affects muscle strength by increased size of anatomical dimension and by the impact of aging in itself. One extra year before puberty increases the strength by 5% to 10% of the average strength of the same group, a gain that at least partly is attributed to the maturation of the central nervous system (Sale 1992). Furthermore, the development of the child's sexual maturity is of special importance for the development of his or her muscle strength (as discussed subsequently). As a matter of fact, about one third of the increase in body height occurs between the ages of 6 and 20, but during the same period, four fifths of the development of strength takes place.

Knussmann and Weder (1995), in a study of 165 12- to 14-year-old boys and girls, found that age affects physiological measures, including grasping strength, primarily through its influence on height and weight. In a study of the correlation of the trunk muscle strength with age in 5- to 18-year-old boys and girls, Sinaki et al. (1996) concluded that age, height, and weight are all important predictors of trunk strength in children.

Backman et al. (1995) measured isometric muscle strength and muscle endurance in normal persons aged 17 to 70 years. The mean strength of women was about 65% to 70% that of men, but when the results were related to weight, the differences almost disappeared. Both men and women seemed to have the greatest muscle strength at the age of about 17 to 18 years. The strength was rather constant up to the age of about 40 years, after which a discrete decline was seen up to about 60, from where the decline was more obvious. Muscular endurance showed great individual variability, but no decrease in endurance was seen in older ages.

## Sex

An increase in muscular strength is definitely a consequence of growth. In young children, up to the age of 12 years, there is no significant difference in strength between girls and boys, although there is a trend for boys to be stronger. After this age boys become continuously stronger for some years, whereas the girls do not improve much in muscle strength (Tanner 1989). The most likely explanation of this difference in development is the greater secretion of the hormone testosterone in the male. In a nonobese woman, the muscle mass is 25% to 35% of the body weight, but in a man it is 40% to 45%.

Is there any difference in the strength per unit muscle mass in males and females? Applying ultrasonic measurements, Ikai and Fukunaga (1968) noticed almost the same maximal isometric strength of the elbow flexors in female and male subjects, aged 12 to 20 years (approximately $60 \ N \cdot cm^{-2}$). Today, the measurement of the cross-sectional area of a muscle group can be performed by **computed tomography (CT).** Applying this method and analyzing tissue samples secured by muscle biopsies, researchers have obtained the following results: (1) The number of muscle fibers in a given muscle group is the same in bodybuilders and in female and male physical education students; and (2) the maximal tension developed per unit of cross-sectional area does not differ between the female and male students or the bodybuilders (figure 8.17). Therefore, the difference in maximal strength is apparently mainly explained by a difference in size of the individual muscle fiber. Costill et al. (1976) presented data on fiber size in untrained women and men and, on average, the cross-sectional area for slow-twitch fibers in women was approximately 70% of the men's size; for fast-twitch fibers it was about 85%. The scatter is very large, however.

In a study of 23 men and 29 women about 80 years of age, Danneskjold-Samsøe et al. (1984) found that muscle strength was significantly lower in women than in men in all muscle groups except for plantar and dorsal flexors of the foot.

In absolute figures, women's maximal strength in isometric or dynamic contractions of the leg muscles (e.g., the knee flexors or extensors) is, on average, 65% to 75% of men's maximum. For trunk muscles, the values are 60% to 70%. In elbow flexion and extension, women are at a greater disadvantage, because they can attain only about 50% of the men's maximum. The cross-sectional area of arm muscles is not in proportion to women's body size. One may speculate that women's leg muscles are relatively well trained because women have less muscle mass in relation to their body weight than do men (Maughan 1984; Nygaard 1981; Schantz et al. 1983).

Glenmark et al. (1994) studied the relationship between muscle strength and muscle fiber type in 55 men and 26 women at the ages of 16 and 27 years. In the women, the researchers found that the more type II fibers, the stronger the subject. A positive

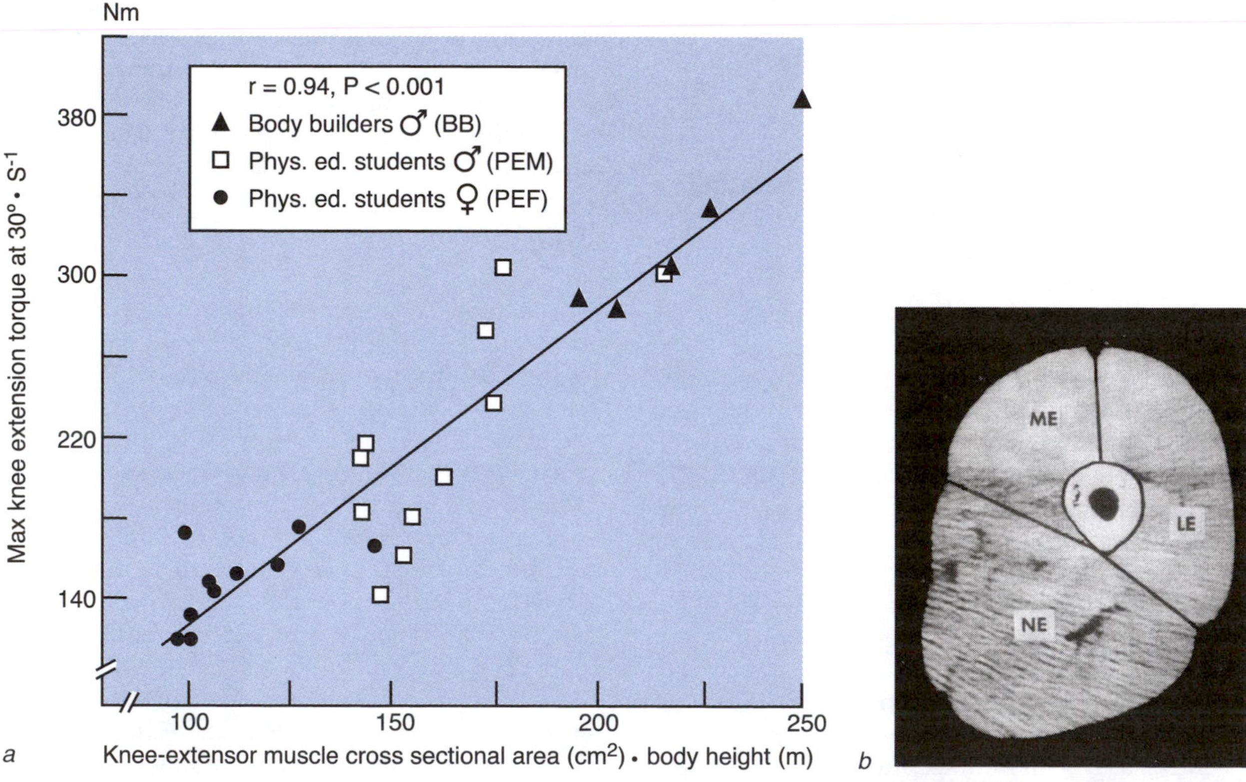


**Figure 8.17** *(a)* Relationship between muscle cross-sectional area times body height (proportional to the lever arm) in bodybuilders (BB) and physical education students, females (PEF) and males (PEM). *(b)* Muscle cross-sectional image of the left thigh, obtained through computed tomography scanning. The medial extensor (ME), lateral extensor (LE), and nonextensor (NE) muscle groups as well as bone are outlined. The contraction was isokinetic at an angular velocity of 30 B · $s^{-1}$. The total area of ME plus LE was for BB 125.0 $cm^2$, for PEM 88.3 $cm^2$, and for PEF 66.5 $cm^2$ (i.e., 75% of the figure for their male colleagues). The mean muscle fiber area was 8,400 $mm^2$, 6,200 $mm^2$, and 4,400 $mm^2$ respectively.

From Schantz et al. 1983.

correlation between strength and the level of physical activity during leisure time was revealed in the women at both ages. The positive correlation between strength and type II fibers in the 16-year-old men had disappeared at age 27. No systematic relationship between strength and the level of physical activity was seen in the men at 16 or at 27 years of age. This might indicate that women are more dependent on physical activity than adult men to develop strength. Differences between individuals in the internal muscle architecture, in limb lengths, and in joint structures are important factors capable of influencing strength (Maughan 1984).

## Summary

The maximal strength per unit of cross-sectional area of the skeletal muscle is about the same in women and men, irrespective of age. In the beginning of a strength training program, an increase in strength can be observed without any change in the cross-sectional area of the muscles involved. A more extensive recruitment of motor units in the trained person explains this finding.

The aging person faces a reduction in the number of motoneurons and therefore a decline in muscle mass. An inevitable consequence of this is a decrease in muscular strength so that, on average, the 65-year-old person can develop 75% to 80% of the strength attained when he or she was young. Because women have a smaller muscular mass than men, it is not surprising that their maximal muscular strength is lower. In women, the maximal leg muscle strength is about 70% of that of the men, but in the case of elbow flexors, the maximal strength of the women is, on average, only about 50% of men's. Evidently, there is no sex or age difference in the "quality" of the skeletal muscles. Therefore, the muscle mass and the ability of the central nervous system to recruit it mainly determine the potential for developing strength.

# Performance in Sports—Women Versus Men

In many sporting events, women and men compete under similar environmental conditions, and a comparison of the results is justified when we evaluate fitness as related to sex. In running events, the world records for women are, on the average, 10% behind the men. They are closest in the 100-m run (8.5%); in the marathon, women are 12 % slower. In long jump, the difference is 25%, in speed skating about 8%, in bicycling just below 12%, and in swimming 6% to 10%. The smaller difference in some swimming events may be explained by a higher swimming efficiency in women (see chapter 16).

The most logical method of studying possible sex differences in, for example, marathon runners is to compare performance-matched subjects of both sexes. Helgerud, Ingjer, and Strømme (1990) found that men and women with equal performance ability over the marathon distance (about 3 h and 20 min) had approximately the same maximal aerobic power (about 60 $ml \cdot kg^{-1} \cdot min^{-1}$). For both sexes, the anaerobic threshold was reached at an exercise intensity of about 83% of maximal aerobic power, or 88% to 90% of maximal heart rate. The women's running economy was poorer; that is, their oxygen uptake during running at a standard submaximal speed was higher. The heart rate, respiratory exchange ratio, and blood lactate concentration also confirmed that a given running speed resulted in higher physiological strain for the women (figure 8.18). The percentage utilization of $\dot{V}O_2max$ at the average marathon running speed was higher for the women. For both sexes, the oxygen uptake at average speed was 93% to 94% of the oxygen uptake corresponding to the anaerobic threshold. Answers to a questionnaire showed that the women's training programs over the last 2 months before the marathon included almost twice as many kilometers of running per week compared with the men (60 and 33 km, respectively). The better state of training of the women also was confirmed by a 10% higher $\dot{V}O_2max$ in relation to lean body mass compared with the men. Thus, the difference between performance-matched male and female marathon runners seemed primarily to be found in running economy and amount of training. The poorer running economy was probably compensated by the higher $\dot{V}O_2max$. Such a higher utilization level in the women would be expected on the basis of their more extensive endurance training before the marathon (Helgerud 1994).

It has been postulated that women's achievements in sports will gradually approach, and even catch up with, those attained by male athletes, particularly in endurance events (Dyer 1984). So far this has not happened. In events that have been on the Olympic program for some 40 years, the sex differences in world records have remained relatively constant. We therefore still face the question whether there are real biological differences in physical performances between the sexes. Alternative

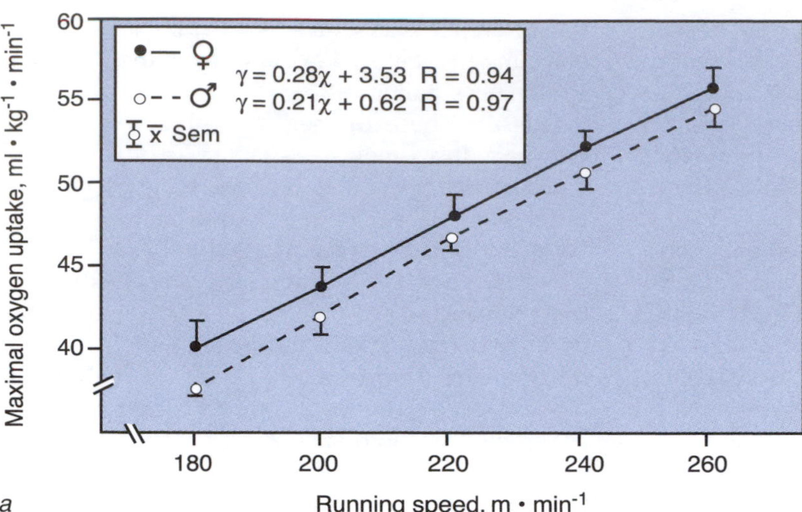


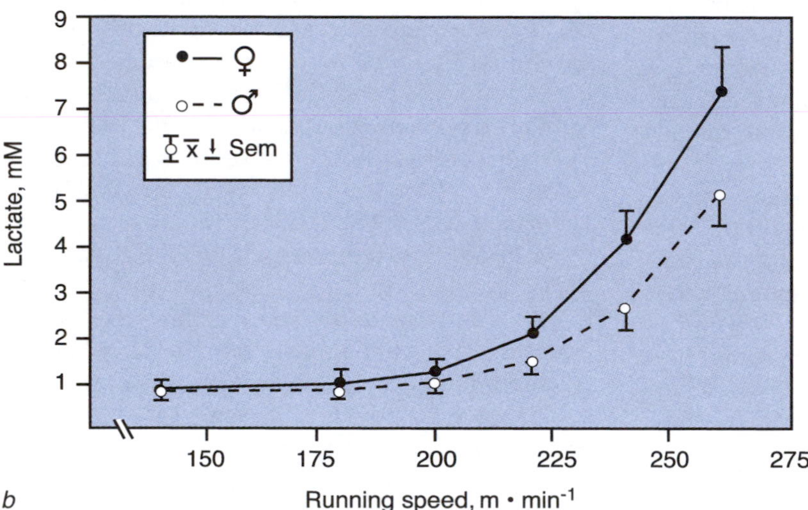


**Figure 8.18** *(a)* Average oxygen uptake in relation to running speed (treadmill at 1° inclination); the equations were derived from linear regressions. *(b)* Average blood lactate concentration in relation to running speed (treadmill at 1° inclination).

From Helgerud, Ingjer, and Strømme 1990.

explanations of the observed superiority of males in sports might be related to such factors as sociological or cultural traditions and biases, levels of selection, training, motivation, and competitiveness.

Wilmore (1981) quoted unpublished results by Grimditch and Sockolov, who "investigated why females perform so poorly in the softball throw. Postulating this difference to be the result of insufficient practice and experience, they recruited over 200 males and females, three to twenty years of age, to throw the softball for distance with both the dominant and nondominant arm. As they had predicted, there was absolutely no difference between males and females throwing with the nondominant arm up to the age of ten to twelve years, a pattern identical to that in the other motor tasks, whereas, with the dominant arm, the males were able to throw over twice the distance of the females at all ages. Thus, the softball throw for distance using the dominant arm appears to be biased by the previous experience and practice of the male."

Boys apparently practice ball throwing much more than do girls. The cause of this difference in behavior is still an open question. P.A. Silva et al. (1984) followed 954 girls and boys from birth to the age of 7 years and submitted them to a "basic motor ability test." At this age, the boys gained significantly higher scores than did the girls on the gross motor measurements, particularly in tasks requiring strength where shoulder girdle and arm muscles were involved, whereas the girls gained significantly higher scores in tasks requiring finer motor involvement. Gross motor performance factor was defined by long jump, agility run, target throwing, push-ups, and face-down to standing subtest; the fine motor performance factor was measured by the tapping board, balance, bead stringing, and hamstring stretch subtests. The authors suggested that different interests and therefore practice may explain the observed sex differences.

Apparently women's performances in sporting events are closer to men's results than mirrored in the laboratory tests. The female top runner has a relative body fat content similar to the male runner, that is, below 10%, in some cases down to 6% (Wilmore 1979). A smaller amount of fat to carry when running is certainly an advantage. For most women, according to Lebrun (1994), there is no significant effect of the phase of the menstrual cycle or the use of oral contraceptives on athletic performance.

## Genetics of Physical Performance

To attain top performance in a given event, one has to have the basic physiological endowment. In addition, one has to have the mental strength to endure the training involved and the determination to mobilize all of one's resources to win. The question is whether, or to what extent, the basic requirements that determine performance in a given event are based on genetic endowment.

Bouchard, Malina, and Pérusse (1997) reviewed this question extensively. They predicted that the exercise sciences will be strongly influenced by advances in human genetics and molecular biology. So far, the fairly extensive studies of the genotypic contribution to performance phenotypes are mainly of the genetic epidemiology type and are limited largely to comparisons of twins, siblings, and parent–offspring pairs, and they are of limited use in determining the genetic contribution to performance phenotypes.

As pointed out by Bouchard, Malina, and Pérusse (1997), estimates of the genetic contribution to individual differences in strength and motor tasks are based largely on younger subjects and twins. The available data suggest a genetic effect, but it seems to be moderate. Gene sharing seems to be associated with covariation between biological relatives in submaximal and maximal aerobic performances. Results of path analyses suggest that variation observed in performance phenotypes is associated mainly with nontransmissible environmental factors and that the contribution of the genotype is at best moderate.

Because lean body mass and muscle strength are both associated with bone mineral density, which is known to be under genetic control, Arden and Spector (1997), in a classic twin study, examined the size of the genetic component of both muscle strength and lean body mass and to what degree they account for the genetic component of bone mineral density. The authors concluded that grip strength, leg strength, and lean body mass have a moderate genetic component, and that the genetic component to muscle bulk and strength accounts for little of the genetic component to bone mineral density.

In the case of cardiorespiratory fitness phenotypes, genetic factors play a significant role in explaining interindividual differences (e.g., in heart size and cardiac function) (Malina and Bouchard 1991).

In a study of parent–offspring similarity in motor performance, Cratty (1960) compared the performances in running long jump and 100-yd dash of 24 college-age men with the performances of their fathers when the latter were of college age 34 years earlier. Father–son correlations were .86 for the running long jump and .59 for the 100-yd dash. This shows that fathers and sons attained fairly similar performances in these speed and power tasks when they were about the same age.

In the Framingham Children's Study (Moore et al. 1991) it was found that when both parents were active, the children were almost six times more likely to be active than children of two inactive parents. According to Mero, Jaakkola, and Komi (1991), heredity may partly affect the selection of sporting events.

In their review of the genetic effect on aerobic performance, Bouchard, Malina, and Pérusse (1997) pointed to the fact that intraclass correlations for **monozygotic twins** reach about .6 to .7, whereas those for **dizygotic twins** reach about .3 to .5. Generally speaking, the available twin studies are of unequal quantitative value, and estimated heritability ranges from near zero to more than 90%.

In a study by Simoneau et al. (1986), interclass correlations for anaerobic performance per unit body weight were much higher in monozygotic than in dizygotic twins. Bouchard et al. (1992) found the estimated heritability for short-term anaerobic performance assessed by total power output in 10 s per unit of body weight to be >50%.

One question of interest is the role of genetic factors in the response to training. In the case of aerobic training, there may be as much as nine times more variance between pairs of twins than within pairs of twins in the response of aerobic power to standardized training (Bouchard et al. 1990).

Data dealing with anaerobic training and strength training are limited. In a study of the effect of 15 weeks of high-intensity intermittent training of 14 pairs of monozygotic twins, Simoneau et al. (1986) found that the training response to long-term anaerobic performance (i.e., 90 s power output) was largely determined by genetic factors. The response to strength training, however, seems to be independent of the genotype (Thibault et al. 1986).

# Chapter 9

# Evaluation of Physical Performance on the Basis of Tests

In this chapter, we shall deal with an important aspect of work physiology, namely testing physical performance ability. This includes both the actual measurement of parameters reflecting basic physiological functions and the prediction of performance on the basis of data obtained at rest or submaximal exercise. Physical performance ability commonly is referred to as **physical fitness,** which is an umbrella concept covering a series of qualities related to how well an individual performs physical activity. Because most of the so-called fitness tests, including evaluation of flexibility, agility, speed, power, balance, and skill, are related to special gymnastic or athletic performance, they are not really suitable for an analysis of basic physiological functions. Practice and training in the performance of the actual test greatly influence the results.

### For Additional Reading

For reviews of commonly used physical fitness tests, the reader is referred to Vandewalle, Pérès, and Monod (1987), Cureton and Warren (1990), and the guidelines for exercise testing and prescription recommended by the American College of Sports Medicine (1995).

## Tests of Maximal Aerobic Power

In laboratory experiments, three methods of producing standard work rates have been mainly applied: running on a treadmill, exercising on a cycle ergometer, and using a so-called **step test.** In general, the exercise should involve large muscle groups, and the measurement of oxygen uptake should be initiated when the exercise has lasted a few minutes to permit the oxygen uptake to reach its maximum. Ideally, a definite plateau should be reached despite an increase in the rate of exercise. However, such a plateau is often difficult to establish in practice, as discussed in chapter 8.

### Type of Exercise

A critical question is which type of exercise will yield the individual's highest oxygen uptake. Consequently, a number of different procedures have been evaluated for applying a proper workload when assessing maximal aerobic power. The values obtained when running on a horizontal treadmill are approximately the same as when bicycling in the upright position, whereas running on a treadmill with 3° or greater uphill inclination gives 5% to 11% higher values. These are old observations that still hold true (Hermansen and Saltin 1969; Kamon and Pandolf 1972; McArdle, Katch, and Pechar 1973). However, the determination of maximal aerobic power during uphill treadmill running still might be inadequate when one is testing specifically trained athletes, particularly those whose endurance fitness to a large extent depends on the muscle groups of the upper extremities (i.e., cross-country skiers and rowers). Also in sport activities in which the performance mainly depends on the legs, the recruitment of muscle fibers and consequently the metabolic demand can be different even though the same muscle groups are involved in the movements (i.e., race cycling, speedskating, running).

Strømme, Ingjer, and Meen (1977) compared the attainment of maximal oxygen uptake in a group of 37 male and female athletes (cross-country skiers, rowers, and cyclists) during uphill running on the treadmill and during maximal performance of their specific sport activity. As demonstrated in figure 9.1, nearly all the subjects reached higher levels of

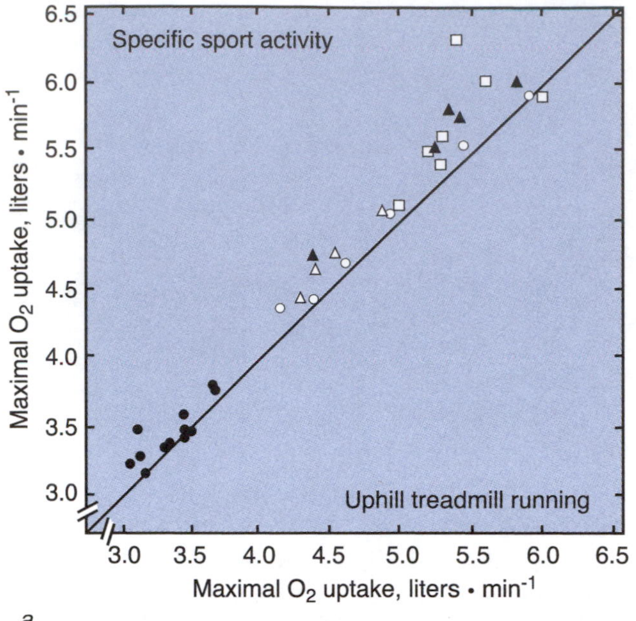


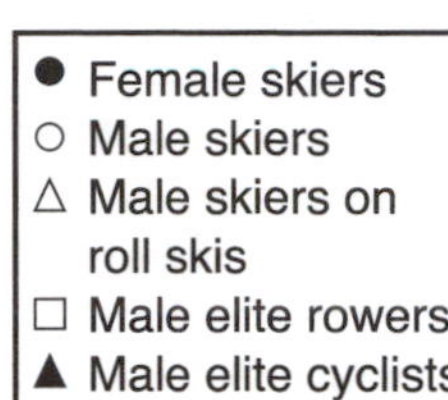


**Figure 9.1** *(a)* Individual values for maximal oxygen uptake ($L \cdot min^{-1}$) obtained during standard treadmill procedure (uphill running) and during maximal performance of specific sport activity of the subjects. *(b)* Individual values for maximal aerobic power ($ml \cdot kg^{-1} \cdot min^{-1}$) obtained during standard treadmill procedure (uphill running) and during maximal performance of specific sport activity of the subjects.

From Strømme et al. 1977.

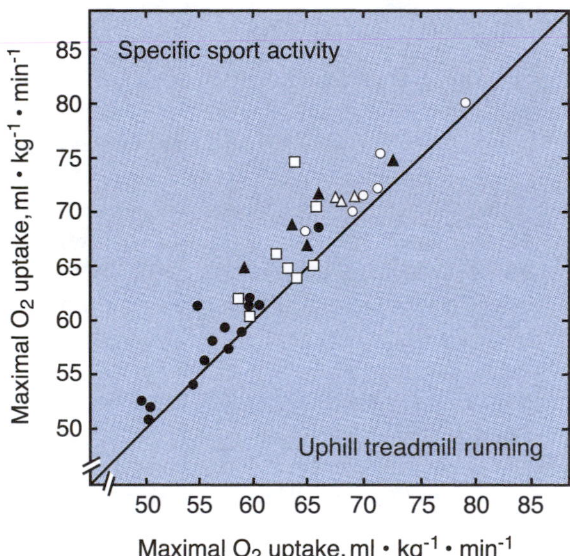


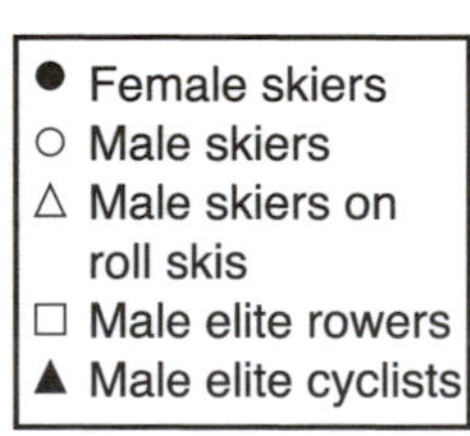


oxygen uptake when tested during their sport activity. For the skiers, the differences in oxygen uptake between uphill skiing and running were 2.9% and 3.1% for females and males, respectively. Rowers were tested on a lake in single sculler, double sculler, or two- or four-shelled boats, whereas the cyclists were tested during uphill bicycling on the treadmill. The rowers and cyclists obtained a difference of 4.2% and 5.6%, respectively. The largest individual differences between the two test procedures were 12.2%, 5.4%, 14.3%, and 7.9% for female and male cross-country skiers, rowers and cyclists, respectively.

Also, "ski-walking" on a motor-driven treadmill with an inclination of 12° at speeds between 60 and 160 $m \cdot min^{-1}$ (the subject walking with slightly bent knees and using ski poles as in cross-country skiing) has been shown to produce a significantly higher maximal oxygen uptake than maximal uphill treadmill running (Hermansen 1973).

These data emphasize that in the evaluation of maximal aerobic power of athletes, it is important to select a work situation that allows optimal use of the specifically trained muscle fibers. This means that the test preferably should be as similar to the subject's sport activity as possible, under the assumption

**Table 9.1** Maximal Oxygen Uptake Attained in Various Types of Exercise in Ordinary and Specially Trained Subjects

| Type of exercise | Ordinary subjects | Specially trained subjects |
|---|---|---|
| Running uphill | 100 | 100 |
| Arms and legs | 100 | 100–115 |
| Running horizontally | 95–98 | |
| Cycling, upright | 92–96 | 100–108 |
| Cycling, supine | 82–85 | |
| One leg, upright | 65–70 | 75–80 |
| Arms | 65–70 | 105–115 |
| Step test | 97 | |
| Rowing | 100 | 100–115 |
| Skiing | 100 | 100–112 |
| Swimming | 85 | 100 |

All values are expressed as percentage of the maximal oxygen uptake attained at uphill treadmill running, which represents the 100% reference value. For literature references, see Table 1 in Åstrand 1976.

that a large muscle mass is engaged during the performance. Table 9.1 summarizes the mean values for maximal oxygen uptake in various types of exercise in "normal" subjects as well as in individuals trained for special activities.

During cycling, the subject often can experience a feeling of local fatigue or a sensation of pain in the thighs or knees, which can be disturbing. This discomfort can cause the effort to be interrupted before the oxygen-transporting organs have been fully taxed. Particularly for persons who have never ridden a bicycle before, a maximal test on the cycle ergometer may be an unsuitable method to assess maximal oxygen uptake. The exercise position is of critical importance. The subject should sit almost vertically over the pedals. The seat should be high enough so that the leg is almost completely stretched when the pedal is in its lowest position. Motivation and stimulation of the subject are especially important in the case of cycling. When a person runs on the treadmill, it is, so to speak, a matter of all or nothing: the subject is forced to follow the speed of the belt or jump off. On the cycle ergometer it is usually possible to continue to exercise at a reduced rate in most types of cycle ergometers. Figure 9.2 summarizes results of studies by Hermansen and Saltin (1969). Despite the 7% difference in maximal oxygen uptake between running and cycling, the maximal pulmonary ventilation, heart rate, and blood lactate concentration were not significantly different in the two procedures.

Normally, a test of maximal aerobic power starts with a submaximal rate of exercise that serves as a warm-up. After this, the load can be increased in one of several ways: First, the load can be increased immediately to a level that in preliminary experiments has been found to represent the predicted maximal load for the subject; this level is maintained for 3 to 6 min. For the second method, the load can be increased stepwise with several submaximal, maximal, or "supermaximal" loads, the subject exercising for 5 to 6 min at each work rate, with or without resting periods between each load. In the third method, the load can be increased stepwise every or every other minute until exhaustion.

It is often of interest to obtain steady-state conditions when measuring oxygen uptake, pulse rate, and ventilation at submaximal work rates. This requires an exercise period of at least 5 min. The more or less continuously increasing rate of exercise in the third method is a quick method to reveal the subject's maximal oxygen uptake. However, because steady-state conditions are not attained at changing work rates, this procedure does not provide reliable information about how the oxygen transport problem is solved at different levels of physical effort, a type of information that is of considerable interest.

In examining patients, the observer can on one day apply method 3 to assess the patient's maximal performance and then another day apply a submaximal rate of exercise at a given percentage of her or his maximal oxygen uptake. During steady-state conditions, various measurements then can be taken (see N.L. Jones and Campbell 1982).

## *Summary*

If the objective is to determine the individual's maximal aerobic power, running uphill or combined arm

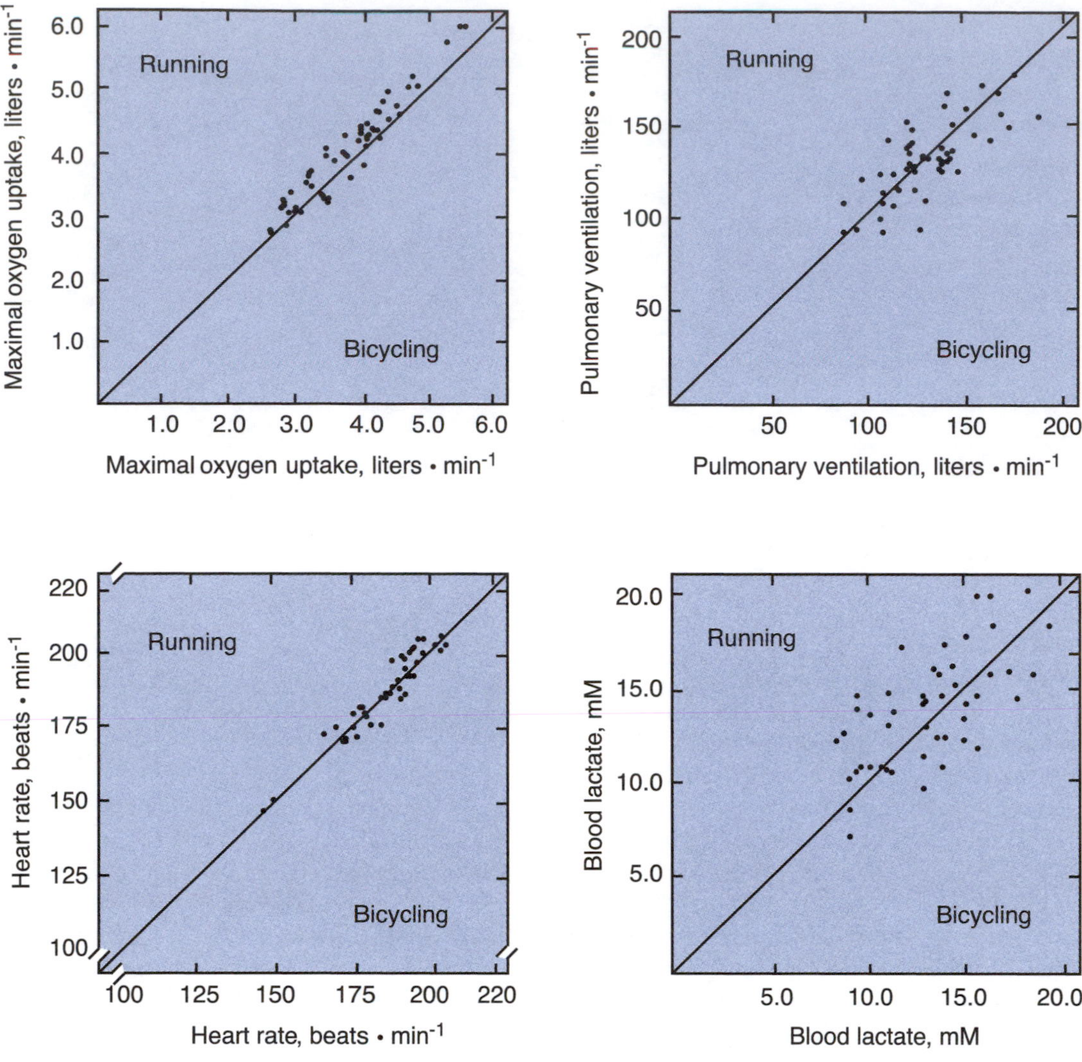


**Figure 9.2** Individual values for some functions studied in 55 subjects during maximal running uphill on a treadmill ( 3°) and work on a cycle er gometer (50 rev · min$^{-1}$) in a sitting position for about 5 min. Line of identity is drawn. A trend is noted for a somewhat higher maximal oxygen uptake during running compared with cycling, but pulmonary ventilation, heart rate, and blood lactic acid concentration were not different. A pedaling rate of 60 rev · min$^{-1}$ may give a higher maximal oxygen uptake than occurs with a pedaling rate of 50 rev · min$^{-1}$.

From Hermansen and Saltin 1969.

plus leg exercise is in general slightly superior to cycling or performing a step test. With athletes, it becomes important to select a work situation that allows optimal use of the specifically trained muscle fibers. This means that the test preferably should be as similar to the subject's sport activity as possible, under the assumption that a large muscle mass is engaged during the performance.

## Measurement of Oxygen Uptake

The classic method for determining oxygen uptake, the Douglas bag method, rests on a very secure foundation. It is theoretically sound, and it is well tested under a wide variety of circumstances. In all its relative simplicity, it is unsurpassable in accuracy. A disadvantage with the method is that the subject is somewhat hampered by the equipment required for the collection of the expired air. This limits the subject's freedom of movement. Furthermore, it merely provides an average figure for the oxygen uptake of, say, 60 s depending on the length of the time in which the expired air is collected. Figure 8.2 gives examples of how the original, but today less used, Douglas bag method can be applied when studying the aerobic energy output during exercise. Today's equipment allows for breath-to-breath analysis from the first second.

## *Treadmill Experiments*

The treadmill has an advantage when (1) the aim is to determine the highest oxygen uptake the subject can reach in a laboratory test, (2) it is of interest to measure the time the subject can endure a maximal trial, and (3) in studies on children. The work rate is dependent on the subject's body weight, a fact that can prove disadvantageous in longitudinal studies. Because the energy output per kilogram and kilometers per hour is more variable than it is at a given power on the cycle ergometer, the oxygen uptake should, whenever possible, be measured during walking and during running on the treadmill (coefficient of variation is about 15%; I. Åstrand 1960; Mahadeva, Passmore, and Woolf 1953). This requirement limits application of the treadmill. Older individuals may have some difficulty in walking on the treadmill. Providing a handrail for support will make the metabolic demand still more unpredictable. The energy cost of running (jogging) at a low speed is much higher than when walking (see chapter 16, figure 16.6). Therefore, it may be a speed at which some subjects prefer to walk (e.g., those with long legs) but others will prefer to run. The energy demand will be quite different and it cannot be predicted from the speed alone. One example is presented in figure 9.3. Note that at a given speed and slope, the oxygen uptake for this particular subject could vary from 2.2 to 3.2 L · min$^{-1}$ depending on technique—with or without support, walking or jogging.

It is evident from the preceding discussion that there is an advantage in keeping the speed of the belt constant and increasing the work rate by elevating the slope. The initial energy demand of an optimal inclination depends on the subject's aerobic fitness. Unfortunately, this fitness is impossible to assess from examinations of the subject when resting. It is more predictable during a submaximal exercise test but still is subject to misjudgment. It is, in a way, a trial-and error-situation. An optimal exercise time to establish a subject's maximal aerobic power is 5 to 10 min, the warm-up period excluded.

Table 9.2 presents one protocol that has been successfully applied. It is a running protocol developed for relatively well-trained people and athletes. To select a suitable starting speed and slope for a maximal run on the treadmill, the subject's maximal oxygen uptake (ml · kg$^{-1}$ · min$^{-1}$) is first predicted from the heart rate at a submaximal rate of exercise on treadmill or cycle ergometer (see subsequent list), and then the data presented in table 9.2 are applied.

1. Starting at the given initial speed and incline, the incline is increased by 1.5° (2.67%) every 3rd min, keeping the speed constant. Under these conditions, not even top athletes are able to run for more than 7 min. Experience has shown that it is advisable to give women a somewhat lower relative starting work rate than men (see table 9.2). Immediately preceding the maximal run, the subject is given a 10-min warm-up on the treadmill at a speed and slope corresponding to 50% of the selected starting work rate. The subjects are not allowed to hold onto the treadmill railing during the run.

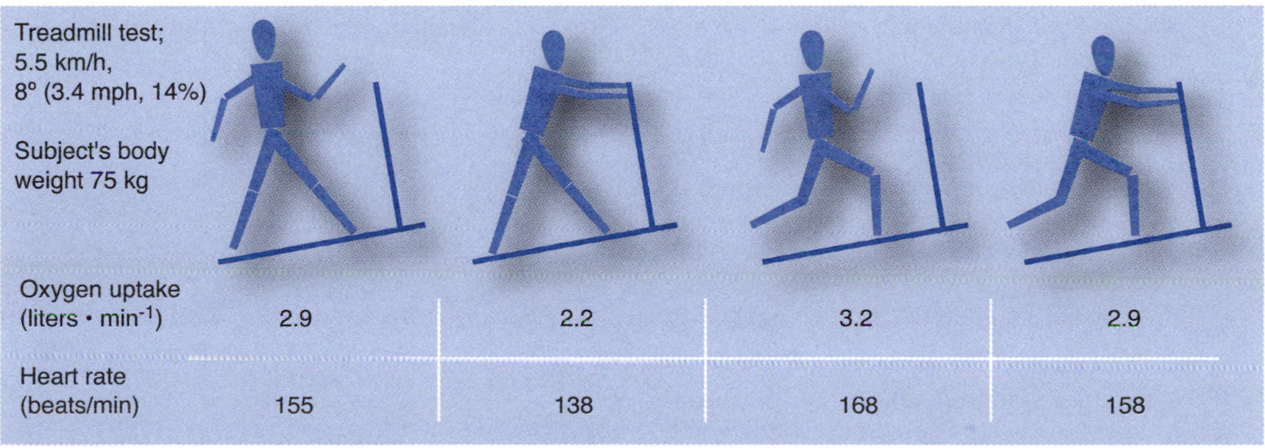


**Figure 9.3** Factors that influence the oxygen uptake when testing a subject on the treadmill at a given slope and speed.
From P.-O. Åstrand 1982.

## Table 9.2 Starting Work Rate Used for the Maximal Run on the Treadmill

| Predicted max $\dot{V}O_2$ ml · kg⁻¹ min⁻¹ | Starting speed and inclination for max run on treadmill* | | | | | | | |
|---|---|---|---|---|---|---|---|---|
| | ♂ Speed uphill | | | | ♀ Speed uphill | | | |
| | km · hr⁻¹ | Degree | mph | % | km · hr⁻¹ | Degree | mph | % |
| <40 | 10.0 | 3.0 | 6.2 | 5.25 | 10.0 | 1.5 | 6.2 | 2.52 |
| 40–50 | 12.5 | 3.0 | 7.8 | 5.25 | 10.0 | 3.0 | 6.2 | 5.25 |
| 55–75 | 15.0 | 3.0 | 9.3 | 5.25 | 12.5 | 3.0 | 7.8 | 5.25 |
| >75 | | | | | | | | |
| A | 15.0 | 4.5 | 9.3 | 7.9 | | | | |
| B | 17.5 | 3.0 | 10.9 | 5.25 | | | | |
| C | 20.0 | 1.5 | 12.5 | 2.62 | | | | |

*A = cross-country skiers, skaters, etc.; B = cross country runners, orientation runners, etc.; C = track runners.

Source: *Saltin and Åstrand: J. Appl. Physiol.,* 23:353, 967.

## CLASSIC STUDY

A widely used method of establishing an individual's maximal aerobic power was described by Bruno Balke in 1954. The principle of his protocol has survived the test of time. The treadmill speed is kept constant in a standard test procedure at around 5 km · $h^{-1}$ (3.0–3.5 mph). The slope of the treadmill is increased every minute or every other minute in steps of 2.5% (1.4°). With a speed of 4.8 km · $h^{-1}$ (3 mph) the increase in energy demand is about 3.5 ml · $kg^{-1}$ · $min^{-1}$ per step, which is one **metabolic unit (MET;** the approximate oxygen uptake at rest). Starting at 4 METs, at a slope of 2.5%, for example, energy demand increases to 5 METs at 5%, 7 METs at 10%, and 15 METs at 30%. One drawback with this protocol is that it takes a long time for fit subjects to complete the test. In such cases, a running speed may be chosen instead of walking.

Balke 1954.

2. Another procedure that in our experience has proved satisfactory is to let the subject run in bouts of 3 min, with 4- to 5-min rest periods between each running bout, keeping the treadmill inclination constant at 3° while increasing the speed by 15 m · $min^{-1}$ each time. It is advisable at first to allow the subject to become accustomed to treadmill running and, in particular, to learn how to jump on and off the running treadmill with ease. The actual test should be started at a running speed that is not far from the subject's maximal level yet no greater than a rate that the subject will feel confident of keeping up with for several minutes.

The protocol developed by Bruce (1971) is extensively used in the United States, particularly when testing patients for diagnostic purposes. In this test, speed and slope are increased every third minute (see table 9.3). Figure 9.3 illustrates how the subject's individual "technique" markedly modifies the oxygen uptake and therefore the heart rate at stage III in the Bruce test. If the oxygen uptake is not measured, its prediction from the subject's weight and the speed and inclination of the treadmill evidently has

## Table 9.3 Robert Bruce's Treadmill Protocol

| Stage | I | II | III | IV | V |
|---|---|---|---|---|---|
| Min | 3 | 3 | 3 | 3 | 3 |
| Speed | | | | | |
| mph | 1.7 | 2.5 | 3.4 | 4.2 | 5.0 |
| km · hr⁻¹ | 2.7 | 4.0 | 5.5 | 6.8 | 8.0 |
| Grade | | | | | |
| percent | 10 | 12 | 14 | 16 | 18 |
| degree | 5.7 | 6.9 | 8.0 | 9.1 | 10.3 |
| METs* | 4 | 6–7 | 8–9 | 15–16 | 21 |

*1 MET, a metabolic unit, is 3.5 ml $O_2$ uptake · $kg^{-1}$ · $m^{-1}$

its limitation at stages where walking or jogging is the alternative.

Running on a treadmill at a given speed and slope has the same efficiency as running on a track with a similar surface and similar conditions, neglecting air resistance. There is no sex difference in the energy demand.

Formulas are available for calculating the energy cost of walking and running (Bobbert 1960; Margaria et al. 1963; Passmore and Durnin 1955; Van Baak 1979; Van der Walt and Wyndham 1973).

## Cycle Ergometer Experiments

The term *bicycle ergometer* traditionally has been used, but most ergometers have just one wheel, so that *cycle ergometer* is a more appropriate term.

In many cases, the cycle ergometer is the preferred instrument for use in routine studies of physical power and adaptation to exercise. Energy output can be predicted with greater accuracy in cycling than in any other type of exercise. Within limits, the mechanical efficiency is independent of body weight. This is a definite advantage in studies that require repeated examinations over the years. The work rate can, however, be selected simply according to the subject's gross body weight and calculated lean body mass (e.g., 1 or 2 W · kg$^{-1}$ initially). A cycle ergometer operated with a mechanical brake is inexpensive. It is easy to move from place to place and is not dependent on the availability of electrical power. Because the subject on the cycle ergometer exercises in a sitting or lying position with arms and chest relatively immobile, it is simple to obtain good **electrocardiogram (ECG)** tracings and to perform studies with indwelling catheters. During submaximal exercise, a pedal frequency of 40 to 50 rev · min$^{-1}$ produces the lowest oxygen uptake (i.e., the greatest mechanical efficiency) and therefore also a relatively low pulse rate (Eckermann and Millahn 1967).

The variation in oxygen uptake with different pedal frequencies at a standard work rate should be kept in mind if the oxygen uptake is not measured but merely calculated from the rate of exercise used. Cycle ergometers producing a constant load, even with relatively large variations in pedal frequency, have certain advantages. Nevertheless, oxygen uptake is not strictly determined by the external power but varies with the pedal frequency. Respiration also is affected by the pedal frequency. Usually the subject is asked to try to maintain a certain pedal frequency, such as 50 or 60 rev · min$^{-1}$. Acoustic signals, such as those produced by a metronome, are easier to follow than visual signals. It is by no means essential that the chosen pedal frequency be optimal, as long as the mechanical efficiency is known.

In heavy or maximal exercise, the optimal pedal frequency is higher (e.g., the use of lower gear when traveling uphill during bicycle racing) (P.-O. Åstrand, 1953). Sixty rev · min$^{-1}$ will produce a somewhat higher maximal oxygen uptake than 50 rev · min$^{-1}$ (Hermansen and Saltin 1969; McKay and Banister 1976). Thus, one should change from 50 rev · min$^{-1}$ to 60 rev · min$^{-1}$ in the case of heavy work rate, if the objective is to measure maximal oxygen uptake. The type of cycle ergometer that produces a constant power regardless of pedal frequency has the following drawback: During maximal exercise when the subject is tired and is unable to maintain the tempo, the force increases for each pedal revolution, and the subject may be forced to discontinue exercise before it is possible to complete critical measurements. With the use of a cycle ergometer in which the work rate varies with the pedal frequency, the power decreases when the tempo no longer can be maintained, but the subject can in any case continue. It is important to be able to record the pedal frequency continuously if the work rate is critical for the experiment.

Inbar et al. (1983) drew attention to the importance of the crank-length of the cycle ergometer for obtaining maximal leg power output during tests of short duration, in the case of subject populations that are heterogeneous in body size. They found, however, that for a homogeneous population, the conventional 17.5-cm crank is close to the calculated optimum for power production.

A quick and quite effective procedure to measure a subject's maximal aerobic power is the following: The initial power selected should be relatively low and should be increased gradually until a heart rate is reached around 140 beats · min$^{-1}$ for subjects less than 50 years of age and approximately 120 beats · min$^{-1}$ for older subjects. A subjective **rating of perceived exertion (RPE)** will most likely be around 13 to 14 when evaluated with a scale from 6 to 20 that was developed by Borg (1998; table 9.4).

This is the warm-up period. The steady-state heart rate is recorded. The maximal oxygen uptake is then predicted from heart rate and the power setting on the ergometer, for example, using the Åstrand and Åstrand **nomogram** (figure 9.4; explained in more detail later). A power is then chosen that will require an oxygen uptake approximately 10% to 20% higher than the predicted maximum, using table 9.5. If the subject at the end of the first minute on this selected power has difficulty keeping up the pedaling rate and starts to hyperventilate markedly, the power is lowered slightly to allow the subject to continue for a total of about 3 min. If, on

**Table 9.4** The 15-Grade Scale for Ratings of Perceived Exertion, the RPE Scale

| Score | Subjective rating |
|---|---|
| 6 | No exertion at all |
| 7 | Extremely light |
| 8 | |
| 9 | Very light |
| 10 | |
| 11 | Light |
| 12 | |
| 13 | Somewhat hard |
| 14 | |
| 15 | Hard (heavy) |
| 16 | |
| 17 | Very hard |
| 18 | |
| 19 | Extremely hard |
| 20 | Maximal exertion |

Reprinted, by permission, from G. Borg, 1998, *Borg's Perceived Exertion and Pain Scales* (Champaign, IL: Human Kinetics), 47. © Gunnar Borg, 1970, 1985, 1994, 1998.

the other hand, after a minute or two the subject appears to have more strength left than originally predicted, the work rate is slightly increased.

This "quick method" requires a considerable amount of experience on the part of the investigator to attain satisfactory results. As discussed in chapter 8, the peak value of blood lactate mirrors fairly well the severity of the load on the aerobic and anaerobic processes and the degree of exhaustion, as does the respiratory quotient (R). Blood for the determination of lactate concentration can be obtained easily from the fingertip. The hand should be prewarmed and washed in hot water. Blood samples should be secured within 1 min after the end of the exercise and again after about 3, 6, and 10 min.

Many types of cycle ergometers can be adapted to or are specially built for arm exercise. They are useful tools when testing individuals who cannot perform leg exercise. Table 9.1 illustrates that subjects not particularly trained in arm exercise can attain approximately 70% of the maximal oxygen uptake in leg exercise when exercising with the arms. Also, heart rate and arterial blood pressure are significantly higher at a given oxygen uptake and cardiac output in arm exercise compared with leg exercise.

## Step Test

In field studies, a step test may be the only realistic test alternative. It is more difficult, however, to standardize and offers limited possibilities for varying the demand on the oxygen transporting system. However, equipment is available that provides automatic adjustment of the stepping height from 2 up to 50 cm. The rate of stepping is established with a metronome, and the stepping cadence has four counts: up, up, down, down. At maximal effort, the stepping pace must be high, and there is always the risk that the subject will stumble when approaching a maximal rate of work. Besides, in the untrained subject, the aftermath is sore muscles. It is even more difficult to perform recordings such as ECG on exercising subjects during a step test than is the case during the treadmill test. Gradational step tests have been described by Hettinger and Rodahl (1960), Maritz et al. (1961), Nagle, Balke, and Naughton (1965), Kasch et al. (1966), and Shephard (1966). The introduction of electronic stepping ergometers allows a graded exercise test to be carried out in a manner similar to that on a treadmill (Howley, Colacino, and Swenson 1992).

### FOR ADDITIONAL READING

For further details concerning step ergometry, see Ben-Ezra and Verstraete (1991) and Howley and Franks (1992).

## Summary

Ideally, any test of maximal oxygen uptake should meet at least the following general requirements: (1) The exercise must involve large muscle groups, (2) the work rate must be measurable and reproducible, (3) the test conditions must be such that the results are comparable and reproducible, (4) the test must be tolerated by all healthy individuals, and (5) the mechanical efficiency (skill) required to perform the task should be as uniform as possible in the population to be tested.

The magnitude of the external power can be expressed exactly, and it can be reproduced with a high degree of accuracy in tests using the cycle ergometer although with less accuracy in exercise on the treadmill. However, the use of both cycle ergometer and treadmill is advisable. We have good experience with using the cycle ergometer for submaximal exercise testing but prefer the treadmill test for maximal exercise testing. The running should be uphill with an inclination of 3° or more. On a cycle ergometer, the subject should sit in a position directly above the pedals. The seat should be sufficiently high, and the pedal frequency should be 50 to 60 rev $\cdot$ min$^{-1}$; the higher frequency is particularly advisable for subjects with limited muscular strength. The work rate should be se-

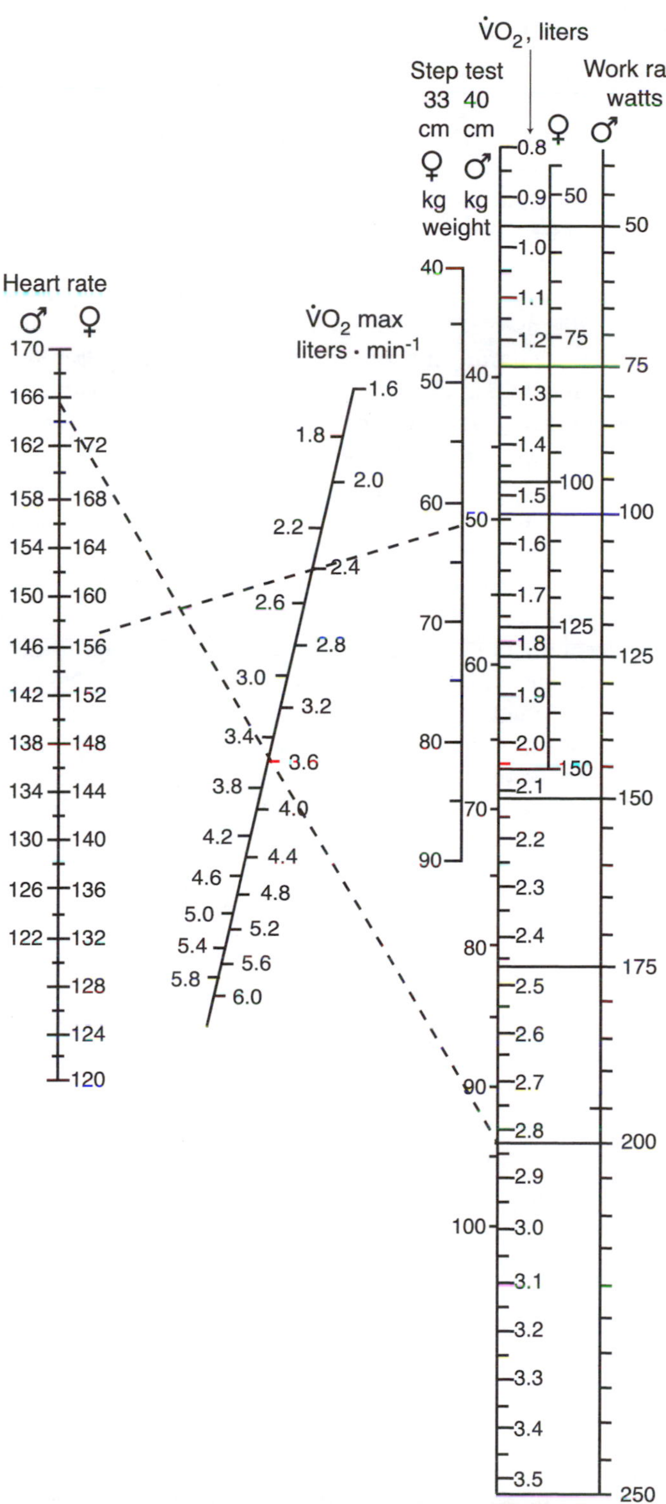


**Figure 9.4** The adjusted nomogram for calculation of maximal oxygen uptake from submaximal pulse rate and oxygen uptake values (cycling, running, or walking, and step test). In tests without direct oxygen uptake measurement, it can be estimated by reading horizontally from the "body weight" scale (step test) or "work rate" scale (cycle test) to the "oxygen uptake" scale. The point on the oxygen uptake scale ($\dot{V}O_2$, liters) is then connected with the corresponding point on the pulse rate scale, and the predicted maximal oxygen uptake is read on the middle scale. A female subject (62 kg) reaches a heart rate of 156 at step test, predicted maximum $\dot{V}O_2 = 2.4 \text{ L} \cdot \text{min}^{-1}$. A male subject reaches a heart rate of 166 at cycling test on a work rate of 200 W, predicted $\dot{V}O_2\text{max} = 3.6 \text{ L} \cdot \text{min}^{-1}$ (exemplified by dotted lines).

From I. Åstrand 1960.

lected so that the subject can proceed for at least 3 min. Repeatedly, it has been emphasized that an objective criterion for an attained maximal oxygen uptake is a final "leveling off" of the oxygen uptake despite an increasing rate of exercise, that is, a failure of a higher work rate to increase the aerobic metabolic rate significantly. But in many cases, this criterion is difficult to attain. The heart rate attained at the end of an exercise performed for the purpose of measuring the subject's maximal oxygen uptake is a poor basis for evaluating whether the exercise actually represented a maximal load on the oxygen

**Table 9.5 Oxygen Uptake as Related to Work Rate**

| Work rate | | Oxygen uptake, $L \cdot min^{-1}$ |
|---|---|---|
| W | $kpm \cdot min^{-1}$ | |
| 50 | 300 | .9 |
| 100 | 600 | 1.5 |
| 150 | 900 | 2.1 |
| 200 | 1200 | 2.8 |
| 250 | 1500 | 3.5 |
| 300 | 1800 | 4.2 |
| 350 | 2100 | 5.0 |
| 400 | 2400 | 5.7 |

transporting system. It is quite common to use an average maximal heart rate for a given age group as a reference when evaluating whether a subject is exercising at a level demanding maximal aerobic power. It was mentioned in chapter 5 that within a group of 100 subjects having an average maximal heart rate of 180 beats · min $^{-1}$, two to three persons will stop at a maximal heart rate of 160 or below, and two to three subjects may exceed a heart rate of 200 (standard deviation = ±10 beats · $min^{-1}$).

## Prediction From Data Obtained at Rest or Submaximal Exercise

Although the aerobic power in terms of maximal oxygen uptake can be determined with a reasonable degree of accuracy, the method is rather time consuming. It requires fairly complicated laboratory procedures and demands a high degree of cooperation from the subject. Although this is the method of choice for any scientific investigation, it is by no means a routine method that can be conveniently applied in a physician's office.

The practicing physician (especially the industrial physician), the coach, the physical therapist, or anyone else interested in physical performance is nevertheless often faced with the need to assess a person's circulatory fitness. This requirement has created the need for a simple test for such an evaluation of an individual, based on a submaximal "stress" test. In treating older individuals, as well as certain other patients, the physician may be reluctant to expose the patient to the risk of an exhausting maximal work rate. This is true whether one is considering job placement, fitness for continued employment, or retirement.

### *Rest*

No objective measurements made on resting individuals will reveal their capacity for physical exercise or their maximal aerobic power. Even a simple questionnaire may provide more useful information than can be obtained from measurements made at rest. A low heart rate at rest, a large heart size, or similar parameters may indicate a high aerobic power, but may, on the other hand, be a symptom of disease.

A significant correlation between any parameter and maximal oxygen uptake indicates a direct or indirect dependence. Figure 10.5 (lower part), in chapter 10, illustrates such a relationship. The range of the observed values is of decisive importance for the numerical value of the correlation coefficient, but for an evaluation of the individual case, the standard deviation from the regression line is the critical factor. The deviation may be large (i.e., a prediction of one parameter from the other is very uncertain) despite a high correlation between the parameters in question. From the data presented in figure 10.5 (lower part), we find that the correlation coefficient between maximal oxygen uptake and total amount of hemoglobin ($Hb_T$) is as high as .970, but the standard deviation of Hb weight is still as high as 10.5% at an oxygen uptake of 2.6 L · $min^{-1}$. The figure shows that one girl with 350 g of $Hb_T$ can transport up to 2.7 L · min $^{-1}$, but another girl with the same $Hb_T$ reaches an oxygen uptake of only 1.9 L · $min^{-1}$. This is to be expected statistically, because 350 g of Hb represents an oxygen uptake of 2.25 L · $min^{-1}$ with such a standard deviation that 95 of 100 subjects of a similar group are expected to fall within the range of 1.8 to 2.7 L · $min^{-1}$ and only five subjects will lie outside these values.

In conclusion, it may be stated that the correlation between parameters is of interest when analyzing biological interactions, but for an evaluation of one individual on the basis of indirect methods, the standard deviation from the regression line indicates the accuracy of the prediction.

### *A Simple Submaximal Cycle Ergometer Test*

It has been repeatedly pointed out that there is a high correlation between cardiac output during exercise and oxygen uptake in liters per minute. Because cardiac output is the product of heart rate and stroke volume, it is evident that a low heart rate at a given oxygen uptake is most likely associated with a large stroke volume. In chapter 5, it was also pointed out that such a heart functions with a higher efficiency than a heart that produces a given cardiac output with a high heart rate and small stroke volume. This physiological fact forms an important basis for submitting people to submaximal exercise tests. When exercise is performed on a cycle ergometer, there are relatively small individual variations in mechanical efficiency. Therefore, the oxygen uptake can be predicted from the external power to which the subject is subjected (see table 9.5). Figure 9.5 illustrates the effect of training on the heart rate at two work rates. The oxygen uptake at a given power was constant throughout the study (100 W, 1.5 L · min$^{-1}$; 150 W, 2.1 L · min$^{-1}$). Most likely, the cardiac output did not change and the decline in heart rates is associated with an increase in stroke volume.

In our experience, the submaximal exercise test is a very useful tool for evaluating whether a training program has improved an individual's circulatory capacity. This test has been widely applied in top athletes, in trained and untrained adults, and in children. In this test, the individual is his or her own control; the individual's own performances are compared in repeated tests over months or years. In the simple test on the cycle ergometer, counting the heart rate is all that is needed for the evaluation. A gradually decreasing heart rate at a standard work rate as the training progresses stimulates the individual's efforts to continue to improve his or her circulatory capacity further or to maintain a given heart rate level.

## *How Is the "Normal" Test Conducted?*

When the test is conducted outside medical institutions, the subject usually is just asked about her or his health. The risks involved are almost negligible. In a retrospective study involving 370,000 tests conducted over 2 years, primarily submaximal but also maximal, no serious complications were reported (Jonsson 1981).

The subject should feel well and be free of infection. Several hours should elapse between the last meal and the test. The subject should not engage in any physical activity heavier than the test load the last few hours before the test. Smoking should not be allowed during the 2 h immediately before the test. If possible, the room temperature should be 18 to 20° C (64.4 to 68° F), and the room should be adequately ventilated. If the temperature is higher, an electric fan may be used.

For trained, active individuals, the risk of overstraining in connection with a test is very slight. For women, a suitable power is 75 to 100 W, and for men 100 to 150 W. If the heart rate exceeds about 130 beats · min$^{-1}$, the load can be considered adequate and the test can be discontinued after 6 min. If the heart rate is slower than about 130 beats · min$^{-1}$ after some minutes of exercise, the power should be increased by 50 W. If the purpose is a multistage test, the work rate can be increased by 50 W every 6 min for as long as the heart rate remains below about 150 beats · min$^{-1}$. The final exercise period can be continued for 6 min even if the heart rate exceeds 150 beats · min$^{-1}$.

For persons who are expected to be physically unfit, persons who are completely untrained, and older individuals, lower work rates should be chosen and an initial load of 50 W might be suitable. A convenient pedaling frequency is 50 rev · min$^{-1}$ paced by a metronome set at 100. In the event of pressure or pain in the chest or marked shortness of breath or distress, the test must be discontinued immediately.

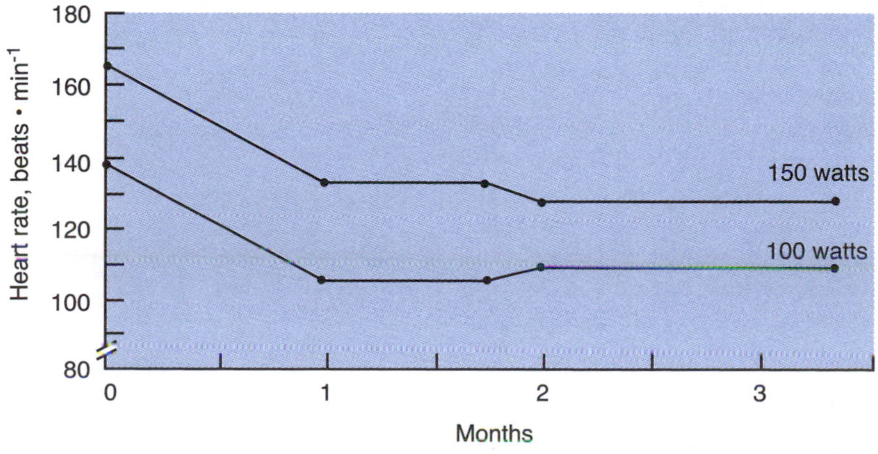


**Figure 9.5** Decrease in the heart rate tested at two fixed work rates in the course of 3 1/2 months of training.

Provided that the exercise is not too heavy, respiration and circulation will increase during the first 2 to 3 min, after which a steady state is attained. The increase in pulse rate can be established easily by counting the pulse once every minute. In any case, after 4 to 5 min of exercise, the pulse rate generally has reached a steady state. As a rule, a period of about 5 to 6 min is therefore sufficient for the pulse rate to adapt to the task being performed. The pulse rate should then be taken every minute, and the mean value of the pulse rate at the 5th and 6th min is taken as the heart rate response to the exercise demand. If the difference between these last two pulse rates exceeds 5 beats · min$^{-1}$, the exercise time should be prolonged one or more minutes until a constant level is reached.

The pulse rate can be obtained by using a stethoscope on the chest wall or by using surface electrodes that transmit the signal to an electrocardiograph or to a monitor that displays the pulse rate directly. If measured by **palpation,** the pulse rate is most easily felt over the carotid artery just below the mandibular angle. In this case, care must be taken, because too much finger pressure can slow down the pulse rate by eliciting the baroreceptor reflex. For the inexperienced, it is rather difficult to count the pulse rate; the subject is in motion, and the pulse rate may vary. Practice under experienced supervision is important. The pulse rate may be measured during the last 15 to 20 s of every minute. The most exact value is obtained by taking the time for 30 pulse beats. A person trained in taking pulse rates, whether by palpation or by auscultation, obtains values that are in close agreement with those obtained by ECG recordings, provided the pulse rate is not irregular.

It has proven to be useful to ask the subject about the subjective rating of exertion, using Borg's RPE scale, at the end of each exercise state (see table 9.4). There are exceptions, but in many situations the heart rate mirrors the physical strain experienced subjectively, although the heart rate level at a given grade varies individually (Ekblom and Goldbarg 1971). The questionnaire can be modified to serve as a rating of general fatigue, leg fatigue, and dyspnea. The decline in heart rate as a consequence of physical training as illustrated in figure 9.5 is also accompanied by a reduced perceived exertion when the subject is exposed to a standard rate of exercise.

## FOR ADDITIONAL READING

For further discussion concerning the prediction of maximal oxygen uptake from submaximal exercise testing, see Hartung et al. (1993).

## Can Maximal Oxygen Uptake Be Predicted Accurately From Data Recorded During a Submaximal Exercise Test?

The answer is definitely no, but the question warrants some further comments. For a long time there has been a need for a test by which one could select people for physically demanding tasks. One example is the classic Harvard fitness step test (see R.E. Johnson, Brouha, and Darling 1942). In this and in some other tests, the heart rate is counted during early recovery from a standardized exercise test. In the same individual, the correlation coefficient between the heart rate during submaximal exercise and the heart rate 1 to 1-1/2 min after the exercise is high. In one study, the correlation coefficient was .96 with a deviation from the regression line of 5%. For a group of subjects, the $r$ value for the steady-state heart rate and the recovery rate was reduced to .77 and the deviation increased to 10% (Ryhming 1953). These data indicate that the recovery heart rate gives only a rough idea of the heart rate attained during exercise if the results are compiled from different subjects.

Most modern circulatory exercise tests are based on a linear increase in heart rate with increasing oxygen uptake, or work rate, as illustrated by figure 9.6. If the maximal heart rate were fixed in *Homo sapiens*, from a few measurements of the heart rate during a multistage submaximal exercise protocol one could extrapolate and predict the subject's oxygen uptake at the maximal heart rate. As discussed in chapter 5, however, there are marked individual

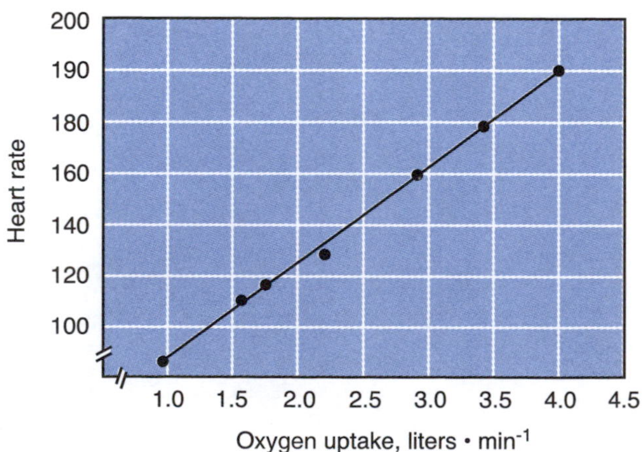


**Figure 9.6** The relationship between heart rate and oxygen uptake on the cycle ergometer.

variations in maximal heart rate within a given age group, and it decreases with age. Therefore, an extrapolation to an age-predicted maximal heart rate may give quite erroneous results. Let us examine in more detail the reliability of predicting the maximal aerobic power and potential of the oxygen-transporting system from the submaximal heart rate response to a given rate of exercise.

1. The linear increase in heart rate with increase in oxygen uptake is a typical feature. There are many exceptions, however. In some cases, the oxygen uptake increases relatively more than the heart rate as the work rate becomes very heavy (subject B in figure 9.7). One possible explanation of this phenomenon is that an efficient redistribution of blood, giving the exercising muscles an appropriate share of the cardiac output, is not brought about until the very heavy work rates are reached. The consequence is that in this subject, the maximal oxygen uptake will be underestimated by an extrapolation from the heart rate response to submaximal loads.
2. The maximal heart rate declines with age (see figure 5.22). Therefore, if old and young subjects are included in the same study, the circulatory capacity of the older subjects will be consistently overestimated compared with the younger subjects. By introducing an age factor, a correction can be made. However, the standard deviation for maximal heart rate within an age group is about ±10 beats · min$^{-1}$; thus, 50% of the tested subjects will be more or less overestimated and the remainder underestimated. In figure 9.7, subject A was assumed to have a maximal heart rate of 195, maximal oxygen uptake = 3.5 L · min$^{-1}$. If the maximal heart rate was 170, the maximal oxygen uptake would be only 2.9 L · min$^{-1}$. An extrapolation of the heart rate to 215 for a subject with the same slope for the relationship between heart rate and oxygen uptake as subject A would give an oxygen uptake of 4.0 L · min $^{-1}$.
3. In cases where the oxygen uptake is predicted from the work rate, assuming a fixed mechanical efficiency, it should be kept in mind that the mechanical efficiency can vary by ±6% (cycle ergometer). In figure 9.7, 150 W is indicated with a mean oxygen uptake of 2.1 L · min$^{-1}$. An oxygen uptake as low as 1.9 or as high as 2.3 L · min$^{-1}$ at the same work rate would not be unusual, however. The consequence is that in a subject with a low mechanical efficiency (whose oxygen uptake at the submaximal rate of exercise is relatively high), maximal oxygen uptake would be predicted to be lower than it really is, because the heart rate is influenced by the extra oxygen transport. The oxygen uptake at a given submaximal rate of exercise on a cycle ergometer or treadmill is very constant, even if the exercise is performed under different conditions, such as with the subject exposed to hypoxia or hyperoxia (chapter 14), a hot environment, or dehydration (see Rowell 1993). The variations in mechanical efficiency in running and cycling in the course of a training period are usually small or insignificant.
4. The last factor to be considered is based on figure 5.19. The cardiac output is not strictly related to the oxygen uptake but shows individual variations. To predict maximal oxygen uptake from heart rate at a submaximal rate of exercise, this variation does not matter. The oxygen uptake (measured or predicted) per heartbeat is actually evaluated during the test. When the maximal heart rate is considered, the maximal oxygen uptake is

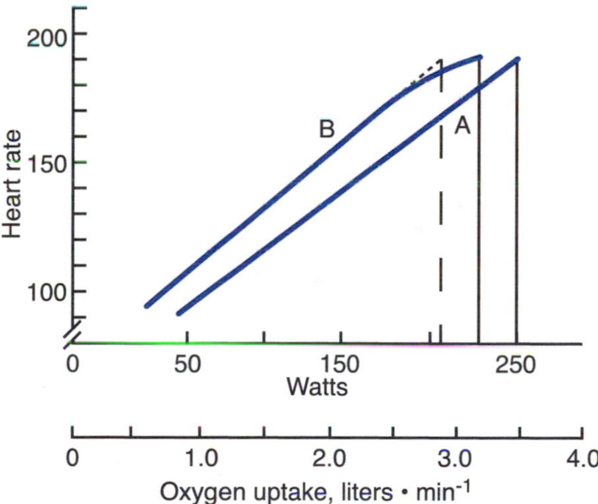


**Figure 9.7** The increase in heart rate with increasing work rate (and oxygen uptake) is linear within a wide range. In some subjects (subject B), the oxygen uptake increases relatively more than the heart rate as the work rate becomes very heavy. The prediction of maximal oxygen uptake by extrapolating to the subject's presumed maximal heart rate (195 in this case) suggests a maximum of 2.9 L · min$^{-1}$ (dotted line), but the actual maximal aerobic power is 3.2 L · min$^{-1}$. The individual's maximal heart rate is therefore a critical factor in an extrapolation.

"calculated." However, if the test is conducted to evaluate cardiac performance (e.g., stroke volume), the individual variation in cardiac output and arteriovenous oxygen difference must be taken into consideration.

Under standardized conditions, the variation from day to day in heart rate at a given oxygen uptake is less than 5 beats · min$^{-1}$, provided the state of training is the same. The mean value for a group of subjects undergoing repeated tests remained exactly at the same level under these conditions.

However, a number of situations can markedly increase the heart rate at a given submaximal rate of exercise, without being accompanied by a reduction in the individual's maximal oxygen uptake. The following are examples:

1. Dehydration during heavy exercise or during exposure to heat (chapter 13)
2. Prolonged heavy exercise (chapters 12 and 13)
3. Exercise with a muscle mass involved which is less than in, for instance, running and cycling (chapter 5)
4. Exercise after ingestion of alcohol (chapter 14)

In some of these situations, the potential for top physical performance is actually reduced. The excessive heart rate response to a given submaximal work rate is often a better criterion for reduced physical fitness than the person's maximal oxygen uptake. Also, fear, excitement, and related emotional stress can markedly elevate heart rate at a submaximal work rate without affecting either maximal oxygen uptake or performance capacity. The heavier the work rate, however, the less pronounced is this nervous effect on the heart rate. It is usually recommended that the test load should be high enough to bring the heart rate up to, or above, 150 beats · min$^{-1}$ in the case of younger subjects.

In some instances, the heart rate at a standard work rate is unchanged while the maximal oxygen uptake and the performance capacity are reduced (e.g., after acclimatization for a certain period at high altitude or during semistarvation). These examples illustrate further the danger of drawing conclusions concerning maximal oxygen uptake from heart rate data obtained during submaximal tests.

One method of predicting an individual's maximal oxygen uptake by use of a nomogram (P.-O. Åstrand and Ryhming 1954) has been critically examined over the years in a number of studies. Therefore, it may be justifiable to devote some attention to this particular method. The original data were based on 86 female and male students in physical education. Not unexpectedly, the scatter of the heart rates at a given oxygen uptake was considerable (figure 9.8). Note that in this group of relatively well-trained

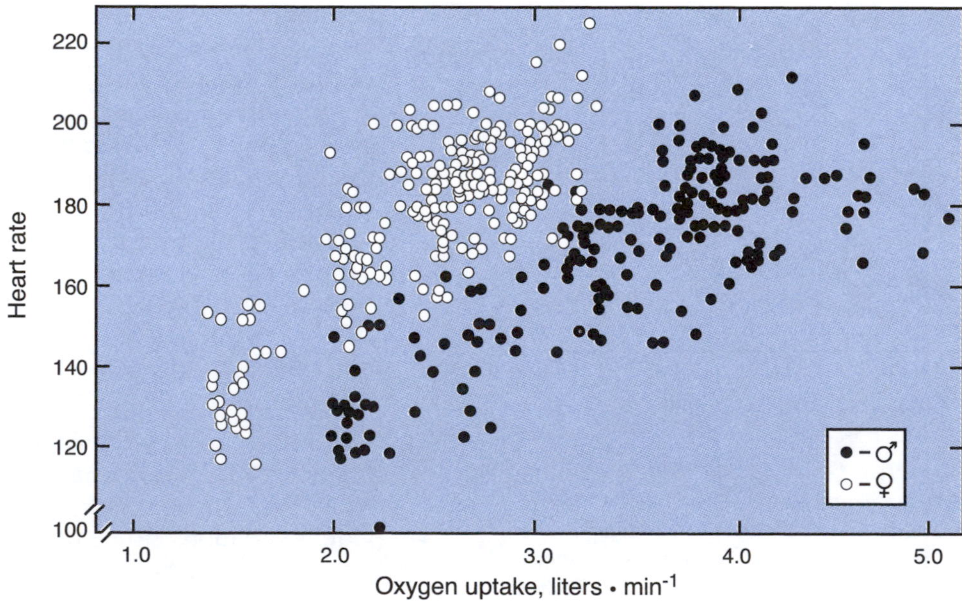


**Figure 9.8** Heart rates in relation to oxygen uptake for 86 adult female and male subjects. Maximal as well as submaximal values are represented.

From P.-O. Åstrand 1952.

students, heart rates from 140 to 220 beats · min$^{-1}$ were recorded at an oxygen uptake of 3.0 L · min$^{-1}$. A heart rate of 180 beats · min$^{-1}$ represented an oxygen uptake of only 2.0 L · min$^{-1}$ for some female subjects and as much as 5.0 L · min$^{-1}$ for some male subjects. One major explanation of this scatter was the differences in the subjects' maximal aerobic power. When young men were exercising at 50% of their maximal oxygen uptake, their heart rates were on the average 128 beats · min$^{-1}$; for women, the heart rates were 138. When the subjects were exercising against a heavier power demanding an oxygen uptake of 70% of their maximal aerobic power, the average heart rates were 154 for males and 164 for females, with a standard deviation of 8 to 9 beats · min$^{-1}$. Based on these data, a nomogram was constructed (see figure 9.4). From the steady-state value of the heart rate and oxygen uptake, the maximal oxygen uptake can be predicted. If the oxygen uptake is not measured during the exercise, it can be predicted from the work rate during cycling and from body weight in the case of a particular step test. This nomogram is not based on any sophisticated theory, and the individual's maximal heart rate originally was not considered in the construction of the nomogram. The predicted maximal oxygen uptake of persons over approximately 30 years of age was, in most cases, overestimated. This could be explained by the reduction in maximal heart rate with age (see figure 5.22). A correction factor for age and also for the individual's maximal heart rate, when known, was introduced (see table 9.6; I. Åstrand 1960).

The standard error of the method for such a prediction of maximal oxygen uptake from submaximal exercise tests was, in the studies behind the nomogram, 10% in relatively well-trained individuals of the same age but up to 15% in moderately trained individuals of different ages when the age correction factor of maximal oxygen uptake was applied. Untrained persons often are underestimated, and extremely well-trained athletes often are overestimated. The consequence of a standard error of 15% is as follows: With a maximal aerobic power predicted to 3.0 L · min$^{-1}$, the actual oxygen uptake for five of 100 subjects is then less than 2.1 or higher than 3.9 L · min$^{-1}$. It is important to keep in mind this limitation in accuracy; this drawback holds true for any submaximal cardiopulmonary test described so far. The validity of the nomogram has been tested in other laboratories. In some cases, there has been good agreement between the actually measured and predicted maximal oxygen uptake from the nomogram (Kavanagh and Shephard 1976; figure 9.9). In other studies, the subject's maximal oxygen uptake was found to be underestimated when the nomogram was applied (Chase, Graves, and Rowell 1966; Harrison, Brown, and Cochrane 1980). Siconolfi et al. (1982) suggested the following equations as a modification of the nomogram:

**Table 9.6** Factor to Be Used for Correction of Predicted Maximal Oxygen Uptake

| Age | Factor | Max heart rate | Factor |
|---|---|---|---|
| 15 | 1.10 | 210 | 1.12 |
| 25 | 1.00 | 200 | 1.00 |
| 35 | .87 | 190 | .93 |
| 40 | .83 | 180 | .83 |
| 45 | .78 | 170 | .75 |
| 50 | .75 | 160 | .69 |
| 55 | .71 | 150 | .64 |
| 60 | .68 | | |
| 65 | .65 | | |

(1) When the subject is over thirty to thirty-five years of age or (2) when the subjects' maximal heart rate is known, the actual factor should be multiplied by the value that is obtained from the nomogram, Fig. 9.4.

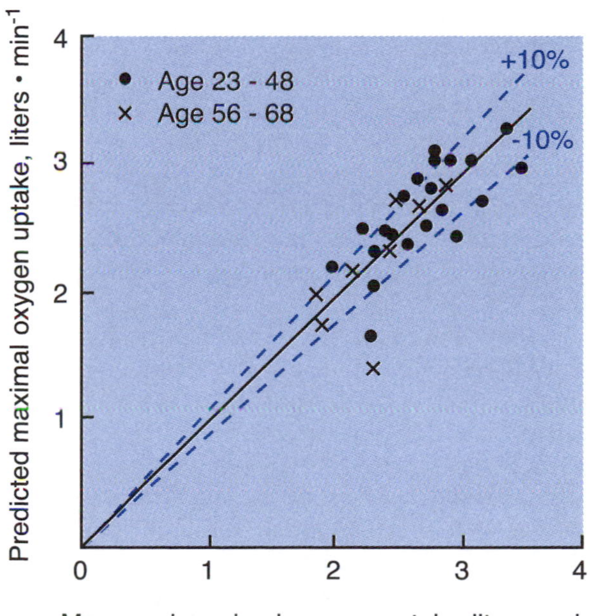


**Figure 9.9** Estimated maximal oxygen uptake calculated from the Åstrand nomogram in relation to measured maximal oxygen uptake in 22 male subjects, ages 23 to 48 (●) and nine male subjects, ages 56 to 68 (×). Broken lines denote a deviation of ±10% from the "ideal line."

From K. Rodahl and Issekutz 1962.

For males:

$$y = .348(X1) - .035(X2) + 3.011 \quad (1)$$

($r = .86$; standard error of the estimate = .359 L · min$^{-1}$)

For females:

$$y = .302(X1) - .019(X2) + 1.593 \quad (2)$$

($r = .97$; standard error of the estimate = .199 L · min$^{-1}$) where $y$ is the $\dot{V}O_2$ (L · min$^{-1}$), $X1$ is the $\dot{V}O_2$ (L · min$^{-1}$) from the nomogram (figure 9.4; not corrected for age), and $X2$ is the age (years).

Maritz et al. (1961) and Wyndham et al. (1966) applied a similar principle in their predictions. Their subjects performed four submaximal exercise loads (usually step tests). From measurements of heart rate and oxygen uptake, a straight line was fitted to the four pairs of plots and extrapolated to a mean maximal heart rate for the population in question. Margaria, Aghemo, and Rovelli (1965) also introduced a nomogram based on similar concepts. Their test was modified by Harrison, Brown, and Cochrane (1980) by adapting the step test, using a step the height of which was adjusted to fit the length of the subject's leg. (In the Margaria test it is a constant 40 cm step height, 15 steps · min$^{-1}$ for 5 min and 25 steps · min$^{-1}$ for another 5 min.)

It would be reasonable to assume that accuracy would vary from one group of subjects to another, with regard to both the difference between predicted and measured maximal oxygen uptake and the scatter of the data. Regarding the considerable error of the method and the fact that it is applied, at best, only as a screening test, a consistent difference between measured and predicted maximal oxygen uptake of a few 100 ml · min$^{-1}$ is of minor importance.

In some test procedures, the **physical work capacity (PWC)** is evaluated from data on work rate, oxygen uptake, or **oxygen pulse** at a given heart rate, such as 170 or 150 beats · min$^{-1}$ (Sjöstrand 1947). The methodological error must necessarily be about the same in these tests as in the application of the Åstrand–Ryhming nomogram (I. Åstrand, 1960). The calculated $PWC_{170}$ is related to the maximal stroke volume of the heart, but it is not a measure of effectiveness or maximal power. In any such estimate, the maximal heart rate must be considered. If oxygen uptake is measured during the submaximal test, it is illogical to disregard individual variation in mechanical efficiency by expressing the "capacity" in watts or kilopond-meters per minute at a heart rate of 170. However, the most important error is introduced when scientists compare or evaluate individuals of different ages without correcting for the decline in maximal heart rate with age. The mean value for heart rate at a given submaximal oxygen uptake is the same for individuals of the same sex and state of training regardless of age (from 25 up to at least 70 years of age; I. Åstrand 1960). By definition, therefore, the calculated $PWC_{170}$ or oxygen pulse at a given rate of exercise is the same. The real performance capacity, however, declines with age. Furthermore, the subjective feeling of strain is higher at a given heart rate in the older individual.

Studies also have confirmed that there is a low correlation between the oxygen uptake or work rate achieved at a heart rate of 170 or 150 beats · min$^{-1}$ and the measured maximal oxygen uptake, cardiac output, heart size, or blood volume in individuals from 20 to 70 years of age (Strandell 1964).

The conclusion is that an evaluation of the maximal effect of the oxygen-transport system, based on studies at submaximal work rates or oxygen uptake, should be performed with the utmost caution, especially when persons of different age groups are considered. Figure 10.7, in chapter 10, presents mean values for various functional parameters in relation to age. It is evident that the decline in physical performance is not related to a similar change in heart size, blood volume, or heart rate during a standard work rate.

## Submaximal or Maximal Test?

Particularly in the clinical examination of patients or during a health checkup, it is often important to include an exercise test to examine the cardiovascular system under functional stress. Usually ECG and blood pressure monitoring, specific objective changes, signs, and symptoms will guide the tester as to when the safety of the test subject is threatened. Some clinicians are reluctant to push the subject to a maximal effort, and the test is terminated at a predetermined end point, mostly based on the function of the heart. One such criterion for termination of the test is a heart rate of 200 minus the subject's age in years. The rationale behind these criteria is a simple formula, not far from the truth, assuming the maximal heart rate to be 220 minus the subject's age. For a discussion of this concept, see Bovens et al. (1993). By this procedure, the risk of submitting a subject to a maximal performance is relatively small. As mentioned, the standard deviation in maximal heart rate is approximately ±10 beats · min$^{-1}$. Another cutoff point for automatic termination of a test is 85% of the predicted maximal heart rate based on the subject's age (Pollock, Wilmore, and Fox 1978). This level of exercise demands approximately 80% of the subject's maximal oxygen uptake (figure 9.10).

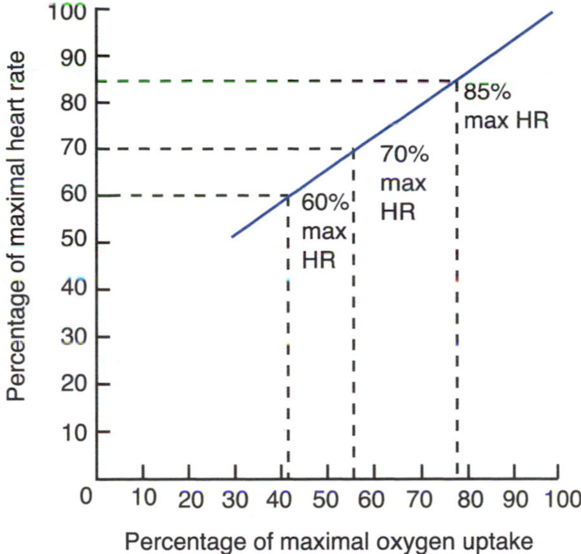


**Figure 9.10** As illustrated in figure 9.6, the heart rate (HR) during exercise (e.g., when walking, running, or cycling) increases for the average person linearly with the oxygen uptake. This figure illustrates that the relationship is not strictly on a percentage basis. When the heart rate is 85% of the maximum, the oxygen uptake is slightly below 80% of the maximal aerobic power.

From Pollock, Wilmore, and Fox 1978.

The inevitable problem when using a submaximal exercise test with a fixed heart rate to predict a person's cardiovascular fitness is that a heart rate of 160 may be maximal for one person but 40 beats · $min^{-1}$ or more below the maximum for another one. Considering the wide scatter in maximal heart rates reported in the literature (see Åstrand et al. 1973; Cooper et al. 1977), it is quite meaningless to be too sophisticated in the choice of the target heart rate.

There is a definite trend, supported by Bruce's experience (1971), to continue a test until the subject has alarming symptoms or has to stop because of fatigue. This end point is termed the functional maximum, but the measured oxygen uptake is not necessarily the subject's maximal oxygen uptake. It has been observed that the degree of ST-segment depression (a sign of insufficient coronary perfusion), if present in the ECG, will increase with higher cardiac output (and oxygen uptake) up to maximum. Studies conducted at the Institute of Aerobics Research in Dallas, Texas, showed that of the 7,059 males who were subject to a maximal exercise test, approximately 15% were considered abnormal or questionable. Among the 552 abnormal tests, 34% did not become abnormal until after the heart rate exceeded 85% of the predicted maximal heart rate (Pollock, Wilmore, and Fox 1978). However, more false-positive cases may appear in a maximal test, whereas some false-negative cases probably result from a submaximal test. If the purpose of a test is to predict the maximal oxygen uptake because the methods to measure oxygen uptake are not available, a multistage submaximal test followed by a maximal spurt can be useful. The steady-state heart rates are plotted against the predicted submaximal oxygen uptakes. (When the subject is using the cycle ergometer, the data in table 9.5 can be applied.) An interpolation to the subject's measured maximal heart rate, determined at the end of about a 3-min spurt, gives a more reliable maximal aerobic power than extrapolation to the maximal heart rate typical for the person's age.

If equipment is not available for recording or monitoring heart rate, then the clinician should measure the length of time for 30 heartbeats with a stopwatch, counting the pulse at the carotid artery by palpation or counting the heart rate with the use of a stethoscope at the apex of the heart. It is advisable to learn how to count the heart rate manually, because any recording system can fail unexpectedly. If the heart rate is needed in connection with sport activities (e.g., during training), taking the length of time for 10 heartbeats, starting the counting immediately after the exercise, gives a good picture of the heart rate during the last part of the exercise.

Cooper (1982) presented tables with fitness categories from "very poor" to "superior" based on performance in a 12-min running test, 1.5-mile run test, 3-mile walking test, or 12-min swimming or cycling test. The main purpose of these tests is to follow one's own score over the years.

In Canada, a "safe, simple, self-administered fitness test, the purpose of which would be motivational rather than to accurately evaluate fitness" has been developed, the Canadian Home Fitness Test. After answering the questions in a Physical Activity Readiness Questionnaire, the person can evaluate whether she or he can safely go through the test. The subject climbs the bottom two steps of a standard staircase, step height 20.3 cm (8 in.), at a rhythm that is chosen according to sex and age, and after a 3-min warm-up, the person does one or two 3-min bouts of stepping depending on the heart rate taken from 5 to 15 s during recovery after each stage. From the duration of the stepping and the heart rate, one obtains a fitness score. The individual then can compare the fitness level from time to time by repeating the test with the same pace. The stepping rate is chosen so that the demand on the oxygen transporting system is approximately 70% in an average person of the same age and sex. This test is an interesting approach to reaching a large population in health and fitness education (Shephard, Bailey, and Mirwald 1976; Shephard, Thomas, and Weller 1991).

## *Blood Lactate, Respiratory Quotient*

By securing blood samples after the exercise and during early recovery, one can determine peak lactate concentration. Concentrations exceeding 8 to 9 mM support the assumption that the subject's aerobic power was taxed to its maximum.

Chapter 8 discussed the rate of exercise in relation to the person's maximal oxygen uptake when the blood lactate begins to accumulate (the anaerobic threshold concept). An individual who is not very well trained can exercise at about 50% of her or his maximal aerobic power without a noticeable increase in blood lactate concentration. Therefore, if this person can exercise with a work rate demanding an oxygen uptake of 1.5 L · min $^{-1}$ with no significant increase in the blood lactate, the maximal oxygen uptake is probably some 3 L · min$^{-1}$ or higher.

A person involved in manual labor who is free to set the pace normally works with an energy output that is approximately within 40% of the maximal aerobic power. If such a person can exercise for at least 5 min at a rate requiring an oxygen uptake of 2.5 L · min$^{-1}$, that person can most likely exercise under steady-state conditions for hours with an oxygen uptake of at least 1 L · min$^{-1}$. Another approach would be to let the person exercise on a cycle ergometer or a treadmill with the mentioned oxygen demand. If she or he does not develop clinical symptoms or signs, does not feel fatigued, experiences no increase in blood lactate concentration, and has a low heart rate (<120 beats · min$^{-1}$), then that person can probably safely take part in an activity that demands an oxygen uptake of 1 L · min$^{-1}$.

During heavy exercise of short duration (up to 5 min), the R exceeds 1.0. Issekutz, Birkhead, and Rodahl et al. (1962) showed that the R value obtained during standardized exercise conditions can be used to predict maximal aerobic power. It was noted that ΔR (exercise R – .75) increased logarithmically with the exercise rate, and maximal oxygen uptake was reached at ΔR = .40. This observation offers the possibility of predicting a person's maximal oxygen uptake from the measurement of R during a single 5-min cycle ergometer test at a submaximal work rate. An accumulation of blood lactate is correlated with an increase in pulmonary ventilation beyond the demand of a sufficient oxygen supply to exercising muscles. More or less by definition, such a situation induces hyperventilation, which decreases the body's carbon dioxide content. The carbon dioxide/oxygen ratio in the expired air will eventually exceed 1.0 as the exercise approaches maximum. This fact, along with determination of blood lactate concentration, can be used as a rough method for assessing the effectiveness of an individual's oxygen-transporting system during standardized exercise. Such measurements are particularly useful when one is testing the aerobic exercise potential of patients who take medications (e.g., β-blockers) or have irregular heart rates. In such cases, the heart rate attained during submaximal exercise is a poor guide in evaluating the function of the oxygen-transporting system.

## *End Point for Termination of a Test*

Especially with patients who have established or suspected cardiovascular disease, it is important to have some predetermined rules as a guide for deciding when to terminate an exercise test. Pollock, Wilmore, and Fox (1978) listed the following indications for discontinuing the test:

1. Failure of the monitoring system
2. Progressive angina (chest pain)
3. Two millimeter horizontal or down-sloping ST-segment depression on the ECG (as long as a patient has no symptoms, some clinicians do not use this criterion as a reason for terminating the test)
4. Sustained supraventricular tachycardia
5. Ventricular tachycardia
6. Exercise-induced left or right bundle branch block
7. Any significant decrease (10 mmHg, 1.3 kPa) in systolic blood pressure
8. Light-headedness, confusion, **ataxia,** pallor, **cyanosis,** nausea, or any sign of peripheral circulatory insufficiency
9. Inappropriate bradycardia (slow heart rate)
10. Excessively high blood pressure: systolic greater than 260 mmHg (36 kPa), diastolic greater than 120 mmHg (16 kPa)
11. Presence of dangerous dysrhythmias (irregular heartbeats) such as frequent premature ventricular contractions and multifocal premature ventricular contractions (premature beats that are triggered from more than one area of the heart)

Heart rate, blood pressure, and ECG monitoring should continue during exercise and recovery, at least up to 5 min after the end of exercise. Emergency equipment and qualified personnel should be available.

### For Additional Reading

For further facts about the principles and practice of stress testing, see Ellestad (1995).

## *Evaluation of Test Results*

An evaluation of cardiac performance should be based on total oxygen uptake (in $L \cdot min^{-1}$), because this is correlated with cardiac output, myocardial oxygen consumption, and blood flow. In a group of individuals with different body weights, the oxygen uptake corrected for weight ($ml \cdot kg^{-1} \cdot min^{-1}$) is unrelated to the actual load on their hearts. In longitudinal studies, the subject's body weight can vary considerably. The milliliter per kilogram figure for maximal oxygen uptake is a good predictor of the subject's potential to move and lift the body, but again it may not mirror the cardiac performance. For example, a subject with a body weight of 80 kg has a maximal heart rate 180, $\dot{V}O_2max$ of 2.0 $L \cdot min^{-1}$, cardiac output of 14 $L \cdot min^{-1}$, and stroke volume of 78 ml. His maximal oxygen uptake is expressed as $ml \cdot kg^{-1} \cdot min^{-1}$ is 25, equivalent to 7 METs. He is retested after having lost 10 kg in body weight. His weight is now 70 kg, maximal heart rate 180, as before, $\dot{V}O_2$ 2.0 $L \cdot min^{-1}$, cardiac output 14 $L \cdot min^{-1}$, and stroke volume 78 ml, but his measured maximal MET value is now 8 and his maximal oxygen uptake 29 $ml \cdot kg^{-1} \cdot min^{-1}$, a difference of 14% from the previous test, merely because of the weight loss. Yet the cardiac performance is unchanged.

Maximal oxygen uptake and, for that matter, the subject's response to a submaximal or maximal exercise test actually are of limited value in judging the subject's potential for running, skiing, and swimming. A superb technique, high motivation, and a good state of training, giving the potential to exercise close to maximal oxygen uptake for long periods of time, can significantly compensate for an inadequate maximal aerobic power. Actually, some marathon runners of world elite class have quite a modest maximal oxygen uptake, below 70 $ml \cdot kg^{-1} \cdot min^{-1}$. On the other hand, they have a very good running technique and can exercise for hours at a high percentage of their maximal oxygen uptake. In one study of marathon runners, the speed that was attained at the anaerobic threshold or the onset of blood lactate accumulation was highly correlated with their marathon time ($r > .90$; Sjödin and Svedenhag 1985). This is logical, because the threshold is dependent on the running economy, the maximal oxygen uptake, and the ability to exercise close to this maximum for hours.

## Summary: Tests of Maximal Aerobic Power

Every type of exercise is, in a sense, a unique situation. However, all forms of muscular exercise increase the metabolic rate, and therefore it is of particular interest to be able to analyze the involvement of the oxygen-transporting system. Oxygen uptake provides an accurate measure of aerobic power, and it is highly related to the cardiac output. The $\dot{V}O_2max$ is, under standardized conditions, a highly reproducible measure of the individual's aerobic fitness. However, it is subject to variations under certain conditions (i.e., after prolonged inactivity, after training, or as a consequence of cardiac disease). The main factor behind such variations in $\dot{V}O_2max$ is a proportional change in stroke volume. Therefore, a recording of the heart rate during exercise at a given oxygen uptake will reflect these variations in longitudinal studies. Generally speaking, a high heart rate usually is associated with a low stroke volume. However, from this information it is not possible to tell whether this exercise response is caused by genetic factors, lack of training, impaired heart function, or other factors. A multistage exercise test on a treadmill or cycle ergometer will indicate the rate of work an individual is able to tolerate without symptoms or ECG abnormalities. To predict the subject's ability to move his or her body, the maximal oxygen uptake per kilogram of body weight should be calculated. However, an evaluation of cardiac performance should be based on total oxygen transport ($\dot{V}O_2$ in $L \cdot min^{-1}$), because this is correlated with cardiac output, myocardial oxygen consumption, and blood flow. Variations in body fat content are not followed by similar changes in the dimensions of muscles, heart, and blood volume and in the demands for local blood flow. In other words, in a heterogeneous group of individuals, the $\dot{V}O_2 \cdot kg^{-1} \cdot min^{-1}$ value is unrelated to the actual load on their hearts. One good measure of cardiac performance is the ratio of oxygen uptake/heart rate, that is, the oxygen pulse.

No test protocol is ideal for all situations. It is recommended that the tester adapt the initial rate of work and the increment in work rate to the assumed maximal aerobic power of the person to be tested. The exercise time at each stage should be at least 3 min if possible; the larger the increments in rate of work, the longer the time at each stage. If the main purpose of the exercise test is to establish the $\dot{V}O_2max$ or symptom-limited exercise tolerance, one can apply a non-steady-state protocol with 1 to 2 min at each stage.

For the investigator who is willing to accept the small but definite risk involved, a multistage test carried to symptom-limited work rate or to maximal power will provide the clearest results, particularly for differentiating between normal subjects and coronary heart disease patients.

An alternative is to terminate the multistage exercise test at a heart rate close to 200 minus the subjects' age (in years). For most individuals, this means a submaximal test. The third alternative is to simulate on a cycle ergometer or a treadmill the metabolic rate of the subject's job or recreational activities. Notes about perceived exertion, objective measurements of blood pressure, blood lactate concentration, and recording of the ECG will give essential information about the subject's potential to meet those demands.

Repeated single-stage or multistage tests give excellent indication of variations in physical fitness. (Changes in body weight must be considered in a treadmill test.)

## Evaluation of Anaerobic Power

An athlete's physical fitness cannot be assessed by maximal oxygen uptake alone. Anaerobic metabolism, speed, strength, and maximal power are also determining factors in many types of athletic performance.

Theoretical as well as practical aspects of the anaerobic energy yield during exercise at various work rates are discussed in chapter 8. Unfortunately, at present we do not have any satisfactory methods to measure anaerobic power with a desirable degree of accuracy. It is true that if the mechanical efficiency in cycle ergometer exercise is constant, the energy requirement during exhausting exercise can be predicted by extrapolation. By measuring the oxygen uptake continuously during the entire exercise period, the investigator can calculate oxygen deficit and use it as a measure of anaerobic energy yield. The problem is, however, that in most events or types of exercise, one cannot estimate the total energy demand with sufficient accuracy when dealing with maximal or near-maximal effort. Consequently, anaerobic power cannot be determined. An increase in lactate concentration in the muscle or in the blood will not give a quantitative measurement of the total lactate production, because we do not know the total volume of body fluid into which the lactate is dissolved. This is a handicap when considering how to train anaerobic power most effectively, because we are unable to measure anaerobic power objectively.

However, there are tests that to some extent reflect the maximal anaerobic power of a subject. Laboratory tests with 30 to 60 s sprint bouts on a cycle ergometer have been the subject of increasing interest. One of them, the **Wingate anaerobic test,** which was introduced in the early 1970s, consists of an exhaustive cycling exercise against a resistance that is related to the subject's body weight. The subject is asked to make as many revolutions as possible for a total of 30 s, and the number of half-revolutions is recorded at 5-s intervals. The peak mechanical power, the average power, and the decline in this power from the peak recording down to the last 5 s period are calculated. There are two physiological bases for these measurements: (1) The maximal 5-s power is assumed to be correlated with the power generated by the intramuscular high-energy phosphate compounds, adenosine triphosphate (ATP) and phosphocreatine (PC), and (2) the 30-s average power output should reflect the anaerobic capacity. There may be a high correlation between those two factors, but a 30-s bout of exercise will not exhaust the anaerobic capacity. Since the introduction of the Wingate test, a number of modifications to the original protocol have been proposed (see Bar-Or 1987; Ingen Schenau, Jacobs, and de Konig 1991).

Gastin et al. (1991) developed a test that involves 60 s of maximal exercise and uses a variable resistance loading on the cycle ergometer. According to the authors, the variable resistance design permits the measurement of both peak anaerobic power (peak ATP–PC system power) and maximal anaerobic power (glycolytic) over 60 s duration.

When we discussed the force–velocity curve for a maximally contracting muscle (in connection with figure 3.34), we emphasized that the selected force/speed of contraction in relation to the maximal potential was critical. Actually, to achieve optimal human power output, cycling force and velocity should be equal and close to one third of their maximal values (Davies et al. 1982). Sargeant, Hoinville, and Young (1991) noticed that the velocity for the greatest mechanical power output was 110 rev $\cdot$ min$^{-1}$ (mean value over a complete revolution of 840 W in their male subjects). Bergh (1985) suggested a braking force of 10% of the subject's body weight, which is very close to the optimum reported by Nadeau and Brassard (1983), 50 N for their female and 70 N for their male subjects. Usually, the maximal power when tested on a cycle ergometer is attained during the first 2 to 3 s of the exercise. Without a sophisticated apparatus, it is difficult to measure the total force, that is, the force exerted to accelerate the mass of the wheel and to overcome the braking force. However, Bergh (1985) found a close relation between this total force per second and the peak velocity of the wheel or pedals. The latter is much easier to measure.

The simplest apparatus is a mechanically braked, calibrated cycle ergometer with a speedometer (e.g., a pendulum-braked or "weight" ergometer). The peak velocity is quite simple to read on the speedometer, and the number of revolutions can be counted and timed. However, it is certainly an improvement to modify a regular cycle ergometer by adding toe clips, installing dual microswitches on the cranks, and using racing handlebars and a reinforced and lengthened seat stem.

In this and similar all-out tests on a cycle ergometer, the mean coefficient of variance in data for such parameters as peak torque, mechanical power, average power, ride time, fatigability, and peak blood lactate concentration is reported to be 5% to 6% under optimal conditions (Coggan and Costill 1984; J.A. Evans and Quinney 1981). Significant correlations have been noted between the 30-s results and performance in running shorter distances, such as 40 m up to 300 m ($r$ = .7–.8; see Shephard 1982, p. 66). As already emphasized, such a correlation coefficient is not high enough for a good prediction of one individual's running ability from a cycle ergometer test.

Bosco, Terjung, and Greenleaf (1983) proposed a test for measuring mechanical power during a series of vertical rebounce jumps (e.g., in four successive 15-s periods). The sum of the flight times is measured with a digital timer. On the basis of this information and the number of jumps measured, their formula provides the mechanical power per unit mass. In the authors' hands, the reproducibility was high ($r$ = .95). Because of the simplicity of the test, it may be applied in both the laboratory and the field.

Vandewalle, Pérès, and Monod (1987) extensively reviewed the standard anaerobic exercise tests, concluding that if only one anaerobic test can be performed within the testing session, it is better to measure the maximal anaerobic power than maximal anaerobic capacity. They advised using a force–velocity test rather than the other maximal anaerobic power tests because the accuracy of the measure is probably greater and it also gives information on the force and velocity components of maximal power.

### *Summary*

The method of measuring maximal power output during a 30-s cycle ergometer test is assumed to reflect the maximal muscular strength of the engaged muscle groups and the power of the high-energy phosphate compounds. The average power reflects the anaerobic capacity but it is not a measure of this capacity. The decline in power during the test (e.g., measured at 5 s intervals) provides a fatigue curve under these standardized conditions.

# Muscular Strength

It is almost universally accepted that the maximal force potential of a muscle depends on its physiological cross section, commonly defined as a cross section perpendicular to the long axis of the muscle fibers. It is more correct, however, to say that the maximal force potential is the combined cross sections of all fascicles of the muscle. Of course, this means that a pennate muscle is stronger than a parallel-fibered muscle of the same volume. So far, this is the textbook version of the story, but other factors complicate the picture.

## The Anatomical Basis for Muscular Strength

As discussed in chapter 3, the force along the line of pull of a muscle is reduced by a factor equal to the cosine of its pennation angle. This means that the increase in pennation inherent in the process of hypertrophy in a pennate muscle will reduce the increase in force along the line of pull of the muscle (Kawakami, Abe, and Fukunaga 1993). In addition, the increase of the angle of pennation during shortening will influence the length–tension relationship on the whole muscle level (Kaufman, An, and Chao 1989).

Furthermore, the notion that muscle strength depends on the number of sarcomeres in parallel (see, e.g., the simplified model presented by Edgerton et al. 1986) derives from the idea that the contractile force developed by a myofibril is transmitted serially to its ends and eventually to the tendons through myotendinous junctions. There are indications that we may have to modify this view. As discussed in the section "Force Transmission From Muscle Fiber to Tendon" in chapter 3, the demonstration of muscle fibers that do not span the entire length of the fascicle they are part of shows that serial transmission of force can no longer be regarded as the only mode. This is not least attributable to the fact that these fibers have tapered ends, leaving room for fewer myofibrils in the ends than in the middle part. The possibility that part of the force is transmitted through lateral connections to the cell membrane at each Z-disk level raises the question whether sarcomeres, albeit morphologically in series, act in parallel as well through their individual attachments to the cell membrane and the endomysium (see figure 3.6).

If this assumption holds true, a long muscle fiber may be stronger than a shorter muscle fiber with the same diameter, all other factors being equal. In fact, it has been shown that long muscle fascicles may confer an advantage in activities requiring high power output. Kumagai et al. (2000) found that sprinters with personal-best 100-m time in the range of 10 to 10.9 s had significantly longer muscle fascicles in the lateral vastus and gastrocnemius muscles than sprinters with personal-best times in the 11- to 11.7-s range. This cannot be taken as support of the preceding theoretical considerations, however, because what counts in sprinting is power rather than static strength. Consequently, it is entirely possible that the advantage of longer fascicles can be explained by the higher external velocity of shortening attained because of more sarcomeres in series, despite identical velocity of shortening at the sarcomere level. In other words, longer fascicles may compensate in part for the loss of force attributable to high intrinsic velocity of shortening (see the section in chapter 3, "Force in Relation to Speed of Shortening: Force–Velocity Relationship").

## Requirements for Maximal Force Development

There is evidence that a maximal voluntary muscle effort, in most situations and with unconditioned subjects, does not engage all the motor units of the active muscle maximally. This can be demonstrated experimentally by applying a vibratory stimulus to the muscle during a presumed maximal contraction or by the **twitch interpolation** technique (Gandevia, Enoka et al. 1995; Vøllestad 1997). Theoretically, this inability to engage the muscle maximally may be due to insufficient excitation or excessive inhibition of the motoneurons in question, resulting from supraspinal influence or reflex activity. In a specific situation, say an emergency, and most probably also as an effect of training, net excitation increases, and the full force potential of the muscle can be revealed.

Many athletes shout in the critical stage of an effort, which may be advantageous in the light of the following experiment. Ikai and Steinhaus (1961) conducted experiments in which the subjects made a maximal arm flexion every minute during a 30-min period. The investigators found that the firing of a 22-caliber gun 2 to 10 s before a pull could significantly increase the exerted maximal strength. Shouting, hypnosis, and amphetamine also tend to improve performance compared with controls. The positive effect on strength was noticeable in untrained subjects but slight or absent in well-trained athletes such as weightlifters. Ikai and Steinhaus cited Pavlov's statement that "any unusual sensory experience or excitement may inhibit inhibitions" (i.e., disinhibition). They emphasize that their findings "support the thesis that in every voluntarily executed, all-out maximal effort, psychological rather than physiological factors determine the limits of performance." Furthermore, they claimed that, "because such psychological factors are readily modified, the implications of this position gravely challenge all estimates of fitness and training effects based on testing programs that involve measures of all-out or maximal performance."

Although we only can speculate about the mechanisms responsible for these effects, they may explain the many anecdotal reports indicating that individuals in a stressful situation perform better than otherwise; they become exceptionally powerful. We know, however, that in controlled experiments, epinephrine and norepinephrine, the levels of which are enhanced during stress, increase both excitability and contractility above normal values.

The thesis that central factors are of decisive significance for the development of strength is also supported by the observation that strength can increase without a proportional hypertrophy of the muscles. McDonagh, Hayward, and Davies (1983) reported that a 5-week training producing a 20% increase in the force of a maximal voluntary isometric contraction was not combined with any modification of the electrically evoked twitch and tetanus maximal force. They concluded that the improvement was based on factors other than the force-generating potential of the muscle fibers themselves. As reviewed by Sale (1992), other observations indicate that during the first weeks of strength training, the improvement is not associated with an increase in the cross-sectional area of the muscle groups involved. Later there is a gradual increase in both strength and the muscle cross-sectional area, particularly of the type II fibers. These events are further discussed in chapter 11.

## Measuring Muscle Strength

Muscle strength is a very complicated parameter. Given the preceding information, it is not surprising that even in well-standardized measurements of muscle strength, the standard deviation of the results obtained in repeated tests on the same subjects can be ±10% or even higher. Furthermore, there is no doubt that the more complex a contraction, the more difficult it is to measure accurately (Simonson and Lind 1971).

From studies on individuals of different body size, age, and sex, Asmussen, Hansen, and Lammert (1965) and O. Lambert (1965) concluded that the correlation between symmetrical muscle groups (right and left) is quite high ($r$ = .8). This depends, however, on the type of muscular activity habitually performed by the individual. Tennis players have been reported to have less symmetrical strength profile than wrestlers (Sward, Svensson, and Zetterberg 1990). Between flexors and extensors of the same extremity the correlation is fairly high, but between muscles from different parts of the body the correlation is rather low ($r$ = ~.4 or less). Again, the prevailing type of physical activity should be taken into account. Therefore, these authors concluded, the overall muscle strength of an individual should not be evaluated from measurements in one single muscle group, such as the finger flexors in a handgrip, but from the application of a battery of well-standardized, selected tests of muscular strength. The correlation between maximal dynamic and isometric strength recorded in these studies was .8 but can depend on the individual's muscle fiber type distribution and the actual speed of shortening of the dynamic contraction. Thus, an extrapolation from one type of testing to another has its limitations.

The value of comparing concentric strength in antagonistic muscle groups has been debated from a functional point of view. In antagonistic pairs, one muscle works concentrically while the other works eccentrically to control the movement. Consequently, comparing concentric strength in one muscle group with eccentric strength in the antagonistic group, while correcting for the influence of gravity, may give a more correct picture of the muscular balance over the joint in question (Aagaard et al. 1995).

Many investigators have speculated whether the maximal force produced with both left and right limbs involved simultaneously is simply the sum of the individually measured forces (i.e., left and right limb performance). The literature provides conflicting data on this point. Some studies indicate an inhibition on the left side of the body's performance when the right side is simultaneously involved and vice versa (Coyle et al. 1981; Ohtsuki 1981). Others have noticed that the two-limb strength exceeded the sum of the right and left forces measured separately. One side of the body enhanced performance on the other side (Wawrzinoszek and Kramer 1984). Such discrepancies may be attributable to differences in the training status of the experimental subjects. It appears logical that when one arm is not enough to accomplish a task, that arm should not be disadvantaged when the other arm is brought into use to assist it. However, other points should be considered: With two arms (or legs) pushing at maximum simultaneously, the strain on the back, for instance, might be so great that an inhibition is warranted.

When an evaluation of maximal dynamic strength development is essential, the dependence of the force–velocity relationship represents a problem. For this reason, isokinetic instruments such as the Cybex apparatus have been developed by which the speed of contraction can be accurately controlled. In a way, this tests an "artificial" contraction because movements are not normally isokinetic. Whether this has any practical implications for the interpretation of the data is difficult to assess.

In the isokinetic strength test, one measures the muscular force times the lever arm, that is, the torque (in newton-meters). If it is of interest to compare data obtained on subjects of different body size, a measure of the force may be desirable. Because the length of the lever is very difficult to measure, one may divide the newton-meter value by the subject's body height, assuming that it is proportional to the muscle's lever.

When discussing tests of anaerobic power, we mentioned that a vertical jump and a "sprint" on the cycle ergometer can be applied as a test of knee extensor muscle strength. A "fatigue test" also has been developed for the isokinetic muscle contraction. The subject is instructed to perform 50 repeated contractions with maximal effort without rest in between. The average peak torque for the last three contractions is given in percentage of the highest peak torque recorded (or the mean of the first three contractions). The decline in torque is taken as a fatigue index (Thorstensson 1976).

The maximal muscular force can be measured in various types of exercise with resistance weight-training machines (e.g., bench press, "curl," "squat"), but for obvious reasons such activities are very difficult to standardize. The same is true for the pull-up and push-up.

As will be discussed in chapter 11, there is also specificity in the effect of strength training; the cause of this specificity is more likely related to events in the central nervous system than in the muscles. To increase muscle strength for a particular activity, the best training is that same activity. The gain in strength attained through other procedures may be comparatively modest when the individual is tested in the particular activity in question. The explanation for this is that a gain in strength after a particular training program is attributable not only to changes in the muscle tissue but also to modifications in the central nervous system. In a different procedure, this effect on the central nervous system is absent, although the same muscles may be active, but in a different pattern of movement. Even slight changes in a movement may reduce the load on one muscle and increase it on another.

## Summary: Muscular Strength

Basic muscle physiology holds that the maximal force potential of a muscle depends on the combined cross-sectional area of all its fascicles—its physiological cross section. Although this means that a pennate muscle is stronger than a parallel-fibered muscle of the same volume, the reverse side of the coin is that the pennate muscle's force along the line of pull is reduced by a factor equal to the cosine of its angle of pennation. An increase in the angle of pennation during muscle contraction, or even after hypertrophy, will affect the force along the line of pull accordingly.

Many people, especially untrained ones, are unable to engage all motor units maximally in a voluntary contraction but may experience unexpected strength in critical situations, possibly because of the enhanced levels of adrenaline and noradrenaline. As mentioned in chapter 4, plateau potentials induced by these monoamines increase the excitability of motoneurons (D.F. Collins, Burke, and Gandevia 2002; Hounsgaard 2002), thus increasing the individual's ability to recruit all relevant motor units maximally. This phenomenon may be related to the athlete's habit of shouting at the critical stage of an effort, although training by itself increases the ability to recruit all motor units maximally.

Measuring muscle strength is not as simple as it may seem, and rather large variations have been found in repeated, standardized experiments. An individual's overall muscular strength should not be evaluated from measurements in a single muscle group but rather from a battery of well-standardized tests, selected according to the purpose of the evaluation, be it in relation to a job or the ability to lead an independent daily life.

# Specific Tests for Specific Performance

For a number of reasons, it would be useful to be able to assess, by specific tests, the extent to which a person has the basic requirements for the development of superior performance in a specific athletic event. Any such testing, however, should start with a physical examination to provide information concerning the athlete's ability to participate safely in sports. As pointed out by Kibler, Chandler, and Maddux (1989), by collecting information specific to the musculoskeletal system in addition to that from a general medical examination, we can gain information that improves performance and helps prevent certain injuries.

By examining more than 2,000 athletes from a variety of sports from the junior high school to the college level using specific tests for flexibility, strength, and endurance, Kibler, Chandler, and Maddux (1989) found that females were significantly more flexible than males on all flexibility measurements. The males were significantly stronger than the females on all strength measurements. Their results indicate that the adaptations of the musculoskeletal system are sport specific and depend on the part of the body that is placed under stress. The following examples are meant to point out the potential of using sport-specific tests to evaluate athletic performance.

Seeking a specific test for performance in endurance running, Duggan and Tebutt (1990) found in a study of 11 nonendurance athletes that the onset of blood lactate accumulation may be a valid and reproducible predictor of 4-km run performance. They also found, however, that this test is no better than the "La 12" test (i.e., blood lactate at 12 km $\cdot$ $h^{-1}$), which is easier to perform and less traumatic for the subject, because it requires less blood sampling.

According to Thissen-Milder and Mayhew (1991), it is possible to select and classify high school volleyball players by general and specific tests. They tested 40 high school volleyball players during the first week of practice by six general and four specific motor performance tests. The specific tests included the overhead volley, forearm pass, wall spike, and self-bump/set test. The general tests included height, weight, percent body fat, agility run, vertical jump, and two flexibility maneuvers. Varsity players were significantly better in vertical jump, agility, and all specific ball-handling tests than freshmen and junior varsity players. The combination of forearm pass, overhead volley, vertical jump, and weight correctly classified 68% of the players to their team level. The combination of height, weight, and shoulder flexibility correctly classified 78% of the starters and nonstarters.

Rundell (1996b) examined the physiological consequences of the low "sitting" posture of speed skaters. Seven male short-track speed skaters performed running in-line skating in an upright position and in-line skating in the usual sitting posture, on a motor-driven treadmill. In addition, two 1,000-m time trials on ice were performed to assess the relationship between performance and laboratory measurements. The results showed that peak $\dot{V}O_2$ was significantly lower when skaters were in the sitting position than when skating upright or when running. At equivalent speeds, submaximal oxygen uptake was lower for skating in the sitting position, and blood lactate was significantly higher. Peak $\dot{V}O_2$ when skaters were in the sitting position was strongly related to the 1,000-m time trials on ice. It was suggested that the depressed $\dot{V}O_2$ and higher blood lactate during skating in the sitting position are related to decreased knee and trunk angle.

Mygind, Larsson, and Klausen (1991) studied six Danish male cross-country skiers. Their maximal oxygen uptake was measured while they ran on a treadmill and used an upper body ski ergometer incorporating the double-poling technique. Maximal oxygen uptake during treadmill running and double-poling was correlated with performance, expressed as a ranking score during 10 ski races. The maximal oxygen uptake measured during double-poling, when using the ski ergometer, was significantly correlated with performance. It was concluded that the upper body ski ergometer can be used to evaluate elite cross-country skiers. See also Wisløff and Helgerud (1998).

To succeed in the biathlon, one has to ski fast and shoot accurately. Rundell and Bacharach (1995) attempted to determine whether physiological laboratory test results relate to success in the biathlon. Their tests included treadmill run, double-pole lactate profile, $\dot{V}O_2$peak tests, and a double-pole peak power test. National Point Rank, racing ski time, and shooting percentage from the 1993 World Team Trials and laboratory test results from 11 males and 10 females were examined. Of the athletes tested, six males and six females were top 10 U.S. ranked. Racing ski time was significantly related to National Points Rank, both in males and females. Shooting percentage and National Points Rank were related for the females. Maximum run time during the $\dot{V}O_2$peak test was the only parameter related to National Points Rank or racing ski time for males. For females, National Points Rank was related to running $\dot{V}O_2$peak and to the double-pole peak power test. Double-pole and running $\dot{V}O_2$peak were related to shooting percentage for females. This study suggests that shooting percentage is more important to National Points Rank for females than for males. According to the authors, sex-specific tests might better predict success in elite biathlon skiers.

# Chapter 10

# Body Dimensions and Muscular Exercise

The thrill of watching athletic competitions is partially caused by the fact that it is impossible to predict who is going to win. It is impossible from appearance alone to tell who is an athletic champion. On the other hand, it is often possible to exclude those who obviously cannot reach top results in certain sport events, such as shot putting, rowing, and American football. A tiny individual hardly has a chance in these events. This chapter deals with the effect of body dimensions on physical activity, including the human resources in relation to muscular exercise from a biological viewpoint. If we compare animals of different size, it is evident that certain dimensions and functional capacities are determined by fundamental mechanical necessities. In addition, it may be a matter of biological adaptation. **Allometry** is the term used to describe the effect of size on bodily proportions and functions (R.M. Alexander 1971).

## Scaling of Body Dimensions

If we take two geometrically similar cubes of different size, the relationship between the surface and the volume of the two cubes can easily be calculated when only the scale factor between the sides of the two cubes is known. Let one cube have equal sides with length $l_1$ and another cube have equal sides with length $l_2$. Then the ratio or scale factor $L$ can be found as

$$L = l_1 / l_2 = L{:}1. \tag{1}$$

Now, regarding one side of each cube, their surface areas are $l_1 \cdot l_1$, and $l_2 \cdot l_2$, respectively. Considering all six sides of each cube, the total surface ratio is

$$\begin{aligned} \text{Surface ratio} &= (6 \cdot l_1 \cdot l_1) : (6 \cdot l_2 \cdot l_2) \\ &= (l_1 / l_2) \cdot (l_1 / l_2) = L \cdot L \\ &= L^2 = L^{2/1} = L_2{:}1. \end{aligned} \tag{2}$$

Similarly, for the volume ratios, we get

$$\begin{aligned} \text{Volume ratio} &= (l_1 \cdot l_1 \cdot l_1) : (l_2 \cdot l_2 \cdot l_2) \\ &= (l_1 / l_2) \cdot (l_1 / l_2) \cdot (l_1 \cdot l_2) \\ &= L \cdot L \cdot L = L^3 = L^{3/1} = L_3{:}1. \end{aligned} \tag{3}$$

We see that the area and volume ratios are known once the ratio or scale factor ($L$) is known. The importance of such a single scaling factor ($L$) becomes evident when establishing grounds for comparison of differently sized individuals:

Imagine two identical individuals—$I_1$ and $I_2$. All lengths (of arms, legs, fingers) in $I_1$ are exactly equal to the corresponding lengths in $I_2$. Now, let all linear dimensions in $I_2$ be multiplied by a single scaling factor. If this scaling factor is >1, all measures in $I_2$ will be enlarged. The scaling factor used to enlarge $I_2$ is nothing else than the scaling factor introduced previously. This is easily realized when regarding some of the ratios between $I_1$ and $I_2$. Take height, for example. If the height of $I_1$ is $h_1$, and the height of $I_2$ is $h_2$, then the ratio is

$$\text{Height ratio} = h_2 / h_1. \tag{4}$$

But, because $h_2 = h_1 \cdot L$, we get

$$\text{Height ratio} = h_2 / h_1 = h_1 \cdot L / h_1 = L. \tag{5}$$

The same ratio is found between all linear measures of $I_1$ and $I_2$. Furthermore, it follows from the preceding that the area and volume ratios between $I_1$ and $I_2$ are $L^2$ and $L^3$, respectively. We understand that the scaling factor $L$ represents a lot of information about two qualitatively similar individuals of different size.

Why talk about such hypothetical and nonexistent individuals $I_1$ and $I_2$? The point is that if two real individuals are sufficiently similar, they might by approximation be viewed as our $I_1$ and $I_2$. The benefit is the simple relations during comparison. Whether such a simplified point of view is legitimate depends on a lot of factors and will sometimes be acceptable and sometimes not. Precision is the crucial point.

If we compare two boys, one 180 cm high and the other 120 cm high, the scale factor will be such that all lengths, levers, and muscular contractions during a specific motion will be related as 180:120 or as 1.5:1 (figure 10.1). Cross sections of, for instance, a muscle, the aorta, a bone, the trachea, the alveolar surface, or the surface of the body are then related as $180^2:120^2$, or $1.5^2:1^2$ (i.e., 2.25:1). Volumes such as lung volumes, blood volumes, or heart volumes should similarly be related as $180^3:120^3$, or $1.5^3:1^3$ (i.e., 3.375:1).

# Functional Significance of Body Dimensions

It is generally accepted that the maximal force a muscle can develop (*F*) is proportional to ($\propto$) its cross-sectional area. The difference between physiological and anatomical cross section does not matter in this context. It therefore would be expected that the 1.5 times taller body should be able to produce a 2.25 times larger muscular force; that is, he should be able to lift a 2.25 times heavier weight. The advantage of the 1.5 times longer levers (*a* in figure 10.1) for the taller boy's muscles to work on is offset by the fact that the weight to be lifted also has a 1.5 times longer lever (*A* in figure 10.1).

## Force and Work

The magnitude of the work (*W*) to be performed by a muscle is determined by the developed force, or more correctly, by the force component in the direction of the movement—both of which are proportional to $L^2$—times the distance over which the force moves an object (e.g., the part of a limb distal to the joint where the movement takes place), that is proportional to *L*. Consequently, *W* is proportional to $L^2 \cdot L = L^3$. Thus, the work that the larger boy in our example should be able to perform with a specific muscle or muscle group is 3.375 times larger than would be expected from the smaller boy.

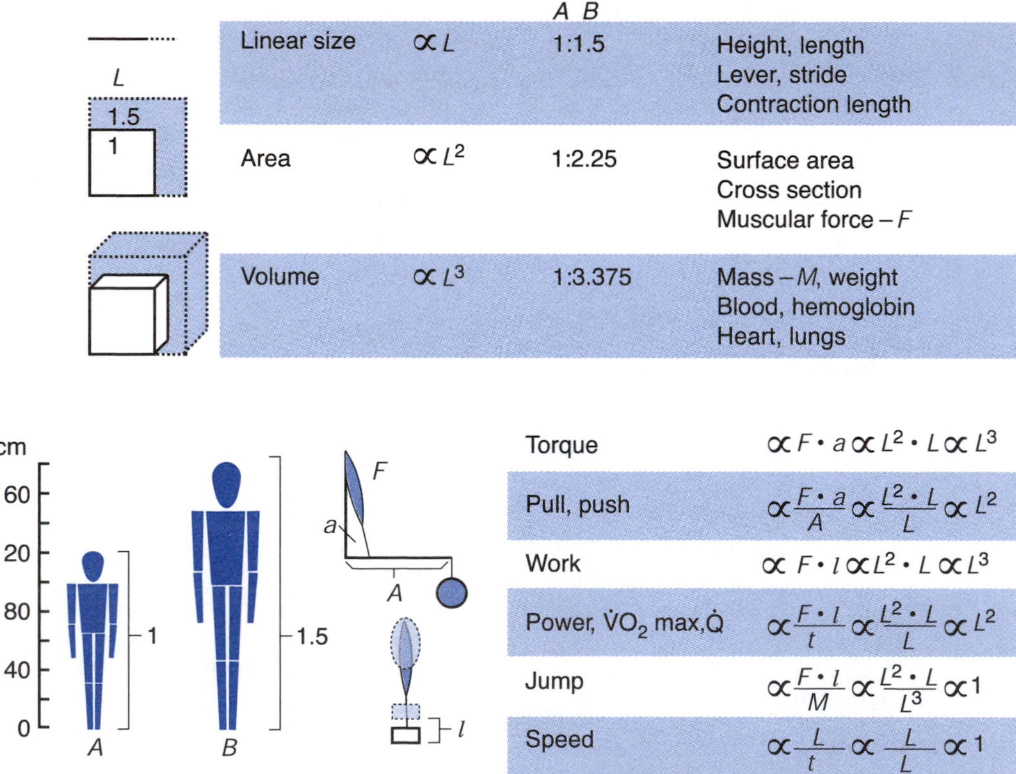


**Figure 10.1** Schematic illustration of the influence of dimensions on some static and dynamic functions in geometrically similar individuals. *A* and *B* represent two persons with body height 120 and 180 cm, respectively. See text for explanation.

Adapted from Asmussen and Christensen 1967.

If it is a matter of lifting one's own body with a mass ($M$), as in chinning the bar, then the muscular force times the muscular lever must be larger than the body weight times the body lever. For the sake of simplicity, we disregard movements in all joints of the body, except the elbow joints, making the axis of rotation of the elbow joints the only movement axes. Consequently, the body lever is the distance between the body line of gravity and the elbow axis of rotation, and the muscular lever is the distance between the vertical force component and the elbow axis of rotation. Force ($F$) is also equal to mass ($M$) times acceleration ($a$), which means that

$$a = F/M \propto L^2/L^3 = 1/L = 1 \cdot L^{-1}. \quad (6)$$

Consequently, taller and heavier persons are handicapped when accelerating their body mass.

## Frequency

Periodic events repeat themselves after a time $T$; that is, it takes $T$ seconds for one period to occur. A characteristic of such events is the frequency $f$, which is defined as

$$f = 1/T \text{ Hz}. \quad (7)$$

In other words, a frequency expresses the number of repeated events per second (Hz = $s^{-1}$). Biological examples are heartbeats, action potentials, and the steps of a walking person. Because the time scale is proportional to the length scale (Döbeln 1966), this can be expressed as follows (for details, see Döbeln 1966):

$$f \propto 1 \cdot t^{-1}, \text{ and accordingly, } f \propto 1 \cdot L^{-1} = L^{-1}. \quad (8)$$

According to this reasoning, we might expect that the frequency of limb motion should vary as an inverse function of limb length. As pointed out by A.V. Hill (1950), this is generally the case. A hummingbird moves its wings about 75 to 100 times per second while flying forward, a sparrow some 15 times per second, and a stork only two or three times. These frequencies are roughly in inverse proportion to the linear size of the birds. "If the sparrow's muscles were as slow as the stork's, it would be unable to fly. If the stork's muscles were as fast as the hummingbird's, it would be exhausted very quickly" (A.V Hill 1950).

So far, we have no indication that the maximal specific force of a contracting voluntary muscle varies with the size of the body, but the speed of contraction and hence the maximal speed of shortening vary enormously between different muscles and different animals. The balance between muscle strength, length of levers, and speed of shortening is very delicate. If humans had muscles contracting as fast as those that move the wings of a hummingbird, they would soon break the bones and tear the muscles and tendons. It is possible to swing a rod made of fragile material back and forth if it is short and relatively thick. But if it is long and with a thickness just proportional to $L^2$, the inertia will cause it to break if moved at the same speed. For the same reason, a small motor may run at higher revolutions per minute than a large one, and the strength of the material is an important factor determining the maximal speed. For similar reasons, a smaller creature can tolerate a greater speed of movement than a larger one. In fact, the margin of safety is quite small, so that occasionally muscles tear, tendons break, and bones splinter during unusual strain, such as during strenuous athletic events. Without altering the general design, it would be extremely hazardous for the athlete's locomotor organs if the muscles could be altered to allow him or her to run suddenly, say, 25% faster.

## Locomotion

Diamond (1983), who questioned why evolution never evolved the wheel, compared the energy cost of transporting the body by locomotion in terms of energy cost per gram per kilometer for a variety of creatures, from a cockroach to a human being. He found that transport costs are lower for swimming than flying and lower for flying than running, independent of the number of legs. Apart from maybe a whale or a large fish, the most efficient known vehicle is a human on a bicycle, requiring only one-fifth of the energy of the same human walking. According to Denny (1980), the greatest cost of locomotion ever measured is that of a slug crawling.

## Running Speed

The speed that can be attained in moving the body is determined by the length of stride—which is proportional to $L$—and the number of movements per unit of time—which is proportional to $L^{-1}$, among other things. Thus, for similar animals of different size, the maximal speed is proportional to $L \cdot L^{-1} = 1$, which means that the speed is the same. Short limbs with short strides move more rapidly and therefore can cover as much ground as do longer limbs moving more slowly. As stated, however, taller and heavier persons are handicapped when accelerating their body mass to attain that running speed.

It is well known that "athletic animals" of different size can achieve approximately the same maximal speed. A blue whale of 100 tons and a dolphin of 80 kg attain the same steady-state speed of about 15 knots and a maximal speed of about 20 knots (37 $km \cdot h^{-1}$). The speeds of a whippet, a greyhound, and a racehorse, very similar in general design, are nearly the same, or about 65 $km \cdot h^{-1}$ (40 mph; A.V. Hill 1950). Gazelles and antelopes with wide variation in size are all able to reach a maximal speed of about 80 $km \cdot h^{-1}$ (50 mph).

Since A.V. Hill (1950) presented his prediction of how performance can be expected to change as a function of body size, experimental evidence has shown that it is not quite true that all quadrupeds have the same maximal speed, that stride frequency is inversely proportional to limb length, or that the rate of oxygen uptake strictly increases with the cube of running speed (Heglund et al. 1974; Schmidt-Nielsen 1972; C.R. Taylor, Schmidt-Nielsen, and Rabb 1970). A major reason for this is the lack of geometric similarity across species (Heglund et al. 1974), violating an important part of the conditions for comparison.

## Jumping

In the broad jump and high jump, it is also a question of the maximal muscular force and power that can be developed and the distance that the muscle can shorten before the body leaves the ground. Therefore, we expect the performance to be proportional to the muscle cross-sectional area (which is proportional to $L^2$), proportional to its shortening distance (which is proportional to $L$), and inversely related to body mass (which means that it is proportional to $L^{-3}$). Because $L^2 \cdot L \cdot L^{-3} = 1$ (see figure 10.1), a small and a large animal should be equally able to lift their own centers of gravity. In broad jumping, kangaroos, jackrabbits, horses, mule deer, and impala antelopes actually seem to be equals to the record-holding human (A.V Hill 1950). Borelli drew similar conclusions some 300 years ago. In high jumping, in which the aim is to lift the body as high as possible, the larger animal has an advantage, however, because its center of gravity before the jump is already at a higher level (A.V Hill 1950), but at the same time it is at a disadvantage in accelerating its larger body mass.

We might expect that the 180 cm tall body would perform better in high jump than the 120 cm tall body, which it actually does. Their ability to move their centers of gravity vertically, on the other hand, should not be different, nor should their maximal speeds. The outstanding high jumper is without exception a tall individual (Khosla and McBroom 1985). This person is usually not geometrically similar to the average individual but is long-legged (Fuster, Jerez, and Ortega 1998), with a body weight that is not proportional to $L^3$, but lower. Again, deviations from our assumed similarity are seen.

## Maximal Running Speed for Children

In an analysis of speed in children of different size (figure 10.2), Asmussen divided them into age groups and plotted maximal speed, calculated from their personal best time in the 50 to 100 m dash, in relation to body height. From the age of about 10, the body proportions remain approximately the same. We therefore can consider the children represented in figure 10.2 as geometrically similar (figure 10.3).

For 11- to 12-year-old boys, there is no significant variation in speed with body size, as expected from the preceding discussion. The somewhat better performance of the 12-year-olds over the 11-year-old boys may be attributable to maturity of the neuromuscular function, improving the coordination. Even better are 14-year-old boys, but one also finds that the taller boys can run faster than the shorter boys. There is a further improvement in coordination with age, but this is also probably attributable to sexual maturity. Their male sex hormones may have increased their muscular strength. The smaller 14-year-old boys may not yet have reached puberty, in contrast to the taller boys. In the 18-year-old group, there is again hardly any variation in the results despite a large difference in body height. At this age, all the boys have passed puberty and are sexually mature.

In girls, there is an increase in maximal speed up to the age of 14, but from then on there is no further improvement. The results are not influenced by the size of the girls in any of the age groups, which supports the assumption that the superiority of the taller 14-year-old boys is attributable to the effect of male sex hormones. Slender body structure is favorable to long-distance running (Katić, 1996).

This independence of maximal speed with body height is in contrast to the greater muscle strength in taller children. From an anatomic viewpoint, this is just what would be expected: The muscle force should increase in proportion to $L^2$, but the speed should be independent. As already discussed, the results obtained for boys of different size do not strictly follow the results predicted from body dimensions. Apparently, biological factors can modify muscular dynamics. We have considered age factors as well as sexual maturity, which is particularly relevant in boys.

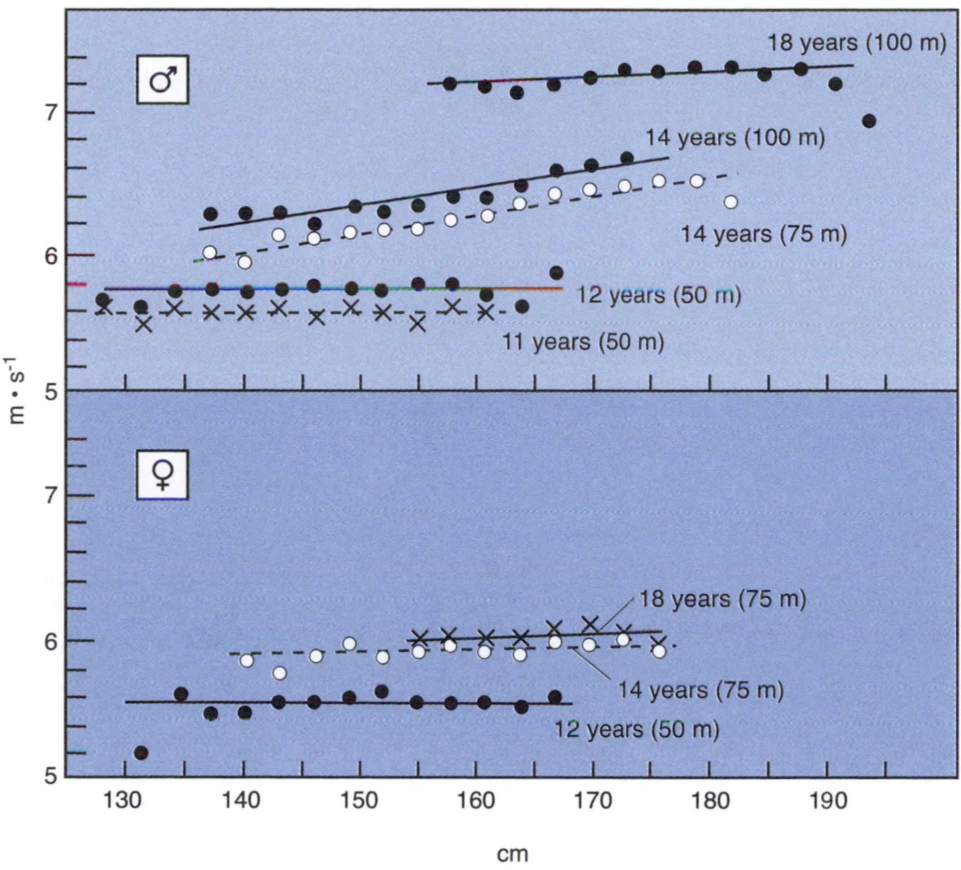


**Figure 10.2** Maximal speed in relation to body height for girls and boys of different age. Based on almost 100,000 subjects. See text for explanation.

Adapted from Asmussen and Christensen 1967.

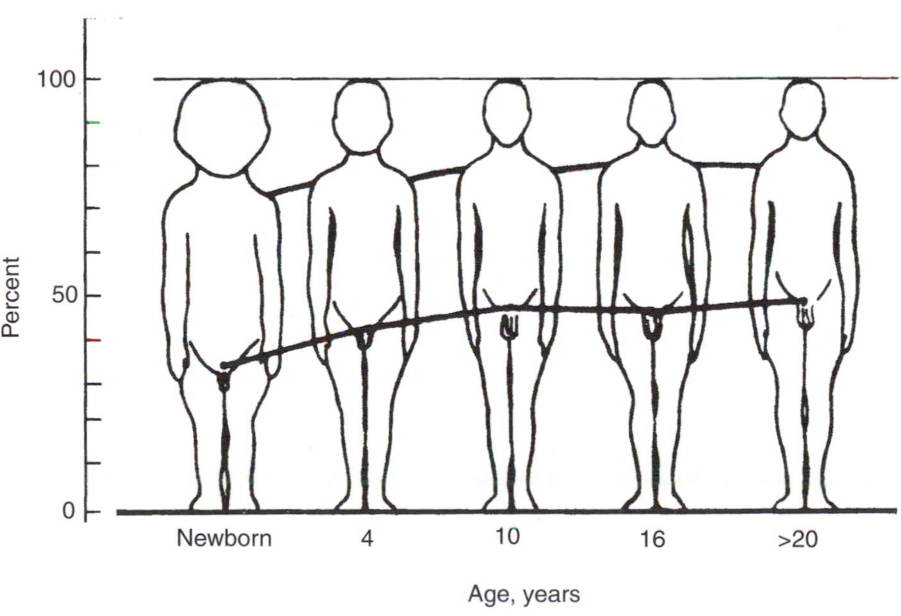


**Figure 10.3** Body proportions at different ages. Note that from the age of 10 years there is no marked change in proportions.

From Asmussen and Christensen 1967.

## CLASSIC STUDY

An extensive series of experimental studies of physical work capacity in relation to sex and age was carried out at the Swedish College of Physical Education some 50 years ago and provided a foundation for understanding some of the basic aspects of physical working capacity in relation to these variables.
P.-O. Åstrand 1952.

## Kinetic Energy and External Work

According to A.V. Hill (1950), the work performed during a single movement, calculated per unit of body weight, in producing and using kinetic energy in the limbs, is the same in large and small animals, and the external work done in overcoming the resistance of air or water is proportional to the square of the linear dimensions (i.e., the surface area). The same is the case regarding the muscular force used to overcome the resistance. Therefore, the effect of this resistance should be the same in similar animals of different size. In running uphill at a given speed, however, the effect of the slope of the hill is inversely proportional to the linear size $L$: The smaller the animal, the faster it should be when running uphill.

We concluded earlier that maximal speed is independent of an animal's size. If one animal is 1,000 times as heavy as another (proportional to $L^3$), each linear movement that is proportional to $L$ will be 10 times larger than in the smaller animal. However, the larger animal has to take only one tenth the number of steps, each step taking 10 times as long, to attain the same speed as the smaller animal. Because the work per movement and per unit of body weight is the same in the two animals, it follows that it will take roughly 10 times as long for the larger animal to become exhausted during a maximal run.

## Energy Supply

It is obviously important that the power output permitted by the mechanical design be matched by an equivalent supply of chemical energy. As mentioned previously, work (being the product of force applied in the direction of the movement and the distance moved) is proportional to $L^2 \cdot L = L^3$, or to $M$. The total energy output therefore should be related to the mass of the muscles and the body weight in similar animals. The power, or work output per unit of time, must then be proportional to $L^3 \cdot t^{-1}$, that is, to $L^3 \cdot L^{-1} = L^2$ ($\propto M^{2/3}$; see Brody 1945; Döbeln 1956a; Heusner 1982).

It is well established that the basal metabolic rate in animals with large differences in body size, from the mouse to the elephant, follows this prediction. The resting oxygen uptake is actually proportional to $M^{.74}$ rather than to $M^{2/3}$, but this difference is surprisingly small considering the wide variation in size, shape, and other factors (Brody 1945; Heusner 1982). This relation tells us that smaller animals must be more metabolically active per unit of body weight than larger ones. In proportion to its weight, a mouse has to eat 50 times more than a horse to maintain its basal metabolism.

Theoretically speaking, it is expected that the maximal oxygen uptake should be proportional to $L^2$ (see figure 10.1) or $M^{2/3}$. This holds extremely well for trained athletes, as is evident from figure 10.4. The lower panel of this figure illustrates that the maximal oxygen uptake, expressed as ml · min$^{-1}$ · kg$^{-2/3}$, is not related to body weight and therefore can be used as a meaningful fitness index instead of the conventional method of expressing maximal oxygen uptake as ml · min$^{-1}$ · kg$^{-1}$, which penalizes heavy individuals. Because maximal cardiac output and pulmonary ventilation are also volumes per unit of time, they should be proportional to $L^3 \cdot L^{-1} = L^2$ or $M^{2/3}$. Taylor et al. (1981) noted that in animals ranging in body weight ($w$) from 7 to 260 kg, the maximal oxygen uptake measured during treadmill running scaled roughly to $w^{0.75}$. The estimated diffusing capacity of the lung, however, scaled approximately to $w^{1.0}$ (Weibel et al. 1981).

It thus appears that the larger animals require a larger pulmonary diffusion capacity to transfer oxygen at a given rate from the air to the blood than do smaller animals. There is another approach with which to analyze this relationship, giving the same result: Cardiac output ($\dot{Q}$) is the product of the frequency of heartbeats and stroke volume. Frequency is proportional to $L^{-1}$ (see Döbeln 1966) and stroke volume to $L^3$, and therefore

$$\dot{Q} \propto L^{-1} \cdot L^3 = L^2. \quad (9)$$

Similarly, pulmonary ventilation is the product of respiratory frequency and tidal air:

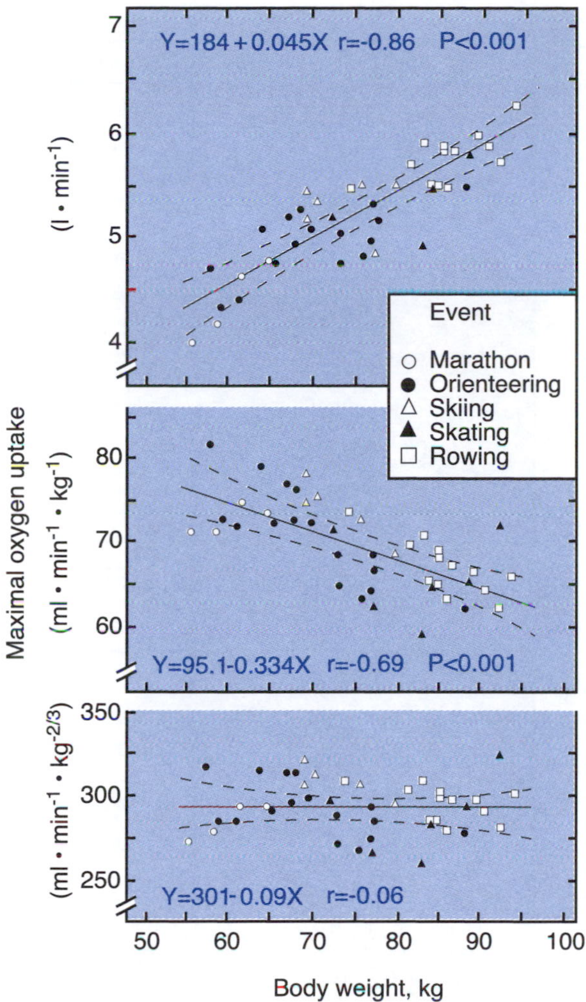


**Figure 10.4** Maximal oxygen uptake in a group of Norwegian top athletes trained in different events, expressed as $L \cdot min^{-1}$, $ml \cdot min^{-1} \cdot kg^{-1}$, and $ml \cdot min^{-1} \cdot kg^{-2/3}$.
Courtesy of O. Vaage and L. Hermansen.

$$\dot{V}_E \propto L^{-1} \cdot L^3 = L^2. \quad (10)$$

With pulmonary ventilation proportional to $\dot{V}O_2$ and with the production of carbon dioxide proportional to $\dot{V}O_2$, we should expect the alveolar partial pressure of carbon dioxide to be the same in different mammals, which it is. On the basis of measurements in subjects 7 to 30 years of age (P.-O. Åstrand 1952) it can be calculated that the vital capacity is proportional to $L^{3.1}$ in males and $L^{3.0}$ in females and, thus, very close to the expected $L^3$. We therefore can conclude that children have lung volumes that are proportionally dimensional to their body size.

In heavy exercise, heat production is great and related to $\dot{V}O_2$, which in turn (according to the formula in figure 10.1), is related to $L^2$ (i.e., to the surface of the body, from which most of the excess heat is lost, at least in humans). There is also heat loss via the expired air, increasing within limits, in direct proportion to $\dot{V}O_2$. This seems, however, to be in conflict with the usual rule that heat production is related to body volume, whereas heat loss is related to body surface. Furthermore, for the male subjects 8 to 18 years of age studied by P.-O. Åstrand (1952), it can be calculated that the maximal attainable oxygen uptake is proportional to $L^{2.9}$ and not, as expected, to $L^{2.0}$.

For a group of 65 young female and male subjects, Döbeln (1956b) calculated the value of the exponent $b$ in the equation maximal oxygen uptake = $k \cdot$ (body weight – adipose tissue)$n$ and found that $n$ = 0.71 (maximal oxygen uptake predicted from the Åstrand and Åstrand nomogram; adipose tissue calculated by means of hydrostatic weighing).

## Maximal Aerobic Power in Children

As stated, it appears that children's maximal oxygen uptake is not as high as expected for their size, and compared with adults, they do not have the aerobic power to handle their weight. It is also significant that the 8-year-old boy could increase his basal metabolic rate only 9.4 times during maximal running for 5 min, whereas the 17-year-old boy could attain an aerobic power that was 13.5 times the basal power. Therefore, the child has less in the way of a power reserve than the adult. In addition, the young subjects had a significantly higher oxygen uptake per kilogram of body weight than the older boys and adults when running at a given submaximal speed on a treadmill (P.-O. Åstrand 1952). These two factors together may explain the fact that children have difficulty in following their parents' speed, even if maximal oxygen uptake per kilogram body weight is the same. The children's lower efficiency can be explained partially by their high stride frequency, which is expensive in terms of energy use per unit of time.

Because the oxygen is transported by the hemoglobin (Hb), it is of interest to compare the children's total amount of Hb ($Hb_T$) with their body size and maximal oxygen uptake. In most of the subjects discussed previously, the total Hb was determined. In similar animals of different size, $Hb_T$ should be proportional to $M$. Per kilogram of body weight, however, the younger boys had only 78% of the amount of Hb of the older boys. Thus, the amount of Hb was definitely not proportional to body size. Assuming that the maximal oxygen uptake is proportional to $Hb_T$, we can calculate the maximal $\dot{V}O_2$ in children from the equation $\dot{V}O_2 = a \cdot Hb_T^{2/3}$ and use a value for $a$ calculated from the data on older boys. It is found that the child's maximal oxygen uptake should be 1.92 L · min$^{-1}$ if the child's Hb is as effective in transporting oxygen as the adult. This calculated value is not far from the determined 1.75 L · min$^{-1}$. If we use the exponent .74, the calculated maximal aerobic power in 7- to 9-year-old children will be 1.78 L · min$^{-1}$, or very close to real maximum.

The sample of subjects selected for these analyses was limited to 21 individuals, but it was a homogenous group. They were nonobese and in the same state of training. The results support the assumption that the maximal oxygen uptake in children and young adults is proportional to the muscular strength and to $Hb_T^{.76}$, or roughly to $Hb_T^{2/3}$, but not to $M^{2/3}$.

Rowland, Vanderburgh, and Cunningham (1997a) measured maximal oxygen uptake in 20 children annually over 5 years and found significant sex differences when $\dot{V}O_2$max was related to lean body mass. They found that factors other than body size affect the development of $\dot{V}O_2$max in children, and that sex differences exist in $\dot{V}O_2$max during childhood that are independent of body composition.

It can be concluded that children are physically handicapped compared with adults. When related to a child's body dimensions, the muscular strength of the child is low and so are his or her maximal oxygen uptake and other parameters of importance for the oxygen transport. Furthermore, the mechanical efficiency of children is often inferior to that of adults.

## Maximal Aerobic Power in Women

For female subjects, 8 to 16 years of age, the $\dot{V}O_2$max is proportional to $L^{2.5}$. In light of the previous discussion, the noted discrepancy from the theoretically expected $L^2$ is not surprising.

Women have approximately the same maximal oxygen uptake per kilogram of fat-free body mass as men, but it should be higher in women because of their smaller size (Döbeln 1956a, 1956b). The lower Hb concentration in women may explain why they cannot fully use their cardiac output for oxygen transport.

Figure 10.5 presents data on 227 children and young adults that stress these points further. There is a very high correlation between maximal oxygen uptake and body weight for male, nonobese subjects (upper panel). The lower maximal oxygen uptake for female subjects above 40 kg of body

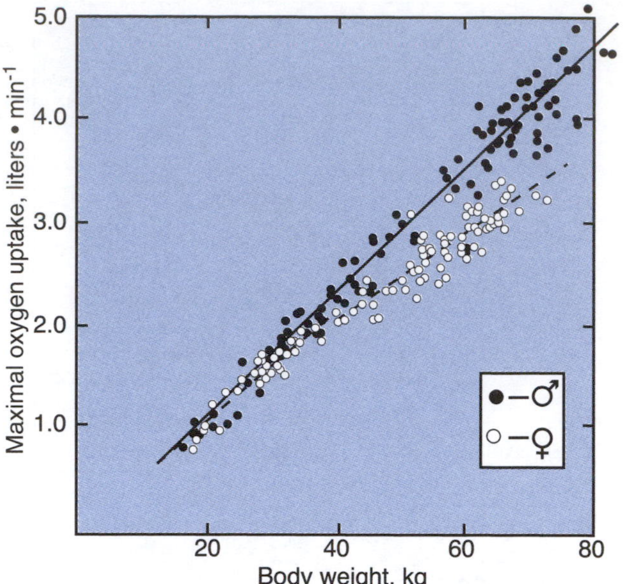


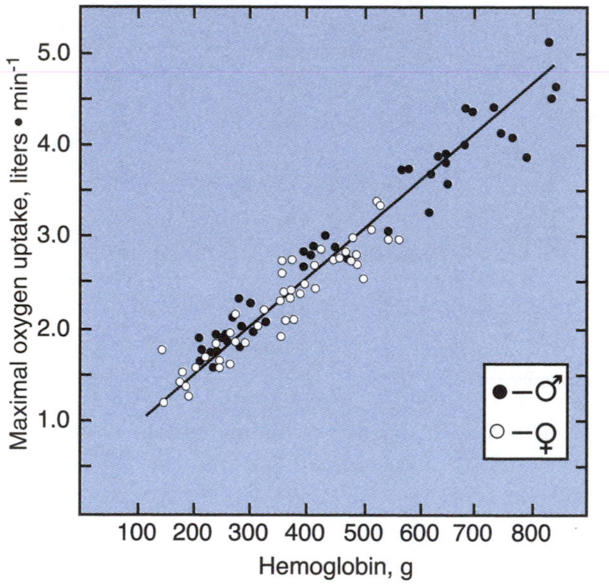


**Figure 10.5** Upper panel: Maximal oxygen uptake measured during cycling or running in 227 female and male subjects 4 to 33 years of age, in relation to body weight. For male subjects, the exponent $b$ = .76 in the equation maximal $\dot{V}O_2 = a \cdot Mb$, and $y = -.108 + .060x$ ($r$ = .980 ± .004; deviation from regression line = 7.5%). Lower panel: Maximal oxygen uptake for 94 of the same subjects, age 7 to 30 years, in relation to total amount of hemoglobin (Hb). In the equation, maximal $\dot{V}O_2 = a \cdot Hb^b$, the exponent $b$ = .76. (For all subjects, $r$ = .970 ± .006; two standard deviations within shadowed area. To determine total amount of Hb, a carbon monoxide method was used and the absolute values are questionable.) All subjects were fairly well trained, and none of them was overweight.

Adapted from P.-O. Åstrand 1952.

weight (age about 14 years) is largely explained by their higher content of adipose tissue. The lower concentration of Hb in women also contributes to the observed sex difference. When one relates the total amount of Hb to the maximal oxygen uptake, the difference between regression lines for the female and male subjects is insignificant (lower panel). The exponent $b$ in the equation $y = a \cdot x^b$ (i.e., maximal $\dot{V}O_2 = a \cdot Hb_T{}^b$) is .76.

Although sex differences have to be taken into account, body dimensions have predictive value also in women. After studying 670 Estonian female students 18 to 24 years of age, Kaarma et al. (1996) concluded that when studying the physical development of women and their individual body characteristics, body height and mass should always be starting parameters and taken together.

## Maximal Cardiac Output

According to theoretical analysis, maximal $\dot{V}O_2$ should be proportional to $L^2$ (or $M^{2/3}$), but as stated earlier, experimental data on children and adults do not support this assumption, because the exponent is closer to 3. Several factors contribute to this discrepancy. In addition to the possibility that the calculations are based on invalid assumptions, there are also reasons to believe that biological factors have modified the aerobic power in the subjects studied. Because $\dot{V}O_2$ must be related to cardiac output, it should be emphasized that if the total output of the heart per minute was proportional to the body weight ($L^3$), the blood velocity in the aorta would have to be so great in the largest mammals that the heart would be faced with an impossible task (A.V. Hill 1950; Hoesslin 1888). It is therefore more likely that the cardiac output is proportional, not to the body mass, but to $M^{2/3}$ (or $L^2$), like the basal metabolism. However, Rowland, Popowski, and Ferrone (1997b), in a study of 15 boys (mean age 11 years) and 16 men (mean age 31 years), found no significant differences in resting, submaximal, or peak values of stroke volume, aortic peak velocity, or systolic ejection time related to body size between the boys and the men.

The blood pressure is independent of body size, as pointed out by Döbeln (1956b), because pressure is force per cross-sectional area, or proportional to $L^2 \cdot L^{-2} = 1$. In hearts that work against the same blood pressure and are anatomically uniform in the sense that the coronary blood flow is a given percentage of the total blood flow, the maximal linear velocity of the blood in the aortic ostium during the period of expulsion is the same regardless of the size of the heart. This means that the maximal cardiac output, and consequently the maximal aerobic power of the entire animal, is proportional to the cross-sectional area of the aortic ostium. This area is found to be proportional to $L^2$ (or actually $M^{0.72}$; A.J. Clark 1927). Therefore, in uniformly built organisms, the maximal aerobic power is proportional to body weight raised to the 2/3 power (Döbeln 1956b).

## Heart Weight, Oxygen Pulse, and Heart Rate

Figure 10.6 illustrates how the heart weight is directly proportional to the body weight in mammals the size of a mouse up to the size of a horse (heart weight = .0066 · body weight .98; Adolph 1949). In other words, over the full range of mammalian size, the heart weight is a constant fraction of the body weight. However, an animal capable of severe exercise has a heart ratio greater than .6 (heart ratio = heart weight · 100 · body weight$^{-1}$), whereas animals incapable of heavy, steady exercise have ratios less than .6 (A.J. Clark 1927; Tenney 1967). (Compare the data on hare and rabbit in figure 10.6.) For the presented parameters, the best correlation and least deviation from the regression line are noticed between heart weight and Hb weight. A similar picture is demonstrated for the oxygen pulse (i.e., oxygen uptake per heartbeat at rest) and its relation to body weight (oxygen pulse = .061 · body weight$^{.99}$), blood volume, and Hb, respectively. Thus, the oxygen pulse is a relative measure of the stroke volume.

As already mentioned, the heart rate is likely to be proportional to $L^{-1}$ (Lambert and Teissier 1927). Therefore, the larger animal should have a lower heart rate at rest and possibly also during exercise than the smaller animal. The resting heart rate of the 25-g mouse is about 700 beats · min$^{-1}$ and of the 3,000-kg elephant is 25 beats · min$^{-1}$. This difference is not a biological adaptation but a mechanical necessity.

### For Additional Reading

For a more profound treatment of dimensional analysis and the theory of biological similarity, see Günther (1975).

Günther (1975) claimed that the statistical analysis of the experimental data can be represented most conveniently by means of the logarithmic equivalent of Huxley's allometric equation: $y = a \cdot W^b$, where $y$ is any function definable in terms of the MLT system ($M$ = mass, $L$ = length, $T$ = time), $a$ is an empirical parameter, $W$ is body weight, and $b$ is

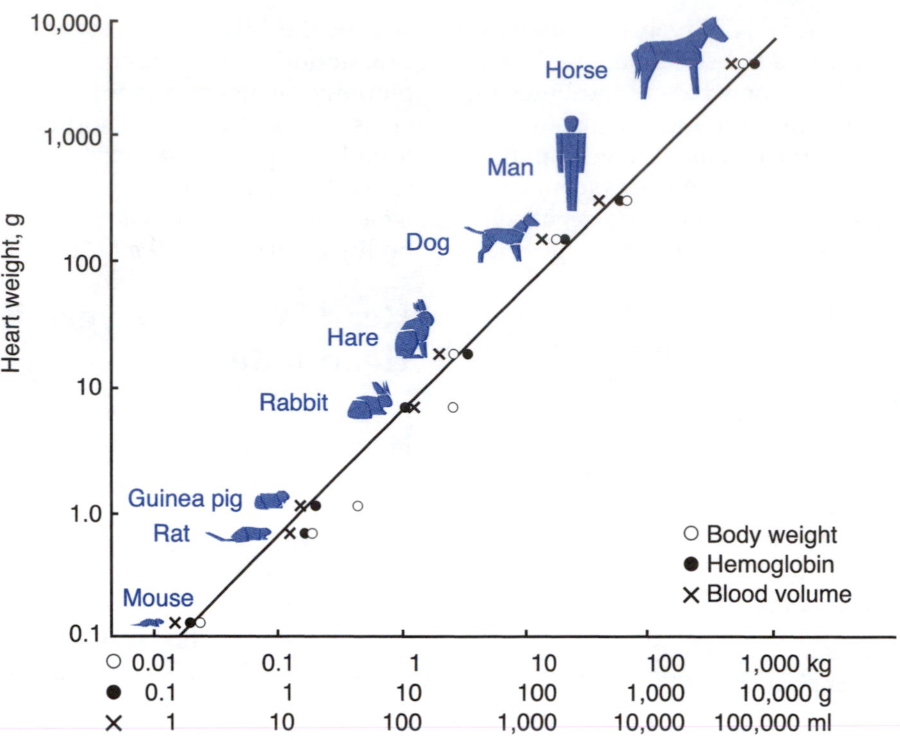


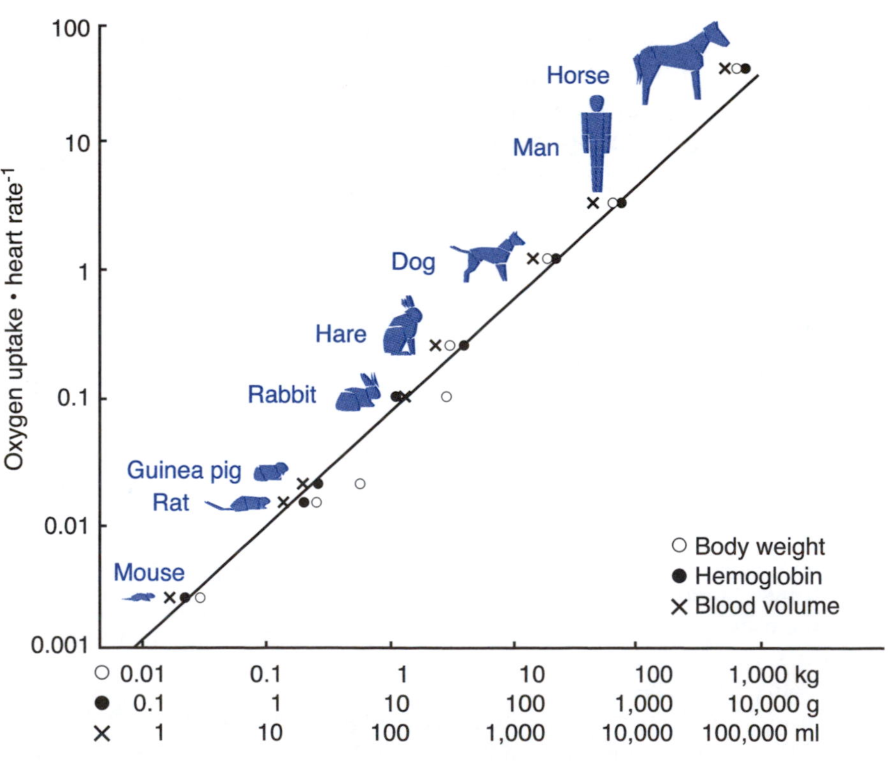


**Figure 10.6** Heart weight (upper panel) and oxygen uptake per heartbeat (oxygen pulse) in relation to body weight, total hemoglobin weight, and blood volume in various mammals.

Data from Sjöstrand 1961.

reduced exponent, the numerical value of which can be obtained from the log–log plot of the experimental data, because the logarithmic expression is a straight line ($\log y = \log a + b \cdot \log W$).

## Secular Increase in Dimension

We now consider a few additional applications of the effect of dimensions on human performance. Because, in many countries, humans have grown taller in recent generations, some improvement in athletic performance is to be expected. This steady secular increase in growth is typical of countries with a satisfactory nutritional status (figure 10.7). The importance of nutrition is demonstrated clearly by the decline in body height for children of all ages shown during the Second World War from 1940 to 1945 (Brundtland, Liestøl, and Walløe 1980). Today, besides the increase in such bodily dimensions as height and weight at all ages from birth to adulthood, the maturation of certain physiological functions, notably those connected with sexual maturity, is also accelerated. There has been a steady decrease in age of menarche, from about 17 years of age in 1840 to 13-1/2 years of age in 1960 (Tanner 1962). A similar trend of earlier maturation of boys is also apparent from the available data; boys now reach their maximal height at an earlier age than they did a generation ago. The influence of sexual maturity on performance is evident from figure 10.2. The rapid increase in height and weight that is accompanied by puberty generally occurs earlier in girls than in boys but is at any rate subject to marked individual variation (Tanner 1962). Thus, the peak height velocity in extreme cases can occur in a 9-year-old girl and in a 16-year-old boy (Lindgren 1978; see figure 8.12). Because this adolescent growth spurt profoundly affects physical performance, it may be unfair, if not harmful, to group children athletically in classes according to age.

Peplowski (1990) investigated the secular trends in body height and body weight in 3,579 Polish female medical students aged 18 to 20 years from 1966 to 1986. Body height showed a tendency to increase regularly in the course of time, and the body weight tended to decrease constantly. The subjects' somatotypes were found to vary according to the social environment. The students coming from professional or intellectual milieus were taller

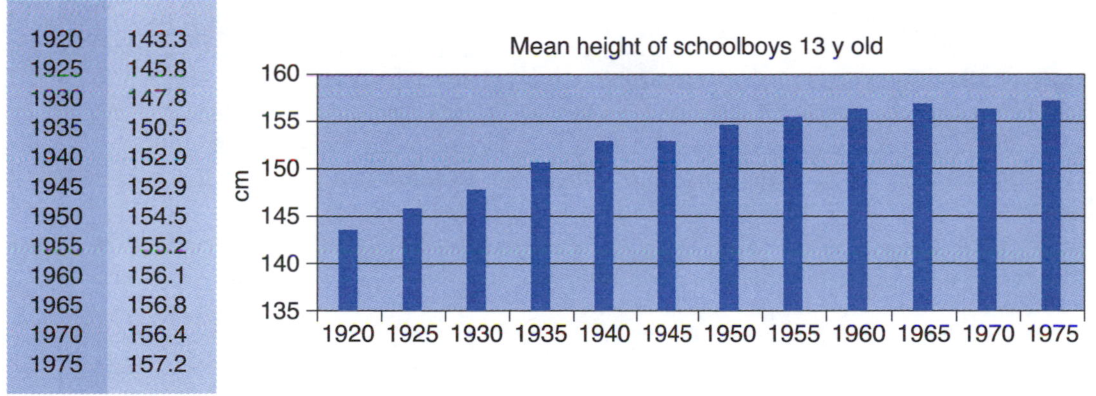


**Figure 10.7** The mean height of 13-year-old schoolboys in Oslo, Norway increased 13.9 cm between 1920 and 1975. The major part of the increase took place between 1920 and 1940, but the increase suffered a temporary stop during World War II, which affected Norway during the years 1940–45. A corresponding increase in body height, including the setback during the war, took place in all age groups attending elementary school, both boys and girls. Some age groups even suffered a small temporary decrease in body height during the war. On the whole, this resulted in an average 8-year-old boy or girl in 1975 having almost the same height as an average 10-year-old in 1920.

Based on data from Brundtland, Liestøl, and Walløe 1980.

and more slender than their peers who had working-class backgrounds or peasant origins.

In a study of a large random population sample of 200,000 children and youth 7 to 19 years of age, Przeveda (1994) observed that the levels of physical fitness increased in succeeding generations, but that the rate of fitness increase was smaller than the secular trend in body height. He also found that the structure of physical fitness in young people was changing; that is, the speed and nimbleness types were more common than strength types. He suggested that these findings indicated that human locomotor skills are now better adjusted to present-day conditions of life.

If the height of an athlete is 184 cm (the mean height of the participants in the decathlon in Rome in 1960), and the height of an athlete 30 years before was 176 cm (mean height of a decathlon athlete in 1930), their heights will compare as 1.06:1 and their muscle strength as 1.13:1 (Asmussen 1964). This means that, because of different dimensions, the average top athlete now is 6% taller than the top athlete of 30 years ago, but his or her muscular strength should be 13% greater than 30 years ago. Therefore, the maximal work that the muscles of the decathlon athlete could perform should be 20% greater than 30 years ago, because the maximal work a muscle can produce is the product of its maximal force and the distance it can shorten. When the size of the oxygen-transporting organs is the limiting factor, the taller athlete consequently should be able to deliver 13% more oxygen to the muscles per unit of time than the smaller athlete.

In such events as javelin or shot putting, the increase in bodily dimensions can influence achievements in two ways: In the first place, the athlete's strength increases in proportion to the second power of the person's height. This will tend to improve the results, particularly because the weight of the equipment is constant and not varied with the weight of the thrower. Second, the greater height from which the javelin and shot start their flight will cause them to travel further. These two factors, and the first one in particular, result in better records and may partially account for the improvements in records that have taken place. Anyway, there is clearly a good physiological basis for the selection of tall throwers.

This discussion is included to demonstrate that in some events, part (but only part) of the improvement in results over the years are attributable to the athletes' dimensional change. Khosla (1968) found that the winners in different throwing events in the Olympic Games in Rome and Tokyo were on the average taller and heavier than their competitors. Similarly, the winners in jumping and running, with the exception of 10,000-m and marathon running, were taller. This, he claims, is unfair to people who are less tall, and suggests classification of the competitors according to height and weight in events in which body size influences the results.

## Old Age

Figure 10.8a summarizes data from the literature on different biological measures in adults from 20 up to 60 years of age. It is evident that there is no strict interrelation of these functions based on dimensions alone. The body height is maintained constant, but body weight, heart weight, and heart volume increase with age. Blood volume and total amount of Hb are not markedly changed. Heart rate at a given submaximal work rate is the same in the old and the young; that is, the oxygen transport per heartbeat (oxygen pulse) is constant. However, maximal oxygen uptake, heart rate, stroke volume, pulmonary ventilation, and muscular strength decrease significantly with age.

As figure 10.8b shows, the urinary elimination of creatinine is reduced by about 30% from the age of 30 to age 75, which indicates a reduction of total muscle mass of about the same order. Concomitantly with this muscular atrophy, muscle strength declines correspondingly (i.e., about a 30% decrease in the strength of the back muscles, as indicated in figure 10.8b). Evidently, endurance-based physical exercise is of little value in maintaining muscle strength and speed of contraction in old age (Harridge, Magnusson, and Saltin 1997).

Himann et al. (1988) examined age-related changes in the speed of walking in 289 males and 149 females aged 19 to 102 years. They found that the walking speed was associated with height before 62 years and with height and age after 62 years.

# Conclusion

In the case of biological phenomena, we express the physical basic units of length, surface, mass, and time on the basis of one single unit: length. This means that all units such as pressure, temperature, and energy are expressed as derivatives of the basic unit. Table 10.1 is prepared on this basis.

The conclusions that can be drawn from table 10.1 obviously do not solve any biological problems. It may, however, facilitate the correct formulation of biological problems. If we know that a fully

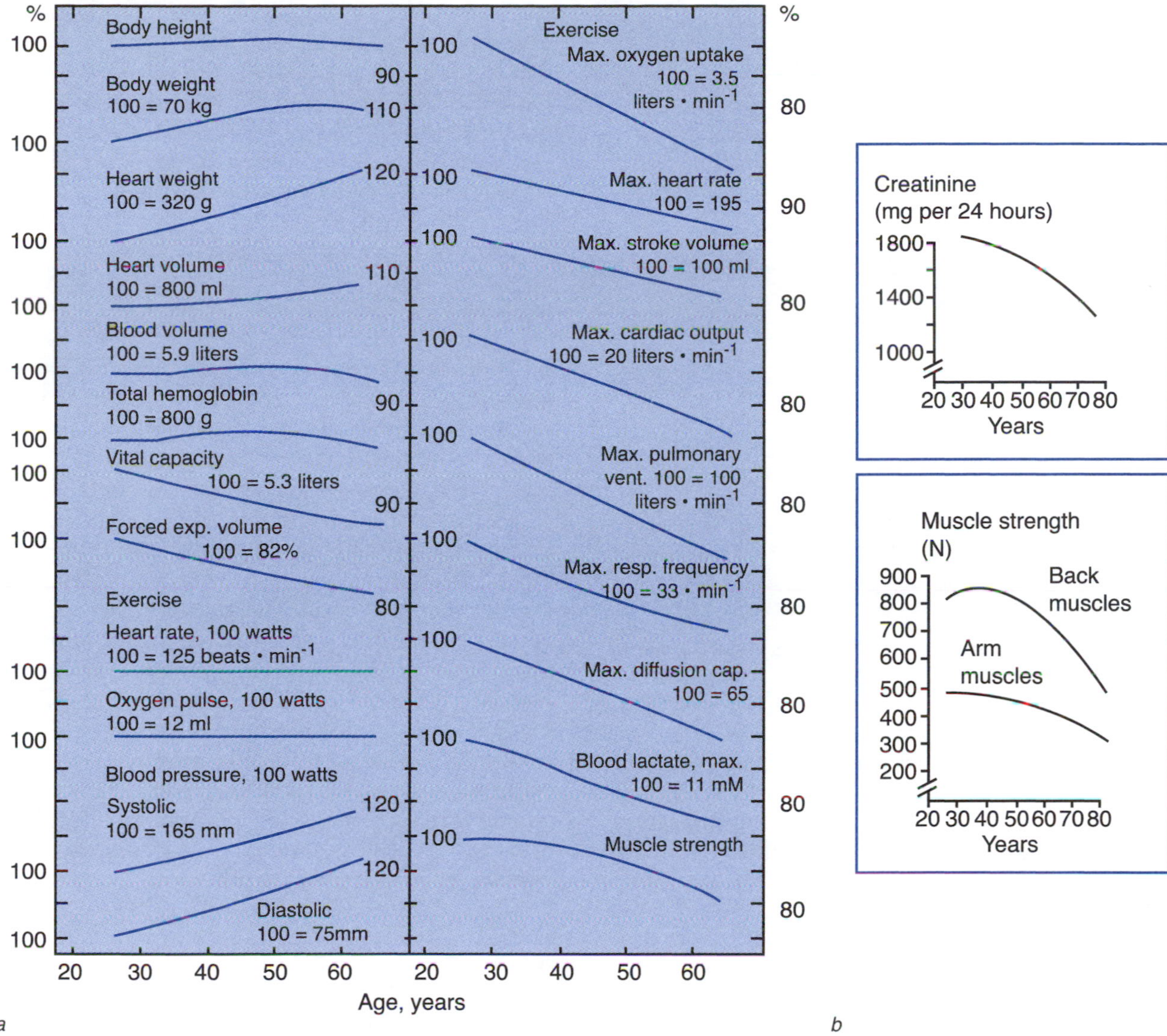


**Figure 10.8** *(a)* Variation in some biological measures with age. Data have been collected from various studies including healthy, male individuals. For data on the same function, only one study was consulted. The values for the 25-year-old subjects = 100%; for the older ages, the mean values are expressed as a percentage of that. The mean values cannot be considered as "normal" values, but their trends illustrate the effect of aging. Note that the heart rate and oxygen pulse at a given work rate (100 W or 600 kpm · min$^{-1}$, oxygen uptake about 1.5 L · min$^{-1}$) are identical throughout the age covered, but the maximal oxygen uptake, heart rate, and cardiac output decline with age. The data on cardiac output and stroke volume are based on relatively few observations and are therefore less certain. *(b)* Variation in 24-h urinary excretion of **creatinine** as an index of total muscle mass and muscle strength.

Data from E. Asmussen (unpublished data).

grown man of 70 kg on average has a maximal pulse of 195 beats · min$^{-1}$ and that a child of 35 kg has a maximal pulse of 210 to 220 beats · min$^{-1}$, the question is not why this latter value is higher than 195, but why it is less than 245. The latter value of 245 is what might be expected from a purely dimensional consideration.

In this chapter we have given a number of examples indicating that biological dimensions in animals of different sizes in many cases are mechanically meaningful and desirable, providing a reasonable efficiency of the movements. In general, the links involved in the chain of oxygen supply and energy output adjust remarkably, so that none of the

**Table 10.1** Dimensions in Physics and Physiology

| Quantity | Dimension: Physical | Physiological |
|---|---|---|
| Length | $L$ | $L$ |
| Mass | $M$ | $L^3$ |
| Time | $t$ | $L$ |
| Surface | $L^2$ | $L^2$ |
| Volume | $L^3$ | $L^3$ |
| Density | $L^{-3}M$ | $L^0$ |
| Velocity | $Lt^{-1}$ | $L^0$ |
| Frequence | $t^{-1}$ | $L^{-1}$ |
| Flow | $L^3t^{-1}$ | $L^0$ |
| Acceleration | $Lt^{-2}$ | $L^{-1}$ |
| Force | $LMt^{-2}$ | $L^2$ |
| Pressure | $L^{-1}Mt^{-2}$ | $L^0$ |
| Temperature* | $L^2t^{-2}$ | $L^0$ |
| Energy | $L^2Mt^{-2}$ | $L^3$ |
| Power | $L^2Mt^{-3}$ | $L^2$ |

*Physical dimension from J.C. Georgian, The Temperature Scale. *Nature*, 201:595, 1964.

Source: By courtesy of W. von Döbeln.

individual links is much stronger or weaker than necessary. On the other hand, examples show that organisms of different size are not in all respects similar and uniform. There are deviations from the general trend. Deviations sometimes are of a physical nature; in larger animals, for example, the weight of the skeleton is relatively high. It is also known that training can markedly improve physical performance. This improvement is sometimes accompanied by changes in organic dimensions, but this does not always occur. The performance of children is lower than expected from their dimensions, which clearly indicates that biological factors are involved. Women and older individuals have a relatively low maximal oxygen uptake compared with 25-year-old men.

It is very fruitful, however, to consider whether differences in performance of animals of different size, including children and adults, can be explained by purely dimensional factors. If such a consideration should fail to account for the difference, biological adaptations would appear probable.

As stated at the beginning of this chapter, it is usually not possible to tell who will be the best athlete without testing his or her ability. The top skier or runner in endurance events has a maximal aerobic power that is about twice that of ordinary nonathletes of similar age (6.0 L · min$^{-1}$ compared with 3.0 L · min$^{-1}$). Yet the top athlete's body height is not 250 cm, which it would have to be if body dimensions alone were to determine oxygen-transporting ability (maximal $\dot{V}O_2 \propto L^2$). In fact, these top athletes are of average height, and some of their dimensions are very similar to those of an average person, whereas others are different.

# Chapter 11

# Physical Training

In this chapter we consider the physiological basis for the development of a training program and the biological long-term effects of different levels of physical activity, as well as the present state of knowledge about extra- and intracellular signals that mediate the training effects.

When we discuss training and training principles, it is essential to keep in mind the specific purpose of the training. As a rule, the best training is achieved simply by carrying out the activity for which one is training. As is evident from chapter 8, physical training can influence a number of the factors that constitute physical performance capacity; that is, it can cause not merely changes in muscle strength and maximal oxygen uptake but also structural and functional changes in a number of organ systems as well as psychological changes. It is, however, quite obvious that the large maximal aerobic power that is characteristic of the top endurance athlete to a large extent depends on endowed biological advantages. Thus, a person with a maximal oxygen uptake of 45 ml · $kg^{-1}$ · $min^{-1}$ cannot, under any circumstance, no matter how well trained, attain a maximal oxygen uptake of 80 ml · $kg^{-1}$ · $min^{-1}$, which is required for Olympic medals in certain sport events (figure 11.1). It may not be quite fair, but it is nevertheless a fact that one's genetic makeup is important for athletic achievement (Bouchard et al. 1999; Myerson et al. 1999).

## Training Principles

Physical training entails exposing the organism to a training load or work stress of sufficient intensity, duration, and frequency to produce a noticeable or measurable training effect, that is, to improve the functions for which one is training. To achieve such a training effect, it is necessary to expose the organism to an overload (i.e., a stress) that is larger than the one regularly encountered during everyday life. It is a common conception in training environments that "to build up, one must first break down." Admittedly, exposure to the training stress is associated with some catabolic processes, such as breakdown of glycogen, followed by an overshoot or anabolic response that causes an increased deposition of the molecules that were mobilized or broken down during training. As to the effect on other cellular components, this is at best an imprecise statement.

Today, the molecular mechanisms involved in training responses have started to emerge, but the picture is still far from complete. As a basis for studying the training process, however, one can safely state that all cells and tissues of the body, regardless of the presence or absence of training, are subject to some kind of continuous exchange and remodeling. On the cellular level, molecules have a restricted lifetime and are constantly replaced by new molecules of the same kind or by another isoform of the same molecule if so demanded by the current activity level. On the tissue level, as exemplified by the bone remodeling process (see chapter 7), cells and extracellular matrix are removed and replaced all the time. Both increased activity (training) and decreased activity (detraining) affect the body by shifting the balance between breakdown and replacement. A major challenge in exercise physiology today is to reveal the signaling pathways regulating this balance.

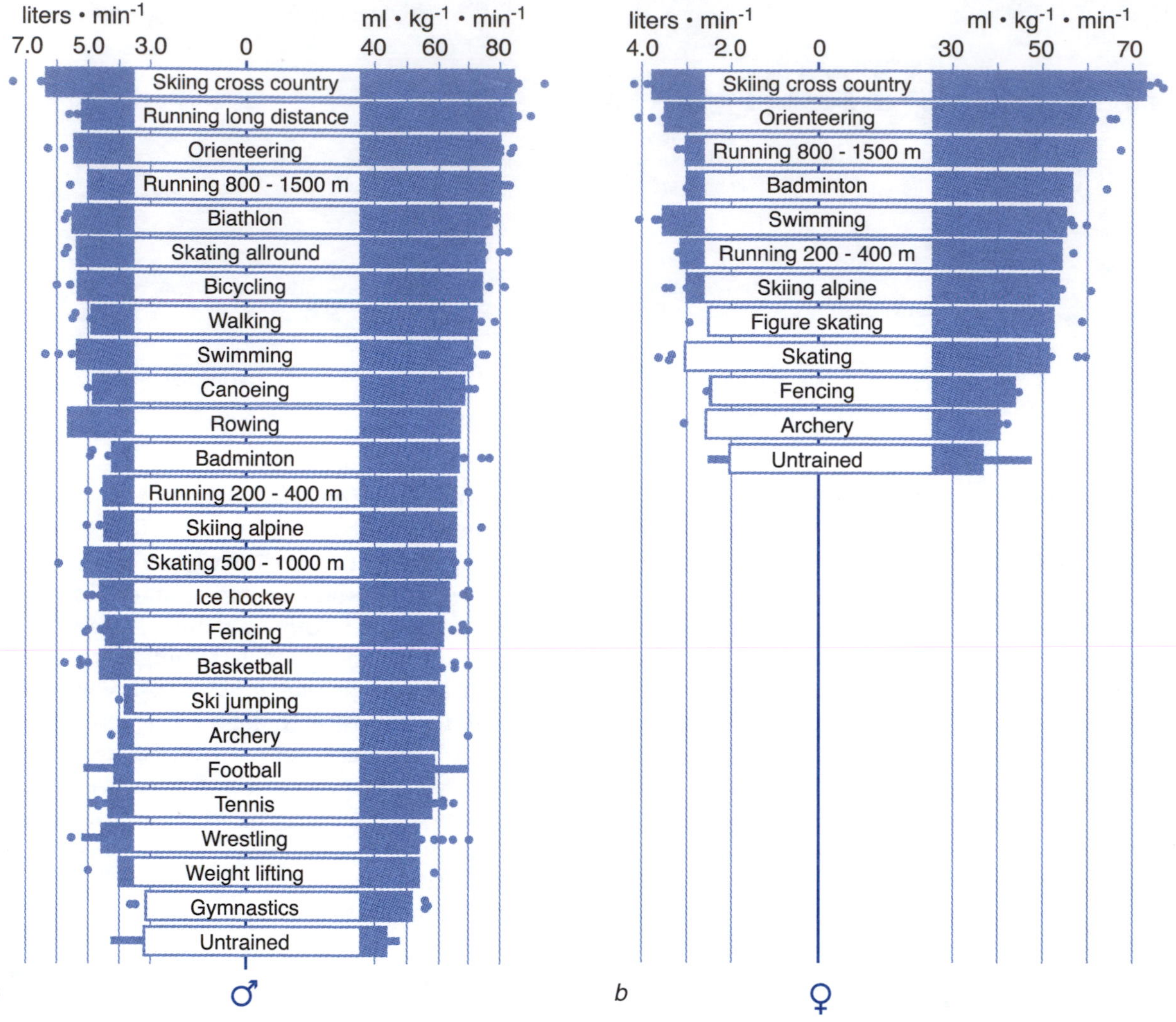


**Figure 11.1** Average maximal oxygen uptake in liters per minute (left part), and in milliliter per kilogram body weight (right part) for male *(a)* and female *(b)* Swedish national teams in different sports. Dots indicate individual values higher than the mean value. Note the difference in scale between *a* and *b*.

Compiled from various sources by Ulf Bergh.

## CLASSIC STUDY

Christensen showed that regular training with a given standard exercise rate gradually lowers the heart rate and that the intensity necessary to produce an effect has to be increased as performance improves in the course of training. The fitter the person, the more it will take to improve that fitness.

Christensen 1931b.

Training increases the maximal oxygen uptake as well as the percentage of it that can be taxed during a workout (figure 11.2). Consequently, the intensity of the load required to produce an effect increases as the performance is improved in the course of training. The training load is therefore relative to the level of fitness of the individual. The fitter a person is, the more it will take to improve that fitness.

Finally, it becomes a matter of time and motivation to continue when the elite athlete has to devote several hours a day to training. The need to gradually increase training load with improved performance, in the case of the effect of heart rate, was demonstrated as early as in 1931 by Christensen (1931b). He observed that regular training with a given standard exercise rate gradually lowered the heart rate. Further training did not modify this heart rate response. After a period of training on a heavier load, the original standard rate of exercise then could be performed with a still lower heart rate. This general principle is apparent during training of a number of functions: An adaptation to a given load takes place, and to achieve further improvement, the training intensity has to be increased. This principle has been elucidated by several studies summarized in other sections of this chapter. There is, however, no linear relationship between amount of training and the training effect. For instance, 2 h of training per week may increase maximal oxygen uptake, say, by 0.4 L · $min^{-1}$. If the training is twice as much, that is, 4 h per week, the increase in oxygen uptake will not be twice as large (.8 L · $min^{-1}$) but possibly .5 to .6 L · $min^{-1}$.

The major effects in a strength-training program are triggered by the first seconds of the first maximal contraction at each session. The gain is most pronounced when the subject is in a low state of fitness, and most of the gain is noticed during the first 5 weeks of a daily training program (Atha 1981), most probably attributable to improved motor control by the central nervous system (Moritani 1992; Sale 1992).

To improve the aerobic power in an average sedentary person, it appears that a training intensity in excess of approximately 50% of the individual's maximal aerobic power, which roughly corresponds to the rate of exercise causing the person to be slightly out of breath, is sufficient to produce a significant improvement (see Kilbom 1971; Pollock 1973). It commonly is recommended that the exercise rate during training should gradually be increased to 70% to 80% of the individual's maximal aerobic power, that is, to approximately 75% to 85% of her or his maximal heart rate (see figure 9.10).

Older individuals (above 50 years of age) may be less trainable than younger ones. But some effect of training is noticed even at very old age

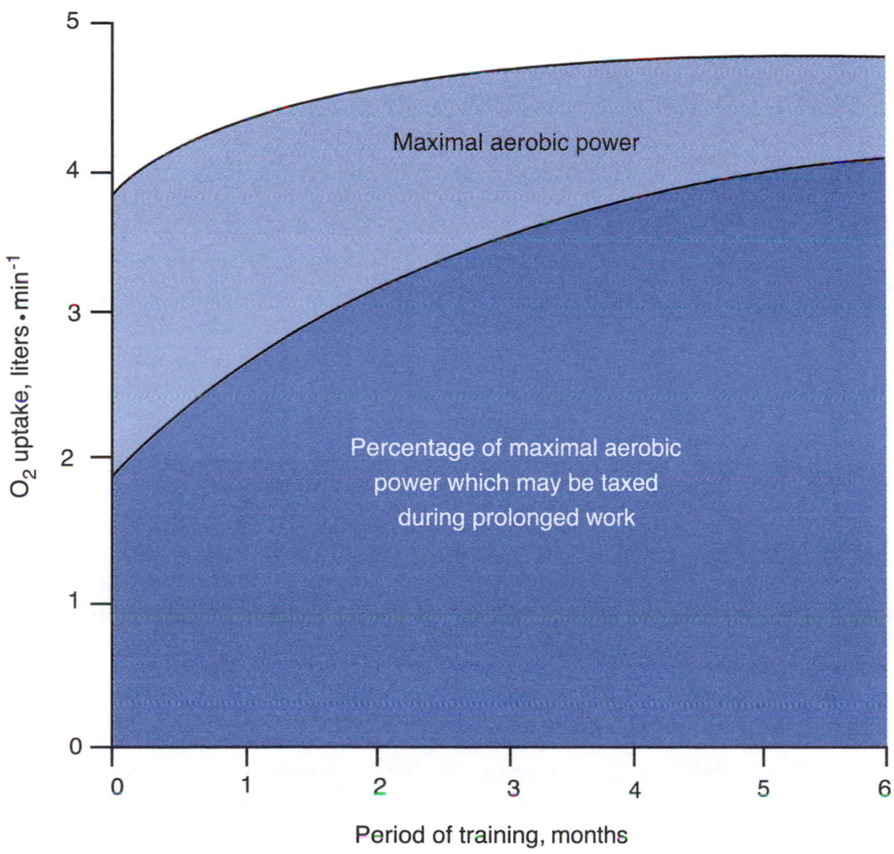


**Figure 11.2** Training increases maximal oxygen uptake. With training, a subject is able to tax a higher percentage of his or her maximal oxygen uptake during prolonged work.

(Ahmaidi et al. 1998; Grimby et al. 1992; Hurley and Hagberg 1998; Seals et al. 1984; Singh et al. 1999).

A certain amount of training of the oxygen-transporting organs is necessary for all categories of athletes, regardless of the nature of the athletic event. Thus, the individual will be better able to cope with the special training required for the event. Furthermore, even the warm-up before the event requires a certain amount of fitness. The general training for all individuals, irrespective of profession, age, and sex, should include the following:

1. Training of the oxygen-transporting function to improve maximal aerobic power and endurance
2. Improvement of muscle strength including the abdominal and respiratory muscles (Boutellier 1998)
3. Training aimed at maintaining joint mobility, enhancing the metabolism of the articular cartilage, and developing improved coordination (chapter 4)

In addition, low energy consumers should be stimulated to increase their metabolism through regular exercise, eventually becoming high energy consumers (chapter 12).

As far as the oxygen-transporting function is concerned, a distinction should be made between factors involved primarily in the heart and central circulation and factors involved in the peripheral circulation. In regard to the central circulation, the training is effective and less strenuous if as large a muscle mass as possible is engaged in the training. In the case of the peripheral circulation, it is a matter of training the muscles that will be engaged in performing the type of event or activity, where an improvement is desired. Many support the assumption that the central circulation limits the maximal oxygen uptake in activities engaging large muscle groups. On the other hand, peripheral factors may be limiting factors in the case of endurance, defined as time to exhaustion at submaximal levels of oxygen uptake (see chapters 5 and 8; Blomqvist and Saltin 1983).

It has often been discussed whether physical training is most effective when the exercise is accomplished continuously or intermittently, that is, with periods of more intensive muscular activity followed by periods of mild exercise or rest. In the following discussion, we summarize results from a few studies in which the physiological effects of these types of exercise are compared.

One subject was made to accomplish a certain amount of work (635 kJ) in the course of 1 h. This could be done either by uninterrupted exercise with a load of 175 W or by intermittent exercise with a heavier load, interrupted by rest periods at regular intervals. The double work rate (350 W) was chosen; thus, the required amount of work (635 kJ) could be accomplished by 30 min of exercise within the span of 1 h. Exercising continuously without any rest periods, the subject could tolerate this high work rate for only 9 min, at the end of which he was completely exhausted. If, instead, he exercised for 30 s, rested for 30 s, exercised for 30 s, and so on, he could complete the work with moderate exertion. The longer the activity periods, the more exhausting the exercise appeared, even though the rest periods were correspondingly increased. Some of the results of these studies are summarized in table 11.1. It appeared that with exercise periods of 3 min interrupted by 3-min rest periods, the load on the oxygen-transporting organs was maximal (oxygen uptake, 4.60 $L \cdot min^{-1}$, heart rate 188), and the degree of exertion was particularly high, as shown by a blood lactate concentration of 13.2 mM (I. Åstrand et al. 1960a).

At least one conclusion can be drawn from these classic experiments: For the purpose of taxing the oxygen-transporting organs maximally, exercise periods of a few minutes' duration represent an effective type of activity. An example of the effect of intermittent exercise of this type is presented in figure 11.3. Table 11.2 presents data on one of the subjects in a study by Christensen, Hedman, and Saltin (1960) of intermittent treadmill running using shorter exercise periods. It is striking that when the subject ran for 10 s followed by a 5- s pause, he could run for 20 min during the 30-min period at a high speed without undue fatigue and with a low blood lactic acid concentration. At the end of each running period, the load on the oxygen-transporting system was maximal, or 5.6 $L \cdot min^{-1}$. On average, the oxygen uptake when the subject was running was 5.1 L, whereas the oxygen requirement was 7.3 L per exercise minute. In other words, there is an apparent deficit of about 46 kJ ($[7.3 - 5.1]L \cdot 21\ kJ \cdot L^{-1}$). It is noticeable, however, that both the oxygen uptake (4.9 $L \cdot min^{-1}$) and pulmonary ventilation (142 $L \cdot min^{-1}$) during resting intervals are elevated almost to the average value during exercise, indicating coverage of an oxygen debt. Admittedly, the blood lactate concentration is not very high, but in both cases where the resting periods are shorter than the exercise periods, the blood lactate is clearly higher than in cases where rest and exercise periods are balanced. The results indicate clearly that the dura-

**Table 11.1** Data on One Subject Performing 635 KJ (64,800 kpm) on a Cycle Ergometer Within 1 Hour With Different Procedures

| Type of exercise | | Oxygen uptake L · hr$^{-1}$ | Oxygen uptake L · min$^{-1}$ | Pulmonary ventilation, L · min$^{-1}$ | Heart rate, beats · min$^{-1}$ | Blood lactic acid, mM |
|---|---|---|---|---|---|---|
| *Continuous* | | | | | | |
| 175 watts | | 148 | 2.44 | 49 | 134 | 1.3 |
| 350 watts* | | | 4.60 | 124 | 190 | 16.5 |
| *Intermittent* | | | | | | |
| 350 watts | | | | | | |
| Exercise | Rest | | | | | |
| 1/2 min | 1/2 min | 154 | 2.90† | 63† | 150 | 2.2 |
| 1 min | 1 min | 152 | 2.93† | 65† | 167 | 5.0 |
| 2 min | 2 min | 160 | 4.40 | 95 | 178 | 10.5 |
| 3 min | 3 min | 163 | 4.60 | 107 | 188 | 13.2 |

*Could be performed for only 9 min.

†Measured during 1/2 min.

Source: I. Astrand et al. 1960.

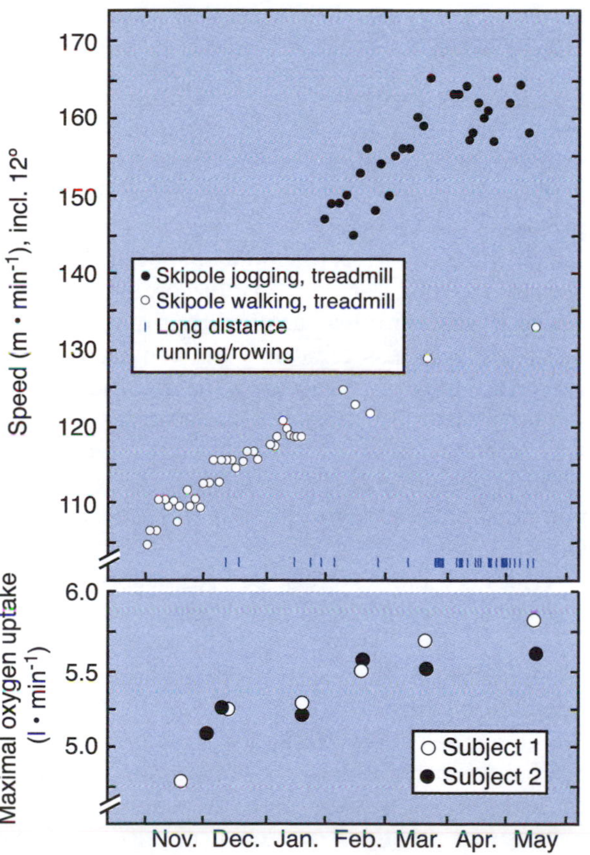


**Figure 11.3** The effect of ski-pole walking and jogging on a treadmill (inclination 12°) on the maximal oxygen uptake in two athletes. Two athletes trained on the treadmill (inclination, 12°) almost daily for more than half a year. Walking with ski-poles, they started at a speed of about 100 m · min$^{-1}$, but later the treadmill speed was increased to more than 140 m · min$^{-1}$ to attain maximal exertion. During the first period, they walked on the treadmill for 4 to 5 min three to five times in succession, with 5- to 9-min rest pauses between each exercise period. During the subsequent period, they alternated on different days between this very strenuous uphill treadmill walking and running in short bouts of about 20 s, interspersed with about 10-s rests for about 8 min. The whole 8-min sequence was repeated after 5- to 9-min rest periods and altogether was repeated two to five times each day. This training was replaced by long-distance running or rowing for 40 to 150 min each time for about 30 days, as indicated by ', showing the effect of this program on the maximal oxygen uptake of the two subjects.

Courtesy of O. Vaage.

tion and spacing of exercise and resting periods are rather critical with respect to the peak load on the oxygen-transport system. If the resting period of 5 s is prolonged to 10 s (still running for 10 s), the peak oxygen uptake observed will be reduced from 5.6 to 4.7 L · min$^{-1}$, probably because less oxygen debt is carried over to the next exercise period, as indicated by the lower lactate and lower resting oxygen uptake. The same kind of reasoning can be applied to running at the same speed for 15 s, then resting for 15 s, which did not bring the oxygen uptake (5.3 L · min$^{-1}$) to a maximum.

**Table 11.2** Data on One Subject During Intermittent Running for 30 Min at 20 Km · $H^{-1}$ on a Treadmill Between Varied Running Periods

| Periods exercise–rest, s | Distance, m | Oxygen uptake, L · $min^{-1}$ | | | Pulmonary ventilation, L · $min^{-1}$ | | | Blood lactate, mM |
|---|---|---|---|---|---|---|---|---|
| | | Exercise | | | Exercise | | | |
| | | Highest | Average | Rest | Highest | Average | Rest | |
| 5–5 | 5,000 | . . . | 4.3 | 4.5 | . . . | 101 | 101 | 2.5 |
| 5–10 | 3,330 | . . . | 3.4 | 3.0 | . . . | 81 | 77 | 1.8 |
| 10–5 | 6,670 | 5.6 | 5.1 | 4.9 | 157 | 142 | 140 | 4.8 |
| 10–10 | 5,000 | 4.7 | 4.4 | 3.8 | 109 | 104 | 95 | 2.2 |
| 15–10 | 6,000 | 5.3 | 5.0 | 4.5 | 140 | 130 | 144 | 5.6 |
| 15–15 | 5,000 | 5.3 | 4.6 | 3.8 | 110 | 90 | 95 | 2.3 |
| 15–30 | 3,330 | 3.9 | 3.6 | 2.8 | 96 | 79 | 64 | 1.8 |

The subject was standing beside the treadmill. During continuous running, he could proceed for 4.0 min, covering a distance of about 1,300 m. Oxygen uptake: 5.6 liters · $min^{-1}$; pulmonary ventilation: 158 liters · $min^{-1}$; blood lactic acid concentration: 16.5 mM.

Source: Data from Christensen et al. 1960.

Figure 11.4 presents another example of how critical the exercise intensity is for the load on the oxygen-transporting organs in intermittent exercise with activity periods of short duration. During running at a speed of 22.75 km · $h^{-1}$, for 20 s, alternating with 10 s rest, the oxygen uptake becomes maximal. If the speed is reduced to 22.0 km · $h^{-1}$, the oxygen uptake is reduced to about 90% of the maximum. On the other hand, the subject can continue for about 60 min at the lower speed as against only 25 min at the higher speed of 22.75 km · $h^{-1}$. It is an important but unsolved question which type of training is the more effective: to maintain a level representing 90% of the maximal oxygen uptake for 40 min, or to tax 100% of the oxygen uptake capacity for about 16 min. Billat et al. (2000) speculated that the time spent at $\dot{V}O_2$max is an important parameter when training to improve $\dot{V}O_2$max. They showed that it was possible, because of the $\dot{V}O_2$ slow component, to reach $\dot{V}O_2$max by running at 50% of $V_{\dot{V}O_2max}$ (the velocity needed to reach $\dot{V}O_2$max by conventional, incremental running velocity increases), but the time spent at $\dot{V}O_2$max was considerably shorter than during an intermittent protocol alternating between 100% and 50% of $\dot{V}O_2$max, 30 s each. The lactate response was insignificantly lower in the intermittent protocol than in the continuous one. No training effect was reported, and as long as the molecular signals leading to the training response are unknown, the importance of the time spent at $\dot{V}O_2$max remains theoretical.

## Summary

A series of studies has shown that maximal oxygen uptake (and cardiac output) can be attained in connection with repeated periods of exercise of very high intensity and of as short duration as 10 to 15 s, provided the rest periods between each burst of activity are very short (of equal or shorter duration than the exercise periods). In more prolonged exercise of several minutes duration, the duration of the rest periods is less critical. If the activity periods exceed about 10 min or so, a high level of motivation is required to attain maximal oxygen uptakes. In the case of continuous activity, the high intensity required to severely tax the oxygen-transporting system must alternate with periods of reduced intensity.

With very short exercise periods of about 30 s or less, a very severe load can be imposed on both muscles and oxygen-transporting organs without engaging anaerobic processes that lead to any significant elevation of the blood lactate. It is thus possible to select the proper load, exercise, and rest periods in such a manner that the main demand is centered on (1) muscle strength, without a major increase in the total oxygen uptake; (2) aerobic processes, without significantly mobilizing anaerobic processes; (3) anaerobic processes, without maximally taxing the oxygen-transporting organs; and (4) both aerobic and anaerobic processes simultaneously. The alternatives 2 and 4 do not entail maximal taxation of muscle strength; alternative 3

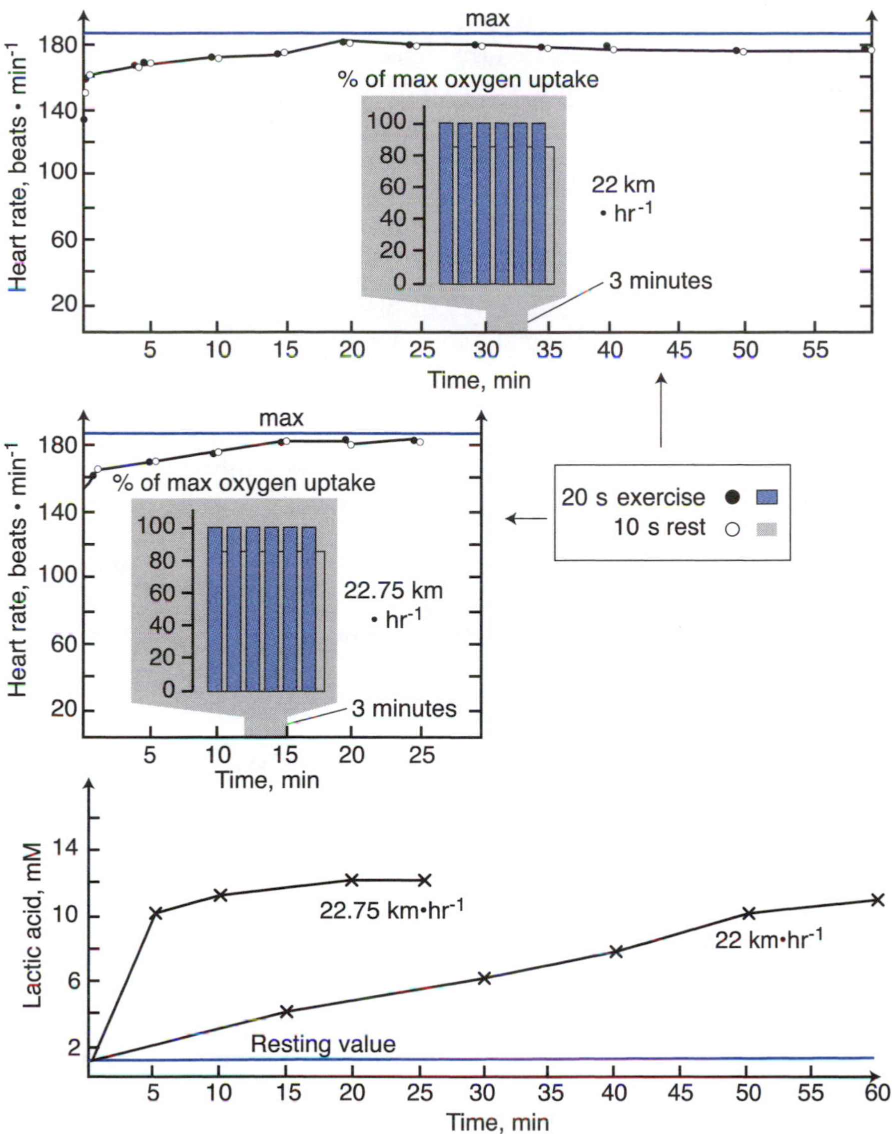


**Figure 11.4** Oxygen uptake, heart rate, and blood lactate concentration during a training program, running on the treadmill, with short exercise and rest periods (20 and 10 s, respectively) at speeds of 22 km · h$^{-1}$ (upper panel) and 22.75 km · h$^{-1}$ (lower panel). Note that the heart rate and oxygen uptake are not maximal at 22 km · h$^{-1}$, but that they are at 22.75 km · h$^{-1}$. Also, total exercise time is reduced to half in the latter case. The lactic acid concentration in the blood had reached about the same level in both cases when the exercise had to be discontinued.

From Karlsson, Hermansen, et al. 1967.

does not necessarily require maximal strength. In the following we consider how these principles can be applied.

## Strength Training

In this section, we discuss in some detail the various long-term training effects indicated in table 11.1. By long-term effects, we mean changes that will require a certain amount of time (weeks, months, or years) to be developed. Furthermore, available training methods, based on scientific data or experience, for achieving such effects, are discussed. At the end of the section, we review our present state of knowledge regarding molecular mechanisms and signaling pathways involved in the training response.

### The Effect of Training on Muscle Maximal Force

In chapters 3 and 4 we pointed out that there is a high correlation between the cross-sectional area of a muscle and its potential to produce force, first documented by Ikai and Fukunaga (1970). However, factors other than the cross-sectional area of a muscle are significant for its maximal strength. Muscle fibers may be attached to the tendons in a manner representing a mechanical disadvantage (e.g., those with pennate fiber arrangements)

(Kawakami, Abe, and Fukunaga 1993). Furthermore, the ability of the CNS to recruit all motor units in the muscle maximally is important. This leads us to ask how much of the training effect on muscle strength is attributable to neural adaptations and how much is attributable to adaptive changes in the skeletal muscle.

A number of reviews discuss various aspects of strength training (Atha 1981; Carpinelli and Otto 1998; Komi 1992; Kraemer, Fleck, and Evans 1996; Sale 1987), but many questions are still unresolved. A high tension must be produced for strength development. It is, therefore, of interest to recall the physiological events illustrated by figure 3.33. The highest tension that a muscle can develop is obtained in an eccentric contraction (lengthening the muscle). An intermediate position is held by the isometric contraction, and the maximal obtainable tension is progressively reduced the faster a concentric contraction is performed. As discussed in chapter 3, the terminology is eccentric, concentric, and isometric contractions, the former two also known as dynamic contractions. Under normal conditions, the muscles do not perform isotonic ("same tension") contractions, because the demand on the muscles involved when moving a given external resistance varies with the mechanical design. For example, when an athlete is lifting a weight kept in the hand by an elbow flexion, the muscle's lever changes during the movement and the demand on the flexor muscles varies. With reference to figure 3.33, it is evident that in tests of dynamic strength before and after a period of training, it is of vital importance to control the speed of shortening. The best way to do that is by using isokinetic equipment, but what is actually achieved with such equipment is a constant angular velocity of a limb segment and not necessarily a constant velocity of shortening of the muscles involved. This is particularly evident when it is a question of two-joint muscles and muscles that follow twisted paths. Furthermore, the centers of rotation of joint and apparatus cannot be kept coincidental. With the possible exception of swimming, isokinetic exercise does not simulate natural movements, and therefore the value of isokinetic training is difficult to evaluate. For the testing of maximal dynamic strength and endurance, however, the isokinetic testing apparatus is very useful.

The following text is limited to a general discussion of physiological effects of strength training.

Strength can be defined as the ability to develop force against a resistance in a single contraction of restricted duration (Atha 1981). When training with weights in dynamic contractions, one talks about **nRM** load, which is the number of repetitions maximum. The weight is so chosen that it can be lifted $n$ times in good style, but it is too heavy to lift $n + 1$ times. Generally, it can be said that if 1RM is 100%, 3RM is 90% to 95%, 6RM approximately 85%, 10 RM 75%, and 15 RM about 65% of the 1RM load. With interspersed intervals of rest, such a set may be repeated a number of times.

Although all fiber types can be recruited in a maximal effort in a slow contraction against high resistance as well as in a fast contraction, resistance being less, the power contributed by type I muscle fibers is negligible at higher shortening velocities (Faulkner, Claflin, and McCully 1986). With the force–velocity curve in mind, we also should point out that it is misleading to express the demand on maximal dynamic strength in running, swimming, and jumping as a percentage of maximal isometric strength. At a high velocity of shortening, a "submaximal" force can, in this respect, be maximum or at least close to maximum.

## The Effect of Training on Muscle Mass and Phenotype

It is an indisputable fact that muscle mass increases after working at intensities exceeding 60% to 70% of their force-generating capacity (MacDougall 1992). The mechanism behind this increase is, however, not equally clear. Theoretically, it may be attributable to each of the following factors or to a combination of them: an increase in fiber cross-sectional area or fiber length (hypertrophy), or an increase in fiber number (hyperplasia). Experimental evidence so far seems to indicate that all three factors are involved, but to a varying degree, depending on the type of training or experimental intervention and on the type of muscle and species studied.

#### FOR ADDITIONAL READING

For a review of the role of myostatin in muscle growth and repair, see Sharma et al. (2001).

### *Strategies Commonly Used to Resolve the Question of Hypertrophy Versus Hyperplasia*

Two main approaches have been used to count the number of muscle fibers in a muscle. Most common is counting the number of muscle fiber profiles in a cross section of the muscle. Less common and far more laborious is counting of muscle fibers after nitric acid digestion (Gollnick et al. 1981). Both methods have their pitfalls. Counting in a cross section is based on an assumption that all fibers run the entire length of the muscle, which is often not the case (Roy and Edgerton 1992; Trotter, Richmond, and Purslow 1995). In such cases, a growth in length of muscle fibers attributable to the experimental intervention can bring more muscle fibers into the plane of section and thus create an impression of an increased fiber number, despite the lack of any de novo fiber formation. Counting after nitric acid digestion, on the other hand, is apt to miss small fibers (Antonio and Gonyea 1993), which can be easier to see after the training if they have increased in size during the experiment. This also can create a false impression of hyperplasia. Despite this, direct counting after nitric acid digestion is the most widely accepted technique for fiber quantification (Antonio and Gonyea 1993). For obvious ethical reasons, however, this method cannot be used for studies of human muscles, except in postmortem situations, where too little or nothing is known about the training status of the deceased.

Since, for reasons given previously, most attempts to solve the question of hypertrophy versus hyperplasia in muscle have been performed in experimental animals, one also has to face the problems inherent in extrapolating the data to humans. These problems are even greater in training studies than in other studies because of the almost impossible task of constructing a training situation that mimics a training program used by humans (Baar et al. 1999). This is especially true for the stretch overload model often used to incur muscle hyperplasia in quail anterior latissimus dorsi muscle. Admittedly, this model results in a more than 50% increase in fiber number in addition to increases in fiber cross-sectional area as well as length after 30 days of constant stretch (Antonio and Gonyea 1993). The constant stretch used in this model is, however, very different from training stimuli used by humans, and in addition, the quail anterior latissimus dorsi is a tonic (nontwitch) muscle.

## FOR ADDITIONAL READING

For a discussion of animal models in the study of exercise induced muscle enlargement, see Timson (1990).

In animal experiments, stretch is also known to induce the formation of more sarcomeres in series, which also may be regarded as hypertrophy in the broader sense because it entails an increase in size of individual cells. It has been hypothesized that the force–length properties adapt to the functional requirements imposed on chronically trained muscles of athletes (Herzog 2000). According to Herzog's reasoning, muscle fibers in the rectus femoris muscle of runners would have more sarcomeres in series than corresponding muscle fibers in cyclists (figure 11.5), and actual measurements of moment–length relationships seem to support this assumption. The intermittent stretch during the eccentric phase of running, as opposed to the almost total lack of eccentric work during cycling, may be a relevant explanation for the difference (Herzog 2000; see also Rassier, MacIntosh, and Herzog 1999).

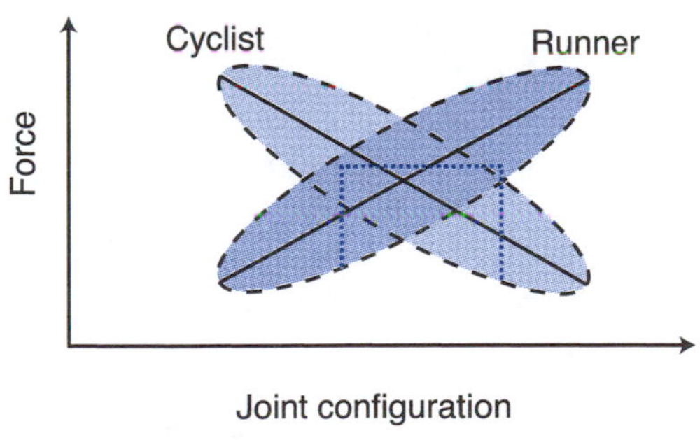


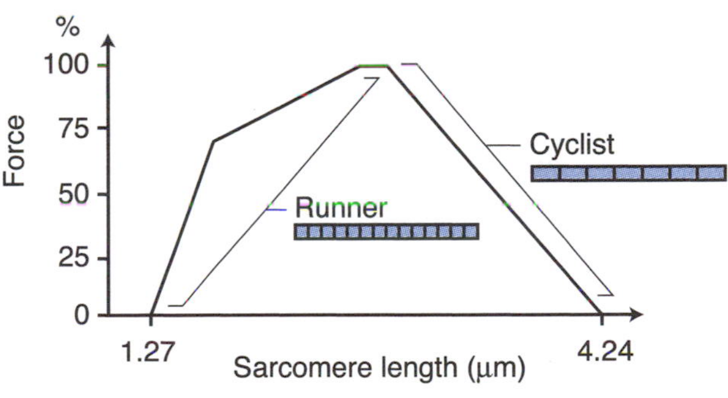


**Figure 11.5** Expected force–length relationship of the rectus femoris muscle of elite cyclists and runners (left panel) and the corresponding schematic explanation of how these differences in the force–length relationships might be explained (right panel). The idea is that adaptation of the force–length relationship occurs by an increase or decrease in the number of sarcomeres that are aligned in series within a muscle fiber. Runners would be expected to have a larger number of sarcomeres in series than cyclist. Consequently, for a given fiber length, individual sarcomere lengths would be smaller for runners than for cyclists. In this way, it could be perceived that the rectus femoris of runners works on the ascending part, whereas that of cyclists works on the descending part of the force–length curve within the anatomical range of motion.

For natural reasons, most morphological studies of human muscles have been based on biopsies, which, because of regional variations within the muscle, are far from representative of the muscle as a whole. Even in studies where an experienced investigator makes every effort to obtain a sample of muscle from the same area and same depth before and after the training intervention, this is a potential source of error. Nevertheless, investigators can use muscle biopsies to indirectly estimate muscle fiber number by using the following formula: Total muscle fiber number = muscle cross-sectional area (corrected for noncontractile tissue)/mean muscle fiber area (Alway et al. 1989). Like counting of muscle fibers in a cross section of the muscle, this method is based on the assumption that all fibers run the entire length of the muscle, which, as we know, is not always true.

## *Hypertrophy in Human Muscles*

Most human studies show that increased cross-sectional area is fully able to account for the increase in muscle mass observed after training (MacDougall 1992), but there are exceptions. In a study of a group of young, previously healthy, right-handed men who had suffered sudden accidental death, Sjöström et al. (1991) found evidence of activity-induced differences in muscle size and muscle fiber number between right and left tibialis anterior muscle, the fiber number on the left side being 10% higher than on the right side, allegedly attributable to asymmetrical use.

Without entering into a lengthy discussion of hypertrophy versus hyperplasia of skeletal muscle, it seems safe to conclude that, in humans, hypertrophy is the more important way of increasing muscle mass. As shown by the aforementioned study by Sjöström et al., hyperplasia cannot be ruled out, but it seems to be the result of certain types of perturbations, such as stretch (Antonio and Gonyea 1993), rather than a general feature of human muscle adaptation to increased demands.

Abe et al. (2000) reported regional differences in muscle adaptation following high-intensity dynamic resistance training, increases in muscle thickness being larger and occurring earlier in upper body muscles than in the lower body. They offered no explanation, however, but it is reasonable to suspect that natural differences in the pretraining status may have contributed to the result. The lack of any further hypertrophy after 24 weeks of heavy resistance training in highly competitive body builders (Alway et al. 1992) supports this assumption. Thus, the influence of the pretraining status on a strength training response could be analogous to its influence on the metabolic response (Tesch 1992a).

In its ability to adapt its size to physiological demands, the skeletal muscle is a very plastic tissue. High-resistance, slow-velocity training can thus induce a hypertrophy of both type I and type II fibers because both fiber types are engaged (Sale and MacDougall 1981). In cross-sectional studies, the ratio between the area of type II and type I fibers in weightlifters, powerlifters, and bodybuilders is about 1.6, whereas it is 1.1 to 1.2 in normal, active people and about 1.0 in sprinters (Saltin and Gollnick 1983). Endurance training, on the other hand, may selectively enlarge the type I fibers. There are reports, however, that muscle groups that have been subjected to intensive endurance training have normal or relatively small type I fiber areas (Jansson and Kaijser 1977). Typical strength training will stimulate neither capillary growth nor any development of mitochondrial enzyme systems. With an increase in the muscle fiber size, there is actually an increase in the fiber area that each capillary must supply with oxygen and substrates. Thus, Chilibeck, Syrotuik, and Bell (1999) showed that the muscle hypertrophy associated with strength training reduced the density of regionally distributed mitochondria, as indicated by the reduction in the activity of succinate dehydrogenase (SDH).

These factors may explain the relatively poor maximal aerobic power and endurance in muscles exclusively trained for strength. In contrast, endurance training will increase capillary density and mitochondrial volume, which certainly can enhance endurance. The energy demand of a muscle activated at maximal or near-maximal force is mainly covered by adenosine triphosphate (ATP) and phosphocreatine (PC). Neither strength training nor endurance training affects the ATP and PC concentrations significantly. If the emphasis is on high-repetition training, however, even strength training has been reported to induce capillary proliferation (Tesch 1992a).

Muscle hypertrophy is often evaluated on the basis of computed tomography or magnetic resonance imaging of the muscles in question. In this connection, a word of caution is appropriate because it has been claimed that the initial response for the muscle fiber is to increase its cross-sectional area at the expense of the extracellular space, without any increase in the total muscle girth (Goldspink 1992). The increase in radiological density of muscles reported after strength training (D.A. Jones and Rutherford 1987) supports this.

To a certain extent, strength training modifies the fiber composition in the skeletal muscle. The proportion of type I fibers is largely unaffected by strength training, but the proportion of type IIA fibers increases at the expense of IIX (formerly IIB, see chapter 3) fibers (Kadi and Thornell 1999; Kraemer, Fleck, and Evans 1996; Tesch 1992a). This is in accordance with increased training in general, and there is reason to regard type IIX fibers as a pool of nonrecruited fibers (Kraemer, Fleck, and Evans 1996), which, when recruited for high threshold type of activities, undergo transformation to type IIA fibers (figure 11.6). There is also reason to believe that there is an increase, at least temporarily, in hybrid fibers (see discussion in chapter 3). Most studies fail to report any changes in the percentage of type I fibers, but because a training-induced increase in myosin heavy chain (MyHC) I has been reported in the trapezius muscle of women (Kadi and Thornell 1999), this may depend on the type of muscle and its pretraining status.

## The Effect of Disuse

Disuse of a skeletal muscle (e.g., attributable to immobilization or joint fixation) will result in muscle atrophy and a reduced cross-sectional area of both type I and type II fibers. Actually, immobilization of a leg following injury could, after some months, reduce the cross-sectional area of both fiber types by some 40% to 45% (Sargeant et al. 1977). There is also reduced activity of the oxidative enzymes (Booth and Gollnick 1983). It is not surprising that prolonged exposure to a weightless environment in space also will result in a loss of muscle mass. It was mentioned earlier that stretching of a muscle could initiate a hypertrophy. It is logical, therefore, that immobilizing a joint in such a manner that the muscles are shortened below their resting length may enhance the atrophy more than expected by inactivity alone.

Experimental protocols aimed at revealing the effect of inactivity on muscle include bed rest or detraining (human), hind-limb suspension (rat), and space flight (human and rat). They all seem to point in the same direction. In rat soleus muscle, hind-limb suspension leads to a shift toward faster isoforms of myofibrillar proteins (Campione et al. 1993; Guezennec, Gilson, and Serrurier 1990), that is, more or less the opposite of the normal development seen with increasing body weight and activity in growing rats (Dahl and Aas 1981). The muscle

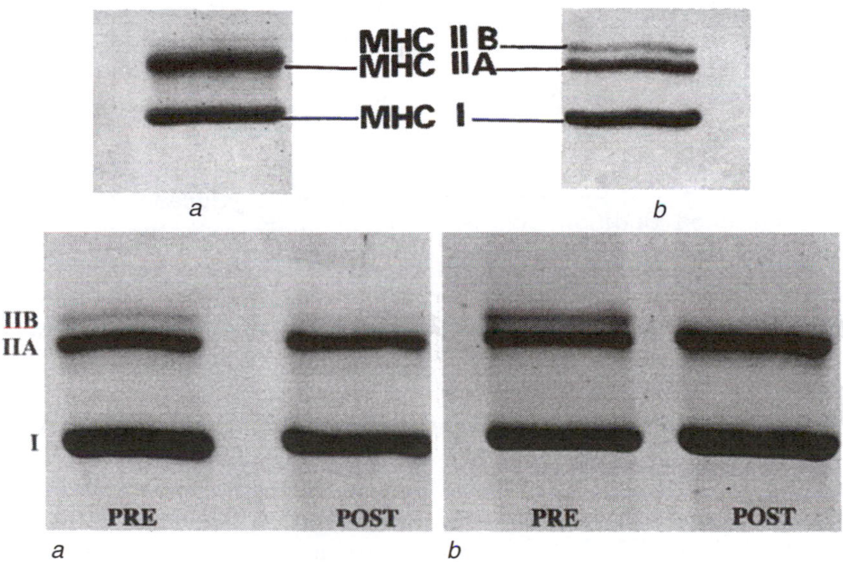


**Figure 11.6** Upper panel: Sodium dodecyl sulfate polyacrylamide gel electrophoresis electrophoretic separation of myosin heavy chains (MyHCs) in the trapezius muscle from an elite powerlifter (A) and a control subject (B). Note that MyHC IIB in this figure is called MyHC IIX in the present book, and that there is less of this isomyosin in the powerlifter than in the control person. Lower panel: MyHCs in female trapezius muscles before (pre) and after (post) training 1 h three times a week for 10 weeks. *(a)* before and after strength training; *(b)* before and after endurance training. Note that the MyHC IIB (IIX) band present before training has disappeared after training.

Upper panel reprinted, by permission, from *Histochemistry and Cell Biology*, Cellular adaptation of the trapezius muscle in strength-trained athletes, Kadi, et al., 111, 189, 1999, © Springer-Verlag.

Lower panel reprinted, by permission, from *Histochemistry and Cell Biology*, Training affects myosin heavy chain phenotype in the trapezius muscle of women, F. Kadi and L.E. Thornell, 112, 73, 1999, © Springer-Verlag.

atrophy resulting from microgravity has been shown to be attributable to changes in the pretranslational regulation of contractile protein gene expression (Thomason et al. 1992). Sedentary students subjected to prolonged bed rest showed decreased isometric muscle strength both in antigravity and nonantigravity muscles, and the decrease was larger than could be accounted for by the accompanying reduction in muscle mass, indicating reductions in neural factors as well (Y. Suzuki et al. 1994). Less easily explained are in vitro measurements of unloaded shortening velocity of segments of human soleus fibers after 17 days of space flight (Widrick et al. 1999). The flight-induced reduction in peak calcium ion-activated force, attributed to reduced cross-sectional area and specific force, was accompanied by an increase in unloaded shortening velocity, despite a lack of change in myosin isoform expression. The increased velocity, tentatively attributed to increased myofilament lattice spacing, was to a certain extent able to offset the postflight loss in fiber absolute peak power. For a discussion of the effect of space flight on the neuromuscular unit, see Roy, Baldwin, and Edgerton (1996).

Patients confined to bed rest or with a joint immobilized by a cast can counteract muscular atrophy by subjecting the muscles to submaximal isometric contractions of a few seconds in duration, repeated a couple of times per day. The same applies to astronauts during prolonged space flights.

A study by MacDougall et al. (1980) illustrates the magnitude of the effects of use and disuse on the skeletal muscle mass. Seven healthy male subjects were observed under controlled conditions, following 5 to 6 months of heavy resistance training, and then again after 5 to 6 weeks of immobilization in elbow casts. Cross-sectional fiber areas were calculated from sections of needle biopsies taken from triceps brachii. Training resulted roughly in a 100% increase in maximal elbow extension strength. Both fast-twitch and slow-twitch fiber areas increased significantly (by about 40% and 30%, respectively). Immobilization reduced strength by 40%; the reduction in fiber areas was slightly above 30% for fast-twitch and 25% for slow-twitch fibers.

### For Additional Reading

For a review of our current knowledge of the molecular events underlying disuse atrophy, see Kandarian and Stevenson (2002).

## The Role of Satellite Cells

Muscle fibers are postmitotic cells, unable to reenter the cell cycle. The only cells in the myogenic lineage still able to divide and serve as a myoblast reserve are the satellite cells (Bischoff 1994; Moss and Leblond 1971).

### For Additional Reading

For techniques to identify satellite cells in microscopic sections, see Lawson-Smith and McGeachie (1998), Kadi et al. (1999), and Putman, Düsterhöft, and Pette (1999).

As discussed in chapter 3, satellite cells serve two main purposes, muscle fiber repair and muscle growth, be it hypertrophy or hyperplasia. In hypertrophy, the point is that the nucleus/cytoplasm ratio is kept more or less constant (Kadi et al. 1999; Yan 2000), meaning that satellite cells fuse with the parent muscle fiber during hypertrophy, increasing the myonuclear density. On the other hand, the number of myonuclei has been reported to decline during inactivity and muscle fiber atrophy (D.L. Allen et al. 1995), but because muscle fiber atrophy without concomitant decrease in myonuclear number has been reported in hypothyroid rat muscle (Putman, Düsterhöft, and Pette 1999), this may depend on the reason for the atrophy.

Nuclei that are removed are not recoverable. Because we know that cells, at least in culture, can go through a limited number of mitotic cycles only, a theoretical concern may arise. Is there a limit to the number of training and detraining cycles an individual can go through without compromising his or her potential for muscle repair and growth? In rat muscle, hind-limb immobilization (plaster casts) resulted in a significant loss of gastrocnemius muscle mass and a reduced proliferative potential of resident satellite cells after just one bout of immobilization. Remarkably, the satellite cell proliferative potential did not recover during the first 9 weeks after immobilization but needed an additional 2 weeks with insulin-like growth factor (IGF)-1 supplementation to do so (Chakravarthy, Davis, and Booth 2000). So far, therefore, we have no indications that lack of satellite cell proliferative potential represents a threat to muscle cell growth and repair, at least in the presence of IGF-1 and in the absence of muscle disease. Apparently, satellite cells have an enormous capacity for growth and replication.

### For Additional Reading

For further discussion of the role of satellite cells, see Bischoff (1994). The possible role of IGF-1 and MGF in the prevention of muscle atrophy in old age is discussed by Hameed, Harridge, and Goldspink (2002).

There are, however, indications that we may have to revise our view of satellite cells and maybe stem cells in general. De Angelis et al. (1999) reported that clonable skeletal myogenic cells are present in the embryonic dorsal aorta of mouse embryos, raising the possibility that a subset of muscle satellite cells may derive from the vascular system. Furthermore, a population of putative stem cells isolated directly from skeletal muscle has been reported to be able to efficiently reconstitute the hematopoietic compartment as well as participate in muscle regeneration after intravenous injection in mice. These studies provide evidence that pluripotent stem cells may be present within adult skeletal muscle (Seale and Rudnicki 2000), but so far the practical implications of this remain subject to speculations.

## The Role of the Central Nervous System

When someone starts a strength-training program, there can be, during the first weeks, a 20% to 40% increase in strength without any noticeable increase in the cross-sectional area of the muscles involved. In the classic study by Ikai and Fukunaga (1970), they noticed that three maximal 10-s isometric contractions repeated daily for about 100 days increased isometric strength approximately 90% with only about a 25% increase in the cross-sectional area of the muscles involved. One explanation for this finding is a more efficient activation of the motor units, illustrated by an increase in electromyographic activity during a maximal effort. An increase in the quantity of recorded electromyographic activity, often called integrated **electromyography**, indicates that more motor units have been recruited, that motor units are firing at higher rates, or that some combination of the two has occurred (Sale 1992). Another factor that can contribute to the initial strength gain without a corresponding increase in total muscle size is an increase in muscle fiber girth at the expense of extracellular spaces (Goldspink 1992). Understandably, this may lead investigators to overestimate the role of neural adaptation.

In cases where both hypertrophy and neural factors contribute, there will be a discrepancy between the increase in maximal force and the increase in the integrated electromyographic activity after training, and the relative contribution of hypertrophy and neural factors can be assessed as shown in figure 11.7. A host of studies have demonstrated that, in the early phases of a heavy resistance training program, increased activation of muscle is the more important contributor to the observed strength increases. Even in competitive Olympic weightlifters, increases in strength and power have

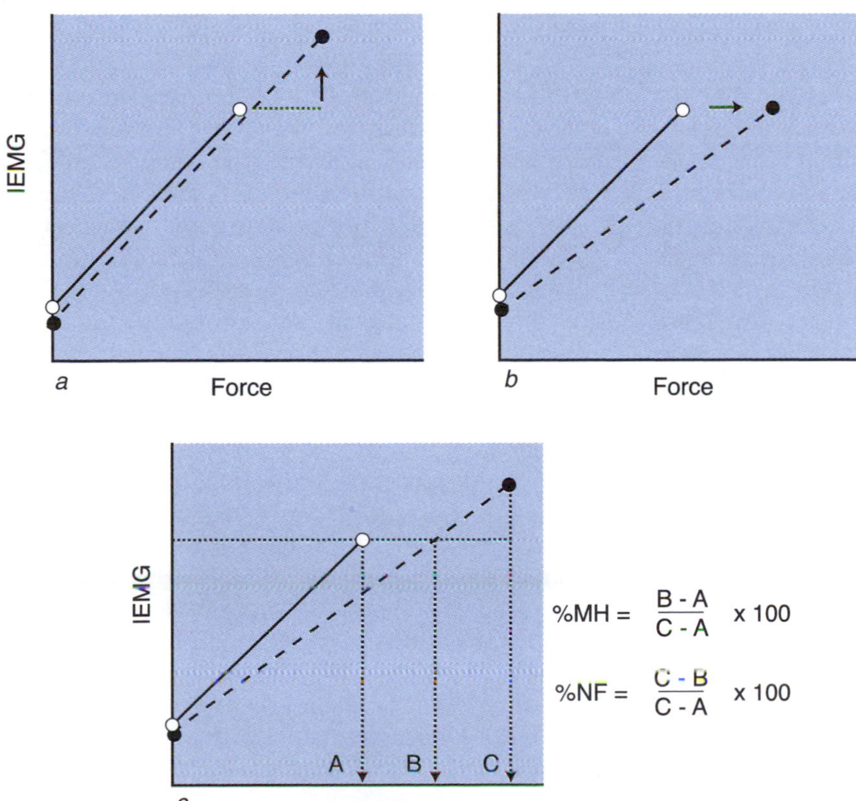


**Figure 11.7** Schema for evaluating the percentage contributions of neural factors (NF) and muscle hypertrophy (MH) to strength gain. If the strength gain is brought about by neural factors such as learning to disinhibit, then we would expect to see proportional increases in integrated electromyography (IEMG) and force, as seen in *(a)*. If, on the other hand, the strength gain were entirely attributable to muscle hypertrophy, we would expect the results shown in *(b)*. Here the force per fiber (or per unit activation) is increased by virtue of the hypertrophy without any increase in the IEMG. *(c)* Method to evaluate the contribution from the two mechanisms.

Reprinted, by permission, from T. Moritani, 1992, Time course of adaptations during strength and power training. In *Strength and power in sport*, edited by P.V. Komi (Oxford: Blackwell Science Ltd.), 268.

been described during 2 years of training, despite minimal changes in muscle fiber size (Kraemer, Fleck, and Evans 1996).

Increased electromyographic activity can also be seen in the contralateral limb in cases of unilateral training. This **cross education** effect can only be explained by neural factors (Moritani 1992b), probably at the spinal level (S. Zhou 2000). Even more convincing evidence for neural adaptation is the demonstration of increased strength after contractions that were imagined only, not actually performed, and the effect of bilateral limb training on the so-called bilateral deficit: Normally, the maximal force that a muscle can exert decreases if homologous muscles in the contralateral limb are activated at the same time. This phenomenon has been reported to be absent or even replaced by a bilateral facilitation in subjects participating in activities requiring simultaneous activation of homologous muscles, such as weightlifting or rowing.

## For Additional Reading

For further discussion of the role of neural adaptation, see Moritani (1992), Enoka (1997), and Bawa (2002).

As noted in chapters 3 and 4, the real working units in the neuromuscular system are the motor units. Consequently, improved performance of a muscle can be attributable to improved performance of individual motor units or improved collaboration between motor units. Improved performance of individual motor units in turn can be attributable to changes in the impulse pattern delivered by the motoneuron: An increased or more optimal impulse frequency may ensure maximal force development and increased rate of force increase, but the muscle fiber composition of the muscle in question may determine the ultimate potential for such adaptations (Häkkinen, Komi, and Alen 1985). Maximum firing frequency has been reported to increase after training and to decrease following periods of disuse (Sale 1992). It also has been shown to be higher during maximal contractions in well-trained older weightlifters than in age-matched controls (Leong et al. 1999). Improved collaboration of motor units may be caused by increased synchronization of synergistic motor units or decreased activity of antagonistic units (Moritani 1992). The most likely functional role of motor unit synchronization is to increase the rate of force development during rapid contractions (Semmler 2002). The role of the CNS in all aspects of muscle strength is quite obvious, because the muscles are the slaves of their motoneurons. They can do nothing unless they receive stimulation via their motor nerves.

Both the increase in strength observed without any noticeable muscle hypertrophy in the initial phase of strength training and the large variation seen in strength measurements from test to test in the same subjects can be explained by changes in the level of inhibition on the relevant motoneurons caused by inhibitory interneurons. These interneurons in turn can be inhibited (a so-called disinhibition, see figure 4.3) as a result of training. The less the inhibition on the motoneurons, the more muscle units can contract at tetanus frequency, eventually becoming synchronized. This can occur if the subject is startled or is in situations involving danger or competition, and it results in an instantaneous increase in muscular strength (Ikai and Steinhaus 1961).

The fact that the effect of strength training is related primarily to the practiced activity (Morrissey et al. 1995) is another example that supports the role of neural factors as a moderator of strength. Isometric training of the elbow flexors with the elbow joint at a 90° angle will increase the strength at that joint position, but the effect is less pronounced when the strength is measured in a 45° or 135° angle (Clarke 1973).

How can this particular specificity in training be explained? As mentioned, joint angles determine the muscle length, alter the mechanical advantages of the system through which the muscles act, and modify the relative contribution of each of the synergic muscles. In addition, reflex facilitation contributes as much as 30% of the excitation of the motoneurons during an isometric contraction. Because this reflex facilitation presumably is attributable to muscle spindle afferents (Gandevia, Herbert, and Leeper 1998), and spindle activity varies with muscle length, differences in reflex facilitation between joint angles are to be expected. All aspects considered, both in the case of muscle and in the case of CNS, the activation pattern is not identical when the joint is at a 90° angle and when it is at a 45° or 135° position.

Actually, there is a special nervous circuit for every single muscle activity and movement. Eccentric contractions, for example, require special activation strategies by the CNS, as shown by Nardone, Romano, and Schiepatti (1989). For discussion, see Enoka (1996). Therefore, it is not surprising that some of the effects of training and practice modify the behavior of the CNS. The technique that is used can be improved, and this learning process involves the CNS. Many strength tests and training activities include factors involving technique. A considerable confusion has emerged from the fact that some investigators have used the same activity for both testing and for training, whereas others have applied a test situation that did not correspond to the actual training.

How are the observed changes in motoneuron activity reflected in their morphology? For obvious reasons, this has only been studied in animal models. The size of motoneuron cell bodies has been studied in rats subjected to functional overload, endurance training (Roy, Ishihara et al. 1999), hindlimb suspension, or space flight (Jiang et al. 1992), but only minimal changes have been found. Apparently, spinal motoneurons, unlike muscle fibers, are highly stable over a wide range of levels of chronic neuromuscular activity (Roy, Ishihara et al. 1999). A 16-week period of endurance training, on the other hand, increased the content of **calcitonin gene-related peptide (CGRP)** in rat lumbar motoneuron cell bodies and axons (Gharakhanlou, Chadan, and Gardiner 1999). Even acute bouts of downhill running have been reported to increase numbers of CGRP-positive motoneurons (Homonko and Theriault 2000). Because CGRP is thought to be involved in morphological and functional adaptations taking place at the neuromuscular junction (Gharakhanlou, Chadan, and Gardiner 1999), this phenomenon may be causally related to the increased size of motor endplates seen after endurance training (Wærhaug, Dahl, and Kardel 1992).

Changes in putative molecular signals like CGRP as a result of changes in the level of physical activity are reported at a slowly increasing pace and contribute to an increased understanding of the adaptive processes in the CNS. Thus, Carro et al. (2000) reported that in rats subjected to 1 h of treadmill running, circulating IGF-1 is able to enter the cerebrospinal fluid and be taken up by specific groups of neurons throughout the CNS. Most importantly, blockade of IGF-1 uptake abolished not only the neuronal accumulation of IGF-1 but also the increased expression of the immediate early gene product **c-fos** induced by the unimpeded IGF-1 uptake. Increased levels of c-fos are known to be associated with neuronal activation (Grassi-Zucconi et al. 1993), and IGF-1 injection significantly increased firing frequency in groups of neurons shown to accumulate IGF-1 (Carro et al. 2000). In addition to their other functions, receptor tyrosine kinase ligands, such as IGF-1, have been found to modulate calcium channels (L.A. Blair and Marshall 1997). So far it is puzzling, however, how an intracellular accumulation of IGF-1 (Carro et al. 2000) corresponds with its action on receptor tyrosine kinase receptors on the extracellular side of the cell membrane (L.A. Blair and Marshall 1997) and, apparently, we have to await further studies to understand its functions.

## Mechanical Efficiency, Technique

In activities that are relatively uncomplicated technically, such as walking, running, or bicycling, there is a very slight increase in efficiency with training, but this increase is less than the variability among individuals. There is no definite difference in the consumption of energy per kilogram of body weight and per kilometer in differently trained runners as groups (Åstrand 1956; B. Sjödin and Svedenhag 1985). There is a wide variation in running economy even in elite marathon runners, with no significant correlation between running economy and performance time within a narrow racing time range. There are reports of world-class runners with maximal oxygen uptake below 70 ml · kg$^{-1}$ · min$^{-1}$, but they are also characterized by an extremely efficient running economy. It is impossible to say whether this is a training effect or whether it is attributable to inherent qualities (B. Sjödin and Svedenhag 1985), or both. In endurance athletes, 9 weeks of simultaneous explosive strength and endurance training improved running economy and 5K time without changing their $\dot{V}O_2$max (Paavolainen et al. 1999). In the general population, on the other hand, inherent differences in running economy are important when it comes to who is and who is not participating in recreational physical activity or training (Ingjer and Dahl 1979).

On a cycle ergometer, both Olympic medal bicyclists and untrained persons have almost the same mechanical efficiency at submaximal work rates. In Eskimos, totally unfamiliar with the use of a bicycle, however, Vokac and Rodahl (1977) found a distinctly higher oxygen uptake at a given work rate on the cycle ergometer than they found in white nonnatives. The more complicated the exercise, the larger the individual variations in mechanical efficiency and the larger the improvement with training. The mechanical efficiency of a person performing heavy exercise might be overestimated if the assessment is based on measurement of oxygen uptake, because anaerobic processes may have contributed to the energy yield.

In many achievements, the aim of the training is not primarily to reduce the energy expenditure during the event in question but to improve "at all costs," for example, by increasing the developed power. Thus, Lauru (1957) described an athlete performing a broad jump from a force platform.

Before training, the push against the platform during the jump reached 1,000 N. During the training, the push increased to 1,400 N, at least partially because of an improved coordination and a smoother sequence of motions.

In the end, technique and skill acquisition rest on the proper recruitment and activation of motor units. When a muscle is used repeatedly in the same contraction mode, its recruitment strategy is changed. The recruitment in the initial segment of the force generation cycle becomes slower and more prolonged, allowing a more precise and accurate control of the force increments (Bernardi et al. 1996).

Motor learning is so specific in nature that one cannot speak of a general learning ability with reference to motor coordination in skilled movements (see chapter 4). Everyone has had personal experience with learning techniques and has experienced how skill and dexterity are maintained for years without practice (e.g., the strokes in swimming, the tennis serve, balance on the bicycle, the grip on a guitar). But the level of endurance is lost without practice. However, extreme skill can be maintained by training up to a very old age. Andrés Segovia gave brilliant concerts at the age of 91, and Arthur Rubinstein played Chopin perfectly at the age of 88.

## Which Is the More Effective Strength Training Program?

From the preceding discussion, it should be evident that this question is unsolvable from a physiological point of view. A point of major importance is the purpose of the training. Careful selection of training program variables to simulate sport-specific movements is necessary to optimize training responses (Kraemer, Duncan, and Volek 1998). Another point is the person's potential to respond to the training, which in turn depends on genetic factors as well as on his or her previous training status, as discussed previously. Many studies have been designed to compare the effect of isometric to dynamic muscle training, with different resistances and a different number of repetitions per session, per day, or per week. In dynamic activities, variations in the speed of shortening in eccentric or concentric contractions have been compared. Last, mixed programs have been evaluated. In his extensive analysis of the literature, Atha (1981) concluded that a high tension, not necessarily fatiguing, is the stimulus that increases strength, but it is not clear which of the three contractions—eccentric, isometric, or concentric—is the most effective. In his 1973 review, Clarke came to a similar conclusion. Specific programs have produced a strength improvement with individual variations from 0% up to 100%. At present, this is an area in which we must base our comments on a mixture of scientific evidence and belief. The number of sets required to induce increases in strength and hypertrophy appears to be among the more controversial elements of any strength training program. Although the prevalent belief is that multiple sets—at least three—are better than single sets, there is little scientific evidence and no physiological basis for this (Carpinelli and Otto 1998).

### FOR ADDITIONAL READING

A summary of advantages and disadvantages of different strength and power training methods, especially in a clinical setting, is found in Grimby (1992). For strength training for specific purposes, see Garhammer and Takano (1992), Schmidtbleicher (1992), and Tesch (1992b).

### *Isometric Strength Training*

Isometric strength training can be conducted with simple equipment. It takes approximately 4 s to reach maximal tension, with some variation between different muscle groups. It is a common recommendation that the contraction should be maintained for about 6 s. Increasing the number of repetitions is advantageous for increasing strength, but the advantage is not proportional to the number of repetitions. A program of isometric contractions repeated five times, three times a week is an efficient program because it achieves significant gains at a minimal cost. As mentioned previously, isometric training improving the strength in one joint position does not always mean improved strength in other joint positions. Thus, to develop a more general strength, the training should include a variety of joint positions. The superiority of daily training can be illustrated by Hollmann and Hettinger's observation (1980) that the effect of every-second-day training was 80% of that of daily training. This type of training does not enhance aerobic power and endurance, but there are some transfer effects to dynamic strength tests. Isometric contractions are effective in preventing substantial loss of muscle mass and muscle function during periods of recovery from injury with joint immobilization.

### *Dynamic Strength Training*

When testing the effect of the number of repetitions in one training session on muscle strength, the majority of the experts are of the opinion that the 5RM to 6RM loads probably represent the optimum and are more effective than 2RM and 10RM. At any rate, because the 6RM load is about 85% of the 1RM load, it shows that the load does not have to be maximal. Individual athletes and coaches have their favorite programs. The trend has been that weightlifters and powerlifters include fewer repetitions, from 1RM to 5RM, and bodybuilders prefer 8 to 12 repetitions with 3 to 5 and approximately 10 sets, respectively, for each activity. Alternatively, upper limb, trunk, and lower limb muscles are exercised two to three times per week.

For dynamic strength training, special equipment and apparatuses are available, such as barbells, dumbbells, and pulleys. Most sophisticated are the isokinetic machines, which can secure a maximal demand on torque over the entire functional range of joint movements, the more advanced ones even both concentrically and eccentrically. So far, there is no scientific evidence showing that isokinetic training is superior to dynamic or isometric strength-training methods with regard to hypertrophy. On the contrary, there is an ambiguity in the literature regarding muscle hypertrophy after isokinetic training. Some studies report hypertrophy (e.g., Coyle et al. 1981; Higbie et al. 1996; Housh et al. 1992), whereas others fail to find any sign of hypertrophy (in the sense of increased fiber girth), although gains in strength are obtained (Akima et al. 1999; Côté et al. 1988). See the discussion of hypertrophy versus hyperplasia earlier in this chapter.

There is, however, a specificity with regard to type of training. After eccentric isokinetic training, average knee extension torque increased more in the eccentric mode than in the concentric, whereas it was the other way around after concentric training (Higbie et al. 1996). The resulting hypertrophy was also larger after eccentric training. A high muscular tension is important for the development of strength, and with a high speed of contraction the tension drops. From this viewpoint, one should expect that a program concentrating on eccentric muscle exercises should be superior to other programs. Actually, eccentric training is effective in increasing muscle strength, but it has been debated whether it is more effective than isometric or concentric training (Atha 1981). It can, however, result in a growth in length of muscle fibers because of the stretch imposed on the muscle fibers. (For a discussion of possible functional consequences of longer muscle fibers, see chapter 3.) Again, maximal loading does not necessarily elicit maximal strength gains. It is a disadvantage with eccentric exercises that they are apt to produce sore muscles in untrained persons. Second, with heavy loads exceeding 1RM, these exercises cannot be handled in an isometric or concentric muscle contraction, which may create a potentially risky situation.

A combination of different training methods seems to have some advantage over single exercise methods in developing strength and muscle hypertrophy. On the basis of their studies, Häkkinen and Komi (1981) suggested a combination of concentric and eccentric exercises for an optimal training of maximal force. To go one step further, it has been suggested that a complete strength training program should devote some 15% of the entire training period to eccentric training, 10% to isometric training, and 75% to concentric training.

The specificity of training complicates, as already mentioned, a comparison of different methods, particularly isometric versus dynamic training programs (see Duchateau and Hainaut 1984). It is understandable that the coach or athlete tries to develop strength-training exercises that in their pattern of movement come as close as possible to the strength demands of the event in question. In many ways, the best form of training for any given activity is to perform that activity. Track-and-field athletes and soccer, football, volleyball, and basketball players often train repetitive jumps with a bounce-loading of the leg muscles, including jumps down from a bench before takeoff to increase the load. Swimmers try to swim at high speed with their normal stroke against a pull of weights or other forces attached to their waists arranged in such a way that the drag is exerted horizontally on the swimmer, just to mention a few examples. As an example of the effect of simultaneous explosive strength and endurance training on physical performance, Paavolainen et al. (1999) showed that strength training improved the 5K running time in well-trained endurance athletes without changing their $\dot{V}O_2$max, by improving running economy and muscle power.

Most authors agree that there is a significant component of velocity specificity in strength training. Training with high-velocity movements increases high-velocity strength relatively more than low-velocity strength, and vice versa. In accordance with this, Häkkinen, Komi, and Alen (1985) and Häkkinen, Alen, and Komi (1985) showed that slow, heavily loaded strength training enhanced maximal strength, with minimal change in the ability to develop force rapidly, whereas explosive **plyometric training**, in the form of drop jumps, increased the rate of force development, with little effect on maximal strength. For further discussion, see Moritani (1992) and Sale (1992). The velocity specificity does not seem to apply to eccentric movements, however, because Aagaard et al. (1996) found eccentric strength gains over the whole range of eccentric velocities in the hamstrings, although the training consisted of high-resistance concentric knee extensions only.

## *Plyometric Training*

Plyometrics is a form of power training that consists of an eccentric contraction followed by a concentric contraction (i.e., a vigorous stretch-shortening cycle; see chapter 3). It has been claimed to be particularly useful for explosive, power-based sports such as basketball, volleyball, throwing, jumping, and sprinting. Theoretically, the advantage of plyometric training may reside in adaptations both in the CNS and in the muscle. The neural adaptation may be in the reaction to high stretch loads. In drop jumps from a height of 110 cm, an untrained person responds with a period of reduced electromyographic activity in the muscles that are stretched during the eccentric phase, whereas a trained jumper responds with facilitation and increased activation of the muscles (figure 11.8). In the muscle itself, the high tension is a stimulus to increased strength, and, in addition, the stretch imposed on the muscle fibers can increase the fiber length, which besides possibly increasing strength (see discussion in chapter 3) can result in higher maximal velocity of shortening on the whole-fiber level. In turn, this may allow a lower velocity of shortening on the sarcomere level (intrinsic velocity of shortening) in the longer muscle fibers compared with shorter ones. Assuming the same velocity of shortening on the whole-muscle level, this possibly can alleviate the negative effect of high intrinsic velocity of shortening on the power output of the muscle.

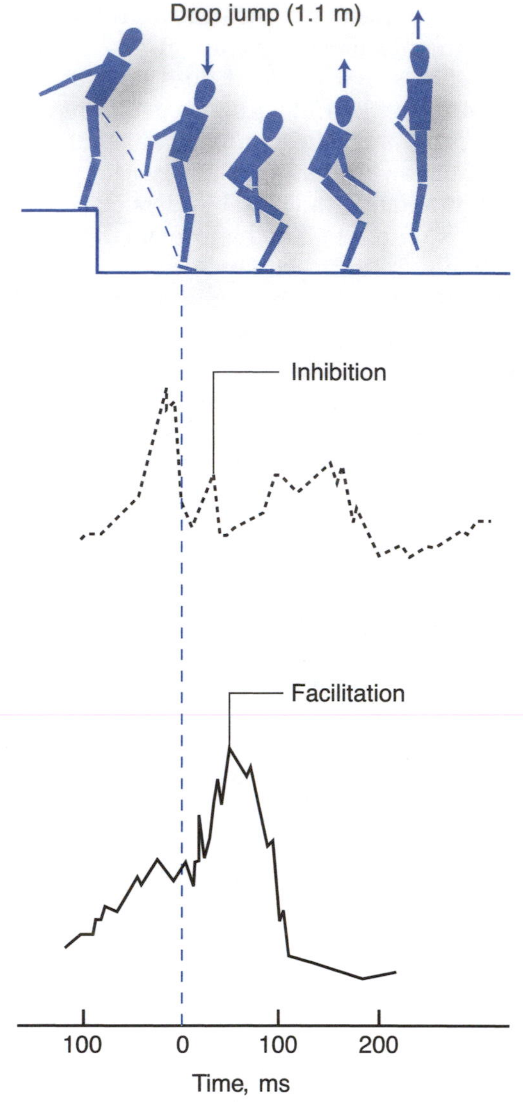


**Figure 11.8** Electromyographic recordings from the gastrocnemius muscle during drop jumps in an untrained subject (dashed line) and a trained jumper (solid line). During the eccentric phase of high stretch load, immediately following ground contact (indicated by vertical, dashed line, at time 0), the untrained subject responded with a period of inhibition. In contrast, the trained jumper responded with a period of facilitation, reflecting a neural adaptation.

Reprinted, by permission, from D.G. Sale, 1992, Neural adaptation to strength training. In *Strength and power in sport*, edited by P.V. Komi (Oxford: Blackwell Science Ltd.), 253.

## Summary

The major determinant of muscle strength is the cross-sectional area of the muscle. However, other factors modify the strength, such as neural factors, the internal muscle architecture, limb length, and joint structure. The muscle fiber composition is not critical in isometric contractions, but when contractions are performed at high velocities, a high percentage of fast-twitch fibers is advantageous. In the initial stages of a training program, strength increases more rapidly than the muscle mass. Gradually, there is a muscle hypertrophy, more so if the training load is at maximum or close to maximum. The increase in muscle mass is attributable to an increase in the cross-sectional area of the individual muscle fibers by the formation of more myofibrils. The number of muscle fibers does not seem to change, but the percentage of type IIA increases at the expense of type IIX.

It is not clear which type of strength training is the more effective, but there is specificity in the training effects. Therefore, it is an advantage if strength-training activities can be designed to fit the individual's specific needs. For obvious reasons, the load should be heavier than the demands of "normal" life (i.e., an overload), and the resistance should be progressively increased.

There is a low correlation between strength, speed, and endurance. The highest loads do not produce the largest gains; 5RM to 6RM has been suggested as optimum in dynamic exercises. It is recommended that eccentric, isometric, and concentric exercises are included, for instance in the following proportions: 15%, 10%, and 75%, respectively, of the time spent on each mode of exercise. Fast contractions can enhance performance in powerful and fast activities.

For the ordinary fitness seeker, a few isometric contractions at near-maximal effort, lasting at least some seconds to allow all muscle fibers in the active muscle to be recruited, make a simple and effective program if repeated about three times per week. The individual should remember, however, that isometric training has position-specific effects.

In addition, maximal or near-maximal efforts in isometric and slow dynamic muscle contractions are associated with a relatively high blood pressure and heart rate, particularly if a large muscle mass is involved. This extra load on the cardiovascular system can be hazardous for patients with cardiovascular diseases. For the recovering patient or recreational athlete, it is not of vital importance to include sophisticated programs and isometric exercises, because submaximal loads can be quite effective in developing strength.

## The Molecular Biology of the Strength-Training Response

Skeletal muscle is a dynamic tissue, capable of adapting to changing functional demands by altering its phenotypic profile, its mass, or both. Despite the fact that the underlying molecular mechanisms were a major topic of research during the last two decades of the 20th century, the overall picture is still far from clear. Part of the reason for this is the large number of different signals elicited by the training. Some of these signals are possibly redundant, whereas others may act synergistically and still others may counteract each other (M.T. Hamilton and Booth 2000). A major question has been and still is how the intracellular signals differ between various types of training adaptations. After a single bout of either high frequency electrical stimulation, low frequency electrical stimulation, or a running exercise protocol in the rat, Nader and Esser (2001) were able to reveal differences in the postexercise intracellular signals. High frequency electrical stimulation, which is a growth-inducing stimulus, resulted in a prolonged increase on $p70^{S6k}$ and a transient increase in protein kinase B phosphorylation, while only ERK phosphorylation was significantly elevated six hours after the running exercise. Undoubtedly, a lot of work is still ahead of us before we fully understand the molecular mechanisms behind the training response and how they relate to variations in the trainability of individuals. There is no reason to believe, however, that the riddle of the training response will be solved by exercise physiologists alone, the point being that the molecular mechanisms involved in cellular adaptation are not unique to muscle cells. On the contrary, it is becoming more and more obvious that such adaptive mechanisms and corresponding signaling pathways are found in a variety of cells, and that they are invoked in many situations, both physiological and pathological, many of which may be alien to an exercise physiologist. In other words, being a small part only of the combined fields of biological research, exercise physiology may have much to gain by looking over the fence to larger and more richly funded fields (e.g., cardiovascular and cancer research) as they uncover the basic cellular mechanisms. Once these mechanisms are uncovered, it may be relatively easy to ascertain whether the same mechanisms are involved in a given training response. The dark side of the emerging knowledge of these molecular mechanisms is their possible abuse in some kind of molecular doping or in genetic screening of children to identify possible future world-class athletes.

### For Additional Reading

For discussion of possible use of molecular biology for cheating, see Goldspink (1992) and J.L. Andersen, Schjerling, and Saltin (2000). For possible molecular and genetic strategies to study the adaptation process, see D.L. Allen, Harrison, and Leinwand (2002). Further details about the subcellular organization of signaling complexes can be found in Donelson Smith and Scott (2001).

### Changes in Protein Synthesis and Breakdown

Body protein is in a continuous state of turnover, with new proteins being synthesized and old ones degraded. In resting humans, body protein turns over at a rate of approximately 3 to 4 g of protein per kilogram of weight per day. Although muscle protein is about 50% of whole-body protein, the latter has a more rapid turnover, and the muscle protein turnover has been estimated to account for 25% of the whole-body turnover, or in the range of 1 g per kilogram of weight per day (Houston 1999).

Basically, there are two ways in which a muscle cell—or any cell, for that matter—can increase its protein content. It can either increase the rate at which proteins are synthesized or decrease the rate at which they are broken down. In a steady state, these two opposite processes are in balance, and the half-life of contractile proteins has been estimated to be in the range of 7 to 15 days, with the half-life of cytosolic proteins even shorter (Goldspink, 1992), although differences between species are expected.

Exercise has a profound effect on muscle protein turnover, but the effect seems to vary between the two main fiber types, although the end result is increased protein content in both instances. In fast fibers, the synthesis rate is increased, whereas in slow fibers the degradation rate is decreased (Goldspink 1992). It has been claimed that an acute bout of exercise is followed by a decreased rate of protein synthesis lasting 2 to 8 h, after which the rate is increased above normal (Viru 1994), but others have demonstrated increases in muscle protein synthesis extending 4 to 24 h postexercise (Chesley et al. 1992).

The control of protein synthesis can take place at different sites on the road from gene to protein. The first possible control point is transcriptional control, where deoxyribonucleic acid (DNA) is copied to messenger ribonucleic acid (mRNA); the second is during RNA processing, where alternative splicing may give rise to different isoforms of the same protein. Together, they are called **pretranslational control** and determine the type and amount of mRNA available for protein synthesis. Only pretranslational mechanisms are able to change protein isoforms. Pretranslational control means that the amount of a specific mRNA is rate limiting. Consequently, it also includes control of the rate of mRNA degradation. **Translational control** implies that the type of mRNA in question is present in excess and that some other factor limits the use of the mRNA in protein synthesis. **Posttranslational control** is the term used for all steps leading from the birth of the new protein molecule until it is fully functional in the right place within or outside the cell.

The relative importance of pretranslational, translational, and posttranslational controls varies, depending on the stage of the training and the type of muscle (Booth 1989). Initiation, elongation, and termination of translation are possible control points, but initiation has been shown to be rate limiting in overall protein synthesis. **Eukaryotic initiation factor-2 (eIF-2)** is thought to regulate general protein synthesis, whereas other proteins (4E-BP and $p70^{S6k}$) mediate growth-related protein synthesis. The kinase activity of $p70^{S6k}$ is controlled by a series of phosphorylation steps, and there is a good correlation between the activation of $p70^{S6k}$ and the long-term increase in muscle mass, making $p70^{S6k}$ phosphorylation a possible good molecular marker for muscle hypertrophy (Baar and Esser 1999; Nader and Esser 2001).

Chesley et al. (1992) found that a single bout of heavy resistance exercise increased muscle protein synthesis in the biceps for up to 24 h postexercise and that the increase was attributable to posttranscriptional events. This is supported by data reported by Welle, Bhatt, and Thornton (1999), suggesting that the stimulation of myofibrillar synthesis by 1 week of resistance exercise is mediated by more efficient translation of mRNA. In a more chronic training situation with changes in muscle phenotype, the pretranslational mechanisms become more important (Booth 1989).

In the recovery period after resistance exercise, both muscle protein synthesis and breakdown are accelerated (Houston 1999), but the control of protein breakdown is less well known (Eble et al. 1999). In general, proteins are degraded by protein-degrading enzymes called **proteases,** which are found partly in the cytosol and partly in lysosomes. Muscle protein degradation seems to be mediated by different cytosolic proteases, among which is the multisubunit protease called **proteasome** (Bochtler et al. 1999; Eble et al. 1999). Proteins destined to be degraded by proteasomes are labeled by the attachment of a highly conserved, small protein called **ubiquitin.** For the protein molecule to be degraded, it is like a "kiss of death." Once ubiquinated, the protein is recognized and degraded by the proteasome. The attachment of ubiquitin has been considered to be the rate-limiting step, but proteasome function also is influenced by regulatory proteins, and both the level of proteasome protein and proteasome activity have been shown to be up-regulated by continuous electrical stimulation of rabbit skeletal muscle. It is believed that this up-regulation is part of the process leading to a change of phenotype in stimulated muscle (Ordway et al., 2000). The signals leading to such up-regulation are less clear. The **ubiquitin–proteasome system** has been shown to be up-regulated in acutely diabetic rats, seemingly attributable to the joint effects of glucocorticoids and low insulin levels (Mitch et al. 1999). In addition, different cytokines may be involved. Thus, rats treated with recombinant interleukin (IL)-1α showed increased total and myofibrillar breakdown rates in the extensor digitorum longus muscle (EDL), and this effect was not mediated by glucocorticoids (Zamir et al. 1991).

### For Additional Reading

For more detailed information about proteasomes, see Bochtler et al. (1999).

Although the ubiquitin–proteasome system is the major pathway for muscle protein breakdown (Mitch et al., 1999), **calpain,** a calcium-activated neutral protease, also seems to participate. Its activity is increased and its subcellular distribution is changed following exercise (Arthur, Booker, and Belcastro 1999; Belcastro 1993).

## The Muscle Cell Is Responsive to Mechanical Stimuli

It has been known for some time that both local and systemic factors control tissue growth. A major systemic growth signal is IGF-1, whereas insulin itself both stimulates protein synthesis (Alberts et al. 1994) and inhibits its degradation (Mitch et al. 1999). For an overview and discussion of the effects of IGF-1 on muscle adaptation, see G.R. Adams (1998). In addition, many cell types, including muscle cells, bone cells, and fibroblasts, respond to local mechanical factors, and many authors refer to them as mechanocytes (Goldspink 1998). Although several cell types react to the same stimuli, the downstream gene expression is likely to vary in a cell type–specific manner (Baar et al., 1999). Both in vivo and in vitro experiments have shown that mechanical forces can alter the regulation of skeletal muscle genes. Specific regulatory regions of these genes have been identified, which, at least partially, regulate the gene's response to mechanical forces, but the linkage of load-activated signaling cascades with specific transcription factors and/or translation-related protein targets has not yet been firmly established in skeletal muscle (Carson and Wei 2000).

Although details are still lacking, the cytoskeletal lattice, including **integrin** and dystrophin (see figure 3.6), is now acknowledged as a possible key element in load-activated growth signals. Geoffrey Goldspink's group in England has cloned a **splice variant** of IGF-1, which they have called **mechano growth factor (MGF),** and which is expressed in muscle only when it is subjected to activity (Yang et al. 1996). In contrast to normal muscle, MGF is not detectable in dystrophic mdx muscles after stretch with or without electrical stimulation, indicating that the dystrophin cytoskeletal complex is critically involved in the mechanotransduction mechanism (Goldspink 1999).

The integrins are membrane-associated proteins involved in the interaction between cells and their immediate surroundings, be it extracellular matrix or other cells. By serving as a link between the extracellular matrix and the cytoskeleton of the cell, they are obvious candidate components of the signaling process by which the cell reacts to mechanical stimuli. They are now thought to be a primary sensor for relaying physical or mechanical signals from the surrounding environment into the interior of the cell (Carson and Wei 2000).

### For Additional Reading

For a general survey of integrins and their emerging role in signal transduction, see M.A. Schwartz, Schaller, and Ginsberg (1995), Burridge and Chrzanowska-Wodnicka (1996), and Carson and Wei (2000).

Stretch-activated ion channels are obvious candidates to be discussed when mechanical forces are transduced into biochemical signals (Baar et al. 1999), but the absence of growth signals in mdx mice, as discussed previously, renders it unlikely that ion channels are able to transduce mechanical forces into growth signals on their own. A cooperative role seems far more likely.

## Possible Signaling Pathways

So far, our knowledge of intracellular signaling pathways in muscle hypertrophy is rather patchy, but some main features are emerging. The recruitment of satellite cells is a crucial element in muscle hypertrophy, and the first question is how these dormant cells are brought back into the cell cycle. Both mechanical factors and growth factors are thought to contribute. Focal adhesion kinase (FAK) activity (discussed later) can stimulate satellite cell proliferation and differentiation, and an increased amount of FAK protein appears to correlate with satellite cell activation (Carson and Wei 2000). It also generally is accepted that growth factors, IGF-1 and others, are important determinants of satellite cell fate (Yan 2000), but this process must necessarily be finely tuned, because after exit from mitosis, some satellite cells apparently return to the quiescent G0 phase, whereas others fuse with the parent muscle fiber and contribute to the hypertrophy.

## FOR ADDITIONAL READING

For further details about the interplay between cell cycle regulation and hypertrophy, see Carson and Wei (2000) and Yan (2000). Possible crosstalk between cAMP and MAP kinase signaling is discussed by Stork and Schmitt (2002).

The number of focal adhesion complexes in muscle cells also known as costameres (Patel and Lieber 1997; see also chapter 3) has been shown to be increased in chronically stretched muscle. Their formation and subsequent tyrosine phosphorylation of signal transduction proteins are among the earliest and most critical components in the transduction of mechanical signaling. A vast array of signaling molecules and cascades have been connected to integrin signaling (figure 11.9), including FAK, **protein kinase C, mitogen activated protein kinase (MAPK)**, phosphatidylinositol-3 (PI-3) kinase, **Ras,** and **Rho** (Carson and Wei 2000).

FAK is recruited to focal adhesion complexes and is thought to have an essential role in their formation. It is a possible site for the integration of growth factor and integrin signaling, because the effect of IGF-1 has been shown to be mediated by specific binding to membrane-associated receptors of the tyrosine kinase family (Semsarian, Sutrave, et al. 1999). The downstream effects of this binding are still elusive, mainly because of the complexity of the signaling pathways involved. The orderly, linear pathways that only a few years ago appeared to link the cell surface to the nucleus have evolved into an array of intersecting and overlapping cascades (Finkel 1999).

A possible way to handle this complexity is, for the time being, to start with the "reductionist approach" used by D.M. Cox, Quinn, and McDermott (2000). As far as changes in gene expression are concerned, the end point of the signaling cascades is transcription and/or activation of transcription factors acting on muscle specific genes. Two families of genes encoding such muscle-specific transcription factors are the MyoD family and the myocyte enhancer-binding factor 2 (MEF2) family. The MyoD family, also known as myogenic determination factors, belongs to the helix-loop-helix type of transcription factors and consists of four members: MyoD, Myf-5, myogenin, and MRF-4. They all contain a highly conserved basic helix-loop-helix domain, which mediates transcriptional activity by dimerizing with a so-called E-protein and binding to a six-base consensus enhancer sequence called the **E-box**. The E-box is found upstream (see chapter 2) of the promoter region of most muscle-specific genes. In particular, it is found upstream of each of the MyoD genes, allowing for positive feedback (D.M. Cox, Quinn, and McDermott 2000).

The MEF2 family consists of nuclear phosphoproteins belonging to the MADS superfamily of DNA binding proteins (D.M. Cox, Quinn, and McDermott 2000). Their binding site is found in the control region of many muscle-specific genes, like myosin heavy chain, actin, troponin C, and others. They are able to act in synergy with members of the MyoD family in the activating muscle-specific genes (figure 11.10), thus emphasizing the role of combinatorial control of gene expression, although com-

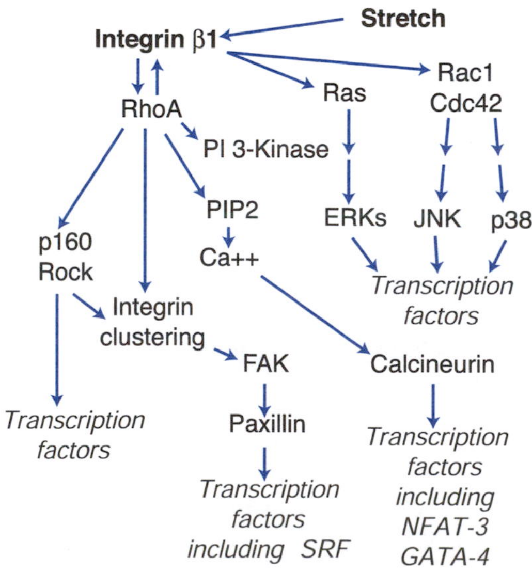


**Figure 11.9** Signaling pathways that can be activated in skeletal muscle by integrin-mediated signaling. An overload stimulus potentially can activate many different signaling cascades in skeletal muscle, but how these different pathways are integrated into a growth response is not yet completely understood. Integrin/RhoA signaling in skeletal muscle is an excellent candidate for integrating at least some of these events for the induction of hypertrophy. FAK = focal adhesion kinase; PI-3 kinase = phosphatidylinositol-3 kinase; JNK = c-jun NH2-terminal kinase; PIP2 = L-α-phosphatidylinositol 4,5-diphosphate.

Reprinted, by permission, from J.A. Carson and L. Wei, 2000, "Integrin signaling's potential for mediating gene expression in hypertrophying skeletal muscle," *Journal of Applied Physiology* 88:337-343.

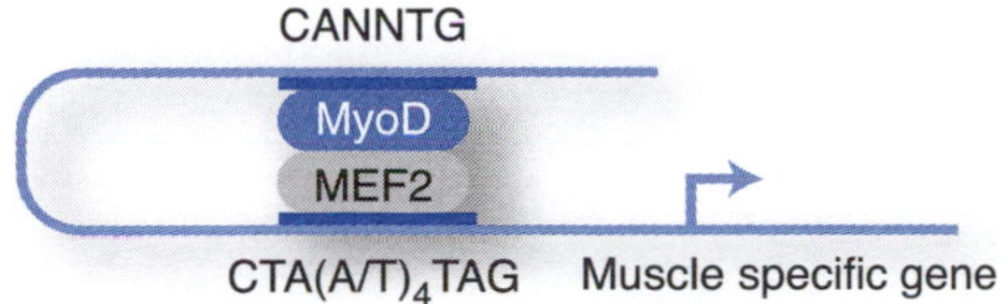


**Figure 11.10** Direct interactions between myocyte enhancer-binding factor 2 and MyoD transcription factors can synergistically transactivate muscle-specific genes by binding to their respective DNA binding sequences, as indicated in the figure, and to each other. This provides a mutually reinforcing network of increased activation for muscle-specific genes during differentiation.

Reprinted, by permission, from D.M. Cox, Z.A. Quinn, and J.C. McDermott, 2000, "Cell signaling and regulation of muscle-specific gene expression by myocyte enhancer-binding factor 2," *Exercise and Sport Sciences Reviews* 28:33-38.

binatorial control apparently involves a larger number of interacting transcription factors.

To a large extent, the signaling pathways from cell surface to the genes are MAPK pathways. There are three different classes of MAPKs: the **extracellular signal-regulated kinases (ERKs),** the **c-jun NH2-terminal kinases/stress-activated protein kinases (JNK/SAPKs),** and the **p38 MAPKs** (D.M. Cox, Quinn, and McDermott 2000). As indicated by their names, they originally were believed to be responsive to different kinds of stimuli (figure 11.11), but more recent data have shown that they are less selective, adding to the complexity of the system. The MAPKs act via the regulation of several transcription factors initiating the expression of a variety of immediate and delayed early genes (Widegren et al. 2000). The most prominent and extensively studied immediate early genes (IEGs) are the **fos** and **jun** gene families. Hetero- and homodimers of fos and jun proteins form the transcription complex **AP-1**, and different combinations of fos and jun proteins seem to determine the transcriptional specificity and activity of the AP-1 complex (Puntschart et al. 1998). As little as 30 min of running in untrained subjects resulted in a transient up-regulation of both c-fos (10- to 20-fold) and c-jun (threefold) mRNA. Surprisingly, the induction did not follow a fiber type–specific pattern as would be expected from known recruitment schemes but showed a rather patchy pattern (Puntschart et al. 1998). Figure 11.12 presents an overview of the different phenomena linking loading of a muscle cell to increased protein synthesis.

## For Additional Reading

For further details about the emerging links between cell signaling and gene expression, see Carson and Wei (2000) and D.M. Cox, Quinn, and McDermott (2000).

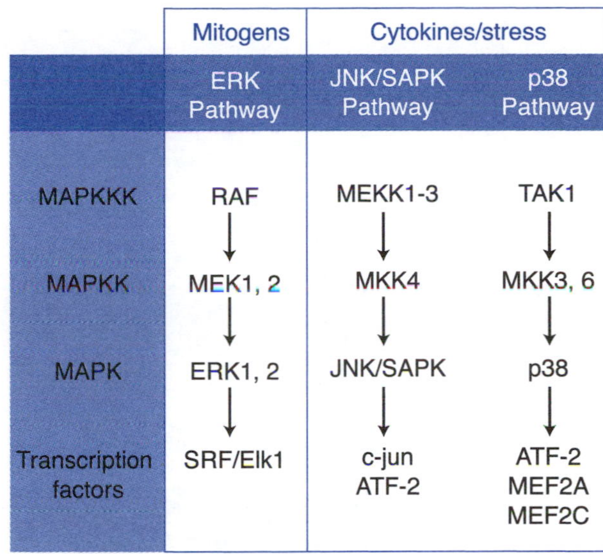


**Figure 11.11** The activation of receptor tyrosine kinases by external stimuli such as mitogens, cytokines, and stress activates a kinase cascade, the generic components of which are listed to the left side of the figure (MAPKKK, etc.). Ultimately, this cascade converges on the MAPKs, specific types of which are listed under the three pathways, These MAPKs then phosphorylate and modify the activity of specific target proteins, such as transcription factors. The potential for cross talk between pathways allows for extremely complex responses to a variety of extracellular stimuli. MAPK = mitogen activated protein kinase; MAPKK = MAPK kinase; MAPKKK = MAPK kinase kinase.

Reprinted, by permission, from D.M. Cox, Z.A. Quinn, and J.C. McDermott, 2000, "Cell signaling and regulation of muscle-specific gene expression by myocyte enhancer-binding factor 2," *Exercise and Sport Sciences Reviews* 28:33-38.

The calcium ion is a ubiquitous intracellular signal (see Berridge, Bootman, and Lipp 1998) with a well-known role in muscle cells, but its role is not restricted to excitation–contraction coupling. Calcineurin (CaN) is a calcium–calmodulin-sensitive serine–threonine phosphatase, which, in addition to being involved in MyHC phenotype switching (see chapter 3), also seems to be able to mediate skeletal muscle hypertrophy. IGF-1 or insulin activates CaN by mobilizing intracellular calcium and induce hypertrophy.

## For Additional Reading

See L.A. Blair and Marshall (1997) for an outline of the effect of IFG-1 on calcium channels and Baar et al. (1999) for a discussion of the variety of effects induced by increased levels of cytosolic calcium.

The hypertrophic response is suppressed by CaN inhibitors but not by inhibitors of the MAPK or the PI-3-kinase pathways (Semsarian, Wu, et al., 1999).

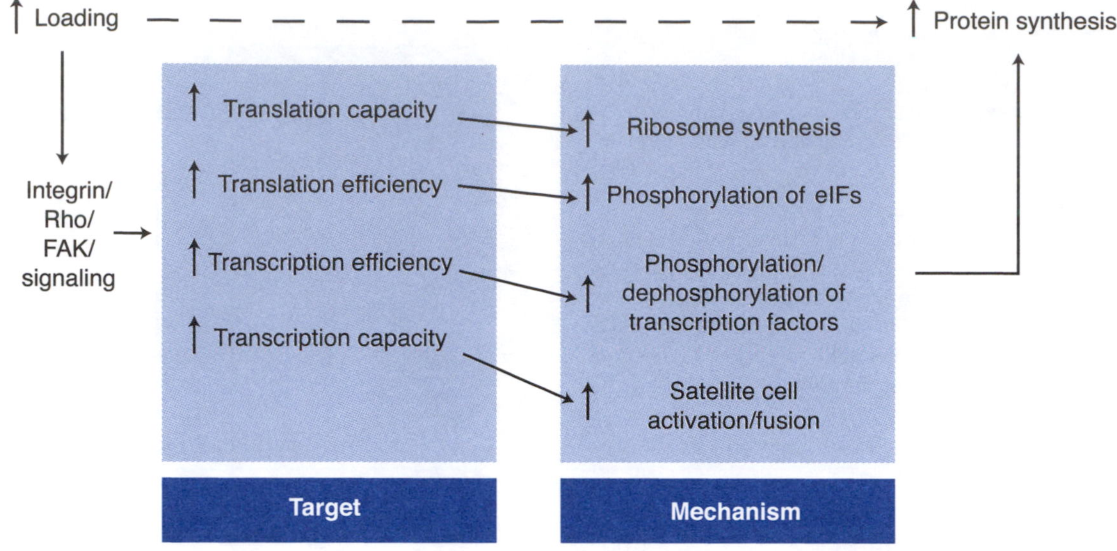


**Figure 11.12** Several levels of cellular regulation appear critical for overload-induced increases in skeletal muscle mass. These points of regulation may have varying degrees of importance for inducing protein synthesis during the time course of overload-induced growth, especially in animal models of extreme hypertrophy. It is critically important to define how different cellular events are integrated to allow for a sequential growth response in skeletal muscle. eIFs = eukaryotic initiation factors.

Reprinted, by permission, from J.A. Carson and L. Wei, 2000, "Integrin signaling's potential for mediating gene expression in hypertrophying skeletal muscle," *Journal of Applied Physiology* 88:337-343.

So far it remains puzzling that CaN seems able to induce apparently opposite adaptations in muscle cells. On one hand it is able to invoke a slow fiber program as shown by Chin et al. (1998). On the other it is reported to induce hypertrophy and switching to a glycolytic metabolism (Semsarian, Wu, et al. 1999). The answer probably lies in the modulatory effect of combinatorial control (see, e.g., Treisman 1996). This is also supported by Naya et al. (2000), who suggested that signals in addition to CaN are necessary to induce muscle hypertrophy.

This is probably analogous to the effects of anabolic steroids. Today, we regard the family of intracellular steroid receptors as ligand-regulated transcription factors, which upon binding to their appropriate hormone or hormone agonist become able to bind to DNA and act in concert with other transcription factors.

So far, we have treated the regulation of muscle mass as the result of the presence or absence of positive growth signals, but several studies show that there is a negative regulation as well. A member of the transforming growth factor-β family of secreted growth and differentiation factors called **growth/differentiation factor 8 (GDF-8)** or **myostatin** has been shown to be a negative regulator of muscle mass (McPherron, Lawler, and Lee 1997; Zhu et al. 2000). Individual muscles of myostatin-deficient mice have been reported to be two to three times larger than the corresponding muscles in wild type animals (McPherron, Lawler, and Lee 1997). How this relates to human muscle hypertrophy or atrophy is, however, unknown.

## Metabolic Adaptations in Muscle

Although strength adaptations primarily alter the instantaneous maximal force, metabolic adaptations confer an increased protection against muscular fatigue. This does not mean, however, that strength adaptations are without effect on muscle fatigability. Because most demands on our muscles in daily work are relatively constant (e.g., lifting an object, turning a wheel), an increased strength means that these tasks are performed at a lower percentage of maximal force than before training, with a corresponding increase in endurance. In addition, low-resistance strength training with many repetitions can elicit favorable effects on local enzyme systems and capillary density.

The metabolic demands during muscular work are basically a demand for resynthesis of ATP, which can take place by means of a complete breakdown of glucose to carbon dioxide ($CO_2$) and water ($H_2O$), or by an incomplete breakdown to pyruvic and lactic acid (see chapter 2). The capacity of ATP resynthesis by incomplete and complete metabolism of glucose is called anaerobic and aerobic power, respectively.

## Training of Anaerobic Power

As discussed in chapter 8, no methods are available to accurately measure anaerobic power or capacity. Therefore, it is not possible to evaluate objectively whether a specific anaerobic training program is effective.

### For Additional Reading

The problems associated with measurement of anaerobic power are discussed by S. Green and Dawson (1993) and Bulbulian, Jeong, and Murphy (1996).

It is a common finding that the peak blood lactate concentration is higher in a track athlete after that athlete has finished an important competition than after an all-out test on the treadmill. What, then, are the limiting factors in efforts with maximal demands on the anaerobic energy yielding system? Saltin and Gollnick (1983) concluded that lactate per se is not responsible for the termination of exercise, nor are the enzymes involved in the anaerobic metabolism significantly inhibited. It is a problem, however, that these conclusions were drawn on the basis of blood lactate measurements, which do not always reflect the intracellular lactate concentration. It is known now that lactate transport through the muscle cell membrane depends on the presence of monocarboxylate transporters (MCTs; see chapters 3 and 8) and that the amount and isoform of MCT present depend on the level of training (Bonen 2000; Dubouchaud et al. 2000).

If, however, one can become adapted to a low pH and increase one's tolerance to it by exposing the tissue to the anaerobic metabolites, the following program should be effective: maximal effort for about 1 min, followed by 4 to 5 min of rest, then a further period of 1 min of maximal effort, followed by 4 to 5 min of rest, and so on. At the end of four or five such exercise periods, a highly motivated runner can gradually attain lactate concentrations in the skeletal muscle in excess of 30 mM, lactate concentrations of 20 mM or more in the blood, and an arterial pH approaching 7.0 or lower. To produce major changes in the cellular environment (e.g., affecting the respiratory function and causing dyspnea), large muscle groups should be engaged in activities such as running. Training of anaerobic power is important for many groups of athletes. Because this form of training is psychologically exhaustive, it should not be introduced until a month or two prior to the competitive season. Such strenuous training is not recommended for average persons.

At a given submaximal oxygen uptake, the content of lactic acid in the blood is lower in a trained subject than in an untrained one, as shown by a series of studies (figure 11.13). This can be interpreted as an expression of a more effective oxygen transport and an increased ability to use it in oxidative metabolism,

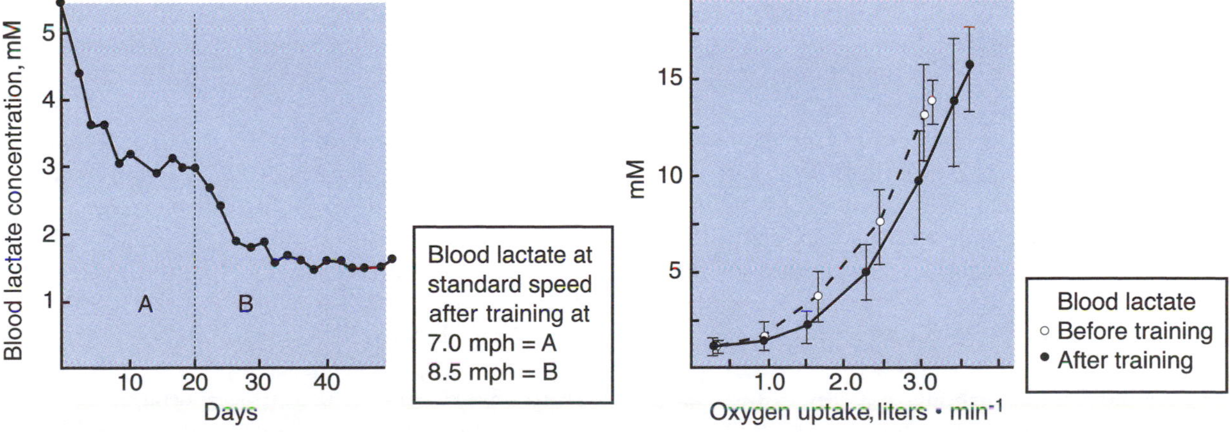


**Figure 11.13** Left panel: Progressive decrease of blood lactate for a standard amount of exercise: running on a treadmill at 7 mph for 10 min. During the first 20 days *(A)*, training consisted of running daily on the treadmill for 20 min at 7 mph. A steady level of blood lactic acid was reached at around 3 mM. During the following 30 days *(B)*, training was increased to running at 8.5 mph for 15 min daily. Blood lactic acid decreased further, and a new steady state was reached around 1.5 mM after the standard test. Right panel: Peak blood lactate concentration in relation to oxygen uptake up to maximum, before (dotted line) and after (solid line) 16 weeks of physical training. Mean values from eight subjects exercising on a cycle ergometer. Vertical bars indicate ±1 standard deviation.

Left panel from H.T. Edwards, Brouha, and Johnson 1939; right panel from Ekblom et al. 1968.

leading to a diminished anaerobic energy yield. Although an onset of blood lactate accumulation can be observed at an oxygen uptake corresponding to 50% to 60% of the maximal oxygen uptake in an untrained individual, this percentage can be elevated to 70% to 80% or even higher in a well-trained individual (see chapter 8, and Denis et al. 1982; Hurley et al. 1984). Coyle et al. (1983) observed that their well-trained ischemic heart disease patients could run at 100% of their maximal oxygen uptake before the blood lactate increased 1 mM above baseline. A similar increase in lactate concentration was already noticed in healthy, well-trained subjects when they were running at about 85% of their maximal oxygen uptake.

Strength training with maximal or near-maximal efforts for approximately 5-s periods will not significantly affect either the resting level of ATP and PC or the activity of the enzymes involved in the anaerobic and aerobic energy yield. Sprint-type interval training (e.g., repeated 200-m dashes) will increase the activity in "anaerobic" enzymes, including phosphofructokinase (PFK) and lactate dehydrogenase (A.D. Roberts, Billeter, and Howald 1982). During maximal effort, children usually attain relatively low muscle lactate concentrations (e.g., approximately 9 mM · $kg^{-1}$ of wet muscle in 11- to 13-year-old boys; B.O. Eriksson 1972). They have also much lower PFK activity than adults. After a period of training that includes sprint and endurance training, the PFK activity as well as the lactate concentration can increase (e.g., to about 14 mM in the aforementioned study). PFK is a rate-limiting enzyme, which means that it is important for regulating the flow of substrate through the glycolytic pathway. The potential of the non-rate-limiting enzymes, on the other hand, is luxuriously high even in a sedentary person's skeletal muscles and yet increases further during training. One may speculate why this is so, but a simple answer is that the potential of the non-rate-limiting steps has to be able to handle the flux coming out of the flux-generating (rate-limiting) steps, whether the latter work at full capacity during maximal work or at a lower level during periods of rest.

It has been reported that elite athletes in sports with great demands on anaerobic power and capacity have a significantly higher buffer capacity in the skeletal muscles than more sedentary subjects (Sahlin and Henriksson 1984).

There are large individual differences in the peak lactate concentration in skeletal muscles and blood after maximal exercise lasting some minutes. This maximum tends to be higher in well-trained individuals.

### For Additional Reading

For a review of lactate production and clearance in exercise and the effect of training, see Stallknecht, Vissing, and Galbo (1998).

### *Summary*

The most significant finding is that training of aerobic power and endurance lowers the lactate concentration in exercising muscles as well as in the blood, both at a given rate of exercise and at a given percentage of the maximal oxygen uptake. Therefore, the intensity at which a continuous accumulation of lactate starts will be shifted upward with training. An enhanced oxygen supply may be one factor behind this change, but an inhibition by the improved potential for an aerobic energy yield in the mitochondria is also a possible mechanism.

## The Dallas Study

Before going into a detailed discussion of the long-term effects of training aimed at improving maximal oxygen uptake and endurance, we will present one study that has become classic in its design because of its broad illustration of the effects of prolonged bed rest on the oxygen-transporting system, followed by a specific training program in the same subjects.

Saltin et al. (1968) carried out extensive studies on the effect of a 50-day period of physical training following a 20-day period of bed rest in five male subjects, aged 19 to 21, three of whom had previously been sedentary and two of whom had been physically active. The training program was rather intensive and continuously supervised. The weekly schedule included two workouts daily, Monday through Friday; on Saturdays there was only one workout, and Sundays were free. The workouts consisted of both interval and continuous exercise, mainly outdoor running of from 4 up to 11 km (2.5–7 miles). In the interval exercise, the speed was chosen so that the oxygen demand during 2- to 5-min periods of running was at or near the individual's maximal oxygen uptake (figure 11.14). During the continuous running, usually for more than 20 min, oxygen uptake varied from 65% to 90% of the subject's maximal value. One of the subjects who had not trained before covered an average of 64 km · $week^{-1}$ (40 miles), and one of the previously trained subjects covered about 80 km · $week^{-1}$ (50 miles).

After bed rest, the maximal oxygen uptake had dropped from an average of 3.3 before bed rest to 2.4 L · $min^{-1}$, a 27% decrease (figure 11.15). The stroke volume during supine exercise on a cycle ergometer

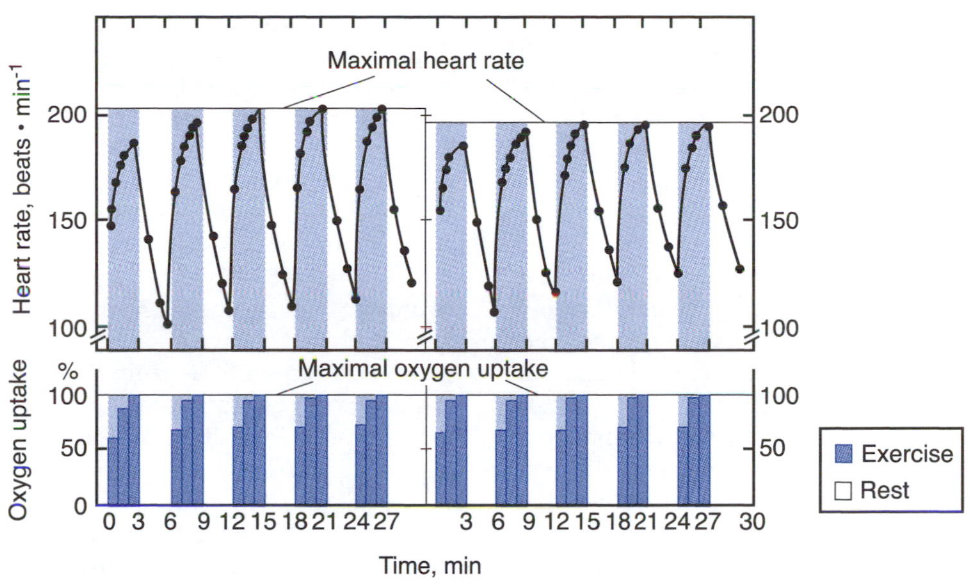


**Figure 11.14** Heart rate and oxygen uptake recorded in two subjects during training with alternating 3 min running and 3 min rest. The efforts were not maximal, but the oxygen uptake reached maximal values, as did the heart rate.

From Saltin et al. 1968.

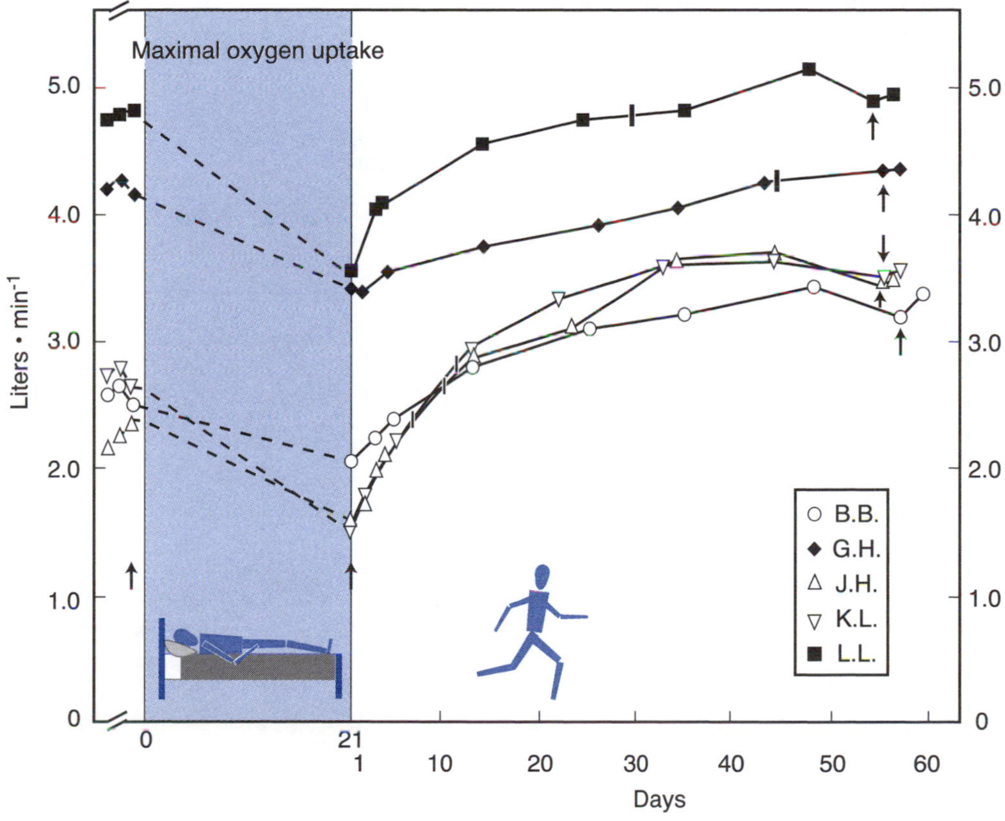


**Figure 11.15** Changes in maximal oxygen uptake, measured during running on a motor-driven treadmill, before and after bed rest and at various intervals during training; individual data from five subjects. Arrows indicate circulatory studies. Heavy vertical bars mark the time during the training period at which the maximal oxygen uptake had returned to the control value before bed rest.

From Saltin et al. 1968.

at 100 W (oxygen uptake, 1.5 L · $min^{-1}$) had decreased from 116 to 88 ml, or about 25%. The heart rate increased from 129 to 154 beats · $min^{-1}$. The cardiac output at this standard load fell from 14.4 to 12.4 L · $min^{-1}$. Thus, the arteriovenous oxygen [(a-$\bar{v}$)$O_2$] difference was somewhat increased. Also, during treadmill running at submaximal loads, there was a reduction in cardiac output (15%) and stroke volume (30%; figure 11.16). An oxygen uptake that originally could be attained with a heart rate of 145 required a heart rate of 180 beats · $min^{-1}$ after bed rest. The cardiac output during maximal treadmill exercise fell from 20.0 to 14.8 L · $min^{-1}$ (a 26% reduction). It should be recalled that the maximal oxygen uptake was now 27% less. Because the oxygen content of arterial blood did not change, the maximal (a-$\bar{v}$)$O_2$ difference was, therefore, not modified by the bed rest. The maximal heart rate was not altered, so the decrease in maximal cardiac output was attributable to a reduction of stroke volume (see figure 11.16).

In the previously sedentary subjects, the subsequent physical training increased the maximal oxygen uptake to 3.41 L · $min^{-1}$, an increase of 33% in relation to the value attained before bed rest, but in relation to the value after bed rest the improvement was 100%, or from 1.74 to an average of 3.41 L · $min^{-1}$. In the previously active subjects, the end result was 4.65 L · $min^{-1}$, compared with 4.48 before bed rest, an improvement of only 4% (see figure 11.15). In relation to the level after bed rest, however, the increase was 34% (from 3.48 to 4.65 L · $min^{-1}$), illustrating how critical the level of physical activity before a training regime is in evaluating the effectiveness of a training program to improve maximal aerobic power. Figure 11.15 shows that the three sedentary subjects exceeded their control values as early as about 10 days after the commencement of the training. The two previously active subjects required 30 to 40 days to achieve the noted improvement.

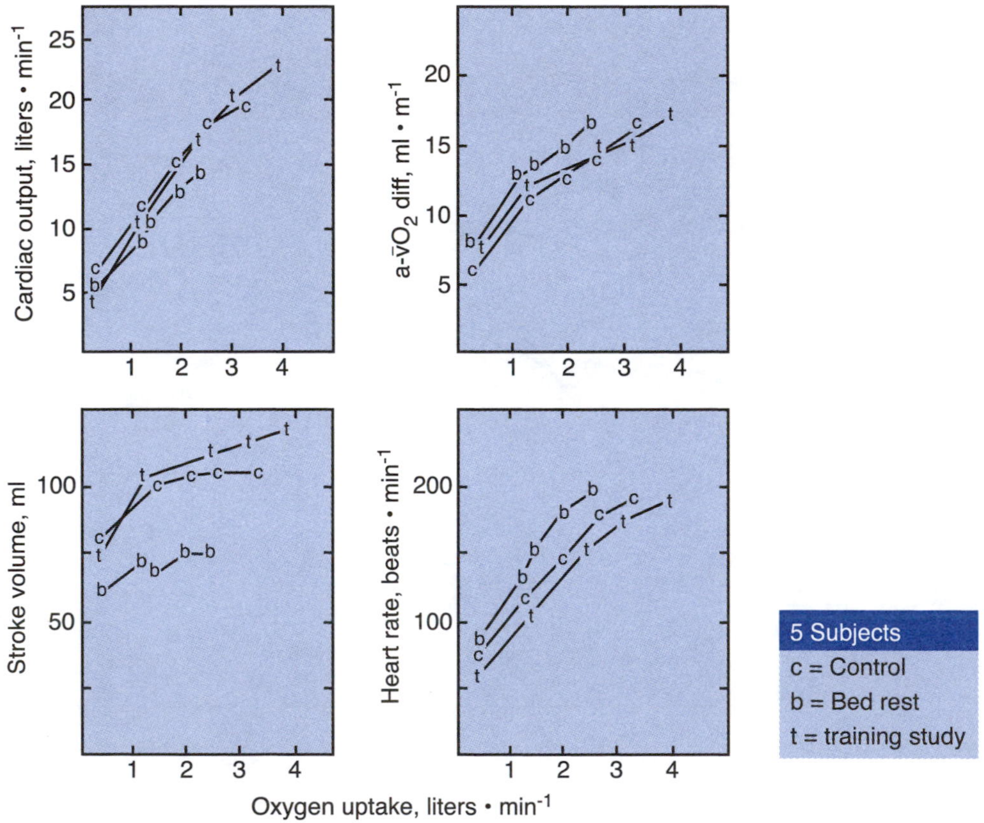


**Figure 11.16** Mean values of cardiac output, arteriovenous oxygen difference [(a-$\bar{v}$)$O_2$ diff.], stroke volume, and heart rate in relation to oxygen uptake during running at submaximal and maximal intensity before *(c)* and after *(b)* a 20-day period of bed rest, and again after a 50-day training period *(t)*.

Adapted from Saltin et al. 1968.

Figure 11.17 illustrates the variation in maximal oxygen uptake for the three normally inactive subjects. The higher maximum noticed during their normal sedentary life, compared with the more extreme inactivity when immobilized in bed, was attributable to a higher cardiac output. The further improvement in maximal oxygen uptake with intensive training was partially attributable to a still higher cardiac output and partly to an increased $(a-\bar{v})O_2$ difference by a more complete extraction of oxygen from the blood in the tissues.

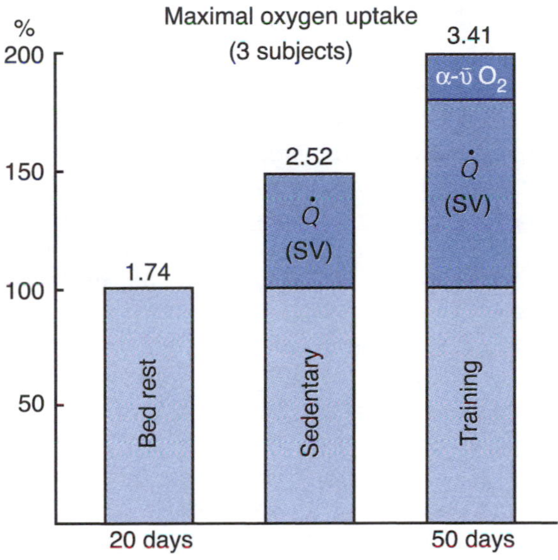


**Figure 11.17** Maximal oxygen uptake during treadmill running for three subjects after bed rest (100%), when they were habitually sedentary, and after intensive training. The higher oxygen uptake under sedentary conditions compared with bed rest is attributable to an increased maximal cardiac output ($\dot{Q}$). The further increase after training is possible because of a further increase in maximal cardiac output and arteriovenous oxygen $[(a-\bar{v})O_2]$ difference. The maximal heart rate was the same throughout the experiment; therefore, the increased cardiac output was attributable to a larger stroke volume (SV).

Data from Saltin et al. 1968.

The training increased the stroke volume and decreased the heart rate at submaximal work rates (see figure 11.16). However, maximal heart rate was not modified by the training. The maximal cardiac output in the sedentary subjects decreased from 17.2 to 12.3 L · min$^{-1}$ during bed rest. After the training, it rose to 20.2 L · min$^{-1}$. The $(a-\bar{v})O_2$ difference, in the same experiments, was 147, 149, and 170 ml · L$^{-1}$ of blood, respectively. The stroke volume fell from 90 to 62 ml during bed rest, but after training it increased to 105 ml. The changes in heart rate at various levels of oxygen uptake caused by inactivity and activity are obvious. However, when the heart rate was related to the oxygen uptake in percentage of the maximum, the heart rate response remained the same in the different conditions.

The mean value for the heart volume in the three sedentary subjects was 740 ml. After bed rest, it was 690 ml. It increased to 810 ml at the end of the training, a 17% increase. The increase in stroke volume was 69%; however, this was partly attributable to decreased peripheral resistance (see chapter 5).

Blood volume decreased significantly during bed rest (from 5.0 to 4.7 L in the five subjects, i.e., a 7% reduction). The decrease in plasma volume was slightly more pronounced than in the red cell mass. During training, the plasma volume and red cell mass increased again, in most subjects above the control values.

The lean body mass changed from 66.3 to 65.3 kg after bed rest and increased to 67.0 after training. There were no significant changes in the ultrastructure of the quadriceps muscle during the experimental period. The basal heart rate recorded during sleep was on the average 51.3 beats · min$^{-1}$ after bed rest, and it decreased to 39.7 at the end of the training.

## Training of Aerobic Power

Physical activity ranging from repeated exercise periods of a few seconds' duration up to hours of continuous activity may place a major load on the oxygen-transporting organs and thereby can induce a training effect, provided the exercise load is sufficiently high. Practical experience has shown that exercise with large muscle groups for 3 to 5 min, followed by rest or light physical activity for an equal length of time, then a further exercise period, as required by the individual's ambition and the objective of the training, is an effective method of training. The load does not have to be maximal during the activity periods, and it is not necessary to be exhausted when the exercise is discontinued, because maximal oxygen uptake can be reached in the absence of exhaustion. Mild exercise such as jogging between the heavier bursts of activity is advantageous, because this eliminates lactic acid faster than complete rest (see chapter 8). It has been shown experimentally that the cardiac output and blood pressure reach their highest values at a load that produces the maximal oxygen uptake. During supramaximal exercise, on the other hand, oxygen uptake, cardiac output, and stroke volume attain lower values than they do at a slightly lower work rate. There is no evidence that it is important to engage the anaerobic processes to any extreme degree to train the aerobic motor power.

The justification for a submaximal tempo in the optimal training of the oxygen-transporting system is further supported by figure 11.18. For six subjects, an individual speed was determined that brought them to complete exhaustion at the end of the 4th min of running. On other days, the speed of the treadmill was decreased stepwise by .5 to 1 km · h$^{-1}$ without changing the total distance of the run. A reduction in speed by as much as 3 km · h$^{-1}$ for some of the subjects did not reduce the oxygen uptake. Therefore, because maximal oxygen uptake can be attained at a submaximal speed, this lower speed may be sufficient and probably optimal as a training stimulus. A highly motivated individual and someone with a high anaerobic capacity will show a wide plateau. Others may just be able, or willing, to push themselves to the point where the maximal oxygen uptake is reached, or not even that far. It is thus impossible to delineate a border value where the greatest load on the oxygen-transporting organs is attained. In the case of healthy young persons, the speed of running can be reduced to about 80% of the maximum, which can be maintained for 3 to 5 min. If, in other words, the distance that can be covered by running, swimming, or bicycling in a matter of, say, 3.0 min is covered, instead, in about 3.5 min, the demand on the aerobic processes remains the same. Therefore, the stopwatch, in such cases, should be used to maintain a reduced tempo, not to stimulate the trainee to achieve more in terms of better timing. Figure 11.14 illustrates how this type of training can be arranged and what effect it has on heart rate and oxygen uptake. In exercise in the upright position, the maximal stroke volume is attained during and not after the exercise. It is a misconception to believe that the advantage of interval training is that frequent recovery periods as such should effectively train the central circulation.

The heart rate can be used to determine whether the load is maximal or nearly maximal. It should not differ more than 10 beats · min$^{-1}$ from the individual's maximal heart rate assessed during controlled laboratory experiments. A person accustomed to heavy exercise also can sense when the pulmonary ventilation has reached the steep part of the slope on the curve relating pulmonary ventilation to oxygen uptake (see figure 6.9). This type of training is certainly far more pleasant than when the tempo is higher. Thus, one should definitely distinguish between training of the aerobic and the anaerobic processes. In competitive events, which require a superior effect in both these processes, the training obviously should also include this combination.

The type of training described for the oxygen-transporting system, with a submaximal tempo for periods of 3 to 5 min, can indeed increase the maximal oxygen uptake. In most of the earlier studies of the effects of training, the intensity during the training sessions included maximal efforts. In the 1970s, Kilbom (1971) and Pollock (1973) conducted studies showing that a demand exceeding 50% of the maximal oxygen uptake, repeated some 30 min three times a week, was enough to elevate the maximal oxygen uptake in previously untrained individuals. When previously sedentary subjects trained at 50% of the maximal oxygen uptake, a 5% to 10% increase in this $\dot{V}O_2$max was found. When subjects train at 70% to 80% of maximal oxygen uptake, the improvement is on average 15%. There are, however, large individual variations, at least partly because of variations in the initial level of fitness.

The problem is to teach a person how to find the exercise intensity that demands 70% to 80% of her or

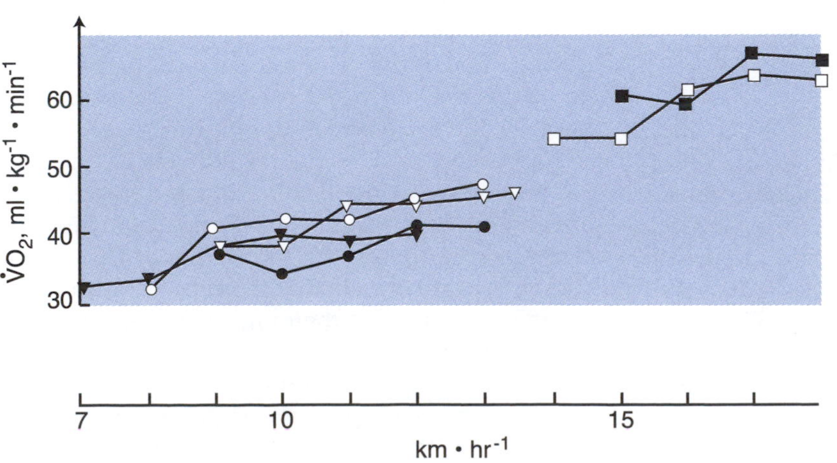


**Figure 11.18** Individual data on oxygen uptake in relation to speed for two female (△, ▲) and four male subjects. The treadmill was set at 3° inclination. The highest speed for each subject could be maintained for just 4.0 min.

From Karlsson et al. 1967b.

his maximal aerobic power with no sophisticated technical aids available. If the person is very breathless, the intensity is too high, and if a fluent conversation is possible, the intensity is too low. R.J. Chow and Wilmore (1984) noticed that about 50% of their subjects learned to pace themselves quite accurately by applying ratings of perceived exertion (Borg 1998). If perceived exertion is used to establish exercise intensity, the treadmill, according to Zeni, Hoffman, and Clifford (1996), is the optimal indoor exercise machine for aerobic training.

Brisk walking, rope skipping, and stair climbing can significantly improve aerobic power (Ilmarinen et al. 1979; Jones and Rodahl 1962; Ogawa et al. 1974). The ability to exercise for prolonged periods of time, using the largest possible percentage of the maximal oxygen uptake, probably can be developed just by exercising continuously during long periods of time (endurance training). The capacity to store glycogen in the muscles and the ability to mobilize and to use free fatty acids (FFAs) play a major role in prolonged exercise. Endurance athletes devote a great deal of time to distance training. They may run or ski up to 250 to 350 km (150–220 miles) weekly. No studies are available with which to objectively evaluate the effects of such training quantities, but it can take up to about 48 h to refill emptied glycogen stores (Piehl 1974). This is an additional complication for those who have the ambition to train daily for several hours at high rates of exercise. Over the years, top athletes have become increasingly professional in the sense that they are employed full time in their sports and are thus free to devote a great deal of time to training.

Analyses of the exercising muscle groups have revealed that the slow-twitch fibers are the first to lose their glycogen content in prolonged submaximal activity (Vøllestad, Vaage, and Hermansen 1984). However, as these fibers are depleted of their glycogen stores, fast-twitch fibers are apparently recruited, type IIX fibers being the last reserve. When the subjects are exercising at rates exceeding 100% of maximal $\dot{V}O_2$, both fiber types appear to be continuously involved (Gollnick, Piehl, and Saltin 1974). Thus, in heavy interval training, all fibers in the activated muscle groups are active simultaneously. As pointed out in chapter 8, the fast-twitch fibers may be forced to depend primarily on an anaerobic energy yield because the oxygen supply and/or the ability of the fibers to use oxygen for oxidative metabolism is not sufficient to cover their energy demand. The timing of these events depends on initial glycogen stores, rate of work, and the exercising person's willpower. At any rate, the biochemical and morphological adaptations to chronic exercise will be restricted to the fibers that have been active. The only possible exception to this regards capillaries. If the number of capillaries around one particular muscle fiber increases because of training, this will also benefit its nearest neighbors. The question is, however, whether the untrained muscle fibers are able to use the increased supply of oxygen and nutrients. One can speculate that intensive interval training provides a balanced adaptation of both slow- and fast-twitch fibers to a combined aerobic and anaerobic stress situation, as in a 5,000 -or 10,000-m race. Exclusive distance training will promote the fast-twitch fibers' aerobic metabolism. Actually, the aerobic potential including the enzymes of the citric acid cycle and the electron-transport chain of both type IIA and type IIX fibers, which eventually turn into type IIA, can attain the same level as in the type I fibers (Saltin and Gollnick 1983).

In the case of patients and completely untrained individuals, it is out of the question to prescribe an accelerated tempo in connection with the training of circulation. A previously bedridden patient should be satisfied with a training load that commences by elevating the heart rate by about 30 beats · $min^{-1}$ above the resting value (to about 100 beats · $min^{-1}$). For the habitually sedentary individual, an elevation of the heart rate by about 60 beats · $min^{-1}$ is a suitable initial intensity. The principle of intermittent exercise is also valid for these individuals. With daily training periods of from 15 to 30 min in duration, or even with only a few training periods per week, the tempo can gradually be increased. The individual's health, age, and interest will determine how strenuous the training should be. As discussed earlier, it is not necessary, and in some cases not even desirable, to attain maximal oxygen uptake and cardiac output during the training. For patients and previously sedentary individuals, endurance training can be accomplished by such activities as walking or bicycling. As mentioned, large muscle groups must be engaged when training the central circulation. Figure 11.1 is of interest in this connection. In a number of top Swedish athletes belonging to the National Team, the maximal oxygen uptake was determined by laboratory experiments, mostly on the treadmill. It is evident that the athletes who had the highest maximal oxygen uptakes had selected events that placed heavy demands on their aerobic power, which means that these events also represent an excellent form of training of the oxygen-transporting system. Ball games fall rather low on the scale (see also chapter 16). This can be explained by the fact that each period of activity with a high tempo is interrupted frequently by periods of reduced tempo (see table 11.1). Most competitive calisthenics entail exercise periods up to 1 min. Needless to say, calisthenics can be performed in such a manner that they entail an excellent training effect of the aerobic power (straddle jump, sequences of movements engaging large muscle groups).

## Circuit Training

Circuit training (R.E. Morgan and Adamson, 1962) entails a series of activities performed one after the other. At the end of the last activity, one starts from the beginning again and carries on until the entire series has been repeated several times. By a preliminary test, in which many of the activities are performed at maximal exertion, the number of repetitions of each activity is determined. The advantage with circuit training is that every individual undergoes a program adjusted to his or her level of fitness, and that it may be accomplished in a limited space. The disadvantage is that an untrained individual is exposed to tests requiring maximal exertion. Persons may follow their own improvement by recording the time required for the series of repetitions and endeavoring to shorten the time required. At the end of a few weeks of training, a new test is performed on the number of repetitions of each activity the individual can manage (e.g., push-ups). This type of training produces a high degree of motivation. It has been found that a correctly planned program varying the involvement of large and small muscle groups and mixing static and dynamic exercise does not produce maximal oxygen uptake measured on the cycle ergometer but rather produces only about 80% of the maximal oxygen uptake (Hedman 1960). Despite this, the heart rate is almost maximal, the lactic acid concentration in the blood is very high, and the degree of exertion is considerable. Circuit training can be included in a training program, not only for athletes (especially those who fall within the lower part of figure 11.1) but also for schoolchildren for the sake of variation and for experience. It also can be applied very effectively in a strength-training program.

## The Effect of Endurance Training on Muscle Fiber Phenotype

Continuous electrical stimulation of a nerve innervating an animal's fast-twitch muscle with a low frequency (10 Hz) gradually transforms the muscle toward characteristics typical for slow-twitch fibers (Lømo, Westgaard, and Dahl 1974; Pette 1984; Salmons and Henriksson 1981). The capillary density and the activity of oxidative enzymes increase during the first week. Some 3 to 6 weeks after the onset of nerve stimulation, a change in the contractile characteristics of the fibers is seen, attributable to a shift of contractile protein isoforms. This shift of isoforms starts as early as after a few days of stimulation, as shown by a shift in the pattern of MyHC mRNA (Salmons 1994) but takes longer to manifest itself at the functional level. When the electrical stimulation is discontinued, the fibers gradually transform back to their original type II characteristics. This illustrates the plasticity of the skeletal muscle. But the plasticity goes further than this. Even within type II fibers, there are transitions between subtypes. In the young adult rat, chronic low-frequency stimulation results in fiber type transitions in the order of type IIB → type IID/X → type IIA (Skorjanc, Traub, and Pette 1998).

Even though the activity imposed on the muscle by chronic stimulation is very simple compared with the activity during physical training, muscles adapt to these two experimental strategies in a strikingly similar way (Salmons 1994). Thus, rats subjected to treadmill running at ~75% of maximal oxygen uptake for 10 weeks showed a shift in their hind-limb MyHC isoform pattern from MyHC IIB toward MyHC IID/X and MyHC IIA. Even increases in MyHC I were found (Demirel et al. 1999). It is to be expected, however, that variations in the degree of fiber type transformation will occur, at least in some muscles, because differences in the degree of recruitment have been demonstrated for muscle units in different parts of the same muscle, even though they belong to the same muscle fiber type (De Ruiter et al. 1996).

In humans, the possibility of a fiber type transition from type I to type II or vice versa was a matter of controversy for many years. Larsson and Ansved (1985) wrote, "The majority of longitudinal training studies in man have, to date, provided no evidence for a fiber type interconversion between type 1 and type 2 fibers when classified by their myofibrillar adenosine triphosphatase (ATPase) stainability" (p. 714). In the same article, however, they were able to show a significant decrease in the proportion of type I fibers after detraining. Similar findings were reported by Häggmark, Eriksson, and Jansson (1986), who followed the fiber type pattern of several athletes referred for surgery after having suffered knee ligament injuries. In one of their patients, a world-class competitive cross-country skier, the proportion of type I fibers in the lateral vastus dropped from 81% at the time of surgery to 58% after 6 weeks of immobilization in a plaster cast. After retraining, the percentage of type I fibers increased to the original level.

A major problem in the search for evidence of fiber type transformations between type I and type II has been the inability of the ATPase method to reveal details in the MyHC content (see discussion in chapter 3). With the advent of immunohistochemical techniques based on monoclonal antibodies and of high-resolution polyacrylamide gel electrophoresis able to separate MyHC isoforms in fragments of single human muscle fibers, this has radically changed. These techniques have revealed that hybrid muscle fibers are more common than previously thought and that they are especially prevalent after changes in the level of physical activity, both increases and decreases. By using these techniques, M.-Y. Zhou et al. (1995) showed that the percentage of type I fibers in the human vastus lateralis muscle decreased significantly after as little as 5 to 11 days in space. Kadi and Thornell (1999) found significant changes in the fiber type profile in the upper part of the trapezius muscle of women after training. After strength training, the proportion of MyHC IIA increased and the amount of MyHC IIX and MyHC I decreased. According to Williamson et al. (2001), the increase in type IIA fibers after resistance training is mainly due to a decrease in single-fiber hybrid proportions.

In the endurance group, Kadi and Thornell (1999) found a significant decrease in the amount of MyHC IIX. According to O'Neill et al. (1999), a significant decline in the amount of MyHC IIX mRNA can be seen after only 1 week of endurance training. Unlike other types of training, sprint training seems to result in a bidirectional fiber type transformation, increasing the proportion of MyHC IIA at the expense of both MyHC IIX and MyHC I (J.L. Andersen, Klitgaard, and Saltin 1994; Harridge et al. 1998).

## Endurance

Available data indicate that the maximal oxygen uptake in $L \cdot min^{-1}$ or $ml \cdot kg^{-1} \cdot min^{-1}$, depending on the task, is of decisive importance in strenuous exercise lasting approximately 5 to 30 min. In prolonged energy-demanding activities, the maximal aerobic power is less important for performance, whereas peripheral factors are more important for success. When the energy demand exceeds approximately 70% of the maximal aerobic power, the enhanced FFA metabolism attributable to training seems to be important because a given glycogen depot then can last longer (chapter 12). Endurance training of respiratory muscles may prolong intense constant-intensity exercise and reduce blood lactate concentrations during exercise (Spengler et al. 1999). Strength training in general improves neuromuscular function, prevents injury, and is a prerequisite for effective endurance training (Henriksson and Tesch 1999). Hardman and Hudson (1994) showed that regular brisk walking can improve endurance fitness and increase high-density lipoprotein cholesterol concentration in sedentary women.

For many years it was commonly accepted that respiratory capacity is not an exercise-limiting factor. Today, there is growing evidence that this may not be entirely true. Thus, Boutellier (1998) found that isolated respiratory training significantly increased endurance in both sedentary and physically active subjects. Most notably, the sensation of breathlessness was gone.

### *Summary*

One effect of interval/endurance training is that the trained individual can tax a larger percentage of her or his maximal oxygen uptake. The reason for this is not obvious. A more effective oxygen supply to the exercising muscle attributable to an enlargement of the vascular bed, an enhanced diffusion, a higher enzymatic potential for increased use of FFAs as a substrate, increased glycogen content, and a higher psychological fatigue threshold may play a role. A higher oxygen uptake can be maintained without any accumulation of lactate. With training, individuals become more accustomed to the exertion and are more willing to push themselves closer to their limit. Achievements in endurance events can be improved beyond those that are mirrored in the observed changes in the maximal aerobic power. From a practical standpoint, it is logical to list four components of a rational training program aimed at developing the different types of power:

1. Bursts of intense activity lasting only a few seconds may develop muscle strength and stronger tendons and ligaments.
2. Intense activity for about 1 min, repeated after about 4 min of rest or mild exercise, may develop anaerobic power.
3. Activity with large muscles involved, at less than maximal intensity, for about 3 to 5 min, repeated after rest or mild exercise of similar duration may develop aerobic power.
4. Activity of submaximal intensity lasting as long as 30 min or more may develop endurance, that is, the ability to tax a larger percentage of the individual's maximal aerobic power.

## Recovery After Exercise

Very few longitudinal studies have examined the payment of oxygen debt, resting level of heart rate, cardiac output, and temperature before and after training. The fast recovery of athletes after muscular exercise, in contrast, is well established. Hartley and Saltin (1968) observed that untrained subjects who (1) exercised for 6 min at an oxygen uptake of about 40% of their maximal aerobic power, (2) then exercised for 6 min at a load of about 70% of their maximum, (3) then rested 10 min, and (4) repeated the 40% load, now had the same oxygen uptake and cardiac output as under (1), but they had a significantly higher heart rate, in many cases as much as 20 beats · min$^{-1}$ higher, and consequently they also had a smaller stroke volume. Following training, this increase in heart rate was much less, and well-trained athletes actually can accomplish the entire procedure with the results from (1) and (4) being almost identical. This type of testing appears quite promising as a method of assessing the level of physical fitness or the level of habitual physical exercise. Recovery after exercise, however, involves far more than restoration of respiration frequency, heart rate, and cardiac output. As mentioned earlier, it may take 48 h before intramuscular glycogen stores are back to resting levels after exercise of long duration (Piehl 1974).

### For Additional Reading

For a discussion of the problem of overtraining and recovery, see Work Kenttä and Hassmén (1998). The diagnosis of overtraining was discussed by Urhausen and Kindermann (2002).

## Specificity of Training

When discussing muscular strength, we pointed out that specific training may not improve strength in other types of activities. In certain situations, there is a similar lack of transfer of the training effects of the oxygen-transporting system. Above all, the adaptations in the skeletal muscles, just described, are limited to the muscles—and even to the muscle units—actually engaged in the training.

The most readily available indication of a training effect on the maximal aerobic power is the heart rate response to a given submaximal rate of oxygen uptake. As mentioned, an increased maximal aerobic power is almost always associated with a reduced heart rate and vice versa. Three examples are presented to illustrate the specificity of the heart rate response to training:

1. When a subject trains one leg, the increase in maximal oxygen uptake in that type of exercise is larger than when the person performs the exercise with both legs simultaneously. The decrease in heart rate is also more pronounced in the one-leg than in the two-leg experiment.

2. When both legs are trained separately, there is a significant decrease in the heart rate during the submaximal one-leg test (i.e., by 11%), but during the two-leg exercise the change is an insignificant 2% decrease (Klausen et al. 1982).

3. Clausen (1976) summarized data from studies on healthy, young subjects in whom the circulatory response to combined arm and leg exercise was assessed after a period of training of either arms or legs alone. Arm training markedly reduced heart rate during exercise with the trained arm muscles from 137 to 118 beats · min$^{-1}$. During exercise with the nontrained leg muscles, a much less pronounced decrease in heart rate was seen (from 132 to 124 beats · min$^{-1}$). After leg training, however, the decrease in heart rate was about the same in the test with the trained leg muscles (from 135 to 122) as with the untrained arm muscles (from 127 to 112 beats · min$^{-1}$). It seems logical that there is more of a transfer effect on the central hemodynamics after training involving large muscle groups, compared with training of small muscle groups. Also in these situations, the most prominent training effects are evident when the test is performed with the trained limbs.

Figure 11.19 gives an example of data on the maximal oxygen uptake of a champion swimmer when swimming in a swimming flume and when running on a treadmill. For about 4 years, his maximal oxygen uptake was the same when running, but varying oxygen uptake values were measured during maximal swimming, attributable to variations in the training of his arms. Peaks were noted when the swimmer was most successful (winning two Olympic gold medals in 200- and 400-m medley, 1972). Two identical twin sisters reached similar maximal oxygen uptake when running (3.6 L · min$^{-1}$), but the sibling who had been training for competitive swimming attained the same maximal oxygen uptake of 3.6 L · min$^{-1}$ when swimming, which was a 30% higher maximal oxygen uptake than her twin sister, who was not trained for competitive swimming. When swimming with arms only, the swim-trained sister could reach a 50% higher maximal oxygen uptake (Holmér and Åstrand, 1972; see also Magel et al. 1975).

The conclusions are as follows:

1. A treadmill test is not a good predictor of performance in other types of activities.
2. To use aerobic potential in an optimal way in a given activity, one must train in that activity.

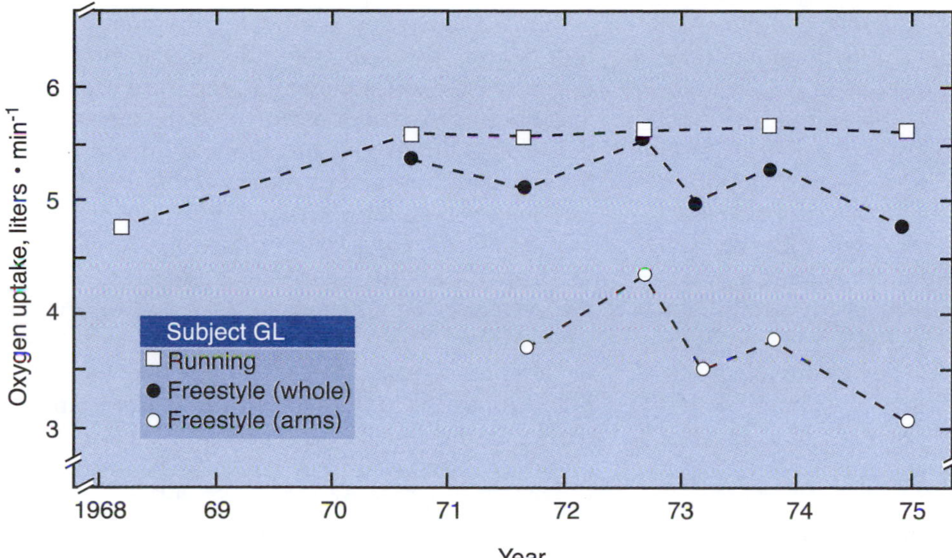


**Figure 11.19** Oxygen uptake during swimming and running at maximal intensities by one world-class swimmer over a 6-year period. Note the unchanged oxygen uptake during running from 1970, whereas oxygen uptake during swimming with the whole stroke and with only the arms varied considerably during the seasons.

Adapted from Holmér 1974b.

It is often claimed that strength training and endurance training are incompatible, in particular that endurance training interferes with and even inhibits a strength-training response, but so far, our understanding of the nature of this inhibition and the mechanisms responsible for it is limited. Two main hypotheses have been forwarded, however. The so-called chronic hypothesis claims that the muscle cells are unable to adapt morphologically by increasing the number of sarcomeres in parallel and metabolically by increasing the oxidative capacity at the same time. The acute hypothesis, on the other hand, blames some kind of residual fatigue after the endurance component. According to M.A. Collins and Snow (1993), however, the sequence of training does not matter, which speaks against this explanation.

### For Additional Reading

For discussion of the compatibility of strength and endurance training, see M.A. Collins and Snow (1993) and Leveritt et al. (1999).

There is reason to mention that the result of concurrent endurance and strength training can depend on many factors other than the sequence of training, such as sex, previous training status, and the duration, frequency, and mode of training (Docherty and Sporer 2000). In addition, the type of strength training used in many of these studies has been characterized as being "somewhat unusual" (McCarthy et al. 1995). Thus, in previously sedentary males, McCarthy et al. (1995) obtained substantial concurrent and compatible increases in peak aerobic power and strength after combined strength and endurance training three times per week for 10 weeks.

Regarding the effect of strength training on muscular endurance, Chilibeck, Syrotuik, and Bell (1999) reported that the muscle fiber hypertrophy after strength training reduced the density of mitochondria, which might indicate a reduced aerobic power. On the other hand, it might be speculated whether the increased maximal force after strength training would enable a muscle to overcome a given external resistance at a lower percentage of its maximal force, thus increasing its endurance. Despite a significant increase in 1RM concentric squat strength after 12 weeks of resistance training in endurance-trained female cyclists, however, Bishop et al. (1999) found no improvement in cycle endurance.

## The Effect of Detraining and Inactivity

The adaptations in the oxygen-transporting system to regular exercise of various intensity, duration, and frequency are reversible, with the exception of heart size, which in many individuals remains enlarged. Bed rest is an extreme form of inactivity, and the Dallas study provides a good illustration of its negative effects on maximal oxygen uptake (see figure 11.15) and other functions; however, the activity of oxidative enzymes may be even more affected by reduced activity (detraining) than is the maximal oxygen uptake (Henriksson and Reitman 1977), leading to an even larger reduction in endurance. The half-life of the oxidative enzymes seems to be approximately 1 week, whereas the half-life of the glycolytic enzymes is much shorter, from about 30 min up to a few days, at least in experimental animals (Illg and Pette 1979).

Individuals who have been involved in intensive training for a long time are, on the other hand, reported to be more resistant to a loss in muscle mitochondria and capillary density than are subjects who have trained for only a few months (Coyle et al. 1984).

In all cases, regular training is important. Within a month it is possible to develop a reasonable level of fitness and strength, but these qualities are lost when the training is discontinued. Fortunately, less effort is required to maintain a certain level of fitness than to develop this level in the first place. With nonathletes, it therefore might be worthwhile, when time permits, to devote more time to training, and then, later on, to maintain the acquired level of fitness by only a few training periods per week.

The athlete often has to train many different functions (aerobic and anaerobic power, strength, endurance, technique). It may be practical and sometimes necessary, because of time constraints, to concentrate on some of these functions at certain periods. This particular type of training has to be continued, however, even though it may be at a reduced intensity, when the athlete concentrates on the next function. Otherwise, the individual will lose the improvement attained. The problem is to decide how intensively the different functions have to be trained to maintain a satisfactory level.

The keen competition of today necessitates year-round training. For a variety of reasons this is necessary, especially with regard to the oxygen-transporting system. This is particularly important for the aging athlete. The reason why many 30- to 35-year-old athletes continue to rank among the world elite in endurance events is often attributable to relatively hard training during all seasons of the year. The older athlete may be what is known as "hard to train," compared with younger athletes, and the person whose fitness has been allowed to deteriorate can have considerable difficulty regaining the former level of training. The seasonal variations in maximal oxygen uptake can be pronounced. Figure 11.20 presents data on three top cross-country skiers.

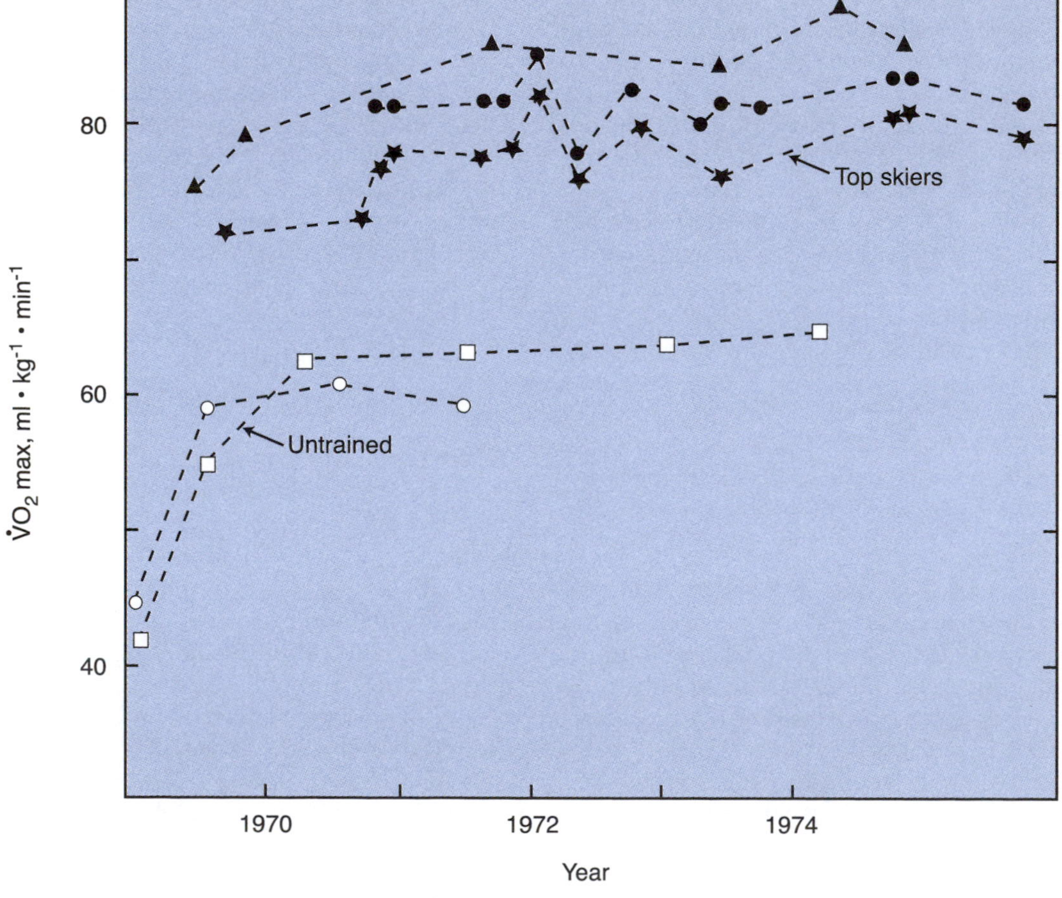


**Figure 11.20** Data on maximal oxygen uptake in three internationally successful cross-country skiers and two untrained subjects who started intensive physical training in 1969.

Courtesy of U. Bergh and B. Ekblom.

They more or less appear to have reached a ceiling, and the intensive daily training they undergo over the years will not markedly affect the maximum. Included are data on two subjects who began intensive training in 1969. Their improvement in maximal oxygen uptake during the first year is quite remarkable, but later their improvement leveled off.

## Adaptations in the Oxygen-Transporting System

Endurance training increases both the blood volume and the total hemoglobin (Hb) so that the Hb concentration usually is maintained constant. Plasma volume increases after a few days of training, whereas the expansion of erythrocyte volume takes longer. Cross-sectional studies show that there is a close relationship between $\dot{V}O_2$max on one hand and blood volume and total amount of Hb on the other (Harrison 1985). There are also indications that the capillary-muscle fiber interphase is regulated in direct proportion to fiber mitochondrial volume or maximal $O_2$ demand. For discussion, see Mathieu-Costello and Hepple (2002).

Acute plasma volume expansion increases the stroke volume during submaximal and maximal exercise in well-trained individuals (Gledhill 1985; Kanstrup and Ekblom 1982; Krip et al. 1997). Because maximal heart rate also remains unchanged, maximal cardiac output is increased. This increase in well-trained athletes is just about enough to compensate for the reduced Hb concentration and thereby reduced oxygen content of arterial blood, so that the $\dot{V}O_2$max is mainly unchanged compared with control experiments. However, in untrained or moderately trained individuals, a corresponding plasma volume expansion can increase maximal cardiac output more than what is needed to compensate for the reduction in Hb concentration during maximal exercise and, thus, increase $\dot{V}O_2$max (Ekblom and Berglund 1991).

It is well known that well-trained endurance athletes often have a lower Hb concentration than nonathletes (Hunding et al., 1981). Why isn't this relative anemia compensated for? Citing the training-related increase in erythrocyte diphosphoglycerate (2,3-DPG), which shifts the oxygen dissociation curve of the Hb to the right (Hespel et al. 1988), thereby increasing the delivery of oxygen to the tissue, Hallberg and Magnusson (1984) launched the following hypothesis: The sensor responsible for the erythropoietin level, regulating the rate of red cell production, will receive the same information about the oxygen delivery potential of the arterial blood of the athletes as if, at a normal level of 2,3-DPG, their Hb concentrations were normal. In other words, the regulatory mechanisms are cheated. The increased level of 2,3-DPG therefore would be associated with a reduction in the production of erythropoietin and thereby would induce a lower Hb concentration. These events would explain the development of a "sports anemia" even if the supply of iron is sufficient.

## Cardiovascular Adaptations to Training

Like skeletal muscle, the heart is adaptable to variations in the individual's physical activity. It is common for athletes in endurance events to have a large heart volume (figure 11.21). In their summary of eight longitudinal studies of endurance training in sedentary individuals, Peronnet et al. (1982) reported a 2.5% increase in left ventricular end-diastolic volume at rest. It is a very modest increase, but expressed in volume changes it is a 16% increase. The average gain in maximal oxygen uptake was actually 17%. On the basis of cross-sectional studies in both female and male endurance trained athletes, Stolt et al. (2000) found that total heart volume was generally 15% to 25% larger than in sedentary size-matched controls, with morphological differences seen in both the ventricles and atria.

The increase in maximal cardiac output following endurance training results from a larger stroke volume, whereas maximal heart rate is unchanged or even slightly reduced. Although heart size is a function of total body size as well as genetic factors, the higher stroke volume elicited by endurance training is attributed to enlargement of cardiac chamber size and to expansion of total blood volume (Pelliccia et al. 1991). The cardiac hypertrophy depends on the type of sport carried out. There are mainly two types of myocardial hypertrophy. In weightlifters and other strength-training athletes, heart wall thickness is increased with only minor increases in heart cavity diameters, whereas endurance athletes have increased heart volume and cavity diameter with a proportional increase in wall muscle thickness (Huston, Puffer, and Rodney 1985). The ratio of wall thickness to cavity diameter is unchanged in the endurance-trained individual but increased as a result of strength training (Morganroth et al. 1975). Deconditioning from elite sport reduces cardiac size and volume toward, if not back to, what is normal for age and sex (B.J. Maron et al. 1993). The cardiac morphology of the female athlete heart is

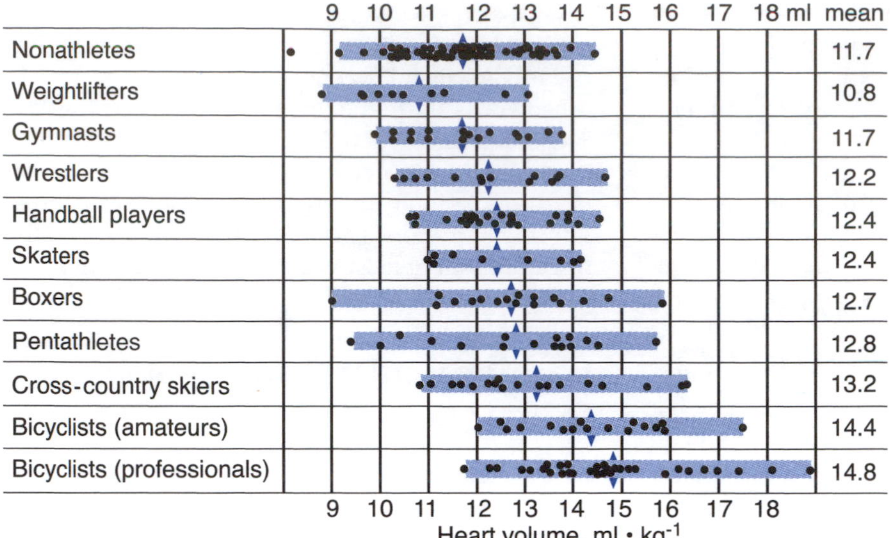


**Figure 11.21** Heart volume per kilogram of body weight for members of different German national teams and a group of untrained individuals of the same age. Each dot represents one subject. Note the large individual variations within each group.

From Roskamm 1967.

the same as in males, but the dimensions are in general smaller (Pelliccia et al., 1996).

In addition to producing structural adaptations, endurance training produces functional improvements in cardiac performance during exercise. Most notable is a more rapid early and peak ventricular filling as well as peak filling rate during diastole. An enlarged blood volume, greater ventricular compliance and distensibility, and a faster and more complete ventricular relaxation are important factors allowing stroke volume to increase even at high heart rates during exercise. Improved myocardial relaxation allows for a more rapid lowering of ventricular pressure, optimizing the left atrial/ventricular pressure gradient for enhanced filling (Brandao et al. 1993; Gledhill, Cox, and Jamnik 1994). The muscle pump facilitates venous return. As a result of an enlarged end-diastolic volume, left ventricular systolic performance is improved mainly by way of the Frank-Starling mechanism (B.D. Levine 1993).

The enhanced diastolic filling and reduced afterload, attributable to lower peripheral resistance, ensure that stroke volume is maintained or even progressively increased from submaximal through maximal exercise, compared with the heart of a sedentary person, in which stroke volume plateaus at submaximal intensities and may even decrease as maximal exertion is approached (Gledhill, Cox, and Jamnik 1994; Spina et al. 1992).

Animal experiments have demonstrated an enhanced vascularization of the heart muscle as a consequence of training, in both capillary density (Hudlicka 1982) and the coronary vascular bed as a whole (C.G. Blomqvist and Saltin 1983). Studies in animals also have shown that endurance training can increase maximal coronary perfusion per unit mass of the myocardium (Laughlin, Overholser, and Bhatte 1989). In a comparison of the cross-sectional area of proximal coronary arteries from endurance-trained and sedentary humans, it has been suggested that coronary vascular volume is increased by training (Haskell et al. 1993). However, it remains unresolved in humans whether endurance training increases coronary vascular dimensions beyond the vascular proliferation that accompanies normal training-induced cardiac hypertrophy.

There is evidence that exercise training elicits changes in vascular tone leading to an optimized distribution of myocardial blood flow, whereby more capillaries are recruited without a change in capillary density (Heiss et al. 1976; Laughlin, Overholser, and Bhatte 1989). This is probably attributable to specific endothelium-mediated vasodilation. Results from animal studies suggest that increased endothelial cell **nitric oxide synthase** (eNOS), an enzyme that synthesizes nitric oxide (NO) from L-arginine, contributes to such an adaptation (Sessa et al. 1994; Woodman et al. 1997).

It has long been established that a distinct cardiovascular adaptation to endurance training is a lowering of the heart rate at rest and during submaximal exercise. Thus, Hoogerwerf (1929) found a mean heart rate of 50 beats · $min^{-1}$ in 260 athletes participating in the Amsterdam Olympic Games in 1928, the lowest value being 30 beats · $min^{-1}$. Observations in Scandinavian cross-country skiers have revealed resting heart rates as low as 28 beats · min$^{-1}$ (P.-O. Åstrand, unpublished results).

The lowering of resting and submaximal heart rate is mediated by alterations in the autonomic nervous system, as well as by changes in the intrinsic automaticity of the sinus node and right atrial myocytes (Ekblom et al. 1973; Lewis et al. 1980). During exercise in the trained individual, a given increase in cardiac output requires less increase in heart rate attributable to the maintenance of a larger stroke volume. Studies focusing on autonomic and endocrine responses to training indicate that heart rate is reduced during submaximal exercise in the trained person because of a lower intrinsic heart rate, a reduction in sympathetic activity and circulating catecholamines, and a greater parasympathetic (vagal) tone (Lewis et al. 1980; Tulppo et al. 1998). Tulppo et al. found that higher levels of physical fitness were associated with an augmentation of cardiac vagal function during exercise, whereas aging results in more evident impairment of vagal function at rest. The lower sympathetic activity to the heart at a given submaximal work rate stems, in part, from diminished reflex signals originating from skeletal muscle attributable to less metabolite accumulation and attenuated discharge of **metaboreceptors** (Mostoufi-Moab et al. 1998).

The mechanisms underlying the training-induced increase in vagal tone are thought to be increased activation of the cardiac baroreceptors in response to the enlargement of blood volume and ventricular filling (Convertino, Mack, and Nadel 1991; B.D. Levine 1993), as well as change in opioid modulation of parasympathetic activity (Angelopoulos et al. 1995). It is not fully resolved whether a lowering of intrinsic heart rate is a true adaptation to endurance training, but it appears that an intensive and lengthy training period is necessary for this adaptation (Bonaduce et al. 1998).

Many patients are treated with β-adrenergic blockers. How do they respond to physical training? Most studies performed on patients, healthy individuals, and animals report similar improvement in maximal oxygen uptake in subjects on chronic β-adrenergic blockage as in nontreated controls. The hemodynamic responses to milder exercises were also similar in the two groups, but at higher rates of exercise the reduction in heart rate after training seems to be less pronounced in the subjects on medication (Svedenhag et al. 1984; Vanhees, Fagard, and Amery 1982). On the basis of experiments on rats, Mullin and associates (1984) came to the same conclusion. Apparently, the stimulation of cardiac β receptors or the magnitude of the heart rate increase as such during the training sessions is not essential for the development of training bradycardia (Nylander 1985).

Robinson and Harmon (1941) documented that a period of training could increase subjects' maximal oxygen uptake. Since then, numerous studies have been devoted to the effect of training programs of different intensity, frequency, and duration on the oxygen transporting system. In sedentary people, low-intensity training with sessions lasting about 30 min, repeated three times per week, and demanding approximately 50% of the maximal oxygen uptake can increase the maximal oxygen uptake 5% to 10% after 6 to 12 weeks. With more intensive training demanding 70% to 80% or more of the maximal aerobic power, a 10% to 20% improvement is a common finding. As we pointed out when discussing the Dallas study, a given training program could increase the maximal oxygen uptake from 4% to 100%, depending on the pretraining conditions. In addition, there are large individual differences in the response to training (Lortie et al. 1984; Pollock 1973). When we examine improvements in percentage of maximal oxygen uptake, it is not surprising that those individuals who start with a relatively low level of fitness improve the most. However, natural endowments set a final ceiling for the improvement (figure 11.20).

Studies including individuals up to 70 years of age give similar figures for training-induced increases in maximal oxygen uptake as in young and middle-aged populations (Seals et al. 1984). The normal decline in maximal oxygen uptake with age beyond 20 years apparently can be modified by regular training, corresponding to a rejuvenation by some 15 to 20 years (see figure 8.13). In the group of former physical education students included in figure 8.13, restudied 20 years later, a decline in maximal oxygen uptake was noted without exception, although most of them were still very active physically. Data from repeated measurements another 12 years later showed, surprisingly, no further decline in mean maximal oxygen uptake in the 27 women who took part in the study and just a few percent reduction in the 26 men.

Few studies are available analyzing the effect of interval and distance training in prepubescent children. One problem is to separate the effects of growth and of training, which makes control groups important. When young boys engaged in strenuous training programs (such as ice hockey) were compared with groups who did not participate in regular training programs, the maximal oxygen uptake (L · min$^{-1}$) was not significantly different (P. Hamilton and Andrew 1976). This was also the case in longitudinal studies (Lussier and Buskirk 1977; Shephard

1982). Kobayashi et al. (1978, p. 666) noted that "a remarkable increase in aerobic power was not observed in trained boys before the age of peak height growth velocity" (which is related to the onset of puberty, see chapter 8, figure 8.12). More studies are needed with different training intensity, duration, and frequency to provide more comprehensive knowledge of the potential to improve aerobic power by training in children.

The maximal oxygen uptake during one-leg exercise on a cycle ergometer is normally 75% of the maximal oxygen uptake attained when cycling with both legs. Training one leg increases the maximal oxygen uptake in the one-leg exercise some 15% to 20%, but there is no change with the untrained leg. The two-leg maximum is only 5% to 10% higher after the period of training. The maximal oxygen uptake during exercise with the trained leg is actually 85% to 90% of the two-leg maximum (C.G. Blomqvist and Saltin 1983; C.T.M. Davies and Sargeant 1974). The conclusion is that the training effect of one-leg exercise is mainly peripheral and confined to that particular leg, but in two-leg exercise the central cardiovascular system limits the oxygen uptake (as discussed subsequently).

What central circulatory adjustments can explain the training-induced increase in maximal oxygen uptake? According to the **Fick principle,** increased cardiac output and/or $(a-\bar{v})O_2$ difference will increase the maximal oxygen uptake. In longitudinal studies of sedentary young men, an increase in maximal cardiac output and systemic $(a-\bar{v})O_2$ difference both contributed to the increased $\dot{V}O_2max$, often at a 1:1 ratio. With prolonged training resulting in a further increase in the maximal aerobic power, an increment in the cardiac output is the cause of the increase. It should be recalled that there is no significant difference in the maximal $(a-\bar{v})O_2$ difference between well-trained athletes with very high maximal aerobic power and trained subjects with much lower $\dot{V}O_2max$. Within each sex, the individual variations in maximal stroke volume are much larger than differences in $(a-\bar{v})O_2$ difference.

There are few longitudinal studies of the effect of training on the central circulation in women and older men. Kilbom (1971) reported no increase in $(a-\bar{v})O_2$ difference in women. For older men, there are similar observations (C.G. Blomqvist and Saltin 1983), but an increase in $(a-\bar{v})O_2$ difference has been noted (Seals et al. 1984).

The cardiac output during submaximal exercise at a given rate of work does not change as a result of training. Well-trained endurance athletes attain the same cardiac output as sedentary individuals. As mentioned, the heart rate, at a given oxygen uptake, is reduced. Whether the primary event is the increase in stroke volume or decrease in heart rate is at present impossible to say, as discussed previously. The larger posttraining left ventricular end-diastolic volume is a consequence of the prolonged diastole.

There is general agreement that endurance training slightly reduces resting blood pressure (Fagard 1995; Kelley and Tran 1995). In addition, long-term exercise training has the beneficial effect of preventing the normal age-related increase in blood pressure. A pressure-lowering effect of endurance training has been shown to occur within 6 days after initiation of an exercise program (Meredith et al. 1990). Reduced adrenal medullary catecholamine output during exercise at a given absolute work rate may be important for the blood pressure–lowering effect of training, as well as changes in sympathetic and renal dopaminergic activity.

The reduction in resting diastolic blood pressure with training is significantly related to the increase in exercise capacity, which suggests that high-intensity training is important. Because the mean blood pressure is the product of the cardiac output times the peripheral resistance, this resistance must be reduced in proportion to the increase in cardiac output. During exercise at a given submaximal load, blood pressure and vascular resistance are reduced after endurance training. This adaptation is associated with reduced sympathetic activation and lower circulating catecholamines. At high exercise intensities and at maximal exercise, blood pressure is generally similar before and after training. Yet a given blood pressure is achieved by a lower vascular resistance and a higher cardiac output in the endurance trained. Thus, well-trained athletes with cardiac output around 40 $L \cdot min^{-1}$ during maximal exercise can have the same blood pressure as less active subjects with much lower maximal oxygen uptake.

## Pulmonary Adaptation to Training

As first systematically studied by Ringqvist (1966), the respiratory muscles show considerable variation in strength and endurance, in relation to stature, age, and sex, with fairly obvious implications in the capacity to achieve and maintain ventilation during exercise and also in the sense of effort experienced in breathing. Subjects with stronger respiratory muscles achieve higher tidal volumes (and thus lower breathing frequencies) and experience less dyspnea than those with weaker muscles (A.L. Hamilton et al. 1995).

In terms of dimensions, the maximal breathing capacity is a function of the total lung volume and the maximal flow rates in inspiration and expiration; volume is related to thoracic volume and flow rates to airway cross-sectional area. For a given stature and weight, both volume and maximal flow tend to be larger in athletes, but studies in twins suggest that this is based on genetics and that training has little influence (Weber, Kartodihardjo, and Klissouras 1976). Within these constraints, athletes use a larger volume by being able to achieve a smaller end-expiratory and larger end-inspiratory volume. They also use larger flow rates in both inspiration and expiration. Indeed, some athletes are capable of using virtually all their maximal flow-volume loop during exercise (G. Grimby, Saltin, and Wilhelmsen 1971; B.D. Johnson et al. 1991). It seems likely that this is because of stronger and more fatigue-resistant respiratory muscles. In older subjects, there is a loss of elastic recoil that reduces flow at low lung volumes, prevents the subjects from reducing end-expiratory volume, and contributes to an increase in respiratory effort in older athletes (B.D. Johnson, Saupe, and Dempsey 1992).

Athletes breathe slower and deeper than nonathletes, and a slower breathing rate is one of the effects of training. The volumes and flows determine the tidal volume and frequency of breathing during exercise; all other matters being equal, larger tidal volumes and slower frequencies lead to greater efficiency in breathing. Some subjects entrain breathing frequency with their pedaling or running cadence (Bechbache and Duffin, 1977), but there is scope for wide variation in such responses (e.g., four strides per breath vs. three strides per breath), so that the entrainment never dominates the pattern. Although the use of a given pattern of breathing is often assumed to be a self-optimizing response to minimize the oxygen cost of breathing, it seems more likely that patterns are adopted consciously or unconsciously to minimize the sense of effort in breathing (Casan et al. 1997).

Respiratory muscle oxygen consumption was once thought to increase disproportionately at high levels of ventilation and thus to limit maximal oxygen intake, but more recent estimates suggest that patterns of breathing are adopted mainly to minimize dyspnea. Because exercise is a voluntary activity, conscious humans stop exercising when the sense of excessive effort and weakness in exercising muscle or of dyspnea becomes intolerable. A number of sensations related to breathing can be discriminated and scaled (Killian, Jones, and Campbell 1999), including inspiratory muscle tension and displacement, a sense of "satiety" or appropriateness related to increases in arterial partial pressure of carbon dioxide (Casan et al. 1997; Manning and Schwartzstein 1999), and a sense of effort related mainly to central outgoing command (Gandevia 1998).

Both effort and dyspnea are less in trained compared with untrained individuals, enabling them to maintain higher power for longer. The factors accounting for these differences are numerous and interactive. The reasons for dyspnea are mainly that the pressure generated by the respiratory muscles is relatively less in subjects with large lungs (because of lower elasticity and resistance) and strong respiratory muscles.

Resistance training is known to improve inspiratory muscle strength and endurance (Pardy et al. 1981). This has important implications in aging athletes in whom there is the normal decline in lung elasticity; end-expiratory lung volume cannot be reduced to the same extent as in the young, forcing the inspiratory muscle to carry a greater proportion of the respiratory work during exercise (B.D. Johnson et al., 1991). B.D. Johnson et al. found that reductions in elastic recoil and increases in end-expiratory volume paralleled reductions in $\dot{V}O_2$max in an older (69 ± 1 years) population; such subjects are likely to have been limited by dyspnea.

During submaximal exercise of relatively low intensity, the pulmonary ventilation per liter of oxygen consumed does not change materially with training. Apparently, the depth of respiration is increased somewhat, associated with a corresponding reduction in the respiratory rate. During heavier exercise, the ventilation per liter of oxygen uptake is reduced but reaches a higher level during maximal exercise. Figure 6.9 helps explain these findings. During light exercise, the level of the pulmonary ventilation is primarily determined by the $CO_2$ production, which is directly related to oxygen utilization. During heavier exercise, the pH is also altered, primarily by the end-products from the glycogenolysis, which causes a relatively steeper increase in the pulmonary ventilation. Because the blood lactate concentration is generally lower during submaximal exercise following training, the respiratory drive is reduced and the result is lower pulmonary ventilation. The higher maximal ventilation is partially caused by the increased maximal aerobic power, leading to an increased $CO_2$ production, and is also attributable in part to the higher maximal lactic acid level. Thus, the untrained individual and individuals with low maximal aerobic power fall within the left side of the shaded area in figure 6.9, whereas training moves the curve downward to the right.

## *Summary*

The magnitude of the increase in maximal oxygen uptake during 2 to 3 months of training, 30 min each session, three times per week, is in the order of 10% to 20%, but with large individual variations. Sedentary unfit persons will improve most. The frequent recommendation that the training sessions should be conducted three times per week is explained partly by the fact that this regimen has been predominantly used in research. Furthermore, as mentioned in the introduction, training six times per week is not twice as effective as training three times per week in improving aerobic power; there is an element of cost–benefit philosophy in recommending training of the nonathlete. Both interval training and more continuous exercise are effective.

There is a dual basis for an increased maximal oxygen uptake: an increased maximal cardiac output and an increased $(a\text{-}\bar{v})O_2$ difference. In younger subjects with initially low maximal oxygen uptake, there is an equal increase in cardiac output and $(a\text{-}\bar{v})O_2$ difference during the first 2 to 3 months. With continued intensive training, a further increase in maximal aerobic power is explained by an additional increased cardiac output. Because the maximal heart rate does not change significantly during training, the elevated maximal cardiac output is exclusively attributable to an increase in stroke volume.

During submaximal exercise at a given aerobic power, the cardiac output does not change, the heart rate is lowered, and stroke volume increases. There is also a reduced heart rate at rest. The systemic arterial blood pressure is moderately reduced or unaffected by the training. Different stages of physical fitness have a minor effect on the pulmonary function.

Endurance training increases heart volume and heart muscle mass involving all chambers, whereas strength training increases an athlete's heart wall thickness with only minor increases in heart cavity diameters. There is a moderate increase in the plasma volume and total amount of Hb, and Hb concentration is maintained or in some cases reduced. Because of the reduced heart rate, there is a diminished demand on the oxygen consumption and thereby on the blood flow through the heart muscle at a given cardiac output. There are good indications that the coronary vascular bed is increased, at least at the capillary level.

Training-related reductions in ventilation closely parallel reductions in both $\dot{V}_{CO_2}$ and plasma lactate concentrations. Although the process is to some extent genetically determined, athletes use a greater ventilation volume by being able to achieve a smaller end-expiratory and larger end-inspiratory volume, mainly because of stronger and more fatigue-resistant respiratory muscles. Resistance training improves inspiratory muscle strength and endurance. Subjects with stronger respiratory muscles achieve higher tidal volumes (and thus lower breathing frequencies) and experience less dyspnea than those with weaker muscles, thus enabling them to maintain higher power for longer.

## Peripheral Adaptation to Training

Peripheral adaptation to aerobic training is particularly evident in the trained skeletal muscle. The muscle is a very plastic tissue, also with regard to aerobic training.

Training the muscles of one limb while the contralateral limb serves as a control is a useful model for studying the effect of physical training. The advantage of this model is that both the trained and the control tissues in the posttraining experiments are exposed to the same supply of circulating substrates, hormones, and other signal molecules. In exercise involving a large muscle mass, the volume of oxygen offered to the skeletal muscles by the central circulation is a limiting factor. Drawing on experiments that included maximal dynamic exercise involving just the quadriceps muscles in knee extensions, Andersen and Saltin (1985) concluded that the potential for oxygen uptake was $0.3\ L \cdot kg^{-1}$ of tissue $\cdot\ min^{-1}$ and that the capacity of the skeletal muscles to use the blood flow exceeded the capacity of the heart to provide that flow by a factor of two to three. Therefore, a locally improved aerobic performance would be of no advantage in a "whole-body exercise" unless the central circulation was improved.

Another model is to train each leg separately and then to analyze the situation during maximal one-leg and two-leg exercise. In the first case, it was found that the maximal oxygen uptake increased by 19% and the cardiac output by 16%. In the two-leg exercise, the improvement was not so large, 11% and 11%, respectively (Klausen et al. 1982). The mean systemic blood pressure decreased in all exercise situations by approximately 10 mmHg (1.3 kP). One interesting finding was made: The left leg was exercised at a high work rate and continued to do so when the right leg started a similar exercise. Then, the blood flow in the left leg decreased because of an increased vascular resistance in the leg. As a consequence, a more pronounced metabolic acidosis developed. Evidently, there is a sympathetic vasoconstriction even in maximally contracting muscles. These data are important as a basis for discussing the effect of training on the regulation of regional blood flow. Before proceeding with a more detailed discussion, let us list some of the effects of endurance training on the skeletal muscle. There is an increase in the following:

- Capillary density
- Mitochondrial volume, with an increase of the enzymes involved in the citric acid cycle and the electron-transport chain
- Use of FFAs at an energy demand related to the muscle's maximal aerobic power
- Potential to store glycogen
- Local $(a-\bar{v})O_2$ difference at submaximal rates of exercise
- Maximal flow rate through the muscle
- Possibly also myoglobin concentration, although this remains controversial

The capillary density in skeletal muscle increases during periods of endurance training (figure 11.22). It is not known whether there are mechanical factors connected with increased blood flow, metabolic factors connected with a greater demand for oxygen, and/or accumulation of metabolites that trigger the capillary growth. In contrast, training with the aim of improving maximal muscular strength does not increase capillary density. Because of an increase in the muscle fiber area, strength training actually can reduce the capillary density (Saltin and Gollnick 1983; Tesch, Thorsson, and Kaiser 1984), in the same way as it can reduce mitochondrial density (Chilibeck, Syrotuik, and Bell 1999). An increase in capillary density reduces the distance between the blood and the cell interior, which enhances the exchange of gases, substrates, and metabolites. The surface area available for this exchange will increase. With more capillaries in a given tissue volume, more blood can be accommodated in this vascular bed per unit of time; the mean transit time is reduced, which allows a more complete exchange

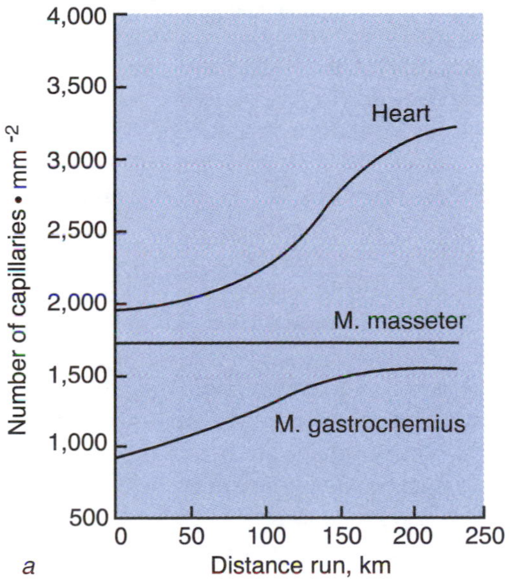


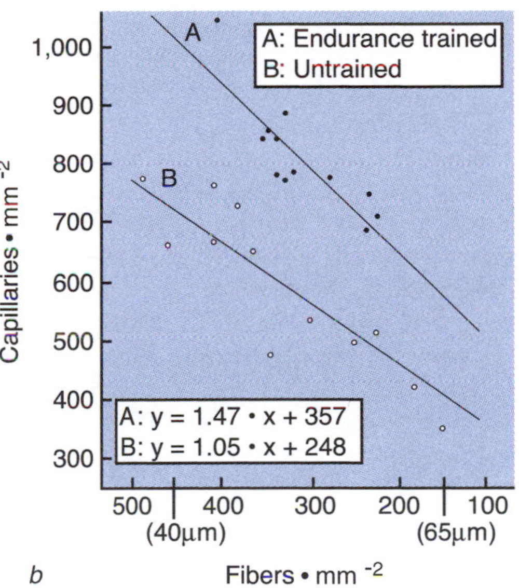


**Figure 11.22** Increased vascularization of muscle as a result of training. *(a)* Capillary density in the heart and two skeletal muscles of guinea pigs in the course of daily training on a treadmill at a speed that was gradually increased to 60 m · min$^{-1}$, running distance about 1,800 m each time. Animals were analyzed after different training times. There was no change in capillary density in the masseter, a masticatory muscle not involved in the exercise. Adapted from Petrén (1936). *(b)* The relation between capillary density (capillaries per square millimeter) and the fiber area (fibers per square millimeter). Each point represents values from one person, untrained (○) or endurance trained (●). Two diameter values are indicated on the abscissa to show the relationship between fiber size and number of fibers per square millimeter. In both trained and untrained subjects, there is a clear reduction in capillary density with increasing fiber size. An endurance-trained person may even have fewer capillaries per square millimeter than an untrained person if his or her fiber diameters are sufficiently different.

*(b)* Courtesy of P. Brodal, F. Ingjer, and L. Hermansen.

of material between the blood and the interstitial fluid. Saltin (1985) pointed out that the primary advantage of the high capillary density in highly trained skeletal muscle is probably that it allows for an adequate mean transit time at high flow rates, thereby promoting the exchange of material. An increase in the myoglobin content of the trained muscle will enhance the diffusion of oxygen. There is actually a high correlation between the capillary density in the skeletal muscles and their maximal oxygen uptake. Thus Brodal, Ingjer, and Hermansen (1977) found that the mean number of capillaries per fiber in the vastus lateralis muscle was 32% higher in a trained than in an untrained group of individuals. The difference was highly significant ($p < .001$) and was almost of the same order of magnitude as the difference in maximal oxygen uptake between the two groups (i.e., 40%). The number of capillaries per square millimeter was 39% higher in the trained group than in the untrained group, the difference being highly significant ($p < .001$).

In a subsequent longitudinal study, Ingjer (1979b) obtained muscle biopsies from seven women before and after a 24-week systematic endurance training program that caused an average increase in $\dot{V}O_2max$ of 25%. A corresponding increase in the number of capillaries occurred. The number of capillaries per muscle fiber increased on the average from 1.39 to 1.79 (29%), and the number of capillaries per square millimeter increased on the average from 348 to 438. Thus, it appears that a number of new capillaries had been formed in the muscle during the training period, evidently as the result of the training itself. Ingjer observed no increase in the cross-sectional area of the muscle fibers, indicating that endurance training of this kind does not stimulate an increased thickness of the muscle fiber.

One of the most important metabolic effects of training of the oxygen-transporting system is an increased capacity of the oxidative pathways. This is reflected by a large increase in the number and size of mitochondria in the trained skeletal muscles (Howald 1982; Saltin and Gollnick 1983). There is a more or less parallel increase in the activities of the oxidative enzymes, including those of the citric acid cycle and electron transfer chain, and a slightly less increase in 3-hydroxy-acyl-CoA-dehydrogenase (HAD), the enzyme involved in the β-oxidation of the FFAs. At one time, it was assumed that the enzyme activity limited the maximal aerobic power. Experimental evidence seems to indicate, however, that the supply of enzymes is abundant, considering the limited volume of oxygen that can be offered during maximal exercise. Thus, Henriksson and Reitman (1977) noticed a 19% increase in maximal aerobic power in 13 subjects who had trained for endurance for 8 to 10 weeks. The activities of SDH and cytochrome oxidase in the vastus lateralis muscle had increased 32% and 35%, respectively. Within 2 weeks posttraining, the cytochrome oxidase activity had returned to the pretraining level and after 6 weeks the SDH activity was back to the control level. In contrast, the maximal oxygen uptake was still 16% above the pretraining level. The authors concluded that an enhancement of the oxidative potential in skeletal muscle is not necessary for a high maximal oxygen uptake. A major problem with their conclusion, however, is that the muscle biopsies represent a small and relatively superficial part of the quadriceps muscle only, while the major part of the oxidative potential resides in the deeper part of the muscle.

SDH, involved in the Krebs cycle, has been very popular in studies of the dynamics of chronic exposure to activity and inactivity. Part of its popularity is attributable to the fact that it is easy to assay and to visualize histochemically. It does not catalyze a nonequilibrium reaction, however, and its value as a measure of the maximum capacity of the Krebs cycle is doubtful (Newsholme and Leech 1984). As a semiquantitative measure, an increase in its activity can be taken as an indication that a certain training program has been effective. Henriksson, Salmons, and Lowry (personal communication) found that 10 weeks of continuous stimulation (10 Hz) of a muscle triggered a much more dramatic increase in enzyme activity than did physical training.

It is an old observation that the trained individual can cover more of the energy demand during exercise at a given rate of oxygen uptake with FFA and triglycerides as substrate and thereby save glycogen and glucose (see chapter 12). Training one leg and using the other as control, Henriksson and Reitman (1977) showed conclusively that the degree of FFA used was higher in the trained leg than in the untrained leg, indicating a difference in preference for substrate. This observation has since been frequently confirmed (Costill et al. 1979; Kiens et al. 1993; Saltin and Gollnick 1983).

Our knowledge of the regulatory mechanisms behind changes in mitochondrial number and enzyme activity is restricted. This lack of basic knowledge is an obstacle to further progress (Essig 1996). From rat experiments, Henriksson et al. (1985) concluded that neither adrenomedullary hormones nor local sympathetic nerves are necessary for the posttraining increase in muscle mitochondrial enzymes. Mitochondria are never made de novo but arise by growth and division of existing mitochondria (Alberts et al. 1994; Essig 1996). Essig even contended that muscle mitochondria may exist as a single reticulum rather than as separate organelles. Mitochondrial biogenesis may be the result of repeated bouts of endurance exercise and requires a coordinated expression of both nuclear and mitochondrial genes. After their synthesis, mitochondrial proteins originating from the nuclear genome have to be imported into the mitochondria, but the details of the mechanisms involved in mitochondrial biogenesis are so far enigmatic.

## For Additional Reading

For an overview and discussion of exercise-induced mitochondriogenesis, see Essig (1996). Tonkonogi and Sahlin (2002) offer an update on mitochondrial adaptation to exercise.

How, then, can the enhanced fat metabolism in a trained muscle be explained? It is not possible to answer this question completely either, but some factors might contribute. It was mentioned that both an increased capillary bed and a reduced mean transit time of blood favor a transport of substrates. The physiological site of the action of lipoprotein lipase (LPL) is in the luminal surface of the capillary epithelium. With an increased capillary bed, more binding sites for LPL are available. This enzyme has a much higher activity in a trained muscle compared with an untrained muscle; in fact, it may be as much as a 40% increase (Kiens and Lithell 1989). The degradation of triglyceride-rich particles largely depends on the activity of LPL. More FFA may be available for the exercising muscles, but in addition the LPL activity will transfer more surface material into high-density lipoprotein (HDL) and transfer it into the plasma compartment. This mechanism could explain the elevated plasma HDL concentration in endurance-trained individuals (Kiens, Lithell, and Vessby 1984).

Inside the cell, an increased activity of HAD may enhance β-oxidation of FFA. Eventually, an increased concentration of acetyl-CoA from the FFA may reduce the demand for acetyl-CoA from pyruvate, thereby reducing the breakdown of glycogen. A reduced flux through the anaerobic pathway can be brought about by an increase in the ratio between ATP and adenosine diphosphate, which are an inhibitor and a stimulator of flux-generating enzymes in the glycolytic pathway, respectively (Newsholme and Leech 1984). More than enough FFA is offered to the exercising muscle, and only a small percentage of the offered substrate (some 5%) actually is taken up by the muscle (Saltin and Gollnick 1983). At any rate, in the endurance-trained muscle, an increased FFA metabolism saves glycogen, which will enhance performance in prolonged exercise.

The glycogen content of a muscle is very much affected by diet. Endurance training more or less empties the glycogen stores in the muscle fibers, and the diet of hard-training athletes is often rich in carbohydrates. These two factors alone might explain the slightly increased glycogen stores in trained muscle. In the trained muscle, the activity of glycogen synthetase and the glycogen branching enzyme increases, which could explain the enhanced potential for storing glycogen in trained muscles (Holloszy and Booth 1976).

With endurance training, a smaller fraction of the cardiac output is distributed to muscles activated at a given rate of oxygen demand. A plausible explanation is that an increased capillary bed allows a larger blood volume to be exposed to the skeletal muscles per unit of time, securing an exchange of gases, substrates, metabolites, and heat energy. Such a modification of the blood distribution would allow for a more liberal distribution of blood flow to the liver, kidneys, and skin, which would be an advantage, particularly during prolonged exercise. During maximal exercise, the trained muscle receives more blood per unit of time than untrained muscle. The increase in peak muscle blood flow appears to be achieved by enhanced endothelium-dependent dilation (EDD) in the muscle that increases its vasodilator capacity in parallel with expanded oxidative capacity. Accordingly, the increase in cardiac output can occur without any rise in arterial pressure. An enhanced peak hyperemic blood flow appears to be an early adaptation to regular exercise (Laughlin, Overholser, and Bhatte 1989; Sinoway et al. 1987). Clarkson et al. (1999) showed a nearly 80% increase in flow-mediated EDD of the brachial artery after 10 weeks of aerobic and anaerobic training. Furthermore, a high correlation between maximal aerobic power and peripheral vasodilator capacity, measured by vascular conductance, has been demonstrated (W.H. Martin, et al. 1990; Snell et al. 1987). Kingwell et al. (1996) found an ~30% greater reduction in forearm vascular resistance to an endothelium-dependent stimulus in endurance athletes compared with sedentary subjects. This reduction was directly related to maximal aerobic power. Also in endurance-trained older people, a significantly greater EDD, compared with age-matched sedentary subjects, has been observed

(Rywik et al. 1999). Additionally, Rinder, Spina, and Ehsani (2000) found that abnormal EDD discovered in older, otherwise healthy individuals could be attenuated with long-term endurance training. They also noted a significant and reasonably good correlation between maximal aerobic power and EDD.

The mechanisms behind the enhanced endothelial function associated with physical training could involve exercise-induced increases in shear stress and pulsatile flow. According to Niebauer and Cooke (1996), chronic increases in blood flow induced by training affect EDD by modulating the expression of eNOS. It has been shown that endothelium-derived NO influences vascular tone in the periods between exercise bouts. In animal studies, reactivity to stimuli that mediate their effects via NO is increased by training in coronary circulation, as mentioned previously in this chapter (Sessa et al. 1994; Woodman et al. 1997). Human studies have produced evidence for a role of NO in the regulation of muscle blood flow (Duffy et al. 1999; Rådegran and Saltin 1999). NOS exists in several isoforms, one of which is eNOS. Another isoform, called neuronal NOS (nNOS), is located in the sarcolemma and cytosol of human skeletal muscle fibers, in apparent association with mitochondria (Frandsen, Lopez-Figueroa, and Hellsten 1996). Frandsen, Lopez-Figueroa, and Hellsten (2000) showed that endurance training can increase the amount of eNOS in parallel with an increase in capillaries in human muscle, while the nNOS levels remain unaltered.

### FOR ADDITIONAL READING

For further discussion of microcirculatory adaptation to metabolic demand in skeletal muscle, see Hepple (2000).

## Hormonal Responses to Exercise

Humans, like all higher animals, react to almost any kind of threat by mobilizing their resources for a physical effort. In 1929, Walter Cannon popularized the concept of the "fight or flight" response, later named the stress reaction. The sight or awareness of any kind of challenge automatically evokes nervous and hormonal reactions, mobilizing the bodily resources for a physical response (Cannon 1929). The purpose of these reactions is primarily to prepare the organism for a physical effort to cope with a threatening or demanding situation, as was the case when the Stone Age hunter suddenly encountered a bear in the wilderness. Immediately, the inherent physiological regulatory mechanisms were brought into action to meet the needs of the body, be it a fight with the bear or a run for safety.

Impulses from motor centers in the brain and working muscles elicit hormonal responses that primarily depend on the physical activity the situation calls for and the intensity and duration of the work. In general, the hormonal responses to exercise are positively correlated to the work intensity as long as this is expressed in relative terms, such as a percentage of maximal oxygen uptake (Galbo 1995). As a result of the hormonal responses, the pulmonary ventilation increases, as do the heart rate and cardiac output, increasing the oxygen uptake and the oxygen delivery to the muscles. Much of the circulating blood volume is shifted from the gut and skin to the muscles. Stored energy is released in the form of glucose and FFAs, now being made available as substrates for the working muscles.

### Catecholamines

Probably, the most important immediate response to physical effort is an increased activation of the sympathoadrenergic system. Nervous stimuli from the CNS to the adrenal medulla cause the hormone adrenaline to be released into the blood. Stimuli reaching the nerve endings in postganglionic sympathetic nerve fibers release the hormone noradrenaline. These two hormones (called catecholamines because they have a six-carbon catechol ring and an amine group) have a number of different physiological effects. Acting on β-adrenergic receptors in the heart, noradrenaline and adrenaline increase the heart rate and the ventricular contractility (the strength of contraction at any given end-diastolic volume). Acting on α-adrenergic receptors on arteriolar smooth muscle cells, both hormones cause vasoconstriction in organs such as the stomach, intestines, and skin. The blood flow to the muscles, however, is increased mainly by the vasodilator effect of circulating adrenaline on β-adrenergic receptors, which dominates in the arterioles of skeletal muscles. Because of the increase in cardiac output and peripheral vasoconstriction, the blood pressure is increased.

Adrenaline acts primarily by stimulating the breakdown of glycogen in the liver and the muscles and of **lipolysis** in adipose and muscular tissue. The liver glycogen is quickly broken down to glucose, which is released to the blood, while glucose resulting from the breakdown of muscle glycogen remains trapped in the individual muscle cell. In adipose tissue, adrenaline and noradrenaline stimulate lipolysis, increasing the concentration of FFAs in the blood. FFAs serve as the main fuel for cells such as those of the heart muscle (Richter and Sutton 1994; Wahrenberg et al. 1987; Yeaman 1990). At a constant work rate, the plasma catecholamine level increases continuously up to the time of exhaustion.

## Insulin and Glucagon

During exercise of increasing intensity and duration, the plasma concentration of insulin progressively decreases, probably because of α-adrenergic inhibition of insulin secretion rather than an increase in insulin clearance (Galbo 1992). This makes the liver more sensitive to the effects of adrenaline and glucagon, thus promoting hepatic glucose output. Hepatic glucose production starts to increase within few minutes after the start of exercise and is directly affected by the relative work rate.

The α-adrenergic inhibition of insulin secretion may explain why the hypoglycemia that follows preexercise sugar intake is not observed when the sugar is consumed during exercise, for instance, during the warm-up period. Furthermore, the increase in sympathoadrenal activity and the decrease in the plasma insulin concentration seem to be the major determinants of lipolysis during exercise (Galbo 1995). Because insulin and muscle contractions have an additive effect on glucose transport in muscle, the decrease in plasma insulin concentration, besides increasing lipolysis, also could be important for limiting muscle glucose uptake and thereby preventing hypoglycemia and delaying exhaustion of hepatic glycogen stores.

The plasma glucagon concentration decreases during brief intensive exercise but starts to increase again if the exercise is continued. The main stimulus for the increased glucagon secretion during prolonged exercise seems to be the decrease in plasma glucose concentration (Galbo 1995). Probably, glucagon plays a significant role in increasing hepatic glucose production during exercise.

Evidently, mechanisms other than changes in plasma insulin and glucagon levels are involved in controlling glucose output during exercise, because a marked increase in hepatic glucose production can take place in the absence of changes in these hormone levels (Galbo 1995).

## Adrenocorticotropic Hormone, Cortisol, and Growth Hormone

The recognition of a threat or a challenge activates nerve cells in the hypothalamus, causing release of peptides known as "releasing factors" or hypophysiotropic hormones. Through blood vessels connecting the hypothalamus with the anterior lobe of the hypophysis (anterior pituitary), the different hypophysiotropic hormones are transported to the hypophysis, where they immediately trigger the production and release into the blood of their respective hormones, including adrenocorticotropic hormone (ACTH, corticotropin) and growth hormone (somatotropin). By the systemic blood circulation, these hormones are in a matter of seconds brought to cells all over the body, changing the behavior of cells that possess the appropriate hormone receptors.

In the adrenal cortex, ACTH stimulates the synthesis and release of the hormone cortisol, which via the blood is spread to all parts of the body. Mental stress immediately before physical activity can as much as double the basal plasma concentration of ACTH and cortisol. Cortisol exerts a profound effect on a number of cellular functions, enabling the organism to mobilize all its resources to deal with a stressful situation. In the liver, it stimulates the breakdown of proteins to amino acids. The carbon skeletons of the amino acids can enter the citric acid cycle, or they can be used for gluconeogenesis, causing an increased blood sugar concentration. At the same time, the protein synthesis in most of the cells in the body is significantly restrained to save energy.

Cortisol makes the smooth muscle cells in arterioles and veins more reactive to adrenaline and noradrenaline. This enhances the vasoconstriction of blood vessels in organs of lesser importance in the body's general reaction to the strain, such as the skin, stomach, and intestines. In turn, this secures as much blood as possible to skeletal muscles, liver, brain, heart, and adrenals. Cortisol also enhances the performance of cardiac muscle cells and can hamper inflammatory reactions in the tissues. Furthermore, it has a protective effect on cellular membranes by counteracting the harmful effects of a shortage of oxygen.

Of special interest is the fact that cortisol is subject to **circadian** rhythmic changes. The serum cortisol concentration is at its lowest level between midnight and 2-4 A.M., followed by a rapid increase during the later part of the morning sleep. The highest cortisol concentration for the entire 24-h period is found just before waking up in the morning. This normal circadian pattern can be disturbed by stress. Serious mental stress, especially conflicts that are not under the subject's own control and that lead to chronic anxiety, can cause a marked and prolonged increase in cortisol production.

In a study of environmental effects on serum cortisol in 577 Russian coal miners in Spitsbergen, Bojko (1997) observed a significant increase in serum cortisol during the transition from polar winter darkness to spring daylight. A corresponding decrease was observed during the transition from daylight to darkness in the fall. This was taken as an indication that seasonal changes in cortisol secretion are mainly determined by changes in daylight.

The effects of growth hormone on the carbohydrate and lipid metabolism are somewhat similar to those of cortisol. Growth hormone increases gluconeogenesis by the liver and renders the adipose tissue more responsive to lipolytic stimuli. It also reduces the ability of insulin to cause glucose uptake by muscle cells and adipocytes and neutralizes the ability of cortisol to inhibit the synthesis of protein in the cells.

## Vasopressin and Aldosterone

Vasopressin, also known as antidiuretic hormone, is produced by hypothalamic neurons and released by the posterior lobe of the hypophysis in response to a physical challenge or stressful situation. It also is secreted in other areas of the brain, serving as a neurotransmitter or neuromodulator. Vasopressin helps control water excretion by the kidneys as well as blood pressure. From the hypophysis, vasopressin is transported by the blood to the kidneys, where it enhances water reabsorption in the collecting ducts by stimulating the insertion into the luminal membrane of special protein molecules, known as aquaporins, which function as water channels (Beitz 1999). This reduces the urine volume and saves body water.

In the case of a sudden decrease in plasma volume (e.g., by loss of blood), the **juxtaglomerular cells** located in the walls of the afferent arterioles in the kidneys will react to the drop in pressure by secreting the enzyme renin. Once in the bloodstream, renin triggers the splitting of the polypeptide angiotensin I from the plasma protein angiotensinogen. Mediated by an enzyme known as angiotensin-converting enzyme, angiotensin I is transformed into the active agent of the renin–angiotensin system, angiotensin II. The two most important effects of angiotensin II are constriction of arterioles and stimulation of the adrenal cortex to produce the steroid hormone aldosterone.

Aldosterone has a profound effect on the sodium reabsorption by the cortical collecting ducts of the kidneys, thus causing retention of fluid in the body. Essentially, all the sodium reaching the collecting ducts is reabsorbed when the plasma concentration of aldosterone is high. Aldosterone induces the synthesis of proteins that function as sodium channels and proteins that constitute the $Na^+$–$K^+$ ATPase pumps. Thus, aldosterone also can increase the potassium excretion in the kidneys.

## Factors That Modulate Hormonal Responses to Exercise

The state of the organism before exercise can affect the hormonal changes, the responsiveness of the fuel depots, and the capacity of the recruited muscle fibers to metabolize the different substrates. This state includes state of health, physical fitness level, training regimen used, nutritional state, body fluid homeostasis, and phase of menstrual cycle.

Training modifies the hormonal responses at a given work rate. For example, during short-term severe exercise, the sympathoadrenergic system is more activated in anaerobically than in aerobically trained men (Strobel et al. 1999). Furthermore, the exercise-induced increases in both plasma glucagon and growth hormone concentrations are less in trained than in untrained subjects at the same relative work rate. This effect is attained after a few weeks of vigorous endurance training. Likewise, after a period of physical training, the catecholamine response to a certain relative work rate can be reduced. On the other hand, the decrease in plasma insulin concentration during prolonged exercise is less pronounced in trained than in untrained individuals. It also has been found that training increases insulin action on glucose uptake in skeletal muscle identically in young and aged subjects (21–64 years; Dela et al. 1996), whereas bed rest (7 days) decreases the insulin effect (Mikines et al. 1991).

Endurance training enhances catecholamine-stimulated lipolysis in adipose tissue. Furthermore, in trained subjects, both the sensitivity of hepatic glucose-producing mechanisms to stimulation by catecholamines and the insulin-mediated inhibition of glucose production are enhanced. Insulin-mediated stimulation of whole-body glucose disposal is also higher in trained than in untrained individuals. This is mainly attributable to an increase in GLUT4 protein in trained muscle. Training also can alter the secretory capacity of the endocrine glands. Thus, an augmented adrenal medullary secretory capacity has been found in elite athletes who have performed endurance training for several years (Galbo 1995).

With regard to nutritional state, the glucose, FFA, and lactate as well as other blood-borne molecules can modulate hormonal responses. High plasma concentrations of glucose and FFA inhibit the mobilization of extra- and intramuscular energy stores. Lactate inhibits FFA release from adipose tissue. A decline in plasma glucose concentration during prolonged exercise will increase plasma adrenaline concentration to restore the blood glucose level.

Plasma volume can be acutely diminished by 7% to 10% during exercise, initially because of fluid shifts, but later also because of evaporation of sweat (Richter and Sutton 1994). This will tend to increase the plasma concentrations of all hormones.

In women, the catecholamine response to exercise is more marked in the follicular phase (the first half of the menstrual cycle) than in the luteal phase (the second half; Sutton et al., 1980). Furthermore, both the type of physical activity and environmental factors like ambient temperature can influence the hormonal response to exercise. Thus, both the aldosterone and cortisol levels increase less when exercise of a certain intensity is carried out with two legs than when it is performed with one leg only (Galbo 1992). Likewise, the increase in plasma noradrenaline level relative to the increase in heart rate and oxygen uptake is larger during isometric than during dynamic exercise. When a woman exercises in the heat, the noradrenaline response is increased more than during normal conditions (Rowell, Brengelmann, and Freund 1987).

### For Additional Reading

For a more extensive discussion of the effects of exercise on the endocrine system, see Richter and Sutton (1994) and Galbo (1995).

## Summary

When exposed to a challenge, humans react by mobilizing the body's resources for a physical effort. The pulmonary ventilation increases, as do the heart rate and cardiac output, increasing the oxygen uptake and the oxygen delivery to the muscles. Much of the circulating blood volume is shifted from the gut and skin to the muscles. Stored energy is released in the form of glucose and FFAs, which are made available as substrates for the working muscles.

In general, the hormonal responses to exercise are positively correlated to work intensity, as long as this is expressed in relative terms.

Probably the most important, immediate response to physical effort is an increased activation of the sympathoadrenergic system, resulting in the release of adrenaline and noradrenaline. These hormones increase heart rate and ventricular contractility; shift the blood flow from the skin, stomach, and intestines to the muscles; and mobilize fuel for cellular metabolism by stimulating breakdown of glycogen in liver and muscles and lipolysis in adipose tissue.

During exercise of increasing intensity and duration, the plasma concentration of insulin progressively decreases, making the liver more sensitive to the effects of adrenaline and glucagon, thus promoting hepatic glucose output. The plasma glucagon concentration decreases during brief intensive exercise but starts to increase again if the exercise is continued. Glucagon probably plays a significant role in increasing hepatic glucose production.

In the adrenal cortex, ACTH stimulates the synthesis and release of cortisol, which exerts a profound effect on a number of cellular functions. In the liver, ACTH stimulates the breakdown of proteins to amino acids. The carbon skeletons of the amino acids can enter the citric acid cycle, or they can be used for gluconeogenesis, thus increasing blood sugar concentration. At the same time, the protein synthesis in most of the cells in the body is significantly restrained to save energy.

Vasopressin helps control water excretion by the kidneys and blood pressure, whereas aldosterone profoundly affects the sodium reabsorption by the cortical-collecting ducts of the kidneys, thus causing retention of fluid in the body.

The state of the organism before exercise can affect hormonal changes, as well as the responsiveness of the fuel depots and the capacity of the recruited muscle fibers to metabolize the different substrates. This includes state of health, physical fitness level, training regimen used, nutritional state, body fluid homeostasis, and phase of menstrual cycle.

# Other Training-Induced Changes in the Body

Skeletal muscle accounts for a considerable fraction of total body weight, and it is not surprising that its level of activity has a profound effect on many aspects of bodily functions, to the extent that a sedentary life tends to result in lifestyle diseases, some of which have reached epidemic dimensions in Western countries.

## Changes in Body Composition

During inactivity or habitual training, the body weight can remain relatively constant, although physical inactivity often produces a gradual weight gain. During intensive training, the density of the body increases whereas the skinfold thickness decreases (Parizkova 1977; Wilmore 1982).

An increased body density (i.e., with a .01 unit) indicates a reduction in fat content. This is supported by the reduced skinfold thickness observed. At the same time, there is a certain proportional increase in muscle mass and blood volume. In the case of fat-free body mass, however, body weight varies only a few kilograms with different states of training. It is therefore evident that change in body weight alone is an inadequate index of possible alterations in body composition. Body composition changes with increasing age in humans, in that lean body mass decreases while fat mass increases. Regular physical training, however, can prevent these changes (Horber et al. 1996).

## Blood Lipids and Lipoproteins

A number of studies have examined the possible relationship between regular physical activity and blood lipid levels. Many of them have investigated the effect of a specific training program on the blood cholesterol level. The reason for this interest in blood lipid profile is, of course, the association between unfavorable lipid profiles and cardiovascular disease. Low-density lipoproteins (LDLs) have been associated with the deposition of cholesterol in the blood vessels, whereas HDLs have the opposite effect. High values of HDL are therefore considered to be desirable, and low levels of HDL are considered to be a risk factor in the development of coronary artery disease. Most data support the conclusion that endurance training improves the plasma lipoprotein profile, increasing HDL levels and the HDL/total cholesterol ratio.

Strength training, on the other hand, does not seem to have a similar effect (for review, see Hurley and Hagberg 1998).

### For Additional Reading

For further discussion of the effect of endurance training on plasma lipoproteins and relevant references, see Paffenbarger and Hyde (1980), Kiens, Lithell, and Vessby (1984), Marrugat et al. (1996), E. Bergström, Hernell, and Persson (1997), Mackinnon and Hubinger (1999), and Tolfrey, Jones, and Campbell (2000).

## Psychological Aspects

It is generally assumed that physical fitness has psychological correlates. The concept of the body–mind relationship is age-old. Psychosomatic research has indicated that physical changes result from continued psychological states; it seems logical to assume the reverse, that psychological changes result from physical states such as fitness.

The prevalence of mental disorders is high in most nations. In the United States and Norway, about one of every four adults suffers from a diagnosable mental disorder every year. About 50% of the population will be affected by a mental disorder during their lifetime. The most common forms are mood and anxiety disorders and abuse of or dependence on drugs or alcohol (Kessler et al. 1994). In addition to human suffering, the economic burden is great. In the United States, the annual costs of the most common disorders, anxiety, and mood disorders equal the costs of coronary heart disease.

The most common forms of treatment are medication and various forms of psychotherapy. These are not always effective, the costs are high, and medication can have unpleasant side effects. Hence, development of self-help strategies is important.

The most well-established psychological effect associated with exercise is a feeling of well-being after the exercise is finished. Exercise-related increases in self-esteem are documented in controlled trials. Many people experience more creative, problem-solving thinking during exercise and improved sleep after exercise, but there is no firm scientific documentation to support this. Patients with mental disorders are in bad physical condition compared with the general population (Martinsen et al. 1989).

Exercise as a treatment method is best documented for mild to moderate forms of depression, where several controlled studies support the effectiveness of exercise intervention (Ernst, Rand, and Stevinson 1998). The effect is of the same magnitude as that associated with psychotherapy and antidepressant medication. In chronic fatigue syndrome, graded exercise has been shown to be effective in two controlled trials (Wearden et al. 1998).

A significant reduction in state anxiety following exercise is well documented, although studies on trait anxiety have given inconsistent results. Among clinical anxiety disorders, panic and generalized anxiety disorder can be ameliorated by exercise (O'Connor et al. 2000). It also has therapeutic potential in alcohol abuse and dependence, but firm scientific documentation does not exist (Dishman 1997; W.P. Morgan 1997).

It appears that humans tend to become inactive after they reach puberty. In the past, physical activity was a part of one's daily life and work; this is still true in some countries and occupations today. Inasmuch as the need for physical activity in connection with most daily occupations is almost nonexistent, it is necessary to devote some leisure time to physical activity to maintain an optimal body function. Effective physical training does not have to be very stressful. Often the ideal solution is to acquire a hobby that involves some degree of physical activity, even though the training may not be very intensive. Most people can be motivated to exercise, provided proper facilities are available (Dunn and Blair 1997).

For nonathletes, physical training may not be sufficiently enjoyable to make them exercise merely for fun, without being motivated by the prospects of health benefits. Thus, people must accept the fact that exercise is necessary to maintain good health.

The most typical physical activity of our ancestors was brisk walking of rather long distances in their search for food. The driving force was hunger. Most likely, none of them were exercising in the present meaning of the word. It is an interesting observation that dog owners are quite convinced that the dog, also a mammal, needs exercise. It is tempting to suggest, "Go out and walk your dog even if you do not have one."

# Genetic Factors in the Training Response

There is considerable evidence of interindividual differences in the response to training. Bouchard et al. (1992) quoted several endurance and high-intensity intermittent training studies with pairs of dizygotic (nonidentical) and monozygotic (identical) twins. They concluded that the trainability of phenotypes governing aerobic and anaerobic performances is, to a large extent, determined by so far unknown genetic factors, and that the number of genes involved may be in the hundreds. In a report from the HERITAGE study, Bouchard et al. (1999) presented confirming evidence for the presence of family lines in the trainability of $\dot{V}O_2$max. There was 2.5 times more variance between families than within families for the adjusted $\dot{V}O_2$max response. Model-fitting analytic procedure yielded a maximal heritability of 47%, with a significant maternal component, which supports a role for mitochondrial DNA. Data concerning the role of genetic factors in strength training are limited, but so far it seems that the response to strength training is independent of genotype.

The search for genetic markers of the variation in trainability is still in its infancy, and, understandably, the mitochondrial genome has been a focus of interest. So far, however, nothing conclusive has emerged. For further details, see Bouchard, Malina, and Pérusse (1997). The relative importance of genes and environment for muscle phenotype was discussed earlier in this chapter and in chapter 3.

# Unwanted Effects of Training

Even though training is basically beneficial, things may happen in the wake of a strenuous workout or during an ambitious training program that impair performance for some time. Although this is an old experience, the mechanisms involved are not well understood.

## Muscle Soreness

If an untrained individual suddenly performs vigorous exercise, the active muscles can become painful, especially after eccentric exercise. The muscular soreness is accompanied by stiffness, tenderness, and reduced muscle strength. The symptoms usually appear from a few hours to a day after the exercise and can be most pronounced during the second day. Because of the time lag, the condition is commonly referred to as **delayed onset muscle soreness (DOMS).** The symptoms gradually fade away so that the muscles become free of symptoms after 4 to 6 days. The symptoms are paralleled by substantial mechanical damage of the myofibrillar structures, especially the Z-disks (Fridén, Kjorell, and Thornell 1984; Fridén, Sjöström, and Ekblom 1981), and in the system of intermediary filaments, as evidenced by the disruption of orderly arrangement of sarcomeres in neighboring myofibrils.

An increased level of creatine kinase (CK) in the blood is considered the best indicator of muscle damage, because CK and other muscle-specific intracellular proteins only escape into the bloodstream in the presence of sarcolemmal disruption or increased permeability. A temporal relationship between DOMS, elevated plasma CK, and signs of muscle inflammation has been claimed, and it has been speculated whether prostaglandins or other inflammatory mediators are involved. Nonsteroidal anti-inflammatory drugs, however, do not affect the leakage of CK, force deficit, or muscular soreness (Bourgeois et al. 1999). This does not exclude the presence of inflammation, because a series of circulating proinflammatory and inflammation-responsive cytokines are found after strenuous exercise. In particular, a parallel increase in IL-6 and CK has been reported after eccentric exercise (Pedersen et al. 1998). The temporal relationship between IL-6 and DOMS does not seem to imply a causal relationship, however, because 3 weeks of eccentric training abolished DOMS and the leakage of myoglobin without any effect on plasma IL-6 levels (Croisier et al. 1999). The parallel disappearance of DOMS and sarcolemmal damage after training, as evidenced by the absence of leakage of intracellular proteins, is more interesting. Extracellular sodium ion ($K^+$) concentration is more than likely to increase during periods of membrane leakage, and $K^+$ is a known activator of **nociceptors** (O'Connor and Cook 1999). This can explain at least part of the pain and, through reflex inhibition, maybe also the decrease in force, although Radák et al. (1999) claimed that an increase in muscular NO content may cause a decrease in force.

It has been realized for a long time (Hough 1902) that DOMS does not occur if the muscle is accustomed to eccentric exercise. This has been referred to as the **repeated bout effect** (McHugh et al. 1999; Nosaka and Clarkson 1995). In a way, the repair of the damaged tissue results in a stronger muscle that is much less susceptible to further injuries, even if the subsequent exercise is much more severe, but the mechanism behind this effect is so far unknown. D.L. Morgan (1990) hypothesized that it is attributable to an incorporation of additional sarcomeres in series. For discussion, see McHugh et al. (1999). In contrast to eccentric training, however, a period of concentric training has been claimed to increase the susceptibility to eccentric damage, possibly by reducing the number of sarcomeres in series, thus increasing the possibility that weak sarcomeres will be overextended (Whitehead et al. 1998).

Despite this, Hurley et al. (1995) showed that middle-aged and older men can safely participate in a total-body strength-training program that is intense enough to substantially increase muscle strength and hypertrophy without promoting muscle soreness or significant muscle cell disruption.

### For Additional Reading

For a review of theories underlying the symptoms of DOMS and the protection incurred by previous activity, see Cleak and Eston (1992), Lieber and Fridén (1999), and Koh (2002).

## Muscle Cramps

Muscle cramps are sudden, involuntary, intense, and painful contractions elicited by motor neuron hyperexcitability. It has been suggested that cramps during muscular work are the result of fluid electrolyte imbalance resulting from excessive sweat loss. However, individuals in occupations that require chronic use of small muscle groups but do not elicit profuse sweating, such as musicians, often experience cramps. Besides, muscle cramps can be experienced during pregnancy and at night in the elderly. The only factor common to these situations is the presenting symptoms. Thus, despite the common occurrence of muscle cramps, the exact cause remains unknown (Miles and Clarkson 1994).

There is a high incidence of muscle cramps among athletes. Cramps are listed among the most frequent complaints for marathon participants. Thus, according to Maughan (1986), 18% of runners reported an attack of muscle cramp that occurred after approximately 35 km had been covered. These subjects were not different from the other participants in terms of training status, racing performance, or serum electrolyte concentrations either before or after the marathon.

In a review of this problem, Schwellnus, Derman, and Noakes (1997) reported that an electromyographic study of runners during exercise-associated muscle cramps revealed that baseline activity is increased during spasms of cramping and that a reduction in baseline electromyographic activity correlates well with clinical recovery. During acute exercise-associated muscle cramps, the electromyographic activity is high, and passive stretching reduces electromyographic activity possibly because it relieves the cramp by invoking the inverse stretch reflex (Schwellnus 1999). Schwellnus, Derman, and Noakes hypothesized that muscle cramps are caused by sustained abnormal spinal reflex activity, secondary to muscle fatigue.

## Sport-Related Hematuria and Proteinuria

Hematuria in athletes is not a new problem. It dates back to at least 1713, when the Italian physician Bernadini Ramazzini, known as the father of occupational medicine, reported that runners sometimes pass bloody urine (Eichner 1990). **Hematuria,** usually termed microscopic hematuria or microhematuria, is defined according to the number of red blood cells (RBC) per high-power field (HPF). The most commonly accepted limit of normal hematuria is 3 RBC/HPF or 1,000 RBC $\cdot$ ml$^{-1}$ of urine (J.M. Sutton 1990). Although microhematuria is quite common in the general population, it is even more common in athletes after exercise. According to Eichner (1990), routine clinical detection methods show that marathons cause microhematuria in up to 20% of the runners, whereas ultramarathons cause it in 50% to 70% of the participants.

Exercise causes **proteinuria** too, which is the presence of protein in the urine. In fact, proteinuria is almost inevitable during strenuous or prolonged heavy exercise. Proteinuria is documented in sports like rowing, running, swimming, cycling, skiing, boxing, lacrosse, football, and baseball. Like microhematuria, proteinuria is transient and benign. The urine is generally protein-free 1 to 2 days after the exercise (Eichner 1990).

In a study of the prevalence of exercise-induced hematuria and proteinuria in 70 healthy male runners, McInnis et al. (1998) found that exercise-related changes in renal function were associated with weight-bearing exercise intensity rather than non-weight-bearing exercise duration. They concluded that both hematuria and proteinuria appear to be intensity related. Alteration of renal function appeared to be the mechanism responsible for the hematuria and proteinuria observed in their study. Specifically, the mechanism appears to be related to a decrease in renal plasma flow caused by vasoconstriction. Actually, transient and benign hematuria and proteinuria constitute an "athletic pseudonephritis" as early named by Gardner (1956).

### For Additional Reading

For details concerning kidney function during exercise and sport-related hematuria, see Poortmans and Vanderstraeten (1994), and G.R. Jones and Newhouse (1997).

## Overtraining

The purpose of athletic training is to improve performance and to make the athlete capable of peak performance when it counts. A large amount of intensive training is needed to push performance capacity to its upper limits, and yet it appears that athletes often are inclined to do too much (Kuipers 1998). This can lead to insufficient recovery and muscle fatigue, the so-called overtraining syndrome or staleness, probably in part related to insufficient metabolic recovery and the body's inability to cope with the total amount of stress. The overtraining syndrome is featured by premature fatigue, decreased performance, mood changes, emotional instability, and reduced motivation, but unfortunately no specific tests are available. It also has been considered unfortunate that the term *overtraining* implies causation. Therefore, the term **unexplained underperformance syndrome (UPS)** has been suggested (Budgett et al. 2000).

### For Additional Reading

For a discussion of the symptoms and markers of overtraining, see Gleeson (1998), Hartmann and Mester (2000), and Urhausen and Kindermann (2002).

Israel (1958) described two clinical forms of overtraining: a sympathetic and a parasympathetic form, the former predominant in team sports and sprint events, the latter in endurance athletes. An early and quickly reversible phase of overtraining is often called **overreaching** (figure 11.23). Most probably it is related to insufficient metabolic recovery (Kuipers 1998). Our understanding of the more severe forms of overtraining is so far very patchy, but many theories exist. The terms *sympathetic* and *parasympathetic* in connection with overtraining relate to a theory of autonomic imbalance as a cause of the symptoms (figure 11.24), whereas the branched-chain amino acid hypothesis postulates that the symptoms are caused by an increased entry of the amino acid tryptophan into the brain and a resulting increased production of 5-hydroxytryptamine (5-HT, or serotonin). Animal studies seem to support the 5-HT hypothesis, but human studies show contradictory results. Snyder (1998) examined the effect of sufficient quantities of carbohydrate consumption and concluded that other mechanisms than reduced muscle glycogen levels (the glycogen depletion theory) must be responsible for the symptoms of overtraining. So far, we have to admit that the mechanisms behind the overtraining syndrome are elusive, and while waiting for a breakthrough in the research, we have to concentrate on how to avoid its symptoms. Foster (1998) suggested that simply monitoring the characteristics of training can allow the athlete to achieve the goals of training while minimizing undesired training outcomes.

### For Additional Reading

For details and discussion of possible mechanisms behind overtraining, see M. Lehmann et al. (1997, 1998) and Gastmann and Lehmann (1998).

## Training and Oxygen Radicals

It is well established that mitochondrial respiration in vitro, and most likely also in vivo, generates free oxygen radicals as a by-product of aerobic metabolism. Such radicals, especially hydroxyl radicals, can initiate lipid peroxidation and thus result in cell injuries. Reactive oxygen species are more likely to occur in type I muscle fibers than in type II fibers, because of their higher oxidative potential, and their production has been reported to increase in old age (Bejma and Ji 1999). Failure to remove these

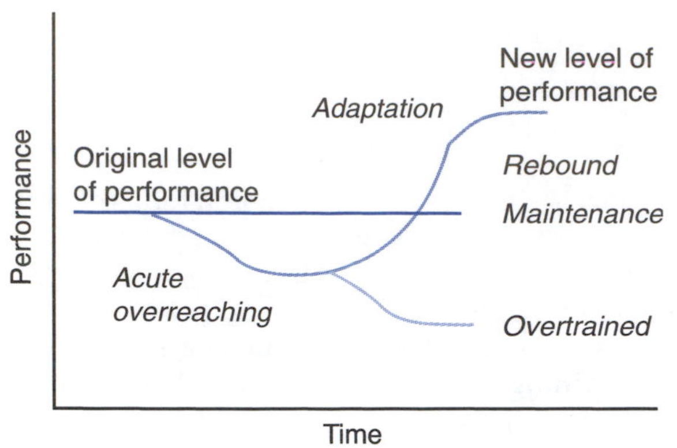


**Figure 11.23** Conceptual model of overreaching, recovery, supercompensation, and overtraining.

Reprinted, by permission, from M. Gleeson, 1998, "Overtraining and stress response," *Sports Exercise and Injury* 4:62-68.

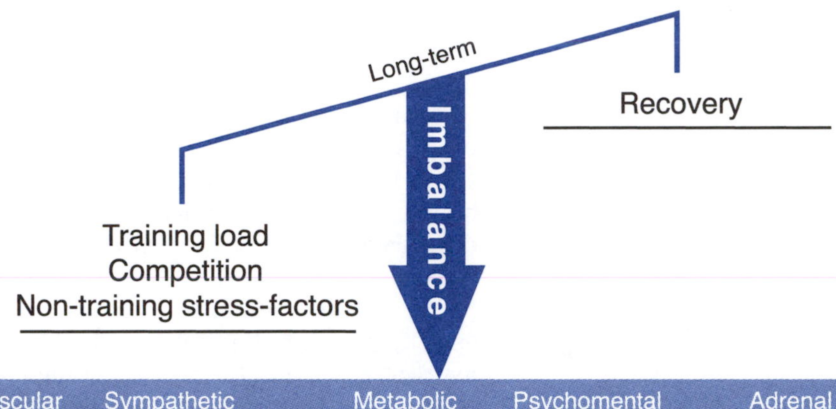


**Figure 11.24** Schematic overview of the genesis of overtraining syndrome in endurance sports related to long-term, high-volume overtraining, as far as known at present.

Adapted from Lehmann et al. 1998.

oxidants during exercise can significantly damage cellular molecules (Powers, Ji, and Leewenburgh 1999).

In the course of evolution, aerobic cells, such as muscle fibers, have been equipped with different scavenger systems against these radicals (Chance, Sies, and Boveris 1979; Del Maestro 1980). Such antioxidant systems include the primary antioxidant enzymes superoxide dismutase, glutathione peroxidase, and catalase as well as important nonenzymatic antioxidants like glutathione and the inducible **stress protein** known as **heat shock protein 72** (Essig and Nosek 1997; Y.F. Liu et al. 1999; Powers, Li, and Leewenburgh 1999).

### For Additional Reading

For a discussion of exercise and oxidative stress, see Ji (1995). Details about the role of stress proteins in physical activity can be found in Locke and Noble (1995 and 2002), and Locke (1997), and a discussion of the effect of nutritional antioxidant supplementation is found in Kanter (1995).

## Tightness of Muscles and the Effect of Stretching

Reduced range of motion can result from both training and inactivity, although substantial individual differences exist. The cause of joint movement constraints can be neurogenic or myogenic or can reside in the joint itself or in tissues surrounding the joint (Hutton 1992). Because a fully functional range of motion is a prerequisite for optimal motor function, several stretching techniques have been developed, but their scientific basis is still controversial.

Different stretching techniques have been developed: ballistic stretch, slow stretch, and variants of contract–relax stretching. Basically, any form of sustained or repetitive stretch that places a joint toward its maximum range of movement will enhance its range of movement over time (Hutton 1992). In theory, however, the stretching techniques aim at changing the neurogenic and myogenic components of the movement constraint. Contract–relax stretching seems to be the most effective form of stretching to increase the range of motion in a joint. A common technique is to start with a maximal isometric contraction of the involved muscle group for 4 to 6 s, to completely relax at least 2 s, then to conclude with a slow passive extension of the joint as far as possible without causing pain, and to maintain this maximally extended position for 8 to 20 s. This sequence is repeated five to six times for each muscle group.

### For Additional Reading

For discussion of the neuromuscular basis of stretching, see Etnyre and Abraham (1988) and Hutton (1992).

Stretching does not improve muscle strength, but it can prevent the quite common reduction of range of motion in joints engaged in training programs, which may persist for at least 24 h after the training session (Wiktorsson-Möller et al. 1983). There are also reports that stretching can prevent muscle and tendon injuries, but a typical stretching protocol performed during preexercise warm-up did not produce any clinically meaningful reductions in the risk of exercise-related injury in army recruits (Pope et al. 2000).

## Contraindications for Physical Training

In some cases, physical activity is contraindicated. More often than not, however, the individual can reduce the level of activity or training rather than completely abolishing it. The practicing physician is often faced with the question of whether a patient can continue the training, and if not, when the patient can begin again. At times it can be quite difficult for the physician to decide how to advise the patient. A sensible recommendation would depend on (in addition to a careful medical examination) a comprehensive knowledge of the physiological demand that the training or activity in question places on the patient, as well as some insight into the pathophysiology of the patient's condition.

Because of uncertainty and to be on the safe side, doctors often recommend prolonged rest or inactivity or ask the patient to "take it easy" without specifying precisely what is meant by that expression. In consequence, the conscientious patient may end up being more inactive than the physician had intended. In the case of an athlete engaged in active competitive sports, the consequence of such prolonged inactivity can be the loss of an entire season. In the case of a nonathlete engaged in regular fitness training, such a prolonged interruption of training actually can end his or her training habit because the patient never gets around to taking it up again. There are a few medical conditions, however, that do require an interruption of the training program, either partially or completely, depending on the nature of the ailment.

Upper respiratory infections with sore throat and nasal congestion are common both in competitive athletes and in the general public. As long as the patient has no fever, it is not necessary to discontinue training. A few days' break in training may be all that is necessary. In the presence of fever, however, one has to be more cautious, because the person's resting pulse as well as the work pulse at a standard work rate is almost invariably elevated. Generally, people should refrain from all training in the presence of fever. When afebrile, the patient should start gently and increase the intensity of training gradually, increasing to full intensity in about a week. One of the reasons why such great caution is recommended in the case of flulike symptoms in athletes is the fear of myocarditis as a possible complication (Jokl 1964). The actual risk of developing myocarditis under such conditions is still an open question, however. A few cases of sudden death in young athletes have been attributed to possible myocarditis, but few documented cases have been reported in the literature. Nonetheless, it is advisable to have this possibility in mind when treating young athletes.

Lower respiratory infections, such as acute bronchitis and pneumonia, require cessation of training until the principal signs and symptoms have returned to normal. Preferably, x-ray examination of the chest should be made before training is resumed.

Acute **nephritis** requires an immediate end of all types of strenuous physical training. The training should be discontinued until urine analysis reveals normal findings and the blood pressure has returned to normal. This disease, as a rule, seriously affects the patient's fitness, and it will take a long time for the patient to regain his or her original level of fitness. Therefore, retraining should be slow and gradual.

Patients with chronic nephritis should stay away from competitive training. In particular, they should refrain from participating in prolonged strenuous physical exertion, such as cross-country ski races covering long distances. Anuria (cessation of urine production) has been observed in presumably healthy individuals with no known history of kidney disease after such extremely strenuous athletic events (Refsum and Strømme 1974). Moderate physical activity, on the other hand, may be beneficial for patients with chronic nephritis, like everyone else. Patients with urinary tract infections, such as cystitis or pyelitis, should not participate in physical training during the acute phase of the illness.

Acute hepatitis is definitely a contraindication for physical training. Because this disease, in general, has a rather protracted course, it is necessary to accept a prolonged break in the training program. The training should not be resumed until the hepatomegalia (swelling of the liver) has disappeared and liver function tests are back to normal. In these cases, it is particularly important that retraining starts very gently and that the patient is regularly checked by the physician during the retraining period. This also applies to patients with infectious mononucleosis. Such patients often have an enlarged spleen. Because of the danger of a possible rupture of the spleen, the patient should not resume physical training, at any rate not in the case of contact sports, until the enlarged spleen has returned to normal. Careful jogging or similar training may be permitted, however.

Last, but not least, retrosternal and other types of chest pain should be subjected to a thorough examination, including electrocardiography and blood tests, before the patient engages in any kind of physical exercise.

CHAPTER 12

# NUTRITION AND PHYSICAL PERFORMANCE

The question of what an athlete should eat to achieve superior performance is as old as the recorded history of organized sports. The practice of consuming large quantities of meat to replenish the supposed loss of muscular substances during heavy exercise was first recorded in Greece during the 5th century B.C. (Christophe and Mayer 1958). Later, the discovery of vitamins offered a new and promising area of experimentation based on the assumption that if a little of something is good, a great deal of it must be much better (Hecker 1984). Researchers have made repeated claims over the last several decades that physical exercise performance can be significantly improved by special diets or dietary supplements (Fenn et al. 1983; Karlsson 1997; Simonson 1951). However, according to Mayer and Bullen (1960) the concept that any well-balanced diet is all that athletes actually require for peak performance has not been superseded. It is the purpose of this chapter to examine, in the light of more recent evidence, whether this is still true.

## Nutrition in General

Evidently, human beings are genetically endowed with the potential to eat, digest, and metabolize both plant and animal foods in a variety of combinations; humans can live on an almost exclusive animal diet, a vegetable diet, or a combination of the two. The earliest humans consumed a considerable amount of meat. This is evident from the findings of large accumulations of animal remains where they lived, and from the tools they used, mainly designed for the processing of game. However, shortly before the introduction of agriculture and animal husbandry, there was a shift away from big game hunting toward a broader spectrum of food sources. This is apparent from the remains of fish, shellfish, and small game, as well as tools suitable for processing plant foods, such as grindstones and mortars. With the introduction of agriculture, the proportion of meat in the diet declined drastically, whereas vegetable foods eventually accounted for some 90% of the diet. However, since the industrial revolution, the animal protein content of Western diets has once more increased (Eaton and Konner 1985).

From the available evidence, Eaton and Konner (1985) concluded that the Paleolithic people generally ate much more protein and far less fat than we do. Their diet contained more essential fatty acids and much higher ratios of polyunsaturated to saturated fats than our diet, although their cholesterol intake was high. Evidently their calcium intake far exceeded even the highest estimates of minimal daily requirement. Meat and entrails from game animals provided them with high iron intakes. Their intake of dietary fiber was much higher than ours, while their sodium intake was remarkably low. By all indications their vitamin intake greatly exceeded ours.

On the whole, one is left with the impression that the diet of our Paleolithic ancestors was superior to ours in terms of promoting health. This impression is further strengthened by the observation that coronary heart disease, hypertension, and type 2 diabetes were relatively unknown among the few surviving hunter-gatherer populations whose way of life and eating habits most closely resembled those of the preagricultural human beings (Eaton and Konner 1985). Even among the more isolated natives living in Alaska 50 years ago, hypertension and coronary heart disease were extremely rare (K. Rodahl 1953).

Today, our knowledge of the number of essential nutrients is rather comprehensive. Patients have been living and maintaining good nutritional conditions and health for years, exclusively on intravenous administration of approximately 45 different nutrients. Actually, one patient has been living on total parenteral nutrition for more than 15 years (D. Hallberg et al. 1982).

We need food as building blocks for our tissues, and we need food for energy. The most important building material is protein. It is essential for building new cells and tissues and for replacing parts of old cells that constantly are being broken down. There are different kinds of proteins, depending on their amino acid composition. Plant proteins are not identical with proteins of animal origin. The tissues of our body need some amino acids that we cannot live without because the body is unable to synthesize them. These vital substances are known as the essential amino acids. Meat, fish, eggs, milk, and cheese contain proteins with about the proper composition of amino acids. They are therefore the most suitable sources of protein.

An adult needs about 1 g of protein of proper composition per kilogram of body weight per day. This corresponds to about 70 g for a full-grown person. Athletes may have a somewhat greater protein requirement during periods of intense training. Those involved in strength training might need to consume as much as 1.6 to 1.7 g of protein $\cdot$ $kg^{-1}$ $\cdot$ $day^{-1}$, whereas those undergoing endurance training, especially that involving muscle strength, might need about 1.2 to 1.4 g $\cdot$ $kg^{-1}$ $\cdot$ $day^{-1}$ (P.W.R. Lemon 1998). Longitudinal studies are needed to confirm these recommendations. However, in developed countries, this amount of protein is found in the diets of even sedentary people. Concern over insufficient protein intake is therefore generally unfounded, as long as energy intake is adequate to maintain body weight.

Minerals are also necessary to maintain body structures, and we need a number of vitamins for a variety of catalytic processes in our biological machinery. However, food is also the fuel for the body machinery. Some of the ingested food is metabolized and used as soon as it is resorbed. But most of it is stored temporarily, mainly in the form of fat. This stored energy then can be mobilized as needed.

In developing the ideal energy stores for the mobile animal, such as the human, nature has to meet a number of requirements. For reasons of portability, each molecule should carry a large amount of energy per unit weight. The material should be fitted into various oddly shaped spaces and compartments of the body. It should possess storage stability and, at the same time, must be readily available and capable of being rapidly converted into oxidizable substrate when needed.

Fat or triglycerides (triacylglycerol) meet these requirements remarkably well. They are high in energy content, and they are stable, yet readily mobilized. The amount of energy held per unit weight of any molecule depends on its content of oxidizable carbon and hydrogen. Oxygen in a molecule of stored energy merely adds dead weight, because oxygen atoms can be obtained from the air as needed. Fat, therefore, approaches maximal storage efficiency because it contains as much as 90% carbon and hydrogen and has an energy density of 39 kJ $\cdot$ $g^{-1}$ (9.3 kcal). It is much superior in storage efficiency to carbohydrates, which contain only 49% carbon and hydrogen and have an energy density of only 17 kJ $\cdot$ $g^{-1}$ (4.1 kcal). The difference is even greater when one considers that fat is deposited in droplets, whereas carbohydrate is deposited together with an appreciable amount of water (in mammals, 2.7 g of water per gram of dried glycogen) (Weis-Fogh 1967). In other words, hydration dilutes the energy intensity of glycogen to about 4 kJ $\cdot$ $g^{-1}$ (l kcal). Thus, carbohydrate is a rather inferior material as an energy store in terms of portability. The energy value of l g of adipose tissue, which does not consist of pure fat, is about 25 to 29 kJ (6 to 7 kcal). On the other hand, there is an almost 10% higher energy yield per liter of oxygen used when carbohydrate is combusted than when protein and fat are burned.

The different animal species have, through adaptation, met their own peculiar needs. In migrating fishes like the salmon and the eel, and in birds, fat constitutes the main source of energy. Weis-Fogh (1967) pointed out that owing to impaired weight economy, carbohydrate cannot sustain a bird in flight for more than a few hours, and the recorded endurance of 1 to 3 days observed in some typical migrants therefore depends almost exclusively on the use of fat mobilized from stored triglycerides.

## Digestion

The digestion of food starts in the oral cavity, where the food is broken up into smaller particles during chewing and is mixed with saliva. The secretion of saliva, which is brought about by reflexes, is greatest when food is dry and gritty. The saliva contains enzymes, which split starch into maltose. This process is more complete the longer the food is chewed.

When the chewed food has attained the proper consistency (bolus), it is gathered by the tongue and pushed backward toward the pharynx. The swallowing reflex automatically provides for the swallowing of the bolus, whereby the food is moved by **peristalsis** down the esophagus into the stomach. The stomach is not a relaxed bag but a slightly constricted muscular container. This container encompasses the stomach content by an even, gentle pressure and is expanded as the stomach is filled. This tension of the stomach wall apparently is connected with the sensation of satiety.

As the food is dissolved by the gastric juice secreted from numerous glands in the stomach mucosa, it is ejected into the duodenum. The gastric secretion is regulated by both neurogenic and hormonal mechanisms. The secretory glands can be excited by smell and taste and possibly also by the mere thought of food. Fatty food, on the other hand, suppresses the production of gastric juice. The gastric juice causes the breakdown of proteins and the emulsification of fat.

The digestion of the food (chyme) in the small intestine is accomplished by the intestinal juice, which contains several digestive enzymes capable of breaking down carbohydrates, fat, and proteins into molecules to be transported through the intestinal wall into the blood. The remainder is moved along by the intestinal peristalsis into the large intestine and eventually expelled as feces.

The nutrient molecules resorbed through the intestinal wall are partially used at once as building blocks for the cells and tissues of the body and as fuel. The rest is stored, partly in the form of glycogen, but for the greater part as fat in the adipose tissues around the guts and under the skin, and for a small part also in the muscles as **intramuscular triglycerides (IMTGs).** Endurance exercise leads to a partial depletion of IMTGs (Björntorp 1991). An increase of this fat store thus would increase substrate availability. In fact, IMTG storage is found to be larger in endurance-trained individuals (Brouns, Saris et al. 1989).

A simplified schematic presentation of the actions of the different digestive enzymes is summarized in figure 12.1.

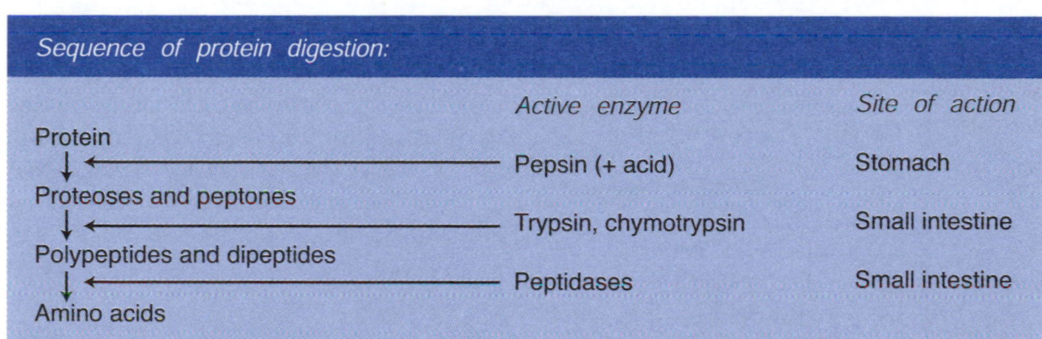


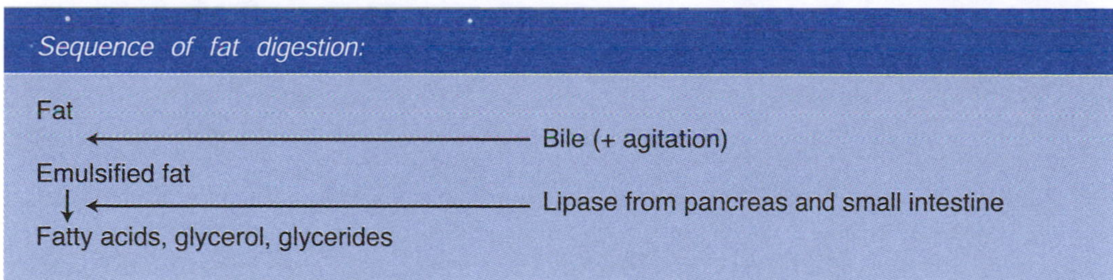


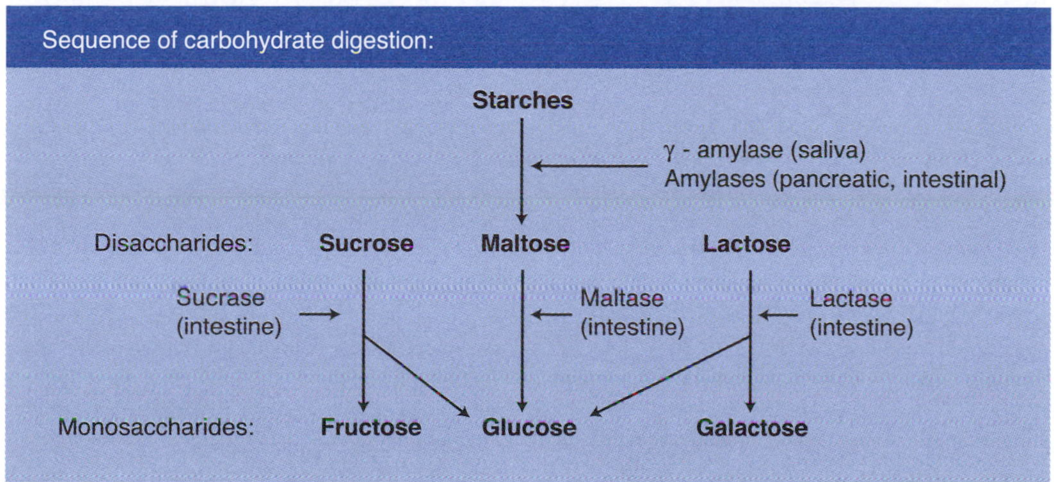


**Figure 12.1** A simplified schematic presentation of the actions of the different digestive enzymes.

# Energy Metabolism and the Factors Governing the Selection of Fuel for Muscular Work

Of the various nutrients in the food we eat, only the carbohydrate, fat, and protein can yield energy for muscular exercise. Alcohol (ethanol), which contains 29 kJ (7 kcal) per gram, cannot be used either directly or indirectly by the muscle (Schürch et al. 1982). The three energy sources, however, do not contribute equally to the energy-yielding processes in the muscle cell. The fact that nitrogen excretion is not significantly increased during muscular exercise in the fed individual (Cathcart and Burnett 1926; Crittenden 1904) has been taken as an indication that protein is not used as a fuel to any appreciable extent as long as the energy supply is adequate (Chauveau 1896; Krogh and Lindhard 1920; Pettenkofer and Voit 1866).

However, several newer studies show that regular exercise increases protein requirements. For endurance exercise, this increased need is probably attributable to an increased amino acid oxidation and perhaps increases in muscle protein synthetic rate to repair those parts of the cells that are being broken down (P.W.R. Lemon 1998). The branched chain amino acids in particular might contribute to the energy liberation (in the glucose-alanine cycle). This is particularly true when an individual is undernourished or fasting and during prolonged heavy exercise lasting 4 to 6 h or more (Dohm et al. 1985; Refsum, Gjessing, and Strømme 1979). Apparently, exercising muscle increasingly uses amino acids for fuel as carbohydrate availability decreases. As a result, dietary protein needs are likely to be greatest if exercise occurs when carbohydrate stores are limited (P.W.R. Lemon and Mullin 1980; Wagenmakers et al. 1991). But normally, protein is only a minor source of energy during exercise (Calles-Eskandon et al. 1984; Wolfe et al. 1984). From a teleological point of view, it would indeed be a catastrophe if the skeletal muscle fibers, like cannibals, consumed themselves; thus, inhibition of protein breakdown during exercise is a necessary provision to maintain an intact muscle machinery. It is known that some nitrogen is eliminated during exercise in the sweat (P.W.R. Lemon and Mullin 1980). For this reason, nitrogen excretion by the kidneys does not fully reflect the total rate of protein deamination.

## Carbohydrate Versus Fat Mobilization

From the previously discussed facts, it should be clear that the choice of fuel for the exercising muscle is mainly limited to carbohydrate and fat. The percentage participation of these two fuels in the energy metabolism usually is assessed by determining the nonprotein respiratory quotient (R), which is the ratio of carbon dioxide volume produced to oxygen volume utilized. An estimation of the amount of oxygen ($ml \cdot min^{-1}$) used for the oxidation of fat can be obtained from the following formula:

$$(1 - R) \cdot .03^{-1} \cdot O_2 \quad (1)$$

In this calculation, the R is handled as a "nonprotein R." The fact that the energy value of oxygen varies with R is not taken into consideration, but this represents, at the most, a 7% difference.

The percentage participation of the two major fuels in the energy metabolism depends mainly on the following factors:

1. *Type of exercise*: whether it is (a) continuous or intermittent, (b) brief or prolonged, and (c) light or heavy in relation to the maximal aerobic power of the engaged muscle groups
2. *State of physical training*: whether the individual is untrained or well-trained
3. *Diet*: whether it is high or low in carbohydrates

### *Type of Exercise*

At rest, fat and carbohydrate contribute about equally to the energy supply, as they also do during light or moderate exercise in the fasting individual. As the exercise progresses, fat contributes in an increasing amount to the energy yield (figure 12.2). During moderately heavy exercise (in fasting subjects) that can be endured for 4 to 6 h (including rest pauses), as much as 60% to 70% of the energy can be derived from fat at the end of the exercise period (Pruett 1971). Thus, the longer the exercise lasts, the smaller the percentage contribution of carbohydrate to energy metabolism in the fasting subject.

With increasing exercise intensity, there is a gradual change toward a proportionally greater share of energy yield from carbohydrates (figure 12.3). This has certain advantages in terms of energy yield. One liter of oxygen can, in the respiration chain, oxidize glycogen and yield energy for a regeneration of about 6.5 mol of adenosine triphosphate (ATP). When fatty acids are oxidized, the ATP formation is reduced to 5.6 mol · $L^{-1}$ of oxygen consumed.

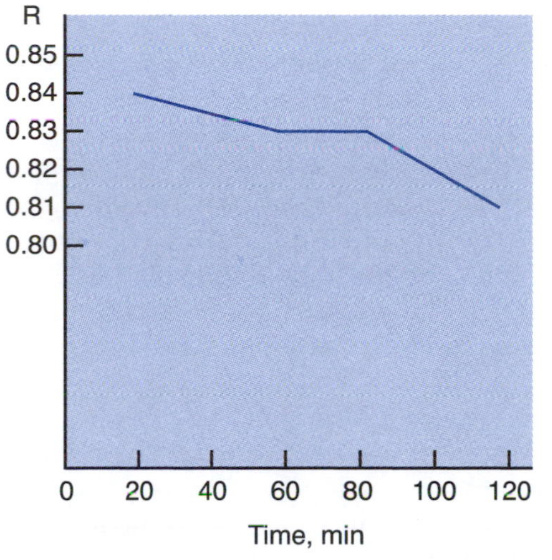


**Figure 12.2** Respiratory quotient (R) in a subject exercising at 175 W for 120 min while living on a normal diet.
Adapted from Christensen and Hansen 1939.

On the other hand, the adequacy of the oxygen supply to the exercising muscle cell is of prime importance because the oxidation of free fatty acids (FFA) depends on oxygen as the hydrogen acceptor. Thus, an inadequate oxygen supply more or less restricts the usable fuel to carbohydrate, of which there are limited stores. In addition, anaerobic oxidation of carbohydrate uses only about 8% of its energy compared with an aerobic oxidation of this fuel.

At extremely heavy or near-maximal rates of exercise, glycogen is the major source of energy for muscular exercise (Gollnick 1985). Under such conditions, when the oxygen supply is lacking, the carbohydrate oxidation proceeds only as far as the formation of lactate. The accumulation of lactic acid in the muscle is associated with an impaired function of the muscle cells. Furthermore, the increased lactate in the blood may inhibit the mobilization of FFA (Fredholm 1969). This in turn suppresses fat metabolism further by limiting the supply of FFA substrate to the muscle cell, although a study by Ahlborg, Hagenfeldt, and Wahren (1976) suggested an augmented removal of FFA from the plasma pool in humans following lactate infusion.

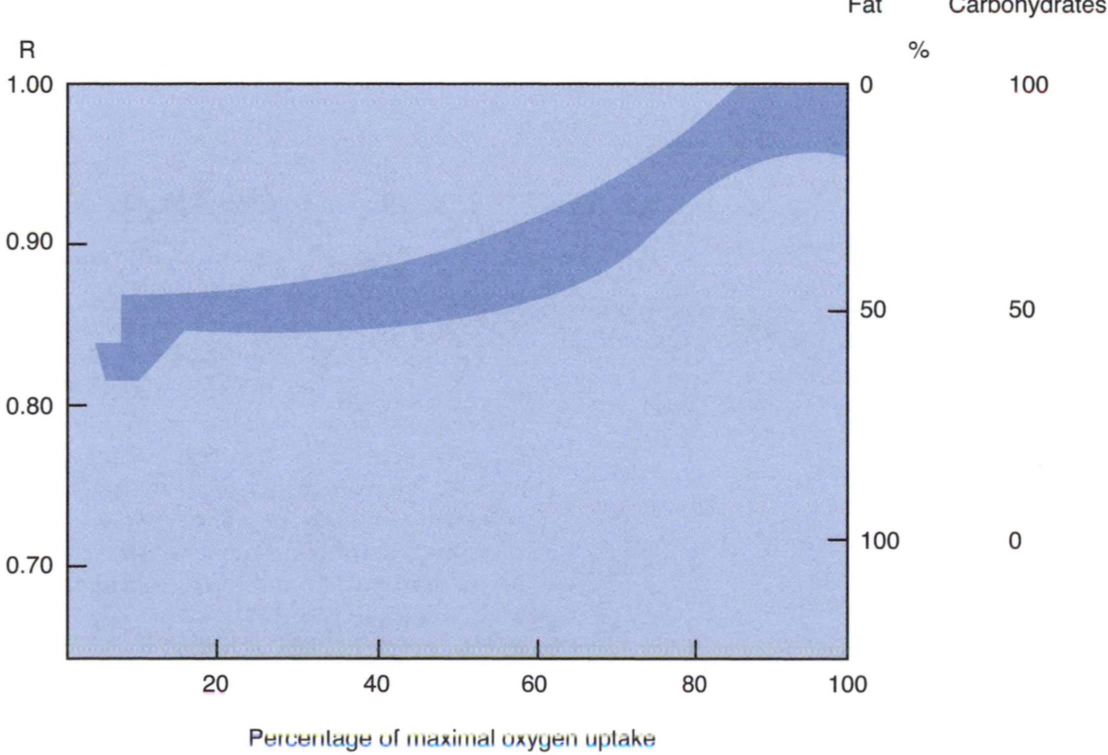


**Figure 12.3** Respiratory quotient at rest and during exercise related to the oxygen uptake in percentage of the subject's maximal oxygen uptake. To the right, the percentage contribution to the energy yield of fat and carbohydrate. Prolonged exercise, endurance training, and diet can markedly modify the metabolic response.

## *State of Training*

Because the ability to use fat as a fuel depends on the oxygen-transporting capacity, the choice of fuel for the exercising muscle depends on the work rate in relation to the individual's maximal oxygen uptake. The greater the maximal oxygen uptake, the greater the percentage contribution of fat to the energy metabolism at a given work rate. Because training increases the maximal oxygen uptake, it also increases the ability to use fat as a source of muscular energy during certain types of activity (Costill et al. 1979; Jeukendrup, Saris, and Wagenmakers 1998; Rahkila et al. 1980).

In prolonged exercise, it is a distinct advantage for the fasting individual to be able to use fat as a source of muscular energy, because the fat stores are infinitely larger than carbohydrate stores. Stored fat amounts perhaps to some 400 MJ (almost 105 kcal) or more in a well-fed individual of average size. The available energy in the form of stored ATP and phosphocreatine is only a few kilojoules, sufficient for only seconds of strenuous physical effort. Stored glycogen amounts to some 8 MJ (2,000 kcal).

The fact that physical training, which increases the maximal oxygen uptake, also increases the individual's facility for fat utilization was shown by Issekutz et al. (1965) in dogs. A trained and an untrained dog performed the same work rate on a treadmill, which the untrained dog could endure for only 30 min. In the untrained dog, the blood lactate rose to 8 mM, whereas in the trained dog, it was only 3 mM. In the trained dog, the utilization of FFA rose during the experiment, whereas in the untrained dog it declined. One reason for this difference is that training enhances the oxygen supply to the active muscle cells, so that the exercise, in the case of the trained dog, can be performed to a greater extent aerobically. Consequently, less lactic acid is formed in the trained dog. The unfit dog, at the same work rate, produces more lactic acid, which inhibits the FFA release from the adipose tissue. The result is that the plasma FFA drops. A decreased plasma FFA always means decreased turnover rate or decreased rate of utilization of FFA.

## *Diet*

By obtaining small pieces of muscle tissue (10–20 mg) using the needle biopsy technique (Bergström 1962) during various stages of different dietary regimens or during exercise, scientists can follow the variation in glycogen content of the muscles. For an individual on a normal mixed diet, the glycogen content in the quadriceps femoris muscle ranges from about 50 to 100 mmol $\cdot$ kg$^{-1}$ of wet muscle (Hultman 1967). In the deltoid muscle, the glycogen content is significantly lower, or about 50 mmol $\cdot$ kg$^{-1}$ of wet muscle. Figure 12.5 presents data from experiments in which biopsies were taken every 20th min during exercise on a cycle ergometer, with an average oxygen uptake of 77% of the individual's maximal aerobic power. Two groups of subjects participated in the study; 10 were trained, whereas 10 were untrained and had a lower maximal oxygen uptake. Therefore, the "77% load" corresponds to a mean oxygen uptake of 3.4 L $\cdot$ min$^{-1}$ for the trained and only 2.8 L $\cdot$ min$^{-1}$ for the untrained subjects. After about 90 min, the exercise had to be terminated because of the subjects' exhaustion. The glycogen content was then in the order of 5 mmol $\cdot$ kg $^{-1}$ of wet muscle. However, the biopsies were taken only from the lateral and superficial portion of the quadriceps femoris muscle, and the analyses do not necessarily reveal events in other portions of the active muscles. The respiratory quotient was higher in the untrained group (about .95) than in the trained subjects (about .90), indicating that FFA metabolism plays a larger role in the trained men. The actual combustion of glycogen was about 155 mmol $\cdot$ min$^{-1}$ in both groups, despite the difference in energy expenditure (3.4 vs. 2.8 L of oxygen per minute). At the end of exercise, the blood sugar level was still 4.4 mM.

Adaptation to a diet rich in fat and poor in carbohydrates results in lower muscle glycogen content and higher rate of fat oxidation during exercise. Thus, the effect of such an adaptation could be a sparing of muscle glycogen. Because muscle glycogen storage is coupled to endurance performance, it is possible that adaptation to a high-fat diet potentially could enhance endurance performance. However, the attainment of an optimal performance ability depends primarily on the quality and quantity of the training regimen. When exercise intensity is increased, there is an increased need for carbohydrates. Since the consumption of a carbohydrate-poor diet decreases the glycogen storage in both muscle and liver, the training intensity may be compromised in those who stick to a fat-rich diet (see Helge 2002).

# Glycogen–Glucose Homeostasis

Despite the reduced carbohydrate use during prolonged moderate exercise, the availability of glycogen may nevertheless be a factor limiting endurance because of the limited stores. At work rates up to some 75% of the individual's maximal oxygen

## Classic Study

Christensen and Hansen (1939), in their classic experiments, examined the participation of fat and carbohydrate in energy metabolism on the basis of the respiratory quotient during physical exercise of different intensities. In subjects on a normal diet, engaged in exercise of such an intensity that the metabolic processes were essentially aerobic, these authors found that about 50% to 60% of the energy was supplied by fat. In prolonged, standardized aerobic exercise of up to 3 h duration, an increased participation of fat was observed, supplying up to 70% of the energy. In heavy exercise, on the other hand, where anaerobic metabolic processes were involved, their findings indicated a major participation of carbohydrates.

Subjects on an extremely high-fat diet for several days, in which less than 5% of the energy intake was derived from carbohydrates, were only able to exercise at a given intensity for about 1 h. Throughout the entire exercise period, 70% to 99% of the energy was obtained from fat combustion. Although the capacity for prolonged heavy exercise was reduced significantly, the subjects nevertheless were able to carry on this exercise for 1 h while using fat almost exclusively as a source of fuel (figure 12.4).

In subjects on a very high-carbohydrate diet, where 90% of the food energy was derived from carbohydrates, the standard load could be performed for a much longer time, up to 4 h. Initially, fat contributed only 25% to 30% to the metabolic fuel, compared with more than 70% when the diet was rich in fat. Gradually, the contribution from fat combustion increased, and it was about 60% at the end of the exercise. Thus, the subjects were able to exercise about three times as long on this very high-carbohydrate diet than on the very high-fat diet (figure 12.4).

Christensen and Hansen 1939.

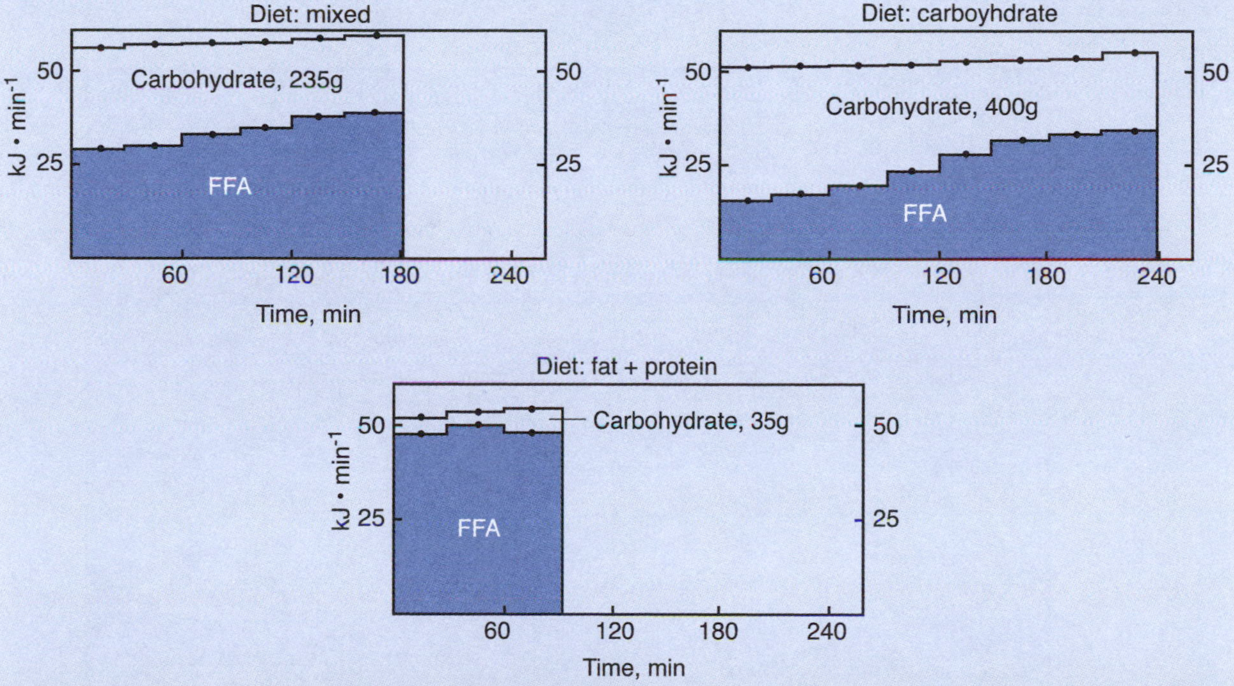


**Figure 12.4** Increase in free fatty acid (FFA) metabolism in prolonged exercise. One well-trained subject exercised on a cycle ergometer at 183 W after a mixed diet and then at 176 W after a carbohydrate-rich diet for 3 days. In another experiment, a 176-W bout was preceded by 3 days of fat and protein intake, excluding carbohydrates from the diet. The subject exercised until exhausted. The total energy output was calculated from the measured oxygen uptake and respiratory quotient (R) during 15-min periods; the energy yield from carbohydrate and free fatty acids (FFA), respectively, was estimated from the R values. The calculated total carbohydrate consumption (g) is presented. Note how exercise time and the diet affect the choice of substrate. At a given rate of exercise, the endurance time varied from 93 to 240 min depending on the diet. (The subject's maximal oxygen uptake was not determined.)

From Christensen and Hansen 1939.

uptake, performed with 15 min rest each hour, and tolerated for as long as 3 to 6 h, the hepatic glucose output may be the limiting factor, causing a decrease in the blood sugar. This leads to central nervous system symptoms typical of hypoglycemia (dizziness, partial blackout, nausea, and confusion) in fasting subjects (Pruett 1971; figure 12.6) without depleting the muscle glycogen depots. K. Rodahl, Miller, and Issekutz (1964) showed that fed subjects could complete 6 h of exercise (15 min rest each hour) at a load corresponding to approximately 60% of their maximal oxygen uptake without difficulty. Their blood sugar level remained more or less unchanged. However, when fasting, the same subjects could barely complete the exercise period and had a marked decrease in their blood sugar levels (figure 12.7).

The central nervous system with its low glycogen content depends to a very great extent on the blood sugar. It has been calculated that in humans, approximately 60% of the hepatic glucose output serves the brain metabolism (Reichard et al. 1961). It therefore seems essential that some sort of barrier should exist to prevent the blood glucose from freely entering the muscle cells and being used in their metabolism, because this might lead to a too rapid decrease in blood glucose levels, resulting in

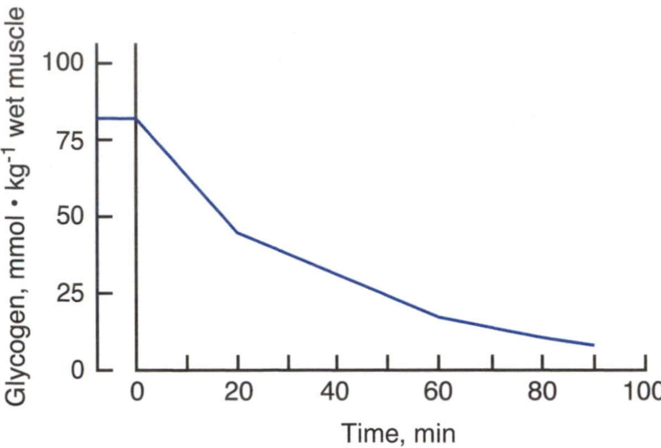


**Figure 12.5** Average values for glycogen content, expressed in glucose units, in needle-biopsy specimens from the lateral portion of the quadriceps muscle taken before and at intervals during exercise until exhaustion, in a group of 20 subjects. Oxygen uptake averaged 77% of the maximal aerobic power.

Adapted from Hermansen, Hultman, and Saltin 1967.

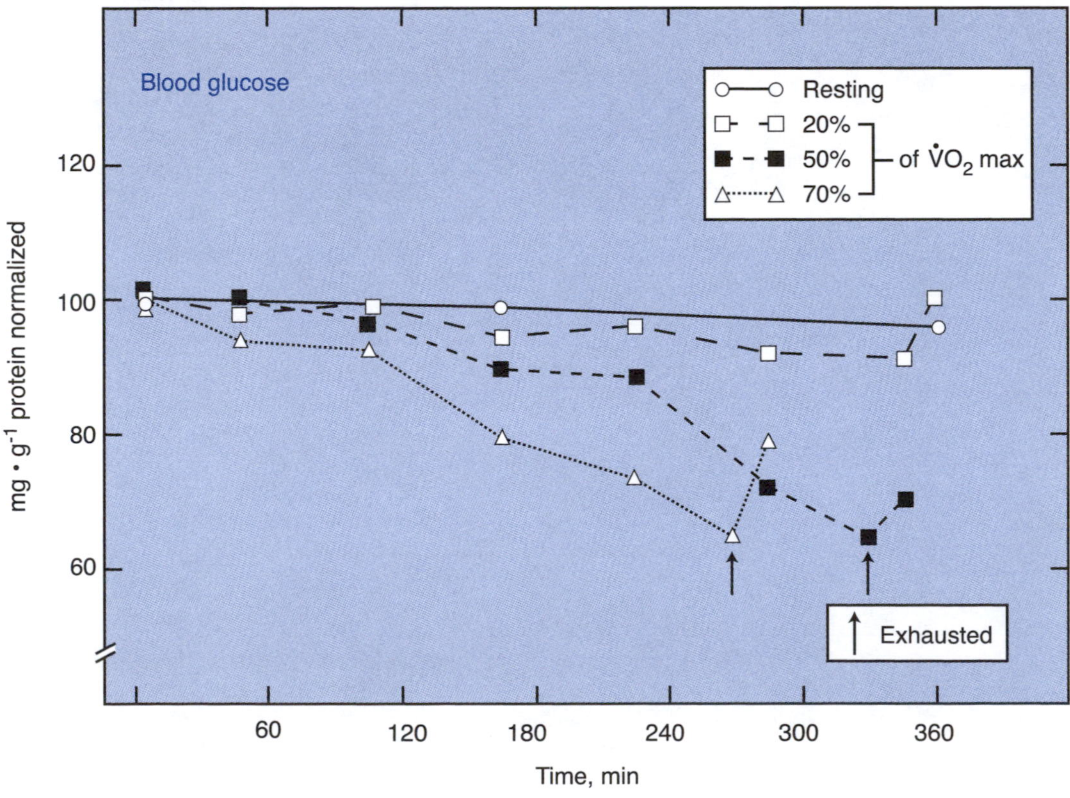


**Figure 12.6** Effect of rest and three levels of exercise on blood glucose concentrations in seven subjects consuming a standard diet.

From Pruett 1971.

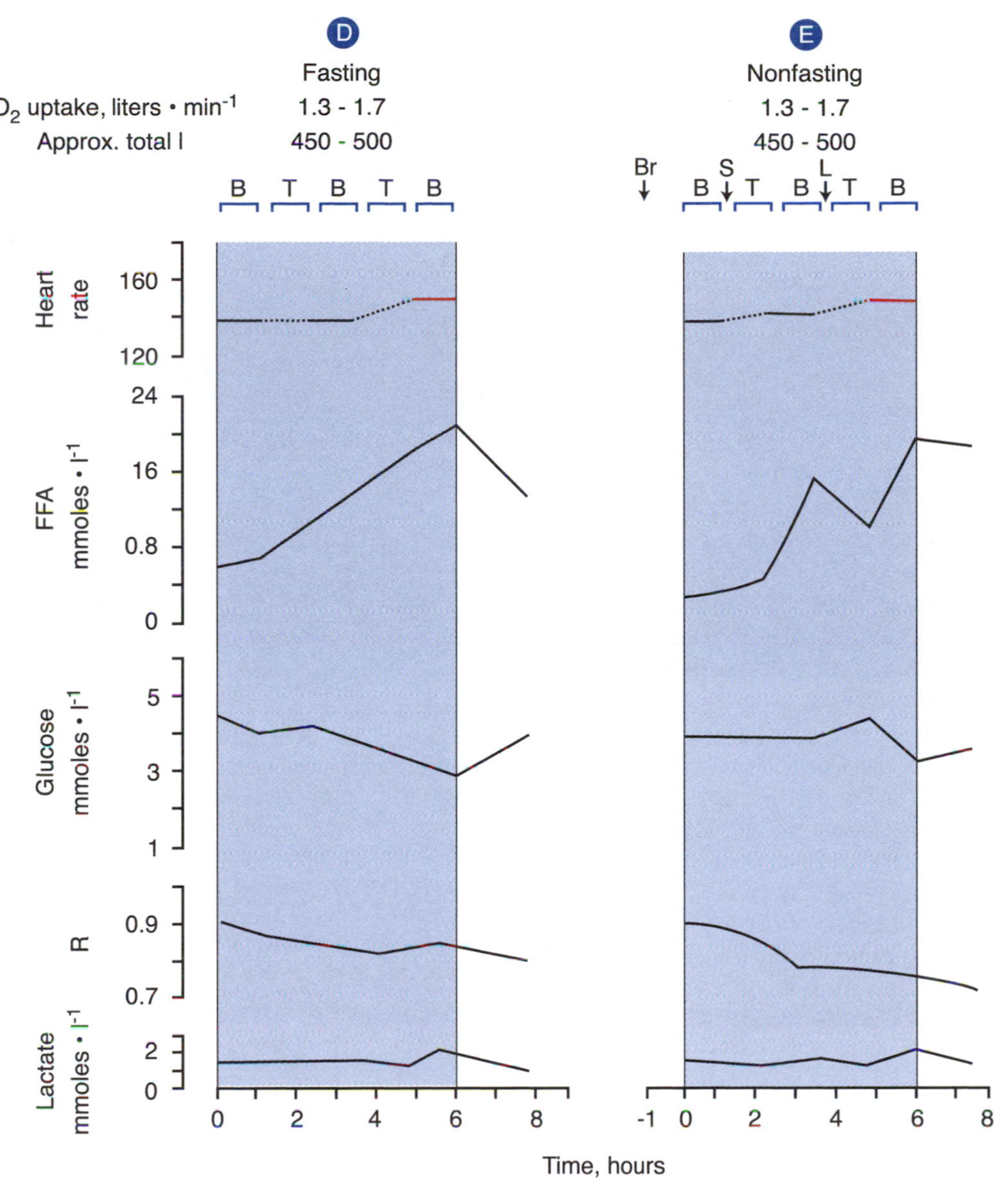


**Figure 12.7** Effect of prolonged exercise. D = fasting subjects; B = exercise on the cycle ergometer at 100 W for 60 min; T = walk on the treadmill with a slope 8.6% at a speed 5.6 km · h$^{-1}$ (3.5 mph) for 60 min; E = nonfasting subjects; Br = breakfast; S = snack; L = lunch; FFA = free fatty acid.

From K. Rodahl, Miller, and Issekutz 1964.

severe symptoms of hypoglycemia. In this respect, it should be pointed out that the permeability of the cell membrane for glucose depends on the plasma insulin concentration, which decreases parallel with the blood sugar during prolonged moderately heavy exercise (Pruett 1971). This can reduce glucose uptake by the exercising muscle cell. Furthermore, enzymes are necessary for the uptake of glucose across the membrane, and at least one such enzyme, hexokinase, is inhibited by products from the breakdown of glycogen (Hultman 1967). Therefore, stored muscle glycogen is a more readily available substrate for the energy metabolism in the exercising muscle cell than exogenous glucose. This is an advantage for the central nervous system, which might otherwise compete with the muscles for blood glucose and suffer from hypoglycemia as a consequence. However, during prolonged exercise the exercising muscles use a significant amount of glucose.

This is partly balanced by an enhanced hepatic **gluconeogenesis** from various glucose precursors, such as alanine and glycerol. Eventually, glucagon may have a stimulatory effect on the hepatic uptake of glucose precursors and on gluconeogenesis. Wahren et al. (1975) estimated the total glucose output from the liver during 4 h of exercise (at 30% of the subject's maximal oxygen output) to be about 75 g. The glucose production in the liver was estimated to be 15 to 20 g. Since the total glycogen content in the liver amounts to 75 to 90 g, the situation can be critical in prolonged heavy exercise. According to Issekutz (1981), the relatively small decline in plasma glucose and insulin in the exercising organism seems to play a major role in increasing hepatic glucose production and in limiting the glucose uptake of exercising muscle during prolonged activity. This mechanism maintains glucose homeostasis as long as possible and guarantees that both liver and muscle glycogen participate in about equal proportions in the elevated glycolysis of the exercising muscle.

At work rates that are above 75% of the individual's aerobic power and that can be tolerated for only about 1-1/2 h at the most, a significant depletion of the muscle glycogen stores can limit endurance. Under these conditions, there is usually no decrease in blood sugar, because the duration of the exercise period is too short to deplete hepatic glycogen stores to any significant extent, whereas the muscle glycogen stores can be markedly reduced indeed. Normally, some glucose from the blood enters the muscle cell at all times. However, when the muscle glycogen store is nearly depleted, an increasing amount of glucose evidently enters the muscle cells from the blood.

Bergström and Hultman (1966) performed an experiment in which one subject exercised with his left leg and the other subject simultaneously exercised with his right leg on the same cycle ergometer. After several hours of exercise, the exercising leg of each individual was almost depleted of glycogen whereas the resting leg still had a normal glycogen content. Feeding the subjects a carbohydrate-rich diet on the following days did not markedly influence the depots of the resting limb, but in the previously exercised leg the glycogen content increased rapidly until the values were about twice as high as those in the nonexercised leg (figure 12.8). This experiment shows that exercise with glycogen depletion enhances the resynthesis of glycogen. It also shows that the factor must operate locally in the exercising muscle (Bergström, Hultman, and Roch-Norlund 1972).

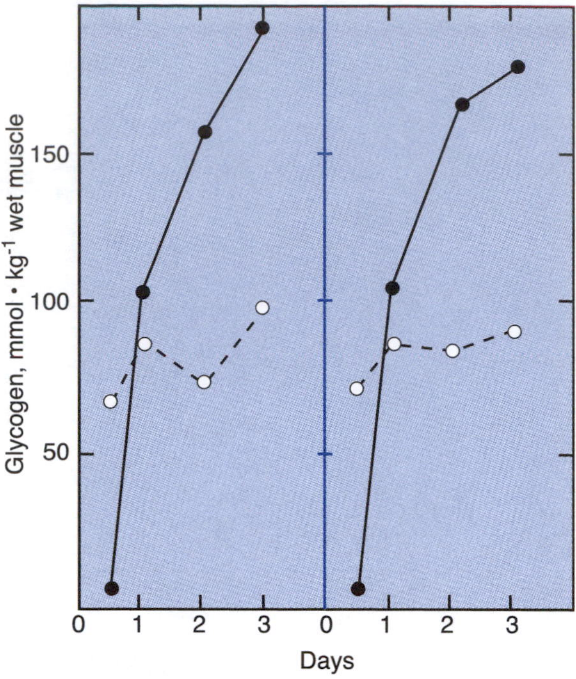


**Figure 12.8** Two subjects were exercised on the same cycle ergometer, one on each side exercising with one leg while the other leg rested (dashed line). After the subjects exercised to exhaustion, their glycogen content was analyzed in specimens from the lateral portion of the quadriceps muscle. Thereafter, a carbohydrate-rich diet was consumed for 3 days. Note that the glycogen content increased markedly in the leg that previously had been emptied of its glycogen content.

Adapted from Bergström and Hultman 1966.

## Carbohydrate Loading

The available evidence suggests that at work rates exceeding about 75% of the individual's maximal oxygen uptake, the initial glycogen content in the skeletal muscles determines the individual's ability to sustain such exercise for more than an hour. Thus, Bergström et al. (1967) showed that at a work rate of about 75% of the maximal oxygen uptake, the larger the initial muscle glycogen stores, the longer the subject could continue to exercise at this load (figure 12.9). After a normal mixed diet that provided an initial glycogen content of about 100 mmol · kg$^{-1}$ of wet muscle, the subject could tolerate the 75% work rate for 115 min. After the subject spent 3 days on an extreme fat and protein diet, the glycogen concentration was reduced to about 35 mmol · kg$^{-1}$ of wet muscle and the standard load could be performed for only about 60 min. After 3 days on a carbohydrate-rich diet, the subject's glycogen content increased, to 200 mmol · kg$^{-1}$ of wet muscle, and the time on the

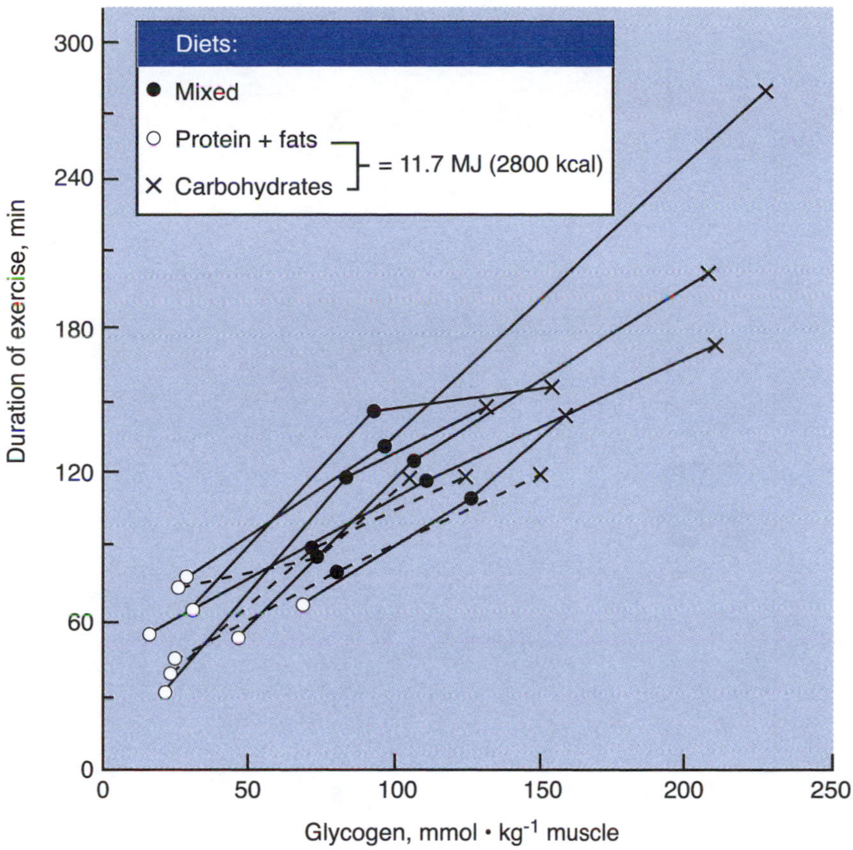


**Figure 12.9** Relation between initial glycogen content in the quadriceps muscle in nine subjects who had been on different diets, and maximal exercise time when they exercised at a given work rate demanding 75% of maximal aerobic power. Broken lines denote subjects who had been on a carbohydrate diet before the fat-plus-protein diet.

From Bergström et al. 1967.

75% work rate was prolonged to about 170 min on the average. It was further observed that the most pronounced effect was obtained if the glycogen depots were first emptied by heavy prolonged exercise and then maintained at low levels by giving the subject a diet low in carbohydrate, followed by a few days with a diet rich in carbohydrates (figure 12.10). With this procedure, the glycogen content could exceed 200 mmol · $kg^{-1}$ of wet muscle and the heavy load could be tolerated for longer periods, in some subjects for more than 4 h. The total muscle glycogen content under these conditions could exceed 700 g.

In prolonged athletic events in which the work rates exceed about 75% of the maximal oxygen uptake, not only endurance but also the ability to maintain high running speed is affected by the initial muscle glycogen content. This is schematically illustrated in figure 12.11. A group of subjects participated in two 30K cross country running races, on the first occasion after their normal mixed diet and on the second occasion after a few days on an extremely high-carbohydrate diet after previously emptying the glycogen depots. At several points on the track, the running time was recorded. It was observed that the lower the initial muscle glycogen content, the lower the subjects' ability to maintain a high running speed toward the end of the race. However, even in the subjects with the lowest initial muscle glycogen content, the speed was maintained during the first hour of the race. This shows that a high muscle glycogen content did not enable the subjects to attain a higher speed at the beginning of the race any more than did a low initial glycogen level.

Under such conditions of very intense or near-maximal effort of sufficient duration, the draining of the muscle glycogen apparently is high enough to significantly deplete muscle glycogen stores. In this event, then, the cause of exhaustion is located in the exercising muscle tissue. However, even at the point of such exhaustion, the muscle is still using significant amounts of glycogen or glucose as judged by the respiratory quotient.

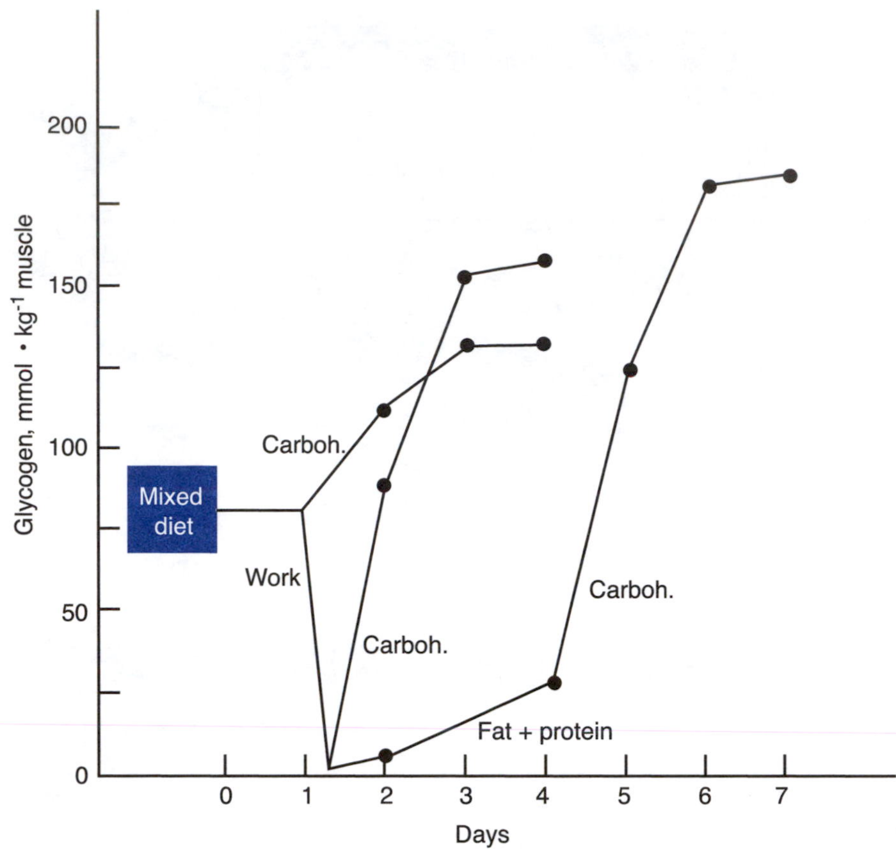


**Figure 12.10** Different possibilities for increasing the muscle glycogen content. For further explanation, see text.

From Saltin and Hermansen 1967.

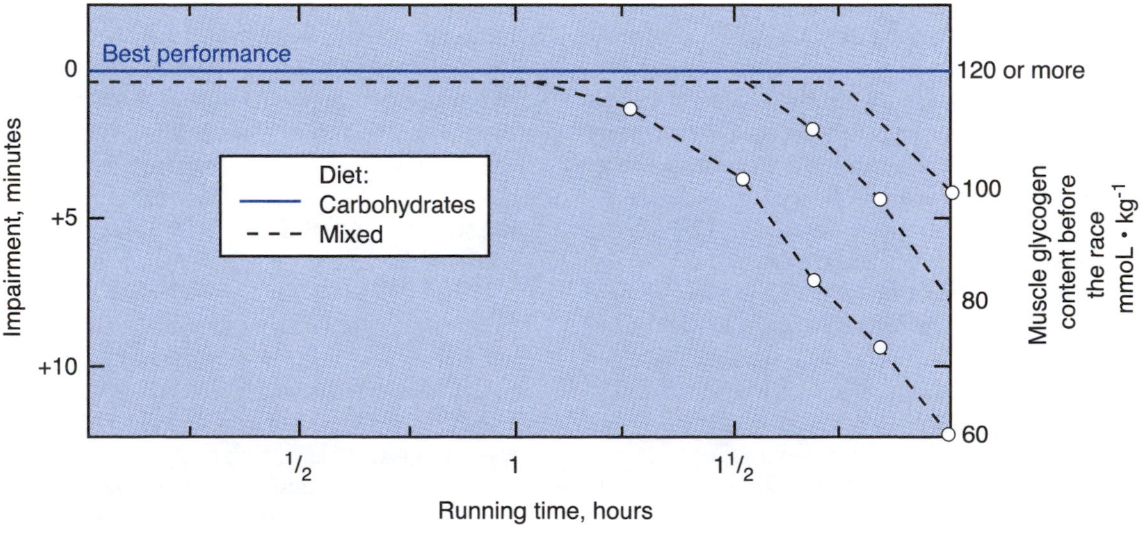


**Figure 12.11** Schematic illustration of the importance of a high glycogen content in the muscle before a 30K race (running). The lower the initial glycogen store, the slower the speed at the end of the race compared with the race performed when the muscle glycogen content was 120 mmol · kg$^{-1}$ of muscle or more at the start of the race. For the first hour, however, no difference in speed was observed.

Courtesy of B. Saltin.

There is one drawback with the high glycogen storage: We mentioned that each gram of glycogen is stored together with about 2.7 g of water. With a glycogen storage of 700 g, body water then increases by about 2 kg. Therefore, in activities in which the body weight has to be lifted, an excessive glycogen store should be avoided.

It is thus clear that under certain conditions, that is, during heavy exercise equivalent to about 75% or more of the maximal oxygen uptake, the level of the muscle glycogen stores can significantly affect performance after about 1 h. This can be influenced by the diet (see figure 12.10).

Since the classic method of "carbohydrate loading" was introduced by Bergström et al. (1967), as previously described, various dietary methods have been developed to increase the muscle and liver glycogen levels prior to exercise. The reason was that the "classic regimen," with 3 days of extreme low-carbohydrate diet combined with exhaustive exercise, frequently led to irritability and dizziness caused by low plasma glucose levels. Furthermore, many believed that such a regimen disrupted "peaking" for the competition. W. Sherman et al. (1981) showed that a gradual reduction in the intensity and duration of daily workouts and an increase in the carbohydrate content of the diet to 70% of the daily energy intake during the last 3 days before the competition was as effective as the traditional method. Endurance performance, determined by constant pace running to exhaustion, also is improved simply by dietary carbohydrate loading for 2 to 3 days before competition (Brewer, Williams, and Patton 1988).

# Regulatory Mechanisms

The use of metabolic fuel is regulated by the interplay of a number of factors, both physiological and biochemical in nature (figure 12.12).

First, a high-carbohydrate diet favors a higher participation of carbohydrate in energy metabolism, as first described by Christensen and Hansen (1939). This was confirmed by Issekutz, Birkhead, and Rodahl (1963), Bergström et al. (1967), and Pruett (1971). Issekutz, Birkhead, and Rodahl (1963) found that the carbohydrate intake rather than the amount of dietary fat determines whether the preferred fuel is FFA or carbohydrate. Ingestion of 100 g of glucose immediately before exercise shifts exercise metabolism toward carbohydrate and correspondingly reduces FFA metabolism. The reason is probably that the carbohydrate intake increases the production of insulin, which inhibits FFA release and stimulates esterification in adipose tissue. Hence, FFA oxidation is reduced. Wirth et al. (1981) showed

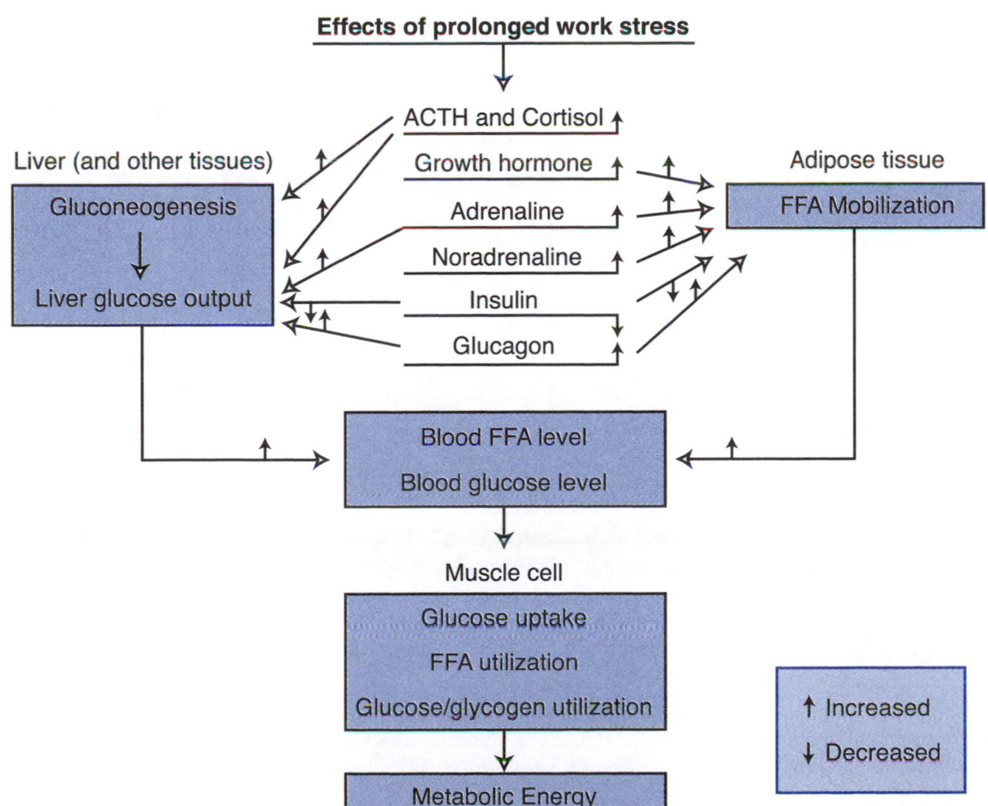


**Figure 12.12** Schematic representation of the effect of exercise on hormone secretion and the concomitant effect of those hormones on fat and carbohydrate metabolism. ACTH = adrenocorticotropic hormone; FFA = free fatty acid.

From Pruett 1971.

that basal plasma insulin concentrations appear to be lower in athletes than in nonathletes, which is attributable to reduced insulin secretion and increased sensitivity. During exercise, however, insulin secretion is diminished independent of the training state. After exercise, active recovery at low intensity can maintain a low insulin level (Wigernæs, Strømme, and Höstmark 2000).

Second, the use of FFA during muscular exercise is determined partly by the level of plasma FFA. During exercise, an increased plasma FFA concentration implies an increased rate of FFA utilization. At any given plasma FFA level, the FFA turnover rate is about twice as high during exercise as during rest (Issekutz et al. 1964; Paul 1975). As pointed out in chapter 11, one-leg training increased utilization of FFA in the trained leg compared with the untrained leg during a standardized two-leg exercise. Because the arterial concentration of FFA was identical in the two legs, it appears that local factors affected by training also influence FFA utilization.

The plasma FFA level during exercise is determined by the combined effects of several factors:

1. Noradrenaline is a most powerful stimulator of FFA mobilization. Even small increases in noradrenaline cause a marked increase in plasma FFA levels associated with a corresponding increase in FFA turnover (Issekutz 1964; Rodahl and Issekutz 1965). During both short, intense and prolonged, moderately heavy exercise, there is a considerable increase in catecholamine production (figure 12.13), which increases FFA mobilization and utilization.

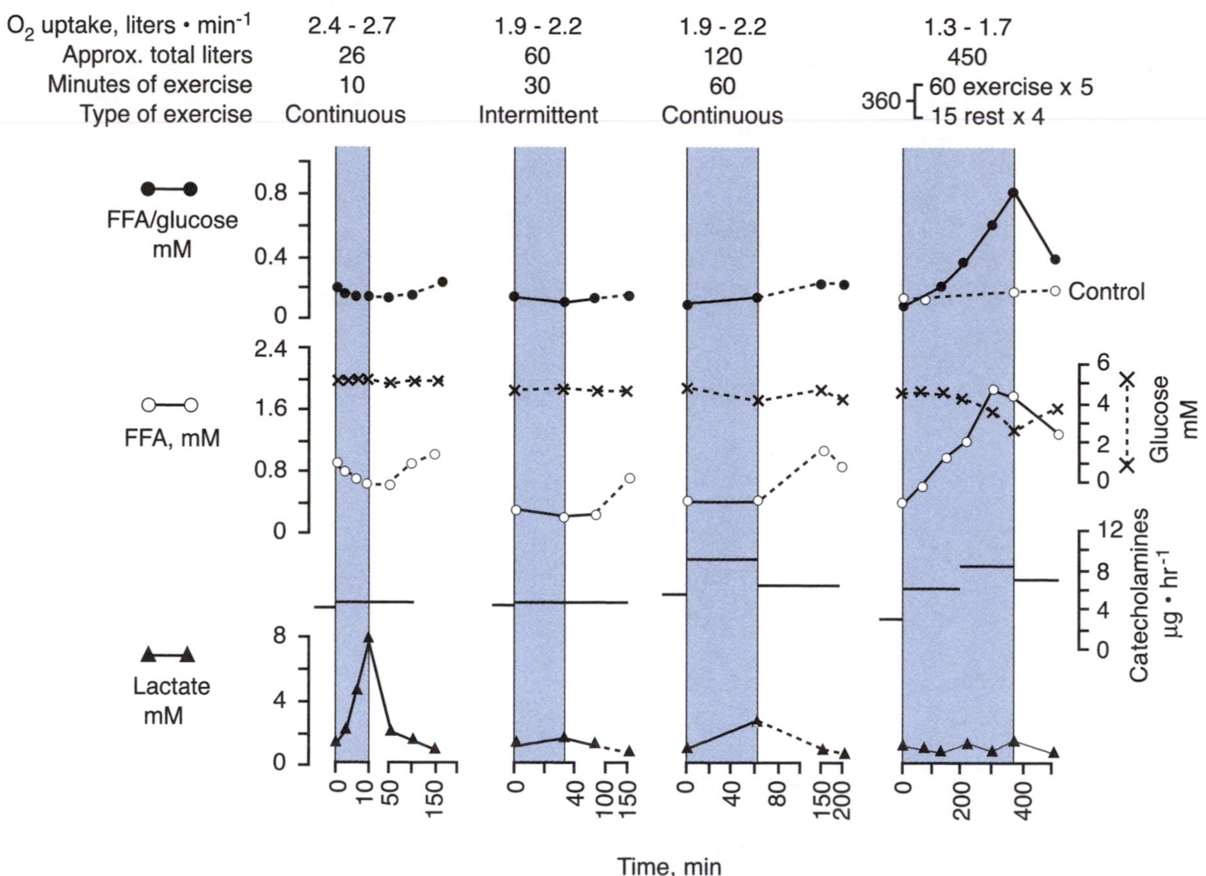


**Figure 12.13** Effects of exercise of different intensity and duration in normal fasting subjects in relation to blood lactate levels, urinary catecholamine excretion, plasma free fatty acid (FFA), and blood glucose: Heavy exercise on the cycle ergometer for 10 min; intermittent treadmill run of 5 s exercise, 5 s rest, slope 8.6%, speed 12 km · h$^{-1}$, 7.5 mph for 30 min; continuous exercise on the cycle ergometer at 150 W for 60 min; and alternating exercise on the cycle ergometer at 100 W for 60 min and exercise on the treadmill with a slope of 8.6% at a speed of 5.9 km · h$^{-1}$, 3.5 mph for 60 min for a total period of 360 min or until exhaustion. Note the increase in plasma FFA and the decrease in blood glucose during prolonged exercise and the marked increase in blood lactate level during heavy short exercise, while the blood lactate is essentially unchanged during the prolonged exercise. Note also the increase in plasma FFA following the cessation of short, intense exercise.

From Rodahl et al. 1964.

2. The lactate accumulated during strenuous muscular exercise suppresses the mobilization of FFA from the adipose tissues. The possibility that lactate might function as a physiological inhibitor of FFA mobilization in severe exercise was first suggested by Issekutz and Miller (1962), who observed an inverse correlation between FFA and lactate levels in the blood (figure 12.14). H.I. Miller et al. (1964) showed that lactate infusion decreased the inflow of FFA into the plasma in depancreatized dogs. Fredholm (1969) induced FFA mobilization in isolated adipose tissue preparations by sympathetic nerve stimulation, which could be counteracted by lactate infusion in physiological concentrations. From his experiments, it appears that an increased reesterification of FFA is the major mechanism underlying the lactate effect. This view is supported by the findings of Issekutz, Shaw, and Issekutz (1975). In prolonged, moderately heavy exercise, the lactate level is not significantly increased, but the catecholamine level is (see figure 12.13). This may explain the observed increase of FFA mobilization under such conditions. In contrast, the accumulation of lactate during short, intense exercise may prevent an increase in plasma FFA during the exercise period itself but permits a subsequent increase after exercise when the lactate level has dropped (see figure 12.13; Rodahl, Miller, and Issekutz 1964). After cessation of intense or near-maximal exercise, the increase in plasma FFA levels is very marked and can last for several hours, or until the next meal is taken (figure 12.15). Such a postexercise increase in FFA levels also can occur following intermittent exercise (Bahr et al. 1991; Ranallo and Rhodes 1998; Rodahl, Miller, and Issekutz 1964). Glucose supply shortly after strenuous exercise reverses the increase in FFA (El-Sayed et al. 1996).

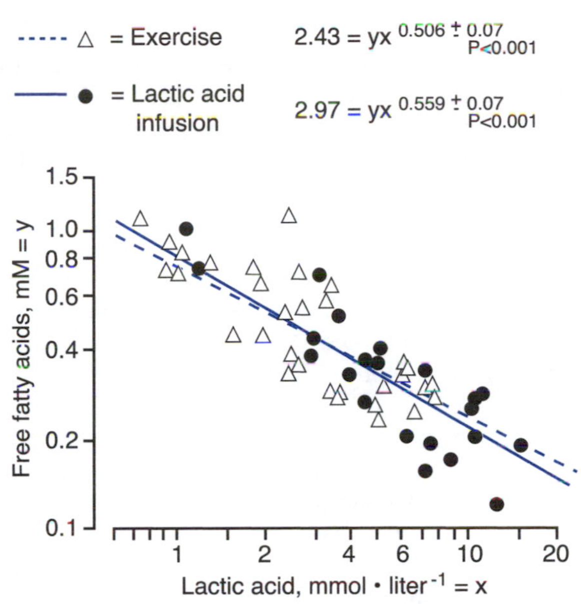


**Figure 12.14** Treadmill experiment on a dog (9.5 kg). Interrelationship between plasma free fatty acid level and blood lactic acid concentration.

From Issekutz and Miller 1962.

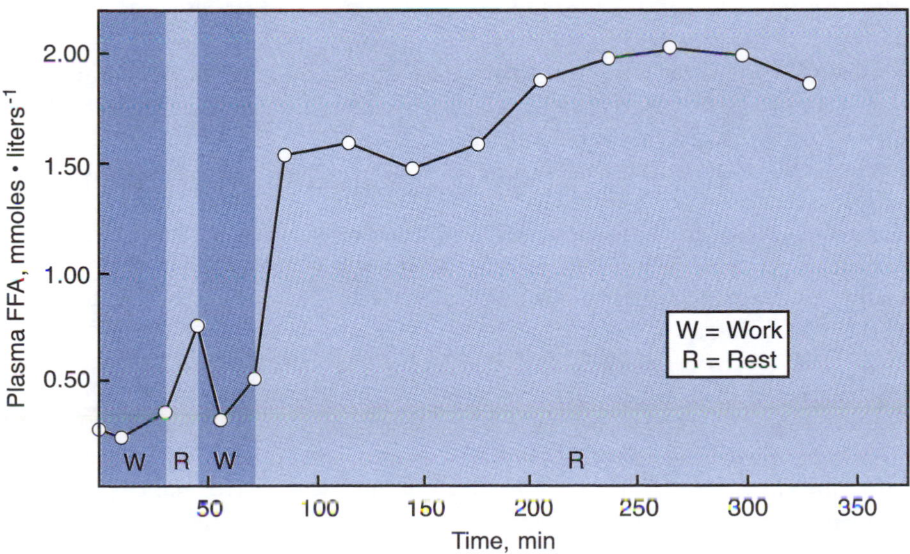


**Figure 12.15** The plasma free fatty acid (FFA) response to two exhausting exercise bouts on the cycle ergometer of 87% $\dot{V}O_2$max.

From Pruett 1971.

3. A number of hormones other than catecholamines are affected by physical exercise. As already mentioned, insulin suppresses FFA mobilization and augments esterifications. However, insulin is decreased during prolonged exercise (Pruett 1971; Wigernæs, Strømme, and Höstmark 2000). Insulin, therefore, cannot prevent the observed increase in plasma FFA concentration during prolonged exercise. It is known that the plasma growth hormone concentration is increased during exercise (Hartley 1975). This increase can affect both glucose and FFA metabolism. The administration of growth hormone is reported to cause an increase in plasma FFA lasting several hours (Grunt et al. 1967).

Cortisol, which is produced during both physical and mental stress, profoundly influences carbohydrate metabolism. It increases the catabolic breakdown of proteins and the formation of carbohydrates from protein, which in turn can enter the metabolic pool.

An important effect of training is increased oxidation of FFA and reduced energy yield from glycogen. This change in fuel utilization is evident not only at a given metabolic rate but also when an individual exercises at a given percentage of maximal aerobic power. This adaptation is certainly a glycogen-saving mechanism. A reduced lactate production in the trained individual could be one factor behind this difference in metabolic pattern. The modification in the mitochondrial enzyme profile induced in trained muscles also could be of importance. Furthermore, a reduction in the diffusion distance between capillaries and the interior of the muscle cells will facilitate FFA uptake. An increased oxidation of FFA will elevate the concentration of citrate, which in turn can inhibit the activity of the enzyme phosphofructokinase, thereby retarding glycogenolysis and lactate formation. All this means that the endurance-trained individual can exercise closer to his or her maximal aerobic power for longer periods of time than the untrained person. Another interesting effect of physical training is that the intracellular muscle fat, which is stored as triglycerides in the form of small fat droplets located near the mitochondria, increases in endurance-trained individuals. It is also known that prolonged strenuous exercise can deplete intramuscular fat. Thus, an increase in this fat store would improve the substrate availability of the muscle cells. However, in proportion to the total body fat content, the intramuscular fat content is rather small (Brouns, Saris et al. 1989).

### For Additional Reading

For a more comprehensive discussion of fat metabolism during exercise and the hormonal regulation of substrate metabolism in general during exercise, see Newsholme, Calder, and Yaqoob (1993), Galbo (1995), Jeukendrup, Saris, and Wagenmakers (1998), and Spriet (2002). For a review of the effects of exercise on the neurobiological control of food intake and energy expenditure, the reader is referred to Richard (1995).

# Food for the Athlete

In general, a varied, well-balanced diet in adequate amounts is all that is necessary from a nutritional point of view for the body to function optimally, and such a diet provides a biological basis for top performance. It is true that the requirements for protein and certain minerals can be somewhat increased in athletes. However, the total food intake in most athletes also is increased, often by more than 4 MJ · day$^{-1}$ (1,200 kcal) in connection with regular training and competition. The greatly increased energy expenditure automatically increases the appetite, resulting in the ingestion of more food. If athletes' diets are balanced, the athletes will automatically take in more of these nutrients as they increase food intake. This is true provided that their diet is not too high in fat (i.e., that no more than 35–40% of the energy is derived from fat) and that the content of refined sugar is relatively low, with the consumption of milk, fish, meat, vegetables, fruit, berries, and grain products being comparatively high. Such a diet ensures an adequate intake of proteins, minerals, and vitamins. This diet holds for athletes in general, whether they are engaged in strength or endurance training and whether they are engaged in competitions or not. As shown in figure 12.16, this was convincingly demonstrated by Erp-Baart (1992).

## Vitamins and Minerals

It is well known that vitamin or mineral deficiency impairs performance capacity, but it takes a long time to develop such deficiency in an individual living on a deficient diet. For this reason, an individual can be without any vitamin intake for a week or so without any detectable detrimental effect on work capacity (K. Rodahl 1960). As long as vitamin status is adequate, additional vitamin intake does not improve performance (Mayer and Bullen 1960; Van der Beek 1991). According to the American College of Sports Medicine (2000), the current **Recommended Daily Allowance (RDAs)** and **Dietary Reference Intakes (DRIs)** are appropriate for most

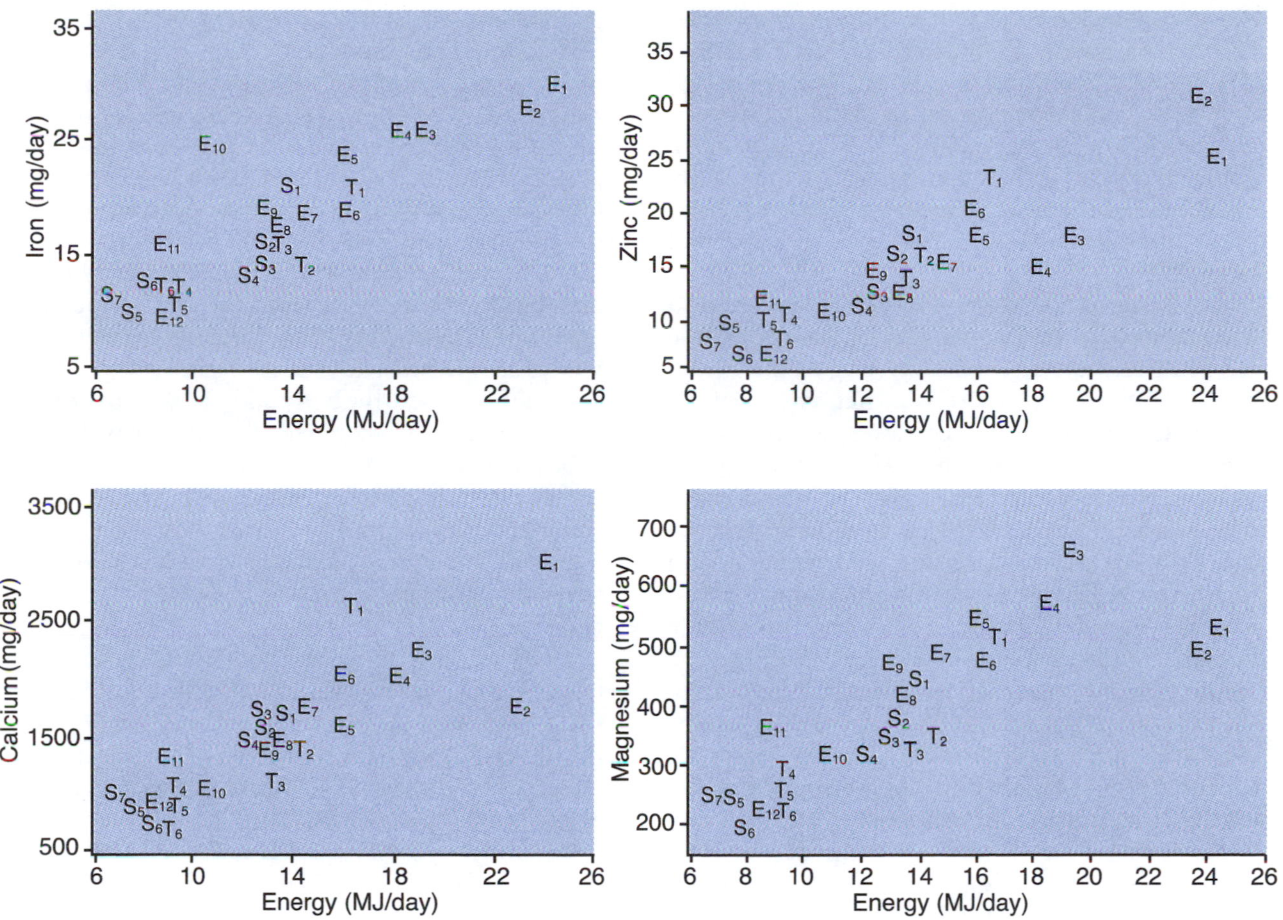


**Figure 12.16** Magnesium, zinc, iron, and calcium intakes in athletes increase with higher energy intakes. E = endurance athletes, S = strength athletes, T = team sport athletes.

Reprinted, from "Die Ernahrungsbedurfnisse von Sportlern," F. Brouns, 78 (33), 80 (34), 86 (36), 88 (37), 1993, © Springer-Verlag. English translation © John Wiley & Sons Limited. Reproduced by permission.

athletes. However, athletes at times subject themselves to stringent dietary regimens, often combined with dehydration, in an endeavor to lose weight and thus to obtain a lower weight classification (e.g., in sports like wrestling, boxing, judo, rowing, and horseracing). Such athletes and those who have chronically low energy intake to achieve low body fat (e.g., in activities like ballet, gymnastics, and aerobics) are at considerable risk of marginal nutrition and consequently of poor vitamin and mineral status. They may need to use a multivitamin and mineral supplement to improve overall micronutrient status. Supplementation with single micronutrients is discouraged unless medical, nutritional, or public health reasons are present, such as iron to treat iron-deficiency anemia or folic acid to prevent birth defects (American College of Sports Medicine 2000).

Particular attention has been focused on the possible beneficial effect of the B-complex vitamins, the reason being that riboflavin, niacin, thiamin, vitamin $B_6$, pantothenic acid, and biotin are involved in energy production during exercise, whereas vitamin $B_{12}$ and folate are needed for production of erythrocytes and protein synthesis. According to Manore and Thompson (2000), exercise increases the need for the B-complex vitamins in athletes perhaps up to twice the current RDAs. However, this increased need generally can be met by the higher energy intake of most athletes. In the case of the water-soluble vitamins, the ingestion of large quantities of vitamin pills is a rather expensive way of increasing the vitamin content of the urine, which serves no useful purpose.

Iron intake is subject to considerable interest because a low iron store is one of the most prevalent nutritional problems. A high percentage of female athletes, especially those involved in endurance exercise, have been found to have some degree of iron deficiency (Clement et al. 1987; Weaver and

Rajaram 1992). Because iron is required for enzymes involved in energy production and for the formation of hemoglobin and myoglobin, iron deficiency reduces aerobic power. Iron depletion usually is caused by low energy intake, avoidance of meat, fish, and poultry (which contain iron in the readily available heme form), vegetarian diets that have poor iron bioavailability, and losses in menstrual blood.

It is hard to conceive how decreased **serum iron** (iron bound to **transferrin**) levels can affect the oxygen-transporting system as long as the hemoglobin content of the blood is normal, as it usually is in athletes. However, the decrease in serum iron is a forerunner for a reduction in the blood hemoglobin content. Before the decrease in serum iron, changes in iron storage (in liver, spleen, and bone marrow) usually have occurred, as mirrored by the serum **ferritin** level. Ferritin binds iron and stores it within the cells, especially in the liver. A small amount of ferritin circulates in the blood, and because the serum ferritin concentration reflects the amount of stored iron, measurement of serum ferritin can be used to estimate iron stores (1 μg ferritin · $L^{-1}$ = 8 mg of stored iron; Manore and Thompson 2000). Follow-up tests are recommended for athletes who experience fatigue, pallor, performance decline, heavy menstrual bleeding, have low energy intake, or are vegetarians, and they should be performed regularly to assess iron status. Because reversal of iron deficiency anemia can require 3 to 6 months, the nutrition intervention should begin as soon as low serum ferritin is detected. LaManca and Haymes (1993) studied the effects of repletion of iron stores on aerobic power, endurance performance, and blood lactate concentration in 20 female athletes with low plasma ferritin concentrations (< 20 ng · $ml^{-1}$). They found that 82 mg of elemental iron per day for 8 weeks significantly improved the iron status and concluded that reversal of mild anemia was associated with higher aerobic power and lower blood lactate level following submaximal exercise.

### For Additional Reading

For further reading about vitamin and mineral status in athletes, see Beard and Tobin (2000) and Fogelholm (1995).

## Hydration

It is well known that work capacity is impaired with progressive dehydration (McConell et al. 1997; Montain and Coyle 1992) and that dehydration increases the risk of developing heat injury (Noakes 1993). Sweat losses during exercise depend mainly on ambient temperature, humidity, exercise intensity, and acclimation status and can exceed 2.5 L · $h^{-1}$ (Sawka and Pandolf 1990). During competitive cross-country orientation, a mean fluid loss of 3.0 L during the 90-min duration of the competition has been observed (Saltin 1964). Even in winter time, fluid loss of more than 7 L has been observed during long-distance ski racing lasting between 5.0 and 6.5 h (Refsum and Strømme 1974). Sweat loss of a similar magnitude has been reported for many industrial operations (K. Rodahl 1994). Because sweating is a vital process which, during heavy muscular activity or in the heat, cools the body and prevents overheating, it is generally recognized that the lost water has to be replaced, preferably at the same rate at which it is lost (see chapter 13). Sweetened carbohydrate-containing drinks may be the best way of maintaining fluid and energy balance during prolonged, strenuous physical exercise (Noakes 1993; Saris et al. 1989).

Daily water requirement represents the amount necessary to cover sweat loss and **insensible perspiration** losses via breathing and skin and to supply the kidney with the fluid needed for excretion of metabolic end-products. Although it is difficult to assess the daily minimum amount of fluid needed by the body, an intake level of 1 ml · $kcal^{-1}$ (4.2 kJ) of energy expenditure has been recommended by the National Research Council (1989).

Athletes should avoid weight loss through dehydration. The amount of fluid ingested by athletes is normally much less than can be tolerated (Maughan and Leiper 1999). During prolonged efforts in a hot climate, adequate fluid intake is essential. A dehydration corresponding to a loss of body water in excess of 1.5% to 2% of the body weight should be avoided, not only to perform optimally but also to avoid exercise-induced nausea and epigastric cramps, which have been shown to be associated with more severe dehydration (van Nieuwenhoven at al. 2000).

Athletes should always be well hydrated when starting exercise. In addition to consuming ample amounts of fluid in the 24 h before an exercise session, athletes should drink 400 to 600 ml of fluid 2 to 3 h before exercise, especially in warm weather. Such a practice optimizes hydration status while allowing sufficient time for any excess fluid to be excreted before exercise begins.

### For Additional Reading

For further details and recommendations about exercise and fluid replacement, see Maughan (1991), American College of Sports Medicine (1996), and Kovacs and Brouns (1997).

## Fuel During Exercise

Carbohydrate feeding during prolonged exercise improves performance, in particular if the carbohydrates are ingested throughout exercise (McConell, Kloot, and Hargreaves 1996). Even late in exercise, a single carbohydrate feeding can supply sufficient carbohydrate to restore euglycemia and increase carbohydrate oxidation, thereby delaying fatigue (Coggan and Coyle 1989). Not only is physical performance ability improved by carbohydrate feeding. Collardeau et al. (2001) showed enhanced complex cognitive performance measured at the end of a 100-min run as a consequence of carbohydrate ingestion before and every 15 min throughout the exercise. Oral administration of about 225 ml of a 5% glucose solution, given every 15 min, was sufficient to prevent hypoglycemia in subjects exercising at 70% of their maximal oxygen uptake for 3 h (Staff and Nilsson 1971). By giving naturally labeled carbon-13 glucose to subjects exercising at 50% $\dot{V}O_2$max, Pirnay et al. (1977) demonstrated that carbon-13 can be detected in the expired carbon dioxide as early as 15 min after the oral intake, indicating a rapid resorption and utilization of exogenous glucose during such levels of exercise. Even at higher exercise intensity (80% $\dot{V}O_2$max), Bonen et al. (1981) observed that a considerable portion of glucose ingested during exercise was metabolized.

Other fuels might participate as sources of energy, such as amino acids, acetoacetate, and 3-hydroxybutyrate, but they play an insignificant role, at any rate in well-fed individuals engaged in intense muscular exercise at high levels of energy expenditure (McGilvery 1975; Suminski et al. 1997). During prolonged fasting, however, ketone bodies can replace glucose oxidation in the brain (Owen et al. 1967).

When we discuss the dietary requirements of athletes, it is necessary to distinguish between events of very short duration, which mainly involve technique and muscular strength, and events lasting up to several hours, which mainly require endurance. In the case of the endurance events, it is necessary from a nutritional standpoint to distinguish between events lasting less than 1 h and events of significantly longer duration.

### *Events Lasting Less Than 1 Hour*

In very intense physical exertion or athletic events lasting less than 1 h, the available supply of stored energy fuel is generally ample to cover the need. Under such conditions, a special diet is unnecessary. For instance, in a study in which well-trained cyclists increased the carbohydrate content of their diet for 3 days from 6 to 9 g · $kg^{-1}$ of body mass, there was only a modest increase in muscle glycogen content (Hawley, Palmer, and Noakes 1997). The authors concluded that additional carbohydrate provides no performance benefit for athletes who compete in intense continuous events lasting 1 h or less. Similar findings were reported by Pitsiladis and Maughan (1999) and Lynch, Galloway, and Nimmo (2000).

Because the digestion of a meal redistributes blood from the muscles to the gut, physical exercise shortly after a meal will result in a competition between the gut and the exercising muscles for the available blood supply. Heavy muscular exercise therefore should be avoided immediately following a heavy meal. As a general rule, a meal should not be ingested less than 2-1/2 h before an athletic event. Furthermore, the last meal before the event should be light. It should consist only of ingredients that the individual knows from experience can be tolerated well.

### *Events Lasting for Several Hours*

The primary cause of fatigue in exercise lasting more than 1 h (but not more than 4–5 h) is usually the depletion of the body's carbohydrate reserves. Thus, in the case of athletic events involving large muscle groups at high work rates for periods exceeding an hour or so, it is advisable to ingest ample quantities of carbohydrates several days preceding the events to fill the glycogen depots (see figure 12.11). Here, however, it is more important, relatively speaking, to ingest carbohydrates during the actual event to supplement the hepatic glucose output.

The individual should avoid heavy muscular activity, which might deplete the existing glycogen depots prior to the event. On the other hand, it is not advisable to live on a very high-carbohydrate diet regularly, because this would condition the metabolic processes to use carbohydrate fuel rather than FFA, as already explained (Durnin 1982; Wilmore and Freund 1984).

The availability of water and carbohydrates can be limited by gastric emptying and intestinal absorption. Gastric emptying of liquids is slowed by the addition of carbohydrate in proportion to the carbohydrate concentration and osmolality of the solution. This can produce abdominal discomfort during exercise.

With increasing glucose concentration, the rate of fluid delivery to the small intestine also is decreased, although the rate of glucose delivery can be increased. Being a passive process, the absorption of water across the gut membrane in the small intestine is stimulated by the active absorption of glucose and sodium (Maughan 1991). The composition of fluids to be used will depend on the relative needs to replace water and to provide energy. If rehydration is a priority, the solution should contain some carbohydrate and sodium and preferably should not exceed isotonicity. This will require the glucose concentration to be low (20–30 g · $L^{-1}$) and the sodium content to be relatively high (up to 60 mmol · $L^{-1}$). If energy provision is most important, a solution containing glucose polymers in concentrations of 150 to 200 g · $L^{-1}$ is preferred. To minimize the limitation imposed by the rate of gastric emptying, the volume of fluid in the stomach should be kept as high as is comfortable by frequent ingestion of small amounts of fluid (Maughan 1991). According to the American College of Sports Medicine (1996), optimal hydration can be facilitated by drinking 150 to 350 ml of fluid at 15- to 20-min intervals, beginning at the start of the activity. Beverages containing carbohydrate in concentrations of 4% to 8% and sodium in amounts between .5 and .7 g · $L^{-1}$ are recommended. Including sodium in fluid replacement beverages also can prevent **hyponatremia** (abnormally low blood sodium level caused by an excess of water) in susceptible people. Most athletes who drink more fluid than they lose simply excrete the excess fluid as urine. However, in some individuals the fluid is retained (Speedy et al. 1999; Vrigens and Rehrer 1999).

The rate of absorption of glucose, water, and various minerals in the gastrointestinal tract is not affected by exercise, at least not by loads demanding less than about 70% of the maximal oxygen uptake. In Fordtran and Saltin's classic experiments (1967), at least 50 g of glucose was emptied from the stomach during 1 h of heavy exercise. This amount corresponded to one fourth to one half of the carbohydrates required by the body during this period. The data of Fordtran and Saltin (1967) suggest that gastric emptying and intestinal absorption of saline solutions could be rapid enough to replace all the losses of sweat incurred during heavy exercise, even in hot environments.

Ingestion of easily absorbable carbohydrates before exercise rapidly increases blood glucose and insulin. When exercise is started, this can result in a rebound hypoglycemia and consequently decreased performance. However, such hypoglycemia can be prevented if the carbohydrate intake occurs when the subjects is physically active (Brouns, Rehrer et al. 1989). Thus, if the period of exertion during the event is very long, it might be advantageous to consume moderate amounts of sugar, preferably in the form of a flavored glucose solution, just before or during the warm-up.

Following are two examples of how the energy and fluid requirements are covered during long-lasting strenuous physical exertion:

In the case of bicycling, Gabel, Aldous, and Edgington (1995) assessed the food and fluid intake of two elite male cyclists participating in an endurance race lasting 15 to 18 h each day for 10 days, covering 2,050 miles. The average daily energy intake was 475 kJ (113 kcal) · $kg^{-1}$ · $day^{-1}$. Energy percentages of protein, carbohydrate and fat were 11%, 62%, and 27%, respectively, with 44% of the carbohydrate coming from simple sugars, cookies, sweetened drinks, and candy. Total fluid intake averaged 10.5 L per day, with an average of 620 ml · $h^{-1}$ of cycling time.

Eden and Abernethy (1994) recorded the food and fluid intake of a male ultraendurance runner throughout a 1,005-km race completed over 9 days. The average daily energy intake was 25 MJ (6,000 kcal) with 62% from carbohydrate, 27% from fat, and 11% from protein. The carbohydrate intake was 16.8 g · $kg^{-1}$ · $day^{-1}$, the protein intake was 2.9 g · $kg^{-1}$ · $day^{-1}$, and the water intake was 11 L per day. Food and fluid were consumed in small amounts every 15 to 20 min to maintain blood glucose levels and adequate hydration.

### FOR ADDITIONAL READING

For nutritional strategies and basic guidelines to minimize fatigue during prolonged exercise, see N. Clark, Tobin, and Ellis (1992) and Dennis, Noakes, and Hawley (1997).

## After Exercise

After exercise, it is important to ingest adequate amounts of water and energy to ensure rapid recovery. Consuming water up to 150% of the weight lost during physical activity may be necessary to replace the water lost and to cover the obligatory urine production (Shirreffs et al. 1996). To reduce the diuresis that often occurs when only plain water is consumed, it is recommended that the athlete include sodium either in or with fluids ingested after exercise. Food items and condiments high in sodium include various soups, sausages, cured meat and fish, salt herring, pizza, cheeses, pretzels, pickles, ketchup, and soy sauce. By maintaining plasma osmolality, sodium intake also stimulates the thirst and thereby the desire to drink (Maughan and Leiper 1995; Maughan, Leiper, and Shirreffs 1996).

The timing of nourishment after exercise termination seems to be important for the rate of glycogen resynthesis. Thus, Ivy et al. (1988) observed that carbohydrate supplements provided within the first several minutes of completion of exercise more rapidly replaced muscle glycogen compared with the same supplement given 2 h after exercise. A combined effect of insulin and contraction-stimulated glucose transport via GLUT4 transporters might be the mechanism responsible for this observation (Coderre et al. 1995). Later, Tarnopolsky et al. (1997) demonstrated that isoenergetic carbohydrate and carbohydrate–protein–fat supplements given early after endurance exercise increased glycogen resynthesis to practically the same extent for both male and female athletes.

In case of glycogen depletion after exercise, a carbohydrate intake of 1.5 g · $kg^{-1}$ of body weight during the first 30 min and again every 2 h for 4 to 6 h will be adequate to replace the glycogen stores within 24 h. Because protein consumed after exercise provides amino acids for the building and repair of muscle tissue, it is recommended that athletes consume a mixed meal providing carbohydrates, protein, and fat soon after strenuous physical exertion (American College of Sports Medicine, 2000).

# Physical Activity, Food Intake, and Body Weight

The energy requirement is essentially a question of energy balance or energy intake versus energy expenditure. Any excess intake of food energy over and above the daily need will be stored as fat. The result is weight gain.

## Energy Balance

The energy requirement is directly proportional to body size and degree of physical activity. As a rough guide, sedentary, middle-aged people need about 150 kJ (35 kcal) per kilogram of body weight per day. This corresponds to about 10.5 MJ (2,500 kcal) per day for a 70-kg person. On growing older, the person is apt to be less active and therefore to experience a gradual decline in energy expenditure and a decline in the resting metabolic rate. Consequently, if food intake continues to be about the same as before, the person will no longer spend what is taken in. The difference will be deposited as stored energy in the fat depots. Thus, if the person takes in some 1.5 MJ (350 kcal) more than is spent each day (the amount of energy contained in a piece of apple pie), 10.5 MJ (3,500 kcal) will be stored in 10 days. This means that the individual will have deposited about half a kilogram of fat in the body, because there are approximately 29 MJ (7,000 kcal) in each kilogram of stored fat. This uptake, if continued day after day, will result in a gain of about 18 kg in body weight in a year. Simply omitting a single piece of apple pie from the daily diet could have prevented this weight gain. Or, if the person had kept up his or her former level of physical activity, the slice of apple pie could have been enjoyed every day without weight gain.

It is a common observation that athletes such as runners and javelin throwers engaged in intense training need about 200 kJ (50 kcal) per kilogram of body weight per day. Thus, a javelin thrower weighing about 90 kg consumes about 19 MJ (4,500 kcal) per day. A 70-kg runner consumes between 12.5 and 14.5 MJ (3,000 and 3,500 kcal) daily. The higher energy intake, compared with that of most sedentary individuals, is caused by the fact that athletes, owing to their daily 1- to 3-h training activities, expend at least 4 MJ (1,000 kcal) more per day. Jogging, for example, involves an energy expenditure of more than 1.7 MJ (400 kcal) per hour, whereas running a marathon costs about 10.5 to 12.5 MJ (2,500–3,000 kcal; (Newsholme and Leech 1984). Depending on the running time, this means an energy expenditure of about 3 MJ (750 kcal) per hour in a recreational athlete and 6.3 MJ (1,500 kcal) per hour in the best runners. Participation in a professional bicycle race, such as the Tour de France, represents an energy expenditure of more than 27 MJ (6,500 kcal) per day. This figure can increase to near 38 MJ (9,000 kcal) per day when the bicyclist crosses a mountain pass (Saris et al. 1989).

Sjödin et al. (1994) found a close match between energy intake and energy expenditure in elite cross-country skiers during a week of training (figure 12.17). Energy intake was calculated from weighed dietary records, and doubly labeled water was used to simultaneously measure energy turnover. Average daily energy intake ranged from 15.7 MJ (3,738 kcal) to 20.4 MJ (4,857 kcal) per day in the females and from 25.7 MJ (6,119 kcal) to 36.0 MJ (8,571 kcal) per day in the males. However, if energy intake over separate 24-h periods was compared with corresponding data for training, no significant relationship was found. This indicates that the skiers were not in energy balance during shorter periods.

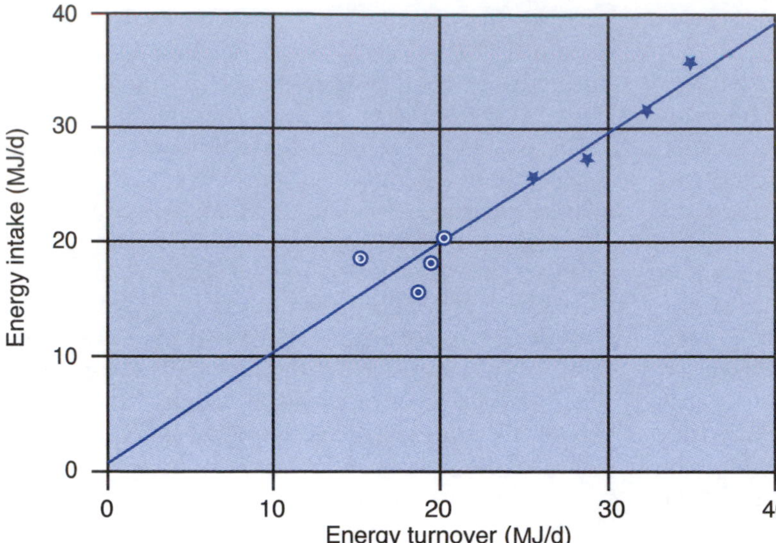


**Figure 12.17** The relationship between energy turnover and energy intake. ◉ = females; ★ = males. The equation for the fitted line was $y = 1.3 + 0.95x$ ($r = .96$, $p = .0001$).

Reprinted, by permission, from M.A. Sjödin et al., 1994, "Energy balance in cross-country skiers: A study using double labeled water," *Medicine and Science in Sports and Exercise* 26 (6): 720-724.

About one third to one half of the total daily energy expenditure is used to maintain the basic metabolic rate (BMR). This BMR can vary greatly from one individual to another. A BMR 10% below or above the normal average is not uncommon. In individuals with a daily total energy expenditure of about 10.5 MJ (2,500 kcal), a difference in BMR of 20% can mean a difference of 2.1 MJ (500 kcal) per day in overall energy metabolism merely attributable to a basic difference in BMR. This could explain why some persons remain slim although they eat more than some heavy persons who are equally active. Such small differences in BMR, 10% below and 10% above the average, are clinically classified as normal, and a BMR 10% higher than the average will mean an additional annual energy expenditure of 375 MJ (90,000 kcal). Because there are about 29 MJ (7,000 kcal) in each kilogram of stored fat, this energy expenditure is equivalent of close to 13 kg of body fat in a year.

On a short-term basis, on the other hand, only physical activity can cause major differences in energy expenditure, in that exercise easily can cause a marked increase in energy expenditure over the resting value. However, physical activity of a more moderate degree, spread out over the whole day, also counts in the long run. There are great individual differences in an individual's attitude to activity. Some instinctively seek it, whereas others avoid it. Some remain seated while others get up and move about. These are indeed small differences but in the course of the day, they may add up to several hundred kilojoules, for it takes only some 15 steps to expend 4 kJ (1 kcal).

Mayer et al. (1956) observed an industrial population in India engaged in a very wide range of physical activity, from tailors and clerks to laborers who carried heavy loads for several hours a day. The diet was quite uniform for each individual and showed little variety within groups and from group to group. The sedentary individuals had the highest energy intakes and the highest body weight. The light workers had lower energy intakes and a low body weight. The groups engaged in medium-heavy and very heavy work had increasing energy intakes, but their body weights were the same. It thus appears that sedentary individuals are apt to eat more than they need and to become obese. The remarkable thing is that the group engaged in light exercise did not eat more than the sedentary workers; in fact, they ate less. Essentially similar findings were reported by Woo et al. (1982) and Bar-Or (1993).

### *Summary*

Within a wide range of energy expenditures through various degrees of physical activity, there is an accurate balance between energy output and intake so that the body weight is maintained constant. If the daily activity is very intensive and prolonged, the spontaneous energy intake is often less than the output, and body weight is reduced as a result. A daily energy expenditure below a threshold level often leads to an energy surplus and consequent obesity. In this case, satiety is not reached until more energy has been taken in than has been expanded.

## Regulation of Food Intake

The exact mechanism by which persons regulate the amount and kind of food they need is not completely understood. As in many control systems, scientists have extrapolated from animal studies (C.L. Hamilton 1965; Mayer and Thomas 1967). However, the concept that the hypothalamus contains a feeding center responsible for the urge to eat or the initiation of feeding, as well as a satiety center capable of exerting inhibitory control over this feeding center, is considered to be faulty (Kandel, Schwartz, and Jessell 2000). It appears that the brain is not organized into discrete centers that control specific functions but that these functions are performed by neural circuits distributed among several structures in the brain.

Mayer and Thomas (1967) originally pointed out that there is some sort of short-term feedback:

1. Information concerning the nutritive value of ingested foods is relayed from gastric sensors to the hypothalamic regulation system by way of neural pathways, humoral pathways, or both.
2. Food intake is also intimately related to the regulation of blood glucose. Information concerning the availability of glucose to the cells mediated via glucose receptors is more important than the actual blood glucose concentration. However, hypoglycemia is almost always followed by hunger sensations.
3. The control of heat exchange is connected in a complex manner with the center that regulates food intake. An elevation of the body temperature inhibits the sensation of hunger (Andersson 1967). This may explain the poor appetite experienced during periods of intense physical activity and the poor appetite often noticed in patients with fever. All this information is integrated with a myriad of other exteroceptive and **interoceptive** inputs at a particular moment, determining the balance of appetite and satiety and the initiation, continuance, or termination of the feeding response.

Cumulative errors in this system could, in turn, be corrected through a long-term feedback, or a lipostatic regulation, which is an inhibition of food intake whenever sufficient energy is derived from the mobilization of surplus body fat. How the adjustment of feeding is modified by the state of the fat stores is not clear. However, after a period of weight gain, food intake often is reduced and the body weight becomes stabilized at a new level.

As pointed out by Garrow (1978) at a symposium on energy balance, humans are not rats. He observed, however, that the rat has an astonishing ability to regulate its energy intake over long periods of time, but this is true only if the food available is monotonous, if the rat has been fed ad libitum with this diet, and if flavoring agents, especially sweeteners, are excluded. These conditions are not relevant for the human species. For primitive humans, there was available a relatively small selection of naturally occurring animal and plant foods. Appetite, which at one time might have been a reliable guide to a correctly balanced diet, is now merely a sensation that can be manipulated in many ways by food manufacturers. In experiments, subjects have been over- or underfed by various methods (e.g., by supplement of food by intragastric tube, or by meals with differing energy density but indistinguishable in taste and volume). The correlation between the state of energy balance and the voluntary energy intake is very low, and the individual's ability to regulate the food intake, at least over a short term, is very poor.

### For Additional Reading

For more complete discussions of body energy balance and food intake, see Le Magnen (1983), Wood, Terry, and Haskell (1985), and M.W. Schwartz and Seeley (1997).

## "Ideal" Body Weight

The body weight and shape are determined largely by the skeletal size, because a certain amount of muscle and other tissues usually goes along with a certain amount of bone. Therefore, the so-called ideal body weight can be modified to some extent by enlargement of the muscles by training, especially such training as weightlifting; a person therefore can be overweight without being obese. However, in most cases, any excess weight above the ideal body weight represents accumulated body fat.

Graphs and tables, which usually are based on height and sex, provide the ideal body weight. However, these norms are rather inaccurate and are meant only as a general guide. Thus, the need for reliable methods of assessing body composition is apparent. Several methods are available, usually based on either a two- or three-component model. The two-component model divides the body into fat mass (all the lipids within the body) and **fat-free mass (FFM)**, whereas the three-component model divides the body into fat mass and two components of FFM (bone mineral and lean tissue). The two-component model typically uses hydrostatic weighing (**hydrodensitometry**) or plethysmography, and the three-component model uses **dual-energy x-ray absorptiometry (DXA)** measurements. However, measurement techniques required for these models are not readily available to most athletes. More common methods used to assess body composition in field or clinical settings are anthropometric measurements, including various diameters (e.g., the radioulnar diameter), skinfold thickness, **bioelectrical impedance analysis (BIA),** and near-infrared interactance (American College of Sports Medicine 2000). With carefully applied skinfold or BIA methods, it is possible to determine relative body fat percentage with an error of 3% to 4% and to estimate FFM within 2.5 to 3.5 kg (Houtkooper 2000).

Body composition, particularly in athletes, is a better guide for determining the desirable weight than the standard height–weight–age tables because of the high proportion of muscular tissue. Observations indicate that an increase in physical activity can profoundly affect body composition, increasing the protein/fat ratio, but the body weight can remain the same. However, instead of setting a specific body fat percentage goal for an individual athlete, coaches and clinicians should recommend a range of target percentage of body fat values (American College of Sports Medicine 2000). On the other hand, for practical purposes, obesity is most conveniently diagnosed and judged by the application of simple height–weight methods.

## Overweight and Obesity

Physically inactive lifestyle and obesity are two of the most prevalent risk factors for common chronic diseases. They are both recognized as major risk factors for cardiovascular disease, non-insulin-dependent diabetes mellitus (type 2 diabetes), hypertension, and other debilitating conditions (Bouchard 2000; Wickelgren 1998). According to the World Health Organization (1998), data from almost all industrialized countries, and even those from the Third World, show that a growing proportion of children and adults are overweight (**body mass index, BMI** = 25.0–29.9) or obese (BMI > 30.0). About 50% of the adults in the United States, Canada, and some of the Western European countries have a BMI above 25 kg $\cdot$ m$^{-2}$. Furthermore, the prevalence of obesity in childhood and adolescence has more than doubled since the early 1960s (Troiano et al. 1995).

If energy intake exceeds energy output, the excess energy will be stored mainly as adipose tissue. If this state of affairs is maintained over a period of time, it will lead to obesity. As Mayer and Thomas (1967) pointed out, obesity is often the result of too little physical activity rather than overeating. This is an old observation. Greene (1939) studied more than 200 overweight adult patients in whom the onset of obesity could be traced to a sudden decrease in activity. Furthermore, in studies of the relationship between body weight and physical activity in children, both Bruch (1940) and Bronstein et al. (1942) found that a great majority of the obese children spent most of their leisure time in sedentary activities. M.L. Johnson, Burke, and Mayer (1956) systematically compared energy intake and activity in paired groups of obese and normal-weight schoolgirls. They found that the time spent by the obese groups in sports or any other sort of exercise was less than half that spent by the lean girls. Energy intakes were generally larger in the nonobese girls than in the obese, and it was concluded that physical inactivity was more important than overeating in the development of obesity. When these schoolgirls attended a summer camp, they all (both obese and nonobese), almost without exception, lost weight under a program of enforced strenuous activity despite simultaneous increased food intake.

Chirico and Stunkard (1960) roughly estimated the degree of physical activity of obese and normal subjects of similar occupation and social status. They were asked to wear a pedometer for recording the number of steps throughout the day. The nonobese subjects were about twice as active as the obese ones. Durnin (1967) also noticed that overweight individuals are less physically active than those who are nonobese. According to Fogelholm et al. (1996), the prevalence of overweight men and women in Finland was, respectively, 39% and 33% in 1982 and 43% and 34% in 1992, although the reported daily energy intake was lower in 1992 than in 1982 for both men and women. Between 1982 and 1992, energy expenditure during work and during moving to and from work was significantly decreased. Energy expenditure during leisure time increased in both men and women but was evidently not enough to counterbalance the decreased habitual daily energy expenditure during work.

Earlier studies distinguished between two types of obesity: one type in which there is an increased number of fat cells (**hyperplasia obesity**), and another in which there is a normal number of fat cells, but each fat cell has an increased content of triglycerides (**hypertrophy obesity;** Björntorp 1974). The total number of fat cells in adipose tissue in women is greater than in men. In transectional studies, the cell size seems to increase with an increase in body fat but levels off at about 30 kg of body fat. The fat cell number, on the other hand, increases with the increase in body fat over the whole range studied (up to 90 kg of body fat). Children already obese before the age of 1 year have a larger number of fat cells than children with a similar degree of obesity of later onset. This supports the hypothesis that during the first year of life, the adipose cells are particularly susceptible to multiplication (**hyperplasia;** Brook 1972). There may be a second critical period for such a hyperplasia between 9 and 13 years of age (Salans, Cushman, and Weismann 1973). Otherwise, short-term longitudinal studies and transectional data on obese subjects of different ages do not indicate that the adipose cell number increases with age and obesity. Therefore, subjects with adult-onset obesity seem to have a normal number of fat cells but an enlarged fat cell size. When fat is lost, there is apparently no loss of fat cells.

Thus, it can be argued that the small fat cells of the reduced obese patients are particularly avid to replenish their fat content, and as such the individual is forever prey to rapid weight gain (Garrow 1978). Björntorp (1974) also emphasized that patients with hyperplasia obesity of early debut and an increased body cell mass seem to be difficult to treat with ordinary dietary and exercise procedures. When treated with a conventional low-energy diet, they seem to fail to lose weight after reaching a certain fat cell size. Obese patients who have lost weight by a restricted energy intake are very prone to regain weight. Therefore, one must try at the earliest stage of life to prevent obesity. Vinten and Galbo (1983) demonstrated that the total number of fat cells is not changed by training.

Rognum, Rodahl, and Opstad (1982) observed a 2.7-kg mean body fat loss and a marked reduction in average fat cell size (from .34 to .24 μg) in 12 well-trained men after 5 days of almost continuous, strenuous combat exercise combined with marked energy deficiency. No significant changes were found in the total number of fat cells. The decrease in fat cell size was most pronounced in the gluteal subcutaneous region, followed by the abdominal region, but no significant decrease in fat cell size was observed in samples from the subcutaneous fat tissue of the thigh. This finding could indicate that gluteal fat depots are most important for energy supply.

An individual's risk of developing obesity is increased when she or he has relatives who are obese. This indicates the existence of obesity genes. Furthermore, 40% to 70% of the variation in obesity-related phenotypes, such as BMI, sum of skinfold thickness, fat mass, and leptin levels, is heritable (Comuzzie and Allison 1998).

### For Additional Reading

For extended reading about physical activity and treatment of obesity, the reader is referred to Buemann and Tremblay (1996) and Bouchard (2000).

### *Summary*

It appears conclusively that obesity results to a large extent from reduced activity with maintenance of an "old-fashioned" appetite center set for an energy expenditure well above the one typical for a sedentary individual. This is true for children as well as for adults. The reason for their different attitudes toward physical exertion is not clear. When studying very obese individuals on the cycle ergometer, authors have noted that obese subjects complain of fatigue and feel exhausted at low work loads, when the blood lactic acid levels are not significantly elevated and are easily tolerated by normal individuals.

It is particularly important to stimulate young individuals to exercise regularly, because such activity will in the long run effectively counteract obesity by keeping the individual within the range where spontaneous energy intake is properly regulated by energy output. Obesity in infancy can increase the number of fat cells, causing predisposition for subsequent overweight. The treatment of obesity is particularly difficult in patients with an increased number of fat cells.

## Slimming

Numerous slimming diets are available, but most of them are not based on physiological principles. The most lenient way to reduce weight involves allowing adequate time for the measures to take effect. A weight loss of more than .5 to 1.0 kg per week is not recommended. The obese person's diet should be critically examined and an attempt should be made to eliminate about 1 MJ (about 240 kcal) per day, for example, by substituting artificial sweeteners for sugar and low-fat milk for whole milk and by eliminating butter and visible fat. Furthermore, a 2-km walk per day would add 400 kJ (100 kcal) to the expenditure, the result being a total net reduction in fatty tissue equivalent to about 1.5 MJ (360 kcal) per day. A combination of a small dietary restriction and an increased energy expenditure equivalent to some 2 MJ (500 kcal) per day will mean the loss of half a kilogram in body weight per week.

It is a common experience that when a slimming diet is instituted, there is often a sudden drop in body weight, followed by a more gradual decline. This initial drop, which indeed can be gratifying and encouraging, can be attributed largely to loss of body water that is regained later on. During the first day or two of energy restriction, stored glycogen is mobilized to cover the energy requirement. Because each gram of glycogen is stored together with almost 3 g of water, the mobilized glycogen liberates this water, which is eliminated and may thus account for the initial drop in body weight. When the person returns to an adequate diet, glycogen is again stored, together with the required amount of water. This, then, would account for the rapid transient weight gain when the individual changes from energy restriction to adequate diet. It is a mistake to avoid carbohydrates completely, because muscle and nerve cells need them in their metabolism. They are particularly important for anyone who is physically active.

In the treatment of overweight and obesity, habitual physical activity has profound effects on body composition and the use of nutrients. Habitual physical activity also affects maintenance of and any increases in skeletal muscle mass, resulting in increased resting metabolic rate and enhanced capacity for lipid oxidation during rest and exercise. Regular exercise also can or limit the loss of lean tissue (FFM) during weight loss regimens. Increased physical activity induces a number of favorable changes in the metabolism of lipoproteins: Serum triglycerides are lowered by the increased lipolytic activity, the high-density lipoprotein concentration increases, and the concentration of small, dense, low-density lipoprotein decreases. In addition, the enhanced metabolic capacity of skeletal muscle (metabolic fitness) favorably influences risk factors such as insulin resistance and hypertension.

Because regular physical activity has favorable effects on several of the comorbidities of obesity, particularly those pertaining to cardiovascular disease and type 2 diabetes, it is not surprising that mortality rates seem to be lower in overweight and moderately obese individuals who are physically fit compared with the unfit (Blair and Brodney 1999; Wickelgren 1998). Thus, in the treatment of overweight and obese persons, one perhaps should focus more on habitual physical activity level than on body weight itself. For most of those who wish to reduce body weight, it is recommended that they combine regular physical activity with somewhat reduced energy intake, in particular a reduction in high-fat food. Emphasis should be on promoting relatively low-intensity, long-duration physical activity that can be conveniently incorporated into daily life.

## FOR ADDITIONAL READING

For further discussion about the effectiveness of traditional dietary and exercise interventions for weight loss and the impact on longevity, see W.C. Miller (1999) and Gaesser (1999).

CHAPTER 13

# TEMPERATURE REGULATION

The protected human being can tolerate variations in environmental temperature between –50 °C and 100 °C. But a person can tolerate a variation of only about 4 °C in deep body temperature without impairment of optimal physical and mental performance. Changes in body temperature affect cellular structures, enzyme systems, and numerous temperature-dependent chemical reactions and physical processes that take place in the body. The maximal limits that the living cell can tolerate range from about –1 °C at one end of the scale, when the ice crystals formed during freezing break the cell apart, to thermal heat coagulation of vital proteins in the cell at about 45 °C at the other end of the scale. Only for shorter periods of time can cells tolerate an internal temperature exceeding 41 °C. In fact, many animals, including humans, live their entire lives only a few degrees removed from their thermal death point.

The hot end of the scale is more problematic than the cold end, because people can protect themselves more easily against overcooling than overheating. Consequently, the controlling mechanism for temperature regulation is particularly geared to protect the body tissues against overheating (Hardy 1967).

The purpose of this chapter is to describe the different factors that help regulate body temperature; to explain the effects of climate, physical activity, and acclimatization; and to discuss the limits of tolerance to heat and cold.

## Heat Balance

If the heat content of the body is to remain constant, heat production and heat gain must equal heat loss, according to the following equation: $M \pm R \pm C - E = 0$, where $M$ = metabolic heat production, $R$ = radiant heat exchange (positive if the environment is hotter than the skin temperature, but negative if the temperature of the environment is lower than that of the skin), $C$ = convective heat exchange (positive if the air temperature is higher than that of the skin, negative if the reverse), and $E$ = evaporative heat loss. This equation is valid only for conditions when the body temperature is constant. If the body temperature varies, a correction must be introduced, and the following equation is applicable: $M \pm S \pm R \pm C - E = 0$, where $S$ = storage of body heat. This was first derived by Winslow, Gagge, and Herrington (1939). $S$ is positive if the body heat content is decreasing and negative if the heat content increases. The specific heat of most tissues is about .83. Conductive heat exchange *(K)* is in most conditions negligible but increases in importance during such activities as swimming, because water has a heat-removing capacity that is some 20 times that of air.

One very important function of the blood circulation is to transport heat, to cool or to heat various tissues as needed, and to carry excess body heat from the interior of the body to the body surface, that is, the skin. The blood is very effective in this function, because it has a high heat capacity (.9), which means that the blood can carry a great deal of heat with only a moderate increase in temperature. Conductance of the tissue ($kJ \cdot m^{-2} \cdot h^{-1} \cdot {}^{\circ}C^{-1}$) is the amount of heat given off per square meter body surface per hour and per degree temperature difference between the interior of the body and its surroundings. When the skin blood flow increases, skin temperature increases and its conductance increases. When the skin blood flow is reduced, skin temperature decreases and the conductance is reduced, that is, the insulating value of the skin increases. Interestingly, increased nasal air flow during exercise possibly could act as a loss avenue contributing to selective brain cooling (M.D. White and Cabanac 1995b).

This control of body temperature, the balance between overcooling and overheating, is the role of temperature regulation. This regulation endeavors to keep the temperature of certain tissues, such as the brain, heart, and gut, relatively constant. Within the body, the temperature is by no means uniform. The greatest gradient is found between the "shell" (the skin) and the "core" (deep central areas including heart, lungs, abdominal organs, and brain). Thus, the average cerebral temperature of a resting subject is considerably higher than the tympanic temperature (Mariak, Bondyra, and Pickarska 1993). The temperature of the core can be as much as 20 °C higher than that of the shell, but the ideal difference between shell and core is about 4 °C at rest. Even within the core, the temperature varies from one place to another. This complicates the calculation of the heat content of the body and makes it difficult to study temperature regulation. Evidently the term *body temperature* is a misnomer. The maintenance of a normal body temperature is actually quite compatible with considerable gains or losses of heat. At issue is which temperatures are being regulated.

## Methods of Assessing Heat Balance

Deep body temperature (core temperature) can be measured with the aid of mercury thermometers, **thermocouples,** or **thermistors.** The classic site of measurement is the rectum. Because the temperature in the rectum ($T_r$) varies with distance from the anus, it is customarily measured at a depth of 5 to 8 cm. This rectal temperature is, in a resting individual, slightly higher than the temperature of the arterial blood; it is about the same as the liver temperature but slightly lower (.2–.5 °C) than the part of the brain where the thermal regulatory center is located. During physical exertion or exposure to heat, the temperature of this part of the brain increases more rapidly than does the rectal temperature. The temperature increase or decline in the brain and in the rectum are of the same magnitude, however. The rectal temperature is therefore appropriate for assessing changes in the deep body temperature, provided the measurement is made under steady-state conditions (i.e., after some 30–40 min). It has been found that the eardrum temperature is a fairly good indication of brain temperature. This can be obtained by placing a thermocouple, introduced through the ear, against the eardrum (Benzinger and Taylor 1963; Gass and Gass 1998). However, the eardrum temperature is not quite identical with the temperature in the thermoregulatory center. Furthermore, it varies with the ambient air temperature (Houdas and Ring 1982). In a study of 33 pregnant women, Yeo et al. (1995) found that auditory canal temperature measured with a tympanic membrane thermometer correlated significantly with rectal temperature measured with a glass mercury thermometer. Another alternative is to measure the temperature in the esophagus, which is relatively accessible for such measurements. Although the temperature of the esophagus is not identical with any of the previously mentioned core temperatures, it generally changes parallel to those temperatures (B. Nielsen 1969; Saltin and Hermansen 1966).

In exercise and heat studies, the oral temperature measurement has its limitation (Strydom et al. 1966), but it can be used as a practical index of deep body temperature. Mairiaux, Sagot, and Candas (1983) showed that sublingual oral temperature variations were highly correlated with esophageal temperature variations under steady-state conditions and under air temperature variations. Under these conditions, oral temperature represented a better estimate of esophageal temperature than did rectal temperature. The reverse was the case under work-rate variations, however. The skin temperature is measured with a **radiometer** or by placing thermocouples or thermistors on the skin at certain locations. For clinical purposes, infrared **thermography** can be used, measuring the infrared radiation emitted by the skin surface (Houdas and Ring 1982). The mean skin temperature ($T_s$) is calculated by assigning certain factors to each measurement in proportion to the fraction of the body's total surface area represented by each specific area, as originally done by Hardy and Dubois (1938):

| | |
|---|---|
| Head | .07 |
| Arms | .14 |
| Hands | .05 |
| Feet | .07 |
| Legs | .13 |
| Thighs | .19 |
| Trunk | .35 |
| Total | 1.00 |

A simple weighing formula for computing the mean skin temperature from observations of four areas of the body was developed by Ramanathan (1964). The formula, which can give values identical to the elaborate Hardy-Dubois formula, is as follows:

$$T_s = .3\,T_{chest} + .3\,T_{arm} + .2\,T_{thigh} + .2\,T_{leg} \quad (1)$$

Furthermore, Ramanathan (1964) confirmed Teichner's earlier observation (Teichner 1958) that the temperature on the surface of the medial thigh is a satisfactory measure of the mean skin temperature under most conditions. In fact, the mean difference of the temperature values for the medial thigh temperature and the mean skin temperature calculated according to the Hardy-Dubois formula in 39 observations was only .17 °C. However, this may not be the case in a very cold environment. When assessing heat loss from the body surface, the clinician should also keep in mind that a posture change can change the effective body surface area available for heat exchange with the environment (Raja and Nicol 1997).

The following equation can be used to calculate the heat content of the body:

$$\text{Heat content} = .83\,W\,(.65\,T_r + .35\,T_s) \quad (2)$$

where *W* represents body weight, .83 is the specific heat of the body, and .65 and .35 are the factors assigned to the rectal and the mean skin temperatures, respectively (Burton 1935). The specific heat of the body was long assumed to be .83, but it varies with the individual's body composition and can range from .70 up to .85 (Hardy, Gagge, and Stolwijk 1970). It is also clear that the heat content of the body during exercise in a hot environment cannot be determined from any fixed ratio of $T_s$ and $T_r$ (Wyndham 1973).

The metabolic rate, or the magnitude of heat production, is assessed by the measurement of oxygen uptake. The volume of 1 L of oxygen consumed corresponds to approximately 20 kJ (4.8 kcal). Human calorimeters, large enough to measure the rate of heat production by direct calorimetry, have been constructed and are used in certain laboratories.

The evaporative heat loss (*E*) plays a major role in cooling the skin and the blood during exposure to heat. At normal skin temperature, the evaporation of 1 L of sweat requires 2.4 MJ (580 kcal). The magnitude of the sweat loss can be estimated simply by weighing the subject nude or dressed in dry clothing, before and after the experiment, and by weighing food and fluid ingested and stools and urine voided during the period of observation. Furthermore, weight loss attributable to respiratory gas exchange should be included in the calculation and can be accounted for as follows (Snellen 1966):

$$C_{ge} = \dot{V}O_2\,(1.977.\mathrm{R} - 1.429) \quad (3)$$

where $C_{ge}$ = weight loss in grams per minute attributable to respiratory gas exchange.

$$\dot{V}O_2 = \text{oxygen uptake in liters per minute STPD} \quad (4)$$

where R = respiratory quotient, 1.977 = weight in grams of 1 L of carbon dioxide STPD, and 1.429 = weight in grams of 1 L of oxygen STPD.

Such measurements are valid for calculating heat balance only as long as all the sweat produced during the experiment is actually evaporated. On the other hand, whether the produced sweat is evaporated or part of it has run off the body, the sweat rate indicates the magnitude of the heat stress.

The air temperature affects convective heat loss or gain (C) and is most conveniently measured with the usual mercury thermometer. If the thermometer is exposed to radiation, it should be shielded. A piece of tinfoil can be used, but care should be taken to allow free air passage around the thermometer.

The humidity of the air can be measured with a sling psychrometer or an electronic device for measuring humidity. The rate of evaporation of the produced sweat depends greatly on the humidity of the air.

The air movement, which may be measured by a hot-wire animometer, affects both convective heat exchange *(C)* and evaporative heat loss *(E)*.

The radiant heat exchange *(R)* depends on the temperature difference between the individual and the surroundings. This can be assessed by the values obtained from a mercury thermometer placed in a hollow, spherical black copper container with a diameter of 15 cm (a globe thermometer). Because of rapidly changing radiation intensities typical for many industrial operations, it is often almost impossible to accurately measure the intensity of the radiation.

Afferent impulses from thermoreceptors in the skin, which respond to very rapid temperature changes, can signal peripheral thermal disturbances long before the central core temperature has been affected. Such impulses, integrated with various sensory input from the rest of the body, result in feelings of thermal comfort or discomfort (Hardy, Stolwijk, and Gagge 1971). More specifically, the sensation of thermal comfort appears to result from the interaction between signals evoking temperature sensation, input signals for temperature regulation, and sensations arising from thermoregulatory activities associated with skin blood flow, sweating, and shivering (Hardy, Stolwijk, and Gagge 1971). Generally speaking, the state of thermal neutrality or thermal comfort is characterized by core temperature of 36.6 to 37.1 °C and skin temperatures between 32 and 35.5 °C (Precht et al. 1973). Apparently, comfort temperature is subject to some degree of adaptation or accustomation. Neither aging nor maximal work capacity appears to affect thermal comfort or cold sensation (Falk et al. 1994).

### For Additional Reading

For an evaluation of miniature data loggers for body temperature measurement during sporting activities, see Fuller et al. (1999).

## Magnitude of Metabolic Rate

Human beings can be considered tropical animals inasmuch as they require an ambient temperature of 28 to 30 °C to maintain thermal balance at rest, in the nude. That is, without clothing and shelter, a human could only survive in a narrow zone along the equator. The oxygen uptake under these conditions is about .20 to .30 $L \cdot min^{-1}$. It is slightly higher when the body size is larger. This corresponds to a production of 250 to 380 kJ (60–90 kcal) per hour, or 70 to 100 W. This energy is the by-product of metabolic processes that are essential for the maintenance of life. This produced heat makes up for the heat lost through convection *(C)*, radiation *(R)*, and evaporation *(E)*. Under these conditions, $C + R$ accounts for about 75% of the heat loss, and $E$ accounts for only 25%. Heat loss through the lungs, through the saturation of the air with water vapor during respiration, accounts for about two fifths of $E$. Not all the rest of $E$ is attributable to the evaporation of sweat, because part of the water loss through the skin occurs without the involvement of the sweat glands, the so-called insensible perspiration. The total water loss through the skin amounts to a minimum of .5 $L \cdot day^{-1}$.

Muscular exercise is associated with an increase in metabolic rate. Because the mechanical efficiency (the ratio of external work to the extra energy used) can vary from 0% to 50% depending on the kind of exercise, at least 50% of the energy used is converted into heat. During short exercise periods of 5 to 10 min duration, well-trained athletes can attain an oxygen uptake of more than 6 $L \cdot min^{-1}$ (2,000 W) and during prolonged exercise as much as 4 to 5 $L \cdot min^{-1}$ or even more. The amount of heat thus produced during 1 h theoretically could increase the body temperature of a 70-kg individual from 37 °C to about 60 °C if the excess heat were not dissipated.

# Effect of Climate

The ideal room temperature is about 20 °C for clothed individuals who are sitting or standing still. The more active the individual, the lower the room temperature should be. Thus, when performing heavy physical work, a person may prefer a room temperature of 15 °C or even lower.

For a nude resting individual, the ideal ambient temperature is about 28 to 30 °C. Under such conditions, the mean skin temperature is about 33 °C, and the temperature of the core is about 37 °C. The temperature gradient from core to skin is then adequate to transfer the excess heat from the metabolically active tissues to the surroundings. Of the total amount of blood (the circulating blood volume) pumped by the heart, about 5% flows through the blood vessels of the skin.

## Cold

If the ambient temperature decreases, the temperature difference between the skin and the environment is increased; this increases the heat loss through convection and radiation. A reduced heat flow to the skin would gradually lower the skin temperature. This would reduce the temperature gradient between the skin and the environment. Actually, the conductance of the tissue is reduced, partly because vasoconstriction of the skin's blood vessels reduces the blood flow and partly because the blood in the veins of the extremities is redirected from the superficial to the deep veins. This vasoconstriction is most pronounced in the extremity (Pendergast 1988; Vangaard 1975). Because of the proximity of the deep veins to the arteries, a heat exchange occurs (figure 13.1). Because of this system of countercurrent, in a subject exposed to an ambient temperature of 9 °C, the blood leaving the heart will have a temperature of about 37 °C. As it flows through the arm, it is gradually cooled so that by the time it reaches the hand, it can have dropped to about 21 °C. The returning venous blood absorbs a considerable part of the heat as the blood flows through the arm. In other words, cooling of arterial blood flowing through the arteries of the limbs depends on the rewarming of cold blood returning in adjacent veins from more distal areas. Thus, a cooling of the body core is prevented (Bazett et al. 1948; Schmidt-Nielsen 1963). In a hot environment, on the other hand, the blood from the limbs returns primarily through superficial veins, thus further cooling the blood.

The major effect of the vasoconstriction of the skin is a sudden displacement of the blood volume from the skin to the central circulation, as evidenced by a sudden increase in central blood volume and a redistribution of blood from superficial to deep veins (Rowell 1993).

The heat exchange between arterial and venous blood is still more important for arctic animals. Sea birds swim in the icy sea, and the large surfaces of their bare feet are exposed to the cooling effect of the water. However, little body heat is lost because their feet cool down to the temperature of the water, thus reducing the loss of metabolic heat (Irving 1967). Warm feet of a gull or a duck standing on snow or ice would cause it to melt, and soon the feet would be frozen solid to the ground where they stood! Hogs, naked as a human in a cold environment, and narwhal, walrus, or seals in arctic waters can prevent the loss of heat by having a very cold skin. These animals have a considerable layer of subcutaneous fatty tissue as an effective insulation when their blood vessels constrict. It is an interesting observation that fats in the peripheral regions have a lower melting point than those in the warmer internal tissues; if they did not, the peripheral tissues and legs would become too inflexible in cold weather (Irving 1967).

Thickly furred animals can use their bare extremities to release excess heat from the body. Heat also can be dissipated by evaporation from the mouth and tongue during panting.

Through peripheral vasoconstriction, a sixfold increase in the insulating capacity of the skin and subcutaneous tissues is possible (Burton 1963). This vascular constriction is particularly active in the fingers and toes. It has been estimated that the blood flow through the fingers can vary by 100-fold or more, from .2 to 120 ml of blood $\cdot$ min$^{-1}$ $\cdot$ 100 g$^{-1}$ of tissue (Robinson 1963). The disadvantage of this vasoconstriction is that the temperature in the peripheral tissues can approach that of the environment (figure 13.2). For this reason, one is apt to suffer cold fingers and toes. The blood vessels of the head are far less subject to active vasoconstriction.

Another protective mechanism to maintain heat balance is an increase of the metabolic rate, mediated through muscle activity in the form of shivering by a reflex mechanism. Shivering consists of a synchronous activation of practically all muscle groups; antagonists are made to contract against one another. Because the mechanical efficiency of shivering is 0%, the heat production is relatively high, and the metabolic rate can increase to two to five times that of resting metabolic rates. At rest, cold exposure shifts substrate utilization to a greater use of carbohydrate representing the main fuel for shivering thermogenesis (Martineau and Jacobs 1988; Vallerand and Jacobs 1992). Dynamic muscular exercise, on the other hand, easily can increase the metabolic rate 10- to 20-fold.

Prior physical exertion can predispose a person to greater heat loss and a larger decline in core temperature when subsequently exposed to cold air (Castellani et al. 1999, 2001).

Whether a significant increase in metabolic heat production can occur in humans in response to cold exposure by the so-called nonshivering thermogenesis is still an open question. According to Houdas and Ring (1982), nonshivering thermogenesis can play a role in human neonates, possibly via increased metabolic activity of their stores of **brown fat**, which are absent in adults. Therefore, brown fat cannot play any role in adult thermogenesis. Nor does it appear likely that catecholamines or thyroid hormones play a significant thermogenic role in humans exposed to cold under ordinary circumstances, which leaves shivering or physical activity as the only ways of increasing metabolic heat production in the unprotected individual exposed to cold.

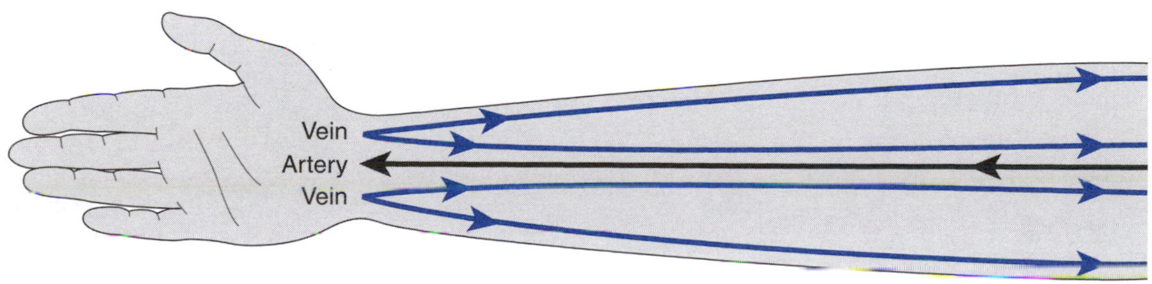


**Figure 13.1** Schematic illustration of the two possible ways that venous blood flow from the hand, the superficial veins, or the deeper ones anatomically close to the arteries can enable a heat exchange.

Adapted from Åstrand and Rodahl 1984.

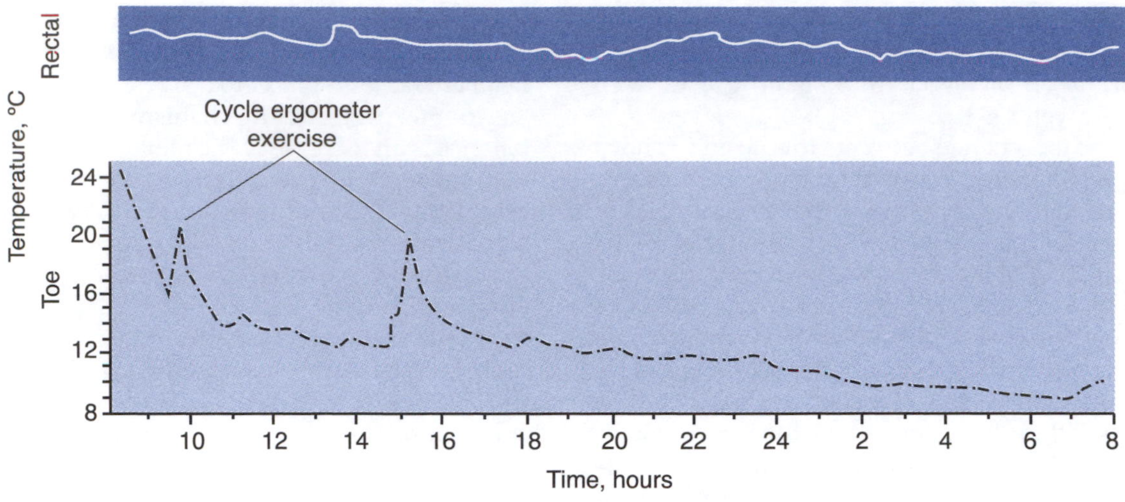


**Figure 13.2** Rectal and toe temperatures in a nude subject exposed to 8 °C continuously for 3 days. By the end of 24 h, the skin temperature of the big toe had dropped to about 8 °C, the temperature of the ambient air.

Although the resources for maintaining the core temperature are quite effective, this maintenance takes place to some extent at the expense of the peripheral tissues, the shell. Local cold injury can be the result in extreme conditions. During prolonged severe cold exposure, even the core temperature can drop. Under certain circumstances, especially when exposed to cold water, obese individuals may be better off in the cold than lean individuals because of the insulating value of the adipose tissue (Golden, Hampton, et al. 1979; Holmér and Bergh 1974; Keatinge 1978). An unclothed individual of average body build will be helpless from hypothermia after 20 to 30 min in water at 5 °C and after 1-1/2 to 2 h in water at 15 °C. With thick conventional clothing, these times can be substantially prolonged. Keatinge (1978) pointed out that one should not exercise in cold water in an attempt to keep warm; it will have the reverse effect. In one study on swimming in water at 18 °C, the oxygen uptake was elevated by approximately .5 L $\cdot$ min$^{-1}$ compared with swimming at the same speed in warmer water. For lean subjects, however, there was a significant drop in esophageal temperature. The maximal oxygen uptake and heart rate were markedly reduced. The effect of exposure to moderately cold water for 20 to 30 min thus will seriously impair physical performance, and swimming can be hazardous (Bergh 1980; M. Davies et al. 1975; Golden, Hampton, and Smith 1979b; Holmér and Bergh 1974).

Vangaard (1975) measured the skin temperature of the hand in a subject suddenly brought from a room temperature of 28 °C into a cold chamber with a temperature of 9 °C. He found that the decrease in skin temperature as a result of the sudden cold exposure was identical to that observed in the same subject when the arterial blood flow to the hand was arrested by an inflated cuff placed around the arm. From this, it appears that sudden cold exposure almost completely shuts down the circulation to the exposed hand. Evidently, every effort is made to maintain the core temperature at the expense of the shell temperature and especially that of the hands and feet. Because manual dexterity is impaired with decreasing hand temperature, unprotected exposure to cold can reduce performance and increase the risk of accidents (Anttonen and Virokannas 1994).

Young and Lee (1997) noted that cold exposure elicited a slightly smaller increase in metabolic heat production, and the cutaneous and the vasoconstrictor responses to cold were less responsive in old than in young men. However, older women appeared to defend core temperature during cold exposure as well as, or better than, younger women. These observations led the authors to suggest that preventable changes in body composition and physical fitness rather than aging per se may contribute to impaired thermoregulatory responses to cold observed in older individuals (Young and Lee 1997).

The speed of nervous impulses and the sensitivity of the receptors are affected by the temperature of the tissues. At about 5 °C, the skin receptors for pressure and touch do not react on stimulation. The execution of coordinated motions depends on the inflow from these receptors to the central nervous system. The numbness in the cold results from this lack of sensitivity of the skin receptors. Irving (1966) reported that the skin at a temperature of 20 °C was only one sixth as sensitive as at 35 °C; that is, an impact on the skin had to be six times greater to be felt at the lower skin temperature. The muscle spindles show increased sensitivity at moderately lowered muscle temperatures, but at 27 °C the activity in response to a standardized stimulus is reduced to 50%; at a temperature of 15 to 20 °C, it is completely abolished (Stuart et al. 1963). This phenomenon also contributes to the difficulty of performing fine coordinated movements in the cold. It may be partially responsible for an increased accident rate in certain types of manual work operations in cold environments (see also Ellis 1982; Tanaka et al. 1983).

Wyon (1982) found that lowering the ambient temperature from 24 to 18 °C or colder significantly impaired manual dexterity. This can be explained partly by changes in nervous conduction velocity. Vangaard (1982) investigated the relationship between local temperature and nervous conduction velocity in a peripheral motor nerve in subjects exposed to a minor cold stress. The decrease in conduction velocity was 15 $m \cdot s^{-1}$ for each 10 °C decrease in temperature. At a local temperature of 8 to 10 °C, a complete nervous block was established. This may explain the common finding that a local cooling in the extremities can be accompanied by a rapid onset of physical impairment. Bergh (1980) found that physical performance is reduced with reducing core temperature. The temperature effect amounted to 4% to 6% $\cdot\ °C^{-1}$ in maximal exercise of less than 3 min duration, and 8% $\cdot\ °C^{-1}$ at 3 to 8 min of exercise. Maximal dynamic muscle strength as well as maximal instantaneous anaerobic power also is reduced with decreasing muscle temperature (Ferretti 1992). A.M. Sjødin et al. (1996) showed that even a small increase in cooling during endurance exercise increases energy expenditure, which may be a relevant problem in winter sports.

C.T.M. Davies and Young (1982) studied the effect of heating and cooling by immersion in water on the electrically evoked mechanical and contractile properties of the triceps surae in five healthy male subjects. An enhanced muscle temperature decreased the time to peak tension and half relaxation time but did not affect supramaximal twitch and tetanic tension and maximal voluntary contraction. Cooling increased the time to peak tension. Furthermore, cooling of the leg muscles markedly reduced the peak power output during cycling and jumping.

### For Additional Reading

For further details concerning the physiology of exercise in the cold, see Doubt (1991) and Beelen and Sargeant (1991).

Exposing the face to a cold stimulus slows the heart rate and increases peripheral vasoconstriction (LeBlanc et al. 1978). Similarly, directing cold wind at the faces of subjects during exercise significantly lowered heart rate (Riggs et al. 1981, 1983). The heart rate response to facial cooling during exercise is probably mediated via a central thermoregulatory response. Facial cooling by an air temperature of –20 °C can significantly elevate both systolic and diastolic blood pressure (Stroud 1991). Exposure to cold stress in general significantly increases heart rate, blood pressure, and cardiac output (Vogelaere et al. 1992).

### *Summary*

When a resting person is exposed to a cold environment, two main mechanisms prevent a decrease in body temperature:

1. a reduction in the peripheral blood flow with a secondary decrease in the skin temperature (reducing the heat loss by radiation and convection); and
2. an increase in the metabolic heat production by shivering. In a hypothermic person, the combined effect of reduced efficiency and reduced maximal aerobic power will impair physical performance.

## Heat

When the nude, resting individual is exposed to heat (when the ambient temperature exceeds 28 °C), or during muscular activity, the heat content of the body tends to increase. Under such conditions, the blood vessels of the skin dilate, venous return in the extremities takes place through superficial veins, and the conductance of the tissue increases. In the comfort zone, the skin blood flow amounts to about 5% of the cardiac minute volume; in extreme heat, it can increase to 20% or more. The increased heat flow to the skin increases the skin temperature. If the temperature of the surroundings is lower than that of the skin, heat loss is facilitated through $C + R$. If the heat load is sufficiently large, the sweat glands are activated, and as the produced sweat is evaporated, the skin is cooled. It has been calculated that there are at least 2 million sweat glands in the skin of an average individual. Recruitment of sweat glands from different areas of the body is not consistent between individuals (Nadel et al. 1971). The activity of individual sweat glands follows a cyclic pattern. The sweat starts to drop off the skin when the sweat intensity has reached about one third the maximal evaporative capacity (Kerslake 1963).

The individual difference in the capacity for sweating is considerable; some people have no sweat glands at all. At any rate, sweat gland function decreases with old age (Inoue 1996). As a person becomes accustomed to heat, the amount of sweat produced in response to a standard heat stress increases. A person may produce several liters of sweat per hour (Torii 1995). Workers exposed to intense heat may lose as much as 6 to 7 L of sweat in the course of the working day. Sweat loss up to 12 L in 24 h has been reported (Leithead and Lind 1964).

During prolonged exposure to a hot environment, there is a gradual reduction in the sweat rate, even if the body water loss is replaced at the same rate. This decline in sweat rate is greater in humid than in dry heat. Ahlman and Karvonen (1961) reported that exercise could again induce sweating after the sweating had ceased during repeated thermal stimuli in a sauna bath. The suppression in sweating is related to the wetting of the skin. Drying the skin with a towel at regular intervals or increasing air velocity around an exercising subject will enhance the sweating rate. When water is evaporated from the skin surface, the solutes are left behind. An increase in osmotic pressure on the skin surface seems to increase the sweat secretion rate (Nadel and Stolwijk 1973).

In a study of local sweat rates at the chest and thigh in athletes during bicycle exercise, Yamouchi et al. (1997) observed that long-term physical training leads to improved circulatory heat transfer to the skin and to a more graded nervous control of sweat expulsion and tends to reduce the rate of sweating. At a given exercise intensity, subjects with a higher aerobic capacity maintain their body temperature with a lower sweating rate than subjects with a lower aerobic capacity (Yoshida et al. 1995).

Frascarolo, Schutz, and Jequier (1992), in their study of sweating response in women exposed to warm environmental conditions, found that the internal temperature set-point for the onset of sweating shifted to a higher value during the luteal phase compared with the follicular phase of the menstrual cycle. Furthermore, it has been shown that the administration of synthetic progestins in oral contraceptives shifts the threshold upward for heat loss responses, resulting in higher body core temperature both at rest and during exercise (Rogers and Baker 1997).

Sweat contains different salts, notably sodium chloride (NaCl), in varying concentrations, and excessive sweating therefore can cause a considerable loss of salt.

Exposing resting subjects to direct whole-body heating causes an increase in skin temperature that is accompanied by an almost immediate increase in heart rate and cardiac output and a decrease in total peripheral resistance and splanchnic blood flow. Cardiac output commonly increases 50% to75% (Rowell 1993). The excess cardiac output is diverted to the skin. This increased skin blood flow is supplemented further by a reduction in splanchnic and renal blood flow. Evidently, muscle blood flow is not increased (Rowell 1993). As the skin temperature rises, cutaneous resistance vessels relax and cutaneous venous pressure and volume increase until a new level of wall tension is reached. This volume displacement is augmented by splanchnic vasoconstriction, which reduces distending pressure in the splanchnic veins, allowing them to empty passively. In this way, blood can be displaced from central to cutaneous venous beds (Rowell 1993). The skin blood flow will increase by local heating. This blood flow can be elevated further if the whole-body skin temperature or core temperature increases. Keeping the local skin temperature high will not abolish the skin vasoconstriction response to lower body negative pressure (simulating the effect of a mild hemorrhage by pooling blood in the leg veins). Thus, local factors and reflex influences to a skin area can interact to modify the degree but not the pattern of the skin vasomotor response (J.M. Johnson, Brengelmann, and Rowell 1976).

Exposure to dry, hot air in the form of the Finnish sauna bath was studied by Eisalo (1956). Both in healthy and in hypertensive subjects, he found that the cardiac output increased by an average of 73% (65% in the hypertensive group), the mean circulation time decreased by almost 60%, and the pulse rate increased by more than 60%. There was a slight but significant decrease in systolic blood pressure in healthy subjects, but in the hypertensive subjects it decreased by 29 mmHg (3.9 kPa) on the average. Twenty minutes after the sauna, the mean decrease was 54 mmHg (7.2 kPa). The diastolic blood pressure remained practically unchanged in the healthy subjects, whereas it decreased significantly in the hypertensive subjects. There was a statistically highly significant decrease in peripheral resistance in both groups of subjects.

## FOR ADDITIONAL READING

For a discussion of the effects of individual characteristics on human responses to heat stress, see Havenith et al. (1998). For practical strategies to minimize the adverse effects of heat on performance, see Terrados and Maughan (1995).

## *Summary*

The resting person exposed to a hot environment experiences a vasodilation in the skin, making an increased heat transfer from "core" to "shell" possible, and perhaps an activation of the sweat glands, with the evaporation of the sweat taking heat from the body and causing an evaporative heat loss.

## Effect of Exercise

Because the mechanical efficiency of the human body is mostly below 25%, more than 75% of the total energy used is converted into heat. The greater the exercise intensity, the greater the total amount of heat produced. This excess heat has to be removed and dissipated to prevent overheating and hyperthermia.

### Maintenance of Thermal Balance

Figure 13.3 shows an example of how the thermal balance is maintained during muscular exercise of different intensity over 1 h. The body temperature increases during work, and this temperature elevation was interpreted some time ago as the result of an active regulation (Berggren and Christensen 1950; Christensen 1931b; M. Nielsen 1938). Apparently, the core temperature increases to establish a gradient for heat flow from core to shell and to stimulate sweating (B. Nielsen, 1980). The difference between energy output and heat production in figure 13.3 is an expression of mechanical efficiency (about 23%), and the difference between heat production and total heat loss is a consequence of the elevated body temperature. Note that the convective and radiative heat losses are almost constant despite the large variations in heat production. Evaporation takes care of the extra heat loss as the rate of exercise increases.

Figure 13.4 shows how the rectal temperature, measured after about 45 min of exercise on the cycle

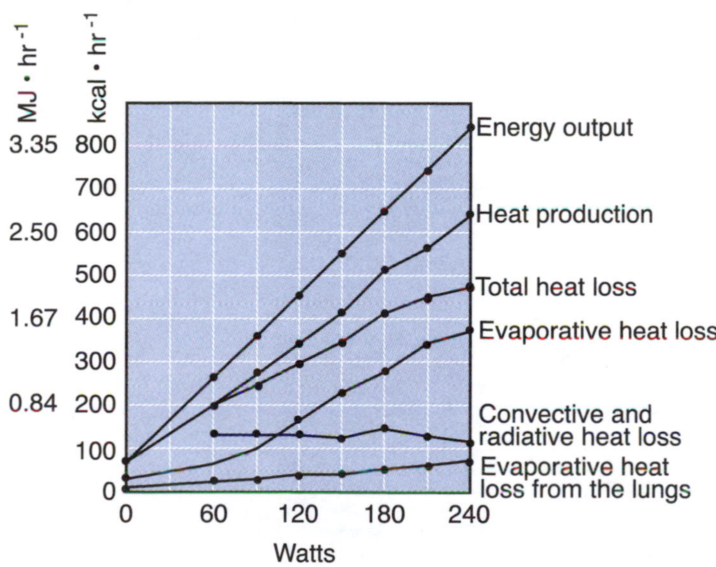


**Figure 13.3** Heat exchange at rest and during increasing exercise intensities in a nude subject at a room temperature of 21 °C.

From M. Nielsen 1938.

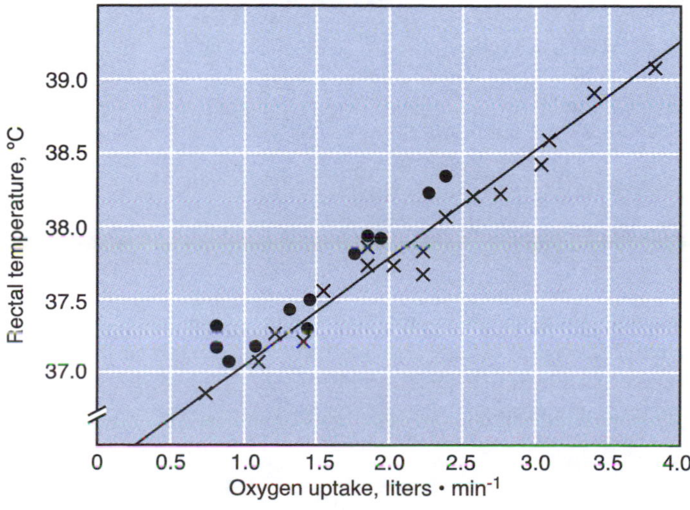


**Figure 13.4** The relationship between oxygen uptake and body temperature in exercise with the legs (×) and exercise with the arms (●).

From M. Nielsen 1938.

ergometer, increases linearly with oxygen uptake, at least up to an energy demand of about 75% of the individual's maximal aerobic power. At higher metabolic rates, the increase in core temperature seems to be curvilinearly related to oxygen uptake (C.T.M. Davies, Brotherhood, and Zeidfard 1976). This end temperature does not depend on the absolute magnitude of the energy output but on the level of metabolism relative to the individual's maximal aerobic power. A subject with a maximal oxygen uptake of 2.0 L · min$^{-1}$ attains a body temperature of about 38 °C during an exercise load, which demands an oxygen uptake of 1.0 L · min$^{-1}$ (i.e., 50% of the person's maximal aerobic power). A subject with a maximum oxygen uptake of 5.0 L · min$^{-1}$ can expend 2-1/2 times more energy (oxygen uptake of 2.5 L · min$^{-1}$) without the body temperature exceeding 38 °C (I. Åstrand 1960).

Figure 13.5 shows the relationship between muscle, rectal, and esophageal temperatures, measured simultaneously, and the oxygen uptake as a percentage of the individual's maximal oxygen uptake. As expected, the highest temperatures are seen in the exercising muscles where most of the heat is produced.

During maximal exercise, the rectal temperature can exceed 40 °C and the muscle temperature 41 °C without causing any discomfort for the exercising person.

## Cardiovascular Response

During prolonged light exercise in a hot environment, the heart rate increases markedly while cardiac output increases more gradually for 30 to 40 min despite a progressive decrease in stroke volume. During prolonged moderate to heavy muscular exercise, the heart rate does not reflect changes in cardiac output or in skin blood flow. Actually, stroke volume decreases while cardiac output is maintained by increased heart rate. During graded exercise, submaximal cardiac output is maintained by increased heart rate despite reduced stroke volume (Rowell 1993). Y. Suzuki (1980) concluded that any

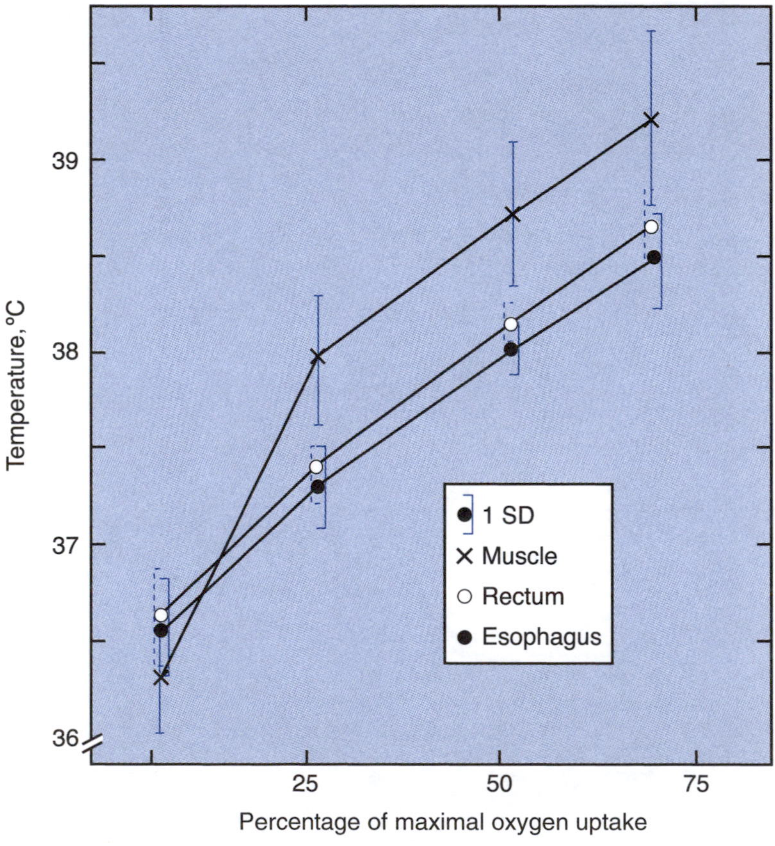


**Figure 13.5** Average temperature measured simultaneously in the esophagus, the rectum, and the exercising muscle in relation to the oxygen uptake in percentage of the individual's maximal oxygen uptake. Seven subjects were exercising for 60 min on a cycle ergometer. To the left, data obtained at rest. SD = standard deviation.

From Saltin and Hermansen 1966.

## CLASSIC STUDY

In his classic study, published in 1938, the Danish physiologist Marius Nielsen (1903–2000) let a subject perform a certain amount of work, 150 W, at ambient temperatures varying between 5 and 36 °C. After 30 to 40 min of exercise, the rectal temperature was the same, regardless of the room temperature. Because the mechanical efficiency was constant, the heat dissipation had to be the same in all these experiments. Figure 13.6 shows how radiation and convection *(R + C)* accounted for about 70% of the heat loss at the lower ambient temperatures. The skin temperature was then 21 °C. In the experiment carried out at the highest ambient temperature, the skin temperature was 35 °C and the body absorbed heat from the environment (i.e., 36 °C air temperature). This was completely counteracted by an increasing evaporative heat loss. In the cold, the subject evaporated 150 g of sweat; in the heat he evaporated 700 g. The greatest variation in the rectal temperature at the end of the work period in all these experiments was only .11 °C.

M. Nielsen 1938.

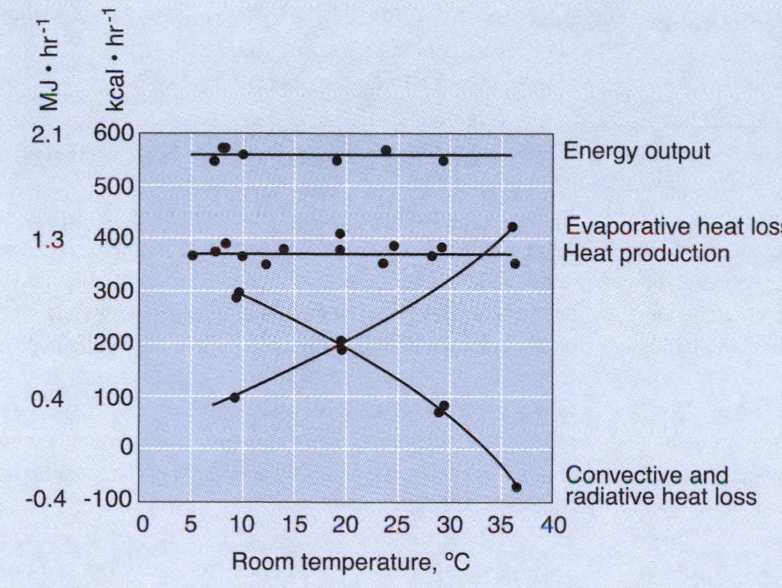


**Figure 13.6** Heat exchange during exercise (150 W) at different room temperatures in a nude subject.

Modified from M. Nielsen 1938.

changes in cardiocirculatory function that limit performance in hot environments are attributable to the decreased stroke volume (i.e., heart filling) caused by alterations in the peripheral blood flow. In 20 to 25 min of heavy exercise in the heat, the relative vasoconstriction helps maintain an adequate stroke volume, preventing a decrease in cardiac output (Nadel et al. 1979). Thus, in this case, circulatory regulation appears to have precedence over temperature regulation (see also M.F. Roberts and Wenger 1980).

During submaximal exercise in a hot environment, there is evidently a different blood flow distribution than with exposure to a normal climate. At a given cardiac output, more blood flows through the skin and less through the active muscle mass. This might explain the higher blood lactate concentration observed in the warm environment (Dimri et al. 1980; Wyndham 1973).

M.D. White and Cabanac (1995b) suggested that the nasal cavity acts as a heat exchanger in selective brain cooling in exercising humans. When the nostrils of their subjects exercising in a warm room were physically dilated, both tympanic temperature and forehead skin blood flow were significantly lower than during the control conditions. This was supported by their finding that body warming significantly increased nasal mucosal blood flow (White and Cabanac 1995a).

The circulatory changes associated with heat exposure are exaggerated by **hypohydration.** Nadel, Fortney, and Wenger (1980) studied the influence of hydration state on circulatory controls at an ambient temperature of 35 °C in four relatively fit subjects during ergometer exercise (55% $\dot{V}O_2$max). They found that hypohydration resulted in a significantly reduced cardiac output during exercise, attributable to a further reduction in stroke volume without corresponding elevation in heart rate. A marked reduction in skin blood flow maintained an already compromised venous return. The internal temperature (esophageal temperature) threshold for cutaneous vasodilation was elevated by .42 °C in hypohydrated conditions.

After conducting experiments that involved blood withdrawal and blood reinfusion in six subjects exercising at 60% $\dot{V}O_2$ in a 35 °C environment, Fortney et al. (198la) concluded that cardiac stroke volume and cutaneous blood flow vary in proportion to changes in absolute blood volume. They also observed that the increase in body temperature during exercise was significantly greater in hypovolemia but was not significantly reduced following volume expansion (i.e., blood reinfusion). The same authors showed that the blood volume significantly affects body fluid and sweating responses. The reduction in blood volume during exercise was significantly smaller under hypovolemic conditions and significantly greater during hypervolemic than during control conditions. Blood volume reduction also significantly altered the control of sweating rate, tending to reduce the sweating rate relative to the esophageal temperature. Thus, both the body fluid and sweating responses during hypovolemia conserve circulating blood volume during exercise.

Gonzalez-Alonso et al. (1999) showed that high internal body temperature per se causes fatigue in trained individuals during prolonged exercise. Time to exhaustion in their subjects was inversely related to the initial temperature and directly related to the rate of heat storage. In line with this, J. Booth, Marino, and Ward (1997) demonstrated that whole-body water immersion precooling improved running performance in hot, humid conditions and resulted in less thermoregulatory strain.

## Summary

Muscular exercise can increase the heat production from 10 to 20 times the heat production at rest. During exercise in a "neutral" environment, body temperature increases up to a maximum of about 40 °C or slightly higher at maximal work rates. Body temperature is not related to the absolute heat production but to the relative work rate, that is, the actual oxygen uptake in relation to the individual's maximal aerobic power; at a 50% load, the deep body temperature is about 38 °C. Within a wide range, the deep body temperature at rest and during exercise is not affected by the environmental temperature but the skin temperature is. In a given environment, the sweating rate depends mainly on the actual heat production and not primarily on the skin or rectal temperature.

# Physiology of Temperature Regulation

Various aspects of the physiology of temperature regulation have been presented by Hardy, Stolwijk, and Gagge (1970), Cabanac (1975), Nadel, Fortney, and Wenger (1980), Fortney and Vroman (1985), Nielsen et al. (1990), J.M. Johnson (1992), and Webb (1993). In this section we shall attempt to summarize some of the concepts covered.

## Temperature-Sensitive Receptors

In the hypothalamus and the adjacent preoptic region, as shown in animal experiments, there are nerve cells which by local heating and cooling elicit the same reactions that occur during exposure to heat or cold (Hammel 1965; Hardy 1967). These cells belong to the temperature regulatory center, which is connected via nervous pathways with receptors in the skin, the central nervous system, and possibly elsewhere in the body, such as in the deep leg veins, the muscles, the abdomen, and the spinal cord (Hensel 1974). These receptors consist of a net of fine nerve endings that are activated specifically by heat or cold stimuli (Hensel 1981; Zotterman 1959). They are especially sensitive to rapid changes in temperature and are highly susceptible to adaptation. In the heat receptors, the maximal frequency of the impulses occurs in a steady-state condition between 38 and 43 °C; in the cold receptors, the maximal impulse frequency occurs at 15 to 34 °C (Zotterman 1959). At temperatures above 45 °C, the cold receptors again can be activated. This could explain the paradoxical cold sensation experienced at the first contact with very hot water. The receptors register not only temperature changes but also temperature levels, particularly if the skin temperature is below 32 °C in the case of the cold receptors, and above 37 °C in the case of the heat receptors (Kenshalo, Nafe, and Brooks 1961). The number of active receptors determines to some extent the sensation of temperature. The fact that temperature sensation is relative is best illustrated by the simple experiment of putting one finger in warm water and another finger in cold water. When both fingers are then simultaneously put into lukewarm water, the finger that previously was exposed to cold water will sense the lukewarm water as warm; the other finger will interpret the water as cold.

For the regulation of body temperature, the hypothalamic temperature regulatory center and the temperature-sensitive receptors in the skin play a dominating role (Smiles, Elizondo, and Barney 1976). It appears that several temperature sensors are capable of triggering defense reactions independently to maintain thermal balance and that the response is a function of several inputs combined at the same time. According to Cabanac (1975), the temperature sensors in the spinal cord are at least one quarter to one half as sensitive as the hypothalamic temperature sensors. Warm stimulation of the spinal cord in animals is followed by all of the warmth defense reactions (skin vasodilation, reduced heat production, increase in evaporative heat loss) proportional to the spinal cord temperature. Cooling the spinal cord in unanesthetized dogs induces shivering, peripheral vasoconstriction, and **piloerection** proportional to the magnitude of the stimulus. Nevertheless, it appears that the hypothalamus remains the main center of temperature regulation, because studies of the effects of spinal cord lesions have shown that the spinal network does not possess complete thermoregulatory capability. On the basis of the available evidence, Cabanac (1975) concluded that it is probably more accurate to consider the control system for the regulation of body temperature as consisting of a number of networks operating independently of one another.

Greenleaf (1979) has discussed the hypothesis that fluid and electrolyte changes influence thermal regulation within the finer control boundaries, assuming that the control of mammalian thermoregulation has two separate components: a broad control that operates when core temperature deviates widely from its normal range and a fine control that operates between these two limits.

### For Additional Reading

For details concerning neuroendocrine control of thermogenesis, see J.E. Silva (1995), Arancibia et al. (1998), and Frank et al. (1996).

## Thermoregulatory Centers

The anterior hypothalamus and the preoptic region are sensitive to changes in the local temperature. Many neurons have been observed to increase their discharge rate when heated; only a few cells increase their discharge frequency when the local tissue temperature is lowered (Hardy 1967). Preoptic heating in animals exposed to a neutral environment can induce vasodilation in the skin and, eventually, panting and sweating. With the animal in a cool environment, such a local heating can inhibit the normal response of shivering and vasoconstriction of peripheral blood vessels. The depression of the metabolic rate causes the body temperature to decrease.

An intact posterior hypothalamus is required to induce the reactions to a cold environment, namely shivering and an increase of metabolic rate, and to restrict flow of heat to the skin by a vasoconstriction of skin blood vessels. This posterior center, however, is essentially insensitive to local temperature changes. The function of this area is largely coordinating, and it receives, rather than generates, temperature signals. Afferent impulses from cold receptors in the skin seem to be the main drive for this center.

The two thermoregulatory centers are, in a way, connected so that the response to stimulation of the anterior center includes stimulation of sweating but inhibition of shivering and vasoconstriction. Conversely, the action of a stimulation of the posterior center involves stimulation of vasoconstriction and an increase of heat production but a simultaneous inhibition of responses to heat. The final common pathways include not only the motor pathways of the synaptic and somatic systems but also blood humoral transmissions. Cold can, via the thermoregulatory center in the hypothalamus, affect the pituitary gland and the release of hormones, which in turn act on their target organs to release thyrotropic and adrenal hormones, increasing the heat production in the tissues.

There are functional connections between the central control of body temperature and the areas regulating water and food intake (Andersson 1967). According to Montain, Latzka, and Sawka(1995), both hypohydration level and exercise intensity produce independent effects on the control of thermoregulatory sweating. McCaffrey et al. (1979) demonstrated that under steady-state conditions, cutaneous thermoreceptors influence sweating rate. C.T.M. Davies (1980) observed that during exercise, the integrating and modulating effect of skin temperature from different regions of the body is responsible for the control of sweat loss, under conditions of constant central thermal drive. That peripheral inputs are a major factor in thermoregulatory processes also under non-steady-state conditions was demonstrated by Libert, Candas, and Vogt (1979) in subjects exposed to rapid or slow alterations in air and wall temperatures (28–45 °C). They concluded that cutaneous receptors produce a positive and a negative rate component within the central thermal integrator. In nude, resting men exposed to environmental temperatures changing from neutral (28 °C) to warm (50 °C), Libert, Candas, and Vogt (1978) observed that sweating started before appreciable variation in rectal temperature, suggesting a peripheral control of the onset of sweating.

Under steady-state conditions, the central drive for sweating in humans generally has been described as a function of internal and mean skin temperature signals (Stolwijk 1970; figure 13.7). The experiments of Libert et al. (1982) during thermal transients, however, showed the insufficiency of explanations based on a simple additive function of core and skin temperature. During such thermal transients, there appears to be an interplay of additive and multiplicative controls. The multiplier effect appears to be exhibited not only in controlling local sweat rate by local skin temperature but also in making up the central drive for sweating.

B. Nielsen (1969) noticed a remarkable similarity in the thermoregulatory responses to passive heating by **diathermy** and to active heating by muscular exercise. Thus, at the same level of heat production, rectal and skin temperatures and estimated skin blood flow were increased to the same level in the two kinds of experiments. Therefore, she concluded, a work factor of nervous origin (from mechanoreceptors or cortical irradiation) could operate on the thermoregulatory centers at the start of exercise but hardly in the later phase of exercise. In intermittent exercise with the same heat production over a given period of time, as in continuous exercise (with different intensity during the periods of activity), B. Nielsen (1969) found the same body temperature and sweat rate, which also should exclude nervous impulses related to the severity of exercise as the "work factor." She proposed that a chemical factor, liberated during exercise proportionate to the engagement of the aerobic processes, is responsible for the temperature resetting.

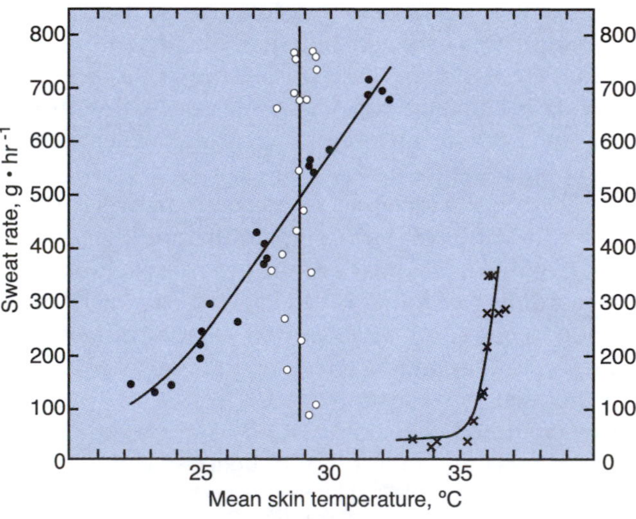


**Figure 13.7** Steady-state values of sweat rate plotted against the corresponding values of mean skin temperature. ○ = exercise intensity from 90 to 235 W at constant environmental temperature of 20 °C; ● = constant exercise intensity (150 W) at environmental temperatures from 5 to 30 °C; × = experiments at rest at environmental temperatures from 25 to 44 °C.

From B. Nielsen 1969.

## For Additional Reading

For further discussions about thermoregulation during exercise, see Nielsen et al. (1990) and Rowell (1993).

# Summary

Physiological thermoregulation is controlled by nervous centers in the anterior preoptic region of the hypothalamus. The central thermoregulation is quite neurochemically complicated and is still far from understood. A decrease in temperature activates cold receptors in the skin and certain neurons in the hypothalamus and upper part of the spinal cord. These further stimulate heat production by activating the nervous pathways. Body temperature is maintained by vasoconstriction and shivering.

It appears that both a decrease in skin temperature and a decrease in core temperature can independently elicit a thermoregulatory increase in the oxygen uptake, and hence heat production, in humans. It still appears likely, however, that both core and skin temperatures have a combined influence on our subjective and physiological reactions to temperature changes in our immediate environment.

When a person is exposed to heat, sweating and the dilation of the skin blood vessels do not run parallel. The produced sweat in itself can cause a vasodilation; on the other hand, the local skin temperature can affect the diameter of the blood vessels in that area. At high evaporative sweat rates, the skin can cool, causing vasoconstriction. Heating or cooling of the skin of the head increases or decreases, respectively, sweating over the rest of the body.

Other impulses to the temperature regulatory center exist. Fever is caused when the "thermostats" are set at a higher level. The reasons for the diurnal variations in body temperature are unknown.

A great deal is known about the various thermosensitive elements such as cold and hot receptors and deep receptors, which sense temperature changes in different parts of the body core and shell. Similarly, a considerable amount of knowledge is available concerning the effector elements that can contribute to heat production or heat loss as required. What is still missing, however, is an understanding of the integrative mechanisms in the human central nervous system proper that transform afferent inputs into efferent action resulting in mandatory adjustments such as piloerection, shivering, sweat secretion, vasoconstriction or vasodilation, endocrine gland stimulation, or changes in cardiopulmonary function.

## Acclimatization

Continuous or repeated exposure to heat, and possibly also to cold, causes a gradual adjustment or acclimatization, resulting in a better tolerance of the temperature stress in question. Many plants prepare for the winter by increasing their carbohydrate content. Certain types of trees can tolerate air temperatures below –40 °C during the winter, but during the summer, they fail to survive air temperatures below –3 °C. Certain insects accumulate the "antifreeze" glycerol in the fall, which enables them to survive cold (Dill 1964).

### Heat

After a few days of exposure to a hot environment, the individual is able to tolerate the heat much better than when first exposed. This improvement in heat tolerance is associated with increased sweat production, a lowered skin and body temperature, and a reduced heart rate (Robinson et al. 1943; Kuno 1956; C.G. Armstrong and Kenney 1993; Bass 1963; Febbraio et al. 1994; Libert, Candas, and Vogt 1983; Sato et al. 1990). An example is illustrated in figure 13.8. Usually the skin blood flow is reduced after acclimatization; in one experiment it declined from 2.6 to 1.5 $L \cdot m^{-2} \cdot min^{-1}$, that is, to about 60% of the original value (Bass 1963). The increased sweat rate provides for more effective cooling of the skin through the evaporative heat loss, and the resultant lowered skin temperature provides for better cooling of the blood flowing through the skin. Sweat

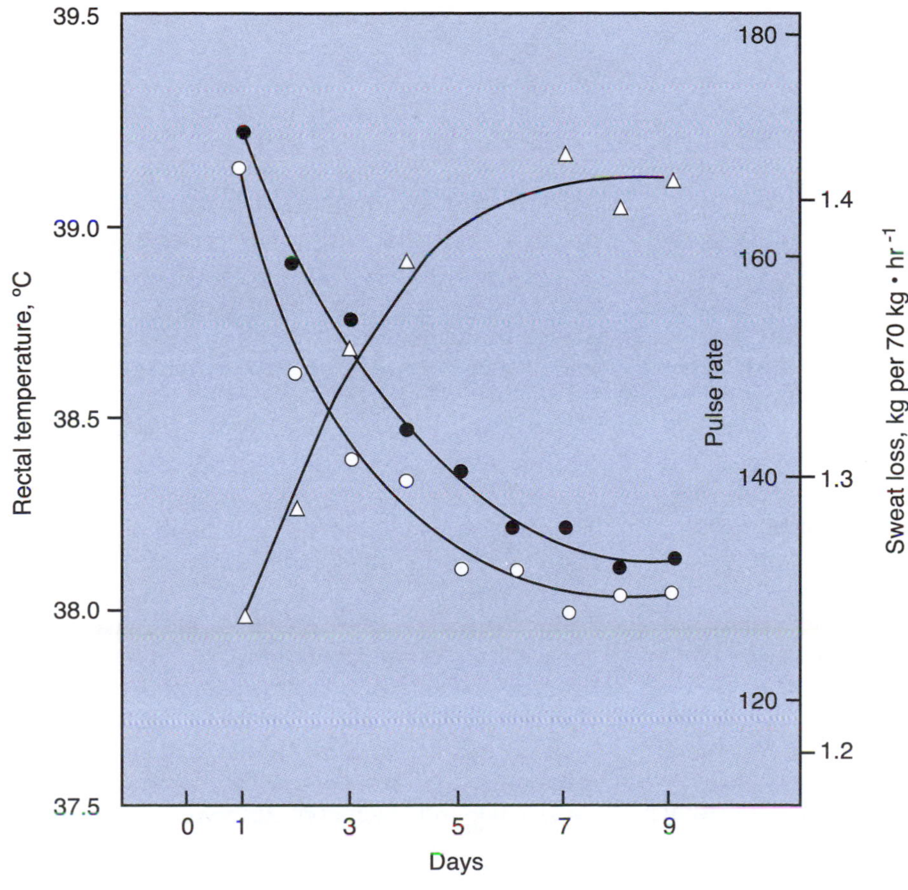


**Figure 13.8** Mean rectal temperature (●), heart rate (○), and sweat loss (△) in a group of men during a 9-day acclimatization to heat. Each day they exercised for 100 min at a rate of energy expenditure of 1.3 MJ (300 kcal) $\cdot h^{-1}$ in a hot climate (48.9 °C dry bulb and 26.7 °C wet bulb temperature).

Adapted from Lind and Bass 1963.

production can increase as much as 100% (Leithead and Lind 1964).

It is possible, then, to demonstrate objectively physiological alterations in response to prolonged exposure to heat. It is also possible to explain why the heat tolerance gradually increases. In acute experiments, the skin blood flow increases at the expense of the blood flow through other tissues, although this may not always be the case (Rowell 1983). As the individual becomes acclimatized, normal distribution of the blood flow is once more established. Within 4 to 7 days of exposure to a hot environment, most of the changes have taken place, and at the end of 12 to 14 days, the acclimatization is complete. Even a relatively short daily heat exposure will have some effect (Cotter, Patterson, and Taylor 1997), but exercising in the heat for 1 h per week has no heat acclimatization effect (Barnett and Maughan 1993).

The effect of heat acclimatization persists several weeks following heat exposure, although some impairment in heat tolerance can be detected a few days following cessation of exposure, such as after a long weekend, especially if the individual is fatigued and alcohol has been consumed (C.G. Williams, Wyndham, and Morrison 1967).

During acute exposure to heat and during the first phase of acclimatization, the heart rate during a standardized prolonged exercise is increased and the stroke volume reduced. The cardiac output remains relatively constant. Arterial blood pressure remains essentially unaltered (Rowell 1983). The mechanisms for these adjustments are not revealed.

Measurements of total blood volume during acclimatization to heat have produced conflicting results, but most studies seem to indicate an increase in blood volume (Harrison 1986; Rowell 1983). Senay, Mitchell, and Wyndham (1976) suggested that the most critical event in the first phase of heat acclimatization is expansion of the plasma volume. However, Fortney and Senay (1979), studying the responses of nine females to moderate exercise in a cool and a hot environment, during a 4-week training program and during heat acclimatization, found that training significantly expanded resting plasma volume and total protein content, whereas acclimatization caused increased sweat rates and decreased skin temperature. Similarly, Harrison et al. (1981) suggested that the expansion of the intravascular volume with heat acclimatization should be regarded merely as a secondary effect of exercise and not as a primary adaptive response to heat. They maintained that hemoconcentration is the usual response to cycling exercise in the heat, whereas Senay (1978) observed a significant **hemodilution** in acclimatized exercising men (30% $\dot{V}O_2$ max) during exposure to heat (33.8 °C).

Seasonal differences in acclimatization to heat have been reported in humans (Shapiro, Hubbard, et al., 1981). Acclimatized humans in the winter season exhibited lower sweat rates than in the summer. During winter acclimatization, the blood volume expansion was attributable to expansion in both the plasma and cellular compartments, whereas in the summer the acclimatization resulted in only plasma volume expansion.

Oddershede and Elizondo (1980) performed a complete body fluid compartment analysis during resting heat acclimatization in rhesus monkeys. Their results suggest that heat acclimatization in primates is characterized by a protein and fluid shift from the interstitial fluid compartment to the cardiovascular and the intracellular compartments.

### *State of Training*

Some authors (Avellini et al. 1982; Edholm and Weiner 1981) have maintained that physical activity, as such, may not be necessary to achieve heat acclimatization but may contribute to the extent that it increases body temperature. On the other hand, Inbar et al. (1981) and Smorawinski and Grucza (1994) reported physiological changes compatible with heat acclimatization by mere physical conditioning in a neutral climate.

It appears that a trained individual is better able to adjust to heat than one who is untrained. Physical training enhances the sweating mechanism at a given level of central sweating drive. The increased metabolic rate during training raises a high thermoregulatory demand, and apparently this demand increases the peripheral sensitivity of the sweat glands to the central sweating drive. For an individual who increases his or her maximal aerobic power by training, a given rate of exercise will require a lower percentage of that person's maximal oxygen uptake. Concomitantly, the core temperature will decrease during a standardized exercise. The increased capability for heat dissipation behind this adaptation is attributable to an enhanced sweating response (Shvartz et al. 1979). In addition, heat acclimatization further enhances the sweating response at a given level of central sweating drive by lowering the zero point of the central nervous system drive for sweating. In other words, physical training seems to increase the slope of the sweating rate–core temperature curves; that is, the activity of the sweat glands increases at a given core temperature. In contrast, acclimatization to heat seems to lower the threshold core temperature at which sweating starts (it moves the curves "to the left" without changing the slope (Nadel et al. 1979). For an optimal heat acclimatization, simultaneous exposure to both heat and exercise is recommended.

### For Additional Reading

For further discussion about the interactions of physical training and heat acclimation, see Dawson (1994), Aoyagi, McLellan, and Shephard (1997), and B. Nielsen (1998).

## Cold

Animals habitually exposed to cold develop an effective protection in the form of fur and an effective heat exchange system in their peripheral blood vessels. In certain types of seals, the skin can be cooled to 0 °C without an increase in oxygen uptake. In the human, lowering the skin temperature by a few degrees can double the metabolic rate. Thus, with respect to reaction to cold, the human takes a path somewhat different from that followed by arctic animals.

Originally, Budd (1962) claimed evidence of general acclimatization to cold in men who wintered in Antarctica. Thirty years later, Savourey, Vallerand, and Bittel (1992) studied a group of eight volunteers submitted to a cold test before and after a ski journey across Greenland lasting for 3 weeks at temperatures of –20 to –30 °C. They observed that the arctic journey lowered the rectal temperature and the mean skin temperature but did not change metabolic heat production.

The metabolic rate of "cold-acclimatized" individuals often is elevated when they are exposed nude to a standardized cold stress, even though shivering is said to be less pronounced. Nevertheless, the metabolic rate is unchanged at normal room temperature. It is possible that hormones, notably noradrenaline, play a role in elevating the metabolic rate, but the mechanism is unclear. The natives of Central Australia and those of the Kalahari desert, wearing little or no clothing, can be exposed to night temperatures of about 0 °C or below. They do not shiver even though their core and skin temperatures keep decreasing throughout the night, and they can sleep, while control subjects are "fighting" against the cold environment by shivering and cannot sleep. There are similar observations of other primitive populations and also of women divers of the Korean peninsula who do not increase their heat production by shivering to the same extent as individuals who have not been exposed to cold for longer periods of time The essential feature of cold adaptation seems to be reduced shivering. The energy saved is not very important, but this effect has other advantages. Shivering is disturbing and uncomfortable. "Shivering is the first line of defense but nobody likes it!" (LeBlanc, 1975).

Radomski and Boutelier (1982), after observing hormonal and thermal responses, suggested that a certain degree of cold resistance can be induced rapidly in humans by short, repeated exposures to an intense cold stress. This was confirmed by Tochihara et al. (1995). This cold resistance persists for a significant period of time after the last exposure.

When a person, whether an arctic native or otherwise, allows his or her hands to be repeatedly exposed to cold for about 1/2 h daily for a few weeks, this cold stress increases the blood flow through the hands, so that they remain warmer and are not so apt to become numb when exposed to cold. This is termed local acclimatization to cold. Although it inevitably will cause a greater amount of heat to be lost from the hands, it will improve the ability of the hand and fingers to perform work of a precise nature in the cold (see Enander, Sköldström, and Holmér 1980; Hellström 1965; LeBlanc 1975; Naidu and Sachdeva 1993; Nelms and Soper 1962; Strømme, Andersen, and Elsner 1963).

Although it is difficult to demonstrate definite evidence of general physiological acclimatization to cold in humans, such acclimatization can be produced in animals exposed to severe cold. If rats are placed in a cold-chamber kept at 5 °C for as long as 6 weeks, their metabolic rates will increase twofold. They will double their food intake, and their thyroid function will increase. The main feature of the cold-acclimatized rat is its ability to maintain a high rate of heat production. This ability is absent in the nonacclimatized rat, which is unable to survive in the cold.

### *Effect of Physical Training and Diet*

The interaction between physical training or fitness and cold acclimatization was investigated by Strømme and Hammel (1967). They found that the metabolic response to cold exposure was significantly higher in exercise-trained rats compared with physically inactive animals. This was mainly attributed to an improved shivering capacity. Later, Östman and Sjöstrand (1975) provided evidence that the adaptive changes following physical training include increased sensitivity of tissues to the vascular and metabolic action of noradrenaline. They demonstrated that chronically exercised rats had a lower demand for noradrenaline in both warm and cold environments and had greater tissue reserves of catecholamines together with greater capacity for synthesis of catecholamines. The consequences of this are obvious: The animal develops better tolerance to conditions that require increased transmitter secretion to maintain homeostasis, as during exposure to severe cold.

These observations in animals are somewhat in line with the findings of Bittel et al. (1988) in 17 male humans submitted to various acute cold temperatures (lying nude in a climatic chamber for 2 h at an ambient temperature of 10, 5, or 1 °C). For all their subjects, there was a direct relationship between physical fitness level and metabolic heat production, mean skin temperature and conductance, and mean skin temperature at the onset of shivering. The predominance of thermogenic or insulative reactions depended on the intensity of the cold stress: Insulative reactions were preferential at 10 °C, or even at 5 °C, whereas colder ambient temperature (1 °C) triggered metabolic heat production, reactions that were closely related to the subject's physical fitness level. The authors concluded that fit subjects have more efficient thermoregulatory abilities against cold stress than unfit subjects, certainly because of an improved sensitivity of the thermoregulatory system.

### For Additional Reading

For further discussion about metabolic adaptations to exercise in the cold, see Shephard (1993).

A person's food intake is not appreciably larger in a cold environment (Rodahl 1963), nor is the rate of heat production. It is true that the arctic natives' rate of heat production is higher than that of nonnatives, but much of this is probably attributable to the specific dynamic effect of diet, because an arctic native living on the nonnative's diet does not have any higher rate of heat production than the nonnative (Rodahl 1952). When nonnatives move to the arctic, their metabolic rate is no higher than it was at home. Healthy arctic natives do not show increased thyroid function compared with nonnatives (Rodahl and Bang 1957). The most probable reason is that in clothed individuals habitually exposed to cold environments, the degree of cold exposure is not severe enough to cause an appreciable increase in metabolism. Without increased metabolism, there is no need for increased food intake. However, M. Levine et al. (1995) observed significant seasonal variations in thyroid function in young male infantry soldiers assigned to the interior of Alaska.

## *Severe Cold Stress*

When the human is exposed to cold stress that is far more severe than that normally encountered by the clothed individual living in the arctic, certain physiological changes occur that can be interpreted as hormonally induced adjustments to cold. When young men dressed only in shorts and sneakers were confined continuously to cold-chamber temperatures of 8 °C for 3 to 10 days, they responded by violent shivering, which lasted more or less continuously night and day throughout the experiment, even at night when they slept under a blanket. They soon learned to continue to sleep despite the shivering. As a result, the increased heat production attributable to the shivering was maintained even during sleep, so that the body temperature remained normal and the subjects slept relatively comfortably. Occasionally, however, the same subjects, for some reason, failed to keep up the vigorous shivering during the night while they slept. Consequently, their body temperature continued to drop and approached dangerously low levels. Under these conditions, the subjects occasionally objected to being disturbed and wanted to be left in peace to continue to sleep. If this were allowed, the subject might continue to cool and, conceivably, might eventually die from hypothermia. It thus appears that an exposed person sleeping in the cold actually can freeze to death if the body's rate of cooling is slow and gradual, so that violent shivering is not produced, which would wake up the person (K. Rodahl et al. 1962).

These subjects in the climatic chamber doubled their metabolic rates because of constant shivering. Their resting heart rate was markedly elevated, and the nitrogen loss in the urine was greatly increased. These changes were interpreted as being the result of hormonal responses brought about by the cold stress (Issekutz, Rodahl, and Birkhead 1962). The cold exposure of these subjects in the climatic chamber (8 °C) was far greater than the cold exposure of the arctic native or any clothed group of individuals living in the arctic. Consequently, all the climatic chamber subjects suffered ischemic cold injury of their feet, although the temperature never approached the freezing temperature of the tissue.

When nutritional deficiency was superimposed on the cold stress in these subjects, physical performance was markedly impaired. After subjects endured 9 days of constant exposure to 8 °C in the nude, maximal oxygen uptake showed no deterioration when the individuals were fed an adequate diet consisting of 12.5 MJ (3,000 kcal) and 70 g of protein. When the energy intake was reduced to 6.3

MJ (1,500 kcal), when the protein intake was reduced to 4 g · day$^{-1}$, or when both energy and protein intakes were reduced, physical performance capacity deteriorated markedly within 3 to 5 days. The effect was most pronounced in the treadmill running time; maximal oxygen uptake in some of the subjects was less affected.

Jansky et al. (1996) showed that repeated exposure of young athletes to cold water (14 °C, 1 h three times per week for 4–6 weeks) altered the regulation of thermal homeostasis. Central and peripheral body temperatures at rest and during cold immersion were lowered. The metabolic response to cold was delayed, and the subjects demonstrated a lowered cold sensation. Because of these physiological changes, about 20% of the "acclimatized" subjects' total heat production was saved during one cold water immersion. Maximal aerobic and anaerobic performances were not altered. Changes in cold sensation and in the regulation of cold thermogenesis were first noticed after four cold water immersions and persisted for at least 2 weeks after termination of the cold exposures.

### *Summary*

Resting humans acclimatize to heat by adjusting the sweating rate and core temperature threshold for sweating and by increasing blood volume. Changes accompanying acclimatization to exercise in hot, dry conditions include reduction in heart rate, skin temperature, and core temperature. Sweating begins at a lower body temperature. Most of the changes occur during the first days of exposure. Apparently, cardiac output is not changed, but plasma volume can be increased.

Whether a clothed human exposed to cold climate really becomes acclimatized to cold, or to what extent, is still an open question. Local acclimatization, however, can occur in the hands of individuals regularly exposed to cold. Such localized cold stress can increase blood flow through the hands, which will improve the ability of the hand and fingers to perform work of a precise nature in the cold. Physically fit subjects seem to have more efficient thermoregulatory abilities against cold stress than unfit subjects, because of an improved sensitivity of the thermoregulatory system.

# Limits of Tolerance

During exposure to severe heat or cold, there is a distinct possibility that health and, in some cases, even life will be threatened. In this section, therefore, we shall briefly address some health issues related to physical activity in harsh environments, including variation in tolerance limits with respect to age and sex.

## Hypothermia

**Hypothermia** is a condition usually characterized by body temperatures below 35 °C. During the initial stages of hypothermia, the patient shivers and gradually becomes disoriented, apathetic, and hallucinatory or can become aggressive, excited, or even euphoric. As the rectal temperature gradually drops below 34 °C, the patient can appear distant and stuporous, he or she can be unconscious, respiration is shallow, and the pulse is weak. Cardiac arrhythmia can develop. Reflexes are reduced, and the pupils are dilated. Finally, the patient reaches the paralytic stage as the rectal temperature drops below 30 °C. The skin is cold, no pulse can be detected, the pupils are dilated, there are no reflexes, and there are no heart sounds.

If the patient cannot be brought to a hospital safely and the treatment must be performed under field conditions, slow rewarming should be applied (.5 °C · h$^{-1}$). The danger of rapid rewarming in the field is a further decrease in deep body temperature attributable to the return of cool venous blood from the skin and extremities to the core. This temperature decrease can be fatal as a result of cardiac arrhythmia (ventricular fibrillation), which requires defibrillation.

Even at more moderate degrees of body cooling, the muscle temperature can be significantly lowered, causing muscular weakness, impaired neuromuscular function, and reduced endurance. This can be the underlying cause of some of the fatal accidents among climbers and cross-country skiers.

## Cold Injury

Cold injury can occur during common winter sports such as skiing and in long-distance runners competing in cold, windy conditions. Local cold injury can occur in the exposed parts of the body such as the face, hands, and feet, either because of the freezing of tissue and formation of ice crystals (i.e., frostbite) or by vasoconstriction that deprives the exposed parts of blood circulation, leading to ischemic cold injury. In the case of frostbite, it has been customary to distinguish between first-degree frostbite, involving merely the superficial layer of the skin; second-degree frostbite, with the formation of exudate-containing blisters in the skin; and third-degree frostbite, involving the freezing of the deeper tissues as well, including subcutaneous and muscle tissues. The pathological consequences of the freezing can, to some extent, depend on the speed of the cooling. In rapid cooling, the ice crystals formed can break the cells and cause tissue destruction and necrosis. During slow cooling, ice crystals are formed in the tissues, but because solutes are excluded in the freezing process, the osmotic pressure in the extracellular fluid increases, pulling fluid out of the cell, with exudate formation as a result. This in itself will cause cell damage, augmented by the concomitant vasoconstriction with increased venous pressure, reduced capillary blood flow, blood cell aggregation, thrombosis, and necrosis.

The treatment of local cold injury consists of local rewarming in the case of first-degree frostbite. In second-degree frostbite, blisters should be left intact; third-degree frostbite requires hospitalization. At any rate, the patient should be brought into shelter, tight clothing should be loosened, the patient should be kept warm and given hot drinks, and if possible the patient should be active to produce internal heat. The treatment should be continued until the return of normal color and feeling in the affected parts, and the patient should be kept under surveillance. If normal functions are not restored within 30 min, the patient should be hospitalized.

Local cold injury of the eyes (i.e., transitory epithelial damage of the cornea with the formation of corneal edema and blurred vision) has been observed in cross-country skiers competing at very low temperatures (Kolstad and Opsahl 1979). Similar afflictions were reported among the early aviators, racing cyclists, speed skaters, and natives of the arctic (for references, see Kolstad and Opsahl 1979). In skiers, this particular type of injury usually is seen only during competitive ski races, not during training. It is especially apt to occur during long-distance races at temperatures below –15 °C combined with wind. The symptoms usually develop toward the end of the race and consist of impaired or blurred vision evidently caused by pathological changes in the lower segment of the cornea. According to Kolstad and Opsahl (1979), who first described this phenomenon, the cause could be an impaired blinking reflex and an incomplete closure of the eyelids during blinking because the surface temperature of the cornea has decreased. As a result of the reduced blinking, the thin tear film covering the cornea and nourishing it will not be maintained. This could explain the observed degeneration of the epithelium of the unprotected part of the cornea. The damage is transitory, however, and usually heals completely within 24 h. A possible prevention is to protect the eyelids and the cornea by wearing suitable headgear, perhaps in combination with the use of contact lenses.

### For Additional Reading

For a comprehensive review of cold injuries, see *International Journal of Circumpolar Health* (2000).

## Failure to Tolerate Heat

The vastly increasing number of participants in long-distance running races under hot conditions has markedly increased heat casualties and heat illnesses, such as heat exhaustion, heat syncope, and heat stroke, often in combination with **dehydration**, especially in marathon runners (J.R. Sutton 1984). Brain function is particularly vulnerable to heat (M.A. Baker 1982). Tolerance to elevated deep body temperature is extended if the brain is kept cool (Carithers and Seagrave 1976).

The most serious consequence of exposure to intense heat is heat stroke, which can be fatal. It is caused by a sudden collapse of temperature regulation leading to a marked increase in body heat content. The rectal temperature can be 41 °C or higher, and the skin is hot and dry. The person has tachycardia and hypotension, metabolic acidosis, **disseminated** intravascular coagulation, and occasionally renal failure. The victim is confused or unconscious. This form of temperature regulatory failure is rare. The risk is higher in nonacclimatized than in acclimatized individuals. Obese persons and older individuals are most susceptible. The treatment is rapid cooling (e.g., by pouring cold water over the victim or applying ice packs) until the rectal temperature has dropped below 39 °C.

Because heat stroke often is associated with peripheral circulatory collapse, the oral temperature of the victim may not necessarily be very high, whereas the rectal temperature always is: This emphasizes the importance of measuring rectal temperature in long-distance runners who collapse (J.R Sutton 1984).

Another type of temperature regulation failure is the so-called anhidrotic heat exhaustion. The victim can have a body temperature of 38 to 40 °C and may sweat very little or not at all. He or she feels very tired, may be out of breath, and has tachycardia. The main trouble is reduced sweat production. When the patient stops exercising and is removed to a cool place, this condition rapidly improves.

A third type of serious disturbance caused by heat exposure is excessive loss of fluid and salt, usually because of failure to replace fluid and salts lost through sweating. After several weeks of exposure, the patient eventually can experience cramps, the so-called *miner's cramps*, which in rare cases can be fatal. Intravenous administration of NaCl will promptly relieve the cramps.

Heat syncope is a less serious affliction caused by heat exposure. This is caused primarily by an unfavorable blood distribution. A large proportion of the blood volume is distributed to the peripheral vessels, especially in the lower extremities as the result of prolonged standing, or by a reduction in blood volume attributable to dehydration. The result is decreased blood pressure and inadequate oxygen supply to the brain, which can lead to unconsciousness. If the victim is placed in the horizontal position, preferably with the legs elevated, he or she will quickly regain consciousness. This type of heat collapse is one of the body's built-in safety mechanisms.

### For Additional Reading

For literature on acute and chronic occupational heat illness, see Dukes-Dobos (1981). For advice about keeping athletes safe in hot weather, see Sandor (1997) and Sparling and Millard-Stafford (1999).

## Age

C.T.M. Davies (1981) showed that thermal responses of children are quantitatively different from young adults, evaporative sweat loss being lower in children and skin temperature higher for the same environmental conditions compared with adults. It is generally believed that children cannot tolerate hot environments as well as adults and that the greatest risk of heat sickness for children is heat exhaustion (i.e., cardiovascular instability; Armstrong and Maresh 1995).

Although the experimental data are limited, earlier evidence suggests that heat tolerance is reduced in older individuals (Leithead and Lind 1964; Lind et al. 1970; Robinson 1963). They start to sweat later than do young individuals. Following heat exposure, it takes longer for their body temperature to return to normal levels. Older people react with a higher peripheral blood flow, but their maximal capacity is probably lower. In one study, it was found that 70% of all individuals who suffered heat stroke were over 60 years of age (Minard and Copman 1963). On the other hand, C.T.M. Davies (1979), who conducted 1-h treadmill exercise experiments on subjects 18 to 65 years of age in a moderate environment, observed no evidence for differences in thermoregulatory function that could be ascribed to sex or age.

Furthermore, studies by Drinkwater et al. (1982) revealed that in healthy older women, aging does not diminish the functional capacity of the sweating mechanism to cope with heat stress while resting.

After reviewing the literature on heat tolerance, Pandolf (1997) concluded that the work heat tolerance of habitually active and aerobically trained middle-aged men was the same as, or better than, that of younger individuals. On the other hand, the elderly may be more susceptible to hypo- or hyperthermia than young adults (G.S. Anderson, Memeilly, and Mekjavic 1996).

## Sex

The available evidence shows that women require lower evaporative cooling both in hot and wet environments and in hot and dry environments (Shapiro, Pandolf, et al. 1981). Women have a lower tissue conductance in cold and a higher tissue conductance in heat than do men. This indicates a greater variation in the peripheral reaction to climatic stress in women. This fact does not appear to be important for the performance of work, however. From studies of active men versus active women during acclimatization to dry heat, Horstman and Christensen (1982) concluded that active women performed exercise of equal relative intensity in dry heat as well as active men. Ventilatory, metabolic, and cardiovascular differences between sexes were minimal. Frye and Kamon (1983) observed no differences in sweating efficiency between sexes in the dry heat, but the women maintained a significantly higher sweating efficiency than the men in the humid heat. In both environments, the men recruited a significantly lower percentage of their available sweat glands than did the women.

Physical fitness is an important factor when men and women are compared in the heat. When fitness levels are similar, the previously reported sex-related differences in response to an acute heat exposure seem to disappear (Avellini, Kamon, and Krajewski 1980; B. Nielsen 1980). Physical training also improves the circulatory potential. Thereby the trained individual can maintain a cardiac output sufficient to meet metabolic requirements and the demand for peripheral blood flow for a longer period of time than untrained people (Drinkwater et al. 1976).

## Mental Work Capacity

An evaluation of mental or intellectual performance during exposure to heat or cold is hampered by subjective variations and lack of suitable objective testing methods. As a rule, a deterioration is observed when the room temperature exceeds 30 °C if the individual is acclimatized to heat. For the unacclimatized, clothed individual, the upper limit for optimal function is about 25 °C. The observed deterioration in performance capacity refers to precise manipulation requiring dexterity and coordination; ability to observe irregular, faint optical signs; the ability to remain alert during prolonged, monotonous tasks; and the ability to make quick decisions. During a 3-h drilling operation, the best results were achieved at 29 °C, but at a room temperature of 33 °C, the performance was reduced to 75%; at 35.5 °C, to 50%; and at 37 °C, to 25%. A high level of motivation can to some extent counteract the detrimental effect of the climate (Pepler, 1963).

Wyon, Andersen, and Lundqvist (1979) examined the effect of moderate heat stress (up to 29 °C) on mental performance in 17-year-old boys and girls. They were subject to rising air temperature conditions, typical of occupied classrooms in the range of 20 to 29 °C. Sentence comprehension was significantly reduced by intermediate levels of heat stress in the third hour. A multiplication task was performed significantly more slowly in the heat by male subjects, showing a minimum at 28 °C. Recognition memory showed a maximum at 26 °C, decreasing significantly at temperatures below and above.

# Water Balance

Reasonable figures for the daily water loss are as follows: from gastrointestinal tract, 200 ml; respiratory tract, 400 ml; skin, 500 ml; kidneys, 1,500 ml; total, 2,600 ml. This loss is balanced by an intake as follows: fluid, 1,300 ml; water in the food, 1,000 ml; water liberated during the oxidation in the cells, 300 ml; total, 2,600 ml. However, the water loss can increase considerably when the individual exercises or is exposed to a hot environment.

Water loss through the respiratory tract varies roughly with the pulmonary ventilation (dryness and temperature of the inspired air have some influence). The ventilation varies within a wide range directly with the production of carbon dioxide, and this production is, in turn, proportional to the metabolic rate. Therefore, the water volume from oxidation, proportional to the metabolic rate, equals, by coincidence, roughly the water loss through the respiratory tract.

During very heavy exercise, glycogen is the preferred fuel. About 2.7 g of water is stored together with each gram of glycogen, and this water becomes free as the glycogen is combusted. If during such heavy exercise, 5 MJ (1,200 kcal) is totally metabolized, 80%, or 4 MJ (960 kcal), can be derived from glycogen. The liberated volume of water (including the water of oxidation) will be close to 800 ml. If the mechanical efficiency is about 25%, 3.8 MJ (900 kcal) of the 5 MJ (1,200 kcal) should be dissipated as heat if the body temperature is maintained unchanged. An exclusive evaporative heat loss demands the evaporation of about 1,500 ml of water to eliminate 3.8 MJ (900 kcal). Under these conditions, only approximately half the necessary water volume must be taken from body "stores," because the rest apparently is liberated in the processes producing the heat. More sweat can be secreted than is evaporated from the skin. On the other hand, radiative and convective heat exchange can reduce the demand on the evaporative heat loss. When the glycogen depots are again restored, extra water is certainly needed.

## Thirst

In adults, about 70% of the lean body weight is water, so there is a substantial buffer to cover water losses over limited periods of time. However, in the long run, water intake must balance water loss by the several routes mentioned. Hypothalamus and adjacent preoptic regions play the essential role in the thirst mechanism (Andersson 1967; Stevenson 1965). Osmoreceptors react to an increase in the osmolarity of the intracellular fluid. Any change in the internal environment leading to cellular hypohydration will elicit thirst. As originally shown by Verney (1947), a second effect of an increase in body fluid osmolarity is an increased secretion of antidiuretic hormone (ADH) from the neurohypophysis (posterior pituitary gland), an effect mediated from the same center. The kidneys must excrete a minimal amount of water as a vehicle to eliminate solids. When water is in excess in the body, little or no ADH is brought to the kidneys and more water is excreted. A water deficit will, as stated, stimulate ADH secretion, increasing the reabsorption of water by increasing the water permeability of the wall of the collecting ducts in the kidneys. It is a common observation that the volume of urine is reduced when sweating is profuse.

Montain et al. (1997) examined the separate and combined effect of hypohydration level and exercise intensity on aldosterone and ADH (arginine vasopressin) responses during exercise-induced heat stress. They found that aldosterone and ADH increased in a graded manner with hypohydration level and this effect persisted during a following exercise-induced heat stress. The higher the exercise intensity, the higher were the increases. The authors also observed that ADH responses were closely coupled to osmolality.

Andersson (1967) reported that injection of minute amounts of hypertonic saline into, and electrical stimulation within, the anterior parts of the hypothalamus can elicit excessive drinking in the goat. Later C.J. Thompson et al. (1986) demonstrated the relation of plasma osmolality to the intensity of thirst by infusion of hypertonic saline in humans (figure 13.9). Similar stimulations not only induce drinking but inhibit feeding. The intracellular fluid volume in the specific cells of the hypothalamus could be the crucial factor in thirst. The osmotic pressure across the cell membrane will, of course, influence this volume. Volume receptors (stretch receptors) in the walls of the atria and great veins can detect and reflexively adjust variations in the volume of intravascular fluid (see Goetz, Bond, and Bloxham 1975; Takamata et al. 1994; Zimmerman and Blaine 1981).

Decreasing the extracellular fluid volume also stimulates thirst. Thus, hemorrhage causes increased drinking even though there is no change in the osmolality of the plasma. This effect is mediated in part via the renin–angiotensin system.

Sensation of oral–pharyngeal dryness can elicit the urge to drink, but this reflex is not essential in maintaining a normal water intake. When one drinks, there is a temporary relief from thirst. This negative feedback operates partly from the oral–pharyngeal level, but it is also induced from stomach distension. The rapid relief of thirst after water intake is also explained by a normalization of blood osmolarity. Not only does water move out of the gastrointestinal tract, but salts diffuse in the opposite direction along a concentration gradient.

The salt content of the sweat is less than that of the blood, and sweat loss therefore increases the salt concentration of the blood. As discussed earlier, increased salt concentration of the body fluids leads to the sensation of thirst and reduced urine volume. Certain studies have indicated that the salt content of sweat can be reduced as the result of acclimatization to heat, which should increase

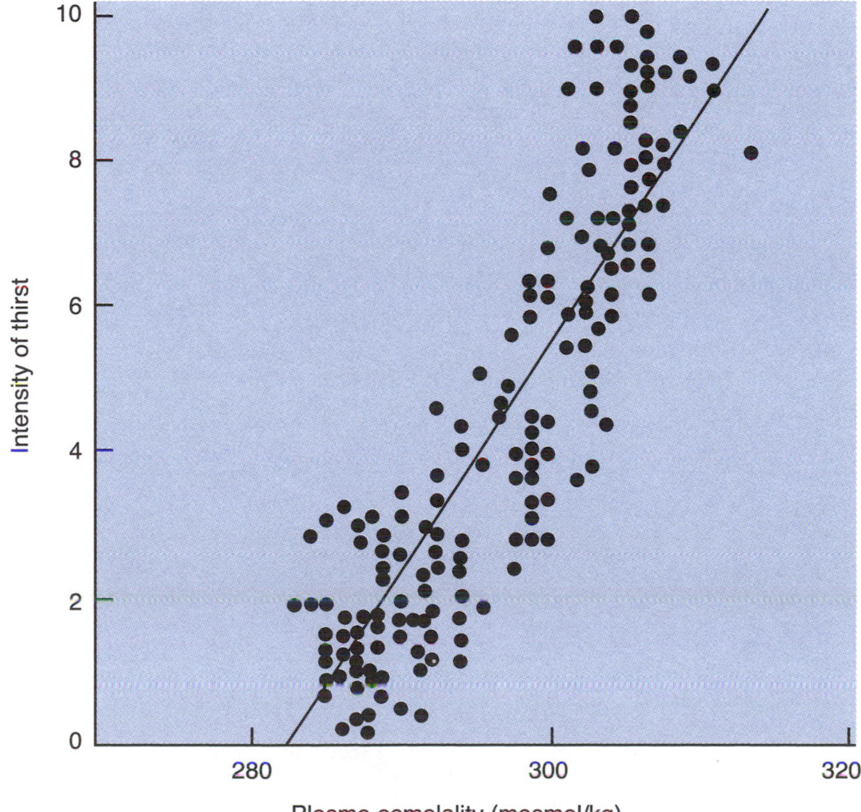


**Figure 13.9** Relation of plasma osmolality to thirst in healthy adult humans during infusion of hypertonic saline. The intensity of thirst is measured on a special analog scale.

Reproduced with permission from C.J. Thompson, et al., (1986), (Clinical Science), (71), (651-656). © The Biochemical Society and the Medical Research Society.

osmolarity and increase thirst at a given sweat loss (Robinson 1963). The acclimatized individual is better able to maintain the fluid balance than one who is not acclimatized. Here again, experience may be an important factor, because a water deficit will impair physical performance (Sawka and Greenleaf 1992).

It is commonly observed that a voluntary water intake does not necessarily cover the water loss induced by excessive sweating (Adolph 1947; Leithead and Lind 1964; Pitts, Johnson, and Consolazio 1944; K. Rodahl 1994). The risk of voluntary hypohydration is greatest in an individual unaccustomed to heat. The risk is also greater when a large portion of the food consists of dried or dehydrated rations, because a considerable volume of the daily water intake comes normally with regular meals.

### *Summary*

Osmometric, volumetric, and thermal excitations appear to be nature's signals that feed information into the control system for water intake, located mainly in the anterior hypothalamus. Sweat loss increases the osmotic pressure of the body fluids and thereby elicits the urge to drink. However, the sensation of thirst does not always "force" the individual to cover the water loss, particularly not when this loss is pronounced because of profuse sweating or because the individual does not eat normal meals containing a large amount of water.

## Water Deficit

We have concluded that high sweat rates with excessive loss of body fluids can cause a deficit of body water (hypohydration or dehydration). Mild to severe dehydration commonly occurs among athletes even when fluid is readily available (Murray 1995). The regulation of body temperature has priority over the regulation of body water. Therefore, a hypohydration can be driven very far and in fact can threaten life if the environment is very hot and water is not available.

Hypohydration causes greater heat storage in the body, which reduces the heat strain tolerance. As pointed out by Sawka (1992), the increased heat storage is mediated by reduced sweating rate and reduced skin blood flow for a given core temperature. The displacement of the blood to the skin makes it difficult to maintain central venous pressure and an adequate cardiac output to simultaneously support metabolism and thermoregulation during exercise-induced heat stress.

Prolonged exposure to heat or prolonged exercise certainly causes hypohydration. In both situations, a decrease in plasma volume has been noticed. Costill and Fink (1974) observed a 16% to 18% reduction in plasma volume at a hypohydration equivalent to a 4% decrease in body weight. The red cells shrink during hypohydration, so changes in hematocrit are not a reliable measure of changes in plasma volume (Costill et al. 1974; Harrison, Edwards, and Leitch 1975). During exercise or exposure to heat stress, protein moves from the interstitial spaces, particularly in skeletal muscles, to the vascular volume. The increase in plasma protein concentration will raise plasma oncotic (osmotic) pressure, thereby helping to maintain blood volume by reducing water loss and enhancing water gain (Harrison, Edwards, and Leitch 1975; Senay 1972). As suggested by Senay (1972), heat acclimatization may increase the ability to shift protein and fluid from the interstitial to the intravascular volume, which could improve the efficiency of the cardiovascular system. Irrespective of the cause of sweating, hypohydration is associated with a decrease in stroke volume during exercise, a concomitant increase in heart rate during submaximal exercise (figure 13.10), and decreased endurance (i.e., the time that a standardized maximal exercise load can be tolerated; figure 13.11). It also has been found that rectal temperatures are significantly higher in the dehydrated subject (figure 13.12), and the increase is related to the weight loss incurred (Gisolfi and Copping 1993; K. Rodahl 1994). The excessive increase in core temperature with hypohydration is

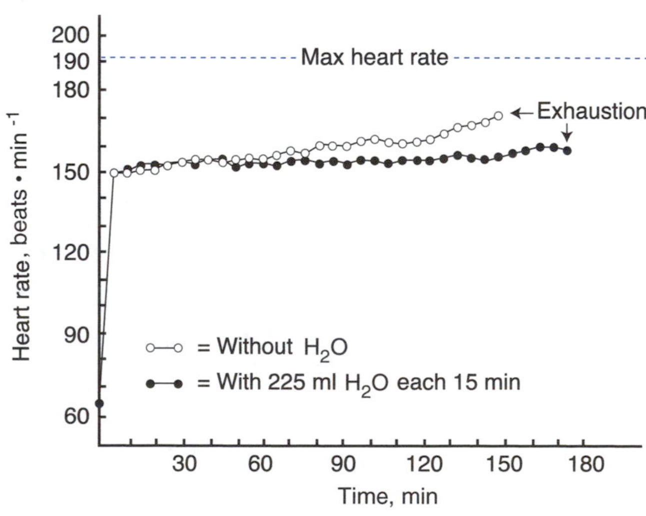


**Figure 13.10** Mean heart rates in subjects running on the treadmill at 70% of the maximal oxygen uptake until exhaustion, with and without water.

From Staff and Nilsson 1971.

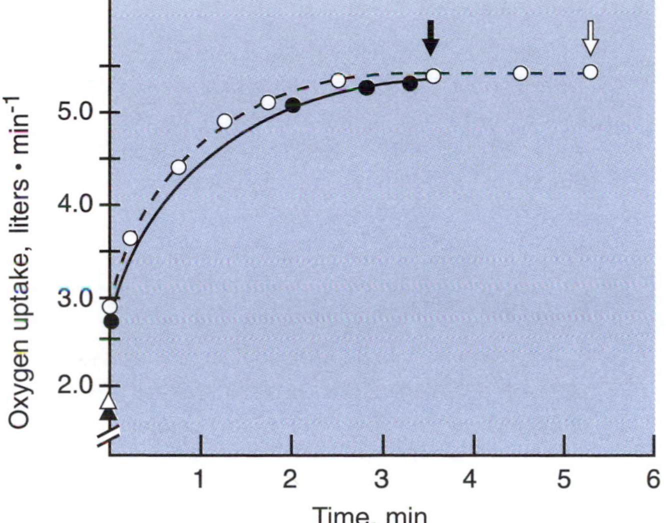


**Figure 13.11** Oxygen uptake (in liters per minute) at an exercise load that could be tolerated for 5 1/2 min during normal conditions (○) but only for 3 1/2 min after hypohydration (●). Arrows indicate maximal exercise time.

From Saltin 1964.

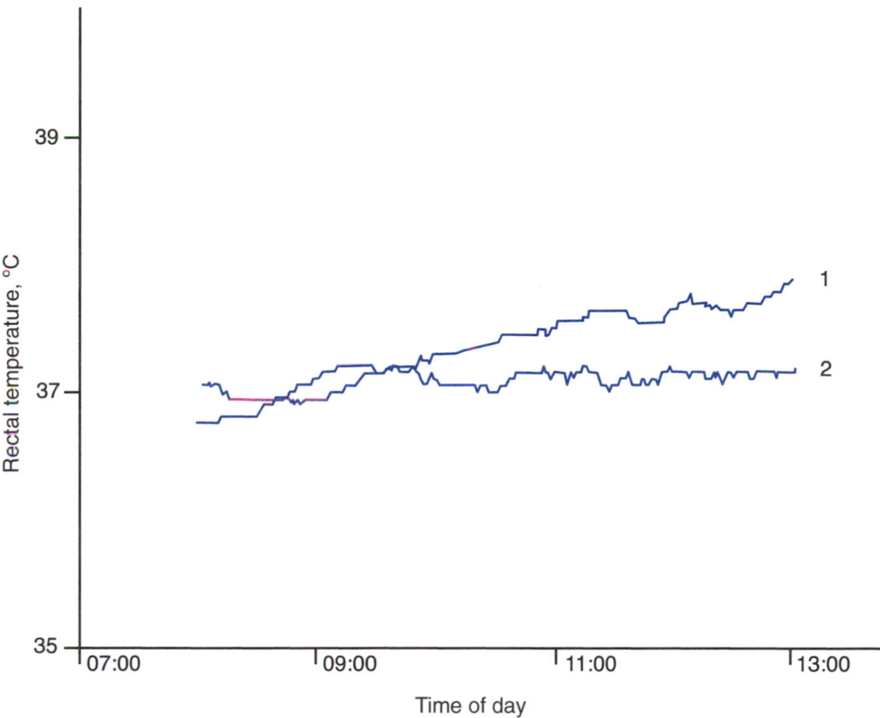


**Figure 13.12** The effect of adequate fluid intake on the rectal temperature of an operator in a glass bottle production plant: (1) without fluid; (2) with fluid.

From K. Rodahl 1994.

probably attributable to inadequate sweating (Greenleaf and Castle 1971).

Studies in a Norwegian cement factory where the ambient temperature was fairly high even in winter showed that the heart rates of the workers at a standardized submaximal work rate were significantly lower (more than 10 beats · min$^{-1}$) at the end of the work shift when 2 L of fluid was taken in the course of the shift, compared with the days when no fluid was taken between meals. These findings have been confirmed in other types of Norwegian industry (K. Rodahl 1989, 1994).

The maximal isometric strength after a water deficit is reported to be unaffected (Saltin 1964) or slightly decreased in connection with progressive hypohydration (Bosco, Terjung, and Greenleaf 1968). In any case, a reduced water content within the muscle cell and a disturbed electrolyte balance easily can influence the muscle cell's ability to contract and its susceptibility to metabolites. The reduction in work performance is more marked if a water deficit is caused by extended heavy exercise than after exposure to a hot environment without exercise.

As originally demonstrated by Adolph (1947), dehydration causes a reduced tilt-table tolerance. A person who normally could tolerate a prolonged 45° head-up tilt with a heart rate of 90 beats · $min^{-1}$ fainted within 7.5 min after a fluid loss corresponding to 3% of his body weight. Following a fluid loss corresponding to 6% of his body weight, he fainted within 1.5 min. His heart rate before fainting in both conditions was 115 and 135, respectively. A low stroke volume was a characteristic finding.

Dehydration also affects crew members of high-performance aircraft. Nunneley and Stribley (1979) showed that dehydration tends to lower G-tolerance and increase the variability of response to heat.

According to Bar-Or et al. (1980), exercising children become progressively dehydrated when not forced to drink.

Well-trained subjects are less affected in their performance by hypohydration than are untrained subjects (Buskirk, Iampietro, and Bass 1958; Saltin 1964). Acclimatization to heat does not seem to protect people from the deteriorating effect of hypohydration.

The simplest method of determining whether the fluid intake has been adequate is by weighing the individual under standard conditions. Even a reduction in body weight of 1% to 2% can lead to a deterioration in physical performance (Adolph 1947; Gisolfi and Copping 1993; Ladell 1955; Pitts, Johnson, and Consolazio 1944; Saltin 1964). However, the degree of body hypohydration is overestimated from measurements of body weight when large amounts of glycogen have been metabolized, which will deliver a "surplus" of water.

Excessive fluid loss can occur in cold environments. Lennquist (1972) showed that cold-induced diuresis can persist for several days and can lead to a considerable fluid deficit, accompanied by hemoconcentration and reduction in blood volume. The cold-induced increase in osmolar excretion largely could be accounted for by significant increases in the excretion of sodium, chloride, and calcium. The increased excretion of sodium was dependent on a reduced tubular sodium reabsorption. According to Lennquist (1972), the releasing mechanism of cold diuresis lies in the renal tubules and not in an increased glomerular filtration rate. Wallenberg (1974) reported that the tubular sodium reabsorption in a cold-exposed person is influenced by plasma osmotic pressure and changes in arterial blood pressure. Cold-induced suppression of distal tubular sodium reabsorption could be abolished almost completely by an albumin infusion of .4 to .5 g · $kg^{-1}$.

### *Summary*

An individual tolerates heavy physical exercise less well if subjected to a water deficit, even if the water loss is only about 1% of the body weight. At a submaximal work rate, the heart rate is increased, the stroke volume is reduced, and the body temperature is higher than normal. Drinking water to satiety may not fully compensate for a water loss.

## Rehydration

Complete restoration of fluid balance is an essential part of recovery after physical exertion. This is especially so in hot, humid conditions when the sweat loss is sufficiently high to cause a significant loss of electrolytes from the body, which have to be replaced, as well as the lost water. Fluid intake attenuates the exercise-induced increase in heart rate and core temperature and decreases in stroke volume, cardiac output, and skin blood flow, mainly by restoring the blood volume. Although fluid and carbohydrate individually enhance exercise performance, the combination of both increases performance to the greatest extent. Thus, Gonzalez-Alonso, Heaps, and Coyle (1992) showed that the ingestion of a 6% carbohydrate electrolyte solution was significantly more effective than water in restoring body weight and blood volume during a 2-h rehydration period following exercise-induced dehydration. Furthermore, ingesting a 6% carbohydrate–electrolyte beverage following prolonged, constant-pace running improved endurance capacity 4 h later compared with a group ingesting equal volumes of a placebo solution (Fallowfield, Williams, and Singh 1995).

Using **deuterium (D2O)**-labeled beverages, Koulman et al. (1997) showed that ingestion of a 6% glucose–electrolyte solution after passive heating (which resulted in 2% loss of body mass) followed by 1 h of exercise in the heat caused the proportion of ingested water to be twice as high in sweat as it was in urine. He also observed that the plasma volume was completely restored and the drifts of heart rate and rectal temperature were less marked than during other trials (mineral water, a 6% maltodextrin solution, and a 6% maltodextrin–electrolyte solution). These results suggest that rehydration with a 6% glucose–electrolyte solution was more efficient, probably because of an internal redistribution of water, resulting in less transfer of plasma water into urine. Gisolfi and Copping (1993) found that consuming 1 L of water during a 1.5- to 2.5-h run at 75% $\dot{V}O_2max$ was more effective in preventing a marked increase in rectal temperature than drinking an equal volume of water 30 min before the run.

In a study of the effect of drink flavor and NaCl on voluntary drinking and hydration in boys exercising in the heat, Wilk and Bar-Or (1996) observed that whereas flavoring of water reduced children's voluntary dehydration, further addition of 6% carbohydrates and 18 mmol · $L^{-1}$ of NaCl prevented it altogether. Furthermore, Meyer et al. (1994) observed that the magnitude of rehydration in children was statistically greater with drinks made of grape and orange than of water and apple. (See also Bar-Or and Wilk, 1996, and Wilk et al., 1998.)

After studying water balance and physical performance in very cold climate, Rintamaki et al. (1995) recommend a continuous maintenance of water balance by using a beverage with a fluid temperature of 25 to 30 °C and with a carbohydrate content less than 7%.

When solid food is consumed during recovery, this normally replaces the electrolytes lost in the sweat (mainly sodium), so that water alone can be adequate for rehydration purposes (Bergeron et al. 1995; Maughan, Leiper, and Shirreffs 1996). When the intake of solid food is not appropriate, as might be the case between exercise bouts, including electrolytes in rehydration beverages is essential (Maughan and Shirreffs 1997).

According to Shirreffs and Maughan (1997), there appears to be no difference in recovery from dehydration whether the rehydration beverage is alcohol free or contains up to 2% alcohol, but drinks containing 4% alcohol tend to delay the recovery process.

### For Additional Reading

For further discussion about replacement of water and sodium losses and guidelines for rehydration strategies during training and competition, see Luetkemeier, Coles, and Askew (1997), Burke and Hawley (1997), and Shirreffs and Maughan (2000).

## Working Environment: Practical Application

It is not feasible to quote exact permissible temperature limitations for the working environment. These limitations depend on the different climatic factors, the nature and type of work, the severity of the work rate, and the duration of the work, and clothing and other measures of protection. Finally, there are wide individual variations in the tolerance of climatic stress, besides the effect of acclimatization. Nevertheless, it is desirable at times to be able to objectively assess the environmental conditions at individual work places, especially as a point of reference when changes are being made, to determine to what extent such changes actually represent improvement.

For such assessment of the thermal heat load in industrial situations, several so-called heat stress indexes have been introduced. With such indexes, one tries to predict the heat load on the individual. The indexes usually are based on data from physical measurements of air temperature **(dry bulb)**, radiant temperature (globe), relative humidity **(wet bulb)**, and air velocity. A measurement of the individual's metabolic rate is also often included; that is, the energy demand of the work is measured or estimated (I. Åstrand et al. 1975; Kerslake 1972). The so-called wet bulb globe temperature index, however, is rather complicated and is time-consuming to record. In the case of a predominantly dry, radiant heat environment, the much simpler and faster reacting wet globe thermometer or Botsball, devised by Botsford (1971), records similar values and the results are interchangeable with a simple formula (Ciriello and Snook 1977).

J.E. Ramsey and Chai (1983) critically examined the inherent variability in heat stress decision rules. Their approach consisted of pairing wet bulb globe temperature values with the corresponding values of human heat exchange calculated from the heat stress index equation (Belding 1973). They suggest that simplified rules, when combined with appropriate administrative and work practices, can provide protection to the worker that is comparable to seemingly more precise methods.

Solving the problems associated with industrial heat exposure involves a combination of practical measures such as shielding the heat source; reducing each period of heat exposure to about 20 min, interspersed with brief 10-min cooling-off breaks; eliminating strenuous physical work close to the heat source when possible, introducing mechanical aids to take the place of manual labor wherever possible; and monitoring the industrial process properly at all times to avoid the need for drastic measures that cause excessive exposure to heat stress.

Figure 13.13 illustrates how the radiant heat can be reduced from 5.4 MJ (1,300 kcal) · $h^{-1}$ to about 54 kJ (13 kcal) · $h^{-1}$ by placing an aluminum shield between the worker and the heat source, which in this particular case had a temperature of 188 °C. A sheet of fiberboard covered with tinfoil provides an inexpensive protection against radiation. It is important that the surface of the shield be kept clean. If the shield has to be transparent, substances should be used that will reflect infrared light, such as glass.

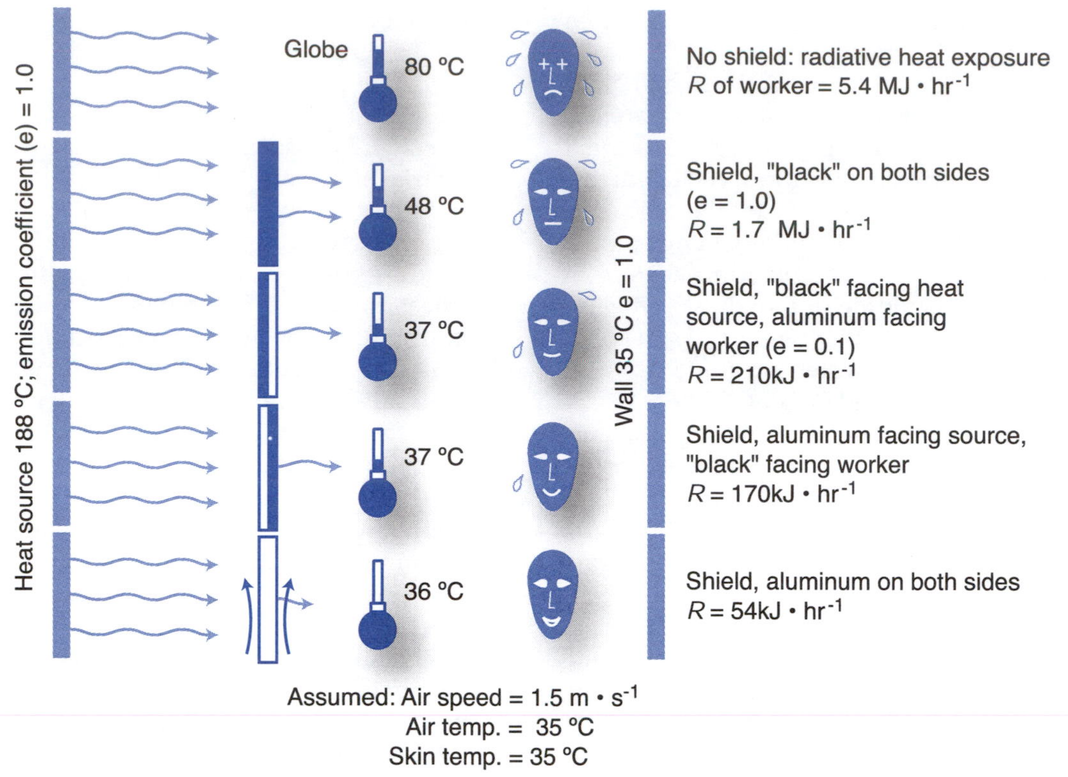


**Figure 13.13** Effect of shields with various surface emissivities in reducing radiant heat load.
Adapted from Hertig and Belding 1963.

A key point is to avoid prolonged intense heat exposure, which will lead to profuse sweating. This will reduce the fluid loss and thereby reduce the need for fluid intake correspondingly. At any rate, however, the fluid loss should be replaced as it is lost, by ingesting either plain water, or in the case of prolonged work, carbohydrate–electrolyte solutions (Galloway et al. 1997). The aim should be for workers to leave their workplace fully hydrated, able to enjoy their leisure time.

In laboratory experiments, Kamon (1979) studied different exercise cycles in six heat-acclimatized male subjects exercising in hot ambient conditions. Using the end points of stable heart rate and rectal temperature and setting a maximum rectal temperature of 38 °C, the authors found that a schedule of exercise and rest periods of 20 min was adequate for hot and dry ambient conditions, with a dry bulb temperature of 50 °C and a wet bulb temperature of 25 °C, irrespective of whether the resting took place under the same ambient conditions as for the exercising conditions or at a neutral ambient condition (dry bulb temperature 23 °C, wet bulb temperature 16 °C). The work rate was 40% of $\dot{V}O_2$max.

Essentially similar observations were made in a Norwegian magnesium plant with a work schedule of 20 min work interspaced with a 10-min cool-down at prevailing outside ambient temperatures (figure 13.14).

Heat exposure in itself represents an extra load on the blood circulation. Exhaustion occurs much sooner during heavy physical exercise in the heat because the blood, in addition to carrying oxygen to the exercising muscle, also has to carry heat from the interior of the body to the skin. This represents an extra burden on the heart, which has to pump that much harder. This is demonstrated in table 13.1, which shows the difference in heart rate in a subject performing the same work in a hot environment and in a cool one. The stress of heat and the hydrostatic factors in prolonged standing work are added to the stress of work itself. According to Maw, Boutcher, and Taylor (1993), subjects feel worse, perceive work to be harder, and experience greater thermal sensation in the hot condition compared with neutral and cool conditions.

C.G. Williams, Wyndham, and Morrison (1967) observed no difference in maximal oxygen uptake in subjects working in the heat and at comfortable temperature. At submaximal work rates, however, they found that the major change in hemodynamics in the heat was a decrease in stroke volume

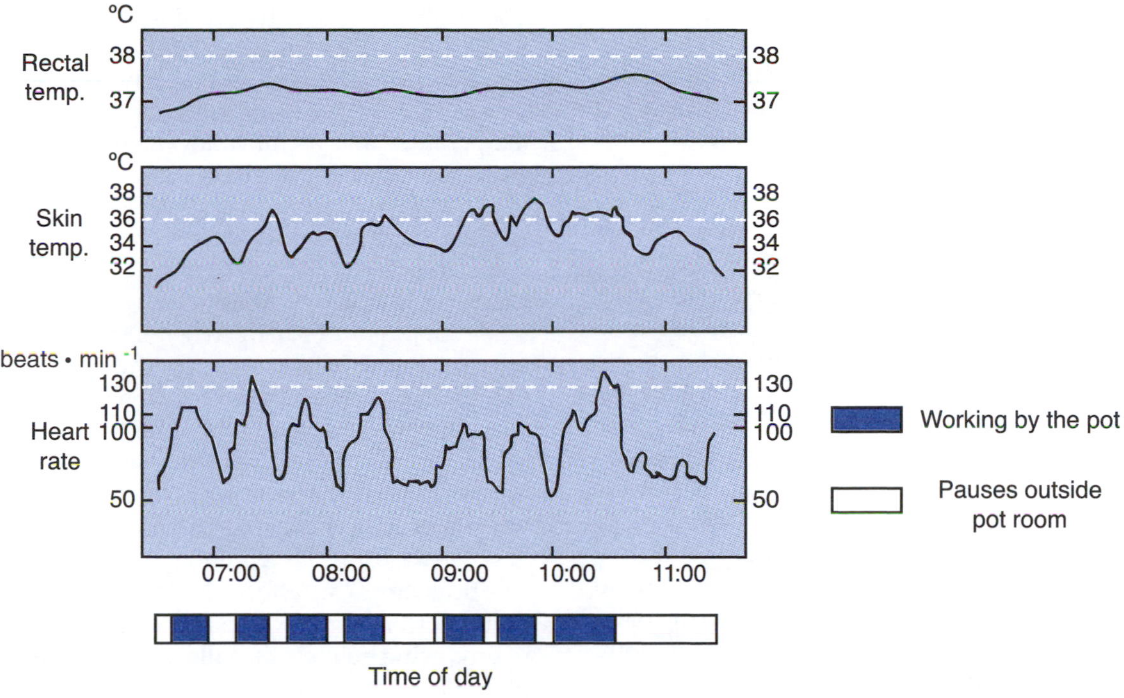


**Figure 13.14** Worker in a magnesium plant, slag removal, revised work schedule; approximately 20 min work, 10 min cool-downs.

## Table 13.1 Effect of Environmental Temperature on Human Response to Standard Exercise on a Cycle Ergometer for 45 Min

| Environmental temp. | Heart rate | Rectal temp. 0°C | Mean skin temp. 0°C | $O_2$ uptake $1 \cdot min^{-1}$ | Weight loss kg. | Weight loss % of body weight |
|---|---|---|---|---|---|---|
| Cool | 104 | 37.7 | 32.8 | 1.5 | .25 | .3 |
| Hot: 40–50°C + radiation | 166 | 38.8 | 37.6 | 1.5 | 1.15 | 1.6 |

At the same exercise rate, the temperature difference between core and shell is 4.9°C in the cool environment, but only .9°C in the heat. This necessitates a much greater skin blood flow in the heat. Hence, the markedly elevated heart rate in the heat.

and an increase in heart rate. Neither cardiac output nor arteriovenous difference was significantly altered compared with comfortable conditions. These authors also demonstrated a larger lactate production during exercise in the heat compared with exercise in a neutral environment. This finding can be explained by the reduced muscle blood flow.

Heat stress also can be a significant problem for pilots during low-level flights in hot climates, especially in fighter aircraft that impose high task loads and repetitive maneuvering forces (Nunneley and Flick 1981). Pilots noted lowered G-tolerance and increased general fatigue on the hotter flights. Weight losses up to 2.3% were observed.

In the armed forces and in certain industries, an efficient and rapid method of acclimatizing a large number of people is often of practical importance. Daily exposure to a hot environment for about 1 h will result in some acclimatization within a week. However, studies by Wyndham (1973) to establish the minimal number of days required for acclimatization (to work in gold mines in South Africa) showed that a person cannot be acclimatized adequately for a normal shift of 6 to 8 h in less than 4 h per day and in less than 8 to 9 days. They applied a step test with a gradual increase in rate of exercise up to an oxygen uptake of 1.4 L · min$^{-1}$ combined with heat stress conditions (air temperature 31.7 °C, 100% relative humidity).

## Head Protection

Brain function appears to be especially vulnerable to heat (M.A. Baker 1982). It is therefore of particular interest to note that a rather unique, selected cooling of the brain is possible because of a special vascular arrangement in the head (Cabanac 1986). The temperature inside the head (i.e., in the brain) does not rise as high as in the rest of the body (figure 13.15). When hyperthermia is caused by prolonged exercise such as running, the face is exposed to increased convection, which augments the cooling by evaporation of sweat. Bald individuals could have the additional advantage of heat loss via sweating from the hairless scalp, because sweating from a bald scalp can be as great or greater than that of the forehead (for references, see Cabanac 1986).

Cabanac and Brinnel (1988) showed that during light hyperthermia, the evaporation rate on the bald scalp is two to three times higher than on the hairy scalp. They claimed that this supports the hypothesis that male baldness is a thermoregulatory compensation for the growth of a beard in adults.

The use of helmets can cause heating of the head. The effect of using bicycle helmets in six subjects exercising in the laboratory, and in four subjects bicycling in the field, was examined in terms of head skin temperature and psychomotor performance. The results confirmed the fact that the head is subject to its own temperature regulation, more or less separate from the rest of the body. The head skin temperature was markedly higher than the leg skin temperature and significantly higher when riders used a helmet than when they did not use a helmet (K. Rodahl, Bjørklund, et al. 1992; figure 13.16). The relative humidity under an open helmet was 55% to 70% compared with 100% under a closed helmet. The head skin temperature was significantly lower when subjects wore an open helmet than when they wore a closed helmet. The psychomotor tests revealed an improvement in reaction time following exercise both with and without helmet, probably as a result of an increased head temperature. The number of errors made, however, was significantly higher when a helmet was used. These observations suggest that wearing helmets under conditions of excessive heat stress can affect certain aspects of psychomotor performance, which points to the desirability of producing helmets that allow sufficient exchange of air inside the helmets.

Sheffield-Moore et al. (1997) studied the effect of wearing a helmet on body temperature and perceived heat sensation in 10 males and four females who were cycling in a hot and dry environment and a hot and humid environment. Their results indicate that using a cycling helmet under the condi-

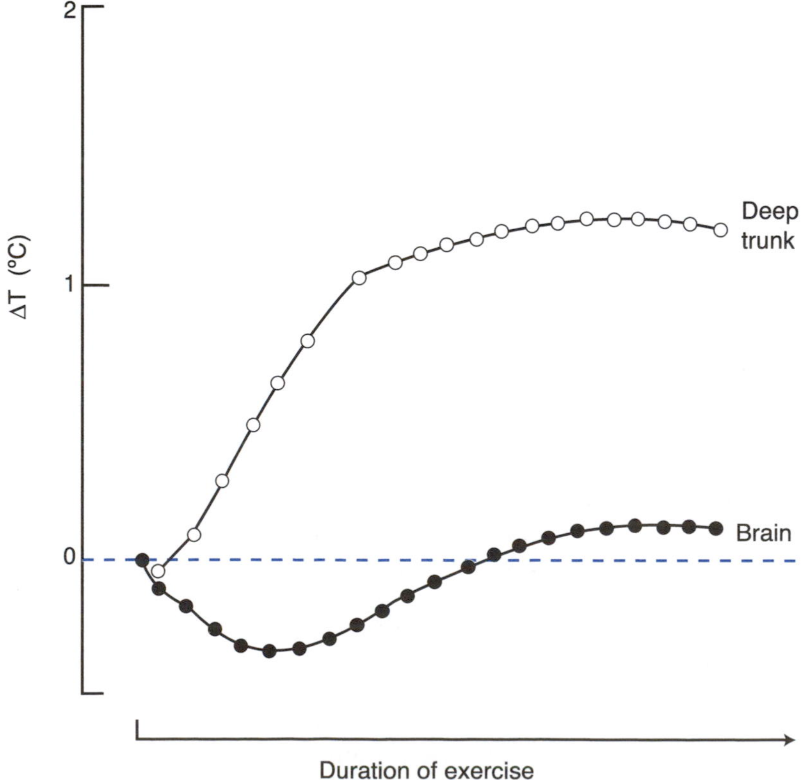


**Figure 13.15** Time course of changes in deep trunk and deep intracranial temperatures during exercise in human subjects. $\Delta T$ = temperature change.

Modified from Cabanac 1986.

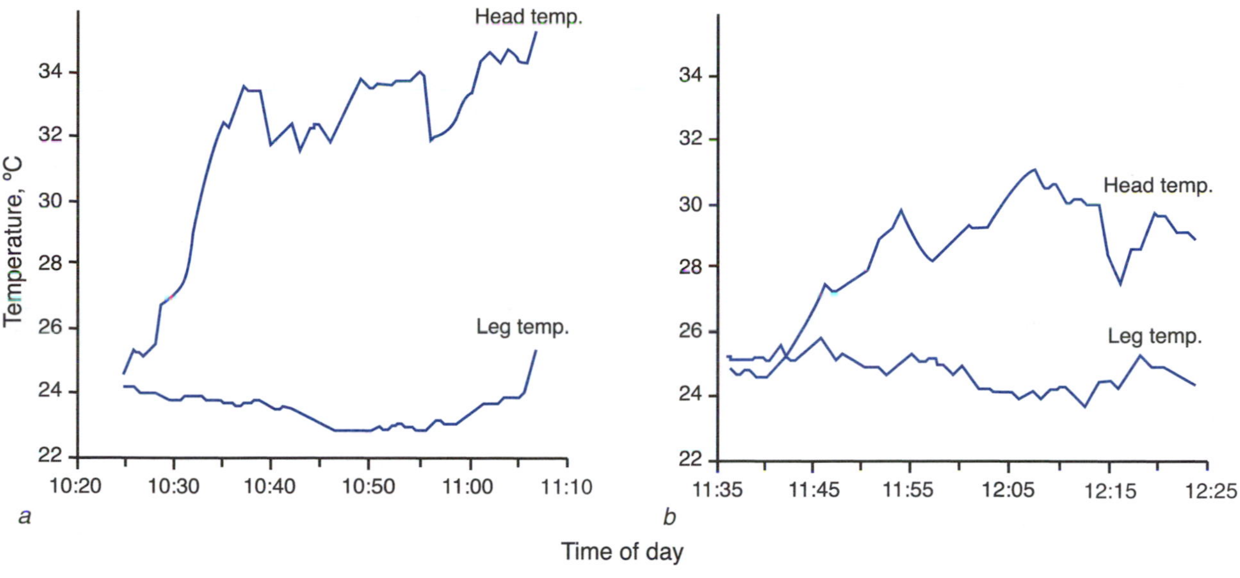


**Figure 13.16** The effect of using a closed bicycle helmet on the forehead temperature in a subject during a 60-min outdoor bicycle ride *(a)* with helmet and *(b)* without helmet.

K. Rodahl, Bjørklund et al. 1992.

tions of the study did not cause the subject to become more hyperthermic or increase perceived heat sensation.

## Clothing

In a moist, hot climate where the temperature of the environment is lower than that of the skin, it is advisable to wear as little clothing as possible. If the ambient temperature is higher than that of the skin, clothing can protect the individual from the radiant heat of the environment. Loose-fitting clothing that permits free circulation of air between the skin and the clothing is preferable. Workers habitually exposed to intense heat have learned to dress in heavy clothing for protection. This allows some of the radiant heat to be absorbed in the clothing a distance away from the skin. However, it also impairs evaporative heat loss.

The problem of clothing in the cold when heavy physical activity is alternated with rest periods is even more complicated, because there is no single item of clothing capable of both protecting against cold at rest and facilitating heat dissipation during heavy work. The conventional method is to unbutton the coat during work and to button it up during inactivity.

The surface of the hands represents about 5% of the total surface area of the body. In a nude individual, about 10% of the heat produced can be eliminated through the hands. In a clothed individual, up to 20% of the heat produced can be eliminated through the hands (Day 1949).

To remain in heat balance, a person sleeping outdoors at –40 °C needs protective clothing with an insulation value of about 12 **Clo units.** (A Clo unit equals the amount of insulation provided by the clothing required to maintain in comfort a resting–sitting human adult male whose metabolic rate is approximately 50 kcal · m$^2$ of body surface per hour, when the environmental temperature is 70 °F and humidity is less than 50%.) However, when the same individual is physically active, moving about or walking, only the equivalent of about +4 Clo units is needed. This is because the person's body heat production is now at least three times greater than it was when he or she was sleeping, the reason being the increased metabolic rate associated with the increased physical activity (figure 13.17). This requirement is adequately met by the original double-layer caribou clothing of the arctic native. Two layers of caribou fur, amounting to a thickness of 75 mm, have a total insulation value of about 12 Clo units (figure 13.18). This is adequate to maintain heat balance under practically any condition likely to be encountered in the Arctic. Temperature measurements inside such clothing, taken in the field, have confirmed that the person's body inside the clothing is indeed comfortably warm. There is, therefore, some truth to the old statement that arctic natives, by virtue of their clothing, are really surrounded by a tropical climate. The ordinary uniform usually worn by military personnel

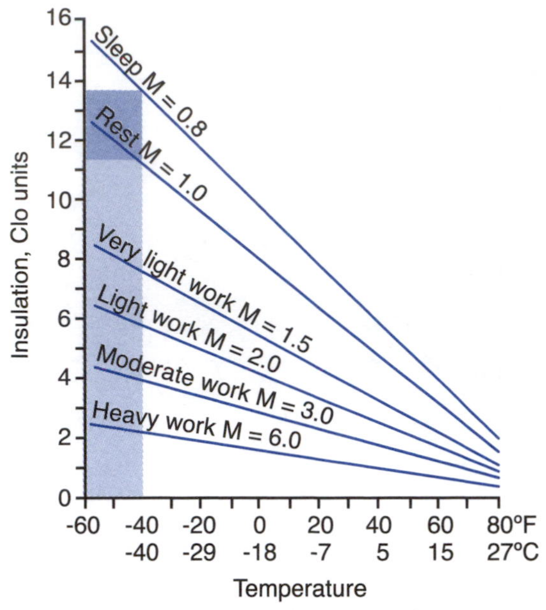


**Figure 13.17** Insulating requirements in the cold when a subject is protected from the wind during different rates of heat production. A Clo unit is the thermal insulation that will maintain a resting human indefinitely comfortable in an environment of 21 °C, relative humidity less than 50%, and air movement 6 m · min$^{-1}$. M = metabolic unit. One M equals the metabolic rate of a resting human, 3.5 ml of oxygen · kg$^{-1}$.

From Burton and Edholm 1969.

in the arctic, on the other hand, has an insulation value of only 4 Clo units, which is only one third that of the arctic native's clothing. The arctic uniform offers adequate protection for an active person at temperatures as low as –40 °C, but the person would be in negative heat balance if inactive (see figure 13.17). Temperatures below –40 °C occur on an average of about 2 days per month in the winter in the interior of Alaska.

The insulation value of most materials is proportional to the amount of air that is trapped within the material itself, because air is such a superb insulator. In the case of fur, air is trapped in the space between each hair, but the superior insulation quality of caribou fur, over and above other furs, lies in the fact that the caribou hair is hollow and contains trapped air inside each hair as well as in the spaces between them.

As a rough estimate, the insulating value of most clothing is approximately 1.6 Clo · cm$^{-1}$. Because the insulating value is primarily a function of the amount of trapped air in the clothing, the type of fiber used is of less importance.

Holmér (1983) reviewed the role of clothing for human heat exchange with the environment, with particular emphasis on how to objectively evaluate the protection offered by a clothing system against the thermal stress of the environment and how to compare differences between garments. The heat

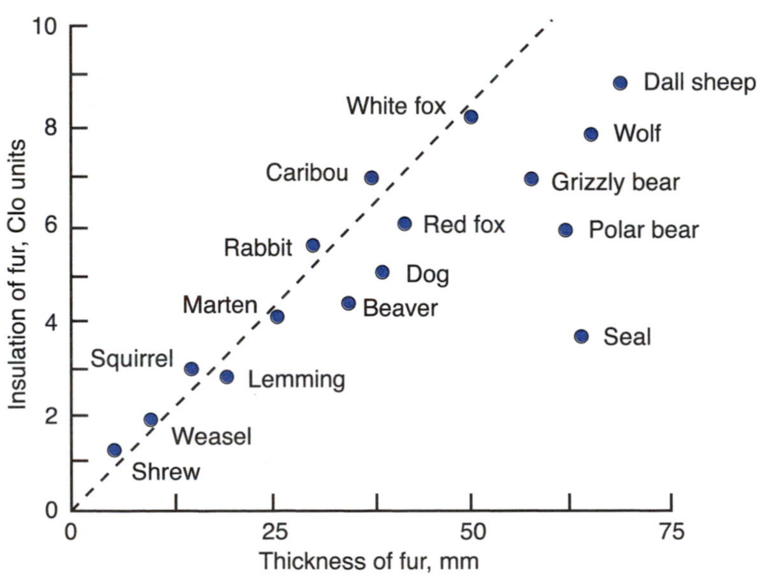


**Figure 13.18** Insulating value of different furs.

Adapted from Scholander et al. 1950.

exchange with the environment is expressed in terms of the following heat exchange equation:

$$S = M \pm R \pm C - E \quad (5)$$

where $S$ = storage of body heat, $M$ = metabolic heat production, $R$ = radiant heat exchange, $C$ = convective and conductive heat exchange, and $E$ = evaporative heat loss. $S$ can be calculated by measuring core and mean skin temperature. (A rate of heat storage of about 40 W · m$^2$ is equivalent to an increase in mean body temperature of about 1 °C · h$^{-1}$.) Of these factors, the heat content of the body can be calculated as described previously in this chapter. Metabolic heat production can be assessed by measuring oxygen uptake. Evaporative heat loss can be estimated by measuring the changes in body weight. The only remaining unknown variables are the $R$ and $C$, that is, heat transmitted via the skin surface to the environment, which roughly ranges from 80% to 90% of all heat transferred from the body. This heat must pass through the clothing layers, covering up to 95% of the body surface. This takes place as dry heat transfer and as humid heat transfer.

The dry heat transfer, $H_{dry}$, accounts for the heat transported by convection, radiation, and conduction and can be calculated with the following equation:

$$H_{dry} = \frac{(T_s - T_a)}{R_c} \quad (6)$$

where $T_s$ = mean skin temperature, $T_a$ = ambient air temperature, and $R_c$ = resistance of clothing and air layers to dry heat transfer, expressed in Clo units.

The humid heat loss ($H_{humid}$) accounts for the heat of sweat evaporated at the skin surface and transported as vapor through clothing and air layers.

$$H_{humid} = \frac{P_s - P_a}{R_c} \quad (7)$$

where $P_s$ = the actual average water vapor pressure at the skin surface, $P_a$ = the ambient water vapor pressure, expressed in pascals, and $R_c$ = the resistance to evaporative heat transfer by clothing and air layers.

Thermal properties of clothing systems can be determined by the guarded hot plate technique, by the thermal manikin or "copper man," or by means of direct or partitioned calorimetry (Holmér and Elnäs 1981).

## FOR ADDITIONAL READING

For further discussion of the determination of thermal properties of clothing, including gloves, see Holmér (1992) and Anttonen (1993).

The insulating value of clothing can be compared simply by measuring heat production (oxygen uptake) and heat loss as evidenced by changes in stored body heat (by applying the formula and procedures for obtaining rectal and mean skin temperatures described earlier in this chapter) in normal subjects under controlled climatic-chamber conditions. An example of such a study is presented in figure 13.19, in which similar garments made of nylon pile and wool pile were compared in paired experiments with subjects at rest for 1 h and during 2 h of fairly strenuous physical activity (treadmill walking at 100 m · min$^{-1}$, 5 °incline) followed by a 2-h rest in a climatic chamber at –20 °C. Evaporative weight loss was determined by weighing the subjects in the nude before and after the experiment. The accumulation of moisture in the experimental clothing was assessed by weighing the garments before and after the experiment on a scale with an accuracy of ±10 g. In this study, no significant difference could be detected between the two types of garments in terms of thermal insulation or in the ability of the two types of fabric to allow free escape of moisture produced by sweating during the physical activity (K. Rodahl, Giere et al. 1974).

The type of clothing worn in daily life plays an important role in the seasonal warm acclimatization of thermal responses in women (Li, Tokura, and Midorikawa 1995). Women wearing knee-length skirts improve their heat tolerance with the advance of the hot seasons, compared with women wearing full-length slacks (X. Li and Tokura 1996). Similar findings were reported by Xiuxian and Hiromi (1996). In the case of cold tolerance, Li, Tokura, and Midorikawa (1994) showed that women leaving their legs uncovered by wearing skirts for 3 months during a fairly cold period of the year improved their ability to tolerate cold.

Kim and Tokura (1995) studied the effect of the menstrual cycle on dressing behavior in the cold. Their findings indicate the establishment of a higher set-point in core temperature during the luteal compared with the follicular phase of the menstrual cycle. Most subjects dressed with thicker clothing in the luteal phase.

Kwon et al. (1998) showed that the hydrophilic properties of fabrics have physiological significance for reducing heat strain during exercise, especially under the influence of wind.

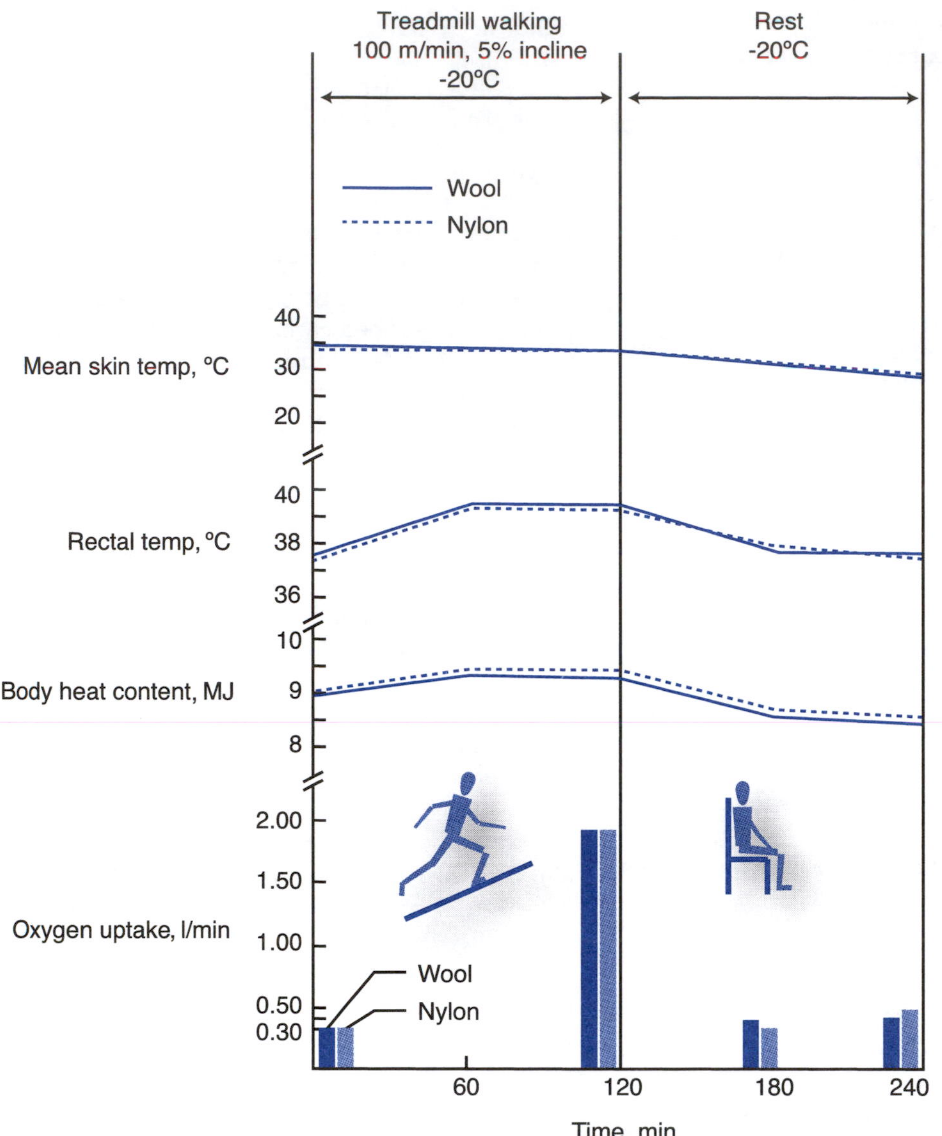


**Figure 13.19** A comparison between similar garments made of nylon pile and of wool pile during 2 h of rest in a climatic chamber at –20 °C. There was no statistically significant difference in oxygen uptake, skin or rectal temperature, or stored heat resulting from the two types of garments worn. Figures represent means of five subjects.

From K. Rodahl, Giere et al. 1974.

## For Additional Reading

For further discussion of the thermal responses affected by different underwear materials during light exercise, see Rissanen et al. (1994) and Ha, Tokura, and Yamashita (1995).

## Microclimate

Preferred environmental temperatures range from 17 °C up to 31 °C, depending on the climate and the clothing worn. However, 21 to 24 °C represents the comfort zone for a large majority of individuals. Edholm and Weiner (1981) concluded that the most effective environmental temperature for mental effort is 28 °C in terms of number of signals missed in a particular set of experiments.

Evidently the temperature of the face is critical, because according to Crawshaw et al. (1975), the skin of the face is four times as sensitive to the sensation of heat as is the skin of the thigh per unit of surface area.

Providing an optimal working environment may involve creating local microclimates by cooling or heating the clothing, by providing environmental suits to be worn under special circumstances, or by enclosing the work area in a suitable artificially made environment that will provide an effective climatic control. Radiant heaters can be used, and exposure suits may be applicable. With any solution, there will be certain complications. In the ideal climate, the temperature of the skin is about 33 °C, but not uniformly so. The feet are normally colder than the trunk, and a person may accept a greater lowering of the temperature of the feet without feeling cold. A bath with a water temperature of 33 °C feels cold. In order for the bather to feel comfortable, the water temperature has to be about 35 °C. However, such a water temperature does not produce temperature equilibrium; it will increase the body temperature. As Burton (1963) said, humans were not constructed to spend much time in water.

Heating the floor is an unphysiological manner of regulating room temperature because the receptors in the skin of the feet exert a relatively dominating influence on the temperature-regulating center. An induced vasodilation of the feet increases heat loss. Despite warm feet, the subjects of one study eventually became cold. Given these findings, it might appear unphysiological to heat our dwellings and factories by keeping the floor hotter than the room air (Burton 1963).

It is thus possible, by improper clothing or by local heating or cooling of limited skin areas, to upset the normal physiological temperature regulation. Local heating of hands and feet, for example, can cause shivering and sweating at the same time. Even if the air temperature is high, heat loss by outgoing radiation to cold surfaces, like a cold window, can cause a most unpleasant sensation, commonly referred to as draft. Because of radiative heat loss to the night sky, a person exposed, unshielded, in the arctic actually can be exposed to a cold stress 10 to 20 °colder than that indicated by the air thermometer. In a small enclosed area, such as the cabin of an aircraft or a car, it is difficult to satisfy the requirement for adequate ventilation, hot or cold, without causing certain areas of the body to be too hot or too cold. It is evident that much money, as well as many heartbeats and much sweating, could be saved by proper planning of lecture halls, offices, and factories, taking into account all of the factors that constitute the optimal climate. Even so, however, it is not possible at all times to satisfy the individual comfort requirements of all of those who work in large open rooms or offices, because of individual differences in temperature sensation, metabolic rate, and clothing.

The secret of success of an experienced arctic traveler or hunter lies in the ability to avoid the extreme cold. The arctic native's dwelling, whether it be a skin tent, a peat-covered house, or a log cabin, is comfortably warm at all times. The temperature is kept around 21 °C (70 °F) in the day and can drop to about 10 °C (50 °F) during the night, when the sleeping person is well covered by fur. Although, in the summer, the arctic native can spend as much as 9 h or more out of doors, the average amount of time spent outside in the winter is only 1 to 4 h. Furthermore, experience has taught arctic dwellers to take advantage of the characteristic temperature distribution in their environment, produced by the so-called temperature inversion during the winter. The coldest spot is at the surface of the snow, and especially in a depression in the terrain, such as a riverbed, where cold, heavy air is trapped. A few feet up the hillside, the temperature is usually many degrees warmer. Although the temperature at the actual snow surface can be as cold as –50 °C, the temperature under the snow cover, in the narrow air space between the ground and the snow crust, is usually maintained at –9 to –6 °C throughout the winter. By taking advantage of this characteristic temperature distribution, the experienced traveler can successfully escape the extreme degrees of cold stress.

## Warm-Up for Physical Performance

It is a relatively old observation that physical performance is improved following warm-up (Asmussen and Böje 1945; Högberg and Ljunggren 1947; Muido 1946; Simonson, Teslenko, and Gorkin 1936). Högberg and Ljunggren (1947) examined the effect of warm-up in the form of running at moderate speed combined with calisthenics on the speed of running 100, 400, or 800 m in well-trained athletes. The authors compared this effect with the effect of heating the body passively in a sauna bath for 20 min before the race and found that the beneficial effect of passively elevating the body temperature was much less than that of elevating the body temperature by a warm-up through physical exercise. In the 100-m dash, the improvement after a proper warm-up was on the order of .5 to .6 s, corresponding to 3% to 4 %, compared with the results without any warm-up. In the 400-m race, the improvement amounted to 1.5 to 3.0 s, corresponding to 3% to 6%. In the 800-m race, the improvement was 4 to 6 s, or 2.5% to 5.0%. Thus, the percentage improvement was roughly the same at all distances examined. Similar results were obtained in swimming by Muido (1946).

The benefit of the higher temperature during exercise lies in the fact that the metabolic processes in the cell can proceed at a higher rate, because these processes are temperature-dependent. For each degree of temperature increase, the metabolic rate of the cell increases by about 13%. At the higher temperature, the exchange of oxygen from the blood to the tissues is also much more rapid. Furthermore, the nerve messages travel faster at higher temperatures. At the temperature of the human body, which is much higher than that of a frog, the nerve messages go up to eight times as fast as those of the frog (A.V. Hill 1927). Thus, there is a very good reason for a person to keep the body temperature up, even at considerable expense, to be able to move more quickly. This is also the reason why athletes have discovered that it pays to warm up before an athletic event.

Ingjer and Strømme (1979) studied the effects of active, passive, or no warm-up on the physiological response to a maximal aerobic workload, consisting of running uphill on a treadmill for 4 min. The increase in body temperature during the active and passive warm-up procedures was controlled, so that the temperature during each procedure reached the same level before the subject was exposed to maximal exercise. On average, the rectal temperature rose to 38.3 °C. The standard work resulted in a significantly higher oxygen uptake, lower lactate concentration, and higher pH when the work was preceded by active warm-up compared with passive or no warm-up (figure 13.20). The authors concluded that the physiological effects of a thorough active warm-up can substantially benefit athletic performance.

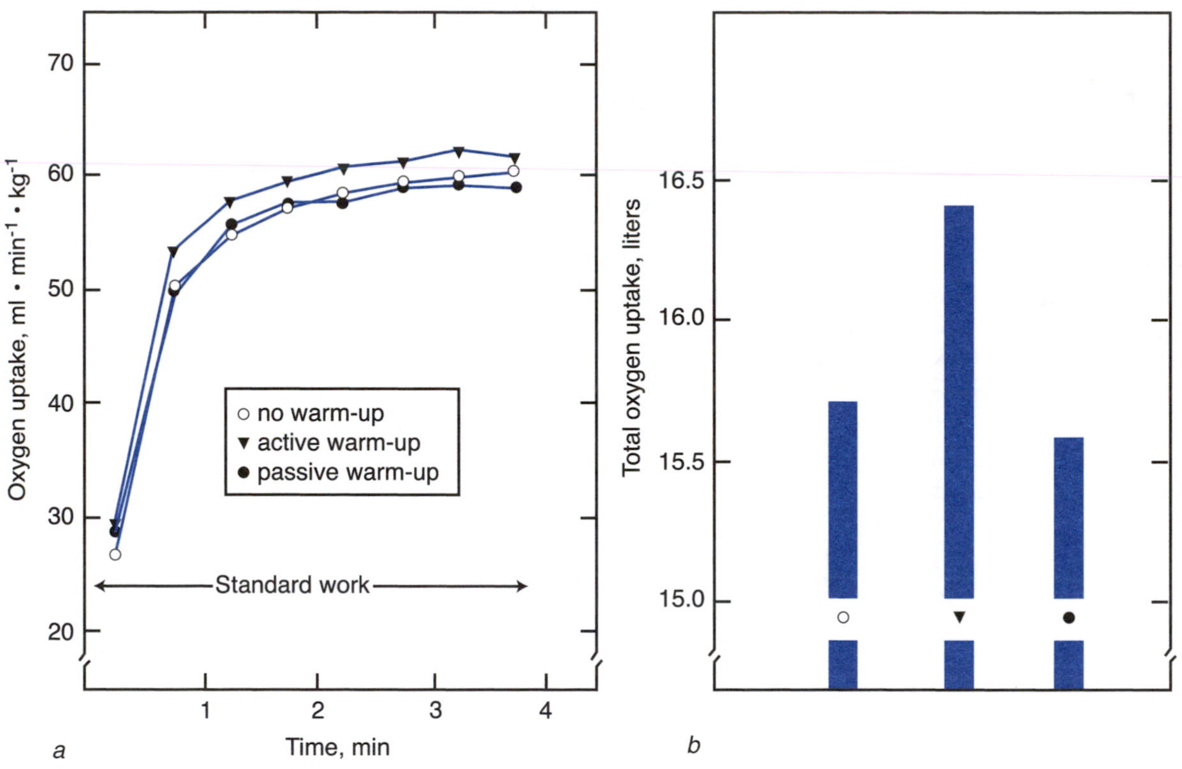


**Figure 13.20** *(a)* Mean values for oxygen uptake during standard work following active, passive, and no warm-up procedures. *(b)* Mean values for the total oxygen cost of standard work following the same three warm-up procedures.
From Ingjer and Strømme 1979.

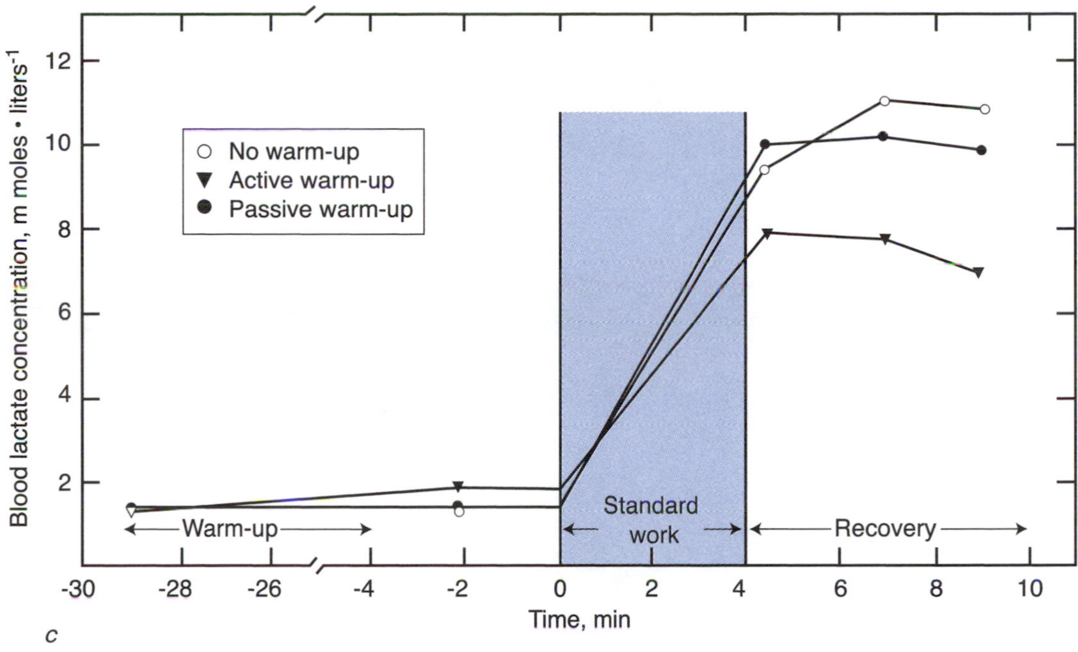


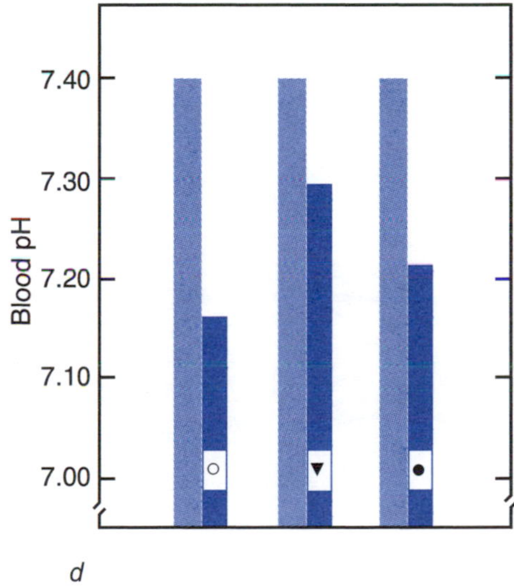


**Figure 13.20** *(c)* Mean values for lactate concentration in blood, sampled before and after the warm-up procedures and after standard work following the active, passive, and no warm-up experimental conditions. *(d)* Mean blood pH values before the warm-up procedures and 3 min after standard work following the active, passive, and no warm-up experimental conditions.

From Ingjer and Strømme 1979.

# CHAPTER 14

# FACTORS AFFECTING PERFORMANCE

Physical performance is affected not only by genes, age, sex, and state of training but also by biological rhythm, sleep, nutrition, environment, and a number of other factors such as use of caffeine, nicotine, alcohol, and various drugs.

Figure 8.1 is a schematic presentation of factors influencing physical performance ability. Most of these factors have been discussed rather extensively from various viewpoints and in different connections. Thus, the nature of exercise and somatic factors were examined in chapters 8 and 9, training adaptation and psychic factors in chapter 11, and nutritional aspects in chapter 12. As far as the effect of the environment is concerned, the consequences of high gas pressure were explored in chapter 6, and the effect of temperature and water balance was considered in chapter 13. In this chapter we discuss the effects of high altitude and the influence of use of tobacco, alcohol, and caffeine.

## High Altitude

The decision to hold the 1968 Olympic Games in Mexico City at an altitude of 2,300 m created a special interest in the effects of altitude on physical performance. Several symposia were held and reviews and books written to discuss various problems related to physical performance and training at high altitude (D.M. Bailey and Davies 1997; Berglund 1992; Jackson and Sharkey 1988; Jokl and Jokl 1968; Sutton et al. 1983; Ward, Milledge, and West 1995; J.B. West 1981; Wilber 2001; Wolski, McKenzie, and Wenger 1996).

### CLASSIC STUDY

In 1878, the French physiologist Paul Bert (1833–86) recognized that the detrimental effects of high altitude were attributable to the diminished partial pressure of oxygen at reduced barometric pressure ($P_{bar}$). Bert, who began his career as an engineering student, became a physician and then a physiologist. His scientific activity was very diverse, but his main interest was respiratory physiology. As a gifted designer of complex apparatuses, he was able to construct an altitude chamber in which he could study the effect of varying air pressure on human subjects. In his famous book *La Pression barométrique* (1878), he convincingly described the results of hundreds of his own experiments, discussing not only their scientific significance but also their practical applications. Because Paul Bert's main achievements concern the effects of $P_{bar}$ on life, he is considered the creator of barophysiology. He proved the essential role of oxygen for short-term survival on high mountains and in balloons and was named "the father of aviation medicine."

Bert 1878.

## Physics

The $P_{bar}$ at a given altitude depends on the weight of the air column over the point in question. Its pressure and density decrease almost linearly with altitude.

Table 14.1 presents $P_{bar}$ and the oxygen pressure of the inspired air (tracheal air; $PO_2$) at various altitudes. With a constant oxygen concentration of 20.94% of the dry air, the oxygen pressure of the inspired air in the trachea, saturated with water vapor, can easily be calculated from the following formula:

$$PO_2 = (P_{bar} - 47) \cdot 20.94 \cdot 100^{-1} \quad (1)$$

This means that if a person attempts to breathe at an altitude of about 19,000 m, where the barometric pressure is 47 mmHg (6.3 kPa), there should be nothing but water molecules in the trachea!

The oxygen tension of the alveolar air, and thereby also the oxygen tension of the arterial blood, is determined by the magnitude of the pulmonary ventilation in addition to the composition and pressure of the inspired air. The critical alveolar $PO_2$ at which an unacclimatized person loses consciousness within a few minutes in acute exposure to hypoxia is 30 mmHg (4.0 kPa), with minimal individual variations (Christensen and Krogh 1936). This limit is set at an altitude of slightly more than 7,000 m. Down to this low $PO_2$, the demand of the nerve cells for oxygen apparently can be maintained (Noell 1944). The well-acclimatized individual can spend hours at an altitude above 8,000 m breathing the ambient air.

The reduced density of the air at high altitudes affects the mechanics of breathing. Part of the work of breathing is expended in moving the air against the resistance of the airways. The resistance is relatively high when the flow is turbulent, as it may be at high rates of airflow. Therefore, the influence of reduced density is more noticeable in hyperpnea, during heavy exercise, or in flow-dependent pulmonary function tests. The maximal breathing capacity is considerably higher at high altitude than at sea level, because the net effect of the reduced resistance to air flow at lowered $P_{bar}$ is a diminished respiratory work to move a given volume of air in and out of the lungs (S. Miles 1957; Ulvedal et al. 1963). On several occasions, pulmonary ventilations of 200 $L \cdot min^{-1}$ have been measured during maximal exercise at high altitude.

Another effect of the reduced density of air at a low $P_{bar}$ is a diminished external air resistance. The air resistance changes with the wind speed raised to the second power. The external work therefore is reduced at high altitude in sprint-type activities, events involving jumping or throwing, bicycling, speed skating, and alpine skiing with high velocities. Dickinson, Piddington, and Brain (1966) stated, on the basis of ballistic calculations, that at an altitude corresponding to Mexico City, one might expect an improvement of 6 cm in shot putting, 53 cm

**Table 14.1** Barometric Pressure (Standard Atmosphere) at Various Altitudes and the Pressure of Oxygen After the Inspired Gas Has Been Saturated With Water Vapor at 37 °C (Tracheal Air)*

| Altitude | | | | | | Altitude | | | | | |
|---|---|---|---|---|---|---|---|---|---|---|---|
| m | ft | Pressure, mmHg | kPa | $Po_2$ tracheal air, mmHg | kPa | m | ft | Pressure, mmHg | kPa | $Po_2$ tracheal air, mmHg | kPa |
| 0 | 0 | 760 | 101.3 | 149 | 19.9 | 5,500 | 18,050 | 379 | 50.5 | 69 | 9.2 |
| 500 | 1,640 | 716 | 95.5 | 140 | 18.7 | 6,000 | 19,690 | 354 | 47.2 | 64 | 8.5 |
| 1,000 | 3,280 | 674 | 89.9 | 131 | 17.5 | 6,500 | 21,330 | 330 | 44.0 | 59 | 7.9 |
| 1,500 | 4,920 | 634 | 84.5 | 123 | 16.4 | 7,000 | 22,970 | 308 | 41.1 | 55 | 7.3 |
| 2,000 | 6,560 | 596 | 79.5 | 115 | 15.3 | 7,500 | 24,610 | 287 | 38.3 | 50 | 6.7 |
| 2,500 | 8,200 | 560 | 74.7 | 107 | 14.3 | 8,000 | 26,250 | 267 | 35.6 | 46 | 6.1 |
| 3,000 | 9,840 | 526 | 70.1 | 100 | 13.3 | 8,500 | 27,890 | 248 | 33.1 | 42 | 5.6 |
| 3,500 | 11,840 | 493 | 65.7 | 93 | 12.4 | 9,000 | 29,530 | 230 | 30.7 | 38 | 5.1 |
| 4,000 | 13,120 | 462 | 61.6 | 87 | 11.6 | 9,500 | 31,170 | 214 | 28.5 | 35 | 4.7 |
| 4,500 | 14,650 | 433 | 57.7 | 81 | 10.8 | 10,000 | 32,800 | 198 | 26.4 | 32 | 4.3 |
| 5,000 | 16,400 | 405 | 54.0 | 75 | 10.0 | 19,215 | 63,000 | 47 | 6.3 | 0 | .0 |

*Based on dry conditions for average temperature at altitude when the temperature at sea level is 15 °C and the barometric pressure is 760 mmHg (101.3 kPa).

in hammer throwing, 69 cm in javelin throwing, and 162 cm in discus throwing. Also, the force of gravity is reduced with the distance from the earth's center. The former long-standing (23 years) world record in long jumping (8.90 m) by Bob Beaman in the Mexico City Olympic Games should be noted.

The air temperature is, on the whole, lower the higher the altitude. With a mean annual temperature of 15 °C at sea level, the air temperature decreases linearly by 6.5 °C · 1,000 $m^{-1}$ to about 11,000 m. The air also becomes increasingly dry with increasing altitude. Therefore, the water loss via the respiratory tract is higher at high altitudes than at sea level. If much work is performed at high altitude, this loss can result in hypohydration and soreness and dryness in the throat.

Finally, the solar radiation is more intense at high altitude, and the ultraviolet radiation can cause sunburn or snow blindness.

### For Additional Reading

The interested reader is referred to Erslev (1980) and Dejours and Dejours (1992).

## Physical Performance

The reduced performance at high altitudes is well established; it is already evident at an altitude of about 1,200 m for heavy exercise that engages large muscle groups for about 2 min or longer. As an example of the physical strain experienced during the conquest of very high mountain peaks, Somervell (1925) reported the following:

> *It may be of interest to record some personal observations which I made while climbing in the neighborhood of 27,000 to 28,000 feet. The heart rate during the actual motion upwards was found to be 160–180 beats per minute, sometimes even more, regular in rhythm. The respiratory frequency was about 50 to 55 per minute while climbing. Approaching 28,000 feet I found that for every single step forward and upward, seven to ten complete respirations were required.*

E.F. Norton (1925) wrote that at an altitude of 8,500 m, he required 1 h to climb a distance of 35 m even though the climb was not particularly difficult.

Sutton et al. (1983) made an interesting analysis of the remarkable ascent of Mount Everest by Habeler and Messner without the use of supplementary oxygen. Although Habeler and Messner were exceptional athletes, they had almost reached their limits as they were climbing the last 48 m to the summit, collapsing into the snow every few steps and barely crawling on. Although they proved that it is indeed possible to climb Mount Everest without supplementary oxygen under fortunate circumstances, a decrease in the ambient $P_{bar}$ would perhaps have been enough to make it impossible. It can only be done at the expense of extreme hyperventilation and respiratory alkalosis (observed arterial partial pressure of carbon dioxide, 7.5 mmHg–1 kPa; arterial partial pressure of oxygen, 35 mmHg–4.6 kPa, arterial pH 7.7; Hackett et al. 1983; J.B. West 1983). Schoene (1982) reported that a group of climbers who successfully climbed to at least 7,800 m had significantly higher ventilatory responses to hypoxia and hypercapnia than a group of outstanding long-distance runners. He suggests that this factor may be selective for this elite group of athletes.

In competitions in Mexico City (altitude 2,300 m), the same or better performance in running distance up to 400 m has been reported. In 1,500 m running, an impairment of about 3% has been observed, and in 5,000 and 10,000 m, an impairment of roughly 8% compared with sea level. Similarly, in swimming, impairments of about 2% to 3% and 6% to 8% for 100 m and 400 m or longer, respectively, have been registered (A.B. Craig, 1969; Shephard, 1973).

What then, is the physiological explanation for such observations?

## Limiting Factors

The initial effect of exposure to very high altitude often includes so-called acute mountain sickness (AMS). AMS is associated with various symptoms such as headache, vomiting, physical and mental fatigue, interrupted sleep, and digestive disorders. In rare cases, previously healthy persons can develop **pulmonary edema** within a few hours after exposure to high altitude. It has been suggested that AMS is the result of abnormalities in the handling of body water (Hackett et al. 1981), possibly attributable to a hormonal constellation that favors greater water and salt retention (Bärtsch et al. 1991). J.B. West and Mathieu-Costello (1992) cited strong evidence that the mechanism behind high-altitude pulmonary edema is stress failure of pulmonary capillaries with escape of high molecular weight proteins and blood cells into the alveolar space.

### For Additional Reading

For further discussions about the implications of pulmonary hemodynamics for the development of high-altitude pulmonary edema, see Gibbs (1999). A general review about medicine and mechanisms in altitude sickness was prepared by Coote (1995).

If one disregards altitudes above 3,000 m, it is evident that aerobic power is affected by a reduced oxygen pressure in the inspiratory air. Studies have shown that exercise during acute exposure to high altitude increases the blood lactate concentration at lighter work rates than is the case at sea level. At a fixed rate of exercise, the lactate level is higher, but the maximal concentration attained during exhaustive exercise is roughly the same as at sea level (Asmussen, Döbeln, and Nielsen 1948; P.-O. Åstrand 1954; Brooks et al. 1991; Buskirk et al. 1967; H.T. Edwards 1936; Hermansen and Saltin 1967; Stenberg, Ekblom, and Messin 1966; Terrados et al. 1988;). Figure 14.1 gives an example of how the blood lactate concentration is affected during exercise at sea level, at 2,300 m, and at 4,000 m simulated altitude. In this subject, however, the lactate concentration was the same when it was related to the relative, instead of the absolute, oxygen uptake. This indicates that anaerobic processes are brought into play at a relatively lower work rate at high altitude. On the other hand, maximal anaerobic power, at least the part of it that is determined by glycogenolysis, is not affected. The fact that the maximal oxygen debt is the same following maximal exercise at sea level and at altitude also points in the same direction (Buskirk et al. 1967; Saltin 1966).

It is difficult to explain the higher blood level of lactic acid at a standard submaximal work rate at high altitude in view of the fact that the oxygen uptake is the same as at sea level (Asmussen and Chiodi 1941; P.-O. Åstrand 1954; Christensen and Forbes 1937; Pugh 1964). Some studies have indicated a slower increase in oxygen uptake at the beginning of exercise; that is, the oxygen deficit is enlarged. Therefore, the demand of the anaerobic energy yield is correspondingly increased (Raynaud et al. 1974).

It can be concluded that the maximal aerobic power is directly affected during exercise under conditions of reduced oxygen pressure in the in-

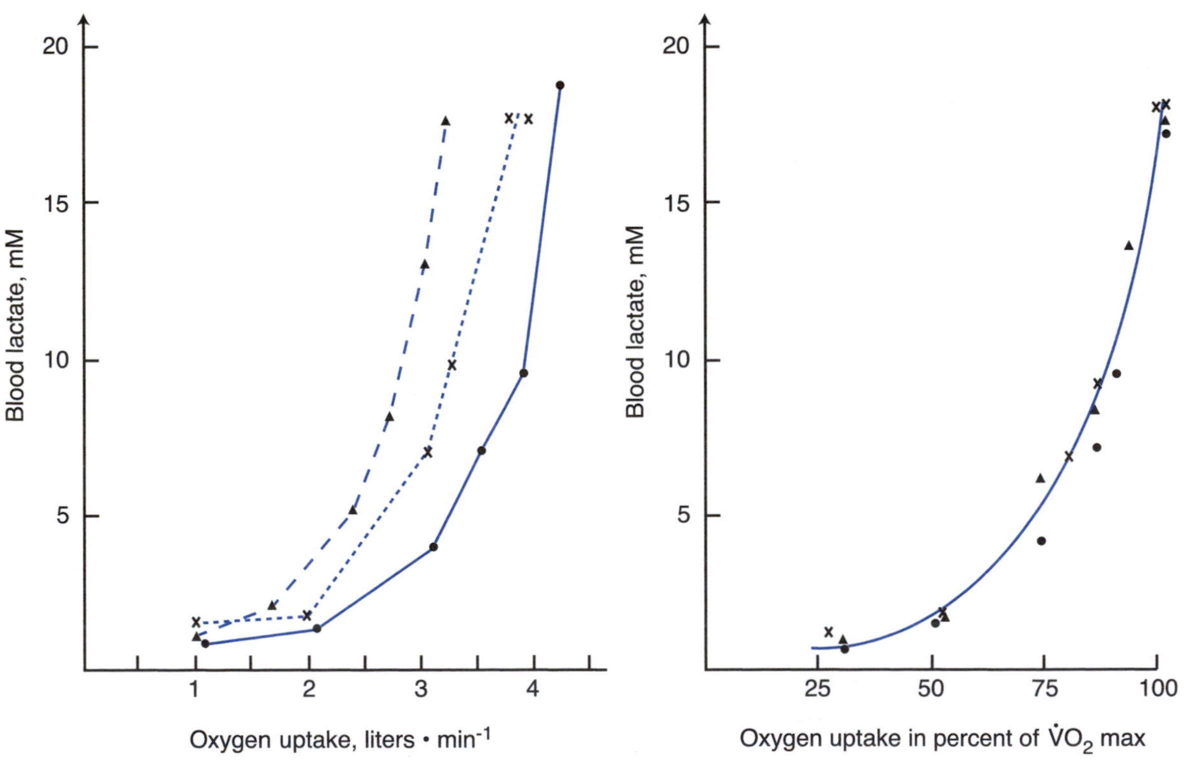


**Figure 14.1** Blood lactate concentrations in one subject during exercise at sea level (●) and acute exposure to 2,300 (×) and 4,000 m (▲) (760, 580, and 462 mm Hg, or 101, 77, and 66 kPa, respectively). Exercise times for all submaximal work rates were 10 min, and all blood samples were taken between the 9th and 10th min. Exercise times for the maximal work rates were 4 to 5 min, and peak values for lactic acid are given. In the left panel, the absolute oxygen uptake is on the abscissa; on the right, oxygen uptake is in percentage of the maximum.

From Hermansen and Saltin 1967.

spired air. We shall therefore summarize how the different steps in oxygen transport from the air to the mitochondria of the exercising muscle cell are affected during exposure to high altitude.

## Oxygen Transport

Figure 14.2 indicates the pressure levels at different distances along the transport chain from atmospheric air to the ultimate destination, the mitochondria. In the given example, the pressure gradient for oxygen between air and mixed venous blood at rest at sea level is about 110 mmHg (14.7 kPa), that is, 150 – 40 mmHg (20 – 5.3 kPa). When one goes to an altitude of 5,500 m, $PO_2$ = 70 mmHg (9.3 kPa), which is about the highest altitude to which people can become acclimatized and at which they can live and work for months and years (Rahn 1966), the pressure drop for the same oxygen delivered is reduced to about 50 mmHg (6.7 kPa). Close to the mitochondria, the oxygen pressure can be 10 mmHg (1.3 kPa) at sea level and about 5 mmHg (0.7 kPa) at 5,400 m. This pressure is still adequate to provide optimal conditions for oxidative enzyme reactions. Many factors explain the increased "conductance" to keep the tissue oxygen pressure at this almost constant level. We shall first consider acute hypoxia.

1. The pulmonary ventilation at a given oxygen uptake is markedly elevated (figure 14.3). In this subject, the ventilation at an oxygen uptake of 4.0 L · $min^{-1}$ was 80 L · $min^{-1}$ when pure oxygen was inhaled, 105 L · $min^{-1}$ at sea level when the subject was breathing air, 140 L · $min^{-1}$ at 2,000 m, and 160 L · $min^{-1}$ at 3,000 m, or twice as high as when breathing oxygen. Even at sea level, there exists a hypoxic drive that is evident when this subject's oxygen uptake exceeds 1.5 L · $min^{-1}$. This hypoxic hyperpnea is elicited through the reflex pathway originating in the chemoreceptors. Because the production of carbon dioxide ($CO_2$) is roughly the same at a given oxygen uptake, this hyperventilation will inevitably wash out $CO_2$ from the blood into the inspired air, the dissolved $CO_2$ of the blood being more affected than the bicarbonate. The secondary effect of the hyperpnea therefore will be an increase in the pH of the blood (i.e., an uncompensated respiratory alkalosis). The reduced partial pressure

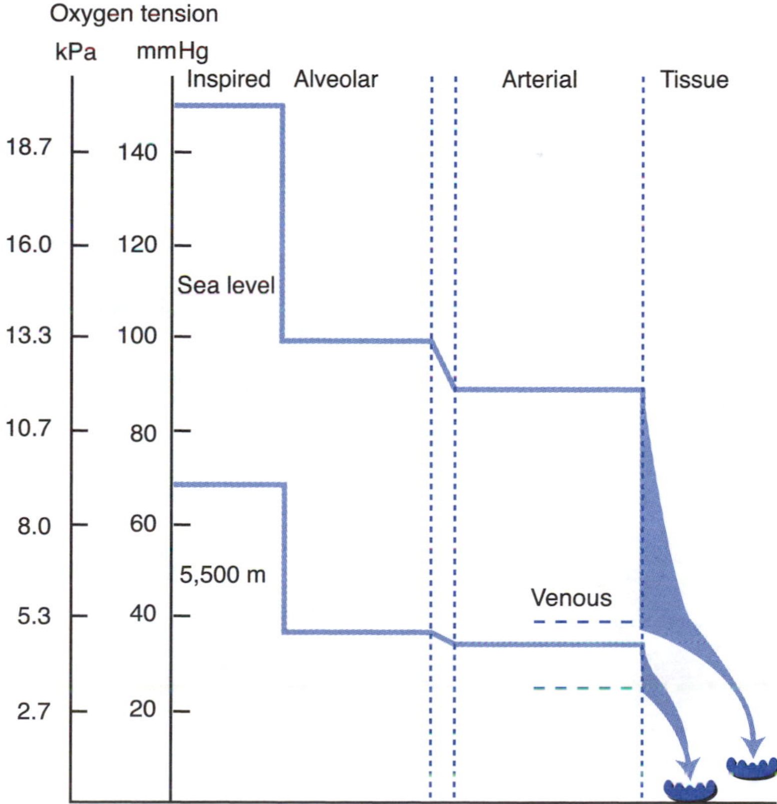


**Figure 14.2** A comparison of the oxygen cascade from inspired air to tissue in the human at sea level and at 5,500 m (18,000 ft).

Modified from Rahn 1966.

of $CO_2$ ($PCO_2$) and elevated pH of the arterial blood exert an inhibitory influence on the respiratory center. On the other hand, the earlier accumulation of lactic acid in the blood decreases the pH.

At sea level, the alveolar $PCO_2$ was about 38 mmHg (5.1 kPa) in the experiment, bringing the pulmonary ventilation to 80 L · min$^{-1}$ (see figure 14.3). The reduction in ventilation when the subject was breathing oxygen brought the alveolar $PCO_2$ to a somewhat higher level; at a simulated altitude of 4,000 m, the alveolar $PCO_2$ was reduced to 28 mm Hg (3.7 kPa) during exercise, with a similar pulmonary ventilation of 80 L · min$^{-1}$ (Dejours, Kellogg, and Pace 1963; Dempsey et al. 1972). Thus, the hypoxic drive during exercise at high altitude must be stronger than reflected in the magnitude of the pulmonary ventilation. If the alveolar and arterial $PCO_2$ are maintained at about 40 mmHg (5.3 kPa) by the addition of a proper volume of $CO_2$ to the inspired air, the pulmonary ventilation during exercise in hypoxic conditions will be still higher than illustrated in figure 14.3. The ventilatory response must be viewed as a physiological compromise, with the call for an adequate oxygen supply matched against the need to maintain the acid-base balance as normal as possible. Because of the great increase in ventilation during exercise at high altitude and a given oxygen uptake, the alveolar $PO_2$ is higher than normal. This obviously facilitates the diffusion of oxygen to the blood in the pulmonary capillaries. The maximal pulmonary ventilation during exercise at high altitude is the same as, or higher than, at sea level (Grover and Reeves 1967; Saltin 1966; Stenberg, Ekblom, and Messin 1966).

2. The diffusing capacity is reported to be unchanged in people who have moved from sea level to higher altitudes (West et al. 1962) or is slightly increased (Guleria et al. 1971; Reeves et al. 1969). The alveolar–arterial $PO_2$ gradient is greater for a given oxygen uptake compared with sea-level determinations (Cruz, Hartey, and Vogel 1975).

3. During submaximal exercise with reduced oxygen pressure in the inspired air, the lower oxygen saturation is compensated for by an increased cardiac output (Hartley, Vogel, and Landowne 1973; Stenberg, Ekblom, and Messin 1966). This and other effects of high altitude on oxygen transport in humans are illustrated in figure 14.4. The increase in the cardiac output is brought about by an increased heart rate; the stroke volume even can be reduced (McManus et al. 1974). The arterial blood pressure is largely unchanged. The lower arterial $CO_2$ pressure (hypocapnia) causes a venoconstriction that can preserve the cardiac output by increasing the central venous volume and cardiac filling pressure (Cruz et al. 1976).

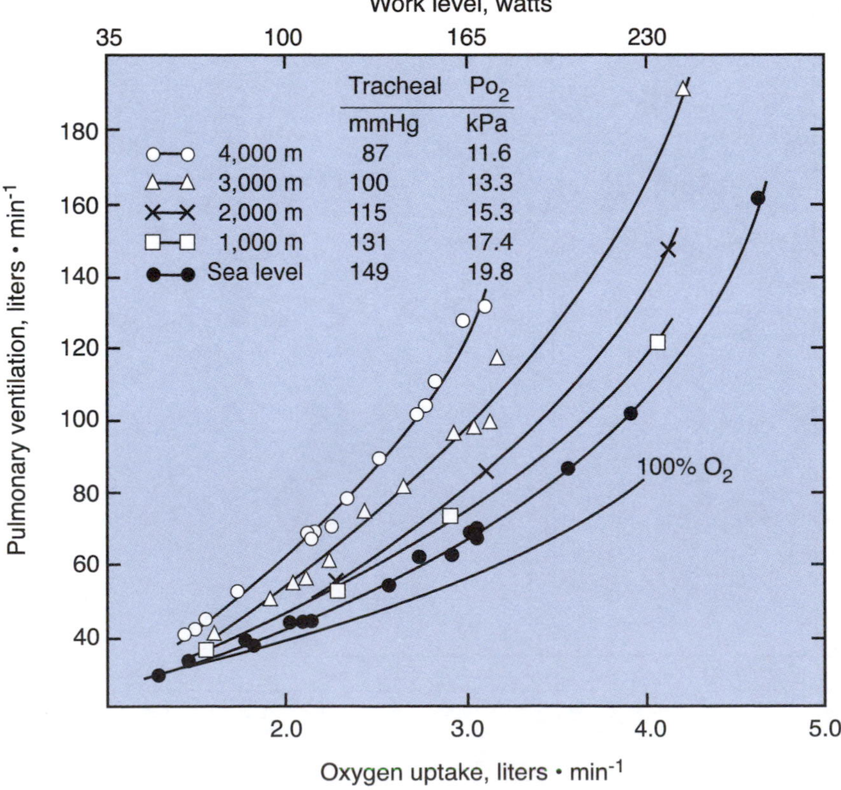


**Figure 14.3** Pulmonary ventilation (BTPS) in relation to oxygen uptake in one subject at different work rates when breathing oxygen or air during exposure to various simulated altitudes. Note the high pulmonary ventilation of 190 L · min$^{-1}$ reached during exercise at 3,000 m altitude.

From P.-O. Åstrand 1954.

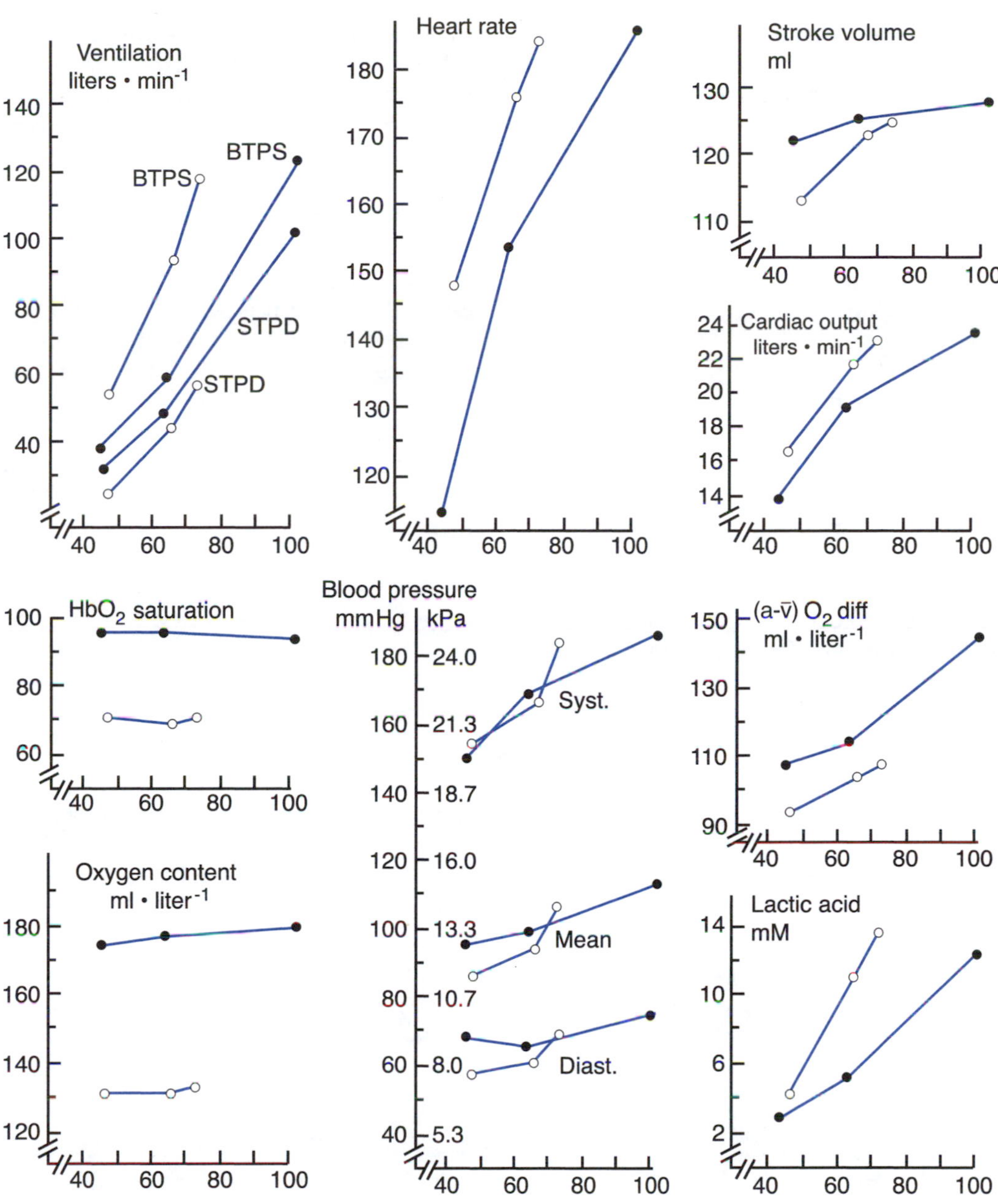


**Figure 14.4** Mean values of six subjects studied at two submaximal work rates and one maximal work rate (cycle ergometer) at sea level (●) and when acutely exposed to simulated altitude of 4,000 m in a decompression chamber (○). Abscissa = oxygen uptake in percentage of the maximum attained at sea level.

From Stenberg, Ekblom, and Messin 1966.

The previously mentioned hypocapnia during acute lack of oxygen produces a shift in the hemoglobin (Hb) dissociation curve (see figure 5.20), which means a net advantage for oxygen transport because of a higher arterial saturation. The arterial oxygen content is definitely reduced at high altitudes, however, and the arteriovenous oxygen difference decreases.

Observed maximal values for heart rate, cardiac output, and stroke volume are the same at an altitude of 4,000 m (acute exposure) as at sea level (Stenberg, Ekblom, and Messin 1966). Apparently, the lack of oxygen is not of such a magnitude to reduce the pumping capacity of the heart muscle, despite the fact that the arterial $PO_2$ is estimated to be lower than 50 mmHg (6.7 kPa). In a comparable study, Blomqvist and Stenberg (1965) showed that there were no electrocardiographic signs of myocardial ischemia when their subjects were performing maximal exercise at the same simulated altitude of 4,000 m. Stenberg, Ekblom, and Messin (1966) concluded that the reduced maximal oxygen uptake in moderate acute hypoxia compared with normoxia is related closely to the reduction of arterial oxygen content. During maximal exercise at hypoxia, the oxygen uptake was on an average 72%, the arterial oxygen content was 74%, and the cardiac output was 100% of the values attained at sea level. In other words, the maximal oxygen uptake was highly correlated with the volume of oxygen offered to the tissue (arterial oxygen content times maximal cardiac output). During maximal exercise at sea level, almost all the oxygen is extracted from the blood passing the active muscles, so that there is nothing more to gain in this respect at acute exposure to high altitude.

The quantitative effect on oxygen transport during maximal exercise at different altitudes is illustrated in figure 14.5. At the altitude of Mexico City, 2,300 m, the reduction is on the order of 15%; at 4,000 m, it is about 30% (from 4.24 to 3.07 $L \cdot min^{-1}$ in the study of Stenberg et al., 1966). The considerable scatter of the data observed when different studies are compared can be explained by several facts: (1) The effect of the reduced oxygen pressure on the maximal aerobic power is different in different individuals; (2) different techniques have been applied, especially concerning criteria for ascertaining that the maximal oxygen uptake has been reached; and (3) conceivably, persons with a high maximal aerobic power are more affected than persons with a lower maximal aerobic power, in that the diffusing capacity could be more critical for the former (M.R. Edwards et al. 2000; Hopkins et al, 1998). The progressive decrease in arterial oxygen saturation as the exercise level increases at high altitude, despite an increasing alveolar tension, and the resulting large alveolar–arterial oxygen differences, can be explained by diffusion limitations of the lung

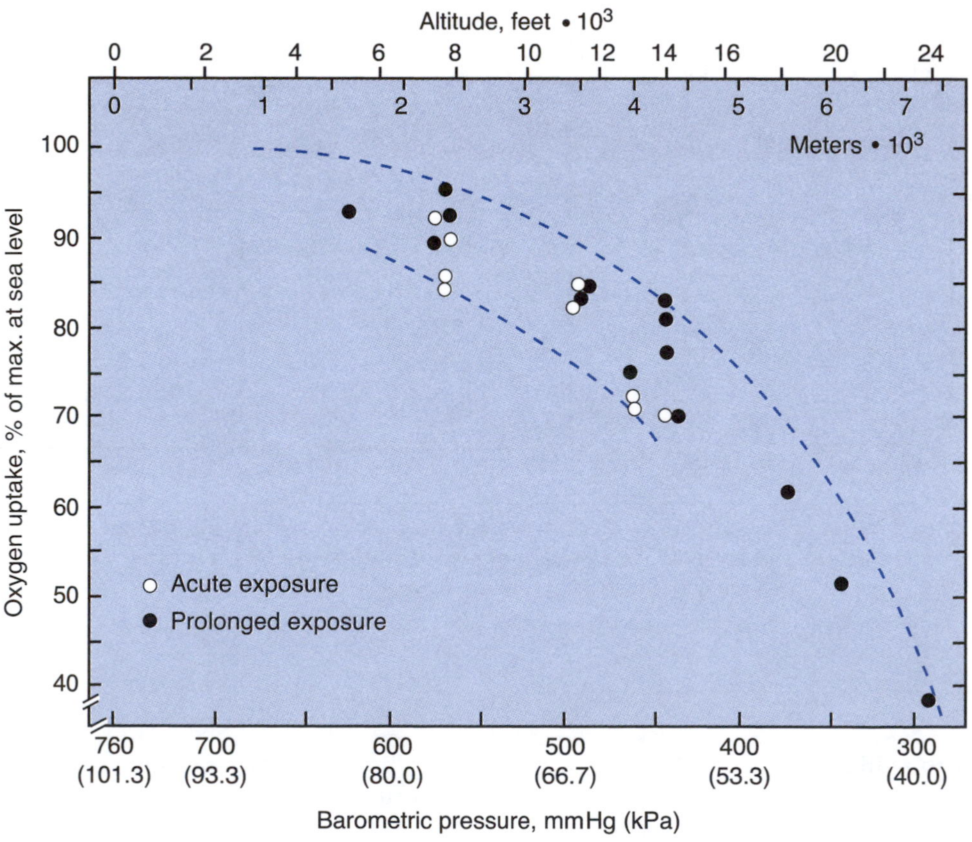


**Figure 14.5** Reduction in maximal oxygen uptake in relation to altitude barometric pressure. ○ = experiment in the acute hypoxic stage; ● = data obtained after various periods of acclimatization. In principle, the maximal aerobic power during acute exposure to reduced oxygen pressure decreases at the lower part, within the dotted lines; during acclimatization, it is shifted toward the upper part of the field.

Data from Balke 1960; Buskirk et al. 1967; Hansen et al. 1967; Pugh 1964; Roskamm et al. 1968; Saltin 1967; Stenberg, Ekblom, and Messin 1966.

(Chapman, Emery, and Stager 1999; Dempsey and Wagner 1999).

## For Additional Reading

For discussions concerning the limitations to $\dot{V}O_2$max and exercise performance ability at sea level and at high altitude, see Fulco, Rock, and Cymerman (1998) and Wagner (2000).

## *Summary*

Acute exposure to a reduced oxygen pressure in the inspired air during exercise is associated with hyperpnea in excess of that at sea-level conditions for the same energy requirement. The cardiac output also increases out of proportion to the oxygen uptake. These factors, combined with the displacement of the physiological range of the oxygen dissociation curve to its steep part, enhance oxygen transport. However, these compensatory responses cannot fully compensate for the reduced oxygen pressure. The maximal oxygen uptake is reduced, and the importance of the anaerobic energy yield increases.

# Adaptation to High Altitude

We now discuss the effect of a prolonged stay at high altitude, that is, acclimatization to reduced oxygen pressure in the inspired air. It is customary to distinguish between short-term adaptation, when the person is at high altitude for days, weeks, or a few months, and long-term adaptation, when the stay at high altitude lasts for years.

The first few days of exposure to reduced oxygen pressure entails a further increase in pulmonary ventilation at a given work rate. This hyperpnea will further increase the $PO_2$ and reduce the $PCO_2$ of the alveolar air. This ventilatory response is illustrated in figure 14.6, which shows data obtained at sea level, during acclimatization for 4 weeks to an altitude of 4,300 m (14,250 ft), and again at sea level during the reacclimatization. Four subjects were

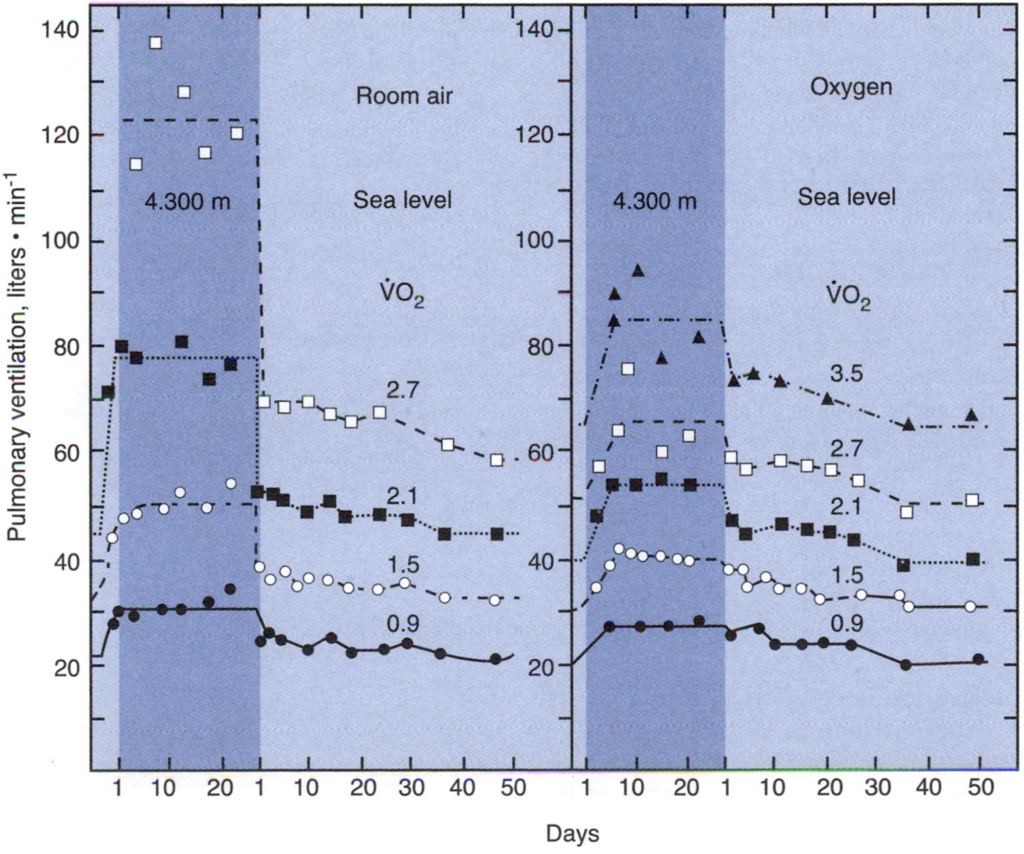


**Figure 14.6** Pulmonary ventilation in one subject exercising on a cycle ergometer at sea level (baseline to the left), during a 4-week sojourn at an altitude of 4,300 m (shaded area), and again after return to sea level including almost 50 days of observation time. Oxygen uptake is indicated on the respective lines. Left panel, subject breathing room air; right panel, subject breathing oxygen during exercise.

Unpublished data obtained by I. Åstrand and P.-O. Åstrand at White Mountain Research Station, California, 1957.

studied, but the figure presents the data on only one of the subjects. However, he represents the normal reaction reasonably well. Three comments should be made: First, it is evident that within a week at high altitude, a new level for pulmonary ventilation is attained, exceeding the value noticed in acute exposure to the same degree of hypoxia. The prolonged exposure to this hypoxia caused a 40% to 100% increase in pulmonary ventilation compared with the sea-level controls, the increase being more pronounced the heavier the work rates. At the end of the stay, the ventilation at an oxygen uptake of 2.7 L · $min^{-1}$ (200 W) was about 120 L · $min^{-1}$ when the subject was breathing air, compared with 60 L · $min^{-1}$ at sea level. The alveolar $PCO_2$ was 24 mmHg (3.2 kPa) at 4,300 m, compared with 40 mmHg (5.3 kPa) at sea level. Second, even when oxygen is inhaled during exercise at high altitude, blocking the peripheral chemoreceptor drive, there is a gradual increase in pulmonary ventilation (see figure 14.6). In the example chosen, the ventilation in the control experiment at sea level was 50 L · $min^{-1}$, but at the end of the 4-week sojourn at high altitude, it was raised to 65 L · $min^{-1}$ during the standard exercise. Third, when the subjects returned to sea level following exposure to altitude, it took several weeks before the control level was attained (Buskirk et al. 1967; H.V. Forster et al. 1971). The return of the alveolar $PCO_2$ to control levels paralleled the shift in pulmonary ventilation.

The extra increase in ventilation in connection with exercise at high altitude must be attributed to a hypoxic drive via the peripheral chemoreceptors. This hypoxic drive is prevalent even during chronic hypoxia, at least for a considerable period of time (P.-O. Åstrand 1954; Dejours, Kellogg, and Pace 1963). The induced alkalosis on sudden exposure to low $PO_2$, which reduces the central chemoreceptor drive, becomes gradually compensated by a proportionate decrease in the blood bicarbonate, and normal pH is restored in the acclimatized person. Prolonged alteration in arterial $PCO_2$ in either direction tends to alter the renal acid-base excretion, which tends to slowly return the arterial hydrogen ion ($H^+$) toward normalcy. As the alkalosis is reduced (pH is lowered), there is a further increase in pulmonary ventilation, as illustrated in figure 14.6. However, a restoration of cerebrospinal fluid $H^+$ concentration does not sufficiently explain the hyperventilation obtained on sojourn to high altitude (Dempsey et al. 1972).

Evidence of a real change in the regulation of the body $PCO_2$ developing during the first week after ascent from sea level to high altitudes is presented in figure 14.7. The respiratory response to $CO_2$ was tested by having subjects breathe $CO_2$–oxygen mixtures during exercise at a load of 100 W (oxygen uptake, 1.5 L · $min^{-1}$). There was a marked shift to the left so that, after acclimatization, a given pulmonary ventilation was attained at 15 to 20 mmHg (2.0–2.7 kPa) lower alveolar $PCO_2$ compared with controls at sea level. The difference in the response to $CO_2$ can be illustrated by the following example: At sea level, a pulmonary ventilation of 35 L · $min^{-1}$ was obtained when the end-expiratory $PCO_2$ was 45 mmHg (6.0 kPa). After 1 week at 4,300 m, the same $PCO_2$ was recorded when ventilation was as high as 83 L · $min^{-1}$.

After subjects returned to sea level, the $CO_2$ response curve gradually returned to the control level. This slow reacclimatization shows a pattern parallel with the one illustrated in figure 14.6. When the hypoxic drive was reduced by the increased oxygen pressure in the inspired air, the pulmonary ventilation was reduced. This produced a reduced washout of $CO_2$ and an uncompensated acidosis with an increased central chemoreceptor drive as a consequence. Millhorn, Eldridge, and Waldrop (1980) suggested that the observed long-lasting facilitation of respiration, mediated endogenously by the central neurotransmitter serotonin, explains the slow secondary in-

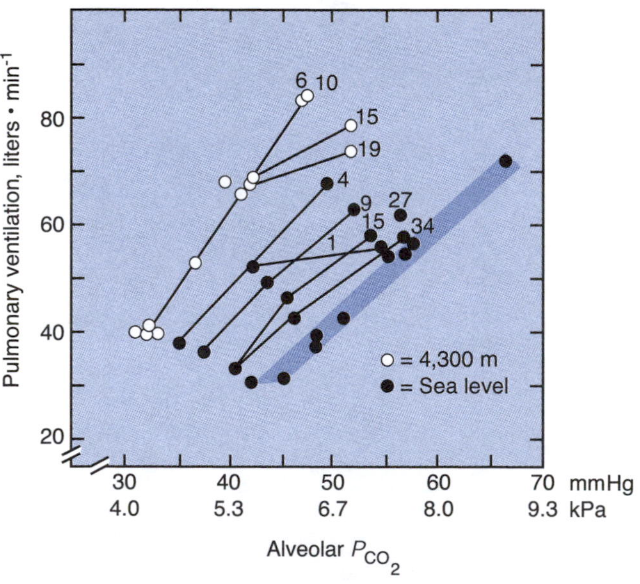


**Figure 14.7** Ventilatory response during a standard work rate (100 W) to inhalation of various carbon dioxide–oxygen mixtures at sea level (shaded area), during a prolonged sojourn at altitude of 4,300 m (○), and at various intervals during the reacclimatization to sea-level conditions (●). Figures at top of lines denote the day for the experiment after arrival at altitude and sea level.

Åstrand and Åstrand (unpublished data).

crease in ventilation during acclimatization to high-altitude hypoxia.

Pugh (1964) carried out hemodynamic studies including maximal exercise at very high altitudes. They reported that a prolonged sojourn at various altitudes brought the cardiac output at a given work rate down to the level typical for the same work rate performed at sea level. However, the maximal cardiac output was markedly reduced, and after several months' stay at 5,800 m, the values were 16 to 17 L · min$^{-1}$ compared with 22 to 25 L · min$^{-1}$ at sea level. The reduction of cardiac output was a combined effect of a lowered stroke volume and maximal heart rate (reduced from 192 down to 135 beats · min$^{-1}$). This study confirmed the data by Christensen and Forbes (1937).

A number of measurements of cardiac output during exercise at altitudes between 3,000 and 4,300 m have been made with exposure up to a few weeks (Alexander et al. 1967; Hartley et al. 1967; Klausen 1966; Vogel, Hansen, and Harris 1967; Vogel et al. 1974). The results indicate that after a few days, the minute volume of the heart during submaximal exercise is already reduced, compared with the cardiac output during acute exposure to the hypoxic condition, and that minute volume returns gradually to values typical for sea-level conditions or can even become subnormal. During maximal exercise, the cardiac output is reduced. A reduced stroke volume appears to be the primary reason for the reduced cardiac output; the lowering of the heart rate, in any case during maximal exercise, is a more inconsistent finding. Figure 14.4 shows that the stroke volume during light exercise is reduced at acute exposure to hypoxia. Grover, Lufschanowski, and Alexander (1976) ruled out myocardial hypoxia as a basis for the decrease in stroke volume.

Horstman, Weiskopf, and Jackson (1980) studied the aerobic power at sea level and during a 2-week stay at 4,300 m in nine subjects. After 2 weeks at this altitude, the maximal oxygen uptake and systemic oxygen transport increased by 10%, compared with the first day at altitude. The increased systemic oxygen transport resulted from a 19% increase in arterial oxygen content, despite a 9% decrease in maximal cardiac output. They concluded that a relative **polycythemia** was a major contributor to the increased maximal aerobic power.

An example of the heart rate response to fixed work rates was presented by Åstrand and Åstrand (1958; see figure 14.8). During acute exposure to a tracheal oxygen tension of about 85 mm Hg (i.e. 11.3 kPa at an altitude of 4,300 m), the heart rate was 15 to 30 beats higher per minute than at sea level. When the hypoxia was prolonged, heart rate gradually decreased at a given oxygen uptake. In the later stage of acclimatization, the heart rate attained during lower levels of exercise decreased to the same range as those recorded at sea level. At the heavier loads, however, the heart rate was even lower than in experiments with high tracheal $PO_2$. In this subject, the normal maximal heart rate was about 190 beats · min$^{-1}$. At the high altitude, it gradually declined to 135. When the subject was allowed to breathe 100% oxygen during almost maximal exercise, the heart rate increased within seconds by as much as 25 beats · min$^{-1}$.

A decline in maximal heart rate at prolonged exposure to altitudes exceeding 3,000 m also has been reported by Christensen and Forbes (1937), Cerretelli and Margaria (1961), Pugh (1964), and Cerretelli (1976). Similar to Åstrand and Åstrand (1958), Pugh (1964) noticed that a maximal heart rate of 130 to 150 beats · min$^{-1}$ was elevated to almost sea-level values when the subjects were breathing oxygen. An increased parasympathetic activity during prolonged exposure to hypoxia could partly explain the reduction in maximal exercise heart rate.

Within the first few days at high altitude, the Hb concentration in the blood increases, but this increase is mainly attributable to a hemoconcentration secondary to a decrease in the plasma volume (Buskirk et al. 1967; Merino 1950; Surks, Chinn, and

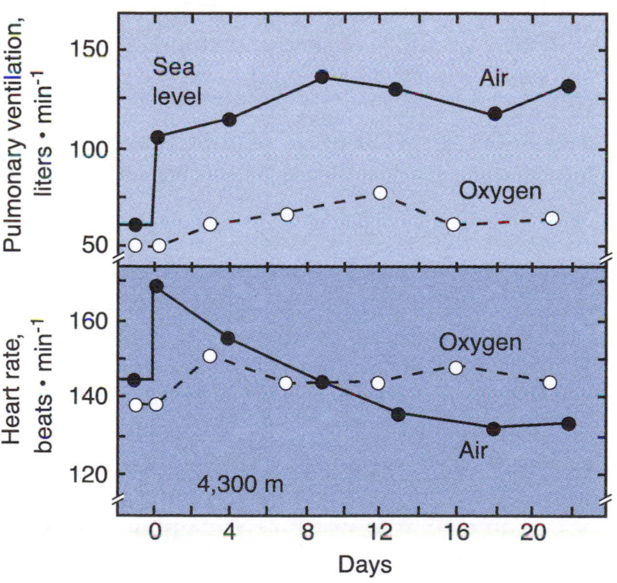


**Figure 14.8** Pulmonary ventilation and heart rate during exercise on a cycle ergometer, work rate 200 W at sea level, and during a 22-day sojourn at an altitude of 4,300 m. On some days, room air was inhaled during exercise; on other days, pure oxygen was taken.

Adapted from Åstrand and Åstrand 1958.

Matoush 1966). Gradually, the increased erythropoiesis brings Hb content to high levels so that the oxygen content per liter of arterial blood can be the same in the acclimatized person at 4,500 m as it is in a person at sea level (Christensen and Forbes 1937; Hurtado, Merino, and Delgado 1945; Mairbäurl et al. 1986; Reynafarje 1967). At an altitude of 4,500 m, the native residents of Morococha in Peru had an Hb concentration that averaged 208 g · $L^{-1}$ of blood (Hurtado, Merino, and Delgado 1945). As already mentioned, Horstman, Weiskopf, and Jackson (1980) emphasized that relative polycythemia is a major contributor to an increased maximal aerobic power during a prolonged high-altitude sojourn.

Ingjer and Myhre (1992), investigating elite male cross-country skiers for 3 weeks at an altitude of 1,900 m, found that those skiers who had the lowest initial Hb concentration and hematocrit (Hct) values experienced the greatest increases. Moreover, the same skiers showed the greatest reduction in blood lactate increase produced by a standard submaximal exercise test. A significant correlation was found between the magnitude of increase in Hb concentration and Hct and the difference in the lactate response to the standard test before and after altitude training. Although maximal aerobic power remained unchanged, the lower lactate response reflects an improved ability to exercise at higher submaximal work loads after the altitude training, possibly because of an improved buffer capacity of the blood. The authors suggest that individuals with low initial Hb concentrations and Hct improve their aerobic performance ability most during high-altitude training.

As a consequence of an increased hemoconcentration during acclimatization to high altitude, the oxygen offered to the tissues per liter of arterial blood can be the same in the individual living at high altitude as in the resident at sea level. However, these values of oxygen are really not comparable. The oxygen pressure gradient between blood and tissue is most important for the final transfer of oxygen to the mitochondria, and this gradient is reduced during hypoxia. However, the gradual decline in cardiac output during prolonged exposure to a low oxygen pressure in the ambient air can be explained partially by the concomitant rise in the oxygen-combining capacity of the blood. The increased viscosity of the blood with the elevated hematocrit necessitates an increased cardiac work at a given cardiac output, but the net effect of the hematologic response to prolonged hypoxia in terms of work of the heart cannot be evaluated at present. In any case, the increase in Hb concentration and the mentioned shift in the operational range to the steeper slope of the oxygen-dissociation curve contribute greatly to the gradually increased oxygen conductance within the body at altitude. During chronic hypoxia, there is also an increase in the 2,3-diphosphoglycerate (2,3-DPG) concentration in the blood, which enhances the unloading of oxygen to the tissues (see chapter 6). However, the opposite effect on the oxygen dissociation curve of reduced $PCO_2$ can cancel out the 2,3-DPG effect (Morpurgo et al. 1972).

Increased capillary density (i.e., capillary number per square millimeter of muscle tissue) after a period of acclimatization to high altitudes has been reported. An increase in the number of capillaries reduces the distance between the capillary and the most distant cells within its tissue cylinder. A relatively low oxygen tension in the capillaries therefore still provides the oxygen supply to these distant cells. However, in cases of exposure to severe altitude, such as expeditions to the Himalayas, the capillary density might be increased not because of capillary neoformation but because of a reduction in muscle fiber size as a consequence of reduced muscle mass (Hoppeler et al. 1990; MacDougall et al. 1991). Prolonged exposure to extreme hypoxia also decreases the muscle oxidative capacity, probably via loss of mitochondria (Howald et al. 1990). In contrast to these results, training studies carried out in moderate hypoxia or under severe, but intermittent hypoxia reveal a different effect on muscle tissue. Thus, in a study involving single-leg bicycling, Terrados et al. (1990) found a significantly larger increase in citrate synthase activity in the leg trained in moderate hypoxia (2,300 m) than the leg trained in normoxia. Capillary to fiber ratio and mean fiber area were both slightly increased after training in hypoxia, but not in normoxia. Furthermore, the myoglobin content increased only in the leg trained under hypobaric conditions. This last finding confirms an early observation by Reynafarje (1962), who reported that the myoglobin content in the skeletal muscles increases during altitude adaptation; this will favorably affect the oxygen transport. Desplanches et al. (1993) presented data from a study in which the subjects were asked to perform bipedal ergometer training (2 h · $day^{-1}$, 6 days · $week^{-1}$) in severe hypoxia (approximately 5,500 m) for 3 weeks. The morphometric analysis of muscle biopsy samples showed that training in hypoxia, in contrast to similar training (same absolute work load) in normoxia, significantly increased the muscle fiber area, capillary to fiber ratio, and total mitochondrial volume.

Mizuno et al. (1990) studied a group of cross-country skiers before and after a 2-week stay at altitude (2,100 m) and training (2,700 m). On return to sea level, the maximal aerobic power was the same as the prealtitude value, whereas increases in oxygen deficit (29%) and in short-term running performance (17%) were observed. Buffer capacity (in m. gastrocnemius and m. triceps brachii) was increased by 6%, providing evidence that the improvement in short-term exercise performance is attributable to increased muscle buffer capacity. The net effect of acclimatization to high altitude is a gradual improvement in oxygen availability to the active muscles and thereby an improvement in physical performance during endurance events. The well-trained individual is not acclimatized to high altitude any sooner or any more effectively than the untrained individual. The earlier the age when one works at high altitudes, the greater the influence of the environment in modifying the expression of inherited potential. The optimum time of exposure seems to be during growth and development. Also, there is a considerable individual variation in response to hypoxic conditions. This can be illustrated by the following example: In a group of eight international top athletes, the reduction in maximal oxygen uptake at the altitude of 2,300 m averaged 16% (ranging from 9% to 22%); after 19 days at this altitude, the reduction averaged 11% with a range of 6% to 16% (Saltin 1967).

### *Summary*

With prolonged exposure to reduced oxygen pressure in the inspired air, compensatory devices such as the following are acquired more slowly:

1. A further increase in pulmonary ventilation
2. Increased Hb concentration in the blood
3. Morphological and functional changes in the tissues (increased formation of capillaries, increased myoglobin content, modified enzyme activity)
4. Increased buffer capacity

The initially observed increase in cardiac output during submaximal exercise is replaced by a gradual decline to, or even below, the values observed at sea level. During submaximal as well as maximal exercise, the stroke volume becomes reduced. If the sojourn has been at very high altitude, 4,000 m or higher, the maximal heart rate can be reduced compared with sea-level values. All these adaptive changes are reversible, but it may take several weeks before the values return to sea-level values in the case of sea-level dwellers who have stayed at high altitudes for a month or more.

## Practical Applications

To attain top achievement at altitudes of 1,700 to 2,000 m or higher in activities requiring maximal aerobic power, an acclimatization period of at least 2 to 3 weeks commonly is recommended. At lower altitudes, the time required is probably less. A longer exposure to high altitude would probably be beneficial from a physiological point of view, but this advantage must be considered against possible psychological, social, and economic factors.

Practical experience has shown, especially in the case of cross-country skiers, that it is sensible to take it easy during the initial period of exposure to high altitude. If the effort is too intense, a considerably longer recovery time will be required than is needed at sea level.

The athlete must become accustomed to the fact that the subjective feeling of fatigue is different, which must be reflected in the choice of tactics. One is forced to accept a slower tempo, and the intensity and duration of training activities must be reduced. Swimmers discover that they can remain under water for a shorter time after turning than they normally do, and they must adapt their swimming strokes to a different breathing rhythm. The ability to tolerate an intense tempo for long periods of time at high altitudes is different from one individual to the next. This complicates the selection of the athletes for a team. There are examples of outstanding athletes in long-distance events at sea level who consistently fail at high altitude (Smith and Sharkey 1984). In competitions at high altitude, collapse from unknown causes occurs more frequently than at sea level. This fact, however, does not appear to represent an increased health hazard. Ingestion of additional fluid and carbohydrates may be required during exposure to high altitude combined with heavy physical exercise. Butterfield (1999) reviewed the evidence for increased requirements for nutrients at high altitude and made recommendations regarding appropriate intakes.

Athletes competing in events where technique is of prime importance or in events primarily involving anaerobic metabolic processes can arrive at more or less the time of the competition if the altitude is not so high that mountain sickness is expected.

The use of high-altitude training to improve sea-level endurance performance is widely practiced by athletes. However, the proposed benefits from the traditional model where the subjects lived and trained at moderate altitude are controversial. Although acclimatization to high altitude can improve oxygen delivery and utilization, as discussed previously, hypoxia at altitude limits training intensity, which in elite athletes can result in relative deconditioning. Scientific evidence confirming a positive effect on performance ability at sea level is rather scarce. Lack of appropriate controls, few subjects, and incomplete characterization of athletic performance have complicated the interpretation of the majority of reports published since the 1960s (for references, see Levine and Stray-Gundersen 1997).

During the 1980s, groups of endurance athletes, among others cross-country skiers and bicyclists, started to experiment with a somewhat different approach to altitude training. Although they stayed at high altitude, parts of the training were conducted at considerably lower altitude. In the early 1990s this strategy became the object of scientific investigations and was introduced as the "living high–training low" model (Levine and Stray-Gundersen 1992). The rationale behind the concept was that if athletes could live at moderate altitude (2,000–3,000 m) but train at low altitude (1,500 m or lower), they could acquire the physiological advantages of high-altitude acclimatization for increasing the oxygen transport capacity and, at the same time, be able to train at exercise intensities similar to those at sea level. Parallel with the live high–train low approach, several novel devices and modalities for altitude training were introduced. These include **hypobaric hypoxia** (reduced $P_{bar}$), normobaric hypoxia via nitrogen dilution (reduction in the number of oxygen molecules in a given unit of air), hypoxic sleeping devices, and supplemental oxygen (simulating normoxic conditions during high-intensity training at altitude).

Since the introduction of the living high–training low model, a number of investigations have been carried out to determine its efficacy. In their later studies, Levine and Stray-Gundersen (1997) showed that 4 weeks of living at 2,500 m and training around 1,300 m (HL group) resulted in a 1.4% improvement in 5,000-m running performance immediately on subjects' return to sea level and up to 3 weeks thereafter. This was not seen in a similar population of college runners who had lived and trained either at 2,500 m (HH group) or at sea level (LL group). Both HL and HH groups increased red cell mass and maximal aerobic power, whereas there were no changes in the LL group. However, there was considerable variation with respect to both the increase in cell mass and the performance improvement. It is speculated that some individuals increase red cell mass while living at altitudes of 2,000 m, whereas others need to live close to 3,000 m to increase red cell mass. The group data indicated that living at 2,500 m for 4 weeks increased mean red cell mass by 9%, which increased aerobic power by 5%. On basis of this study and other observations (Chapman, Stray-Gundersen, and Levine 1998), Stray-Gundersen and Levine (1999) concluded that if a healthy endurance athlete with normal iron stores lives high enough (2,000–3,000 m) for long enough (3–4 weeks) to produce a robust and sustained increase in erythropoietin, he or she will achieve an increase in red cell mass. This increase will increase maximal aerobic power, which, as long as training intensity is maintained, will improve subsequent sea-level performance.

In contrast to these findings, Ashenden, Gore, Martin et al. (1999) reported no effect on red cell production in male endurance athletes who spent 23 nights (8–10 h per night) at simulated altitude of 3,000 m in a "nitrogen house" while training at 600 m. The authors argued that it is more appropriate to use sensitive measures of hematological stimulation, such as automated analysis of reticulocyte parameters, because these techniques have greater resolution with which to detect small changes in erythropoietic stimuli than gross measures of red cell mass or plasma volume.

The effect of the living high–training low model on cardiac function was investigated by Y. Liu et al. (1998). A group of 21 well-trained triathletes were divided into a control group (LL), living and training at sea level, and a HL group, living at approximately 2,000 m and training at sea level. Echocardiographic recordings (Doppler) revealed reduced left ventricular end-systolic diameter and increased stroke volume, suggesting improved systolic function in the HL group. This was underlined by incremented left ventricular contractility, which might be associated with increased β-adrenergic receptor expression or an enhanced myocardial energy utilization.

The use of intermittent hypobaric hypoxia also can induce altitude acclimatization processes that improve aerobic performance ability. Rodriguez et al. (1999) exposed 17 subjects to intermittent hypoxia in a hypobaric chamber over 9 days at simulated altitude, which was progressively increased from 4,000 to 5,500 m in sessions ranging from 3 to 5 h $\cdot$ day$^{-1}$. They found significant increases in red blood cell count, packed cell volume, and Hb concentration, implying an increase in the oxygen transport capacity. Significant differences in exercise blood lactate kinetics and heart rate also were observed. The lactate versus exercise load curve shifted to the right and heart rate decreased, thus revealing an improvement of aerobic endurance.

To study the effects of living high–training low on anaerobic performance at sea level, Nummela and Rusko (2000) let eight 400-m runners live for 10 days in normobaric hypoxia in an altitude house (oxygen content = 15.8%) and train outdoors in ambient normoxia at sea level (HL group). Before and within 1 week after the altitude exposure, a 400-m race and running tests to determine maximum speed and the speed at different submaximal blood lactate concentrations were performed. During the same period, ten 400-m runners lived and trained at sea level and were subjected to identical test procedures (LL group). The results indicated that the HL group but not the LL group improved 400-m race time during the experimental period. The speeds at blood lactate concentrations of 5 to 13 mmol · $L^{-1}$ tended to increase in the HL group, but the responses were significant only at 5 and 7 mmol · $L^{-1}$. Furthermore, resting blood pH was increased in six of the eight subjects in the HL group. The authors suggested that changes in the acid-base balance and lactate metabolism might be responsible for the improvement in sprint performance.

### For Additional Reading

For reviews concerning physiological implications of altitude training for endurance performance at sea level, see Hahn (1991), Rusko (1996), Bailey and Davies (1997), and Wilber (2001). For details about training at altitude in practice, when to use altitude relative to target competitions, and specific training examples, see Dick (1992) and Bailey et al. (1998).

# Tobacco Use

The effect of cigarette smoking on work capacity is well known. More than a half century ago Juurup and Muido (1946) demonstrated that during submaximal cycle exercise at a fixed oxygen uptake, the heart rate was 10 to 20 beats · $min^{-1}$ higher when the work test was preceded by smoking one or two cigarettes. The difference in heart rate between smokers and nonsmokers was greater the higher the work rate.

## Cardiovascular and Respiratory Effects

Nicotine has pronounced acute effects on the cardiovascular system. It causes vasoconstriction, decreases peripheral circulation, and increases heart rate and blood pressure, secretion of antidiuretic hormone and catecholamines, and levels of glucagon, insulin, and cortisol (Omvik 1996).

Tobacco smoke contains up to 4% by volume of carbon monoxide, which has an affinity to Hb approximately 225 times that of oxygen. Therefore, the presence of even small amounts of carbon monoxide can noticeably reduce the oxygen-transporting capacity of the blood. Carbon monoxide also hinders the unloading of oxygen in the tissues by shifting the oxyhemoglobin dissociation curve to the left, thus unloading less oxygen at any given oxygen tension (King, Dodd, and Cain 1987). The combination of reduced arterial oxygen content and increased oxygen affinity for Hb (decreased oxygen loading) can cause the myoglobin to desaturate to levels that reduce the driving force for oxygen into the muscle cells, thereby limiting mitochondrial adenosine triphosphate (ATP) production (Schaffartzik et al. 1993).

Because the inhalation of cigarette smoke can, within seconds, cause a two- to threefold increase in airway resistance that can last 10 to 30 min (Da Silva and Hamosh 1973), the oxygen cost of ventilation is increased (Koike et al. 1991). This contributes to the reduced maximal aerobic power and exercise performance following smoking. In addition to this acute effect, smoking also causes a more chronic swelling of the mucous membranes of the airways, which increases airway resistance. At rest when the pulmonary ventilation is less than 10 L · $min^{-1}$, the increased airway resistance is not noticeable, however. When the demand on respiration is elevated, the increased respiratory resistance caused by smoking may be noticeable. A reduced pulmonary ventilation capacity causes a smaller volume of oxygen to reach the alveoli, which impairs gas exchange.

### For Additional Reading

For further details about the effect of cigarette smoking and passive smoking on the oxygen transport system and exercise performance, see Morton and Holmik (1985), McMurray et al. (1985), Celermajer et al. (1996), and McDonough and Moffatt (1999).

The term *passive smoking* is used when a person breathes air contaminated by tobacco smoke. Long-term exposure to environmental tobacco smoke increases the risk of lung cancer and heart disease (Gottlieb 1999). In particular, passive smoking is associated with dose-related impairment of endothelium-dependent dilation in healthy young adults, suggesting early arterial damage (Celermajer et al. 1996).

### For Additional Reading

For further reading about the detrimental effects of smoking on respiratory and cardiovascular functions and the mortality rate from smoking worldwide, see C.B. Sherman (1992), McBride (1992), Rigotti and Pasternak (1996), and Peto et al. (1996).

## Metabolic Effects

Wahren et al. (1983) examined the metabolic response to acute tobacco smoking in 11 healthy, habitual smokers. The consumption of three cigarettes increased oxygen uptake by 11%, which was sustained for more than an hour. Venous concentrations of noradrenaline and adrenaline rose in some subjects by as much as 50%, whereas insulin and glucagon levels were unchanged. Free fatty acid (FFA) concentrations increased during and after tobacco smoking by 50% to 125%, whereas glucose levels remained unchanged.

In one subject who smoked 20 cigarettes during 11 h, the investigators made 24-h continuous respiratory exchange measurements using a metabolic chamber. His energy expenditure was found to be 14% greater than on the following day when he did not smoke, food intake and physical activity being the same. Wahren et al. (1983) concluded that smoking significantly raises energy expenditure. This could explain why smokers often gain weight when giving up smoking (Moffatt and Owens 1991).

Brundin (1980) found that the exercise-induced increase in blood temperature was considerably augmented after smoking. This effect was paralleled by, and probably caused by, an increased heat production reflected by an increased oxygen uptake. During long-term exercise, cardiac output and heart rate increased and the stroke volume decreased in response to smoking.

Furthermore, it is known that smokers develop increased dependence on blood glucose as a fuel during rest and sustained submaximal exercise. Long-term exposure to nicotine and catecholamines caused by tobacco use is probably responsible for this alteration in carbohydrate utilization (Colberg et al. 1994). Together, these factors mentioned can contribute to earlier fatigue in smokers compared with nonsmokers who exercise. Similar effects on exercise tolerance are noted in those who inhale environmental tobacco smoke (McDonough and Moffatt 1999). In line with this, it has been shown that women who undergo a vigorous exercise training program and quit smoking demonstrate improved exercise performance over those who continue to smoke (Albrecht et al. 1998).

### For Additional Reading

For a review of the effects of smoking on exercise performance, see Huie (1996).

## Smokeless Tobacco

Smokeless tobacco, also called *oral tobacco,* comes in the form of leaf ("chewing tobacco") and moist snuff ("dipping tobacco"). Although few athletes consider smoking to be part of the game or a performance enhancer, many do hold such beliefs about smokeless tobacco (Wichmann and Martin 1994).

Nicotine is absorbed easily by the oral mucosa and distributed extensively throughout the body. Because, in general, snuff is kept in the oral cavity during several hours per day, the nicotine intake can be higher in habitual snuff users than in smokers (Holm et al. 1992). Furthermore, the use of biological markers has shown that certain tobacco-specific **carcinogenic** nitrosamines are substantially higher in snuff users compared with smokers (Carmella et al. 1990). The use of smokeless tobacco can lead to strong physical dependence. As a drug, nicotine is comparable to cocaine and other highly addictive substances. Smokeless tobacco users have the same types of problems that smokers do in stopping their use. They will experience withdrawal symptoms such as headache, gastrointestinal discomfort, sleeping problems, irritability, anxiety, aggressiveness, and craving for tobacco (Christen, McDonald, and Olson 1990).

Snuff-induced lesions in oral soft and hard tissues have been described in several reports, including gingival gum recession, periodontal bone destruction, leukoplakia and other mucosal changes, and tooth abrasion (Christen et al. 1989; Ernster et al. 1990). Bad breath, discolored teeth and dental restorations, and decreased ability to taste bitter, salty, and sweet substances are other oral effects of smokeless tobacco (Christen et al. 1989).

Although the rate of nicotine absorption through smokeless tobacco use is more gradual than through cigarette smoking, the acute cardiovascular effects are quite marked, including peripheral vasoconstriction and increased heart rate and blood pressure (Squires et al. 1984; Van Duser and Raven 1992). Van Duser and Raven (1992) demonstrated an increased lactate concentration and lowered stroke volume at rest and during exercise at 60% and 85% of $\dot{V}O_2$max when subjects used smokeless tobacco. An increased hypertension rate among users of smokeless tobacco was reported by Schroeder and Chen (1985) and Bolinder, Ahlborg, and Lindell (1992).

Case reports have indicated a negative effect of smokeless tobacco on muscular performance. Symptoms like muscle aches, fatigue, twitches, and loss of strength were observed by Patten (1984) in a 20-year-old man using snuff and chewing tobacco. The symptoms disappeared when the subject stopped using all tobacco products, and he gradually returned to normal strength and activity during a 3-month period. As an experiment he decided to restart using snuff to see whether the condition would return. Within 2 weeks he again had muscle aches, fatigue, and twitches. He again returned to normal after the snuff was stopped. His condition remained normal for at least 3 years of follow-up. Relatively few data are available on the effects of smokeless tobacco on strength. However, Escher et al. (1998) compared various parameters of strength and reaction time in regular smokeless tobacco users while using and abstaining from tobacco. The investigators concluded that smokeless tobacco did not affect reaction time but decreased maximum voluntary force and maximum rate of force generation.

Persson et al. (2000) demonstrated an association between the use of snuff and type 2 diabetes. They also provided evidence that habitual smoking and snuff use negatively affect insulin secretion.

### For Additional Reading

For further details about the health implications of smokeless tobacco, see Pershagen (1996).

## Delayed Recovery After Injury

The association between tobacco use and delayed tissue healing (muscle, bone, and wound repair) is well recognized in clinical practice. Nicotine, being a strong vasoconstrictor, reduces nutritional blood flow to the tissue in question, impairing the healing processes. Nicotine also increases platelet adhesiveness, raising the risk of thrombotic microvascular occlusion and tissue ischemia. In addition, proliferation of red blood cells, fibroblasts, and macrophages is reduced by nicotine. Carbon monoxide reduces the oxygen transport, whereas hydrogen cyanide inhibits the enzyme systems necessary for oxidative metabolism. Slower healing has been observed clinically in tobacco users with wounds resulting from trauma, disease, or surgical procedures (Silverstein 1992). Studies of rabbits exposed to nicotine have demonstrated that bone union, bone-graft **revascularization,** resorption of granulation tissue, bone formation, and remodeling are all considerably slower than in unexposed rabbits. Inhibition of early revascularization by nicotine is proposed as the pathophysiological mechanism by which tobacco use adversely affects the healing processes (Daftari et al. 1994; Haverstock and Mandracchia 1998). The reduced capacity of tissue healing for tobacco users should be of particular concern in athletes who have been injured and want to be back in action as soon as possible.

### For Additional Reading

For further reading about the effect of smoking on bone mineral density, see Hollenbach et al. (1993) and Hopper and Seeman (1994).

# Alcohol and Exercise

Although alcohol (ethyl alcohol, ethanol) is a drug, it also can be classified as a nutrient because it provides close to 30 kJ (7 kcal) per gram. The absorption is relatively rapid, about 20% and 80% of the alcohol ingested being taken up from the stomach and the intestines, respectively. The majority is metabolized by the liver at a rate of 8 to 10 $g \cdot h^{-1}$ (M.H. Williams 1991). Alcohol is distributed to all cells, being diluted by the water content of the body. Needless to say, the effects of alcohol consumption depend on the blood alcohol concentration, which usually is expressed in grams per deciliter ($g \cdot dl^{-1}$) or in mg %. The question of dosage in the case of alcohol consumption in connection with athletic performance is a most difficult one. The tolerance varies greatly from one individual to another and probably also from time to time in the same individual.

## Central Nervous and Neuromuscular Function

Most of the ergogenic benefits of alcohol have been related to its **anxiolytic** and antitremor effects and potency to increase self-confidence and decrease sensitivity to pain. Such effects can enhance performance in certain sport events, such as reduction of tremor in precision shooting sports like pistol shooting, riflery, and archery (Koller and Biary 1984). Some individuals taking part in competitive events such as target shooting maintain that they achieve better results following moderate alcohol consumption. They feel more relaxed. It is conceivable that the depressing influence of alcohol on certain central nervous system centers can cause routine procedures to progress normally without the disturbing effect caused by the anxiety of competition. According to Borg, Domserius, and Kaijser (1990), moderate doses of alcohol do not alter the rating of perceived exertion during exercise.

Psychomotor performance can be tested using tasks such as simple and complex reaction time, balance ability, hand–eye coordination, and gross motor coordination.

It is well established that alcohol can temporarily impair coordination. The performance of rather simple movements is sometimes used to test whether an individual is under the influence of alcohol (such walking on a straight line or touching the tip of the nose with the index finger with the eyes closed). A precise assessment in borderline cases is impossible, however. The deleterious effect of alcohol is particularly noticeable in tasks requiring divided attention (N. Brewer and Sandow 1980).

Research findings are somewhat inconsistent regarding the influence of blood alcohol concentrations below .05 g · $dl^{-1}$ on psychomotor performance. However, several investigators have reported decrements in various performance tasks, both simple and complex. At blood alcohol levels above .05 g · $dl^{-1}$ most complex psychomotor tasks are adversely affected, and the effect is dose related, with serious impairment occurring at blood levels ~.10 g · $dl^{-1}$. Above this level, almost all behavioral skills are impaired by alcohol. Although the molecular mechanisms by which alcohol affects the brain are not well understood, there is a deleterious effect on central processing capacity and consequently a decrease in the ability to process information, that is, to interpret stimuli and to produce an appropriate response (M.H. Williams 1991).

Ikai and Steinhaus (1961) noted that the maximal isometric muscle strength could be improved in some cases, especially with untrained subjects, after moderate alcohol consumption. They theorized that this results from a depressing influence of alcohol on the central inhibition of impulse traffic in the nerve fibers to the skeletal muscles during maximal effort. The result is an increased impulse activity and increased strength. However, newer research does not support an ergogenic effect of alcohol in events characterized by maximal force development for short periods of time. In a review, M.H. Williams (1991) noted no effects of alcohol on tests of static strength, isometric muscular endurance, or laboratory and field tests characterized by energy outputs comparable to a 100-m dash. In five studies, including laboratory tests of strength and power and field tests of speed ranging from 80 to 400 meters, alcohol actually impaired performance.

## Metabolic Function

Alcohol is not used as a major energy source during prolonged aerobic exercise (Schürch et al. 1982). However, the oxidation of alcohol is associated with metabolic changes in a variety of tissues, particularly the liver, which can inhibit splanchnic uptake of lactate and glycerol (Krebs 1968). The reduced utilization of these precursors decreases gluconeogenesis and accompanying hypoglycemia. Juhlin-Dannfelt et al. (1977) studied the influence of ethanol administration on splanchnic and leg metabolism during rest and continuous bicycle exercise. They concluded that ethanol reduces glucose output from the liver after prolonged (180 min) exercise, interferes with hepatic uptake of lactate, and can reduce the stimulatory effect of prolonged exercise on glucose uptake by the exercising legs. However, ethanol did not interfere with lipolysis and FFA utilization during the exercise.

Blomqvist, Saltin, and Mitchell (1970) studied eight young male subjects during cycle exercise at two submaximal and one maximal work rate before and after oral alcohol intake producing blood levels of 90 to 200 mg percent. Oxygen uptake and heart rate were determined. In four of the subjects, cardiac output, stroke volume, and intra-arterial pressure were also measured. During maximal exercise with an oxygen uptake of about 3 L · $min^{-1}$ and a cardiac output of about 21 L · $min^{-1}$, no difference was observed when the results before and after alcohol consumption were compared. Maximal heart rate, stroke volume, arteriovenous oxygen difference, and calculated peripheral resistance also were unaffected. During submaximal exercise, on the other hand, the heart rate was on average 12 to 14 beats higher per minute in the alcohol experiments ($p < .01$). In the latter case, the cardiac output was greater whereas the stroke volume was unaffected. The oxygen uptake during submaximal exercise was slightly higher after alcohol, but the arteriovenous oxygen difference was nevertheless reduced (i.e., the cardiac output was more elevated than would be expected from the increased oxygen uptake). At rest and during submaximal exercise, the calculated total peripheral resistance was reduced.

With higher alcohol consumption, giving a blood alcohol concentration of .20 g · $dl^{-1}$, the heart function can be impaired, as suggested by a 6% reduction in left ventricular ejection fraction during an aerobic exercise task (Kelbaek et al.,1985). In addition, alcohol is dehydrating, which reduces aerobic performance ability.

## Conclusion

The American College of Sports Medicine (1982) came to the following conclusions after conducting a comprehensive analysis of alcohol and exercise:

1. The acute ingestion of alcohol has a deleterious effect on many psychomotor skills.
2. Alcohol consumption does not substantially influence physiological function crucial to physical performance.

3. Alcohol ingestion will not improve muscular work capacity and can decrease performance levels.
4. Alcohol can impair temperature regulation during prolonged exercise in a cold environment.

### For Additional Reading

For reviews of the acute and chronic effects of alcohol on exercise and sport performance, see O'Brien (1993) and Gutgesell and Canterbury (1999).

## Caffeine

Caffeine is a naturally occurring trimethylxantine that is present in many different plant species. It is found in a range of food products, with coffee being the main source. One cup of coffee (250 ml) provides about 100 mg of caffeine. Substantial amounts are also present in tea, chocolate, and various soft drinks, especially colas (12 oz contains 38–54 mg).

The use of caffeine among athletes has been popular over the years, probably because of reports of its ergogenic benefits and because of public acceptance, availability, and relative inexpensiveness of the drug. The legality of caffeine use in sport events has been controversial, with a suggestion as early as 1939 that it should be prohibited. The International Olympic Committee (IOC) classified caffeine as a doping agent in 1962, removed it from its list of banned substances in 1972, and currently has classified it under Section II as a restricted drug (positive at >12 mg · $L^{-1}$) in urine. It is unlikely that the IOC limit of 12 mg · $L^{-1}$ in urine would be exceeded under any circumstances by casual caffeine ingestion. However, before attempting to use caffeine as an ergogenic aid, athletes and coaches would be wise to consult expertise within sports medicine. For further discussion, see Tarnopolsky (1994).

### Central Nervous and Neuromuscular Function

Caffeine is a central nervous system stimulant and is known to have most apparent effects on cognitive functions during fatigue by increasing alertness and repelling drowsiness. It has been suggested that caffeine exerts its actions, at least partly, through central neural processes involved in activity and fatigue. Central dopaminergic and cholinergic deficits can play an important role in the increased perception of effort to maintain a certain workload (Davis and Bailey 1997). Caffeine antagonizes the central adenosine receptor system, which increases the activity of dopamine, noradrenaline, and acetylcholine (Ach) (Ashton 1987). Hence, caffeine, by reversing central dopaminergic and cholinergic deficiencies, can reduce the perception of effort and fatigue and thus increase the work output and improve reaction time, precision, and coordination. These effects are most pronounced during and after strenuous physical exertion, in which concentration, reaction time, and technical and tactical skills have a major influence on performance. Hogervorst et al. (1999) investigated the effect of different dosages of caffeine (0–5 mg · $kg^{-1}$) on cognitive performance in a double-blind, crossover study with trained male athletes. The results demonstrated that a low and medium dose of caffeine improved complex cognitive functions after strenuous exercise. These functions could be improved with a relatively low dose of caffeine (in the range of 150–225 mg · $L^{-1}$), and the effects were not further enhanced by increasing the dose of caffeine (320 mg · $L^{-1}$).

The effect of moderate (300 mg) and large doses (600 mg) of caffeine on simple reaction time and movement time was demonstrated by Jacobson and Edgley (1987) using a double-blind format. They noted significant improvements in the 300-mg group, whereas no significant effect was found between the pre- and posttest measurements in the 600-mg group. A concern with this study was that the investigators did not account for body weight when administering the caffeine.

Kalmar and Cafarelli (1999) studied the effect of caffeine (6 mg · $kg^{-1}$) on maximal force production and endurance of human quadriceps muscle. The subjects were able to hold a 50% maximal voluntary contraction for an average of 66 s in the absence of caffeine. After caffeine administration, the time to fatigue increased significantly by approximately 26 %. It is difficult to determine whether changes in the contractile properties of muscle or central mechanisms, such as motivation, are responsible for the enhanced ability to sustain a submaximal voluntary contraction. Although the effects of caffeine on skeletal muscle calcium ion ($Ca^{2+}$) availability at the dosage used in this study are not documented, it is possible that caffeine diminishes the decline in $Ca^{2+}$ that may occur in fatigued muscle.

Jacobsen et al. (1992) examined the effect of caffeine (7 mg · $kg^{-1}$) on strength and power of the knee extensors and flexors in 20 elite male athletes. They concluded that caffeine can improve some strength parameters in highly resistance-trained males, but that differences in subject fiber type, motivation, and caffeine sensitivity need to be elucidated.

MacIntosh and Wright (1995), who investigated the effect of caffeine (6 mg · $kg^{-1}$) administered 2-1/2 h prior to a 1,500-m swim (an event that is completed in less than 25 min), found that the subjects swam significantly faster with caffeine than without.

Several other studies have been conducted, and many of them have failed to show any ergogenic effect of caffeine on muscle strength and fatigue, work output, and speed. The contradictory results might be attributable to differences among experimental protocols, like dosages and timing of caffeine ingestion, fitness levels of subjects, habituation and tolerance to caffeine, and interactions with diet.

Although moderate doses of caffeine (200–300 mg) can elevate mood and improve psychomotor and intellectual performance, larger doses can be **anxiogenic,** producing symptoms such as **insomnia,** headache, irritability, tremor, nausea, and diarrhea. Sensitivity to these effects varies among individuals, depending on tolerance, rate of absorption, metabolism, and age. Furthermore, the half-life of caffeine can be doubled by the use of oral contraceptives and is also increased in pregnancy (for references, see Ashton 1987). Furthermore, J. George et al. (1986) showed that alcohol in amounts commonly consumed (50 g · day$^{-1}$) is a strong inhibitor of caffeine metabolism, prolonging caffeine half-life by approximately 70%. Elevated serum caffeine levels can disturb sleep patterns and in severe cases lead to caffeine toxicity.

## Metabolic Function

It appears that caffeine can enhance physical performance during prolonged submaximal exercise. Thus, Costill, Dalsky, and Fink (1978) found an approximate 20% improvement in time to exhaustion after the ingestion of 330 mg of caffeine 1 h prior to a cycling exercise at an intensity of about 80% $\dot{V}O_2$max. The subjects also perceived the effort to be less for the caffeine trial compared with the placebo. Ivy et al. (1979) demonstrated that 250 mg of caffeine given 1 h before and another dose of 250 mg given during a 2-h cycling bout improved the work output by 7.4% compared with the placebo trial. The explanation for the enhanced performance ability could be that caffeine increases lipolysis and consequently spares muscle glycogen by increasing FFA oxidation. This theory was strengthened by Essig, Costill, and Van Handel (1980), who examined the glycogen concentration in muscle biopsy samples taken pre- and postexercise in caffeine and placebo trials. They observed that the reduction in muscle glycogen after 30 min of cycling at 69% of $\dot{V}O_2$max was 42% less in the caffeine trial compared with the placebo trial. In line with this, Spriet et al. (1992) showed that net glycogenolysis was reduced by about 55% in a caffeine trial during the first 15 min of exercise at 80% $\dot{V}O_2$max, compared with the placebo trial, but it remained similar thereafter. These investigators observed that caffeine ingestion increased levels of muscle citrate and the ratio of acetyl-coenzyme A to coenzyme A sulf-hydryl , and they suggested that this might inhibit glycolysis in the early stage of exercise, thereby sparing muscle glycogen.

In line with this, D.E. Anderson and Hickey (1994), who examined the effect of caffeine ingestion on the metabolic and catecholamine responses to exercise in a cold environment, found that the combination of caffeine and cold increased plasma adrenaline and lipid metabolism but decreased carbohydrate metabolism.

Dodd et al. (1991) investigated the effects of caffeine (5 mg · kg$^{-1}$) given to subjects who were recreationally active but either caffeine habituated or not (caffeine naive). The caffeine-habituated subjects showed an increase in blood FFA levels at the start of cycle exercise (1 h after caffeine ingestion), followed by a decrease. In contrast, the blood FFA levels remained elevated throughout the exercise for the caffeine-naive subjects. The authors suggested that caffeine tolerance could explain the different metabolic responses among the subjects.

Collomp et al. (1991) studied the effect of caffeine ingestion (5 mg · kg$^{-1}$) on performance and metabolic responses during supramaximal exercise by using the Wingate Anaerobic Test. They found that caffeine, essentially via adrenaline, evidently modified glycolytic metabolism but failed to improve performance. Conversely, Anselme et al. (1992) found that caffeine increased maximal anaerobic power and blood lactate concentration. Furthermore, Doherty (1998) concluded that caffeine ingestion (5 mg · kg$^{-1}$) can be an effective ergogenic aid for short-term, supramaximal running performance and can increase the maximal accumulated oxygen deficit. However, these results did not appear to be related to an increased lactate concentration.

### FOR ADDITIONAL READING

For further references and discussion concerning caffeine as an ergogenic aid, see Clarkson (1993), Nehlig and Debry (1994), and Spriet (1995). An extended account of the use and abuse of performance-enhancing substances is found in Bahrke and Yesalis (2002).

# Chapter 15

# Fatigue

Fatigue is a very common and important constraint in manual labor and physical activity, but the processes involved are still not fully understood, despite a century of research. This is obviously attributable to the fact that fatigue is a very complex conception, involving both psychological and a host of physiological factors. Feelings of fatigue without preceding exercise are not uncommon, and a given power output can be perceived as heavier today than it was yesterday. General fatigue can be a symptom of disease. It also can be psychological in nature, often associated with lack of motivation, lack of interest, and low reserve capacity. None of these problems will be dealt with here, because they fall outside the scope of this book. We shall deal only with exercise-induced fatigue that has a physiological basis. Such physiological fatigue, thought to be a warning mechanism preventing overstraining of the organism or part of it, can be general and systematic or it can be local and, as a rule, muscular in nature.

## General Physical Fatigue and Its Symptoms

Everybody has experienced the feeling of exercise fatigue, but when it comes to explaining fatigue in physiological terms, there is not even a consensus about which organ in the body is the main culprit. Obviously, this is because fatigue in general is a multifactorial condition and because the relative importance of individual factors depends on the type of work performed and the condition of the subject. A popular concept in the exercise sciences holds that fatigue develops when the cardiorespiratory system is no longer able to cover the oxygen demand incurred by the muscular work, but this concept has been met by severe criticism.

Physical fatigue can be defined as a state of disturbed homeostasis attributable to work and to work environment (Christensen 1960). This can give rise to subjective as well as objective symptoms, but so far very little is known about the nature of this disturbed homeostasis. Obviously, the decrease in blood sugar observed in a fasting subject engaged in prolonged submaximal work can lead to a feeling of general fatigue as one of the symptoms of hypoglycemia, and lactic acid accumulated in muscles engaged in intense work can cause local fatigue.

The subjective symptoms of fatigue range from a slight feeling of tiredness to complete exhaustion. Attempts have been made to relate these subjective feelings to objective physiological criteria such as the accumulation of lactate in the blood. Although such a relationship often is observed in connection with strenuous athletic events, this relationship is not usually present in prolonged light or moderate work. Subjective feelings of fatigue usually occur at the end of an 8-h workday when the average work load exceeds 30% to 40% of the individual's maximal aerobic power, and certainly when the load exceeds 50% of the maximal aerobic power. Subjective feelings of fatigue also can be associated with hypohydration in heat-exposed workers who do not consume enough water during their working shift.

Over the years, the word *fatigue* has had so many different connotations that it has been suggested to abandon it entirely (Petajan 1996). Most authors, however, seem to use the words *fatigue* and *muscle fatigue* interchangeably, which is why we focus on the latter in the following. We define muscle fatigue as "any exercise-induced reduction in the maximal capacity to generate force or power output" (table 15.1; Vøllestad 1997).

**Table 15.1 Terminology and Definitions**

| Term | Definition |
|---|---|
| Maximal voluntary contraction force | The force generated with feedback and encouragement, when the subject believes it is a maximal effort |
| Maximal evocable force | The force generated by a muscle or group of muscles when additional electrical stimulation does not augment force |
| Maximal power output | The power generated with feedback and encouragement when the subject believes it is a maximal effort |
| Muscle fatigue | Any exercise-induced reduction in the capacity to generate force or power output |
| Central fatigue | Any exercise-induced reduction in maximal voluntary contraction force that is not accompanied by the same reduction in maximal evocable force |

### For Additional Reading

For a broader discussion of exercise fatigue, see Noakes (2000).

## Muscle Fatigue

One of the difficulties in understanding muscle fatigue lies in the fact that there is no single, unifying cause of muscle fatigue that is valid for all kinds of physical activity, be it in daily life or in an experimental setting. Fatigue can result from simultaneous failure at a number of sites, but for any particular task one site could be primarily responsible for the loss of force-generating capacity (Binder-Macleod and Russ 1999). Consequently, the literature reveals several different opinions regarding the mechanisms involved, reflecting the wide range of exercise models, protocols, and methods used to study muscle fatigue (Noakes 2000; Vøllestad 1997).

In the 19th century, the prevailing view was that fatigue was caused by processes in the central nervous system. Mosso (1892) has been quoted as advocating this view, but actually he clearly demonstrated that muscle fatigue could be entirely peripheral. Mosso was the first to record the decrease in muscle force by means of an "ergograph" during fatiguing contractions in humans, and he found a comparable decline in force both during voluntary contractions and during direct stimulation of the muscle (figure 15.1). "One must conclude," he wrote, "that the mental factor does not exercise a preponderating influence, and the fatigue may even be a peripheral phenomenon." Despite this, the view that fatigue was central in origin prevailed until the middle of the 20th century, when Merton (1954) and Næss and Storm-Mathisen (1955) showed that processes within the muscle could be the only mechanism. "Fatigue is peripheral," Merton wrote, "for when strength fails, electrical stimulation of the motor nerve cannot restore it" (p. 564).

a

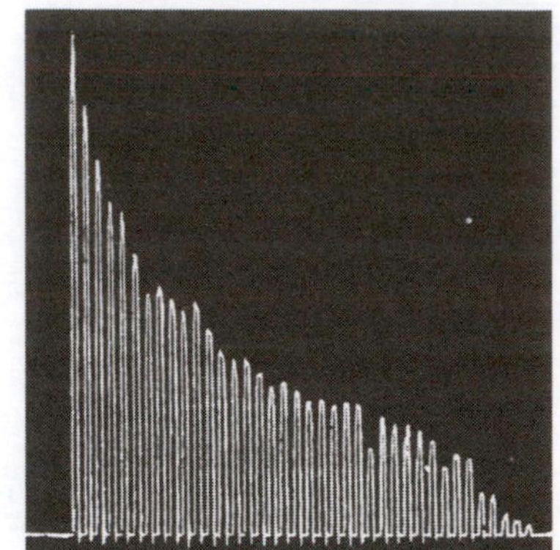

b

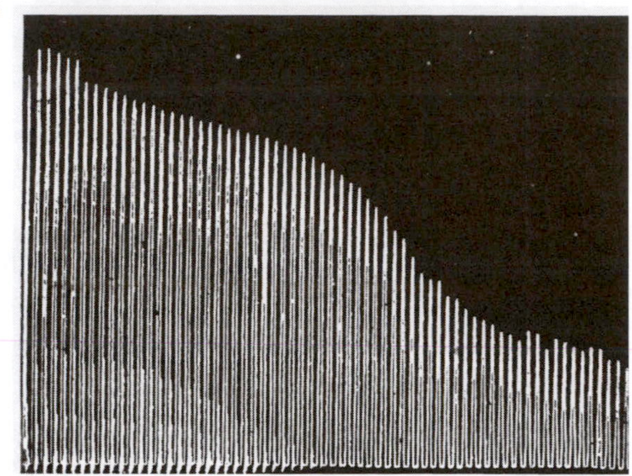

**Figure 15.1** Mosso (1892) showed convincingly that muscle fatigue could be entirely peripheral. The two pictures show comparable fatigue curves when the experimental subject (one of his junior assistants) lifts 3 kg repeatedly with his middle finger. The decreasing height of the consecutive recordings from the left to the right shows the decline in force as the number of repetitions increases *(a)* by means of voluntary contraction and *(b)* by stimulation of the median nerve. In the text he commented on the two curves as follows: "The middle finger lifts 3 kg in figure 13 [our panel b]. The steady reduction in height takes place in the same way as in figure 8 [our panel a], where the muscle was made to contract voluntarily. When the personal type of fatigue shows so little difference when the will is excluded, it must mean that the influence of the psyche is small, or that the fatigue mainly depends on peripheral conditions" (p. 101).

From Mosso 1892.

## CLASSIC STUDY

The study by Mosso (1892), which was first published in Italian but later translated into German, is the first systematic study of muscle fatigue. Unfortunately, the German edition is printed in gothic letters, which makes it inaccessible for many readers. This may be the reason why Mosso has been quoted as advocating the view that muscle fatigue was caused by processes within the central nervous system, whereas, in fact, he clearly showed and stated that muscle fatigue could be entirely attributable to processes in the muscle.

Mosso 1892.

In the 1970s, Asmussen and Mazin (1978) reintroduced a central component as a factor in muscle fatigue, but the relative contribution of central factors is not yet settled and probably never will be, because their importance is bound to vary with the task at hand. Admittedly, the central command undergoes definite changes during the development of fatigue, but as we discuss later, part of these changes may be compensatory mechanisms rather than factors causing the fatigue.

### FOR ADDITIONAL READING

For a historical view on muscle fatigue, see R.H.T. Edwards, Toescu, and Gibson (1995).

Muscle fatigue inevitably decreases performance in activities requiring muscular force, and most people experience an increasing difficulty in maintaining a certain activity level. If that level is high, or more specifically, higher than what is defined as "the habitual level of physical activity" for the individual, the problems arise early and increase rapidly. If the activity level is lower, the problems come later and are more protracted. Although the symptoms of fatigue can appear late during the fatiguing activity, the processes underlying the fatigue begin to develop as soon as the activity begins (J.L. Taylor, Butler, and Gandevia 2000).

Understandably, everyday muscle fatigue is not suitable as a basis for a scientific definition of the term, although loss of the ability to reach a certain target force can be useful, provided that the target force is precisely defined. The most easily defined target force is maximal voluntary isometric force, and many investigators therefore define fatigue as "a loss of maximal force generating capacity." See also Vøllestad (1997).

The chain of command from volition to measurable force is a long one (figure 15.2), and during exercise, changes occur at all steps in force production from upstream of the motor cortex to the myofibril. In addition, many of these changes influence earlier steps in the chain, thus complicating the picture. A schematic view of the chain of command with possible feedback loops is shown in figure 15.3. Admittedly, most causes of muscle fatigue are located beyond the neuromuscular junction (Gandevia, Enoka, et al. 1995a), but processes within the nervous system also contribute. In everyday life, for example, volition can be a limiting factor, as recognized by A.V. Hill (1926): "If one took a patient from the hospital and made him work till he could barely move, one could never be sure that he had driven

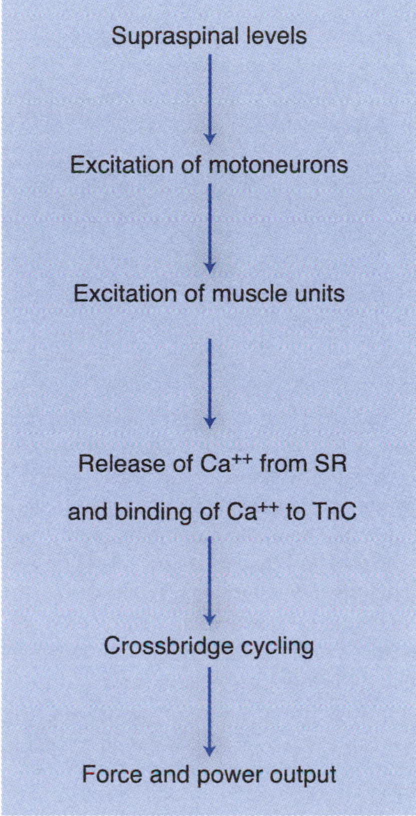


**Figure 15.2** The chain of command during voluntary activation of a skeletal muscle. Fatigue can be caused by deficient function at any of the steps. SR = sarcoplasmic reticulum; **TnC** = troponin C.

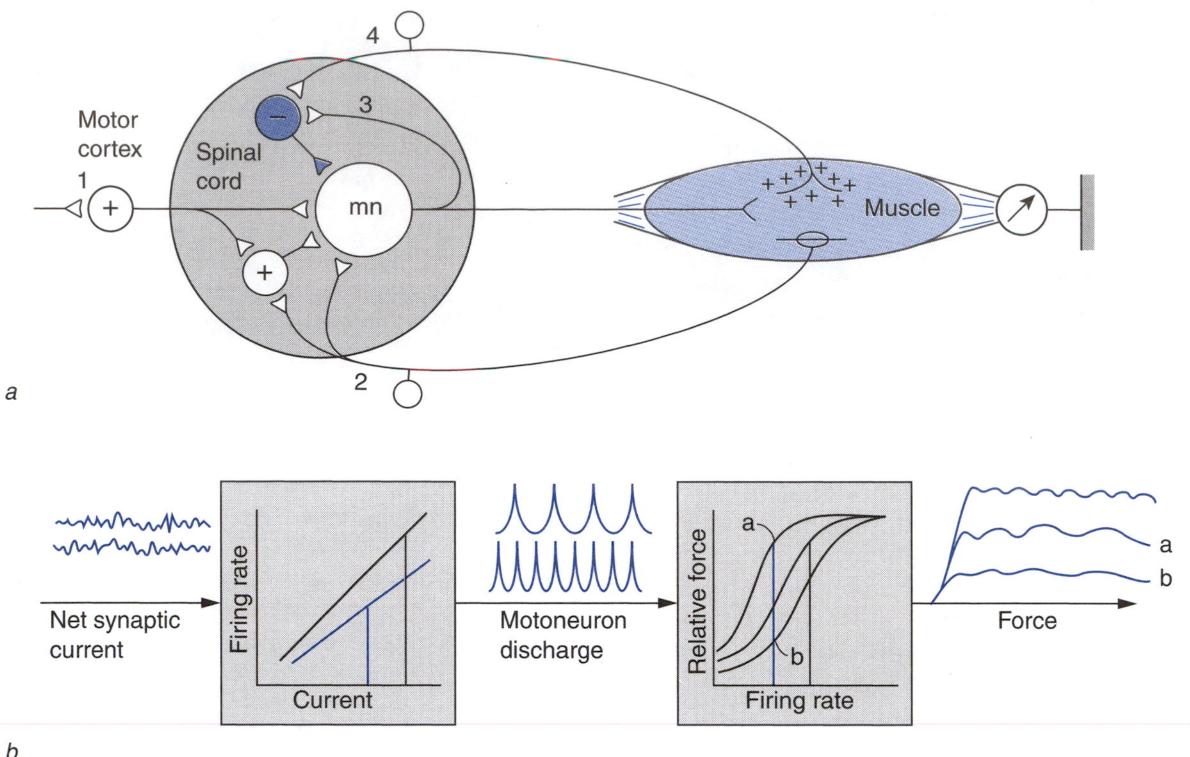


**Figure 15.3** *(a)* Schematic view of the chain of command from the motor cortex to the muscle, including some of the neural pathways thought to decrease excitatory drive to the motoneurons (mn) during fatigue. Reduced excitation from supraspinal levels (1), reduced input from spindle afferents (2), recurrent inhibition via Renshaw cells (3), and increased activity in metabolite-sensitive group III and IV muscle afferents (4) combine to reduce the net excitation of motoneurons. *(b)* Key changes in motoneurons and muscle fibers from before (black lines) to after fatiguing contractions (colored lines). With fatigue, the firing rate induced by a given synaptic current (left box) is reduced as indicated by the change from upper, black line to lower, blue line. The resulting reduction in firing rate is accentuated by the fatigue-induced decrease in synaptic current. The relationship between firing rate and resulting force is shown by a sigmoid curve (right box). During fatigue, slowing of contractile characteristics shifts the curve to the left (a), which potentiates the relationship. Consequently, a reduced firing rate as indicated by a change from the right vertical, black line to the left, blue line because of fatigue is without effect on the force output ("muscular wisdom"). If, instead, the frequency/force curve is shifted to the right because of fatigue (b), a reduced firing rate will substantially lower force output.

A.J. Fuglevand, *The Neuroscientist* 2:203-206, © 1996 by Sage Publications, Inc. Adapted by permission of Sage Publications, Inc.

himself to his limit. . . . With young athletic people one may be sure that they really have gone all out."

In most experimental situations, fatigue has been assessed in static contractions, engaging a restricted group of muscles acting on one single joint. Much less attention has been paid to the evaluation of fatigue during dynamic contractions involving larger groups of muscles. This is understandable because several theoretical and methodological difficulties limit the feasibility of such evaluation during complex, multijoint movements (Lewis and Fulco 1998). Despite this, there is a need to include dynamic work in our definition of muscle fatigue.

### For Additional Reading

See Lewis and Fulco (1998) for discussion of the problems inherent in the study of muscle fatigue during dynamic contractions involving larger muscle groups.

## Central and Peripheral Fatigue

In the fatigue literature, the distinction between central and peripheral fatigue is very common, but as we shall see, it is not always easy to distinguish between the two. A convenient definition of central fatigue could be "any force decline caused by a reduction in the firing frequency of the motoneurons involved." This would include all kinds of reduced excitatory drive to the motoneurons, regardless of their origin, but unfortunately it does not include failure in the propagation of the action potentials from the motoneuron to the muscle fibers. This possibility has to be considered because the main method for assessing central fatigue is the twitch interpolation technique, where a single or double supramaximal electrical stimulus is delivered to the muscle (or to the nerve) during a fatigu-

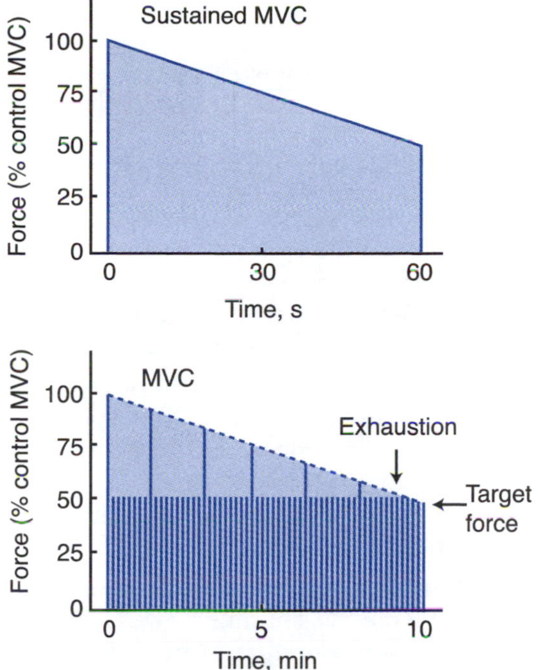


**Figure 15.4** Upper panel: Schematic demonstration of the decline in force during a sustained maximal voluntary contraction (MVC). Lower panel: There is no decline in force during repeated contractions at 45% of MVC until after about 10 min, but the interpolated twitches (the progressively shorter vertical lines) reveal that fatigue starts to develop from the first contractions.

Reprinted from Journal of Neuroscience Methods, 74, N.K. Vøllestad, Measurement of human muscle fatigue, 219-227, Copyright 1997, with permission from Elsevier Science.

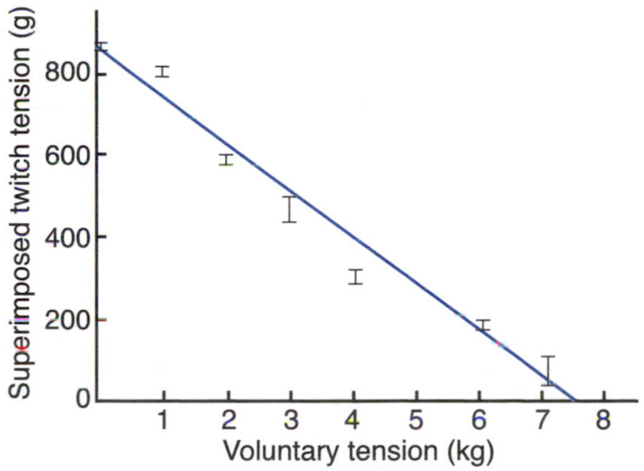


**Figure 15.5** In rested muscle, the superimposed (interpolated) twitch tension (shown as "superimposed twitch tension" in the figure) decreases as voluntary force approaches maximum, showing that it is a measure of the maximal muscle force potential not evoked by submaximal voluntary activation. In fatigued muscle, the superimposed twitch tension will show the part of the muscle force potential not evoked by the nervous system, despite maximal voluntary activation.

Adapted, by permission, from P.A. Merton, 1990, "Voluntary strength and fatigue," *Journal of Physiology* 123:559.

ing contraction (figure 15.4). The force increment caused by the stimulus shows the part of the muscle force potential not evoked by the nervous system and represents the force deficit attributable to central fatigue (figure 15.5). In table 15.1, the **maximal evocable force** is defined as "the force generated by a muscle or group of muscles when additional electrical stimulation does not augment force." In accordance with this, central fatigue can be defined as "any exercise-induced reduction in maximal voluntary contraction force, which is not accompanied by the same reduction in maximal evocable force" (Vøllestad 1997, p. 220). In other words, there is an insufficient neural activation of the muscle, insufficient in the sense that it is unable to make all its muscle units do their best. Peripheral fatigue, on the other hand, can be defined as "a force or power deficit that occurs despite optimal activation of the muscle fibers by their motoneurons"; that is, the muscle fibers are unable to respond maximally, despite adequate stimulation from the nervous system. Fatigue-promoting factors related to the neuromuscular junction or terminal branches of the motor axons are categorized as peripheral or central depending on where the interpolated twitch stimulus is applied. If the nerve is stimulated, they will appear as peripheral. If the muscle is stimulated directly, they will appear to be central.

Thus, although a distinction between central and peripheral fatigue seems logical at first sight, such a distinction in many ways is arbitrary, and the terms *central* and *peripheral* are not used in a strictly anatomical meaning of the words. The reason for this is that the definition of central fatigue is an operational one (see table 15.1). In addition, it is evident that several sites that can give rise to "central fatigue" definitely are located in the periphery. We return to this later.

## Reduced Excitation of the Motoneurons

Several factors contribute to reduced excitation of spinal motoneurons. Basically, any reduction in the excitatory drive and/or increase in inhibitory influence acting on the motoneurons, whether from supraspinal levels, from the segmental circuitry, or from the periphery, will reduce firing rates. As shown in figure 3.10, reduced firing frequencies can mean incomplete activation of and reduced force in the motor units involved. (See, however, the discussion of "muscle wisdom" later in this chapter.)

### *Reduced Supraspinal Drive to the Motoneurons*

Motivation is a word that easily comes to one's mind when speaking about central fatigue, but it is not easy to define motivation in neurophysiological terms, although changes in central enkephalinergic, dopaminergic, and serotonergic systems have been mentioned as possible mechanisms (Gandevia 1998). It also has been argued that motivation should be considered separately as part of psychological fatigue rather than as a cause of central fatigue (Brooke 1990). The feeling that it is difficult to start again after having taken a short break from strenuous activity could be a related phenomenon. It feels as if the momentum is gone, but the feeling is difficult to explain in physiological terms, although reduced levels of catecholamines could play a role (see chapter 4).

Not only acute reductions in activity but also long-term reductions, such as during limb immobilization, have been reported to disproportionately reduce maximal voluntary force compared with twitch and tetanic forces (Duchateau and Hainaut 1987). This may be taken to indicate that voluntary muscle activity is necessary to retain the ability to produce high levels of drive to the motoneuron pool (Gandevia, Allen, and McKenzie 1995). In a way, the reduced drive following a period of inactivity can be regarded as the inverse of neural adaptation to strength training (see chapters 4 and 11). A similar explanation may be offered for the greater prevalence of central fatigue in older subjects reported by Bilodeau et al. (2001).

Changes in the human motor cortex have been shown to accompany central fatigue. Experiments with **transcranial magnetic stimulation (TMS)** in healthy humans have shown decreased intracortical facilitation after fatiguing exercise, and the changes were confined to the areas of the motor cortex involved in the fatiguing contractions. The mechanisms behind these changes could not be established, although events upstream of the motor cortex have been considered to be possible candidates (Tergau et al. 2000), such as supraspinal effects of group III and IV muscle afferents (Gandevia 1998). As alluded to previously, this means that elements of central fatigue could originate in the periphery. For discussion, see Gandevia, Allen, and McKenzie (1995), Gandevia (1998), and J.L. Taylor, Butler, and Gandevia (2000). It has been argued, however, that changes in cortical excitability *per se* are not necessarily the direct cause of central fatigue (Gandevia, Allen, and McKenzie 1995).

Although little is known about the neurophysiological or biochemical mechanisms behind central fatigue, some theories exist, one of which is the so-called branched-chain amino acid (BCAA) theory. The main points of this theory are as follows:

1. BCAAs and tryptophan enter the brain via the same amino acid carrier, and competition between the two types of amino acid for entry into the brain can occur.
2. In the brain, tryptophan is converted to the neurotransmitter 5-hydroxytryptamine (5-HT), also known as serotonin, which is known to promote sleep and tiredness, a fact that makes it a possible fatigue-promoting substance.
3. During physical activity, BCAAs are taken up by muscle, shifting the balance between the plasma concentrations of tryptophan and BCAAs in the direction of tryptophan. This can increase the uptake of tryptophan in the brain and the subsequent formation of 5-HT (Newsholme and Blomstrand 1995).

So far, however, this remains a theory because experiments involving ingestion of tryptophan and BCAAs during sustained exercise have failed to show a performance effect (Van Hall et al. 1995).

#### FOR ADDITIONAL READING

For further discussion of the BCAA theory, see Newsholme and Blomstrand (1995) and Van Hall et al. (1995). The possible involvement of 5-HT in the beneficial effect of exercise on depression is discussed by Dunn and Dishman (1991).

More recent theories suggest that both hyperthermia and the cytokine interleukin (IL)-6, which is released from the brain during prolonged exercise, may induce central fatigue (Nybo and Nielsen 2001; Nybo, Nielsen, Pedersen et al. 2002).

### *Changes in the Input to the Motoneurons From the Periphery*

In conscious humans, fatiguing muscle contractions are accompanied by a decrease in the firing rates of $\alpha$ motoneurons. This has been interpreted as a mechanism to ensure proper matching between motoneuron firing rate and muscle unit force output and is commonly referred to as **muscle wisdom** (Marsden, Meadows, and Merton 1983). The mechanisms underlying muscle wisdom are, however, controversial.

### FOR ADDITIONAL READING

For discussion of muscle wisdom, see Windhorst and Boorman (1995), who also discuss the role of segmental motor circuitry in muscle fatigue. The applicability of the muscle wisdom concept in different contraction types is discussed by Garland and Gossen (2002).

During a fatiguing maximal voluntary contraction, the electromyographic activity declines roughly in parallel with the loss of force. This decline has been attributed to reflex inhibition of the motoneuron pool (Woods, Furbush, and Bigland-Ritchie 1987). Many reflex inputs can reduce motoneuronal output, but the contribution by the different kinds of afferents is to a large extent unknown (Garland and Kaufman 1995).

Oligosynaptic spinal pathways are likely contributors, possibly mediated by group III (corresponding to group Aδ) or IV (corresponding to group C) muscle afferents, which are known to be activated by muscle contractions (Gandevia, Allen, and McKenzie 1995). Group IV contains unmyelinated nerve fibers responsive to pain stimuli, whereas group III fibers are thin, myelinated fibers responsive to pain and temperature stimuli. Neuropharmacological research has shown that some group III and IV muscle afferents—the so-called **capsaicin-sensitive nerves**—are responsive to protons as well. There is a good correlation between tissue pH and afferent discharge when the pH falls to 6.6 or less (Bevan and Geppetti 1994), and the resulting inhibition of motoneurons can be regarded as a protective mechanism. There is reason to believe, however, that the processes associated with the fatigue-related reflex inhibition are too complex to be explained by small-diameter afferents responding to the byproducts of muscle contraction (Walton et al. 2002). Depending on the situation, fatigue also can slow muscle contractile properties, shifting the force–frequency curve to the left (see figure 15.3b). If there was not a parallel decline in motoneuronal firing rate, the activation of motor units would become supratetanic, and rate coding as a means of force modulation would be ineffective (Bigland-Ritchie 1981). This is the reason why the phenomenon has been named muscle wisdom.

Muscle spindles are the main source of afferent excitation from the periphery, and spindle afferent activity declines during muscle contraction (Hagbarth and Macefield 1995). Blockade of spindle afferents reduces motoneuron firing rates, indicating that fusimotor-mediated spindle facilitation is necessary to reach peak rates (Gandevia, Allen, and McKenzie 1995). Although spindle afferent activity declines during static contractions, resulting in a disfacilitation of α motoneurons (figure 15.6), the situation could be different during dynamic contractions. During concentric work, fusimotor drive is insufficient to compensate for muscle shortening, and spindle afferents become silent. In eccentric contractions, on the other hand, the stretch sensitivity of muscle spindles increases, which increases the activity in spindle afferents and increases muscle stiffness (Gandevia 1998).

There are conflicting results regarding the effect of fatigue on stretch reflex sensitivity. Gollhofer et al. (1987) revealed both increased and decreased reflex amplitudes, whereas Hortobagyi, Lambert, and Kroll (1991) found an increase and Nicol et al. (1996) found a decrease in stretch reflex sensitivity. The reason for this may be that the effect depends on the fatigue level and the type of task requested (Nicol et al., 1996).

## Factors Affecting Generation of an Endplate Potential

In normal rested muscle, each action potential launched from the motoneuron results in a corresponding action potential and a twitch in all muscle fibers in the muscle unit. During fatiguing contractions, this may not always be so. Changes in the muscle fibers may make them unable to react properly to the impulse trains from their motoneurons, but this inability also can result from faulty transmission of the nerve signals from the motoneuron to the muscle fiber. Such failure could be a propagation failure, preventing the signals from reaching the motor endplate, or it could be located in the motor endplate itself.

### *Motor Axon Propagation Failure—Branch Failure*

It has been known for quite some time that action potentials can fail to propagate along each branch of a motor axon (Barron and Matthews 1935), and that the failure is likely to occur at axonal branching points. Experiments have suggested that shifts in the ionic balance over the axon membrane could be responsible (figure 15.7). Such ionic perturbations are more likely to occur in smaller axons, because the surface-to-volume ratio is larger in thin axons than in thicker ones, and they are more likely to occur during high-frequency stimulation (see discussion by Sieck and Prakash 1995). Consequently, branch failure is believed to start peripherally and spread gradually in a centripetal direction.

The most peripheral branch points are the terminal branchings of the axon as it approaches each

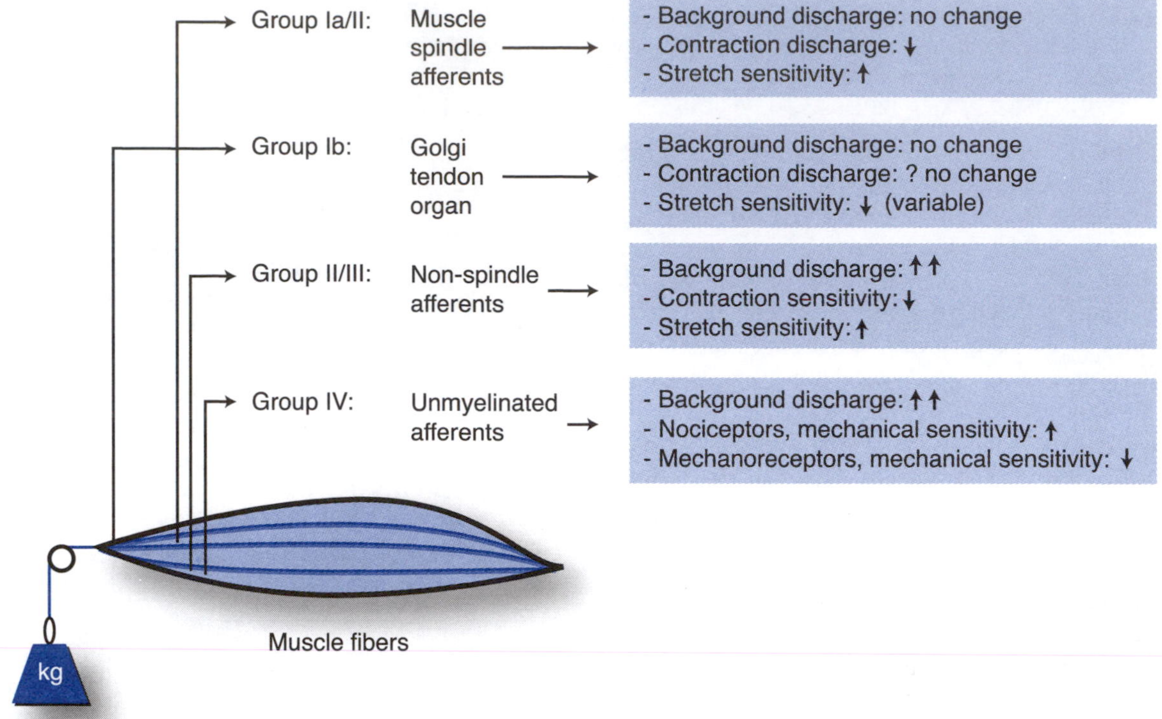


**Figure 15.6** Schematic representation of likely changes in muscle afferent input to the spinal cord during sustained maximal voluntary contractions of a limb muscle.

Reprinted, by permission, from S.C. Gandevia, 1998, "Neural control in human muscle fatigue," *Acta Physiologica Scandinavica* 162:275-283.

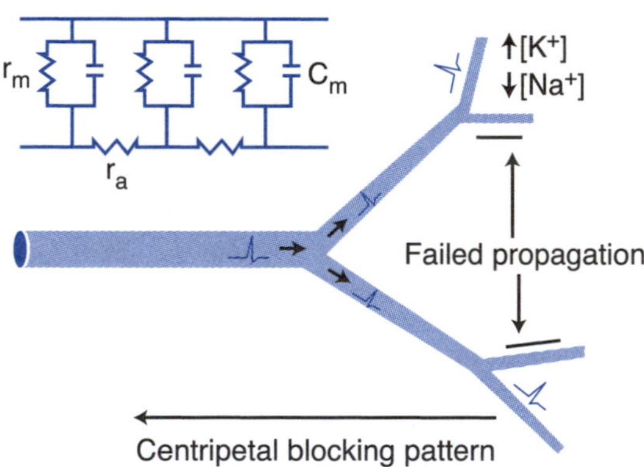


**Figure 15.7** Potential mechanisms underlying axonal branch point propagation failure include the following: changes in axon geometry, such as higher axial resistivity ($r_a$) and lower membrane capacitance ($C_m$) of daughter branches, and increased membrane refractoriness at smaller branches attributable to perineural accumulation of potassium ion ($K^+$) caused by repetitive stimulation. As a result, axonal propagation failure will start in small, distal branches and spread in a centripetal direction. $r_m$ = membrane resistivity; For further details, see original publication.

Reprinted, by permission, from G.C. Sieck and Y.S. Prakash, 1995, "Fatigue at the neuromuscular junction—Branch point vs presynaptic vs postsynaptic mechanisms," *Advances in Experimental Medicine and Biology* 384:90.

muscle fiber (cf. figure 3.9). For each individual axonal branch, the effect of the ionic perturbations on action potential propagation is probably transient, meaning that an affected synaptic terminal within the motor endplate is not completely silenced, but that occasional action potentials are missing, decreasing the firing rate for that particular terminal at the level of the muscle fiber and reducing the safety factor of the motor endplate in question. We shall return to this when discussing fatigue at the neuromuscular junction.

A branch failure where the axon divides into separate branches for each muscle fiber will inevitably reduce impulse frequency to the muscle fibers innervated by the affected branches and will correspondingly reduce the force contribution of those muscle fibers (cf. figures 3.10 and 15.3). Experimental evidence for such failure has been reported (Sieck and Prakash 1995).

## *Fatigue at the Neuromuscular Junction*

The successful transmission of an action potential in a motor endplate depends on the size of the combined EPSPs (excitatory postsynaptic potentials, see chapter 4) of all the synaptic terminals of the endplate in relation to the threshold for muscle fiber action potential generation. Normally, the combined EPSPs are well above the threshold, which is why the relation between the two has been named the safety factor of the endplate. Morphologically, the safety factor is reflected in the size of the endplate in relation to the diameter of the muscle fiber, but other differences also contribute (Wærhaug and Lømo 1994).

Theoretically, a transmission failure at the neuromuscular junction may be the result of each of the following factors or a combination of them: a reduced number of discharging synaptic terminals, a reduced number of synaptic vesicles in the discharging terminals, a reduced amount of acetylcholine (ACh) in the vesicles, or a reduced sensitivity of the ACh receptors (AChRs) in the postsynaptic membrane (figure 15.8).

A reduced number of discharging synaptic terminals in a motor endplate can result from a branch failure within the terminal arborization of the axon, as outlined previously. To a certain extent, this can be handled by the safety factor. The safety factor for type II muscle fibers has been reported to be larger than for type I fibers, but for both fiber types it is large enough to ensure reliable neuromuscular transmission under nonfatigued conditions. On repetitive stimulation, however, the safety factor for type I fibers remains unchanged, whereas that of type II fibers rapidly declines (Gertler and Robbins 1978). The reason for this difference is so far speculative. It is also unknown how and to what extent the ionic perturbations created by repeated muscle fiber action potentials influence propagation in the neighboring terminal branches of the axon.

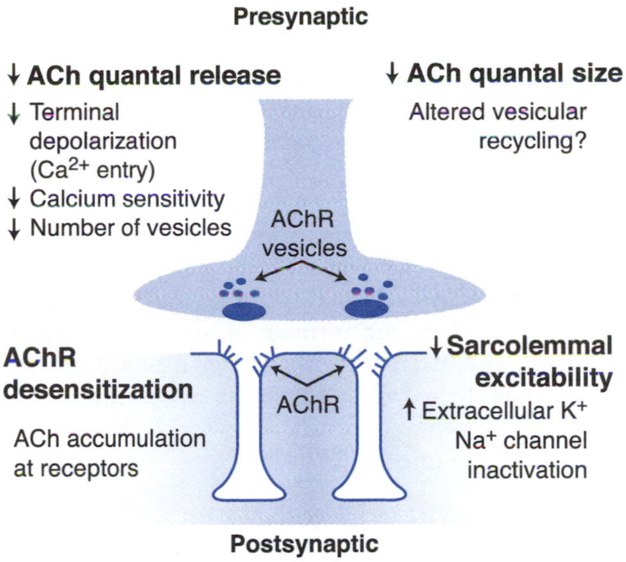

During postnatal growth, both muscle fibers and motor endplates increase in size, the latter relatively more than the former, thus increasing the size of the endplates in relation to muscle fiber size (Wærhaug 1992). In accordance with this, newborn rats have been found to be more susceptible to neuromuscular transmission failure than adult animals (Sieck and Prakash 1995). In line with this, the relative increase in the size of endplates seen in animals subjected to training (Wærhaug, Dahl, and Kardel 1992) can be taken as an indication that training confers resistance to neuromuscular transmission failure. For natural reasons, corresponding findings in humans are lacking.

A reduced release of ACh from each active synaptic terminal may be attributable to a reduction in the number of synaptic vesicles released per action potential or to a reduced quantal size (i.e., a reduced number of ACh molecules per vesicle). Several authors have shown that the number of synaptic vesicles is reduced by repetitive stimulation (Korneliussen, Barstad, and Lilleheil 1972; Kurihara

**Figure 15.8** Possible sites where neuromuscular transmission failure may occur. See text for details. ACh = acetylcholine.

Reprinted, by permission, from G.C. Sieck and Y.S. Prakash, 1995, "Fatigue at the neuromuxcular junction—Branch point vs presynaptic vs postsynaptic mechanisms," *Advances In Experimental Medicine and Biology* 384:92. © Kluwer Academic/Plenum Publishers.

and Brooks 1975). This has been suggested to be attributable to impaired vesicle recycling (Lentz and Chester 1982), whereas Boyne, Bohan, and Williams (1975) concluded that both reduced vesicle number and reduced vesicle content contribute to transmission failure.

Signal transduction depends just as much on the state and number of the appropriate receptor as on the signal molecules. In the case of neuromuscular transmission, a desensitization of the AChRs has been shown to occur during prolonged exposure to ACh (Katz and Thesleff 1957). Such desensitization is believed to involve a transition of ACh-regulated ion channels to a nonconducting state in the continued presence of ACh.

### FOR ADDITIONAL READING

For further discussion of neuromuscular transmission failure, see Sieck and Prakash (1995).

## Summary

Muscle fatigue can be categorized as central or peripheral, but it should be acknowledged that the words *central* and *peripheral* are not used in the anatomically correct way. The main reasons for this are that the definition of central fatigue is operational, and that factors leading to central fatigue may have their origin in the periphery. Central fatigue is said to be present when maximum voluntary force is less than maximal evocable force. The interpolated twitch technique is the main method to separate central from peripheral fatigue.

Transcranial magnetic stimulation has revealed the presence of decreased intracortical facilitation after fatiguing exercise, and the changes are confined to the somatotopically relevant parts of the motor cortex. The changes, however, could be secondary to changes in afferent activity to the motor cortex (e.g., supraspinal effects of increased activity in group III and IV muscle afferents, known to be sensitive to ionic and metabolic perturbations in the muscles).

For normal, rested muscle, the launching of an action potential from the motoneuron is the point of no return in muscle control. Thus, in nonfatigued muscle, any change in the balance between excitatory and inhibitory inputs to the motoneurons, in favor of the latter, will reduce the impulse frequency to the muscle units and tend to lower their force output. This can be attributed to reduced supraspinal drive or reduced muscle spindle afferent activity. Theoretically, failure of propagation of the nerve action potential (branch failure) and neuromuscular transmission failure also can give rise to central fatigue, but the practical importance of this is probably limited, at least in daily life.

# Fatigue Attributable to Processes Beyond the Neuromuscular Junction—Peripheral Fatigue

This is the kind of fatigue that is not alleviated by direct stimulation of the muscle at its motor point. The muscle fibers are unable to respond as well as in a nonfatigued situation, despite optimal stimulation by its motor nerve. This kind of fatigue can be caused by factors affecting the propagation of muscle action potentials, factors affecting the release of calcium from the sarcoplasmic reticulum, or factors affecting the response of individual myofibrils to increased levels of cytosolic calcium.

## Failure in the Propagation of the Muscle Fiber Action Potential—High-Frequency Fatigue

Although the firing threshold of the muscle fiber membrane is exceeded, there can be a failure in the propagation of the muscle fiber action potential along the surface membrane and into the T-tubules. During fatigue, the level of excitation and the conduction velocity are reduced along the membrane of the muscle fiber (Arendt-Nielsen, Mills, and Forster 1989; Juel 1988; Milner-Brown and Miller 1986). At least part of this is caused by an increase in interstitial [K+] (potassium ion concentration) (Sejersted and Sjøgaard 2000; Sjøgaard and McComas 1995). Furthermore, the integrated electromyographic activity tends to increase during fatiguing contractions at constant force because of recruitment of motor units with higher threshold and firing rates. This counteracts the decreasing force output from low threshold units (Bigland-Ritchie and Woods 1984).

Indirect evidence indicates that failure in the propagation of muscle action potentials along the T-tubules plays an important role (Bianchi and Narayan 1982), and this has been attributed to accumulation of $K^+$ in the T-tubule lumen (D.A. Jones 1996; Sjøgaard and McComas 1995). Theoretically, the T-tubules would be expected to be the more susceptible to ionic perturbations because of repetitive stimulation. First, the lumen is very small in relation to the T-tubule membrane. Second, the density of $Na^+$–$K^+$ ATPase molecules (the sodium–potassium pump) has been claimed to be lower in the T-tubule membrane than in the surface membrane (Fambrough et al. 1987; Venosa and Horowicz 1981), and the more important effects of an increased extracellular $K^+$ concentration have been considered likely to involve excitation–contraction coupling via the T-tubules, rather than the excitability of the surface membrane (Sjøgaard and McComas 1995). It is a problem, however, that a great deal of the information about the density of ion channels and ion transporters in the T-tubule membrane comes from studies of amphibian muscles, and it is not known whether the information is valid for mammalian muscles as well (Sejersted and Sjøgaard 2000). Even if substantial perturbations of the ion balance across the T-tubule membrane really occur, it is unclear to what extent these changes affect tubular function (Sejersted and Sjøgaard 2000). However, human studies have provided support for the possible role of T-tubules in high-frequency fatigue (HFF), because a kind of fatigue resembling HFF results from stimulation of human tibialis anterior muscle under ischemic conditions. Muscles held at shortened length during stimulation fatigued more than muscles held at optimum length. Both groups recovered substantially after cessation of stimulation, although still ischemic, but the muscles fatigued in the shortened position showed further recovery when returned to optimum length (Sacco et al. 1994). These results are consistent with restricted movement and consequent accumulation of $K^+$ in the lumen of T-tubules, possibly because of narrowing of their openings to the surface of the fibers when in the shortened position (D.A. Jones 1996).

Indirect support for a propagation failure in the T-tubules comes from the demonstration in freeze-clamped fatigued muscle fibers that centrally located myofibrils had a wavy appearance, indicating that they were less activated and thus longer than more peripherally located myofibrils (Garcia et al. 1991; Howell and Oetliker 1987).

As indicated by its name, HFF is induced by high-frequency stimulation, and it has been questioned whether HFF is a physiologically relevant and "normal" type of fatigue, mainly because the frequencies necessary to induce HFF are higher than those thought to occur during normal sustained isometric contractions (D.A. Jones 1996). On a theoretical basis, it has been calculated that $Na^+$–$K^+$ ATPase is able to keep pace with the influx of sodium ions and efflux of potassium at excitation frequencies up to around 55 Hz (Nielsen and Clausen 2000).

### For Additional Reading

For a more general discussion of the role of $K^+$ and other ionic perturbations in muscle fatigue, see Sjøgaard (1990), Nielsen and Clausen (2000), and Sejersted and Sjøgaard (2000). Regarding the use of microdialysis to measure muscle interstitial $K^+$ concentration during rest and exercise, see S. Green et al. (1999).

## Fatigue at the T-Tubule–Sarcoplasmic Reticulum Junction—Low-Frequency Fatigue

In 1977, Richard Edwards and coworkers (R.H.T. Edwards et al. 1977) described a type of muscle fatigue that they called low-frequency fatigue (LFF). It was characterized by being most pronounced at low frequencies (figure 15.9) and by having a very

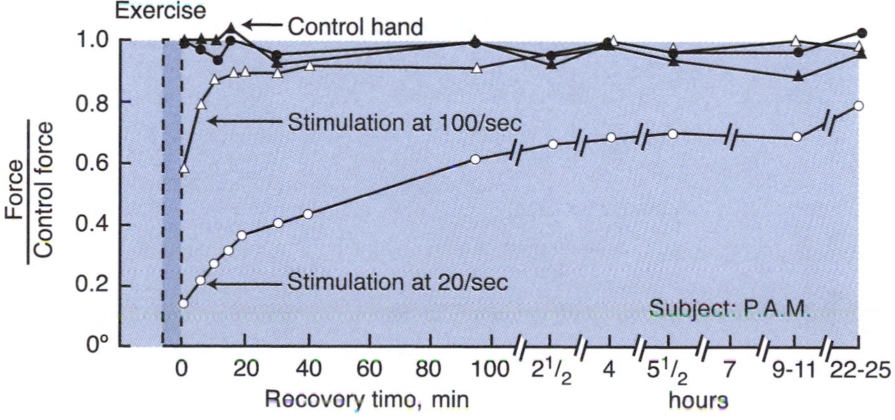


**Figure 15.9** Recovery from fatigue in human m. adductor pollicis caused by motor nerve stimulation at 20 Hz for 50 s alternating with 50-s recovery intervals. Muscle blood supply was arrested during stimulation but not during recovery intervals. The force during high-frequency stimulation (100 Hz for 1 s) increased in the course of minutes to values approaching those in the control hand, whereas the force during 20 Hz stimulation for 1 s took many hours to recover.

Adapted, by permission, from R.H.T. Edwards, et al., 1990, "Fatigue of long duration in human skeletal muscle after exercise," *Journal of Physiology* 272:772.

slow recovery—in fact taking hours, maybe days, despite absence of any signs of metabolic or electrical disturbance in the muscle (Baker et al. 1993). This is why "long-lasting fatigue" has been suggested as an alternative name (Sejersted and Sjøgaard 2000). The slow recovery is in sharp contrast to the very fast recovery, in the order of minutes, seen after HFF.

It has been acknowledged for a long time (see D.A. Jones 1996 for a brief history of LFF) that reduced levels of cytosolic calcium could explain LFF, but it could not be excluded that a reduced calcium sensitivity of troponin was the cause. This point was impossible to resolve until sufficiently sensitive methods for measuring intracellular calcium became available. The slow recovery indicated that some kind of repair process involving protein synthesis was necessary, and this suspicion was strengthened by the fact that LFF is common after eccentric contractions (Newham et al. 1983), a type of muscular activity known to cause substantial damage (see chapter 11).

Over the years, many different methods, each with advantages and disadvantages, have been used to measure the cytosolic calcium concentration. For an overview, see Westerblad et al. (2000). Westerblad, Duty, and Allen (1993) were able to show that LFF was accompanied by a uniform reduction of cytosolic calcium across the fatigued muscle fiber, whereas no change was seen in calcium sensitivity or in maximal calcium-activated tension, thus settling the previously mentioned uncertainty. It remained unknown, however, what was responsible for the reduced calcium levels. Furthermore, although decreased cytosolic calcium has been shown during LFF, it may not be the only cause; a redistribution of sarcomere lengths also has been suspected to contribute to LFF (D.A. Jones 1996). Binder-Macleod and Russ (1999) presented evidence in support of two factors contributing to LFF: one rapidly recovering, metabolite-dependent mechanism, and a slow-developing, slow-recovering factor, which is not the result of any metabolite buildup. Because its effect is transient only, the question is whether the rapidly recovering factor should be regarded a genuine part of the mechanism behind the characteristically long-lasting LFF.

Among the mechanisms considered likely to explain the reduced cytosolic calcium ion ($Ca^{2+}$) concentrations during LFF are elevated $Ca^{2+}$ concentrations—a theory often referred to as the calcium paradox (Westerblad et al. 2000). Other possibilities include perturbations of the metabolic microenvironment or subtle structural derangement within the triadic space, and modification of key molecules by reactive oxygen species. The difference in fatigue at high and low frequencies can be explained by

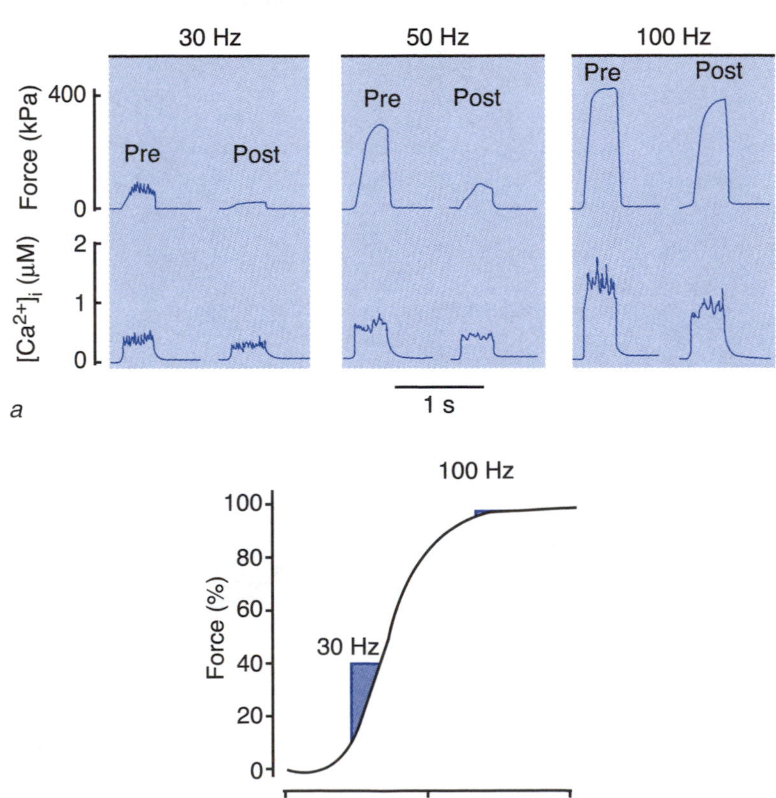


**Figure 15.10** *(a)* Force and cytosolic calcium concentration [$Ca^{++}$] from single mouse muscle fibers before (pre) and after (post) induction of low-frequency fatigue. In the fatigued state, cytosolic [$Ca^{++}$] is reduced at all frequencies, while the force is reduced only at the lower frequencies. *(b)* The force/$Ca^{++}$ concentration curve explains why the decrease in cytosolic [$Ca^{++}$], as seen during low-frequency fatigue, causes a more pronounced force reduction at low frequencies than at high frequencies.

Reprinted, by permission, from *European Journal of Applied Physiology*, Functional significance of $Ca^{2+}$ in long-lasting fatigue of skeletal muscle, H. Westerblad, et al., 83, 170, 2000, © Springer-Verlag.

figure 15.10. At low frequencies, a small reduction in cytosolic calcium concentration greatly reduces force output because we are on the steep part of the curve. At high frequencies, on the other hand, the change in force output for a comparable reduction in calcium concentration is negligible. For further details and discussion, see Favero (1999) and Westerblad et al. (2000).

## Myofibrillar Fatigue

Fatigue caused by inability of the muscle fiber to react properly to elevated cytosolic $Ca^{2+}$ concentration is called myofibrillar fatigue. In such cases, the force/$Ca^{2+}$ concentration curve is shifted to the right, but this has been shown only for type I fibers (figure 15.11). Under these circumstances, caffeine-induced elevation of cytosolic $Ca^{2+}$ concentration is unable to increase force (figure 15.12).

In normal, rested muscle, the force per sarcomere is proportional to the number of active cross-bridges. Theoretically, a reduced force output despite unchanged sarcomere length and cytosolic $Ca^{2+}$ concentration can be attributable to a reduced number of active cross-bridges or a reduced force per cross-bridge. The factors leading to such changes are metabolic in origin and therefore are discussed under that heading.

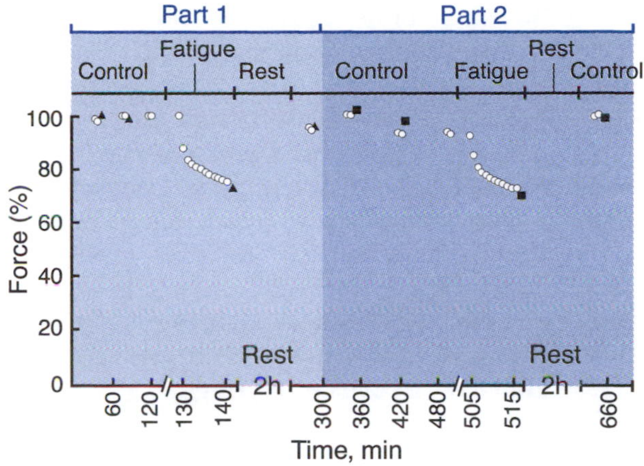


**Figure 15.12** Relative tetanic force in control and moderately fatigued single frog muscle fibers. ○ = isometric tetanus; ▲ = contracture responses to 15 mM caffeine; ■ = contracture responses to 3 mM caffeine. Note the lack of effect of caffeine during fatigue.

Reprinted, by permission, from K.A.P Edman, 1995, "Myofibrillar fatigue versus failure of activation," *Advances in Experimental Medicine and Biology* 384:32. © Kluwer Academic/Plenum Publishers.

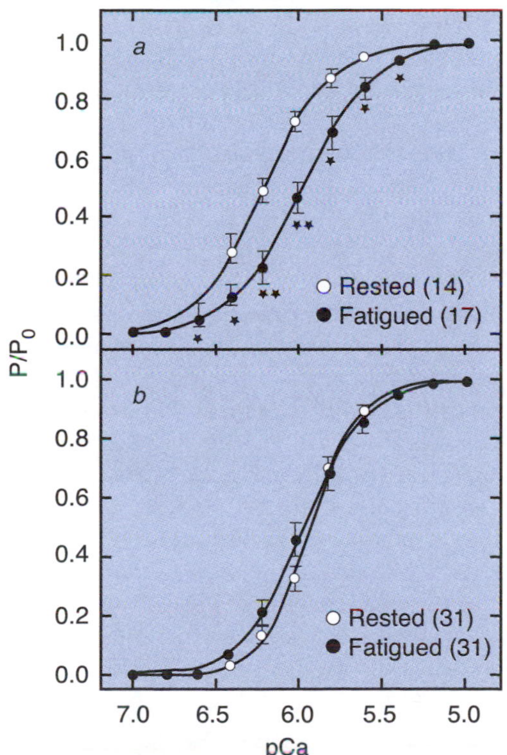


**Figure 15.11** The relation between force, expressed as P/Po, and calcium concentration, expressed as $pCa^{++}$, in soleus *(a)* and extensor digitorum longus (EDL) muscle fibers *(b)*, before and after induction of fatigue. The curve is shifted to the right for soleus fibers, but not for EDL.

Reprinted, by permission, from D. Danieli-Betto et al., "Effects of fatigue on sarcoplasmic reticulum and myofibrillar properties of rat single muscle fibers," *Journal of Applied Physiology* 89:891-898.

## Summary

Peripheral fatigue is the kind of fatigue that is not alleviated by direct stimulation of the muscle by the twitch interpolation technique. It can be caused by factors affecting the propagation of muscle action potentials along the surface membrane and into the T-tubules, factors affecting the release of calcium from the sarcoplasmic reticulum, or factors affecting the response of individual myofibrils to increased levels of cytosolic calcium.

Propagation failure is mostly seen after unphysiologically high stimulation frequencies and seems to be of limited importance in daily life. Such HFF is rapidly reversed when the fatiguing activity is terminated, and it is believed to be caused by ionic perturbations, in particular a loss of $K^+$ from the cytosol to the extracellular space. Ionic perturbations in the T-tubule are often claimed to be responsible for failure of T-tubule function, but direct evidence is lacking.

Failure in the calcium release process gives rise to a characteristic force deficit, which is more pronounced at low than at high stimulation frequencies, hence the name LFF. In contrast to HFF, LFF shows a very slow recovery, indicating that molecules critically involved in the calcium release process are seriously damaged and in need of replacement. The mechanism behind this process is not known yet, but elevated levels of calcium are implicated, a phenomenon referred to as the calcium paradox.

Inability of the individual myofibrils to respond properly to increased levels of cytosolic calcium is called myofibrillar fatigue. It can be caused by a decreased number of active cross-bridges or a decreased force per active cross-bridge.

## The Role of Metabolic Factors in Muscle Fatigue

In lay opinion, metabolic factors play a rather prominent role in muscle fatigue, in particular the accumulation of lactate because of anaerobic processes in the muscles. In addition, the feeling of fatigue often is blamed on a depletion of energy stores, as indicated by the expression "feeling empty." To a certain extent and under given circumstances, both accumulation of lactate and depletion of energy stores are of importance, although not at all as often as believed. In fact, the role of metabolic factors in muscle fatigue is still debated. When present, however, metabolic factors affect muscle function in several ways, by compromising adenosine triphosphate (ATP) supply or by interfering with excitation–contraction coupling or cross-bridge dynamics (Vøllestad 1995).

Historically, the needle biopsy technique (Bergström 1962) has been a major method for studying metabolic factors in muscle fatigue, but its time resolution is poor, and the samples of muscle tissue obtained by this technique are small and few. Furthermore, unless single fiber analysis is possible, the end result is a mean of individual fiber values. Consequently, metabolic measurements based on the biopsy technique should be interpreted with caution (Vøllestad 1995).

In phosphorus-31 nuclear **magnetic resonance spectroscopy (MRS),** the time resolution is far better, but there are still disadvantages. First, the fact that the activity has to be performed within the restricted space of the apparatus restricts the experimental design. Second, the spatial resolution is lower than one could want. It is impossible to restrict the sample volume. The method does not recognize borders between individual muscles and can pick up signals from more than one muscle or less than one entire muscle (Miller et al. 1995), depending on the local anatomy. The main advantage of the method is that it provides continuous biochemical data, mainly about phosphorus compounds during the contractile task. Figure 15.13 shows that the phosphocreatine (PCr) peak and the peak for inorganic phosphate ($P_i$) gradually decrease and increases, respectively, in an almost reciprocal way during the contraction. During the subsequent recovery, these changes are gradually reversed. We shall return to this later.

### The Role of Lactate

The association between fatigue and lactic acid has been known since the beginning of the 20th century (Fletcher and Hopkins 1907). At work rates above 50% to 60% of maximal aerobic capacity, blood lactate concentration increases and generally averages 10 to 20 mM following short-duration maximal exercise. Following intermittent maximal exercise, the increase can be substantially higher. Thus, Osnes and Hermansen (1972) reported blood lactate concentrations as high as 32 mM after intermittent 40- to 60-s maximal exercise bouts. In an accompanying article (Hermansen and Osnes, 1972), however, they showed that whereas muscle pH decreased to a certain level only, blood pH continued to decrease following repeated intermittent bouts of all-out exercise. The obvious conclusion is that blood lactate concentration does not accurately reflect the conditions in the muscle. In general, however, mean muscle lactate concentration is in the order of 10 mM higher than blood lactate (Fitts 1994), and the intracellular concentration in the least oxidative fibers is probably even higher. This provides a concentration gradient that serves as the driving force for the diffusion of lactate from the muscle cell via muscle interstitial fluid to the blood. For an update on lactate transporters (monocarboxylate transporters), see chapters 8 and 11.

Thus, it is beyond doubt that high lactate levels occur during fatiguing activities, but the question is whether they cause fatigue. The answer is possibly a cautious yes, but the relative importance of lactate depends heavily on the type of activity performed. If the activity requires a high glycolytic rate, the decline in force is highly correlated with the increase in muscle lactate (Fitts 1994). On the other hand, if glycolysis is compromised because of myophosphorylase deficiency, fatigue develops faster than in normal subjects even though no acidosis occurs (D.L. Allen et al. 1995). This is the ultimate proof that lactic acidosis is not an indispensable factor in muscle fatigue.

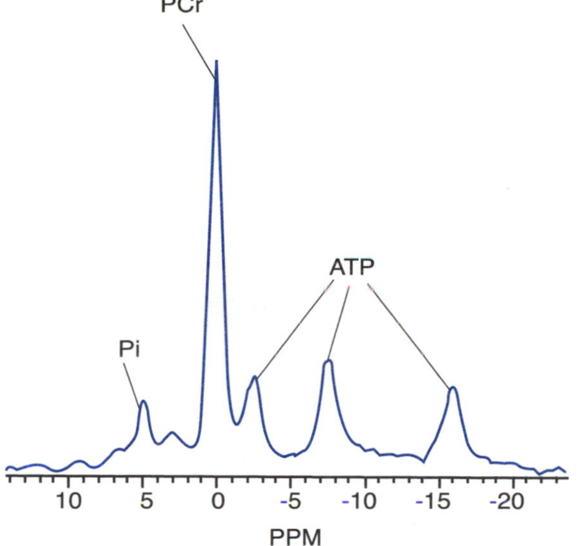


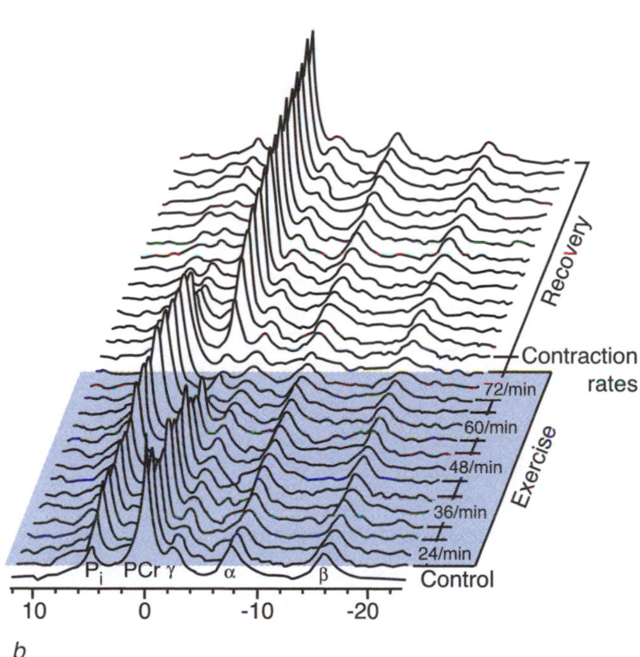


**Figure 15.13** *(a)* Phosphorus-31 magnetic resonance spectrum from human m. tibialis anterior. The peaks originating from inorganic phosphate ($P_i$), phosphocreatine (PCr), and the three peaks of ATP are clearly seen. *(b)* Stacked plot of nuclear magnetic resonance spectra obtained during exercise and recovery. Note almost inverse changes in $P_i$ and PCr peaks.

*(a)* Reprinted, by permission, from R.G. Miller, et al., 1995, "Mechanism of human muscle fatigue—Quantitating the contribution of metabolic factors and activation impairment," *Advances in Experimental Medicine and Biology* 384:196. © Kluwer Academic/Plenum Publishers.

*(b)* Reprinted, by permission, from R.G. Miller, et al., 1995, "The fatigue of rapid repetitive movements," *Neurology* 43:755-761.

It has become increasingly clear, however, that the association between lactic acid and fatigue is mainly indirect and—if present—is mediated through the resulting decrease in pH. The lactate anion itself is first and foremost a source of energy not yet transferred to ATP, and it is the only important means of energy transfer between individual muscle cells (cf. figure 2.10).

In theory, acidic pH can affect muscle force production in several ways. To start with, optimum pH of cytosolic proteins is generally around resting pH values, and lowering the pH to values seen during fatigue affects key proteins to varying degrees, depending on their pH susceptibility. This is particularly evident in the case of enzymes, and even small changes in cytosolic pH can have a large impact on enzyme activity and thus on cellular metabolism (D.L. Allen et al. 1995). Phosphorylase, adenylate cyclase, and phosphofructokinase are inhibited, directly or indirectly, by low pH (Chasiotis 1983).

Low pH also affects intracellular membrane proteins, most importantly the calcium release channel in the sarcoplasmic reticulum (the ryanodine receptor, see chapter 3) and the $Ca^{2+}$ ATPase, responsible for the calcium reuptake process. Each contributes to characteristic features of muscle fatigue, the former to a reduction in cytosolic calcium and the latter to a slowing of relaxation.

The association of low intracellular pH and fatigue, however, is based on experiments at unphysiologically low temperatures, and studies at temperatures only slightly lower than physiological indicate that pH is less important than we used to believe (Pate et al. 1995).

In fatigued muscle, the osmotic pressure increases, causing water to enter muscle cells. Such increased water content is highly correlated with increases in lactate content (Bergström, Guarnieri, and Hultman 1971; Sahlin et al. 1976, 1978), although not necessarily caused by it. Theoretically, such swelling could affect the lateral spacing of contractile filaments and reduce peak force, but this does not seem to be a major cause of fatigue (Fitts 1994). It also has been discussed whether the low pH during fatigue affects the binding of $Ca^{2+}$ to troponin, but this remains a controversial point. It has been convincingly demonstrated, however, that the force/p$Ca^{2+}$ curve is shifted to the right during fatigue, meaning that—for one reason or another—the level of cytosolic calcium needed to obtain a certain force level is higher during fatigue than in the rested muscle (Fitts 1994). A direct effect of low pH on the contractile proteins also has been implicated (Fabiato and Fabiato 1978), possibly involving a reduction of the number of cross-bridges and/or a decrease in the force per cross-bridge. The number of attached cross-bridges can be estimated by measuring fiber instantaneous stiffness (figure 15.14), which shows that the primary effect of low pH is decreased force per cross-bridge (Edman 1995; Edman and Lou 1990). Differences could exist between fiber types, however, because low pH has been claimed to decrease the number of active cross-bridges in fast- but not in slow-twitch muscle fibers (Metzger and Moss 1990).

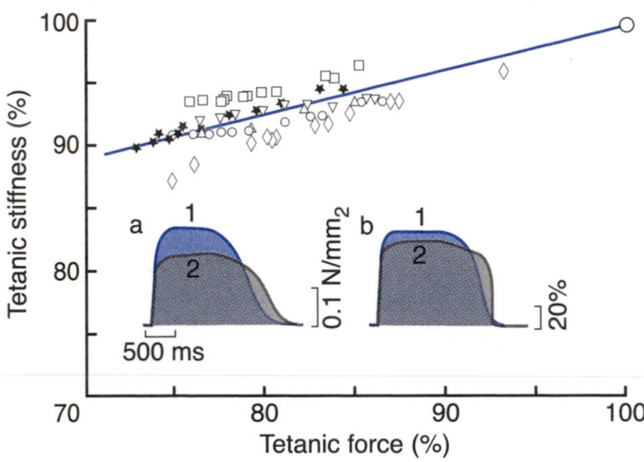


**Figure 15.14** The relation between maximum tetanic stiffness and maximum tetanic force in control and during developing fatigue in six single frog muscle fibers. Data are normalized with respect to maximum force and maximum stiffness under control conditions in the respective fibers. Values for individual fibers are shown by identical symbols. Both regression line and inserts show that the reductions in force (a) are larger than the reductions in stiffness (b) during fatigue (2) in relation to control (1).

Reprinted, by permission, from K.A. Edman and F. Lou, 1990, "Changes in force and stiffness induced by fatigue and intracellular acidification in frog muscle fibers," *Journal of Physiology* 424:133-149.

Fatigue is often accompanied by a reduced maximal shortening speed. Some experiments with skinned fibers have indicated that this may be entirely attributable to the acidic pH (Cooke et al. 1988; Metzger and Moss 1987), whereas other experiments indicate that additional factors are equally important (Westerblad and Lännergren 1988). Again, the disagreement may be attributable to differences between species and fiber types (D.L. Allen et al. 1995). In general, slowing of contractile characteristics seems to depend on the degree of metabolic perturbation. In the absence of increased levels of lactate or $P_i$, no slowing is seen. On the contrary, an increased speed is seen. Thus, in human muscle, low-force isometric contractions have been shown to gradually decrease half relaxation time, whereas contractile speed was unaltered (Vøllestad, Sejersted, and Saugen 1997). If present, the slowing of relaxation could be caused by an inhibition of the calcium reuptake mechanism in the sarcoplasmic reticulum or slowing of the detachment rate of cross-bridges.

### For Additional Reading

For further details and discussion of the role of lactic acid in muscle fatigue, see Fitts (1994).

## The Role of Energy Stores

Limitation in energy supply is a classic hypothesis in muscle fatigue, and depending on the type of activity, both depletion of muscle phosphagens and depletion of muscle glycogen can be limiting factors. Within 10 s of exercise, the maximal power output decreases considerably and coincides with depletion of PCr (Sahlin, Tonkonogi, and Söderland 1998). During recovery, maximal force is restored with a similar time course as the resynthesis of PCr (figure 15.15). In accordance with this, cell PCr has been regarded as a potential limiting factor, but it remains a matter of discussion whether the relation between low PCr concentration and fatigue is a causative one (Fitts 1994). It has been claimed, however, that dietary supplementation of creatine increases performance in high-intensity exercise (Sahlin, Tonkonogi, and Söderlund 1998; Skare, Skadberg, and Wisnes 2001; Valvik, unpublished data), but the results so far are rather conflicting. The mechanism behind the alleged effect is also elusive, but increased muscle glycogen content (Op't et al., 2001; Wojtaszewski et al. 1999) and increased growth hormone secretion (Schedel et al. 2000) have been reported. We should emphasize that the evaluation of a possible effect of dietary supplements depends critically on the purity of the supplement in question.

Several studies have shown that there is a temporal correlation between depletion of muscle intracellular glycogen and exhaustion, but again the question of causal relationship is unresolved. At work rates between 65% and 86% of $\dot{V}O_2$max, muscle fatigue is highly correlated with depletion of muscle glycogen. In the 1960s it was shown that resting muscle glycogen and time to exhaustion (figure 15.16) could be increased by a diet rich in carbohydrates (Bergström et al. 1967; Karlsson and Saltin 1971).

The rate of glycogen depletion differs between muscle fiber types. In submaximal exercise (figure 15.17), glycogen depletion progresses in the se-

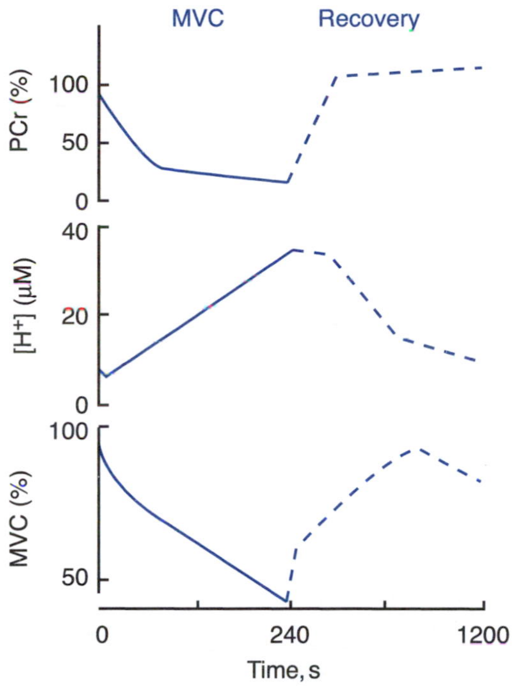


**Figure 15.15** Changes in phosphocreatine (PCr), hydrogen ion concentration [$H^+$], and maximal voluntary contraction (MVC) during and after 4 min of sustained maximal contraction.

M. DeGroot, et al., 1993, "Dissociation of [H+] from fatigue in human muscle detected by high time resolution 31P-NMR," *Muscle and Nerve* 16:91-98. Reprinted by permission of John Wiley & Sons, Inc.

quence type I, type IIA, and IIX (formerly IIB, see chapter 3), and at the point of exhaustion the glycogen content of type I and IIA is reduced to less than 10% of the resting values (Vøllestad, Vaage, and Hermansen 1984). In high-intensity exercise, on the other hand, type IIX fibers have been shown to be the first to lose glycogen (Gollnick et al. 1973), but glycogen depletion is not a major factor in fatigue in this type of activity. P.-O. Åstrand et al. (1986) showed that exercise sufficiently intense to produce fatigue within 5 min depletes muscle glycogen by less than 50%. It has been argued that maintaining blood glucose concentration is important to delay muscular exhaustion (figure 15.18) but is unable to offset it. In the end, exhaustion occurs despite maintained plasma glucose.

## For Additional Reading

For further discussion of the role of energy depletion in muscle fatigue, see Fitts (1994).

# The Role of Inorganic Phosphate

Data from both in vivo and in vitro experiments support the hypothesis that increased concentrations of $P_i$ contribute to muscle fatigue, but there is no consensus concerning the species of inorganic phosphate involved, that is, whether total $P_i$ or the diprotonated form $H_2PO_4^-$ is the more important

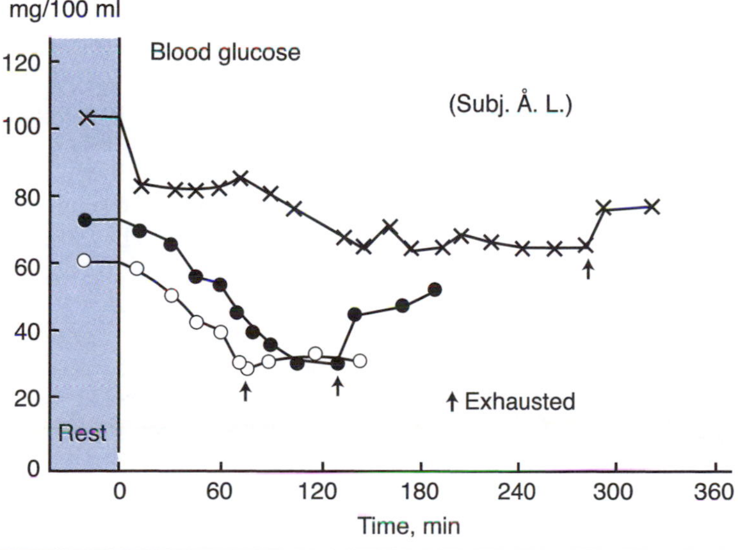


**Figure 15.16** The effect of different diets (× = carbohydrate diet; ● = mixed diet; ○ = fat + protein diet) on blood glucose and time to exhaustion.

Reprinted, by permission, from J. Bergström et al., 1967, "Diet, muscle glycogen and physical performance," *Acta Physiologica Scandinavica* 71:146.

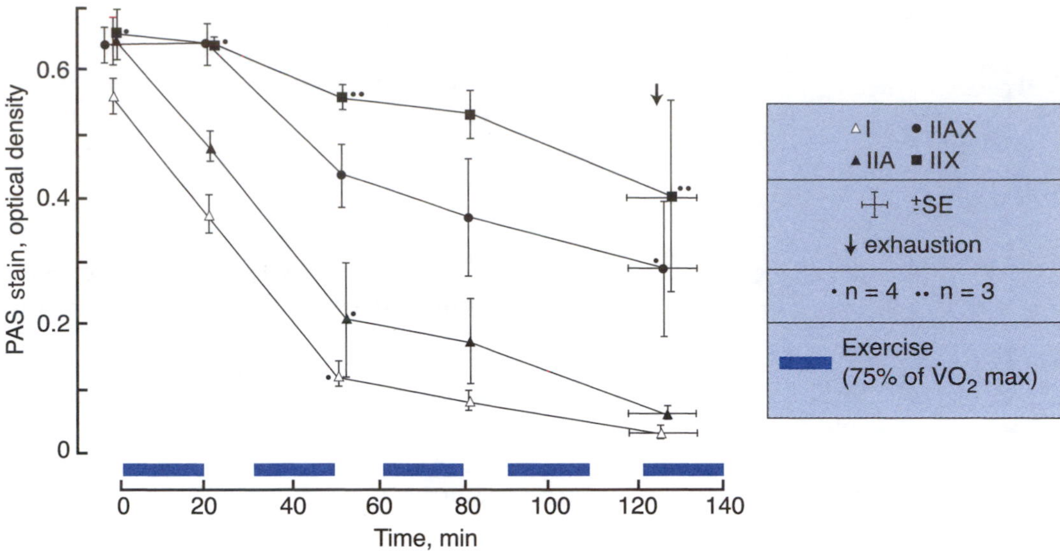


**Figure 15.17** Glycogen content, measured as optical density of periodic acid Schiff (PAS) stain, of different muscle fiber types at rest and during bouts of exhaustive exercise (75% of $\dot{V}O_2$max). The point of exhaustion coincided with an almost complete depletion of PAS stain in type I and type IIA fibers.

Reprinted, by permission, from N.K. Vøllestad, O. Vaage, and L. Hermansen, 1984, "Muscle glycogen depletion patterns in type I and subgroups of type II fibres during prolonged severe exercise in man," *Acta Physiologica Scandinavica* 122:437.

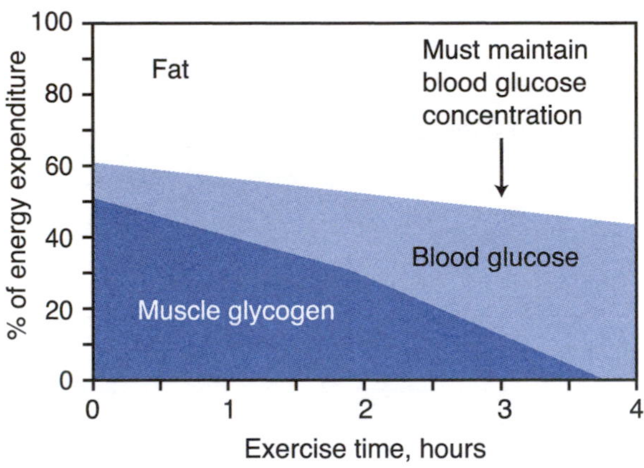


**Figure 15.18** Sources of energy during prolonged exercise at 70% of $\dot{V}O_2$max. Blood glucose becomes increasingly important as muscle glycogen is gradually depleted.

Reprinted, from E.R. Coyle, 1991, Carbohydrate metabolism and fatigue. In *Muscle fatigue: biochemical and physiological aspects* edited by G. Atlan, L. Beliveau, and P. Bouissou (Paris: Massin et Cie), 153-164. Reprinted by permission from E.R. Coyle.

(Fitts 1994). Experiments with skinned fibers seem to indicate that it may depend on the type of muscle fiber. The effect of $P_i$ is most pronounced in fast-twitch fibers where the decline in force correlates well with the level of $H_2PO_4^-$ (Nosek et al. 1990). The mechanism behind the effect of $P_i$ is elusive, but it has been shown that high $P_i$ shifts the force/pCa$^{++}$ curve to the right (cf. figure 15.10), thus exaggerating the effect of decreased cytosolic calcium during fatigue.

## For Additional Reading

For further discussion of the role of $P_i$ in muscle fatigue, see Fitts (1994) and Miller et al. (1995).

# The Role of Static Contractions

Fatigue in static exercise produces a sensation of discomfort and sometimes even pain. The disposition to subdue the feeling of fatigue is very different among individuals. Cooperative and well-motivated subjects can maintain a muscular contraction to the point of muscular fatigue, whereas others terminate the activation before reaching that point. The reason for the discomfort and fatigue during static contractions is thought to be the resultant muscle ischemia. With increasing contractile force, the intramuscular pressure approaches and eventually exceeds the perfusion pressure, thereby lowering and ultimately stopping muscle blood flow. There seems to be a linear relationship between intramuscular pressure and percentage of maximal isometric contraction force, but the slope of the relationship is likely to vary between muscles and individuals. Partial occlusion of human quadriceps muscle blood flow is likely to occur at 20% to 25% of maximal isometric contraction, and occlusion may be close to complete at 50% of maximal voluntary contraction (MVC; Fulco et al. 1995; Sejersted et al. 1984). According to R.H. Edwards, Hill, and McDonnell (1972), this can occur at even lower percentages.

Figure 15.19 summarizes data on maximal holding time in sustained isometric contractions at various levels of forces related to the maximal force of contraction (MFC). The maximal force can be maintained for only a few seconds, 50% of MFC can be maintained for about 1 min, but at 15% and below, it has been claimed that an isometric contraction could be maintained for more than 10 min and even up to hours (Rohmert 1968; Simonson and Lind 1971), although other studies have indicated that the upper limit for isometric contraction maintained for an "indefinite" time is below 10% of MFC (Björksten and Jonsson 1977). One must expect variation depending on the muscle groups studied, fiber types, and individual variations in local anatomy, maximal strength, and ability to endure discomfort. There are also indications that stronger subjects are less able to maintain isometric contractions at the same percentage of their maximal strength than are weaker subjects (Mundale 1970; Thorstensson 1976). From a mechanical point of view, an impediment of blood flow should be more dependent on the absolute than on the relative muscle forces; that is, the stronger subject will be handicapped by an impaired circulation at a force that is relatively low in percentage of the maximum. This has also been a common explanation for the female advantage often reported in fatigue resistance (Hicks, Kent-Braun, and Ditor 2001; Hunter and Enoka 2002).

Simonson and Lind (1971) showed that at 5% and 10% of MVC, the forearm blood flow increased to a steady state and dropped immediately after exercise. As mentioned previously, contractions at that level can be held for a very long time and the energy yield is most likely aerobic. At 20% to 30% of MVC, the blood flow increased steadily during the activity and increased further immediately after the end of contraction. Apparently, there had been a "blood flow debt," and the muscle fibers had had to pay part of the energy cost by anaerobic processes. Forces exceeding 30% of MVC decreased the blood flow and brought it to a complete arrest at about 70% of MVC.

In repeated isometric exercise, the combination of force and frequency of repetitions determines the length of time that the exercise can be endured. Figure 15.20 shows results from experiments in which the subjects performed rhythmic, maximal isometric contractions on a dynamometer in pace with a metronome (Molbech 1963). The force gradually decreased because of fatigue but then finally leveled off at a value that could be maintained for a

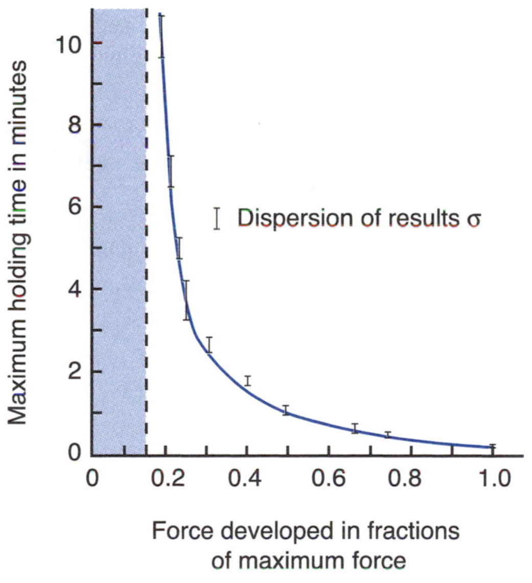


**Figure 15.19** Maximal work time plotted versus force in isometric contractions. Average of results obtained in studies of different muscle groups. Fifty percent of maximal load can be maintained for just 1 min, whereas loads less than 15% of maximum can be maintained almost "indefinitely."

Reprinted, by permission, from W. Rohmert, 1968, "Die Beziehung Zwischen Kraft und Ausdauer bei Statisher Muskelarbeit. Schriftenreihe Arbeitsmedizin," *Arbeitshygiene* 22:118.

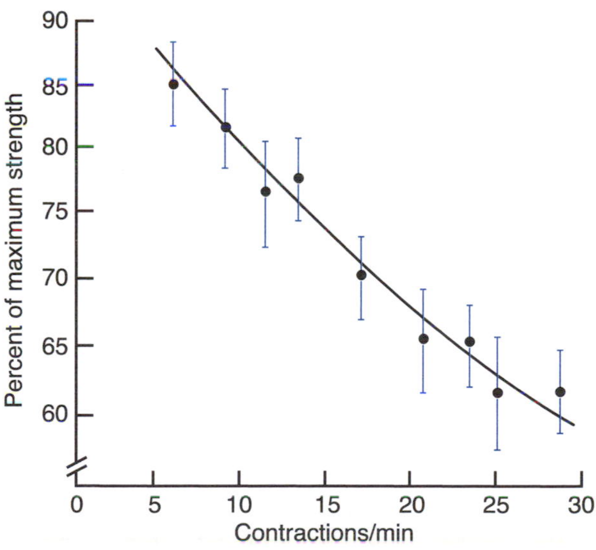


**Figure 15.20** Percentage of maximum isometric force that can be maintained in a steady state during rhythmic contractions, in relation to the number of contractions per minute. Points are averages ± standard error for finger muscles, hand muscles, arm muscles, and leg muscles combined.

From Molbech 1963.

long time. With 10 contractions per minute, about 80% of the maximal isometric strength could be applied without impairment. With 30 contractions per minute, the force was reduced to 60%. The values seemed to be independent of the size of the activated muscle group. The capacity to exercise isometrically appears to be optimal when the ratio between period of contraction time and period of rest is 1:2.

With the blood flow occluded by a cuff just before and during a vigorous contraction, there was no difference in the initial forces developed compared with the controls. This was the case with contraction forces from 60% to 70% and higher in Lind's studies; the blood flow was then occluded anyway. At lower forces, an occlusion of the local blood flow reduced the duration of maximal isometric contraction. It is not surprising that the reduction of endurance caused by external occlusion of blood flow is much more pronounced at lower forces than at forces closer to MVC. The normal occlusion of blood flow can be both a result of "nipping" of the arteries between moving and nonmoving tissues and an effect on the capillary flow attributable to the increased intramuscular pressure.

With a blood flow below the level required for an adequate supply of oxygen and removal of carbon dioxide, metabolites, and heat, there must inevitably be a shortage of oxygen and an accumulation of metabolites and heat. According to Ahlborg, Hendler, and Felig (1972), the lactate accumulation at exhaustion is maximal between 30% to 60% of MVC. The accumulation rate is linear with respect to contraction forces above 20% of MVC.

## Fatigability of Different Fiber Types

There are typical differences in the fatigability of human muscle fiber types. Type I fibers are the more fatigue resistant, type IIX the least, and type IIA intermediate. This is demonstrated very clearly by figure 15.21, which shows the correlation between percentage of type IIX fibers (formerly named type IIB or FTb fibers, see chapter 3) in human quadriceps muscle and peak torque decline after 50 maximal knee extensions (Tesch 1980a). There are rather large variations in peak torque decline for comparable percentages of IIX, however, especially for the lower percentages. This turns out to be attributable at least in part to differences in the percentage of type I fibers. The open circles in figure 15.21 denote individuals with less than 35% type II fibers, that is, with more than 65% type I. Part of the variation also

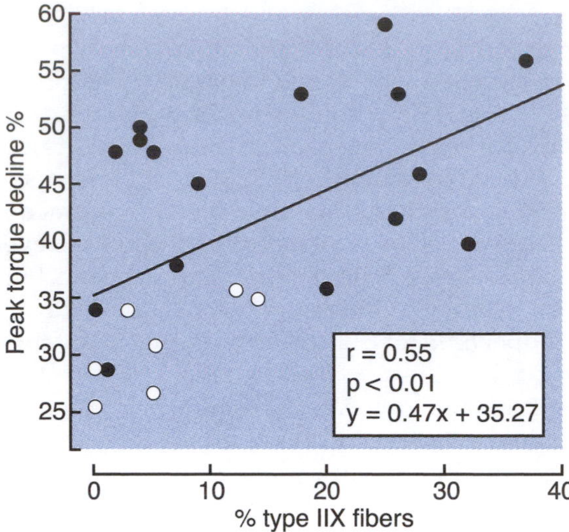


**Figure 15.21** The relationship between the percentage of type IIX fibers in m. vastus lateralis and the decline in peak torque after 50 maximal knee extensions. • = individuals with less than 35% type II fibers (type IIA and IIX combined).

Adapted, by permission, from P.A. Tesch, 1980, "Muscle fatigue in man with special reference to lactate accumulation during short term intense exercise," *Acta Physiologica Scandinavica* Supplementum 480:80.

could be attributable to the fact that differences between "identical" fiber types in different individuals are even larger than differences between different fiber types within one muscle. This is caused by differences in individuals' training status and can even apply to differences between muscles in different parts of the body in one individual, depending on his or her training level.

Basically, the different fatigue resistance of muscle fiber types derives from three characteristics: contractile economy, oxidative capacity, and relative training status, which among other things is subsequent to their rank in the recruitment hierarchy (cf. chapters 3 and 4). The contractile economy of a muscle fiber is based on its rate of cross-bridge cycling, which in turn is reflected in its ATPase activity and speed of shortening (Bárány 1967). This is an important reason why type I fibers are more economic than type II fibers, but differences in the energy consumption of active transport processes, like the $Ca^{2+}$ ATPase of the sarcoplasmic reticulum, also contribute.

The relative oxidative potential of the different muscle fiber types is type I > type IIA > type IIX, which also contributes to the differences in fatigue resistance between the muscle fiber types. Because the oxidative potential of a muscle fiber is highly trainable, the fatigue resistance of low-threshold type I muscle fibers may be attributable in part to their recruitment history. The size principle (Henneman, Somjen, and Carpenter 1965) governs the recruitment of motor units during isometric and concentric work. In these cases, the recruitment hierarchy can be likened to a flight of stairs, which has to be climbed from the bottom every time, regardless of how high you have to climb. Consequently, the lower steps (the low threshold units) will always be the heavier worn (the better trained). To take this analogy a little further, one might say that the steps will be worn as far up the stairs as it is climbed on a regular basis. In the case of the recruitment hierarchy, one can define a corresponding level that can be called the habitual recruitment level of the individual and muscles in question (figure 15.22). As long as an activity can take place without recruitment of motor units above that level, the fatigue resistance is reasonably good, but as soon as additional recruitment is necessary, fatigue develops rather quickly. Consequently, to improve fatigue resistance in daily life, the logical thing to do is to climb higher in the recruitment hierarchy on a sufficiently regular basis. The improvements in daily functioning of older adults after strength training (Fiatarone et al. 1990, 1994) support this statement.

## Summary

Metabolic factors can affect muscle function in a number of ways, by compromising ATP supply or by interfering with excitation–contraction coupling or with cross-bridge dynamics. The needle biopsy technique has been a major tool in studying the role of metabolic factors in muscle fatigue, but its time resolution is poor, and the samples obtained are small and not as representative of the whole muscle as one could wish. Magnetic resonance spectroscopy (MRS) has a far better time resolution but still has its disadvantages. Apart from the need for special equipment and a suboptimal spatial resolution, the fatiguing exercise has to be performed within the restricted space allowed by the equipment.

Lactate traditionally has been regarded as a major culprit in muscle fatigue, but to the extent this is really the case, the role of lactate anions is mainly indirect, through their accompanying protons, which lower muscle pH. There is an indisputable increase in lactate concentration in muscle during fatiguing activity, but its role as a causative factor in muscle fatigue is still debated, although low pH can negatively affect key muscle proteins.

Depletion of energy stores is another popular explanation of muscle fatigue, but although depletion of glycogen can be easily demonstrated during fatiguing activity, it remains to be proven that it is the cause of the fatigue. The content of PCr is clearly diminished during fatiguing contractions, with a concurrent increase in $P_i$. $P_i$ affects muscle function in a number of negative ways. ATP is not completely depleted from the muscle until after death, when the muscle enters a state of rigor mortis.

Sustained static contractions elevate intramuscular pressure, occluding the blood supply at 20% to 25% of maximal contractile force, in some muscles maybe even at lower percentages. It has been argued that force levels below 15% of MVC can be sustained "indefinitely," but even such low levels

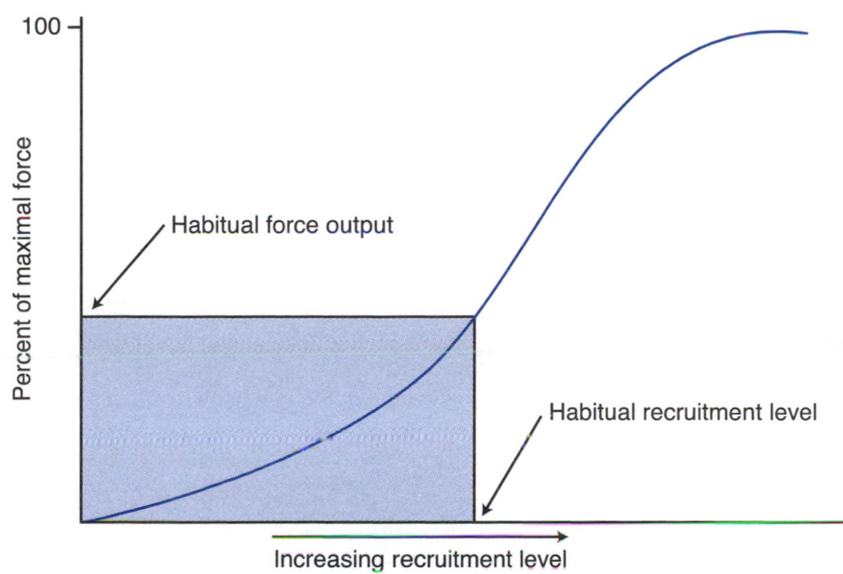


**Figure 15.22** Highly schematic representation of the relation between recruitment level and force output. The habitual force output can be defined as the force output obtained on a sufficiently regular basis to confer reasonable muscular endurance, and the habitual recruitment level as the recruitment level necessary to provide that force.

cause fatigue at the muscle unit level, where the relative force level is substantially higher.

Type I is the most fatigue-resistant muscle fiber type, type IIA follows, and type IIX is most easily fatigued. This difference is based on three characteristics: muscle fiber contractile economy, oxidative capacity, and relative training status, subsequent to their rank in the recruitment hierarchy.

## Measurement of Human Muscle Fatigue

Because muscle fatigue is defined as "an activity-induced reduction in maximal muscle force or power," a logical measure of muscle fatigue is the fraction of maximal muscle force or power left at the end of the fatiguing activity. This fraction often is referred to as the **fatigue index** (Burke et al. 1973; Milner-Brown et al. 1986). The peak torque decline given as a percentage of original torque (cf. figure 15.21) also can be used. Most fatigue estimates in dynamic work report the end result only, but the fatigue profile of Karatzaferi, Giakas, and Ball (1999) describes the temporal development of fatigue during cycle ergometry.

Electromyography also can be used to estimate the degree of fatigue. Because the electromyographic activity depends on summation of individual motor unit action potentials, it can provide an estimate of fatigue attributable to changes upstream of the muscle action potential (reduced motoneuron activation) or to changes in the propagation of the muscle action potential itself (high-frequency fatigue) but cannot estimate fatigue caused by factors located downstream of the muscle action potential (Moxham et al. 1982; Sandercock et al. 1985). The use of electromyography for fatigue measurements, however, is restricted to isometric contractions, because length changes will alter the relationship between electromyographic activity and activation (Basmajian and De Luca 1985).

### For Additional Reading

For further discussion of methods to quantify muscle fatigue, see Vøllestad (1997). The use of the twitch interpolation technique during submaximal contractions is discussed by Bülow et al. (1995).

## Muscle Fatigue in Everyday Life

Understandably, most studies of muscle fatigue have been performed in laboratory settings, often vastly different from situations encountered in real life. The challenge is, therefore, to relate the various factors contributing to experimental fatigue to the fatigue encountered in the workplace or at the athletic field.

A comprehensive study of physiological evidence of fatigue was carried out in four mechanical factory workers who were engaged in the production of automobile drive shafts (Rodahl 1994). In addition to assessing maximal working capacity, the investigator made a series of pertinent physiological measurements at the beginning and at the end of each work shift, 5 days running. The results failed to disclose any objective signs of fatigue, either local or general, at the end of a normal working shift or at the end of a normal working week. The subjects taxed, on the average, 25% of their maximal work capacity when working, and this relative workload was the same at the beginning and at the end of the day. Grip strength of the hand used to lift the drive shafts was higher at the end of the working day than at the beginning and increased from day to day. Whether this could be the result of a training effect can be discussed, but the results certainly do not indicate any deterioration of muscle strength in the involved muscle groups as a consequence of overloading or fatigue.

### Nonbiomechanical Muscle Activity and Fatigue

Muscular complaints are a frequent cause of occupational malfunction, often leading to lengthy absences from work and eventually to premature retirement (Hägg 2000). This happens despite a decreasing biomechanical workload in most occupations, but in many cases, an increasing mental workload. An increasing quest for efficiency, speed, and high degree of precision leads to what has become known as nonbiomechanical muscle activity or task-irrelevant muscle activity (Wærsted 2000). Such nonbiomechanical muscle activity has been reported to be more frequent in the upper trapezius than in other parts of the body (Wærsted and Westgaard 1996). Consequently, three different needs must be met by muscle activity during a given manual task: muscle power to perform the task, stabilization of the shoulder girdle, and nonbiomechanical muscle activity (Wærsted 2000). The need for high speed and/or precise movements increases the two latter factors significantly, although the power needed in the task itself may be negligible. Actually, long-lasting, low-level static contractions of neck and shoulder muscles are widespread in modern society, not only in occupational work (Hägg 2000).

It has been well documented that anxious subjects have more nonbiomechanical muscle activity than others (Hazlett, McLeod, and Hoehn-Saric 1994), and athletes in shooting sports have higher trapezius electromyographic activity during competition than during training (Helin et al. 1988). Such nonbiomechanical muscle activity is typically low level, rarely exceeding 5% of electromyographic activity during maximal contraction, and according to figure 15.19 should be sustained almost indefinitely without fatigue. Several points need to be discussed in this context. First, figure 15.19 concerns isometric contractions of up to 10 min duration and may not be valid for more extended periods of activity, even though the activity is more intermittent. Thus, it is acknowledged now that myalgia can occur even at very low levels of muscular activity, where muscle circulation should be adequate, at least on a macro level (Veiersted and Westgaard 1993). Such myalgia is believed to be caused by sensitization of muscle nociceptors (group III and IV muscle afferents) following local metabolic changes in fatigued low-threshold muscle units (Sejersted and Vøllestad 1993). Second, differences are likely to exist between individuals and maybe even between individual muscles in the same individual as to what percentage of maximal force they are able to sustain without fatigue. Surely, differences exist between individuals with regard to what extent they show short periods of low-level electromyographic activity, called **electromyographic gaps,** in muscles engaged in long-term, low-level activity (Veiersted, Westgaard, and Andersen 1990). Veiersted, Westgaard, and Andersen showed that such electromyographic gaps were fewer in a group of chocolate packers who reported neck and shoulder complaints than in individuals without complaints, and Hägg and Åström (1997) reported a significantly shorter accumulated electromyographic gap time in medical secretaries with complaints compared with secretaries without complaints.

A rotation of activity between motor units has been postulated as a means of reducing muscle fatigue during low-level contractions, but its importance has been debated and varies between muscles (B.R. Jensen, Pilegaard, and Sjøgaard 2000). In cases where there is a low degree of rotation between motor units, extended low-force contractions can overload certain motor units (Hägg 1991; Sejersted and Vøllestad 1993). Such units have been named **Cinderella units** (Hägg 1991), alluding to the fairy tale by the Grimm Brothers, where Cinderella had to do all the hard work while her stepsisters were excused. Even though the force output is low at the whole-muscle level, the output at the individual muscle unit level is substantially higher. It has been calculated, for example, that during a shoulder abduction torque amounting to 12% of maximum, the force at the level of a working muscle unit is on the order of 50% of its maximal force (B.R. Jensen, Pilegaard, and Sjøgaard 2000).

Computer work and cashier work are considered to be typical examples of occupations conferring high mental loads and nonbiomechanical muscle activity, often leading to complaints centering on shoulder, neck, and forearm ("mouse arm"). Interestingly, low-frequency fatigue has been shown to occur in human forearm muscles after using a computer mouse (B.R. Jensen, Laursen, and Ratkevicius 1999). A very important point in this context is the typical slow recovery of LFF, which makes it likely that the muscles do not recover from LFF fully between work shifts (Westerblad et al. 2000). Using a mouse muscle, Chin, Balnave, and Allen (1997) showed that LFF obtained in successive fatigue bouts is indeed additive. Because such low-level muscle activity is highly prevalent, LFF could be responsible for a large part of the muscular complaints of modern humans. Failure in the propagation of muscle action potentials (HFF), including propagation into the T-tubules on the other hand, seems to be unimportant in daily life. For further discussion, see Westerblad et al. (2000).

## The Importance of Maximal Force

In both static and dynamic muscular work, endurance is related to the developed force expressed as a percentage of the maximal force that the muscle can develop. Thus, in isometric muscular contraction, a 50% load can be maintained for about 1 min, whereas the contraction can be maintained almost indefinitely as long as the load is significantly less than 15% of the maximal force of the muscle. Similar relationships exist in the case of rhythmic contractions. Thus, if the load corresponds to about 80% of maximal strength, only 10 contractions per minute can be maintained in a steady state, but if the load is reduced to about 60% of maximal strength, a contraction rate of $30 \cdot min^{-1}$ can be maintained. Therefore, the stronger the muscles, the heavier load they can endure without developing muscular fatigue. But, in any situation, the load and the rate of contraction have to be adjusted according to the strength of the muscle to avoid muscular fatigue.

In a study of 29 female industrial sewing-machine operators, C. Jensen, Vasseljen, and Westgaard (1993) showed that those with the highest frequency of shoulder and neck troubles had significantly lower muscle strength than those with the lowest frequency of shoulder and neck troubles. The investigators also concluded that repetitive precision work, such as operating sewing machines, imposes a high static load on the shoulder muscles, and electromyographic signs of muscle fatigue develop during the work day.

In a study of the microcirculation in the upper trapezius muscle during varying levels of static contraction, Larsson, Cai, and Öberg (1993) concluded that recruitment of motor units for increased muscle tension to maintain a static shoulder load necessitated an increased muscle blood flow. This was, however, insufficient to prevent local fatigue even with low-grade static contraction. They suggested that specific muscle training to increase capillary density is required to improve capacity to perform work tasks characterized by static load of the trapezius muscle.

Hagberg (1981a) observed a rapid decrease of endurance time at contraction levels above 15% of maximum voluntary contraction for both sustained isometric and dynamic exercise involving elbow flexion. This value could be too high for prolonged work, for other field studies have indicated that loads below 10% of the maximal voluntary strength of a muscle produce symptoms of local muscular fatigue in the course of a working day. Malmqvist et al. (1981) showed that fatigue can occur even when the force exerted is small and that a repetitive task has a greater tendency to produce localized muscle fatigue than a more varied one, even if the latter is heavier.

Another very common experience is the gradually increasing pain in the arms when one carries heavy luggage, a task that requires contraction at a substantial percentage of maximal isometric force. After a few seconds of rest, with the luggage on the ground, the pain disappears. One obvious possible cause of the trouble is an impaired blood flow to the active arm muscles. According to Simonson and Lind (1971), the blood flow in forearm muscles is suboptimal at 20% to 30% of MVC. To a certain extent, premature fatigue in forearm muscles while carrying heavy objects can be offset or at least delayed by training to increase the muscles' MVC, although the perfusion of a muscle depends more on the contractile force in absolute terms than in percentage of MVC. In the workplace, regularly handling heavy objects automatically provides necessary training. In other cases, especially in the case of senior citizens, more active intervention is necessary. As shown by a number of studies (see, e.g., Fiatarone et al. 1994; G. Grimby et al. 1992), training improves muscular function in elderly persons and thus extends their period of independent living.

In activities including frequent dynamic concentric contractions, the energy output for a given force is relatively high. According to Asmussen (1973), this type of activity probably can be performed for long periods of time only if the developed force does not exceed 10% of the maximal isometric force. This kind of activity is found in a series of manual tasks in the workplace, such as painting, cleaning, or spading, and in rowing and cycling (see chapter 16). As usual, fatigue is most prevalent in less experienced performers because of their low maximal power and insufficient technique. With practice, both will improve via muscular and neural adaptation (see chapter 11).

## Nonmuscular Symptoms and Signs of Fatigue

I. Åstrand (1960) observed an increase in heart rate in subjects working at a load corresponding to about 50% of the individual's maximal oxygen uptake during a period of about 8 h. However, because these experiments were carried out during the day (morning and afternoon), it is unknown whether these changes were caused by, or partially caused by, the development of fatigue, or whether they were the result of the normal circadian rhythms, which increase heart rate and rectal temperature.

Volle et al. (1979) examined fatigue induced in workers subject to a compressed work week, that is, 40 h in 4 days, compared with the usual 40 h in 5 days. The study subjects were two groups of workers from two different factories that manufactured similar products, one group practicing the 4-day week and the second the 5-day week. The data did not reveal any significant difference between the two groups in terms of reaction time, heart rate, blood pressure, body temperature, oxygen uptake, or carbon dioxide output measured before and after the first and the last day of the week. The only difference observed was that the critical flicker fusion frequency and the right-hand strength were significantly more impaired in the 4-day week group.

The results of these studies show that physical fatigue during submaximal work is a most difficult parameter to assess objectively by measuring physiological parameters.

## Summary

In addition to demonstrating muscle activity necessary to stabilize the shoulder girdle and provide the necessary force to handle the object at hand during a manual task, most people also show some task-irrelevant muscle activity, often called nonbiomechanical muscle activity, especially in the upper part of the trapezius muscle. The amount of such nonbiomechanical muscle activity increases with the quest for speed and accuracy in occupational work but is also widespread in the general public. Although such nonbiomechanical muscle activity is usually low level, it seems reasonable to assume that it confers lasting fatigue to the muscles involved. Comparable force levels, leading to long-lasting low-frequency fatigue, are found in forearm muscles after working with a computer mouse. Because of its slow recovery, daily bouts of LFF can add up, eventually leading to an almost permanent muscle malfunction.

A strong muscle is more able to sustain prolonged exposure to heavy workloads than a weaker muscle. Strength training therefore can increase endurance by allowing the muscle to engage in a given task at a lower percentage of its MVC.

# CHAPTER 16

# APPLIED SPORT PHYSIOLOGY

The development of lightweight electronic instruments and devices capable of recording and transmitting impulses by telemetry, or by direct recording with the aid of portable miniature recorders, has made it possible to study a variety of physiological functions in the person exposed to different types of work stress, including athletic events (Ikegami et al. 1988; Kawakami et al. 1992).These studies have produced a considerable amount of data concerning physiological characteristics of the individual athlete as well as physiological requirements of the specific athletic event.

With the use of electromyography, it is possible to evaluate the engagement of individual muscle groups. Electromyography furnishes information pertaining to:

1. which muscle or which parts of a muscle are activated.
2. the chronological order of participation of the respective muscles in the activity.
3. the degree and duration of the contraction of the respective muscles in each movement.

Such studies also facilitate the development of an individual muscle training program.

The kinematic analysis describes the geometric form of a movement. To arrive at an idea of the forces that produce movement (i.e., a kinetic analysis), investigators can use force-platforms with strain gauges as force-sensitive devices. It is typical for an elite athlete, such as a champion golf player, to repeat precisely, again and again, a certain motion or force, whereby the path of movement, the force developed, and the electromyogram are practically identical each time (Carlsöö 1967).

### FOR ADDITIONAL READING

For a historical review of the research leading to our present understanding of skeletal muscle adaptation to exercise, see C.L. Hamilton and Booth (2000).

## Analysis of Specific Athletic Events

Such information as briefly described provides a foundation for selecting athletes, evaluating technique and methods of training, and monitoring training progress. In this chapter we present data from various athletic events, including walking, running, bicycling, swimming, speedskating, cross-country skiing, Alpine skiing, canoeing, rowing, and ball games, to exemplify the meaning and purpose of applied sport physiology.

An attempt to schematically present the major components of physical or athletic performance in general is shown in figure 16.1.

## Walking

The energy cost of walking varies within wide limits, not only among individuals but also in the same individual, depending on the circumstances. It certainly depends on total body weight, including clothing, speed of walking, type of surface, and gradient (figure 16.2). The freely chosen step rate requires the least oxygen uptake at any given speed (Zarrugh and Radcliffe 1978). If the subject is forced to use any other step rate at the same speed, the oxygen cost will be higher than that required to maintain that speed at the subject's own freely chosen step rate.

## Physical performance

| | Function | Structural basis | | Biochemical processes involved | Modifying factors |
|---|---|---|---|---|---|
| Locomotor organs | 1. Muscular strength | Motor unit: | a) Muscle cell-contractile elements | Contraction-tension (static, dynamic) | Genetic endowment; sex; age<br>Training; use-disuse |
| | | | | Energy metabolism: chemical energy mech. work<br>Aerobic-anaerobic processes, enzymat. react.<br>Fuels: carbohydrate, fat; Nutritional intake, storage | Training<br>Diet |
| | | | b) Motoneuron, synapses, endplate | Excitation, impulse propagation, membrane depolarization | Psychic factors<br>Training |
| | 2. Joint mobility | Skeleton: | a) Muscular attachment | | |
| | | | b) Skeletal levers | | |
| | | | c) Joints and ligaments, articul. cartilage, synovia, bursae | | Training of joint mobility |
| | 3. Coordination | Neuromuscular apparatus: | Afferent efferent pathways, senses | Propagation of nerve impulses, facilitation, inhibition, regulation | Training<br>Psychic factors<br>Drugs |
| Service organs | 4. Endurance | Oxygen transport organs: | a) Pulmonary ventilation | | |
| | | | b) $O_2$ binding capacity of the blood (Hb, blood vol., etc.) | | Training, altitude, air pollution, smoking, iron intake, etc. |
| | | | c) Cardiac output: stroke volume, heart rate | | Genetic endowment, state of health, training, environment |
| | | | d) a-$\bar{v}O_2$ diff: local milieu in muscle cell (shift of the dissociation curve, etc.)<br>Venous return (muscle pump)<br>Negative pressure in the thorax<br>Redistribution of the blood volume to the muscle | | |
| | | | e) Fluid balance | | Fluid intake, fluid loss, heat, etc. |
| Central nervous system | 5. Will to win | Reticular formation, etc. | | | Psychic factors<br>Attitude |

**Figure 16.1** Physical performance. $(a-\bar{v})O_2$ diff. = arteriovenous oxygen difference.

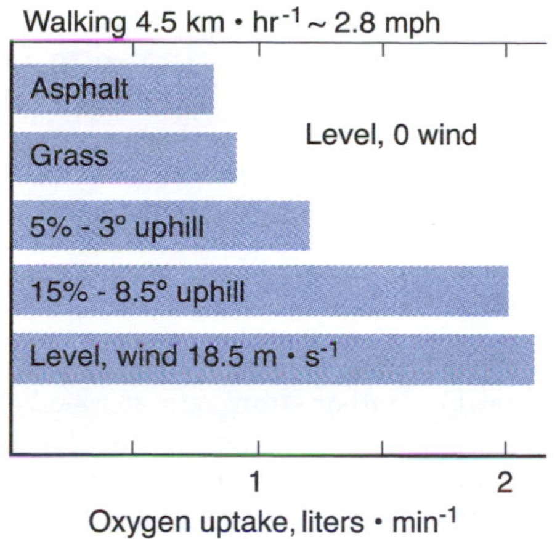


**Figure 16.2** The energy expenditure of walking under different conditions.

From Pugh 1971, 1976.

C.R. Taylor and Heglund (1982) emphasized the importance of energy storage and recovery in the elastic elements of the extremities during locomotion. Their observations demonstrate that the storage of energy in elastic elements occurs in one part of the stride and that this energy is recovered as useful work in another.

A fair amount of data have accumulated from different countries on the energy cost of walking on level ground, and generally these data are in good agreement. Figure 16.3, based on data from Passmore and Durnin (1955), shows the combined effects of varying speed and varying body weight on the energy expenditure of walking.

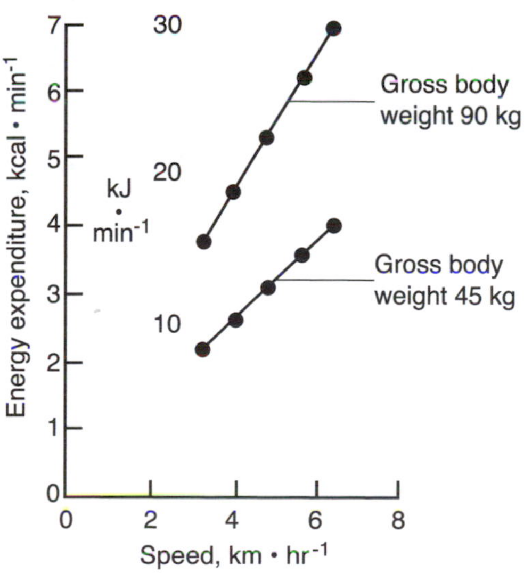


**Figure 16.3** Effect of speed (km · h$^{-1}$) and gross body weight (kg) on energy expenditure (kcal · min$^{-1}$ or kJ · min$^{-1}$) of walking.

From Passmore and Durnin 1955.

A classic, comprehensive, treadmill-grade walking study was carried out by Margaria (1938), who found that going down a slope of 1 in 10 at varying speeds involved an energy expenditure of up to 25% less than walking on level ground. However, on very steep declines, particularly at low speeds, energy expenditure can be considerably higher than when walking on level ground.

M.J. Gordon et al. (1983) compared load carrying with just walking on a treadmill. They showed that both heart rate and the rating of perceived exertion increased linearly with increases in power output, but added load carriage brought about substantially larger increases in heart rate and rating of perceived exertion than did unloaded walking for equivalent increases in power. They suggested that these differential responses are related to differences in muscular fatigue and biomechanical action.

Holewijn, Heus, and Wammes (1992) determined the effect of wearing heavy footwear on physiological responses in five male and five female subjects walking on a treadmill. The investigators found that the mass of footwear resulted in an increase in energy expenditure that was 1.9 to 4.7 times greater than that of an additional kilogram of body mass, depending on sex and walking speed.

The type of surface affects the energy cost of walking, which ranges from 23 kJ (5.5 kcal) · min$^{-1}$ on an asphalt road to 31 kJ (7.5 kcal) · min$^{-1}$ on a plowed field for a 70-kg man walking at a speed of about 5.5 km · h$^{-1}$ (Granati and Busca 1945). When a person is walking at speeds greater than 3 km · h$^{-1}$, the energy cost of locomotion per unit of distance is about 1.8 times greater on sand than on compact terrain (Zamparo et al. 1992).

Walking upstairs may represent an energy expenditure of as much as 42 kJ (10 kcal) · min$^{-1}$ for a 75-kg person (Passmore, Thomson, and Warnock 1952). Going downstairs involves only about one third the energy used in going upstairs. Going, and especially running, upstairs thus represent fairly heavy exercise, and it is therefore not surprising that so many people find this activity very tiring. On the other hand, stair-climbing is effective in improving physical fitness.

The use of sticks or so-called power poles, which are specially constructed, rubber tipped poles designed for use during walking, simulates the arm motion of cross-country skiing. This increases the muscle mass used during walking. Porcari et al. (1997), who investigated the potential increases in exercise intensity and energy cost associated with the use of walking poles in 32 volunteers, found that walking with poles resulted in an average of 23% higher $\dot{V}O_2$ and 16% higher heart rate responses compared with walking without poles on a treadmill. The investigators concluded that the use of poles can increase the intensity of walking at a given speed and provides additional training benefits to walkers.

Tripping over an obstacle while walking is a common cause of falls in the elderly. In a study of 16 young and 16 old healthy adults, Chen et al. (1996) observed that divided attention degraded obstacle avoidance abilities of the old significantly more than it did in the young.

After the Olympic Games in London, 1948, the winner of the 10K walking event, John Mikaelsson, was studied when walking on a treadmill. When he simulated the race by adjusting the speed of the treadmill so that the walking speed during the race was attained (13.3 km · $h^{-1}$ = 10 km in 45 min 13.2 s), his measured oxygen uptake was 4.0 L · $min^{-1}$, or 58 ml · $kg^{-1}$ · $min^{-1}$ (P.-O. Åstrand, unpublished results). The oxygen uptake during competitive walking seems to be approximately 75% of the maximum; uphill it even exceeds that level.

Peters et al. (1999) examined the occurrence of gastrointestinal symptoms among 79 men and 76 women, 30 to 49 years old, during long-distance walking during four consecutive days, covering a total distance of 203 km for men and 164 km for women. The results showed that 24% of the subjects experienced one or more symptoms, nausea, headache, and flatulence being the most frequent complaints. However, a relationship between gastrointestinal symptoms and age, sex, training status, or walking speed could not be found.

R.L. Thompson and Hayward (1996) monitored thermal, metabolic, and motor responses of 18 male subjects while they walked for 5 h at an air temperature of 5° C. During the last 4 h of the walk the subjects were continuously exposed to rain and wind. They found that the rain caused substantial cold stress, evidenced by a 40% increase in heat production attributable to shivering and significant loss of strength and manual dexterity.

## Running

In adults, individual variations in energy expenditure are fairly small at submaximal speeds. Under these conditions, oxygen uptake per kilogram of body weight is the same regardless of sex or athletic rating (P.-O. Åstrand 1956). On the other hand, the oxygen uptake per kilogram of body weight is higher for children than for adults when they both run at a certain speed (P.-O. Åstrand 1952). The running economy is poorer in children than in adults because children have a higher resting metabolic rate, greater ventilatory equivalents for oxygen, and disadvantageous stride rates and stride lengths (Krahenbuhl and Williams 1992).

Ariens et al. (1997), studying the development of running economy at three different slopes in teenage to adult males and females (13–27 years old), found that females had significantly better running economy than males at all ages measured and for all three slopes. In a study of the total metabolic costs of running at self-selected, comfortable speeds in males and females, Bhambhani and Singh (1985) observed that the metabolic costs in kcal · $kg^{-1}$ · $km^{-1}$ were significantly higher in females than in males. Similarly, in a study of performance-matched marathon runners, Helgerud, Ingjer, and Strømme (1990) found that the females' running economy was poorer; that is, their oxygen uptake during running at a standard submaximal speed was higher than that of performance-matched males. The oxygen uptake during running depends on the stride length, as is evident from figure 16.4. The subject represented in

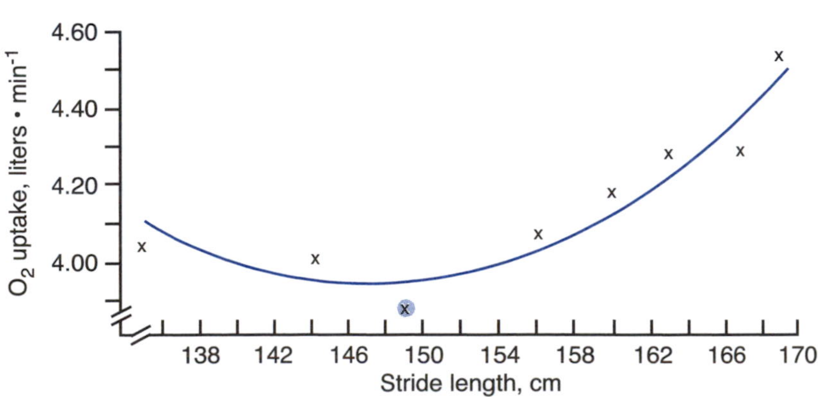


**Figure 16.4** Oxygen uptake during running at a speed of 16 km · $h^{-1}$ with different lengths of stride. The encircled cross represents the freely chosen length of stride.

From Högberg 1952.

the figure ran at a given speed paced by a metronome, and oxygen uptake was measured at steady state. In some experiments, he was free to choose the stride frequency. In general, the stride length that is natural for the individual is also the most economical one. The energy cost of running is increased greatly with a further increase of stride length. This fact has considerable practical significance. In long-distance running, economy and energy efficiency are important, so that it becomes essential to maintain the stride length most efficient for the individual. In short-distance running, such as the 100-m dash, on the other hand, where speed is more important than economy, the one who can run with rapid, long strides has an advantage and can afford to disregard energy cost for the limited period of time involved.

The well-known Finnish runner Nurmi had an unusually long stride length. It was thought that this was the key to his success, and others attempted to copy his style. This, however, reduced their efficiency and performance. It appears that the long stride was natural for Nurmi, and for him this represented the most economical style. This example clearly shows the danger of generalization on the basis of single cases.

Otherwise, an increase in the speed of running is brought about primarily by an increase in the stride length. When an experienced 800-m runner ran on the treadmill at different speeds from 8 up to 30 km · h$^{-1}$, the length of the stride increased more or less rectilinearly from about 80 to 220 cm, whereas the stride frequency increased from only about 170 to 230 steps · min$^{-1}$ (Högberg 1952).

There is very little difference in the measured values for energy expenditure per meter in elite runners, indicating that any influence of running technique on energy expenditure must be small. Thus, in well-trained elite runners running at a speed of 20 km · h$^{-1}$ on level ground, the oxygen uptake varied only between 67 and 71 ml · kg$^{-1}$ · min$^{-1}$ (Karlsson et al., 1972; Margaria, Aghemo, and Pinera Limas 1975).

The energy expenditure is greatly increased when the athlete runs against the wind (figure 16.5). According to C.T.M. Davies (1980), the energy cost of overcoming air resistance on a calm day outdoors is about 8% for sprinting (10 m · s$^{-1}$), 4% for middle-distance (6 m · s$^{-1}$), and 2% for marathon (5 m · s$^{-1}$) running.

As long as the speed is kept below a certain level, which varies with the individual, it is more economical to walk than to run (figure 16.6). In elite walkers, this level is higher than in less experienced walkers. It costs the same amount of energy to run at a rate of 14 km · h$^{-1}$ (8.7 mph) as it does to walk at a rate of only 10 km · h$^{-1}$ (6.2 mph).

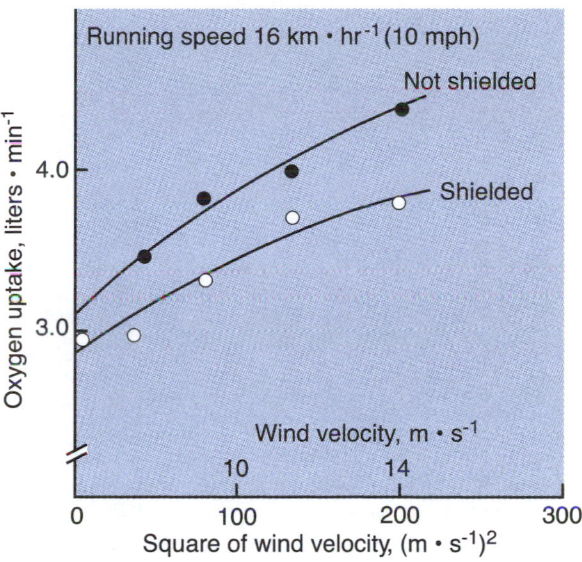


**Figure 16.5** The effect of increased air resistance attributable to wind on the energy expenditure of running. The curve "Not shielded" was obtained when the subject was running alone on the treadmill and "Shielded" when he was about 1 m behind another runner. The wind was achieved by a big fan. An extrapolation of the data indicated that running 1,500 m at about 4 min speed on a track in calm air close behind another runner can save up to 6% of the energy cost. Running behind and to one side of another runner gives a gain of about 1 s per lap.

From Pugh 1971.

When a person runs within a wide range of speeds, the energy demand per kilometer is practically the same (jogging at lower speeds, running at higher speeds). Walking at lower speeds is less costly, but at higher speeds the energy cost approaches and even exceeds that of running (Böje 1944; Margaria et al. 1963). As a general rule, the energy cost of jogging or running is approximately 4 kJ · kg$^{-1}$ · km$^{-1}$ (1 kcal), but for walking at 4 to 5 km · h$^{-1}$ the energy demand is only half the figure, or 2 kJ · kg$^{-1}$ · min$^{-1}$ (.5 kcal). As pointed out, the surface, wind, and slope will modify these figures. At about 4 km · h$^{-1}$, the work done at each step to lift the center of mass of the body equals the work done to increase its forward speed. The total mechanical energy involved (potential plus kinetic) is at this speed at a minimum, as is the energy cost (Cavagna, Thys, and Zamboni 1976). When a person runs on sand, the energy cost of locomotion per unit of distance is about 1.8 times greater than when running on compact terrain (Zamparo et al. 1992).

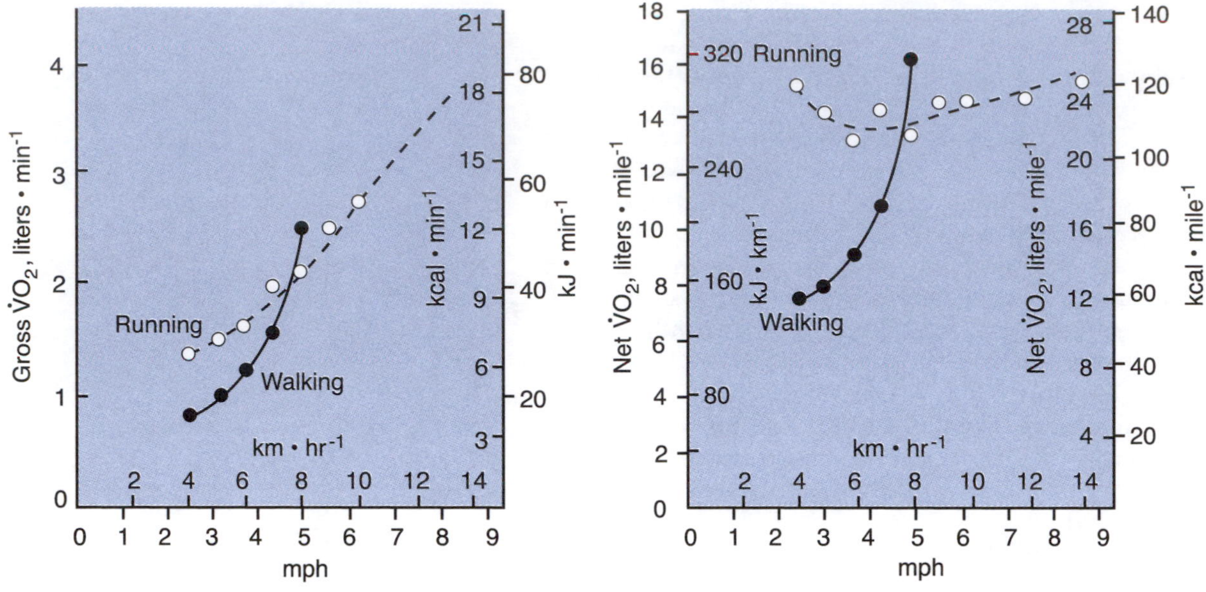


**Figure 16.6** Energy cost of walking and running at different speeds. To calculate the net energy cost (right panel), the oxygen uptake on the sitting subject was subtracted from the oxygen uptake measured during walking and running. The body weight of the subject was 75 kg.

For an athlete to attain a level of achievement equivalent to the world elite in middle- and long-distance running, a maximal aerobic power close to, or preferably above, 80 ml $\cdot$ kg$^{-1}$ $\cdot$ min$^{-1}$ is necessary. An athlete with a high maximal oxygen uptake has the advantage of being better able to tolerate bouts of forced tempo than the competitor with a lower aerobic power in that he or she does not have to use the anaerobic energy yield to the same extent during the bouts of increased tempo. Ideally, the pace should be selected corresponding to the intensity when the oxygen uptake, following a linear increase with the increasing intensity, begins to level off.

According to our present state of knowledge, a top runner who has had a proper warm-up attains his or her maximal oxygen uptake after about 45 s. Accordingly, it can be assumed that in a 400-m dash lasting 40 to 50 s, the runner taxes his or her anaerobic capacity maximally. In these cases, blood lactate concentrations of 20 to 25 mM have been measured. At shorter distances (i.e., 100- to 200-m races), it is estimated that some 90% of the energy is derived from anaerobic processes. At distances exceeding 400 m, aerobic metabolic processes assume an increasingly important role. In an 800-m race, aerobic metabolic processes account for about 40% of the total energy used, and in a 1,500-m race, 65%.

Ingjer and Dahl (1979) compared the dropouts from an endurance training program with those who continued. The investigators found differences in muscle fiber composition and in the economy of the subjects' running technique. In contrast to the rest of the group, the dropouts did not show any improvement in maximal aerobic power during the training period. It was suggested that constitutional factors are important to the ability and willingness of an individual to pursue an endurance training program successfully.

## Bicycling

Bicycling deserves our attention, not only as a popular sport but also because of the use of cycle ergometers in the study of the exercise physiology. Cycling has been the subject of extensive studies, and a considerable body of information is available (for references, see Chavarren and Calbet 1999; Faria 1992; Hartung et al. 1993; Sjøgaard 1984; Whitt and Wilson 1974).

Basically, the bicycle is a very efficient method of locomotion. The energy cost of bicycling is only about one fifth that of walking, yet the speed can be five times greater. Improvements have been made aerodynamically by more streamlined construction, the use of racing suits, and streamlined helmets to reduce drag. But above all, the training of the bicyclist has improved vastly to meet the requirements of this demanding sport, which includes races ranging from a 200-m sprint to the 5,000-km Tour de France.

The average aerobic power of most elite cyclists ranges between 70 and 74 ml · kg$^{-1}$ · min$^{-1}$. The quadriceps muscle is recruited proportionally to the changes in oxygen uptake during cycling. The saddle height significantly affects muscle involvement. Mechanical efficiency ranging from 19.6% to 28.8% for cycling at pedal frequencies between 40 and 100 rev · min$^{-1}$ has been reported. Data on the efficiency of changing the pedal rate are conflicting, but there appears to be a significant advantage in using a high pedaling rate at high power output. Evidently, the body posture during cycling is crucial for optimal performance (for references, see Faria 1992). P.-O. Åstrand (1953) studied the effect of different pedal crank lengths (16–20 cm) on mechanical efficiency when subjects rode a bicycle placed on a horizontal treadmill (speed 20 km · h$^{-1}$, slope 2°). He found no difference in efficiency, whereas Faria (1992) suggested that the crank length on most commercially available bicycles and ergometers is probably too long for most cyclists.

The air resistance, which increases as the square of speed, is the dominant factor a cyclist must overcome. Because a larger cyclist has a greater surface area than a smaller cyclist, the former experiences a higher absolute air resistance. However, the larger cyclist usually has a lower ratio of surface area to body weight and consequently a lower relative air resistance for his or her muscle mass to overcome. Thus, the larger cyclist would be expected to have a lower ratio of oxygen uptake to body weight than the smaller cyclist at any given speed. This was proven to be true by Swain et al. (1987), who photographically determined the frontal area of cyclists in a racing posture and found that the large cyclists had a 16% lower frontal area/body weight ratio than the small cyclists. The investigators concluded that large cyclists are at a distinct advantage, in terms of the ratio of oxygen requirement to body weight, while cycling on level roads, and that this advantage is principally attributable to their lower frontal area/body weight ratio.

### For Additional Reading

For further discussion and detailed treatment of the ergogenics of bicycling, see Kyle (1991).

## Swimming

Swimming engages practically all muscle groups of the body. It is therefore not surprising that very high oxygen uptakes have been obtained in swimmers (Åstrand, Engström, et al. 1963; Holmér 1974a). In male swimmers of good world standard, a maximum of around 6 L · min$^{-1}$ has been measured.

Because the specific gravity of the body is not much different from that of water, the weight of the body submerged in water is reduced to a few kilograms. Especially the obese individual can keep afloat with very little energy expenditure. Swimming therefore is an easy task when performed at a low level of intensity. For this reason, swimming and various exercises performed in water are a very common form of training for physically handicapped individuals.

The functional demands of competitive swimming were evaluated in 22 female swimmers on the basis of the relationship between oxygen uptake during swimming at competitive speed and maximal oxygen uptake during work on a cycle ergometer (Åstrand, Hallbäck, et al. 1963). A very high correlation was observed, but the oxygen uptake during swimming averaged only 92.5% of the maximal oxygen uptake reached during cycling. However, five girls reached a higher value in the former case. The high correlation is also evident from the fact that the blood lactate concentration after swimming was of the same order of magnitude as after maximal cycling (10.3 and 10.5 mM, respectively). Similar results were reported by P.-O. Åstrand and Saltin (1961a and Holmér (1974b): 12.8 mM after maximal swimming, and 12.9 mM after running for male swimmers. The quotient of pulmonary ventilation to oxygen uptake was significantly lower during maximal swimming than during cycling (27.7 and 35.5, respectively). The reason for this relative hypoventilation could be the different mechanical conditions of breathing. The water pressure on the thorax makes respiration more difficult. Furthermore, breathing is not as free during swimming as in most other types of exercise, in that the respiration during competitive swimming is synchronized with the swimming strokes. Holmér I., E.M. Stein et al. (1974) also noted a similar difference in the ratio between maximal pulmonary ventilation and maximal oxygen uptake when comparing the data obtained during swimming and running (29.8 and 37.4, respectively, for five subjects). However, despite the relative hypoventilation in the swimming subjects compared with running, the arterial oxygen pressure and content were the same in the two types of exercise.

World swimming records have been attained by girls at an increasingly younger age. Chapter 8 presented an analysis of their physiological possibilities. It was concluded that girls, by the age of 13 to 14 years, have almost reached the maximal power of their aerobic processes. During puberty, the organism responds more strongly to training. In the aforementioned study of female swimmers (P.-O. Åstrand, Engström, et al., 1963), the swimmers had significantly greater functional dimensions than girls who had not taken part in competitive sport or undergone any special physical training. Vital capacity and heart volume were highly correlated to the maximal oxygen uptake.

Thus, young girls exhibit a very high energy power. It has been shown that women during breaststroke swimming at a certain speed have a lower oxygen uptake (greater mechanical efficiency) than men. This is explained by the fact that the lower specific gravity in women, because of their greater fat content, reduces the effort required to keep the body floating. However, considerable individual variations in technique are typical for swimming (Holmér 1974a).

Metabolic and cardiorespiratory responses of swimmers were compared by Holmér and Gullstrand (1980) during a so-called hypoxic swim training session, in which the swimmer breathed less than usual, and during a training session performing normal breathing. Their results do not support the view that hypoxic swim training has an advantage over normal swim training.

The energy requirement of different types of swimming is best mirrored in the record tables, which show that crawling is the most economical type of swimming as long as the swimmer masters the proper technique. It is possible to study the different swimming techniques in detail in a specially constructed swimming flume. In the flume, the water is made to flow at different speeds through a channel, thus providing opportunity for studies similar to those on walking and running on a treadmill.

Holmér (1974a) studied the physiological responses to swimming in 87 subjects of varying states of skill and training, compared with running and cycling. He found that oxygen uptake during swimming at a given submaximal speed depended on the amount of swimming training, body dimensions, swimming style, and technique. Thus, oxygen uptake at a given submaximal speed was higher for untrained than for trained swimmers and for tall subjects than for short subjects. The arm stroke has the highest efficiency, not the leg kick. In freestyle, the maximal speed with arm strokes was almost the same (1.31 m $\cdot$ s$^{-1}$) as for the whole stroke (1.34 m $\cdot$ s$^{-1}$) but at a significantly lower oxygen uptake. One might speculate that over longer distances, the leg kicks should be de-emphasized because they waste oxygen. In breaststroke, on the other hand, the leg kicks are probably as important or more important than the arm strokes. The mechanical efficiency in elite swimmers was 6% to 7% in freestyle and 4% to 6% in breaststroke. Most costly from the standpoint of energy expenditure is the butterfly stroke. (For details, see di Prampero et al. 1974; Holmér 1974b; McLean and Hinrichs 1998.)

The increase in oxygen uptake with increasing swimming speed was linear or slightly exponential (Holmér 1974b). Maximal oxygen uptake during swimming was, for elite swimmers, 6% to 7% lower than during running and approximately the same as during cycling. For subjects untrained in swimming, their maximal oxygen uptake during swimming was on the average 80% of the running maximum. When the body was submerged, vital capacity was reduced by 10% and the expiratory reserve volume was less than 1 L compared with 2.5 L in air. The increase in tidal volume in water was exclusively achieved by the use of the inspiratory reserve volume. Heart rate, cardiac output, and stroke volume during submaximal swimming were of the same magnitude and increased with increasing speed in approximately the same way as during running. Heart rate was significantly lower in maximal swimming than in maximal running. The mean intra-arterial blood pressure at submaximal as well as maximal rates of work was higher in swimming than in running.

Breaststroke and butterfly swimming required 1 to 2 L $\cdot$ min$^{-1}$ higher oxygen uptake at a given submaximal speed than did freestyle and backstroke. As mentioned, in elite swimmers the mechanical efficiency was 4% to 6% in breaststroke and 6% to 7% in freestyle. Maximal oxygen uptake measured during swimming varied as a consequence of swimming training, whereas the maximal oxygen uptake measured during treadmill running remained relatively unchanged. Therefore, maximal oxygen uptake measured during running is not representative of swimming performance, and training to increase a swimmer's maximal oxygen uptake should consist of swimming, to the greatest amount possible.

Bonen et al. (1980) compared the maximal oxygen uptake obtained in swimmers during tethered swimming, free swimming, and flume swimming and found essentially identical results. They also compared maximal oxygen uptake obtained during arm ergometer exercise and during swimming and found large differences. Thus, predicting swimming maximal oxygen uptake from arm-ergometer data would yield considerable errors, as in the case of treadmill data (see figure 11.15).

H. Tanaka and Seals (1997) performed a 5-year (1991–95) retrospective analysis of top freestyle performance times from the U.S. Masters Swimming Championships. Both men's and women's endurance swimming performance (1,500 m) declined linearly from ages 35 to 40 to approximately 70 years of age. From then on, performance declined exponentially. The rate and magnitude of the declines in both short- and long-duration swimming performance with age were significantly greater in women than in men.

## Speedskating

Ekblom, Hermansen, and Saltin (1967) studied speedskating athletes when running on a treadmill as well as when skating at submaximal and maximal speeds. Table 16.1 summarizes some of the maximal data attained in the two types of activities by five subjects. The average value for maximal oxygen uptake was 5.49 L · min$^{-1}$ during treadmill running, compared with 4.85 L · min$^{-1}$ during skating, which was a difference of about 12%. Otherwise, the achieved maximal values were rather uniform. The extra burden imposed by the equipment that the skaters had to carry for the collection of the expired air might have caused the oxygen uptake values to be higher than they would have been when skating without equipment. The values for blood lactate and heart rate attained in connection with skating competitions were similar to those obtained during the determination of maximal oxygen uptake.

When the ice condition is excellent, the maximal oxygen uptake expressed in liters per minute is probably of greater practical importance than the maximal oxygen uptake expressed as milliliters of oxygen per kilogram of body weight per minute. During speedskating, the body's center of gravity moves relatively parallel to the ice surface without the marked vertical movements made during running. The potential to perform work is related from a dimensional point of view to the body mass ($L^3$), but the air resistance is proportional to the body surface area ($L^2$). Therefore, a better performance can be expected from a larger subject, and more so at higher speeds (di Prampero et al. 1976). When the ice surface is soft, it is a drawback for the skater to be heavy, because weight increases the friction. The results from the study mentioned (see table 16.1) show that the elite long-distance skaters had a maximal aerobic power of about 5.5 L · min$^{-1}$. The 500-m race and especially the 1,500-m race (skating time slightly under 2.0 min) impose particularly high demands on anaerobic power. The peak blood lactate concentration for three of the skaters at the end of the race averaged as follows: 500 m, 13.6 mM; 1,500 m, 17.3 mM; 5,000 m, 15.1 mM; 10.000 m, 13.3 mM. The technique is also different in the shorter distances compared with the longer ones. The skaters who place first in the 500-m race, and to some extent also in the 1,500-m race, infrequently win the 10,000-m race, thus illustrating that the physiological requirements are different in the two types of races. The skaters referred to in table 16.1 are all typical long-distance skaters.

Figure 16.7 presents oxygen uptake during skating at different speeds at two different skating rinks. As is evident, the curve is not linear. An increase in the speed from 4 to 6 m · s$^{-1}$ requires an additional .7 L · min$^{-1}$, whereas the increase from 8 to 10 m · s$^{-1}$

**Table 16.1** Maximal Data Attained During Running and Speedskating in Five Elite Long-Distance Skaters

| | | Maximal values | | | | | | | | | |
|---|---|---|---|---|---|---|---|---|---|---|---|
| | | Oxygen uptake | | | | Pulmonary ventilation, L · min$^{-1}$ | | Blood lactates, mM | | Heart rate, beats · min$^{-1}$ | |
| | | L · min$^{-1}$ | | ml · kg$^{-1}$ · min$^{-1}$ | | | | | | | |
| Subject | Best time 10,000 m | R* | S* | R | S | R | S | R | S | R | S |
| 1 | 15:20.6 | 5.77 | 5.48 | 79.0 | 75.1 | 184 | 183 | 20.3 | 19.9 | 186 | 186 |
| 2 | 15:47.0 | 5.70 | 4.80 | 79.2 | 66.7 | 172 | 139 | 17.0 | 18.0 | 188 | 186 |
| 3 | 15:23.6 | 5.39 | 4.69 | 71.9 | 62.5 | 137 | 159 | 15.4 | 15.4 | 185 | 185 |
| 4 | 17:08.0 | 5.39 | 4.89 | 64.9 | 58.9 | 141 | 141 | 17.4 | 17.4 | 194 | 190 |
| 5 | 16:19.4 | 5.20 | 4.38 | 76.5 | 64.4 | 138 | 121 | 16.8 | 16.6 | 171 | 171 |
| | | | | | | | | | | | |
| value | | 5.49 | 4.85 | 74.3 | 65.5 | 154 | 149 | 17.4 | 17.4 | 185 | 184 |

*R = Running; S = Speed skating

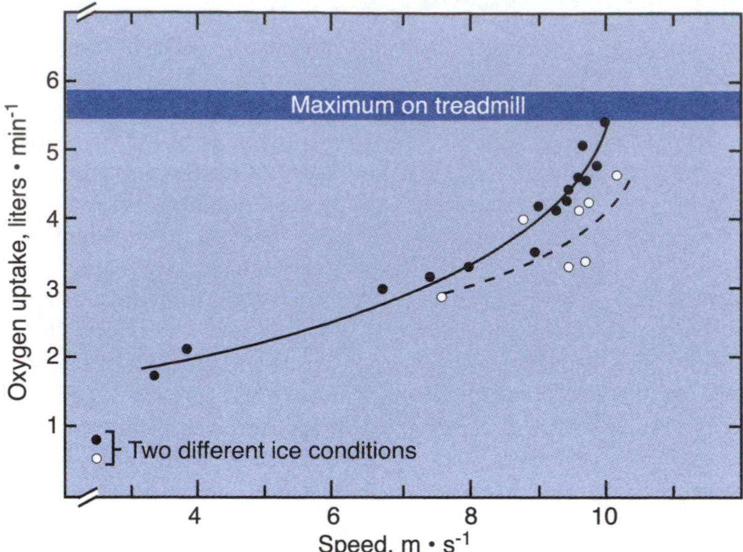


**Figure 16.7** Oxygen uptake during speedskating at different speeds. The different symbols represent experiments performed at different speedskating rinks. In excellent ice conditions as well as at high altitude (reduced atmospheric pressure = reduced air resistance), the curve is shifted to the right; that is, the same expenditure of energy gives greater speed.

From Ekblom, Hermansen, and Saltin 1967.

necessitates an increase in the aerobic power of 2.0 $L \cdot min^{-1}$. The authors of this study did not consider a possible contribution from anaerobic processes. The explanation must be sought primarily in the increased air resistance at the higher speed. Air resistance increases with the speed raised to the second power and accounts for a large part of the energy expenditure at high speeds. From experiments in a wind tunnel it can be calculated that at a speed of 10 $m \cdot s^{-1}$, 70% of the external work is devoted to overcoming the air resistance, whereas the remaining 30% is required to overcome the friction of the ice. Thus, a more ideal aerodynamic profile would to a large extent reduce the air resistance. The speed-skating style, with the arms held on the back, undoubtedly has developed from experience. It is also known that the type of clothing is very important in regard to air resistance.

To reduce the negative effect of air resistance and headwind on the energy cost, one can skate directly behind a competitor, a technique called drafting. The effects of drafting during short-track speedskating were examined by Rundell (1996b), who observed significantly lower heart rate and lactate responses during drafting. Sprint performance was also significantly better following drafting. Some skaters were more effective drafters than others.

According to de Koning et al. (1995), who described the start in speedskating, from running to gliding, the mechanics of the first strokes of a sprint differ considerably from the mechanics of strokes later on. The first push-offs take place against fixed location on the ice. In these "runninglike" push-offs, the contribution of the "rotational" velocity component in the forward direction is larger than the "extension" velocity component. Later on, the strokes are characterized by a gliding push-off, in which the "extension" velocity component increases. In these gliding push-offs, no direct relation exists between forward velocity of the skater and the extension of the joints. This allows skaters to obtain much higher velocities than can be obtained during running. In a study of speedskaters during the 1988 Winter Olympic Games, de Boer and Nilsen (1989) found that the higher work per stroke of the faster skaters was correlated with a longer gliding phase and a more horizontally directed push-off.

The increase in the speed of the skater is perhaps best illustrated by comparing the world records of the 10,000 meter events from 1913 through the Olympic Games in 2002 (see table 16.2).

The style of speedskating appears rather rigorous, with many muscle groups involved in static contraction. This fact explains why the maximal oxygen uptake is lower during skating than during running. Figure 16.8 also shows that the blood lactate concentration at a given oxygen uptake is considerably higher during skating than cycling. From a training point of view, it appears essential that the speedskater allow the muscle groups engaged in skating to become accustomed to tolerating a high lactic acid concentration.

The sitting posture of speedskating can compromise blood flow to the working thigh muscles, limiting oxygen uptake. Rundell (1996a) examined this problem in seven short-track speedskaters who performed running, in-line skating upright, and in-line skating in a sitting position on a motor-driven

**Table 16.2** World Records for the 10,000 Meter Speedskating Events Through the Olympic Games of 2002

| Year | Record holder | Time (min:s) | Average speed (km · $hr^{-1}$) |
|---|---|---|---|
| 1913 | Oscar Mathisen | 17:22.6 | 34.53 |
| 1952 | Hjalmar Andersen | 16:32.6 | 36.27 |
| 1960 | Knut Johannessen | 15:46.6 | 38.03 |
| 1976 | Sten Stensen | 14:38.8 | 41.00 |
| 1980 | Eric Heiden | 14:28.1 | 41.47 |
| 1984 | Igor Malkov | 14:21.5 | 41.79 |
| 1988 | Tomas Gustafson | 13:48.2 | 43.47 |
| 1994 | Johann Koss | 13:30.5 | 44.42 |
| 1998 | Gianni Romme | 13:08.7 | 45.64 |
| 2002 | Jochem Uytdehaage | 12:58.9 | 46.22 |

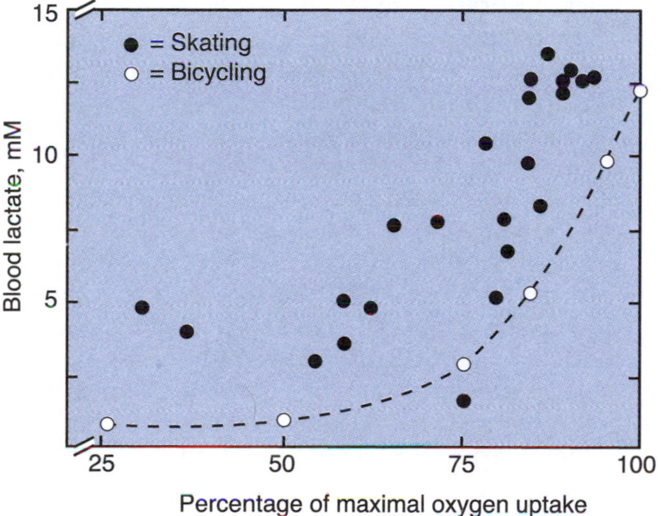


**Figure 16.8** Blood lactate concentration at different work rates expressed in percentage of maximal oxygen uptake during speedskating and bicycling. The bicycling data are from well-trained noncyclists. The speedskating data are obtained from well-trained speedskaters. At corresponding oxygen uptakes, more lactic acid is produced during speedskating than during bicycling.

From Ekblom, Hermansen, and Saltin 1967.

treadmill. Peak $\dot{V}O_2$ was significantly lower in the sitting position compared with skating in the upright position and with running. At equivalent speeds, submaximal oxygen uptake was significantly lower and blood lactate was significantly higher in the sitting position. These findings were confirmed by Rundell, Nioka, and Chance (1997). In any case, the task of balancing the body weight on one leg in the sitting position places a heavy burden on some pelvic muscles. In the upright position, this balancing act is taken care of by the gluteus medius muscle, but in the sitting position of a skater, this muscle is largely unable to abduct the leg. Instead, a group of pelvic muscles, consisting of m. piriformis, m. obturatorius internus, and m. gemelli, is brought into action (Samuelsen, Høiseth, and Dahl 1993). Unless they are properly trained and stretched, these muscles can hamper peak performance.

The maximal isometric and dynamic muscle strength of skaters is not particularly great. It has been observed, however, that the leg muscles of the skaters who are the best sprinters and specialize in 500-m skating are much stronger than those of long-distance skaters.

In a study of force-generation capacity of knee extensor muscles in speedskaters, Kanchisa et al. (1996) found that the speedskaters have a higher muscle performance in a repetitive maximal contraction task than in a single contraction compared with untrained subjects.

The technique in speedskating requires that plantar flexion is largely suppressed during the gliding push-off to prevent the tip of the blade from scratching through the ice. To allow the skater to maintain contact with the ice until the knee is fully extended, a new skate, the "klapskate," has been developed, which permits the shoe to rotate relative to the blade in a hinge between the shoe and the blade. Van Ingen Schenau et al. (1996) showed in a group of Dutch skaters that those using the new skate improved their personal best times by 6.2% on average compared with 2.3% in the control group, which is a significant difference. In a more recent study comparing the use of klapskates and conventional skates, Houdijk et al. (2000) also found that the use of the klapskate increases gross efficiency in speedskating.

## Cross-Country Skiing

Because more muscle groups generally are engaged in skiing than in walking (the arms are used to pull and push on the ski poles), the overall energy expenditure involved in transporting the body on skis from one place to another is as high as, or higher than, the energy expenditure when moving the body the same distance on foot. In fact, the energy expenditure of skiing, especially uphill, occasionally is surprisingly high (Strømme, Ingjer, and Meen 1977).

Because skiing engages about all of the major muscle groups in the body, cross-country skiing is an excellent method of training for physical fitness and dynamic muscular endurance. For the same reason, elite cross-country skiers generally have exceedingly high maximal oxygen uptakes. Thus, a maximal oxygen uptake of 7.4 L · min$^{-1}$ has been reported (Bergh 1974). Evidently, cross-country skiers have the highest maximal aerobic power ever recorded: 93 ml · kg$^{-1}$ · min$^{-1}$ for a male Olympic champion, and 80 ml · kg$^{-1}$ · min$^{-1}$ for a female skier (P.-O Åstrand unpublished results). Rusko (1987) investigated the influences of growth, training, and various training methods by analyzing long-term training effects in 129 young cross-country and biathlon skiers. The results indicated that maximal aerobic power and heart volume increased between 15 and 20 years of age and the most significant changes in heart volume were observed between 16 and 18 years of age. International-level skiers were able to increase their maximal aerobic power and heart volume also after 20 years of age. Anaerobic threshold (AT) increased like maximal oxygen uptake, but when expressed as percentage of maximal oxygen uptake, the AT was similar in every age group over 16 years of age. Training at the intensity of AT or somewhat higher was most effective in improving maximal aerobic power. Low-intensity distance training was more effective in improving AT.

An additional advantage of cross-country skiing is that the work rate varies greatly with the changing features of the terrain, with very high loads when climbing uphill and lighter loads when sliding downhill (figure 16.9). The workload also can be reduced by drafting. Bilodeau, Roy, and Boulay (1995) compared the heart rate response during cross-country skiing for a leading skier, as well for a skier in a drafting situation 3 m behind the leader. Their results revealed that the heart rate was significantly lower when drafting, as opposed to leading. The heart rate differences between the leading and the drafting situations were larger for smaller skiers drafting behind larger skiers.

Heart rate recording during competitive skiing (see figure 16.9) shows that heart rate reaches maximal levels during uphill skiing; drops, at the most, to some 20 beats · min$^{-1}$ below maximal values during downhill skiing; and stays slightly below maximal values during skiing on fairly long stretches of level ground. It appears that top competitive cross-country skiers need to tax some 85% or more of their maximal oxygen uptake during a race.

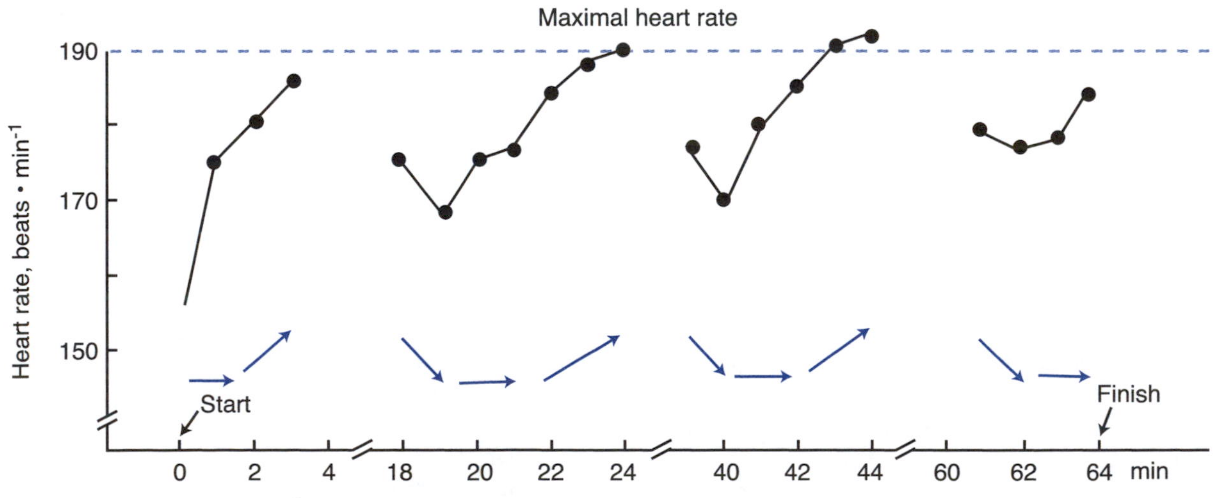


**Figure 16.9** Upper notations: Heart rate of the winner in a 21-km cross-country ski race. Lower notations: Arrows indicate terrain profile: → = level ground; ↗ = uphill; ↘ = downhill.

From Bergh 1974.

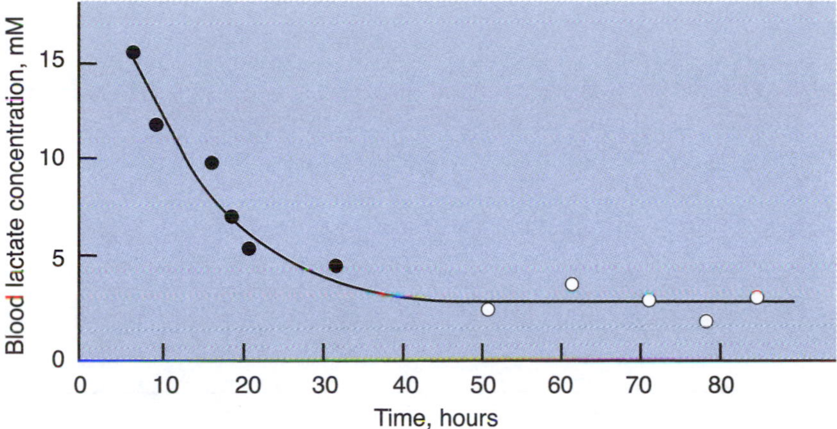


**Figure 16.10** Blood lactate concentration at the end of races of distances from 10 to 85 km. ● = mean values; ○ = individual values.

From Åstrand, Hällback, et al. 1963.

Figure 16.10 shows that the longer the duration of the race, the lower the blood lactate concentration measured at the end of the race (Åstrand, Hällback, et al., 1963). It is thus evident that a high anaerobic capacity is important in the shorter distance, 5K to 10K races, and particularly in the relay races on skis where the tempo is very uneven.

Millerhagen, Kelly, and Murphy (1983), after studying combined arm and leg exercise, suggested that skiing could be made more economical through effective use of a well-developed upper body. In an investigation of 15 female cross-country skiers, using a double-pooling ski ergometer, Hoff, Helgerud, and Wislöff (1999) found that maximal strength training in the upper body improved the double-poling performance by improved work economy. The work economy was improved by a reduction in relative workload (one repetition maximum, or 1RM) and time to peak force while double pooling. Time to peak force and reduction in relative force used while working on the ski ergometer correlated both with time to exhaustion and double-pooling economy. The results showed considerable potential for improvement in 1RM and time to peak force, even when the strength training is superimposed on a relatively large volume of endurance training.

Roller-skis have been widely used by competitive skiers as a training device when there is no snow on the ground. It appears, on the basis of a comparison between the oxygen uptake during maximal roller-skiing and maximal treadmill running, that prolonged roller-skiing taxes the oxygen-transporting system as much as, or more than, running. Thus, roller-skiing and running conceivably are of equal value as a method of endurance training (figure 16.11).

## For Additional Reading

For a comparison of injuries in classic and skating cross-country ski techniques, see Butcher and Brannen (1998).

# Alpine Skiing

One of the first studies of elite Alpine skiers was carried out by Agnevik et al. (1969 unpublished results), who tested the subjects both in the laboratory and during regular international competitions. The maximal oxygen uptake of one of the foremost Alpine skiers in the world at that time measured on the cycle ergometer was 3.9 L · min$^{-1}$ or 66 ml · kg$^{-1}$ · min$^{-1}$; his maximal heart rate was 207 beats · min$^{-1}$ (figure 16.12a). With the aid of telemetry, his heart rate was recorded before and during a competitive slalom race. Because of technical difficulties, his heart rate could be followed only during the end of the race in both giant slalom and downhill skiing. Figure 16.12a shows a heart rate of more than 160 beats · min$^{-1}$ at the start, which no doubt was attributable to emotional factors and some degree of nervousness at the start of the race. The heart rate quickly rose to more than 200 beats · min$^{-1}$, to the same maximal heart rate that was obtained at a maximal load on the cycle ergometer. The same heart rate was recorded at the end of the other competitive events. The figure also shows how the heart rate increased at "supermaximal" work rates on the cycle ergometer. In this case, the increase

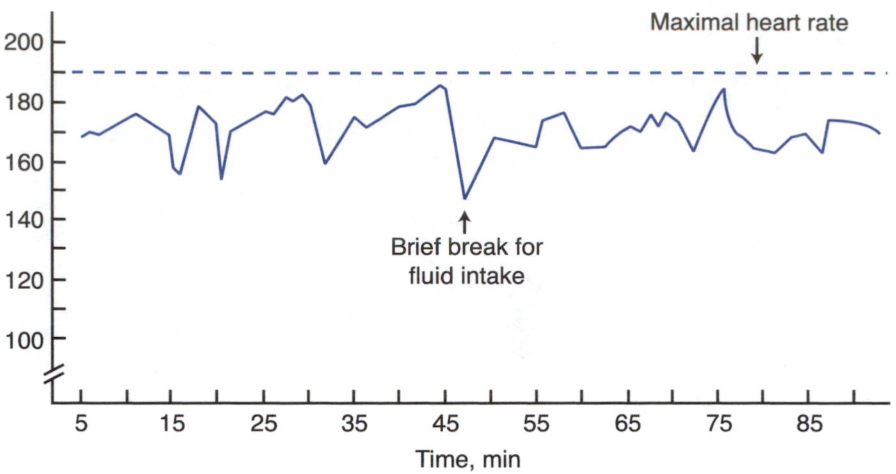


**Figure 16.11** Heart rate during 90 min of roller-skiing on a hilly road. In this subject, a heart rate of 180 beats · $min^{-1}$ corresponds to 90–93% of his maximal oxygen uptake.
From Bergh 1974.

occurred more slowly than during skiing competitions, a difference that must be attributed to psychic factors. The day following the slalom race, some of the participants covered exactly the same track, but then the oxygen uptake was determined as well. The subjects completed the run in almost exactly the same time as they had during the actual race. Figure 16.12b presents an example of the data collected. The oxygen uptake in this particular subject was 3.3 L · $min^{-1}$ or 87% of his maximal oxygen uptake measured in laboratory experiments. In another subject, the oxygen uptake was 3.9 L · $min^{-1}$ or 78% of his maximal aerobic power. On this occasion, the recorded heart rates were submaximal. Figure 16.12b shows that the heart rate measured lies on the regression line that was obtained for the heart rate–oxygen uptake relationship during tests on the cycle ergometer. This observation supports the assumption that the competition itself represents an extra psychic stress that elevates heart rate. It should be added that the winner of the World Cup in Alpine skiing in 1976 and 1977 had a maximal oxygen uptake of 5.2 L · $min^{-1}$, or 70 ml · $kg^{-1}$ · $min^{-1}$, when running on the treadmill.

Figure 16.12c presents mean values for peak blood lactic acid in connection with competitions, maximal work on the cycle ergometer, and running. The high lactic acid levels during skiing, despite a relatively short exercise time, are explained by the assumption that certain muscle groups are engaged in intense static exercise. The best skier attained in his special event the highest lactic acid levels of the entire group. Finally, it is evident from figure 16.12d that the Alpine skiers are characterized by a great isometric strength in the stretch muscles of the legs. The static muscular strength of the leg extensor muscles was very high on the skiers, on an average 2,900 N. Athletes from other sports events studied had mean values at or below 2,500 N. The skiers had even greater muscle strength than a group of weightlifters. Various studies (E. Eriksson, Nygaard, and Saltin 1977) have confirmed these results. The observed high values of leg strength in Alpine skiers are not surprising, because the force the Alpine skier must develop during actual skiing can reach several thousand newtons. It also has been shown that the electromyographic activity of the knee extensors reaches near-maximal levels during the course of a turn (Berg and Eiken 1999).

A relatively wide scatter in muscle fiber composition of the quadriceps muscle of Alpine skiers has been noted, with a mean value of 43% slow-twitch (type I) fibers (E. Eriksson, Nygaard, and Saltin 1977).

In addition to these basic requirements, the technique obviously determines the level of achievement. From a training standpoint, it is important to master a good technique despite a high lactic acid concentration in the muscles (Tesch 1995).

Watanabe and Ohtsuki (1977) performed wind-tunnel experiments to measure aerodynamic forces acting on an Alpine skier running down a slope at top speed and to clarify the relation of these forces to the skier's postural changes. The aerodynamic advantage of the so-called egg-shaped posture in Alpine skiing was confirmed. However, even at the

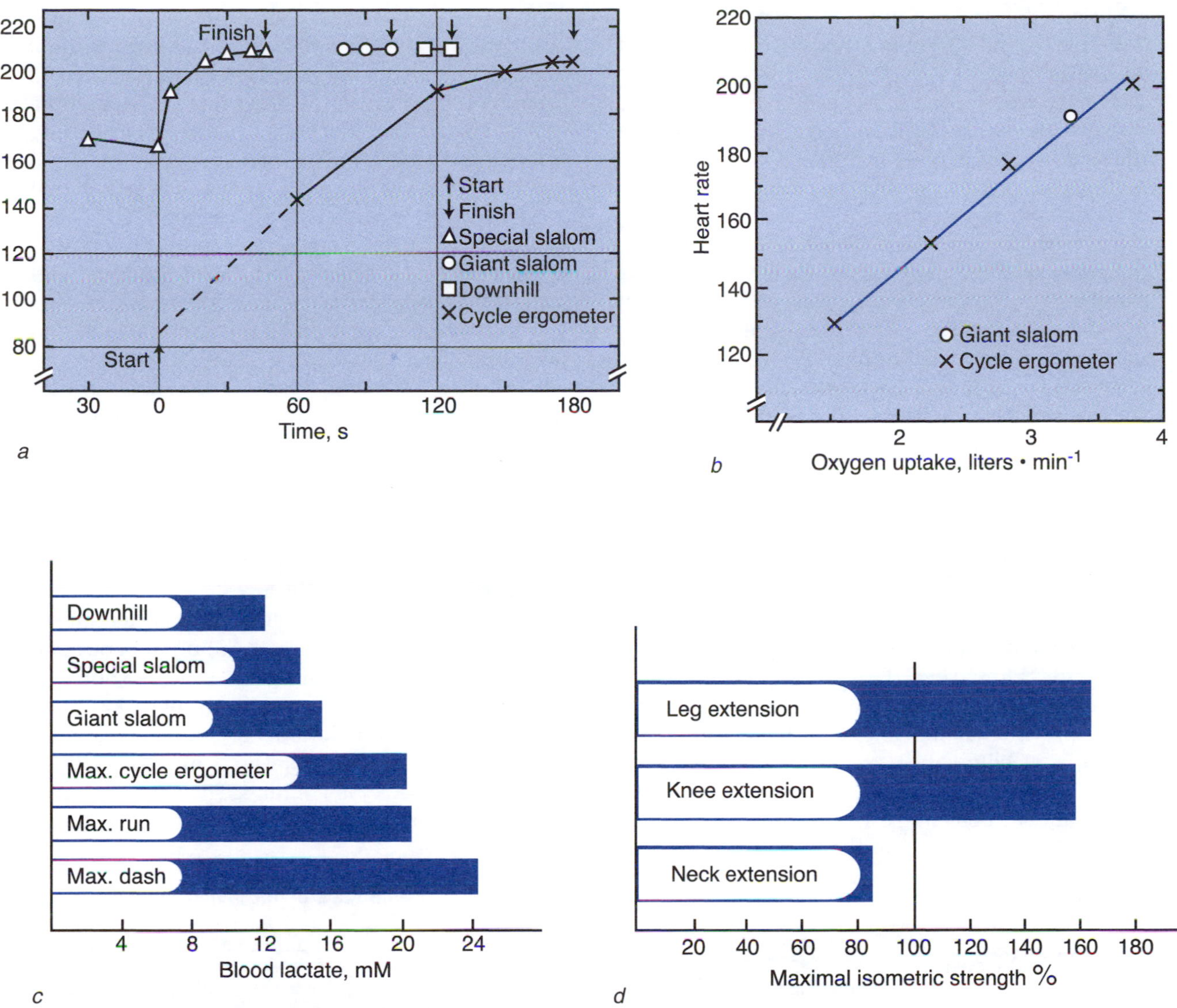


**Figure 16.12** Data on male top athletes in Alpine skiing. *(a)* Heart rate before and during competitions in Alpine skiing and during maximal exercise on a cycle ergometer. Note the high heart rate before start and the rapid increase in heart rate after start. *(b)* Heart rate in relation to oxygen uptake during cycling and giant slalom the day after actual race on the same course. *(c)* Peak blood lactate concentration after various activities; mean values of three to five determinations. Max. run = 3 × 1,000 m at maximum speed with a few minutes of rest in between; Max. dash = 5 × 50 s in a similar manner. *(d)* Maximal isometric muscle strength; 100% = data obtained on a group of recruits.

From Agnevik et al. 1969.

lowest posture, lateral extension of the arm substantially increased drag, which was compatible with the increase caused by raising the trunk.

### For Additional Reading

For further details concerning the physiological requirements of Alpine skiing, see R.E. Andersen and Montgomery (1988) and Bacharach and Duvillard (1995). For a critical evaluation of the effectiveness of ski bindings and their professional adjustment for preventing Alpine skiing injuries, see Finch and Kelsall (1998).

## Canoeing

Canoeing is rather unique in that it is about the only competitive sport in which mainly the arm and trunk muscles are engaged in endurance efforts. In addition, the same individual often competes in 500-, 1,000-, and 10,000-m races, lasting from about 1.45 to 45 min.

For the success of the canoeist, maximal aerobic power is more critical than body weight, because the slightly greater resistance caused by the friction of the canoe in the water by a few kilograms of extra body weight is insignificant. It is therefore more meaningful to express the canoeist's aerobic power in terms of liters of oxygen per minute than in milliliters of oxygen per kilogram of body weight per minute. According to Tesch et al. (1976), the mean maximal oxygen uptake of an elite group of canoeists, including two world champions, was 5.40 L · min$^{-1}$ (range 4.7– 6.1) or 68 ml · kg$^{-1}$ · min$^{-1}$ (range 64–75). Figure 16.13 indicates that some canoeists use all their aerobic power when paddling the canoe.

Although the major part of the canoeist's work appears to be performed by the arms, trunk muscles are indispensable in stabilizing the upper body during each stroke. A high maximal oxygen uptake in arm exercise therefore is a major requirement for the canoeist. Tesch et al. (1974) compared maximal oxygen uptake during leg exercise (cycle ergometer exercise) with maximal oxygen uptake during arm cranking in elite canoeists and found very high values for arm exercise, both in absolute terms (L · min$^{-1}$) and in percentage of the maximal oxygen uptake measured during leg exercise. Male junior and senior canoeists combined had a mean maximal oxygen uptake during exercise with the legs of 5.13 L · min$^{-1}$ compared with 4.45 L · min$^{-1}$ in arm exercise (87% of that in leg exercise) and 4.48 L · min$^{-1}$ during canoeing. In elite female canoeists, the maximal oxygen uptake during arm exercise was 90% of that obtained during exercise with the legs. The corresponding figure for a group of weightlifters was 78%, and 71% for physical education students. Figure 16.14 compares maximal oxygen uptake in elite canoeists during leg exercise (cycle ergometer) and arm cranking, and during 500-m, 1,000-m, and 10,000-m canoe racing. One of the canoeists, who had started systematic canoe training at a very early age, showed during repeated tests consistently maximal oxygen uptakes in arm exercise that were as high as, or higher than, in leg exercise. The fact that he started canoe training at an early age could explain, at least in part, his exceptionally high aerobic work capacity with the arms.

In a review of the physiological characteristics of elite kayak paddlers, Tesch (1983) pointed out that compared with other groups of athletes known to exhibit great upper body muscle strength, kayakers had high values for shoulder strength, endurance, and anaerobic capacity. Total body maximal oxygen uptake averaged 5.36 L · min$^{-1}$. The values for arm cranking and paddling were 4.30 and 4.67 L · min$^{-1}$.

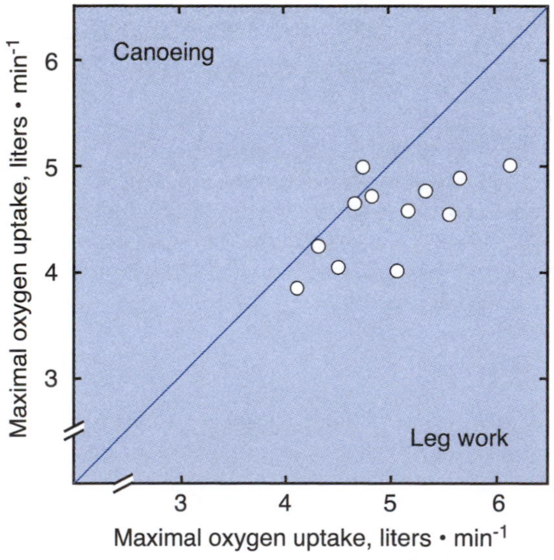


**Figure 16.13** Relationship between maximal oxygen uptake during canoeing and during exercise with the legs in elite canoeists.

From Tesch et al. 1974.

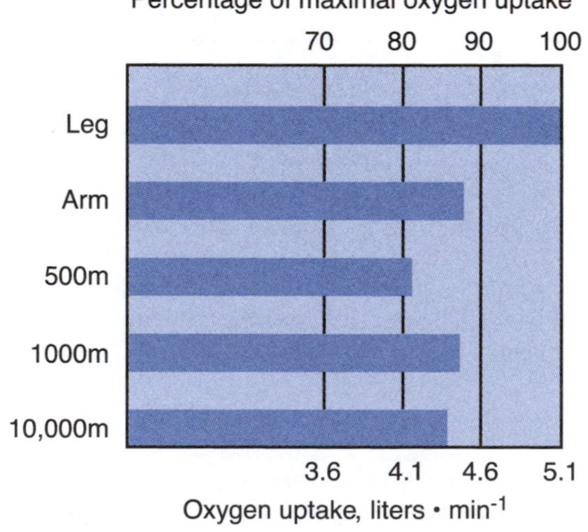


**Figure 16.14** Oxygen uptake (liters per minute and percentage of the maximal oxygen uptake measured during leg exercise) in elite canoeists (junior and senior) during the final minutes of 500-m, 1,000-m, and 10,000-m races. The maximal oxygen uptake in arm exercise is included for comparison.

From Tesch et al. 1974.

High blood lactate levels were observed under training conditions and at the end of a race.

For comparison in a group of Greenland native kayakers/hunters studied by Vokac and Rodahl (1977), the maximal oxygen uptake in arm cranking averaged 89% (81–104%) of that of cycling on a cycle ergometer. This is approximately the same rate between maximal oxygen uptakes in arm and leg exercise as found by Vrijens et al. (1975) in various nonnative subjects, including highly trained athletes specialized in kayak paddling.

Studies of Swedish elite canoeists (Tesch et al. 1976) indicate that canoeists generally are unable to tax their maximal oxygen uptake fully during the 500-m race. In 1,000-m canoe races, almost maximal heart rates have been recorded (189 beats · $min^{-1}$ compared with a maximum of 195) during the last half of the race. In two Swedish canoeists studied during a 10,000-m race lasting 45 min, the mean heart rate during the entire race was kept at a level corresponding to 96% and 98% of the maximal heart rate. In one of them, the heart rate was maximal during the last 2,000 m of the race.

According to Tesch et al. (1976), the higher the maintained speed during the race, the greater the lactate levels in the blood, the highest values being obtained at the end of the 500-m race (figure 16.15). Under comparable circumstances, blood lactate values are higher in senior than in junior elite canoeists and higher in the final heat than in the same canoeist's semifinal or trial heat, perhaps indicating the positive relationship between degree of exertion and blood lactate levels.

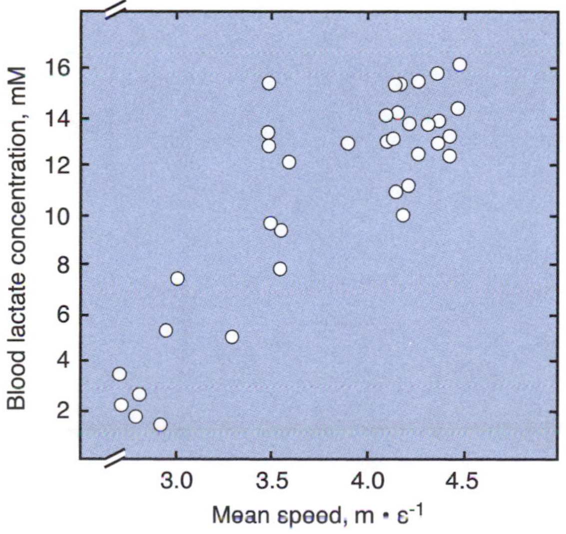


**Figure 16.15** Blood lactate levels (mM) in elite canoeists, related to mean speed (m · $sec^{-1}$).

From Tesch et al. 1974.

Whereas the maximal oxygen uptake measured during maximal leg exercise tended to remain more or less unchanged throughout the year in elite canoeists, the maximal oxygen uptake measured during arm cranking was, on the average, 8% higher during the active canoeing season than during the winter (Tesch et al. 1976).

Tenosynovitis is a common injury in long-distance canoeists. This was studied by du Toit et al. (1999), who found that development of tenosynovitis is not related to the equipment used but is probably caused by difficult paddling conditions, in particular uneven surface conditions that alter paddling style. They concluded that important factors include level of fitness and the ability to balance a less stable canoe, thereby maintaining optimum paddling style without repeated eccentric loading of the forearm tendons to limit hyperextension of the wrist during the pushing phase of the stroke.

## Rowing

Rowing is divided into two related disciplines: sweep rowing and sculling. In the sweep boats, each oarsman uses a single oar, whereas in sculling he or she uses two smaller oars (sculls). In contrast to traditional rowing boats, the rowing is made more efficient with the use of a sliding seat, thereby adding leg extension to the work performed by the arms and upper body. In sweep rowing, the boats include, two, four, or eight persons. Pairs and fours are rowed both with and without coxswain, whereas a coxswain is always present in the eights. Sculls are rowed without a coxswain and include single, double, and quadruple boats.

The relative anaerobic contribution to a male rower's performance during a 2-km race has been estimated to about 20% (Secher 1993). The relatively recent inclusion of a 2-km race for females in rowing competitions triggered Pripstein et al. (1999) to study the aerobic/anaerobic energy contributions in female rowers during a simulated 2-km race on a rowing ergometer. The average time taken for the rowers to complete the distance was 7.5 min, and the anaerobic system accounted for 12% of the rower's total energy production during the race. Secher (1973, 1983, 1993) compiled the winning times in international rowing championships from 1893 to 1985. The results are presented in figures 16.16a and 16.16b for eights and single sculls, respectively. As is evident from the figures, there is considerable scatter of the data because of varying wind and water conditions. Because these external conditions change very quickly (in a matter of minutes), it is quite meaningless to compare individual results from one race to another even when the time interval is as short as 1 h. Mean results of races from several regattas, on the other hand, indicate the general level of performance for any 1 year. On this basis, the mean improvement in rowing performance is 0.66 s per year, corresponding to a 1.6% increase in speed. Secher (1973) explained this improvement as the result of a combination of factors: the increasing body size and consequently an increase in the maximal aerobic power of the general population; better selection and training of rowers; better rigging of the boats, allowing better technique and greater mechanical efficiency; and better boats, causing less water resistance.

By actually measuring the oxygen uptake of the rower while rowing at submaximal intensity during near steady-state conditions, Secher (1983) observed that the mean oxygen uptake during rowing increases with velocity to the power 2.4. From their data, di Prampero et al. (1971) concluded that the mechanical power output necessary to maintain the boat progression as well as the energy expenditure appears to increase as the 3.2 power function of the average speed.

The maximal aerobic power of the rower seems to be a limiting factor in rowing performance. This supposition is strengthened by the finding of a positive correlation between the average oxygen uptake of the crew and their placing in international championships (figure 16.17). The mean maximal oxygen uptake was 6.1 L · min$^{-1}$ for the crew taking first place, 5.7 L · min$^{-1}$ for the crew taking sixth place, and 5.1 L · min$^{-1}$ for the crew attaining 12th place. Furthermore, a comparison of the mean maximal oxygen uptakes of individual international champion oarsmen shows the following figures for the maximal oxygen uptakes in liters per minute: German, 5.9 ($n$ = 5); Norwegian, 5.8 ($n$ = 6); Danish, 5.7 ($n$ = 4) (Vaage and Secher, unpublished data). The mean maximal oxygen uptake of a group of 20 top rowers (members of national teams) was 5.0 L · min$^{-1}$ compared with 4.3 L · min$^{-1}$ in a group of 44 beginners (Secher et al. 1983). In a group of 21 highly conditioned oarsmen, Clark, Hagerman, and Gelfand (1983) observed an average maximal oxygen uptake of 6.6 L · min$^{-1}$.

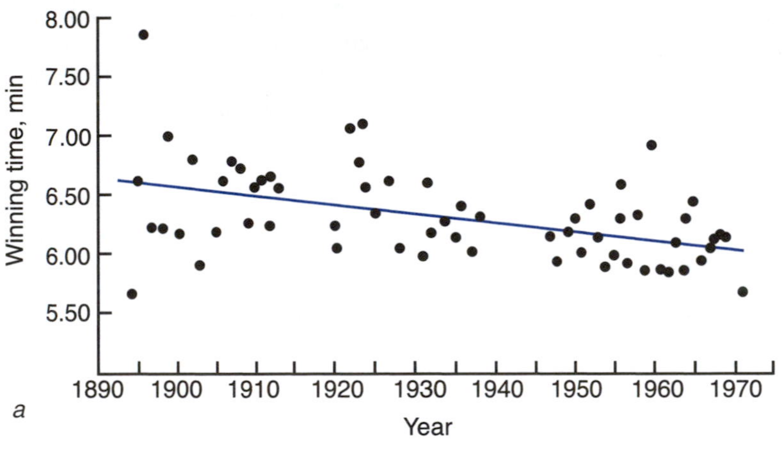


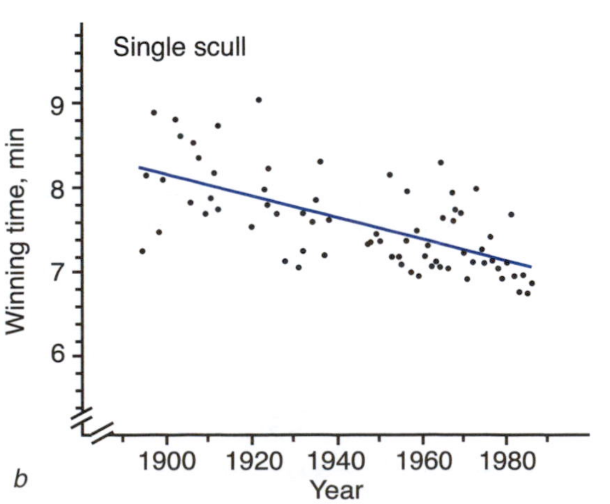


**Figure 16.16** *(a)* Improvements of results in international rowing championships. *(b)* Results obtained by Fédération Internationale des Sociétés d'Aviron (FISA) regatta winning single scull over 2,000 m from 1893 to 1986. The regression line is shown.

From T. Reilly et al., 1990, *Physiology of sports* (London: Chapman and Hall), 261. Reprinted by permission of Routledge, Inc., part of The Taylor & Francis Group.

So far, we have considered performance in rowing championships, expressed in terms of mean speed during the entire race. However, a closer analysis of the speed attained at different points during the race shows a general speed pattern starting with an initial very high speed followed by a decline to a steady level after the first quarter of the race, and then increasing again in a final spurt. Secher (1993) measured oxygen uptake during simulated racing conditions on a cycle ergometer leading to about the same state of exhaustion in all the subjects at the end of 6 min. Oxygen uptake increased more rapidly during the simulated racing experiment than during the control experiment when the speed was kept constant during the entire 6-min period. The author concluded that the greater expenditure early in the simulated race was at least partially covered by the more rapid attainment of maximal oxygen uptake. This appears to provide a physiological basis for the procedure generally practiced during racing of this kind. It also supplies the scientific basis for the development of an optimal speed profile for any type of race. In this connection, the heart rate, generally speaking, attains its peak value more rapidly during a real race than during a training race (figure 16.18).

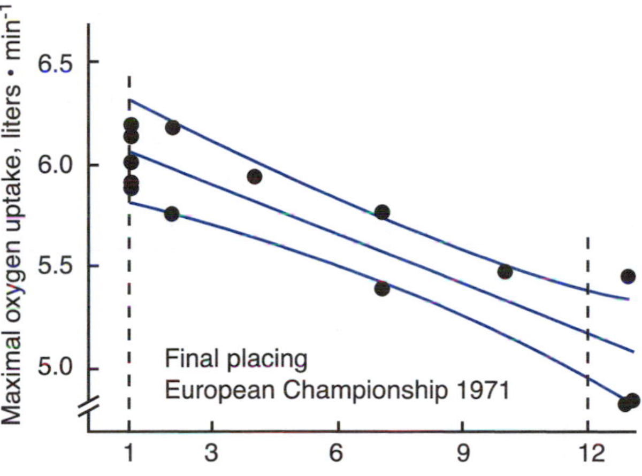


**Figure 16.17** Relationship between the maximal oxygen uptake of individual Scandinavian rowers and their final placing in the European championship in 1971; the higher the maximal oxygen uptake, expressed in liters per minute, the better the performance. The outer lines indicate 95% confidence limit.

From Secher, Vaage, and Jackson 1976 (unpublished results).

The best rowers tend to be heavy. This can be explained by the positive correlation between body weight and maximal oxygen uptake, and the fact that there is no statistically significant correlation between maximal oxygen uptake in milliliters per kilogram of body weight and rowing performance (Vaage, unpublished data). In addition, there is a positive correlation between rowing strength and body weight. This means that maximal oxygen uptake divided by body weight is not a good indicator of rowing performance (Secher 1993). It also appears that the larger body mass is associated with a larger anaerobic capacity. We therefore can conclude that the three attributes of greater maximal oxygen uptake, greater rowing strength, and greater anaerobic capacity together make a superior rower, and that these attributes more than compensate for the higher water resistance caused by the greater body weight.

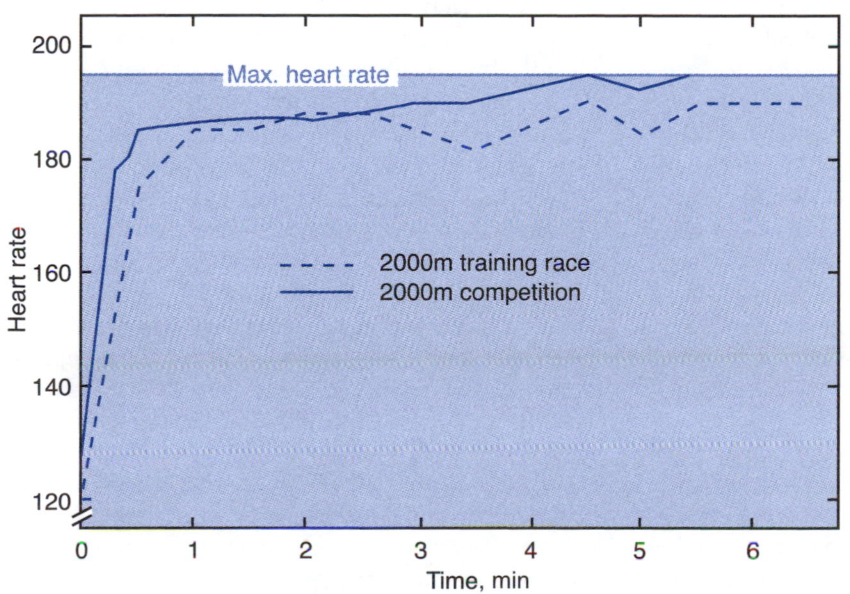


**Figure 16.18** Heart rate of a rower during a real race and during a training race.

Courtesy of E.D.R. Pruett.

## Ball Games

Generally speaking, most types of ball games represent more or less intermittent work with frequent, short bursts of physical effort interspersed with brief pauses. For this reason, various ball games have not, in the past, required the same level of physical endurance or aerobic power in the players as required in long-distance runners, cross-country skiers, or athletes engaged in other events requiring continuous, long-lasting effort of near maximal intensity.

### *Soccer*

In the past, studies of superior individual soccer players or national teams have, with some exceptions, shown maximal oxygen uptakes significantly below the levels found in the endurance athletes just mentioned. Thus, 30 to 35 years ago, maximal aerobic power for Swedish top soccer players was reported to be in the range of 50 to 69 ml · kg$^{-1}$ · min$^{-1}$. However, current figures for players are in the range of 60 to 70 ml · kg$^{-1}$ · min$^{-1}$. A minimum of 65 ml · kg$^{-1}$ · min$^{-1}$ has been suggested as desirable in top players. Differences in maximal aerobic power with playing position are today relatively small; the highest figures are seen in midfielders, and outside fullbacks have higher values than center-backs (Ekblom 1986; Shephard 1999). Further improvements in the level of play in soccer require greater emphasis on optimizing the athletes' functional strength and endurance capacity. Superior technical and tactical ability in soccer can be consistently demonstrated throughout the course of a 90-min game only by players with a high level of endurance and strength (Wisløff, Helgerud, and Hoff 1998).

The resting heart rate of top soccer players is typically 48 to 52 beats · min$^{-1}$, whereas figures for maximal heart rate are similar to those for the same age group in the general population: 187 to 193. Echocardiographic studies have revealed a moderate enlargement of the left ventricular cavity and heart volume above 13 ml · kg$^{-1}$. Resting stroke volume is typically about 100 ml, and a high peak oxygen pulse (25–29 ml · beat$^{-1}$) suggests that a large stroke volume is maintained during maximal effort. Hemoglobin concentrations are generally normal, but ferritin levels are sometimes low, suggesting that one should be aware of latent iron deficiency (Shephard 1999).

Figure 16.19 presents the heart rate in one top player during a major soccer match. The heart rate, with the exception of a few brief occasions, remained well below the maximal heart rate for this player and reflected a fairly regular pattern, with periods of high effort interspersed with brief pauses, during which the heart rate dropped as much as 50 beats · min$^{-1}$. His mean heart rate during the entire match (90 min) was 175 beats · min$^{-1}$, whereas his maximal heart rate was 189 beats · min$^{-1}$. Similar studies have been made in other players, and although the mean heart rate differs somewhat from one player to another during the match, the alternating pattern with changing heart rate is a fairly common feature found in most soccer players studied.

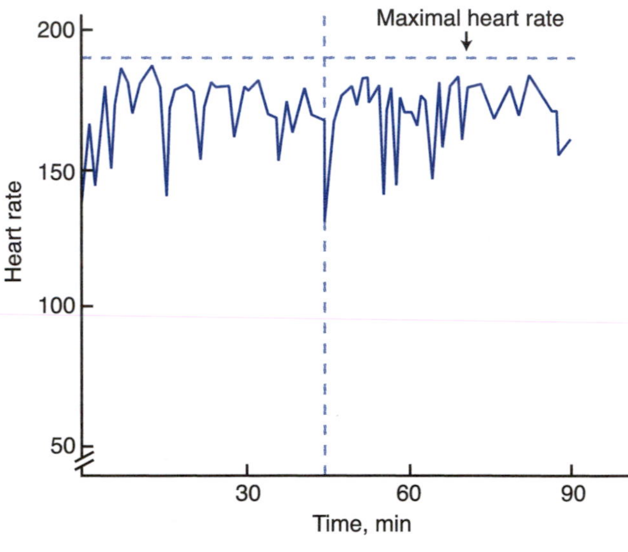


**Figure 16.19** Heart rate in a soccer player on the national team during an important match.

From Agnevik 1970b.

Soccer players cover a total distance of about approximately 10 km per game, of which 8% to 18% is at the highest individual speed. At higher levels of competition, a greater percentage of the game is performed at maximal speed. The average aerobic energy yield during a game is around 80% of the individual maximum. Registration of blood lactate during a soccer match has shown peak concentrations of 9.5 mM · L$^{-1}$ at the end of the first half and 7.2 mM · L$^{-1}$ at the end of the second half (Ekblom 1986). Many other studies, however, have reported blood lactate concentrations of 4 to 6 mM · L$^{-1}$ (Shephard 1999). Obviously, much depends on when the sample is collected. For instance, blood lactate is increased markedly during rapid dribbling and can be metabolized during intervals of less intensive play. Because of the higher energy yield, most players have empty muscle glycogen stores at the end of the match. Generally, the level of activity in the second half of the game is reduced compared with the first. There is some evidence that increased aerobic fitness helps to counteract this (Tumilty 1993).

There has been an explosion in women's soccer over the last 20 years. According to Mandelbaum and Putukian (1999), 22% of the world's soccer players are women, and in the United States, women represent 43% of all soccer players. Although the same types of traumatic soccer injuries are found both in male and female players, there are notable differences. Whereas ankle sprains are as frequent in female as in male soccer players, knee injuries occur more often in female players.

### For Additional Reading

For further reading of the biology and medicine of soccer, see Bangsbo (1994) and Shephard (1999). For physiological reviews of American football, see J. Brewer and Davis (1995) and Pincivero and Bompa (1997).

## Golf

A golfer plays the game with a combination of mind and physical body. Steinberg, Frehlich, and Tennant (1995) investigated the relationship between eye and hand dominance and the relative positioning of the ball with respect to the subject's eyes on putting performance. They found that pure dextral (right-eyed and right-handed) golfers demonstrated less absolute and variable errors in their putting performance when they focused their eyes midway between the ball and their feet than when they positioned their eyes directly over the ball.

The golf swing involves a complex sequence of body movements that adjust according to the demands of a given shot. According to Fujimoto-Kanatani (1996), who analyzed the swing patterns of 13 elite professional golfers, the critical element during the swing is the constraint of the right knee and hip during the take-away phase. During take-away, the lower body rotates about the right leg while the upper body rotates about a point near midshoulders. Kao et al. (1995) studied the role of the scapular muscles in the golf swing in 15 competitive male golfers and demonstrated the importance of the scapular muscles in the golf swing and the need for specific strengthening exercises.

Although golf has not been considered to be a sport associated with injuries, epidemiological studies have revealed that back and elbow injuries are most common in male amateur golfers, whereas lower back and wrist injuries occur in the male professional golf players. Female professional golfers are most likely to suffer injuries to the wrist (McCarroll 1996). Such injuries are best avoided by a good strength and flexibility program, proper warm-up exercises, and good stroke mechanics (Kohn 1996).

### For Additional Reading

For a systematic description of the mind–body connection of the golf swing, see Selicki and Segall (1996).

## Tennis

Tennis is a physically demanding sport that requires considerable neuromuscular skills. As pointed out by Chandler (1995), the junior player should begin the physical development for this sport with a sound program for physical fitness, including flexibility, cardiorespiratory endurance, and muscular strength and endurance. This should be followed by the development of athletic fitness and sport-specific movement training as well as injury prevention.

To obtain a more specific understanding of muscle function during the tennis volley under different ball placement and speed conditions, J.W. Chow et al. (1999) examined the activity of selected superficial muscles of the stroking arm and shoulder and muscles related to postural support during the volley in seven skilled tennis players. This was done under 18 experimental conditions, including variations in lateral contact location (forehand and backhand), ball contact height (high, middle, and low), and ball speed (fast, medium and low). A ball machine was modified so that the subjects could not predict the ball trajectory before it was released from the machine. Surface electromyographic techniques were used to determine muscle activity, and the critical instants of a volley were determined by using two force platforms and two high-speed video cameras. The investigators found that muscle activity in general increased with increasing ball speed. The extensor carpi radialis was more active than the flexor carpi radialis during both the forehand and backhand volleys, suggesting the importance of wrist extension/abduction and grip strength. The increase in electromyographic levels in the forearm muscles shortly before the ball impact indicated that the subjects did not tighten their grip and wrist until moments before ball impact. Both antero-middle and postero-middle deltoids were active in most stroke phases.

The Leuven Tennis Performance Test is a recently developed test to measure stroke performance in matchlike conditions in elite tennis players. Vergauwen et al. (1998) critically evaluated the test as to its value for research purposes. They concluded that this test is an accurate, reliable, and valid instrument for evaluating stroke quality in high-level tennis players.

Kannus et al. (1995) studied the difference in bone mineral content between the playing and the nonplaying arms in Finnish female national-level tennis players. The authors found that bones of the playing extremity clearly benefited from active tennis in that it increased the mineral mass. According to the authors, the benefit of playing tennis (or squash) is about two times greater if females start playing at or before, rather than after, menarche. In a study of former competitive male tennis players, Kontulainen et al. (1999) found that, on the average, the players' exercise-induced bone mineral gain in the playing arm did not disappear with time despite decreased playing activity. High bone mineral density also has been observed in elite professional volleyball players, both male (Calbet, Herrera, and Rodriguez 1999) and female (Alfredson et al. 1998).

## For Additional Reading

For a review of common sports injuries in young tennis players, see Bylak and Hutchinson (1998), and for a review of the biomechanics of tennis elbow specifically, see Roetert et al. (1995).

## *Table Tennis*

In an early study by Lundin (1973), the maximal oxygen uptakes in seven male elite table tennis players, including three world champions, were found to be 4.42 L · min$^{-1}$ (3.6–5.1 L) corresponding to 65.0 ml · kg$^{-1}$ · min$^{-1}$ on average. According to Lundin, there were only small changes in maximal oxygen uptake in the same player from month to month, possibly because the elite table tennis player is engaged in systematic year-round training.

The heart rate during important matches varied considerably. In some cases, the heart rate was maintained close to the maximal level (figure 16.20). In other cases, the heart rate dropped far below the maximal level, as in figure 16.21, when the losing player appeared to have lost interest in the game. In general, however, Lundin (1973) concluded that among top Swedish table tennis players, the heart rate during a match was, on the whole, 20 to 30 beats · min$^{-1}$ below the maximal level. During the actual playing, the blood lactate concentrations were around 2 to 3 mM with peak values up to 5 mM. Similar concentrations were observed in the muscles. Measurements of the actual oxygen uptake during a simulated table tennis match in seven elite players showed that, on the average, they taxed slightly more than 70% of their maximal aerobic power, corresponding to roughly 50 ml · kg$^{-1}$ · min$^{-1}$. In four players, the proportion of slow- and fast-twitch fibers in the deltoid and quadriceps muscles varied markedly, with a mean value of approximately 45% fast-twitch (type II) fibers.

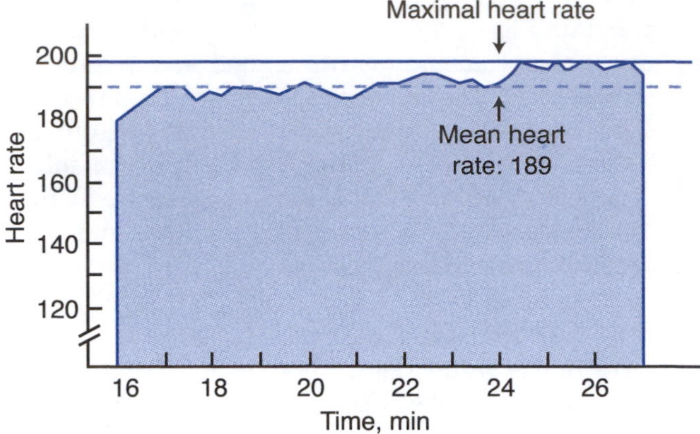


**Figure 16.20** Heart rate in an elite table tennis player during a match.

From Lundin 1973.

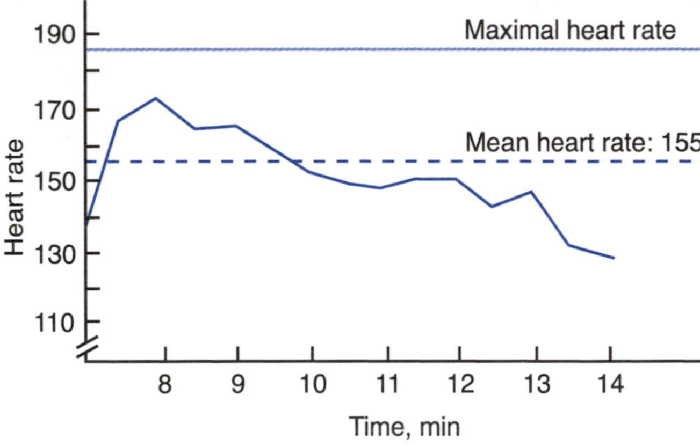


**Figure 16.21** Heart rate in an elite table tennis player during a losing match.

From Lundin 1973.

## *Badminton*

An early study of a group of Swedish top badminton players (Agnevik 1970a) revealed that the mean maximal oxygen uptake of the female players was 2.57 (2.3–2.7) L · min$^{-1}$ or 46.9 ml · kg$^{-1}$ · min$^{-1}$. For the male players, the corresponding figures were 3.74 (3.0–4.4) L · min$^{-1}$, or 55.6 ml · kg$^{-1}$ · min$^{-1}$. This showed that the maximal oxygen uptake of these badminton players, at that time, was quite low compared with other elite athletes in endurance events.

Measurements of the oxygen uptake during a simulated badminton match demonstrated oxygen uptakes up to 3.9 L · min$^{-1}$ for men (as against 3.9 L · min$^{-1}$ during maximal exercise on the cycle ergometer) and 2.6 L · min$^{-1}$ for women (as against 2.6 L · min$^{-1}$ during maximal cycling), showing that during a hard game the players actually tax all their aerobic power maximally. This is in line with a study of Indian national male badminton players by Majumdar et al. (1997) demonstrating lactate values around 8 to 10 mM and heart rates ranging from 82% to 100% of maximum during badminton matches.

Recordings of heart rates during important international matches (Agnevik 1970a) showed that the players attained near-maximal values in single matches, whereas doubles and mixed doubles matches clearly showed lower heart rates (figure 16.22). Blood lactate values around 12 to 13 mM were obtained. In a study of 10 junior national-level female badminton players, 13 to 14 years old, who had 3 to 6 years of training, Ghosh et al. (1990) observed that heart rates were higher in the second and third games than in the first, but there was no difference in blood lactate concentration.

## *Bandy*

Häkkinen and Sinnemaki (1990) examined maximal oxygen uptake, anaerobic power, and neuromuscular performance characteristics in bandy players of

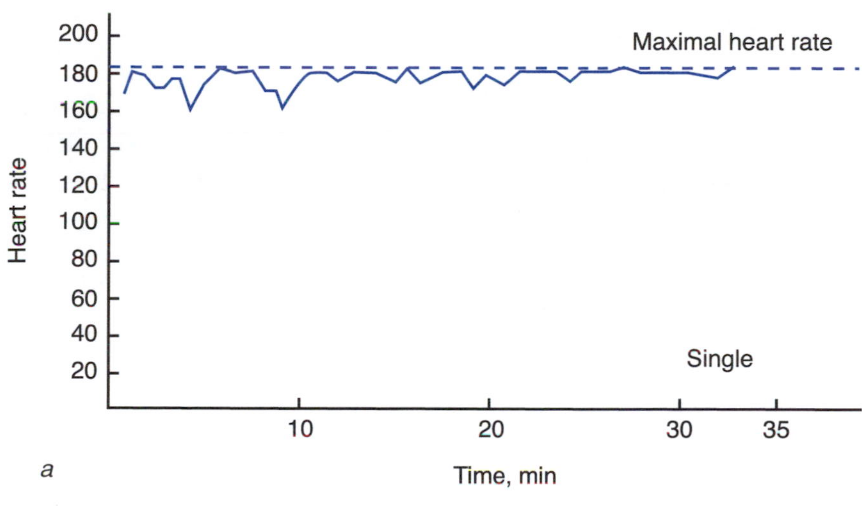


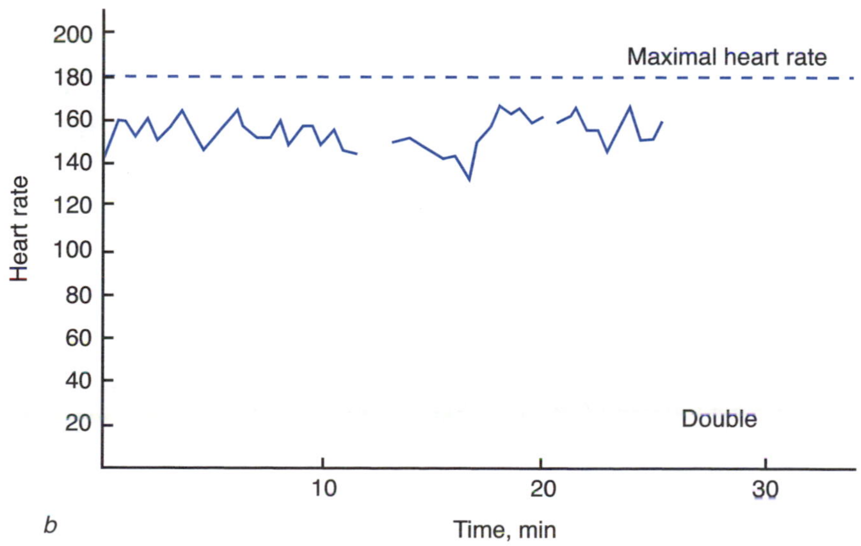


**Figure 16.22** Heart rates recorded in an elite badminton player during a match. Upper panel: singles; lower panel: doubles.

From Agnevik 1970a.

two different competitive levels. In the total subject sample, the individual values for $\dot{V}O_2max$ had the highest correlation with the individual's playing ability in bandy. The elite bandy players differed from those players of a lower competitive level especially with regard to maximum oxygen uptake and also maximal anaerobic power, whereas only minor differences were observed between the groups in various neuromuscular performance characteristics.

## *Ice Hockey*

Ice hockey is a very intense, intermittent sport. The typical player performs for 15 to 20 min of a 60-min game. Each shift lasts from 30 to 80 s with 4 to 5 min of recovery between shifts. These high-intensity bursts with rapid changes in velocity and duration and frequent body contact require a wide variety of motor skills and a high level of fitness to compete successfully at an elite level (Quinney 1990).

Forsberg et al. (1974) found that the mean maximal oxygen uptakes of the Swedish elite ice hockey team during the 1973–74 season was 4.9 L · min$^{-1}$ (4.5–5.5) or 65 ml · kg$^{-1}$ · min$^{-1}$ (61–69). Slightly lower values (56 ± 4 ml · kg$^{-1}$ · min$^{-1}$) were observed in a study of Hungarian top league ice hockey players (Heller et al. 1991).

According to Forsberg et al. (1974), the oxygen uptake during simulated matches demonstrated that the players at maximal intensity used at the most 85% to 90% of their maximal oxygen uptakes measured during running on the treadmill.

Heart rates recorded during important international ice hockey matches showed mean heart rates of about 180 beats · min$^{-1}$, with peaks reaching maximal heart rate values (204 beats · min$^{-1}$). Heart rates were recorded and blood lactate concentrations were measured in a Swedish forward player during all three periods in the international championship match against Soviet Russia in 1974. As is evident from figure 16.23, the more intense the play, the higher the blood lactate level. Again, the intermittent nature of activity is reflected in the heart rate record shown in the upper panel of figure 16.23. Similar findings were reported by H. Green et al. (1976) in Canadian ice hockey players and by Boyle et al. (1994), who found that the mean heart rate in nine international male hockey players wearing a telemetric heart rate monitor was 159 ± 8 beats · min$^{-1}$ during competition.

### FOR ADDITIONAL READING

For an extensive discussion of the physiology of ice hockey, see Montgomery (1988).

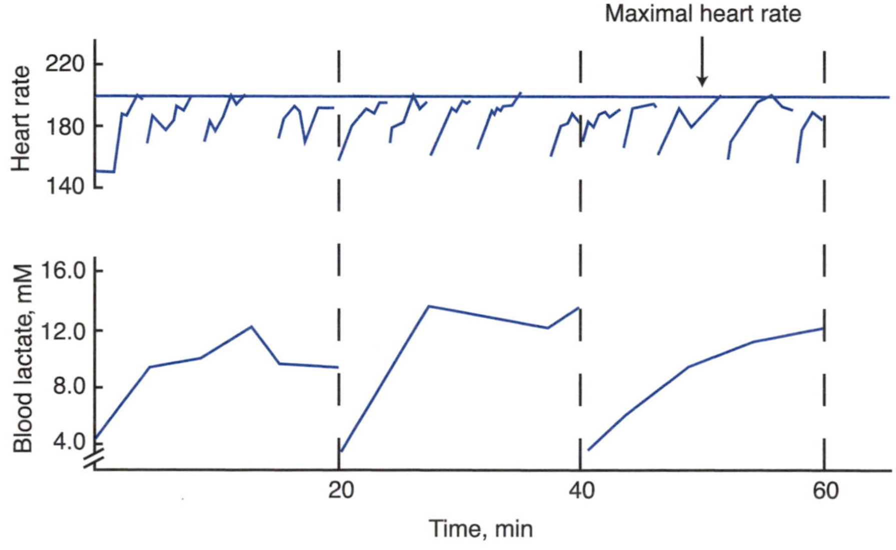


**Figure 16.23** Relationship between heart rate and blood lactate concentration in an ice hockey player during an important match.

From Forsberg et al. 1974.

# Chapter 17

# Applied Work Physiology

In this chapter, which deals with occupational work, we have deliberately used the term *work*, meaning job, manual labor, or mental pursuit, instead of the term *exercise* as used throughout the rest of the book.

The main objective of the work physiologist is to enable working individuals to accomplish their tasks without undue fatigue so that at the end of the working day, they are left with sufficient vigor to enjoy their leisure time.

With the development of mechanization, automation, and many work-saving devices, modern technology has eliminated much heavy physical work. Nevertheless, some heavy physical work, at least occasionally, is still obligatory in a number of occupations, such as commercial fishing, agriculture, forestry, construction, transportation, and many service occupations. In some cases, when the workload appears excessively high, it is evident that the only way the task can be accomplished is by intermittent work (i.e., brief work periods interspersed with short periods of rest). However, the tendency is to eliminate physical strain, while at the same time the need for increased output and greater efficiency through rationalization and automation has accelerated the tempo in most industrial operations. The short working week has resulted, on the whole, in higher work intensity. The outcome is increasing nervous tension and mounting emotional stress. Consequently, the greatest problem in many industrial operations today is not the physical load but rather the mental stress and unfavorable working environment. Furthermore, static muscular work, such as that of computer operators, is an increasing problem in many occupations.

The task facing the work physiologist in the field is to assess the strain imposed on the worker by the total stress of the work and the working environment. Because practical experience has shown that one cannot tax more than some 30% to 40% of one's maximal aerobic power during an 8-h working day without developing subjective or objective symptoms of fatigue, one of the most obvious problems is to determine the ratio between workload and work capacity. If the burden placed on the worker is too high in relation to the person's capacity for sustained physical work, fatigue invariably will develop. This is true whether the work in question involves the entire body (large muscle groups) or only part of it (small muscle groups). A basic task for the work physiologist therefore is to measure the rate at which work is being done and match this rate with the worker's ability to perform the work. In this chapter we shall deal with the applied physiology of work, both manual and mental, and the effect of the working environment on working performance.

## Factors Affecting Work Performance

The relationship between work rate and work capacity is affected by a complicated interplay of many factors, internal as well as external, which must be taken into consideration (figure 17.1).

The ability to perform physical work depends on the ability of the muscle cell to transform chemical energy in the food into mechanical energy for muscular work. This in turn depends on the capacity of the service functions that deliver fuel and oxygen to the working muscle, that is, on the nutritional state, nature, and quality of the food ingested; frequency

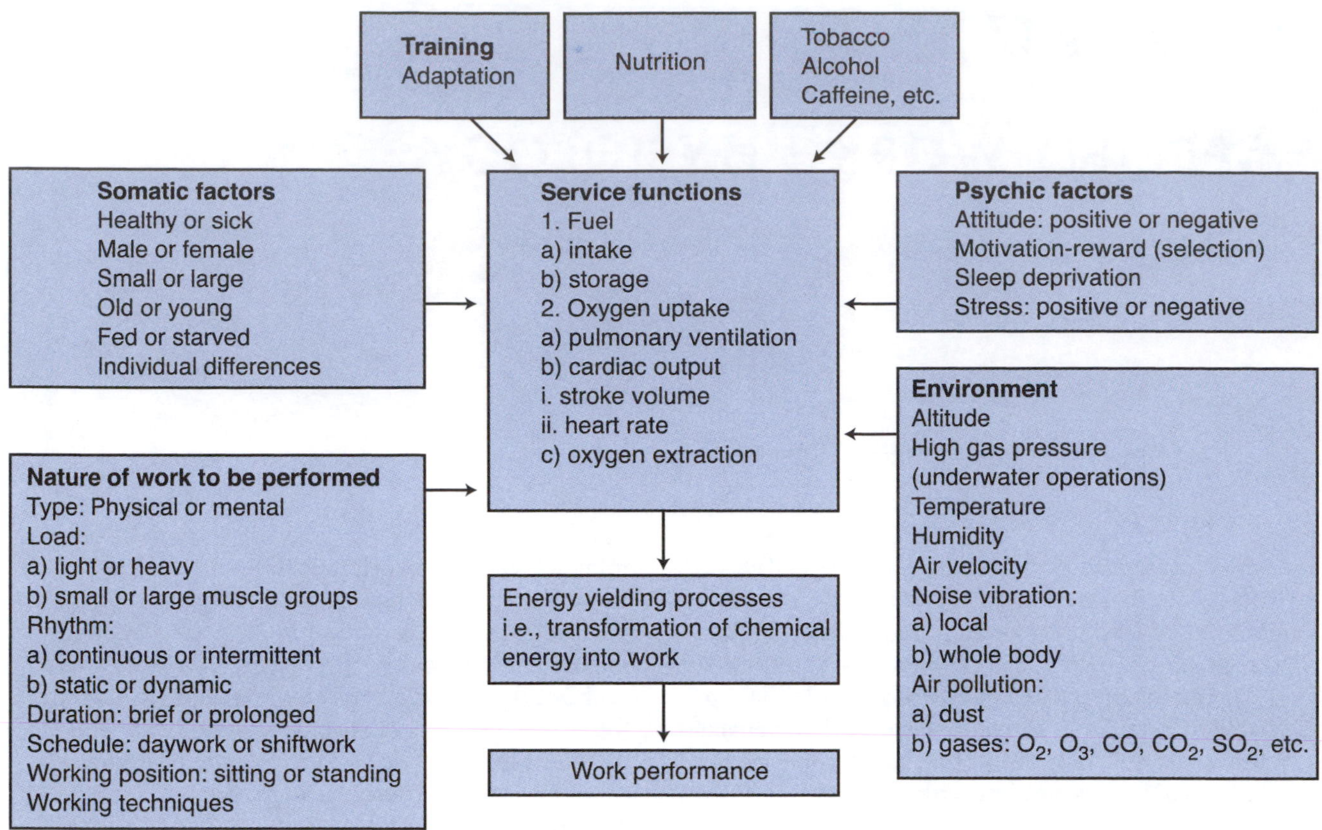


**Figure 17.1** Factors affecting work performance.

of meals; oxygen uptake including pulmonary ventilation, cardiac output, and oxygen extraction; and the nervous and hormonal mechanisms that regulate these functions.

Many of these functions depend on sex, age, body dimensions, and state of health. In addition, physical performance is influenced, to a significant extent, by psychological factors, notably motivation, attitude to work, and the will to mobilize one's resources to accomplish the task in question. Several of these factors are affected by training and adaptation.

Physical performance also is greatly influenced, directly or indirectly, by factors in the external environment. Thus, air pollution affects physical performance directly by increasing airway resistance and thereby pulmonary ventilation and indirectly by causing ill health. The same applies to tobacco use. Alcohol also can have a negative effect on performance (Kahn and Cooper 1990; Price and Hicks 1979). Noise is a stress that not only damages hearing but also elevates heart rate and affects other physiological parameters that reduce physical performance. Cold weather, if severe, can reduce physical performance because of numbness of the hands or lowered body temperature. But it also can involve the hobbling effect of bulky clothing and the slowing down of ordinary, simple functions because of snow and ice. Heat, if intense, can greatly reduce endurance because more of the circulating blood volume must be devoted to transporting heat rather than transporting oxygen and because sweating can lead to dehydration. High gas pressures, encountered in underwater operations in connection with modern offshore oil exploration, present relatively new and rather unique problems for the work physiologist. Work in space represents another set of problems concerning the environmental effects on physical performance capacity. In fact, significant muscle fiber atrophy has been noted after only 5 days in space (F.W. Booth and Criswell 1997).

Finally, the nature of the work to be performed, apart from work intensity and duration, is of decisive importance when we consider an individual's capacity to endure prolonged work stress. Because all life functions generally consist of rhythmic, dynamic muscular work in which work and rest, muscle contraction and relaxation, are interspersed at more or less regular, fairly short intervals, the ideal way to perform physical work is to perform it dynamically, with brief work periods interrupted by brief pauses. This routine provides some rest during the actual work period so the worker can avoid fatigue and exhaustion. Similarly, the work position is also important in that working in a standing position represents a greater circulatory strain than does working in a sitting position. Conversely, working in a standing position that permits the worker to move about and thereby vary the load on individual muscle groups and facilitate circulation at times is preferable. The working technique is of major importance in conserving energy and providing varied use of different muscle groups. The monotony of a working operation is a stress for some individuals but a relief for others who can carry on the work more or less automatically while thinking about something else. In any case, the tempo of work performance is very important and can impose a type of stress that in some instances is unbearable or harmful to the individual. Finally, the work schedule, including shift work, is a problem requiring increasing attention in modern industry.

**These definitions of terms and units will be helpful in this chapter:**

**power:** Ability to work, the faculty of performing work; capacity for performance; the rate of transfer of energy.

**capacity:** Ability; maximum power output.

**load:** The burden placed on the worker; the rate at which work is being done at any time.

**units:** To facilitate the transition from old units to the new SDI system (Le Systéme International d'Unités), both the conventional and the new units are used.

- The new unit for force ($F$) is newton (N). $F = m \times a$, where $m$ = mass and $a$ = acceleration. $1\ \text{N} = 1\ \text{kg} \times \text{m} \times \text{s}^{-2}$ = the force which gives the mass of 1 kg an acceleration of $1\ \text{m} \times \text{s}^{-2}$. In the old system, the commonly used unit for force is kilopond (1 kp is the force acting on the mass of 1 kg at normal acceleration of gravity); 1 kp = 9.80665 N or approximately 10 N.
- The unit for work ($W$) or energy is derived from the equation $W = F \times L$ = force times distance. The unit for force is N and the unit for distance is m (meter); therefore, the unit for work or energy is N $\times$ m = joule (J). 1,000 J = 1 kJ; 1,000 kJ = 1 MJ; 1 cal = 4.1868 J; 1 kcal = 4.1868 kJ.
- For power (work/time), the following units apply:

$$1\ W = 1\ \text{J} \times \text{s}^{-1} = 6.12\ \text{kpm} \times \text{min}^{-1}$$
$$9.81\ W = 1\ \text{kpm} \times \text{s}^{-1}$$
$$16.35\ W = 100\ \text{kpm} \times \text{min}^{-1}$$
$$1\ \text{hp (horsepower)} = 736\ \text{W} = 75\ \text{kpm} \times \text{s}^{-1} = 4500\ \text{kpm} \times \text{min}^{-1}$$

# Assessment of Workload in Relation to Work Capacity

Because maximal oxygen uptake varies greatly from one person to another, a workload that is fairly easy for one worker can be quite exhausting for another. Suppose two men are to perform the same task, such as carrying a heavy load uphill, requiring an energy expenditure of about 2 L of oxygen per minute. One of the individuals has a maximal oxygen uptake of $6.0\ \text{L} \cdot \text{min}^{-1}$, the other $2.0\ \text{L} \cdot \text{min}^{-1}$. In the first case, the individual is merely taxing 30% of his aerobic power. Consequently, he can work all day without fatigue, as is normally true when the workload is less than about 40% of the individual's maximal aerobic power. Furthermore, he can continue to cover more than half his energy expenditure from the oxidation of fat. The second man, on the other hand, is taxing his aerobic power maximally and can carry on for only a few minutes, during which time he is compelled to rely on his carbohydrate stores as a major source of metabolic fuel.

Expression of the workload, as such, in absolute values (liters of oxygen uptake per minute) therefore can be quite meaningless. Instead, it should be expressed in percentage of the individual's maximal aerobic power. This means that the ratio between load and power should be assessed individually; that is, the individual's maximal oxygen uptake has to be determined and his or her rate of work has to be assessed. In general, the same principle also applies to the muscle groups that are performing the work in question, because only a certain percentage of the maximal muscle strength can be taxed without developing muscular fatigue.

## Assessment of Maximal Aerobic Power and Physical Workload

A person's maximal aerobic power can be determined by direct measurement of the individual's maximal oxygen uptake or estimated on the basis of data obtained from submaximal tests. The physical workload can be assessed either by measurement of the oxygen uptake during the actual work operation or by indirect estimation of the oxygen uptake on the basis of the heart rate recorded during performance of the work.

## Measuring the Oxygen Uptake in a Typical Work Situation

Since the validity of using oxygen uptake as a basis for measuring energy expenditure has been established, this indirect calorimetry has been used to determine the energy cost of a great variety of human activities. Figure 17.2 presents oxygen uptakes based on average values for various loads on a cycle ergometer, compared with equivalent activities. With the development of highly portable devices for collecting expired air under field conditions and rapid methods of analyzing the oxygen and carbon dioxide content of the air samples, a vast body of knowledge of the energy cost of physical work has accumulated.

In field studies, the classical method was to collect expired air in Douglas bags. Other methods are now available by which the volume of the expired air is measured with flow meters, and aliquot samples of the expired air are collected in a small rubber bladder. The expired air then can be analyzed for oxygen and carbon dioxide by using a conventional gas-analysis technique or electronic oxygen and carbon dioxide analyzers. If accuracy is not too critical (±10%), it is sufficient to analyze only the oxygen content by a portable oxygen analyzer. Various methods have been described and discussed over the years, for example, by Consolazio, Johnson, and Pecora (1963), Durnin and Passmore (1967), Morehouse (1972), Kamon (1974), and Harrison,

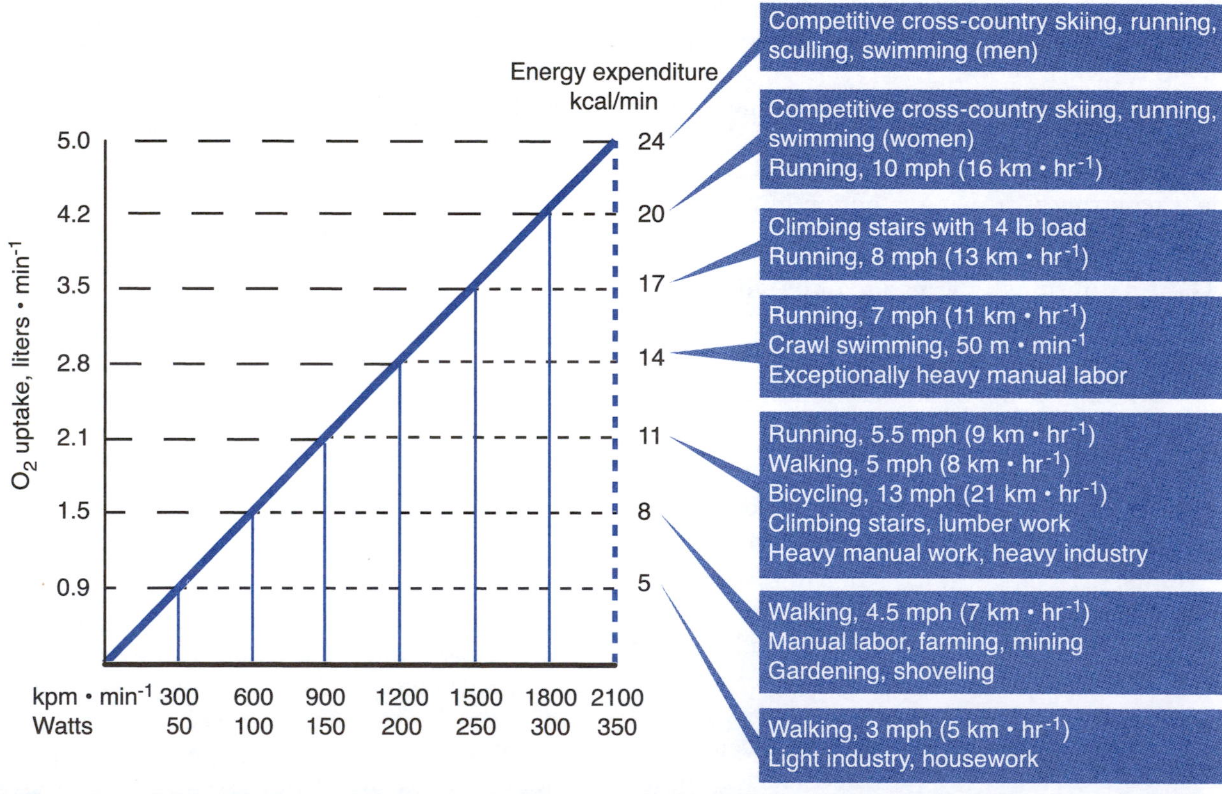


**Figure 17.2** Cycle ergometer work and equivalent physical activity.

Brown, and Belyavin (1982). G.D. Wilson and Sklenka (1983) designed a modified portable system for measuring energy cost during particularly intense, dynamic activity. Novitsky et al. (1995) compared the oxygen uptake values collected with a new portable, indirect calorimeter (AeroSport Teem 100 Metabolic Analysis System) against a more traditional, large calorimeter system and found that it produces valid data at rest and at low to moderate work rates.

An example of measured oxygen uptake in typical commercial fishing operations is given in figure 17.3. It is evident that the measured oxygen uptake during work performance represents the energy expenditure only at the time when the expired air sample was collected and may not represent the work performed during the whole working day. Furthermore, the test subject usually is affected by the investigation, causing the test situation to be atypical. The test equipment is apt to affect heart rate, pulmonary ventilation, and oxygen uptake and can hamper the actual work operation.

In contrast to this, indirect assessment of workload via continuously recorded heart rate reveals a general picture of the overall activity level during the entire working day. Moreover, by using time–activity records for each subject, collected by an observer during the whole working day, the investigator can separate the different activities with respect to heart rate. Thus, the indirect assessment of workload based on the recorded work pulse is preferable in many work situations.

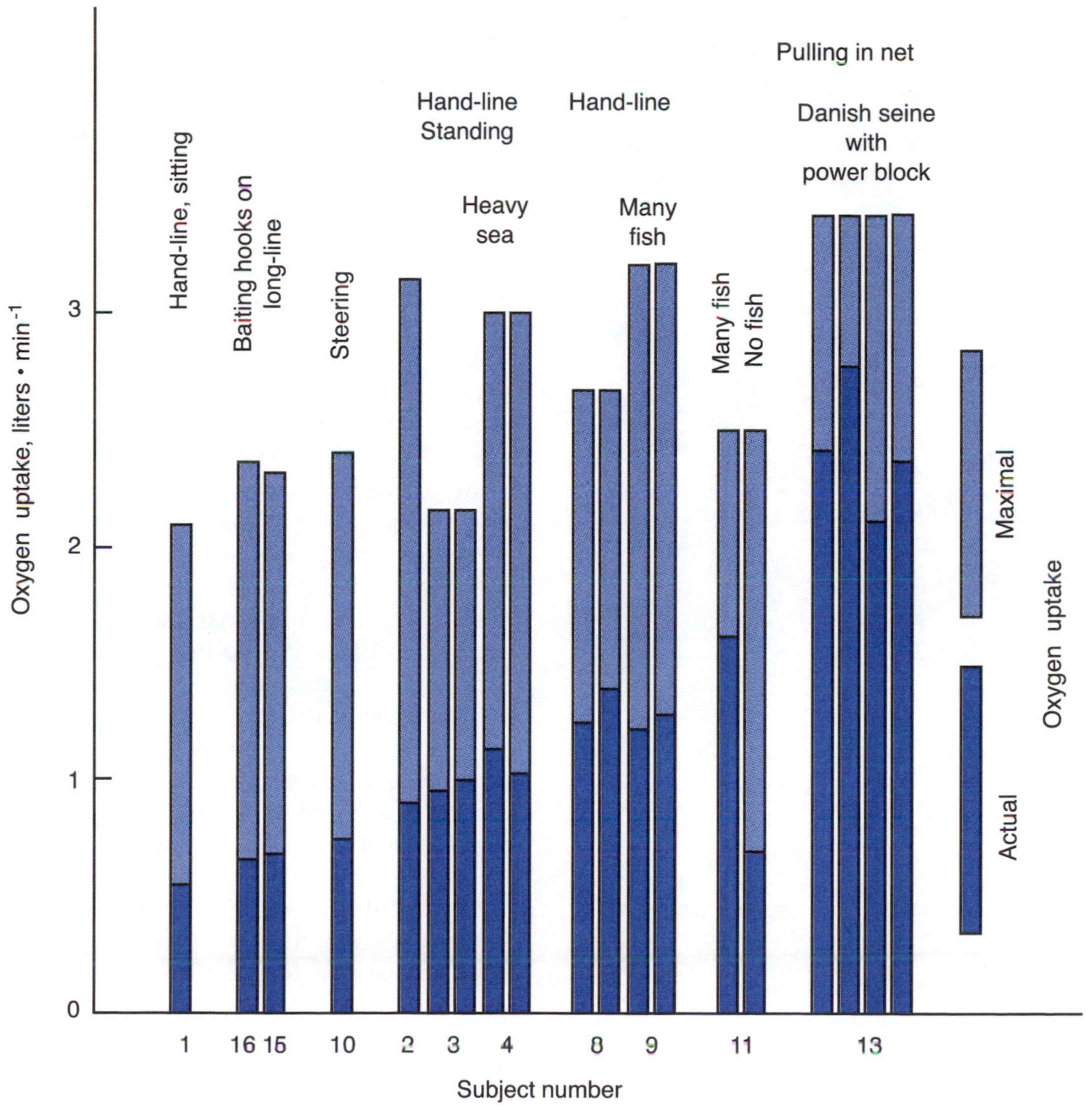


**Figure 17.3** Measured oxygen uptake in typical fishing operations.

From I. Åstrand, Fugelli et al. 1973.

## Indirect Assessment by Recording the Heart Rate During Work

In a given person, there is generally a linear relationship between oxygen uptake and heart rate. Therefore, the heart rate, under certain standardized conditions, can be used to estimate workload, if the workload–heart rate relationship has been established for the individual in question, if roughly the same large muscle groups are engaged in the work in both cases, and if environmental temperature, emotional stress, and other factors are the same. A heart rate of 130 in the average worker roughly corresponds to about 50% of the individual's aerobic power.

The individual's circulatory response to work engaging large muscle groups can be measured on a cycle ergometer. Starting at a low load such as 50 W, the load is increased stepwise every 6 min, usually by 50 W, until a heart rate of about 150 beats $\cdot$ min$^{-1}$ is reached (figure 17.4). By using the resulting line representing the relationship between the individual's workload and heart rate, to the investigator can estimate the workload from the heart rate recorded during a specific work situation in the field (figures 17.5 and 17.6 show typical examples).

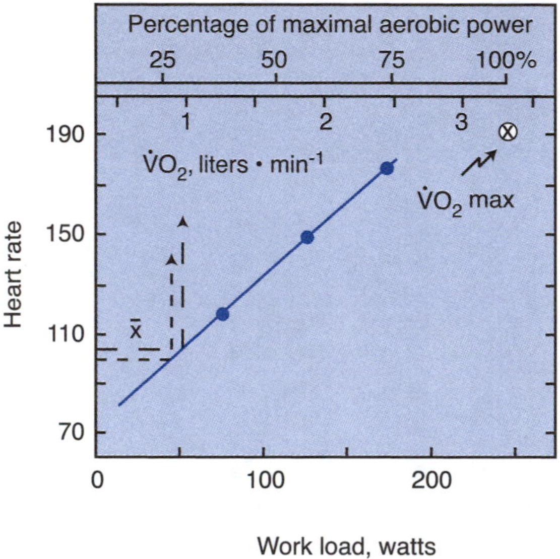


**Figure 17.4** The individual relationship between the heart rate at different submaximal workloads and the predicted corresponding oxygen uptake (I. Åstrand 1967b) is established graphically (scale $\dot{V}O_2$). The measured maximal oxygen uptake ($\dot{V}O_2$max) is used to construct another parallel scale, which shows the load expressed in percentage of the individual's maximal aerobic power. The weighted mean ($\bar{x}$) of the continuous recording of the heart rate is then used to assess the approximate average oxygen uptake during work as well as the load expressed as percentage of the maximal aerobic power.

Heart rate is easily recorded in the field with a portable miniature battery-operated recorder, which can be carried on the worker to record the parameters in question at the workplace and while the actual work is being performed. For this reason, the recorders have to be robust, but light, and they have to be able to store the recorded data in internal memory for instantaneous display on a portable computer screen. This will allow all investigators to observe and discuss the recorded results.

A variety of such recorders, or loggers, are available on the market, such as Ramlog EI 9000 portable data logger (a.b.i. Data, Brussels, Belgium), the Vitalog Pocket-Polygraph (Vitalog Monitoring, Inc., Palo Alto, CA), and the AMS-1000 Ambulatory Monitoring System (Consumer Sensory Products, Inc., Palo Alto, CA). The Squirrel meter/logger produced by Grant Instruments Ltd. (Cambridge, UK) is capable of logging any parameter convertible to voltage or electric current (figure 7.7). It has a large number of inputs and a high degree of flexibility. It is compact, light, and very simple to operate. It is set up similar to a digital watch, and only three push buttons are used in connection with the liquid crystal display. The function of each button is actually printed on the logger, making it easy to use for nontechnical personnel. It can record several parameters simultaneously, related to both the working environment and the physiological responses to some of the environmental parameters, such as body temperature. It does this by using different sensors already available that are compatible with the logger. It is a combined meter and data logger; that is, it can measure and immediately display data such as ambient and body temperature, environmental gasses, and heart rate. At the same time, it can store the recorded data in its internal memory for immediate or subsequent transfer to a small, portable battery-operated personal computer (PC), a battery-operated portable printer in the field, or a conventional office PC for data analysis, printout, or graphic display. Both the conventional 12-bit Squirrel and a special 8-bit version are equipped with separate channels for recording heart rate, based on the principle of counting the number of R-waves per minute. The device is supplied with filters to exclude artifacts or interference. In controlled laboratory tests, it has been found to be quite accurate and reliable. The 12-bit Squirrel meter/logger weighs about 1 kg and can be carried in a leather carrying case by a shoulder strap or by a belt around the waist (figure 17.7).

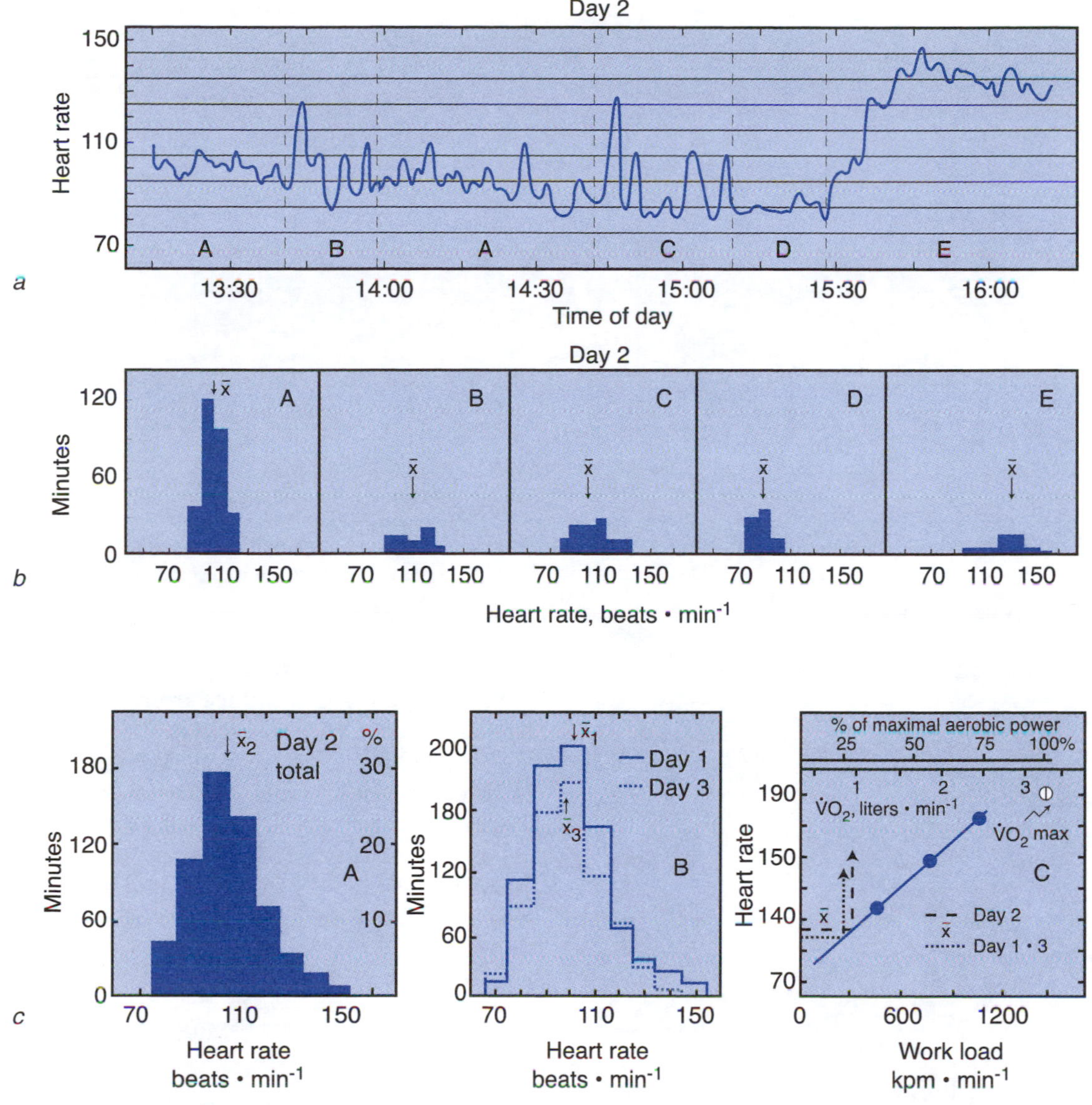


**Figure 17.5** Example of analysis of heart rate recorded at sea in a 21-year-old net-fisherman. *(a)* Heart rate curve in the latter part of the second day observation period. Activities: A = arranging and putting out net; B = bleeding and cleaning the fish; C = other unspecified activities; D = resting; E = unloading the catch at the pier. *(b)* Heart rate distribution curves and weighted means ($\bar{x}$) of the previously listed five types of activity throughout the whole observation period. *(c)* A = heart rate distribution curve and weighted mean ($\bar{x}_2$) for the whole observation period of the second day (06.45–16.50 h). B = corresponding results from the first and third day of observation. C = relationship between workload, heart rate, estimated oxygen uptake ($\dot{V}O_2$), and percentage of maximal aerobic power in the subject as assessed in the ergometer test. Estimation of average oxygen uptake and percentage of maximal aerobic power during the 3 days of observation on board by using the weighted means ($\bar{x}$) of the heart rate from sections A and B of *(c)*.

From K. Rodahl, Vokac, et al. 1974.

The recorded heart rate, coupled with time–activity records, shows the degree and variations of the circulatory strain. By comparing the individual's heart rate during work in the field with the heart rate response to known, increasing workloads on a cycle ergometer, the investigator can convert heart rate into the approximate oxygen uptake (see figure 17.5). The heart rate curves thus obtained can be replayed and transcribed on recording paper and graphically evaluated (I. Åstrand, Fugelli et al., 1973) or evaluated by computer analysis (Rodahl, Vokac, et al. 1974).

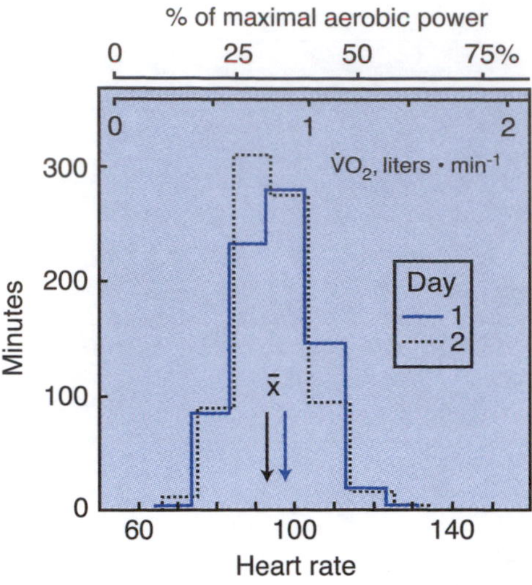


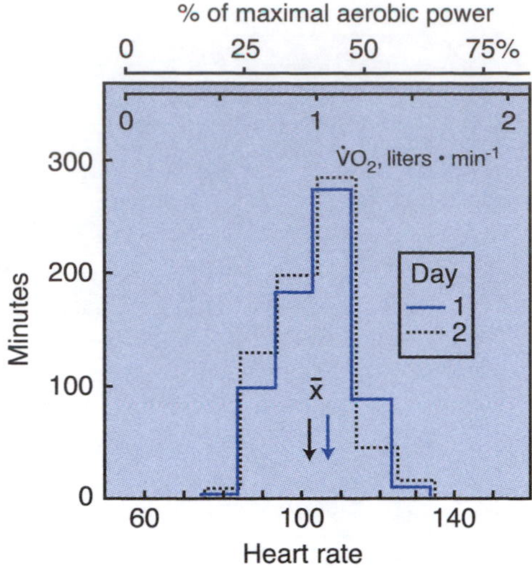


**Figure 17.6** Weighted means ($\bar{x}$) and distribution curves of the heart rate and corresponding workload in two subjects during consecutive days of observation. Left: 41-year-old long-line fisherman. Right: 56-year-old catch handler on land.
From K. Rodahl, Vokac, et al. 1974.

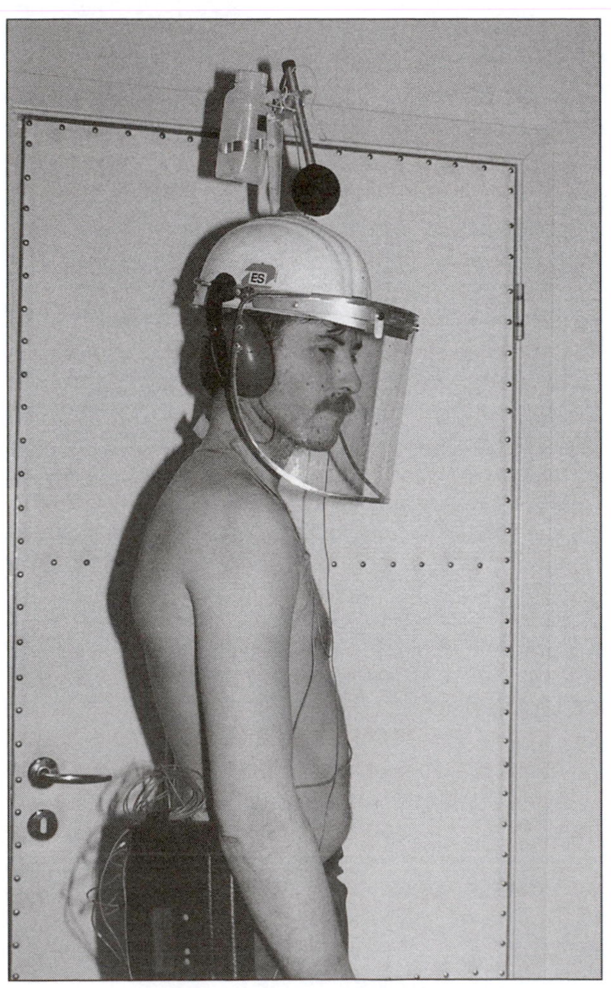

**Figure 17.7** Subject wearing Squirrel apparatus.

Estimating oxygen uptake from recorded heart rate can be subject to considerable inaccuracy. However, in a field study by Rodahl, Vokac, et al. (1974), 24 direct measurements of oxygen uptake were compared with the oxygen uptake calculated from the simultaneously recorded heart rate in six fishermen. The calculated values deviated from the measured ones in both directions by no more than ±15% (figure 17.8).

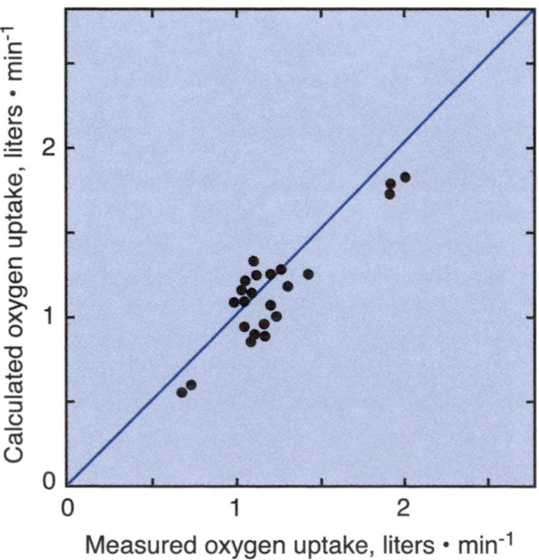


**Figure 17.8** Relationship between oxygen uptake measured by the Douglas bag method and calculated from the simultaneously recorded heart rate. There were twenty-four observations in six subjects during the various fishing operations.
From Rodahl, Vokac, et al. 1974.

The reproducibility of the results from day to day in such field studies over a 3-day period was examined in connection with a study of fishermen by Rodahl, Vokac, et al. (1974). They found a remarkable reproducibility of the day-to-day results in the same individual doing the same work. The weighted mean heart rates were practically the same for all 3 days, and the distribution curves, when superimposed, showed the same shape (see figure 17.6).

It is thus clear that recorded heart rate in the field, compared with the heart rate at known workloads on the cycle ergometer, can be used to estimate work when the work operation involves the use of the same large muscle groups as are used in the cycle work. This comparison is not feasible in work situations where mostly small muscle groups are involved, such as in arm work, because it is well established that the heart rate is higher with arm work than with leg work at the same workload. Although this is indeed the case in prolonged work lasting 5 min or more, Vokac et al. (1975) showed that the discrepancy in heart rate between arm and leg work at the same workload is not so marked at the onset of the work but increases as the work proceeds (figure 17.9). Because most ordinary work operations involve a dynamic type of work with a rhythmic alteration between muscular contraction and relaxation, in which each period of work effort is rather brief, it appears that using recorded heart rate to estimate workload is acceptable even in many work situations involving arm work or the use of small muscle groups.

Because the heart rate response to the same work load varies individually, the circulatory strain is best expressed as percentage of the heart rate reserve of the subject, the heart rate reserve being the difference between maximal heart rate and resting heart rate. To overcome the time-dependent variability, the heart rate results can be presented as weighted means, which are calculated from the individuals' values multiplied by the fraction of the sum of their durations (Rodahl, Vokac, et al. 1974).

Verma et al. (1979) suggested that a combination of minute ventilation and heart rate is a better predictor of energy expenditure than either of the two variables used singly.

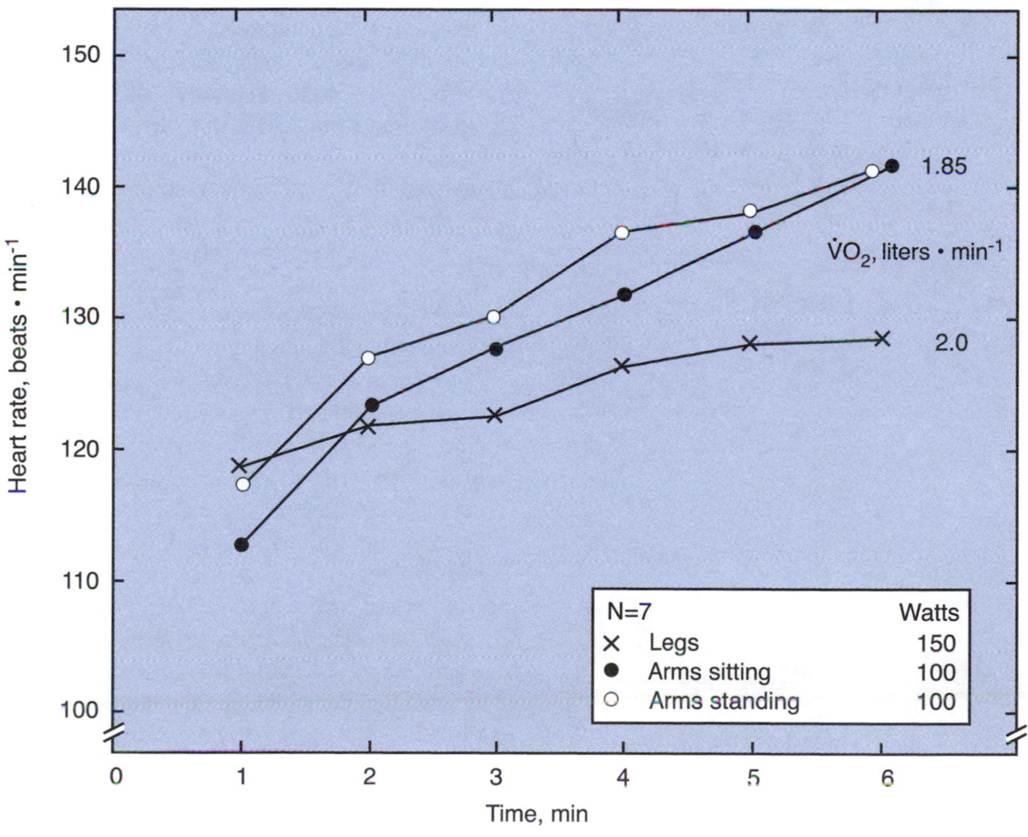


**Figure 17.9** Heart rates in the first 6 min of cycling at 900 kpm · min$^{-1}$ (150 W) (**x**) and of arm cranking at 600 kpm · min$^{-1}$ (100 W) in sitting (•) and standing (○) positions.

From Vokac et al. 1975.

## Assessment of the Load Exerted on Specific Muscles

Although oxygen uptake and heart rate can be used to assess the magnitude of muscular work in general, these parameters are not suitable for monitoring localized muscular loads (Malmqvist et al. 1981). Under certain conditions, however, the recording of heart rate is a quite sensitive index of muscular force and even responds significantly to minor changes in posture (Hanson and Jones 1970).

Electromyographic recordings, on the other hand, permit load measurements on single muscles as well as groups of muscles. The contraction of muscle fibers is associated with small electrical changes, or oscillations, that can be picked up by suitable surface electrodes placed on the skin over the muscle in question. The instrument picking up the electrical oscillations is known as an electromyograph, and the display of the myoelectrical oscillations is called an electromyogram (EMG). The greater the load on the muscle, the larger the EMG signals. In more sophisticated quantitative electromyography, the generated electrical activity is measured by rectifying and integrating the action potentials recorded from surface electrodes. These integrated action potentials are linearly or semilinearly related to the muscular force that is being exerted.

It is well established that the EMG reflects the magnitude of the muscle engagement (figure 17.10) and can be used to measure the exerted force in percentage of the maximal voluntary muscle strength.

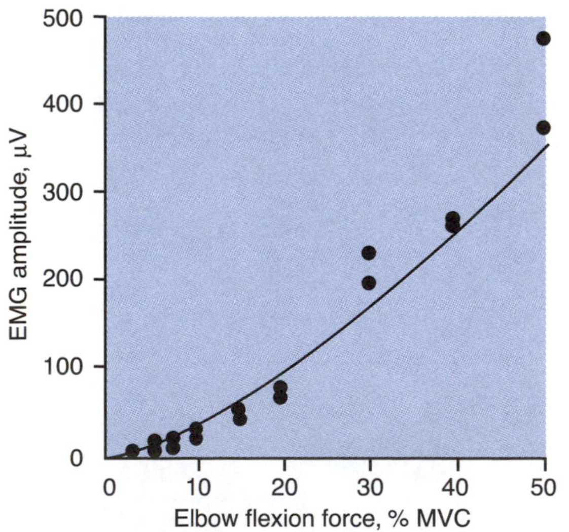


**Figure 17.10** Relation between electromyogram (EMG) activity from the biceps brachii and elbow flexion force as a percentage of the maximal voluntary contraction (% MVC).

From Hagberg 1981b.

Bobet and Norman (1982) showed that the average EMG provides useful information even concerning whole-body tasks but that its use is less straightforward in this respect than in the investigation of more constrained or simplified tasks.

### For Additional Reading

For a detailed discussion of muscular endurance and surface electromyogram in isometric and dynamic exercise, see Hagberg (1981a).

## Assessment of Static Muscular Work

Most people are unable to tolerate prolonged, intense, static muscular work as a regular occupation without ill effects. These effects include muscular stiffness, pain, tension, and even symptoms and signs of neuromuscular disorders, and they represent a major cause of absenteeism in many industrialized countries (Tola et al. 1988; Westgaard et al. 1984).

Integrated electromyographic recordings allow the investigator to measure the load on single muscles as well as groups of muscles. It is well established that the electromyography reflects the magnitude of the muscle engagement and can be used to measure the exerted forces in percentage of the maximal voluntary muscle strength (Caldwell, Jamison, Lee 1993; Jensen et al. 1993; K. Rodahl 1989, 1994). This can be done with a pocket-sized, battery-operated electromyograph, such as the Squirrel-compatible Myolog, produced by Aleph One Ltd. (Cambridge, UK). It uses three disposable skin electrodes to pick up signals from the muscles in question. The Myolog is plugged into the Squirrel logger, and both instruments are attached to a belt and carried by the subject. This combination with a Squirrel provides an ambulatory system that can be used for logging the tension in certain muscle groups over extended periods, even 24 h.

To relate the amplitude of the electromyographic signals to known levels of muscular tension, a routine calibration procedure is introduced at the beginning and again at the end of the recording period, after the electrodes are in place and the logging system is started. This can be done by asking the subject to relax in a resting position, followed by loading the muscle group in question with a known weight (figure 17.11). For further details, see K. Rodahl (1994).

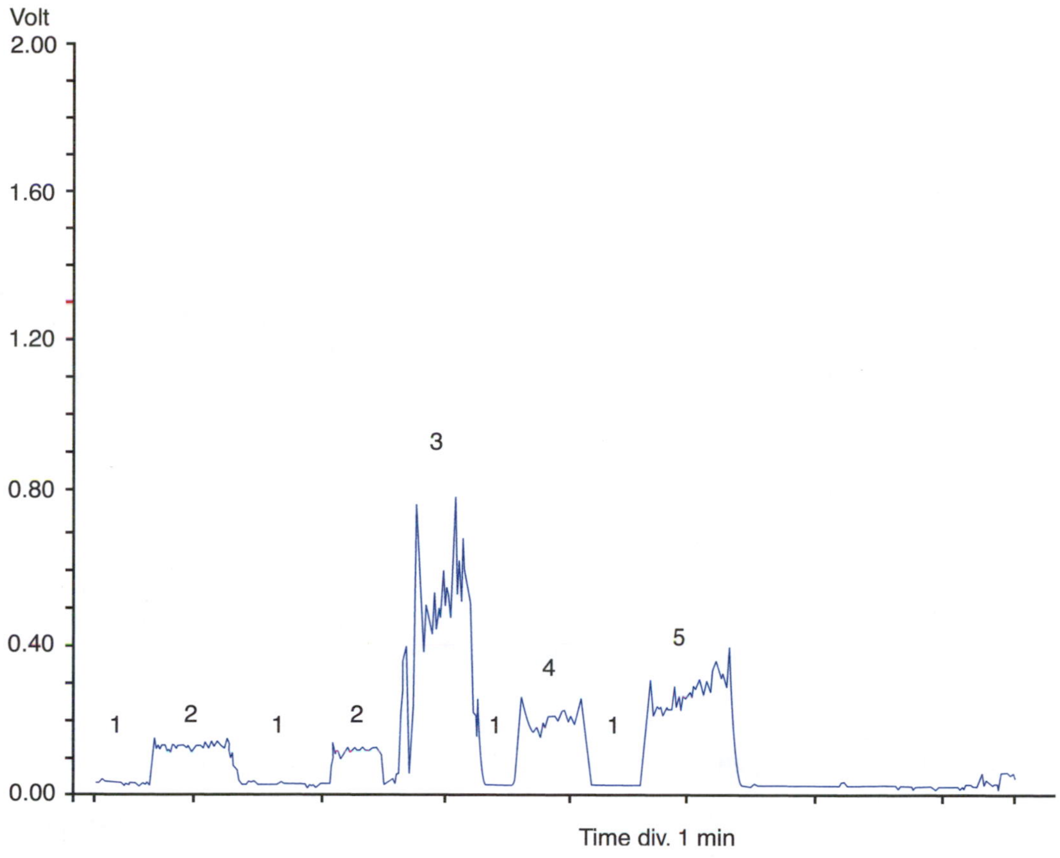


**Figure 17.11** Electromyographic calibration (the right trapezius muscle): (1) relaxing; (2) arm stretched out horizontally; (3) lifting 10 kg; (4) lifting 2 kg placed at elbow; (5) lifting 2 kg placed at wrist.

A major advantage with this ambulatory electromyographic logging system is that it can be used to visualize the consequences of the practical working procedures on the load imposed on the muscles involved. This is evident from a number of studies made in modern industrial operations, as well as office work, including the use of PCs with and without the use of a mouse. Such recordings then can be used to implement improvements (figures 17.12 and 17.13).

"Tennis elbow" (lateral humeral epicondylalgia) is an extremely common complaint, not only in tennis players but also in computer operators and manual workers engaged in static-arm muscular work (Laursen, Jensen, and Ratkevicius 2001). The pain appears to be localized in the muscular attachment at the epicondyle and occurs when the muscles pull the tissues in and around the muscle attachment (K. Rodahl, 1994). Application of this EMG–logging combination in a group of PC-mouse operators, among whom there had been a noticeable increase in shoulder and arm complaints, revealed that there was no relationship between the state of tension in the muscles used during work and the complaints. In all the subjects examined, whether they had symptoms or not, merely grasping or holding the mouse in the hand while resting the hand or the arm on the table, without moving it, and without pressing the button, significantly increased the electromyographic amplitude, recorded from the extensor digitorum muscle, as is evident from figure 17.14 (subject without symptoms) and figure 17.15 (subject with symptoms). The subjects, whether they had symptoms or not, were able to relax the muscles examined when asked to do so.

### For Additional Reading

For a series of articles on motor control and the mechanisms of muscle disorders in computer users, see Sjøgaard, Kadefors, and Lundberg (2000).

## Assessment of the Organism's Response to the Total Stress of Work

The total stress imposed on the organism by a given work situation (physical as well as psychological) generally is reflected by nervous and hormonal

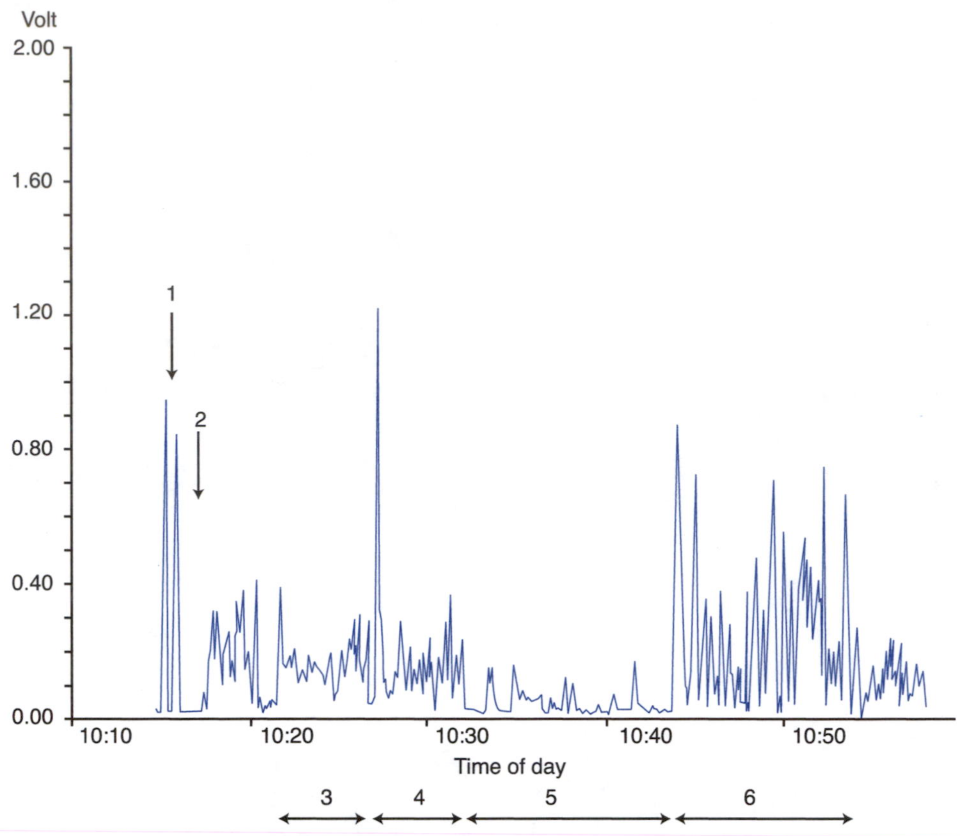


**Figure 17.12** Electromyographic amplitude of the right trapezius muscle in a 37-year-old female computer operator who had just completed a 5-month sick leave because of intense aches and pains in her back, shoulder, arm, and wrist, considered to be related to her work involving the use of the computer mouse: (1) calibration; (2) relaxing; (3) typing on the computer using both hands; (4) using the mouse with arm support; (5) using the mouse while resting the arm on the table; (6) doing a variety of jobs.

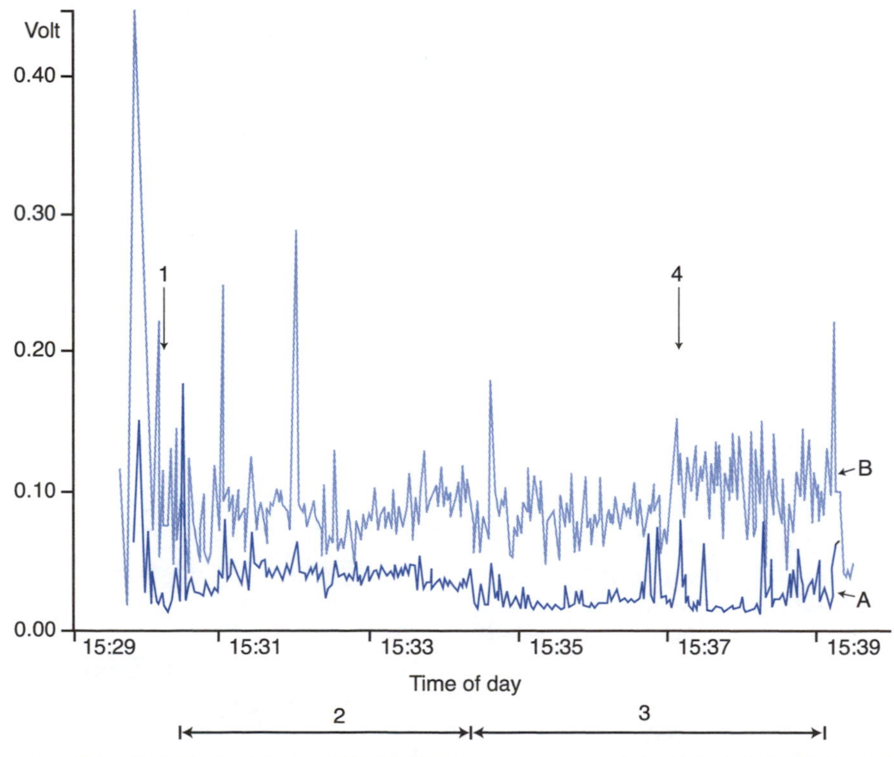


**Figure 17.13** Simultaneous electromyographic recording from the right trapezius muscle (A) and from the right extensor digitorum muscle (B) in the same computer operator as shown in figure 17.12: (1) relaxing; (2) working with the mouse, resting the hand on the mouse; (3) working with the mouse while resting the arm on the table; (4) taking a telephone call.

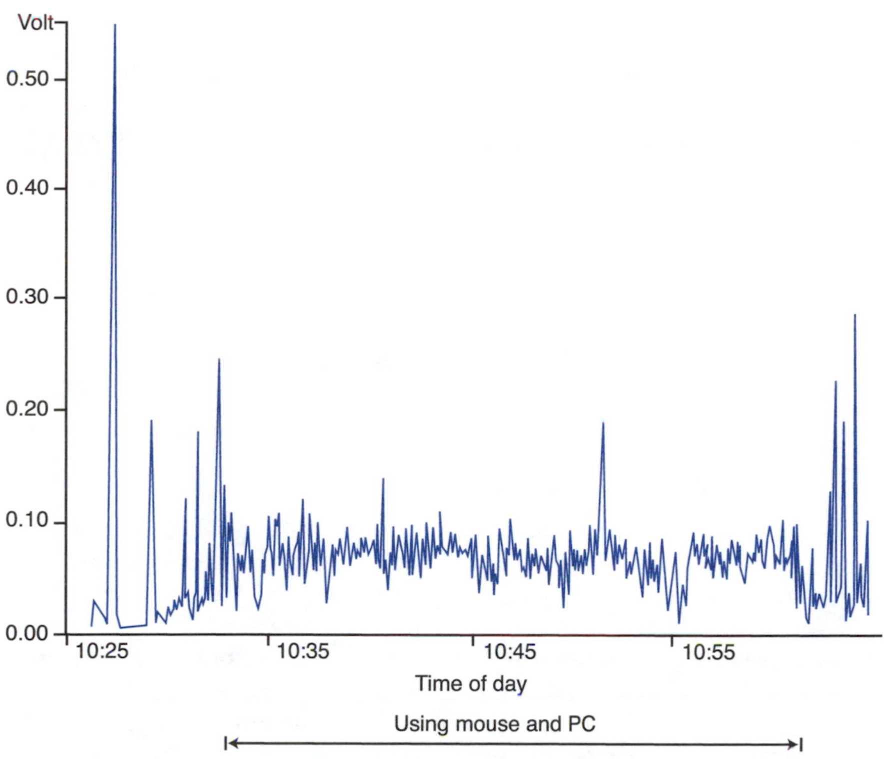


**Figure 17.14** Electromyogram recorded from the right extensor digitorum muscle in a computer operator using a mouse with her right hand. This subject had no symptoms of neuromuscular complaints despite the persistent, static muscle tension.

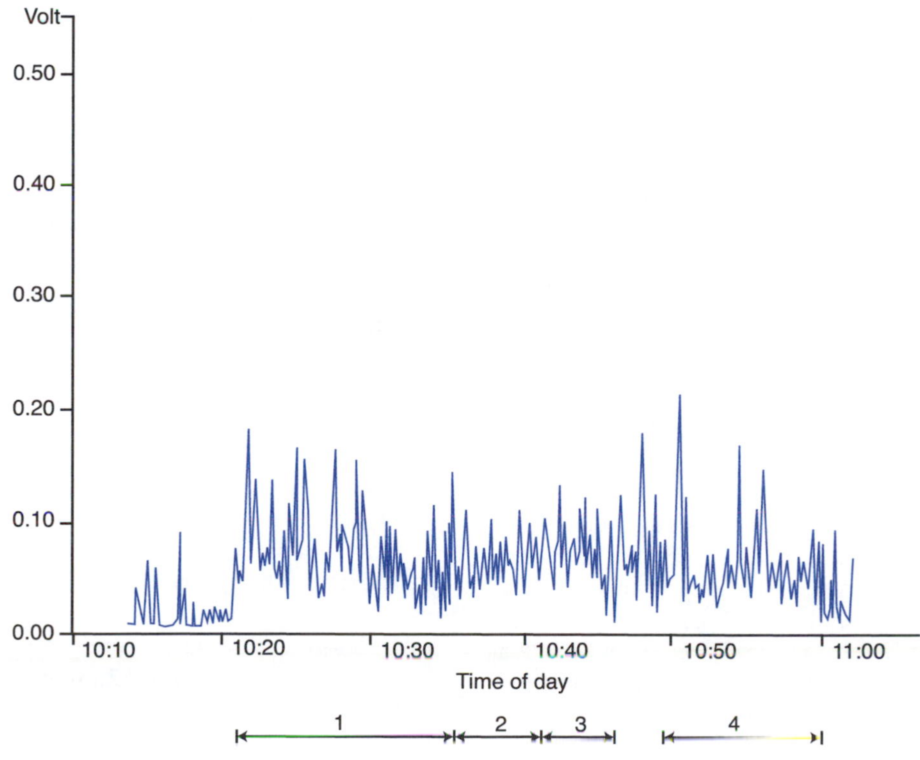


**Figure 17.15** Electromyogram recorded from the right extensor digitorum muscle in a personal computer (PC) operator using a mouse with her right hand. This operator had symptoms of neuromuscular complaints: (1) using PC and mouse without arm support; (2) using PC and mouse while resting the distal end of the arm; (3) moving the mouse with the first finger instead of the thumb; (4) using movable arm support while using the mouse.

reactions, more or less proportional to the degree of stress.

This stress concept is based on the fact that human beings, like all higher animals, react to almost any kind of threat by mobilizing their resources for a physical effort, that is, by the so-called stress reaction (Cannon 1929). The pulmonary ventilation increases, as do the heart rate and cardiac output, increasing the oxygen uptake and the oxygen delivery to the muscles in readiness for work. Much of the circulating blood volume is shifted from the gut and skin to the muscles. Because of the effort of some of the stress hormones liberated as a consequence of the stress mechanism, stored energy is released in the form of glucose and free fatty acids, which are made available as substrates for the working muscles. At the same time, the excreted stress hormone cortisol profoundly affects a number of cellular functions, including cell membrane permeability (Francis 1979).

## Classic Study

Walter Bradford Cannon (1871–1945), who became George Higgins Professor of Physiology at the Harvard Medical School in 1906 only 6 years after his graduation from that institution, was the first to point out that the sight or awareness of any kind of threat automatically evokes nervous and hormonal reactions, mobilizing the bodily resources for a physical response to the threat. According to Cannon, the purpose of all these changes, brought about by the stress reaction, is primarily to prepare the organism to cope with the threatening situation, as was the case when a Stone Age hunter suddenly encountered a threatening bear in the wilderness. As soon as the threatened individual engaged him- or herself in the physical activity that the situation called for, be it a fight with the bear or a run for safety, the inherent physiological regulatory mechanisms were adjusted to meet the body's needs in terms of physical performance. Similarly, as soon as the effort was over, the same regulatory mechanisms brought the excited organism back to a state of rest and recuperation.

Cannon 1929. See also Heard 1972.

Our problem is that any stressful situation in our civilized existence can trigger the same stress reaction as in our primitive ancestor confronted with a wild beast, but we are without a physical outlet for the stress response. This can elevate blood pressure (Mundal et al. 1990).

### *Hormonal Response*

It is well known that the total stress response of the organism is reflected by the sympathoadrenal activity. This can by assessed roughly by measuring urinary excretion of adrenaline and noradrenaline by the method described by von Euler and Lishajko (1961) and modified by Andersson et al. (1974), using the resting night urine as base value. Expressed in nanograms per minute, it can serve as a measure of occupational stress. In one study, a nearly 10-fold increase in adrenaline excretion and about a fourfold increase in noradrenaline excretion were observed during the workday, compared with the excretion of coastal fishermen during the night (I. Åstrand, Fugelli et al. 1973). An even greater increase in noradrenaline excretion was observed in war college cadets during a strenuous battle course including marked sleep deprivation (Holmboe, Bell, and Norman 1975).

An increase in both adrenaline and noradrenaline can be attributable to a number of single factors or a combination of them. Generally, both the circadian rhythm (Fröberg et al. 1972) and the change of body posture from the recumbent to the standing position (Sundin 1956) increase the catecholamine excretion during the day. This increase is enhanced markedly by the effect of physical exertion (von Euler and Hellner 1952), cold (Lamke et al. 1972), and emotional factors (Levi 1967; von Euler 1964). The level of plasma catecholamines increases with both the duration and the severity of muscular exertion (Banister and Griffiths 1972).

It is generally agreed that urinary catecholamine excretion can be used to quantitate stress response in a laboratory situation and in simulated work situations (S. Cox et al. 1982). It also can be used in field studies when the collection of urinary samples is carefully supervised and controlled. Jenner, Reynolds, and Harrison (1980) reported the analysis of catecholamine excretion rates of men living in 12 villages north of Oxford. Their results showed differences according to day of week and occupation. Workday adrenaline excretion rates were higher than rest-day rates, rates for nonmanual occupations were higher than for manual occupations, and rates for professional and managerial occupations were higher than for other occupations.

The fact remains, however, that it is at best very difficult to obtain complete urinary collections; the volume of residual urine is a source of error, and it is difficult to obtain samples more frequently than every 2 h or so. This means that acute stress situations of brief duration may not be reflected in the urinary sample, which shows the average value for as much as a couple of hours. In addition, there are considerable individual differences in urinary catecholamine and noradrenaline elimination, even among persons exposed to the same level of work stress (figure 17.16). Extensive studies of urinary catecholamine elimination in a large number of workers at a variety of workplaces, which were carried out at the Norwegian Institute of Work Physiology, failed to show any significant elevation of excretion rates in most industries. Finally, Follenius et al. (1980) showed that although exposure to high levels of noise in the working environment can be stressful for those who are exposed, high levels of noise do not necessarily induce any significant endocrinological changes. For these reasons, this method is not recommended to assess levels of work stress. Instead, continuously recorded heart rate can be used routinely to examine autonomous nervous system response to stress, both physically and mentally.

## *Nervous Response*

An increased tonus of the sympathetic system, brought about by emotional as well as physical stress, will give rise to an accelerated heart rate, which consequently serves as an index of stress response. It is thus well established that the heart rate increases linearly with increasing physical workload, provided there is no major change in the subject's emotional state (see figure 17.4). Also, the heart rate varies markedly as a consequence of emotional stress in an individual who is at rest or subject to a constant light workload (figure 17.17). Hitchen, Brodie, and Harness (1980) found that during a demanding mental task, the heart rate increased by 15%.

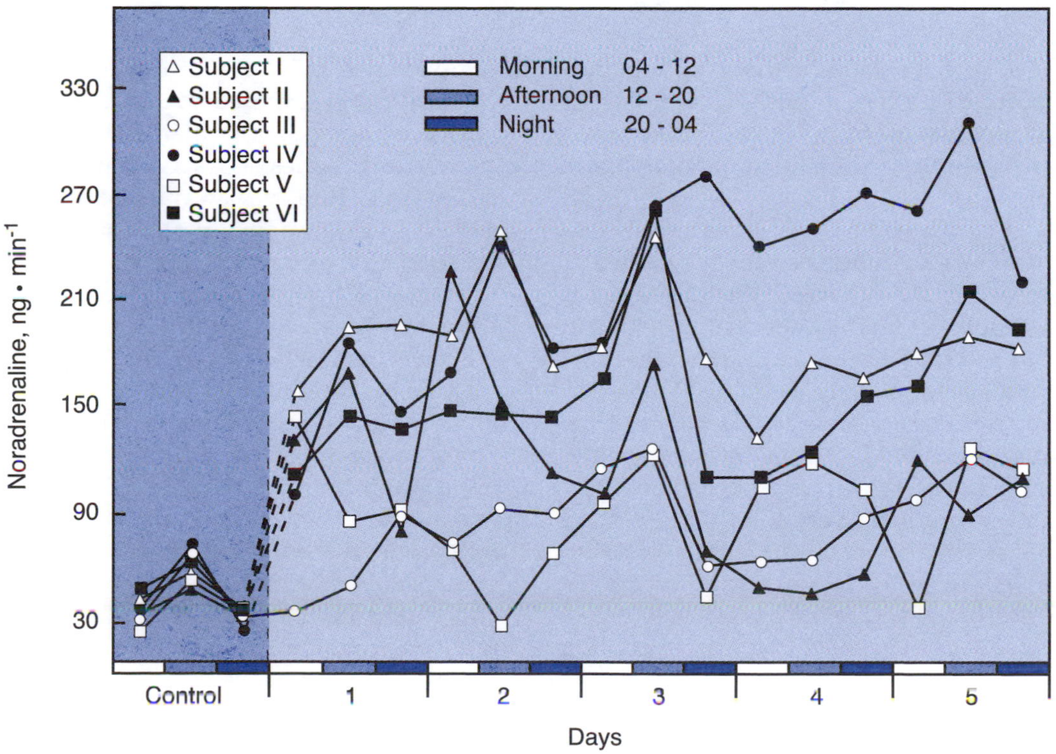


**Figure 17.16** Urinary excretion of noradrenaline in war academy cadets during simulated battle course.

From Holmboe, Bell, and Norman 1975.

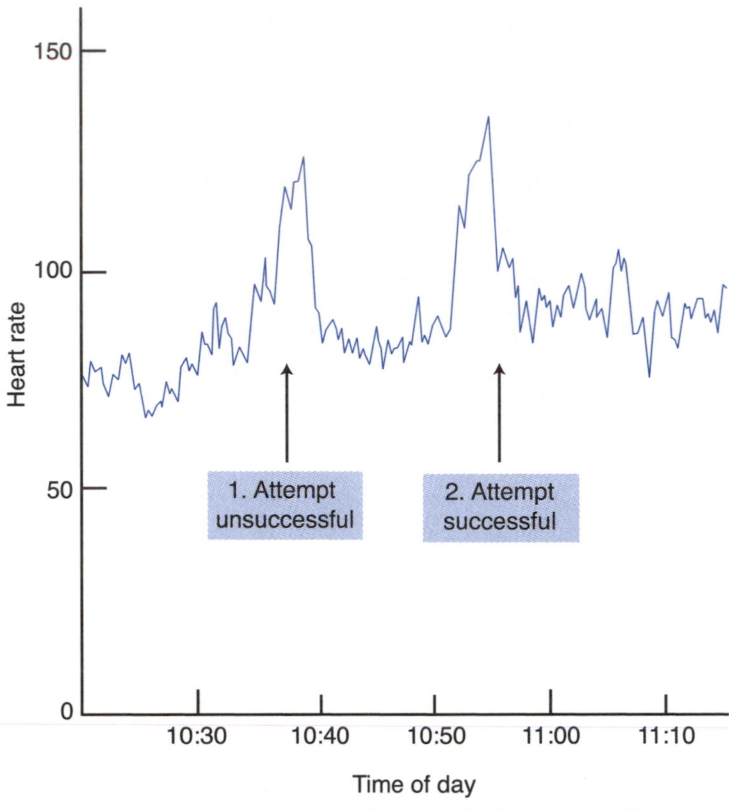


**Figure 17.17** Heart rate response in commercial pilot landing a twin-Otter aircraft at a small airfield under difficult weather conditions. He made one unsuccessful attempt but succeeded to land on the second attempt. He was not performing any actual physical work during the landing.

Combined with a detailed activity log kept by a trained observer, the continuously recorded heart rate indeed provides useful information concerning a subject's response to stressful situations (figure 17.18).

From a practical point of view, continuously logging the heart rate is a most useful way of visualizing the physiological response to the total stress of work, as is evident from the examples presented in figures 17.19 and 17.20. Figure 17.19 shows the heart rate of a young actor playing the role of Karlson in the children's play *Karlson on the Roof*. Part of the time he was suspended by a rope from the ceiling, pretending to hover in the air, while continuously gesticulating and talking very loudly. His resting heart rate was on the average 48, but his working heart rate was close to 200 on several occasions. Figure 17.20 shows the heart rate of a subject crossing the inland ice of Greenland on skis with a dog team (Westby, Rodahl, and Rodahl 1997).

A quantitative numerical analysis of the recorded data, supplemented by visual analysis of the replayed heart rate curves (see figure 17.5), permits a comprehensive and dynamic evaluation of the circulatory strain imposed by workloads of varying intensity. Using a computer makes it possible to analyze large series of observations with respect to the mean values, peak values, time distribution, and the occurrence and duration of excessively high heart rates (e.g., more than 50% of the heart rate reserve, which is the maximal heart rate minus the heart rate at rest). Because the heart rate response to the same workload varies with individuals, the circulatory strain is best expressed as a percentage of the subject's heart rate reserve (I. Åstrand 1960). The results can be presented more conveniently and more graphically by converting the recorded heart rate individually into the corresponding estimated oxygen uptake (I. Åstrand 1967b). Estimated oxygen uptake serves, then, as a measure of workload, and, expressed in percentage of the subject's maximal aerobic power, it indicates the relative degree of the exertion in the same way as percentage of heart rate reserve. In most cases, the reliability of the conversion is adequate for all practical purposes of field investigation (Rodahl, Vokac, et al. 1974).

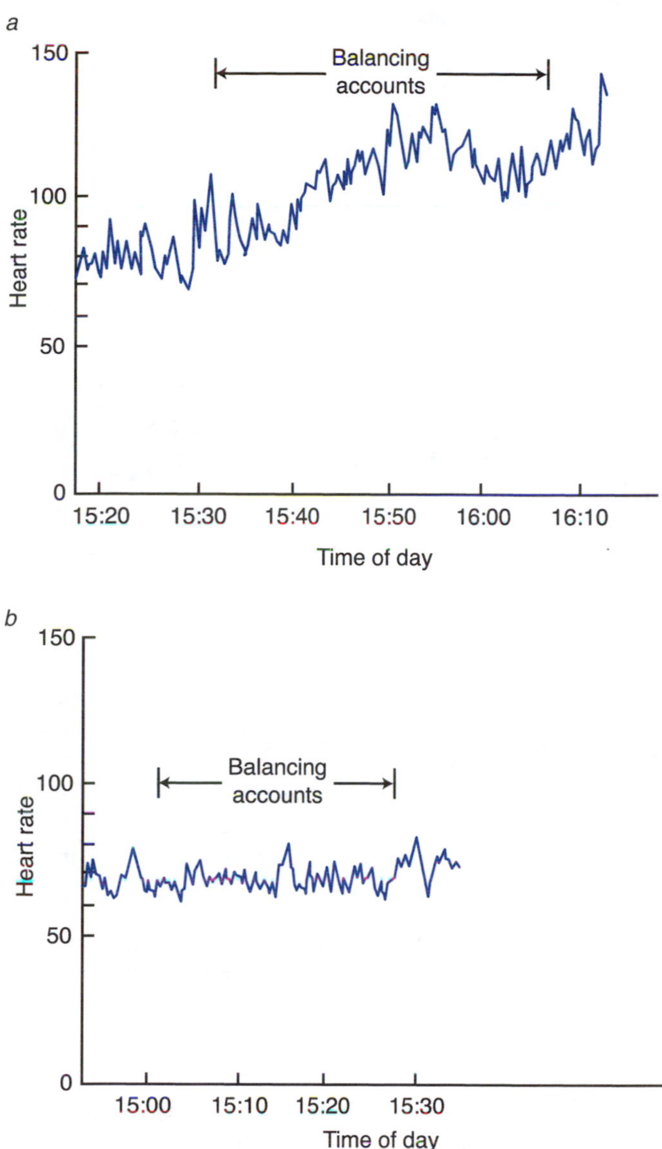


**Figure 17.18** Stress reaction of an inexperienced bank clerk involved in balancing accounts at the end of the day *(a)*, compared with an experienced clerk *(b)*.

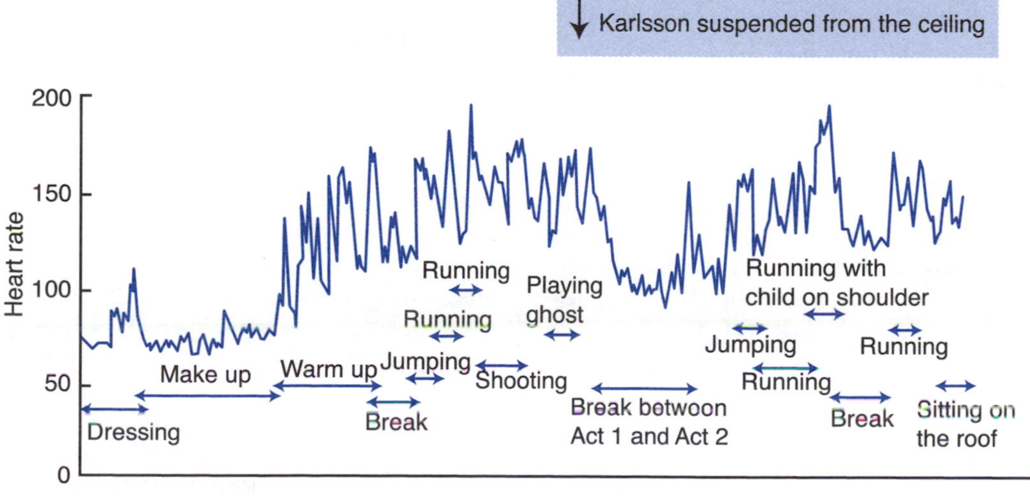


**Figure 17.19** The heart rate reaction to a combination of physical and mental stress in an actor, acting part of the time suspended by a rope from the ceiling, pretending to be flying, in the play *Karlsson on the Roof*.

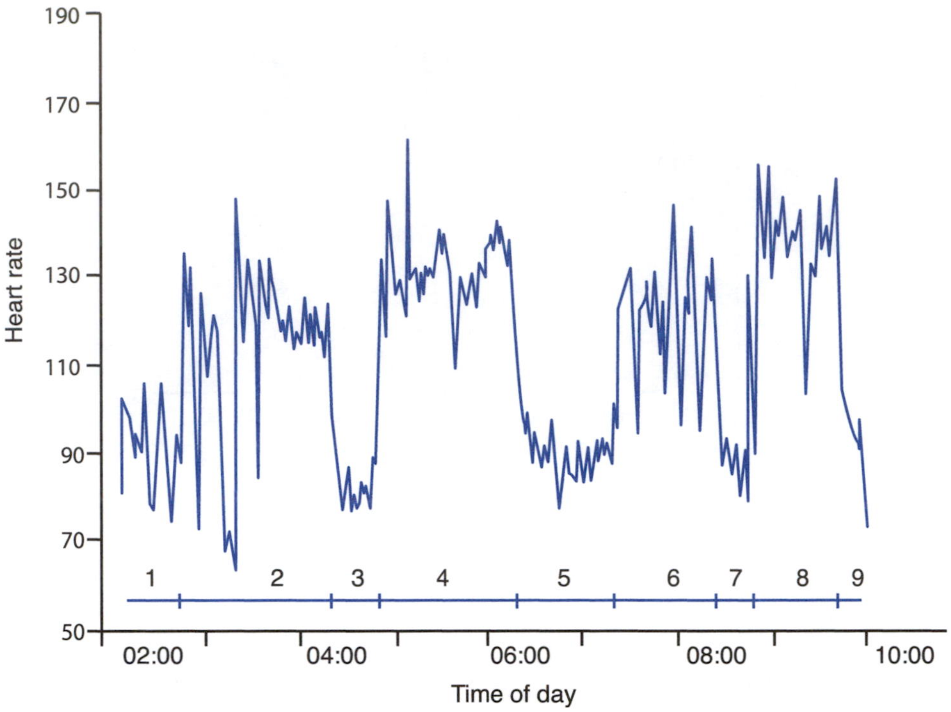


**Figure 17.20** Continuous recording of the heart rate of a subject crossing the Greenland icecap on skis, with a dog team, from west to east. 1 = preparing the dog team, packing the dogsled. 2 = first stage (89 min); some trouble with the dog team; skiing behind the dogsled. 3 = rest (31 min); drinking water and chocolate. 4 = second stage (93 min); skiing ahead of the team. 5 = rest (53 min); drinking soup, water, chocolate; eating biscuits. 6 = third stage (67 min); skiing behind the sled. 7 = rest (18 min); drinking water and chocolate. 8 = fourth stage (58 min); skiing ahead of the team. 9 = making camp, putting up tent. Distance covered: 42 km in 5 h 7 min. Total break periods: 1 h 42 min. Altitude: 2,500 m above sea level.

From Westby, Rodahl, and Rodahl 1997.

# Energy Expenditure of Work, Rest, and Leisure

Evidently, the human being is not ideally suited to be a source of mechanical power and in this respect cannot compete with modern mechanical devices, such as a bulldozer or a truck. The power output of an average person engaged in prolonged work over an 8-h working day may amount to little more than .1 hp (1 hp = 750 W). A horse yields at least seven times that amount, and an ordinary farm tractor, 70 hp. To a large extent, modern life has eliminated the need for heavy physical exertion. Yet humans must be physically active, or they will deteriorate. The solution, therefore, is to include some physical activity during leisure time, to provide for varied use of the locomotive system, and to select a proper ratio between work and rest to supply adequate recuperation during work.

## Classification of Work

In most instances, at least in the Western world with its advanced technology, excessively heavy work easily can be eliminated with technical aids; it is merely a matter of cost and priority. Establishing limits for permissible physical workloads is therefore of limited practical value. Of far greater importance to the worker today is the manner in which the work is being performed, the opportunity to influence the working situation and to govern one's own rate of work, the safety and general atmosphere of the working environment, and the arrangement of work shifts. In most jobs in modern industry, the worker or operator is able to adjust the rate of work according to his or her personal capacity. However, there are some exceptions, as when the work is performed by a team. Here, the weak have to keep up with the strong. In such teamwork, older workers, who generally are slower and have a reduced physical working capacity, may be hard-pressed to keep pace with the younger team members. In any event, great individual differences exist in physical working capacity, and practical experience has indicated that a workload taxing 30% to 40% of the individual's maximal oxygen uptake is a reasonable average upper limit for physical work performed regularly over an 8-h working day. Similarly, no more than 40% of maximal muscle strength should be applied in repetitious muscular work in which the time of each muscular contraction is about one half the time of each period of relaxation.

The physiological and psychological effects of a given energy output (per 1 min, per 8 h, per day) are determined by the individual's maximal aerobic power, size of the engaged muscle mass, working position, whether work is intermittent at a high rate or continuous at a lower intensity, and environmental conditions. In general, a person's subjective experience of a particular workload or work rate is more closely related to heart rate than to oxygen uptake during performance of the work, because the heart rate, in addition to reflecting the actual workload, also reflects emotional factors, heat, and the size of the engaged muscle groups.

An identification of prolonged physical work, classified as to severity of workload and to cardiovascular response, may be of some use (table 17.1). These figures refer to average individuals 20 to 30 years of age and can be used only as general guidelines in view of the vast individual variations in ability to perform physical work.

## Daily Rates of Energy Expenditure

An estimate of daily energy expenditure can be made by several methods: (1) the 24-h recording of heart rate by portable miniature recorders, as described earlier in this chapter; (2) estimation based on time–activity data and measurements of the energy cost of all pertinent activity; and (3) assessment of daily food intake (by food weighing) required to maintain body weight. All three of these methods are about equally accurate and reliable, with an error of no more than about 15% (K. Rodahl 1960).

As expected, there is a wide individual variation in energy output depending on occupation, leisure activity, and attitude or individual proneness to physical activity in general. The range of daily rates of energy expenditure is from 1,340 up to 5,000 kcal (5.63–21.00 MJ; Durnin and Passmore 1967; K. Rodahl 1960). About 2,900 kcal · day$^{-1}$ (12 MJ · day$^{-1}$) is a reasonable expenditure for a man who is not engaged in heavy manual labor but who is regularly active during leisure time. A reasonable figure for a woman would be about 2,100 kcal · day$^{-1}$ (9 MJ · day$^{-1}$; table 17.2). It is possible, however, that these figures are too high because of the increasing inactivity and the sedentary life led by large segments of the population.

The effect of environmental temperature, as such, on metabolic work is rather small (Consolazio, Johnson, and Pecora 1963). Gray et al. (1951) reported variations in metabolism up to 4% in men at a fixed cycle ergometer workload at ambient temperatures of –15 °C and 32 °C. It appears that any increase in the energy cost of work in a cold environment is primarily attributable to the additional energy cost of bodily movement in bulky clothing through difficult terrain and snow. Nelson et al. (1948) found that metabolic heat production for a given amount of work remained unchanged in three men walking in seven hot environments between 32 °C and 49 °C.

In the majority of professional activities (including office work), housework, light industry, laboratory and hospital work, and retail and distribution trade, the energy output is less than 5 kcal (21 kJ) · min$^{-1}$ (or less than 1 L of oxygen · min$^{-1}$). In the building industry, agriculture, the iron and steel industries, and the armed services, many jobs occasionally demand a energy expenditure of up to 7.5 kcal (31.5 kJ) · min $^{-1}$ or even higher, particularly if mechanical aids are few and prefabricated materials are used to only a small extent. Still higher energy demands are found in fishing, forestry, mining, and dock labor, where figures up to or exceeding 10 kcal (42 kJ) · min$^{-1}$ have been reported.

The energy expenditure in recreational activity naturally covers the whole range from near-resting values up to full use of aerobic and anaerobic processes, depending on the type of activity and the degree of vigor with which it is pursued.

**Table 17.1** Prolonged Physical Work Classified As to Severity of Workload and to Cardiovascular Response

| Workload | Oxygen uptake, L · min$^{-1}$ | Heart rate, beats · min$^{-1}$ |
|---|---|---|
| Light work | up to 0.5 | up to 90 |
| Moderate work | 0.5-1.0 | 90–110 |
| Heavy work | 1.0-1.5 | 110–130 |
| Very heavy work | 1.5-2.0 | 130–150 |
| Extremely heavy work | over 2.0 | 150–170 |

**Table 17.2 Energy Expenditure**

| | | Men | | | Women | | |
|---|---|---|---|---|---|---|---|
| | Time, h | Rate, kcal · min$^{-1}$ | Total kcal | Total MJ | Rate, kcal · min$^{-1}$ | Total kcal | Total MJ |
| Sleeping, lying (1) | 8 | 1.1 | 540 | 2.3 | 1.0 | 480 | 2.0 |
| Sitting (2) | 6 | 1.5 | 540 | 2.3 | 1.1 | 420 | 1.8 |
| Standing (3) | 6 | 2.5 | 900 | 3.8 | 1.5 | 540 | 2.3 |
| Walking (4) | 2 | 3.0 | 360 | 1.5 | 2.5 | 300 | 1.3 |
| Other (5) | 2 | 4.5 | 540 | 2.3 | 3.0 | 360 | 1.5 |
| Total | 24 | | 2,880 | 12.1 | | 2,100 | 8.8 |

(1) Essentially basal metabolic rate plus some allowance for turning over or getting up or lying down.

(2) Includes normal activity carried on while sitting, e.g. reading, driving an automobile, eating, playing cards, and desk or bench work.

(3) Includes normal indoor activities while standing and walking spasmodically in limited areas, e.g. performing personal toilet, moving from one room to another.

(4) Includes purposeful walking, largely outdoors, e.g. home to commuting station to work site, and other comparable activities.

(5) Includes spasmodic activities in occasional sport exercises, limited stair climbing, or occupational activities involving light physical work. This category may include weekend swimming, golf, tennis, or picnic using 5 to 20 kcal · min$^{-1}$ for limited times.

Various attempts have been made to establish maximal permissible limits for daily energy output for people working at the same task year-round (Banister and Brown 1968). G. Lehmann (1953) suggested 4,800 kcal (20 MJ) · day$^{-1}$ as the limit. Subtracting about 2,300 kcal (9.6 MJ) for basal metabolism, eating and various basic necessities, leisure activity, and travel to and from work, 2,500 kcal (10.5 MJ) is left for the actual 8 h of work. Banister and Brown (1968) considered 2,000 kcal (8.4 MJ) a more suitable load for heavy workers, giving an average rate of energy expenditure of 4.2 kcal (18 kJ) · min$^{-1}$.

Establishing such norms may be quite meaningless, however, in view of the large individual differences in physical work capacity or fitness. Furthermore, the level of activity in many industrial tasks is actually self-regulatory in that the rate of work and the spacing of rest pauses are set by the individual's level of physical fitness. In fact, in some cases, such as the older commercial fishermen studied by K. Rodahl, Vokac, et al. (1974), the only way it is possible for a person to endure workloads close to the permissible physiological limits, day after day, year after year, is by working intermittently, with periods of high work intensity interspersed with frequent, brief rest periods.

In terms of strain imposed on the worker, the peak load of the task is more important than the mean energy expenditure. A steel worker may expend 10 kcal (42 kJ) · min$^{-1}$ during 1 h of shoveling gravel or dolomite, but during the rest of the 8-h shift, energy output may be only 2 to 2.5 kcal (8–10 kJ) · min$^{-1}$. The 8-h energy expenditure is then 600 + 900 = 1,500 kcal (6.3 MJ). A worker with a job that demands a consistent rate of work with no peak loads can attain a higher 8-h energy expenditure (e.g., 480 · 4 = 1,920 kcal, or 8.0 MJ), without requiring as strong a physique as the steel worker. Heavy work or awkward working positions often hamper the recruitment for certain types of work, even though these factors operate for only short periods of time. The same applies to many types of automated industrial operations where monotony or lack of personal influence on the process can be boring.

## *Sleeping*

In the past, it has been difficult to obtain accurate values for energy expenditure in a sleeping subject because of the discomfort and restriction imposed by noseclip and mouthpiece or the technical problem of leaks when a face mask is used. The use of the open-circuit method with a plastic hood covering the subject's entire head offers many distinct advantages when making metabolic measurements in a resting or sleeping individual. With this method, repeated samples can be taken throughout the observation period without disturbing the subject in any way. Generally, one third of the 24-h period is spent in bed, sleeping or resting. The energy spent in this way accounts for about one tenth to one fourth of the daily expenditure.

The metabolism, as a rule, falls below the basal metabolic rate (BMR) when the fasting subject is asleep. BMR is the rate of energy metabolism in a resting individual 14 to 18 h after eating. At the start of the night's sleep, however, the metabolism is above the basal level because of the specific dynamic action effect of the evening meal. These two factors thus cancel each other out, so that the energy expenditure throughout the night is not far from BMR. In any case, a deviation of 10% above or below the normal basal level represents an error of less than about 3% of the total 24-h energy expenditure. The BMR value therefore can be taken as a measure of the metabolic rate of a subject in bed, asleep or awake.

The metabolic rate of a subject going to sleep is affected by his or her mental state. Figure 17.21 provides data for an aircraft pilot worried about getting to work in time the next day because of falling snow and unplowed roads. In comparison, figure 17.22 shows data for his copilot going to sleep in a hotel within walking distance from the airport.

## FOR ADDITIONAL READING

For a comprehensive review of the physiology of sleep and wakefulness and how physical fitness influences sleep, see Nicholson and Stone (1982) and S.M. Shapiro et al. (1984).

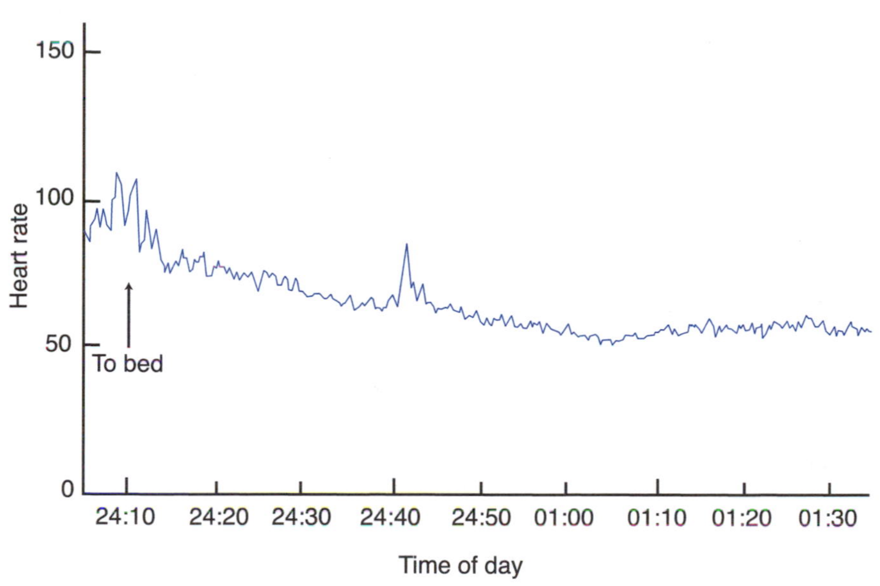


**Figure 17.21** Heart rate in a commercial pilot trying to sleep, who is worried about getting to work on time the next day because of snowfall in the night.

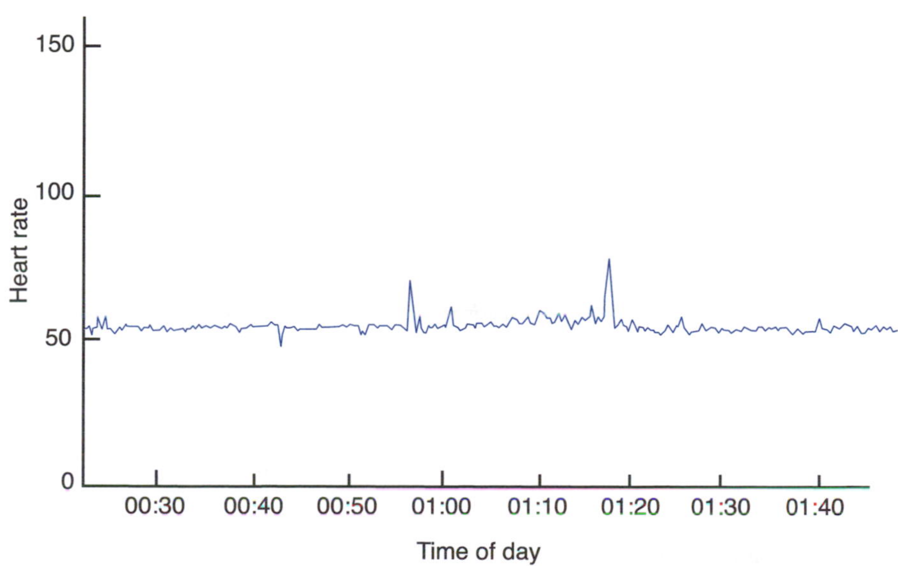


**Figure 17.22** Heart rate in a commercial pilot (copilot to the subject in figure 17.21), who spent the night prior to takeoff sleeping in a hotel near the airport.

## Sedentary Work

The energy expenditure of mental work, including office work, is not materially different from that of sitting or standing unless such occupations involve a great deal of physical activity, such as walking, bending, or opening drawers. Passmore and Durnin (1955) listed a mean value of 1.6 kcal (6.7 kJ) · min$^{-1}$ for miscellaneous office work while sitting and 1.8 kcal (7.5 kJ) · min$^{-1}$ while standing. Figure 17.23 shows the heart rate of a female bank clerk getting up at fairly regular intervals to collect papers and forms.

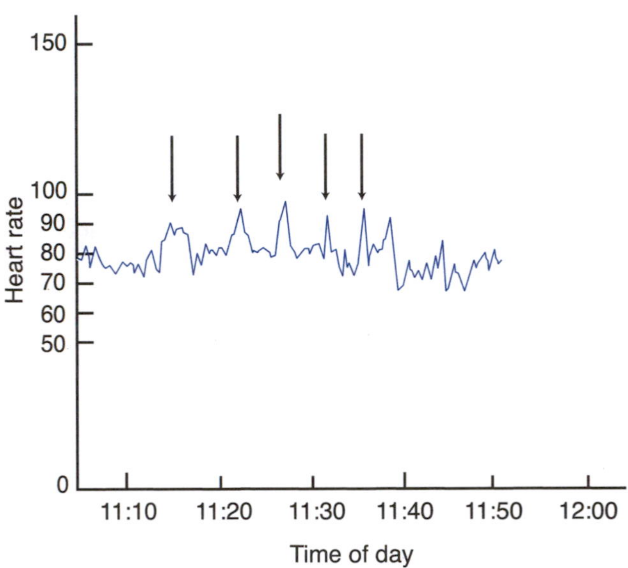


**Figure 17.23** The rhythmic work routine of a bank clerk who gets up at fairly regular intervals to collect papers and forms (indicated by arrows). The result is a dynamic work rhythm providing change of body position and circulatory variations.

Although the central nervous system is extremely sensitive to lack of oxygen and lowered oxygen tension, and the brain uses a substantial part of the total oxygen uptake of the body at rest, it is well established that mental work requires only an insignificant increase in oxygen uptake, at least as long as the mental effort is not associated with markedly increased muscular tension or emotional stress (Benedict and Carpenter 1909). Benedict and Benedict (1933) observed no substantial difference in metabolism at rest and during mental effort in six subjects engaged in 15-min periods of arithmetic exercises.

## Housework

Domestic work involves many tasks that are classified as fairly heavy physical work (Oberai, Dhillon, and Miglani 1983), although modern equipment has contributed greatly to making housekeeping easier. This is all relative, however, because performance of physical work depends not only on the severity of the workload but also on the individual's physical work capacity. In a study of stress at sea (K. Rodahl 1989), the heaviest physical load, relatively speaking, was observed in women working in the mess and cleaning cabins (figure 17.24).

Torgén, Nygård, and Kilbom (1995) investigated the physical workload, physical capacity, physical strain, and perceived health among elderly Swedish aids in home-care service, 45 to 65 years old. They concluded that home-care work is characterized by long periods of standing and walking and that postures potentially harmful for the lower back and shoulders occurred frequently.

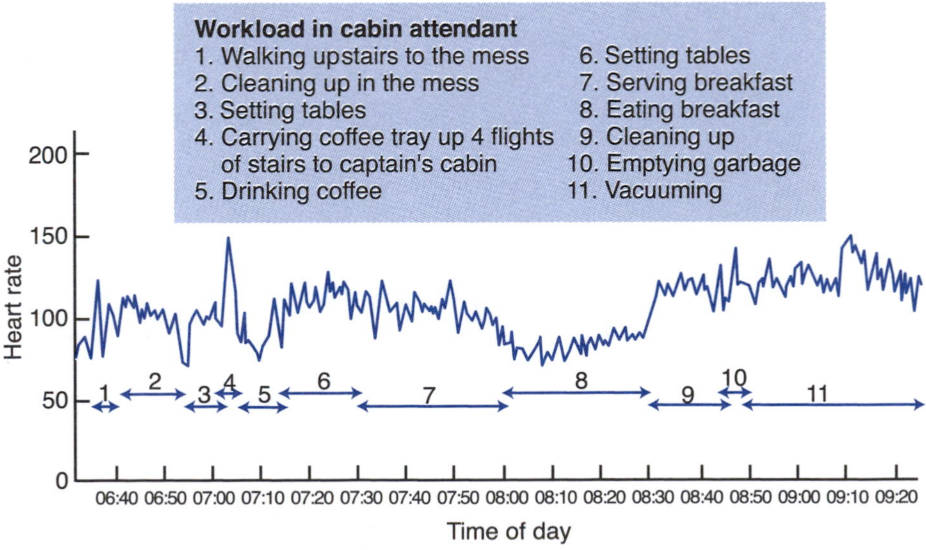


**Figure 17.24** Physical workload in a ship's cabin attendant.

## Light Industry

A fair amount of data are available regarding the energy cost of different kinds of manual labor and industrial tasks. With the development of automation, the physical workload of the industrial worker on the whole has been greatly reduced. This is evident from an early study by Kagan et al. (1928), who compared energy expenditure of men assembling machinery entirely by hand with those using a conveyer system. In the former case, the energy expenditure varied from 5.2 to 6.4 kcal · $min^{-1}$ (22–27 kJ · $min^{-1}$); in the latter case, it varied between 1.8 and 4.7 kcal · $min^{-1}$ (7.5–20 kJ · $min^{-1}$). According to Passmore and Durnin (1955) and Durnin and Passmore (1967), a wide variety of industrial activities, classified by them as light industry, demanded energy expenditure rates between 2 to 5 kcal · $min^{-1}$ (8–21 kJ · $min^{-1}$) for men and 1.5 to 4 kcal · $min^{-1}$ (6–17 kJ · $min^{-1}$) for women.

## Manual Labor

A number of studies have shown that the energy expended during the performance of similar types of work varies greatly, depending on the technique used in accomplishing the work. This is even true for relatively simple activities, such as carrying a load, depending on how the activity is performed. Thus, already in 1924 it was demonstrated that when a person carried a load, the energy expenditure was minimal when a yoke across the shoulder was used; energy expenditure was maximal when the load was carried on the hip under the arm (Bedale 1924). The energy cost is less when the load is kept close to the center of the body (Soule, Pandolf, and Goldman 1978). When a person carries a load, the energy expenditure increases markedly with increased speed of walking and by the increasing weight of the load (Holewijn 1990).

Garg and Saxena (1980) showed that handles have a profound effect on maximal acceptable weights to be lifted. The subject's own weight, in addition to that of the load, also should be taken into account when assessing the optimal load to be carried (Pierrynowski, Winter, and Norman 1981). O.M. Evans et al. (1983) recorded heart rate and EMG in seven young men who held and carried loads of 15, 20, 25, 30, and 40 kg until exhaustion. The mean maximum time for load holding was found to be hyperbolically related to the mass of the load, the mean maximum time for load carrying being shorter than that for load holding. The heart rate at exhaustion was linearly related to the load and was greater when the load was carried than when held. The physiological response to lifting a load depends on the manner in which it is being lifted. Kumar (1980) measured intra-abdominal pressure by telemetry and the activity of the erector spinae and external oblique muscles by electromyography in male volunteers lifting 10 kg from ground to knee, hip, and shoulder levels in the sagittal, lateral, and oblique planes. He found a significant difference between the responses in these three planes, the sagittal plane activities evoking least response. The intra-abdominal pressures and the electromyographic activity of the erector spinae and the external oblique activity were significantly correlated in each of the three planes. According to Cresswell and Thorstensson (1994), the intra-abdominal pressure is less during lowering than during lifting at any given force.

The importance of using proper technique when lifting is evident from the studies of Ekholm, Arborelius, and Németh (1982), who showed that the compressive forces on the lower back (the L5–S1 disks) were about 4,400 N when a subject lifted 12.8 kg with straight knees, compared with 3,400 N when the subject lifted with bent knees, straight back, and the burden between the knees and close to the pelvis. However, this only applies as long as the load is kept close to the body.

The energy cost of climbing up and down stairs with a load is 11 times that of walking on level ground with the same load (Crowden 1941; Pimental and Pandolf 1979). In a study of the energy cost of transporting a load with the aid of a wheelbarrow, J.E. Hansson (1968) showed that the energy expenditure is higher the smaller and softer the wheel and that a two-wheel cart is more efficient than a wheelbarrow with a single wheel.

Madeleine, Voigt, and Arendt-Nielsen (1998) examined the subjective and physiological responses to prolonged light, repetitive, manual work while subjects stood on a soft, polyurethane standard mat and on a hard, aluminum casting surface for 2 h. Standing on a soft surface caused a lower intensity of discomfort. Standing on a hard surface enhanced swelling of the shank, increased electromyographic activity of the lower leg, and led to more pronounced muscle fatigue.

The efficiency of a particular work operation can be estimated roughly by measuring the energy cost based on oxygen uptake or heart rate. The influence of the choice of tools on the work rate and work effort was studied in a group of men engaged in nailing impregnated paper under standardized conditions (J.E. Hansson 1968). The rate attained in nailing when they used a stapler was three to four times higher than when they used an ordinary hammer. There were no significant differences, however, in oxygen uptake per minute, quality of work, or estimated degree of fatigue when the subjects used the different tools. I. Åstrand, Guharay, and Wahren (1968) studied the energy cost of hammer nailing at three different heights: at bench level, into a wall at head level, and into a ceiling above the head. The number of hammer strokes did not differ significantly in the three situations. However, the number of nails driven per minute was lower when subjects were nailing into the wall (10.6 nails · min$^{-1}$) than when nailing into the bench (14.6 nails · min$^{-1}$) and still lower for nailing into the ceiling (4.5 nails · min$^{-1}$), indicating that the strokes become less powerful or were less well aimed when subjects were nailing into the wall and ceiling than when nailing into the bench. Nailing in the three positions resulted in an oxygen uptake of about 1.0 L · min$^{-1}$ in each case. The 11 subjects, all of whom were skilled carpenters, also performed leg work on a cycle ergometer with the same oxygen uptake. Nailing into the wall or ceiling elevated the heart rate more than nailing at bench level; the heart rate during cycle exercise at a comparable oxygen uptake was lower (or 102 beats · min$^{-1}$) than for all types of nailing (130 beats · min$^{-1}$ when nailing into the ceiling). It is interesting to note that leg exercise with an oxygen uptake of 1.4 L · min$^{-1}$ gave approximately the same increase in heart rate as nailing with an oxygen uptake of only 1.0 L · min$^{-1}$. This finding refers to prolonged continuous work in a more or less steady-state situation and not to intermittent work. The intra-arterially measured blood pressures during nailing were higher than during leg exercise at a given oxygen uptake, the difference being most pronounced between nailing into the ceiling and cycle work ($p < .01$).

Studies of coal miners in different countries have shown a good general agreement for work with pick and shovel. It appears that the energy expenditure of shoveling ranges from 6 to 7 kcal (25–29 kJ) · min$^{-1}$. In a German study (G. Lehmann, Müller, and Spitzer 1950), the mean gross energy expenditure during actual coal mining was 5 kcal (21 kJ) · min$^{-1}$, and the mean expenditure of energy per minute for the total time spent underground was 3.5 kcal (15 kJ). According to Passmore and Durnin (1955), walking to and from the coal face in a stooped position requires as much as 10 kcal (42 kJ) · min$^{-1}$. It appears that in despite increased mechanization, mining is still hard physical work. Studies in the iron and steel industry (Christensen 1953; G. Lehmann, Müller, and Spitzer 1950; Paleneiano et al. 1996) show wide variations in energy expenditure, so that generalizations cannot be made.

Several studies have verified the general impression that farming is hard work, at least in the busy season, especially when mechanization is not available (Durnin and Passmore 1967). Milking by hand requires about 4.5 kcal (19 kJ) · min$^{-1}$ compared with 3.5 kcal (15 kJ) · min$^{-1}$ for machine milking. Horse-drawn plowing uses 6 kcal (25 kJ) · min$^{-1}$, whereas tractor plowing requires less than 5 kcal (21 kJ) · min$^{-1}$ (Hettinger and Wirths 1953a, 1953b).

In Norwegian coastal fishermen, the average energy expenditure of all activities on board during a whole day at sea was 34% to 39% of the fisherman's maximal aerobic power, with occasional peaks of up to 80% (K. Rodahl, Vokac, et al. 1974). The heart rate exceeded 50% of the fisherman's heart rate reserve for 9% to 23% of the time. The most strenuous activities were pulling in the seine with a power block (oxygen uptake up to 2.7 L · min$^{-1}$) and unloading the catch, taxing the subjects by more than 50% of their maximal aerobic power for two thirds of the duration of these activities. In general, it appears that the energy expenditure of commercial fishermen during the active fishing season reaches levels as high as about 5,000 kcal (21 MJ) · day$^{-1}$.

It is generally recognized that lumber work involves heavy energy expenditure. In fact, lumbering is probably the hardest form of physical work, requiring sometimes as much as 6,000 kcal (25 MJ) · day$^{-1}$ (Lundgren 1946). J.E. Hansson (1965) noted that lumberjacks with very high earnings were characterized by a particularly high maximal aerobic power compared with that of the average earner. With regard to muscle strength and precision in a variety of standardized work operations, there was no significant difference between the two categories. Because of his higher aerobic power, the top worker could attain a higher work output, did not take as long breaks, and became less tired at a given energy output compared with his less productive colleague. It appears that workers involved in manual labor who are more or less free to set their own pace normally work with an energy output less than about 40% of their individual maximal aerobic power (I. Åstrand 1967b).

## For Additional Reading

For a comprehensive discussion of energy expenditure in household, occupational, recreational, and sports activities in relation to body mass, see Katch, Katch, and McArdle (2000).

## Sex-Related Aspects of Firefighter Work Stress

It has been claimed that women should be allowed to serve as firefighters on equal terms with men, despite the fact that fire fighting involves exposure to extreme levels of work stress and requires great muscle strength and physical work capacity. Fire fighting involves carrying the weight of fire-protective clothing and self-contained breathing apparatuses amounting to more than 25 kg (Louhevaara et al. 1995), in addition to hoses and ladders, as well as carrying or pulling unconscious victims up and down stairs under extremely hot and smoky conditions.

Despite this, there are a number of active female firefighters in the United States (Murphy et al. 1994), while in Scandinavia there are rather few. The fitness test of firefighters used in Sweden includes moving a 98-kg doll 30 m across a rough floor while wearing heavy protective equipment. It includes taking down ladders from the truck and carrying them together with a motor saw weighing 11.5 kg. The ladder has to be raised and climbed and then taken down and carried back to the truck. Two bags with water hoses, weighing 26 kg each, have to be carried 100 m; one of them has to be taken down one flight (10 steps) and up again and back to the starting point. In a selection of eight applicants picked out of 85, only one passed all the tests in the final trial.

If female applicants compete with male applicants on equal terms on the basis of the physical tests used in Sweden, most likely only male applicants would be appointed because the strength and physical work capacity of the average male are significantly superior to the average female. In fact, there is almost no overlap of the mean figures for strength and maximal oxygen uptake between males and females.

In an earlier study of eight professional firefighters in Canada, Romet and Frim (1987) found that the most demanding activity, building search and victim rescue, resulted in an average heart rate of 153 beats · min$^{-1}$ and a mean rectal temperature increase of 1.3 °C. The least demanding activity was that of the crew captain directing the fire fighting, which resulted in an average heart rate of only 122 beats · min$^{-1}$and a rectal temperature increase of .3 °C. On the basis of these data, the work of the firefighter crew captain most likely would be within the limits of the work capacity of the average woman.

# Working in the Heat

Environmental temperature affects human performance (see figure 17.1). At body temperatures substantially higher than the optimal levels, both physical and mental performance deteriorate because of the complicated interplay of physiological and pathophysiological processes. Prolonged heat stress leads to loss of body fluid (hypohydration), which in itself impairs performance, especially endurance. In addition, prolonged heat strain impairs mental and psychomotor functions, thereby affecting performance. It is therefore of considerable practical importance to be able to assess the magnitude of thermal stress in the working environment and the worker's physiological reaction to it, to ensure optimal conditions for health and productivity.

One of the most commonly used methods of assessing environmental heat exposure at different places of work is the wet bulb globe temperature (WBGT) index, which is based on the measurement of air temperature (dry bulb), radiant temperature (globe), and relative humidity (wet bulb). This can be metered and logged on the standard version of the Squirrel meter/logger. This, however, occupies three of the temperature channels of the logger. Furthermore, the WBGT index is rather complicated and is time consuming to record. In the case of a predominantly dry, radiant heat environment, the much simpler and fast reacting wet globe thermometer (WGT) index or Botsball, devised by Botsford (1971), records similar values, and the results are interchangeable with the aid of a simple formula (Beshir, Ramsey, and Burford 1982; Ciriello and Snook 1977; Dernedde 1992).

The Botsball thermometer combines air temperature, humidity, wind, and thermal radiation into a single reading, expressing the thermal stress of the environment. The instrument is quite simple and easily adapted to industrial use. It consists of a small, 60-mm, hollow black globe, covered with a double layer of black cloth, which is continuously moistened by water seeping from a reservoir tube attached to the globe. The temperature sensor attached to the Squirrel passes through a plastic tube along the center line of the water reservoir and into the globe, thus sensing its temperature. When placed in a hot area, the globe is heated by the surrounding air and by radiant heat from any hot surfaces in the surroundings. It is cooled by the evaporation of water from the globe surface, depending on air humidity and movement. The wet globe reaches an equilibrium temperature when these heating and cooling effects come into balance, which usually takes about 5 min. Furthermore, the recording of the Botsball requires only one of the temperature channels of the Squirrel. The remaining channels can be used to record physiological response to the heat stress by recording changes in body heat content assessed by rectal temperature (one channel) and mean skin temperature (indicated by the midthigh temperature (one channel). For references, see K. Rodahl and Guthe (1988). Additional channels can be used for temperature parameters of special concern, such as radiant temperature (black ball temperature), eardrum temperature (as an indication of central body temperature), or skin temperature (e.g., under the foot of a worker standing on a hot floor in front of a furnace). With the multichannel Squirrel loggers, a number of other pertinent parameters in addition to temperature, such as heart rate (as an indication of work stress), muscular tension, and concentration of dust and different gases in the ambient air, can be continuously and simultaneously recorded on the same logger.

## Continuous Recording of Environmental Temperature Indexes

The Squirrel logger, as already indicated, is well suited for the continuous recording of both the WBGT and WGT indexes. The temperature probes for the Squirrel have been compared with a standard mercury thermometer and found to be quite accurate. The WBGT has three probes, occupying three channels in the Squirrel (1 = dry bulb temperature; 2 = globe temperature; 3= wet bulb temperature). The WBGT index can be calculated by the following formula:

$$\text{WBGT} = (\text{Channel } 1 \times 0.1) + (\text{Channel } 2 \times 0.2) + (\text{Channel } 3 \times 0.7) \quad (1)$$

The reading of the Botsball (WGT) can be converted to WBGT by the following formula:

$$\text{WBGT} = (\text{WGT2} + 0.0212) + (\text{WGT} \times 0.192) + 9.5 \quad (2)$$

The converted readings of the WGT have been compared with the WBGT readings, the two instruments being placed side by side in different places in and around the laboratory. The mean difference between the two sets of readings was .5 °C, which is quite satisfactory. A similar comparison between the WBGT and WGT converted into WBGT was made in a ferroalloy plant in the north of Norway in the month of July. Here, under actual field conditions with changing radiant heat and considerable air movement, the main difference was 1.5 °C; the difference decreased as the temperature increased (range of difference .2–2.9°C). This degree of difference is quite acceptable given that the environmental temperature in the plant can fluctuate by as much as 10 °C from hour to hour in the same place.

Dernedde (1992) showed that the WBGT can be predicted from the Botsball (BB in the following equation) temperature index and the ambient water vapor pressure ($P_a$), measured with a psychrometer, according to the following equation:

$$\text{WBGT} = 1.5157\,(\text{BB}) + .0112\,(\text{BB} \times P_a) - 0.7379\,(P_a) - 2.5591 \quad (3)$$

Another advantage of the Botsball thermometer is that it can be mounted on a helmet and thus carried by the person, allowing the environmental temperature to be recorded at the worker's actual location at any time. The only drawback with this arrangement is that walking briskly causes the WBGT to drop a couple of degrees as air motion increases evaporation from the globe surface, hence cooling the probe. The ideal solution is probably to use the Botsball thermometer attached to the worker initially to survey the temperature distribution in the working area, and then to place the Botsball thermometer stationary at the location where the temperature is most extreme or most representative.

## Assessment of Heat Stress

Environmental heat stress can be assessed with the aid of the Squirrel logger by using a WBGT assembly or a Botsball WGT instrument. However, the environment is one thing, but the human reaction to that environment is of far greater importance. For this reason, greater emphasis should be placed on assessing human response to the heat stress encountered by workers while performing their everyday work at their actual workplaces. Without such information, any discussion of safe or upper limits of exposure is meaningless.

A significant correlation was found by Kiladze and Afanas'eva (1993) between the WBGT index, average skin temperature, and heat sensation in a study of microclimate in Russian industry.

To assess the physiological reaction to heat loads imposed by the environment, the body heat content *(S)* is a most meaningful index of body heat gain or body heat loss. It can be calculated by the following formula:

$$S \text{ (in kcal)} = .83 \times \text{(body weight in kg)} \times \text{(rectal temperature} \times .65) + \text{(mean skin temperature} \times .35) \quad (4)$$

## *Deep Body Temperature and Skin Temperature*

Rectal temperature mirrors the core temperature and, by itself, reflects body heat gain or body heat loss. Under such conditions, rectal temperature can be used as an index of heat stress, although the investigator must remember that it is a slow-reacting parameter. In most cases of industrial heat exposure, it takes some 45 min for the rectal temperature to reach a plateau. This emphasizes the importance of continuous recording of rectal temperature by portable electronic loggers. This, however, requires cooperative subjects as well as patience and persuasion by the investigator. However, when correctly carried out, the continuous recording of rectal temperature in workers in the course of their normal work and leisure is indeed a most revealing parameter in terms of thermal strain.

Ideally, the temperature sensor should be placed no less than 8 cm into the rectum to achieve stable readings representing deep body temperature. This can be done by supplying the rectal temperature sensor with a taped knob 8 cm from the end, inserting the sensor far enough into the rectum for the knob to be placed inside the rectal sphincter, and taping the wire to the subject's lower back to prevent it from being pulled out of the rectum. For hygienic reasons, the sensor can be placed inside a thin disposable plastic envelope before insertion.

In some cases, it might be simpler to record the temperature inside the ear as an index of deep body temperature, replacing rectal temperature in the body heat content equation. For basic research purposes, it has been customary to measure the tympanic temperature as an expression of deep body temperature close to the location of the thermoregulatory center in the brain, by placing a thin thermocouple deep into the ear touching the lower part of the eardrum (Benzinger and Taylor 1963). When the investigator is making field measurements in working individuals at their industrial workplaces, taking tympanic temperature might not be any easier than using a rectal thermometer. Manipulating the temperature sensor into the ear duct and placing the sensor so it touches the eardrum require professional skill. In addition, it can be painful to the subject, even when a very thin and flexible thermistor wire is used. In any case, it is necessary to plug the external opening of the ear duct to prevent the ambient air from interfering with the readings. For applied studies or less sophisticated research projects, Keatinge and Sloan (1975) suggested a combination of an aural canal sensor and servo-controlled heating of the outer ear to attain temperature readings very close to eardrum temperatures.

For a rough estimation of deep body temperature, the much simpler oral (sublingual) temperature can be used, providing the sensor or thermocouple is kept in place under the tongue and the subject keeps her or his mouth completely closed. Oral temperatures recorded in this manner are, in most cases, about .5 °C lower than rectal temperature.

Skin temperature can be measured relatively easily with commercially available skin temperature sensors connected to the Squirrel. Skin temperature can be used as an index of local heat exposure but cannot by itself be used as an indication of general heat stress or thermal balance. Proper shielding of the temperature sensor under conditions of intense radiating heat is essential to obtain real values of skin temperatures. The skin temperature recorded on the inside of the thigh closely resembles mean skin temperature under most normal conditions (Ramanathan 1964), so this value can be used in the body heat content equation previously referenced.

## The Effect of Heat Stress on Heart Rate

It is well known that heat stress places an additional load on the cardiovascular system. This is evidenced by an elevated heart rate at the same workload in a hot environment versus room temperature. Thus, when investigators recorded the heart rates of workers operating in the pot room in an aluminum-production plant, in some cases the heart rates reflected the environmental temperature (Botsball temperature index). In other words, they fluctuated synchronously (figure 17.25).

To determine the effect of heat stress per se, heart rate was recorded in the laboratory at room temperature (mean Botsball temperature 15.4–16.4 °C) at rest and at two different workloads on a cycle ergometer and at a selected location in the pot room (mean Botsball temperature 22.0–23.1°C). Each subject was studied on two different days. The first day, the subject was studied at normal room temperature in the laboratory in the morning. The subject rested in a sitting position for 15 min. This was followed by 10 min of cycle ergometer exercise at a load of 300 kpm · min$^{-1}$, then a 5-min rest in a sitting position, and, finally, a 10-min cycle ergometer exercise at a load of 600 kpm · min$^{-1}$ followed by 60 min

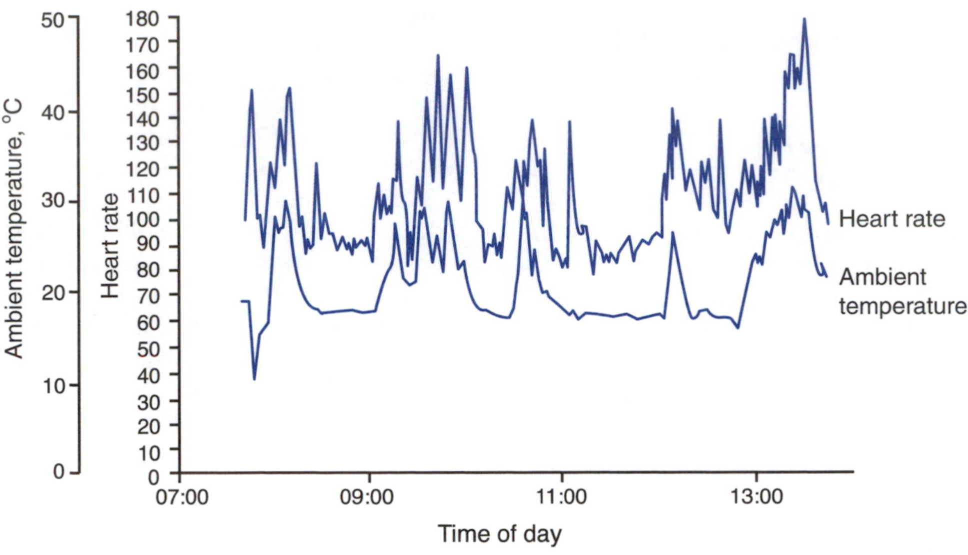


**Figure 17.25** Relationship between ambient temperature (Botsball) and heart rate in a 34-year-old female operator in an aluminum production plant.

of rest at normal room temperature. The subject then went to a selected place in front of a pot, where he or she remained seated in a chair for 45 min for the body temperature to become adjusted to the prevailing ambient temperature. Following this, the program was identical to that followed at room temperature in the laboratory. The second day, the same subject went through the program in reverse order; that is, he or she started in the pot room and finished in the laboratory. This was done to adjust the possible effects of circadian rhythmic changes in body temperatures and heart rate.

The mean difference in the heart rate in the laboratory (room temperature) and in the pot room (heat stress) is shown in table 17.3.

The results of this study show that the heat stress alone which the operator experiences when working at a rate of 300 and 600 kpm · min$^{-1}$ (50 and 100 W · min$^{-1}$) in front of the pot represents an additional load on the cardiovascular system of 20 to 25 beats · min$^{-1}$, corresponding to an approximate 20% increase in the workload. This should be taken into account to prevent undue fatigue in workers exposed to excessive heat.

**Table 17.3** The Mean Difference in Heart Rate in the Laboratory (Room Temperature) and in the Pot Room (Heat Stress)

| Location | Rest | 300 kpm · min$^{-1}$ | 600 kpm · min$^{-1}$ |
|---|---|---|---|
| Pot room | 79 | 119 | 140 |
| Laboratory | 74 | 100 | 115 |
| Difference | 5 | 19 | 25 |

From Nes et al. 1990.

While logging the physiological effects of intensive heat exposure in a glass factory (Rodahl and Guthe 1991; K. Rodahl, Guthe, and Klüwer 1991), the investigators observed that the heart rate reacted surprisingly quickly to the ambient temperature and more or less oscillated synchronously with the temperature surrounding the subject. This very rapid increase in the heart rate was not caused by physical activity, because the subject was standing quietly in front of the heat source. Therefore, the increase could only be caused by heat stress. This poses the question of what physiological mechanism is behind the elevated heart rate. The answer to this question is essential to counteract the effect.

There are a number of indications that the head and especially the face play a key role in the body's reaction to heat stress. Riggs et al. (1981) showed that cooling the face in physically active subjects caused a decrease in heart rate without any changes in blood pressure or rectal temperature. Furthermore, in a study of one of the subjects in the glass factory referred to previously (K. Rodahl, Guthe, and Klüwer 1991), the subject's heart rate and ambient and skin temperatures were recorded in repeated 10-min periods, while the subject stood inactively 1.5 m away from the production line of red-hot bottles. Between each exposure period, the subject spent 5 min inside a cooled control room to change headgear. The heart rate was recorded with the aid of an 8-bit Eltek Special Squirrel with a built-in heart rate counter (Rodahl, Leuba, et al. 1992).

In the first case (figure 17.26), the subject was exposed bareheaded, dressed in normal work clothing. In addition, the trunk was shielded by a reflective aluminum-foil apron as well as a reflective jacket. This was done to determine the influence of the trunk skin temperature on heart rate. The investigators also examined the effect of cooling the face around the mouth and nose with the aid of ice bags taped to the face. Also examined were the effect of cooling the area around the ears and, finally, the effect of inhaling cooled air by using a rubber tube connected to a bag containing ice-cooled air (K. Rodahl, Guthe, and Klüwer 1992).

The heart rate of this subject, standing inactive in front of a heat source, fluctuated synchronously with the ambient temperature—the hotter the environment, the higher the heart rate (figure 17.26). Changes in the heart rate were extremely rapid. Figure 17.27 shows that the heart rate increased in the heat-exposed subject even though the body was protected by reflective clothing that prevents an increase in the body skin temperature. However, if the protective clothing was removed from the body, skin temperature rose and a further increase in the heart rate was observed. It appears from figure 17.28 that cooling the face around the mouth and nose, in addition to trunk shielding, did not reduce the elevated heart rate in the heat-exposed subject, although it reduced the face skin temperature. Neither did cooling the area around the ears, placing a visor in front of the helmet, or covering the outside of the helmet with aluminum foil seem to have any effect. The attempt to have the subject inhale cooled air while being exposed to the same heat source gave inconclusive results.

The effect of added fluid intake on heart rate and body temperature was investigated in one operator at a glass bottle production plant. The operator went 1 day with and 1 day without fluid intake (250 ml of water every 30 min) and was exposed to a 10-min standard workload of 600 kpm · min$^{-1}$ on a cycle ergometer at normal room temperature, before and after the regular work shift. Without fluid, he had a

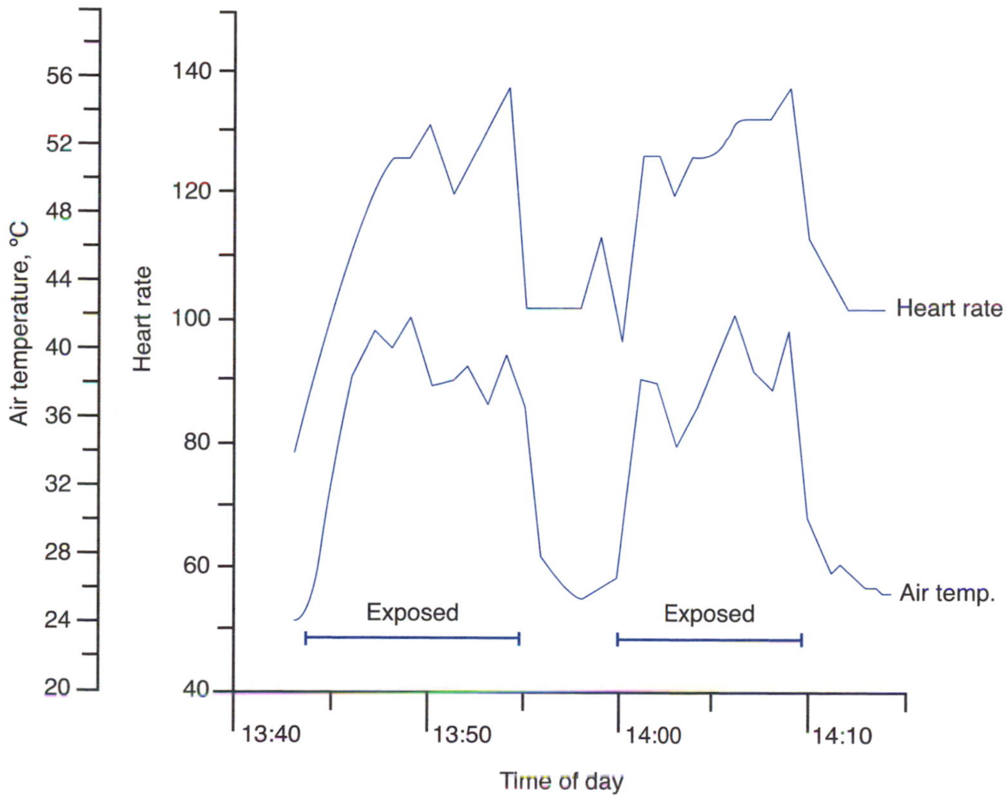


**Figure 17.26** The heart rate increases synchronously with the air temperature in a subject exposed twice for 10 min each time in front of a glass bottle production line (1.5 m away), interspersed with a 5-min stay in a control room at normal room temperature. The subject was bareheaded, dressed in normal work clothing with an aluminum-foil reflective jacket and apron.

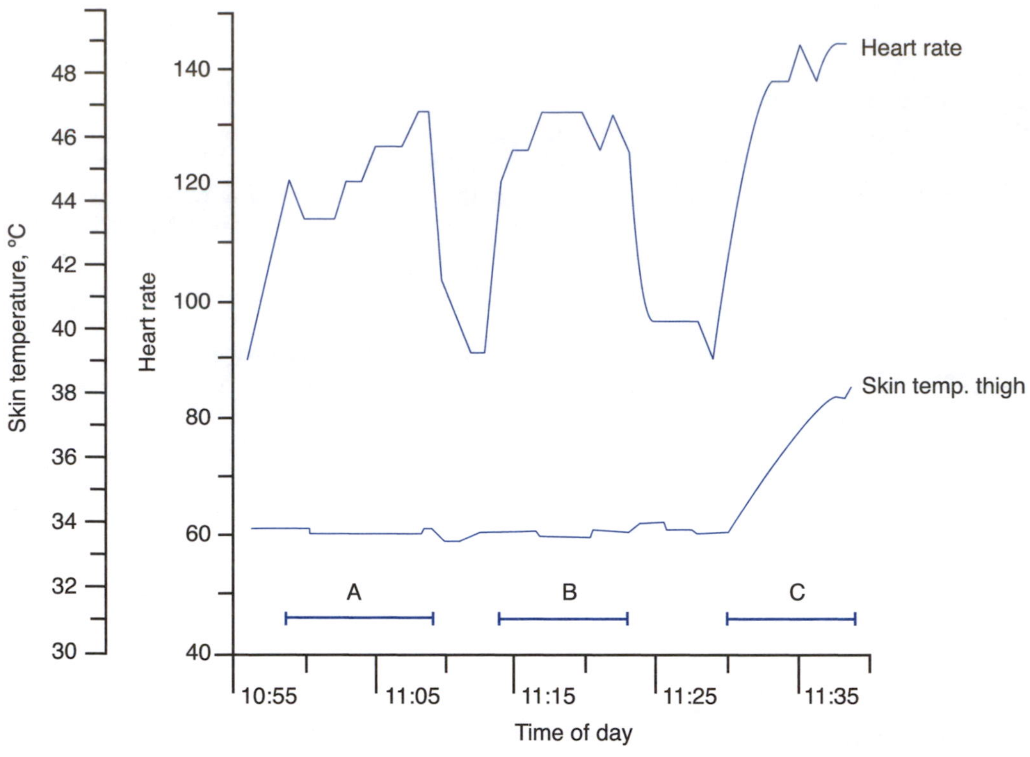


**Figure 17.27** Heart rate and thigh skin temperature in a subject exposed to the heat of a glass bottle production line (1.5 m away), interspersed with a 5-min stay in a control room at normal room temperature. A = bareheaded, dressed in work clothes and aluminum reflective jacket and apron; B = dressed as (A) but with an ice bag strapped in front of his mouth and nose; C = bareheaded, dressed in work clothes only.

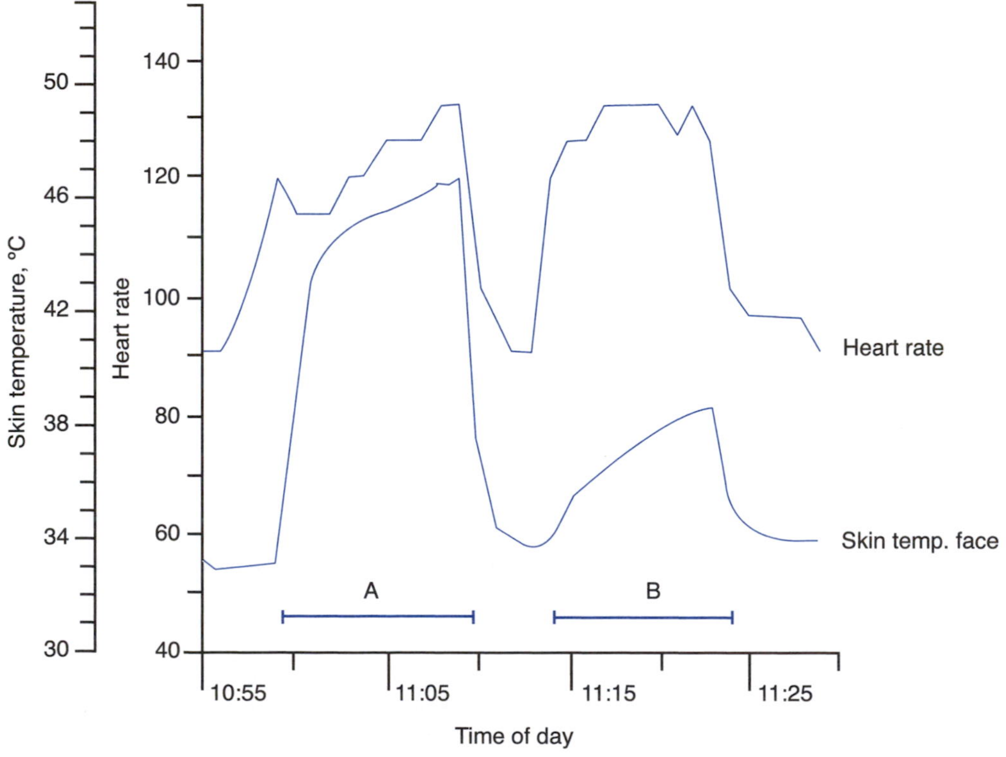


**Figure 17.28** Heart rate and face skin temperature in a subject exposed bareheaded to the heat of a glass bottle production line (1.5 m away), interspersed with a 5-min break at normal room temperature. A = dressed in normal work clothing and an aluminum reflective jacket and apron; B = dressed as (A) but with an ice bag strapped in front of his mouth and nose.

net fluid loss attributable to sweating of 1.9 L, and his mean heart rate during the ergometer exercise was 5 beats · $min^{-1}$ higher after than before the work shift. On the day with fluid, his fluid intake exceeded his fluid loss in terms of sweating, and his mean heart rate during the ergometer exercise was 13 beats · $min^{-1}$ lower after than before the work shift. The mean rectal temperature during the 10-min work test was .6 °C higher after than before the work shift in the case where no fluid was taken versus only .3 °C higher when adequate fluid was consumed. The minute-to-minute changes in rectal temperature with and without fluid are shown in figure 17.29.

## Heat Stress in an Aluminum Production Plant

Nes, Karstensen, and Rodahl (1990) systematically assessed the heat stress to which the pot room operators in an aluminum production plant were exposed and the effect of the exposure on the operators. The operators were studied during a total of 38 work shifts, during both the winter and the summer. The physiological reaction to the heat stress in terms of body temperature was recorded with temperature sensors connected to the 12-bit Squirrel logger. The logger was shielded against the magnetic field in the pot room by being kept in a fitted steel box and carried on a belt on the subject's back. The stored data were transferred to a PC and the results displayed immediately after the observation period to those involved in the study. One advantage of this immediate display of results was that the subjects could see the effects of different activities in terms of specific heat stress and the direct relationship between cause and effect. The fluid loss in terms of sweating was determined by weighing the subject before and after the work shift.

The results showed considerable differences in heat stress at the different plant operations, as well as variations in heat stress from one day to the next and at different seasons of the year. It was evident that burner cleaning represented by far the most severe heat stress, with Botsball temperatures as high as 35 °C and radiant temperatures exceeding 60 °C. The rectal temperature exceeded 38 °C in all the subjects studied, in some cases with peaks over 39 °C (figure 17.30). The skin temperature oscillated with the radiant temperature exceeding 40 °C and, in some cases, reached peaks close to the pain threshold (figure 17.31).

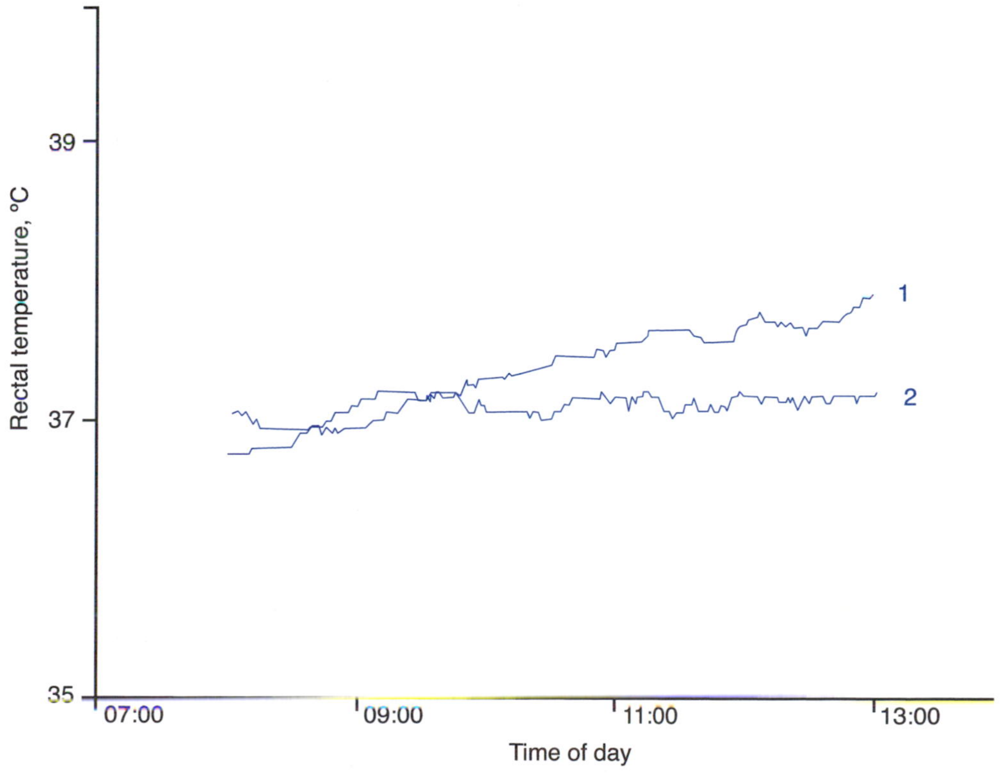


**Figure 17.29** The effect of adequate fluid intake on the rectal temperature of an operator in a glass bottle production plant. 1 = without fluid; 2 = with fluid.

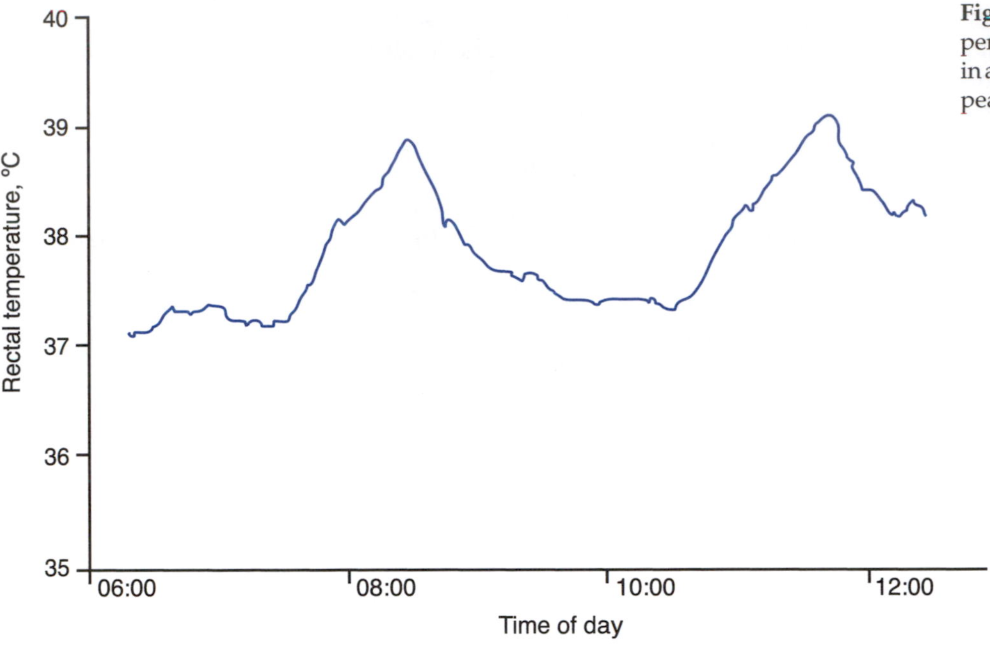


**Figure 17.30** Rectal temperature in a burner cleaner in an aluminum plant, with peaks exceeding 39 °C.

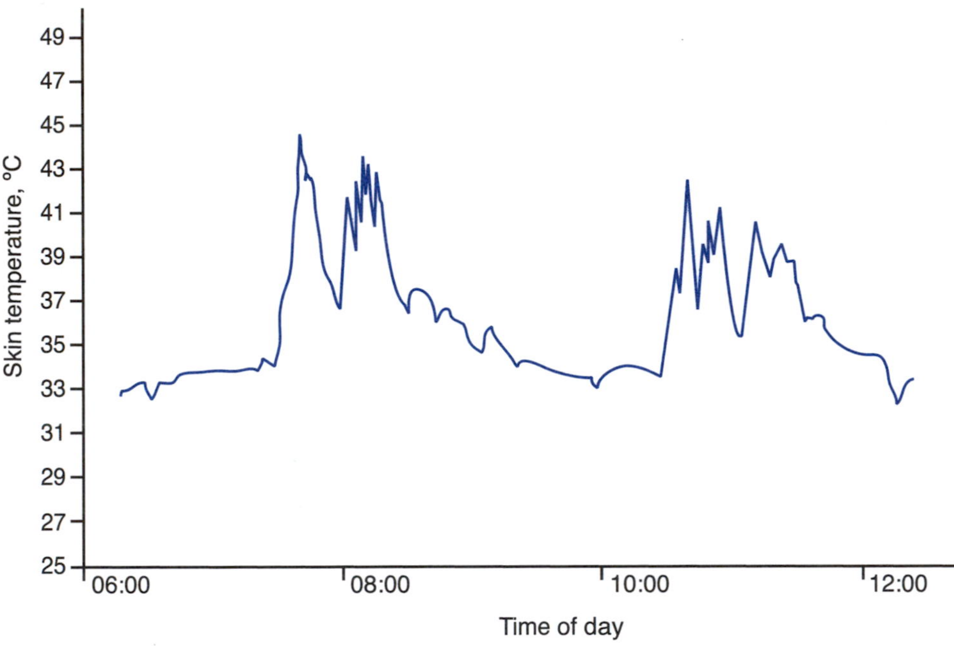


**Figure 17.31** Skin temperature on the inside thigh of a burner cleaner in an aluminum plant, approaching the pain threshold.

Next in terms of heat stress was gas manifold changing, where in one case the skin temperature under the foot reached about 42 °C (figure 17.32). In another case involving continuous exposure for more than 2 h, the rectal temperature rose gradually from 37 °C to more than 38 °C despite a moderate Botsball temperature from 15 to about 25 °C.

The weighing of fluid loss and hypohydration attributable to sweating reflected, on the whole, the magnitude of the heat stress as evidenced by the Botsball and deep body temperatures. The mean values showed that those who were most exposed to heat stress also had the greatest fluid loss by sweating. The fluid loss in the same subjects (examined both during summer and winter) was significantly lower during the winter, but even in the winter the fluid loss exceeded the level that produces symptoms of hypohydration,

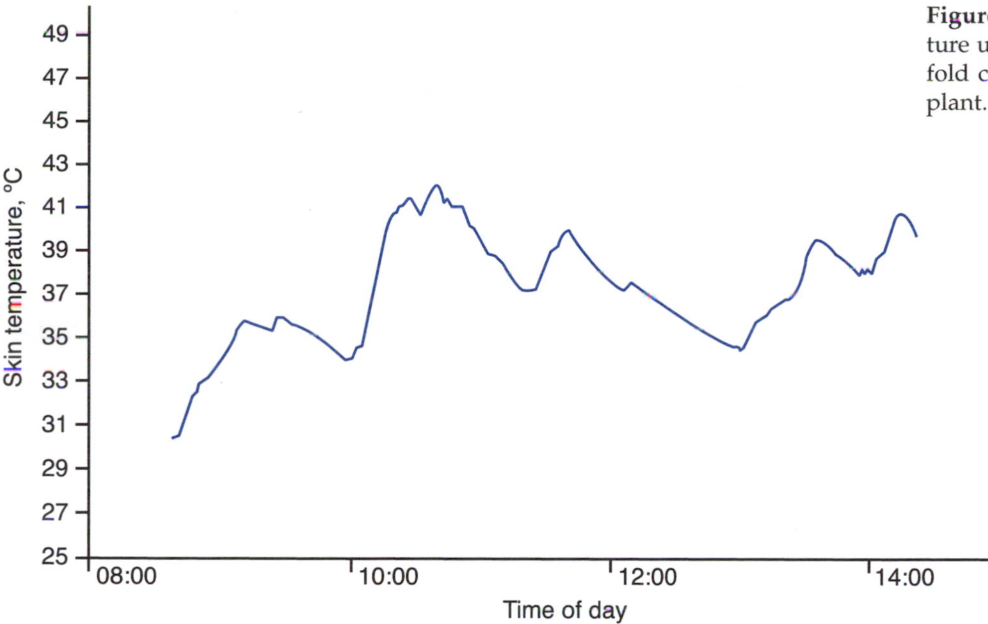


**Figure 17.32** Skin temperature under the foot of a manifold changer in an aluminum plant.

with fatigue, reduced stamina, and reduced alertness.

In some subjects, a net deficiency in fluid intake of 3.5 L during a work shift was recorded. This represents a significant degree of hypohydration and emphasizes the need for operators to drink water regularly during the work shift, regardless of whether they feel thirsty. This is because thirst is a slow-reacting indicator of the body's state of fluid balance. In addition, sweat loss can be significantly reduced by reducing each period of heat exposure, for example, 20 min, interspersed by frequent 10-min cooling-off periods to prevent an excessive increase of the internal body temperature. Heat strain can be reduced significantly by using heat-resistant and heat-reflective clothing, including reflective aprons. The need for effective protective garments is particularly stressed by the finding of very high skin temperatures both on the legs and under the feet, which merely confirmed operators' subjective reports. The skin temperatures can be reduced by using reflective shields in front of the legs and heat-resistant insulating insoles and perhaps even covering the boots with heat-reflective material, as well as using heat-resistant gloves or mittens.

## For Additional Reading

For further details concerning industrial heat stress, including industry producing steel, ferroalloys, aluminum, magnesium, silicocarbide, and paper, as well as studies of sailors on ships operating in tropical waters, see K. Rodahl (1989). For a discussion of aging and heat tolerance, see Pandolf (1991). For a review of the effect of chemical protective clothing on work tolerance in the heat, see M.K. White, Hodous, and Vercruyssen (1991).

# Working in the Cold

Working in the cold is primarily a matter of maintaining thermal balance, because both energy metabolism and neuromuscular functions are temperature dependent. Although local acclimatization to cold is well established and is of considerable practical benefit, general acclimatization to cold (if indeed possible) is at best of limited practical value compared with experience and environmental protection.

Some of the major problems associated with performing physical work in the cold are consequences of the hobbling effect of protective clothing as well as the obstructive effects of snow and ice and the chilling effect of wind. In a study in Finland, Anttonen and Virokannas (1994) showed that in outdoor work during the winter, cold stress frequently reduced working ability by some 70%, at least for a short period. The most common problem during light outdoor work in the cold is cooling of the extremities and the face (Virokannas 1996).

Because of permafrost, the temperature in the Spitsbergen coal mines is quite low, being –2 to –4 °C year-round. Because of the geological conditions—almost horizontal sedimentary layers and coal seams only 70 to 110 cm thick—the miners have to work lying on the ground. To get to the coal face, the workers have to crawl several hundred yards. The work is performed in a lying, half-sitting, or squatting position for two sessions of approximately 3 h in each shift period. As a consequence, the workers have always complained of difficulty keeping their feet warm.

Alm and Rodahl (unpublished data) assessed actual work stress in four of the miners. They were studied for 24-h periods, both during work in the mine and during time off and sleep. The investigators assessed the following:

1. Maximal work capacity based on recording of heart rate during submaximal cycle ergometer exercise
2. Physical workload based on continuous recording of heart rate
3. Thermal stress based on continuous recording of rectal and skin temperature
4. General stress response, based on the analysis of urinary catecholamine elimination

The estimated physical workload in the mine corresponded, on the average, to about 30% to 40% of the miners' maximal work capacity. This is considerably higher than workloads commonly encountered in most industries, which seldom exceed 25% of the maximal work capacity. Figure 17.33 shows that this type of mining operation imposes some rather unique types of stress; for example, at the onset of the work shift, the miner crawls along the narrow passage, dragging a box containing 50 kg of dynamite tied to his leg, causing his heart rate to approach 165 beats · min$^{-1}$. The workload of these coal miners (figure 17.34) is comparable to that of coastal fishermen. The levels of urinary catecholamine elimination of the fishermen equaled the levels observed in the coal miners.

The rectal temperature ranged from 37.5 to 38.5 °C during work. Despite the high rectal temperature, the skin temperature of the thigh dropped (figure 17.35). Thus, the observation supported the miners' complaints of cold feet, a problem that under existing circumstances could only be remedied by using properly insulated trousers and boots.

## For Additional Reading

For a review of different methods to evaluate the needs for occupational cold protective clothing and the thermal insulation of textile materials and clothing, see Anttonen (1993).

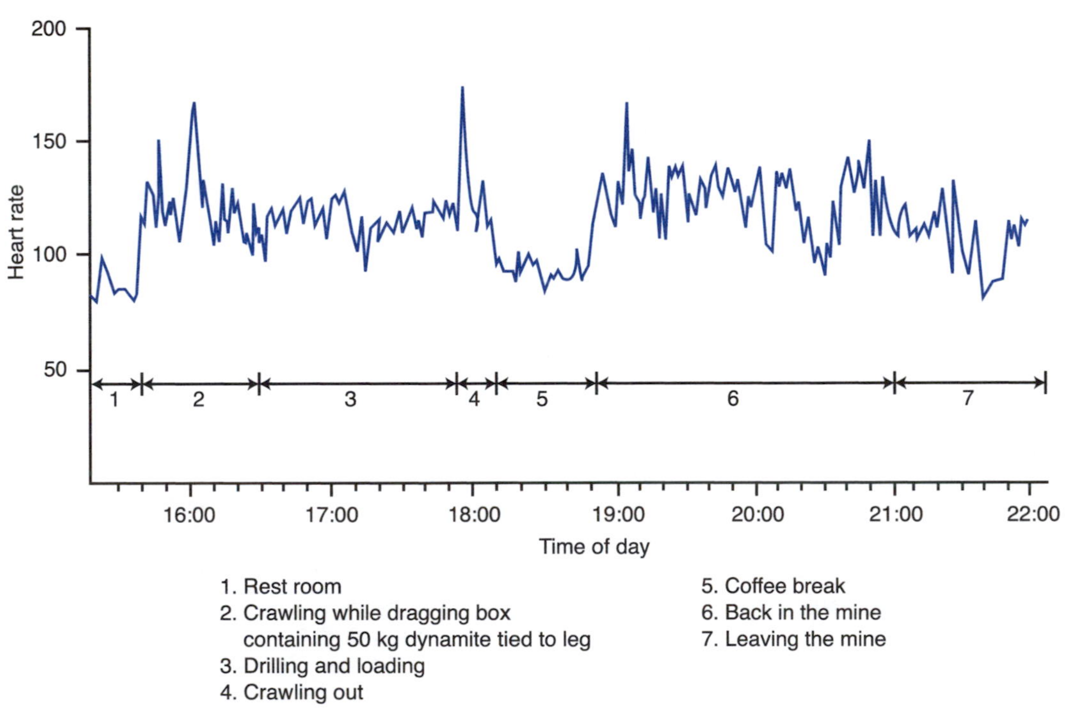


**Figure 17.33** Heart rate in a coal miner in Spitsbergen during an afternoon shift.

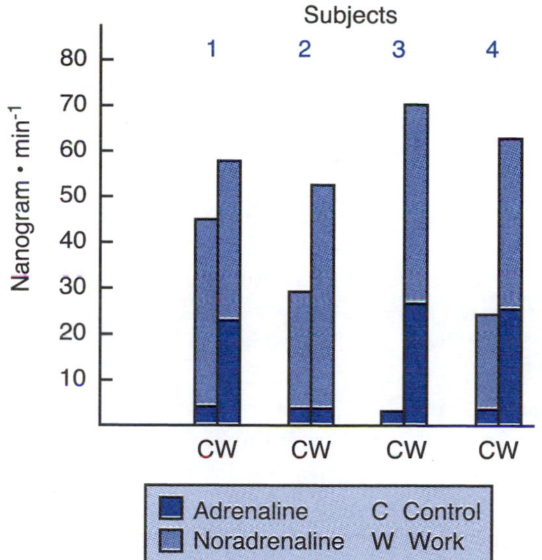


**Figure 17.34** Mean urinary catecholamine elimination (adrenaline and noradrenaline) in four Spitsbergen coal miners during work, compared with night values.

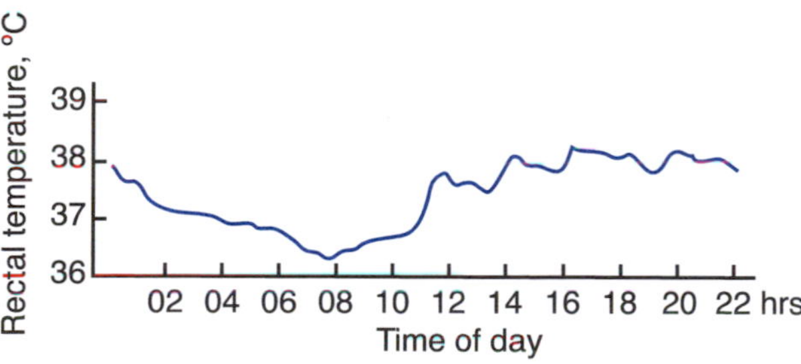


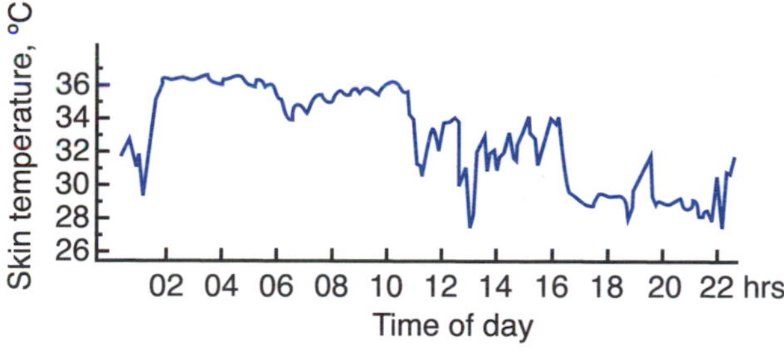


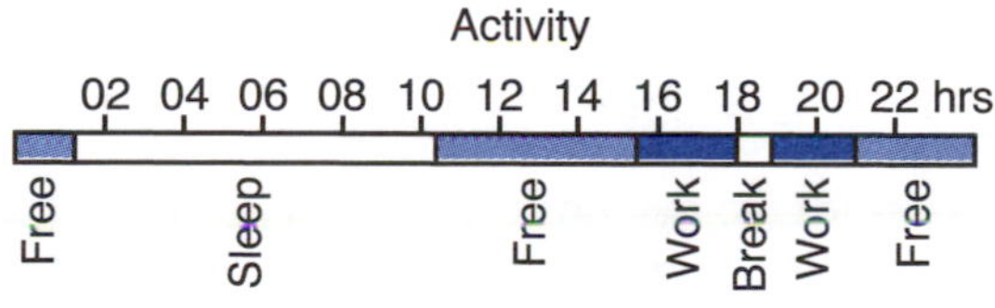


**Figure 17.35** Body temperature (rectal temperature and skin temperature recorded at the medial side of the right thigh) in a coal miner in Spitsbergen during work and time off.

# Vibration

Vibration is characterized by its frequency, acceleration, and direction. It can affect the whole body (whole-body vibration) or parts of it, such as the hands (segmental vibration). Vibration tolerance usually is defined as the level of vibration that impairs performance of a motor or visual task.

The pathophysiological consequences of segmental vibration (e.g., working with vibrating tools such as pneumatic drills and motor-driven hand saws) are fairly well known. In the case of chain saw operators, a considerable reduction in the prevalence of vibration syndrome has been observed since 1976 because of a marked reduction in the acceleration level of chain saws and improvements in general working conditions (H. Suzuki 1994).

Segmental vibration, especially of the hands when using power tools, varies according to the type of power tool being used, in that the weight, size, and design of the tool determine its vibration characteristics (Abrams and Suggs 1977). As far as segmental vibrations of the hands are concerned, frequencies in the range of 25 to 150 Hz and accelerations from 1.5 to 80 g are most likely to lead to problems such as Raynaud's phenomenon, or white fingers (i.e., circulatory disturbances of the fingers causing numbness, pain, and loss of strength; Ramsey, 1975). The patient may be unable to perform fine, manipulative tasks, particularly in cold environments.

Exposure to local vibration can decrease skin temperature (Iwata 1974), which reduces blood supply to the fingers (Makarenko 1969) and increases in plasma adrenaline and noradrenaline concentrations (Krasavina, Volkova, and Gornik 1977; Miyashati et al. 1983).

The effect of whole-body vibration on the worker is far less understood (Seidel et al. 1980; Ullsperger and Seidel 1980). In the case of whole-body vibration, low frequency vibration is particularly undesirable. According to Rodgers and Eggleton (1983), vibration is slightly better tolerated while the worker is standing than while sitting, and women, on the whole, experience greater discomfort than men exposed to the same level of vibration.

Vibrations with frequencies in the .1- to 20-Hz range, and accelerations generally less than .2 g, predominantly vertical, are common in most construction equipment, trucks, and buses (1 g = approximately 10 m $\times$ $s^{-2}$). Exposure to this degree of whole-body vibration continuously for more than 8 h is associated with considerable discomfort and objective physiological reactions. Vibration in the 2-

to 20-Hz range at 1 *g* acceleration can cause abdominal pain, nausea, dizziness, muscle tension, chest pain, and shortness of breath (Rodgers and Eggleton 1983).

After reviewing the literature on duration effects of whole-body vibration on comfort, Kjellberg and Wickström (1985) concluded that the time dependency proposed by the International Standard for whole-body vibration, ISO 2631, appears to overestimate the importance of exposure time for the strength of the effects.

### For Additional Reading

For an extensive review of the problem of hand–arm vibration, see Starck and Pyykkø (1986).

## Noise

In most cases, a standard sound-level meter can be used to evaluate noise. It can be used to identify the sound pressure level in decibels on the A scale, that is, dB(A). It also can be used to assess the potential harmful effect of the noise in terms of hearing loss and communication interference.

When there is a need to characterize noise according to frequency bands, precision sound meters equipped with filters are used. When the sound level is measured in each band, the relative proportion of high, medium, and low frequencies can be determined. This can be useful in determining the effectiveness of various noise-control measures and in identifying specific parts of noise-producing equipment that need to be sound-shielded.

To record an individual worker's noise exposure during the entire work shift or work day, portable noise dosimeters, which offer direct reading display and output to printers and computers, are available.

According to Lang, Fouriaud, and Jacquinet-Salord (1992), prolonged occupational noise exposure is a risk factor for high blood pressure. Lesnik and Makowiec-Dabrowska (1989) found that noise of a 70 dB(A) level increased cardiac ejection volume and arterial blood pressure, especially diastolic, in a study of women during 4 h of monotonous work. Umemura, Honda, and Kikuchi (1992) observed an elevation of heart rate during mental work in subjects exposed to a background noise level of 70 dB(A) but almost no elevation of the heart rate when exposed to music of the same A-weighted sound pressure level.

Hartikainen et al. (1994) concluded from a study of pregnant women that working in high noise exposure is a risk during pregnancy.

### For Additional Reading

For practical guidelines for the use of noise measurement equipment, see Rodgers and Eggleton (1983).

## Circadian Rhythms and Performance

In human beings, a variety of physiological functions, such as heart rate, oxygen uptake, rectal temperature, and urinary excretion of potassium and catecholamines, show distinct rhythmic changes in the course of a 24-h period, with the values falling to their lowest level during the night (low dip around 4 A.M.) and rising during the day, reaching their peak in the afternoon (Smolander et al. 1993). This phenomenon is known as circadian rhythm and is thought to be regulated by several separately operating biological clocks. It occurs in most individuals, although there are apparently a few exceptions; some individuals show reversed rhythms, the rectal temperature, for example, being highest at night (Folk 1974).

These rhythmic changes in physiological functions are associated with changes in performance. This relationship appears to exist especially in the case of rectal temperatures and performance. In general, the lowest performance is observed early in the morning (about 4 A.M.). Thus, the delay in answering calls by switchboard operators on night shift was twice as long between 2 and 4 A.M. as during the daytime (Colquhoun 1971). A similar relationship may exist also in the case of athletic performance. Thus, A. Rodahl, O'Brien, and Firth (1976), studying the performance of top swimmers who competed under comparable conditions early in the morning and late in the evening, found that the swimmers performed significantly better in the evening than in the morning ($p < .001$).

These findings show that circadian rhythm must be considered when we interpret results from prolonged physiological experiments and perform fitness tests in athletes at different times of the day.

The effect of crossing time zones on circadian rhythm was studied in two officers on board a supertanker en route from South Africa to South America (K. Rodahl 1989). During the voyage, the ship's clock was put back 1 h every other night. In the first mate, who worked 4-h shifts and had 8 h free, and in the chief engineer, who only worked during the day, the rectal temperature was continuously recorded every other 24-h period for 2 weeks. In the officer working regular shifts, there appeared to be a slight shift to the left in the circadian rhythm when the temperature records for the first and the last day of the voyage were compared (figure 17.36). This suggested a possible lag in the rhythm that disappeared during the subsequent days when the ship remained at the same latitude on the east coast of Brazil. In the chief engineer, who worked during the day only and was free to sleep regularly during the entire night, no such phase shift was observed.

### For Additional Reading

For a review of the basic aspects of mammalian circadian rhythm, see Illnerova (1991).

## Shift Work

The fact that human beings are "day animals" and that some of their basic physiological functions associated with their performance capacities are subject to circadian rhythmic changes suggests that humans may not be suited for night work. Nonetheless, shift work has been practiced for generations in one form or another. Yet, little precise information is available about the effects of shift work on physiological functions or physical performance, and there is no general agreement as to what type of shift work or work schedule is preferred. Most of the available information refers to clinical, social, or psychological aspects of shift work.

A review of the literature indicates that the health of shift workers in general is good despite such complaints as loss of sleep, disturbance of appetite and digestion, and a high rate of stomach ulcers (Fujita et al. 1993). The social and domestic effects of shift work represent greater problems than do the physiological effects (Alward-Monk 1990). The results of studies pertaining to the effects on productivity are conflicting, as are results concerning accident rates. Absenteeism because of illness appears to be lower among shift workers than among day workers. It has been suggested that the physiological and biological effects are probably related to circadian rhythms rather than to work schedule. To what extent such circadian rhythms are related to health, performance, and a feeling of well-being is still undetermined. According to Harma et al. (1990), physiological adjustment to night work is not influenced by age.

Systematic studies of men engaged in rotating shift work and in continuous night work (Vokac and Rodahl 1974, 1975) indicate that shift work places a physiological strain on the organism. It causes a desynchronization between functions such as body temperature and the biological clocks governing these functions. These studies show that there are considerable individual differences in the reaction to shift work, supporting the general experience that not everyone is equally suited for such work. (Some individuals consistently show relatively high values for urinary catecholamine elimination during shift work, whereas others have consistently low values.) As judged by catecholamine excretion, the greatest strain occurs when the worker, after several free days, starts work on a night shift. The results of these studies indicate that it is preferable, from a physiological standpoint, to distribute the free days more evenly throughout the entire shift cycle, that is, to alternate between work and free time regularly instead of assigning several consecutive free days.

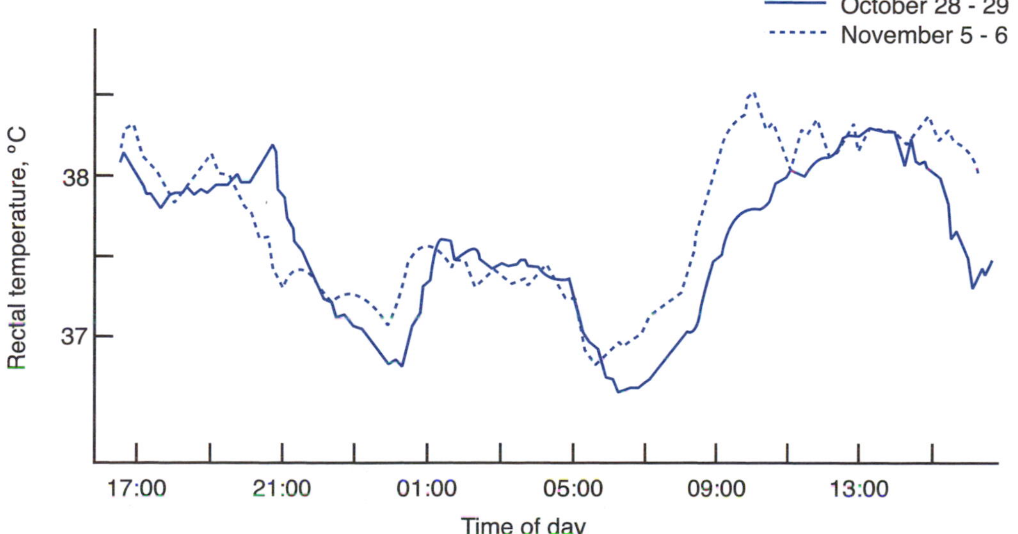


**Figure 17.36** Rectal temperature (24 h) of the first officer on board a ship en route from South Africa to Brazil, recorded at the beginning and at the end of the voyage. The ship's clock was put back 1 h every other night. The record indicates a slight shift in the circadian rhythm to the left.

From K. Rodahl 1989.

A study of continuous night work showed that at the onset, body temperature and work pulse fell during the course of the night as if the subject were sleeping, although he was working (figure 17.37). It takes several weeks for this normal rhythm to be reverted (i.e., for an increase in body temperature during the course of night work). In view of this, it appears unrealistic to keep shift workers on continuous night work for prolonged periods in order to obtain the benefit of the reverted physiological reactions, because such a reversion takes too long to occur and is lost when interrupted by a single day. Bright light has been used to adjust the biological clock following changes of circadian rhythms during shift work (Eastman, Liv, and Fogg 1995; Wetterberg 1994).

Disturbances in circadian rhythms can cause considerable problems for those who have to travel by air from one continent to another to conduct business, to take part in political negotiations, or to participate in athletic competitions. It is an open question whether the indisposition or functional disturbances experienced after such intercontinental flights are in fact caused by disturbed circadian rhythms, loss of sleep, or both. It is a common experience, however, that by being able to sleep during such travel, if necessary by using sleep-producing drugs, the individual can maintain a reasonable functional capacity despite the rapid shifts from one time to another (Rutenfranz and Colquhoun 1978, 1979). According to Shinkai, Watanabe, and Kurokawa (1993), salivary cortisol appears to be a useful measure of monitoring circadian rhythm variation in healthy individuals during shift work.

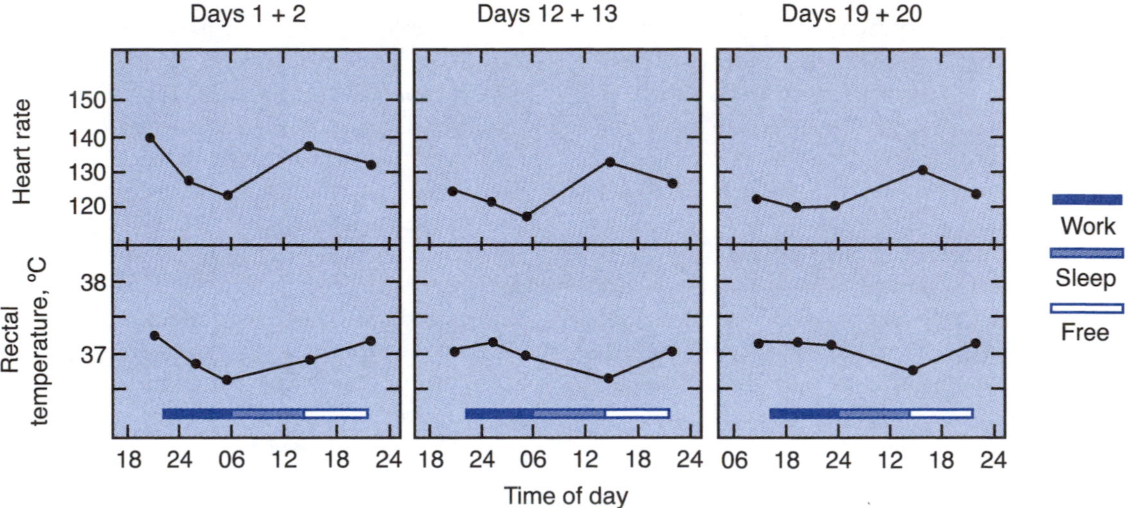


**Figure 17.37** Mean heart rate at a fixed submaximal work load and mean rectal temperature in steel mill workers during three weeks' continuous night shifts.

Adapted from Vokac and Rodahl 1975.

# Chapter 18

# Physical Activity and Health

Humans are built for physical activity. The last century has witnessed a revolution in the lives of millions of people, whereby mechanical implements have taken over tasks earlier performed by human power. We have created a new world with a different environment and different living conditions from those of our ancestors. Overall these changes have been to our benefit, but they also have created problems. Many of these problems stem from the fact that humans were built for the Stone Age but must now fit into the present, extremely technical world. Basically, infants born today have the same physical equipment as the children of our prehistoric ancestors. The vast changes have occurred in the outside world, not in the human. We are "constructed" for physical activity, and regular activity is essential for our optimal functioning and health. Therefore, a portion of our leisure time should be devoted to active recreation and training.

In this final chapter we briefly discuss the health consequences of physical inactivity, cover the role of physical activity in preventing various diseases and disorders, and present some concrete physiological and medical data showing the need for regular physical activity for adults, children, adolescents, older adults, and pregnant women.

## Health Effects of Physical Inactivity

Physical activity is a comprehensive concept. It encompasses many other terms related to physical exertion, such as work, sports, exercise, recreation, play, training, workouts, physical education, athletics, and gymnastics. Physical activity can be defined as any body movement that significantly increases energy expenditure in relation to the resting state. According to this definition, people who are sedentary both at work and in their leisure time, sitting or lying down a lot and largely using motorized devices and transportation, would be characterized as physically inactive.

The consequences of extreme physical inactivity or immobilization have been well studied under controlled conditions and have been shown to consist of major physiological changes, especially in the circulation. In fact, the cardiovascular responses to immobilization are similar to those caused by exposure to weightlessness (Sandler and Vernikos 1986). In both cases, there is a loss of orthostatic tolerance, unfavorable changes in fluid and electrolyte balance, and a marked reduction in maximal aerobic power.

During prolonged immobilization and space flights, skeletal bone structure is demineralized through the accelerated urinary excretion of calcium, eventually leading to bone fragility. Studies of the effect of prolonged bed rest on calcium metabolism have shown that the increased urinary calcium excretion that occurs during prolonged confinement to bed is not attributable to physical inactivity per se but to the absence of longitudinal pressure on the long bones (K. Rodahl et al. 1966). Also in people who sit or lie down most of the time, calcium begins to disappear from the bones because they too seldom have any weight to bear. The loss of calcium can lead to osteoporosis. This risk increases with age, particularly in postmenopausal women.

Muscles that are not stimulated sufficiently through regular use are weakened; this applies not least to the heart muscle. Joint function is reduced, and stiffness gradually increases because of joint capsule deterioration and the shortening of adjacent muscle groups. In weight-bearing joints such as the hips and knees, the cartilage becomes thinner and more vulnerable to strain and injury. Blood circulation in the muscles and bones becomes impaired, with possible leg pain as a result. Physical inactivity also is linked to digestive problems, reduced metabolism, the development of overweight and obesity, high blood pressure, reduced glucose tolerance, increased plasma lipid levels with the development of unfavorable blood cholesterol profile, and mental disorders such as depression.

### For Additional Reading

For comprehensive documentation of the various subjects related to physical inactivity/activity and health consequences, see Strømme et al. (1982), Bouchard, Shephard, and Stephens. (1994), and U.S. Department of Health and Human Services (1996).

# Physical Fitness for Everyday Life

The various organs and organ functions adjust to the demands placed on them. Some feeling of fatigue is inevitable after hours of physically heavy work. The well-trained individual has a broad margin of safety between his or her maximal power or capacity on the one hand and what is being demanded of him or her physically on the other.

One important aim of regular physical activity is to achieve a physical condition and degree of fitness that are well above those required for the routine job. If performing a job occasionally requires the heart to pump out 10 L of blood per minute at a rate of 120 beats $\cdot$ min$^{-1}$, it is an evident advantage if it has been trained to attain an output of 15 L $\cdot$ min$^{-1}$ at a rate of, say, 150. This is essentially what is meant by being fit for everyday life.

A number of studies have shown that physical inactivity in itself is a risk factor contributing to coronary heart disease, hypertension, overweight and obesity, type 2 (non-insulin-dependent) diabetes, colon and breast cancer, and early death, and that regular physical activity can prevent disease (S.N. Blair et al. 1989, 1995, 1996; Carpenter et al. 1999; Ekelund et al. 1988; Farrell et al. 1998; Friedenreich et al. 1998; N.F. Gordon and Scott 1991; Lakka et al. 1994; Paffenbarger et al. 1993; Thune and Lund 1996). Regular physical activity also improves mental health (W.P. Morgan 1997) and is vital for the musculoskeletal system (Dalsky et al. 1988; Skargren and Øberg 1996).

The question of how much and what type of exercise are needed to obtain these results is still a subject of debate. It is recommended, however, that people of all ages include a minimum of 30 min of physical activity of moderate intensity, such as brisk walking, on most if not all the days of the week. The activity need not necessarily be continuous, but could, for instance, be split up into three sections of 10 min. The intensity should be sufficiently high to cause a moderate degree of breathlessness, which represents a level of 50% to 60% of one's maximal capacity and which has been shown to produce a training effect. In the case of most people, the more vigorous and prolonged the activity, the greater are the health benefits, provided that the increments in activity level are not too abrupt (P.-O. Åstrand 1997).

## Active Recreation

By active recreation is meant a stimulating hobby or activity that involves some muscular activity, as opposed to passive recreation which, although it may be stimulating, lacks any marked demand on the circulation or locomotor organs. Watching television, reading a book, playing cards or chess, stamp collecting, and listening to music are all examples of passive recreation. It is extremely important to have something enjoyable to look forward to every week, and possibly every day, and passive recreations are in this respect very valuable. However, they should be supplemented by active recreation, preferably outdoors. Good examples of active recreation are gardening, botany, and of course sports, including fishing, hunting and mountaineering, and walking. Important beneficial effects of active recreation, depending on the type of activity chosen, are improved strength and endurance and increased range of movement.

Most individuals are physically lazy from the age of puberty on. Therefore, training and active recreation must be offered in a positive manner. The cost of simple sports grounds, possibly with changing rooms, showers, and sauna, is low compared with the cost of building and running a hospital.

# Physical Activity in the Prevention of Various Diseases and Disorders

The rapid increasing burden of noncommunicable diseases is a key determinant of global public health. According to the World Health Organization, in 1999 these diseases contributed to approximately 60% of deaths in the world and 43% of the global burden of disease. About half of these deaths are attributable to cardiovascular disease. In the following we shall take a closer look at the relationship between habitual physical activity and cardiovascular disease, type 2 diabetes, and some common musculoskeletal disorders, including backache, osteoporosis, joint stiffness, and pain. (The issue of overweight and obesity in relation to diet and physical activity is discussed in chapter 12.)

## Cardiovascular Disease

As early as 1949–50, Morris et al. (1953) found that bus conductors on double-decker vehicles had less coronary heart disease than other bus drivers and that the same was true for postal carriers compared with clerks and telephone operators. These observations were strengthened when Paffenbarger and Hale (1975) reported an inverse association between physical activity and coronary heart disease mortality from a 22-year follow-up study of 6,351 San Francisco longshoremen aged 35 to 74.

A number of more recent studies have convincingly shown an inverse relationship between physical activity/physical fitness and cardiovascular disease (Ekelund et al. 1988; G. Erikssen et al. 1998; Hein, Suadicani, and Gyntelberg 1992; I. Lee, Hsieh, and Paffenbarger 1995; Leon et al. 1987; Morris et al. 1980; Paffenbarger et al. 1986; Sandvik et al. 1993). For example, Sandvik et al. (1993) found that the risk for an early death from cardiovascular disease was significantly less in those who had the highest level of physical fitness compared with those who had the lowest level (figure 18.1).

Paffenbarger, Hyde, and Wing (1990) presented data from a longitudinal study of approximately 17,000 Harvard alumni. Men expending 2,000 kcal (8.4 MJ) weekly in walking, stair climbing, and sport play were at 39% lower risk of developing coronary heart disease than less active classmates. These results indicate that sedentary adults who increase physical activity even relatively late in life can reduce their relative risk of cardiac heart disease and all-cause mortality to levels similar to those who were always physically active.

Overweight and obesity also increase the risk of cardiovascular mortality. However, because regular physical activity has favorable effects on several of the comorbidities of obesity, particularly those pertaining to cardiovascular disease and type 2 diabetes, it is not surprising that mortality rates seem to be lower in overweight and moderately obese individuals who are physically fit compared with the unfit (S.N. Blair and Brodney 1999; C.D. Lee et al. 1999; Wickelgren 1998). Thus, in treating overweight and obese persons, the clinician should focus more on the habitual physical activity level than on the body weight itself.

For most of those who wish to reduce their body weight, it is recommended that they combine regular physical activity with somewhat reduced energy intake, in particular a reduction of fat-rich food (S.A. Anderssen et al. 1995). Emphasis should be on promoting relatively low-intensity, long-duration physical activity that can be conveniently incorporated into daily life. For example, taking a daily walk of about 30 min duration can reduce overweight by as much as half a kilogram per week.

### FOR ADDITIONAL READING

For further discussion about the effect of dietary and exercise interventions on weight loss and the risk of developing cardiovascular disease, see W.C. Miller (1999) and Gaesser (1999).

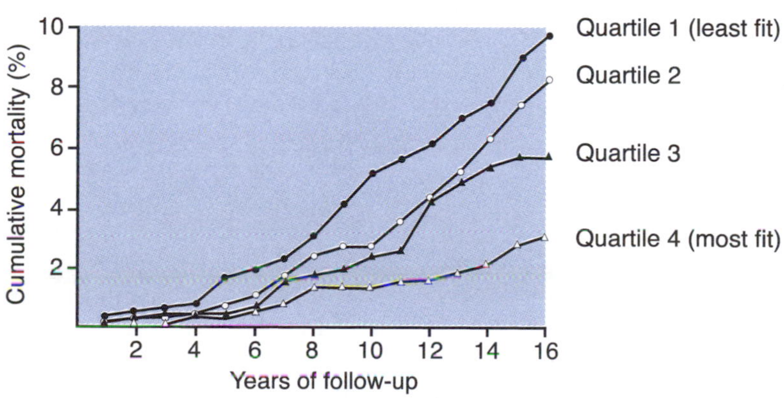


**Figure 18.1** Cumulative age-adjusted mortality from cardiovascular causes over 16 years of follow-up, according to fitness quartile.

Sandvik et al. 1993.

## Type 2 Diabetes

Habitual physical activity has profound effects on body composition and the use of nutrients. Activity helps maintain as well as increase skeletal muscle mass, which increases resting metabolic rate and enhances capacity for lipid oxidation during rest and exercise. Furthermore, increased physical activity induces favorable changes such as increased capillary density and increased number of type IIX muscle fibers. The enhanced metabolic capacity of skeletal muscle (metabolic fitness) will positively influence risk factors such as unfavorable blood lipid profile, hypertension, and insulin resistance. For example, when skeletal muscle has a high capacity for lipid oxidation, more saturated fatty acids are oxidized and more unsaturated fatty acids are incorporated into the phospholipid fraction of the plasma membrane, improving insulin sensitivity (Borkman et al. 1993).

Physical activity increases insulin sensitivity and reduces blood glucose concentration in both type 1 and type 2 diabetes (Koivisto, Yki-Järvinen, and Defronzo 1986; Peirce 1999). According to Yamanouchi et al. (1992), regular physical activity also improves insulin response in older persons (74 years of age) to a level equal to that of young (20 years of age) physically active individuals. It appears that those who are at greatest risk of developing type 2 diabetes benefit the most from regular physical activity.

## Musculoskeletal Disorders

Movement patterns—or motor skills—are developed during childhood, and the basis for the development of many musculoskeletal ailments is laid in these years. Musculoskeletal disorders are a major cause of sick leaves in most countries, and the economic potential represented by reducing the occurrence of these disorders is considerable.

## Backache

Diseases of the spinal column rank are very common. They are responsible for many days of sick leave and thus give rise to economic problems as well as much suffering. When a load is lifted or carried, a reflex mechanism calls the trunk muscles into action to fix the rib cage and compress the abdominal contents. The intracavitary pressures are thereby increased and help support the spine. Such observations emphasize the important role that the trunk muscles have in supporting the spine (Lahad et al. 1994).

Whereas flabby abdominal muscles leave the spine exposed to injurious stress, well-developed abdominal muscles, on the other hand, prevent damage to the spinal column. The trunk muscles probably have no influence on the inevitable changes in the spinal column that come with age, but if these muscles are well developed and trained they can, to a great extent, prevent the symptoms caused by the a weak back. Walking or running upstairs or uphill will train the leg and trunk muscles (van Tulder et al. 1997). Because back troubles are so common, everyone should keep the trunk muscles fit and to know how to lift and carry loads in a way that reduces the force on the spinal column.

Strengthening exercises—targeting the muscles that stabilize the back—reduce the incidence of back problems, particularly in persons who have a history of back problems, but also to a certain degree among those who have not previously experienced such problems (Lahad et al. 1994).

An interesting analysis of the tolerance of the spinal column was published more than 40 years ago by Morris, Lucas, and Bresler (1961). They pointed out that the spinal column has both an intrinsic and extrinsic stability, the former being provided by the alternating rigid and elastic components of the spine bound together by a system of ligaments, and the latter provided by the paraspinal and other trunk muscles. The stability of the ligamentous spine, which can be considered as a modified elastic rod, depends largely on the action of the extrinsic support provided by the trunk muscles. It has been shown that the critical load value at which buckling of the isolated ligamentous spine occurs, fixed at the base, is much less than the weight of the body above the pelvis (Lucas and Bresler 1960).

Morris, Lucas, and Bresler (1961) pointed out that if the nucleus pulposus of the fifth lumbar disk is considered as the fulcrum of movement and if a heavy weight is lifted with the hands, the arms and trunk form a long anterior lever (figure 18.2). The weight being lifted and the weight of the head, arms, and upper part of the trunk are counterbalanced by the contraction of the deep muscles of the back acting through a much shorter lever arm, that is, the distance from the center of the disk to the center of the spinous process. With these factors in mind, the authors computed the force that results when a 77-kg man lifts a 90-kg weight, and they concluded that the force on the lumbosacral disk is about 8,800 N, if the role of the trunk is omitted.

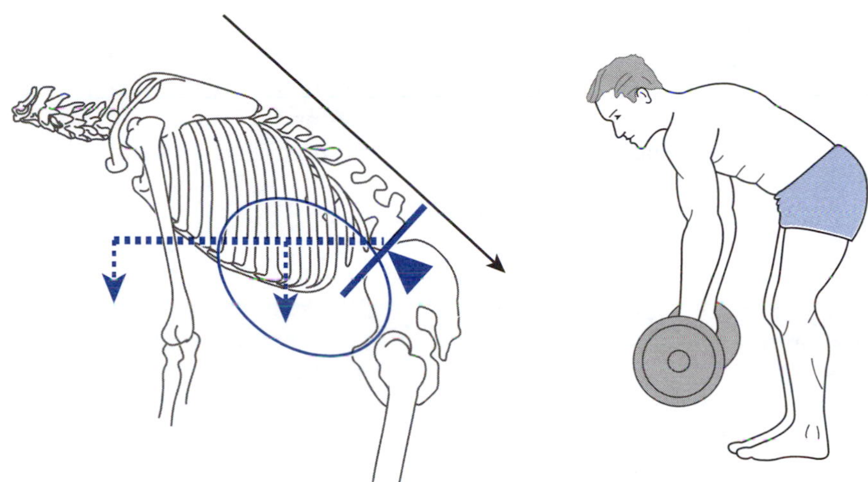

**Figure 18.2** Schematic drawing of a 77-kg man lifting 90 kg. The nucleus pulposus of the fifth lumbar disk is considered as the fulcrum of movement. The arms and trunk form a long anterior lever. The weight being lifted is counterbalanced by the contraction of the deep muscles of the back acting on a much shorter lever (the distance from the center of the disk to the center of the spinous process). If the role of the trunk is omitted, the force on the lumbosacral disk would be about 9,000 N, which is considerably more than segments of the isolated ligamentous spine can withstand without structural failure. Structural failure does not happen because the contracted trunk muscles convert the abdominal and thoracic cavities into semirigid cylinders that relieve the load on the spine itself.

From Morris, Lucas, and Bresler 1961.

## *Osteoporosis*

Osteoporosis is a generic term referring to decreased mass per unit volume (density) of normally mineralized bone. It is the most common skeletal disorder in the world and is second only to arthritis as a leading cause of musculoskeletal morbidity in the elderly (Kaplan 1987).

The relationship between applied load and bone morphology was first recognized in 1683 by Galileo, who observed a direct correlation between body weight and bone size. Today it is well established that weight-bearing activities play an important role in increasing and maintaining the mineral content of bone, and that physical activity, particularly during childhood and youth, contributes to greater maximum bone density in adult life (Grimston, Willows, and Hanley 1993; Kirchner, Lewis, and O'Connor 1996). For adults and the elderly, physical activity retards bone loss and thus can counteract the development of osteoporosis (Kriska et al. 1988). Older persons fall frequently, and many of them suffer fractures as a result. A large percentage of femoral neck fractures are caused by factors that can be influenced. Among the many reversible risk factors for falls are muscle weakness in the limbs, poor balance, and poor overall fitness. These qualities can be improved through regular physical activity, including strengthening exercises, even among those who already have developed osteoporosis (Prior et al. 1996).

Although exercise plays an important role in preventing and treating osteoporosis, one should not rely on exercise alone to maintain bone density. Exercise should be an adjunct to an adequate diet. Calcium balance studies indicate that premenopausal women require approximately 1,000 mg of calcium per day. Calcium requirements of both elderly men and women are likely to be greater (between 1,000 and 1,500 mg per day) because of a possible age-related decrease in the efficiency of intestinal calcium absorption (Santora 1987). For postmenopausal women, estrogen replacement therapy should be considered.

## Joint Stiffness and Pain

Movement is required in many joints to counteract stiffening and impairment of the cartilage. A prolonged static load on a joint will compress the articular cartilage. Movement in the joint, on the other hand, possibly increases the thickness of the cartilage and improves the exchange of nutrients (Panush et al. 1995). Therefore, the limbs should be moved as far as they will go, gently and with no force applied. No pain should be caused by these movements.

A sustained load can have a detrimental effect on the functioning of muscles and joints. In the erect position, there is a tendency for blood to stagnate in the legs. Patients whose cardiac function is impaired and those with varicose veins are particularly prone to such adverse blood distribution, resulting in edema of the lower leg. The contraction of muscles during movement compresses the veins and thus promotes the flow of blood toward the heart. This beneficial effect of dynamic muscular activity on the blood flow leads to the important conclusion that sitting or standing postures should not be too fixed but should permit movement. When driving long distances, a person must stop regularly, get out of the car, and stretch the legs. No place of work should be so well organized that everything is too comfortably within reach; it is good, from a physiological point of view, to get up from time to time, walk a few steps, and climb the odd flight of stairs.

It has been well documented that physical activity is important to reduce functional limitations and improve physical fitness among those suffering from rheumatic conditions (Hansen et al. 1993). There are also indications that physical activity has a disease-modifying effect (Lyngberg et al. 1991).

## The Significance of Physical Activity for Children and Youths

Regular physical activity among children and adolescents is necessary for normal growth and the development of aerobic capacity, muscle strength, flexibility, motor skills, and agility. Physical activity during the formative years strengthens the bones and connective tissues and favorably affects metabolism (Bar-Or 1996). Furthermore, physical activity seems to promote self-esteem in children and adolescents, and children and adolescents who are involved in physical activity seem to experience fewer mental health problems (Calfas and Taylor 1994; Mutrie 1997). There is evidence that neglect of regular physical activity during adolescence cannot be fully compensated later in life (Haywood 1993; Losse et al. 1991).

Therefore, it is very important that children and youth in general be introduced to the principles of training and active recreation, in theory as well as in practice. Good habits and motivation must be developed early in life. It is a pity that the time set aside for compulsory physical education in schools gradually is being reduced in many countries. If more time is considered necessary for mathematics or physics, it should not be provided at the expense of physical education, as has often been the case. Children should have a minimum of 60 min of outdoor play or other physical activity every day in daycare centers and schools at all levels.

From a physiological–medical viewpoint, physical education in schools should accomplish the following:

1. Improve the oxygen-transporting system
2. Generally strengthen the locomotive organs (including the muscles of the back and abdomen)
3. Give instruction on how to lie, sit, stand, walk, lift, and carry
4. Give instruction in technique, tactics, and rules in games and sports to reduce or eliminate accidents
5. Provide physical and psychological recreation and variety
6. Arouse interest in regular physical activity after schooling has been finished

Active commuting to school by walking or bicycling is a potential source of moderate activity that often has been ignored in surveys of physical activity. Suggestive evidence of steep declines in the amount of children's destination walking can be gleaned from national transportation surveys. At the same time, there has been a dramatic increase in the reported use of motorized vehicles, including transporting children. Therefore, active commuting to school should be promoted as an important source of children's physical activity (Tudorlocke, Ainsworth, and Popkin 2001).

## Physical Activity and Aging

Physical inactivity is a key risk factor for increased mortality and morbidity as well as increased functional disability in old age. Regular physical activity is essential for quality of life in old age and reduces the need for assistance. (For references, see American College of Sports Medicine 1998.)

Even very old people benefit from exercise, in terms of both endurance and muscular strength (Fiatarone et al., 1994). Stronger muscles mean a more rapid walking pace, improved balance and ability to use stairs, reduced use of aids, and increased spontaneous physical activity (R.S. Schwartz and Buchner 1999).

Older persons fall frequently, and many suffer fractures when they fall. Among the many reversible risk factors for falls are muscle weakness in the limbs, poor balance, and a poor level of overall

## Classic Study

The Greek physician Hippocrates (460–375 B.C.) proclaimed the view—revolutionary at that time—that disease is a natural process and looked on the patient as an individual whose "constitution" would react to disease in its own way. He stressed the importance of observation and reasoning, rather than mysticism, and made careful and systematic examinations of the patient's condition. Some of his descriptions of disease have hardly been improved to this day. Hippocrates understood the importance of physical activity and stated the following:

> *All parts of the body which have a function, if used in moderation and exercised in labors in which each is accustomed, become thereby healthy, well-developed and age more slowly, but if unused and left idle they become liable to disease, defective in growth, and age quickly.*

This early concept indeed has been confirmed and supported by our present-day experience and scientific investigations.

F. Adams 1891.

fitness. These are qualities that can be improved through regular physical activity, including strengthening exercises.

The effect of habitual regular physical exercise on physiological age was convincingly demonstrated in a study of 221 healthy men 20 to 85 years old, by Nakamura, Moritani, and Kanetaka (1996), using a group of 17 physiological function tests and five physical fitness tests. The results support the view that regular physical exercise can delay physiological changes normally seen with aging and consequently improves quality of life and increases the lifespan.

Age-related loss of skeletal muscle mass not only decreases strength and functional capacity but also is closely linked to age-related loss in bone mineral, basal metabolic rate, and increased body fat content. Through physical exercise and training, especially resistance training, older people can prevent loss in skeletal muscle mass and other age-related ailments, such as non-insulin-dependent diabetes, coronary heart disease, hypertension, osteoporosis, and obesity. According to W.J. Evans (1995) there is no pharmacologic intervention that holds a greater promise of improving health and promoting independence in the elderly than does exercise.

The heel pad (i.e., the thick mass of elastic adipose tissue located beneath the calcaneus bone) is an essential biological shock-absorbing mechanism of the locomotor system. Kinoshita et al. (1996) examined the shock-absorbing characteristics of the heel pad in two groups of active elderly individuals 60 to 67 years and 71 to 86 years of age, and the investigators compared the results to a group of 17- to 30-year-old adults. They concluded that the capacity for shock absorbency of the heel pad declines with age. Consequently, older individuals should be encouraged to choose walking rather than jogging or other high-impact activities, especially on hard surfaces. Attention also should be given to the selection of appropriate shoes and floor surfaces for the elderly.

### For Additional Reading

For concrete exercise training guidelines for the elderly, see Evans (1999).

## Physical Activity and Pregnancy

The age of menarche usually is delayed a few months in athletes compared with nonathletes (Carey 1983). There is a great deal of evidence that menstrual problems diminish in physically active women, but with the increasing participation of women in long-distance running as well as in other types of intensive training programs, the incidence of **amenorrhea** has increased.

Women who are moderately physically active during pregnancy experience easier pregnancies and deliveries, have better self-esteem, gain less weight, and have fewer perinatal complications than women who have not engaged in physical activity during their pregnancy (Artal and Sherman 1999). There is no evidence of increased risk of miscarriage, premature delivery, or growth retardation attributable to moderate physical activity during pregnancy (Lokey 1991).

Although moderate exercise generally is recommended during pregnancy, highly conditioned athletes seem to be able to train safely at more demanding activity levels. Thus, Kardel and Kase (1998) found that a rigorous training program that included both strength and endurance training 6 days a week, continued nearly to term, had no adverse effect on fetal growth in 42 well-trained women.

The typical adjustments to exercise in pregnant women are essentially similar to those observed in nonpregnant controls (Hartman and Bung 1999). Lynch and Nimmo (1998) studied five women using oral contraceptives compared with 10 normally menstruating women performing intermittent exercise. Their results suggest that exercise performance does not vary between the midfolicular phase and the late luteal phase of the menstrual cycle, nor does exercise performance appear to be affected by the number of days using the oral contraceptive.

However, too much exercise can negatively affect the reproductive system in young women, including primary and secondary amenorrhea (Warren and Stiehl 1999). This is thought to be caused by several factors, including low body weight and improper nutrition.

The **female athlete triad,** a term coined by the American College of Sports Medicine in 1992 (Otis et al. 1997), is a common clinical entity among female athletes (J.M. Anderson 1999; R.V. West 1998). It consists of three interrelated problems: eating disorders, amenorrhea, and **osteopenia.**

### FOR ADDITIONAL READING

For a review of amenorrhea among athletes, see Gidwani (1999).

## Conclusions: Recommendations for Physical Activity and Health

- Regular physical activity protects against cardiovascular disease, high blood pressure, type 2 (non-insulin-dependent) diabetes, overweight, obesity, colon cancer, and breast cancer. Physical activity also improves mental health and is vital for the musculoskeletal system and the joints. There is a higher incidence of premature death and greater functional disability among physically inactive persons than among those who are physically active.

- For inactive adults, physical activity of moderate intensity representing a daily energy expenditure corresponding to a 30-min moderate to brisk walk yields substantial health gains. The daily activity can be divided into 5- to 10-min intervals. Any increase in activity beyond this level will yield further health gains.

- Regular physical activity among children and adolescents is necessary for normal growth and development and has great significance for health both during the formative years and later in life. Thus, children should have a minimum of 60 min of outdoor play or other physical activity every day in daycare centers and schools at all levels.

- Physical inactivity is a key risk factor for mortality, morbidity, and reduced functional ability among older people. Regular physical activity is essential for self-reliance and quality of life. Exercise is just as good for older people as for younger people, and the same recommendations apply as to other adults. In addition, older people should do strengthening exercises to counter the development of osteoporosis, muscle loss, and poor balance. Older people are a heterogeneous group, and exercise programs must be adapted to the individual's physical fitness and functioning level.

- Women who are moderately physically active while pregnant experience easier pregnancies and deliveries and have fewer complications than inactive pregnant women. Thus, pregnant women should continue to exercise at an appropriate level. Pregnant women who have not previously been physically active can engage in moderate-intensity exercise during pregnancy.

- The question is frequently raised whether a medical examination is advisable before commencing a training program. Certainly anyone who is doubtful about his or her state of health should consult a physician. In principle, however, there is less risk in activity than in continued inactivity. Because many people can benefit from exercise, we hope that the physician of the future will prescribe specific exercise programs for patients in the same way he or she prescribes drugs or any other form of treatment. In any case, it is exceedingly important to start the program gently and to increase the intensity very gradually over a period of weeks and months.

# Appendix A

# Definition of Units

The SI units (Le Système International d'Unités) are the preferred units of measurement. To facilitate the transition from old units to this system, both the conventional and SI units are often used in the text.

## Basic Units

Length: meter (m)

Mass: kilogram (kg)

Time: second (s)

Quantity: mole (mol)

The *meter* is the length equal to 1,650,763.73 wavelengths in vacuum of the radiation corresponding to the transition from the level $5d_5$ to $2p_{10}$ of the krypton-86 atom.

The *kilogram* is the mass of a particular cylinder of platinum–iridium alloy, International Prototype Kilogram, which is preserved in a vault at Sèvres, France, by the International Bureau of Weights and Measures.

The *second* is the duration of 9,192,631,770 periods of the radiation corresponding to the transition between the two hyperfine levels of the ground state of the cesium-133 atom.

The *mole* is the amount of substance that contains Avogadro's number of particles. Avogadro's number is the number of atoms in exactly .012 kg of carbon-12. In biological contexts, it is often convenient to use the unit millimole, micromole, nanomole, or picomole.

From these basic units, other units can be defined:

*Force (F)* can be defined as that which changes the state of rest or motion of a body, measured by the rate of change of momentum. The unit of force is the newton (N), which is the force producing an acceleration of 1 m per second every second ($1\ m \cdot s^{-2}$) when it acts on 1 kg. Also $F = m \cdot a$, where $m$ is mass in kilograms and $a$ is acceleration in meters per second every second.

In the old system, the commonly used unit for force was kilopond (kp). One kilopond is the force acting on a mass of 1 kg at normal acceleration of gravity: 1 kp = 9.80665 N 10 N.

*Work* is done when a force acts against resistance to produce motion. It is therefore measured by the product of the force *(F)* and the distance moved in the direction of the force *(L)*: Work = $F \cdot L$. The unit of work is the joule (J). One joule of work is done when a force of 1 N moves the point of application through a distance of 1 m; 1,000 J = 1 kJ; 1,000 kJ = 1 MJ; 1 calorie (cal) = 4.186 J; 1 kcal = 4.186 kJ.

*Power* is the rate of doing work and is measured in watts (W); $1\ W = 1\ J \cdot s^{-1}$. An old alternative is kilopond-meters per minute, $kpm \cdot min^{-1}$; $1\ W = 6.12\ kpm \cdot min^{-1}$; $9.81\ W = 1\ kpm \cdot s^{-1}$; $16.35\ W = 100\ kpm \cdot min^{-1}$; 1 hp (horsepower) = $736\ W = 75\ kpm \cdot s^{-1} = 4{,}500\ kpm \cdot min^{-1}$.

*Pressure* is the force per unit area, and it is measured in newtons per square meter ($N \cdot m^{-2}$) or in pascals (Pa): $1\ Pa = 1\ N \cdot m^{-2}$. The practice of expressing pressure in millimeters of mercury (mmHg) or torr in medical contexts (blood pressure, osmotic pressure, and gas pressure) will probably continue for some time. The pressure in millimeters of mercury multiplied by .133 gives the pressure in kilopascals (kPa). Hence, 1 kPa = 7.5 mmHg. Torr is a provisional international standard term to replace mmHg; 1 torr = 1 mmHg.

*Concentrations* can be expressed in moles per liter: $mol \cdot L^{-1}$ or molar (M), $mmol \cdot L^{-1}$ or mM. A 1-molar (1-M) solution contains 1 mol or *g* mol weight of the solute in 1 L of solution. Concentrations can also be expressed in moles per kilogram, for example, moles of lactate per kilogram of muscle tissue, wet weight (ww) or dry weight (dw). A 1-*molal* solution, on the other hand, contains 1 mol or *g* mol weight of the solute in 1 L of *solvent*.

*Frequency* is the rate of oscillation, measured in cycles per second or hertz (Hz): $1\ cycle \cdot s^{-1} = 1\ Hz$.

# Appendix B

# Prefixes for Unit Abbreviations

## Prefixes for Multiples of Units

| Multiple | Prefix | Symbol |
|---|---|---|
| $10^{-1}$ | deci | d |
| $10^{-2}$ | centi | c |
| $10^{-3}$ | milli | m |
| $10^{-6}$ | micro | μ |
| $10^{-9}$ | nano | n |
| $10^{-12}$ | pico | p |
| 10 | deka | da |
| $10^{2}$ | hecto | h |
| $10^{3}$ | kilo | k |
| $10^{6}$ | mega | M |
| $10^{9}$ | giga | G |
| $10^{12}$ | tera | T |

Examples: $m \cdot 10^{-3}$ = mm; $m \cdot 10^{-6}$ = μm; $m \cdot 10^{-9}$ = nm; $g \cdot 10^{-3}$ = mg; $g \cdot 10^{3}$ = kg; $M \cdot 10^{-3}$ = mM (millimolar concentration).

As mentioned above, the following abbreviations are used:

s = second; min = minute

J = joule; kJ = kilojoule; MJ = megajoule

V = volt; mV = millivolt

W = watt

kp = kilopond

## Summary

| Quantity | Unit | Symbol | Other units |
|---|---|---|---|
| Force | newton | N | $(kg \cdot m \cdot s^{-2})$ |
| Work or energy | joule | J | $(kg \cdot m^2 \cdot s^{-2} = N \cdot m)$ |
| Power | watt | W | $(kg \cdot m^2 \cdot s^{-3} = J \cdot s^{-1})$ |
| Pressure | pascal | Pa | $(N \cdot m^{-2})$ |

# Appendix C

# Conversion Tables

## Length, Weight, and Volume

1 cm = .39370 in.

1 m = 3.28 ft

1 km = .62137 mi

1 in. = 2.54 cm

1 ft = 30.480 cm

1 ml = .03381 fl oz (U.S.)

1 L = 1.0567 qt (U.S.)

1 kg = 2.2046 lb

## Power

1 W = .001 kW

1 W = .73756 ft-lb $\cdot$ $s^{-1}$

1 W = $1 \cdot 10^7$ ergs $\cdot$ $s^{-1}$

1 W = .056884 BTU $\cdot$ $min^{-1}$ = 3.41304 BTU $\cdot$ $h^{-1}$ (BTU = British thermal unit)

1 W = .01433 kcal $\cdot$ $min^{-1}$

1 W = $1.341 \cdot 10^{-3}$ hp (horsepower)

1 W = 1 J $\cdot$ $s^{-1}$

1 W = 6.12 kpm $\cdot$ $min^{-1}$

1 kcal $\cdot$ $min^{-1}$ = 69.767 W

1 kcal $\cdot$ $min^{-1}$ = 51.457 ft-lb $\cdot$ $s^{-1}$

1 kcal $\cdot$ $min^{-1}$ = $6.9770 \cdot 10^8$ ergs $\cdot$ $s^{-1}$

1 kcal $\cdot$ $min^{-1}$ = 3.9685 BTU $\cdot$ $min^{-1}$

1 kcal $\cdot$ $min^{-1}$ = .093557 hp

1 hp (horsepower) = 745.7 W

1 hp = 550 ft-lb $\cdot$ $s^{-1}$

1 hp = $7.457 \cdot 10^9$ ergs $\cdot$ $s^{-1}$

1 hp = 42.4176 BTU $\cdot$ $min^{-1}$

1 hp = 10.688 kcal $\cdot$ $min^{-1}$

1 hp = 745.7 J $\cdot$ $s^{-1}$

1 hp = 75 kpm $\cdot$ $s^{-1}$

1 kpm $\cdot$ $min^{-1}$ = .1635 W

## Work and Energy

1 kcal = $4.186 \cdot 10^{10}$ erg

1 kcal = 4,186 J

1 kcal = 3.9680 BTU

1 kcal = 3,087.4 ft-lb

1 kcal = 426.85 kpm

1 kcal = $1.5593 \cdot 10^{-3}$ hp $\cdot$ $h^{-1}$

1 erg = $2.3889 \cdot 10^{-11}$ kcal

1 erg = $1 \cdot 10^{-7}$ J

1 erg = $9.4805 \cdot 10^{-14}$ BTU

1 erg = $7.3756 \cdot 10^{-8}$ ft-lb

1 erg = $1.0197 \cdot 10^{-8}$ kpm

1 erg = $3.7251 \cdot 10^{-14}$ hp $\cdot$ $h^{-1}$

1 J = $2.3889 \cdot 10^{-4}$ kcal

1 J = $1 \cdot 10^7$ erg

1 J = $9.4805 \cdot 10^{-4}$ BTU

1 J = .73756 ft-lb

1 J = .10197 kpm

1 J = $3.7251 \cdot 10^{-7}$ hp $\cdot$ $h^{-1}$

1 BTU (British thermal unit) = .25198 kcal

1 BTU = $1.0548 \cdot 10^{10}$ erg

1 BTU = 1054.8 J

1 BTU = 777.98 ft-lb

1 BTU = 107.56 kpm

1 BTU = $3.9292 \cdot 10^{-4}$ hp $\cdot$ $h^{-1}$

1 ft-lb (foot-pound) = $3.2389 \cdot 10^{-4}$ kcal

1 ft-lb = $1.35582 \cdot 10^7$ erg

1 ft-lb = 1.3558 J

1 ft-lb = $1.2854 \cdot 10^{-3}$ BTU

1 ft-lb = .13825 kpm

1 ft-lb = $5.0505 \cdot 10^{-7}$ hp $\cdot$ $h^{-1}$

1 kg $\cdot$ m (kilogram-meter) = $2.3427 \cdot 10^{-3}$ kcal

1 kg · m = $9.8066 \cdot 10^7$ erg
1 kg · m = 9.8066 J
1 kg · m = $9.2967 \cdot 10^3$ BTU
1 kg · m = 7.2330 ft-lb
1 kg · m = $3.6529 \cdot 10^{-6}$ hp · $h^{-1}$

1 kpm = 9.80665 J

## Pressure

1 ATM[1] (atmosphere) = 1,013.25 mb (millibar)
1 ATM = 101,325 Pa (pascal)
1 ATM = 101,325 N · $m^{-2}$
1 ATM = 1.03323 kg · $cm^{-2}$
1 ATM = 1,033.26 cm $H_2O$ at 4 °C
1 ATM = 760 torr
1 ATM = 760 mmHg at 0 °C
1 ATM = 29.9213 in. Hg at 32 °F

1 torr = $1.31579 \cdot 10^{-3}$ ATA
1 torr = 1.33322 mb
1 torr = 133.322 Pa
1 torr = $1.35951 \cdot 10^{-3}$ kg · $cm^{-2}$
1 torr = 1.35955 cm $H_2O$ at 4 °C
1 torr = 1 mmHg at 0 °C
1 torr = $3.93701 \cdot 10^{-2}$ in. Hg at 32 °F

1 Pa (pascal) = $7.5 \cdot 10^{-3}$ mmHg
1 kPa = $9.9 \cdot 10^{-3}$ ATA
1 kPa = .01 bar

[1] The pressure of 760 mmHg at a density of 13.5951 g · $cm^{-1}$ and an acceleration of $g = 980.655$ cm · $s^{-2}$

## Speed

| km · $h^{-1}$ | mph | m · $s^{-1}$ |
|---|---|---|
| 10 | 6.21 | 2.78 |
| 20 | 12.4 | 5.56 |
| 30 | 18.6 | 8.33 |
| 40 | 24.9 | 11.1 |
| 50 | 31.1 | 13.9 |
| 60 | 37.3 | 16.7 |
| 70 | 43.5 | 19.4 |
| 80 | 49.7 | 22.2 |
| 90 | 55.9 | 25.0 |
| 100 | 62.1 | 27.8 |
| 120 | 74.6 | 33.3 |
| 140 | 87.0 | 38.9 |
| 160 | 99.4 | 44.5 |
| 180 | 112 | 50.0 |
| 200 | 124 | 55.6 |
| 220 | 137 | 61.2 |
| 240 | 149 | 66.7 |
| 260 | 162 | 72.2 |
| 280 | 174 | 77.8 |
| 300 | 187 | 83.4 |
| 320 | 199 | 88.9 |
| 340 | 211 | 94.5 |
| 360 | 224 | 100 |
| 380 | 236 | 106 |
| 400 | 249 | 111 |
| 420 | 261 | 117 |
| 440 | 274 | 122 |
| 460 | 286 | 128 |

# APPENDIX D

# SYMBOLS

| | |
|---|---|
| $\bar{x}$ | dash over any symbol indicates a mean value |
| $\dot{x}$ | dot above any symbol indicates time derivative |

## Respiratory and Hemodynamic Notations

| | |
|---|---|
| V | gas volume |
| $\dot{V}$ | gas volume per unit time (usually $L \cdot min^{-1}$) |
| R or RQ | respiratory exchange ratio (volume $CO_2 \cdot$ volume $O_2^{-1}$) |
| *I* | inspired gas |
| *E* | expired gas |
| *A* | alveolar gas |
| F | fractional concentration in dry gas phase |
| f | respiratory frequency (breath $\cdot$ unit $time^{-1}$) |
| TLC | total lung capacity |
| VC | vital capacity |
| FRC | functional residual capacity |
| RV | residual volume |
| T | tidal gas |
| D | dead space |
| FEV | forced expiratory volume |
| $FEV_{1.0}$ | forced expiratory volume in 1 s |
| MVV | maximal voluntary ventilation |
| $MVV_{40}$ | maximal voluntary ventilation at f = 40 |
| $D_L$ | diffusing capacity of the lungs ($ml \cdot min^{-1} \cdot mmHg^{-1}$) |
| P | gas pressure |
| B or bar | barometric |
| STPD | 0 °C, 760 mmHg (101.3 kPa), dry |
| BTPS | body temperature and pressure, saturated with water vapor |
| ATPD | ambient temperature and pressure, dry |
| ATPS | ambient temperature and pressure, saturated with water vapor |
| Q | blood flow or volume |
| $\dot{Q}$ | blood flow $\cdot$ unit $time^{-1}$ (without other notation, cardiac output; usually $L \cdot min^{-1}$ |
| SV | stroke volume |
| HR | heart rate (usually beats $\cdot$ $min^{-1}$) |
| BV | blood volume |
| $HB_T$ | total amount of hemoglobin in body |
| Hb | hemoglobin |
| Hct | hematocrit |
| BP | blood pressure |
| R | resistance |
| C | concentration in blood phase |
| S | percentage saturation of Hb |
| a | arterial |
| c | capillary |
| v | venous |

# Temperature Notations

| | |
|---|---|
| *T* or *t* | temperature |
| r or re | rectal |
| s | skin |
| e | esophageal |
| m | muscle |
| ty | tympanic |
| M | metabolic energy yield |
| C | convective heat exchange |
| R | radiation heat exchange |
| E | evaporative heat loss |
| S | storage of body heat |
| °C | temperature in degrees Centigrade |
| °F | temperature in degrees Fahrenheit |

# Dimensions

| | |
|---|---|
| *W* | weight |
| *H* | height |
| *L* | length |
| LBM | lean body mass |
| BSA | body surface area |

# Statistical Notations

| | |
|---|---|
| *M* | arithmetic mean |
| *SD* | standard deviation |
| *SE* | standard error of the mean |
| *n* | number of observations |
| *r* | correlation coefficient |
| range | smallest and largest observed value |
| Σ | summation |
| *D* or *d* | difference |
| *p* | probability |
| * | denotes a (probably) significant difference; .05 $p > .01$ |
| ** | denotes a significant difference; .01 $p > .001$ |
| *** | denotes a (highly) significant difference: *p* .001 |

# Examples

| | |
|---|---|
| $V_A$ | volume of alveolar gas |
| $\dot{V}_E$ | expiratory gas volume · $min^{-1}$ |
| $\dot{V}O_2$ | volume of oxygen · $min^{-1}$ (oxygen uptake · $min^{-1}$) |
| $V_T$ | tidal volume |
| $P_A$ | alveolar gas pressure |
| $P_B$ | barometric pressure |
| $F_IO_2$ | fraction concentration of oxygen in inspired gas |
| $P_AO_2$ | alveolar pressure of oxygen |
| $P_aO_2$ | arterial pressure of oxygen |
| $pH_a$ | arterial pH |
| $C_aO_2$ | oxygen content in arterial blood |
| $(a\text{-}\bar{v})O_2$ difference | arteriovenous oxygen difference |
| $T_r$ | rectal temperature |
| $T_{\bar{s}}$ | mean skin temperature |

# Glossary

This is a list of key words and abbreviations with short explanations. For a more extensive explanation of a term in its functional context, see Index.

**2,3-DPG**—See *diphosphoglycerate.*

**5-hydroxytryptamine**—Serotonin, an important neurotransmitter.

**A-band**—The middle part of a sarcomere, containing myosin filaments.

**accessory**—Assisting, auxiliary; an organ or part that supplements a similar organ or part (e.g. accessory inspiratory muscles).

**acetyl CoA**—acetyl coenzyme A.

**acetylcholine (ACh)**—The neurotransmitter in motor endplates and in some cholinergic synapses in the central nervous system.

**acidosis**—An increase in the blood concentration of hydrogen ions above the normal range.

**actin**—A globular protein forming the building blocks of actin filamernts.

**action potential**—A nerve impulse or a corresponding impulse in a muscle fiber.

**active site**—The part of an enzyme molecule that binds the substrate.

**active transport**—Energy-consuming, transmembrane transport.

**adenosine triphosphate (ATP)**—A high energy molecule serving as the ubiquitous energy currency of cells.

**ADH**—See *antidiuretic hormone.*

**ADP**—adenosine diphosphate.

**adrenaline**—The adrenal medullary hormone (also called epinephrine).

**adrenergic**—A synaptic terminal, which uses the neurotransmitter noradrenaline.

**afterload**—The resistance against which blood is expelled from the heart.

**alactic (alactacid)**—Without lactate buildup.

**alactic power**—The non-lactate contribution to power in maximal exercise.

**albumin**—The most abundant group of plasma proteins.

**alkali light chains**—Another name for essential light chains MyLC1 and MyLC3, which form part of the hexameric myosin molecule.

**alkalosis**—A decrease in the blood concentration of hydrogen ions below the normal range.

**allo- (allos)**—A prefix meaning "other."

**allometric scaling**—A mathematical procedure to establish a proper relationship between a body size variable and some other factor of interest (e.g., aerobic capacity, muscular strength, jumping height, etc.)

**allometry**—A way of describing the effect of body size on bodily proportions and functions.

**allosteric regulation**—Regulation of enzyme activity by binding of a regulatory molecule to the enzyme molecule in a place different from the active site.

**allosteric site**—The part of an enzyme molecule, different from the active site, which upon binding of a ligand, the allosteric regulator, confers a change in enzyme activity.

**alpha motoneuron (α-motoneuron)**—A motoneuron innervating extrafusal muscle fibers (i.e., the ordinary muscle fibers).

**alpha–gamma (α–γ) coactivation**—The simultaneous activation of alpha- and gamma-motoneurons.

**alternative splicing**—Splicing of exons in a sequence different from that in the gene. May also imply omission of an exon.

**amenorrhea**—Absence of menstruation.

**amoeboid mobility**—Resembling an amoeba in movement.

**AMP**—Adenosine monophosphate.

**anabolism**—The type of metabolism concerned with the synthesis of larger and more complex molecules from smaller and simpler ones.

**anaerobic threshold**—The onset of anaerobic metabolism during exercise; the oxygen uptake in relation to maximal aerobic power when the blood lactate concentration starts to increase during continuous exercise. The terms lactate threshold, OBLA, and anaerobic threshold are often used interchangeably.

**anhidrosis**—Deficiency or absence of sweat secretion.

**anisomelia**—Unequal limb length.

**ANP**—See *atrial natriuretic peptide.*

**antibody**—Proteins, collectively known as immunoglobulins, produced by B-lymphocytes in response to exposure to an antigen.

**antidiuretic hormone (ADH)**—A peptide hormone secreted by the posterior hypophysis (also called vasopressin).

**antigen**—A molecule or part of a molecule, which is able to induce an immune response. See *epitope.*

**antigravity muscle**—A muscle that opposes the effect of gravity on the body during standing.

**antiport**—Transmembrane transporter, carrying two different kinds of ions in opposite directions at the same time (e.g., a $Na^+/K^+$ antiport).

**anulus fibrosus**—The outer, fibrous part of an intervertebral disc.

**anxiogenic**—Something that increases the level of anxiety.

**anxiolytic**—Something that decreases the level of anxiety.

**AP-1**—Transcription complex consisting of homo and hetero-dimers of c-fos and c-jun proteins.

**apolar**—Hydrophobic, not water soluble.

**aquaporin**—Protein forming transmembrane water channels.

**aspect ratio**—The ratio between length and width of a muscle fiber.

**ataxia**—Incoordination of voluntary muscular action.

**ATP**—See *adenosine triphosphate.*

**ATPase**—An ATP-splitting enzyme. Molecules such as myosin and active transmembrane transporters, which use ATP as their source of energy, have ATPase activity.

**atrial natriuretic peptide(ANP)**—A peptide hormone secreted by the cardiac atria (also called atrial natriuretic factor = ANF).

**autocrine signaling**—Chemical signaling acting on the same cell, which secreted the signal molecule.

**autologous blood reinfusion**—Reinfusion of blood to the same individual it was taken from.

**autonomic nervous system**—The part of the nervous system controlling the internal organs through involuntary and largely unconscious reflex activity.

**autopsy**—A postmortem investigation.

**autoradiography**—A histochemical technique based on radioactive tags.

**axolemma**—The cell membrane of an axon.

**axon terminal**—The slightly dilated terminal part of the axon; synaptic terminal; bouton.

**ballistic movement**—A (rapid) movement where all kinetic energy is supplied initially, like in a shot or a throw.

**baroreceptor**—Stretch receptor in the wall of the heart or blood vessels.

**basal ganglia**—A group of intracerebral nuclei comprising the lentiform nucleus (putamen and globus pallidus) and the caudate nucleus.

**basal metabolic rate**—The rate of energy metabolism in a resting individual 14 to 18 hours after eating.

**basic multicellular unit (BMU)**—A group of osteoclasts and osteoblasts active in bone remodeling.

**bends**—A dull, throbbing, shifting pain of muscles, bones, and joints in decompression sickness.

**bioelectrical impedance analysis (BIA)**—Analysis technique for body composition based on the fact that impedance to electric current flow relates to the quantity of total body water; this in turn relates to FFM, body density, and percentage body fat.

**biopsy**—A piece of tissue obtained from a living organism by a minor surgical procedure.

**blast**—Suffix denoting young cell (e.g., fibroblast, chondroblast, osteoblast).

**BMI**—See *body mass index.*

**BMP**—See *bone morphogenic proteins.*

**BMR**—Basic metabolic rate.

**BMU**—See *basic multicellular unit.*

**body mass index (BMI)**—The body weight (in kg) divided by the square of the height (in m); normal value for the index is 20–25 $kg \cdot m^{-2}$.

**bone morphogenic proteins (BMPs)**—A group of growth factors belonging to the transforming growth factor beta superfamily.

**Borg scale**—A numerical scale (6–20) used to rate perceived exertion; invented by Gunnar Borg.

**bouton**—The presynaptic part of a synapse. See *axon terminal.*

***bouton en passage***—As opposed to "terminal bouton" or axon terminal, which is found at the tip of an axon, this kind of presynaptic structure appears as local swellings, one after the other, along the course of an axon.

**bradycardia**—Slowness of the heart rate.

**bradykinin**—A vasodilator peptide; it also increases capillary permeability and produces edema.

**brown fat**—Fat tissue with a thermogenic function; a source of considerable heat, particularly in infants.

**buffer**—A substance that reduces change of hydrogen ion concentration when either acid or alkali is added to the solution.

**$Ca^{++}$-ATPase**—A transmembrane calcium transporter, a calcium "pump."

**calcineurin (CaN)**—A calcium-calmodulin sensitive serine-threonine phosphatase, involved in intracellular signaling and muscle phenotype switching; also known as protein phosphatase IIB.

**calcitonin-gene-related peptide (CGRP)**—A neuroactive peptide stored in and released from dense core vesicles in motor nerve terminals.

**calcium transient**—The rise and fall of the intracellular calcium concentration (e.g., during a twitch).

**callus**—The first, primitive bone mass bridging the gap between the two parts of a fractured bone.

**calpain**—A calcium-activated neutral protease.

**calsequestrin**—A calcium-binding protein in the lumen of the sarcoplasmic reticulum

**cAMP**—cyclic AMP.

**CaN**—See *calcineurin.*

**capsaicin-sensitive nerves**—A subgroup of group III (Aδ) and IV (C) muscle afferents responsive to protons (and to capsaicin, an active ingredient of chili pepper).

**carbamino compound**—Formed by the combination of carbon dioxide with a free amino group in an amino acid or a protein (e.g., carbaminohemoglobin).

**carbonic anhydrase**—An enzyme, found in erythrocytes and in tissues that catalyzes the reaction $H_2O + CO_2 = H_2CO_3$.

**carcinogenic**—Pertaining to any agent or substance that may produce cancer.

**catabolism**—The type of metabolism concerned with the breakdown of large and complex molecules to smaller ones, often with liberation of energy.

**catch property (of muscle)**—The increased force output of a muscle resulting from increased frequency of the first two or three impulses (doublets or triplets) of an impulse train.

**central sulcus**—The sulcus between the frontal and the parietal lobes of the brain.

**c-fos**—Protein encoded by the early response gene *c-fos*.

**CGRP**—See *calcitonin-gene-related peptide*.

**chemically gated ion channel**—See *ligand gated ion channel*.

**chemoreceptor**—A sensory end organ capable of reacting to a chemical stimulus.

**chloride shift**—The reversible exchange of chloride and bicarbonate ions between erythrocytes and plasma.

**cholinergic**—An axon or synaptic terminal, which uses the neurotransmitter acetylcholine.

**chondroblast**—Young cartilage cell.

**chronotropic action**—Having an effect on cardiac rate, cf. inotropic.

**ciliary neurotrophic factor (CNTF)**—A protein synthesized by Schwann cells, important for the differentiation and survival of many cell types in nerve tissue.

**Cinderella units**—Term used for low threshold motor units prone to become fatigued during low intensity, sustained work.

**circadian pattern**—A pattern that repeats itself on a daily basis, such as the sleep-wake cycle.

**circulus vitiosus**—An increasingly bad situation where unfavorable factors contributing to the situation reciprocally reinforce each other.

**citric acid cycle**—See *tricarboxylic acid cycle (TCA)*.

**c-jun**—Protein encoded by the early response gene *c-jun*.

**clathrin**—A protein attached to the cytoplasmic side of the membrane of coated vesicles.

**climbing fibers**—Important afferent fibers to the cerebellum.

**clo unit**—An unit of the thermal insulation value of clothing.

**clone**—A clone of cells is a subset of a particular cell type with certain genetic properties.

**closed loop mode**—"With feedback."

**CNTF**—See *ciliary neurotrophic factor*.

**coated vesicle**—Intracellular vesicle with a protein coat consisting of clathrin on its cytoplasmic side. Coated vesicles are involved in different kinds of vesicular transport, both between different intracellular compartments, and between cell interior and the extracellular fluid.

**collagen**—The most ubiquitous fibrous protein in the body. A major constituent of all connective tissues, including tendons.

**colloidosmotic pressure**—The pressure exerted by the plasma proteins (also called oncotic pressure).

**colony stimulating factors (CSFs)**—A group of glycoprotein growth factors important for the proliferation and differentiation of many cell types. Although their individual names may indicate a rather restricted group of target cells, they usually act on a variety of cell types, and the specificity in their actions is due to different combinations of factors rather than to one specific CSF.

**combinatorial control**—Regulation of gene expression by different combinations of gene regulatory proteins.

**computed tomography (CT)**—Computer-based X-ray technique generating detailed cross-sectional, two-dimensional radiographic images of different body segments.

**concentric contraction**—A contraction where the force exerted by the muscle is larger than the resistance opposing the contraction, allowing the muscle to shorten.

**connectin**—Another name for titin.

**Cori cycle**—A series of reactions resulting in the conversion of lactic acid (formed from glycogen during exercise) to glucose in the liver, and its subsequent anabolism to glycogen in muscle.

**corticospinal tract**—The only uninterrupted, descending nerve tract from the cerebral cortex to the spinal cord. Also known as the pyramidal tract since it passes through the structure called the pyramid on the ventral side of the medulla oblongata.

**costamere**—Riblike structure, which encircles a muscle fiber at regular intervals along its length. May be visualized by staining for vinculin, one of its constituent proteins.

**creatinine**—The breakdown product of creatine.

**cross education**—The training effect obtained in the contralateral, untrained limb after unilateral training.

**cryosections**—Thin sections of frozen tissue used in histochemistry.

**CSFs**—See *colony stimulating factors*.

**CT**—See *computed tomography*.

**curare**—A mixture of plant toxins used by South American Indians to poison arrowheads for hunting. The poison paralyzes the animal by binding to the nicotinic acetylcholine receptor (AChR) in the motor endplate, thereby preventing the activation of the receptor by ACh.

**cyanosis**—A bluish purple discoloration of the skin, due to excessive amounts of reduced hemoglobin in capillaries.

**cytokine**—Various types of secreted molecules acting as local mediators in many different tissues.

**cytosol**—The part of the cytoplasm that is not organelles.

**$D_2O$**—See *deuterium*.

**dark-field microscopy**—A special form of light microscopy where the object appears bright on a dark background.

**dead space**—The space in the respiratory passages that contains air that does not reach the alveoli during respiration (anatomical dead space).

**dehydration**—The removal of water, as from the body or a tissue; excessive loss of body fluid.

**delayed onset muscle soreness (DOMS)**—Muscle soreness developing one to two days after unaccustomed muscle activity, in particular eccentric work.

**denervation**—The removal or destruction of the physical contact between an axon and its muscle fiber.

**desmin**—A type of intermediate filament, important part of the cytoskeleton.

**deuterium ($D_2O$)**—Water that contains double-weight atoms of hydrogen instead of ordinary hydrogen atoms; also called heavy water.

**diaphysis**—The shaft of a long bone.

**diastole**—The relaxation and dilatation phase of the cardiac cycle.

**diathermy (diathermia)**—The use of oscillating electric current of high frequency to produce local heat in body tissue.

**diffusion**—Passive transport of molecules due to Brownian movements, tending to eliminate concentration differences.

**dihydropyridine receptor**—A voltage-sensitive protein in the T-tubule membrane, important link in excitation-contraction coupling.

**diphosphoglycerate (2,3-DPG)**—A substance, produced and found in the erythrocytes, that shifts the oxygen-hemoglobin dissociation curve to the right.

**disinhibition**—The removal of inhibition by inhibition of an inhibitory neuron.

**disseminated**—Scattered, dispersed, diffused.

**dizygotic twins**—Non-identical twins.

**DNA**—deoxyribonucleic acid.

**DOMS**—See *delayed onset muscle soreness*.

**Douglas bag**—A collecting bag for expired air used in determining oxygen uptake.

**downstream**—Allusion to the direction of flow in a river. May be used for a part of a structure or process, which have a direction (e.g., a gene or a signaling pathway); cf. upstream.

**DRI**—Dietary reference intake.

**dry-bulb temperature**—The actual air temperature, with the exclusion of variations due to radiation or conduction.

**DTNB light chain**—Another name for MyLC2, regulatory light chain, or phosphorylatable light chain.

**dual-energy x-ray absorptiometry(DXA)**—Quantifies fat and nonebone regional lean body mass, including the mineral content of the body's deeper bony structures.

**ductus arteriosus (Botalli)**—A shunt between the pulmonary artery and the aorta that largely short-circuits the pulmonary circulation during intra-uterine life.

**dynamic contraction**—A contraction during which the length of the muscle changes; may be concentric or eccentric.

**dyspnea**—Shortness of breath; difficult or labored breathing.

**early response genes**—Genes that are activated within 15 minutes of growth factor exposure. Many of their products are gene regulatory proteins.

**E-box**—A six-base consensus enhancer sequence in DNA, which binds E-protein and members of the MyoD family of transcription factors.

**E-C coupling**—See *excitation-contraction coupling*.

**eccentric contraction**—A muscle contraction where the force exerted by the muscle is lower than the force opposing the contraction, resulting in a lengthening contraction.

**ECF**—See *extracellular fluid*.

**eIF-2**—See *eukaryotic initiation factor-2*.

**electrocardiography (ECG)**—Recording of the electrical impulses in the heart as they vary during the cardiac cycle.

**electrochemical gradient**—The combined voltage- and concentration gradient across the cell membrane.

**electromyography (EMG)**—Registration of the electrical activity in a muscle.

**EMG gaps**—Short periods of low-level EMG activity during sustained work.

**end-diastolic volume**—The volume of blood in the ventricles at the end of the diastole/beginning of systole.

**endochondral (enchondral) ossification**—Indirect bone formation by gradual replacement of a cartilagineous "bone" by bone tissue.

**endocrine signaling**—Classical hormone signaling.

**endocytosis**—The process by which cells internalize particles and large molecules enwrapped in vesicles.

**endomysium**—The connective tissue surrounding individual muscle fibers.

**endoplasmic reticulum**—Cellular organelle consisting of a system of flattened vesicles. Important for the compartmentalization of cellular activities.

**endosome**—Endocytotic vesicle with content. The result of an endocytotic process.

**enhancer**—Part of a DNA molecule that, upon binding of a gene regulatory protein, confers increased likelihood for expression of the gene in question.

**enteric nervous system**—The largely independent part of the automonic nervous sytem serving the intestinal tract.

**enzyme-linked receptors**—A major group of cell surface receptors, including the insulin receptor and the IGF receptor.

**epimysium**—The connective tissue on the outer surface of a muscle. If not adherent to surrounding tissue, it is also called a fascia.

**epiphyseal disc (plate)**—The cartilagineous zone between the diaphysis and the epiphyses of a long bone; the site of growth in length of long bones.

**epiphysis**—The ends of a long bone. May also mean pineal gland.

**epitope**—Part of an antigenic molecule, which evokes an immune response from a single clone of lymphocytes.

**EPO**—See *erythropoietin*.

**EPOC**—See *excess postexercise oxygen consumption*.

**equilibrium potential**—The membrane potential at which the transmembrane flux of a certain ion is zero due to the opposing effects of its concentration gradient and voltage gradient.

**ER**—endoplasmic reticulum.

**ergogenic**—Tending to increase physical performance, work output.

**ERK**—Extracellular signal-regulated kinase, a subgroup of MAP kinases.

**erythrocyte**—Non-nucleated blood cell containing the oxygen-carrying pigment hemoglobin.

**erythrocythemia (erythrocytosis)**—An elevation above normal of red blood cells.

**erythropoiesis (erythrocytopoiesis)**—The formation and development of erythrocytes.

**erythropoietin (EPO)**—A hormone acting on the bone marrow to produce erythrocytes.

**essential light chains**—Another name for alkali light chains (MyLC1 and MyLC3).

**eukaryotic initiation factor-2 (eIF-2)**—A GTP-binding protein necessary for binding of methionyl initiator t-RNA to the start codon AUG, thus initiating translation.

**excess post-exercise oxygen consumption (EPOC)**—Elevated aerobic metabolism during recovery from exercise, the magnitude depending on the duration and intensity of exercise.

**excitation-contraction coupling**—The sequence of events coupling action potentials to muscle contraction.

**excitatory postsynaptic potential (EPSP)**—The change in membrane potential (depolarization) due to the action of an excitatory synapse.

**excitatory synapse**—A synapse that depolarizes the postsynaptic membrane, thereby increasing the probability of an action potential.

**exocytosis**—The process by which cells externalize particles and large molecules enwrapped in vesicles.

**exon**—Coding sequence of a gene.

**exteroceptive**—Activated by, or pertaining to, stimuli impinging on an organism from the outside.

**extracellular fluid (ECF)**—The fluid outside cells.

**extrafusal muscle fiber**—The ordinary muscle fibers of a muscle.

**facilitated diffusion**—A kind of passive transmembrane transport requiring special carrier molecules in the membrane.

**F-actin**—Filamentous actin, a string of G-actin molecules.

**F-6-P**—Fructose-6-phosphate.

**F-1,6-P**—Fructose-1,6-diphosphate.

**fascia**—An epimysium that is not adherent to the tissue surrounding the muscle, allowing the muscle to move freely during contraction.

**fatigue index**—Index used to quantify muscle fatigue. Several varieties exist (e.g., the relation between maximal force after and before development of fatigue).

**FDPase**—Fructose diphosphatase.

**female athlete triad**—Includes three medical disorders commonly found in female athletes: disordered eating, amenorrhea, and osteopenia.

**fenestration**—Very thin areas in the endothelial cells of some capillaries (fenestra = window).

**ferritin**—The principal storage form of iron in tissues.

**FFA**—Free fatty acid.

**FFM**—Fat free mass.

**fibrin**—A fibrous insoluble protein formed by the proteolytic action of thrombin on fibrinogen.

**fibrinogen**—A protein present in plasma; the soluble precursor of fibrin.

**fibroblast**—Young connective tissue cell, which makes the intercellular substance, including the collagen fibers.

**fibrocartilage**—A hybrid form of cartilage, between hyaline cartilage and connective tissue proper, found in articular discs and menisci.

**Fick equation**—A formula that expresses the relationship between cardiac output, oxygen uptake, and the arteriovenous oxygen difference.

**flux-generating reaction**—A non-equilibrium enzyme reaction, which is saturated with pathway substrate.

**focal adhesion kinase**—Kinase located to focal adhesion complexes, also known as costameres, in muscle fibers.

**functional stretch reflex**—See *long loop stretch reflex*.

**G-actin**—Globular actin molecules, the monomer of actin filaments (F actin).

**G-1-P**—Glucose-1-phosphate.

**G-6-P**—Glucose-6-phosphate.

**gamma (γ) motoneuron** —A motoneuron innervating intrafusal muscle fibers (e.g., the muscle fibers within a muscle spindle).

**ganglion (pl. ganglia)**—A group of nerve cell bodies located outside of the central nervous system.

**GDF-8**—Growth/differentiation factor 8. See *myostatin*.

**gene regulatory protein**—A protein that upon binding to a DNA molecule takes part in the control of expression of the gene in question.

**gene regulatory sequence**—A part of a DNA molecule that binds a gene regulatory protein.

**generator potential**—See *receptor potential.*

**genotype**—The fundamental genetic constitution of an organism comprising genes from both parents.

**globin**—The four polypeptides of a hemoglobin molecule.

**globulin**—A group of plasma proteins.

**glottis**—The two vocal folds and the space between them.

**gluconeogenesis**—The formation of glucose from noncarbohydrate molecules.

**GLUT**—Glucose transporter protein.

**GLUT4**—The insulin and contraction sensitive isoform of GLUT.

**glycocalyx**—Any polysaccharide-containing structure on the external surface of a cell.

**glycogenolysis**—The liberation of glucose from glycogen in the liver or other tissues.

**glycolysis**—The breakdown of glucose (or glycogen) to pyruvic acid (and lactic acid).

**glyconeogensis**—The formation of glycogen by the liver from non-carbohydrate sources (gluconeogenesis designates the formation of glucose).

**goblet cells**—Unicellular mucous glands found in the epithelium of mucous membranes (e.g., in the respiratory passages).

**golgi apparatus**—Intracellular organelle consisting of stacks of flattened vesicles involved in secretory processes.

**golgi tendon organ**—The force sensor of muscle, located in the tendon.

**GP**—Glycogen phosphorylase.

**G-protein**—GTP-binding protein functioning as a molecular on/off switch.

**G-protein-linked receptor**—A major family of cell surface receptors, including adrenergic receptors.

**gray matter**—Nervous tissue dominated by neuronal cell bodies and dendrites.

**GS**—Glycogen synthetase.

**GTP**—Guanosine triphosphate.

**haversian system (osteon)**—A group of concentric bone lamellae surrounding a small artery. Haversian systems are temporary structures found in compact bone, with their long axis parallel to the bone in question.

**H-band**—The central part of the A-band of a sarcomere, which is devoid of actin filaments.

**heat shock proteins (HSPs)**—A large family of highly conserved proteins, originally discovered in cells subjected to heat shock, but subsequently found to be induced by a variety of homeostatic perturbances. See *stress proteins.*

**hematocrit**—The percentage of blood volume that is occupied by erythrocytes.

**hematogenous**—Pertaining to the blood or its constituents.

**hematuria**—The discharge of urine containing blood.

**heme**—A subunit of the hemoglobin molecule containing one atom of iron to which oxygen binds.

**hemoconcentration**—An increase in the concentration of blood cells resulting from a loss of plasma or water from the blood stream.

**hemodilution**—A condition of the blood in which the ratio of blood cells to plasma is reduced.

**hemoglobin (Hb)**—A heme-containing protein that combines reversibly with oxygen.

**Hering-Breuer reflex**—Nervous mechanism that controls respiration by impulses mediated through pulmonary fibers of the vagus nerves.

**hexamer**—A compound molecule consisting of six parts (or monomers).

**high frequency fatigue (HFF)**—A type of muscle fatigue induced by (unphysiologically) high frequency stimulation, causing impaired propagation of action potentials along the surface membrane and into the T-tubules.

**highly conserved**—Term used to characterize proteins that have undergone few changes during phylogenesis, presumably because of their fundamental importance for the life of cells.

**histochemistry**—A collective name for techniques aimed at studying the chemical composition of a tissue.

**HK**—Hexokinase.

**homeostasis**—The maintenance of uniformity or stability in the organism by coordinated physiological processes.

**homonymous**—"With the same name."

**HSP72**—Heat shock protein 72, a highly inducible type of stress protein with MW 72 kD, believed to take part in the rescue of protein molecules damaged by excessive heat or other environmental disturbances.

**hybridoma**—A cell line resulting from the fusion of an antibody-producing lymphocyte and a cancer cell, conferring immortality to the hybrid cell. Used for production of monoclonal antibodies.

**hydrodensitometry**—See *hydrostatic weighing.*

**hydrophilic**—Water soluble, polar.

**hydrophobic**—Water insoluble, apolar.

**hydrostatic weighing**—Underwater weighing (hydrodensitometry). Indirect method to assess body composition; percentage body fat is computed from body density, the ratio of body mass to body volume.

**hypercapnia**—Excessive amount of carbon dioxide in the blood.

**hyperplasia**—Increase in size of an organ due to an increase in the number of cells.

**hyperplasia-obesity**—Obesity involving an increased number of fat cells.

**hyperpnea**—Increase in depth and rate of respiration.

**hypertrophy**—Increase in size of an organ due to increased size of individual cells.

**hypertrophy-obesity**—Obesity involving increased content of triglycerides in each fat cell, while the number of fat cells is normal.

**hyperventilation**—Abnormally rapid, deep breathing; overbreathing.

**hypobaric hypoxia**—Oxygen deficiency caused by lower than normal air pressure (e.g., at high altitude).

**hypocapnia**—Subnormal concentration of carbon dioxide in the blood.

**hypohydration**—See *dehydration.*

**hyponatremia**—Abnormally low blood sodium level.

**hypothalamic**—Pertaining to or involving the hypothalamus.

**hypothermia**—Abnormally low body temperature.

**hypoventilation**—Reduced repiratory effort; reduced alveolar ventilation.

**hypoxia**—Oxygen deficiency.

**I-band**—The lateral parts of a sarcomere, which contains actin but not myosin.

**ICF**—Intracellular fluid.

**IGF**—See *insulin-like growth factors.*

**immediate early gene**—See *early response gene.*

**immunohistochemistry**—A histochemical method to show the localization of specific molecules in a microscopic section with the aid of specific antibodies against the molecule in question.

**immunoperoxidase staining**—An immunohistochemical method based on horseradish peroxidase as visualizing agent.

**IMTG**—Intramuscular triglycerides.

**in vitro**—Literally "in glass." Denotes test tube experiments with living tissue or cells outside the organism.

**inferior colliculus**—Small, round protrusions on the dorsal side of the mesencephalon, one on each side of the midline, serving as centers for auditory reflexes.

**inhibitory postsynaptic potential (IPSP)**—The change in potential (hyperpolarization) due to the action of an inhibitory synapse.

**inhibitory synapse**—A synapse that hyperpolarizes the postsynaptic membrane, thereby decreasing the probability of an action potential.

**innervation**—Denotes the presence of a functional contact between an axon and its target, usually a muscle fiber.

**inotropic action**—Having an effect on muscular contractile force.

**insensible perspiration**—Perspiration that takes place constantly, the fluid being evaporated as fast as excreted.

**in situ hybridization**—A method to show the localization of mRNA molecules in a microscopic section, based on the property of nucleotides to hybridize with its complimentary strand.

**insomnia**—Sleeplessness, disturbed sleep.

**insulin-like growth factors (IGFs)**—A group of growth factors formerly known as somatomedins, produced by liver cells and muscle cells.

**integrins**—Membrane-associated proteins involved in the interaction between cells and their immediate surroundings; possible role in the transduction of mechanical stimuli into intracellular signals.

**intercalated disc**—Transverse thickening occurring at the abutting surfaces of cardiac muscle cells, providing a strong union between the cells.

**intermediary filament**—Tough and durable protein filaments found in most cells, particularly in cells subjected to mechanical stress. The name intermediary refers to their diameter (8–10 nm), which is between that of actin and myosin filaments.

**interoceptive**—Activated by, or pertaining to, sensory information from the body itself.

**interstitial fluid**—The fluid surrounding cells, except blood cells.

**intracellular fluid (ICF)**—The fluid inside cells.

**intrafusal muscle fibers**—The muscle fibers within a muscle spindle.

**intramembranous ossification**—Direct bone formation in connective tissue proper, without a cartilaginous predecessor.

**intron**—Non-coding sequence of a gene. The parts of the primary transcript, which represent introns, are removed before mRNA is transferred to the cytoplasm.

**ion channel–linked receptor**—Family of cell surface receptors that form part of ligand-gated ion channels.

**ionotropic event**—A synaptic event where the immediate effect is a change in membrane conductance for certain ions.

**ischemia**—Local diminution of blood supply; localized tissue anemia.

**isoelectric point**—The pH at which the net electric charge of a particular molecule is zero.

**isoenzyme**—Slightly different molecular variant (isoform) of the same enzyme, with correspondingly different functional properties.

**isokinetic equipment**—Equipment allowing joint movements with a predetermined constant angular velocity.

**isometric contraction**—A muscle contraction without any change in muscle length.

**isotonic contraction**—A muscle contraction without any increase in muscle tension. A true isotonic contraction is only possible during weightlessness.

**JNK/SAPK**—c-jun $NH_2$-terminal kinase/stress-activated protein kinase, a subgroup of MAP kinases.

**juxtaglomerular cells**—Cell located in the wall of the afferent arterioles, close to the renal corpuscles (glomeruli).

**kinase**—An enzyme that phosphorylates another molecule (e.g., a protein kinase).

**Krebs cycle**—Tricarboxylic acid cycle; the common pathway for converting foodstuffs to energy (also called citric acid cycle).

**lactate threshold**—Describes the highest relative work intensity (or oxygen uptake) with less than 1.0 mM per liter increase in blood lactate concentration above the preexercise level.

**lactic (lactacid)**—Associated with elevated blood lactate levels.

**laminar flow**—Streamline, silent flow, as opposed to turbulent flow.

**larynx**—The organ of the voice situated between the trachea and the base of the tongue.

**lateral cerebral fissure (of Sylvius)**—The deep sulcus on the lateral side of the brain, separating the temporal lobe from the frontal lobe and the anterior part of the parietal lobe.

**LDH**—lactate dehydrogenase, the enzyme catalyzing the interconversion of pyruvate and lactate. Several isoenzymes exist.

**leptin**—A multifunctional hormone secreted by fat cells (adipocytes); basically a "stop signal" for food intake.

**leukocyte**—White blood cell.

**LFF**—See *low frequency fatigue*.

**ligand**—Denotes any molecule that binds specifically to a receptor.

**ligand gated ion channel**—An ion channel that opens when its receptor part binds its specific ligand.

**light chain**—See *myosin light chain*.

**lipolysis**—The hydrolysis of fat.

**lipostatic regulation**—Based on the hypothesis that adipose tissue produces a humoral signal that is proportionate to the amount of fat and acts on the hypothalamus to decrease food intake and increase energy output. See *leptin*.

**long loop stretch reflex**—A stretch reflex with a long, polysynaptic reflex arch, possibly involving the cerebral cortex.

**long term depression (LTD)**—Long lasting decreased efficiency in synaptic transmission, due to previous synaptic activity, thought to be important in learning and memory.

**long term potentiation (LTP)**—Long lasting increased efficiency in synaptic transmission, due to previous synaptic activity, thought to be important in learning and memory.

**low frequency fatigue (LFF)**—A type of long lasting fatigue, most pronounced at low stimulation frequencies, thought to be due to reduced levels of cytosolic calcium.

**M1**—See *primary motor cortex*.

**macrophage colony stimulating factor (M-CSF)**—See *colony stimulating factors*.

**MADS** Acronym for the MCM1, Agamous, Deficiens, Serum response factor superfamily of DNA binding phosphoproteins.

**magnetic resonance imaging (MRI)**—A noninvasive method of producing pictures of the interior of the body.

**magnetic resonance spectroscopy (MRS)**—Methods providing information about shifting levels of phosphorus compounds in a tissue (muscle) during activity.

**MAP kinase (MAPK)**—Mitogen activated protein kinase.

**maximal aerobic power**—Maximal oxygen uptake expressed as ml oxygen · kg body weight$^{-1}$ · min$^{-1}$.

**maximal anaerobic power**—The highest mechanical power generated during a certain time interval.

**maximal evocable force**—The force generated by a muscle or a group of muscles when additional stimulation does not augment force.

**M-CSF**—Macrophage colony stimulating factor; see *colony stimulating factors*.

**MCT**—See *monocarboxylate transporter*.

**Mdx mice**—A strain of mice with a mutated dystrophin gene. Mdx mice are used as an animal model of muscular dystrophy.

**mechanical efficiency**—The percentage of total chemical energy expended that contributes to external work, with the remainder lost as heat.

**mechano growth factor (MGF)**—A splice variant of IGF-1 secreted by muscle cells.

**mechanosensory cell (mechanocyte)**—A cell that is able to respond to mechanical stress; often used to characterize bone cells and muscle cells, but most cells more or less have that ability.

**medulla oblongata**—The lower part of the brain stem, between the spinal cord and the pons.

**MEF2**—See *myocyte enhancer-binding factor 2*.

**mesencephalon**—The part of the brain stem between pons and the diencephalon. The ventral part of the mesencephalon is the cerebral peduncles.

**MET**—See *metabolic unit*.

**metabolic pathway**—A series of enzyme-catalyzed reactions, initiated by a flux-generating step and ending with either storage of the product, its loss to the environment, or in a reaction, which precedes another flux-generating step, which represents the beginning of the next metabolic pathway.

**metabolic unit (MET)**—The approximate oxygen uptake at rest. 1 MET equals 3.5 ml · kg$^{-1}$ · min$^{-1}$. METS are defined as multiples of the resting metabolic rate (RMR).

**metaboreceptor**—Receptors reacting to products of the metabolism.

**metabotropic event**—A synaptic event resulting in a modulation of synaptic efficiency.

**MGF**—See *mechano growth factor*.

**metabotropic receptor**—Receptor in the membrane of a nerve cell mediating a metabotropic event.

**milieu interne**—The external and internal fluid environment of cells.

**M-lines**—Three to five lines across the myofibril in the middle of the A-band.

**monocarboxylate transporters (MCTs)**—A family of specific proton-linked transporter isoforms (MCT 1–4) involved in transmembrane facilitated diffusion of monocarboxylates (lactic acid and pyruvic acid).

**monocarboxylic**—Containing one carboxyl group.

**monoclonal antibody**—Antibody directed against one particular epitope only. Monoclonals are produced by a single clone of lymphocytes, hence the name.

**monozygotic twins**—Identical twins.

**mossy fibers**—Important afferent fibers to the cerebellum.

**motoneuron pool**—A group of motoneurons innervating one particular muscle.

**motor endplate**—The synapse between a branch of the motor axon and a muscle fiber. Each endplate contains a cluster of synaptic terminals (boutons).

**motor unit**—A motoneuron, all its axonal branches and the muscle fibers they command.

**mrf4**—See *myogenic determination factors*.

**MRI**—See *magnetic resonance imaging*.

**mRNA**—messenger RNA.

**MRS**—See *magnetic resonance spectroscopy*.

**MTJ**—See *myotendinous junction*.

**muscarinic receptor**—The type of acetylcholine receptor found in the heart, smooth muscle, and many places in the brain.

**muscle regionalization**—Means that different parts of the same muscle may contain different subsets of muscle units and differ functionally.

**muscle spindle**—The length sensor of muscle.

**muscle unit**—The muscle fibers belonging to one motor unit.

**muscle wisdom**—Term used to denote the fact that when the force-frequency curve is shifted to the left due to slowing of contractile properties during the development of fatigue, the stimulation frequency is lowered correspondingly.

**myf5**—See *myogenic determination factors*.

**MyHC**—See *myosin heavy chain*.

**MyLC**—See *myosin light chain*.

**myoblast**—A muscle cell precursor. See also *satellite cell*.

**myocyte enhancer-binding factor 2 (MEF2)**—A family of phosphoprotein transcription factors that act in synergy with transcription factors of the MyoD family.

**MyoD**—See *myogenic determination factors*.

**myofibril**—The contractile element of muscle cells consisting of a chain of serially coupled sarcomeres.

**myofilament**—Actin and myosin filaments, which are the main constituents of a sarcomere.

**myogenic determination factors (MDFs)**—A group of gene regulatory proteins (MyoD, Myf5, mrf4, myogenin) pivotal in the differentiation of muscle cells from naïve mesodermal cells; also known as myogenic regulatory factors (MRFs) or myogenic proteins.

**myogenic**—Of muscular origin.

**myogenin**—See *myogenic determination factors*.

**myosin**—A hexameric motor protein molecule consisting of two myosin heavy chains and four myosin light chains.

**myosin heavy chain (MyHC)**—The larger of the two kinds of subunits constituting the hexameric myosin molecule; consists of a motor-domain head and a rod-like tail.

**myosin light chain (MyLC)**—The smaller of the two kinds of subunits constituting the hexameric myosin molecule. Two light chains are attached close to the head of each of the two MyHCs; one is always MyLC2, the other either MyLC1 or MyLC3.

**myostatin (GDF-8)**—A negative regulator of muscle mass, which belongs to the TGF-β family of growth and differentiation factors.

**myotendinous junction (MTJ)**—The junction between the end of a muscle fiber and the connective tissue of the tendon.

**$Na^+$-$K^+$-ATPase**—A sodium–potassium antiport; "the sodium–potassium pump."

**NAD**—Nicotinamide dinucleotide.

**NADH**—Nicotinamide dinucleotide, the reduced form.

**nebulin**—A filamentous protein lying alongside the actin filament in the sarcomere, thought to act as a ruler to ensure correct length of actin filaments.

**nephritis**—Inflammation of the kidney.

**neuromodulator**—A metabotropic neurotransmitter.

**neuromuscular synapse**—See *motor endplate*.

**neutral fat**—See *triacylglycerol*.

**nicotinic receptor**—The type of acetylcholine receptor found in motor endplates.

**nitric oxide (NO)**—A vasodilator released from endothelial cells.

**nitric oxide synthase (NOS)**—An enzyme that synthesizes NO from L-arginine.

**nociceptor**—Pain receptor.

**nodes of Ranvier**—The short, naked part of the axon between two neighboring segments of myelin.

**nomogram**—A graph on which appear lines for variables in a formula, arranged in such a manner that the value of one variable can be read on the appropriate line from a knowledge of the values of the other variables.

**non-adrenergic-non-cholinergic (NANC) transmission**—Transmission in the autonomic nervous system, notably in the enteric system, which does not use adrenaline or acetylcholine as neurotransmitter.

**noradrenaline**—A hormone formed at sympathetic nerve-endings as a mediator of nerve activity (also called norepinephrine).

**noradrenergic**—An axon or synaptic terminal that uses the neurotransmitter noradrenaline.

**NOS**—See *nitric oxide synthase.*

**noxious**—Harmful, deleterious; poisonous.

**nuclear bag fiber**—A type of intrafusal muscle fiber.

**nuclear chain fiber**—A type of intrafusal muscle fiber.

**nuclear domain**—The part of muscle fiber cytoplasm that is "governed" by one particular myonucleus.

**nucleus pulposus**—The inner, soft part of an intervertebral disc.

**nucleus ruber**—See *red nucleus.*

**OAA**—Oxaloacetate.

**OBLA**—See *onset of blood lactate accumulation.*

**ODF**—See *osteoclast differentiation factor.*

**oligosynaptic**—"With a limited number of synapses" (oligo = few).

**onset of blood lactate accumulation (OBLA)**—Signifies when blood lactate concentration shows a systematic increase. (A threshold value often used is an arterial lactate concentration of 4.0 mM).

**open loop mode**—"Without any feedback."

**OPG**—See *osteoprotegerin.*

**OPG-L (osteoprotegerin ligand) = RANK-L = ODF**—See *osteoclast differentiation factor.*

**opioid**—Resembling an opium alkaloid.

**osteoarthritis**—A degenerative joint disease, to a large extent due to wear and tear.

**osteoblast**—Young bone cell, capable of producing new organic bone intercellular substance.

**osteocalcin**—A molecule that inhibits osteoclast function.

**osteoclast differentiation factor (ODF) = RANK-L = OPG-L**—A cytokine necessary for fusion and differentiation of osteoclast progenitor cells to osteoclasts.

**osteoclast progenitor cell**—A stem cell of the monocyte/macrophage family.

**osteoclast**—The "bone-eating" cell, a multinucleate bone macrophage arising from fusion of osteoclast progenitor cells.

**osteocytes**—Mature bone cell located between bone lamellae.

**osteogenic cell**—The stem cells of bone tissue.

**osteon**—See *haversian system.*

**osteopenia**—Denotes too little bone; bone atrophy. Any condition presenting less bone than normal.

**osteoporosis**—Deossification with absolute decrease in bone tissue, resulting in structural weakness.

**osteoprogenitor cell**—See *osteogenic cell.*

**osteoprotegerin (OPG)**—A soluble cytokine receptor secreted by osteoblasts.

**osteotendinous junction**—The junction between a tendon and bone.

**overreaching**—An early, quickly reversible phase of overtraining.

**oxygen pulse**—Oxygen uptake (transport) per heart beat.

**oxyhemoglobin ($HbO_2$)**—Hemoglobin combined with oxygen.

**p38 MAPK**—A subgroup of MAP kinases.

**PAGE**—See *polyacrylamide gel electrophoresis.*

**palpation**—A technique of investigating an external or superficial part of the body with the fingertips.

**paracrine signaling**—Chemical signaling between neighboring cells.

**parasympathetic nervous system**—The part of the autonomic nervous system responsible for everyday functions.

**parietal pleura**—The portion of the pleura lining the internal surface ("the walls") of the thoracic cavity.

**partial pressure**—The pressure exerted by one gas in a mixture of gases.

**parvalbumin**—Calcium-binding protein in the cytoplasm of muscle fibers.

**PAS staining**—A staining method for glycogen, based on Periodic Acid and Schiff's reagent.

**passive transport**—Transmembrane transport that does not require the use of energy.

**PCR**—Polymerase chain reaction; important laboratory method in molecular biology to increase the number of available copies of a certain string of DNA.

**PCr**—See *phosphocreatine.*

**PDH**—Pyruvate dehydrogenase.

**pennation angle (of a muscle)**—The angle between the line of pull of a muscle and the long axis of its muscle fibers.

**PEP**—Phosphoenol pyruvate.

**PFK**—Phosphofructokinase.

**$P_i$**—Orthophosphate.

**$PP_i$**—Pyrophosphate.

**perikaryon (pl. perikarya)**—Neuronal cell body (literally "the part around the nucleus").

**perimysium**—Connective tissue surrounding bundles of muscle fibers.

**peristalsis**—A progressive wave of contraction seen in tubes provided with longitudinal and transverse muscular fibers (e.g., intestinal peristalsis).

**phagocytosis**—Endocytosis of particulate matter; "cell eating."

**pharynx**—The throat.

**phenotype**—The observable hereditary characteristics of an organism.

**phosphocreatine (PCr)**—Creatine phosphate, a phosphoric acid derivative of creatine that contains an energy-rich phosphate bond.

**phosphorylatable light chain**—Another name for regulatory light chain, DTNB light chain, or MyLC2.

**physical work capacity (PWC)**—The work performed or oxygen uptake per minute at a given heart rate.

**piloerection**—Erection of hairs, resulting in "goose flesh."

**pinocytosis**—Endocytosis of extracellular fluid; "cell drinking."

**plasma skimming**—The phenomenon of plasma flowing into a side branch of a vessel, the erythrocytes continuing along the main channel.

**plasma**—The extracellular fluid of blood.

**plethysmography**—Measurement of volume changes of an extremity or of an organ.

**pleura**—The serous membrane enveloping the lung and lining the internal surface of the thoracic cavity.

**plyometric training**—A form of power training, consisting of an eccentric contraction, followed by a concentric contraction.

**PMA**—See *premotor area.*

**polar**—Hydrophilic, water soluble.

**polyacrylamide gel electrophoresis (PAGE)**—A major method for separation of protein molecules, most often on the basis of their size, sometimes on the basis of size and charge.

**polycythemia (erythrocythemia)**—A condition characterized by an increased number of erythrocytes.

**post**—Prefix meaning "after."

**postganglionic axon**—Autonomic axon connecting an autonomic ganglion and its peripheral target organ.

**postmitotic cell**—A cell that is no longer able to divide.

**postsynaptic cell**—The receiving cell in a synapse.

**postsynaptic potential**—The change in potential, depolarization or hyperpolarization, caused by a synaptic event.

**posttranslational control (of protein synthesis)**—Control that takes place after translation; includes all steps that lead from the birth of a new protein molecule to the fully functional molecule in its correct location.

**posttranslational modification**—Chemical modification of a protein molecule after its synthesis.

**pre-**—Prefix meaning "before."

**precapillary sphincter**—Minute bundles of smooth muscle fibers surrounding the openings of true capillaries.

**precentral gyrus**—The gyrus just in front of the central sulcus, the primary motor cortex.

**preganglionic axon**—Autonomic axon connecting the central nervous system and an autonomic ganglion.

**preload**—The degree to which the myocardium is stretched before it contracts.

**premotor area (PMA)**—The area of the cerebral cortex anterior to the lateral part of the primary motor cortex.

**presynaptic inhibition**—Inhibition of a presynaptic structure (axon terminal).

**pretranslational control (of protein synthesis)**—Control that takes place before translation; includes transcriptional control and RNA processing.

**prevertebral ganglia**—Sympathetic ganglia located outside of the sympathetic trunc.

**primary motor cortex (M1)**—The area of the cerebral cortex that gives rise to corticospinal fibers.

**primary transcript**—The original copy of a gene, made during transcription, that contains transcripts of both exons and introns.

**probe**—An oligonuceotide designed to be complimentary to (part of) an mRNA molecule. Probes usually carry a tag, which may help localize the probe in the tissue.

**promoter**—A specific part of a DNA molecule, located upstream of transcription start, that directs RNA polymerase to attach and start the transcription.

**pronation**—Turning of the forearm until the palm of the hand faces downward.

**proprioceptor**—A receptor conveying information from the musculoskeletal system, relevant for posture and locomotion.

**propriospinal fiber**—Ascending or descending axons connecting different segments of the spinal cord.

**prostaglandins**—A group of important, locally acting signaling molecules, derived from fatty acids, mainly arachidonic acid.

**protease**—A protein-degrading enzyme.

**proteasome**—A multisubunit protease, important for degradation of muscle proteins that have been labeled with ubiquitin.

**protein kinase A**—Protein kinase that is activated by cyclic AMP (cAMP).

**protein kinase C**—Protein kinase that is activated by diacylglycerol and calcium.

**proteinuria**—The presence of protein in the urine (also called albuminuria).

**prothrombin**—An inactive precursor of the enzyme thrombin in blood.

**proton**—A subatomic particle identical with the nucleus of the hydrogen atom.

**pulmonary edema**—An effusion of fluid into the alveoli and interstitial tissue of the lungs, producing severe dyspnea.

**putative transmitter**—A transmitter candidate, not yet confirmed.

**PWC**—See *physical work capacity.*

**pyramidal tract**—See *corticospinal tract.*

**quantal size**—The amount of neurotransmitter contained in one synaptic vesicle

**R**—See *respiratory quotient.*

**radiometer**—Instrument for detection and measuring of radiant energy (normally infrared, visible, or ultraviolet).

**ramp movement**—A (slow) movement where kinetic energy is supplied throughout the duration of the movement, allowing for corrections if necessary.

**RANK**—"Receptor for activation of nuclear factor kappa B," a cytokine receptor on the surface of osteoclast progenitor cells.

**RANK ligand (RANK-L)**—A molecule that binds to RANK.

**rapidly adapting receptor**—A receptor that upon continued stimulation stops responding with a receptor potential, or responds more weakly.

**ras**—Protein belonging to a family of monomeric regulatory GTPases; plays a crucial role in cell signaling, acting as an on-off switch.

**rate coding**—Frequency coding.

**rating of perceived exertion (RPE)**—The exerciser rates on a numerical scale *(Borg scale)* perceived feelings in relation to the exertion level.

**RDA**—Recommended daily allowance.

**readiness potential**—Widespread electrical activity in the cerebral cortex, prior to the activity in the primary motor cortex, preceding a movement.

**receptor mediated endocytosis**—Endocytosis of specific molecules by means of (clathrin-)coated vesicles.

**receptor potential**—The local change in membrane potential resulting from stimulation of a receptor.

**receptor tyrosine kinases**—Important subgroup of enzyme-linked cell surface receptors.

**recruitment**—Denotes the activation and subsequent contraction of muscle fibers.

**recruitment order**—The order of recruitment of motor units when the force output of a muscle is gradually increased.

**recurrent inhibition**—Inhibition of motoneurons by recurrent collaterals acting on inhibitory interneurons (Renshaw cells).

**red nucleus (nucl. ruber)**—Motor nucleus in the mesencephalon projecting to the spinal cord.

**regulatory light chain**—Another name for phosphorylatable light chain, DTNB light chain, or MyLC2.

**reinnervation**—The reestablishment of a functional contact between an axon and its muscle fiber, after a period of denervation.

**Renshaw cell**—Inhibitory interneuron in the spinal cord. See *recurrent inhibition*.

**repeated bout effect**—Refers to the absence of muscle soreness in muscles accustomed to eccentric exercise.

**respiratory quotient (R or RQ)**—The ratio of the amount of carbon dioxide produced to the amount of oxygen consumed.

**respiratory threshold**—Describes the point at which pulmonary ventilation increases disproportionately with oxygen consumption during graded exercise (also called ventilatory threshold).

**responsive element**—See *gene regulatory sequence*.

**reticular formation**—Area in the central parts of the brain stem consisting of functionally heterogeneous nuclei (groups of nerve cell bodies).

**revascularization**—Reestablishment of blood supply, as after destruction of the old vessels due to injury.

**rho**—Protein belonging to a family of monomeric regulatory GTPases, involved in intracellular signaling; closely related to the ras protein.

**ribosome**—Ribonucleoprotein-particle involved in translation.

**RM**—"Repetition maximum," used for quantification of load during strength training; 3RM is the load one particular person is able to lift three times, but not a fourth time.

**RNA polymerase**—Enzyme that joins nucleotides end to end to a polynucleotide string, the RNA chain, during transcription.

**RNA processing**—The sequence of events leading from the primary transcript containing copies of exons and introns alike to the ready-to-use mRNA chain.

**RNA**—Ribonucleic acid.

**rough endoplasmic reticulum**—Endoplasmic reticulum vesicles with ribosomes attached to their cytoplasmic surface.

**RPE**—See *rating of perceived exertion*.

**RQ**—See *respiratory quotient*.

**ryanodine receptor**—The calcium channel in the terminal cisternae of the sarcoplasmic reticulum, which releases calcium ions during E-C coupling.

**safety factor**—The relation between the combined endplate potentials of all synaptic terminals in a motor endplate and the firing threshold of the muscle fiber in question.

**saltatory conduction**—"Jumping" conduction. The kind of impulse conduction found in myelinated axons, where the action potentials seem to "jump" from one node of Ranvier to the next.

**sarcolemma**—The muscle cell membrane.

**sarcomere**—The part of a myofibril lying between two neighboring Z-disks; the basic contractile unit of a striated muscle cell.

**sarcopenia**—Decreased amount of muscle tissue; muscle atrophy.

**sarcoplasm**—The cytoplasm of a muscle cell.

**sarcoplasmic reticulum**—The endoplasmic reticulum of a muscle cell, storing calcium ions.

**satellite cell**—A quiescent myoblast lying just outside the muscle cell membrane, like a satellite, between the cell membrane and the basal lamina in fully developed muscle. An important myoblast reserve in muscle repair and hypertrophy.

**SDH**—Succinate dehydrogenase.

**second wind**—Expression used to denote a period of renewed energy, originally in connection with increased uptake of blood-borne glucose in cases of myophosphorylase deficiency (McArdles disease).

**series elastic component**—Elastic elements in series with the contractile apparatus of muscle cells, partly located in tendinous tissue, partly in the cross-bridges themselves.

**series-fibered muscle**—A muscle with muscle fibers that do not span the entire length between tendon of origin and tendon of insertion.

**serotonin**—See *5-hydroxytryptamine*.

**serum iron**—Iron bound to transferrin in the blood.

**serum**—The cell and protein-free fluid after blood or plasma clots.

**slowly adapting receptor**—A receptor that only slowly loses its ability to respond to continuous stimulation.

**SMA**—See *supplementary motor area*.

**small HSPs**—A family of heat shock proteins with MW in the range of 18-30 kDa believed to be important for cytoprotection during exercise.

**smooth ER**—Smooth endoplasmic reticulum; the part of endoplasmic reticulum that is devoid of ribosomes.

**sodium-potassium pump (sodium pump)**—A membrane transporter (antiport) for sodium and potassium, often called $Na^+/K^+$ ATPase, due to its ATPase activity.

**spatial summation**—The summation of simultaneous synaptic potentials in neighboring synapses.

**spinal animal**—An animal where all connections between the brain and the spinal cord have been cut, leaving spinal reflexes as the only basis of movement.

**spinal level**—A level of the central nervous system consisting first and foremost of the spinal cord, which gives rise to motor axons innervating skeletal muscle directly.

**spirometer**—A device for measuring and recording the amount of air inhaled and exhaled.

**splanchnic**—Pertaining to or supplying the viscera (especially organs within the abdominal cavity).

**splicing**—The joining together of exons during RNA processing.

**squamous epithelial cells**—Thin, flat, scalelike epithelial cells.

**static contraction**—See *isometric contraction*.

**step test**—A method to classify people for aerobic fitness, based on the heart rate during recovery from a standardized stepping exercise.

**stitch**—A sudden, sharp, lancinating pain, often at a costal margin.

**stress proteins**—A large group of cellular proteins involved in the defense against homeostatic perturbations; includes heat shock proteins and glucose regulated proteins.

**stroke volume**—The volume of blood pumped into the aorta per heartbeat.

**SuccCoA**—Succinyl coenzyme A.

**superfamily**—A term increasingly often used to denote a group of related molecules, usually based on sequence homology on the gene level.

**superior colliculus**—Small, round protrusions on the dorsal side of the mesencephalon, one on each side of the midline, serving as centers for visual reflexes.

**supination**—Turning of the forearm until the palm of the hand faces upward.

**supplementary motor area (SMA)**—The area of the cerebral cortex anterior to the medial part of the primary motor cortex.

**supraspinal level**—A level of the central nervous system that is able to influence motor functions only indirectly by influencing motoneurons located in a spinal level.

**surfactant**—A lipid surface-tension-lowering agent.

**sympathetic nervous system**—The part of the autonomic nervous system responsible for emergency reactions.

**sympathetic trunk**—The chain of sympathetic ganglia, one on each side of the vertebral column, that serve to spread sympathetic activity to most parts of the body.

**symport**—Transmembrane transporter, carrying two different kinds of ions in the same direction at the same time.

**synaptic cleft**—The narrow, 200-nm cleft between the axon terminal and the postsynaptic cell.

**synaptic ending**—See *axon terminal*.

**synaptic potential**—See *postsynaptic potential*.

**synovial joint**—A joint where the inside of the articular capsule is lined by a synovial membrane.

**systole**—The contraction phase of the cardiac cycle.

**tachycardia**—Excessive rapidity of the heart rate.

**TBW**—See *total body water*.

**TATA box**—Part of the promoter of a gene, rich in T and A nucleotides.

**TCA**—Tricarboxylic acid.

**telemetry**—The measurement of a property (e.g., heart rate or temperature) and the transmission of the results to a distant receiving station where it is recorded.

**temporal summation**—The summation of sequential synaptic potentials.

**tendon organ**—See *Golgi tendon organ*.

**terminal cisterna**—The dilated, lateral part of a sarcoplasmic reticulum sac, adjoining the T-tubule.

**TGFβ**—See *transforming growth factor beta*.

**thalamus**—The larger part of the diencephalons, situated in the depth of each cerebral hemisphere; important relay center for sensory information to the cerebral cortex.

**thermistor**—An electrical resistance element made of material whose resistance value varies with temperature.

**thermocouple**—A device for measuring temperature; based on two dissimilar metals that form a junction that develops a thermoelectric current when heated.

**thermography**—Recording of infrared radiations spontaneously emanating from the body's surface to provide a thermogram.

**thoroughfare channel**—Blood vessel that directly connects a metarteriole with a venule, short-circuiting the capillary bed.

**thrombin**—An enzyme elaborated in shed blood by proteolytic cleavage of its inactive precursor prothrombin.

**thrombocyte**—Blood platelet.

**thrombosis**—The formation of a thrombus (a blood clot).

**titin**—A giant protein molecule, MW 3000 kD, spanning the half sarcomere from the Z-disc to the middle of the A-band.

**TMS**—See *transcranial magnetic stimulation.*

**TnC**—Troponin C, the calcium binding part of the troponin complex.

**TnI**—Troponin I, the inhibitory part of the troponin complex.

**TnT**—Troponin T, the part of the troponin complex that binds to tropomyosin.

**torque**—The measure of the effectiveness of a force in producing rotation or torsion of a body about an axis; the moment of force.

**total body water (TBW)**—The total water content of the body.

**trabecular bone**—Spongelike bone, with bone marrow in the cavities.

**trachea**—The windpipe.

**transcranial magnetic stimulation (TMS)**—A method to stimulate the cerebral cortex in living humans.

**transcription factor**—See *gene regulatory protein.*

**transcriptional control**—The control of protein synthesis at the level of transcription.

**transferrin**—A plasmaprotein (globulin) concerned with binding and transportation of iron.

**transforming growth factor beta (TGFβ)**—A group of protein growth factors with broad specificity.

**translational control**—The control of protein synthesis at the level of translation.

**transmural pressure** The pressure difference across a wall.

**triacylglycerol (triglycerides)**—Glycerol acylated by three fatty acids; constitutes the major storage form of fat in fat cells (sometimes termed "neutral fats").

**triad**—A T-tubule and two adjoining terminal cisternae of the sarcoplasmic reticulum.

**tricarboxylic acid cycle**—See *Krebs' cycle.*

**tropomyosin**—Regulatory protein attached to actin filaments.

**troponin**—Regulatory protein complex attached to actin filaments; see *TnC, TnI, and TnT.*

**T-tubules**—Transverse tubules.

**Tunica media**—The middle part of the wall of a blood vessel, composed of varying amounts of smooth muscle and elastic tissue.

**turbulent flow**—Non-laminar, non-silent flow, which occurs above a critical velocity and beyond an obstruction.

**twitch interpolation**—A technique used in the assessment of muscle fatigue, where supramaximal stimuli are applied to a muscle or its nerve during a maximal voluntary contraction.

**type IIX muscle fibers**—Muscle fiber type first described in rat muscle, where it is a third type of fast-twitch fiber. Also known as type IID or type IID/X. In man, IIX is what previously was called IIB.

**ubiquitin**—A small, highly conserved protein molecule that attaches to and marks other protein molecules destined for degradation.

**ubiquitin-proteasome system**—See *proteasome.*

**unexplained underperformance syndrome (UPS)**—Term used as an alternative to overtraining since it does not imply any reason for the situation.

**UTR**—Untranslated region. The part of a mRNA molecule not coding for a protein.

**upstream**—Allusion to the direction of flow in a river; may be used for a part of a structure or process, which has a direction (e.g., a gene or a signaling pathway); cf. downstream.

**UTP**—Uridine triphosphate.

**Valsalva maneuver**—Forcible exhalation against the closed glottis, increasing the intrathoracic pressure.

**vasoconstriction**—A decrease in the diameter of a blood vessel (usually an arteriole) resulting in a reduction of blood flow.

**vasodilation**—An increase in the diameter of a blood vessel (usually an arteriole) resulting in an increased blood flow.

**vasomotor**—Pertaining to vasoconstriction and vasodilation.

**vestibular nuclei**—A group of nuclei receiving sensory information from the vestibular apparatus through the 8th cranial nerve.

**visceral nervous system**—Autonomic nervous system.

**visceral pleura**—Pulmonary pleura; the portion of the pleura enveloping and attaching to the surface of the lung itself.

**viscosity**—The resistance that a liquid exhibits to the flow.

**voltage gated ion channel**—An ion channel that opens at a certain change in the membrane potential.

**wet-bulb temperature**—The temperature indicated when a current of air is passed over a thermometer bulb that is enclosed by a wet jacket (evaporation of water from the jacket lowers the temperature).

**white matter**—Nervous tissue dominated by myelinated axons.

**wild type (animal)**—Denotes a "normal" experimental animal, as opposed to an animal in which a gene has been manipulated (knocked out, amplified, or inserted de novo).

**Windkessel vessel**—Pertains to the elastic walls of the arteries, converting pulsatile flow to nearly continuous flow (from German *Windkessel* = elastic reservoir).

**Wingate anaerobic test**—All-out stationary test on either a leg-cycle ergometer or an arm crank with frictional resistance preset at high load and lasting for 30 s.

**Z-disk**—A platelike structure between neighboring sarcomeres in a myofibril; serves as anchoring site for the actin filaments. In electron micrographs it appears as a dark line, which is why it is sometimes called Z-line.

# References

Aagaard, P., and J.L. Andersen. 1998. Correlation between contractile strength and myosin heavy chain isoform composition in human skeletal muscle. *Med. Sci. Sports Exerc.* 30:1217–1222.

Aagaard, P., E.B. Simonsen, M. Trolle, J. Bangsbo, and K. Klausen. 1995. Isokinetic hamstring/quadriceps strength ratio: Influence from joint angular velocity, gravity correction and contraction mode. *Acta Physiol. Scand.* 154:421–427.

———. 1996. Specificity of training velocity and training load on gains in isokinetic knee joint strength. *Acta Physiol. Scand.* 156:123–129.

Abe, T., D.V. DeHoyos, M.L. Pollock, and L. Garzarella. 2000. Time course for strength and muscle thickness changes following upper and lower body resistance training in men and women. *Eur. J. Appl. Physiol. Occup. Physiol.* 81:174–180.

Abrams, C.F., Jr. and C.W. Suggs. 1977. Development of a simulator for use in the measurement of chain saw vibration. *Appl. Ergon.* 8:130–134.

Acker, H., and D.W. Lübbers. 1980. The physiology of chemoreceptors in the carotid body. In *XXIII International Congress of Physiological Sciences,* vol. 14, p. 7. Budapest.

Adams, F. 1891. *The genuine works of Hippocrates.* New York: W. Wood.

Adams, G.R. 1998. Role of insulin-like growth factor I in the regulation of skeletal muscle adaptation to increased loading. *Exerc. Sport Sci. Rev.* 26:31–60.

Adams, K.F., S.M. McAllister, S. El-Ashmaway, S. Atkinson, G. Koch, and D.S. Streps. 1992. Interrelationship between left ventricular volume and output during exercise in healthy subjects. *J. Appl. Physiol.* 75:2097–2104.

Adolph, E.F. 1949. Quantitative relations in the physiological constitution of mammals. *Science* 109:579–585.

Adolph, E.F., and Members of the Rochester Desert Unit. 1947. *Physiology of man in the desert.* New York: Interscience.

Agnevik, G. 1970a. *Badminton.* Idrottsfysiologi Rapport no. 8. Stockholm: Trygg-Hansa.

———. 1970b. *Fotball.* Idrottsfysiologi Rapport no. 7. Stockholm: Trygg-Hansa.

Agre, P., L.S. King, M. Yasui, Wm.B. Guggino, O.P. Ottersen, Y. Fujiyoshi, A. Engel, and S. Nielsen. 2002. Aquaporin water channels – from atomic structure to clinical medicine. *J. Physiol.* 542.1:3–16.

Ahlborg, B., J. Bergström, L.G. Ekelund, G. Guarnieri, R.C. Harris, E. Hultman, and L.O. Nordesjö. 1972. Muscle metabolism during isometric exercise performed at constant force. *J. Appl. Physiol.* 33:224–228.

Ahlborg, G., L. Hagenfeldt, and J. Wahren. 1976. Influence of lactate infusion on glucose and FFA metabolism in man. *Scand. J. Clin. Lab. Invest.* 36:193–201.

Ahlborg, G., R. Hendler, and P. Felig. 1981. Lactate production by resting muscle during and after prolonged exercise: Mechanism of redistribution of muscle glycogen. *Clin. Physiol.* 1:608.

Ahlman, K., and M.J. Karvonen. 1961. Stimulating of sweating by exercise after heat induced "fatigue" of the sweating mechanism. *Acta Physiol. Scand.* 53:381.

Ahmaidi, S., J. Masse-Biron, B. Adam, D. Choquet, M. Freville, J.P. Libert, and C. Prefaut. 1998. Effects of interval training at the ventilatory threshold on clinical and cardiorespiratory responses in elderly humans. *Eur. J. Appl. Physiol.* 78:170–176.

Akima, H., H. Takahashi, S.Y. Kuno, K. Masuda, T. Masuda, H. Shimojo, I. Anno, Y. Itai, and S. Katsuta. 1999. Early phase adaptations of muscle use and strength to isokinetic training. *Med. Sci. Sports Exerc.* 31:588–594.

Akiyama, Y., and Y. Kawakami. 1999. Clinical assessment of the respiratory control system. In *Control of breathing in health and disease,* ed. M.D. Altose and Y. Kawakami, pp. 251–287. New York: Marcel Dekker.

Alam, A.S., A. Gallagher, V. Shankar, M.A. Ghatei, H.K. Datta, C.L. Huang, B.S. Moonga, T.J. Chambers, S.R. Bloom, and M. Zaidi. 1992. Endothelin inhibits osteoclastic bone resorption by a direct effect on cell motility: Implications for the vascular control of bone resorption. *Endocrinology* 130:3617–3624.

Alam, A.S., B.S. Moonga, P.J.R. Bevis, C.L.-L. Huang, and M. Zaidi. 1993. Amylin inhibits bone resorption by a direct effect on the motility of rat osteoclasts. *Exp. Physiol.* 78:183–196.

Alberts, B., D. Brady, A. Johnson, J. Lewis, M. Raff, K. Roberts, and P. Walter. 1998. *Essential cell biology.* New York: Garland.

Alberts, B., D. Brady, J. Lewis, M. Raff, K. Roberts, and J.D. Watson. 1994. *Molecular biology of the cell.* 3rd ed. New York: Garland.

Alberts, B., A. Johnson, J. Lewis, M. Raff, K. Roberts, and P. Walter. 2002. *Molecular biology of the cell.* 4th ed. New York: Garland.

Albrecht, A.E., B.H. Marcus, M. Roberts, D.E. Forman, and A.F. Parisi. 1998. Effect of smoking cessation on exercise performance in female smokers participating in exercise training. *Am. J. Cardiol.* 82:950–955.

Alexander, J.K., L.H. Hartley, M. Modelski, and R.F. Grover. 1967. Reduction of stroke volume during exercise in man following ascent to 3,100 m altitude. *J. Appl. Physiol.* 23:849–858.

Alexander, R.M. 1971. *Size and shape*. Studies in Biology, no. 29. London: Edward Arnold.

Alfredson, H., P. Nordstrom, T. Pietila, and R. Lorentzon. 1998. Long-term loading and regional bone mass of the arm in female volleyball players. *Calcif. Tissue Int.* 62:303–308.

Allen, D.L., B.C. Harrison, and L.A. Leinwand. 2002. Molecular and genetic approaches to studying exercise performance and adaptation. *Exerc. Sport Sci. Rev.* 30:99–105.

Allen, D.L., S.R. Monke, R.J. Talmadge, R.R. Roy, and V.R. Edgerton. 1995. Plasticity of myonuclear number in hypertrophied and atrophied mammalian skeletal muscle fibers. *J. Appl. Physiol.* 78:1969–1976.

Allen, M. 1983. Activity-generated endorphins: A review of their role in sports science. *Can. J. Appl. Sport Sci.* 8:115–133.

Allinger, T.L., M. Epstein, and W. Herzog. 1996. Stability of muscle fibers on the descending limb of the force–length relation: A theoretical consideration. *J. Biomech.* 29:627–633.

Altose, M.D., and Y. Kawakami, eds. 1999. *Control of breathing in health and disease.* New York: Marcel Dekker.

Alward, R.R., and T.H. Monk. 1990. A comparison of rotating-shift and permanent night nurses. *Int. J. Nurs. Stud.* 27:297–302.

Alway, S.E., W.H. Grumbt, W.J. Gonyea, and J. Stray-Gundersen. 1989. Contrasts in muscle and myofibers of elite male and female bodybuilders. *J. Appl. Physiol.* 67:24–31.

Alway, S.E., W.H. Grumbt, J. Stray-Gundersen, and W.J. Gonyea. 1992. Effects of resistance training on elbow flexors of highly competitive bodybuilders. *J. Appl. Physiol.* 72:1512–1521.

American College of Sports Medicine. 1982. Position statement on the use of alcohol in sport. *Med. Sci. Sports Exerc.* 14:ix–x.

———. 1995. *Guidelines for exercise testing and prescription.* 5th ed. Baltimore: Williams and Wilkins.

———. 1996. Exercise and fluid replacement: Position stand. *Med. Sci. Sports Exerc.* 28:i–vii.

———. 1998. Exercise and physical activity for older adults. ACSM Position Stand. *Med. Sci. Sports Exerc.* 30:992–1008.

———. 2000. Nutrition and athletic performance: Joint position statement. *Med. Sci. Sports Exerc.* 32:2130–2145.

Amis, T.C., H.A. Jones, and J.M.B. Hughes. 1984. Effect of posture on inter-regional distribution of pulmonary perfusion and $V_A/Q$-ratios in man. *Respir. Physiol.* 56:169.

Andersen, J.L., T. Gruschy-Knudsen, C. Sandri, L. Larsson, and S. Schiaffino. 1999. Bed rest increases the amount of mismatched fibers in human skeletal muscle. *J. Appl. Physiol.* 86:455–460.

Andersen, J.L., H. Klitgaard, and B. Saltin. 1994. Myosin heavy chain isoforms in single fibres from m. vastus lateralis of sprinters: Influence of training. *Acta Physiol. Scand.* 151:135–142.

Andersen, J.L., and S. Schiaffino. 1997. Mismatch between myosin heavy chain mRNA and protein distribution in human skeletal muscle fibers. *Am. J. Physiol. Cell Physiol.* 272:C1881–C1889.

Andersen, J.L., P. Schjerling, and B. Saltin. 2000. Muscle, genes and athletic performance. *Sci. Am.* 283:48–55.

Andersen, J.L., G. Terzis, and A. Kryger. 1999. Increase in the degree of coexpression of myosin heavy chain isoforms in skeletal muscle fibers of the very old. *Muscle Nerve* 22:449–454.

Andersen, P. 1975. Capillary density in skeletal muscle of man. *Acta Physiol. Scand.* 95:203.

Andersen, P., and B. Saltin. 1985. Maximal perfusion of skeletal muscle in man. *J. Physiol.* 366:233–249.

Andersen, R.E., and D.L. Montgomery. 1988. Physiology of alpine skiing. *Sports Med.* 6:210–221.

Anderson, D.E., and M.S. Hickey. 1994. Effects of caffeine on the metabolic and catecholamine responses to exercise in 5 and 28 degrees C. *Med. Sci. Sports Exerc.* 26:453–458.

Anderson, G.S., G.S. Meneilly, and I.B. Mekjavic. 1996. Passive temperature lability in the elderly. *Eur. J. Appl. Physiol.* 73:278–286.

Anderson, J.M. 1999. The female athlete triad: Disordered eating, amenorrhea and osteoporosis. *Conn. Med.* 63:647–652.

Anderssen, S.A., A. Haaland, I. Hjermann, P. Urdal, K. Gjesdal, and I. Holme. 1995. Oslo Diet and Exercise Study: A one-year randomized intervention trial; effect on hemostatic variables and other coronary risk factors. *Nutr. Metab. Cardiovasc. Dis.* 5:189–200.

Andersson, B. 1967. The thirst mechanism as a link in the regulation of the "milieu intérieur." In *Les concepts de Claude Bernard sur le milieu intérieur,* p. 13. Paris: Masson.

Andersson, B., S. Houmøller, C.-G. Karlson, and S. Svensson. 1974. Analysis of urinary catecholamines: An improved auto-analyzer fluorescence method. *Clin. Chim. Acta* 51:13–28.

Angelopoulos, T.J., B.G. Denys, C. Weikart, S.G. Dasilva, T.J. Michael, and R.J. Robertson. 1995. Endogenous opioids may modulate catecholamine secretion during high intensity exercise. *Eur. J. Appl. Physiol. Occup. Physiol.* 70:195–199.

Anselme, F., K. Collomp, B. Mercies, S. Ahmaiedi, and C. Prefaut. 1992. Caffeine increases maximal anaerobic power and blood lactate concentration. *Eur. J. Appl. Physiol. Occup. Physiol.* 65:188–191.

Antonio, J., and W.J. Gonyea. 1993. Skeletal muscle fiber hyperplasia. *Med. Sci. Sports Exerc.* 25:1333–1345.

Anttonen, H. 1993. Occupational needs and evaluation methods for cold protective clothing. *Arctic Med. Res.* 52 (suppl. 9): 1–76.

Anttonen, H., and H. Virokannas. 1994. Assessment of cold stress in outdoor work. *Arctic Med. Res.* 53:40–48.

Aoyagi, Y., T.M. McLellan, and R.J. Shephard. 1997. Interactions of physical training and heat acclimation: The thermophysiology of exercising in a hot climate. *Sports Med.* 23:173–210.

Aoyagi, Y., and R.J. Shephard. 1992. Aging and muscle function. *Sports Med.* 14:376–396.

Arancibia S., F. Rage, H. Astier, and L. Tapia-Arancibia. 1998. Neuroendocrine and autonomous mechanisms underlying thermoregulation in cold environment. *Neuroendocrinology* 64:257–267.

Archambault, J.M., J.P. Wiley, and R.C. Bray. 1995. Exercise loading of tendons and the development of overuse injuries: A review of current literature. *Sports Med.* 20:77–89.

Arden, N.K., and T.D. Spector. 1997. Genetic influences on muscle strength, lean body mass and bone mineral density: A twin study. *J. Bone Miner. Res.* 12 (12):2076–2081.

Arendt-Nielsen, L., K.R. Mills, and A. Forster. 1989. Changes in muscle fiber conduction velocity, mean power frequency and mean EMG voltage during prolonged submaximal contractions. *Muscle Nerve* 12:493–497.

Ariens, G.A.M., W. van Mechelen, H.C.G. Kemper, and J.W.R. Twisk. 1997. The longitudinal development of running economy in males and females aged between 13 and 27 years: The Amsterdam Growth and Health Study. *Eur. J. Appl. Physiol. Occup. Physiol.* 76:214–220.

Armstrong, C.G., and W.L. Kenney. 1993. Effects of age and acclimation on responses to passive heat exposure. *J. Appl. Physiol.* 75:2162–2167.

Armstrong, L.E., and C.M. Maresh. 1995. Exercise-heat tolerance of children and adolescents. *Pediatr. Exerc. Sci.* 7:239–252.

Armstrong, N., and J. Welsman. 1997. *Young people and physical activity.* Oxford: Oxford University Press.

Arokoski, J.P.A., J.S. Jurvelin, U. Väätäinen, and H.J. Helminen. 2000. Normal and pathological adaptations of articular cartilage to joint loading. *Scand. J. Med. Sci. Sports* 10:186–198.

Artal, R., and C. Sherman. 1999. Exercise during pregnancy: Safe and beneficial for most. *Physician Sportsmed.* 27:51–60.

Arthur, G.D., T.S. Booker, and A.N. Belcastro. 1999. Exercise promotes a subcellular redistribution of calcium-stimulated protease activity in striated muscle. *Can. J. Physiol. Pharmacol.* 77:42–47.

Ashenden, M.J., C.J. Gore, G.P. Dobson, and A.G. Hahn. 1999. "Live high, train low" does not change the total haemoglobin mass of male endurance athletes sleeping at a simulated altitude of 3000 m for 23 nights. *Eur. J. Appl. Physiol.* 80:479–484.

Ashenden, M.J., C.J. Gore, D.T. Martin, G.P. Dobson, and A.G. Hahn. 1999. Effects of a 12-day "live high, train low" camp on reticulocyte production and haemoglobin mass in elite female road cyclists. *Eur. J. Appl. Physiol.* 80:472–478.

Ashton, C.H. 1987. Caffeine and health. *Br. Med. J.* 295:1293–1294.

Asmussen, E. 1964. Growth and athletic performance. *FIEP* 34 (4):22.

———. 1967. Exercise and regulation of ventilation. *Circ. Res.* 20:1–132.

———. 1973. Growth in muscular strength and power. In *Physical activity: Human growth and development,* ed. G.L. Rarick, pp. 60–79. New York: Academic Press.

Asmussen, E., and O. Böje. 1945. Body temperature and capacity for work. *Acta Physiol. Scand.* 10:1.

Asmussen, E., and H. Chiodi. 1941. The effect of hypoxemia on ventilation and circulation in man. *Am. J. Physiol.* 132:426.

Asmussen, E., and E.H. Christensen. 1939. Einfluss der Blutverteilung auf den Kreislauf bei körperlicher Arbeit. *Skand. Arch. Physiol.* 82:185.

———. 1967. *Kompendium i legemsöverelsernes specielle teori.* Copenhagen: Köbenhavns Universitets Fond til Tilvejebringelse af Læremidler.

Asmussen, E., E.H. Christensen, and M. Nielsen. 1939. Pulsfrequenz und Körperstellung. *Scand. Arch. Physiol.* 81:190.

Asmussen, E., W. von Döbeln, and M. Nielsen. 1948. Blood lactate and oxygen debt after exhaustive work at different oxygen tensions. *Acta Physiol. Scand.* 15:57.

Asmussen, E., O. Hansen, and O. Lammert. 1965. *The relation between isometric and dynamic muscle strength in man.* Copenhagen: Communications from the Testing and Observations Institute of the Danish National Association for Infantile Paralysis, vol. 20.

Asmussen, E., and B. Mazin. 1978. A central nervous component in local muscular fatigue. *Eur. J. Appl. Physiol. Occup. Physiol.* 38:9–15.

Åstrand, I. 1960. Aerobic work capacity in men and women with special reference to age. *Acta Physiol. Scand.* 49 (suppl. 169):67.

———. 1967a. Aerobic working capacity in men and women in some professions. *Försvars-medicin* 3:163.

———. 1967b. Degree of strain during building work as related to individual aerobic work capacity. *Ergonomics* 10:293–303.

Åstrand, I., P.-O. Åstrand, E.H. Christensen, and R. Hedman. 1960a. Intermittent muscular work. *Acta Physiol. Scand.* 48:448–453.

———. 1960b. Myohemoglobin as an oxygen-store in man. *Acta Physiol. Scand.* 48:454.

Åstrand, I., P.-O. Åstrand, I. Hallbäck, and Å. Kilbom. 1973. Reduction in maximal oxygen uptake with age. *J. Appl. Physiol.* 35 (5):649–654.

Åstrand, I., P.-O. Åstrand, and K. Rodahl. 1959. Maximal heart rate during work in older men. *J. Appl. Physiol.* 14:562–566.

Åstrand, I., I.O. Axelson, U. Eriksson, and L. Olander. 1975. Heat stress in occupational work. *AMBIO* 3:37.

Åstrand, I., P. Fugelli, C.G. Karlsson, K. Rodahl, and Z. Vokac. 1973. Energy output and work stress in coastal fishing. *Scand. J. Clin. Lab. Invest.* 31:105.

Åstrand, I., A. Guharay, and J. Wahren. 1968. Circulatory responses to arm exercise with different arm positions. *J. Appl. Physiol.* 25:528–532.

Åstrand, P.-O. 1952. *Experimental studies of physical working capacity in relation to sex and age.* Copenhagen: Munksgaard.

———. 1953. Study of bicycle modifications using a motor driven treadmill–bicycle ergometer. *Arbeitsphysiologie* 15:23.

———. 1954. The respiratory activity in man exposed to prolonged hypoxia. *Acta Physiol. Scand.* 30:343.

———. 1956. Human physical fitness with special reference to sex and age. *Physiol. Rev.* 36:307.

———. 1960. Breath holding during and after muscular exercise. *J. Appl. Physiol.* 15:220.

———. 1976. Quantification of exercise capability and evaluation of physical capacity in man. *Prog. Cardiovasc. Dis.* 19:51–67.

———. 1982. How to conduct and evaluate an exercise stress test. In *International approaches to issues in pulmonary disease,* ed. I. Kass, p. 20. International Exchange of Information Rehabilitation, Monograph 18. New York: World Rehabilitation Fund, Inc.

———. 1997. Why exercise? *Adv. Exerc. Sports Physiol.* 3:45–54.

Åstrand, P.-O., and I. Åstrand. 1958. Heart rate during muscular work in man exposed to prolonged hypoxia. *J. Appl. Physiol.* 13:75.

Åstrand, P.-O., U. Bergh, and Å. Kilbom. 1997. A 33-yr follow-up of peak oxygen uptake and related variables of former physical education students. *J. Appl. Physiol.* 82 (6):1844–1852.

Åstrand, P.-O., and E.H. Christensen. 1964. Aerobic work capacity. In *Oxygen in the animal organism,* ed. F. Dickens, E. Neil, and W.F. Widdas, p. 295. New York: Pergamon Press.

Åstrand, P.-O., T.E. Cuddy, B. Saltin, and J. Stenberg. 1964. Cardiac output during submaximal and maximal work. *J. Appl. Physiol.* 19:268.

Åstrand, P.-O., B. Ekblom, R. Messin, B. Saltin, and J. Stenberg. 1965. Intra-arterial blood pressure during exercise with different muscle groups. *J. Appl. Physiol.* 20:253.

Åstrand, P.-O., L. Engström, B. Eriksson, P. Karlberg, I. Nylander, B. Saltin, and C. Thorèn. 1963. Girl swimmers. *Acta Paediatr. Suppl.,* vol. 147.

Åstrand, P.-O., I. Hallbäck, R. Hedman, and B. Saltin. 1963. Blood lactates after prolonged severe exercise. *J. Appl. Physiol.* 18:619.

Åstrand, P.-O., E. Hultman, A. Juhlin-Dannfelt, and G. Reynolds. 1986. Disposal of lactate during and after strenuous exercise in humans. *J. Appl. Physiol.* 61:338–343.

Åstrand, P.-O., and I. Ryhming. 1954. A nomogram for calculation of aerobic capacity (physical fitness) from pulse rate during submaximal work. *J. Appl. Physiol.* 7:218.

Åstrand, P.-O., and B. Saltin. 1961a. Maximal oxygen uptake and heart rate in various types of muscular activity. *J. Appl. Physiol.* 16:977–981.

———. 1961b. Oxygen uptake during the first minutes of heavy muscular exercise. *J. Appl. Physiol.* 16:971–976.

Atha, J. 1981. Strengthening muscle. *Exerc. Sport Sci. Rev.* 9:1–73.

Atkinson, P.J., J.A. Weatherell, and S.M. Weidmann. 1962. Changes in density of human femoral cortex with age. *J. Bone Joint Surg.* 44B:496–502.

Avellini, B.A., E. Kamon, and J.T. Krajewski. 1980. Physiological responses of physically fit men and women to acclimatization to heat. *J. Appl. Physiol.* 49:254.

Avellini, B.A., Y. Shapiro, S.M. Fortney, C.B. Wenger, and K.B. Pandolf. 1982. Effects on heat tolerance of physical training in water and on land. *J. Appl. Physiol.* 53:1291.

Baar, K., E. Blough, B. Dineen, and K. Esser. 1999. Transcriptional regulation in response to exercise. *Exerc. Sport Sci. Rev.* 27:333–379.

Baar, K., and K. Esser. 1999. Phosphorylation of $p70^{S6k}$ correlates with increased skeletal muscle mass following resistance exercise. *Am. J. Physiol. Cell Physiol.* 276:C120–C127.

Bacharach, D.W., and S.P. Duvillard. 1995. Intermediate and long-term anaerobic performance in elite alpine skiers. *Med. Sci. Sports Exerc.* 27:305–309.

Backman, E., V. Johansson, B. Hager, P. Sjöblom, and K.G. Henriksson. 1995. Isometric muscle strength and muscular endurance in normal persons aged between 17 and 70 years. *Scand. J. Rehabil. Med.* 27 (2):109–117.

Bahr, R., A.T. Höstmark, E.A. Newsholme, O. Grönneröd, and O.M. Sejersted. 1991. Effect of exercise on recovery changes in plasma levels of FFA, glycerol, glucose and catecholamines. *Acta Physiol. Scand.* 143:105–115.

Bahr, R., and O.M. Sejersted. 1991. Effect of intensity of exercise on excess postexercise $O_2$ consumption. *Metabolism* 40:836–841.

Bahrke, M., and C. Yesalis. (Eds.) 2002. *Performance-Enhancing Substances in Sport and Exercise*. Champaign, IL: Human Kinetics.

Bailey, D.A., R.A. Faulkner, and H.A. McKay. 1996. Growth, physical activity and bone mineral acquisition. *Exerc. Sport Sci. Rev.* 24:233–266.

Bailey, D.M., and B. Davies. 1997. Physiological implications of altitude training for endurance performance at sea level: A review. *Br. J. Sports Med.* 31:183–190.

Bailey, D.M., B. Davies, L.R.L. Castell, E. Newsholme, and G. Gandy. 1998. Implications of moderate altitude training for sea-level endurance in elite distance runners. *Eur. J. Appl. Physiol.* 78:360–368.

Baker, A.J., K.G. Kostov, R.G. Miller, and M.W. Weiner. 1993. Slow force recovery after long-duration exercise: Metabolic and activation factors in muscle fatigue. *J. Appl. Physiol.* 74 (5):2294–3000.

Baker, M.A. 1982. Brain cooling in endotherms in heat and exercise. *Annu. Rev. Physiol.* 44:85.

Baker, S.K., K.J.A. McCullagh, and A. Bonen. 1998. Training intensity–dependent and tissue-specific increases in lactate uptake and MCT-1 in heart and muscle. *J. Appl. Physiol.* 84 (3):987–994.

Baldwin, K.M., P.J. Campbell, and D.A. Cooke. 1977. Glycogen, lactate and alanine changes in muscle during graded exercise. *J. Appl. Physiol.* 43:288–291.

Balke, B. 1954. Optimale Körperliche Leistungsfähigkeit, ihre Messung und Veränderung infolge Arbeitsermüdung. *Arbeitsphysiol.* 15:311.

Balke, B. 1960. Work capacity at altitude. In *Science and medicine of exercise and sports,* ed. W.R. Johnson, p. 339. New York: Harper and Row.

Balsom, P.D. 1995. High intensity intermittent exercise: Performance and metabolic responses with a very high intensity short duration work periods. Thesis, Karolinska Institute, Stockholm.

Bang, O. 1936. The lactate content of the blood during and after muscular exercise in man. *Skand. Arch. Physiol.* 74 (suppl. 10):51.

Bangsbo, J. 1994. The physiology of soccer. *Acta Physiol. Scand. Suppl.* 619:151.

Banister, E.W., and S.W. Brown. 1968. The relative energy requirements of physical activity. In *Exercise physiology.* New York: Academic Press.

Banister, E.W., and J. Griffiths. 1972. Blood levels of adrenergic amines during exercise. *J. Appl. Physiol.* 33:674–676.

Bär, A., and D. Pette. 1988. Three fast myosin heavy chains in adult rat skeletal muscle. *FEBS Lett.* 235:153–155.

Bárány, M. 1967. ATPase activity of myosin correlated with speed of muscle shortening. *J. Gen. Physiol.* 50 (suppl.):197–218.

Barbour, B., and M. Häusser. 1997. Intersynaptic diffusion of neurotransmitter. *Trends Neurosci.* 20:377–384.

Barcroft, H., and H.J.C. Swan. 1953. *Sympathetic control of human blood vessels.* London: Edward Arnold.

Barnett, A., and R.J. Maughan. 1993. Response of unacclimatized males to repeated weekly bouts of exercise in the heat. *Br. J. Sports Med.* 27 (1):39.

Barney, J.A., T.J. Ebert, L. Groban, P.A. Farrell, C.V. Hughes, and J.J. Smith. 1988. Carotid baroreflex responsiveness in high-fit and sedentary young men. *J. Appl. Physiol.* 65:2190–2194.

Bar-Or, O. 1987. The Wingate anaerobic test: An update on methodology, reliability and validity. *Sports Med.* 4:381–394.

———. 1993. Physical activity and physical training in childhood obesity. *J. Sports Med. Phys. Fitness* 33:323–329.

———, ed. 1996. *The child and adolescent athlete.* Champaign, IL: International Olympic Committee.

Bar-Or, O., R. Dotan, O. Inbar, A. Rothstein, and H. Zonde. 1980. Voluntary hypohydration in 10 to 12-year-old boys. *J. Appl. Physiol.* 48:104.

Bar-Or, O., and B. Wilk. 1996. Water and electrolyte replenishment in the exercising child. *Int. J. Sport Nutr.* 6:93–99.

Barrett, B. 1962. The length and mode of termination of individual muscle fibers in the human sartorius and posterior femoral muscles. *Acta Anat.* (Basel) 48:242–257.

Barron, D.H., and B.H.C. Matthews. 1935. Intermittent conduction in the spinal cord. *J. Physiol.* (Lond.) 85:73–103.

Bartlett, D., Jr. 1979. Effect of hypercapnia and hypoxia on laryngeal resistance to airflow. *Respir. Physiol.* 37:293.

Barton, P.J., and M.E. Buckingham. 1985. The myosin alkali light chain proteins and their genes. *Biochem. J.* 231:249–261.

Bärtsch, P., M. Maggiorini, W. Schobersberger, S. Shaw, W. Rascher, J. Girard, P. Weidmann, and O. Oswald. 1991. Enhanced exercise-induced rise of aldosterone and vasopressin preceding mountain sickness. *J. Appl. Physiol.* 71:136–143.

Basmajian, J.V., and C.J. De Luca. 1985. *Muscles alive: Their functions revealed by electromyography.* 5th ed. Baltimore: Williams and Wilkins.

Bass, E.E. 1963. Thermoregulatory and circulatory adjustments during acclimatization to heat in man. In *Temperature: Its measurement and control in science and industry,* ed. J.D. Hardy, vol. 3, part 3, p. 299. New York: Reinhold.

Basset, C.A. 1962. Current concepts of bone formation. *J. Bone Joint Surg.* 44A:1217–1244.

Bassett, D.R., Jr., P.W. Merrill, F.J. Nagle, J.C. Agre, and R. Sampedro. 1991. Rate of decline in blood lactate after cycling exercise in endurance-trained and untrained subjects. *J. Appl. Physiol.* 70:1816–1820.

Bawa, P. 2002. Neural control of motor output: Can training change it? *Exerc. Sport Sci Rev.* 30:59–63.

Bazett, H.C.L.L., M. Newton, L. Eisenberg, R. Day, and R. Forster. 1948. Temperature changes in blood flowing in arteries and veins in man. *J. Appl. Physiol.* 1:3.

Beard, J., and B. Tobin. 2000. Iron status and exercise. *Am. J. Clin. Nutr.* 72 (2 suppl):5594–5597.

Bechbache, R.R., and J. Duffin. 1977. The entrainment of breathing frequency by exercise rhythm. *J. Physiol.* 272:553–561.

Bedale, E.M. 1924. *Comparison of the energy expenditure of a woman carrying loads in eight different positions.* Medical Research Council Industrial Fatigue Research Board, no. 29.

Beelen, A., and A.J. Sargeant. 1991. Effect of lowered muscle temperature on the physiological response to exercise in men. *Eur. J. Appl. Physiol.* 63:387–392.

Beitz, E. 1999. The mammalian aquaporin water channel family: A promising new drug target. *Curr. Med. Chem.* 6:457–467.

Bejma, J., and L.L. Ji. 1999. Aging and acute exercise enhance free radical generation in rat skeletal muscle. *J. Appl. Physiol.* 87:465–470.

Belcastro, A.N. 1993. Skeletal muscle calcium-activated neutral protease (calpain) with exercise. *J. Appl. Physiol.* 74:1381–1386.

Belcastro, A.N., and A. Bonen. 1975. Lactic acid removal rates during controlled and uncontrolled recovery exercise. *J. Appl. Physiol.* 39:932–936.

Belding, H.S. 1973. *The industrial environment: Its evaluation and control.* Department of Health, Education and Welfare Publ. No. NIOSH. Washington, DC: U.S. Government Printing Office.

Bell, G.I., T. Kayano, J.B. Buse, C.F. Burant, J. Takeda, D. Lin, H. Fukumoto, and S. Seino. 1990. Molecular biology of mammalian glucose transporters. *Diabetes* 13:198–208.

Bemben, M.G. 1998. Age-related alterations in muscular endurance. *Sports Med.* 25:259–269.

Bencowitz, H.Z., P.D. Wagner, and J.B. West. 1982. Effect of change in $P_{50}$ on exercise tolerance at high altitude: A theoretical study. *J. Appl. Physiol. Respir. Environ. Exerc. Physiol.* 53:1487.

Benedict, F.G., and C.G. Benedict. 1933. *Mental effort in relation to gaseous exchange, heart rate and mechanics of respiration.* Publication No. 446. Carnegie Institute. Washington, DC.

Benedict, F.G., and T.M. Carpenter. 1909. *Influence of muscular and mental work on metabolism and efficiency of the human body as a machine.* U.S. Department of Agriculture, Office of Experimental Stations, Bulletin 208. Washington, DC: U.S. Department of Agriculture.

Ben-Ezra, V., and R. Verstraete. 1991. Step ergometry: Is it task-specific training? *Eur. J. Appl. Physiol.* 63:261–264.

Bennett, M.R. 1999. Synapse formation molecules in muscle and autonomic ganglia: The dual constraint hypothesis. *Prog. Neurobiol.* 57:225–287.

Benninghoff, A., and K. Goertler. 1964. *Lehrbuch der Anatomie des Menschen.* Berlin: Urban und Schwarzenberg-München.

Benzinger, T.H., and G.W. Taylor. 1963. Cranial measurements of internal temperature in man. In *Temperature: Its measurement and control in science and industry,* ed. J.D. Hardy, vol. 3, part 3, p. 111. New York: Reinhold.

Berg, H.E., and O. Eiken. 1999. Muscle control in elite alpine skiing. *Med. Sci. Sports Exerc.* 31:1065–1067.

Bergeron, M.F., C.M. Maresh, L.E. Armstrong, J.F. Signorile, J.W. Castellani, R.W. Kenefick, K.E. LaGasse, and D.A. Riebe. 1995. Fluid-electrolyte balance associated with tennis match play in a hot environment. *Int. J. Sport Nutr.* 5:180–193.

Berggren, G., and E.H. Christensen. 1950. Heart rate and body temperature as indices of metabolic rate during work. *Arbeitsphysiologie* 14:255.

Bergh, U. 1974. *Längdlöpning.* Idrottsfysiologi Rapport no. 11. Stockholm: Trygg-Hansa.

———. 1980. Human power at subnormal body temperatures. *Acta Physiol. Scand. Suppl.,* vol. 478.

———. 1985. The cycle ergometer as a tool for muscle strength measurements. In *10th International Congress of Biomechanics, June 15–20.* ed. U. Bergh and H. Robertsson, p. 27. Umeä, Sweden. Arbete och hälsa. 14.

Bergh, U., I.-L. Kanstrup, and B. Ekblom. 1976. Maximal oxygen uptake during exercise with various combinations of arm and leg work. *J. Appl. Physiol.* 41:191–196.

Berglund, B. 1992. High altitude training: Aspects of haematological adaptation. *Sports Med.* 14:289–303.

Bergström, E., O. Hernell, and L.A. Persson. 1997. Endurance running performance in relation to cardiovascular risk indicators in adolescents. *Int. J. Sports Med.* 18:300–307.

Bergström, J. 1962. Muscle electrolytes in man, determined by neutron activation analysis on needle biopsy specimens: A study on normal subjects, kidney patients and patients with chronic diarrhoea. *Scand. J. Clin. Lab Invest.* 14 (suppl. 68).

———. 1975. Percutaneous needle biopsy of skeletal muscle in physiological and clinical research. *Scand. J. Clin. Lab. Invest.* 35:609–616.

Bergström, J., G. Guarnieri, and E. Hultman. 1971. Carbohydrate metabolism and electrolyte changes in human muscle tissue during heavy work. *J. Appl. Physiol.* 30:122–125.

Bergström, J., L. Hermansen, E. Hultman, and B. Saltin. 1967. Diet, muscle glycogen and physical performance. *Acta Physiol. Scand.* 71:140–150.

Bergström, J., and E. Hultman. 1966. Muscle glycogen synthesis after exercise: An enhancing factor localized to the muscle cells in man. *Nature* 210:309–310.

Bergström, J., E. Hultman, and A.F. Roch-Norlund. 1972. Muscle glycogen synthetase in normal subjects. *Scand. J. Clin. Lab. Invest.* 29:231–236.

Bernard, T., M. Giacomoni, O. Gavarry, M. Seymat, and G. Falgairette. 1998. Time-of-day effects in maximal anaerobic leg exercise. *Eur. J. Appl. Physiol.* 77:133–138.

Bernardi, M., M. Solomonow, G. Nguyen, A. Smith, and R. Baratta. 1996. Motor unit recruitment strategy changes with skill acquisition. *Eur. J. Appl. Physiol. Occup. Physiol.* 74:52–59.

Berridge, M.J., M.D. Bootman, and P. Lipp. 1998. Calcium—a life and death signal. *Nature* 395:645–648.

Bert, P. 1878. *La pression barométrique: Recherches de physiologie expérimentale.* Paris: Masson et Cie.

Beshir, M.Y., J.D. Ramsey, and C.L. Burford. 1982. Threshold values for the Botsball: A field study of occupational heat. *Ergonomics* 25:247.

Bevan, S., and P. Geppetti. 1994. Protons: Small stimulants of capsaicin-sensitive sensory nerves. *Trends Neurosci.* 17:509–512.

Bevegård, B.S., and J.T. Shepherd. 1967. Regulation of the circulation during exercise in man. *Physiol. Rev.* 47:178.

Bhambhani, Y., and M. Singh. 1985. Metabolic and cinematographic analysis of walking and running in men and women. *Med. Sci. Sports Exerc.* 17:131–137.

Bianchi, C.P., and S. Narayan. 1982. Muscle fatigue and the role of transverse tubules. *Science* 215:295–296.

Bigland-Ritchie, B. 1981. EMG/force relations and fatigue of human voluntary contractions. *Exerc. Sport Sci. Rev.* 9:75–117.

Bigland-Ritchie, B., A.J. Fuglevand, and C.K. Thomas. 1998. Contractile properties of human motor units: Is man a cat? *Neuroscientist* 4:240–249.

Bigland-Ritchie, B., and J.J. Woods. 1984. Changes in muscle contractile properties and neural control during human muscular fatigue. *Muscle Nerve* 7:691–699.

Billat, V.L., J. Slawinski, V. Bocquet, A. Demarle, L. Lafitte, P. Chassaing, and J.P. Koralsztein. 2000. Intermittent runs at the velocity associated with maximal oxygen uptake enables subjects to remain at maximal oxygen uptake for a longer time than intense but submaximal runs. *Eur. J. Appl. Physiol.* 81:188–196.

Bilodeau, B., B. Roy, and M.R. Boulay. 1995. Effect of drafting on work intensity in classical cross-country skiing. *Int. J. Sports Med.* 16 (3):190–195.

Bilodeau, M., T.K. Henderson, B.E. Nolta, P.J. Porsley, and G.L. Sandfort. 2001. Effect of aging on fatigue characteristics of elbow flexor muscles during sustained submaximal contraction. *J. Appl. Physiol.* 91:2654–2664.

Binder-Macleod, S.A., and D.W. Russ. 1999. Effects of activation frequency and force on low-frequency fatigue in human skeletal muscle. *J. Appl. Physiol.* 86:1337–1346.

Bischoff, R. 1994. The satellite cell and muscle regeneration. In *Myology: Basic and clinical,* ed. A.G. Engel and C. Franzini-Armstrong, pp. 97–118. New York: McGraw-Hill.

Bishop, D., D.G. Jenkins, L.T. Mackinnon, M. McEniery, and M.F. Carey. 1999. The effects of strength training on endurance performance and muscle characteristics. *Med. Sci. Sports Exerc.* 31:886–891.

Bittel, J.H.M., C. Nonotte-Varly, G.H. Livecchi-Gonnot, G.L.M.J. Savourey, and A.M. Hanniquet. 1988. Physical fitness and thermoregulatory reactions in a cold environment in men. *J. Appl. Physiol.* 65:1984–1989.

Björksten, M., and B. Jonsson. 1977. Endurance limit of force in long-term intermittent static contractions. *Scand. J. Work Environ. Health* 3:23–27.

Björntorp, P. 1974. Effects of age, sex and clinical conditions on adipose tissue cellularity in man. *Metabolism* 23:1091–1102.

———. 1991. Importance of fat as a support nutrition for energy: Metabolism of athletes. *J. Sports Sci.* 9 (suppl.):71–76.

Bjurholm, A., A. Kreicbergs, M. Schultzberg, and U.H. Lerner. 1992. Neuroendocrine regulation of cyclic AMP formation in osteoblastic cell lines (UMR-106–01, ROS 17/2.8, MC3T3-E1 and Saos-2) and primary bone cells. *J. Bone Miner. Res.* 7:1011–1019.

Bjurstedt, H., G. Rosenhamer, U. Balldin, and V. Katkov. 1983. Orthostatic reactions during recovery from exhaustive exercise of short duration. *Acta Physiol. Scand.* 119:25.

Blair, L.A., and J. Marshall. 1997. IGF-1 modulates N and L calcium channels in a PI 3-kinase-dependent manner. *Neuron* 19:421–429.

Blair, S.N., and S. Brodney. 1999. Effects of physical activity and obesity on morbidity: Current evidence and research issues. *Med. Sci. Sports Exerc.* 31 (suppl.):646–662.

Blair, S.N., J.B. Kampert, H.W. Kohl, C.E. Barlow, C.A. Macera, R.S. Paffenbarger, et al. 1996. Influences of cardiorespiratory fitness and other precursors on cardiovascular disease and all-cause mortality in men and women. *JAMA* 276:205–210.

Blair, S.N., H.W. Kohl, III, C.E. Barlow, R.S. Paffenbarger, L.W. Gibbons, and C.A. Macera. 1995. Changes in physical fitness and all-cause mortality: A prospective study of healthy and unhealthy men. *JAMA* 273:1093–1098.

Blair, S.N., H.W. Kohl, R.S. Paffenbarger, D.G. Clark, R.H. Cooper, and L.W. Gibbons. 1989. Physical fitness and all-cause mortality: A prospective study of healthy men and women. *JAMA* 262:2395–2401.

Blimkie, C.J., J. Lefevre, G.P. Beunen, R. Renson, J. Dequeker, and P. Van Damme. 1993. Fractures, physical activity and growth velocity in adolescent Belgian boys. *Med. Sci. Sports Exerc.* 25:801–808.

Bliss, T.V., and T. Lømo. 1973. Long-lasting potentiation of synaptic transmission in the dentate area of the anaesthetized rabbit following stimulation of the perforant path. *J. Physiol.* (Lond.) 232:331–356.

Block, J.E., A.L. Friedlander, G.A. Brooks, P. Steiger, H.A. Stubbs, and H.K. Genant. 1989. Determinants of bone density among athletes engaged in weight-bearing and non-weight-bearing activity. *J. Appl. Physiol.* 67:1100–1105.

Blomqvist, C.G., and B. Saltin. 1983. Cardiovascular adaptations to physical training. *Annu. Rev. Physiol.* 45:169–189.

Blomqvist, G., B. Saltin, and J.H. Mitchell. 1970. Acute effects of ethanol ingestion on the response to submaximal and maximal exercise in man. *Circulation* 42:463–470.

Blomqvist, G., and J. Stenberg. 1965. The ECG response to submaximal and maximal exercise during acute hypoxia. In G. Blomqvist, The Frank lead exercise electrocardiogram. *Acta Med. Scand.* 178 (suppl. 440):82.

Blomstrand, E., and B. Ekblom. 1982. The needle biopsy technique for fibre type determination in human skeletal muscles—a methodological study. *Acta Physiol. Scand.* 116:437–442.

Bloom, W., and D.W. Fawcett. 1994. *A textbook of histology.* 12th ed. New York: Chapman and Hall.

Bobbert, A.C. 1960. Energy expenditure in level and grade walking. *J. Appl. Physiol.* 15:1015–1021.

Bobet, J., and R.W. Norman. 1982. Use of the average electro-myogram in design evaluation: Investigation of a whole-body task. *Ergonomics* 25:1155.

Bochtler, M., L. Ditzel, M. Groll, C. Hartmann, and R. Huber. 1999. The proteasome. *Annu. Rev. Biophys. Biomol. Struct.* 28:295–317.

Böje, O. 1944. Energy production, pulmonary ventilation and length of steps in well-trained runners working on a treadmill. *Acta Physiol. Scand.* 7:362.

Bojko, E.R. 1997. Metabolical changes induced by adaptation to circumpolar conditions in Spitsbergen. *Int. J. Circumpolar Health* 56:134–141.

Bolinder, G.M., B.O. Ahlborg, and J.H. Lindell. 1992. Use of smokeless tobacco: Blood pressure elevation and other health hazards found in a large scale population survey. *J. Intern. Med.* 232:327–334.

Bonaduce, D., M. Petretta, V. Cavallaro, C. Apicella, A. Ianniciello, M. Romano, R. Breglio, and F. Marciano. 1998. Intensive training and cardiac autonomic control in high level athletes. *Med. Sci. Sports Exerc.* 30:691–696.

Bonde-Petersen, F., and Y. Suzuki. 1982. Heart contractility at pressure loads induced by ischemia of exercised muscle in humans. *J. Appl. Physiol. Respir. Environ. Exerc. Physiol.* 52:340.

Bonen, A. 2000. Lactate transporters (MCT proteins) in heart and skeletal muscles. *Med. Sci. Sports Exerc.* 32:778–789.

Bonen, A., S.A. Malcolm, R.D. Kilgour, K.P. MacIntyre, and A.N. Belcastro. 1981. Glucose ingestion before and during intense exercise. *J. Appl. Physiol.* 50:766–771.

Bonen, A., B.A. Wilson, M. Yarkony, and A.N. Belcastro. 1980. Maximal oxygen uptake during free, tethered and flume swimming. *J. Appl. Physiol.* 48:232–235.

Booth, F.W. 1989. Application of molecular biology in exercise physiology. *Exerc. Sport Sci. Rev.* 17:1–27.

Booth, F.W., M.V. Chakravarthy, and E.E. Spangenburg. 2002. Exercise and gene expression: Physiological regulation of the human genome through physical activity. J. Physiol. 543.2:399–411.

Booth, F.W., and D.S. Criswell. 1997. Muscular events underlying skeletal muscle atrophy and the development of effective countermeasures. *Int. J. Sports Med.* 18 (suppl. 4):265–269.

Booth, F.W., and P.D. Gollnick. 1983. Effects of disuse on the structure and function of skeletal muscle. *Med. Sci. Sports Exerc.* 15:415–420.

Booth, F.W., S.E. Gordon, C.J. Carlson, and M.T. Hamilton. 2000. Waging war on modern chronic diseases: Primary prevention through exercise biology. *J. Appl. Physiol.* 88:774-787.

Booth, F.W., S.H. Weeden, and B.S. Tseng. 1994. Effect of aging on human skeletal muscle and motor function. *Med. Sci. Sports Exerc.* 26:556–560.

Booth, J., F. Marino, and J.J. Ward. 1997. Improved running performance in hot humid conditions following whole body precooling. *Med. Sci. Sports Exerc.* 29:943–949.

Borg, G.A. 1982. Psychophysical bases of perceived exertion. *Med. Sci. Sports Exc.* 14:377–381.

———. 1998. *Borg's perceived exertion and pain scales.* Champaign, IL: Human Kinetics.

Borg, G., M. Domserius, and L. Kaijser. 1990. Effect of alcohol on perceived exertion in relation to heart rate and blood lactate. *Eur. J. Appl. Physiol. Occup. Physiol.* 60:382–384.

Borkman, M., L.H. Storlien, D.A. Pan, A.B. Jenkins, D.J. Chisholm, and L.V. Campbell. 1993. The relation between insulin sensitivity and the fatty-acid composition of skeletal muscle phospholipids. *N. Engl. J. Med.* 328:238–244.

Bosco, C., P. Luhtanen, and P.V. Komi. 1983. A simple method for measurement of mechanical power in jumping. *Eur. J. Appl. Physiol.* 50:273–282.

Bosco, J.S., R.L. Terjung, and J.E. Greenleaf. 1968. Effects of progressive hypohydration on maximal isometric muscular strength. *J. Sports Med.* 8:81.

Botsford, J.H. 1971. A wet globe thermometer for environmental heat measurements. *Am. Ind. Hyg. Assoc. J.* 32: 1–10.

Bottinelli, R., R. Betto, S. Schiaffino, and C. Reggiani. 1994. Unloaded shortening velocity and myosin heavy chain and alkali light chain isoform composition in rat skeletal muscle fibres. *J. Physiol.* 478:341–349.

Bottinelli, R., M.A. Pellegrino, M. Canepari, R. Rossi, and C. Reggiani. 1999. Specific contributions of various muscle fibre types to human muscle performance: An in vitro study. *J. Electromyogr. Kinesiol.* 9:87–95.

Bouchard, C., ed. 2000. *Physical activity and obesity.* Champaign, IL: Human Kinetics.

Bouchard, C., P. An, T. Rice, J.S. Skinner, J.H. Wilmore, J. Gagnon, L. Pérusse, A.S. Leon, and D.C. Rao. 1999. Familial aggregation of $\dot{V}O_2$max response to exercise training: Results from the HERITAGE Family Study. *J. Appl. Physiol.* 87:1003–1008.

Bouchard, C., M.R. Boulay, F.T. Dionne, L. Pérusse, M.C. Thibault, and J.A. Simoneau. 1990. Genotype, aerobic performance and response to training. In *Children and exercise,* ed. G. Beunen, J. Chesquiere, T. Reybrouch, and A.L. Claessens. Stuttgart, Ferdinand Enke Verlag: Band 4, Schriftenreihe der Hamburg-Mannheimer-Stiftung für Informationsmedizin, pp. 124–135.

Bouchard, C., F.T. Dionne, J.A. Simoneau, and M.R. Boulay. 1992. Genetics of aerobic and anaerobic performances. *Exerc. Sport Sci. Rev.* 20:27–58.

Bouchard, C., and R.M. Malina. 1983. Genetics of physical fitness and motor performance. *Exerc. Sport Sci. Rev.* 11:306–339.

Bouchard, C., R.M. Malina, and L. Pérusse. 1997. *Genetics of fitness and physical performance.* Champaign, IL: Human Kinetics.

Bouchard, C., R.J. Shephard, and T. Stephens, eds. 1994. *Physical activity, fitness, and health: International proceedings and consensus statements.* Champaign, IL: Human Kinetics.

Bourgeois, J., D. MacDougall, J. MacDonald, and M. Tarnopolsky. 1999. Naproxen does not alter indices of muscle damage in resistance-exercise trained men. *Med. Sci. Sports Exerc.* 31:4–9.

Boutellier, U. 1998. Respiratory muscle fitness and exercise endurance in healthy humans. *Med. Sci. Sports Exerc.* 30:1169–1172.

Bove, A.A., and J.C. Davis, eds. 1990. *Diving medicine.* 2nd ed. Baltimore: Williams and Wilkins.

Bovens, A.M.P.M., M.A. van Baak, J.G.P.M. Vrencken, J.A.G. Wijnen, and F.T.J. Verstappen. 1993. Maximal heart rates and plasma lactate concentrations observed in middle-aged men and women during a maximal cycle ergometer test. *Eur. J. Appl. Physiol.* 66:281–284.

Boyle, P.M., C.D.A. Mahoney, and W.F.M. Wallace. 1994. The competitive demands of elite male field hockey. *J. Sports Med. Phys. Fitness* 34:235–241.

Boyne, A.F., T.P. Bohan, and T.H. Williams. 1975. Changes in cholinergic synaptic vesicle populations and the ultrastructure of the nerve terminal membranes of *Narcine brasiliensis* electron organ stimulated to fatigue in vivo. *J. Cell Biol.* 67:814–825.

Bramble, D.M., and D.R. Carrier. 1983. Running and breathing in mammals. *Science* 219:251.

Brandao, M.U., M. Wajngarten, E. Rondon, M.C. Giorgi, F. Hironaka, and C.E. Negrao. 1993. Left ventricular function during dynamic exercise in untrained and moderately trained subjects. *J. Appl. Physiol.* 75:1989–1995.

Brand-Saberi, B., and B. Christ. 1999. Genetic and epigenetic control of muscle development in vertebrates. *Cell Tissue Res.* 296:199–212.

Braunwald, E., E.H. Sonnenblick, J. Ross, Jr., G. Glick, and S.E. Epstein. 1967. An analysis of the cardiac response to exercise. *Circ. Res.* 20–21:44.

Brewer, J., and J. Davis. 1995. Applied physiology of rugby league. *Sports Med.* 20:129–135.

Brewer, J., C. Williams, and A. Patton. 1988. The influence of high carbohydrate diets on endurance running performance. *Eur. J. Appl. Physiol.* 57:698–706.

Brewer, N., and B. Sandow. 1980. Alcohol effects on driver performance under conditions of divided attention. *Ergonomics* 23:185–190.

Brice, A.G., and H.G. Welch. 1983. Metabolic and cardiorespiratory responses to He-$O_2$ breathing during exercise. *J. Appl. Physiol. Respir. Environ. Exerc. Physiol.* 54:387.

British Museum. 1980. *Man's Place in Evolution.* Cambridge, UK: Cambridge University Press.

Brodal, P. 1972. Leddinnervasjon – et forsømt kapittel? *Fysioterapeuten* 39:65–68.

———. 1998. *The central nervous system: Structure and function.* 2nd ed. New York: Oxford University Press.

Brodal, P., H.A. Dahl, and S. Fossum. 1990. *Menneskets anatomi og fysiologi [Human anatomy and physiology.].* Oslo: Cappelens Forlag.

Brodal, P., F. Ingjer, and L. Hermansen. 1977. Capillary supply of skeletal muscle fibers in untrained and endurance-trained men. *Am. J. Physiol.* 232:H705–H712.

Brody, S. 1945. *Bioenergetics and growth, with special reference to the efficiency complex in domestic animals.* New York: Reinhold.

Bronstein, I.P., S. Wexler, A.W. Brown, and L.J. Halpern. 1942. Obesity in childhood. *Am. J. Dis. Child.* 63:238.

Brook, C.G.D. 1972. Evidence for a sensitive period in adipose-cell replication in man. *Lancet* 2:624–627.

Brooke, M.H. and R. Miller. 1990. Fatigue testing. *Muscle Nerve* (13 Suppl.):S35–S37.

Brooke, M.H., and K.K. Kaiser. 1970a. Muscle fiber types: How many and what kind? *Arch. Neurol.* 23:369–379.

———. 1970b. Three "myosin adenosine triphosphatase" systems: The nature of their pH lability and sulfhydryl dependence. *J. Histochem. Cytochem.* 18:670–672.

Brooks, G.A. 1985. Anaerobic threshold: Review of the concept and directions for future research. *Med. Sci. Sports Exerc.* 17:22–31.

———. 2000. Intra- and extra-cellular lactate shuttles. *Med. Sci. Sports Exerc.* 32:790–799.

———. 2002. Lactate shuttle – between but not within cells? *J. Physiol.* 541.2:333.

Brooks, G.A., G.E. Butterfield, R.R. Wolfe, B.M. Groves, R.S. Mazzeo, J.R. Sutton, E.E. Wolfel, and J.T. Reeves. 1991. Decreased reliance on lactate during exercise after acclimatization to 4,300 m. *J. Appl. Physiol.* 71:333–341.

Brooks, G.A., and T.D. Fahey. 1984. *Exercise physiology: Human bioenergetics and its applications.* New York: Wiley.

Brouns, F., N.J. Rehrer, W.H.M. Saris, E. Beckers, P. Menheere, and F. ten Hoor. 1989. Effect of carbohydrate intake during warming-up on the regulation of blood glucose during exercise. *Int. J. Sports Med.* 10 (suppl.1):S68–S75.

Brouns, F., W.H.M. Saris, E. Beckers, H. Adlercreutz, G.J. van der Vusse, H.A. Keizer, H. Kuipers, P. Menheere, A.J. Wagenmakers, and F. ten Hoor. 1989. Metabolic changes induced by sustained exhaustive cycling and diet manipulation. *Int. J. Sports Med.* 10 (suppl. 1):S49–S62.

Brown, W.F., M.J. Strong, and R. Snow. 1988. Methods for estimating numbers of motor units in biceps-brachialis muscles and losses of motor units with aging. *Muscle Nerve* 11:423–432.

Bruce, R.A. 1971. Exercise testing of patients with coronary heart disease. *Ann. Clin. Res.* 3:323–332.

Bruch, H. 1940. Energy expenditure of obese children. *Am. J. Dis. Child.* 60:1082.

Brundin, T. 1980. Effects of tobacco smoking on the blood temperature during exercise. *Acta Physiol. Scand. Suppl.* 479:43–47.

Brundtland, G.H., K. Liestøl, and L. Walløe. 1980. Height, weight and menarchal age of Oslo schoolchildren during the last 60 years. *Ann. Hum. Biol.* 7:307–322.

Buchthal, F., and H. Schmalbruch. 1980. Motor unit of mammalian muscle. *Physiol. Rev.* 60:90–142.

Budd, G.M. 1962. Acclimatization to cold in Antarctica as shown by rectal temperature response to a standard cold stress. *Nature* 193:886.

Budgett, R., E. Newsholme, M. Lehmann, C. Sharp, D. Jones, T. Peto, D. Collins, R. Nerurkar, and P. White. 2000. Redefining the overtraining syndrome as the unexplained underperformance syndrome. *Br. J. Sports Med.* 34:67–68.

Buemann, B., and A. Tremblay. 1996. Effects of exercise training on abdominal obesity and related metabolic complications. *Sports Med.* (Auckland) 21:191–212.

Bulbulian, R., J.W. Jeong, and M. Murphy. 1996. Comparison of anaerobic components of the Wingate and critical power tests in males and females. *Med. Sci. Sports Exerc.* 28:1336–1341.

Buller, A.J., J.C. Eccles, and R.M. Eccles. 1960. Interactions between motor neurons and muscles in respect of the characteristic speeds of their responses. *J. Physiol.* (Lond.) 150:417–439.

Bülow, P.M., J. Nørregaard, J. Mehlsen, and B. Danneskiold-Samsøe. 1995. The twitch interpolation technique for study of fatigue of human quadriceps muscle. *J. Neurosci. Methods* 62:103–109.

Burger, E.H., and J. Klein-Nulend. 1999. Mechanotransduction in bone—role of the lacuno-canalicular network. *FASEB J.* 13 (suppl.):S101–S112.

Burke, L.M. and J.A. Hawley. 1997. Fluid balance in team sports. Guidelines for optimal practices. *Sports Med.* 24:38–54.

Burke, R.E. 1981. Motor unit recruitment: What are the critical factors? In *Motor unit types, recruitment and plasticity in health and disease,* ed. J.E. Desmedt, pp. 61–84. Basel: Karger.

Burke, R.E., D.N. Levine, P. Tsairis, and F.E. Zajac, III. 1973. Physiological types and histochemical profiles in motor units of the cat gastrocnemius. *J. Physiol.* (Lond.) 234:723–748.

Burke, R.E., D.N. Levine, F.E. Zajac, III., P. Tsairis, and W.K. Engel. 1971. Mammalian motor units: Physiological–histochemical correlation in three types in cat gastrocnemius. *Science* 174:709–712.

Burridge, K., and M. Chrzanowska-Wodnicka. 1996. Focal adhesions, contractility and signaling. *Annu. Rev. Cell Dev. Biol.* 12:463–518.

Burton, A.C. 1935. Human calorimetry: II. The average temperature of the tissues of the body. *J. Nutr.* 9:261.

———. 1963. The pattern of response to cold in animals and the evolution of homeothermy. In *Temperature: Its measurement and control in science and industry,* ed. J.D. Hardy, vol. 3, part 3, p. 363. New York: Reinhold.

Burton, A.C., and O.G. Edholm. 1969. *Man in a cold environment.* New York: Hafner.

Busa, W.B., and R. Nuccitelli. 1984. Metabolic regulation via intracellular pH. *Am. J. Physiol.* 246 (15):R409–R438.

Buskirk, E.R., P.F. Iampietro, and D.E. Bass. 1958. Work performance after dehydration: Effects of physical conditioning and heat acclimatization. *J. Appl. Physiol.* 12:189.

Buskirk, E.R., J. Kollias, R.F. Akers, B.K. Prokop, and E. Picón-Reátegui. 1967. Maximal performance at altitude and on return from altitude in conditioned runners. *J. Appl. Physiol.* 23:259–266.

Butcher, J.D., and S.J. Brannen. 1998. Comparison of injuries in classic and skating Nordic ski techniques. *Clin. J. Sport Med.* 8:88–91.

Butterfield, G.E. 1999. Nutrient requirements at high altitude. *Clin. Sports Med.* 18:607–621.

Bye, P.T.P., S.A. Esau, K.R. Walley, P.T. Macklem, and R.L. Pardy. 1984. Ventilatory muscles during exercise in air and oxygen in normal men. *J. Appl. Physiol. Respir. Environ. Exerc. Physiol.* 56:464.

Bye, P.T.P., G.A. Farkas, and C. Roussos. 1983. Respiratory factors limiting exercise. *Annu. Rev. Physiol.* 45:465.

Bylak, J., and M.R. Hutchinson. 1998. Common sports injuries in young tennis players. *Sports Med.* 26:119–132.

Cabanac, M. 1975. Temperature regulation. *Annu. Rev. Physiol.* 37:415.

———. 1986. Keeping a cool head. *News Physiol. Sci.* 1:41.

Cabanac, M., and H. Brinnel. 1988. Beards, baldness and sweat secretion. *Eur. J. Appl. Physiol.* 58:39–46.

Calbet, J.A.L., P.D. Herrera, and L.P. Rodriguez. 1999. High bone mineral density in male elite professional volleyball players. *Osteoporos. Int.* 10:468–474.

Caldwell, G.E., J.C. Jamison, and S. Lee. 1993. Amplitude and frequency measures of surface electromyography during dual task elbow torque production. *Eur. J. Appl. Physiol.* 66:349–356.

Calfas, K.J., and C. Taylor. 1994. Effects of physical activity on psychological variables in adolescents. *Pediatr. Exerc. Sci.* 6:406–423.

Calles-Eskandon, J., J.J. Cunningham, P. Snyder, R. Jacob, G. Huszar, J. Loke, and P. Felig. 1984. Influence of exercise on urea, creatinine and 3-methyl histidine excretion in normal human subjects. *Am. J. Physiol.* 246:E334–E338.

Campbell, E.J.M. 1964. Motor pathways. In *Handbook of physiology,* sec. 3, *Respiration,* ed. W.O. Fenn and H. Rahn, vol. 1, p. 535. Washington, DC: American Physiological Society.

Campione, M., S. Ausoni, C.Y. Guezennec, and S. Schiaffino. 1993. Myosin and troponin changes in rat soleus muscle after hindlimb suspension. *J. Appl. Physiol.* 74:1156–1160.

Cannon, W.B. 1929. *Bodily changes in pain, hunger, fear and rage.* New York: Appleton.

Carey, R.A. 1983. Physiological aspects of women and exercise. In *Exercise medicine,* ed. A.A. Bove and D.T. Lowenthal. New York: Academic Press.

Carithers, R.W., and R.C. Seagrave. 1976. Canine hyperthermia with cerebral protection. *J. Appl. Physiol.* 40:543.

Carlsöö, S. 1967. A kinetic analysis of the golf swing. *J. Sports Med. Phys. Fitness* 7:76–82.

Carmella, S.G., S.S. Kagan, M. Kagan, P.G. Foiles, G. Palladino, A.M. Quart, E. Quart, and S.S. Hecht. 1990. Mass spectrometric analysis of tobacco-specific nitrosamine hemoglobin adducts in snuff dippers, smokers and nonsmokers. *Cancer Res.* 50:5438–5445.

Carpenter, C.L., R.K. Ross, A. Pagani-Hill, and L. Bernstein. 1999. Lifetime exercise activity and breast cancer risk among postmenopausal women. *Br. J. Cancer* 80:1852–1858.

Carpinelli, R.N., and R.M. Otto. 1998. Strength training. Single versus multiple sets. *Sports Med.* 26:73–84.

Carro, E., A. Nuñez, S. Busiguina, and I. Torres-Aleman. 2000. Circulating insulin-like growth factor I mediates effects of exercise on the brain. *J. Neurosci.* 20:2926–2933.

Carson, J.A., and L. Wei. 2000. Integrin signaling's potential for mediating gene expression in hypertrophying skeletal muscle. *J. Appl. Physiol.* 88:337–343.

Caruana-Montaldo, B., K. Gleeson, and C.W. Zwillich. 2000. The control of breathing in clinical practice. *Chest* 117:205–225.

Casaburi, R. 1995. Mechanisms of the reduced ventilatory requirement as a result of exercise training. *Eur. Respir. Rev.* 5:42–46.

Casan, P., C.C. Villafranca, M.C. Kearon, E.J.M. Campbell, and K.J. Killian. 1997. Contribution of respiratory muscle oxygen consumption to breathing limitation and dyspnea. *Can. Respir. J.* 4:101–107.

Castellani, J.W., A.J. Young, D.W. Degroot, D.A. Stulz, B.S. Cadarette, J. Rhi Zamecnik, P.N. Shek, and M.N. Sawka. 2001. Thermoregulation during cold exposure after several days of exhaustive exercise. *J. Appl. Physiol.* 90:939–946.

Castellani, J.W., A.J. Young, J.E. Kain, A. Rouse, and M.N. Sawka. 1999. Thermoregulation during cold exposure: Effects of prior exercise. *J. Appl. Physiol.* 87:247–252.

Cathcart, E.P., and W.A. Burnett. 1926. Influence of muscle work on metabolism in varying conditions of diet. *Proc. R. Soc. (Biol.)* 99:405.

Catterall, W.A. 1997. Modulation of sodium and calcium channels by protein phosphorylation and G proteins. *Adv. Second Messenger Phosphoprotein Res.* 31:159–181.

Cavagna, G.A., H. Thys, and A. Zamboni. 1976. The sources of external work in level walking and running. *J. Physiol.* 262:639–657.

Celermajer, D.S., M.R. Adams, P. Clarkson, J. Robinson, R. McCredie, A. Donald, and J.E. Deanfield. 1996. Passive smoking and impaired endothelium-dependent arterial dilatation in healthy young adults. *N. Engl. J. Med.* 334:150–154.

Cerretelli, P. 1976. Limiting factors to oxygen transport on Mount Everest. *J. Appl. Physiol.* 40:658–667.

Cerretelli, P., and R. Margaria. 1961. Maximum oxygen consumption at altitude. *Int. Z. Angew. Physiol.* 18:460.

Chakravarthy, M.V., B.S. Davis, and F.W. Booth. 2000. IGF-I restores satellite cell proliferative potential in immobilized old skeletal muscle. *J. Appl. Physiol.* 89:1365–1379.

Chamari, K., S. Ahmaidi, C. Fabre, and J. Masse-Biron. 1995. Anaerobic and aerobic peak power output and the force–velocity relationship in endurance-trained athletes: Effects of aging. *Eur. J. Appl. Physiol. Occup. Physiol.* 71 (2/3):230–234.

Chance, B., B. Schoener, and F. Schindler. 1964. The intracellular oxidation-reduction state. In *Oxygen in the animal organism,* ed. F. Dickens and E. Neil, p. 367. New York: Pergamon Press.

Chance, B., H. Sies, and A. Boveris. 1979. Hydroperoxide metabolism in mammalian organs. *Physiol. Rev.* 59:527–605.

Chandler, T.J. 1995. Exercise training for tennis. *Clin. Sports Med.* 14:33–46.

Chapman, R.F., M. Emery, and J.M. Stager. 1999. Degree of arterial desaturation in normoxia influences $\dot{V}O_2$max decline in mild hypoxia. *Med. Sci. Sports Exerc.* 31:658–663.

Chapman, R.F., J. Stray-Gundersen, and B.D. Levine. 1998. Individual variation in response to altitude training. *J. Appl. Physiol.* 85:1448–1456.

Chase, G.A., C. Grave, and L.B. Rowell. 1966. Independence of changes in functional and performance capacities attending prolonged bed rest. *Aerospace Med.* 37:1232–1238.

Chasiotis, D. 1983. The regulation of glycogen phosphorylase and glycogen breakdown in human skeletal muscle. *Acta Physiol. Scand. Suppl.* 518:1–68.

Chatham, J.C. 2002. Lactate: The forgotten fuel. *J. Physiol.* 542(2):333.

Chauveau, A. 1896. Source et nature du potentiel directement utilisé dans le travail musculaire d'après les éxchanges respiratoires, chez l'homme en état d'abstinence. *C.R.A. Sci.* (Paris) 122:1163.

Chavarren, J., and J.A.L. Calbet. 1999. Cycling efficiency and pedalling frequency in road cyclists. *Eur. J. Appl. Physiol.* 80:555–563.

Chen, H.C., A.B. Schultz, J.A. Ashton-Miller, B. Giordani, N.B. Alexander, and K.E. Guire. 1996. Stepping over obstacles: Dividing attention impairs performance of old more than young adults. *J. Gerontol. A Biol. Sci. Med. Sci.* 51:116–122.

Chesley, A., J.D. MacDougall, M.A. Tarnopolsky, S.A. Atkinson, and K. Smith. 1992. Changes in human muscle protein synthesis after resistance exercise. *J. Appl. Physiol.* 73:1383–1388.

Chesnut, C.H. 1993. Bone mass and exercise. *Am. J. Med.* 95:34S–36S.

Chih, C.P., P. Lipton, and E.L. Roberts, Jr. 2001. Do active cerebral neurons really use lactate rather than glucose? *Trends Neurosci.* 24:573–578.

Chilibeck, P.D., D.H. Paterson, D.A. Cunningham, A.W. Taylor, and E.G. Noble. 1997. Muscle capillarization, $O_2$ diffusion distance and $\dot{V}O_2$ kinetics in old and young individuals. *J. Appl. Physiol.* 1:63–69.

Chilibeck, P.D., D.G. Sale, and C.E. Webber. 1995. Exercise and bone mineral density. *Sports Med.* 19:103–122.

Chilibeck, P.D., D.G. Syrotuik, and G.J. Bell. 1999. The effect of strength training on estimates of mitochondrial density and distribution throughout muscle fibres. *Eur. J. Appl. Physiol. Occup. Physiol.* 80:604–609.

Chin, E.R., C.D. Balnave, and D.G. Allen. 1997. Role of intracellular calcium and metabolites in low-frequency fatigue of mouse skeletal muscle. *Am. J. Physiol.* 272:C550–C559.

Chin, E.R., E.N. Olson, J.A. Richardson, Q. Yano, C. Humphries, J.M. Shelton, H. Wu, W.G. Zhu, R. Bassel-Duby, and R.S. Williams. 1998. A calcineurin-dependent transcriptional pathway controls skeletal muscle fiber type. *Genes Dev.* 12:2499–2509.

Chirico, A.M., and A.J. Stunkard. 1960. Physical activity and human obesity. *N. Engl. J. Med.* 263:935.

Chow, J.W., L.G. Carlton, Y.T. Lim, J.H. Shim, W.S. Chae, and A.F. Kuenster. 1999. Muscle activation during the tennis volley. *Med. Sci. Sports Exerc.* 31:846–854.

Chow, R.J., and J.H. Wilmore. 1984. Continuing medical education: The regulation of exercise intensity by rating perceived exertion. *J. Cardiac Rehabil.* 4:382–387.

Christen, A.G., R.K. McDaniel, and J.L. McDonald, Jr. 1990. The smokeless tobacco "time bomb." *Postgrad. Med.* 87:69–74.

Christen, A.G., J.L. McDonald Jr., B.L. Olson, and J.A. Christen. 1989. Smokeless tobacco addiction: A treat to the oral and systemic health of child and adolescent. *Pediatrician* 16:170–177.

Christensen, E.H. 1931a. Beiträge zur Physiologie schwerer körperlicher Arbeit, Die Körpertemperatur während und unmittelbar nach schwerer körperlicher Arbeit. *Arbeitsphysiologie* 4:154.

———. 1931b. Beiträge zur Physiologie schwerer körperlicher Arbeit: Minutenvolumen und Schlagvolumen des Herzens während schwerer körperlicher Arbeit. *Arbeitsphysiologie* 4:453–470.

———. 1953. Physiological valuation of work in the Nykoppa Iron Works. In *Ergonomic Society Symposium on Fatigue*, ed. W.F. Floyd and A.T. Welford, p. 93. London: Lewis.

———. 1960. Muscular work and fatigue. Chap. 9 in *Muscle as a tissue*, ed. K. Rodahl and S.M. Horvath. New York: McGraw Hill.

Christensen, E.H., and W.H. Forbes. 1937. Der Kreislauf in grossen Höhen. *Skand. Arch. Physiol.* 76:75.

Christensen, E.H., and O. Hansen. 1939. Arbeitsfähigkeit und Ehrnährung. *Skand. Arch. Physiol.* 81:160–171.

Christensen, E.H., R. Hedman, and B. Saltin. 1960. Intermittent and continuous running. *Acta Physiol. Scand.* 50:269–286.

Christensen, E.H., and A. Krogh. 1936. Fliegeruntersuchungen: Die Wirkung niedriger $O_2$ Spannung auf Höhenflieger. *Skand. Arch. Physiol.* 73:145.

Christensen, E.H., and M. Nielsen. 1942. Investigation of the circulation in the skin at beginning of muscular work. *Acta Physiol. Scand.* 4:162.

Christie, J., L.M. Sheldahl, F.E. Tristani, K.B. Sagar, M.J. Ptacin, and S. Wann. 1987. Determination of stroke volume and cardiac output during exercise: Comparison of two-dimensional and Doppler echocardiography, Fick oximetry, and thermodilution. *Circulation* 76:539–547.

Christophe, J., and J. Mayer. 1958. Effect of exercise on glucose uptake in rats and men. *J. Appl. Physiol.* 13:269–272.

Ciriello, V.M., and S.H. Snook. 1977. The prediction of WBGT from the Botsball. *Am. Ind. Hyg. Assoc. J.* 38:264–271.

Clark, A.J. 1927. *Comparative physiology of the heart.* Cambridge, England: University Press.

Clark, J.M., F.C. Hagerman, and R. Gelfand. 1983. Breathing patterns during submaximal and maximal exercise in elite oarsmen. *J. Appl. Physiol. Respir. Environ. Exerc. Physiol.* 55:440–446.

Clark, N., J. Tobin, and C. Ellis. 1992. Feeding the ultraendurance athlete: Practical tips and a case study. *J. Am. Diet. Assoc.* 92 (10):1258–1262.

Clarke, D.H. 1973. Adaptations in strength and muscular endurance resulting from exercise. *Exerc. Sport Sci. Rev.* 1:73–102.

Clarkson, P., H.E. Montgomery, M.J. Mullen, A.E. Donald, A.J. Powe, T. Bull, M, Jubb, M. World, and J.E. Deanfield. 1999. Exercise training enhances endothelial function in young men. *J. Am. Coll. Cardiol.* 33:1379–1385.

Clarkson, P.M. 1993. Nutritional ergogenic aids: Caffeine. *Int. J. Sports Nutr.* 3:103–111.

Clausen, J.P. 1976. Circulatory adjustments to dynamic exercise and effect of physical training in normal subjects and in patients with coronary artery disease. *Prog. Cardiovasc. Dis.* 18:459–495.

Cleak, M.J., and R.G. Eston. 1992. Delayed onset muscle soreness: Mechanisms and management. *J. Sports Sci.* 10:325–341.

Clement, D.B., D.R. Lloyd-Smith, J.G. MacIntyre, G.O. Matheson, R. Brock, and M. DuPont. 1987. Iron status in winter Olympic sports. *J. Sports Sci.* 5:261–271.

Cobb, F.R., M.J. Sullivan, M.B. Higginbotham, and D. Kitzman. 2000. Aging: Can the effects be prevented? In *Exercise and circulation in health and disease*, ed. B. Saltin, R. Boushel, N. Secher, and J. Mitchell. Champaign, IL: Human Kinetics.

Coderre, L., K.V. Kandror, G. Vallega, and P.F. Pilch. 1995. Identification and characterization of an exercise-sensitive pool of glucose transporters in skeletal muscles. *J. Biol. Chem.* 46:27584–27588.

Coggan, A.R., and D.L. Costill. 1984. Biological and technological variability of three anaerobic ergometer tests. *Int. J. Sports Med.* 5 (3):142–145.

Coggan, A.R., and E.F. Coyle. 1989. Metabolism and performance following carbohydrate ingestion late in exercise. *Med. Sci. Sports Exerc.* 21:59–65.

Coggan, A.R., R.J. Spina, D.S. King, M.A. Rogers, M. Brown, P.M. Nemeth, and J.O. Holloszy. 1992. Skeletal muscle adaptations to endurance training in 60- to 70-yr-old men and women. *J. Appl. Physiol.* 72:1780–1786.

Colberg, S.R., G.A. Casazza, A. Horning, and G.A. Brooks. 1994. Increased dependence on blood glucose in smokers during rest and sustained exercise. *J. Appl. Physiol.* 76:26–32.

Cole, R., P.C. Sukanek, J.B. Wittenberg, and B.A. Wittenberg. 1982. Mitochondrial function in the presence of myoglobin. *J. Appl. Physiol. Respir. Environ. Exerc. Physiol.* 53:1116.

Collardeau, M., J. Brisswalter, F. Vercruyssen, M. Audiffren, and C. Goubault. 2001. Single and choice reaction time during prolonged exercise in trained subjects: Influence of carbohydrate availability. *Eur. J. Appl. Physiol.* 86:150–156.

Collins, D.F., D. Burke, and S.C. Gandevia. 2002. Sustained contractions produced by plateau-like behaviour in human motoneurons. *J. Physiol.* 538.1:289–301.

Collins, M.A., and T.K. Snow. 1993. Are adaptations to combined endurance and strength training affected by the sequence of training? *J. Sports Sci.* 11:485–491.

Collomp, K., S. Ahmaidi, M. Audran, J.L. Chanal, and C.H. Prefaut. 1991. Effects of caffeine ingestion on performance and anaerobic metabolism during the Wingate test. *Int. J. Sports Med.* 12:439–443.

Colquhoun, W.P. 1971. Circadian variations in mental efficiency. In *Biological rhythms and human performance,* pp. 39–108. London: Academic Press.

Comroe, J.H., Jr. 1966. Some theories of the mechanism of dyspnoea. In *Breathlessness,* ed. J.B. Howell and E.J.M. Campbell, p. 1. Oxford: Blackwell Scientific.

Comuzzie, A.G., and D.B. Allison. 1998. The search for human obesity genes. *Science* 280:1374–1377.

Connett, R.J., T.E.J. Gayeski, and C.R. Honig. 1984. Lactate accumulation in fully aerobic working dog gracilis muscle. *Am. J. Physiol.* 246:H120–H128.

Consolazio, C.F., R.E. Johnson, and L.J. Pecora. 1963. *Physiological measurements of metabolic functions in man.* New York: McGraw-Hill.

Convertino, V.A., G.W. Mack, and E.R. Nadel. 1991. Elevated central venous pressure: A consequence of exercise training-induced hypervolemia? *Am. J. Physiol.* 260:R273–R277.

Cooke, R., K. Franks, G.B. Luciani, and E. Pate. 1988. The inhibition of rabbit skeletal muscle contraction by hydrogen ions and phosphate. *J. Physiol.* 395:77–97.

Cooper, K.H. 1982. *The aerobics program for total well-being.* Toronto: Bantam.

Cooper, K.H., J.G. Purdy, S.R. White, M.L. Pollack, and A.C. Linnerud. 1977. Age-fitness adjusted maximal heart rates. In *The role of exercise in internal medicine,* ed. D. Brunner and E. Jokl, p. 78. Basel: Karger.

Coote, J.H. 1995. Medicine and mechanisms in altitude sickness: Recommendations. *Sports Med.* 20:148–159.

Costill, D.L. 1970. Metabolic response during distance running. *J. Appl. Physiol.* 28:251–255.

Costill, D.L., L. Branam, D. Eddy, and W. Fink. 1974. Alterations in red cell volume following exercise and dehydration. *J. Appl. Physiol.* 37:912.

Costill, D.L., G.P. Dalsky, and W.J. Fink. 1978. Effects of caffeine ingestion on metabolism and exercise performance. *Med. Sci. Sports Exerc.* 10:155–158.

Costill, D.L., J. Daniels, W. Evans, W. Fink, G. Krahenbuhl, and B. Saltin. 1976. Skeletal muscle enzymes and fiber composition in male and female track athletes. *J. Appl. Physiol.* 40:149–154.

Costill, D.L., and W.J. Fink. 1974. Plasma volume changes following exercise and thermal dehydration. *J. Appl. Physiol.* 37:521.

Costill, D.L., W.J. Fink, L.H. Getchell, J.L. Ivy, and F.A. Witzmann. 1979. Lipid metabolism in skeletal muscle of endurance-trained males and females. *J. Appl. Physiol.* 47 (4):787–791.

Côté, C., J.A. Simoneau, P. Lagasse, M. Boulay, M.C. Thibault, M. Marcotte, and C. Bouchard. 1988. Isokinetic strength training protocols: Do they induce skeletal muscle fiber hypertrophy? *Arch. Phys. Med. Rehabil.* 69:281–285.

Cotter, J.D., M.J. Patterson, and N.A. Taylor. 1997. Sweat distribution before and after repeated heat exposure. *Eur. J. Appl. Physiol.* 76:181–186.

Cox, D.M., Z.A. Quinn, and J.C. McDermott. 2000. Cell signaling and the regulation of muscle-specific gene expression by myocyte enhancer–binding factor 2. *Exerc. Sport Sci. Rev.* 28:33–38.

Cox, S., T. Cox, M. Thirlaway, and C. Mackay. 1982. Effects of simulated repetitive work urinary catecholamine excretion. *Ergonomics* 25:1129–1141.

Coyle, E.F. 1991. Carbohydrate metabolism and fatigue. In *Muscle fatigue: Biochemical and physiological aspects,* ed. G. Atlan, L. Beliveau, and P. Bouissou., pp. 153–64. Paris: Masson.

Coyle, E.F., D.C. Feiring, T.C. Rotkis, R.W. Cote, III, F.B. Roby, W. Lee, and J.H. Wilmore. 1981. Specificity of power improvements through slow and fast isokinetic training. *J. Appl. Physiol.* 51:1437–1442.

Coyle, E.F., and M. Hamilton. 1990. Fluid replacement during exercise: Effects on physiological homeostasis and performance. In *Perspectives in exercise science and sports medicine,* ed. C.V. Gisolfi and D.R. Lamb, vol 3: *Fluid homeostasis during exercise,* pp 281–308. Carmel, IN: Benchmark.

Coyle, E.F., W.H. Martin, A.A. Ehsani, J.M. Hagberg, S.A. Bloomfield, D.R. Sinacore, and J.O. Holloszy. 1983. Blood lactate threshold in some well-trained ischemic heart disease patients. *J. Appl. Physiol.* 54:18–23.

Coyle, E.F., W.H. Martin, D.R. Sinacore, M.J. Joyner, J.M. Hagberg, and J.O. Holloszy. 1984. Time course of loss of adaptations after stopping prolonged intense endurance training. *J. Appl. Physiol.* 57:1857–1864.

Craig, A.B. 1969. Olympics, 1968: A post-mortem. *Med. Sci. Sports* 1:177.

———. 1976. Summary of 58 cases of loss of consciousness during underwater swimming and diving. *Med. Sci. Sports* 8:171–175.

Craig, R. 1994. The structure of the contractile filaments. In *Myology: Basic and clinical,* ed. A.G. Engel and C. Franzini-Armstrong, pp. 134–175. New York: McGraw-Hill.

Cratty, B.J. 1960. A comparison of fathers and sons in physical ability. *Res. Q.* 31:12–15.

Crawshaw, L.I., E.R. Nadel, J.A.J. Stolwijk, and B.A. Stanford. 1975. Effect of local cooling on sweating rate and cold sensation. *Pflügers Arch.* 354:19.

Cresswell, A.G., and A. Thorstensson. 1994. Changes in intra-abdominal pressure, trunk muscle activation and force during isokinetic lifting and lowering. *Eur. J. Appl. Physiol.* 68:315–321.

Crittenden, R.H. 1904. *Physiological economy in nutrition.* New York: Frederick A. Stokes.

Croisier, J.L., G. Camus, I. Venneman, G. Deby-Dupont, A. Juchmès-Ferir, M. Lamy, J.M. Crielaard, C. Deby, and J. Duchateau. 1999. Effects of training on exercise-induced muscle damage and interleukin 6 production. *Muscle Nerve* 22:208–212.

Crowden, G.P. 1941. Stair climbing by postmen. *The Post* (Lond.), p. 10, July 26.

Cruz, J.C., C.R. Grover, J.T. Reeves, J.T. Maher, A. Cymerman, and J.C. Denniston. 1976. Sustained venoconstriction in man supplemented with $CO_2$ at high altitude. *J. Appl. Physiol.* 40:96–100.

Cruz, J.C., L.H. Hartley, and J.A. Vogel. 1975. Effect of altitude relocations upon $AaDO_2$ at rest and during exercise. *J. Appl. Physiol.* 39:469–474.

Crystal, R.G., and J.B. West, eds. 1997. *The lung: Scientific foundations.* 2 vols. 2nd ed. Philadelphia: Lippincott-Raven.

Cumming, G.R., L. Hastman, J. McCort, and S. McCullough. 1980. High serum lactates do occur in young children after maximal work. *Int. J. Sports Med.* 1:66–69.

Cureton, K.J., and G.L. Warren. 1990. Criterion-referenced standards for youth health-related fitness test: A tutorial. *Res. Q. Exerc. Sports* 61:7–19.

Czesla, M., G. Mehlhorn, D. Fritzsche, and G. Asmussen. 1997. Cardiomyoplasty improvement of muscle fibre type transformation by an anabolic steroid (metenolone). *J. Mol. Cell Cardiol.* 29:2989–2996.

Daftari, T.K., T.E. Whitesides, Jr., J.G. Heller, A.C. Goodrich, B.E. McCarey, and W.C. Hutton. 1994. Nicotine on the revascularization of bone graft: An experimental study in rabbits. *Spine* 19:904–911.

Dahl, H.A., and O.G. Aas. 1981. The effect of activity on the normal development of rat muscle fibres. In *International series on biomechanics*, ed. A. Morecki, K. Fidelus, K. Kedzior, and A. Wit, vol. 3A, pp. 100–106. Warsaw: PWN and University Park Press.

Dahl, H.A., and E. Rinvik. 1999. *Menneskets funksjonelle anatomi. [Human functional anatomy.]* Oslo: Cappelen Akademisk Forlag.

Dalsky, G.P., K.S. Stocke, A.A. Ehsani. 1988. Weight bearing exercise training and lumbar bone mineral content in postmenopausal women. *Ann. Int. Med.* 108:824–828.

Danbolt, N.C. 2001. Glutamate uptake. *Prog. Neurobiol.* 65:1–105.

Daniel, H., C. Levenes, and F. Crepel. 1998. Cellular mechanisms of cerebellar LTD. *Trends Neurosci.* 21:401–407.

Danieli-Betto, D., E. Germinario, A. Esposito, D. Biral, and R. Betto. 2000. Effects of fatigue on sarcoplasmic reticulum and myofibrillar properties of rat single muscle fibers. *J. Appl. Physiol.* 89:891–898.

Danneskiold-Samsøe, B., V. Kofod, J. Munter, G. Grimby, P. Schnohr, and G. Jensen. 1984. Muscle strength and functional capacity in 78–81-year-old men and women. *Eur. J. Appl. Physiol.* 52:310–314.

Da Silva, A.M.T., and P. Hamosh. 1973. Effect of smoking a single cigarette on the "small airways." *J. Appl. Physiol.* 34:361–365.

Davee, A.M., C.J. Rosen, and R.A. Adler. 1990. Exercise patterns and trabecular bone density in college women. *J. Bone Miner. Res.* 5:245–250.

Davenport, H.W. 1969. *The ABC of acid–base chemistry.* 5th ed. Chicago: University of Chicago Press.

Davies, C.T.M. 1979. Thermoregulation during exercise in relation to sex and age. *Eur. J. Appl. Physiol.* 42:71

———. 1980. Effects of wind resistance on the forward motion of a runner. *J. Appl. Physiol.* 48:702–709.

———. 1981. Thermal responses to exercise in children. *Ergonomics* 24:55.

Davies, C.T.M., J.R. Brotherhood, and E. Zeidifard. 1976. Temperature regulation during severe exercise with some observations on effects of skin wetting. *J. Appl. Physiol.* 41:772.

Davies, C.T.M., and R. Rennie. 1968. Human power output. *Nature.* 217:770.

Davies, C.T.M., and A.J. Sargeant. 1974. Physiological responses to one- and two-leg exercise breathing air 45% oxygen. *J. Appl. Physiol.* 36:142.

Davies, C.T.M., and M.W. Thompson. 1979. Aerobic performance of female marathon and male ultramarathon athletes. *Eur. J. Appl. Physiol.* 41:233–245.

Davies, C.T.M., J. Wemyss-Holden, and K. Young. 1982. Maximal power output during cycling and jumping. *J. Physiol.* 322:43P–44P.

Davies, C.T.M., J. White, and K. Young. 1983. Electrically evoked and voluntary maximal isometric tension in relation to dynamic muscle performance in elderly male subjects, aged 69 years. *Eur. J. Appl. Physiol.* 51:37–43.

Davies, C.T.M., and K. Young. 1982. Maximal power output in relation to the contractile properties of the triceps surae in man. *J. Physiol.* 325:51.

Davies, M., B. Ekblom, U. Bergh, and I.-L. Kanstrup-Jensen. 1975. The effect of hypothermia on submaximal and maximal work performance. *Acta Physiol. Scand.* 95:201.

Davis, J.M., and S.P. Bailey. 1997. Possible mechanisms of central nervous system fatigue during exercise. *Med. Sci. Sports Exerc.* 29:45–57.

Dawson, B. 1994. Exercise training in sweat clothing in cool condition to improve heat tolerance. *Sports Med.* 17:233–244.

Day, R. 1949. Regional heat loss. In *Physiology of heat regulation,* ed. L.W. Newburgh, p. 240. Philadelphia: Saunders.

De Angelis, L., L. Berghella, M. Coletta, L. Lattanzi, M. Zanchi, M.G. Cusella-De Angelis, C. Ponzetto, and G. Cossu. 1999. Skeletal myogenic progenitors originating from embryonic dorsal aorta coexpress endothelial and myogenic markers and contribute to postnatal muscle growth and regeneration. *J. Cell Biol.* 147:869–877.

de Boer, R.W., and K.L. Nilsen. 1989. The gliding and push-off technique of male and female Olympic speed skaters. *Int. J. Sport Biomech.* 5:119–134.

DeGroot, M., B.M. Massie, M. Boska, J. Gober, R.G. Miller, and M.W. Weiner. 1993. Dissociation of $[H^+]$ from fatigue in human muscle detected by high time resolution $^{31}P$-NMR. *Muscle Nerve* 16:91–98.

Dejours, P. 1981. *Principles of comparative respiratory physiology.* 2nd ed. New York: Elsevier, North Holland.

Dejours, P., and S. Dejours. 1992. The effects of barometric pressure according to Paul Bert: The question today. *Int. J. Sports Med.* 13 (suppl. 1):S1–S5.

Dejours, P., R.H. Kellogg, and N. Pace. 1963. Regulation of respiration and heart rate response in exercise during altitude acclimatization. *J. Appl. Physiol.* 18:10.

de Koning, J.J., R. Thomas, M. Berger, G. de Groot, and G.J. Schenau. 1995. The start in speed skating: From running to gliding. *Med. Sci. Sports Exerc.* 27:1703–1708.

Dela, F., K.J. Mikines, J.J. Larsen, and H. Galbo. 1996. Training-induced enhancement of insulin action in human skeletal muscle: The influence of aging. *J. Gerontol. A. Biol. Sci. Med. Sci.* 51:B247–B252.

Del Maestro, R.F. 1980. An approach to free radicals in medicine and biology. *Acta Physiol. Scand. Suppl.* 492:153–168.

Delp, M.D. 1995. Effects of exercise training on endothelium-dependent peripheral vascular responsiveness. *Med. Sci. Sports Exerc.* 27:1152–1157.

De Luca, C.J., and Z. Erim. 1994. Common drive of motor units in regulation of muscle force. *Trends Neurosci.* 17:299–305.

Demirel, H.A., S.K. Powers, H. Naito, M. Hughes, and J.S. Coombes. 1999. Exercise-induced alterations in skeletal muscle myosin heavy chain phenotype: Dose–response relationship. *J. Appl. Physiol.* 86:1002–1008.

Dempsey, J.A., and H.V. Forster. 1982. Mediation of ventilatory adaptations. *Physiol. Rev.* 62:262.

Dempsey, J.A., H.V. Forster, M.L. Birnbaum, W.G. Reddan, J. Thoden, R.F. Grover, and J. Rankin. 1972. Control of exercise hyperpnea under varying durations of exposure to moderate hypoxia. *Respir. Physiol.* 16:213–231.

Dempsey, J.A., P. Hanson, and K. Henderson. 1984. Exercise induced arterial hypoxemia in healthy humans at sea-level. *J. Physiol.* (Lond.) 355:161–175.

Dempsey, J.A., E.H. Vidruk, and G.S. Mitchell. 1985. Pulmonary control systems in exercise: Update. *Fed. Proc.* 44:2260–2270.

Dempsey, J.A., and P.D. Wagner. 1999. Exercise induced arterial hypoxemia. *J. Appl. Physiol.* 87:1997–2006.

Dengler, R., R.B. Stein, and C.K. Thomas. 1988. Axonal conduction velocity and force of single human motor units. *Muscle Nerve* 11:136–145.

Denis, C., R. Fouquet, P. Poty, A. Geyssant, and J.R. Lacour. 1982. Effect of 40 weeks of endurance training on the anaerobic threshold. *Int. J. Sports Med.* 3:208–214.

Dennis, S.C., T.D. Noakes, and J.A. Hawley. 1997. Nutritional strategies to minimize fatigue during prolonged exercise: Fluid, electrolyte and energy replacement. *J. Sports Sci.* 15:305–313.

Denny, M. 1980. Locomotion: The cost of gastropod crawling. *Science* 208:1288–1290.

Dernedde, E. 1992. A correlation of the wet-bulb globe temperature and Botsball heat stress indexes for industry. *Am. Ind. Hyg. Assoc. J.* 53:169–174.

De Ruiter, C.J., P.E.M.H. Habets, A. De Haan, and A.J. Sargeant. 1996. In vivo IIX and IIB fiber recruitment in gastrocnemius muscle of the rat is compartment related. *J. Appl. Physiol.* 81:933–942.

Desmedt, J.E. 1981. The size principle of motoneuron recruitment in ballistic or ramp voluntary contractions in man. *Prog. Clin. Neurophysiol.* 9:97–136.

Desmedt, J.E., and E. Godaux. 1978. Ballistic contractions in fast or slow human muscles: Discharge patterns of single motor units. *J. Physiol.* (Lond.) 285:185–196.

———. 1981. Spinal motoneuron recruitment in man: Rank deordering with direction not with speed of voluntary movement. *Science* 214:933–936.

Desmurget, M., D. Pélisson, Y. Rossetti, and C. Prablanc. 1998. From eye to hand: Planning goal-directed movements. *Neurosci. Biobehav. Rev.* 22:761–788.

Desplanches, D., H. Hoppeler, M.T. Linossier, C. Denis, H. Claassen, D. Dormois, J.R. Lacour, and A. Geyssant. 1993. Effects of training in normoxia and normobaric hypoxia on human muscle ultrastructure. *Pflügers Arch.* 425:263–267.

Diamond, J. 1983. The biology of the wheel. *Nature* 302:572–573.

Dick, F.W. 1992. Training at altitude in practice. *Int. J. Sports Med.* 13:S203–S205.

Dickinson, E.R., M.J. Piddington, and T. Brain. 1966. Project Olympics. *Schw. Zschr. Sportmed.* 14:305–313.

Dietrichson, P., J. Coakley, P.E. Smith, R.D. Griffiths, T.R. Helliwell, and R.H. Edwards. 1987. Conchotome and needle percutaneous biopsy of skeletal muscle. *J. Neurol. Neurosurg. Psychiatry* 50:1461–1477.

Dill, D.B., ed. 1964. *Handbook of physiology,* sec. 4, *Adaptation to the environment.* Washington, DC: American Physiological Society.

DiMarco, A.F., J.R. Romaniuk, C. von Euler, and Y. Yamamoto. 1983. Immediate changes in ventilation and respiratory pattern associated with onset and cessation of locomotion in the cat. *J. Physiol.* 343:1.

Dimri, G.P., M.S. Malhotra, J. Sen Gupta, T. Sampath Kumar, and B.S. Arora. 1980. Alterations in aerobic–anaerobic proportions of metabolism during work in heat. *Eur. J. Appl. Physiol.* 45:43.

di Prampero, P.E. 1981. Energetics of muscular exercise. *Rev. Physiol. Biochem. Pharmacol.* 89:143–222.

di Prampero, P.E., U. Boutellier, and P. Pretsch. 1983. Oxygen deficit and stores at onset of muscular exercise in humans. *J. Appl. Physiol. Respir. Environ. Exerc. Physiol.* 55:146–153.

di Prampero, P.E., and P. Cerritelli. 1969. Maximal muscular power (aerobic and anaerobic) in African natives. *Ergonomics* 12:51–59.

di Prampero, P.E., G. Cortili, F. Celentano, and P. Cerretelli. 1971. Physiological aspects of rowing. *J. Appl. Physiol.* 31:851–857.

di Prampero, P.E., G. Cortili, P. Mognoni, and F. Saibene. 1976. Energy cost of speed skating and efficiency of work against air resistance. *J. Appl. Physiol.* 40:584–591.

di Prampero, P.E., D.R. Pendergast, D.W. Wilson, and D.W. Rennie. 1974. Energetics of swimming in man. *J. Appl. Physiol.* 37:1–5.

Dishman, R.K. 1997. Brain monoamines, exercise and behavioral stress: Animal models. *Med. Sci. Sports Exerc.* 29:63–74.

Döbeln, W. von. 1956a. Human standard and maximal metabolic rate in relation to fat-free body mass. *Acta Physiol. Scand.* 37 (suppl. 126).

———. 1956b. Maximal oxygen intake, body size and total hemoglobin in normal man. *Acta Physiol. Scand.* 38:193–199.

———. 1966. Kroppsstorlek, energiomsättning och kondition. Chap. 10 in *Handbok i ergonomi,* ed. G. Luthman, U. Åberg, and N. Lundgren, pp. 245–253. Stockholm: Almqvist and Wiksell.

Docherty, D., and B. Sporer. 2000. A proposed model for examining the interference phenomenon between concurrent aerobic and strength training. *Sports Med.* 30:385–394.

Dodd, S., S.K. Power, T. Callender, and E. Brooks. 1984. Blood lactate disappearance at various intensities of recovery exercise. *J. Appl. Physiol.* 57:1462–1465.

Dodd, S.L., E. Brooks, S.K. Powers, and R. Tulley. 1991. The effects of caffeine on graded exercise performance in caffeine naive versus habituated subjects. *Eur. J. Appl. Physiol.* 62:424–429.

Doherty, M. 1998. The effects of caffeine on the maximal accumulated oxygen deficit and short-term running performance. *Int. J. Sport Nutr.* 8:95–104.

Dohm, G.L., G.J. Kasperek, E.B. Tapscott, and H.A. Barakat. 1985. Protein metabolism during endurance exercise. *Fed. Proc.* 44 (2):348–352.

Dolmetsch, R.E., R.S. Lewis, C.C. Goodnow, and J.I. Healy. 1997. Differential activation of transcription factors induced by $Ca^{2+}$ response amplitude and duration. *Nature* 386:855–858.

Donelson Smith, F. and J.D. Scott. 2001. Signaling complexes: Junctions on the intracellular information super highway. *Curr. Biol.* 12:R32-R40.

Donovan, C.M., and M.J. Pagliassotti. 2000. Quantitative assessment of pathways for lactate disposal in skeletal muscle fiber types. *Med. Sci. Sports Exerc.* 32:772–777.

Doubt, T.J. 1991. Physiology of exercise in the cold. *Sports Med.* 11:367–381.

Doust, J.H., and J.M. Patrick. 1981. The limitation of exercise ventilation during speech. *Respir. Physiol.* 46:137.

Drinkwater, B.L., J.F. Bedi, A.B. Loucks, S. Roche, and S.M. Horvath. 1982. Sweating sensitivity and capacity of women in relation to age. *J. Appl. Physiol.* 53:671.

Drinkwater, B.L., J.E. Denton, L.C. Kupprat, T.S. Talag, and S.M. Horvath. 1976. Aerobic power as a factor in women's response to work in hot environment. *J. Appl. Physiol.* 41:815.

Drinkwater, B.L., S.M. Horvath, and C.L. Welles. 1975. Aerobic power of females, ages 10 to 68. *J. Gerontol.* 30:385–394.

Dubouchaud, H., G.E. Butterfield, E.E. Wolfel, B.C. Bergman, and G.A. Brooks. 2000. Endurance training, expression and physiology of LDH, MCT1 and MCT4 in human skeletal muscle. *Am. J. Physiol. Endocrinol. Metab.* 278:E571–E579.

Dubowitz, V., and M.H. Brooke. 1973. *Muscle biopsy: A modern approach.* London: Saunders.

Dubowitz, V., and A.G.E. Pearse. 1960. A comparative histochemical study of oxidative enzyme and phosphorylase activity in skeletal muscle. *Histochemie* 2:105–117.

Duchateau, J., and K. Hainaut. 1984. Isometric or dynamic training: Differential effects on mechanical properties of a human muscle. *J. Appl. Physiol.* 56:296–301.

———. 1987. Electrical and mechanical changes in immobilized human muscle. *J. Appl. Physiol.* 62:2168–2173.

Ducy, P., C. Desbois, B. Boyce, G. Pinero, B. Story, C. Dunstan, E. Smith, J. Bonadio, S. Goldstein, C. Gundberg, A. Bradley, and G. Karsenty. 1996. Increased bone formation in osteocalcin-deficient mice. *Nature* 382:448–452.

Ducy, P., T. Schinke, and G. Karsenty. 2000. The osteoblast: A sophisticated fibroblast under central surveillance. *Science* 289:1501–1504.

Duffy, S.J., G. New, B.T. Tran, R.W. Harper, and I.T. Meredith. 1999. Relative contribution of vasodilator prostanoids and NO to metabolic vasodilation in the human forearm. *Am. J. Physiol.* 276:H663–H670.

Duggan, A., and S.D. Tebutt. 1990. Blood lactate at 12 km/h and vOBLA as predictors of run performance in non-endurance athletes. *Int. J. Sports Med.* 11(2):111–115.

Dukes-Dobos, F.N. 198l. Hazards of heat exposure. *Scand. J. Work Environ. Health* 7:73.

Dunn, A.L., and S.N. Blair. 1997. Exercise prescription. In *Physical activity and mental health*, ed. W.P. Morgan. Washington, DC: Taylor and Francis.

Dunn, A.L., and R.K. Dishman. 1991. Exercise and the neurobiology of depression. *Exerc. Sport Sci. Rev.* 19: 41–98.

Dunn, S.E., J.L. Burns, and R.N. Michel. 1999. Calcineurin is required for skeletal muscle hypertrophy. *J. Biol. Chem.* 274:21908–21912.

Durnin, J.V.G.A. 1967. Activity patterns in the community. *Can. Med. Ass. J.* 96:882.

———. 1982. Muscle in sports medicine—nutrition and muscular performance. *Int. J. Sports Med.* 3 (suppl. 1): 52–57.

Durnin, J.V.G.A., and R. Passmore. 1967. *Energy, work and leisure.* London: William Heinemann.

du Toit, P., G. Sole, P. Bowerbank, and T.D. Noakes. 1999. Incidence and causes of tenosynovitis of the wrist extensors in long distance paddle canoeists. *Br. J. Sports Med.* 33:105–109.

Dyer, K. 1984. Catching up the men, Olympics '84. *New Scientist* 103:25–26.

Eastman, C.I., L. Liv, and L.F. Fogg. 1995. Circadian rhythm adaptation to simulated night shift work: Effect of nocturnal bright-light duration. *Sleep* 18:399–407.

Eaton, S.B., and M. Konner. 1985. Paleolithic nutrition. *N. Engl. J. Med.* 312:283.

Eble, D.M., M.L. Spragia, A.G. Ferguson, and A.M. Samarel. 1999. Sarcomeric myosin heavy chain is degraded by the proteasome. *Cell Tissue Res.* 296:541–548.

Eccles, J.C. 1957. *The physiology of nerve cells.* Baltimore: Johns Hopkins.

———. 1973a. The cerebellum as a computer: Patterns in space and time. *J. Physiol.* (Lond.) 229:1–32.

———. 1973b. *The understanding of the brain.* New York: McGraw-Hill.

———. 1982. The synapse: From electrical to chemical transmission. *Annu. Rev. Neurosci.* 5:325–339.

Eckermann, P., and H.P. Millahn. 1967. Der Einfluss der Drehzahl auf die Herzfrequenz und die Sauerstoffaufnahme bei konstanter Leistung am Fahrradergometer. *Int. Z. Angew. Physiol. Einschl. Arbeitsphysiol.* 23:340–344.

Eden, B.D., and P.J. Abernethy. 1994. Nutritional intake during an ultraendurance running race. *Int. J. Sport Nutr.* 4:166–174.

Edgerton, V.R., R.R. Roy, R.J. Gregor, and S. Rugg. 1986. Morphological basis of skeletal muscle power output. In *Human muscle power*, ed. N.L. Jones, N. McCartney, and A.J. McComas, pp. 43–64. Champaign, IL: Human Kinetics.

Edholm, O.G. and Y.S. Weiner. 1981. *The principles and practice of human physiology.* London. Academic Press.

Edman, K.A.P. 1992. Contractile performance of skeletal muscle fibres. In *Strength and power in sport*, ed. P.V. Komi, pp. 96–114. Oxford: Blackwell Science.

———. 1995. Myofibrillar fatigue versus failure of activation. *Adv. Exp. Med. Biol.* 384:29–43.

Edman, K.A.P., and F. Lou. 1990. Changes in force and stiffness induced by fatigue and intracellular acidification in frog muscle fibres. *J. Physiol.* 424:133–149.

Edström, L., and E. Kugelberg. 1968. Histochemical composition, distribution of fibres and fatiguability of single motor units: Anterior tibial muscle of the rat. *J. Neurol. Neurosurg. Psychiatry* 31:424–433.

Edwards, H.T. 1936. Lactic acid in rest and work at high altitude. *Am. J. Physiol.* 116:367.

Edwards, H.T., L. Brouha, and R.T. Johnson. 1939. Effets de l'entrainement sur le taux de l'acide lactique sanguin au cours du travail musculaire. *Travail Humain* 8:1–9.

Edwards, M.R., G.S. Hunte, A.S. Belzberg, A.W. Sheel, D.F. Worsley, and D.C. McKenzie. 2000. Alveolar epithelial integrity in athletes with exercise-induced hypoxemia. *J. Appl. Physiol.* 89:1537–1542.

Edwards, R.H.T., D.K. Hill, D.A. Jones, and P.A. Merton. 1977. Fatigue of long duration in human skeletal muscle after exercise. *J. Physiol.* (Lond.) 272:769–778.

Edwards, R.H.T., D.K. Hill, and M. McDonnell. 1972. Myothermal and intramuscular pressure measurements during isometric contractions of the human quadriceps muscle. *J. Physiol.* 224:58P–59P.

Edwards, R.H.T., V. Toescu, and H. Gibson. 1995. Historical perspective: A framework for interpreting pathobiological ideas on human muscle fatigue. *Adv. Exp. Med. Biol.* 384:481–494.

Effros, R.M., G. Mason, and P. Silverman. 1981. Assymetric distribution of anhydrase in the alveolar–capillary barrier. *J. Appl. Physiol. Respir. Environ. Exerc. Physiol.* 51:190.

Eichner, E.R. 1990. Hematuria—a diagnostic challenge. *Physician Sportsmed.* 18:53–63.

Eisalo, A. 1956. Effects of the Finnish sauna on circulation. Annales Medicinae Experimentalis et Biologicae Finniae. 34:(Suppl. 4).

Eisenberg, E., and G.S. Gordan. 1961. Skeletal dynamics in man measured by nonradioactive strontium. *J. Clin. Invest.* 40:1809–1825.

Ekblom, B. 1969. Effect of physical training on oxygen transport system in man. *Acta Physiol. Scand. Suppl.*, vol. 328.

———. 1986. Applied physiology of soccer. *Sports Med.* 3:50–60.

Ekblom, B., P.-O. Åstrand, B. Saltin, J. Stenberg, and B. Wallström. 1968. Effect of training on circulatory response to exercise. *J. Appl. Physiol.* 24:518–528.

Ekblom, B., and B. Berglund. 1991. Effect of erythropoietin administration on maximal aerobic power in man. *Scand. J. Med. Sci. Sports* 1:125–130.

Ekblom, B., and A.N. Goldbarg. 1971. The influence of physical training and other factors on the subjective rating of perceived exertion. *Acta Physiol. Scand.* 83:399–406.

Ekblom, B., A.N. Goldbarg, and B. Gullbring. 1972. Response to exercise after blood loss and reinfusion. *J. Appl. Physiol.* 33:175.

Ekblom, B., A.N. Goldbarg, Å. Kilbom, and P.-O. Åstrand. 1972. Effects of atropine and propranolol on the oxygen transport system during exercise in man. *Scand. J. Clin. Lab. Invest.* 30:35.

Ekblom, B., and L. Hermansen. 1968. Cardiac output in athletes. *J. Appl. Physiol.* 25:619.

Ekblom, B., L. Hermansen, and B. Saltin. 1967. *Hastighetsåkning på skridsko.* Idrottsfysiologi Rapport no. 5. Stockholm: Framtiden.

Ekblom, B., R. Huot, E.M. Stein, and A.T. Torstensson. 1975. Effect of changes in arterial oxygen content on circulation and physical performance. *J. Appl. Physiol.* 39:71.

Ekblom, B., A. Kilbom, and J. Soltysiak. 1973. Physical training, bradycardia and autonomic nervous system. *Scand. J. Clin. Lab. Invest.* 32:251–256.

Ekblom, B., G. Wilson, and P.-O. Åstrand. 1976. Central circulation during exercise after venesection and reinfusion of red blood cells. *J. Appl. Physiol.* 40:379–383.

Ekelund, L.G., W.L. Haskell, J.L. Johnson, F.S. Whaley, M.H. Criqui, and D.S. Sheps. 1988. Physical fitness as a predictor of cardiovascular mortality in asymptomatic North American men. The Lipid Research Clinic's mortality follow-up study. *N. Engl. J. Med.* 319:1379–1384.

Eken, T. 1999. Tonic motor activity in a physiological extensor muscle. Thesis, University of Oslo.

Eken, T., and O. Kiehn. 1989. Bistable firing properties of soleus motor units in unrestrained rats. *Acta Physiol. Scand.* 136:383–394.

Ekholm, J., U.P. Arborelius, and G. Németh. 1982. The load on the lumbo-sacral joint and trunk muscle activity during lifting. *Ergonomics* 25:145.

Eldridge, L.F., D.E. Millhorn, and T.G. Waldrop. 1981. Exercise hyperpnea and locomotion: Parallel activation from the hypothalamus. *Science* 211 (4484):844.

Ellestad, M. 1995. *Stress testing: Principles and practice.* 4th ed. Philadelphia: F.A. Davis.

Ellis, H.D. 1982. The effects of cold on the performance of serial choice reaction time and various discrete tasks. *Human Factors* 24 (5):589.

El-Sayed, M.S., R.J.M. Angelheart, J.M. Rattu, X. Lin, and T. Reilly. 1996. Effects of active warm-down and carbohydrate feeding on free fatty acid concentrations after prolonged submaximal exercise. *Int. J. Sport Nutr.* 6:337–347.

Enander, A., B. Sköldström, and I. Holmér. 1980. Reaction to hand-cooling in workers occupationally exposed to cold. *Scand. J. Work Environ. Health* 6:58.

Engel, A.G., and C. Franzini-Armstrong, eds. 1994. *Myology: Basic and clinical.* Vol. 1. New York: McGraw-Hill.

English, A.W., S.L. Wolf, and R.L. Segal. 1993. Compartmentalization of muscles and their motor nuclei: The partitioning hypothesis. *Phys. Ther.* 73:857–867.

Ennion, S., J. Sant'ana Pereira, A.J. Sargeant, A. Young, and G. Goldspink. 1995. Characterization of human skeletal muscle fibres according to the myosin heavy chains they express. *J. Muscle Res. Cell Motil.* 16:35–43.

Enoka, R.M. 1996. Eccentric contractions require unique activation strategies by the nervous system. *J. Appl. Physiol.* 81:2339–2346.

———. 1997. Neural adaptations with chronic physical activity. *J. Biomech.* 30:447–455.

Epstein, Y., G. Keren, R. Udassin, and Y. Shapiro. 1981. Way of life as a determinant of physical fitness. *Eur. J. Appl. Physiol.* 47:1–5.

Erikssen, G., K. Liestøl, J. Bjørnholt, E. Thaulow, L. Sandvik, and J. Erikssen. 1998. Changes in physical fitness and changes in mortality. *Lancet* 352:759–762.

Erikssen, J., and K. Rodahl. 1979. Seasonal variation in work performance and heart rate response to exercise. *Eur. J. Appl. Physiol.* 42:133–140.

Eriksson, B.O. 1972. Physical training, oxygen supply and muscle metabolism in 11–13-year-old boys. *Acta Physiol. Scand. Suppl.* 384:1–48.

Eriksson, B.O., A. Lundin, and B. Saltin. 1975. Cardiopulmonary function in former girl swimmers and the effects of physical training. *Scand. J. Clin. Lab. Invest.* 35:135.

Eriksson, E., E. Nygaard, and B. Saltin. 1977. Physiological demands in downhill skiing. *Physician Sportsmed.* 5:29–37.

Ernst, E., J.I. Rand, and C. Stevinson. 1998. Complementary therapies for depression: An overview. *Arch. Gen. Psychiatry* 55:1026–1032.

Ernster, V.L., D.G. Grady, J.C. Greene, M. Walsh, P. Robertson, P. Daniels et al.1990. Smokeless tobacco use and health effects among baseball players. *JAMA* 264:218–224.

Erp-Baart van, A.M.J. 1992. Food habits in athletes. Thesis, University of Nijmegen, The Netherlands.

Erslev, A.J. 1980. Blood and mountains. In *Blood, pure and eloquent,* ed. M.M. Wintrobe. New York: McGraw-Hill.

Escher, S.A., A.M. Tucker, T.M. Lundin, and M.D. Grabiner. 1998. Smokeless tobacco, reaction time and strength in athletes. *Med. Sci. Sports Exerc.* 30:1548–1551.

Essén, B. 1978. Studies on the regulation of metabolism in human skeletal muscle using intermittent exercise as an experimental model. *Acta Physiol. Scand. Suppl.,* vol. 454.

Essig, D., D.L. Costill, and P.J. Van Handel. 1980. Effects of caffeine ingestion on utilization of muscle glycogen and lipid during leg ergometer cycling. *Int. J. Sports Med.* 1:86–90.

Essig, D.A. 1996. Contractile activity–induced mitochondrial biogenesis in skeletal muscle. *Exerc. Sport Sci. Rev.* 24:289–319.

Essig, D.A., and T.M. Nosek. 1997. Muscle fatigue and induction of stress protein genes: A dual function of reactive oxygen species? *Can. J. Appl. Physiol.* 22:409–428.

Etnyre, B.R., and L.D. Abraham. 1988. Antagonist muscle activity during stretching: A paradox re-assessed. *Med. Sci. Sports Exerc.* 20:285–289.

Euler, C. von. 1974. On the role of proprioceptors in perception and execution of motor acts with special reference to breathing. In *Loaded breathing,* ed. L.D. Pengelly, A.S. Rebuck, and E.J.M. Campbell, p. 139. Ontario: Longman Canada.

———. 1983. On the central pattern generator for the basic breathing rhythmicity. *J. Appl. Physiol. Respir. Environ. Exerc. Physiol.* 55:1647.

Euler, U.S. von. 1964. Quantification of stress by catecholamine analysis. *Clin. Pharm. Ther.* 5:398–404.

Euler, U.S. von, and S. Hellner. 1952. Excretion of noradrenaline and adrenaline in muscular work. *Acta Physiol. Scand.* 26:183–191.

Euler, U.S. von, and F. Lishajko. 1961. Improved technique for the fluorimetric estimation of catecholamines. *Acta Physiol. Scand.* 51:348–356.

Evans, B.W., and K.J. Cureton. 1983. Effect of physical conditioning on blood lactate disappearance after supramaximal exercise. *Br. J. Sports Med.* 17:40–45.

Evans, J.A., and H.A. Quinney. 1981. Determination of resistance settings for anaerobic power testing. *Can. J. Appl. Sports Sci.* 6 (2):53–56.

Evans, J.W. 1999. Exercise guidelines for the elderly. *Med. Sci. Sport Exerc.* 31:12–17.

Evans, O.M., Y. Zerbib, M.H. Faria, and H. Monod. 1983. Physiological responses to load holding and load carriage. *Ergonomics* 26:161–171.

Evans, W.J. 1995. Exercise, nutrition and aging. *Clin. Geriatric Med.* 11:725–734.

Evarts, E.V. 1973. Motor cortex reflexes associated with learned movement. *Science* 179:501–503.

Eyzaguirre, C., and S.J. Fidone. 1980. Transduction mechanisms in carotid body: Glomus cells, putative neurotransmitters and nerve endings. *Am. J. Physiol.* 239:C135.

Faber, J.E., P.D. Harris, and F.N. Miller. 1982. Microvascular sensitivity to PGE2 and PGI2 in skeletal muscle of decerebrated rat. *Am. J. Physiol.* 243:H844.

Fabiato, A., and F. Fabiato. 1978. Effects of pH on the myofilaments and the sarcoplasmic reticulum of skinned cells from cardiac and skeletal muscles. *J. Physiol.* 276:233–255.

Fagard, R.H. 1995. The role of exercise in blood pressure control: Supportive evidence. *J. Hypertens.* 13:1223–1227.

Falk, B., O. Bar-Or, J. Smolander, and G. Frost. 1994. Response to rest and exercise in the cold: Effects of age and aerobic fitness. *J. Appl. Physiol.* 76:72–78.

Fallentin, N., K. Jørgensen, and E.B. Simonsen. 1993. Motor unit recruitment during prolonged isometric contractions. *Eur. J. Appl. Physiol.* 67:335–341.

Fallowfield, J.L., C. Williams, and R. Singh. 1995. The influence of ingesting a carbohydrate–electrolyte beverage during four hours of recovery on subsequent endurance capacity. *Int. J. Sport Nutr.* 5:385–399.

Fambrough, D.M., B.A. Wolitzky, M.M. Tamkim, and K. Takeyasu. 1987. Regulation of the sodium pump in excitable cells. *Kidney Int.* 32:S97–S112.

Farfan, H.F. 1973. *Mechanical disorders of the low back.* Philadelphia: Lea and Febiger.

Faria, I.E. 1992. Energy expenditure, aerodynamics and medical problems in cycling: An update. *Sports Med.* 14:43–63.

Farrell, P.A., J.H. Wilmore, E.F. Coyle, J.E. Billing, and D.L. Costill. 1979. Plasma lactate accumulation and distance running performance. *Med. Sci. Sports Exerc.* 11:338–344.

Farrell, S.T., J.B. Kampert, H.W. Kohl, C.E. Barlow, C. Macera, R.S. Paffenbarger, et al. 1998. Influences of cardiorespiratory fitness levels or other predictors on cardiovascular disease mortality in men. *Med. Sci. Sports. Exerc.* 30:899–905.

Faulkner, J.A., and S.V. Brooks. 1995. Muscle fatigue in old animals: Unique aspects of fatigue in elderly humans. *Adv. Exp. Med. Biol.* 384:471–480.

Faulkner, J.A., D.R. Claflin, and K.K. McCully. 1986. Power output of fast and slow fibers from human skeletal muscles. In *Human muscle power,* ed. N.L. Jones, N. McCartney, and A.J. McComas, pp. 81–94. Champaign, IL: Human Kinetics.

Favero, T.G. 1999. Sarcoplasmic reticulum $Ca^{2+}$ release and muscle fatigue. *J. Appl. Physiol.* 87:471–483.

Favier, R., D. Desplanches, J. Frutoso, M. Grandmontagne, and R. Flandrois. 1983. Ventilatory transients during exercise: Peripheral or central control? *Pflügers Arch.* 396:269.

Febbraio, M., R.J. Snow, M. Hargreaves, C.G. Stathis, I.K. Martin, and M.F. Carey. 1994. Muscle metabolism during exercise and heat stress in trained men: Effect of acclimation. *J. Appl. Physiol.* 76:589–597.

Feigl, E.O. 1983. Coronary physiology. *Physiol. Rev.* 63:1.

Fenn, C.E., J.B. Leiper, I.M. Light, and K.O. Maughan. 1983. Effects of the oral administration of fluid, electrolytes and substrate on endurance capacity in man. *J. Physiol.* 341:66P.

Fenn, W.O., and B.S. Marsh. 1935. Muscular force at different speed of shortening. *J. Physiol.* 85:277–297.

Ferrell, W.R., S.C. Gandevia, and D.I. McCloskey. 1987. The role of joint receptors in human kinaesthesia when intramuscular receptors cannot contribute. *J. Physiol.* (Lond.) 386:63–71.

Ferretti, G. 1992. Cold and muscle performance. *Int. J. Sports Med.* 13:S185–S187.

Ferris, D.P., K.L. Liang, and C.T. Farley. 1999. Runners adjust leg stiffness for their first step on a new running surface. *J. Biomech.* 32:787–794.

Ferrus, L., D. Commenges, J. Gire, and P. Varene. 1984. Respiratory water loss as a function of ventilation and environmental factors. *Respir. Physiol.* 56:11.

Fiatarone, M.A., E.C. Marks, N.D. Ryan, C.N. Meredith, L.A. Lipsitz, and W.J. Evans. 1990. High-intensity strength training in nonagenarians: Effects on skeletal muscle. *JAMA* 263:3029–3034.

Fiatarone, M.A., E.F. O'Neill, N.D. Ryan, K.M. Clements, G.R. Solares, M.E. Nelson, S.B. Roberts, J.J. Kehayias, L.A. Lipsitz, and W.J. Evans. 1994. Exercise training and nutritional supplementation for physical frailty in very elderly people. *N. Engl. J. Med.* 330:1769–1775.

Finch, C.F., and H.L. Kelsall. 1998. The effectiveness of ski bindings and their professional adjustment for preventing alpine skiing injuries. *Sports Med.* 25:407–416.

Finkel, T. 1999. Myocyte hypertrophy: The long and winding RhoA'd. *J. Clin. Invest.* 103:1619–1620.

Fischer, K.J. 2000. Biological response to forces acting in the locomotor system. In *Biomechanics and biology of movement*, ed. B.M. Nigg, B.R. MacIntosh, and J. Mester, pp. 307–329. Champaign, IL: Human Kinetics.

Fitts, R.H. 1994. Cellular mechanisms of muscle fatigue. *Physiol. Rev.* 74:49–94.

Fletcher, W.M., and G. Hopkins. 1907. Lactic acid in amphibian muscle. *J. Physiol.* (Lond.) 35:247–309.

Floyd, W.F., and P.H.S. Silver. 1955. The function of the erectores spinae muscles in certain movements and postures in man. *J. Physiol.* (Lond.) 129:184–203.

Fogelholm, M. 1995. Indicators of vitamin and mineral status in athletes' blood: A review. *Int. J. Sport Nutr.* 5:267–284.

Fogelholm, M., S. Mannisto, E. Vartiainen, and P. Pietinen. 1996. Determinants of energy balance and overweight in Finland 1982 and 1992. *Int. J. Obesity* 20:1097–1104.

Folinsbee, L.J., E.S. Wallace, J.A. Bedi, J.A. Gliner, and S.M. Horvath. 1982. Exercise respiratory pattern in elite athletes and sedentary subjects. *Am. Rev. Respir. Dis.* 125:240.

Folk, G.E., Jr. 1974. *Textbook of environmental physiology.* 2nd ed. Philadelphia: Lea and Febiger.

Folkow, B., U. Haglund, M. Jodahl, and O. Lundgren. 1971. Blood flow in the calf muscle of man during heavy rhythmic exercise. *Acta Physiol. Scand.* 81:157.

Folkow, B., and E. Neil. 1971. *Circulation*. London: Oxford University Press.

Follenius, M., G. Brandenberger, C. Lecornu, M. Simeoni, and B. Reinhardt. 1980. Plasma catecholamines and pituitary adrenal hormones in response to noise exposure. *Eur. J. Appl. Physiol.* 43:253–261.

Fonnum, F., J. Storm-Mathisen, and F. Walberg. 1970. Glutamate decarboxylase in inhibitory neurons: A study of the enzyme in Purkinje cell axons and boutons in the cat. *Brain Res.* 20:259–275.

Fordtran, J.S., and B. Saltin. 1967. Gastric emptying and intestinal absorption during prolonged severe exercise. *J. Appl. Physiol.* 23:331–335.

Forsberg, A., B. Hultén, G. Wilson, and J. Karlsson. 1974. *Ishockey.* Idrottsfysiologi Rapport no. 14. Stockholm: Trygg-Hansa.

Forssberg, H., and G. Svartengren. 1983. Hardwired locomotor network in cat revealed by a retained motor pattern to gastrocnemius after muscle transposition. *Neurosci. Lett.* 41:283–288.

Forster, H.V., J.A. Dempsey, M.L. Birnbaum, W.G. Reddan, J. Thoden, R.F. Grover, and J. Rankin. 1971. Effects of chronic exposure to hypoxia in ventilatory response to $CO_2$ and hypoxia. *J. Appl. Physiol.* 31:586–592.

Forster, R.E., and E.D. Crandell. 1976. Pulmonary gas exchange. *Annu. Rev. Physiol.* 38:69.

Fortney, S.M., E.R. Nadel, C.B. Wenger, and J.R. Bove. 1981a. Effect of acute alterations of blood volume on circulatory performance in humans. *J. Appl. Physiol.* 50:292.

———. 1981b. Effect of blood volume on sweating rate and body fluids in exercising humans. *J. Appl. Physiol.* 51:1594.

Fortney, S.M., and L.C. Senay, Jr. 1979. Effect of training and heat acclimation on exercise responses of sedentary females. *J. Appl. Physiol.* 47:978.

Fortney, S.M., and N.B. Vroman. 1985. Exercise, performance and temperature control: Temperature regulation during exercise and implications for sports performance and training. *Sports Med.* 2:8–20.

Foster, C. 1998. Monitoring training in athletes with reference to overtraining syndrome. *Med. Sci. Sports Exerc.* 30:1164–1168.

Francis, K.T. 1979. Psychologic correlates of serum indicators of stress in man: A longitudinal study. *Psychosom. Med.* 41:617–628.

Frandsen, U., L. Hoffner, A. Betak, B. Saltin, J. Bangsbo, and Y. Hellsten. 2000. Endurance training does not alter the level of neuronal nitric oxide synthase in human skeletal muscle. *J. Appl. Physiol.* 89:1033–1038.

Frandsen, U., M. Lopez-Figueroa, and Y. Hellsten. 1996. Localization of nitric oxide synthase in human skeletal muscle. *Biochem. Biophys. Res. Commun.* 227:88–93.

Frank, S.M., N. el-Gamal, S.N. Raja, and P.K. Wu. 1996. Alpha-adrenoceptor mechanisms of thermoregulation during cold challenge in humans. *Clin. Sci.* (Colch.) 91:627–631.

Franzini-Armstrong, C. 1994. The sarcoplasmic reticulum and the transverse tubules. In *Myology: Basic and clinical*, ed. A.G. Engel and C. Franzini-Armstrong, pp. 176–199. New York: McGraw-Hill.

Franzini-Armstrong, C., and A.O. Jorgensen. 1994. Structure and development of E-C coupling units in skeletal muscle. *Annu. Rev. Physiol.* 56:509–534.

Frascarolo, P., F.Y. Schutz, and E. Jequier. 1992. Influence of the menstrual cycle on the sweating response measured by direct calorimetry in women exposed to warm environmental conditions. *Eur. J. Appl. Physiol. Occup. Physiol.* 64:449–454.

Fredholm, B.B. 1969. Inhibition of fatty acid release from adipose tissue by high arterial lactate concentrations. *Acta Physiol. Scand.* 77 (suppl. 330).

Freund, H.J. 1983. Motor unit and muscle activity in voluntary motor control. *Physiol. Rev.* 63:387–436.

Fridén, J., U. Kjorell, and L.E. Thornell. 1984. Delayed muscle soreness and cytoskeletal alterations: An immunocytological study in man. *Int. J. Sports Med.* 5:15–18.

Fridén, J., M. Sjöström, and B. Ekblom. 1981. A morphological study of delayed muscle soreness. *Experientia* 37:506–507.

Friedenreich, C.M., I. Thune, L.A. Brinton, and D. Albanes. 1998. Epidemiologic issues related to the association between physical activity and breast cancer. *Cancer* 83:600–610.

Fröberg, J.C., C.G. Karlsson, L. Levi, and L. Lidberg. 1972. Circadian variations in performance, psychological ratings, catecholamine excretion and diuresis during prolonged sleep deprivation. *Inter. J. Psychobiol.* 2:23–36.

Frokiaer, J., D. Marples, M.A. Knepper, and S. Nielsen. 1998. Pathophysiology of aquaporin-2 in water balance disorders. *Am. J. Med. Sci.* 316:291–299.

Frontera, W.R., V.A. Hughes, R.A. Fielding, M.A. Fiatarone, W.J. Evans, and R. Roubenoff. 2000. Aging of skeletal muscle: A 12-yr longitudinal study. *J. Appl. Physiol.* 88:1321–1326.

Frost, H.M. 1987. Bone "mass" and the "mechanostat": A proposal. *Anat. Rec.* 219:1–9.

———. 2001. From Wolff's law to the Utah paradigm: Insights about bone physiology and its clinical applications. *Anat. Rec.* 262:398–419.

Frye, A.J., and E. Kamon. 1983. Sweating efficiency in acclimated men and women exercising in humid and dry heat. *J. Appl. Physiol.* 54:972.

Fuglevand, A.J. 1996. Neural aspects of fatigue. *Neuroscientist* 2:203–206.

Fujimoto-Kanatani, K. 1996. *Determining the essential elements of golf swings used by elite golfers.* Eugene, OR: Microform Publications, International Institute for Sport and Human Performance, University of Oregon.

Fujita, T., H. Mori, M. Minowa, H. Kimura, J. Tsujishita, K. Kimura, J. Gao, A. Yoshida, T. Morita, and M. Mitsubayashi. 1993. A retrospective cohort study of long-term health effects of shift work. *Nippon Koshu Eisei Zasshi* 40:273–283.

Fukunaga, T., Y. Kawakami, K. Kubo, and H. Kanehisa. 2002. Muscle and tendon interaction during human movement. *Exerc. Sport Sci. Rev.* 30:106–110.

Fukunaga, T., Y. Kawakami, S. Kuno, K. Funato, and S. Fukashiro. 1997. Muscle architecture and function in humans. *J. Biomech.* 30:457–463.

Fulco, C.S., S.F. Lewis, P.N. Frykman, R. Boushel, S. Smith, E.A. Harman, A. Cymerman, and K.B. Pandolf. 1995. Quantitation of progressive muscle fatigue during dynamic leg exercise in humans. *J. Appl. Physiol.* 79:2154–2162.

Fulco, C.S., P.B. Rock, and A. Cymerman. 1998. Maximal and submaximal exercise performance at altitude. *Aviat. Space Environ. Med.* 69:793–801.

Fuller, A., T. Oosthuyse, S.K. Maloney, and D. Mitchell. 1999. Evaluation of miniature data loggers for body temperature measurement during sporting activities. *Eur. J. Appl. Physiol.* 79:341–346.

Fuster, V., A. Jerez, and O. Ortega. 1998. Anthropometry and strength relationship: Male–female differences. *Anthropol. Anz.* 56 (1):49–56.

Gabel, K.A., A. Aldous, and C. Edgington. 1995. Dietary intake of two elite male cyclists during 10-day, 2,050-mile ride. *Int. J. Sport Nutr.* 5:56–61.

Gaesser, G.A. 1999. Thinness and weight loss: Beneficial or detrimental to longevity? *Med. Sci. Sports Exerc.* 31:1118–1128.

Galbo, H. 1992. Exercise physiology: Humoral function. *Sport Sci. Rev.* 1:65–93.

———. 1995. Integrated endocrine responses and exercise. In *Endocrinology*, ed. L.J. Groot et al., vol 3. Philadelphia: Saunders.

Galloway, S.D.R., S.M. Shirreffs, J.B. Leiper, and R.J. Maughan. 1997. Exercise in the heat: Factors limiting exercise capacity and methods for improving heat tolerance. *Sports Exerc. Inj.* 3:19–24.

Gandevia, S.C. 1998. Neural control in human muscle fatigue: Changes in muscle afferents, moto neurones and moto cortical drive. *Acta Physiol. Scand.* 162:275–283.

Gandevia, S.C., G.M. Allen, J.E. Butler, R.B. Gorman, and D.K. McKenzie. 1998. Human respiratory muscles: Sensations, reflexes and fatiguability. *Clin. Exp. Pharmacol. Physiol.* 25:757–763.

Gandevia, S.C., G.M. Allen, and D.K. McKenzie. 1995. Central fatigue: Critical issues, quantification and practical implications. *Adv. Exp. Med. Biol.* 384:281–294.

Gandevia, S.C., R.M. Enoka, A.J. McComas, D.G. Stuart, and C.K. Thomas, eds. 1995. Fatigue: Neural and muscular mechanisms. *Adv. Exp. Med. Biol.*, vol. 384.

Gandevia, S.C., L.A. Hall, D.I. McCloskey, and E.K. Potter. 1983. Proprioceptive sensation at the terminal joint of the middle finger. *J. Physiol.* (Lond.) 335:507–517.

Gandevia, S.C., R.D. Herbert, and J.B. Leeper. 1998. Voluntary activation of human elbow flexor muscles during maximal concentric contractions. *J. Physiol.* 512:595–602.

Ganong, W.F. 1999. *Review of medical physiology.* 19th ed. Stamford, CT: Appleton and Lange.

Garcia, C.K., J.L. Goldstein, R.K. Pathak, G.W. Anderson, and M.S. Brown. 1994. Molecular characterization of a membrane transporter for lactate, pyruvate and other monocarboxylates: Implications for the Cori cycle. *Cell* 76:865–873.

Garcia, M.C., H. Gonzalez-Serratos, J.P. Morgan, C.L. Perreault, and M. Rozycka. 1991. Differential activation of myofibrils during fatigue in phasic skeletal muscle cells. *J. Muscle Res. Cell Motil.* 12:412–424.

Gardner, K.D., Jr. 1956. "Athletic pseudonephritis"—alteration of urine sediment by athletic competition. *JAMA* 161:1613–1617.

Garg, A., and U. Saxena. 1980. Container characteristics and maximum acceptable weight of lift. *Hum. Factors* 22:487.

Garhammer, J., and B. Takano. 1992. Training for weightlifting. In *Strength and power in sport,* ed. P.V. Komi, pp. 357–369. Oxford: Blackwell Science.

Garland, S.J. and E.R. Gossen. 2002. The muscle wisdom hypothesis in human muscle fatigue. *Exerc. Sport Sci. Rev.* 30:45–49.

Garland, S.J., and M.P. Kaufman. 1995. Role of muscle afferents in the inhibition of motoneurons during fatigue. *Adv. Exp. Med. Biol.* 384:271–278.

Garnett, R., and J.A. Stephens. 1981. Changes in the recruitment threshold of motor units produced by cutaneous stimulation in man. *J. Physiol.* (Lond.) 311:463–473.

Garrow, J.S. 1978. *Energy balance and obesity in man.* 2nd ed. New York: Elsevier North-Holland.

Gass, E.M., and G.C. Gass. 1998. Rectal and esophageal temperatures during upper and lower body exercise. *Eur. J. Appl. Physiol.* 78:38–42.

Gastin, P., D. Lawson, M. Hargreaves, M. Carey, and I. Fairweather. 1991. Variable resistance loadings in anaerobic power testing. *Int. J. Sports Med.* 12:513–518.

Gastmann, U.A., and M.J. Lehmann. 1998. Overtraining and the BCAA hypothesis. *Med. Sci. Sports Exerc.* 30:1173–1178.

George, J., T. Murphy, R. Roberts, W.G.E. Cooksley, J.W. Halliday, and L.W. Powell. 1986. Influence of alcohol and caffeine consumption on caffeine elimination. *Clin. Exp. Pharm. Physiol.* 13:731–736.

George, K.P., P.E. Gates, K.M. Birch, and L.M. Campbell. 1999. Left ventricular morphology and function in endurance-trained female athletes. *J. Sports Sci.* 17:633–642.

George, K.P., L.A. Wolfe, and G.W. Burggraf. 1991. The "athletic heart syndrome": A critical review. *Sports Med.* 11:300–331.

Georgian, J.C. 1964. The temperature scale. *Nature* 201:695.

Gerrard, D., and S. Hollings. 1992. Bicarbonate loading. *NZ J. Sports Med.* 20:11–12.

Gerstenblith, G., E.G. Lakatta, and M.L. Weisfeldt. 1976. Age changes in myocardial function and exercise response. *Prog. Cardiovasc. Dis.* 19:1.

Gertler, R.A., and N. Robbins. 1978. Differences in neuromuscular transmission in red and white muscles. *Brain Res.* 142:160–164.

Gharakhanlou, R., S. Chadan, and P. Gardiner. 1999. Increased activity in the form of endurance training increases calcitonin gene-related peptide content in lumbar motoneuron cell bodies and in sciatic nerve in the rat. *Neuroscience* 89:1229–1239.

Ghosh, A.K., P. Mazumdar, A. Goswami, A. Ahuja, and T.P.S. Puri. 1990. Heart rate and blood lactate response in competitive badminton. *Ann. Sports Med.* (New York) 5:85–88.

Gibbs, J.S. 1999. Pulmonary hemodynamics: Implications for high altitude pulmonary edema (HAPE): A review. *Adv. Exp. Med. Biol.* 474:81–91.

Gidwani, G.P. 1999. Amenorrhea in the athlete. *Adolesc. Med.* 10:275–290.

Giori, N.J., G.S. Beaupre, and D.R. Carter. 1993. Cellular shape and pressure may mediate mechanical control of tissue composition in tendons. *J. Orthop. Res.* 11:581–591.

Gisolfi, C., S. Robinson, and E.S. Turrell. 1966. Effects of aerobic work performed during recovery from exhausting work. *J. Appl. Physiol.* 21:1767–1772.

Gisolfi, C.V., and J.R. Copping. 1993. Thermal effects of prolonged treadmill exercise in the heat. *Med. Sci. Sports Exerc.* 25:310–315.

Gladden, L.B. 1991. Net lactate uptake during progressive steady-level contractions in canine muscle. *J. Appl. Physiol.* 71:514–520.

———. 2000. Muscle as a consumer of lactate. *Med. Sci. Sports Exerc.* 32:764–771.

Gledhill, N. 1984. Bicarbonate ingestion and anaerobic performance. *Sports Med.* 1:177–180.

———. 1985. The influence of altered blood volume and oxygen transport capacity on aerobic performance. *Exerc. Sport Sci. Rev.* 13:75–93.

Gledhill, N., D. Cox, and R. Jamnik. 1994. Endurance athletes' stroke volume does not plateau: Major advantage is diastolic function. *Med. Sci. Sports Exerc.* 26:1116–1121.

Gleeson, M. 1998. Overtraining and stress response. *Sports Exerc. Inj.* 4:62–68.

Glenmark, B. 1994. Skeletal muscle fibre types, physical performance, physical activity and attitude to physical activity in women and men: A follow-up from age 16–27. *Acta Physiol. Scand.* 151 (suppl. 623):1–42.

Glenmark, B., G. Hedberg, and E. Jansson. 1992. Changes in muscle fibre type from adolescence to adulthood in women and men. *Acta Physiol. Scand.* 146:251–259.

Glenmark, B., G. Hedberg, L. Kaijser, and E. Jansson. 1994. Muscle strength from adolescence to adulthood: Relationship to muscle fibre types. *Eur. J. Appl. Physiol.* 68 (1):9–19.

Godfrey, S., and E.J.M. Campbell. 1970. The role of afferent impulses from the lung and chest wall in respiratory control and sensation. In *Breathing: Hering-Breuer Centenary Symposium*, ed. R. Porter, p. 219. London: Churchill.

Goetz, K.L., G.C. Bond, and D.D. Bloxham. 1975. Atrial receptors and renal function. *Physiol. Rev.* 55:157.

Golden, F.S.C., I.F.G. Hampton, G.R. Harvey, and A.V. Knibbs. 1979. Shivering intensity in humans during immersion in cold water. *J. Physiol.* (Lond.) 290:48.

Golden, F.S.C., I.F.G. Hampton, and D. Smith. 1979. Cold tolerance in long-distance swimmers. *J. Physiol.* (Lond.) 290:49.

Goldman, R.M., R.S. Tarr, B.G. Pinchuk, and R.E. Kappler. 1985. The effects of oscillating inversion on systemic blood pressure, pulse, intraocular pressure and central retina pressure. *Physician Sportsmed.* 13:93–96.

Goldspink, G. 1992. Cellular and molecular aspects of adaptation in skeletal muscle. In *Strength and power in sport*, ed. P.V. Komi, pp. 211–229. Oxford: Blackwell Science.

———. 1998. Selective gene expression during adaptation of muscle in response to different physiological demands. *Comp. Biochem. Physiol. Comp. Biochem. Molec. Biol.* 120:5–15.

———. 1999. Changes in muscle mass and phenotype and the expression of autocrine and systemic growth factors by muscle in response to stretch and overload. *J. Anat.* 194:323–334.

Gollhofer, A., P.V. Komi, N. Fujitsuka, and M. Miyashita. 1987. Fatigue during stretch–shortening cycle exercises: II. Changes in neuromuscular activation patterns of human skeletal muscle. *Int. J. Sports Med.* 8 (suppl. 1):38–47.

Gollnick, P.D. 1985. Metabolism of substrates: Energy substrate metabolism during exercise and as modified by training. *Fed. Proc.* 44 (2):353–357.

Gollnick, P.D., R.B. Armstrong, W.L. Sembrowich, R.E. Shepherd, and B. Saltin. 1973. Glycogen depletion pattern in human skeletal muscle fibers after heavy exercise. *J. Appl. Physiol.* 34:615–618.

Gollnick, P.D., D. Parsons, and C.R. Oakley. 1983. Differentiation of fiber types in skeletal muscle from the sequential inactivation of myofibrillar actomyosin ATPase during acid preincubation. *Histochemistry* 77:543–555.

Gollnick, P.D., K. Piehl, and B. Saltin. 1974. Selective glycogen depletion pattern in human muscle fibres after exercise of varying intensity and at varying pedalling rates. *J. Physiol.* 241:45–57.

Gollnick, P.D., B.F. Timson, R.L. Moore, and M. Riedy. 1981. Muscular enlargement and number of fibers in skeletal muscles of rats. *J. Appl. Physiol.* 50:936–943.

Gonzalez-Alonso, J., C.L. Heaps, and E.F. Coyle. 1992. Rehydration after exercise with common beverages and water. *Int. J. Sports Med.* 13:399–406.

Gonzalez-Alonso, J., C. Teller, S.L. Andersen, F.B. Jensen, T. Hyldig, and B. Nielsen. 1999. Influence of body temperature on the development of fatigue during prolonged exercise in the heat. *J. Appl. Physiol.* 86:1032–1039.

Gordon, M.J., B.R. Goslin, T. Graham, and J. Hoare. 1983. Comparison between load carriage and grade walking on a treadmill. *Ergonomics* 26:289–298.

Gordon, N.F., and C.B. Scott. 1991. The role of exercise in the primary and secondary prevention of coronary artery disease. *Clin. Sports Med.* 10:87–103.

Gottlieb, S. 1999. Study confirms passive smoking increases coronary heart disease. *Br. Med. J.* 318:891A.

Gowlett, J.A.J., J.W.K. Harris, D. Walton, and B.A. Wood. 1981. Early archeological sites, hominid remains and traces of fire from Chesowanja, Kenya. *Nature* 294:125.

Granati, A., and L. Busca. 1945. II Lavoro della tribiatura. *Boll. Soc. Ital. Biol. Sper.* 20:51.

Granit, R. 1972. Constant errors in the execution and appreciation of movement. *Brain* 95:649–660.

Granit, R., and R.E. Burke. 1973. The control of movement and posture. *Brain Res* 53:1–28.

Grassi-Zucconi, G., M. Menegazzi, A.C. De Prati, A. Bassetti, P. Montagnese, P. Mandile, C. Cosi, C. and M. Bentivoglio. 1993. c-*fos* mRNA is spontaneously induced in the rat brain during the activity period of the circadian cycle. *Eur. J. Neurosci.* 5:1071–1078.

Gray, E.L., C.F. Consolazio, and R.M. Karl. 1951. Nutritional requirements for men at work in cold, temperate and hot environments. *J. Appl. Physiol.* 4:270.

Green, H., P. Bishop, M. Houston, R. McKillop, R. Norman, and P. Stothart. 1976. Time–motion and physiological assessments of ice hockey performance. *J. Appl. Physiol.* 40:159–163.

Green, H., C. Goreham, J. Ouyang, M. Ball-Burnett, and D. Ranney. 1999. Regulation of fiber size, oxidative potential and capillarization in human muscle by resistance exercise. *Am. J. Physiol.* 276:R591–596.

Green, J.S., and S.F. Crouse. 1993. Aging, cardiovascular function and endurance exercise: A review of the literature. *Med. Exerc. Nutr. Health* 2:299–309.

Green, S., J. Bülow, and B. Saltin. 1999. Microdialysis and the measurement of muscle interstitial $K^+$ during rest and exercise in humans. *J. Appl. Physiol.* 87:460–464.

Green, S., and B. Dawson. 1993. Measurement of anaerobic capacities in humans: Definitions, limitations and unsolved problems. *Sports Med.* 15 (5):312–327.

Greene, J.A. 1939. Clinical study of the etiology of obesity. *Ann. Intern. Med.* 12:1797.

Greenleaf, J.E. 1979. Hyperthermia and exercise. In *International review of physiology, environmental physiology III*, ed. D. Robertshaw, vol. 20. Baltimore: University Park.

Greenleaf, J.E., and B.I. Castle. 1971. Exercise temperature regulation in man during hypohydration and hyperhydration. *J. Appl. Physiol.* 30:847.

Gregg, D.E. 1974. The natural history of coronary collateral development. *Circ. Res.* 35:335.

Grillner, S. 1975. Locomotion in vertebrates: Central mechanisms and reflex interaction. *Physiol. Rev.* 55:247–304.

———. 1981. Control of locomotion in bipeds, tetrapods and fish. In *Handbook of physiology: The nervous system*, ed. V. Brooks, p. 1179. Baltimore: Williams and Wilkins.

Grimby, G. 1992. Clinical aspects of strength and power training. In *Strength and power in sport*, ed. P.V. Komi, pp. 338–354. Oxford: Blackwell Science.

Grimby, G., A. Aniansson, M. Hedberg, G.B. Henning, U. Grangard, and H. Kvist. 1992. Training can improve muscle strength and endurance in 78- to 84-yr-old men. *J. Appl. Physiol.* 73:2517–2523.

Grimby, G., and B. Saltin. 1983. The ageing muscle. *Clin. Physiol.* 3:209–218.

Grimby, G., B. Saltin, and L. Wilhelmsen. 1971. Pulmonary flow–volume and pressure–volume relationship during submaximal and maximal exercise in young well-trained men. *Bull. Physiopathol. Respir.* (Nancy) 7:157–172.

Grimby, L. 1984. Firing properties of single human motor units during locomotion. *J. Physiol.* (Lond.) 346:195–202.

Grimby, L., and J. Hannerz. 1976. Disturbances in voluntary recruitment order of low and high frequency motor units on blockades of proprioceptive afferent activity. *Acta Physiol. Scand.* 96:207–216.

Grimston, S.K., N.D. Willows, and D.A. Hanley. 1993. Mechanical loading regime and its relationship to bone mineral density in children. *Med. Sci. Sports Exerc.* 25:1203–1210.

Grover, R.F., R. Lufschanowski, and J.K. Alexander. 1976. Alterations in the coronary circulation of man following ascent to 3,100 m altitude. *J. Appl. Physiol.* 41:832–838.

Grover, R.F., and J.T. Reeves. 1967. Exercise performance of athletes at sea level and 3,100 meters altitude. In *The international symposium on the effects of altitude on physical performance*, ed. R.F. Goddard, p. 80. Chicago: Athletic Institute.

Grunt, J.A., J.F. Crigler, Jr., D. Slone, and J.S. Soeldner. 1967. Changes in serum insulin, blood sugar and free fatty acid levels four hours after administration of human growth hormone in fasting children with short stature. *Yale J. Biol. Med.* 40:68–74.

Guezennec, C.Y., E. Gilson, and B. Serrurier. 1990. Comparative effects of hindlimb suspension and exercise on skeletal muscle myosin isozymes in rats. *Eur. J. Appl. Physiol.* 60:430–435.

Guillet, C., P. Auguste, W. Mayo, P. Kreher, and H. Gascan. 1999. Ciliary neurotrophic factor is a regulator of muscular strength in aging. *J. Neurosci.* 19:1257–1262.

Guleria, J.S., J.N. Parde, P.K. Sethi, and S.B. Roy. 1971. Pulmonary diffusing capacity at high altitude. *J. Appl. Physiol.* 31:536–543.

Gundersen, K. 1998. Determination of muscle contractile properties: The importance of the nerve. *Acta Physiol. Scand.* 162:333–341.

Günther, B. 1975. Dimensional analysis and theory of biological similarity. *Physiol. Rev.* 55:659–699.

Gustafsson, F., S. Ali, B. Hanel, J.C. Toft, and N.H. Secher. 1996. The heart of the senior oarsman: An echocardiographic evaluation. *Med. Sci. Sports. Exerc.* 28:1045–1048.

Gutgesell, M., and R. Canterbury. 1999. Alcohol usage in sport and exercise. *Addiction Biol.* 4:373–383.

Guth, L., and F.J. Samaha. 1970. Procedure for the histochemical demonstration of actomyosin ATPase. *Exp. Neurol.* 28:365–367.

Ha, M., H. Tokura, and Y. Yamashita. 1995. Effects of two kinds of clothing made from hydrophobic and hydrophilic fabrics on local sweating rates at an ambient temperature of 37 degrees C. *Ergonomics* 38:1445–1455.

Haar Romeny, B.M., J.J. van der Gon, and C.C. Gielen. 1984. Relation between location of a motor unit in the human biceps brachii and its critical firing levels for different tasks. *Exp. Neurol.* 85:631–650.

Hackett, P.H., K.H. Maret, J.S. Milledge, R.M. Peters, C.J. Pizzo, J.B. West, and R.M. Winslow. 1983. Physiology of man on the summit of Mt. Everest. *J. Physiol.* 334:99P–100P.

Hackett, P.H., D. Rennie, R.F. Grover, and J.T. Reeves. 1981. Acute mountain sickness and the edemas of high altitude: A common pathogenesis? *Respir. Physiol.* 46:383–390.

Hagbarth, K.E., and V.G. Macefield. 1995. The fusimotor system: Its role in fatigue. *Adv Exp. Med. Biol.* 384:259–270.

Hagberg, J.M., and E.F. Coyle. 1983. Physiological determinants of endurance performance as studied in competitive racewalkers. *Med. Sci. Sports Exerc.* 15:287–289.

Hagberg, J.M., E.F. Coyle, J.E. Carroll, J.M. Miller, W.H. Martin, and M.H. Brooke. 1982. Exercise hyperventilation in patients with McArdle's disease. *J. Appl. Physiol. Respir. Environ. Exerc. Physiol.* 52:991.

Hagberg, M. 1981a. Muscular endurance and surface electromyogram in isometric and dynamic exercise. *J. Appl. Physiol. Respir. Environ. Exerc. Physiol.* 51 (1):1–7.

———. 1981b. On evaluation of local muscular load and fatigue by electromyography. In *Arbete och hälsa*, vol. 24. Arbetarskyddsstyrelsen Vetenskaplig Skriftserie, Stockholm.

Hägg, G.M. 1991. Static work and occupational myalgia: A new explanation model. In *Electromyographical kinesiology*, ed. P. Anderson, D. Hobart, and J. Danoff, pp. 141–144. Amsterdam: Elsevier.

———. 2000. Human muscle fibre abnormalities related to occupational load. *Eur. J. Appl. Physiol.* 83:159–165.

Hägg, G.M., and A. Åström. 1997. Load pattern and pressure pain threshold in the upper trapezius muscle and psychosocial factors in medical secretaries with and without shoulder/neck disorders. *Int. Arch. Occup. Environ. Health* 69:423–432.

Häggmark, T., E. Eriksson, and E. Jansson. 1986. Muscle fiber type changes in human skeletal muscle after injuries and immobilization. *Orthopedics* (Thorofare, NJ) 9:181–185.

Häggquist, G. 1931. *Die Gewebe. 3. Teil: Gewebe und Systeme der Muskulatur*. Berlin: Verlag von Julius Springer.

Hahn, A.G. 1991. The effect of altitude training on athletic performance at sea level: A review. *EXEL* 7:9–23.

Häkkinen, K., M. Alen, and P.V. Komi. 1985. Changes in isometric force- and relaxation-time, electromyographic and muscle fibre characteristics of human skeletal muscle during strength training and detraining. *Acta Physiol. Scand.* 125:573–585.

Häkkinen, K., and P.V. Komi. 1981. Effect of different combined concentric and eccentric muscle work regimes on maximal strength development. *J. Hum. Mov. Stud.* 7:33–44.

Häkkinen, K., P.V. Komi, and M. Alen. 1985. Effect of explosive type strength training on isometric force- and relaxation-time, electromyographic and muscle fibre characteristics of leg extensor muscles. *Acta Physiol. Scand.* 125:587–600.

Häkkinen, K., and P. Sinnemaki. 1990. Maximum oxygen uptake, anaerobic power and neuromuscular performance characteristics in band players of two different competitive levels. *J. Sports Med. Phys. Fitness* 30:67–73.

Halestrap, A.P., and N.T. Price. 1999. The proton-linked monocarboxylate transporter (MCT) family: Structure, function and regulation. *Biochem. J.* 243:281–299.

Hall, L.A., and D.I. McCloskey. 1983. Detections of movements imposed on finger, elbow and shoulder joints. *J. Physiol.* (Lond.) 335:519–533.

Hallberg, D., B. Hallgren, O. Schuberth, and A. Wretlind. 1982. Parental nutritional goals and achievements, part II. *Nutr. Support Serv.* 2:35.

Hallberg, L., and B. Magnusson. 1984. The etiology of "sports anemia": A physiological adaptation of the oxygen-dissociation curve of hemoglobin to an unphysiological exercise load. *Acta Med. Scand.* 216:147–148.

Halpain, S. 2000. Actin and the agile spine: How and why do dendritic spines dance? *Trends Neurosci.* 23:141–146.

Hameed, M., S.D.R. Harridge, and G. Goldspink. 2002. Sarcopenia and hypertrophy: A role for insulin-like growth factor-1 in aged muscle? *Exerc. Sport Sci. Rev.* 30:15–19.

Hamilton, A.L., K.J. Killian, E. Summers, and N.L. Jones. 1995. Muscle strength, symptom intensity and exercise capacity in patients with cardiorespiratory disorders. *Am. J. Respir. Crit. Care Med.* 152:2021–2031.

Hamilton, C.L. 1965. Control of food intake. In *Physiological controls and regulations,* ed. W.S. Yamamoto and J.R. Brobeck, p. 274. Philadelphia: Saunders.

Hamilton, M.T., and F.W. Booth. 2000. Skeletal muscle adaptation to exercise: A century of progress. *J. Appl. Physiol.* 88:327–331.

Hamilton, P., and G.M. Andrew. 1976. Influence of growth and athletic training on heart and lung functions. *Eur. J. Appl. Physiol. Occup. Physiol.* 36:27–38.

Hammel, H.T. 1965. Neurons and temperature regulation. In *Physiological controls and regulations,* ed. W.S. Yamamoto and J.R. Brobeck, p. 71. Philadelphia: Saunders.

———. 1994. How solutes alter water in aqueous solutions. *J. Phys. Chem.* 98:4196–4204.

———. 1995. Roles of colloidal molecules in Starling's hypothesis and in returning interstitial fluid to the vasa recta. *Am. J. Physiol.* 268:H2133–H2145.

———. 1999. Evolving ideas about osmosis and capillary fluid exchange. *FASEB J.* 13:213–231.

———. 2001. Osmotic effects on solvent of solute diffusing in solution. *In. Adv. Res. Physical Chem.* 2:11–33. Global Research Network, Trivandrum, India.

Hammel, H.T., and W.F. Brechue. 2000. Plasma-ISF fluid exchange is driven by diffusion of carbon dioxide and bicarbonate in presence of carbonic anhydrase. *FASEB J.* 14:Abstract 315.3.

Hannerz, J. 1974. Discharge properties of motor units in relation to recruitment order in voluntary contraction. *Acta Physiol. Scand.* 91:374–384.

Hansen, J.E., J.A. Vogel, G.P. Stelter, and C.F. Consolazio. 1967. Oxygen uptake in man during exhaustive work at sea level and high altitude. *J. Appl. Physiol.* 23:511–522.

Hansen, T.M., G. Hansen, A.M. Langaard, and J.O. Rasmussen. 1993. Longterm physical training in rheumatoid arthritis. A randomized trial with different programs and blinded observers. *Scand. J. Rheumatol.* 22:107–112.

Hanson, J.A., and F.P. Jones. 1970. Heart rate and small postural changes in man. *Ergonomics* 13:483–487.

Hansson, H.A., A.M. Engstrom, S. Holm, and A.L. Rosenqvist. 1988. Somatomedin C immunoreactivity in the Achilles tendon varies in a dynamic manner with the mechanical load. *Acta Physiol. Scand.* 134:199–208.

Hansson, J.E. 1965. *Relationship between individual characteristics of the worker and output of work in logging operations.* Studia Forestalia Suecia, no. 29. Stockholm: Skogshögskolan.

———. 1968. *Work physiology as a tool in ergonomics and production engineering.* Al- Rapport 2, Ergonomi och produktionsteknik. Stockholm: National Institute of Occupational Health.

Harber, V.J., and J.R. Sutton. 1984. Endorphins and exercise. *Sports Med.* 1:154–171.

Hardman, A.E., and A. Hudson. 1994. Brisk walking and serum lipid and lipoprotein variables in previously sedentary women: Effect of 12 weeks of regular brisk walking followed by 12 weeks of detraining. *Br. J. Sports Med.* 28:261–266.

Hardy, J.D. 1967. Central and peripheral factors in physiological temperature regulation. In *Les concepts de Claude Bernard suer le milieu intérieur,* p. 247. Paris: Masson.

Hardy, J.D., and E.F. Dubois. 1938. The technique of measuring radiation and convection. *J. Nutr.* 15:461.

Hardy, J.D., A.P. Gagge, and J.A.J. Stolwijk, eds. 1970. *Physiological and behavioral temperature regulation.* Springfield, IL: Charles C Thomas.

Hardy, J.D., J.A.J. Stolwijk, and A.P. Gagge. 1971. In *Comparative physiology of thermoregulation*, ed. G.C. Whittow, vol. 2, p. 327. New York: Academic Press.

Harma, M., P. Knauth, J. Ilmarinen, and H. Ollila. 1990. The relation of age to the adjustment of the circadian rhythms of oral temperature and sleepiness to shift work. *Chrono-Biol. Int.* (England) 7:227–233.

Harridge, S.D., R. Bottinelli, M. Canepari, M. Pellegrino, C. Reggiani, M. Esbjornsson, P.D. Balsom, and B. Saltin. 1998. Sprint training, in vitro and in vivo muscle function and myosin heavy chain expression. *J. Appl. Physiol.* 84:442–449.

Harridge, S., G. Magnusson, and B. Saltin. 1997. Life-long endurance trained elderly men have high aerobic power, but have similar muscle strength to non-active elderly men. *Aging* (Milano) 9 (1–2):80–87.

Harrison, M.H. 1985. Effects on thermal stress and exercise on blood volume in humans. *Physiol. Rev.* 65:149–209.

———. 1986. Heat and exercise: Effects on blood volume. *Sports Med.* 3:214–223.

Harrison, M.H., G.A. Brown, and A.J. Belyavin. 1982. The "Oxylog": An evaluation. *Ergonomics* 25:809–820.

Harrison, M.H., G.A. Brown, and L.A. Cochrane. 1980. Maximal oxygen uptake: Its measurement, application and limitations. *Aviat. Space Environ. Med.* 51 (10):1123–1127.

Harrison, M.H., R.J. Edwards, M.J. Graveney, L.A. Cochrane, and J.A. Davies. 1981. Blood volume and plasma protein responses to heat acclimatization in humans. *J. Appl. Physiol.* 50:597.

Harrison, M.H., R.J. Edwards, and D.R. Leitch. 1975. Effect of exercise and thermal stress on plasma volume. *J. Appl. Physiol.* 39:925–933.

Hartikainen, A.L., M. Sorri, H. Anttonen, R. Tuimala, and E. Laara. 1994. Effect of occupational noise on the course and outcome of pregnancy. *Scand. J. Work Environ. Health* 20:444–450.

Hartley, L.H. 1975. Growth hormone and catecholamine response to exercise in relation to physical training. *Med. Sci. Sports* 7:34–36.

Hartley, L.H., J.K. Alexander, M. Modelski, and R.F. Grover. 1967. Subnormal cardiac output at rest and during exercise in residents at 3,100 m altitude. *J. Appl. Physiol.* 23:839–848.

Hartley, L.H., and B. Saltin. 1968. Reduction of stroke volume and increase in heart rate after a previous heavier submaximal work load. *Scand. J. Clin. Lab. Invest.* 22:217–223.

Hartley, L.H., J.A. Vogel, and M. Landowne. 1973. Central, femoral and brachial circulation during exercise in hypoxia. *J. Appl. Physiol.* 34:87–90.

Hartman, S., and P. Bung. 1999. Physical exercise during pregnancy: Physiological considerations and recommendations. *J. Perinat. Med.* 27:204–215.

Hartmann, U., and J. Mester. 2000. Training and overtraining markers in selected sport events. *Med. Sci. Sports Exerc.* 32:209–215.

Hartung, G.H., L.P. Krock, C.G. Crandall, R.U. Bisson, and L.G. Myhre. 1993. Prediction of maximal oxygen uptake from submaximal exercise testing in aerobically fit and nonfit men. *Aviat. Space Environ. Med.* 64:735–740.

Haskell, W.L., C. Sims, J. Myll, W.M. Bortz, F.G. St. Goar, and E.L. Alderman. 1993. Coronary artery size and dilating capacity in ultradistance runners. *Circulation* 87:1076–1082.

Hauschka, S.D. 1994. The embryonic origin of muscle. In *Myology: Basic and clinical*, ed. A.G. Engel and C. Franzini-Armstrong, pp. 3–73. New York: McGraw-Hill.

Havenith, G., J.M.L. Coenen, L. Kistemaker, and W.L. Kenney. 1998. Relevance of individual characteristics for human heat stress response is dependent on exercise intensity and climate type. *Eur. J. Appl. Physiol.* 77:231–241.

Haverstock, B.D., and V.J. Mandracchia. 1998. Cigarette smoking and bone healing: Implications in foot and ankle surgery. *J. Foot Ankle Surg.* 37:69–74.

Hawley, J.A., G.S. Palmer, and T.D. Noakes. 1997. Effects of 3 days of carbohydrate supplementation on muscle glycogen content and utilisation during a 1-h cycling performance. *Eur. J. Appl. Physiol.* 75:407–412.

Haynes, R.H., and S. Rodbard. 1962. Arterial and arteriolar systems, biophysical principles and physiology. In *Blood vessels and lymphatics*, ed. D.I. Abramson, chapt. 2, p. 26. New York: Academic Press.

Haywood, K.M. 1993. *Life span motor development*. Champaign, IL: Human Kinetics.

Hazlett, R.L., D.R. McLeod, and R. Hoehn-Saric. 1994. Muscle tension in generalized anxiety disorder: Elevated muscle tonus or agitated movement? *Psychophysiology* 31:189–195.

Hebb, C. 1972. Biosynthesis of acetylcholine in nervous tissue. *Physiol. Rev.* 52:918–957.

Hecker, A.L. 1984. Nutritional conditioning for athletic competition. *Clin. Sports Med.* 3:567–582.

Hedman, R. 1960. Fysiologiska synpunkter på cirkelträning. *Tidskrift i Gymnastik* 87:3–13.

Heglund, N.C., C.R. Taylor, and T.A. McMahon. 1974. Scaling stride frequency and gait to animal size: Mice to horses. *Science* 186:1112–1113.

Hein, H.O., P. Suadicani, and F. Gyntelberg. 1992. Physical fitness or physical activity as a predictor of ischemic heart disease? A 17-year follow-up in the Copenhagen Male Study. *J. Intern. Med.* 232:471–479.

Heiss, H.W., J. Barmeyer, K. Wink, G. Hell, F.J. Cerny, J. Keul, and H. Reindell. 1976. Studies on the regulation of myocardial blood flow in man: I. Training effects on blood flow and metabolism of the healthy heart at rest and during standardized heavy exercise. *Basic Res. Cardiol.* 71:658–675.

Helge, J. W. 2002. Long-term fat diet adaptation effects on performance, training capacity, and fat utilization. *Med. Sci. Sports Exerc.* 34 (9):1499–1504.

Helgerud, J. 1994. Maximal oxygen uptake, anaerobic threshold and running economy in women and men with similar performance level in marathon. *Eur. J. Appl. Physiol.* 68:155–161.

Helgerud, J., F. Ingjer, and S.B. Strømme. 1990. Sex differences in performance-matched marathon runners. *Eur. J. Appl. Physiol. Occup. Physiol.* 61:433–439.

Helin, P., K. Kuoppasalmi, J. Laakso, and M. Harkonen. 1988. Human urinary biogenic amines and some physiological responses during situation stress. *Int. J. Psychophysiol.* 6:125–132.

Heller, J., V. Bunc, J. Pesek, R. Diouha, and J. Novotny. 1991. Physical characteristics of ice hockey performance in young and adult players. *Hung. Rev. Sports Med.* 32:174–177.

Hellström, B. 1965. *Local effects of acclimatization to cold in man.* Oslo: Universitetets-forlaget.

Henneman, E., G. Somjen, and D.O. Carpenter. 1965. Excitability and inhibitability of motoneurons of different sizes. *J. Neurophysiol.* 28:599–620.

Hennig, R., and T. Lømo. 1985. Firing patterns of motor units in normal rats. *Nature* 314:164–166.

Henriksson, J., and J.S. Reitman. 1977. Time course of changes in human skeletal muscle succinate dehydrogenase and cytochrome oxidase activities and maximal oxygen uptake with physical activity and inactivity. *Acta Physiol. Scand.* 99:91–97.

Henriksson, J., J. Svedenhag, E.A. Richter, N.J. Christensen, and H. Galbo. 1985. Skeletal muscle and hormonal adaptation to physical training in the rat: Role of the sympathoadrenal system. *Acta Physiol. Scand.* 123:127–138.

Henriksson, J., and P. Tesch. 1999. Current knowledge on muscle training: Endurance and strength yield complementary effects. *Läkartidningen* 96:56–60.

Hensel, H. 1974. Thermoreceptors. *Annu. Rev. Physiol.* 36:233.

———. 1981. *Thermoreception and temperature regulation.* Monographs of the Physiological Society, no. 38. London: Academic Press.

Hentsch, U., and H.-V. Ulmer. 1984. Trainability of underwater breath-holding time. *Int. J. Sports Med.* 5:343–347.

Hepple, R.T. 2000. Skeletal muscle: Microcirculatory adaptation to metabolic demand. *Med. Sci. Sports Exerc.* 32:117–123.

Hepple, R.T., S.L.M. Mackinnon, J.M. Goodman, S.G. Thomas, and M.J. Plyley. 1997. Resistance and aerobic training in older men: Effects on $\dot{V}O_2$ peak and the capillary supply to skeletal muscle. *J. Appl. Physiol.* 82:1305–1310.

Herd, J.A. 1972. The physiology of strong emotions: Cannon's legacy reexamined. Sixteenth Bowditch Lecture. *Physiologist* 15:5–16.

Hermansen, L. 1973. Oxygen transport during exercise in human subjects. *Acta Physiol. Scand. Suppl.* 399:1–104.

Hermansen, L., B. Ekblom, and B. Saltin. 1970. Cardiac output during submaximal and maximal treadmill and bicycle exercise. *J. Appl. Physiol.* 29:82.

Hermansen, L., E. Hultman, and B. Saltin. 1967. Muscle glycogen during prolonged severe exercise. *Acta Physiol. Scand.* 71:129–39.

Hermansen, L., and J.B. Osnes. 1972. Blood and muscle pH after maximal exercise in man. *J. Appl. Physiol.* 32:304–308.

Hermansen, L., and B. Saltin. 1967. Blood lactate concentration during exercise at acute exposure to altitude. In *Exercise at altitude,* ed. R. Margaria, p. 48. Amsterdam: Excerpta Medica Foundation.

———. 1969. Oxygen uptake during maximal treadmill and bicycle exercise. *J. Appl. Physiol.* 26:31–37.

Hermansen, L., and I. Stensvold. 1972. Production and removal of lactate during exercise in man. *Acta Physiol. Scand.* 86:191–201.

Hermansen, L., and O. Vaage. 1977. Lactate disappearance and glycogen synthesis in human muscle after maximal exercise. *Am. J. Physiol.* 2:E422–E429.

Heron, M.I., and F.J. Richmond. 1993. In-series fiber architecture in long human muscles. *J. Morphol.* 216:35–45.

Herrmann, U., and M. Flanders. 1998. Directional tuning of single motor units. *J. Neurosci.* 18:8402–8416.

Hertig, B.A., and H.S. Belding. 1963. Evaluation and control of heat hazards. In *Temperature: Its measurements and control in science and industry,* ed. J.D. Hardy, vol. 3, part 3, p. 347. New York: Reinhold.

Herzog, W. 2000. Muscle activation and movement control. In *Biomechanics and biology of movement,* ed. B.M. Nigg, B.R. MacIntosh, and J. Mester, pp. 179–192. Champaign, IL: Human Kinetics.

Hespel, P., P. Lijnen, R. Fagard, R. Van Hoof, W. Goossens, and A. Amery. 1988. Effects of training on erythrocyte 2,3-diphosphoglycerate in normal men. *Eur. J. Appl. Physiol. Occup. Physiol.* 57:456–461.

Hettinger, T., and K. Rodahl. 1960. Ein Modifizierter Stufentest zur Messung der Belastungsfähigkeit des Kreislaufes. *Deutsche Medizinische Wochenschrift* 14:553.

Hettinger, T., and W. Wirths. 1953a. Der Energieverbrauch beim Hand und Motorpflügen. *Arbeitsphysiologie* 15:41.

———. 1953b. Über die körperliche Beanspruchung beim Hand und Maschinemelken. *Arbeitsphysiologie* 15:103.

Heunks, L.M.A., M.J. Cody, P.C. Geiger, P.N.R. Dekhuijzen, and G.C. Sieck. 2001. Nitric oxide impairs $Ca^{2+}$ activation and slows cross-bridge cycling kinetics in skeletal muscle. *J. Appl. Physiol.* 91 :2233–2239.

Heunks, L.M.A., H.A. Machiels, P.N.R. Dekhuijzen, Y.S. Prakash, and G.C. Sieck. 2001. Nitric oxide affects sarcoplasmic calcium release in skeletal myotubes. *J. Appl. Physiol.* 91:2117–2124.

Heusner, A.A. 1982. Energy metabolism and body size: Is the 0.75 mass exponent of Kleiber's equation a statistical artifact? *Respir. Physiol.* 48:1–12.

Heymans, C., and E. Neil. 1958. *Reflexogenic areas of the cardiovascular system.* London: Churchill.

Hicks, A.L., J. Kent-Braun, and D.S. Ditor. 2001. Sex differences in human skeletal muscle fatigue. *Exerc. Sport Sci. Rev.* 29:109–112.

Higbie, E.J., K.J. Cureton, G.L. Warren, III, and B.M. Prior. 1996. Effects of concentric and eccentric training on muscle strength, cross-sectional area and neural activation. *J. Appl. Physiol.* 81:2173–2181.

Higginbotham, M.B., K.G. Morris, R.S. Williams, P.A. McHale, R.E. Coleman, and F.R. Cobb. 1986. Regulation of stroke volume during submaximal and maximal upright exercise in normal man. *Circ. Res.* 58:281–291.

Hikosaka, O., K. Nakamura, K. Sakai, and H. Nakahara. 2002. Central mechanisms of motor skill learning. *Curr. Opin. Neurobiol.* 12:217–222.

Hill, A.V. 1926. *Muscular activity.* Baltimore: Williams and Wilkins.

———. 1927. *Living machinery.* New York: Harcourt, Brace and World.

———. 1938. The heat of shortening and the dynamic constants of muscle. *Proc. R. Soc. Lond.* B126:136–195.

———. 1950. The dimensions of animals and their muscular dynamics. *Proc. R. Inst. Great Britain* 34:450–471.

Hill, D.W., D.O. Borden, K.M. Darnaby, and D.N. Hendricks. 1994. Aerobic and anaerobic contributions to exhaustive high-intensity exercise after sleep deprivation. *J. Sports Sci.* 12 (5):455–461.

Hill, D.W., D.O. Borden, K.M. Darnaby, D.N. Hendricks, and D.M. Hill. 1992. Effect of time of day on aerobic and anaerobic responses to high-intensity exercise. *Can. J. Sport Sci.* 17 (4):316–319.

Himann, J.E., D.A. Cunningham, P.A. Rechnitzer, and D.H. Paterson. 1988. Age-related changes in speed of walking. *Med. Sci. Sports Exerc.* 20:161–166.

Hirshman, M.F., H. Wallberg-Henriksson, L.J. Wardzala, E.L. Horton, and E.S. Horton. 1988. Acute exercise increases the number of plasma membrane glucose transporters in rat skeletal muscle. *FEBS Lett.* 238:235–239.

Hitchen, M., D.A. Brodie, and J.B. Harness. 1980. Cardiac responses to demanding mental load. *Ergonomics* 23:379–385.

Hoesslin, H. von. 1888. Ueber die Ursache der scheinbaren Abhängigkeit des Umsatzes von der Grösse der Körperoberfläche. *Arch. F. Anat. Physiol. Physiol. Abt.* 323–379.

Hoey, T., Y.L. Sun, K. Williamson, and X. Xu. 1995. Isolation of two new members of the *NF-AT* gene family and functional characterization of the NF-AT proteins. *Immunity* 2:461–472.

Hofbauer, L.C., F. Gori, B.L. Riggs, D.L. Lacey, C.R. Dunstan, T.C. Spelsberg, and S. Khosla. 1999. Stimulation of osteoprotegerin ligand and inhibition of osteoprotegerin production by glucocorticoids in human osteoblastic lineage cells: Potential paracrine mechanisms of glucocorticoid-induced osteoporosis. *Endocrinology* 140:4382–4389.

Hoff, J., J. Helgerud, and U. Wislöff. 1999. Maximal strength training improves work economy in trained female cross-country skiers. *Med. Sci. Sports Exerc.* 31:870–877.

Högberg, P. 1952. How do stride length and stride frequency influence the energy output during running? *Arbeitsphysiologie* 14:437.

Högberg, P., and O. Ljunggren. 1947. Uppvärmningens inverkan på löpprestationerna. *Svensk Idrott,* vol. 40.

Hogervorst, E., W.J. Riedel, E. Kovacs, F. Brouns, and J. Jolles. 1999. Caffeine improves cognitive performance after strenuous physical exercise. *Int. J. Sports Med.* 20:354–361.

Hoh, J.F., S. Hughes, and J.F. Hoy. 1988. Myogenic and neurogenic regulation of myosin gene expression in cat jaw-closing muscles regenerating in fast and slow limb muscle beds. *J. Muscle Res. Cell Motil.* 9:59–72.

Hökfelt, T., O. Johansson, A. Ljungdahl, J.M. Lundberg, and M. Schultzberg. 1980. Peptidergic neurones. *Nature* 284:515–521.

Holewijn, M. 1990. Physiological strain due to load carrying. *Eur. J. Appl. Physiol.* 61:237–245.

Holewijn, M., R. Heus, and L.J.A. Wammes. 1992. Physiological strain due to load carrying in heavy footwear. *Eur. J. Appl. Physiol. Occup. Physiol.* 65:129–134.

Hollenbach, K.A., E. Barrett-Connor, S.L. Edelstein, and T. Holbrook. 1993. Cigarette smoking and bone mineral density in older men and women. *Am. J. Public Health* 83:1265–1270.

Hollmann, W., and T. Hettinger. 1980. *Sportsmedizin-Arbeits-und Trainingsgrundlagen.* 2nd ed. Stuttgart: F.K. Schattauer Verlag.

Holloszy, J.O., and F.W. Booth. 1976. Biochemical adaptations to endurance exercise in muscle. *Annu. Rev. Physiol.* 38:273–291.

Holm, H., M.J. Jarvis, M.A. Russell, and C. Feyerabend. 1992. Nicotine intake and dependence in Swedish snuff takers. *Psychopharmacology* 108:507–511.

Holmboe, J., H. Bell, and N. Norman. 1975. Urinary excretion of catecholamines and steroids in military cadets exposed to prolonged stress. *Försvarsmedicin* 11:183.

Holmdahl, D.E., and B.E. Ingelmark. 1948. Der Bau der Gelenkknorpels unter verschiedenen funktionellen Verhältnissen. *Acta Anat.* (Basel) 6:309–375.

Holmér, I. 1974a. Physiology of swimming man. *Acta Physiol. Scand. Suppl.* 407:1–55.

———. 1974b. Energy cost of arm stroke, leg kick and the whole stroke in competitive swimming styles. *Eur. J. Appl. Physiol.* 33:105–118.

———. 1983. Role of clothing for man's heat exchange with the environment. Proceedings of the International Conference on Protective Clothing Systems, Aug. 23–27, 1981, pp. 31–50, Stockholm, Sweden.

———. 1992. Protective clothing against cold-performance standards as method for preventive measures. *Arct. Med. Res.* 51 (suppl. 7):94–98.

Holmér, I., and P.-O. Åstrand. 1972. Swimming training and maximal oxygen uptake. *J. Appl. Physiol.* 33:510–513.

Holmér, I., and U. Bergh. 1974. Metabolic and thermal response to swimming in water at varying temperatures. *J. Appl. Physiol.* 37:702.

Holmér, I., and S. Elnäs. 1981. Physiological evaluation of the resistance to evaporative heat transfer by clothing. *Ergonomics* 24:63.

Holmér, I., and L. Gullstrand. 1980. Physiological responses to swimming with a controlled frequency of breathing. *Scand. J. Sports Sci.* 2:1–6.

Holmér, I., E.M. Stein, B. Saltin, B. Ekblom, and P.-O. Åstrand. 1974. Hemodynamic and respiratory responses compared in swimming and running. *J. Appl. Physiol.* 37:49.

Homonko, D.A., and E. Theriault. 2000. Downhill running preferentially increases CGRP in fast glycolytic muscle fibers. *J. Appl. Physiol.* 89:1928–1236.

Honda, Y., S. Myojo, S. Hasegawa, T. Hasegawa, and J. Severinghaus. 1979. Decreased exercise hyperpnea in patients with bilateral carotid chemoreceptor resection. *J. Appl. Physiol. Respir. Environ. Exerc. Physiol.* 46:908.

Honig, C.R., and C.L. Odoroff. 1981. Calculated dispersion of capillary transit times: Significance for oxygen exchange. *Am. J. Physiol.* 240:H199.

Hoogerwerf, S. 1929. Elektrokardiographische Untersuchungen der Amsterdamer Olympiakämpfer. *Arbeitsphysiologie* 2:61.

Hopkins, S.R., T.P. Gavin, N.M. Siafakas, L.J. Haseler, I.M. Olfert, H. Wagner, and P.D. Wagner. 1998. Effect of prolonged, heavy exercise on pulmonary gas exchange in athletes. *J. Appl. Physiol.* 85:1523–1532.

Hoppeler, H., and D. Desplanches. 1992. Muscle structural modifications in hypoxia. *Int. J. Sports Med.* 13:S166–S168.

Hoppeler, H., E. Kleinert, C. Schlegel, H. Claassen, H. Howald, and P. Cerretelli. 1990. Muscular exercise at high altitude: II. Morphological adaptation of skeletal muscle to chronic hypoxia. *Int. J. Sports Med.* 11 (suppl.):S3–S9.

Hopper, J.L., and E. Seeman. 1994. The bone density of female twins discordant for tobacco use. *N. Engl. J. Med.* 330:430–431.

Horber, F.F., S.A. Kohler, K. Lippuner, and P. Jaeger. 1996. Effect of regular physical training on age-associated alteration of body composition in men. *Eur. J. Clin. Invest.* 26:279–285.

Horowits, R. 1992. Passive force generation and titin isoforms in mammalian skeletal muscle. *Biophys. J.* 61:392–398.

Horowitz, J.F., L.S. Sidossis, and E.F. Coyle. 1994. High efficiency of type I muscle fibers improves performance. *Int. J. Sports Med.* 15:152–157.

Horstman, D.H., and E. Christensen. 1982. Acclimatization to dry heat: Active men versus active women. *J. Appl. Physiol.* 52 (4):825.

Horstman, D., R. Weiskopf, and R.E. Jackson. 1980. Work capacity during 3-week sojourn at 4,300 m: Effects of relative polycythemia. *J. Appl. Physiol.* 49:311–318.

Hortobagyi, T., N.J. Lambert, and W.P. Kroll. 1991. Voluntary and reflex responses to fatigue with stretch-shortening exercise. *Can. J. Sport Sci.* 16:142–150.

Horwitz, A.F., D.L. Schotland, and C. Franzini-Armstrong. 1994. The plasma membrane of the muscle fiber: Composition and structure. In *Myology: Basic and clinical*, ed. A.G. Engel and C. Franzini-Armstrong, pp. 200–222. New York: McGraw-Hill.

Houdas, Y., and E.F.J. Ring. 1982. *Human body temperature.* New York: Plenum.

Houdijk, H., E.A.M. Heijnsdijk, J.J. de Koning, G. de Groot, and M.F. Bobbert. 2000. Physiological responses that account for the increased power output in speed skating using klapskates. *Eur. J. Appl. Physiol.* 83:283–288.

Hough, T. 1902. Ergographic studies in muscle soreness. *Am. J. Physiol.* 7:76–92.

Houk, J.C. 1979. Regulation of stiffness by skeletomotor reflexes. *Annu. Rev. Physiol.* 41:99–114.

Hounsgaard, J. 2002. Motoneurons do what motoneurons have to do. *J. Physiol.* 538 (1):4.

Housh, D.J., T.J. Housh, G.O. Johnson, and W.K. Chu. 1992. Hypertrophic response to unilateral concentric isokinetic resistance training. *J. Appl. Physiol.* 73:65–70.

Houston, M.E. 1999. Gaining weight: The scientific basis of increasing skeletal muscle mass. *Can. J. Appl. Physiol.* 24:305–316.

Houtkooper, L.B. 2000. Body composition. In *Sport nutrition for health and performance*, ed. M.M. Manore and J.L. Thompson, pp. 199–219. Champaign, IL: Human Kinetics.

Howald, H. 1982. Training-induced morphological and functional changes in skeletal muscle. *Int. J. Sports Med.* 3:1–12.

Howald, H., D. Pette, J.-A. Simoneau, A. Uber, H. Hoppeler, and P. Cerretelli. 1990. Muscular exercise at high altidude: III. Effects of chronic hypoxia on muscle enzymes. *Int. J. Sports Med.* 11 (suppl.):S10–S14.

Howell, J.N., and H. Oetliker. 1987. Effects of repetitive activity, ruthenium red and elevated extracellular calcium on frog skeletal muscle: Implications for T-tubule conduction. *Can. J. Physiol. Pharmacol.* 65:691–696.

Howley, E.T., D.R. Bassett, Jr., and H.G. Welch. 1995. Criteria for maximal oxygen uptake: Review and commentary. *Med. Sci. Sports Exerc.* 27:1292–1301.

Howley, E.T., D.L. Colacino, and T.C. Swensen. 1992. Factors affecting the oxygen cost of stepping on an electronic stepping ergometer. *Med. Sci. Sports Exerc.* 24:1055–1058.

Howley, E.T., and B.D. Franks. 1992. *Health/fitness instructor's handbook.* 2nd ed. Champaign, IL: Human Kinetics.

Hubbard, A.W. 1960. Homokinetics: Muscular function in human movement. In *Science and medicine of exercise and sports,* ed. W.R. Johnson, pp. 7–39. New York: Harper and Row.

Hudlicka, O. 1982. Growth of capillaries in skeletal and cardiac muscle. *Circ. Res.* 50:451–461.

Hughes, E.F., S.C. Turner, and G.A. Brooks. 1982. Effects of glycogen depletion and pedaling speed on "anaerobic threshold." *J. Appl. Physiol. Respir. Environ. Exerc. Physiol.* 52:1598–1607.

Hughes, S.M., M.M. Chi, O.H. Lowry, and K. Gundersen. 1999. Myogenin induces a shift of enzyme activity from glycolytic to oxidative metabolism in muscles of transgenic mice. *J. Cell Biol.* 145:633–642.

Hughson, R.L., C.A. Russel, and M.R. Marshall. 1984. Effect of metropol on cycle and treadmill maximal exercise performance. *J. Cardiac Rehabil.* 4:27.

Huie, M.J. 1996. The effects of smoking on exercise performance. *Sports Med.* 22:355–359.

Huijing, P.A. 1992. Elastic potential of muscle. In *Strength and power in sport,* ed. P.V. Komi, pp. 151–68. Oxford: Blackwell Science.

Huiskes, R., R. Ruimerman, G.H. van Lenthe, and J.D. Janssen. 2000. Effects of mechanical forces on maintenance and adaptation of form in trabecular bone. *Nature* 405:704–706.

Hulett, G. 1903. Beziehung zwischen negativem Druck und osmotischem Druck. *Z. Phys. Chem.* 42:353–368.

Hultborn, H., R. Katz, and R. Mackel. 1988. Distribution of recurrent inhibition within a motor nucleus: II. Amount of recurrent inhibition in motoneurones to fast and slow units. *Acta Physiol. Scand.* 134:363–374.

Hultman, E. 1967. Studies on muscle metabolism of glycogen and active phosphate in man with special reference to exercise and diet. *Scand. J. Clin. Lab. Invest.* 19 (suppl. 94):1–63.

Hultman, E., and H. Sjöholm. 1983. Substrate availability. In *Biochemistry of exercise,* ed. H.G. Knuttgen, J.A. Vogel, and J. Poortmans, pp. 63–75. International Series on Sport Sciences, vol. 13. Champaign, IL: Human Kinetics.

Hunding, A., Jordal, R. and Paulev, P.E. 1981. Runners' anemia and iron deficiency. *Acta Med. Scand.* 209, 315–318.

Hunter, S.K. and R.M. Enoka. 2002. Sex differences in the fatigability of arm muscles depends on absolute force during isometric contractions. *J. Appl. Physiol.* 91:2686–2694.

Hurley, B.F., and J.M. Hagberg. 1998. Optimizing health in older persons: Aerobic or strength training? *Exerc. Sport Sci. Rev.* 26:61–89.

Hurley, B.F., J.M. Hagberg, W.K. Allen, D.R. Seals, J.C. Young, R.W. Cuddihee, and J.O. Holloszy. 1984. Effect of training on blood lactate levels during submaximal exercise. *J. Appl. Physiol. Respir. Environ. Exerc. Physiol.* 56:1260–1264.

Hurley, B.F., R.A. Redmond, R.E. Pratley, M.S. Treuth, M.A. Rogers, and A.P. Goldberg. 1995. Effects of strength training on muscle hypertrophy and muscle cell disruption in older men. *Int. J. Sports Med.* 16:378–384.

Hurtado, A., C. Merino, and E. Delgado. 1945. Influence of anoxemia on the hemopoietic activity. *Arch. Int. Med.* 75:284.

Huston, T.P., J.C. Puffer, and W.M. Rodney. 1985. The athletic heart syndrome. *N. Engl. J. Med.* 313:24–32.

Hutton, R.S. 1992. Neuromuscular basis of stretching exercises. In *Strength and power in sport,* ed. P.V. Komi, pp. 29–38. Oxford: Blackwell Science.

Huxley, A.F. 1974. Muscular contraction. *J. Physiol.* (Lond.) 243:1–43.

Ikai, M., and T. Fukunaga. 1968. Calculation of muscle strength per unit cross-sectional area of human muscle by means of ultrasonic measurement. *Int. Z. Angew. Physiol. Einschl. Arbeitsphysiol.* 26:26–32.

———. 1970. A study on training effect on strength per unit cross-sectional area of muscle by means of ultrasonic measurement. *Int. Z. Angew. Physiol.* 28:173–180.

Ikai, M., and A.H. Steinhaus. 1961. Some factors modifying the expression of human strength. *J. Appl. Physiol.* 16:157–163.

Ikegami, Y., S. Hiruta, H. Ikegami, and M. Miyamura. 1988. Development of a telemetry system for measuring oxygen uptake during sports activities. *Eur. J. Appl. Physiol.* 57:622–626.

Illg, D., and D. Pette. 1979. Turnover rates of hexokinase I, phosphofructokinase, pyruvate kinase and creatine kinase in slow-twitch soleus muscle and heart of the rabbit. *Eur. J. Biochem.* 97:267–273.

Illnerova, H. 1991. Mammalian circadian clock and its resetting. *NIPS Int. Union Physical Sci.* 6:129–134.

Ilmarinen, J., R. Ilmarinen, A. Koskela, O. Korhonen, F. Fardy, T. Partanen, and J. Rutenfranz. 1979. Training effects of stair-climbing during office hours in female employees. *Ergonomics* 22:507–516.

Inbar, O., O. Bar-Or, R. Dotan, and B. Gutin. 1981. Conditioning versus exercise in heat as methods for acclimatizing 8- to 10-year-old boys to dry heat. *J. Appl. Physiol.* 50 (2):406.

Inbar, O., R. Dotan, T. Trousil, and Z. Dvir. 1983. The effect of bicycle crank-length variation upon power performance. *Ergonomics* 26 (12):1139–1146.

Ingelmark, B.E., and R. Ekholm. 1948. A study on variations in the thickness of articular cartilage in association with rest and periodical load. *Uppsala Läkareförenings Förhandlingar* 53:61.

Ingen Schenau, G.J. van., R. Jacobs, and J.J. de Koning. 1991. Can cycle power output predict sprint running performance? *Eur. J. Appl. Physiol.* 63:255–260.

Ingjer, F. 1979a. Capillary supply and mitochondrial content of different skeletal muscle fiber types in untrained and endurance-trained men: A histochemical and ultrastructural study. *Eur. J. Appl. Physiol.* 40:197–209.

———. 1979b. Effects of endurance training on muscle fibre ATP-ase activity, capillary supply and mitochondrial content in man. *J. Physiol.* (Lond.) 294:419–432.

———. 1992. Development of maximal oxygen uptake in young elite male cross-country skiers: A longitudinal study. *J. Sports Sci.* 10:49–63.

Ingjer, F., and P. Brodal. 1978. Capillary supply of skeletal muscle fibers in untrained and endurance-trained women. *Eur. J. Appl. Physiol. Occup. Physiol.* 38:291–299.

Ingjer, F., and H.A. Dahl. 1979. Dropouts from an endurance training program. *Scand. J. Sports Sci.* 1:20–22.

Ingjer, F., and K. Myhre. 1992. Physiological effects of altitude training on elite male cross-country skiers. *J. Sports Sci.* 10:37–47.

Ingjer, F., and S.B. Strømme. 1979. Effects of active, passive or no warm-up on the physiological response to heavy exercise. *Eur. J. Appl. Physiol.* 40:273–282.

Inoue, Y. 1996. Longitudinal effects of age on heat-activated sweat gland density and output in healthy active older men. *Eur. J. Appl. Physiol.* 74:72–77.

*International Journal of Circumpolar Health.* 2000. 59 (2):89–148. (Contains a number of tutorial and historical articles on frostbite.)

Irving, L. 1966. Adaptations to cold. *Sci. Am.* 214:94.

———. 1967. Ecology and thermoregulation. In *Les concepts de Claude Bernard sur le milieu intérieur,* p. 381. Paris: Masson.

Israel, S. 1958. Die Erscheinungsformen des Uebertrainings. *Sportsmedizin* 9:207–209.

Issekutz, B., Jr. 1964. Effect of exercise on the metabolism of plasma free fatty acids. Chap. 11 in *Fat as a tissue,* ed. K. Rodahl and B. Issekutz, Jr. New York: McGraw-Hill.

———. 1981. Effects of glucose infusion on hepatic and muscle glycogenolysis in exercising dogs. *Am. J. Physiol.* 240 (5):E451–E457.

Issekutz, B., Jr., N.C. Birkhead, and K. Rodahl. 1962. Use of respiratory quotients in assessment of aerobic work capacity. *J. Appl. Physiol.* 17:47–50.

———. 1963. Effect of diet on work metabolism. *J. Nutr.* 79:109.

Issekutz, B., Jr., J.J. Blizzard, N.C. Birkhead, and K. Rodahl. 1966. Effect of prolonged bed rest on urinary calcium output. *J. Appl. Physiol.* 21:1013–1020.

Issekutz, B., Jr. and H. Miller. 1962. Plasma free fatty acids during exercise and the effect of lactic acid. *Proc. Soc. Exp. Biol. Med.* 110:237–239.

Issekutz, B., Jr., H.I. Miller, P. Paul, and K. Rodahl. 1964. Source of fat oxidation in exercising dogs. *Am. J. Phys.* 207:583.

———. 1965. Aerobic work capacity and plasma FFA turnover. *J. Appl. Physiol.* 20:293–296.

Issekutz, B., Jr., K. Rodahl, and N.C. Birkhead. 1962. Effect of severe cold stress on the nitrogen balance of men under different dietary conditions. *J. Nutr.* 78:189.

Issekutz, B., Jr., W.A.S. Shaw, and T.B. Issekutz. 1975. Effect of lactate on FFA and glycerol turnover in resting and exercising dogs. *J. Appl. Physiol.* 39 (3):349–352.

Ito, M., Y. Kawakami, Y. Ichinose, S. Fukashiro, and T. Fukunaga. 1998. Nonisometric behavior of fascicles during isometric contractions of a human muscle. *J. Appl. Physiol.* 85:1230–1235.

Ito, M., M. Yoshida, K. Obata, N. Kawai, and M. Udo. 1970. Inhibitory control of intracerebellar nuclei by the Purkinje cell axons. *Exp. Brain Res.* 10:64–80.

Ivy, J.L., D.L. Costill, W.J. Fink, and R.V. Lower. 1979. Influence of caffeine and carbohydrate feedings on endurance performance. *Med. Sci. Sports Exerc.* 11:6–11.

Ivy, J.L., A.L. Katz, C.L. Cutler, W.M. Sherman, and E.F. Coyle. 1988. Muscle glycogen synthesis after exercise: Effect of time of carbohydrate ingestion. *J. Appl. Physiol.* 64:1480–1485.

Iwanaga, K., M. Sakurai, T. Minami, Y. Kato, K. Sairyo, and Y. Kikuchi. 1996. Is the intracellular pH threshold an anaerobic threshold from the view point of intracellular events? A brief review. *Appl. Human Sci.* 15:59–65.

Iwata, H., ed. 1974. Body reaction by vibration. In *Vibration syndrome,* p. 21. Tokyo: Kindaishupann.

Izumo, S., B. Nadal-Ginard, and V. Mahdavi. 1986. All members of the MHC multigene family respond to thyroid hormone in a highly tissue-specific manner. *Science* 231:597–600.

Jackson, C.G.R., and B.J. Sharkey. 1988. Altitude, training and human performance. *Sports Med.* 6:279–284.

Jacobson, B.H., and B.M. Edgley. 1987. Effects of caffeine on simple reaction time and movement time. *Aviat. Space Environ. Med.* 58:1153–1156.

Jacobson, B.H., M.D. Weber, L. Claypool, and L.E. Hunt. 1992. Effect of caffeine on maximal strength and power in elite male athletes. *Br. J. Sports Med.* 26:276–280.

Jami, L. 1992. Golgi tendon organs in mammalian skeletal muscle: Functional properties and central actions. *Physiol. Rev.* 72:623–666.

Jansky, L., H. Janakova, B. Ulicny, P. Sramek, V. Hosek, J. Heller, and J. Parizkova. 1996. Changes in thermal homeostasis in humans due to repeated cold water immersion. *Pflügers Arch.* 432:368–372.

Jansson, E., and L. Kaijser. 1977. Muscle adaptation to extreme endurance training in man. *Acta Physiol. Scand.* 100:315–324.

Janz, K.F., T. L. Burns, J.D. Witt, and L.T. Mahoney. 1998. Longitudinal analysis of scaling $\dot{V}O_2$ for differences in body size during puberty: The Muscatine Study. *Med. Sci. Sports Exerc.* 30 (9):1436–1444.

Jasinskas, C.L., B.A. Wilson, and J. Hoare. 1980. Entrainment of breathing rate to movement frequency during work at two intensities. *Respir. Physiol.* 42:199.

Jenner, D.A., V. Reynolds, and G.A. Harrison. 1980. Catecholamine excretion rates and occupation. *Ergonomics* 23:237–246.

Jensen, B.B., B. Schibye, K. Søgaard, E.B. Simonsen, and G. Sjøgaard. 1993. Shoulder muscle load and muscle fatigue among industrial sewing-machine operators. *Eur. J. Appl. Physiol.* 67:467–475.

Jensen, B.R., B. Laursen, and A. Ratkevicius. 1999. Foreram extensor muscle fatigue in young and elderly subjects induced by four hours of computer mouse work. International Society of Biomechanics XVIIth Congress, Calgary, Canada, Aug. 1999.

Jensen, B.R., M. Pilegaard, and G. Sjøgaard. 2000. Motor unit recruitment and rate coding in response to fatiguing shoulder abductions and subsequent recovery. *Eur. J. Appl. Physiol.* 83:190–199.

Jensen, C., O. Vasseljen, and R.H. Westgaard. 1993. The influence of electrode position on bipolar surface electromyogram recordings of the upper trapezius muscle. *Eur. J. Appl. Physiol.* 67:266–273.

Jensen-Urstad, K., B. Saltin, M. Ericson, N. Storck, and M. Jensen-Urstad. 1997. Pronounced resting bradycardia in male elite runners is associated with high heart rate variability. *Scand. J. Med. Sci. Sports* 7:274–278.

Jeukendrup, A.E., W.H.M. Saris, and A.J.M. Wagenmakers. 1998. Fat metabolism during exercise: A review. Part I. Fatty acid mobilization and muscle metabolism. *Int. J. Sports Med.* 19:231–244. Part II. Regulation of metabolism and the effects of training. *Int. J. Sports Med.* 19:293–302. Part III: Effects of nutritional interventions. *Int. J. Sports Med.* 19:371–379.

Ji, L.L. 1995. Exercise and oxidative stress: Role of the cellular antioxidant systems. *Exerc. Sport Sci. Rev.* 23:135–166.

Jiang, B., R.R. Roy, I.V. Polyakov, I.B. Krasnov, and V.R. Edgerton. 1992. Ventral horn cell responses to spaceflight and hindlimb suspension. *J. Appl. Physiol.* 73:107S-111S.

Johnson, B.D., W.G. Reddan, D.F. Pegelow, K.C. Seow, and J.A. Dempsey. 1991. Flow limitation and regulation of functional residual capacity during exercise in a physically active aging population. *Am. Rev. Respir. Dis.* 143:960–967.

Johnson, B.D., K.W. Saupe, and J.A. Dempsey. 1992. Mechanical constraints on exercise hyperpnea in endurance athletes. *J. Appl. Physiol.* 73:874–886.

Johnson, J.M. 1992. Exercise and the cutaneous circulation. *Exerc. Sport Sci. Rev.* 20:59–98.

Johnson, J.M., G.L. Brengelmann, and L.B. Rowell. 1976. Interactions between local and reflex influences on human forearm skin blood flow. *J. Appl. Physiol.* 41:826.

Johnson, M.L., B.S. Burke, and J. Mayer. 1956. Relative importance of inactivity and overeating in the energy balance of obese high school girls. *Am. J. Clin. Nutr.* 4:37.

Johnson, R.E., L. Brouha, and R.C. Darling. 1942. A test of physical fitness for strenuous exertion. *Rev. Can. Biol.* 1:491.

Jokl, E. 1964. *Heart and sport.* Springfield, IL: Charles C Thomas.

Jokl, E., and P. Jokl, eds. 1968. *Exercise and altitude.* New York: Karger.

Jones, D.A. 1996. High- and low-frequency fatigue revisited. *Acta Physiol. Scand.* 156:265–270.

Jones, D.A., and O.M. Rutherford. 1987. Human muscle strength training: The effects of three different regimens and the nature of the resultant changes. *J. Physiol.* (Lond.) 391:1–11.

Jones, D.M., and K. Rodahl. 1962. Effect of rope skipping on physical work capacity. *Res. Q.* 33:236–238.

Jones, G.R., and I. Newhouse. 1997. Sport-related hematuria: A review. *Clin. J. Sport Med.* 7:119–125.

Jones, N., and R. Ehrsam. 1982. The anaerobic threshold. *Exerc. Sport Sci. Rev.* 10:49–83.

Jones, N.L., and E.J.M. Campbell. 1982. *Clinical exercise testing.* 2nd ed. Philadelphia: Saunders.

Jonsson, B.G. 1981. A statistical evaluation of data gathered in connection with exercise tests. Thesis, Karolinska inst., Stockholm.

Józsa, L., and P. Kannus. 1997. *Human tendons: Anatomy, physiology and pathology.* Champaign, IL: Human Kinetics.

Judex, S., T.S. Gross, R.C. Bray, and R.F. Zernicke. 1997. Adaptation of bone to physiological stimuli. *J. Biomech.* 30:421–429.

Juel, C. 1988. Muscle action potential propagation velocity changes during activity. *Muscle Nerve* 11:714–719.

Juel, C., and A.P. Halestrap. 1999. Lactate transport in skeletal muscle: Role and regulation of the monocarboxylate transporter. *J. Physiol.* 517 (3):633–642.

Juhlin-Dannfelt, A., G. Ahlborg, L. Hagenfeldt, L. Jorfeldt, and P. Felig. 1977. Influence of ethanol on splanchnic and skeletal muscle substrate turnover during prolonged exercise in man. *Am. J. Physiol.* 233:E195–E202.

Jurata, L.W., J.B. Thomas, and S.L. Pfaff. 2000. Transcriptional mechanisms in the development of motor control: Current opinion. *Neurobiology* 10:72–79.

Juurup, A., and L. Muido. 1946. On acute effects of cigarette smoking on oxygen consumption, pulse rate, breathing rate and blood pressure in working organisms. *Acta Physiol. Scand.* 11:48.

Kaarma, H., H. Tapfer, G. Veldre, M. Thetloff, L. Saluste, and J. Peterson. 1996. Some principles to be considered when using young women's anthropometric data. *Biol. Sport* (Warsaw) 13:127–135.

Kadi, F., A. Eriksson, S. Holmner, G.S. Butler-Browne, and L.E. Thornell. 1999. Cellular adaptation of the trapezius muscle in strength-trained athletes. *Histochem. Cell Biol.* 111:189–195.

Kadi, F., and L.E. Thornell. 1999. Training affects myosin heavy chain phenotype in the trapezius muscle of women. *Histochem. Cell Biol.* 112:73–78.

Kagan, E.M., P. Dolgin, P.M. Kaplan, C.O. Linetskaja, J.L. Lubarsky, M.F. Neumann, J.J. Semernin, J.S. Starch, and P. Spilberg. 1928. Physiologische Vergleichs-untersuchung der Hand- und Fleiss- (Conveyor) Arbeit. *Arch. Hyg.* 100:335.

Kakuda, N., M. Nagaoka, and J. Wessberg. 1999. Common modulation of motor unit pairs during slow wrist movement in man. *J. Physiol.* 520:929–940.

Kalmar, J.M., and E. Carafelli. 1999. Effects of caffeine on neuromuscular function. *J. Appl. Physiol.* 87:801–808.

Kamon, E. 1974. Instrumentation for work physiology. *Trans. N.Y. Acad. Sci.* 36 (7):625–639.

———. 1979. Scheduling cycles of work for hot ambient conditions. *Ergonomics* 22:427.

Kamon, E., and K.B. Pandolf. 1972. Maximal aerobic power during laddermill climbing, uphill running and cycling. *J. App. Physiol.* 32:467–473.

Kanchisa, H., I. Nemoto, H. Okuyama, S. Ikegawa, and T. Fukunaga. 1996. Force generation capacity of knee extensor muscles in speed skaters. *Eur. J. Appl. Physiol. Occup. Physiol.* 73:544–551.

Kandarian, S.C. and E.J. Stevenson. 2002. Molecular events in skeletal muscle during disuse atrophy. *Exerc. Sport Sci. Rev.* 30:111–116.

Kandel, E.R. 1979. Small systems of neurons. *Sci. Am.* 241 (3):67.

Kandel, E.R., J.H. Schwartz, and T.M. Jessell. 2000. *Principles of neural science.* 4th ed. New York: McGraw-Hill.

Kannus, P., H. Haapasalo, M. Sankelo, H. Sievanen, M. Pasanen, A. Heinonen, P. Oja, and I. Vuori. 1995. Effect of starting age of physical activity on bone mass in the dominant arm of tennis and squash players. *Ann. Intern. Med.* 123:27–31.

Kanstrup, I.-L., and B. Ekblom. 1978. Influence of age and physical activity on central hemodynamics and lung function in active adults. *J. Appl. Physiol.* 45 (5):709–717.

———. 1982. Acute hypervolemia, cardiac performance and aerobic power during exercise. *J. App. Physiol.* 52:1186–1191.

———. 1984. Blood volume and hemoglobin concentration as determinants of maximal aerobic power. *Med. Sci. Sports Exerc.* 16:256.

Kanstrup, I.-L., J. Marving, N. Gadsbøll, H. Lønborg-Jensen, and P.F. Høilund-Carlsen. 1995. Left ventricle haemodynamics and vaso-active hormones during graded supine exercise in healthy male subjects. *Eur. J. Appl. Physiol.* 72:86–94.

Kanstrup, I.-L., J. Marving, and P.F. Høilund-Carlsen. 1992. Acute plasma expansion: Left ventricular hemodynamics and endocrine function during exercise. *J. Appl. Physiol.* 73:1791–1796.

Kanter, M. 1995. Free radicals and exercise: Effects of nutritional antioxidant supplementation. *Exerc. Sport Sci. Rev.* 23:375–397.

Kao, J.T., M. Pink, F.W. Jobe, and J. Perry. 1995. Electromyographic analysis of the scapular muscles during a golf swing. *Am. J. Sports Med.* 23:19–23.

Kaplan, F.S. 1987. Osteoporosis: Pathophysiology and prevention. *Clin. Symp.* 39:1–32.

Karatzaferi, C., G. Giakas, and D. Ball. 1999. Fatigue profile: A numerical method to examine fatigue in cycle ergometry. *Eur. J. Appl. Physiol. Occup. Physiol.* 80:508–510.

Kardel, K.R., and T. Kase. 1998. Training in pregnant women: Effects on fetal development and birth. *Am. J. Obstet. Gynecol.* 178:280–86.

Karlsson, J. 1997. *Antioxidants and exercise.* Champaign, IL: Human Kinetics.

Karlsson, J., P.-O. Åstrand, and B. Ekblom. 1967. Training of the oxygen transport system in man. *J. Appl. Physiol.* 22:1061–1065.

Karlsson, J., F. Bonde-Petersen, J. Henriksson, and H.G. Knuttgen. 1975. Effects of previous exercise with arms or legs on metabolism and performance in exhaustive exercise. *J. Appl. Physiol.* 38:763–767.

Karlsson, J., L. Hermansen, G. Agnevik, and B. Saltin. 1967. *Energy demand during running.* Idrottsfysiologi Rapport no. 4. Stockholm: Trygg-Hansa.

Karlsson, J., L. Hermansen, G. Agnevik, and B. Saltin. 1972. "Löpning" [Running]. Idrottsfysiologi. Rapport No. 4. Stockholm: Trygg-Hansa.

Karlsson, J., and I. Jacobs. 1982. Onset of blood lactate accumulation during muscular exercise as a threshold concept. *Int. J. Sports Med.* 3:190–201.

Karlsson, J., and B. Saltin. 1971. Diet, muscle glycogen and endurance performance. *J. Appl. Physiol.* 31:203–206.

Kasch, F.W., W.H. Phillips, W.D. Ross, J.E.L. Carter, and J.L. Boyer. 1966. A comparison of maximal oxygen uptake by treadmill and step-test procedures. *J. Appl. Physiol.* 21:1387–1388.

Katić, R. 1996. The influence of morphological characteristics on selected motor variables in boys and in girls. *Biol. Sport* (Warsaw) 13 (1):47–53.

Katz, A., D.L. Costill, D.S. King, M. Hargreaves, and W.J. Fink. 1984. Maximal exercise tolerance after induced alkalosis. *Int. J. Sports Med.* 5:107–110.

Katz, B., and S. Thesleff. 1957. On the factors which determine the amplitude of the "miniature end-plate potential." *J. Physiol.* 137:267–278.

Kaufman, K.R., K.N. An, and E.Y. Chao. 1989. Incorporation of muscle architecture into the muscle length–tension relationship. *J. Biomech.* 22:943–948.

Kavanagh, T., and R.J. Shephard. 1976. Maximum exercise tests on "postcoronary" patients. *J. Appl. Physiol.* 40:611–618.

Kawakami, Y., T. Abe, and T. Fukunaga. 1993. Muscle-fiber pennation angles are greater in hypertrophied than in normal muscles. *J. Appl. Physiol.* 74:2740–2744.

Kawakami, Y., D. Nozaki, A. Matsuo, and T. Fukunaga. 1992. Reliability of measurement of oxygen uptake by a portable telemetric system. *Eur. J. Appl. Physiol.* 65:409–414.

Kearney, R.E., M. Lortie, and R.B. Stein. 1999. Modulation of stretch reflexes during imposed walking movements of the human ankle. *J. Neurophysiol.* 81:2893–2902.

Keatinge, W.R. 1978. *Survival in cold water.* 2nd ed. Oxford: Blackwell.

Keatinge, W.R., and R.E.G. Sloan. 1975. Deep body temperature from aural canal with servo-controlled heating to outer ear. *J. Appl. Physiol.* 38:919–921.

Keele, C.A., E. Neil, and N. Joels. 1982. *Samson Wright's applied physiology.* Oxford: Oxford University Press.

Kelbaek, H., T. Gjorup, I. Brynjolf, N. Christensen, and J. Godtfredsen. 1985. Acute effects of alcohol on left ventricular function in healthy subjects during rest and upright exercise. *Am. J. Cardiol.* 55:164–167.

Kelley, G., and Z.V. Tran. 1995. Aerobic exercise and normotensive adults: A meta-analysis. *Med. Sci. Sports Exerc.* 27:1371–1377.

Kelly, A.M., and Rubinstein, N.A. 1994. The diversity of muscle fiber types and its origin during development. In *Myology: Basic and clinical*, 2nd ed., ed. A.G. Engel and C. Franzini-Armstrong, pp. 119–133. New York: McGraw-Hill.

Kenshalo, D.R., J.P. Nafe, and B. Brooks. 1961. Variations in thermal sensitivity. *Science* 134:104.

കenttä, G., and P. Hassmén. 1998. Overtraining and recovery: A conceptual model. *Sports Med.* 26:1–16.

Kernell, D. 1998. Muscle regionalization. *Can. J. Appl. Physiol.* 23:1–22.

Kerslake, D.M. 1963. Errors arising from the use of mean heat exchange coefficients in calculation of the heat exchanges of a cylindrical body in a transverse wind. In *Temperature: Its measurement and control in science and industry*, ed. J.D. Hardy, vol. 3, part 3, p. 183. New York: Reinhold.

———. 1972. *Monographs of the Physiological Society: The stress of hot environment.* Cambridge: Cambridge University Press.

Kessler, R.C., K.A. McGonagle, S. Zhao, C.B. Nelson, M. Hughes, S. Eshleman, H.U. Wittchen, and K.S. Kendler. 1994. Lifetime and 12-month prevalence of DSM-III-R psychiatric disorders in the United States: Results from the National Comorbidity Survey. *Arch. Gen. Psychiatry* 51:8–19.

Keul, I., D. Keppler, and E. Doll. 1967. Standard bicarbonate, pH, lactate and pyruvate concentrations during and after muscular exercise. *German Med. Monthly* 12:156–158.

Khan, H., and C.L. Cooper. 1990. Mental health, job satisfaction, alcohol intake and occupational stress among dealers in financial markets. *Stress Med.* 6:285–298.

Khan, K., H. McKay, P. Kannus, D. Bailey, J. Wark, and K. Bennell. 2001. *Physical activity and bone health.* Champaign, IL: Human Kinetics.

Khosla, T. 1968. Unfairness of certain events in the Olympic Games. *Br. Med. J.* 4:111–113.

Khosla, T., and V.C. McBroom. 1985. Age, height and weight of female Olympic finalists. *Br. J. Sports Med.* 19:96–99.

Kibler, W.B., T.J. Chandler, and R.E. Maddux. 1989. A musculoskeletal approach to the preparticipation physical examination. *Am. J. Sports Med.* 17 (4):525–531.

Kiens, B., B. Essén-Gustavsson, N.J. Christensen, and B. Saltin. 1993. Skeletal muscle substrate utilization during submaximal exercise in man: Effect of endurance training. *J. Physiol.* 469:459–478.

Kiens, B., and H. Lithell. 1989. Lipoprotein metabolism influenced by training-induced changes in human skeletal muscle. *J. Clin. Invest.* 83:558–564.

Kiens, B., H. Lithell, and B. Vessby. 1984. Further increase in high density lipoprotein in trained males after enhanced training. *Eur. J. Appl. Physiol. Occup. Physiol.* 52:426–430.

Kiladze, N.A., and R.F. Afanas'eva. 1993. Use of the WBGT-index for evaluation of the heating microclimate in industrial conditions. *Med. Tr. Prom. Ekol.* 9–10:26–29.

Kilbom, Å. 1971. Physical training in women. *Scand. J. Clin. Lab. Invest. Suppl.* 119:1–34.

Kilbom, Å., and J. Persson. 1981. Circulatory response to static muscle contractions in three muscle groups. *Clin. Physiol.* 1:215.

Killian, K.J., and E.J.M. Campbell. 1983. Dyspnea and exercise. *Annu. Rev. Physiol.* 45:465.

Killian, K.J., N.L. Jones, and E.J. Campbell. 1999. Control of breathing during exercise. In *Control of breathing in health and disease*, ed. M.D. Altose and Y. Kawakami, pp. 137–162. New York: Marcel Dekker.

Kim, H.E., and H. Tokura. 1995. Effects of the menstrual cycle on dressing behavior in the cold. *Phys. Behav.* 58:699–703.

King, C.E., S.L. Dodd, and S.M. Cain. 1987. $O_2$ delivery to contracting muscle during hypoxic or CO hypoxia. *J. Appl. Physiol.* 63:726–732.

Kingwell, B.A., and G.L. Jennings. 1998. The role of aerobic training in the regulation of vascular tone. *Nutr. Metab. Cardiovasc. Dis.* 8:173–183.

Kingwell, B.A., B. Tran, J.D. Cameron, G.L. Jennings, and A.M. Dart. 1996. Enhanced vasodilation to acetylcholine in athletes is associated with lower plasma cholesterol. *Am. J. Physiol.* 270:H2008–H2013.

Kinoshita, H.P., R. Francis, T. Murase, S. Kawai, and T. Ogawa. 1996. The mechanical properties of the heel pad in elderly adults. *Eur. J. Appl. Physiol.* 73:404–409.

Kirchheim, H.R. 1976. Systemic arterial baroreceptor reflexes. *Physiol. Rev.* 56:100.

Kirchner, E.M., R.D. Lewis, and P.J. O'Connor. 1996. Effect of past gymnastics participation on adult bone mass. *J. Appl. Physiol.* 80:225–232.

Kirkendall, D.T., and W.E.J. Garrett. 1998. The effects of aging and training on skeletal muscle. *Am. J. Sports Med.* 26:598–602.

Kjær, M., H. Langberg, D. Skovgaard, J. Olesen, J. Bülow, M. Krogsgaard, and R. Boushel. 2000. In vivo studies of peritendinous tissue in exercise. *Scand. J. Med. Sci. Sports* 10:326–331.

Kjellberg, Å., and B.-O. Wickström. 1985. Whole-body vibration: Exposure time and acute effects—a review. *Ergonomics* 28:535–544.

Klausen, K. 1966. Cardiac output in man in rest and work during and after acclimatization to 3,800 m. *J. Appl. Physiol.* 21:609–616.

Klausen, K., L.B. Andersen, and I. Pelle. 1981. Adaptive changes in work capacity, skeletal muscle capillarization and enzyme levels during training and detraining. *Acta Physiol. Scand.* 113:9–16.

Klausen, K., N.H. Secher, J.P. Clausen, O. Hartling, and J. Trap-Jensen. 1982. Central and regional circulatory adaptations to one-leg training. *J. Appl. Physiol.* 52:976–983.

Klip, A., T. Ramlai, D.A. Young, and J.O. Holloszy. 1987. Insulin-induced translocation of glucose transporters in rat hindlimb muscles. *FEBS Lett.* 224:224–230.

Klitgaard, H., O. Bergman, R. Betto, G. Salviati, S. Schiaffino, T. Clausen, and B. Saltin. 1990. Co-existence of myosin heavy chain I and IIa isoforms in human skeletal muscle fibres with endurance training. *Eur. J. Physiol.* 416:470–472.

Knabb, R.M., S.W. Ely, A.N. Bacchus, R. Rubio, and R.M. Berne. 1983. Consistent parallel relationships among myocardial oxygen consumption, coronary blood flow and pericardial infusate adenosine concentration with various interventions and β-blockade in the dog. *Circ. Res.* 53:33.

Knussmann, R., and N. Weden. 1995. Indicators of circulation, respiration and muscle strength in puberty. *Z. Morphol. Anthropol.* 81 (1):101–110.

Knuttgen, H.G., and B. Saltin. 1973. Oxygen uptake, muscle high-energy phosphates and lactate in exercise under acute hypoxic conditions in man. *Acta Physiol. Scand.* 87:368–376.

Kobayashi, K. 1982. *Aerobic power of the Japanese.* Tokyo: Kyorin Shoin.

Kobayashi, K., K. Kitamura, M. Miura, H. Sodeyama, Y. Murase, M. Miyashita, and H. Matsui. 1978. Aerobic power as related to body growth and training in Japanese boys: A longitudinal study. *J. Appl. Physiol.* 44:666–672.

Koh, T.J. 2002. Do small heat shock proteins protect skeletal muscle from injury? *Exerc. Sport Sci. Rev.* 30:117–121.

Kohn, H.S. 1996. Prevention and treatment of elbow injuries in golf. *Clin. Sports Med.* 15:65–83.

Koike, A., K. Wasserman, Y. Armon, and D. Weiler-Ravell. 1991. The work-rate dependent effect of carbon monoxide on ventilatory control during exercise. *Respir. Physiol.* 85:169–183.

Koivisto, V.A., H. Yki-Järvinen, and R.A. Defronzo. 1986. Physical training and insulin sensitivity. *Diabetes Metab. Rev.* 1:445–481.

Koller, W.C., and N. Biary. 1984. Effect of alcohol on tremors: Comparison with propranolol. *Neurology* 34:221–222.

Kolstad, A., and R. Opsahl. 1979. Cold injury to corneal epithelium, a cause of bluffed vision in cross-country skiers. *Acta Ophthalmol.* 48:789.

Komi, P.V., ed. 1992. Stretch-shortening cycle. In *Strength and power in sport,* pp. 169–179. Oxford: Blackwell Science.

Komi, P.V., J.H. Viitasalo, M. Havu, A. Thorstensson, B. Sjödin, and J. Karlsson. 1977. Skeletal muscle fibres and muscle enzyme activities in monozygous and dizygous twins of both sexes. *Acta Physiol. Scand.* 100:385–392.

Kondo, T., and M. Raff. 2000. Oligodendrocyte precursor cells reprogrammed to become multipotent CNS stem cells. *Science* 289:1754–1756.

Kong, Y.Y., H. Yoshida, I. Sarosi, H.L. Tan, E. Timms, C. Capparelli, S. Morony, A.J. Oliveira-dos-Santos, G. Van, A. Itie, W. Khoo, A. Wakeham, C.R. Dunstan, D.L. Lacey, T.W. Mak, W.J. Boyle, and J.M. Penninger. 1999. OPGL is a key regulator of osteoclastogenesis, lymphocyte development and lymph-node organogenesis. *Nature* 397:315–323.

Kontulainen, S., P. Kannus, H. Haapasalo, A. Heinonen, H. Sievanen, P. Oja, and I. Vuori. 1999. Changes in bone mineral content with decreased training in competitive young adult tennis players and controls: A prospective 4-yr follow-up. *Med. Sci. Sports Exerc.* 31:646–652.

Korneliussen, H., J.A. Barstad, and G. Lilleheil. 1972. Vesicle hypothesis: Effect of nerve stimulation on the synaptic vesicles of motor endplates. *Experientia* 28:1055–1057.

Koulmann, N., B. Melin, C. Jimenez, A. Charpenet, G. Savourey, and J. Bittel. 1997. Effect of different carbohydrate-electrolyte beverages on the appearance of ingested deuterium in body fluids during moderate exercise by humans in the heat. *Eur. J. Appl. Physiol. Occup. Physiol.* 75:525–531.

Kovacs, E.M.R., and F. Brouns. 1997. Nutrition and physiological aspects of exercise-induced dehydration and rehydration, *Medicina sportiva* (Krakow) 1:27–36.

Kovanen, V. 2002. Intramuscular extracellular matrix: Complex environment of muscle cells. *Exerc. Sport Sci. Rev.* 30:20–25.

Kraemer, W.J., N.D. Duncan, and J.S. Volek. 1998. Resistance training and elite athletes: Adaptations and program considerations. *J. Orthop. Sports Phys. Ther.* 28:110–119.

Kraemer, W.J., S.J. Fleck, and W.J. Evans. 1996. Strength and power training: Physiological mechanisms of adaptation. *Exerc. Sport Sci. Rev.* 24:363–397.

Krahenbuhl, G.S., J.S. Skinner, and V.M. Kohrt. 1985. Developmental aspects of maximal aerobic power in children. *Exerc. Sport Sci. Rev.* 13:503–38.

Krahenbuhl G.S., and T.J. Williams. 1992. Running economy: Changes with age during childhood and adolescence. *Med. Sci. Sports Exerc.* 24:462–66.

Krall, E.A., and B. Dawson-Hughes. 1994. Walking is related to bone density and rates of bone loss. *Am. J. Med.* 96:20–26.

———. 1999. Smoking increases bone loss and decreases intestinal calcium absorption. *J. Bone Miner. Res.* 14:215–220.

Krasavina, T., T. Volkova, and V. Gornik. 1977. Fundamental condition of the sympatho-adrenal system in patients with late stages of vibration disease. *Klin. Med.* (Moscow) 55:77.

Krebs, H. 1964. Gluconeogenesis: The Croonian lecture, 1963. *Proc. R. Soc. London* 159:545.

———. 1968. The effects of ethanol on the metabolic activities of the liver. *Adv. Enzyme Reg.* 6:467–481.

Krip, B., N. Gledhill, V. Jamnik, and D. Warburton. 1997. Effect of alterations in blood volume on cardiac function during maximal exercise. *Med. Sci. Sports Exerc.* 29:1469–1476.

Kriska, A.M., R.B. Sandler, J.A. Cauley, R.E. LaPorte, D.L. Hom, and G. Pambianco. 1988. The assessment of historical physical activity and its relation to adult bone parameters. *Am. J. Epidemiol.* 127:1053–1061.

Krogh, A. 1929. *The anatomy and physiology of capillaries.* New Haven, CT: Yale University Press.

———. 1941. *The comparative physiology of respiratory mechanisms.* Philadelphia: University of Pennsylvania Press.

Krogh, A., and J. Lindhard. 1913. Regulation of respiration and circulation during the initial stages of muscular work. *J. Physiol.* (Lond.) 47:112–136.

———. 1920. Relative value of fat and carbohydrate as source of muscular energy. *Biochem. J.* 14:290.

Krølner, B., E. Tøndevold, B. Toft, B. Berthelsen, and S.P. Nielsen. 1982. Bone mass of the axial and the appendicular skeleton in women with Colles' fracture: Its relation to physical activity. *Clin. Physiol.* 2:147–157.

Krombholz, H. 1997. Physical performance in relation to age, sex, social class and sports activities in kindergarten and elementary school. *Percept. Mot. Skills* 84 (3):1168–1170.

Krüger, M., J. Wright, and K. Wang. 1991. Nebulin as a length regulator of thin filaments of vertebrate skeletal muscles: Correlation of thin filament length, nebulin size and epitope profile. *J. Cell Biol.* 115:97–107.

Kubo, K., H. Kanehisa, M, Ito, and T. Fukunaga. 2001. Effects of isometric training on the elasticity og human tendon structures in vivo. *J. Appl. Physiol.* 91:26–32.

Kubo, K., H. Kanehisa, Y. Kawakami, and T. Fukunaga. 2001. Influences of repetitive muscle contractions with different modes on tendon elasticity in vivo. *J. Appl. Physiol.* 91:277–282.

Kugelberg, E. 1973. Histochemical composition, contraction speed and fatiguability of rat soleus motor units. *J. Neurol. Sci.* 20:177–198.

Kuipers, H. 1998. Training and overtraining: An introduction. *Med. Sci. Sports Exerc.* 30:1137–1139.

Kumagai, K., T. Abe, W.F. Brechue, T. Ryushi, S. Takano, and M. Mizuno. 2000. Sprint performance is related to muscle fascicle length in male 100-m sprinters. *J. Appl. Physiol.* 88:811–816.

Kumar, S. 1980. Physiological responses to weight lifting in different planes. *Ergonomics* 23:987–993.

Kuno, Y. 1956. *Human perspiration.* Springfield, IL: Charles C Thomas.

Kurihara, T., and J.E. Brooks. 1975. The mechanism of neuromuscular fatigue: A study of mammalian muscle using excitation-contraction uncoupling. *Arch. Neurol.* 32:168–174.

Kwon, A., M. Kato, H. Kawamura, Y. Yanai, and H. Tokura. 1998. Physiological significance of hydrophilic and hydrophobic textile materials during intermittent exercise in humans under the influence of warm ambient temperature with and without wind. *Eur. J. Appl. Physiol.* 78:487–493.

Kyle, C.R. 1991. Ergogenics of bicycling. In *Perspectives in exercise science and sports medicine,* vol. 4, *Ergogenics: Enhancement of performance in exercise and sport,* ed. D.R. Lamb and M.H. Williams. Madison, WI: Brown and Benchmark.

Laaneots, T., K. Karelson, and A. Viru. 1996. Relation of aerobic capacity to the stage of sexual maturation in girls. *Biol. Sport* (Warsaw) 13 (2):137–144.

Ladell, W.S.S. 1955. The effects of water and salt intake upon the performance of men working in hot and humid environments. *J. Physiol.* 127:11.

Lahad, A., A.D. Malter, A.O. Berg, and R.A. Deyo. 1994. The effectiveness of four interventions for the prevention of low back pain. *JAMA* 272:205–211.

Lahiri, S., A. Mokashi, E. Mulligan, and T. Nishino. 1981. Comparison of aortic and carotid chemoreceptor responses to hypercapnia and hypoxia. *J. Appl. Physiol. Respir. Environ. Exerc. Physiol.* 51:55.

Lahiri, S., E. Mulligan, T. Nishino, A. Mokashi, and R.O. Davies. 1981. Relative responses of aortic body and carotid body chemoreceptors to carboxyhemoglobinia. *J. Appl. Physiol. Respir. Environ. Exerc. Physiol.* 50:580.

Laidlaw, D.H., M. Bilodeau, and R.M. Enoka. 2000. Steadiness is reduced and motor unit discharge is more variable in old adults. *Muscle Nerve* 23:600–612.

Lakka, T.A., J.M. Venalainen, R. Rauramaa, R. Salonen, J. Tuomilehto, and J.T. Salonen. 1994. Relation of leisure-time physical activity and cardiorespiratory fitness to the risk of acute myocardial infarction in men. *N. Engl. J. Med.* 330:1549–1554.

LaManca, J.J., and E.M. Haymes. 1993. Effects of iron repletion on $\dot{V}O_2$max, endurance and blood lactate in women. *Med. Sci. Sports Exerc.* 25:1386–1392.

Lambert, O. 1965. The relationship between maximum isometric strength and maximum concentric strength at different speeds. *Int. Fed. Phys. Ed. Bull.* 35:13.

Lambert, R., and G. Teissier. 1927. Théorie de la similitude biologique. *Ann. Physiol.* 3:212–246.

Lamke, L.O., S. Lennquist, S.O. Liljedahl, and B. Wedin. 1972. The influence of cold stress on catecholamine excretion and oxygen uptake of normal persons. *Scand. J. Clin. Lab. Invest.* 30:57–62.

Lang, T., C. Fouriaud, and M.C. Jacquinet-Salord. 1992. Length of occupational noise exposure and blood pressure. *Int. Arch. Occup. Environ. Health* 63:369–372.

Langberg, H., J.L. Olesen, C. Gemmer, and M. Kjær. 2002. Substantial elevation of interleukin-6 concentration in peritendinous tissue, in contrast to muscle, following prolonged exercise in humans. *J. Physiol.* 542.3:985–990.

Lanyon, L.E. 1996. Using functional loading to influence bone mass and architecture: Objectives, mechanisms and relationship with estrogen of the mechanically adaptive process in bone. *Bone* 18:37S-43S.

Lanyon, L.E., and J.A. O'Connor. 1980. Adaptation of bone artificially loaded at high and low physiological strain rates. *J. Physiol.* (Lond.) 303:36P.

Larsson, L., and T. Ansved. 1985. Effects of long-term physical training and detraining on enzyme histochemical and functional skeletal muscle characteristic in man. *Muscle Nerve* 8:714–722.

Larsson, S.E., H. Cai and Å. Öberg. 1993. Microcirculation in the upper trapezius muscle during varying levels of static contraction, fatigue and recovery in healthy women – a study using percutaneous Laser-Doppler flowmetry and surface electromyography, *Eur. J. Appl. Physiol.* 66:483–488.

Larsson, L., G. Grimby, and J. Karlsson. 1979. Muscle strength and speed of movement in relation to age and muscle morphology. *J. Appl. Physiol. Respir. Environ. Exerc. Physiol.* 46:451–456.

Larsson, L., and R.L. Moss. 1993. Maximum velocity of shortening in relation to myosin isoform composition in single fibres from human skeletal muscles. *J. Physiol.* 472:595–614.

Larsson, L., B. Sjödin, and J. Karlsson. 1978. Histochemical and biochemical changes in human skeletal muscle with age in sedentary males, age 22–65 years. *Acta Physiol. Scand.* 103:31–39.

Laughlin, M.H., K.A. Overholser, and M.J. Bhatte. 1989. Exercise training increases coronary transport reserve in miniature swine. *J. Appl. Physiol.* 67:1140–1149.

Laursen, B., B. Jensen, and A. Ratkevicius. 2001. Performance and muscle activity during computer mouse tasks in young elderly subjects. *Eur. J. Appl. Physiol.* 84:329–336.

Lauru, L. 1957. Physiological study of motion. *Adv. Manage.* 22:17.

Lavoisier, A.L. 1775. Sur la nature du principe qui se combine avec les métaux pendant leur calcination, et qui en augmente le poids (a). In *Histoire de l'Académie Royale des Sciences, 1775,* pp. 520–526. Paris: Académie Royale des Sciences.

Lawson-Smith, M.J., and J.K. McGeachie. 1998. The identification of myogenic cells in skeletal muscle, with emphasis on the use of tritiated thymidine autoradiography and desmin antibodies. *J. Anat.* 192:161–171.

Layne, J.E., and M.E. Nelson. 1999. The effects of progressive resistance training on bone density: A review. *Med. Sci. Sports Exerc.* 31:25–30.

LeBlanc, J. 1975. *Man in the cold.* American Lecture Series. Springfield, IL: Charles C Thomas.

LeBlanc, J.J., J. Côté, S. Dulac, and F. Dulon-Turcot. 1978. Effect of age, sex and physical fitness on responses to local cooling. *J. Appl. Physiol.* 44:813.

Leblond, C.P., and R.C. Greulich. 1956. Autoradiographic studies of bone formation and growth. In *The biochemistry and physiology of bone,* ed. G.H. Bourne. New York: Academic Press.

Lebrun, C.M. 1994. The effect of the phase of the menstrual cycle and the birth control pill on athletic performance. *Clin. Sports Med.* 13 (2):419–441.

Lee, C.D., S.N. Blair, and A.S. Jackson. 1999. Cardiorespiratory fitness, body composition and all-cause and cardiovascular disease mortality in men. *Am. J. Clin. Nutr.* 69:373–380.

Lee, I., C.C. Hsieh, and R.S. Paffenbarger. 1995. Exercise intensity and longevity in men: The Harvard Alumni Health Study. *JAMA* 273:1179–1184.

Lee, S.C., C.N. Becker, and S.A. Binder-Macleod. 1999. Catchlike-inducing train activation of human muscle during isotonic contractions: Burst modulation. *J. Appl. Physiol.* 87:1758–1767.

Lehmann, G. 1953. *Praktische Arbeitsphysiologie.* Stuttgart: Thieme.

Lehmann, G., E.A. Müller, and H. Spitzer. 1950. Der Kalorienbedarf bei gewerblicher Arbeit. *Arbeitsphysiologie* 14:166.

Lehmann, M.J., C. Foster, H.H. Dickhuth, and U. Gastmann. 1998. Autonomic imbalance hypothesis and overtraining syndrome. *Med. Sci. Sports Exerc.* 30:1140–1145.

Lehmann, M.J., W. Lormes, A. Opitz-Gress, J.M. Steinacker, N. Netzer, C. Foster, and U. Gastmann. 1997. Training and overtraining: An overview and experimental results in endurance sports. *J. Sports Med. Phys. Fit.* 37:7–17.

Lehtonen-Veromaa, M., T. Möttönen, E. Svedström, P. Hakola, O.J. Heinonen, and J. Viikari. 2000. Physical activity and bone mineral acquisition in peripubertal girls. *Scand. J. Med. Sci. Sports* 10:236–243.

Leithead, C.S., and A.R. Lind. 1964. *Heat stress and heat disorders.* London: Cassell.

Le Magnen, J. 1983. Body energy balance and food intake: A neuroendocrine regulatory mechanism. *Physiol. Rev.* 63:314.

Lemon, P.W.R. 1998. Effects of exercise on dietary protein requirements. *Int. J. Sport. Nutr.* 8:426–447.

———. 2000. Energy and nutrient intake for athletic performance. In *Biomechanics and biology of movement,* ed. B.M. Nigg, B.R. MacIntosh, and J. Mester, pp. 103–28. Champaign, IL: Human Kinetics.

Lemon, P.W.R., and J.P. Mullin. 1980. Effects of initial muscle glycogen levels on protein catabolism during exercise. *J. Appl. Physiol.* 48:624–629.

Lemon, R.N., R.S. Johansson, and G. Westling. 1995. Corticospinal control during reach, grasp and precision lift in man. *J. Neurosci.* 15:6145–6156.

Lennquist, S. 1972. Cold induced diuresis. *Scand. J. Urol. Nephrol. Suppl.*, vol. 9.

Lentz, T.L., and J. Chester. 1982. Synaptic vesicle recycling at the neuromuscular junction in the presence of a presynaptic membrane marker. *Neuroscience* 7:9–20.

Leon, A.S., J. Connett, D.R.J. Jacobs, and R. Rauramaa. 1987. Leisure-time physical activity levels and risk of coronary heart disease and death: The Multiple Risk Factor Intervention Trial. *JAMA* 256:2388–2395.

Leong, B., G. Kamen, C. Patten, and J.R. Burke. 1999. Maximal motor unit discharge rates in the quadriceps muscles of older weight lifters. *Med. Sci. Sports Exerc.* 31:1638–1644.

Lesnik, H., and T. Makowiec-Dabrowska. 1989. Hemodynamic reactions to monotonous work performed in silence and in noise of 70 dB (A). *Pol. J. Occup. Med.* 2:51–61.

Leveritt, M., P.J. Abernethy, B.K. Barry, and P.A. Logan. 1999. Concurrent strength and endurance training: A review. *Sports Med.* 28:413–427.

Levi, L. 1967. Sympatho-adrenomedullary responses to emotional stimuli: Methodologic, physiologic and pathologic considerations. In *An introduction to clinical neuroendocrinology*, ed. E. Bajusz, pp. 78–105. New York: Karger.

Levine, B.D. 1993. Regulation of central blood volume and cardiac filling in endurance athletes: The Frank-Starling mechanism as a determinant of orthostatic tolerance. *Med. Sci. Sports Exerc.* 25:727–732.

Levine B.D., and J. Stray-Gundersen. 1992. A practical approach to altitude training: Where to live and train for optimal performance enhancement. *Int. J. Sports Med.* 13 (suppl. 1):S209–S212.

———. 1997. "Living high—training low": Effect of moderate-altitude acclimatization with low-altitude training on performance. *J. Appl. Physiol.* 83:102–112.

Levine, M., L. Duffy, D.C. Moore, and L.A. Matej. 1995. Acclimation of a non-indigenous sub-Arctic population: Seasonal variation in thyroid function in interior Alaska. *Comp. Biochem. Physiol. A. Physiol.* 111:209–214.

Lewin, R. 1980. Evolutionary theory under fire. *Science* 210:883.

———. 1981. Ethiopian stone tools are world's oldest. *Science* 211:806.

Lewis, S.F., and C.S. Fulco. 1998. A new approach to studying muscle fatigue and factors affecting performance during dynamic exercise in humans. *Exerc. Sport Sci. Rev.* 26:91–116.

Lewis, S.F., E. Nylander, P. Gad, and N.H. Areskog. 1980. Non-autonomic component in bradycardia of endurance trained men at rest and during exercise. *Acta Physiol. Scand.* 109:297–305.

Lexell, J. 1997. Muscle capillarization: Morphological and morphometrical analyses of biopsy samples. *Muscle Nerve Suppl.* 5:S110–S112.

Lexell, J., K. Henriksson-Larsén, and M. Sjöström. 1983. Distribution of different fibre types in human skeletal muscles: 2. A study of cross-sections of whole m. vastus lateralis. *Acta Physiol. Scand.* 117:115–122.

Li, M., W.S. Jee, H.Z. Ke, L.Y. Tang, Y.F. Ma, X.G. Liang, and R.B. Setterberg. 1995. Prostaglandin E2 administration prevents bone loss induced by orchidectomy in rats. *J. Bone Miner. Res.* 10:66–73.

Li, X., and H. Tokura. 1996. The effects of two types of clothing on seasonal heat tolerance. *Eur. J. Appl. Physiol.* 72:287–291.

Li, X., H. Tokura, and T. Midorikawa. 1994. The effects of two types of clothing on seasonal cold tolerance. *Eur. J. Appl. Physiol.* 69:498–501.

———. 1995. The effects of two different types of clothing on seasonal warm acclimatization. *Int. J. Biometeorol.* 38:111–115.

Libert, J.P., V. Candas, and J.J. Vogt. 1978. Sweating response in man during transient rises of air temperature. *J. Appl. Physiol.* 44:284.

———. 1979. Effect of rate of change in skin temperature on local sweating rate. *J. Appl. Physiol.* 47:306.

———. 1983. Modifications of sweating responses to thermal transients following heat acclimation. *Eur. J. Appl. Physiol.* 50:235.

Libert, J.P., V. Candas, J.J. Vogt, and P. Mairiaux. 1982. Central and peripheral inputs in sweating regulation during thermal transients. *J. Appl. Physiol.* 52:1147.

Lieber, R.L., and J. Fridén. 1999. Mechanisms of muscle injury after eccentric contraction. *J. Sci. Med. Sport* 2:253–265.

Lind, A.R., and D.E. Bass. 1963. Optimal exposure time for development of acclimatization to heat. *Fed. Proc.* 22:704.

Lind, A.R., P.W. Humphreys, K.J. Collins, K. Foster, and K.F. Sweetland. 1970. Influence of age and daily duration of exposure on responses of men to work in heat. *J. Appl. Physiol.* 28:50.

Linderman, J.K., and K.L. Gosselink. 1994. The effects of sodium bicarbonate ingestion on exercise performance. *Sports Med.* 18:75–80.

Lindgren, G. 1978. Growth of schoolchildren with early, average and late ages of peak height velocity. *Ann. Hum. Biol.* 5:253–267.

Liu, J., J.D.J. Farmer, W.S. Lane, J. Friedman, I. Weissman, and S.L. Schreiber. 1991. Calcineurin is a common target of cyclophilin–cyclosporin A and FKBP-FK506 complexes. *Cell* 66:807–815.

Liu, Y., J.M. Steinacker, C. Dehnert, E. Menhold, S. Baur, W. Lormes, and M. Lehmann. 1998. Effect of "living high—training low" on the cardiac functions at sea level. *Int. J. Sports Med.* 19:380–384.

Liu, Y.F., S. Mayr, A. Opitz-Gress, C. Zeller, W. Lormes, S. Baur, M. Lehmann, and J.M. Steinacker. 1999. Human skeletal muscle HSP70 response to training in highly trained rowers. *J. Appl. Physiol.* 86:101–104.

Livingston, D.J., G.N. LaMar, and W.D. Brown. 1983. Myoglobin diffusion in bovine heart muscle. *Science* 220 (4592):71.

Locke, M. 1997. The cellular stress response to exercise: Role of stress proteins. *Exerc. Sport Sci. Rev.* 25:105–136.

Locke, M., and E.G. Noble. 1995. Stress proteins: The exercise response. *Can. J. Appl. Physiol.* 20:155–167.

Locke, M. and E.G. Noble (eds.) 2002. Exercise and stress response. Boca Raton, FL: CRC Press.

Loke, J., D.A. Mahler, and J.A. Virgulto. 1982. Respiratory muscle fatigue after marathon running. *J. Appl. Physiol. Respir. Environ. Exerc. Physiol.* 52:821.

Lokey, E.A., Z.V. Tran, C.L. Wells, B.C. Myers, and A.C. Tran. 1991. Effects of physical exercise on pregnancy outcomes: A metaanalytic review. *Med. Sci. Sports Exerc.* 23:1234–1239.

Lømo, T., R.H. Westgaard, and H.A. Dahl. 1974. Contractile properties of muscle: Control by pattern of muscle activity in the rat. *Proc. R. Soc. Lond. B. Biol. Sci.* 187:99–103.

Lönnerholm, G. 1982. Pulmonary carbonic anhydrase in the human, monkey and rat. *J. Appl. Physiol. Respir. Environ. Exerc. Physiol.* 52:352.

Lorentzon, M., R. Lorentzon, and P. Nordstrom. 2000. Interleukin-6 gene polymorphism is related to bone mineral density during and after puberty in healthy white males: A cross-sectional and longitudinal study. *J. Bone Miner. Res.* 15:1944–1949.

Lortie, G., J.A. Simoneau, P. Hamel, M.R. Boulay, F. Landry, and C. Bouchard. 1984. Responses of maximal aerobic power and capacity to aerobic training. *Int. J. Sports Med.* 5:232–236.

Losse, A., S.E. Henderson, D. Elliman, D. Hall, E. Knight, and M. Jongmans. 1991. Clumsiness in children: Do they grow out of it? A 10-year follow up study. *Dev. Med. Child. Neurol.* 33:55–68.

Louhevaara, V., R. Ilmarinen, B. Griefahn, C. Kunemund, and H. Makinen. 1995. Maximal physical work performance with European standard based fire-protective clothing system and equipment in relation to individual characteristics. *Eur. J. Appl. Physiol.* 71:223–229.

Lucas, D.B., and B. Bresler. 1960. *Stability of the ligamentous spine.* Technical Report Series 11, no. 40. Berkeley: University of California, Biomechanics Laboratory.

Ludbrook, J. 1983. Reflex control of blood pressure during exercise. *Annu. Rev. Physiol.* 45:213.

Luetkemeier, M.J., M.G. Coles, and E.W. Askew. 1997. Dietary sodium and plasma volume levels with exercise. *Sports Med.* 23:279–286.

Lundgren, N. 1946. Physiological effects of time schedule work on lumbar workers. *Acta Physiol. Scand.* 13 (suppl. 41).

Lundin, A. 1973. *Bordtennis.* Idrettsfysiologi Rapport no. 12. Stockholm: Trygg-Hansa.

Lussier, L., and E.R. Buskirk. 1977. Effects of an endurance training regimen on assessment of work capacity in prepubertal children. *Ann. N.Y. Acad. Sci.* 301:734–747.

Lynch, N.J., S.D.R. Galloway, and M.A. Nimmo. 2000. Effects of moderate dietary manipulation on intermittent exercise performance and metabolism in women. *Eur. J. Appl. Physiol.* 81:197–202.

Lynch, N.J., and M.A. Nimmo. 1998. Effects of menstrual cycle phase and oral contraceptive use on intermittent exercise. *Eur. J. Appl. Physiol.* 78:565–572.

Lyngberg, K., N. Tvede, J. Halkjær-Kristensen, V. Andersen, and B.K. Pedersen. 1991. Physical exercise modulate the cellular immune system in patients with rheumatoid arthritis. *Scand. J. Med. Sci. Sports* 1:167–173.

MacCallum, J.B. 1898. On the histogenesis of the striated muscle fibre and the growth of the human sartorius muscle. *Johns Hopkins Hosp. Bull.* 9:208–215.

MacDougall, J.D. 1992. Hypertrophy or hyperplasia. In *Strength and power in sport*, ed. P.V. Komi, pp. 230–238. Oxford: Blackwell Science.

MacDougall, J.D., G.C. Elder, D.G. Sale, J.R. Moroz, and J.R. Sutton. 1980. Effects of strength training and immobilization on human muscle fibres. *Eur. J. Appl. Physiol.* 43:25–34.

MacDougall, J.D., H.J. Green, J.R. Sutton, G. Coates, A. Cymerman, P. Young, and C.S. Houston. 1991. Operation Everest: II. Structural adaptation in skeletal muscle in response to extreme simulated altitude. *Acta Physiol. Scand.* 142:421–427.

MacIntosh, B.R., and B.M. Wright. 1995. Caffeine ingestion and performance of a 1,500 metre swim. *Can. J. Appl. Physiol.* 20:168–177.

Mackinnon, L.T., and L.M. Hubinger. 1999. Effects of exercise on lipoprotein (a). *Sports Med.* 28:11–24.

Madeleine, P., M. Voigt, and L. Arendt-Nielsen. 1998. Subjective, physiological and biomechanical responses to prolonged manual work performed standing on hard and soft surfaces. *Eur. J. Appl. Physiol.* 77:1–9.

Maffulli, N., and J.B. King. 1992. Effects of physical activity on some components of the skeletal system. *Sports Med.* 13:393–407.

Magel, J.R., G.F. Foglia, W.D. McArdle, B. Gutin, G.S. Pechar, and F.I. Katch. 1975. Specificity of swim training on maximum oxygen uptake. *J. Appl. Physiol.* 38:151–155.

Maggi, C.A. 1995. Neuropeptides as regulators of airway function: Vasoactive intestinal peptide and the tachykinins. *Physiol. Rev.* 75:277.

Mahadeva, K., R. Passmore, and B. Woolf. 1953. Individual variations in the metabolic cost of standardized exercises: The effect of food, age, sex and race. *J. Physiol.* 121:225.

Mairbäurl, H., E. Humpeler, G. Schwaberger, and H. Pessenhofer. 1983. Training-dependent changes of red cell density and erythrocytic oxygen transport.

*J. Appl. Physiol. Respir. Environ. Exerc. Physiol.* 55:1403.

Mairbäurl, H., W. Schobersberger, E. Humpeler, W. Hasibeder, W. Fischer, and E. Raas. 1986. Beneficial effects of exercising at moderate altitude on red cell oxygen transport and on exercise performance. *Pflügers Arch.* 406:594–599.

Mairiaux, P., J.C. Sagot, and V. Candas. 1983. Oral temperature as an index of core temperature during heat transients. *Eur. J. Appl. Physiol.* 50:331.

Majumdar, P., G.L. Khanna, V. Malik, S. Sachdeva, M. Arif, and M. Mandal. 1997. Physiological analysis to quantify training load in badminton. *Br. J. Sports Med.* 31:342–345.

Makarenko, N.A. 1969. Vegetative nervous system in patients with vibration disease. *Gigiena truda i professional'nye zabolevaniia* 13:5.

Malina, R.M., and C. Bouchard. 1991. *Growth, maturation, and physical activity.* Champaign, IL: Human Kinetics.

Malmqvist, R., I. Ekblom, L. Lindström, I. Petersén, and R. Örtengren. 1981. Measurement of localized muscle fatigue in building work. *Ergonomics* 24:695–709.

Mandelbaum, B.R., and M. Putukian. 1999. Medical concerns and specificities in female soccer players. *Sci. Sports* 14:254–260.

Manley, L. 1990. Apnoeic heart rate responses in humans. *Sports Med.* 9:286–310.

Manning, H.L., and R.M. Schwartzstein. 1999. Dyspnea and the control of breathing. In *Control of breathing in health and disease,* ed. M.D. Altose and Y. Kawakami, pp. 105–135. New York: Marcel Dekker.

Manolagas, S.C. 2000. Birth and death of bone cells: Basic regulatory mechanisms and implications for the pathogenesis and treatment of osteoporosis. *Endocr. Rev.* 21:115–137.

Manore, M.M., and J. Thompson. 2000. *Sport nutrition for health and performance.* Champaign, IL: Human Kinetics.

Marcial, J.J., and H.S. Slutsky, eds. 1997. *Physiology of ventilatory support.* New York: Marcel Dekker.

Marcus, R. 1987. Normal and abnormal bone remodelling in man. *Adv. Int. Med.* 38:129–141.

Marcus, R., B. Drinkwater, G. Dalsky, J. Dufek, D. Raab, C. Slemenda, and C. Snow-Harter. 1992. Osteoporosis and exercise in women. *Med. Sci. Sports Exerc.* 24:S301–S307.

Margaria, R. 1938. Sulla fisiologia e specialmente sul consumo energetico, della marcia e della corsa a varie velocita ed inclinazioni del terreno. *Atti dei Lincei* 7:299.

Margaria, R., P. Aghemo, and F. Pinera Limas. 1975. A simple relation between performance in running and maximal aerobic power. *J. Appl. Physiol.* 38:351–352.

Margaria, R., P. Aghemo, and E. Rovelli. 1965. Indirect determination of maximal $O_2$ consumption in man. *J. Appl. Physiol.* 20:1070–1073.

Margaria, R., P. Cerretelli, P. Aghemo, and G. Sassi. 1963. Energy cost of running. *J. Appl. Physiol.* 18:367.

Mariak, Z., Z. Bondyra, and M. Piekarska. 1993. The temperature within the circle of Willis versus tympanic temperature in resting normothermic humans. *Eur. J. Appl. Physiol. Occup. Physiol.* 66:518–520.

Marin, M.G. 1994. Update: Pharmacology of airway secretion. *Pharmacol. Rev.* 46:36.

Maritz, J.S., J.F. Morrison, J. Peter, N.B. Strydom, and C.H. Wyndham. 1961. A practical method of estimating an individual's maximal oxygen intake. *Ergonomics* 4:97.

Marken Lichtenbelt, W.D., M. Fogelholm, R. Ottenheijm, and K.R. Westerterp. 1995. Physical activity, body composition and bone density in ballet dancers. *Br. J. Nutr.* 74:439–451.

Maron, B.J., A. Pelliccia, A. Spataro, and M. Granata. 1993. Reduction in left ventricular wall thickness after deconditioning in highly trained Olympic athletes. *Br. Heart J.* 69:125–128.

Maron, M., S.M. Horvath, J.E. Wilkerson, and J.A. Gliner. 1976. Oxygen uptake measurements during competitive marathon running. *J. Appl. Physiol.* 40:836–838.

Marrugat, J., R. Elosua, M.I. Covas, L. Molina, and J. Rubies-Prat. 1996. Amount and intensity of physical activity, physical fitness and serum lipids in men: The marathon investigators. *Am. J. Epidemiol.* 143:562–569.

Marsden, C.D., J.C. Meadows, and P.A. Merton. 1983. "Muscular wisdom" that minimizes fatigue during prolonged effort in man: Peak rates of motoneuron discharge and slowing of discharge during fatigue. *Adv. Neurol.* 39:169–211.

Marsden, J.F., S.F. Farmer, D.M. Halliday, J.R. Rosenberg, and P. Brown. 1999. The unilateral and bilateral control of motor unit pairs in the first dorsal interosseous and paraspinal muscles in man. *J. Physiol.* 521:553–564.

Marsden, J.F., K.J. Werhahn, P. Ashby, J. Rothwell, S. Noachtar, and P. Brown. 2000. Organization of cortical activities related to movement in humans. *J. Neurosci.* 20:2307–2314.

Marshall, J.M. 1995. Skeletal muscle vasculature and systemic hypoxia. *NIPS* 10:274–280.

Martin, B., J. Edward, J. Morgan, C.W. Zwillich, and J.V. Weil. 1981. Control of breathing during prolonged exercise. *J. Appl. Physiol. Respir. Environ. Exerc. Physiol.* 50:27.

Martin, B., M. Heintzelman, and H. Chen. 1982. Exercise performance after ventilatory work. *J. Appl. Physiol. Respir. Environ. Exerc. Physiol.* 52:1581.

Martin, W.H., III, W.M. Kohrt, M.T. Malley, E. Korte, and S. Stoltz. 1990. Exercise training enhances leg vasodilatory capacity of 65-yr-old men and women. *J. Appl. Physiol.* 69:1804–1809.

Martineau, L., and I. Jacobs. 1988. Muscle glycogen utilization during shivering thermogenesis in humans. *J. Appl. Physiol.* 65:2046–2050.

Martino, M., N. Gledhill, and V. Jamnik. 2002. High $\dot{V}O_2$ max. with no history of training is primarily due to high blood volume. *Med. Sci. Sports Exerc.* 34:966–971.

Martinsen, E.W., J. Strand, G. Paulsson, and J. Kaggestad. 1989. Physical fitness level in patients with anxiety and depressive disorders. *Int. J. Sports Med.* 10:58–61.

Mason, J.M., A.S. Breitbart, M. Barcia, D. Porti, R.G. Pergolizzi, and D.A. Grande. 2000. Cartilage and bone regeneration using gene-enhanced tissue engineering. *Clin. Orthop.* 379 (suppl.):S171–S178.

Mathieu-Costello, O. and R.T. Hepple. 2002. Muscle structural capacity for oxygen flux from capillary to fiber mitochondria. *Exerc. Sport Sci. Rev.* 30:80–84.

Matthews, P.B. 1964. Muscle spindles and their motor control. *Physiol. Rev.* 44:219–288.

———. 1991. The human stretch reflex and the motor cortex. *Trends Neurosci.* 14:87–91.

Maughan, R.J. 1984. Relationship between muscle strength and muscle cross-sectional area. *Sports Med.* 1:263.

———. 1986. Exercise-induced muscle cramp: A prospective biochemical study in marathon runners. *J. Sports Sci.* 4:31–34.

———. 1991. Fluid and electrolyte loss and replacement in exercise. *J. Sports Sci.* 9:117–142.

Maughan, R.J., and J.B. Leiper. 1995. Sodium intake and post-exercise rehydration in man. *Eur. J. Appl. Physiol.* 71:311–319.

———. 1999. Limitations of fluid replacement during exercise. *Can. J. Appl. Physiol.* 24:173–187.

Maughan, R.J., J.B. Leiper, and S.M. Shirreffs. 1996. Restoration of fluid balance after exercise-induced dehydration: Effects of food and fluid intake. *Eur. J. Appl. Physiol. Occup. Physiol.* 73:317–325.

Maughan, R.J., and S.M. Shirreffs. 1997. Recovery from prolonged exercise: Restoration of water and electrolyte balance. *J. Sports Sci.* 15:297–303.

Maw, G.J., S.H. Boutcher, and N.A.S. Taylor. 1993. Ratings of perceived exertion and affect in hot and cool environments. *Eur. J. Appl. Physiol.* 67:174–179.

Mayer, J., and B. Bullen. 1960. Nutrition and athletic performance. *Physiol. Rev.* 40:369.

Mayer, J., P. Roy, and K.P. Mitra. 1956. Relation between caloric intake, body weight and physical work in an industrial male population in West Bengal. *Am. J. Clin. Nutr.* 4:169.

Mayer, J., and D.W. Thomas. 1967. Regulation of food intake and obesity. *Science* 156:328–337.

McArdle, W.D., F.I. Katch, and V.L. Katch. 2001. Exercise physiology. Philadelphia: Lippincott Williams and Wilkins.

McArdle, W.D., F.I. Katch, and G.S. Pechar. 1973. Comparison of continuous and discontinuous treadmill and bicycle tests for max $\dot{V}O_2$. *Med. Sci. Sports* 5:156–160.

McBride, P.E. 1992. The health consequences of smoking: Cardiovascular disease. *Med. Clin. North Am.* 76:333–353.

McCaffrey, T.V., R.D. Wurster, H.K. Jacobs, D.E. Euler, and G.S. Geis. 1979. Role of skin temperature in the control of sweating. *J. Appl. Physiol.* 47:591.

McCarroll, J.R. 1996. The frequency of golf injuries. *Clin. Sports Med.* 15:1–7.

McCarthy, J.P., J.C. Agre, B.K. Graf, M.A. Pozniak, and A.C. Vailas. 1995. Compatibility of adaptive responses with combining strength and endurance training. *Med. Sci. Sports Exerc.* 27:429–436.

McCartney, N., G.J.F. Heigenhauser, and N.L. Jones. 1983. Effects of pH on maximal power output and fatigue during short-term dynamic exercise. *J. Appl. Physiol. Respir. Environ. Exerc. Physiol.* 55:225–229.

McConell, G.K., C.M. Burge, S.L. Skinner, and M. Hargreaves. 1997. Influence of ingested fluid volume on physiological responses during prolonged exercise. *Acta Physiol. Scand.* 160:149–156.

McConell, G., K. Kloot, and M. Hargreaves. 1996. Effect of timing of carbohydrate ingestion on endurance exercise performance. *Med. Sci. Sports Exerc.* 28:1300–1304.

McCormick, D.A. 1995. Motor control: The cerebellar symphony. *Nature* 374:412–413.

McDonagh, M.J., C.M. Hayward, and C.T. Davies. 1983. Isometric training in human elbow flexor muscles: The effects on voluntary and electrically evoked forces. *J. Bone Joint Surg.* 65:355–358.

McDonough, P., and R.J. Moffatt. 1999. Smoking-induced elevations in blood carboxyhaemoglobin levels: Effect on maximal oxygen uptake. *Sports Med.* 27:275–283.

McFadden, E.R., Jr. 1983. Respiratory heat and water exchange: Physiological and clinical implications. *J. Appl. Physiol. Respir. Environ. Exerc. Physiol.* 54:331.

McFadden, E.R., Jr. and R.H. Ingram, Jr. 1983. Exercise induced airway obstruction. *Annu. Rev. Physiol.* 45:453.

McGilvery, R.W. 1975. The use of fuels for muscular work. In *Metabolic adaptation to prolonged physical exercise,* ed. H. Howald and J.R. Poortmans, pp. 12–30. Basel: Birkhäuser Verlag.

McHugh, M.P., D.A. Connolly, R.G. Eston, and G.W. Gleim. 1999. Exercise-induced muscle damage and potential mechanisms for the repeated bout effect. *Sports Med.* 27:157–170.

McInnis, M.D., L.J. Newhouse, S.P. Duvillard, and R. Thayer. 1998. The effect of exercise intensity on hematuria in healthy male runners. *Eur. J. Appl. Physiol.* 79:99–105.

McKay, G.A., and E.W. Banister. 1976. A comparison of maximum oxygen uptake determination by bicycle ergometry at various pedaling frequencies and by treadmill running at various speeds. *Eur. J. Appl. Physiol.* 35:191–200.

McKenna, M.J., G.J.F. Heigenhauser, R.S. McKelvie, J.D. MacDougal, and N.L. Jones. 1997. Sprint training enhances ionic regulation during intense exercise in man. *J. Physiol.* 501:687–702.

McLean, F.C., and M.R. Urist. 1955. *Bone: An introduction to the physiology of skeletal tissue.* Chicago: University of Chicago Press.

McLean, S.P., and R.N. Hinrichs. 1998. Sex differences in the centre of buoyancy location of competitive swimmers. *J. Sports Sci.* 16:373–383.

McManus, B.M., S.M. Horvath, N. Bolduan, and J.C. Miller. 1974. Metabolic and cardio-respiratory responses to long-term work under hypoxic conditions. *J. Appl. Physiol.* 36:177–182.

McMaster, W.C., T. Stoddard, and W. Duncan. 1989. Enhancement of lactate recovery by continuous sub-maximal swimming. *J. Swimming Res.* 5:19–21.

McMurray, R.G., L.H. Lindsay, and D.L. Thompson. 1985. The effects of passive inhalation of sigarette smoke on exercise performance. *Eur. J. Appl. Physiol.* 54:196–200.

McPherron, A.C., A.M. Lawler, and S.J. Lee. 1997. Regulation of skeletal muscle mass in mice by a new TGF-beta superfamily member. *Nature* 387:83–90.

Medbø, J.I., A.C. Mohn, and I. Tabata. 1998. Blood lactate concentration versus anaerobic energy release during exhausting and non-exhausting treadmill running. *Acta kinesiologiae universitatis tartuensis* (Tartu, Estonia) 3:22–37.

Melzack, R., and P.D. Wall. 1965. Pain mechanisms: A new theory. *Science* 150:971–979.

Meredith, I.T., G.L. Jennings, M.D.Esler, E.M. Dewar, A.M. Bruce, V.A. Fazio et al.1990. Time-course of the anti-hypertensive effects of regular endurance exercise in human subjects. *J. Hypertens.* 8:859–866.

Merino, C. 1950. Studies on blood formation and destruction in the polycythemia of high altitude. *Blood* 5:1.

Mero, A., L. Jaakkola, and P.V. Komi. 1991. Relationship between muscle fibre characteristics and physical performance capacity in trained athletic boys. *J. Sports Sci.* 9 (2):161–171.

Merton, P.A. 1954. Voluntary strength and fatigue. *J. Physiol.* (Lond.) 123:553–564.

Metzger, J.M., and R.L. Moss. 1987. Greater hydrogen ion–induced depression of tension and velocity in skinned single fibres of rat fast than slow muscles. *J. Physiol.* 393:727–742.

———. 1990. Effects of tension and stiffness due to reduced pH in mammalian fast- and slow-twitch skinned skeletal muscle fibres. *J. Physiol.* 428:737–750.

Meyer, F., O. Bar-Or, A. Salsberg, and D. Passe. 1994. Hypohydration during exercise in children: Effect of thirst, drink preferences and rehydration. *Int. J. Sport Nutr.* 4:22–35.

Mikines, K.J., E.A. Richter, F. Dela, and H. Galbo. 1991. Seven days of bed rest decrease insulin action on glucose uptake in leg and whole body. *J. Appl. Physiol.* 70:1245–1254.

Miles, D.S., M.N. Sawka, R.M. Glaser, and J.S. Petrofsky. 1983. Plasma volume shifts during progressive arm and leg exercise. *J. Appl. Physiol. Respir. Environ. Exerc. Physiol.* 54:491.

Miles, M.P., and P.M. Clarkson. 1994. Exercise-induced muscle pain, soreness and cramps. *J. Sports Med. Phys. Fitness* 34:203–216.

Miles, S. 1957. The effect of changes in barometric pressure on maximum breathing capacity. *J. Physiol.* 137:85P.

Miller, H.I., B. Issekutz, Jr., P. Paul, and K. Rodahl. 1964. Effect of lactic acid on plasma free fatty acids in pancreatectomized dogs. *Am. J. Physiol.* 207:1226.

Miller, R.G., J.A. Kent-Braun, K.R. Sharma, and M.W. Weiner. 1995. Mechanism of human muscle fatigue: Quantitating the contribution of metabolic factors and activation impairment. *Adv. Exp. Med. Biol.* 384:195–210.

Miller, R.G., R.S. Moussavi, A.T. Green, P.J. Carson, and M.W. Weiner. 1993. The fatigue of rapid repetitive movements. *Neurology* 43:755–761.

Miller, W.C. 1999. How effective are traditional dietary and exercise interventions for weight loss? *Med. Sci. Sports Exerc.* 31:1129–1134.

Millerhagen, J.O., J.M. Kelly, and R.J. Murphy. 1983. A study of combined arm and leg exercise with application to nordic skiing. *Can. J. Appl. Sports Sci.* 8:92–97.

Millhorn, D.E., F.L. Eldridge, and T.G. Waldrop. 1980. Prolonged stimulation of respiration by endogenous central serotonin. *Respir. Physiol.* 42:171–188.

Milner-Brown, H.S., M. Mellenthin, and R.G. Miller. 1986. Quantifying human muscle strength, endurance and fatigue. *Arch. Phys. Med. Rehabil.* 67:530–535.

Milner-Brown, H.S., and R.G. Miller. 1986. Muscle membrane excitation and impulse propagation velocity are reduced during muscle fatigue. *Muscle Nerve* 9:367–374.

Milner-Brown, H.S., R.B. Stein, and R. Yemm. 1973. Changes in firing rate of human motor units during linearly changing voluntary contractions. *J. Physiol.* (Lond.) 230:371–390.

Minajeva, A., C. Neagoe, M. Kulke, and W.A. Linke. 2002. Titin-based contribution to shortening velocity of rabbit skeletal myofibrils. *J. Physiol.* 540.1:177–188.

Minard, D., and L. Copman. 1963. Elevation of body temperature in disease. In *Temperature: Its measurement and control in science and industry*, ed. J.D. Hardy, vol. 3, part 3, p. 253. New York: Reinhold.

Mink, J.W., R.J. Blumenschine, and D.B. Adams. 1981. Ratio of central nervous system to body metabolism in vertebrates: Its constancy and functional bases. *Am. J. Physiol.* 241:R203–R212.

Mitch, W.E., J.L. Bailey, X. Wang, C. Jurkovitz, D. Newby, and S.R. Price. 1999. Evaluation of signals activating ubiquitin-proteasome proteolysis in a model of muscle wasting. *Am. J. Physiol. Cell Physiol.* 276:C1132–C1138.

Mitchell, J.H. 1992. How to recognize "athlete's heart." *Physician Sportsmed.* 20:87–96.

Mitchell, J.H., M.P. Kaufman, and G.A. Iwamoto. 1983. The exercise pressor reflex: Its cardio-vascular effects, afferent mechanisms and central pathways. *Annu. Rev. Physiol.* 45:229.

Mitchell, J.H., B. Schibye, F.C. Payne, III, and B. Saltin. 1981. Response of arterial blood pressure to static exercise in relation to muscle mass, force development and electromyographic activity. *Circ. Res.* 48 (suppl. 1):70.

Miyashita, K., S. Shiomi, N. Itoh, K. Kasamatsu, and H. Iwata. 1983. Epidemiological study of vibration syndrome in response to total handbook-operating time. *Br. J. Ind. Med.* 40:92–98.

Mizuno, M., C. Juel, T. Bro-Rasmussen, E. Mygind, B. Schibye, B. Rasmussen, and B. Saltin. 1990. Limb skeletal muscle adaptation in athletes after training at altitude. *J. Appl. Physiol.* 68:496–502.

Moffatt, R.J., and S.G. Owens. 1991. Cessation from cigarette smoking: Changes in body weight, body composition, resting metabolism and energy consumption. *Metabolism* 40:465–470.

Mohrman, D.E. 1982. Lack of influence of potassium or osmolarity on steady-state exercise hyperemia. *Am. J. Physiol.* 242:H949.

Molbech, S. 1963. *Average percentage force at repeated maximal isometric muscle contractions at different frequencies, 16.* Communications from the Testing and Observations Institute of the Danish National Association for Infantile Paralysis, Copenhagen.

Montain, S.J., and E.F. Coyle. 1992. Influence of graded dehydration on hyperthermia and cardiovascular drift during exercise. *J. Appl. Physiol.* 73:1340–1350.

Montain, S.J., J.E. Laird, W.A. Latzka, and M.N. Sawka. 1997. Aldosterone and vasopressin responses in the heat: Hydration level and exercise. *Med. Sci. Sports Exerc.* 29:661–668.

Montain, S.J., W.A. Latzka, and M.N. Sawka. 1995. Control of thermoregulatory sweating is altered by hydration level and exercise intensity. *J. Appl. Physiol.* 79:1434–1439.

Montgomery, D.L. 1988. Physiology of ice hockey. *Sports Med.* 5:99–126.

Monti, R.J., R.R. Roy, J.A. Hodgson, and V.R. Edgerton. 1999. Transmission of forces within mammalian skeletal muscles. *J. Biomech.* 32:371–380.

Montoye, H.J., E.L. Smith, D.F. Fardon, and E.T. Howley. 1980. Bone mineral in senior tennis players. *Scand. J. Sport Sci.* 2:26–32.

Moore, U., D.A. Lombardi, M.J. White, J.L. Campbell, S.A. Oliveria, and P.C. Ellison. 1991. Influence of parent's physical activity levels on activity levels of young children. *J. Pedriat.* 118:215–219.

Morehouse, L.E. 1972. *Laboratory manual for physiology of exercise.* St. Louis: Mosby.

Morgan, D.L. 1990. New insights into the behavior of muscle during active lengthening. *Biophys. J.* 57:209–221.

Morgan, R.E., and G.T. Adamson. 1962. *Circuit training.* London: Bell.

Morgan, W.P., ed. 1997. *Physical activity and mental health.* Washington, DC: Taylor and Francis.

Morganroth, J., B.J. Maron, W.L. Henry, and S.E. Epstein. 1975. Comparative left ventricular dimensions in trained athletes. *Ann. Intern. Med.* 82:521–524.

Moritani, T. 1992. Time course of adaptations during strength and power training. In *Strength and power in sport,* ed. P.V. Komi, pp. 266–278. Oxford: Blackwell Scientific.

Morpurgo, G., P. Battaglia, N.D. Carter, G. Modiano, and S. Passi. 1972. The Bohr effect and the red cell 2,3-DPG and HB content in Sherpas and Europeans at low and high altitude. *Experientia* 28:1280–1283.

Morris, J.M., M.G. Everitt, R. Pollard, S.P.W. Chave, and A.M. Semmence. 1980. Vigorous exercise in leisuretime: Protection against coronary heart disease. *Lancet* 2:1207–1210.

Morris, J.M., J.A. Heady, P.A.B. Raffle, C.G. Roberts, and J.W. Parks. 1953. Coronary heart disease and physical activity of work. *Lancet* 2:1053–1057.

Morris, J.M., D.R. Lucas, and B. Bresler. 1961. Role of the trunk in the stability of the spine. *J. Bone Joint Surg.* 43A:327.

Morrissey, M.C., E.A. Harman, and M.J. Johnson. 1995. Resistance training modes: Specificity and effectiveness. *Med. Sci. Sports Exerc.* 27:648–660.

Morton, A.R. and E.V. Holmik. 1985. The effects of cigarette smoking on maximal oxygen consumption and selected physiological responses of elite team sportsmen. *Eur. J. Appl. Physiol.* 53:348–352.

Morton, D.P., and R. Callister. 2000. Characteristics and etiology of exercise-related transient abdominal pain. *Med. Sci. Sports Exerc.* 32:432–438.

Mosekilde, L. 1988. Age-related changes in vertebral trabecular bone architecture assessed by a new method. *Bone* 9:247–250.

Moss, F.P., and C.P. Leblond. 1971. Satellite cells as the source of nuclei in muscles of growing rats. *Anat. Rec.* 170:421–435.

Mosso, A. 1892. *Die Ermüdung.* Hirzel Verlag, Leipzig.

Mostoufi-Moab, S., E.J. Widmaier, J.A. Cornett, K. Gray, and L.I. Sinoway. 1998. Forearm training reduces the exercise pressor reflex during ischemic rhythmic handgrip. *J. Appl. Physiol.* 84:277–283.

Moxham, J., R.H. Edwards, M. Aubier, A. De Troyer, G. Farkas, P.T. Macklem, and C. Roussos. 1982. Changes in EMG power spectrum (high-to-low ratio) with force fatigue in humans. *J. Appl. Physiol.* 53:1094–1099.

Muido, L. 1946. The influence of body temperature on performances in swimming. *Acta Physiol. Scand.* 12:102.

Mullin, W.J., R.E. Herrick, V. Valdez, and K.M. Baldwin. 1984. Adaptive responses of rats trained with reductions in exercise heart rate. *J. Appl. Physiol.* 56:1378–1382.

Mundal, R., J. Erikssen, R. Bjørklund, and K. Rodahl. 1990. Elevated blood pressure in air traffic controllers during a period of occupational conflict. *Stress Med.* 6:141–144.

Mundale, M.O. 1970. The relationship of intermittent isometric exercise to fatigue of handgrip. *Arch. Phys. Med. Rehabil.* 51:532–539.

Murphy, S.A., R.D. Beaton, K. Cain, and K. Pike. 1994. Gender differences in fire-fighter job stressors and symptoms of stress. *Women Health* 22:55–69.

Murray, R. 1995. Fluid needs in hot and cold environments. *Int. J. Sport Nutr.* 5:62–73.

Mutrie, N. 1997. The therapeutic effects of exercise on the self. In *The physical self: From motivation to wellbeing,* ed. K. Fox, pp. 506–560. Champaign, IL: Human Kinetics.

Myerson, S., H. Hemingway, R. Budget, J. Martin, S. Humphries, and H. Montgomery. 1999. Human angiotensin I–converting enzyme gene and endurance performance. *J. Appl. Physiol.* 87:1313–1316.

Mygind, E., B. Larsson, and T. Klausen. 1991. Evaluation of a specific test in cross-country skiing. *J. Sports Sci.* 9 (3):249–257.

Nadeau, M., and A. Brassard. 1983. The bicycle ergometer for muscle power testing. *Can. J. Appl. Sports Sci.* 8 (1):41–46.

Nadel, E.R., E. Cafarelli, M.F. Roberts, and C.B. Wenger. 1979. Circulatory regulation during exercise in different ambient temperatures. *J. Appl. Physiol.* 46 (3):430–437.

Nadel, E.R., S.M. Fortney, and C.B. Wenger. 1980. Effect of hydration state on circulatory and thermal regulations. *J. Appl. Physiol.* 49:715.

Nadel, E.R., J.W. Mitchell, B. Saltin, and J.A.J. Stolwijk. 1971. Peripheral modifications to the central drive for sweating. *J. Appl. Physiol.* 31:828.

Nadel, E.R., and J.A.J. Stolwijk. 1973. Effect of skin wettedness on sweat gland response. *J. Appl. Physiol.* 35:689.

Nader, G.A. and K.A.Esser. 2001. Intracellular signaling specificity in skeletal muscle in response to different modes of exercise. *J. Appl. Physiol.* 90:1936–1942.

Næss, K., and A. Storm-Mathisen. 1955. Fatigue of sustained tetanic contractions. *Acta Physiol. Scand.* 34:351–366.

Nagle, F.J., B. Balke, and J.P. Naughton. 1965. Gradational step tests for assessing work capacity. *J. Appl. Physiol.* 20:745–748.

Naidu, M., and V. Sachdeva. 1993. Effect of local cooling on skin temperature and blood flow of men in Antarctica. *Int. J. Biometeorol.* 37:218–221.

Nakamura, E., T. Moritani, and A. Kanetaka. 1996. Effects of habitual physical exercise on physiological age in men aged 20–85 years as estimated using principal component analysis. *Eur. J. Appl. Physiol.* 73:410–418.

Nardone, A., C. Romano, and M. Schieppati. 1989. Selective recruitment of high-threshold human motor units during voluntary isotonic lengthening of active muscles. *J. Physiol.* (Lond.) 409:451–471.

Nashner, L.M. 1976. Adapting reflexes controlling the human posture. *Exp. Brain Res.* 26:59–72.

National Research Council. 1989. *Recommended dietary allowances.* 10th ed. Washington, DC: National Academy Press.

Naya, F.J., B. Mercer, J. Shelton, J.A. Richardson, R.S. Williams, and E.N. Olson. 2000. Stimulation of slow skeletal muscle fiber gene expression by calcineurin in vivo. *J. Biol. Chem.* 275:4545–4548.

Nehlig, A., and G. Debry. 1994. Caffeine and sports activity: A review. *Int. J. Sports Med.* 15:215–223.

Nelms, J.D., and J.G. Soper. 1962. Cold vasodilatation and cold acclimatization in the hands of British fish filleters. *J. Appl. Physiol.* 17:444.

Nelson, N.A., W.B. Shelley, S.M. Horvath, L.W. Eichna, and T.F. Hatch. 1948. Influence of clothing, work and air movement on the thermal exchanges of acclimatized men in various hot environments. *J. Clin. Invest.* 27:209.

Nes, H., R. Karstensen, and K. Rodahl. 1990. *Varmestressundersøkelse ved Elkem aluminium, Mosjøen.* Technical Report.

Newham, D.J., K.R. Mills, B.M. Quigley, and R.H. Edwards. 1983. Pain and fatigue after concentric and eccentric muscle contractions. *Clin. Sci.* (Colch.) 64:55–62.

Newman, E.V., D.B. Dill, H.T. Edwards, and F.A. Webster. 1937. The rate of lactic acid removal in exercise. *Am. J. Physiol.* 118:457.

Newsholme, E.A., and E. Blomstrand. 1995. Tryptophan, 5-hydroxytryptamine and a possible explanation for central fatigue. *Adv. Exp. Med. Biol.* 384:315–320.

Newsholme, E.A., P. Calder, and P. Yaqoob. 1993. The regulatory, informational and immunomodulatory roles of fat fuels. *Am. J. Clin. Nutr.* 57 (suppl.):738–751.

Newsholme, E.A., and A.R. Leech. 1984. *Biochemistry for the medical sciences.* Chichester: Wiley.

Nicholson, A.N., and B.M. Stone. 1982. *Sleep and wakefulness.* Handbook for Flight Medical Officers. AGARDograph no. 270(E). Neuilly-sur-Seine: AGARD.

Nicol, C., P.V. Komi, T. Horita, H. Kyröläinen, and T.E.S. Takala. 1996. Reduced stretch-reflex sensitivity after exhausting stretch-shortening cycle exercise. *Eur. J. Appl. Physiol. Occup. Physiol.* 72:401–409.

Niebauer, J., and J.P. Cooke. 1996. Cardiovascular effects of exercise: Role of endothelial shear stress. *J. Am. Coll. Cardiol.* 28:1652–1660.

Nielsen, B. 1969. Thermoregulation in rest and exercise. *Acta Physiol. Scand. Suppl.*, vol. 323.

———. 1981. Exercise and temperature regulation. In *Contributions to thermal physiology: Satellite of the 28th International Congress of Physiological Sciences, Pecs, Hungary, 1981,* ed. Z. Szelényi and M. Székely, p. 537. Pécs, Hungary.

———. 1998. Heat acclimation: Mechanisms of adaptation to exercise in the heat. *Int. J. Sports Med.* 19 (suppl. 2):154–156.

Nielsen, B., G. Savard, E.A. Richter, M. Hargreaves, and B. Saltin. 1990. Muscle blood flow and muscle metabolism during exercise and heat stress. *J. Appl. Physiol.* 69:1040–1046.

Nielsen, M. 1938. Die R egulation der Körpertemperatur bei Muskelarbeit. *Skand. Arch. Physiol.* 79:193.

Nielsen, O.B., and T. Clausen. 2000. The $Na^+/K(^+)$-pump protects muscle excitability and contractility during exercise. *Exerc. Sport Sci. Rev.* 28:159–164.

Nieman, D.C., D.A. Henson, L.L. Smith, A.C. Utter, D.M. Vinci, J.M. Davis, D.E. Kaminsky, and M. Shute. 2001. Cytokine changes after a marathon race. *J. Appl. Physiol.* 91:109–114.

Niinimaa, W. 1983. Oronasal airway choice during running. *Respir. Physiol.* 53:129.

Noakes, T.D. 1993. Fluid replacement during exercise. *Exerc. Sport Sci. Rev.* 21:297–330.

———. 2000. Physiological models to understand exercise fatigue and the adaptations that predict or enhance athletic performance. *Scand. J. Med. Sci. Sports* 10:123–145.

Noell, W. 1944. Über die Durchblutung und die Sauerstoffversorgung des Gehirns, Vl, Einfluss der Hypoxämie und Anämie. *Arch. Ges. Physiol.* 247:553.

Noirez, P., and A. Ferry. 2000. Effect of anabolic/androgenic steroids on myosin heavy chain expression in hindlimb muscles of male rats. *Eur. J. Appl. Physiol. Occup. Physiol.* 81:155–158.

Norton, E.F. 1925. *The fight for Everest.* London: Edward Arnold.

Norton, K.H., R. Boushel, S. Strange, B. Saltin, and P.B. Raven. 1999. Resetting of the carotid arterial baroreflex during exercise in humans. *J. Appl. Physiol.* 87:332–338.

Norton, K.H., K.M. Gallagher, S.A. Smith, R.G. Querry, R.M. Welch-O'Connor, and P.B. Raven. 1999. Carotid baroreflex function during prolonged exercise. *J. Appl. Physiol.* 87:339–347.

Nosaka, K., and P.M. Clarkson. 1995. Muscle damage following repeated bouts of high force eccentric exercise. *Med. Sci. Sports Exerc.* 27:1263–1269.

Nosek, T.M., J.H. Leal-Cardoso, M. McLaughlin, and R.E. Godt. 1990. Inhibitory influence of phosphate and arsenate on contraction of skinned skeletal and cardiac muscle. *Am. J. Physiol.* 259:C933–C939.

Novitsky, S., K.R. Segal, D. Chatr-Aryamontri, D. Guvakov, and V.L. Katch. 1995. Validity of a new portable indirect calorimeter: The Aero Sport Teem 100. *Eur. J. Appl. Physiol.* 70:462–467.

Noyes, F.R., P.J. Torvik, W.B. Hyde, and J.L. DeLucas. 1974. Biomechanics of ligament failure: II. An analysis of immobilization, exercise and reconditioning effects in primates. *J. Bone Joint Surg.* (Am.) 56:1406–1418.

Nummela, A., and H. Rusko. 2000. Acclimatization to altitude and normaxic training improve 400-m running performance at sea level. *J. Sports Sci.* 18:411–419.

Nunneley, S.A., and C.F. Flick. 1981. Heat stress in the A-10 cockpit: Flights over desert. *Aviat. Space Environ. Med.* 52:513.

Nunneley, S.A., and F. Stribley. 1979. Heat and acute dehydration effects on acceleration response in man. *J. Appl. Physiol.* 47:197.

Nybo, L. and B. Nielsen. 2001. Hyperthermia and central fatigue during prolonged exercise in humans. *J. Appl. Physiol.* 91:1055–1060.

Nybo, L., B. Nielsen, B.K. Pedersen, K. Møller, and N.H. Secher. 2002. Interleukin-6 release from the human brain during prolonged exercise. *J. Physiol.* 542.3:991–995.

Nygaard, E. 1981. Skeletal muscle fiber characteristics in young women. *Acta Physiol. Scand.* 112:299–304.

Nylander, E. 1985. Training-induced bradycardia in rats on cardioselective and non-selective beta receptor blockade. *Acta Physiol. Scand.* 123:147–149.

Oberoi, K., M.K. Dhillon, and S.S. Miglani. 1983. A study of energy expenditure during manual and machine washing of clothes in India. *Ergonomics* 26:375–378.

O'Brien, C.P. 1993. Alcohol and sport: Impact of social drinking on recreational and competitive sports performance. *Sports Med.* 15:71–77.

O'Connor, P.J., and D.B. Cook. 1999. Exercise and pain: The neurobiology, measurement and laboratory study of pain in relation to exercise in humans. *Exerc. Sport Sci. Rev.* 27:119–166.

O'Connor, P.J., J.S. Raglin, and E.W. Martinsen. 2000. Physical activity, anxiety and anxiety disorders. *Int. J. Sport Psychol.* **31**, 136–155.

Oddershede, I.R., and R.S. Elizondo. 1980. Body fluid and hematologic adjustments during resting heat acclimation in rhesus monkey. *J. Appl. Physiol.* 49:431.

Ogawa, S., K. Asano, Y. Furuta, T. Obara, and T. Fujimaki. 1974. The effect of intermittent rope skipping on aerobic work capacity. *Bull. Inst. Sport Sci. Faculty Phys. Ed., Tokyo U. Educ.* 23:1.

Ohira, Y., D.R. Simpson, V.R. Edgerton, G.W. Gardner, and B. Senewiratne. 1983. Characteristics of blood gas in response to iron treatment and exercise in iron-deficient and anemic subjects. *J. Nutr. Sci. Vitaminol.* 29:129.

Ohkuwa, T., H. Itoh, Y. Yamazaki, and Y. Sato. 1995. Salivary and blood lactate after supramaximal exercise in sprinters and long-distance runners. *Scand. J. Med. Sci. Sports* 5 (5):285–290.

Ohtsuki, T. 1981. Decrease in grip strength induced by simultaneous bilateral exertion with reference to finger strength. *Ergonomics* 24:37–48.

Olsson, R.A. 1981. Local factors regulating cardiac and skeletal muscle blood flow. *Annu. Rev. Physiol.* 43:385.

Omvik, P. 1996. How smoking affects blood pressure. *Blood Press.* 5:71–77.

On, A.Y., Z. Colakoglu, S. Herpguler, and R. Akrit. 1997. Local heat effect on sympathetic skin responses after pain of electrical stimulus. *Arch. Phys. Med. Rehabil.* 78:1196–1199.

O'Neill, D.S., D.H. Zheng, W.K. Anderson, G.L. Dohm, and J.A. Houmard. 1999. Effect of endurance exercise on myosin heavy chain gene regulation in human skeletal muscle. *Am. J. Physiol. Reg. Int. Comp. Physiol.* 276:R414–R419.

Op 't, E.B., E.A. Richter, J.C. Henquin, B. Kiens, and P. Hespel. 2001. Effect of creatine supplementation on creatine and glycogen content in rat skeletal muscle. *Acta Physiol. Scand.* 171:169–176.

Ordway, G.A., P.D. Neufer, E.R. Chin, and G.N. DeMartino. 2000. Chronic contractile activity upregulates the proteasome system in rabbit skeletal muscle. *J. Appl. Physiol.* 88:1134–1141.

Osnes, J.B., and L. Hermansen. 1972. Acid-base balance after maximal exercise of short duration. *J. Appl. Physiol.* 32:59–63.

Östman, I., and N.O. Sjöstrand. 1975. Reduced urinary noradrenaline excretion during rest, exercise and cold stress in trained rats: A comparison between physically-trained rats, cold-acclimated rats and warm-acclimated rats. *Acta Physiol. Scand.* 95:209–218.

Otis, C.L., B. Drinkwater, M. Johnson, A. Loucks, and J. Wilmore. 1997. American College of Sports Medicine position stand: The female athlete triad. *Med. Sci. Sports Exerc.* 29:i–ix.

Ounjian, M., R. Roy, E. Eldred, A. Garfinkel, J.R. Payne, A. Armstrong, A.W. Toga, and V.R. Edgerton. 1991. Physiological and developmental implications of motor unit anatomy. *J. Neurobiol.* 22:547–559.

Owen, O.E., A.P. Morgan, H.G. Kemp, J.M. Sullivan, M.G. Herrera, and G.F. Cahill, Jr. 1967. Brain metabolism during fasting. *J. Clin. Invest.* 46:1589–1595.

Paavolainen, L., K. Häkkinen, I. Hämäläinen, A. Nummela, and H. Rusko. 1999. Explosive-strength training improves 5-km running time by improving running economy and muscle power. *J. Appl. Physiol.* 86:1527–1533.

Padykula, H.A., and E. Herman. 1955a. Factors affecting the activity of adenosine triphosphatase and other phosphatases as measured by histochemical techniques. *J. Histochem. Cytochem.* 3:161–169.

———. 1955b. The specificity of the histochemical method for adenosine triphosphatase. *J. Histochem. Cytochem.* 3:170–195.

Paffenbarger, R.S., and W.E. Hale. 1975. Work activity and coronary heart mortality. *N. Engl. J. Med.* 292:545–550.

Paffenbarger, R.S., and R.T. Hyde. 1980. Exercise as protection against heart attack. *N. Engl. J. Med.* 302:1026–1027.

Paffenbarger, R.S., R.T. Hyde, and A.L. Wing. 1990. Physical activity and physical fitness as determinants of health and longevity. In *Exercise, fitness, and health,* ed. C. Bouchard, R.J. Shephard, T.S. Stephens, J.R. Sutton, and B.D. McPherson, pp. 33–48. Champaign, IL: Human Kinetics.

Paffenbarger, R.S., R.T. Hyde, A.L. Wing, and C.C. Hsieh. 1986. Physical activity, all-cause mortality and longevity of college alumni. *N. Engl. J. Med.* 314:605–613.

Paffenbarger, R.S., R.T. Hyde, A.L. Wing, I.M. Lee, D.L. Jung, and J.B. Kampert. 1993. The association of changes in physical activity level and other lifestyle characteristics with mortality among men. *N. Engl. J. Med.* 328:538–545.

Palenciano, L., V. Gonzalez, L.A. Santullano, B. Rodriguez, and M.A. Montoliu. 1996. Cardiac frequency in miners recorded during four to five work shifts. *Eur. J. Appl. Physiol.* 73:369–375.

Palmoski, M.J., and K.D. Brandt. 1981. Running inhibits the reversal of atrophic changes in canine knee cartilage after removal of a leg cast. *Arthritis Rheum.* 24:1329–1337.

Pandolf, K.B. 1991. Aging and heat tolerance at rest or during work. *Exp. Aging Res.* 17:189–204.

———. 1997. Aging and human heat tolerance. *Exp. Aging Res.* 23:69–105.

Pankey, R.B., D.W. Bacharach, and R.A. Gaugler. 1996. Anaerobic power differences in fit women across age. *J. Strength Conditioning Res.* 10 (1):62–64.

Panush, R.S., C.S. Hansson, J.R. Caldwell, S. Longley, J. Stork, and R. Thoburn. 1995. Is running associated with ostearthritis? *Clin. Rheumatol.* 1:35–39.

Pappenheimer, J.R. 1960. Central control of renal circulation. *Physiol. Rev.* 40 (suppl. 4):35.

Pardo, J.V., J.D. Siliciano, and S.W. Craig. 1983. A vinculin-containing cortical lattice in skeletal muscle: Transverse lattice elements ("costameres") mark sites of attachment between myofibrils and sarcolemma. *Proc. Natl. Acad. Sci. USA* 80:1008–1012.

Pardy, R.L., R.N. Rivington, P.J. Despas, and P.T. Macklem. 1981. The effects of inspiratory muscle training on exercise performance in chronic airflow limitation. *Am. Rev. Respir. Dis.* 123:426–433.

Parízková, J. 1977. *Body fat and physical fitness.* The Hague: Nijhoff/Medical Division.

Parry, D.J. 2001. Myosin heavy chain expression and plasticity: Role of myoblast diversity. *Exerc. Sport Sci. Rev.* 29:175–179.

Passmore, R., and J.V.G.A. Durnin. 1955. Human energy expenditure. *Physiol. Rev.* 35:801.

Passmore, R., J.G. Thomson, and G.M. Warnock. 1952. Balance sheet of the estimation of energy intake and energy expenditure as measured by indirect calorimetry. *Br. J. Nutr.* 6:253.

Pate, E., M. Bhimani, K. Franks-Skiba, and R. Cooke. 1995. Reduced effect of pH on skinned rabbit psoas muscle mechanics at high temperatures: Implications for fatigue. *J. Physiol.* 486:689–694.

Pate, R.R., and A. Kriska. 1984. Physiological basis of the sex difference in cardiorespiratory endurance. *Sports Med.* 1:87–98.

Patel, T.J., and R.L. Lieber. 1997. Force transmission in skeletal muscle: From actomyosin to external tendons. *Exerc. Sport Sci. Rev.* 25:321–363.

Paterson, D.H., D.A. Cunningham, J.J. Koval, and C.M. St. Croix. 1999. Aerobic fitness in a population of independently living men and women aged 55–86 years. *Med. Sci. Sports Exerc.* 31:1813–1820.

Patten, B.M. 1984. Neuromuscular disease due to tobacco use. *Texas Med.* 80:47–51.

Paul, P. 1975. Effects of long lasting physical exercise and training on lipid metabolism. In *Metabolic adaptation to prolonged physical exercise,* ed. H. Howald and J.R. Poortman, pp. 156–193. Basel: Birkhäuser Verlag.

Pavlath, G.K., K. Rich, S.G. Webster, and H.M. Blau. 1989. Localization of muscle gene products in nuclear domains. *Nature* 337:570–573.

Pedersen, B.K., K. Ostrowski, T. Rohde, and H. Bruunsgaard. 1998. The cytokine response to strenuous exercise. *Can. J. Physiol. Pharmacol.* 76:505–511.

Pedersen, P.K. 1983. Oxygen uptake kinetics and lactate accumulation in heavy submaximal exercise with normal and high inspired oxygen fractions. In *Biochemistry of exercise*, ed. H.G. Knuttgen, J.A. Vogel, and J. Poortmans, pp. 415–420. International Series on Sport Sciences, vol. 13. Champaign, IL: Human Kinetics.

Peirce, N.S. 1999. Diabetes and exercise. *Br. J. Sports Med.* 33:161–172.

Pelliccia, A., B.J. Maron, F. Culasso, A. Spataro, and G. Caselli. 1996. Athlete's heart in women: Echocardiographic characterization of highly trained elite female athletes. *JAMA* 276:211–215.

Pelliccia, A., B.J. Maron, A. Spataro, M.A. Proschan, and P. Spirito. 1991. The upper limit of physiologic cardiac hypertrophy in highly trained elite athletes. *N. Engl. J. Med.* 324:295–301.

Pendergast, D.R. 1988. The effect of body cooling on oxygen transport during exercise. *Med. Sci. Exerc. Sport* 20:S171–S176.

Penfield, W., and T. Rasmussen. 1950. *The cerebral cortex of man: A clinical study of localization of function.* New York: Macmillan.

Peper, K., R.J. Bradley, and F. Dreyer. 1982. The acetylcholine receptor at the neuromuscular junction. *Physiol. Rev.* 62:1271–1340.

Pepler, R.D. 1963. Performance and well-being in heat. In *Temperature: Its measurement and control in science and industry,* ed. J.D. Hardy, vol. 3, part 3, p. 319. New York: Reinhold.

Peplowski, A. 1990. Secular trends in the physical development level of the Lublin Medical Academy's students examined in years 1966–1986. *Wychowanie fizyczne i sport* (Warsaw) 34 (1):43–50.

Peronnet, F., R.J. Ferguson, H. Perrault, and G. Ricci. 1982. Echocardiographic dimensions determined in normally active college women and in female athletes. *Med. Sci. Sports Exerc.* 14:181–182.

Perrault, H., and R.A. Turcotte. 1994. Exercise-induced cardiac hypertrophy: Fact or fallacy? *Sports Med.* 17:288–308.

Perry, S.V. 1994. Activation of the contractile mechanism by calcium. In *Myology: Basic and clinical,* ed. A.G. Engel and C. Franzini-Armstrong, pp. 529–552. New York: McGraw-Hill.

Pershagen, G. 1996. Smokeless tobacco. *Br. Med. Bull.* 52:50–57.

Persson, P.-G., S. Carlsson, L. Svanström, C.-G. Östenson, S. Efendic, and V. Grill. 2000. Cigarette smoking, oral moist snuff use and glucose intolerance. *J. Intern. Med.* 248:103–110.

Petajan, J.H. 1996. Commentary. *Muscle Nerve Suppl.* 4:S34–S37.

Peters, H.P., M. Zweers, F.J.G. Backx, E. Bol, E.R.H.A. Hendriks, W.L. Mosterd, and V.R. DeVries. 1999. Gastrointestinal symptoms during long-distance walking. *Med. Sci. Sports Exerc.* 31:767–773.

Peto, R., A.D. Lopez, J. Boreham, M. Thun, C. Heath, Jr., and R. Doll. 1996. Mortality from smoking worldwide. *Br. Med. Bull.* 52:12–21.

Petrén, T. 1936. Die totale Anzahl der Blutkapillaren im Herzen und Skelettmuskulatur bei Ruhe und nach langer Muskelübung. *Verhandl. Anatom. Gesellsch. (Suppl. Anat. Anz.)* 81:165.

Pette, D. 1984. J.B. Wolffe memorial lecture: Activity-induced fast to slow transitions in mammalian muscle. *Med. Sci. Sports Exerc.* 16:517–528.

Pette, D., H. Peuker, and R.S. Staron. 1999. The impact of biochemical methods for single muscle fibre analysis. *Acta Physiol. Scand.* 166:261–277.

Pettenkofer, M. von, and C. Voit. 1866. Untersuchungen über dem Stoffverbrauch des normalen Menschen. *Z. Biol.* 2:459.

Piehl, K. 1974. Time course for refilling of glycogen stores in human muscle fibres following exercise-induced glycogen depletion. *Acta Physiol. Scand.* 90:297–302.

Pierrynowski, M.R., D.A. Winter, and R.W. Norman. 1981. Metabolic measures to ascertain the optimal load to be carried by man. *Ergonomics* 24:393–399.

Pimental, N., and K.B. Pandolf. 1979. Energy expenditure while standing or walking slowly uphill or downhill with loads. *Ergonomics* 22:963–973.

Pincivero, D.M., and T.O. Bompa. 1997. A physiological review of American football. *Sports Med.* 23:247–260.

Pirnay, F., M. Lacroix, F. Mosora, A. Luyckx, and P. Lefebre. 1977. Glucose oxidation during prolonged exercise evaluated with naturally labeled $^{13}$C glucose. *J. Appl. Physiol.* 43:258–261.

Pitsiladis, Y.P., and R.J. Maughan. 1999. The effects of alterations in dietary carbohydrate intake on the performance of high-intensity exercise. *Eur. J. Appl. Physiol.* 97:433–442.

Pitts, G.C., R.E. Johnson, and F.C. Consolazio. 1944. Work in the heat as affected by intake of water, salt and glucose. *Am. J. Physiol.* 142:253.

Plum, F. 1970. Neurological integration of behavioural and metabolic control of breathing. In *Breathing: Hering-Breuer Centenary Symposium,* ed. R. Porter, p. 159. London: Churchill.

Poliner, Z.R., G.J. Dehmer, S.E. Lewis, R.W. Parkey, C.G. Blomqvist, and J.T. Willerson. 1980. Left ventricular performance in normal subjects: A comparison of the response to exercise in the upright and supine position. *Circulation* 62:528.

Pollock, M.L. 1973. The quantification of endurance training programs. *Exerc. Sport Sci. Rev.* 1:155–188.

Pollock, M.L., J.H. Wilmore, and S.M. Fox, III. 1978. *Health and fitness through physical activity.* American College of Sports Medicine Series. New York: Wiley.

Poortmans, J.R., and J. Vanderstraeten. 1994. Kidney function during exercise in healthy and diseased humans. *Sports Med.* 18:419–437.

Pope, R.P., R.D. Herbert, J.D. Kirwan, and B.J. Graham. 2000. A randomized trial of preexercise stretching for prevention of lower-limb injury. *Med. Sci. Sports Exerc.* 32:271–277.

Porcari, J.P., T.L. Hendrickson, P.R. Walter, L. Terry, and G. Walsko. 1997. The physiological responses to walking with and without Power Poles on treadmill exercise. *Res. Q. Exerc. Sports* 68:161–166.

Porter, M.M., A.A. Vandervoort, and J. Lexell. 1995. Aging of human muscle: Structure, function and adaptability. *Scand. J. Med. Sci. Sports* 5:129–142.

Powers, S.K. 1988. Incidence of exercise-induced hypoxemia in elite endurance athletes at sea level. *Eur. J. Appl. Physiol.* 58:298–302.

Powers, S.K., L.L. Ji, and C. Leeuwenburgh. 1999. Exercise training-induced alterations in skeletal muscle antioxidant capacity: A brief review. *Med. Sci. Sports Exerc.* 31:987–997.

Prasad, N. 1996. *The relationship between aerobic and anaerobic exercise capacities in pre-pubertal children.* Eugene, OR: Microform Publications, International Institute of Sport and Human Performance, University of Oregon.

Precht, H., J. Christophersen, H. Hensel, and W. Larcker. 1973. *Temperature and life.* Berlin: Springer-Verlag.

Price, D.L., and T.G. Hicks. 1979. The effects of alcohol on performance of a production assembly task. *Ergonomics* 22:37–41.

Prior, J.C., S.I. Barr, R. Chow, and R.A. Faulkner. 1996. Physical activity as therapy for osteoporosis. *Can. Med. Assoc. J.* 155:940–944.

Pripstein, L.P., E.C. Rhodes, D.C. McKenzie, and K.D. Coutts. 1999. Aerobic and anaerobic energy during a 2-km race simulation in female rowers. *Eur. J. Appl. Physiol.* 79:491–494.

Pruett, E.D.R. 1971. Fat and carbohydrate metabolism in exercise and recovery and its dependence upon work load severity. Thesis, Institute of Work Physiology, Oslo.

Przeveda, R. 1994. Is secular trend also present in the physical fitness of youth? In *Physical activity in the lifecycle: The 1993 FIEP World Congress Proceedings,* ed. R. Lidor, D. Ben-Sira, and Z. Artzi, pp. 182–90. Netanya, Israel: Zinman College of Physical Education at Wingate Institute.

Pugh, L.G. 1964. Animals in high altitude: Man above 5,000 meters—mountain exploration. In *Handbook of physiology,* sec. 4, *Adaptation to the environment,* ed. D.B. Dill, p. 861. Washington, DC: American Physiological Society.

———. 1971. The influence of wind resistance in running and walking and the mechanical efficiency of work against horizontal or vertical forces. *J. Physiol.* 213:255–276.

———. 1976. Air resistance in sport. In *Advances in exercise physiology,* ed. E. Jokl, R.L. Anand, and H. Stoboy. Basel: Karger.

Puntschart, A., E. Wey, K. Jostarndt, M. Vogt, M. Wittwer, H.R. Widmer, H. Hoppeler, and R. Billeter. 1998. Expression of *fos* and *jun* genes in human skeletal muscle after exercise. *Am. J. Physiol. Cell Physiol.* 274:C129–C137.

Purvis, J.W., and K.J. Cureton. 1981. Ratings of perceived exertion at the anaerobic threshold. *Ergonomics* 24:295–300.

Putman, C.T., S. Düsterhöft, and D. Pette. 1999. Changes in satellite cell content and myosin isoforms in low-frequency-stimulated fast muscle of hypothyroid rat. *J. Appl. Physiol.* 86:40–51.

Quinney, H.A. 1990. Sport on ice. In *Physiology of sports,* ed. T. Reilly, N. Secher, P. Snell, and C. Williams. London: Spon.

Radák, Z., J. Pucsok, S. Mecseki, T. Csont, and P. Ferdinandy. 1999. Muscle soreness-induced reduction in force generation is accompanied by increased nitric oxide content and DNA damage in human skeletal muscle. *Free Radical Biol. Med.* 26:1059–1063.

Rådegran, G., and B. Saltin. 1999. Nitric oxide in the regulation of vasomotor tone in human skeletal muscle. *Am. J. Physiol. Heart Circ. Physiol.* 276:H1951–H1960.

Radomski, M.W., and C. Boutelier. 1982. Hormone response of normal and intermittent cold-preadapted humans to continuous cold. *J. App. Physiol.* 53:610.

Rahkila, P., J. Soimajärvi, E. Karvinen, and V. Vittko. 1980. Lipid metabolism during exercise: II. Respiratory exchange ratio and muscle glycogen content during 4 hr cycling on bicycle ergometer in two groups of healthy men. *Eur. J. Appl. Physiol.* 44:245–254.

Rahn, H. 1966. Introduction to the study of man at high altitudes: Conductance of $O_2$ from the environment to the tissues. In *Life at high altitudes,* Scientific Publication 140, p. 2. Washington, DC: Pan-American Health Organization, WHO.

Raja, I.A., and F. Nicol. 1997. A technique for recording and analysis of postural changes associated with thermal comfort. *Appl. Ergon.* 28:221–225.

Ramanathan, L.N. 1964. A new weighing system for mean surface temperature of the human body. *J. Appl. Physiol.* 19:531–533.

Ramsay, J.E., and C.P. Chai. 1983. Inherent variability in heat-stress decision rules. *Ergonomics* 26:495.

Ramsey, J.D. 1975. Occupational vibration. In *Occupational medicine: Principles and practical application,* ed. C. Zenz, pp. 553–562. Chicago: Year Book Medical.

Ranallo, R.F., and E.C. Rhodes. 1998. Lipid metabolism during exercise. *Sports Med.* 26:29–42.

Rankinen, T., L. Pérusse, R. Rauramaa, M.A. Rivera, B. Wolfahrt, and C. Bouchard. 2001. The human gene map for performance and health-related fitness phenotypes. Medicine and Science in Sports & Exercise 33:855–867.

Ranvier, L. 1873. Propriétés et structures différentes des muscles rouges et des muscles blancs, chez les lapins et chez des raies. *C.R. Acad. Sci.* (Paris) 77:1030–1034.

Rao, A., C. Luo, and P.G. Hogan. 1997. Transcription factors of the NFAT family: Regulation and function. *Annu. Rev. Immunol.* 15:707–747.

Rassier, D.E., B.R. MacIntosh, and W. Herzog. 1999. Length dependence of active force production in skeletal muscle. *J. Appl. Physiol.* 86:1445–1457.

Rassler, B., and J. Kohl. 2000. Coordination-related changes in the rhythms of breathing and walking in humans. *Eur. J. Appl. Physiol.* 82:280–88.

Raynaud, J., J.P. Martineaud, J. Bordachar, M.C. Tillous, and J. Durand. 1974. Oxygen deficit and debt in submaximal exercise at sea level and high altitude. *J. Appl. Physiol.* 37:43–48.

Reeves, J.T., J. Halpin, J.E. Cohn, and F. Daoud. 1969. Increased alveolar-arterial oxygen difference during simulated high-altitude exposure. *J. Appl. Physiol.* 27:658–661.

Refsum, H.E., L.R. Gjessing, and S.B. Strømme. 1979. Changes in plasma amino acid distribution and urine amino acids excretion during prolonged heavy exercise. *Scand. J. Clin. Lab. Invest.* 39:407–413.

Refsum, H.E., and S.B. Strømme. 1974. Urea and creatinine production and excretion in urine during and after prolonged heavy exercise. *Scand. J. Clin. Lab. Invest.* 33:247–254.

Reichard, G.A., B. Issekutz, Jr., P. Kimbel, R.C. Putnam, N.J. Hochella, and S. Weinhouse. 1961. Blood glucose metabolism in man during muscular work. *J. Appl. Physiol.* 16:1001–1005.

Reindell, H., H. Klepzig, H. Steim, K. Musshoff, H. Roskamm, and E. Schildge. 1960. *Herz Kreislaufkrankheiten und Sport.* Munich: Johann Ambrosius Barth.

Reith, M.E.A., ed. 1997. *Neurotransmitter transporters: Structure, function and regulation.* Totowa, NJ: Humana.

Reynafarje, B. 1962. Myoglobin content and enzymatic activity of muscle and altitude adaptation. *J. Appl. Physiol.* 17:301–305.

Reynafarje, C. 1967. Humoral control of erythropoiesis at altitude. In *Exercise at altitude,* ed. R. Margaria, p. 165. Amsterdam: Excerpta Medica Foundation.

Richard, D. 1995. Exercise and the neurobiological control of food intake and energy expenditure. *Int. J. Obes. Relat. Metab. Disord.* 19:73–79.

Richter, E.A., and J.R. Sutton. 1994. Hormonal adaptation to physical activity. In *Physical activity, fitness, and health,* ed. C. Bouchard, R.J. Shephard, and T. Stephens. Champaign, IL: Human Kinetics.

Riggs, C.E., D.J. Johnson, R.D. Kilgour, and B.J. Konopka. 1983. Metabolic effects of facial cooling in exercise. *Aviat. Space Environ. Med.* 54:22.

Riggs, C.E., D.J. Johnson, B.J. Konopka, and R.D. Kilgour. 1981. Exercise heart rate response to facial cooling. *Eur. J. Appl. Physiol.* 47:323–330.

Rigotti, N.A., and R.C. Pasternak. 1996. Cigarette smoking and coronary heart disease: Risk and management. *Cardiol. Clin.* 14:51–68.

Rinder, M.R., R.J. Spina, and A.A. Ehsani. 2000. Enhanced endothelium-dependent vasodilation in older endurance-trained men. *J. Appl. Physiol.* 88:761–766.

Ringqvist, T. 1966. The ventilatory capacity in healthy subjects: An analysis of causal factors with special reference to the respiratory forces. *Scand. J. Clin. Lab. Invest. Suppl.* 88:5–179.

Rintamäki, H., T. Mäkinen, J. Oksa, and J. Latvala. 1995. Water balance and physical performance in cold. *Arctic Med. Res.* 54 (suppl. 2):32–36.

Rissanen S., M. Hori-Yamagishi, H. Tokura, Y. Tochihara, T. Ohnaka, and K. Tsuzuki. 1994. Thermal responses affected by different underwear materials during light exercise and rest in cold. *Ann. Physiol. Anthropol.* 13:129–136.

Roberg, R.A., J. Chwalbinska-Moneta, J.B. Mitchell, D.D. Pascoe, J. Houmard, and D.L. Costill. 1990. Blood lactate threshold differences between arterialized and venous blood. *Int. J. Sports Med.* 11 (6):446–451.

Roberts, A.D., R. Billeter, and H. Howald. 1982. Anaerobic muscle enzyme changes after interval training. *Int. J. Sports Med.* 3:18–21.

Roberts, M.F., and S.B. Wenger. 1980. Control of skin blood flow during exercise by thermal reflexes and baroflexes. *J. Appl. Physiol.* 48:717.

Robinson, S. 1938. Experimental studies of physical fitness in relation to age. *Arbeitsphysiologie* 10:251.

———. 1963. Circulatory adjustments of men in hot environments. In *Temperature: Its measurement and control in science and industry,* ed. J.D. Hardy, vol. 3, part 3, p. 287. New York: Reinhold.

———. 1980. Physiology of muscular exercise. In *Medical physiology,* ed. V.B. Mountcastle, pp. 1387–1416. St. Louis: Mosby.

Robinson, S., and P.M. Harmon. 1941. The effects of training and of gelatin on certain factors which limit muscular work. *Am. J. Physiol.* 133:161–169.

Robinson, S., E.S. Turell, H.S. Belding, and S.M. Horvath. 1943. Rapid acclimatization to work in hot climates. *Am. J. Physiol.* 140:168.

Rodahl, A., M. O'Brien, and R.G.R. Firth. 1976. Diurnal variation in performance of competitive swimmers. *J. Sports Med. Phys. Fitness* 16:72–76.

Rodahl, K. 1952. Basal metabolism of the Eskimo. *J. Nutr.* 48:359.

———. 1953. *North.* New York: Harper.

———. 1960. *Nutritional requirements under arctic conditions.* Norsk Polarinstitutts Skrifter no. 118. Oslo: Oslo University Press.

———. 1963. *Nutritional requirements in the polar regions.* U.N. Symposium on Health in the Polar Regions, Geneva, 1962. WHO Public Health Paper, vol. 18, p. 97.

———. 1989. *The physiology of work.* London: Taylor and Francis.

———. 1994. *Stress monitoring in the work place.* Boca Raton, FL: Lewis, CRC Press.

Rodahl, K., and G. Bang. 1957. Thyroid activity in men exposed to cold. *Arctic Aeromed. Lab. Tech. Report* 57:36.

Rodahl, K., N.C. Birkhead, J.J. Blizzard, B. Issekutz, Jr., and E.D.R. Pruett. 1966. Physiological changes due to prolonged bed rest. *Nord. Med.* 75:182–186.

Rodahl, K., R.A. Björklund, A.H. Kulsrud, L.D. Klüwer, and T. Guthe. 1992. *Effects of protective helmets on body temperature and psychomotor performance.* Fifth International Conference on Environmental Ergonomics, Maastricht, Nov. 2–6.

Rodahl, K., F.A. Giere, P.H. Staff, and B. Wedin. 1974. A physiological comparison of the protective value of nylon and wool in a cold environment. In *AGARD Report,* no. 620, ed. A. Borg, and J.H. Veghte. Neuilly-sur-Seine: AGARD.

Rodahl, K., and T. Guthe. 1988. Physiological limitations of human performance in hot environments, with particular reference to work in heat-exposed industry. In *Environmental ergonomics,* ed. J.B. Mekjavic, E.W. Banister, and J.B. Morrison. London: Taylor and Francis.

———. 1991. Beskyttelse av hodet mot varmestress. *KIL-amil-dok,* vol. 12.

Rodahl, K., T. Guthe, and L.D. Klüwer. 1991. Bruk av viftehjelm på varme arbeidsplasser: En orienterende undersøkelse. *KIL-amil-dok,* vol. 14.

———. 1992. En orienterende undersøkelse av mekanismen for økningen i hjertefrekvensen ved eksponering til varme. *KIL-amil-dok,* vol. 17.

Rodahl, K., S.M. Horvath, N.C. Birkhead, and B. Issekutz, Jr. 1962. Effects of dietary protein on physical work capacity during severe cold stress. *J. Appl. Physiol.* 17:763.

Rodahl, K., and B. Issekutz, Jr. 1962. Physical performance capacity of the older individual. In *Muscle as a tissue,* ed. K. Rodahl and S.M. Horvath. New York: McGraw-Hill.

———. 1965. Nutritional effects on human performance in the cold. In *Nutritional requirements for survival in the cold and in altitude,* ed. L. Vaughan, p. 7. Fort Wainwright, AK: Arctic Aeromedical Laboratory.

Rodahl, K., L. Leuba, L. Klüwer, and T. Guthe. 1992. Varmestress ved Moss Glassverk: Hodets rolle i den umiddelbare økning av hjertefrekvensen ved plutselig eksponering for varme. *KIL-amil-dok,* vol. 18.

Rodahl, K., H.I. Miller, and B. Issekutz, Jr. 1964. Plasma free fatty acids in exercise. *J. Appl. Physiol.* 19:489–492.

Rodahl, K., Z. Vokac, O. Fugelli, O. Vaage, and S. Mæhlum. 1974. Circulatory strain, estimated energy output and catecholamine excretion in Norwegian coastal fishermen. *Ergonomics* 17:585–602.

Rodan, G.A., and T.J. Martin. 2000. Therapeutic approaches to bone diseases. *Science* 289:1508–1514.

Rodgers, S.H., and E.M. Eggleton, eds. 1983. *Ergonomic design for people at work.* Vol. 1. New York: Van Nostrand Reinhold.

Rodriguez, F.A., H. Casas, M. Casas, T. Pages, R. Rama, A. Ricart, J.L. Ventura, J. Ibanez, and G. Viscor. 1999. Intermittent hypobaric hypoxia stimulates erythropoiesis and improves aerobic capacity. *Med. Sci. Sports Exerc.* 31:264–268.

Roetert, E.P., H. Brody, C.J. Dillman, J.L. Coppel, and J.M. Schultheis. 1995. The biomechanics of tennis elbow: An integrated approach. *Clin. Sports Med.* 14:47–57.

Rogers, S.M., and M.A. Baker. 1997. Thermoregulation during exercise in women who are taking oral contraceptives. *Eur. J. Appl. Physiol. Occup. Physiol.* 75:34–38.

Rognum, T.O., K. Rodahl, and P.K. Opstad. 1982. Regional differences in the lipolytic response of the subcutaneous fat depots to prolonged exercise and severe energy deficiency. *Eur. J. Appl. Physiol.* 49:401–408.

Rohmert, W. 1968. Die Beziehung Zwischen Kraft und Ausdauerbei bei Statischer Muskelarbeit: Schriftenreihe Arbeitsmedizin, Sozialmedizin. *Arbeitshygiene* 22:118. Stuttgart: A.W. Gentner Verlag.

Romanul, F.C.A. 1964. Enzymes in muscle: I. Histochemical studies of enzymes in individual muscle fibers. *Arch. Neurol.* 11, 355–368.

Romet, T.T., and J. Frim. 1987. Physiological responses to fire fighting activities. *Eur. J. Appl. Physiol.* 56:633–638.

Rooney, S.A., S.L. Young, and C.R. Mendelsohn. 1994. Molecular and cellular processing of lung surfactant. *FASEB J.* 8:957.

Roskamm, H. 1967. Optimum patterns of exercise for healthy adults. *Can. Med. Assoc. J.* 96:895–900.

Roskamm, H., L. Samek, H. Weideman, and H. Reindell. 1968. *Leistung und Höhe.* Ludvigshafen am Rheim: Knoll.

Ross, M.H., L.J. Romrell, and G.I. Kaye. 1995. *Histology: A text and atlas.* 3rd ed. Baltimore: Williams and Wilkins.

Rost, R., and W. Hollmann. 1983. Athlete's heart: A review of its historical assessment and new aspects. *Int. J. Sports Med.* 4:147.

Rothe, C.F. 1983. Reflex control of veins and vascular capacitance. *Physiol. Rev.* 63:1281.

Rowell, L.B. 1983. Cardiovascular adjustments to thermal stress. In *Handbook of physiology,* sec. 2, *The cardiovascular system,* vol. 3, *Peripheral circulation and organ blood flow,* part 2, chap. 25, p. 967. Bethesda, MD: American Physiological Society.

———. 1993. *Human cardiovascular control.* New York: Oxford University Press.

Rowell, L.B., G.L. Brengelmann, and P.R. Freund. 1987. Unaltered norepinephrine–heart rate relationship in exercise with exogenous heat. *J. Appl. Physiol.* 62:646–650.

Rowland, T., B. Popowski, and L. Ferrone. 1997. Cardiac responses to maximal upright cycle exercise in healthy boys and men. *Med. Sci. Sports Exerc.* 29:1146–1151.

Rowland, T., P. Vanderburgh, and L. Cunningham. 1997. Body size and the growth of maximal aerobic power in children: A longitudinal analysis. *Ped. Exerc. Sci.* 9:262–274.

Rowland, T.W. 1996. *Developmental physical activity.* Champaign, IL: Human Kinetics.

Roy, R.R., K.M. Baldwin, and V.R. Edgerton. 1996. Response of the neuromuscular unit to spaceflight: What has been learned from the rat model. *Exerc. Sport Sci. Rev.* 24:399–425.

Roy, R.R., and V.R. Edgerton. 1992. Skeletal muscle architecture and performance. In *Strength and power in sport,* ed. P.V. Komi, pp. 115–129. Oxford: Blackwell Science.

Roy, R.R., A. Ishihara, J.A. Kim, M. Lee, K. Fox, and V.R. Edgerton. 1999. Metabolic and morphological stability of motoneurons in response to chronically elevated neuromuscular activity. *Neuroscience* 92:361–366.

Roy, R.R., S.R. Monke, D.L. Allen, and V.R. Edgerton. 1999. Modulation of myonuclear number in functionally overloaded and exercised rat plantaris fibers. *J. Appl. Physiol.* 87:634–642.

Rudjord, T. 1972. Model study of muscle spindles subjected to static fusimotor activation. *Kybernetik* 10:189–200.

Rüegg, J.C. 1992. *Calcium in muscle contraction.* 2nd ed. Berlin: Springer Verlag.

Rüegg, M.A., and J.L. Bixby. 1998. Agrin orchestrates synaptic differentiation at the vertebrate neuromuscular junction. *Trends Neurosci.* 21:22–27.

Rundell, K.W. 1996a. Compromised oxygen uptake in speed skaters during treadmill in-line skating. *Med. Sci. Sports Exerc.* 28 (1):120–127.

———. 1996b. Effects of drafting during short-track speed skating. *Med. Sci. Sports Exerc.* 28:765–771.

Rundell, K.W., and D.W. Bacharach. 1995. Physiological characteristics of top U.S. biathletes. *Med. Sci. Sports Exerc.* 27 (9):1302–1310.

Rundell, K.W., S. Nioka, and B. Chance. 1997. Hemoglobin/myoglobin desaturation during speed skating. *Med. Sci. Sports Exerc.* 29:248–258.

Rusakov, D.A., D.M. Kullmann, and M.G. Stewart. 1999. Hippocampal synapses: Do they talk to their neighbours? *Trends Neurosci.* 22:382–388.

Rusko, H. 1987. The effect of training on aerobic power characteristics of young cross-country skiers. *J. Sports Sci.* 5:273–286.

———. 1996. New aspects of altitude training. *Am. J. Sports Med.* 24:S48–S52.

Rutenfranz, J., and W.P. Colquhoun, eds. 1978. Shiftwork: Theoretical issues and practical problems. *Ergonomics* 21:737.

Rutenfranz, J., and W.P. Colquhoun. 1979. Circadian rhythms in human performance. *Scand. J. Work Environ. Health* 5:167.

Ryhming, I. 1953. A modified Harvard step test for the evaluation of physical fitness. *Arbeitsphysiologie* 15:235.

Rywik, T.M., M.R. Blackman, A.R. Yataco, P.V. Vaitkevicius, R.C. Zink, E.H. Cottrell, J.G. Wright, L.I. Katzel, and J.L. Fleg. 1999. Enhanced endothelial vasoreactivity in endurance-trained older men. *J. Appl. Physiol.* 87:2136–2142.

Sable, D.L., H.L. Brammel, M.W. Sheehan, A.S. Nies, J. Gerber, and D.L. Honvitz. 1982. Attenuation of exercise conditioning by β-adrenergic blockade. *Circulation* 65:679.

Sacco, P., D.B. McIntyre, and D.A. Jones. 1994. Effects of length and stimulation frequency on fatigue of the human tibialis anterior muscle. *J. Appl. Physiol.* 77:1148–1154.

Sahlin, K., A. Alvestrand, R. Brandt, and E. Hultman. 1978. Intracellular pH and bicarbonate concentration in human muscle during recovery from exercise. *J. Appl. Physiol.* 45:474–480.

Sahlin, K., M. Fernström, M. Svensson, and M. Tonkonogi. 2002. No evidence of an intracellular lactate shuttle in rat skeletal muscle. *J. Physiol.* 541.2:569–574.

Sahlin, K., R.C. Harris, B. Nylind, and E. Hultman. 1976. Lactate content and pH in muscle obtained after dynamic exercise. *Pflügers Arch.* 367:143–149.

Sahlin, K., and J. Henriksson. 1984. Buffer capacity and lactate accumulation in skeletal muscle of trained and untrained men. *Acta Physiol. Scand.* 122:331–339.

Sahlin, K., M. Tonkonogi, and K. Söderlund. 1998. Energy supply and muscle fatigue in humans. *Acta Physiol. Scand.* 162:261–266.

Said, S.I. 1982. Metabolic functions of the pulmonary circulation. *Circ. Res.* 50:325.

Sakata, H., M. Taira, M. Kusunoki, A. Murata, and Y. Tanaka. 1997. The TINS Lecture: The parietal association cortex in depth perception and visual control of hand action. *Trends Neurosci.* 20:350–357.

Salamone, L.M., T. Whiteside, D. Friberg, R.S. Epstein, L.H. Kuller, and J.A. Cauley. 1998. Cytokine production and bone mineral density at the lumbar spine and femoral neck in premenopausal women. *Tissue Int.* 63:466–470.

Salans, L.B., S.W. Cushman, and R.E. Weismann. 1973. Studies of human adipose tissue, adipose cell size and number in non-obese and obese patients. *J. Clin. Invest.* 52:929–941.

Sale, D.G. 1987. Influence of exercise and training on motor unit activation. *Exerc. Sport Sci. Rev.* 15:95–151.

———. 1992. Neural adaptation to strength training. In *Strength and power in sport,* ed. P.V. Komi, pp. 249–265. Oxford: Blackwell Science.

Sale, D.G. 2002. Postactivation potentiation: Role in human performance. *Exerc. Sport Sci. Rev.* 30:138–143.

Sale, D.G., and J.D. MacDougall. 1981. Specificity in strength training: A review for the coach and athlete. *Can. J. Appl. Sport Sci.* 6:87–92.

Salmons, S. 1994. Exercise, stimulation and type transformation of skeletal muscle. *Int. J. Sports Med.* 15:136–141.

———. 1995. Muscle. In *Gray's anatomy*, ed. P.L. Williams, pp. 737–900. New York: Churchill Livingstone.

Salmons, S., and J. Henriksson. 1981. The adaptive response of skeletal muscle to increased use. *Muscle Nerve* 4:94–105.

Saltin, B. 1964. Aerobic work capacity and circulation at exercise in man: With special reference to the effect of prolonged exercise and/or heat exposure. *Acta Physiol. Scand.* 62 (suppl. 230):1–52.

———. 1966. Aerobic and anaerobic work capacity at 2,300 meters. *Schweiz. Z. Sportmed.* 14:81–87.

———. 1967. Aerobic and anaerobic work capacity at an altitude of 2,250 meters. In *The International Symposium on the Effects of Altitude on Physical Performance*, ed. R.F. Goddard, p. 97. Chicago: Athletic Institute.

———. 1985. Hemodynamic adaptations to exercise. *Am. J. Cardiol.* 55:42D–47D.

Saltin, B., and P.-O. Åstrand. 1967. Maximal oxygen uptake in athletes. *J. Appl. Physiol.* 23:353–358.

Saltin, B., G. Blomqvist, J.H. Mitchell, R.L. Johnson, K. Wildenthal, and C.B. Chapman. 1968. Response to submaximal and maximal exercise after bed rest and training. *Circulation* 38 (suppl. 7):1–78.

Saltin, B., R. Boushel, N. Secher, and J. Mitchell. 2000. *Exercise and circulation in health and disease.* Champaign, IL: Human Kinetics.

Saltin, B., B. Essen, and P.K. Pedersen. 1976. Intermittent exercise: Its physiology and some practical application. In *Advances in exercise physiology*, ed. E. Jokl, R.L. Anand, and H. Stoboy, pp. 23–51. Basel: Karger.

Saltin, B., and P.D. Gollnick. 1983. Skeletal muscle adaptability: Significance of metabolism and performance. In *Handbook of physiology*, ed. L.D. Peachey, R.H. Adrian, and S.R. Geiger, sec. 10, *Skeletal muscle*, pp. 555–631. Baltimore: Williams and Wilkins.

Saltin, B., and L. Hermansen. 1966. Esophageal, rectal and muscle temperature during exercise. *J. Appl. Physiol.* 21:1757.

———. 1967. Glycogen stores and prolonged severe exercise. In *Nutrition and physical activity*, ed. G. Blix, p. 32. Uppsala, Sweden: Almqvist and Wiksell.

Saltin, B., G. Rådegren, M. Koskolou, R.C. Roach, and J.M. Marchall. 2000. Muscle blood flow and its regulation. In *Exercise and circulation in health and disease*, ed. B. Saltin, R. Boushel, N. Secher, and J. Mitchell. Champaign, IL: Human Kinetics.

Samuelsen, G., A. Høiseth, and H.A. Dahl. 1993. How important is m. piriformis in extension and abduction of the flexed hip joint? 14th Congress of the International Society of Biomechanics, Paris, July 4–8, 1993. Abstracts 2, pp. 1170–1171.

Sandercock, T.G., J.A. Faulkner, J.W. Albers, and P.H. Abbrecht. 1985. Single motor unit and fiber action potentials during fatigue. *J. Appl. Physiol.* 58:1073–1079.

Sandler, H., and J. Vernikos. 1986. *Inactivity: Physiological effects.* Orlando: Academic Press.

Sandor, R.P. 1997. Heat illness: On-site diagnosis and cooling. *Physician Sportsmed.* 25:35–40.

Sandvik, L., J. Erikssen, E. Thaulow, G. Erikssen, R. Mundal, and K. Rodahl. 1993. Physical fitness as a predictor of mortality among healthy, middle-aged Norwegian men. *N. Engl. J. Med.* 328:533–537.

Sant'ana Pereira, J.A.A., A. Wessels, L. Nijtmans, A.F.M. Moorman, and A.J. Sargeant. 1995. New method for the accurate characterization of single human skeletal muscle fibres demonstrates a relation between mATPase and MyHC expression in pure and hybrid fibre types. *J. Muscle Res. Cell Motil.* 16:21–34.

Santora, A.C. 1987. Role of nutrition and exercise in osteoporosis. *Am. J. Med.* 82 (suppl. 1B):73–79.

Sargeant, A.J., C.T. Davies, R.H. Edwards, C. Maunder, and A. Young. 1977. Functional and structural changes after disuse of human muscle. *Clin. Sci. Mol. Med.* 52:337–342.

Sargeant, A.J., E. Hoinville, and A. Young. 1981. Maximum leg force and power output during short-term dynamic exercise. *J. Appl. Physiol. Respir. Environ. Exerc. Physiol.* 51 (5):1175–1182.

Saris, W.H., M.A. van Erp-Baart, F. Brouns, K.R. Westerterp, and F. ten-Hoor. 1989. Study on food intake and energy expenditure during extreme sustained exercise: The Tour de France. *Int. J. Sports Med.* 10:26–31.

Sato, F., M. Owen, R. Matthes, K. Sato, and C.V. Gisolfi. 1990. Functional and morphological changes in the eccrine sweat gland with heat acclimation. *J. Appl. Physiol.* 69:232–236.

Savourey, G., A.L. Vallerand, and J.H.M. Bittel. 1992. General and local cold adaptation after a ski journey in severe arctic environment. *Eur. J. Appl. Physiol. Occup. Physiol.* 64:99–105.

Sawka, M.N. 1988. Body fluid responses and hypohydration during exercise-heat stress. In *Human performance physiology and environmental medicine at terrestrial extremes*, ed. K.B. Pandolf, M.N. Sawka, and R.R. Gonzalez, pp. 227–66. Indianapolis, IN: Benchmark Press.

———. 1992. Physiological consequences of hypohydration: Exercise performance and thermoregulation. *Med. Sci. Sports Exerc.* 24:657–670.

Sawka, M.N., and J.E. Greenleaf. 1992. Current concepts concerning thirst, dehydration and fluid replacement: Overview. *Med. Sci. Sports Exerc.* 24:643–644.

Sawka, M.N., and K.B. Pandolf. 1990. Effects of body water loss in physiological function and exercise performance. In *Perspectives in exercise and science and sports medicine*, ed. D.R. Lamb and C.V. Gisolfi, vol. 3, *Fluid homeostasis during exercise*, pp. 1–38. Carmel, IN: Benchmark Press.

Schaffartzik, W., E.D. Barton, D.C. Poole, K. Tsukimoto, M.C. Hogan, D.E. Bebout et al. 1993. Effect of reduced hemoglobin concentrations on leg oxygen uptake during maximal exercise in humans. *J. Appl. Physiol.* 75:491–498.

Schantz, P., E. Randall-Fox, W. Hutchinson, A. Tyden, and P.-O. Åstrand. 1983. Muscle fiber type distribution, muscle cross-sectional area and maximal voluntary strength in humans. *Acta Physiol. Scand.* 117:219–226.

Schedel, J.M., H. Tanaka, A. Kiyonaga, M. Shindo, and Y. Schutz. 2000. Acute creatine loading enhances human growth hormone secretion. *J. Sports Med. Phys. Fitness* 40:336–342.

Schiaffino, S., L. Gorza, S. Sartore, L. Saggin, S. Ausoni, M. Vianello, K. Gundersen, and T. Lømo. 1989. Three myosin heavy chain isoforms in type 2 skeletal muscle fibres. *J. Muscle Res. Cell Motil.* 10:197–205.

Schiaffino, S., L. Gorza, S. Sartore, L. Saggin, and M. Carli. 1986. Embryonic myosin heavy chain as a differentiation marker of developing human skeletal muscle and rhabdomyosarcoma: A monoclonal antibody study. *Exp. Cell Res.* 163:211–220.

Schiaffino, S., M. Murgia, A.L. Serrano, E. Calabria, and G. Pallafacchina. 1999. How is muscle phenotype controlled by nerve activity? *Ital. J. Neurol Sci.* 20:409–412.

Schiaffino, S., and C. Reggiani. 1994. Myosin isoforms in mammalian skeletal muscle. *J. Appl. Physiol.* 77:493–501.

———. 1996. Molecular diversity of myofibrillar proteins: Gene regulation and functional significance. *Physiol. Rev.* 76:371–423.

Schiavo, G., and G. Stenbeck. 1998. Molecular analysis of neurotransmitter release: Essays. *Biochemistry* 33:29–41.

Schmidt, R.A., and T.D. Lee. 1999. *Motor control and learning: A behavioral emphasis.* 3rd ed. Champaign, IL: Human Kinetics.

Schmidtbleicher, D. 1992. Training for power events. In *Strength and power in sport,* ed. P.V. Komi, pp. 381–395. Oxford: Blackwell Science.

Schmidt-Nielsen, K. 1963. Heat conservation in countercurrent systems. In *Temperature: Its measurement and control in science and industry,* ed. J.D. Hardy, vol. 3, part, 3, p. 143. New York: Reinhold.

———. 1972. Locomotion: Energy cost of swimming, flying and running. *Science* 177:222–228.

Schoene, R.B. 1982. Control of ventilation in climbers to extreme altitude. *J. Appl. Physiol.* 53 (4):886.

Scholander, P.F., H.T. Hammel, H. LeMessurier, E. Hemmingsen, and W. Garey. 1962. Circulatory adjustment in pearl divers. *J. Appl. Physiol.* 17:184.

Scholander, P.F., V. Walters, R. Hook, and L. Irving. 1950. Body insulation of some arctic and tropical mammals and birds. *Biol. Bull.* 99:225.

Schopf, J.W. 1978. The evolution of the earliest cells. *Sci. Am.* 239 (3):85.

Schreiner, A. 1966. *Menneskets anatomi og fysiologi* (Human anatomy and physiology). Oslo: Cappelens Forlag.

Schroeder, K.L., and M.S. Chen. 1985. Smokeless tobacco and blood pressure. *N. Engl. J. Med.* 312:919–923.

Schürch, P.M., J. Radimsky, R. Iffland, and W. Hollman. 1982. The influence of moderate prolonged exercise and a low carbohydrate diet on ethanol elimination and on metabolism. *Eur. J. Appl. Physiol.* 48:407–414.

Schwaller, B., J. Dick, G. Dhoot, S. Carroll, G. Vrbova, P. Nicotera, D. Pette, A. Wyss, H. Bluethmann, W. Hunziker, and M.R. Celio. 1999. Prolonged contraction-relaxation cycle of fast-twitch muscles in parvalbumin knockout mice. *Am. J. Physiol. Cell Physiol.* 276:C395–C403.

Schwartz, M.A., M.D. Schaller, and M.H. Ginsberg. 1995. Integrins: Emerging paradigms of signal transduction. *Annu. Rev. Cell Dev. Biol.* 11:549–599.

Schwartz, M.W., and R.J. Seeley. 1997. The new biology of body weight regulation. *J. Am. Diet Assoc.* 97:54–58.

Schwartz, R.S., and D.M. Buchner. 1999. Exercise in the elderly: Physiologic and functional effects. In *Principles of geriatric medicine and gerontology,* ed. W.R. Hazzard, J.P. Blass, W.H. Ettinger, Jr., J.B. Halter, and J.G. Ouslander. New York: McGraw-Hill.

Schwellnus, M.P. 1999. Skeletal muscle cramps during exercise. *Physician Sportsmed.* 27:109–115.

Schwellnus, M.P., E.W. Derman, and T.D. Noakes. 1997. Aetiology of skeletal muscle "cramps" during exercise: A novel hypothesis. *J. Sports Sci.* 15:277–285.

Sculptoreanu, A., T. Scheuer, and W.A. Catterall. 1993. Voltage-dependent potentiation of L-type $Ca^{2+}$ channels due to phosphorylation by cAMP-dependent protein kinase. *Nature* 364:240–243.

Seale, P., and M.A. Rudnicki. 2000. A new look at the origin, function and "stem-cell" status of muscle satellite cells. *Devel. Biol.* 218:115–124.

Seals, D.R., J.M. Hagberg, B.F. Hurley, A.A. Ehsani, and J.O. Holloszy. 1984. Endurance training in older men and women: I. Cardiovascular responses to exercise. *J. Appl. Physiol.* 57:1024–1029.

Seals, D.R., R.A. Washburn, P.G. Hanson, P.L. Painter, and F.J. Nagle. 1983. Increased cardiovascular response to static contraction of larger muscle groups. *J. Appl. Physiol. Respir. Environ. Exerc. Physiol.* 4:434.

Secher, N. 1973. Development of results in international rowing championships 1893–1971. *Med. Sci. Sports* 5:195–199.

———. 1990. Rowing. In *Physiology of sports,* ed. T. Reilly, N. Secher, P. Snell, and C. Williams, p. 261. London: Spon.

Secher, N.H. 1983. The physiology of rowing. *J. Sports Sci.* 1:23–53.

———. 1993. Physiological and biomechanical aspects of rowing. *Sports Med.* 15:24–42.

Secher, N.H., O. Vaage, K. Jensen, and R.C. Jackson. 1983. Maximal aerobic power in oarsmen. *Eur. J. Appl. Physiol.* 51:155–162.

Seidel, H., R. Bastek, D. Bräuer, C. Buchholz, A. Meister, K.A.-M. Metz, and R. Rothe. 1980. On human response to prolonged repeated whole-body vibration. *Ergonomics* 23:191–211.

Sejersted, O.M., A.R. Hargens, K.R. Kardel, P. Blom, O. Jensen, and L. Hermansen. 1984. Intramuscular fluid pressure during isometric contraction of human skeletal muscle. *J. Appl. Physiol.* 56:287–295.

Sejersted, O.M., and G. Sjøgaard. 2000. Dynamics and consequences of potassium shifts in skeletal muscle and heart during exercise. *Physiol. Rev.* 80:1411–1481.

Sejersted, O.M., and N.K. Vøllestad. 1993. Physiology of muscle fatigue and associated pain. In *Progress in fibromyalgia and myofascial pain*, ed. H. Værøy, and H. Merskey, pp. 41–51. Amsterdam: Elsevier.

Selicki, F.A., and E. Segall. 1996. The mind/body connection of the golf swing. *Clin. Sports Med.* 15:191–201.

Semmler, J.G. 2002. Motor unit synchronization and neuromuscular performance. *Exerc. Sport Sci. Rev.* 30:8–14.

Semsarian, C., P. Sutrave, D.R. Richmond, and R.M. Graham. 1999. Insulin-like growth factor (IGF-I) induces myotube hypertrophy associated with an increase in anaerobic glycolysis in a clonal skeletal-muscle cell model. *Biochem. J.* 339:443–451.

Semsarian, C., M.J. Wu, Y.K. Yu, T. Marciniec, T. Yeoh, D.G. Allen, R.P. Harvey, and R.M. Graham. 1999. Skeletal muscle hypertrophy is mediated by a $Ca^{2+}$-dependent calcineurin signalling pathway. *Nature* 400:576–581.

Senay, L.C., Jr. 1972. Changes in plasma volume and protein content during exposures of working men to various temperatures before and after acclimatization to heat: Separation of the roles of cutaneous and skeletal muscle circulation. *J. Physiol.* 224:61.

———. 1978. Early response of plasma contents on exposure of working men to heat. *J. Appl. Physiol.* 44:166.

Senay, L.C., D. Mitchell, and C.H. Wyndham. 1976. Acclimatization in a hot, humid environment: Body fluid adjustments. *J. Appl. Physiol.* 40:786.

Sessa, W.C., K. Pritchard, N. Seyedi, J. Wang, and T.H. Hintze. 1994. Chronic exercise in dogs increases coronary vascular nitric oxide production and endothelial cell nitric oxide synthase gene expression. *Circ. Res.* 74:349–353.

Shapiro, S.M., P.M. Warren, T. Tringer, S.J. Paxton, I. Oswald, D.C. Flenley, and J.R. Catterall. 1984. Fitness facilitates sleep. *Eur. J. Appl. Physiol.* 53:1–4.

Shapiro, Y., R.W. Hubbard, C.M. Kimbrough, and K.B. Pandolf. 1981. Physiological and hematologic responses to summer and winter dry-heat acclimation. *J. Appl. Physiol.* 50:792.

Shapiro, Y., K.B. Pandolf, B.A. Avellini, N.A. Pimental, and R.F. Goldman. 1981. Heat balance and transfer in men and women exercising in hot-dry and hot-wet conditions. *Ergonomics* 24:375.

Sharma, M., B. Langley, J. Bass, and R. Kambadur. 2001. Myostatin in muscle growth and repair. *Exerc. Sport Sci. Rev.* 29:155–158.

Shaskey, D.J., and G.A. Green. 2000. Sports haematology. *Sports Med.* 29:27–38.

Sheffield-Moore, M., K.R. Short, C.G. Kerr, A.C. Parcell, D.R. Bolster, and D.L. Costill. 1997. Thermoregulatory responses to cycling with and without a helmet. *Med. Sci. Sports Exerc.* 29:755–761.

Shephard, R.J. 1966. The relative merits of the step test, bicycle ergometer and treadmill in the assessment of cardio-respiratory fitness. *Int. Z. Angew. Physiol.* 23:219–230.

———. 1973. Athletic performance at moderate altitudes. *Medicina dello Sport* 26:36.

———. 1982. *Physical activity and growth*. Chicago: Year Book Medical.

———. 1984a. Sleep biorhythms and human performance. *Sports Med.* 1:11.

———. 1984b. Tests of maximum oxygen intake: A critical review. *Sports Med.* 1:99–124.

———. 1993. Metabolic adaptation to exercise in the cold: An update. *Sports Med.* 16:266–289.

———. 1999. Biology and medicine of soccer: An update. *J. Sports Sci.* 17:757–786.

Shephard, R.J., D.A. Bailey, and R.L. Mirwald. 1976. Development of the Canadian home fitness test. *Can. Med. Assoc. J.* 114:675–679.

Shephard, R.J., S. Thomas, and I. Weller. 1991. The Canadian home fitness test: 1991 update. *Sports Med.* 11:358–366.

Shepherd, G.M., and S.D. Eruklar. 1997. Centenary of the synapse: From Sherrington to the molecular biology of the synapse and beyond. *Trends Neurosci.* 20:385–392.

Shepherd, J.T., and P.M. Vanhoutte. 1979. *The human cardiovascular system*. New York: Raven Press.

Sherman, C.B. 1992. The health consequences of cigarette smoking. Pulmonary diseases. *Med. Clin. North Am.* 76:355–375.

Sherman, W., D. Costill, W. Fink, and J. Miller. 1981. Effect of exercise-diet manipulation on muscle glycogen and its subsequent utilization during performance. *Int. J. Sports Med.* 114:114–118.

Sherrington, C.S. 1906. *The integrative action of the nervous system*. Yale University Press.

Shinkai, S., B. Watanabe, Y. Kurokawa, J. Torii. 1993. Salivary cortisol for monitoring circadian rhythm variations in adrenal activity during shift work. *Int. Arch. Occup. Environ. Health* 64:499–502.

Shirreffs, S.M., and R.J. Maughan. 2000. Rehydration and recovery of fluid balance after exercise. *Exerc. Sports Sci. Rev.* 28 (1) :27–32.

Shirreffs, S.M., and R.J. Maughan. 1997. Restoration of fluid balance after exercise-induced dehydration: Effects of alcohol consumption. *J. Appl. Physiol.* 83:1152–1158.

Shirreffs, S.M., A.J. Taylor, J.B. Leiper, and R.J. Maughan. 1996. Post-exercise rehydration in man: Effects of volume consumed and drink sodium content. *Med. Sci. Sports Exerc.* 28:1260–1271.

Shoshan-Barmatz, V., and R.H. Ashley. 1998. The structure, function and cellular regulation of ryanodine-sensitive $Ca^{2+}$ release channels. *Int. Rev. Cytol.* 183:185–270.

Shumway-Cook, A., and M.H. Woollacott. 1995. *Motor control: Theory and practical applications.* Baltimore: Williams and Wilkins.

Shvartz, E., A. Bhattacharya, S.J. Sperinde, P.J. Brock, D. Sciaraffa, and W. van Beaumont. 1979. Sweating responses during heat acclimation and moderate conditioning. *J. Appl. Physiol.* 46:675.

Siconolfi, S.F., E.M. Cullinane, R.A. Carleton, and P.D. Thompson. 1982. Assessing $\dot{V}O_2$ in epidemiologic studies: Modification of the Åstrand-Ryhming test. *Med. Sci. Sports Exc.* 14 (5):335–338.

Sieck, G.C., and Y.S. Prakash. 1995. Fatigue at the neuromuscular junction: Branch point vs presynaptic vs postsynaptic mechanisms. *Adv. Exp. Med. Biol.* 384:83–100.

Silva, J.E. 1995. Thyroid hormone control of thermogenesis and energy balance. *Thyroid* 5:481–492.

Silva, P.A., J. Birkbeck, D.G. Russel, and J. Wilson. 1984. Some biological, development and social correlates of gross and fine motor performance in Dunedin seven year olds. *J. Hum. Mov. Stud.* 10:35.

Silverstein P. 1992. Smoking and wound healing. *Am. J. Med.* 93:22S–24S.

Simon, J., J.L. Young, B. Gutin, D.K. Blood, and R.B. Case. 1983. Lactate accumulation relative to the anaerobic and respiratory compensation thresholds. *J. Appl. Physiol. Respir. Environ. Exerc. Physiol.* 54:13.

Simoneau, J.A., and C. Bouchard. 1995. Genetic determinism of fiber type proportion in human skeletal muscle. *FASEB J.* 9:1091–1095.

Simoneau, J.A., G. Lortie, M.R. Boulay, M. Marcotte, M.C. Thibault, and C. Bouchard. 1986. Inheritance of human skeletal muscle and anaerobic capacity adaptation to high-intensity intermittent training. *Int. J. Sports Med.* 7:167–171.

Simonson, E. 1951. *Influence of nutrition on work performance.* Nutrition Fronts in Public Health, Nutrition Symposium Series no. 3, p. 72. New York: National Vitamin Foundation.

Simonson, E., and A.R. Lind. 1971. Fatigue in static work. In *Physiology of work capacity and fatigue,* ed. E. Simonson, pp. 241–284. Springfield, IL: Charles C Thomas.

Simonson, E., N. Teslenko, and M. Gorkin. 1936. Einfluss von Vorübungen auf die Liestung beim 100 m. Lauf. *Arbeitsphysiologie* 9:152.

Sinaki, M., P.J. Limburg, P.C. Wollan, J.W. Rogers, and P.A. Murtaugh. 1996. Correlation of trunk muscle strength with age in children 5 to 18 years old. *Mayo Clin. Proc.* 71 (11):1047–1054.

Singh, M.A.F., W.J. Ding, T.J. Manfredi, G.S. Solares, E.F. O'Neill, K.M. Clements, N.D. Ryan, J.J. Kehayias, R.A. Fielding, and W.J. Evans. 1999. Insulin-like growth factor I in skeletal muscle after weight-lifting exercise in frail elders. *Am. J. Physiol. Endocrin. Metab.* 277:E135–E143.

Sinoway, L.I., J. Shenberger, J. Wilson, D. McLaughlin, T. Musch, and R. Zelis. 1987. A 30-day forearm work protocol increases maximal forearm blood flow. *J. Appl. Physiol.* 62:1063–1067.

Sjödin, A.M., A.B. Andersson, J.M. Högberg, and K.R. Westerterp. 1994. Energy balance in cross-country skiers: A study using doubly labeled water. *Med. Sci. Sports Exerc.* 26:720–724.

Sjödin, A.M., A. Forslund, P. Webb, and L. Hambraeus. 1996. Mild overcooling increases energy expenditure during endurance exercise. *Scand. J. Med. Sci. Sports* 6:22–25.

Sjödin, B., and J. Svedenhag. 1985. Applied physiology of marathon running. *Sports Med.* 2:83–99.

Sjøgaard, G. 1984. Muscle morphology and metabolic potential in elite road cyclists during a season. *Int. J. Sports Med.* 5:250–254.

———. 1990. Exercise-induced muscle fatigue: The significance of potassium. *Acta Physiol. Scand. Suppl.* 593: 1–63.

Sjøgaard, G., R. Kadefors, and U. Lundberg, eds. 2000. Motor control and mechanisms of muscle disorders in computer users. *Eur. J. Appl. Physiol.* 83 (2–3):99–235.

Sjøgaard, G., and A.J. McComas. 1995. Role of interstitial potassium. *Adv. Exp. Med. Biol.* 384:69–80.

Sjöstrand, T. 1947. Changes in the respiratory organs of workmen at an ore melting works. *Acta Med. Scand. Suppl.* 196:687.

———. 1961. *Relationen Zwischen Bau and Funktion des Kreislaufsystems Unter Pathologischen Bedingungen.* Forum Cardiologicum, Heft 3. Mannheim-Waldhof, Germany: Boehringer and Soehne.

Sjöström, M., J. Lexell, A. Eriksson, and C.C. Taylor. 1991. Evidence of fibre hyperplasia in human skeletal muscles from healthy young men? *Eur. J. Appl. Physiol. Occup. Physiol.* 62:301–304.

Skare, O.C., Ø. Skadberg, and A.R. Wisnes. 2001. Creatine supplementation improves sprint performance in male sprinters. *Scand. J. Med. Sci. Sports* 11:96–102.

Skargren, E. And B.Øberg. 1996. Effects of an exercise program on musculoskeletal symptoms and physical capacity among nursing staff. *Scand. J. Med. Sci. Sports* 6:122–130.

Skoglund, S. 1956. Anatomical and physiological studies of kneejoint-innervation in the cat. *Acta Phys. Scand.* 36 (suppl. 124).

Skorjanc, D., I. Traub, and D. Pette. 1998. Identical responses of fast muscle to sustained activity by low-frequency stimulation in young and aging rats. *J. Appl. Physiol.* 85:437–441.

Smerdu, V., I. Karsch-Mizrachi, M. Campione, L. Leinwand, and S. Schiaffino. 1994. Type IIx myosin heavy chain transcripts are expressed in type IIb fibers of human skeletal muscle. *Am. J. Physiol. Cell Physiol.* 267:C1723–C1728.

Smiles, K., R.S. Elizondo, and C.C. Barney. 1976. Sweating responses during changes of hypothalamic temperatures in the rhesus monkey. *J. Appl. Physiol.* 40:653.

Smith, M.H., and B.J. Sharkey. 1984. Altitude training: Who benefits? *Physician Sportsmed.* 12:48–62.

Smolander, J., M. Harma, A. Lindqvist, P. Kolari, and L.A. Laitinen. 1993. Circadian variation in peripheral blood flow in relation to core temperature at rest. *Eur. J. Appl. Physiol.* 67:192–196.

Smorawinski, J., and R. Grucza. 1994. Effect of endurance training on thermoregulatory reactions to dynamic exercise in man. *Biol. Sport* 11:143–149.

Snell, P.G., W.H. Martin, J.C. Buckey, and C.G. Blomqvist. 1987. Maximal vascular leg conductance in trained and untrained men. *J. Appl. Physiol.* 62:606–610.

Snellen, J.W. 1966. Mean body temperature and the control of thermal sweating. *Acta Physiol. Pharmacol. Neerl.* 14:99.

Snow-Harter, C. and Marcus, R. 1991. Exercise, bone mineral density and osteoporosis. *Exerc. Sport Sci. Rev.* 19, 351–388.

Snyder, A.C. 1998. Overtraining and glycogen depletion hypothesis. *Med. Sci. Sports Exerc.* 30:1146–1150.

Somervell, T.H. 1925. Note on the composition of alveolar air at extreme heights. *J. Physiol.* 60:282.

Sonnenblick, E.H., and C.L. Skelton. 1974. Reconsideration of the ultrastructural basis of cardiac length-tension relations. *Circ. Res.* 35:519.

Sorokin, S.P. 1967. A morphological and cytochemical study of the great alveolar cell. *J. Histochem. Cytochem.* 14:884–893.

Sorrentino, V., and C. Reggiani. 1999. Expression of the ryanodine receptor type 3 in skeletal muscle: A new partner in excitation–contraction coupling? *Trends Cardiovasc. Med.* 9:54–61.

Soule, R.G., K.B. Pandolf, and R.F. Goldman. 1978. Energy expenditure of heavy load carriage. *Ergonomics* 21:373–381.

Sparling, P.B., and K.J. Cureton. 1983. Biological determinants of the sex difference in 12-min run performance. *Med. Sci. Sports Exerc.* 15 (3):218–223.

Sparling, P.B., and M. Millard-Stafford. 1999. Keeping sports participants safe in hot weather. *Physician Sportsmed.* 27:27–34.

Speedy, D.B., T.D. Noakes, I.R. Rogers, J.M.D. Thompson, R.G.D. Campbell, J.A. Kuttner, et al. 1999. Hyponatremia in ultradistance triathletes. *Med. Sci. Sports Exerc.* 31:809–815.

Spengler, C.M., M. Roos, S.M. Laube, and U. Boutellier. 1999. Decreased exercise blood lactate concentrations after respiratory endurance training in humans. *Eur. J. Appl. Physiol. Occup. Physiol.* 79:299–305.

Spina, R.J. 1999. Cardiovascular adaptations to endurance exercise training in older men and women. *Exerc. Sport Sci. Rev.* 27:317–332.

Spina, R.J., T. Ogawa, W.H. Martin, III, A.R. Coggan, J.O. Holloszy, and A.A. Ehsani. 1992. Exercise training prevents decline in stroke volume during exercise in young healthy subjects. *J. Appl. Physiol.* 72:2458–2462.

Spirduso, W.W. 1995. *Physical dimensions of aging.* Champaign, IL: Human Kinetics.

Spirito, P., A. Pelliccia, M.A. Proschan, M. Granata, A. Spataro, P. Bellone et al. 1994. Morphology of the "athlete's heart" assessed by echocardiography in 947 elite athletes representing 27 sports. *Am. J. Cardiol.* 74:802–806.

Spriet, L.L. 1995. Caffeine and performance. *Int. J. Sport Nutr.* (suppl. 5):S84–S99.

Spriet, L.L. 2002. Regulation of skeletal muscle fat oxidation during exercise in humans. *Med. Sci. Sports Exerc.* 34 (9):1477–1484.

Spriet, L.L., N. Gledhill, A.B. Froese, D.R. Wilkes, and E.C. Meyers. 1980. The effect of induced erythrocythemia on central circulation and oxygen transport during maximal exercise. *Med. Sci. Sports Exerc.* 12:122.

Spriet, L.L., D.A. MacLean, D.J. Dyck, E. Hultman, G. Cederblad, and T.E. Graham. 1992. Caffeine ingestion and muscle metabolism during prolonged exercise in humans. *Am. J. Physiol.* 262:E891–E898.

Squires, W.G., Jr., T.A. Brandon, S. Zinkgraf, D. Bonds, G.H. Hartung, T. Murray, A.S. Jackson, and R.R. Miller. 1984. Hemodynamic effects of oral smokeless tobacco in dogs and young adults. *Prev. Med.* 13:195–206.

Staff, P.H., and S. Nilsson. 1971. Fluid and glucose ingestion during prolonged severe physical activity. *Tidsskrift for Den Norske Laegeforening* 16:1235–1238.

Stallknecht, B., J. Vissing, and H. Galbo. 1998. Lactate production and clearance in exercise: Effects of training. A mini-review. *Scand. J. Med. Sci. Sports* 8:127–131.

Starck, J., and I. Pyykkø. 1986. Fourth international symposium on hand-arm vibration. *Scand. J. Work Environ. Health* 12:237–437.

Starling, E.H. 1896. On the absorption of fluids from the connective tissue spaces. *J. Physiol.* (Lond.) 19:312–326.

Staron, R.S., and D. Pette. 1993. The continuum of pure and hybrid myosin heavy chain-based fibre types in rat skeletal muscle. *Histochem.* 100:149–153.

Stefano, G.B., V. Prevot, P. Cadet, and I. Dardik. 2001. Vascular pulsations stimulating nitric oxide release during cyclic exercise may benefit health: A molecular approach. *Int. J. Molec. Med.* 7:119–129.

Stegmann, H., and W. Kindermann. 1982. Comparison of prolonged exercise tests at the individual anaerobic threshold and the fixed anaerobic threshold of 4 mmol $\cdot$ $l^{-1}$ lactate. *Int. J. Sports Med.* 3:105–110.

Stein, R.B. 1974. Peripheral control of movement. *Physiol. Rev.* 54.215–243.

Steinberg, G.M., S.G. Frehlich, and L.K. Tennant. 1995. Dextrality and eye position in putting performance. *Percept. Mot. Skills* 80:635–640.

Stenberg, J., P.-O. Åstrand, B. Ekblom, J. Royce, and B. Saltin. 1967. Hemodynamic response to work with different muscle groups in sitting and supine position. *J. Appl. Physiol.* 22:61–70.

Stenberg, J., B. Ekblom, and R. Messin. 1966. Hemodynamic response to work at simulated altitude, 4,000 m. *J. Appl. Physiol.* 21:1589–1594.

Stevens, C.F. 1979. The neuron. *Sci. Am.* 241:54–65.

Stevens, E.D., and F.G. Carey. 1981. One why of the warmth of warm-bodied fish. *Am. J. Physiol.* 240:R151.

Stevenson, J.A.F. 1965. Control of water exchange. In *Physiological controls and regulations,* ed. W.S. Yamamoto and J.R. Brobeck, p. 253. Philadelphia: Saunders.

Stewart, A.D., and J. Hannan. 2000. Total and regional bone density in male runners, cyclists and controls. *Med. Sci. Sports Exerc.* 32:1373–1377.

Stolt, A., J. Karjalainen, O.J. Heinonen, and U.M. Kujala. 2000. Left ventricular mass, geometry and filling in elite female and male endurance athletes. *Scand. J. Med. Sci. Sports* 10:28–32.

Stolwijk, J.A.J. 1970. Mathematical model of thermoregulation. In *Physiological and behavioral temperature regulation,* ed. J.D. Hardy, A.P. Gagge, and J.A.J. Stolwijk, chap. 48, p. 703. Springfield, IL: Charles C Thomas.

Stork, P.J.S. and J.M. Schmitt. 2002. Crosstalk between cAMP and MAP kinase signaling in the regulation of cell proliferation. *Trends in Cell Biol.* 12:258–266.

Strandell, T. 1964. Circulatory studies on healthy old men with special reference to the limitation of the maximal physical working capacity. *Acta Med. Scand.* 175 (suppl. 414).

Stray-Gundersen, J., and B.D. Levine. 1999. "Living high and training low" can improve sea level performance in endurance athletes. *Br. J. Sports Med.* 33:150–151.

Stringer, C.B. 1990. The emergence of modern humans. *Sci. Am.* 264 (6):68–74.

Strobel, G., B. Friedmann, R. Siebold, and P. Bartsch. 1999. Effect of severe exercise on plasma catecholamines in differently trained athletes. *Med. Sci. Sports Exerc.* 31:560–565.

Strømme, S.B., K.L. Andersen, and R.W. Elsner. 1963. Metabolic and thermal responses to muscular exertion in the cold. *J. Appl. Physiol.* 18:756–763.

Strømme, S.B., H. Frey, O.K. Harlem, O. Stokke, O.D. Vellar, and L.E. Aarö, eds. 1982. Physical activity and health: A documentation. *Scand. J. Soc. Med. Suppl.* 29:5–269.

Strømme, S.B., and H.T. Hammel. 1967. Effects of physical training on tolerance to cold in rats. *J. Appl. Physiol.* 23:815–824.

Strømme, S.B., F. Ingjer, and H.D. Meen. 1977. Assessment of maximal aerobic power in specifically trained athletes. *J. Appl. Physiol. Respir. Environ. Exerc. Physiol.* 42:833–837.

Strømme, S.B., D. Kerem, and R. Elsner. 1970. Diving bradycardia during rest and exercise and its relation to physical fitness. *J. Appl. Physiol.* 28 (5):614–621.

Stroud, M.A. 1991. Effects on energy expenditure of facial cooling during exercise. *Eur. J. Appl. Physiol. Occup. Physiol.* 63:376–380.

Strydom, N.B., C.H. Wyndham, C.G. Williams, J.F. Morrison, G.A.G. Bredell, A.J.S. Benade, and M. von Rahden. 1966. Acclimatization to humid heat and the role of physical conditioning. *J. Appl. Physiol.* 21:636.

Stuart, D.G., E. Eldred, A. Hemingway, and Y. Kawamura. 1963. Neural regulation of the rhythm of shivering. In *Temperature: Its measurement and control in science and industry,* ed. J.D. Hardy, vol. 3, part 3, p. 545. New York: Reinhold.

Suminski, R.R., R.J. Robertson, F.L. Goss, S. Arslanian, J. Kang, S. DaSilva, et al. 1997. Acute effect of amino acid ingestion and resistance exercise on plasma growth hormone concentration in young men. *Int. J. Sport Nutr.* 7:48–60.

Sundin, I. 1956. The influence of body posture on the urinary excretion of adrenaline and noradrenaline. *Acta Med. Scand.* 154 (suppl. 313).

Surks, M.I., K.S. Chinn, and L.O. Matoush. 1966. Alterations in body composition in man after acute exposure to high altitude. *J. Appl. Physiol.* 21:1741–1746.

Sutton, J.M. 1990. Evaluation of hematuria in adults. *JAMA* 263:2475–2480.

Sutton, J.R. 1984. Heat illness. In *Sports medicine,* ed. R.H. Strauss, p. 307. Philadelphia: Saunders.

———. 1992. Limitations to maximal oxygen uptake. *Sports Med.* 13:127–133.

Sutton, J.R., N.L. Jones, L. Griffith, and C.E. Pugh. 1983. Exercise at altitude. *Annu. Rev. Physiol.* 45:427–437.

Sutton, J.R., J.E. Jurkowski, P. Keane, W.H.C. Walker, N.L. Jones, and C.J. Toews. 1980. Plasma catecholamine, insulin, glucose and lactate responses to exercise in relation to the menstrual cycle. *Med. Sci. Sports Exerc.* 12:83–84.

Suzuki, H. 1994. Improvement of chain saw and changes in symptoms of the operators. *Nagoy J. Med. Sci.* (Japan) 57:229–234.

Suzuki, Y. 1980. Human physical performance and cardiocirculatory responses to hot environments during submaximal upright cycling. *Ergonomics* 23:527.

Suzuki, Y., T. Murakami, Y. Haruna, K. Kawakubo, S. Goto, Y. Makita, S. Ikawa, and A. Gunji. 1994. Effects of 10 and 20 days bed rest on leg muscle mass and strength in young subjects. *Acta Physiol. Scand.* 150 (suppl. 616):5–18.

Svedenhag, J., J. Henriksson, A. Juhlin-Dannfelt, and K. Asano. 1984. Beta-adrenergic blockade and training in healthy men: Effects on central circulation. *Acta Physiol. Scand.* 120:77–86.

Swain, D.P., J.R. Coast, P.S. Clifford, M.C. Milliken, and J. Stray-Gundersen. 1987. Influence of body size on oxygen consumption during bicycling. *J. Appl. Physiol.* 62:668–672.

Sward, L., M. Svensson, and C. Zetterberg. 1990. Isometric muscle strength and quantitative electromyography of back muscles in wrestlers and tennis players. *Am. J. Sports Med.* 18:382–386.

Takamata, A., G.W. Mack, C.M. Gillen, and E.R. Nadel. 1994. Sodium appetite, thirst and body fluid regulation in humans during rehydration. *Am. J. Physiol.* 266:R1493–R1502.

Tanaka, H., and D.R. Seals. 1997. Age and gender interactions in physiological functional capacity: Insight from swimming performance. *J. Appl. Physiol.* 82:846–851.

Tanaka, M., Y. Tochihara, S. Yamagaki, T. Ohnaka, and K. Yoshida. 1983. Thermal reactions and manual performance during cold exposure while wearing cold-protective clothing. *Ergonomics* 26:141.

Tanner, J.M. 1962. *Growth of adolescence.* 2nd ed. Oxford: Blackwell Scientific.

———. 1989. *Foetus into man.* London: Castlemead.

Tanner, J.M., R.H. Whitehouse, N. Cameron, W.A. Marshall, M.J.R. Healy, and H. Goldstein. 1983. *Assessment of skeletal maturity and prediction of adult height (TW2 method).* 2nd ed. London: Academic Press.

Tarnopolsky, M.A. 1994. Caffeine and endurance performance. *Sports Med.* 18:109–125.

Tarnopolsky, M.A., M. Bosman, J.R. MacDonald, D. Vandeputte, J. Martin, and B.D. Roy. 1997. Postexercise protein-carbohydrate and carbohydrate supplements increase muscle glycogen in men and women. *J. Appl. Physiol.* 83:1877–1883.

Taylor, A., P.H. Ellaway, and R. Durbaba. 1999. Why are there three types of intrafusal muscle fibers? *Prog. Brain Res.* 123:121–131.

Taylor, C.R., and N.C. Heglund. 1982. Energetics and mechanics of terrestrial locomotion. *Annu. Rev. Physiol.* 44:97.

Taylor, C.R., G.M.O. Maloiy, E.R. Weibel, V.A. Langman, J.M.Z. Kamau, M.J. Seeherman, and N.C. Heglund. 1981. Design of the mammalian respiratory system. III. Scaling maximum aerobic capacity to body mass: Wild and domestic animals. *Respir. Physiol.* 44:25–37.

Taylor, C.R., K. Schmidt-Nielsen, and J.L. Raab. 1970. Scaling and energetic cost of running to body size in mammals. *Am. J. Physiol.* 219:1104–1107.

Taylor, J.A., and S.C. Kandarian. 1994. Advantage of normalizing force production to myofibrillar protein in skeletal muscle cross-sectional area. *J. Appl. Physiol.* 76:974–978.

Taylor, J.L., J.E. Butler, and S.C. Gandevia. 2000. Changes in muscle afferents, motoneurons and motor drive during muscle fatigue. *Eur. J. Appl. Physiol.* 83:106–115.

Teichner, W.H. 1958. Assessment of mean body surface temperature. *J. Appl. Physiol.* 12:169.

Teitelbaum, S.L. 2000. Bone resorption by osteoclasts. *Science* 289:1504–1508.

Tenney, S.M. 1967. Some aspects of the comparative physiology of muscular exercise in mammals. *Circ. Res.* 20:1–7.

Tergau, F., R. Geese, A. Bauer, S. Baur, W. Paulus, and C.D. Reimers. 2000. Motor cortex fatigue in sports measured by transcranial magnetic double stimulation. *Med. Sci. Sports Exerc.* 32:1942–1948.

Terrados, N., E. Jansson, C. Sylvén, and L. Kaijser. 1990. Is hypoxia a stimulus for synthesis of oxidative enzymes and myoglobin? *J. Appl. Physiol.* 68:2369–2372.

Terrados, N., and R.J. Maughan. 1995. Exercise in the heat: Strategies to minimize the adverse effects on performance. *J. Sports Sci.* 13:55–62.

Terrados, N., J. Melichna, C. Sylvén, E. Jansson, and L. Kaijser. 1988. Effects of training at simulated altitude on performance and muscle metabolic capacity in competitive road cyclists. *Eur. J. Appl. Physiol.* 57:203–209.

Tesch, P.A. 1980a. Fatigue pattern in subtypes of human skeletal muscle fibers. *Int. J. Sports Med.* 1:79–81.

———. 1980b. Muscle fatigue in man with special reference to lactate accumulation during short term intense exercise. *Acta Physiol. Scand. Suppl.* 480:1–40.

———. 1983. Physiological characteristics of elite kayak paddlers. *Can. J. Appl. Sports Sci.* 8:87–91.

———. 1992a. Short- and long-term histochemical and biochemical adaptations in muscle. In *Strength and power in sport,* ed. P.V. Komi, pp. 239–248. Oxford: Blackwell Science.

———. 1992b. Training for bodybuilding. In *Strength and power in sport,* ed. P.V. Komi, pp. 370–380. Oxford: Blackwell Science.

———. 1995. Aspects on muscle properties and use in competitive alpine skiing. *Med. Sci. Sports Exerc.* 27:310–314.

Tesch, P.A., K. Piehl, G. Wilson, and J. Karlsson. 1974. *Kanot.* Idrottsfysiologi Rapport no. 13. Stockholm: Trygg-Hansa.

———. 1976. Physiological investigations of Swedish elite canoe competitors. *Med. Sci. Sports* 8:214–218.

Tesch, P.A., A. Thorsson, and P. Kaiser. 1984. Muscle capillary supply and fiber type characteristics in weight and power lifters. *J. Appl. Physiol.* 56:35–38.

Thibault, M.C., J.A. Simoneau, C. Côté, M.R. Boulay, P. Lagassé, M. Marcotte, and C. Bouchard. 1986. Inheritance of human muscle enzyme adaptation to isokinetic strength training. *Hum. Hered.* 36:341–347.

Thissen-Milder, M. and J.L. Mayhew. 1991. Selection and classification of high school volleyball players from performance tests. *J. Sports Med. Phys. Fitness* 31 (3):380–384.

Thomas, C.K., R.S. Johansson, and B. Bigland-Ritchie. 1999. Pattern of pulses that maximize force output from single human thenar motor units. *J. Neurophysiol.* 82:3188–3195.

Thomason, D.B., P.R. Morrison, V. Oganov, E. Ilyina-Kakueva, F.W. Booth, and K.M. Baldwin. 1992. Altered actin and myosin expression in muscle during exposure to microgravity. *J. Appl. Physiol.* 73:90S-93S.

Thompson, C.J., J. Bland, J. Burd, and P.H. Baylis. 1986. The osmotic threshold for thirst and vasopressin release are similar in healthy humans. *Clin. Sci.* (Lond.) 71:651–656.

Thompson, R.L., and J.S. Hayward. 1996. Wet-cold exposure and hypothermia: Thermal and metabolic responses to prolonged exercise in rain. *J. Appl. Physiol.* 81:1128–1137.

Thomson, J.M., J.A. Stone, A.D. Ginsburg, and P. Hamilton. 1982. $O_2$ transport during exercise following blood reinfusion. *J. Appl. Physiol. Respir. Environ. Exerc. Physiol.* 53:1213.

Thorstensson, A. 1976. Muscle strength, fibre types and enzyme activities in man. *Acta Physiol. Scand. Suppl.* 443:7–45.

Thorstensson, A., G. Grimby, and J. Karlsson. 1976. Force-velocity relations and fiber composition in human knee extensor muscles. *J. Appl. Physiol.* 40:12–16.

Thune, I. and E. Lund. 1996. Physical activity and risk of colorectal cancer in men and women. *Br. J. Cancer* 73:1134–1140.

Tidball, J.G. 1991a. Force transmission across muscle cell membranes. *J. Biomech.* 24:43–52.

———. 1991b. Myotendinous junction injury in relation to junction structure and molecular composition. *Exerc. Sport Sci. Rev.* 19:419–445.

Timiras, M.L., and H. Brownstein. 1987. Prevalence of anemia and correlation of hemoglobin with age in geriatric screening clinic population. *J. Am. Geriat. Soc.* 35:639–643.

Timson, B.F. 1990. Evaluation of animal models for the study of exercise-induced muscle enlargement. *J. Appl. Physiol.* 69:1935–1945.

Tipton, C.M., R.D. Matthes, J.A. Maynard, and R.A. Carey. 1975. The influence of physical activity on ligaments and tendons. *Med. Sci. Sports* 7:165–175.

Tochihara, Y., T. Ohnaka, K. Tuzuki, and Y. Nagai. 1995. Effects of repeated exposures to severely cold environments on thermal responses of humans. *Ergonomics* 38:987–995.

Tola, S., H. Rütiimäki, T. Videman, E. Viikari, and K. Hänninen. 1988. Neck and shoulder symptoms among men in machine operating, dynamic physical work and sedentary work. *Scand. J. Work Environ. Health.* 14:299–305.

Tolfrey, K., A.M. Jones, and I.G. Campbell. 2000. The effect of aerobic exercise training on the lipid-lipoprotein profile of children and adolescents. *Sports Med.* 29:99–112.

Tomonaga, M. 1977. Histochemical and ultrastructural changes in senile human skeletal muscle. *J. Am. Geriatr. Soc.* 25:125–131.

Tonkonogi, M. and K. Sahlin. 2002. Physical exercise and mitochondrial function in human skeletal muscle. *Exerc. Sport Sci. Rev.* 30:129–137.

Torgén, M., C.-H. Nygård, and Å. Kilbom. 1995. Physical work load, physical capacity and strain among elderly female aids in homecare service. *Eur. J. Appl. Physiol.* 71:444–452.

Torii, M. 1995. Maximal sweating rate in humans. *J. Hum. Ergol.* 24:137–52.

Traub, P., and R.L. Shoeman. 1994. Intermediate filament proteins: Cytoskeletal elements with gene-regulatory function? *Int. Rev. Cytol.* 154:1–103.

Treisman, R. 1996. Regulation of transcription by MAP kinase cascades. *Curr. Opin. Cell Biol.* 8:205–215.

Trinkhaus, E., and W.W. Howells. 1979. The Neanderthals. *Sci. Am.* 241 (6):94.

Troiano, R.P., K.M. Flegal, R.J. Kuczmarski, S.M. Campbell, and C.L. Johnson. 1995. Overweight prevalence and trends for children and adolescents: The National Health and Nutrition Examination Surveys, 1963 to 1991. *Arch. Pediatr. Adolesc. Med.* 149:1085–1091.

Trost, S., A. Wilcox, and D. Gillis. 1997. The effect of substrate utilization, manipulated by nicotinic acid, on excess postexercise oxygen consumption. *Int. J. Sports Med.* 18:83–88.

Trotter, J.A. 1993. Functional morphology of force transmission in skeletal muscle. A brief review. *Acta Anat.* (Basel) 146:205–222.

Trotter, J.A., F.J. Richmond, and P.P. Purslow. 1995. Functional morphology and motor control of series-fibered muscles. *Exerc. Sport Sci. Rev.* 23:167–213.

Tudorlocke, C., B.E. Ainsworth, and B.M. Popkin. 2001. Active commuting to school—an overlooked source of childrens' physical activity? *Sports Med.* 31:309–313.

Tulppo, M.P., T.H. Makikallio, T. Seppanen, R.T. Laukkanen, and H.V. Huikuri. 1998. Vagal modulation of heart rate during exercise: Effects of age and physical fitness. *Am. J. Physiol.* 274:H424–H429.

Tumilty, D. 1993. Physiological characteristics of elite soccer players. *Sports Med.* 16:80–96.

Turner, R.T. 2000. Invited review: what do we know about the effects of spaceflight on bone? *J. Appl. Physiol.* 89, 840–847.

Ullsperger, P., and H. Seidel. 1980. On auditory evoked potentials and heart rate in man during whole-body vibration. *Eur. J. Appl. Physiol.* 43:1983–1992.

Ulvedal, F., T.E. Morgan, Jr., R.G. Cutler, and B.E. Welch. 1963. Ventilatory capacity during prolonged exposure to simulated altitude without hypoxia. *J. Appl. Physiol.* 18:904–908.

Umemura, M., K. Honda, and Y. Kikuchi. 1992. Influence of noise on heart rate and quality of work in mental work. *Ann. Physiol. Anthropol.* 11:523–532.

Urhausen, A., and W. Kindermann. 1992. Echocardiographic findings in strength- and endurance-trained athletes. *Sports Med.* 13:270–284.

———. 1999. Sports-specific adaptations and differentiation of the athlete's heart. *Sports Med.* 28:237–244.

———. 2002. Diagnosis of overtraining: What tools do we have? *Sports Med.* 32:95–102.

U.S. Department of Health and Human Services. 1996. *Physical activity and health: A report of the Surgeon General*. Atlanta, GA: Centers for Disease Control and Prevention.

Valentine, J.W. 1978. The evolution of multicellular plants and animals. *Sci. Am.* 239 (3):67.

Vallerand, A.L., and I. Jacobs. 1992. Energy metabolism during cold exposure. *Int. J. Sports Med.* 13 (suppl. 1):191–193.

Van Baak, M.A. 1979. *The physiological load during walking, cycling, running and swimming and the cooper exercise programs*. Meppel, The Netherlands: Krips Repro.

Van der Beek, E.J. 1991. Vitamin supplementation and physical exercise performance. *J. Sports Sci.* 9:77–89.

Van der Walt, W.H., and C.H. Wyndham. 1973. An equation for prediction of energy expenditure of walking and running. *J. Appl. Physiol.* 34:559–563.

Vandewalle, H., G. Pérès, and H. Monod. 1987. Standard anaerobic exercise tests. *Sports Med.* 4:268–289.

Van Duser, B.L., and P.B. Raven. 1992. The effect of oral smokeless tobacco on the cardiorespiratory response to exercise. *Med. Sci. Sports Exerc.* 24:389–395.

Vangaard, L. 1975. Physiological reactions to wet-cold. *Aviation Space Environ. Med.* 46:33.

———. 1982. Perifer varmebalanse. In *Arbete i kallt klimat*, ed. I. Holmér and J. Sundell, Arbete och hälsa, Arbetarskyddsverket, vol. l. Stockholm: Arbetslivsinstitutet.

Van Hall, G., J.S. Raaymakers, W.H. Saris, and A.J. Wagenmakers. 1995. Ingestion of branched-chain amino acids and tryptophan during sustained exercise in man: Failure to affect performance. *J. Physiol.* 486:789–794.

Vanhees, L., R. Fagard, and A. Amery. 1982. Influence of beta adrenergic blockade on effects of physical training in patients with ischaemic heart disease. *Br. Heart J.* 48:33–38.

van Ingen Schenau, G.J., G. de Groot, A.W. Scheurs, H. Meester, and J.J. de Koning. 1996. A new skate allowing powerful plantar flexions improves performance. *Med. Sci. Sports Exerc.* 28:531–535.

Van Nieuwenhoven, M.A., B.E.P.J. Vriens, R.-J.M. Brummer, and F. Brouns. 2000. Effect of dehydration on gastrointestinal function at rest and during exercise in humans. *Eur. J. Appl. Physiol.* 83:578–584.

Van Tulder, M.W., B.W. Koes, and L.M. Bouter. 1997. Conservative treatment of acute and chronic nonspecific low back pain: A systemic review of randomized controlled trials of the most common interventions. *Spine* 22:2128–2156.

Veiersted, K.B., and R.H. Westgaard. 1993. Development of trapezius myalgia among female workers performing light manual work. *Scand. J. Work Environ. Health* 19:277–283.

Veiersted, K.B., R.H. Westgaard, and P. Andersen. 1990. Pattern of muscle activity during stereotyped work and its relation to muscle pain. *Int. Arch. Occup. Environ. Health* 62:31–41.

Venosa, R.A., and P. Horowicz. 1981. Density and apparent location of the sodium pump in frog sartorius muscle. *J. Membr. Biol.* 59:225–232.

Vergauwen, L., A.J. Spaepen, J. Lefevre, and P. Hespel. 1998. Evaluation of stroke performance in tennis. *Med. Sci. Sports Exerc.* 30:1281–1288.

Verma, S.S., M.S. Malhotra, and J. Sen Gupta. 1979. Indirect assessment of energy expenditure at different work rates. *Ergonomics* 22:1039–1044.

Verney, E.B. 1947. The antidiuretic hormone and the factors which determine its release. *Proc. Roy. Soc.* 135:25.

Vidal, G. 1984. The oldest eukaryotic cells. *Sci. Am.* 250 (2):32.

Vikstrom, K.L., S.H. Seiler, R.L. Sohn, M. Strauss, A. Weiss, R.E. Welikson, and L.A. Leinwand. 1997. The vertebrate myosin heavy chain: Genetics and assembly properties. *Cell Struct. Func.* 22:123–129.

Viljanen, A., ed. 1982. Reference values for spirometric, pulmonary diffusing capacity and body plethysmographic studies. *Scand. J. Clin. Lab. Inv.* 42 (suppl. 159).

Vinten, J., and H. Galbo. 1983. Effect of physical training on transport and metabolism of glucose in adipocytes. *Am. J. Physiol.* 244:E129–E134.

Virokannas, H. 1996. Thermal responses to light, moderate and heavy daily outdoor work in cold weather. *Eur. J. Appl. Physiol. Occup. Physiol.* 72:483–489.

Vogel, J.A., J.E. Hansen, and C.W. Harris. 1967. Cardiovascular responses in man during exhaustive work at sea level and high altitude. *J. Appl. Physiol.* 23:531–539.

Vogel, J.A., L.H. Hartley, J.C. Cruz, and R.P. Hogan. 1974. Cardiac output during exercise in sea-level residents at sea level and high altitude. *J. Appl. Physiol.* 36:169–172.

Vogel, J.A., M.U. Ramos, and J.F. Patton. 1977. Comparison of aerobic power and muscle strength between men and women entering the U.S. army. *Med. Sci. Sports* 9:58.

Vogelaere, P., G. Deklunder, J. Lecroart, G. Savourey, and J. Bittel. 1992. Factors enhancing cardiac output in resting subjects during cold exposure in air environment. *J. Sports Med. Phys. Fitness* 32:378–386.

Vokac, Z., H. Bell, E. Bautz-Holter, and K. Rodahl. 1975. Oxygen uptake/heart rate relationship in leg and arm exercise, sitting and standing. *J. Appl. Physiol.* 39:54–59.

Vokac, Z., and K. Rodahl. 1974. *A study of continuous night work at the Norwegian steel mill at Mo i Rana*. Oulu: Nordic Council for Arctic Medical Research, Report no. 10.

———. 1975. Field study of rotating and continuous night shifts in a steel mill. In *Experimental studies in shiftwork*, ed. P. Colquhoun, S. Folkard, P. Knaut, and R. Rutenfranz. Opladen, Germany: Westdeutscher Verlag.

———. 1977. *Maximal aerobic power and circulatory strain in Eskimo hunters in Greenland*. Nordic Council for Arctic Medical Research, Report no. 1. Oulu, Finland: Oulu University Press.

Volle, M., G.R. Brisson, M. Pérusse, M. Tanaka, and J. Doyan. 1979. Compressed work-week: Psycho-physiological and physiological repercussions. *Ergonomics* 22:1001–1010.

Vøllestad, N.K. 1995. Metabolic correlates of fatigue from different types of exercise in man. *Adv. Exp. Med. Biol.* 384:185–194.

———. 1997. Measurement of human muscle fatigue. *J. Neurosci. Meth.* 74:219–227.

Vøllestad, N.K., I. Sejersted, and E. Saugen. 1997. Mechanical behavior of skeletal muscle during intermittent voluntary isometric contractions in humans. *J. Appl. Physiol.* 83:1557–1565.

Vøllestad, N.K., O. Vaage, and L. Hermansen. 1984. Muscle glycogen depletion patterns in type I and subgroups of type II fibres during prolonged severe exercise in man. *Acta Physiol. Scand.* 122:433–441.

Vrigens, D.M.G., and N.J. Rehrer. 1999. Sodium-free fluid ingestion decreases plasma sodium during exercise in the heat. *J. Appl. Physiol.* 86:1847–1851.

Vrijens, J., J. Hoekstra, J. Bouckaert, and P. van Tyvanck. 1975. Effects of training on maximal working capacity and haemodynamic response during arm and leg exercise in a group of paddlers. *Eur. J. Appl. Physiol.* 34:113–119.

Wada, M., T. Okumoto, K. Toro, K. Masuda, T. Fukubayashi, K. Kikuchi, S. Niihata, and S. Katsuta. 1996. Expression of hybrid isomyosins in human skeletal muscle. *Am. J. Physiol. Cell Physiol.* 271:C1250–C1255.

Wærhaug, O. 1992. Postnatal development of rat motor nerve terminals. *Anat. Embryol.* (Berl.) 185:115–123.

Wærhaug, O., H.A. Dahl, and K. Kardel. 1992. Different effects of physical training on the morphology of motor nerve terminals in the rat extensor digitorum longus and soleus muscles. *Anat. Embryol.* (Berl.) 186:125–128.

Wærhaug, O., and T. Lømo. 1994. Factors causing different properties at neuromuscular junctions in fast and slow skeletal muscles. *Anat. Embryol.* (Berlin) 190:113–125.

Wærsted, M. 2000. Human muscle activity related to non-biomechanical factors in the workplace. *Eur. J. Appl. Physiol.* 83:151–158.

Wærsted, M., and R.H. Westgaard. 1996. Attention-related muscle activity in different body regions during VDU work with minimal physical activity. *Ergonomics* 39:661–676.

Wagenmakers, A.J.M., E.J. Beckers, F. Brouns, H. Kuipers, P.B. Soeters, G.J. van der Vusse, and W.H.M. Saris. 1991. Carbohydrate supplementation, glycogen depletion and amino acid metabolism during exercise. *Am. J. Physiol.* 260:E883–E890.

Wagner, P.D. 1991. Central and peripheral aspects of oxygen transport and adaptation with exercise. *Sports Med.* 11:133–142.

———. 2000. New ideas on limitations to $\dot{V}O_2$max. *Exerc. Sport Sci. Rev.* 28:10–14.

Wahren, J., P. Felig, L. Hagenfeldt, R. Hendler, and G. Ahlborg. 1975. Splanchnic and leg metabolism of glucose, free fatty acids and amino acids during prolonged exercise in man. In *Metabolic adaptation to prolonged physical exercise,* ed. H. Howald and J.R. Poortmans, p. 144. Basel: Birkhäuser Verlag.

Wahren, J., E. Jequier, K. Acheson, and Y. Schutz. 1983. Influence of cigarette smoking on body oxygen consumption. *Clin. Phys.* 3:91.

Wahrenberg, H., P. Engfeldt, J. Bolinder, and P. Arner. 1987. Acute adaptation in adrenergic control of lipolysis during physical exercise in humans. *Am. J. Physiol.* 253:E383–E390.

Walgenbach, S.C., and D.E. Donald. 1983. Inhibition by carotid baroreflex of exercise-induced increases in arterial pressure. *Circ. Res.* 52:253.

Wallenberg, L.R. 1974. Reduction in cold-induced natriuresis following hyperoncotic albumin infusion in man undergoing water diuresis. *Scand. J. Clin. Lab. Invest.* 34:233.

Walro, J.M., and J. Kucera. 1999. Why adult mammalian intrafusal and extrafusal fibers contain different myosin heavy-chain isoforms. *Trends Neurosci.* 22:180–184. (Erratum published in *Trends Neurosci.* 22:302, 1999.)

Walton, D.M., R.A. Kuchinad, T.D. Ivanova, and S.J. Garland. 2002. Reflex inhibition during muscle fatigue in endurance-trained and sedentary individuals. *Eur. J. Appl. Physiol.* 87:462–468.

Ward, M.P., J.S. Milledge, and J.B. West. 1995. *High altitude medicine and physiology.* 2nd ed. London: Chapman and Hall Medical.

Warren, M.P., and A.L. Stiehl. 1999. Exercise and female adolescents: Effects on the reproductive and skeletal systems. *J. Am. Med. Womens Assoc.* 54:115–120.

Washburn, S.L. 1978. The evolution of man. *Sci. Am.* 239 (3):146.

Wasserman, K. 1994. Coupling of external to cellular respiration during exercise. The wisdom of the body revisited. *Am. J. Physiol.* 266:E519.

Wasserman, K., B.J. Whipp, S.N. Koyal, and W.L. Beaver. 1973. Anaerobic threshold and respiratory gas exchange during exercise. *J. Appl. Physiol.* 35:236–243.

Wasserstrom, J.A. 1998. New evidence for similarities in excitation-contraction coupling in skeletal and cardiac muscle. *Acta Physiol. Scand.* 162:247–252.

Watanabe, K., and T. Ohtsuki. 1977. Postural changes and aerodynamic forces in alpine skiing. *Ergonomics* 20:121–131.

Watson, J.D. 1968. *The double helix.* London: Weidenfeld and Nicolson.

Watson J.D., and F.H.C. Crick. 1953a. Genetic implications of the structure of deoxyribonucleic acid. *Nature* 171:964–967.

———. 1953b. Molecular structure of nucleic acids: A structure for deoxyribonucleic acid. *Nature* 171:737–738.

Wawrzinoszek, M., and H. Kramer. 1984. Experimentelle Untersuchungen zur Maximalkraftentwicklung: Maximale Handslusskräfte bei Ein-und Beidhandbetätigung in zwei unterschiedlichen Arbeidsebenen. *Medizin und Sport* 24:105–107.

Wearden, A.J., R.K. Morriss, R. Mullis, P.L. Strickland, D.J. Pearson, L. Appleby, I.T. Campbell, and J.A. Morris. 1998. Randomised, double-blind, placebo-controlled treatment trial of fluoxetine and graded exercise for chronic fatigue syndrome. *Br. J. Psychiatry* 172:485–490.

Weaver, C.M., and S. Rajaram. 1992. Exercise and iron status. *J. Nutr.* 122:782–787.

Webb, P. 1993. Daily activity and body temperature. *Eur. J. Appl. Physiol. Occup. Physiol.* 66:174–177.

Weber, A. and C. Franzini-Armstrong. 2002. Hugh E. Huxley: Birth of the filament sliding model of muscle contraction. *Trends Cell Biol.* 12:243–245.

Weber, G., W. Kartodihardjo, and V. Klissouras. 1976. Growth and physical training with reference to heredity. *J. Appl. Physiol.* 40:211–215.

Weibel, E.R. 1984. *The pathway for oxygen: Structure and function in the mammalian respiratory system.* Cambridge, MA: Harvard University.

Weibel, E.R., and C.R. Taylor, eds. 1981. Design of the mammalian respiratory system. *Respir. Physiol.* 44:1.

Weibel, E.R., C.R. Taylor, P. Gehr, H. Hoppeler, O. Mathieu, and G.M.O. Maloiy. 1981. Design of the mammalian respiratory system: IX. Functional and structural limits for oxygen flow. *Respir. Physiol.* 44:151–164.

Weiner, J.S. 1971. *Man's natural history.* London: Weidenfeld and Nicolsen.

Weis-Fogh, T. 1967. Metabolism and weight economy in migrating animals, particularly birds and insects. In *Nutrition and physical activity,* ed. G. Blix, pp. 84–97. Uppsala, Sweden: Almqvist and Wiksell.

Weissman, I.L. 2000. Stem cells: Units of development, units of regeneration and units in evolution. *Cell* 100:157–168.

Welle, S., K. Bhatt, and C.A. Thornton. 1999. Stimulation of myofibrillar synthesis by exercise is mediated by more efficient translation of mRNA. *J. Appl. Physiol.* 86:1220–1225.

Welsh, J.P., E.J. Lang, I. Sugihara, and R. Llinás. 1995. Dynamic organization of motor control within the olivocerebellar system. *Nature* 374:453–457.

Welsh, L., and O.M. Rutherford. 1996. Hip bone mineral density is improved by high-impact aerobic exercise in postmenopausal women and men over 50 years. *Eur. J. Appl. Physiol. Occup. Physiol.* 74:511–517.

Welsman J.R., N. Armstrong, B.J. Kirby, G. Parson, and P. Sharpe. 1997. Exercise performance and magnetic resonance imaging-determined thigh muscle volume in children. *Eur. J. Appl. Physiol. Occup. Physiol.* 76 (1):92–97.

Weltman, A., D. Snead, P. Stein, R. Seip, R. Schurrer, R. Rutt, and J. Weltman. 1990. Reliability and validity of a continuous incremental treadmill protocol for the determination of lactate threshold, fixed blood lactate concentrations, and $\dot{V}O_2$max. *Int. J. Sports Med.* 11(1):26–32.

West, J.B., ed. 1981. *High altitude physiology.* Benchmark Papers in Human Physiology, vol. 15. Stroudsburg, PA: Hutchinson Ross.

West, J.B. 1983. Climbing Mt. Everest without oxygen: An analysis of maximal exercise during extreme hypoxia. *Respir. Physiol.* 52:265–279.

———. 1994. *Respiratory physiology: The essentials.* 5th ed. Baltimore: Williams and Wilkins.

West, J.B., S. Lahiri, M.B. Gill, J.S. Milledge, L.G. Pugh, and M.P. Ward. 1962. Arterial oxygen saturation during exercise at high altitude. *J. Appl. Physiol.* 17:617–621.

West, J.B., and O. Mathieu-Costello. 1992. High altitude pulmonary edema is caused by stress failure of pulmonary capillaries. *Int. J. Sports Med.* 13:S54–S58.

West, R.V. 1998. The female athlete: The triad of disordered eating, amenorrhea and osteoporosis. *Sports Med.* 26:63–71.

Westby, S., A. Rodahl, and K. Rodahl. 1997. Kontinuerlig registrering av stress under ekstreme påkjenninger. *Tids. Nor. Lægeforen.* 117 (22):3263–3264.

Westerblad, H., and D.G. Allen. 1991. Changes of myoplasmic calcium concentration during fatigue in single mouse muscle fibers. *J. Gen. Physiol.* 98:615–635.

Westerblad, H., J.D. Bruton, D.G. Allen, and J. Lännergren. 2000. Functional significance of $Ca^{2+}$ in long-lasting fatigue of skeletal muscle. *Eur. J. Appl. Physiol.* 83:166–174.

Westerblad, H., S. Duty, and D.G. Allen. 1993. Intracellular calcium concentration during low-frequency fatigue in isolated single fibers of mouse skeletal muscle. *J. Appl. Physiol.* 75:382–388.

Westerblad, H., and J. Lännergren. 1988. The relation between force and intracellular pH in fatigued, single *Xenopus* muscle fibres. *Acta Physiol. Scand.* 133:83–89.

Westgaard, R.H., and T. Lømo. 1988. Control of contractile properties within adaptive ranges by patterns of impulse activity in the rat. *J. Neurosci.* 8 (12):4415–4426.

Westgaard, R.H., M. Wærsted, T. Jansen, and K. Korsund. 1984. *Belastninger og belastningslidelser hos produksjonsarbeidere ved Helly Hansen A/S.* AFYI Report. Oslo: Institute of Work Physiology.

Westling, G., R.S. Johansson, C.K. Thomas, and B. Bigland-Richie. 1990. Measurement of contractile and electrical properties of single human thenar motor units in response to intraneural motor-axon stimulation. *J. Neurophysiol.* 64:1331–1338.

Wetterberg, L. 1994. Light and biological rhythms. *J. Intern. Med.* 235:5–19.

Whedon, G.D., L. Lutwak, P. Rambaut, M. Whittle, C. Leach, J. Reid, and M. Smith. 1975. *Mineral and nitrogen metabolic studies on Skylab flights and comparison with effects of earth long-term recumbency.* Sofia: Bulgarian Academy of Sciences Press.

Wheeler, M.T., E.C. Snyder, M.N. Patterson, and S.J. Swoap. 1999. An E-box within the MHC IIB gene is bound by MyoD and is required for gene expression in fast muscle. *Am. J. Physiol. Cell Physiol.* 276:C1069–C1078.

Whipp, B.J. 1983. Ventilatory control during exercise in humans. *Annu. Rev. Physiol.* 45:415.

White M.D., and M. Cabanac. 1995a. Nasal mucosal vasodilation in response to passive hyperthermia in humans. *Eur. J. Appl. Physiol. Occup. Physiol.* 70:207–212.

———. 1995b. Physical dilation of the nostrils lowers the thermal strain of exercising humans. *Eur. J. Appl. Physiol. Occup. Physiol.* 70:200–206.

White, M.K., T.K. Hodous, and M. Vercruyssen. 1991. Effects of thermal environment and chemical protective clothing on work tolerance, physiological responses and subjective ratings. *Ergonomics* 34:445–457.

White, M.K., R.B. Martin, R.A. Yeater, R.L. Butcher, and E.L. Radin. 1984. The effects of exercise on the bones of postmenopausal women. *Int. Orthop.* 7:209–214.

Whitehead, N.P., T.J. Allen, D.L. Morgan, and U. Proske. 1998. Damage to human muscle from eccentric exercise after training with concentric exercise. *J. Physiol.* 512:615–620.

Whitelaw, W.A., B. McBride, J. Amar, and K. Corbet. 1981. Respiratory neuromuscular output during breath holding. *J. Appl. Physiol. Respir. Environ. Exerc. Physiol.* 50:435.

Whitt, F.R., and D.G. Wilson. 1974. *Bicycling science, ergonomics and mechanics.* Cambridge: Massachusetts Institute of Technology Press.

Wichmann, S., and D.R. Martin. 1994. Snuffing out smokeless tobacco use. *Physician Sportsmed.* 22:97–110.

Wickelgren, I. 1998. Obesity: How big a problem? *Science* 280:1364–1367.

Widegren, U., C. Wretman, A. Lionikas, G. Hedin, and J. Henriksson. 2000. Influence of exercise intensity on ERK/MAP kinase signalling in human skeletal muscle. *Pflügers Arch.* 441:317–322.

Widrick, J.J., S.T. Knuth, K.M. Norenberg, J.G. Romatowski, J.L. Bain, D.A. Riley, M. Karhanek, S.W. Trappe, T.A. Trappe, D.L. Costill, and R.H. Fitts. 1999. Effect of a 17 day spaceflight on contractile properties of human soleus muscle fibres. *J. Physiol.* 516:915–930.

Wigernæs, I., S.B. Strømme, and A.T. Höstmark. 2000. Active recovery counteracts the post-exercise rise in plasma free fatty acids. *Int. J. Sport Nutr. Exerc. Metab.* 10:404–414.

Wiktorsson-Möller, M., B. Oberg, J. Ekstrand, and J. Gillquist. 1983. Effects of warming up, massage, and stretching on range of motion and muscle strength in the lower extremity. *Am. J. Sports Med.* 11:249–252.

Wilber, R.L. 2001. Current trends in altitude training. *Sports Med.* 31:249–265.

Wilk, B., and O. Bar-Or. 1996. Effect of drink flavor and NaCl on voluntary drinking and hydration in boys exercising in the heat. *J. Appl. Physiol.* 80:1112–1117.

Wilk, B., S. Kriemler, H. Keller, and O. Bar-Or. 1998. Consistency in preventing voluntary dehydration in boys who drink a flavored carbohydrate-NaCl beverage during exercise in the heat. *Int. J. Sport Nutr.* 8:1–9.

Williams, C.G., C.H. Wyndham, and J.F. Morrison. 1967. Rate of loss of acclimatization in summer and winter. *J. Appl. Physiol.* 22:21.

Williams, J.H., S.K. Powers, and M.K. Stuart. 1985. Hemoglobin desaturation in highly trained athletes during heavy exercise. *Med. Sci. Sports Exerc.* 18:168–173.

Williams, M.H. 1991. Alcohol, marijuana and beta blockers. In *Perspectives in exercise science and sports medicine,* ed. D.R. Lamb and M.H. Williams, vol. 4, *Ergogenics: Enhancement of performance in exercise and sport.* Dubuque, IA: Brown and Benchmark.

Williams, P.L., ed. 1995. *Gray's anatomy.* New York: Churchill Livingstone.

Williamson, D.L., P.M. Gallagher, C.C. Carroll, U. Raue, and S.W. Trappe. 2001. Reduction in hybrid single muscle fiber proportions with resistance training in humans. *J. Appl. Physiol.* 91:1955–1961.

Wilmore, J.H. 1979. The application of science to sport: Physiological profiles of male and female athletes. *Can. J. Appl. Sport Sci.* 4:103–115.

———. 1981. Women in sport: An introduction to the physiological aspects. In *Women and sport,* ed. J. Borms, M. Hebbelinck, and M. Venerando, pp. 109–11. Basel: Karger.

———. 1982. *Training for sport and activity.* 2nd ed. Boston: Allyn and Bacon.

Wilmore, J.H., and B.J. Freund. 1984. Nutritional enhancement of athletic performance. *Nutr. Abstr. Rev. Clin. Nutr. Series A* 54:1.

Wilson G.D., and M.P. Sklenka. 1983. A system of measuring energy cost during highly dynamic activities. *J. Sports Med.* 23:155–158.

Wilson, M.C., V.N. Jackson, C. Heddle, N.T. Price, H. Pilegaard, C. Juel, A. Bonen, I. Montgomery, O. F. Hutter, and A.P. Halestrap. 1998. Lactic acid efflux from white skeletal muscle is catalyzed by the monocarboxylate transporter isoform MCT3. *J. Biol. Chem.* 273:15920–15926.

Windhorst, U., and G. Boorman. 1995. Potential role of segmental motor circuitry in muscle fatigue. *Adv. Exp. Med. Biol.* 384:241–258.

Windisch, A., K. Gundersen, M.J. Szabolcs, H. Gruber, and T. Lømo. 1998. Fast to slow transformation of denervated and electrically stimulated rat muscle. *J. Physiol.* 510:623–632.

Winslow, C-E.A., A.P. Gagge, and L.P. Herrington. 1939. Influence of air movement upon heat losses from clothed human body. *Am. J. Physiol.* 127:505.

Wirth, A., C. Diehm, H. Mayer, H. Mörl, I. Vogel, P. Björntorp, and G. Schlierf. 1981. Plasma C-peptide and insulin in trained and untrained subjects. *J. Appl. Physiol.* 50:71–77.

Wisløff, U., and J. Helgerud. 1998. Methods for evaluating peak oxygen uptake and anaerobic threshold in upper body of cross-country skiers. *Med. Sci. Sports Exerc.* 30:963–970.

Wisløff, U., J. Helgerud, and J. Hoff. 1998. Strength and endurance of elite soccer players. *Med. Sci. Sports Exerc.* 30:462–467.

Wittenberg, J.B., and B.A. Wittenberg. 1981. Facilitated oxygen diffusion by oxygen carriers. In *Oxygen and living processes: An interdisciplinary approach,* ed. D.L. Gilbert, p. 177. New York: Springer-Verlag.

Woittiez, R.D., P.A. Huijing, and R.H. Rozendal. 1983. Influence of muscle architecture on the length–force diagram. A model and its verification. *Pflügers Arch.* 397:73–74.

Wojtaszewski, J.F.P., J. Lynge, A.B. Jakobsen, L.J. Goodyear, and E.A. Richter. 1999. Differential regulation of MAP kinase by contraction and insulin in skeletal muscle: Metabolic implications. *Am. J. Physiol. Endocrinol. Metabol.* 277:E724–E732.

Wolfe, R.R., M.H. Wolfe, E.R. Nadel, and J.H. Shaw. 1984. Isotopic determination of amino acid–urea interactions in exercise in humans. *J. Appl. Physiol.* 56:221–229.

Wolpaw, J.R. 1997. The complex structure of a simple memory. *Trends Neurosci.* 20:588–594.

Wolski, L.A., D.C. McKenzie, and H.A. Wenger. 1996. Altitude training for improvements in sea level performance: Is there scientific evidence of benefit? *Sports Med.* 22:251–263.

Woo, R., Y.S. Garrow, and F.X. Pi-Sunyer. 1982. Effect of exercise on spontaneous calorie intake in obesity. *Am. J. Clin. Nutr.* 36:470–477.

Wood, P.D., R.B. Terry, and W.L. Haskell. 1985. Metabolism of substrates: Diet, lipoprotein metabolism and exercise. *Fed. Proc.* 44:358–363.

Woodman, C.R., J.M. Muller, M.H. Laughlin, and E.M. Price. 1997. Induction of nitric oxide synthase mRNA in coronary resistance arteries isolated from exercise-trained pigs. *Am. J. Physiol.* 273:H2575–H2579.

Woods, J.J., F. Furbush, and B. Bigland-Ritchie. 1987. Evidence for a fatigue-induced reflex inhibition of motoneuron firing rates. *J. Neurophysiol.* 58:125–137.

World Health Organization. 1998. *Obesity-preventing and managing the global epidemic.* Report of a WHO consultation on obesity. Geneva: World Health Organization.

Wretling, M.L., B. Gerdle, and K. Henriksson-Larsen. 1987. EMG: A non-invasive method for determination of fibre type proportion. *Acta Physiol. Scand.* 131:627–688.

Wroble, R.R. 2000. Articular cartilage injury and autologous chondrocyte implantation. Which patients might benefit? *Physician Sportsmed.* 28 (11):43–48.

Wyndham, C.H. 1973. The physiology of exercise under heat stress. *Annu. Rev. Physiol.* 35:193–220.

Wyndham, C.H., N.B. Strydom, W.P. Leary, and C.G. Williams. 1966. Studies of the maximum capacity of men for physical effort. *Int. Z. Angew. Physiol.* 22:285–295.

Wyon, D.P. 1982. Kyla och prestation. In *Arbete i kallt klimat,* ed. I. Holmér and J. Sundell, Arbete och hälsa, Arbetarskyddsverket, vol. 1. Stockholm: Arbetlivsinstitutet.

Wyon, D.P., I. Andersen, and G.R. Lundqvist. 1979. The effects of modern heat stress on mental performance. *Scand. J. Work Environ. Health* 5:352.

Xiuxian, L., and T. Hiromi. 1996. The effects of two types of clothing on seasonal heat tolerance. *Eur. J. Appl. Physiol. Occup. Physiol.* 72:287–291.

Yamanouchi, K., H. Nakajima, T. Shinozaki, K. Chikada, K. Kato, Y. Oshida, et al. 1992. Effect of daily physical activity on insulin action in the elderly. *J. Appl. Physiol.* 73:2241–2245.

Yamauchi, M., T. Matsumoto, N. Ohwatari, and M. Kosaka. 1997. Sweating economy by graded control in well-trained athletes. *Pflügers Arch.* 433:675–678.

Yan, Z. 2000. Skeletal muscle adaptation and cell cycle regulation. *Exerc. Sport Sci. Rev.* 28:24–26.

Yang, S.Y., M. Alnaqeeb, H. Simpson, and G. Goldspink. 1996. Cloning and characterization of an IGF-1 isoform expressed in skeletal muscle subjected to stretch. *J. Muscle Res. Cell Motil.* 17:487–495.

Yazawa, S., A. Ikeda, T. Kunieda, S. Ohara, T. Mima, T. Nagamine, W. Taki, J. Kimura, T. Hori, and H. Shibasaki. 2000. Human presupplementary motor area is active before voluntary movement: Subdural recording of Bereitschaftspotential from medial frontal cortex. *Exp. Brain Res.* 131:165–177.

Yeaman, S.J. 1990. Hormone-sensitive lipase: A multipurpose enzyme in lipid metabolism. *Biochim. Biophys. Acta* 1052:128–132.

Yeh, M.P., R.M. Gardner, T.D. Adams, F.G. Yanowitz, and P.O. Crapo. 1983. "Anaerobic threshold": Problems of determination and validation. *J. Appl. Physiol. Respir. Environ. Exerc. Physiol.* 55:1178–1186.

Yeo, S., R.H. Hayashi, J.Y. Wan, and B. Dubler. 1995. Tympanic versus rectal thermometry in pregnant women. *J. Obstet. Gynecol. Neonatal Nurs.* 24:719–724.

Yoo, J.U., I. Mandell, P. Angele, and B. Johnstone. 2000. Chondrogenitor cells and gene therapy. *Clin. Orthop.* 379 (suppl.):S164–S170.

Yoon, S.J., S.H. Seiler, R. Kucherlapati, and L. Leinwand. 1992. Organization of the human skeletal myosin heavy chain gene cluster. *Proc. Natl. Acad. Sci.* 89:12078–12082.

Yoshida, T., S. Nakai, A. Yorimoto, T. Kawabata, and T. Morimoto. 1995. Effect of aerobic capacity on sweat rate and fluid intake during outdoor exercise in the heat. *Eur. J. Appl. Physiol. Occup. Physiol.* 71:235–239.

Young, A.J., and D.T. Lee. 1997. Aging and human cold tolerance. *Exp. Aging Res.* 23:45–67.

Zahalak, G.I. 1997. Can muscle fibers be stable on the descending limbs of their sarcomere length-tension relations? *J. Biomech.* 30:1179–1182.

Zaman, G., A.A. Pitsillides, S.C. Rawlinson, R.F. Suswillo, J.R. Mosley, M.Z. Cheng, L.A. Platts, M. Hukkanen, J.M. Polak, and L.E. Lanyon. 1999. Mechanical strain stimulates nitric oxide production by rapid activation of endothelial nitric oxide synthase in osteocytes. *J. Bone Miner. Res.* 14:1123–1131.

Zamir, O., P.-O. Hasselgren, D. von Allmen, and J.E. Fischer. 1991. The effect of interleukin-1alfa and the glucocorticoid receptor blocker RU 38486 on total and myofibrillar protein breakdown in skeletal muscle. *J. Surg. Res.* 50:579–583.

Zamparo, P., R. Perini, C. Orizio, M. Sacher, and G. Ferretti. 1992. The energy cost of walking or running on sand. *Eur. J. Appl. Physiol. Occup. Physiol.* 65:183–187.

Zarrugh, M.Y., and C.W. Radcliffe. 1978. Predicting metabolic cost of level walking. *Eur. J. Appl. Physiol.* 38:215–223.

Zehr, E.P., and R.B. Stein. 1999 Interaction of the Jendrassik maneuver with segmental presynaptic inhibition. *Exp. Brain Res.* 124:474–480.

Zeni, A.I., M.D. Hoffman, and P.S. Clifford. 1996. Energy expenditure with indoor exercise machines. *JAMA* 275:1424–1427.

Zernicke, R.F., A.C.Vailas, and G.J. Salem. 1990. Biomechanical response of bone to weightlessness. *Exerc. Sport Sci. Rev.* 18, 167–192.

Zhou, B., R.K. Conlee, R. Jensen, G.W. Fellingham, J.D. George, and A.G. Fisher. 2001. Stroke volume does not plateau during graded exercise in elite male distance runners. *Med. Sci. Sports Exerc.* 33:1849–1854.

Zhou, M.-Y., H. Klitgaard, B. Saltin, R.R. Roy, V.R. Edgerton, and P.D. Gollnick. 1995. Myosin heavy chain isoforms of human muscle after short-term spaceflight. *J. Appl. Physiol.* 78:1740–1744.

Zhou, S. 2000. Chronic neural adaptations to unilateral exercise: Mechanisms of cross education. *Exerc. Sport Sci. Rev.* 28:177–184.

Zhu, X., M. Hadhazy, M. Wehling, J.G. Tidball, and E.M. McNally. 2000. Dominant negative myostatin produces hypertrophy without hyperplasia in muscle. *FEBS Lett.* 474:71–75.

Zimmerman, M.B., and E.H. Blaine. 1981. Water intake in hypovolemic sheep: Effects of crushing the left atrial appendage. *Science* 211:489.

Zotterman, Y. 1959. Thermal sensations. In *Handbook of physiology: Neurophysiology,* ed. J. Field, vol. 1, p. 431. Washington, DC: American Physiological Society.

# INDEX

Note: The italicized *f* and *t* following page numbers refer to figures and tables, respectively.

**Q**



**R**

# About the Authors

**Per-Olof Åstrand, MD, PhD,** is a pioneer in the field of work physiology, with his research focusing on the oxygen transport system in humans. Now retired, he made numerous advances in the field, winning many awards and fellowships including the Officer de l'Ordre des Palmes Académiques from France in 1975. Dr. Åstrand has authored approximately 200 publications and has been invited to speak in 55 countries outside Sweden, where he resides today.

**Kaare Rodahl, MD, PhD,** professor emeritus, is the author of numerous scientific publications in the fields of the physiology of exercise and work physiology, nutrition, metabolism, environmental physiology, and stress. He was coauthor with Per-Olof Åstrand of all three previous editions of *Textbook of Work Physiology,* including the last edition in 1986. In 2001, he was listed in *2000 Outstanding Scientists of the 20th Century*.

**Hans A. Dahl, MD,** is professor of anatomy at the Norwegian University of Sport and Physical Education, works at the University of Oslo, Norway, and has been active in the field of muscle biology for most of his professional life. He is author and coauthor of several anatomy and physiology textbooks.

**Sigmund B. Strømme, PhD,** is professor of physiology at the Norwegian University of Sport and Physical Education. He earned his PhD from Yale University in 1967. Since 1968, he has researched and taught at the graduate level with emphasis on exercise and work physiology. He is the author or coauthor of several scientific publications and books on effects of exercise, sport nutrition, and physical activity and health. He was presented the Norwegian Medical Association's award for preventive medicine in 2001.